Microscopy of Semiconducting Materials 1993

Microscopy of Semiconducting Materials 1993

Proceedings of the Royal Microscopical Society Conference
held at Oxford University, 5–8 April 1993

Edited by A G Cullis, A E Staton-Bevan and J L Hutchison

S
M   M
VIII

Institute of Physics Conference Series Number 134
Institute of Physics Publishing, Bristol and Philadelphia

CODEN IPHSAC 134 1–787 (1993)

British Library Cataloguing in Publication Data

A catalogue record for this book is available from the British Library

ISBN 0-7503-0290-9

Library of Congress Cataloging-in-Publication Data are available

Conference Co-Chairmen
        A G Cullis and A E Staton-Bevan
Honorary Editors
        A G Cullis, A E Staton-Bevan and J L Hutchison
Scientific Sponsors
        The Royal Microscopical Society
        The Institute of Physics
        The Materials Research Society

This work relates to Department of the Navy Grant N00014-93-J-9006 issued by the Office of Naval Research European Office. The United States has a royalty-free licence throughout the world in all copyrightable material contained herein.

Published by Institute of Physics Publishing, wholly owned by the Institute of Physics, London
Techno House, Redcliffe Way, Bristol BS1 6NX, UK
US Editorial Office: Institute of Physics Publishing, The Public Ledger Building, Suite 1035, Philadelphia, PA 19106, USA

Printed and bound in Great Britain by Galliard Printers, Great Yarmouth

Preface

This volume contains the invited and contributed papers presented at the conference on the 'Microscopy of Semiconducting Materials' which took place at Oxford University on 5–8 April 1993. The event was organised with sponsorship by the Royal Microscopical Society, the Electron Microscopy and Analysis Group of the Institute of Physics and the Materials Research Society. This conference was the eighth in the series which focuses on the latest international advances in the exploitation of semiconductor microscopy in all its forms. Delegates from 20 countries gave an overview of the work in progress at the state-of-the-art and across the whole field.

The high performance and extended functionality generally required from advanced electronic devices can often only be achieved by the development of novel device structures and, in some cases, new materials systems. In such circumstances, device fabrication will depend upon the highly accurate control of semiconductor growth and processing technologies and this, in turn, requires a suitable in-depth understanding of materials behaviour. It is here that an essential role is played by electron and related forms of microscopy. The latter range from, for example, the high resolution techniques of electron and scanning tunnelling microscopies which reveal materials structures down to the atomic scale, through various specialized methods to the majority of more conventional observation techniques which may exploit a wide variety of imaging signals and so provide the basic information needed for materials analysis. The results of this work are of critical importance for both fundamental research studies and device technology applications. The papers presented at the present conference demonstrate the sophistication of the latest developments which have taken place world-wide.

Each paper in this volume was submitted in camera-ready format: after review by two referees any necessary modifications were carried out prior to publication. The editors are most grateful to the following scientific referees for their rapid and meticulous work:

R M Anderson, A Armigliato, P D Augustus, J L Batstone, H Bender, D Bimberg, G R Booker, O Breitenstein, J Brown, A Cavallini, H Cerva, D Cherns, D J Eaglesham, P Fewster, C Frigeri, D Gerthsen, F Glas, P J Goodhew, A Gustafsson, M A G Halliwell, C Hetherington, V Higgs, D B Holt, R Hull, C J Humphreys, J L Hutchison, A Jakubowicz, P Kidd, R Mallard, E Napchan, A G Norman, C Norman, A Ourmazd, D W Pashley, D D Perovic, J B Pethica, A J Pidduck, J M Poate, F A Ponce, P Pongratz, A Rocher, F M Ross, G Salviati, J W Steeds, K Sumino, J Vanhellemont, A Wilkinson and A C Wright.

The conference received generous financial sponsorship from a number of organizations, and it is a pleasure to acknowledge the contributions of the following:

GEC Marconi Materials Technology Ltd
Office of US Naval Research, European Office
Sharp Laboratories of Europe Ltd
The Royal Society

The organizers would like to thank O D Dosser (DRA), T S Fell and A J Holland (Oxford) for correcting the proof copies of many manuscripts. Special thanks are due to C McConville (RMS) for underpinning the conference organization over an extended period and to D M Handley for efficient secretarial support.

July 1993   **A G Cullis**
A E Staton-Bevan

Contents

† Invited

viii   *Contents*

† Invited

† Invited

† Invited

† Invited

Section 7: Quantum wells and superlattices

† Invited

† Invited

Section 9: X-ray studies

† Invited

Section 10: STM and AFM applications

Section 11: Advanced SEM applications

† Invited

† Invited

Inst. Phys. Conf. Ser. No 134: Section 1
Paper presented at Microsc. Semicond. Mater. Conf., Oxford, 5–8 April 1993

Analysis of the information in transmission electron micrographs

A Ourmazd, P Schwander, C Kisielowski, M Seibt, F H Baumann and Y O Kim

AT&T Bell Laboratories, Holmdel, New Jersey 07733, USA

ABSTRACT: We describe how general lattice images may be used to measure the variation of the potential in crystalline solids in any projection, with no knowledge of the imaging conditions[1]. This approach is applicable to structurally perfect samples, in which interfacial topography, or changes in composition are of interest. We present the first atomic-level topographic map of a Si/SiO_2 interface in plan-view, and the first microscopic compositional map of a $Si/GeSi/Si$ quantum well in cross-section.

1. INTRODUCTION

Lattice images, obtained by Transmission Electron Microscopy (TEM), are routinely used to infer the subsurface microstructure of crystalline materials. In principle, a lattice image is a map of the sample (Coulomb) potential, projected along a zone axis (see, e.g., Spence 1981, Downing et al 1990). In practice, it is difficult to extract quantitative information from lattice images. This stems from two primary reasons. First, electrons are multiply scattered during their passage through crystalline samples of realistic thickness ($\geq 10 \text{Å}$). This results in a complex, highly nonlinear relationship between the sample potential and the characteristics of the lattice image. This relationship changes rapidly with the sample thickness, and thus from point to point over the sample. Second, electromagnetic lenses have severe aberrations. The image details thus depend sensitively on the (contrast) transfer function of the microscope, and hence the lens defocus. As shown in Fig. 1, small changes in imaging conditions substantially alter the dependence of the image characteristics (e.g., intensity) on the projected potential (varied in Fig. 1 by changing the composition). It is not possible to establish a general relationship between the sample potential and the image features. This has led to the development of "image matching" procedures, whereby the sample structure is inferred by visually comparing simulated images of model structures with experimental results. Extraction of information by this procedure, even at the qualitative level, requires accurate knowledge of the imaging conditions (sample thickness, lens defocus, etc.). (Bourret et al 1988, O'Keefe et al 1990) These are difficult to measure, and are often poorly known.

Here, we describe an approach, named QUANTITEM, which *measures* the variation of the potential over the sample from general lattice images of crystalline materials, requiring no knowledge of the imaging conditions. In samples of uniform composition, QUANTITEM can be used to map the topography of buried interfaces in plan-view, with near-atomic resolution and sensitivity. Here, we demonstrate this capability for the Si/SiO_2 interface. We show that QUANTITEM topographic images of such *interfaces* are comparable with those obtained from *surfaces* by the scanning tunnelling microscope. In samples with compositional non-uniformities, QUANTITEM may be used

---

1. For the systems considered here, a unique relationship between the projected potential and the image can be established. A detailed discussion will be presented elsewhere.

to map the compositional variation. We demonstrate this by presenting composition maps across Si/GeSi/Si quantum wells. Unlike chemical mapping, QUANTITEM does not rely on the presence of chemical reflections (Ourmazd et al 1989b, Ourmazd 1993), and is thus applicable to general crystalline materials.

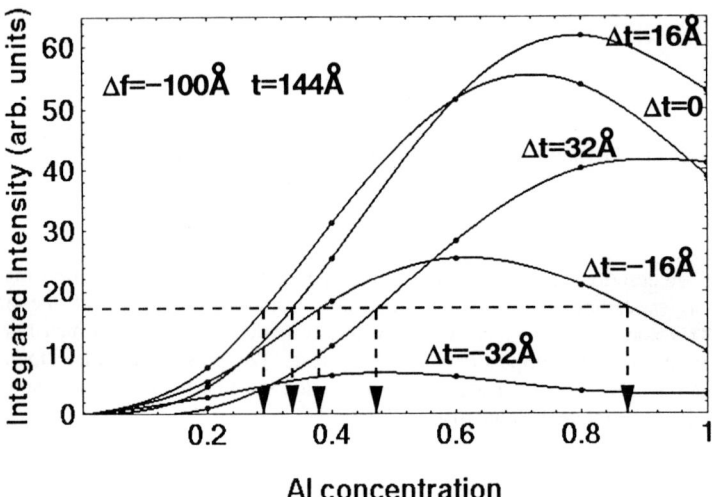

Fig. 1. Plot of image intensity *vs* Al concentration, (which changes the projected potential). The dependence of image characterisitcs on the projected potential changes rapidly with sample thickness. As an example, an intensity of 18 can correspond to an Al concentration of 0.28, 0.33, 0.39, 0.44, or 0.84, as the thickness changes by ±32 Å.

2. PRINCIPLE OF APPROACH

To describe the principle of this approach, it is convenient to represent the information content of an image unit cell in vector notation (Ourmazd et al 1989a). All the available information in a lattice image is contained in the image intensity distribution. The periodicity of the lattice can be used to divide the image into unit cells, within each of which the intensity distribution is digitized. When a unit cell is sampled n×m times, its information content is contained in n×m numbers. We represent these numbers as the components of a (multi-dimensional) vector, whose position and length describe all the available information. (For a discussion of image localization,[2] see Spence (1981) and Marks (1985)).

To measure the variation of the projected potential from a lattice image, one must discover how the image changes with the projected potential, under the particular conditions used to obtain the lattice image under analysis. In vector notation, this requires two steps. First, one needs to determine the path traced by the unit cell image vector as the sample potential varies (Fig. 2). This path changes with imaging conditions, and must be determined afresh for each lattice image. Second, one must determine the rate at which this path is traversed as the sample potential changes. This rate need not be a linear function of the potential change. Determination of the path and the rate at which it is traversed quantify the way that changes in the sample potential affect the image.

The path of the image vector can be directly determined from the experimental image, by plotting the tips of the image unit cell vectors over the region of interest (Fig. 2). Since all TEM

---

2. We have previously shown that under appropriate imaging conditions, the information content of an image unit cell is directly related to the projected potential of a region of the same cross-section in the sample. See: Baumann F H, Bode M, Kim Y O and Ourmazd A 1992 Ultramicroscopy **47** 167.

samples are wedge-shaped, this directly reveals the path described by the image unit cell vector as the projected potential changes. The rate at which this path is traversed can be measured in one of two ways. In the first, one assumes that no particular thickness is favored over the field of view[3]. The density of points (vector tips) along the path is then inversely proportional to the local rate of path traversal. The total length of the path can be calibrated in terms of sample thickness, by recognizing that lattice images vary periodically with the pendellösung oscillations. This calibrates one period of the path in terms of a known change in thickness - the extinction distance.[4]

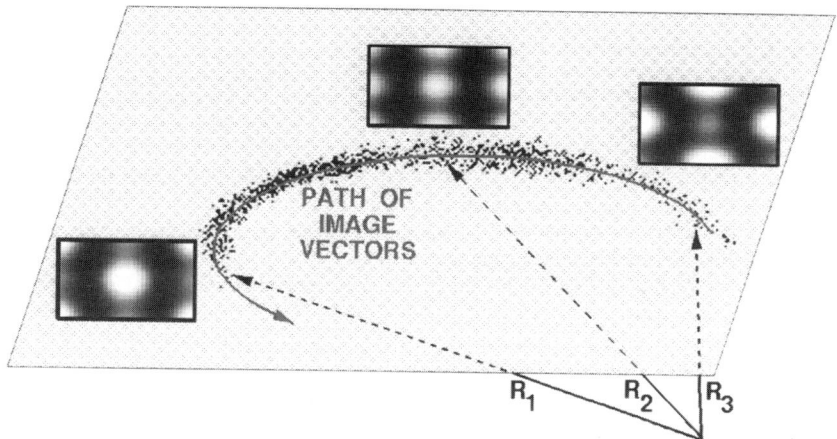

Fig. 2. Simulated lattice image unit cells, and their vector representation $\mathbf{R}^i$, for three different sample thicknesses. The cloud of points represents tips of vectors drawn from an experimental image of a (wedge-shaped) Si sample. The path described by the image vectors quantifies the way changes in the sample projected potential affect the lattice image.

The second approach to measuring the local rate of path traversal requires high signal-to-noise ratios. In the absence of noise, the atomic nature of the sample gives rise to discrete clusters of points along the path, each representing columns with a given number of atoms. Noise tends to smear these clusters into a continuous distribution. Nevertheless, the signal-to-noise ratio is sometimes adequate to reveal the presence of such clustering of points in Fourier transforms of their density, or by autocorrelation techniques. This allows an absolute determination and calibration of the rate at which the path is traversed as a function of projected potential.

The above discussion notwithstanding, we show below that appropriate parameterization of the path can result in a highly linear relationship between the projected potential and the chosen parameter, obviating the need for local calibration of the rate of path traversal.

To summarize, the way that image characteristics change with sample projected potential is represented by the path traced out by the image unit cell vector. This path and the rate at which it is traversed can be determined and calibrated directly from an experimental image, with no knowledge of imaging parameters. The projected potential at each point on the experimental image can then be measured from the position of its image unit cell vector on this path.

We now describe how QUANTITEM may be implemented in practice. The procedure is facilitated by an appropriate choice of reference frame in vector space. We derive a reference frame from the experimental image itself, by extracting a number of "template" vectors from the image. A general unit cell is then expressed in terms of its projections on planes defined by these template vectors. In general, three template vectors suffice, and their choice is not critical. This can be

---

3. This is not as restrictive as assuming a "linear" wedge with a constant slope. However, it does require that, on average, the sample not systematically deviate from a linear wedge.

4. In cases where the pendellösung oscillations cannot be characterized by a single extinction distance, a "local" extinction distance can be used to describe these oscillations.

rationalized by the following argument. In most low-index zone axes, the image consists of three primary elements: the background ($\mathbf{R}^B$); the image due to interference of the central beam with the strongest set of reflections (single-periodic image, $\mathbf{R}^S$); and an image due to interference between these reflections themselves (double-periodic image, $\mathbf{R}^D$). This implies that a general lattice image consists of three large elements:

$$\mathbf{R}^G = a\,\mathbf{R}^B + b\,\mathbf{R}^S + c\,\mathbf{R}^D + \mathbf{r} \quad, \tag{1}$$

where a, b,.. represent numbers, and the residue $\mathbf{r}$ is small. Consider three (template) images $\mathbf{R}_i^T$ (i=1,2,3) extracted from different areas of the lattice image. Since

$$\mathbf{R}_i^T \cong a_i\,\mathbf{R}^B + b_i\,\mathbf{R}^S + c_i\,\mathbf{R}^D , \qquad \mathbf{R}^{B,S,D} = \Sigma_i\ p_i\mathbf{R}_i^T \quad. \tag{2}$$

Substituting for $\mathbf{R}^{B,S,D}$ shows that the general vector $\mathbf{R}^G$ can be written as:

$$\mathbf{R}^G \cong \alpha\,\mathbf{R}_1^T + \beta\,\mathbf{R}_2^T + \gamma\,\mathbf{R}_3^T . \tag{3}$$

Thus a general image can be conveniently expressed in terms of its projections on three template vectors extracted from the image. The primary requirement is that the choice of template vectors $\mathbf{R}^T$ should provide an adequate description of the significant images present. This is easily achieved by extracting a template from areas of the image with distinctly different characteristics.

In general, the path described by the image vector for potential changes of more than half an extinction distance can be well-approximated by an ellipse [5]. Figure 2 shows the tips of experimental image unit cell vectors projected onto the plane defined by the three template vectors $\mathbf{R}_i^T$. We describe the path by fitting an ellipse to the experimental points, and parameterize it in terms of the ellipse phase angle ϕ_e. For the samples we have investigated, this parameterization yields a universal and linear dependence on the sample potential, irrespective of the imaging conditions. Figure 3 is a plot of the variation of the ellipse angle $\Delta\phi_e$ *vs* the projected potential for Si in the <100>, <111> and <110> projections and for Ge_xSi_{1-x} in the <110> projections over the defocus range -100 - -700 Å, and thickness range 80 - 420 Å. These plots were obtained by analyzing simulated images of Si and $Ge_{.25}Si_{.75}$. When present, such a universal relationship obviates the need for fresh measurement and calibration of the rate of path traversal in each lattice image.

3. TOPOGRAPHIC MAPPING: INTERFACIAL ROUGHNESS

By presenting experimental images of the atomic roughness at Si/SiO_2 interfaces in plan-view, we show that QUANTITEM may be used to reveal the topography of buried interfaces with high spatial resolution and sensitivity. Figure 4(a) is a <100> lattice image of a Si sample, after a final rinse in an anisotropic etch (KOH in H_2O) and the formation of a native oxide (~15Å thick on each surface). Figure 5(b) is a QUANTITEM map of the thickness variations in the crystalline part of the $SiO_2/Si/SiO_2$ sample[6], with height representing thickness. Since the sample contains two Si/SiO_2 interfaces, the variations reveal the superimposed roughness of the two Si/SiO_2 interfaces in plan-view. The formation of pyramidal hillocks due to the anisotropic nature of the etch is clear. Such structures are absent when the Si surface is etched isotropically. Quantitative error analysis yields the sensitivity estimates shown in Table I. Figure 4(b) constitutes the first quantitative, high resolution, topographic image of a buried interface in plan-view.

4. COMPOSITION MAPPING

We now describe how QUANTITEM may be used to measure the chemical transition between two regions of known composition in crystalline materials. In the absence of chemical reflections

---

5. For larger thickness variations, the change of defocus due to the wedge shape of the sample can cause significant departures in the path from an ellipse. While convenient, it is not necessary that the path should be an ellipse. Any path can be parameterized, and the parameter related to the projected potential as described above. However, some parameterizations are more convenient than others, primarily because they lead to a more nearly linear relationship with the projected potential.

6. For obvious reasons, QUANTITEM measures only the part of the sample that is crystalline.

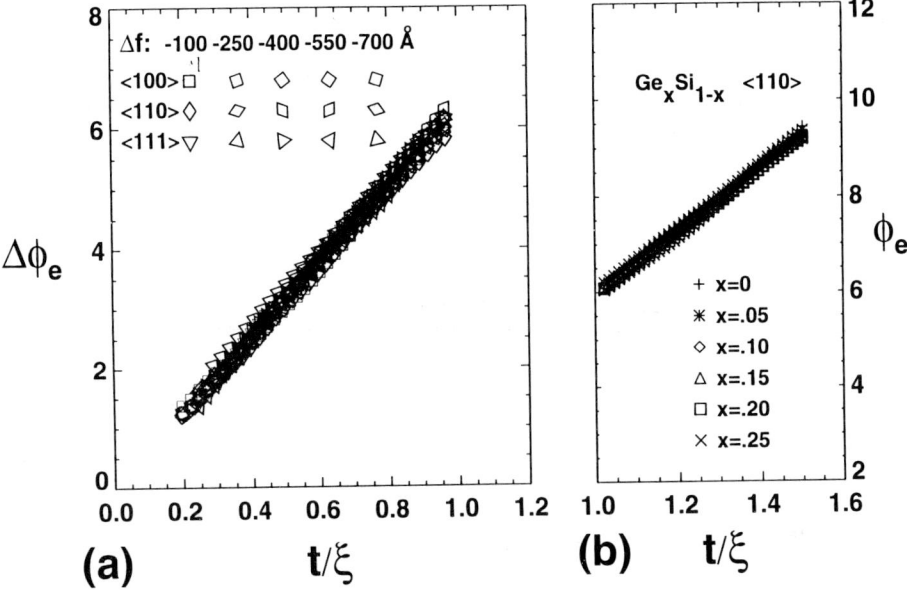

Fig. 3. Variation of ellipse phase angle ϕ_e with sample thickness t, normalized to the extinction distance ξ, for Si (a), and Ge_xSi_{1-x} (b). Note the strong overlap of the points, indicating a universal relation between the variation in ϕ and the projected potential for these systems, irrespective of sample thickness, projection direction, and lens defocus.

(Ourmazd 1993), a lattice image essentially measures the sample projected potential. Changes in sample thickness and composition must therefore be considered on the same footing. It is not possible to neglect changes in sample thickness over the field of view, and assign all changes in the image to compositional variations, for two reasons. First, changes in sample thickness mimic those brought about by changes in composition. Second, in the absence of chemical reflections, small changes in thickness radically alter the dependence of the image characteristics on composition (Fig. 1).

TABLE I

Thickness and chemical sensitivity of QUANTITEM
Best values were obtained by median image filtering of 2 x 2 unit cells

SYSTEM	CELL SIZE	SENSITIVITY TYPICAL	BEST
Si <100>	2.7 x 2.7 $\mathring{A}^2$	15.1 $\mathring{A}$	3.2 $\mathring{A}$
Si <110>	3.8 x 5.4 $\mathring{A}^2$	5.3 $\mathring{A}$	2.0 $\mathring{A}$
Si <111>	2.2 x 3.8 $\mathring{A}^2$	11.0 $\mathring{A}$	3.2 $\mathring{A}$
$Ge_{.25}Si_{.75}$ <110>	3.8 x 5.4 $\mathring{A}^2$	5.4 at% Ge	3.7 at% Ge

Changes in composition have two consequences. First, the path described by the vector can be changed. When present, this can be readily discerned by plotting the experimental vectors in regions of known composition. Second, the extinction distance is altered, which changes the rate at which the path is traversed in each material. QUANTITEM exploits this latter effect to determine the composition of an image unit cell. Consider a target unit cell of unknown thickness and composition, and assume for the moment that its thickness is known. In outline, QUANTITEM proceeds as follows (Fig. 5): (1) it measures the amount by which the ellipse phase angle ϕ_e of the target unit cell is advanced from a reference unit cell in a region of known composition; (2) it subtracts the part $\Delta\phi_e$ due to the thickness change; (3) it ascribes the remainder to changes in the extinction distance, and hence composition. Since the extinction distance can be easily calculated and/or measured, this directly yields the composition of the target unit cell.

To determine the sample thickness at the target unit cell, we map the sample thickness over regions of known composition and fit a two-dimensional model function (surface) to the data, so as to obtain an accurate description of the undulations in the sample thickness. We then infer the sample thickness at the target unit cell by interpolating the model function between the adjoining regions of known composition (Fig. 5(b), inset). Quantitative procedures are used to determine the uncertainty with which the thickness at the target cell has been inferred.

Figure 5(a) is a <110> cross-sectional lattice image of a $Si/Ge_{.25}Si_{.75}/Si$ quantum well. Fig. 5(b) shows the variation of the ellipse phase angle ϕ_e across the image, as determined by QUANTITEM[7]. The variation of ϕ_e over the regions away from the interfaces clearly reveals significant thickness changes, both locally and across the ~150Å field of view. The variation of ϕ_e over the regions away from the interfaces clearly reveals significant thickness changes, both locally and across the ~150Å square field of view. These variations can be reproducd by a model function with a (one-sigma) accuracy of ~3.5Å. As shown in the inset of Fig. 5(b), once the sample thickness at a target cell is determined, its composition is deduced from the part of $\Delta\phi_e$ not due to thickness change.

Figure 6 is a QUANTITEM composition map across a $Si/Ge_{.25}Si_{.75}/Si$ quantum well. The height represents the Ge concentration. This image represents the first quantitative microscopic map of the compositional change across the important Si/Ge_xSi_{1-x} interface, directly revealing its roughness at high resolution. (See Table I.)

We note in passing, that if the composition is known at two points on the sample, QUANTITEM allows an absolute determination of the sample thickness, point by point. This can be understood as follows. The image vector in each material describes its ellipse once every extinction distance $\xi_{1,2}$, where 1,2 refer to the two materials. Since ξ_1 and ξ_2 are different, the two ellipse periods are, in general, incommensurate. This creates a vernier effect between the ϕ_e for the two materials. A given phase difference is then consistent with only a particular absolute thickness for the reference points in each material.

5. PRACTICAL LIMITS

Having outlined the principle of QUANTITEM and demonstrated its implementation, we briefly discuss a few of the factors that determine its practical limits. A more detailed discussion is reserved for a later publication.

5.1 Photographic Nonlinearities

QUANTITEM extracts the projected potential from the details of the image intensity distribution. By analyzing simulated images, we have demonstrated a linear relationship between the ellipse phase angle ϕ_e and the projected potential. We have verified that this linear relationship also holds for experimental images, even when they are recorded on negatives and digitized by

---

7. For this image, the Si and $Ge_{.25}Si_{.75}$ vectors follow the same path to within experimental noise. Changes in composition are reflected in the extinction distance, and hence the rate at which the path is traversed. When the paths differ, ϕ_e is referred to the path for each material.

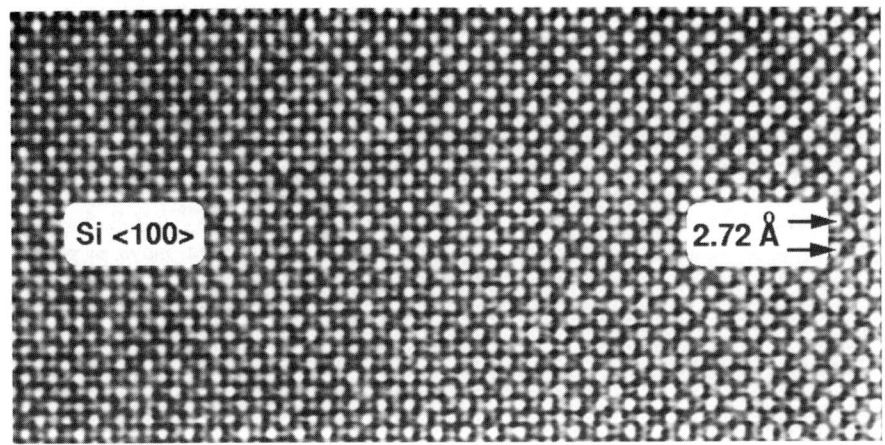

Fig. 4(a). Lattice image of $SiO_2/Si/SiO_2$ sample, viewed in <100> plan-view. The sample was formed by anisotropic etching of Si in KOH, followed by formation of a native oxide. Two Si/SiO_2 interfaces are seen superimposed.

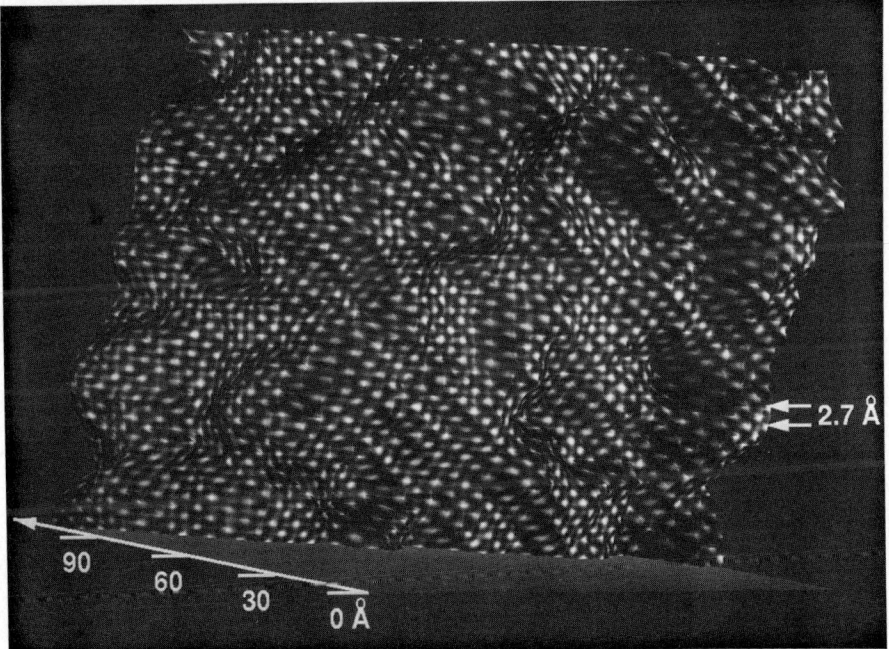

Fig. 4(b). QUANTITEM map of the thickess of crystalline Si sandwitched between the two SiO_2 layers. Height represents sample thickness. This topographic map, deduced from (a) above, directly reveals the superimposed roughness of the two Si/SiO_2 interfaces. Note the pyramidal hillocks produced by the anisotropic etch.

Fig. 5(a). Lattice image of $Si/Ge_{.25}Si_{.75}/Si$ quantum well structure, viewed in <110> cross-section.

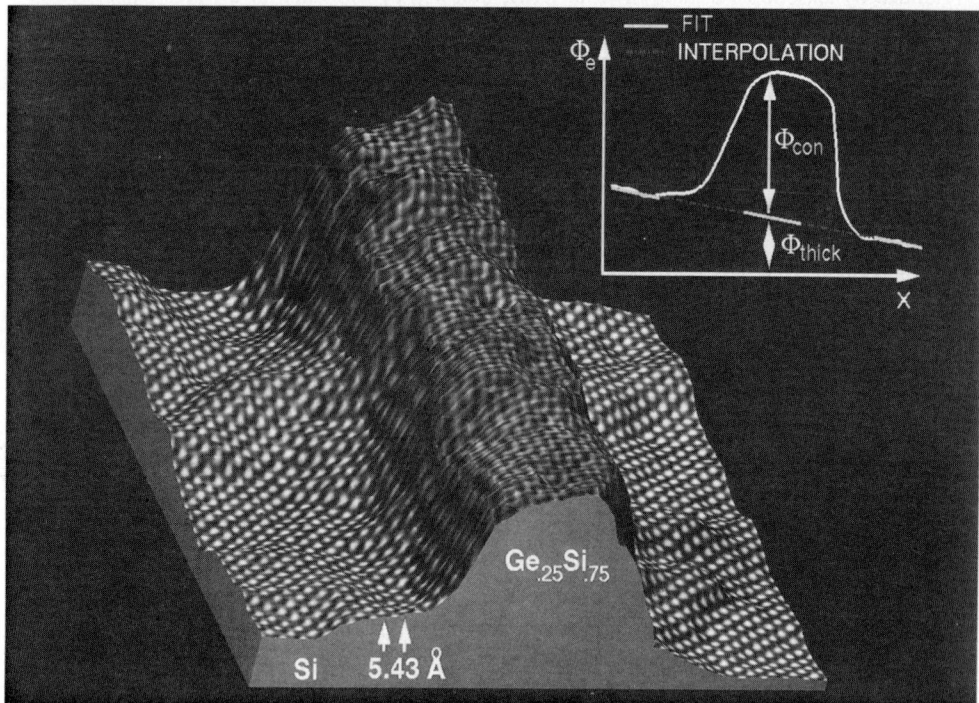

Fig. 5(b). Map of ellipse phase angle ϕ_e across the image shown in (a) above. Note the variations in the Si region, indicating significant thickness changes. Inset: schematic representation of the effect of composition on ϕ_e. The heavier GeSi causes ϕ_e to advance more rapidly. The variation of thickness across the field of view means that part of the change in ϕ_e is due to composition, part due to thickness change.

commercial video systems. First, QUANTITEM analysis of experimental images obtained from cleaved wedges yields wedges of correct constant slope. Second, analysis of simulated images of wedges after convolution with the measured nonlinearity of the (negative + video) system yields the input wedge. Third, QUANTITEM analysis of experimental images recorded directly on an *in situ* CCD camera, or by digitizing exposed negatives yields the same result. These establish that QUANTITEM is robust against recording nonlinearities [8].

5.2 Distortions

A major source of uncertainty in QUANTITEM is the presence of image distortions due to the recording and digitizing instrumentation. Our procedure first corrects pin-cushion distortions, without which, the noise is overwhelming. Second, it resamples the image to remove moiré effects stemming from the often non-commensurate ratio of the unit cell size to the pixel size. Within this procedure each of the unit cells is also centered onto a rectangular grid. This improves the signal-to-noise ratio by a factor of ~2. Additional fine-tuning can be achieved by image filtering with a median filter and by restoration of the exact (e.g., four-fold) symmetry in each unit cell by appropriate averaging over its four quadrants. This improves the signal-to-noise by another factor ~1.3. This remaining noise is due to microbending in the sample, which causes systematic deviation of the data points from one particular path of the image vectors and can only be removed by reduction of the field of view. Reducing the field of view to ~50Å square can improve the signal-to-noise ratio by a factor of ~2. At this level, monolayer changes in sample thickness are resolved (1.9Å for silicon <110>).

5.3 Noise

The presence of amorphous overlayers introduces noise, creating a cloud of points (vector tips) about the path described with sample potential. We reduce this effect by projecting each point onto the path. In this way, only the component of noise that coherently changes the information content of the image unit cell in the direction of the path causes confusion. The ultimate sensitivity of QUANTITEM is limited by residual distortions, and colored noise, which cannot be eliminated by spatial averaging.

5.4 Spatial Resolution

The ultimate (lateral) spatial resolution of QUANTITEM is limited by three factors: spreading of information due to multiple scattering and imperfect lens information transfer[1]; the size of the unit cell analyzed; and noise (Ourmazd et al 1989a). Typical spatial resolutions are summarized in Table I.

6. DISCUSSION AND CONCLUSIONS

We now discuss the more general implications of our work. There can be no single route to quantitative electron microscopy. However, we have described a means for direct measurement of the sample projected potential from general lattice images. This approach is based on the notion that imaging conditions do not need to be individually known. Their combined effect simply produces a relationship between the projected potential and the image features, which can be extracted directly from each experimental image. In view of the complexities of the image formation process, it is remarkable that such a conceptually simple approach can yield valuable information. In particular, the variation of the projected potential in crystals of uniform structure can be directly extracted from general lattice images, with no need for careful control or knowledge of the imaging conditions. Here, we have demonstrated the ability of QUANTITEM to yield high resolution topographic maps of *buried* interfaces in plan-view. This opens the way for the study of a variety of important interfacial reactions at the atomic level, such as surface roughening during oxidation. The ability to map compositional variations in general systems is a significant step toward investigating the relaxation of general multilayered systems and their point defect reactions (Ourmazd 1993). More generally, QUANTITEM constitutes a rapid and robust means of extracting quantitative information from lattice images, which are generally obtained under poorly known conditions. It may thus help to transform high resolution transmission electron microscopy into a practically quantitative tool.

---

8. They also establish that inelastic scattering does not adversely affect analysis by QUANTITEM.

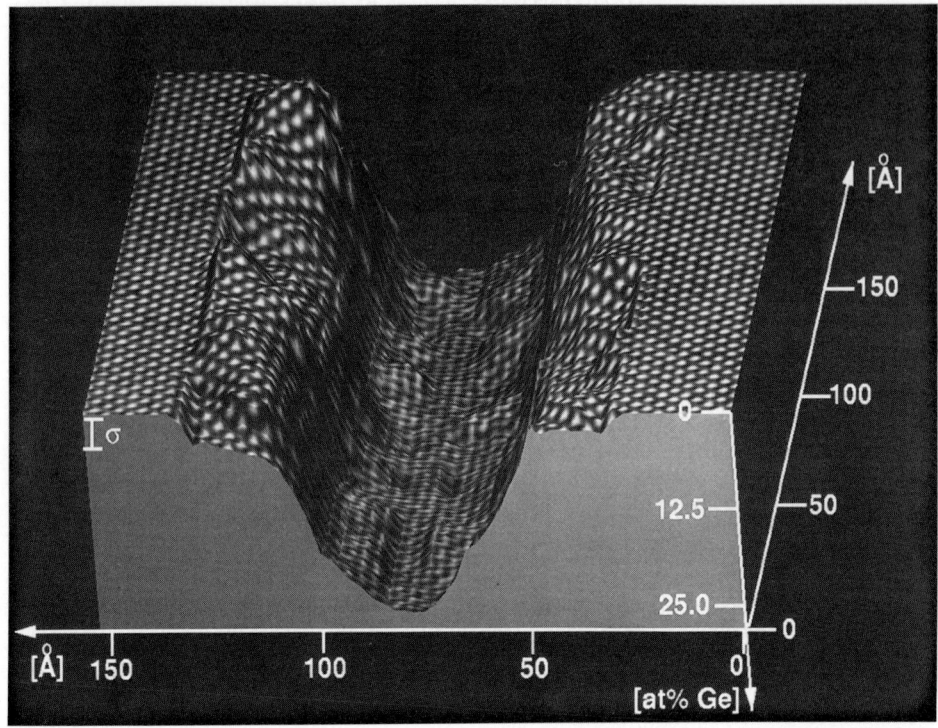

Fig. 6. QUANTITEM composition map deduced from Fig. 5. Height represents Ge concentration, the bar one-sigma accuracy.

ACKNOWLEDGEMENTS

We acknowledge valuable discussions with G Higashi and Y LeCun, expert technical assistance from J A Rentschler and efficient editorial help from C Stiles-Canter. The GeSi/Si samples were kindly provided by J C Bean and J Bevk.

REFERENCES

Bourret A, Rouvière J-L and Penisson J M 1988 Acta Cryst. A **44** 838
Downing K H, Meisheng H, Wenk H-R and O'Keefe M A 1990 Nature **348** 525
Marks L D 1985 Ultramicroscopy **18** 33
O'Keefe M A, Dahmen U and Hetherington C J D 1990 Mat. Res. Soc. Symp. **159** 453
Ourmazd A 1993 Materials Science Reports **9** 201
Ourmazd A, Taylor D W, Bode M and Kim Y O 1989a Science **246** 1571
Ourmazd A, Taylor D W, Cunningham J and Tu C W 1989b Phys Rev Lett **62** 933
Spence J C H 1981 Experimental High Resolution Electron Microscopy (Clarendon Press: Oxford)

Inst. Phys. Conf. Ser. No 134: Section 1
Paper presented at Microsc. Semicond. Mater. Conf., Oxford, 5–8 April 1993

11

Towards a systematic pattern analysis in high resolution electron microscopy: application to quantitative chemical analysis

J-L Rouvière and N Bonnet[1]

CEA - Centre d'Etudes Nucléaires de Grenoble, Département de Recherche Fondamentale sur la Matière Condensée, SP2M, laboratoire Structures, 85X, 38041 Grenoble Cedex, France
[1]Université de Reims, INSERM U314 Laboratoire de Microscopie Electronique, 21 rue Clément Ader, 51100 Reims, France

ABSTRACT: We tested the potential of Multivariate Statistical Analysis (MSA) by applying it to the study of High Resolution Electron Microscopy (HREM) images of the GaAs/AlGaAs system. This pattern analysis could be advantageously applied in HREM as it is a thorough and systematic analysis of all the trends of a set of patterns. It would remove restrictions of actual pattern analysis methods in HREM, such as extraction of one unique parameter and a-priori determination of reference patterns.

1. INTRODUCTION

With the availability of fast computers and inexpensive CCD cameras (and of course good HREM images), numerical analysis of HREM images is becoming a familiar technique, although it is still under development. Schematically, two kinds of information can be extracted from an HREM image. First, the positions of the maxima can be determined. These maxima are generally simply related to the projection of the observed structure and generally give access to translation (Rouvière 1989), rotations between different parts of the samples or more complex local deformation (Loubradou et al, Jouneau 1993, Bierwolf 1993). Second, the whole pattern around each of the maxima can be analyzed. This last method has proved to be very effective in determining chemical profiles in different specific materials (Ourmazd et al 1990, De Jong and Van dick 1990, Thoma and Cerva 1991). However so far, no thorough analysis of the information contained in these patterns has been realized; only one parameter supposed to be the best estimate of the chemical content has been extracted from the pattern analysis. This "blind" estimation could lead to errors.

Here we present and test a tool, the multivariate statistical analysis (MSA), that systematically analyzes a set of elementary cells and could be advantageously applied to the analysis of HREM images and especially to the extraction of the chemical content present in these images. This technique has been already successfully applied to the analysis of biological electron microscopy images (Van Heel et all 1990) and to the analysis of a set of electron energy filtered images (Hannequin and Bonnet 1988), but never to the analysis of HREM images.

2. DESCRIPTION OF THE MULTIVARIATE STATISTICAL ANALYSIS

The multivariate statistical analysis analyzes a set of data and determines all the trends in them. If these data are represented by points in an N dimensional space, MSA determines the minimum set of orthogonal components that best describes the clouds of points.

Here is a quick summary of the different steps to be done in an MSA of an HREM image. First stages are similar to the ones presented by Ourmazd et al (1990) in their algorithm. Maxima positions are first determined. Then, elementary cells are extracted around each maximum. The size of these cells (for instance 0.28nmx0.28nm digitised in 30x30 pixels) determines the lateral resolution. The minimum size value is given by the distance between the maxima (0.28nm in GaAs/AlGaAs sytsem). It is this set of elementary cells, represented by vector intensity I_k (of size 30x30 in the previous example) that constitutes the input of the multivariate analysis. A variance-covariance matrix is formed. It contains all the correlation terms between the intensity vectors and its size is NxN if the set contains N cells (Fig. 1). MSA diagonalises this matrix. Its eigenvectors form a new basis for the representation of the different sources of information in the data set (Trebbia and Bonnet 1990).

The advantages of such an analysis are first that all the elementary cells are analyzed as a whole on the same level, without any a-priori determination of reference cells. It is the analysis itself that determines the constant region that will be used to calibrate the measured values. Second, as all the systematic and relevant trends are analyzed, more than one trend and thus more than one parameter can be evaluated for each elementary cell.

Fig. 1 Variance-covariance matrix (Principal Component Axis -PCA- variant of MSA, see Trebbia and Bonnet 1990 for more details) diagonalised by MSA in the case of a set of N elementary cell of 30x30 pixels, represented by Intensity vector $I_k = \left(i_k^1 \; i_k^2 \; ... \; i_k^{900} \right)$

$$Cov = \begin{pmatrix} I_1.I_1 & I_1.I_2 & .. & I_1.I_N \\ I_2.I_1 & I_2.I_2 & & \\ .. & & & \\ I_N.I_1 & & & I_N.I_N \end{pmatrix}, \; I_1.I_2 = \sum_{m=1}^{900} i_1^m . i_2^m$$

3. APPLICATIONS OF THE MULTIVARIATE STATISTICAL ANALYSIS

We applied the multivariate statistical analysis to two main categories of problems arising from the study of GaAs/AlGaAs interfaces. First, on experimental HREM images in order to test its ability to extract the relevant information from noise and artefacts, and then on simulated images in order to test its ability to separate different physical effects such as thickness, defocus and chemical composition.

3.1 Analysis of an Eperimental GaAs/AlGaAs Interface Image

An experimental image of an $Al_{0.4}Ga_{0.6}As$ /GaAs interface viewed along the [001] axis (Fig. 2) was analyzed. Only two factorial axes (eigenvectors of the variance-covariance matrix, sources of information), representing respectively 55% and 20% of the total variance were considered as having a physical meaning. The information carried by the other axes, was considered to be "noise". The two first eigenvectors of the variance-covariance matrix, also called factorial images, are displayed in the right column of Fig. 2b. Each elementary cell is a linear combination of these factorial images; its coordinates on these axe are displayed on Fig. 2a. as grey maps called coordinate maps. It is the brightest and the darkest regions of these

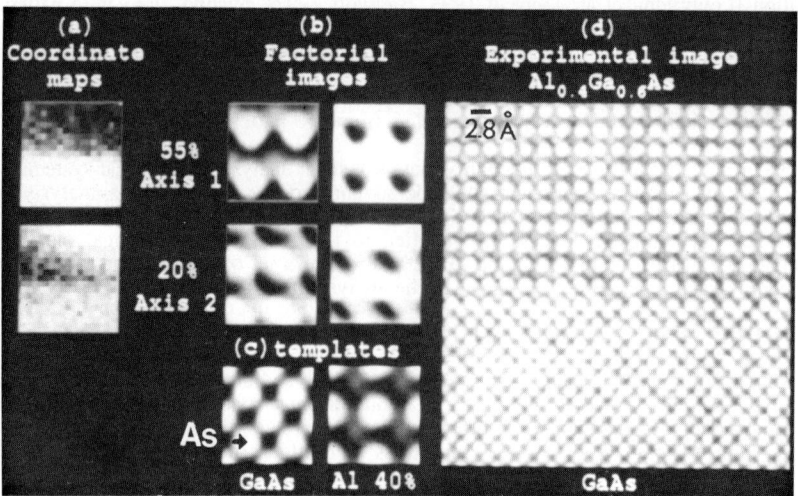

Fig. 2 : a) The elementary cell coordinates on the 2 main axes represented as grey level maps called coordinate maps. b) Factorial images of the 2 main axes (right) and their negative contrasts (left). c) GaAs and $Al_{0.4}Ga_{0.6}As$ template cells that can be compared to factorial images. Four images are put together. d) The originally analyzed HREM image. This picture can be directly compared, although here not on the same scale, to the coordinate maps a).

factorial images that carry information, as it is these that change most when multiplied by positive or negative values (the mean coordinate value is choosen to be zero). In order to better apprehend these areas, the negative contrasts of the factorial images are displayed on Fig. 2b.

The first factorial image (Fig. 2b, axis 1) is characterized by important dark regions at Arsenic positions, surrounded by rather uniform white regions. As has been shown (Ourmazd 1990), it is the brightness of this region that determines the aluminium content of the analyzed cells. The first axis is therefore sensitive to the chemical composition. The second factorial image is not centrosymmetric and cannot be related to "true" physical information. A careful inspection of the data set shows that it is related to a geometrical shift of some cells relative to others. Such an artefact introduced in the procedure at the computer level would not have been perceived when using a global procedure (a procedure extracting one fixed parameter) for quantification. Therefore, individual cell images were carefully realigned and the whole data set was then resubmitted to MSA. The variance percentage increased to 61% for axis 1 and dropped to 5% for axis 2, which was then at the level of "noise" components.

In fact, the coordinate map of the first axis (Fig. 2a) is a chemical map equivalent to the one produced by Ourmazd's algorithm (Ourmazd et al 1990). A projection of this coordinate map perpendicular to the interface gives a chemical profile which, when rescaled, is nearly perfectly identical (values and error bars) to the one obtained by Ourmazd's algorithm (Ourmazd et al 1990) (Fig. 3). Afterwards, you can apprehend why these two methods give a similar result. Both of them are based on a cross correlation method and the cloud of points defined by the cells is well estimated, in "one dimension", by the plane defined by the two template cells of Ourmazd's algorithm. In other words, this plane is a "good estimator" of the axis 1 of MSA.

Fig. 3 : Comparison of a chemical profile as determined by Ourmazd's algorithm and by MSA

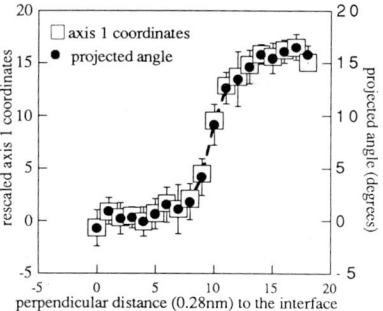

3.2 Analysis of a Set of Simulated Images

A set of simulated images calculated at different defocus D using Stadelmann's software (1987) and representing [001] HREM images of $Al_xGa_{1-x}As$ crystals of different chemical composition x and thickness t, was analyzed as a whole by the multivariate statistical analysis (Fig.4). Defocus and thicknessess were taken around an optimum thickness-defocus condition for a chemical analysis ($5nm \leq D \leq 20nm$, $8nm \leq t \leq 12nm$). Two main axes collecting 66% and 29% of the information were produced by the analysis (Fig. 5).

The chemical information is again mostly contained in the first axis of MSA; except for the extreme values of defocus, the curves obtained at constant defocus and thickness, are nearly horizontal. As in all the previous algorithms, the coordinates on axis 1 are not perfectly linear with composition.

What is interesting is that the second axis carries information on thickness and defocus. Its sensitivity is very low as the simulated images are taken at thickness-defocus conditions where contrast is slowly varying (Fig. 4). Moreover, it is impossible to uniquely correlate the axis 2 coordinates to thickness-defocus values.

However, this analysis proves the sensitivity of the multivariate statistical analysis to detect several uncorrelated coordinates. This ability could be used in determining the defocus and thickness values of experimental images by mixing in the same set, experimental and simulated images. The uncertainty in correlating axis 2 coordinates with thickness-defocus values could be removed because one HREM image contains areas of different thicknesses imaged at the same defocus.

4. CONCLUSION

We have tested the multivariate vectorial analysis in some specific cases. We have systematically explored all the information available in a set of HREM images for a special material without limiting the study to a particular parameter.

We show that indeed the main information present in GaAs images taken around the optimum defocus is linked to the chemical signal and that the detected signal is similar to Ourmazd's one. But we also show that it can detect an artifact (not well centered cells) that could bias the chemical analysis. The method is a little sensitive to thickness and defocus around the optimum condition. In the future, we hope to be able to mix simulated and experimental images in order to get a better interpretation and quantitative evaluation of such physical parameters as thickness-defocus-composition.

At last, as we have mentioned, MSA does not need a-priori determination of template cells. This property could be used in some special problems where localizing reference cells is not as straightforward as in an interface. The case of heavy elements randomly distributed in a matrix is one such situation (Penisson 1992).

Fig. 4 Set of simulated HREM images (sizes 0.28nmx0.28nm), taken at different defocus D, representing $Al_xGa_{1-x}As$ crystals of different chemical composition x and thickness t.
horizontal axis : 21 different compositions x from 0 (left) to 1 (right)
vertical axis : 3 thicknesses (t=83, 99, 105 nm) imaged at 4 defocus values (D=5, 10, 15, 20 nm)

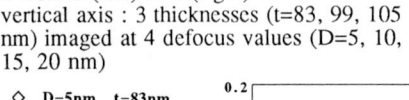

◇   D=5nm   t=83nm
◇   D=5nm   t=99nm
◆   D=5nm   t=105nm
□   D=10nm   t=83nm
▫   D=10nm   t=99nm
■   D=10nm   t=105nm
△   D=15nm   t=83nm
▵   D=15nm   t=99nm
▲   D=15nm   t=105nm
○   D=20nm   t=83nm
○   D=20nm   t=99nm
●   D=20nm   t=105nm

Fig. 5 Coordinates on the 2 main axes, of the simulated images of Fig. 4.

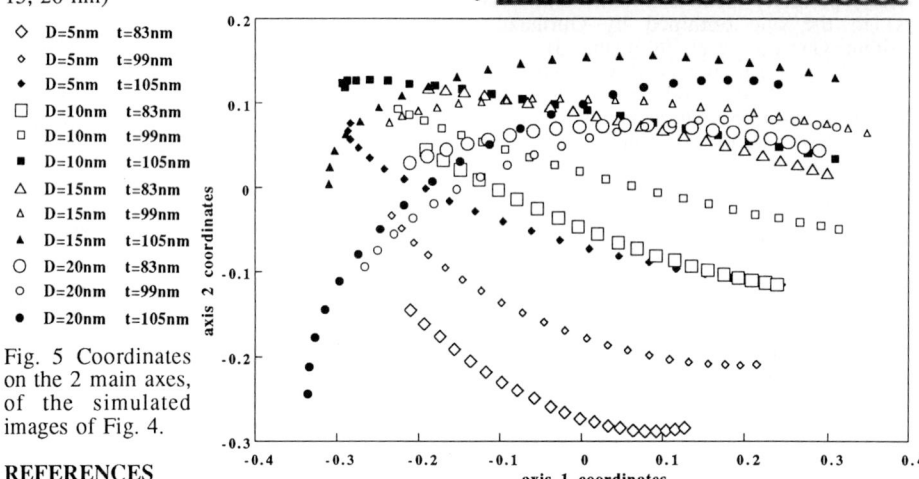

REFERENCES

Bierwolf R, Hohenstein M, Phillip F, Brandt O, Crook GE and Ploog K 1993 Ultramicroscopy (in press)
De Jong AF and Van Dick D 1990 Ultramicroscopy 33 269
Jouneau PH, Tardot A, Feuillet G, Mariette H and Cibert J 1993 (this proceedings volume)
Hannequin P and Bonnet N, Optik 1988 81 6
Loubradou M, Penisson JM and Bonnet R submitted to Phil. Mag. Lett.
Ourmazd A, Baumann F H, Bode M and Kim Y 1990 Ultramicroscopy 34 237
Penisson JM 1992 Electron Microscopy 92 (EUREM 92, Granada) vol I 535
Rouvière JL PhD thesis Grenoble 1989
Stadelmann PA 1987 Ultramicroscopy 21 131
Thoma S, Cerva H 1991 Ultramicroscopy 38 265
Trebbia P and Bonnet N 1990 Ultramicroscopy 34 165
Van Heel M and Frank J 1981 Ultramicroscopy 6 187

ACKNOWLEGMENT
We would like to acknowledge Dr Ourmazd at AT&T Holmdel who provided us with the $Al_{0.4}Ga_{0.6}As/GaAs$ samples and introduced one of us (Jean-Luc Rouvière) to quantitative chemical analysis.

Inst. Phys. Conf. Ser. No 134: Section 1
Paper presented at Microsc. Semicond. Mater. Conf., Oxford, 5–8 April 1993

Compositional and structural analysis of strained Si/SiGe multilayers and interfaces by high resolution transmission electron microscopy

D Stenkamp and W Jäger

Institut für Festkörperforschung, Forschungszentrum KFA Jülich

Postfach 1913, D–5170 Jülich, Germany

ABSTRACT: A method for the quantitative characterization of strained Si/Si_xGe_{1-x} multilayers and interfaces by high-resolution transmission electron microscopy (HRTEM) in [110] and [100] projections is presented. The method relies on the functional relationship between the composition x and the first-order Fourier coefficients of the image intensity, which is quasi-linear and insensitive to strain over a range of imaging conditions for the electron energy of 400 keV considered. By application of a novel image-processing algorithm, which allows a precise measurement of image Fourier coefficients in geometrically distorted lattice images, local composition values x can be determined at near-atomic resolution with an accuracy of $\triangle x \leq \pm 0.1$, and interface sharpness can be detected at the atomic level. Applications of the method to the analysis of strained short-period Si_mGe_n superlattices are presented.

1. INTRODUCTION

Thin strained Si_xGe_{1-x} layers and short-period Si_mGe_n superlattices are of increasing interest in solid state device applications because of their novel electrical and optical properties. These properties strongly depend on the atomic structure of the layer interfaces and therefore require controlled layer growth and a method for compositional and structural interface characterization on an atomic scale. For lattice-matched $GaAs/Al_xGa_{1-x}As$ heterosystems, several approaches have recently demonstrated that local composition values x can be quantitatively obtained from HRTEM lattice images at close-to atomic resolution (Ourmazd et al 1990, de Jong and Van Dyck 1990, Thoma and Cerva 1991). All these methods make special use of "chemical" {200} reflections, which result from the ordered occupancy of sublattices with different atomic species. Since Si_xGe_{1-x} crystallizes as a random alloy with diamond structure, "chemical" reflections like the {200} reflections are kinematically forbidden in this system and these methods are unsuited for a compositional characterisation of lattice-mismatched Si/Si_xGe_{1-x} multilayers. Moreover, image-processing methods are required which are capable of taking into account geometrical image distortions caused by the lattice mismatch. In this paper we describe a method for quantitative compositional and structural analysis of coherent interfaces in strained Si/Si_xGe_{1-x} ($0 \leq x \leq 1$) multilayers with near-atomic resolution and show recent experimental applications of the method.

2. LATTICE IMAGING OF Si_xGe_{1-x} ALLOYS AND Si/Si_xGe_{1-x} INTERFACES

The quantitative characterization of Si/Si_xGe_{1-x} multilayers by HRTEM requires a fundamental understanding of the image contrast behaviour of the individual Si_xGe_{1-x} layers and of the layer interfaces. For this purpose we have extensively studied the image formation process of "bulk" Si_xGe_{1-x} random alloys and of different coherent Si/Si_xGe_{1-x} interface structures for the zone axes $\underline{B} = [110]$ and $\underline{B} = [100]$ by Bloch-wave and multi-slice image simulations at 400 keV and compared with experimental images (Stenkamp and Jäger 1992a, 1992b). The calculations were performed by using the EMS program package (Stadelmann 1987). Here we present a short summary.

The dynamical electron scattering of "bulk" Si_xGe_{1-x} alloys was investigated as a function of composition x for undistorted (cubic) crystals and for tetragonally distorted crystals. For undistorted Si_xGe_{1-x} $(0 \leq x \leq 1)$ the exit wavefunction $\psi(\underline{r})$ is dominated by the interaction of three dominant Bloch waves for $\underline{B} = [110]$ and by two dominant Bloch waves for $\underline{B} = [100]$. Changes in the lattice site occupancy and in the lattice parameter with varying x affect both the amplitudes $U(\underline{g})$ and the phases $\theta(\underline{g})$ of beams contributing to $\psi(\underline{r})$. For thicknesses up to 40 nm, systematic changes in $U(\underline{g})$ and $\theta(\underline{g})$ are observed for x varying from 0 to 1 for both zone axes. For tetragonally distorted Si_xGe_{1-x} alloys, the influence of distortion on the amplitude and phase of low-index beams is found to be negligible for $t < \xi_0/2$ (ξ extinction length) and only small for $t > \xi_0/2$. For $\underline{B} = [110]$ tetragonal lattice distortions maintain the crystal symmetry and keep the number of dominant Bloch waves unchanged. For $\underline{B} = [100]$ the crystal symmetry is reduced, however, and certain diffracted beams strongly depend on an additional third Bloch wave which increases in amplitude with increasing distortion. In order to minimize the sensitivity of the image contrast on distortions, these beams have to be excluded from the subsequent imaging process.

The imaging process of "bulk" Si_xGe_{1-x} alloys was quantitatively investigated by application of the non-linear imaging theory under partially coherent illumination (Ishizuka 1980). The intensity distribution $I(\underline{r})$ in the high-resolution lattice image is given by a Fourier sum $I(\underline{r}) = \sum^g J(\underline{g}) \cos(2\pi \underline{g} \underline{r})$, in which the Fourier coefficients $J(\underline{g})$ depend on the beam amplitudes and phases, on the wave aberration function $\chi(g, \triangle f)$ ($\triangle f$ objective lens defocus) and on damping envelopes. For the case of 5-beam imaging, for which only the undiffracted beam $\underline{g}_0 = (000)$ and the first-order diffracted beams $\underline{g}_1$ ($\underline{g}_1 = \{111\}$ for $\underline{B} = [110]$, $\underline{g}_1 = \{220\}$ for $\underline{B} = [100]$) contribute to the image intensity, it can be shown that $I(\underline{r})$ is dominated by three Fourier contributions: a coefficient $J(\underline{g}_0) = U^2(\underline{g}_0) + 4U^2(\underline{g}_1)$ describing the image background intensity, a first-order linear coefficient $J(\underline{g}_1) \propto 4U(\underline{g}_1)U(\underline{g}_0) \cos(\theta(\underline{g}_1) - \theta(\underline{g}_0) + \chi(g_1, \triangle f))$ describing a basic modulation of the background intensity, and a non-linear coefficient $J(2\underline{g}_1) \propto 2U^2(\underline{g}_1)$ describing a first harmonic. Especially $J(\underline{g}_1)$ is of particular importance since its sign, which depends on the argument of the cosine term in $J(\underline{g}_1)$, determines whether intensity maxima are localized at atomic column ($J(\underline{g}_1) < 0$) or at atomic tunnel positions ($J(\underline{g}_1) > 0$). In thickness regions where the relative beam phases $(\theta(\underline{g}_1) - \theta(\underline{g}_0))$ of Si and Ge differ by 90° or more, pronounced differences in the image contrast can be obtained by a suitable tuning of $\triangle f$ such that $J(\underline{g}_1)$ has the same modulus but opposite signs for Si and Ge. Under such optimized conditions, a systematic reversal from column to tunnel contrast occurs for x varying from 0 to 1, which is the basis of our method for composition determination.

Fig. 1 exemplifies this contrast behaviour by a series of simulated 5-beam images as a function of composition x for (a) $\underline{B} = [110]$ and (b) $\underline{B} = [100]$. Figs. 1a, 1b also show the appearance of a characteristic half-period contrast within the composition region $0.3 < x < 0.7$. This contrast behaviour is caused by the dominating contribution of the coefficient $J(2\underline{g}_1)$ within this composition range where $J(\underline{g}_1)$ is small or zero. A summary of optimized defocus-thickness combinations for 5-beam lattice imaging

zone-axis	transition type Ge → Si	defocus range [nm]	thickness range [nm]
[110]	T → C	0 — 20	6 — 12
	C → T	-60 — -40	6 — 12
[100]	T → C	-38 — -30	0 — 18
	C → T	-60 — -50	0 — 18

Tab. 1: Optimized ranges of defocus and thickness for 5-beam imaging (T: tunnel contrast, C: column contrast).

at 400 keV is given in Tab. 1. A systematic transition between a Si-like contrast and a Ge-like contrast occurs also when additional higher-order beams contribute to the image intensity (Fig. 1c). Tetragonal lattice distortions, which occur in lattice-mismatched heterostructures with coherent interfaces, are found to have a negligible influence on the image contrast under such optimized conditions, especially at defoci near Scherzer defocus ($\triangle f \approx -50$ nm), and leave the basic image morphology unchanged over the full range of experimentally conceivable strain states.

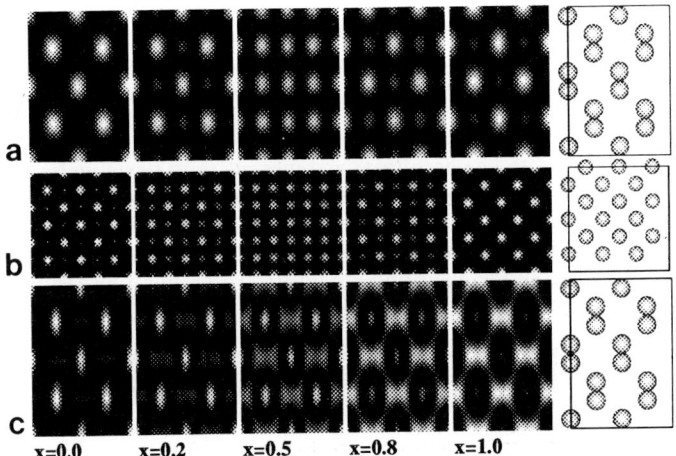

Fig. 1: Simulated 5-beam images as fct. of composition x for (a) $\underline{B} = [110]$ ($\triangle f = -53$ nm, $t = 9.6$ nm) and and (b) $\underline{B} = [100]$ ($\triangle f = -35$ nm, $t = 12$ nm) together with the corresponding projected crystal structure. The influence of higher-order beams is shown by a series of simulated 9-beam [110] images for $\triangle f = -48$ nm and $t = 9$ nm (c).

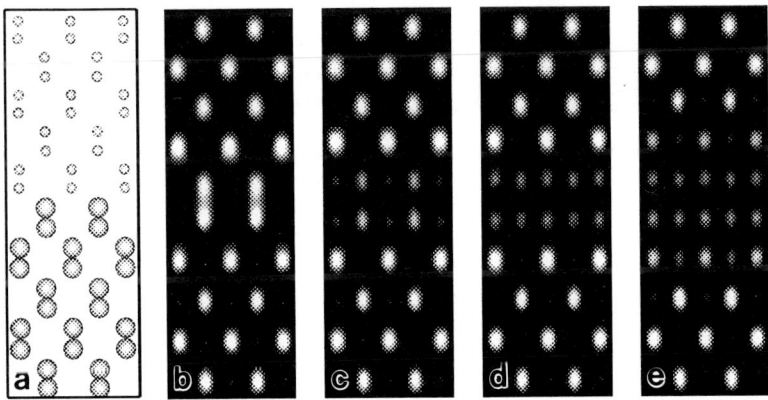

Fig. 2: Simulated 5-beam [110] images for various types of Si/Si$_x$Ge$_{1-x}$ interfaces. (a) Projected atomic interface structure of a sharp Si/Ge interface. (b–e) Images of sharp (b) and diffuse interfaces (c-e) with increasing degree of intermixing ($\triangle f = -55$ nm, $t = 9$ nm).

Lattice image simulations of coherent Si/Si$_x$Ge$_{1-x}$ interfaces were performed for various types of interface structures and for optimized imaging conditions which lead to pronounced contrast differences between layers of different x. Fig. 2 shows as example simulated 5-beam [110] cross-section images of Si/Ge interfaces with a sharp compositional transition (a) and with diffuse transitions extending over 2 (b), 4 (c) and 10 monolayers (d). The image simulations directly show the excellent agreement between the contrast behaviour of Si_xGe_{1-x} alloys at interfaces with the contrast behaviour of "bulk" Si_xGe_{1-x} alloys for the whole range of x. This allows a semi-quantitative determination of the individual layer composition by simply comparing the contrast of each layer with the contrast of "bulk" Si_xGe_{1-x} alloys (Fig. 1). Non-local effects such as Fresnel diffraction were found to have a negligible influence on the contrast behaviour at interfaces under optimized imaging conditions.

3. QUANTITATIVE COMPOSITION DETERMINATION

For the determination of local composition values from lattice images of Si/Si_xGe_{1-x} multilayers and interfaces, we have developed a novel three-step image-processing algorithm which is based on the functional relation between the first-order image Fourier coefficient $J(g_1)$ and x. This relation is found to be quasi-linear and insensitive to strain under optimized imaging conditions. The algorithm allows a precise measurement of the Fourier coefficient $J(g_1)$ within local projected unit cells of digitized lattice images and can be generally applied to the geometrical image correction of lattice images of both strained and unstrained heterostructures. The three process steps of the algorithm are illustrated by Fig. 3 for the example of a sharp interface between layers of pure Si and pure Ge. In a first step, the digitized image (a) is

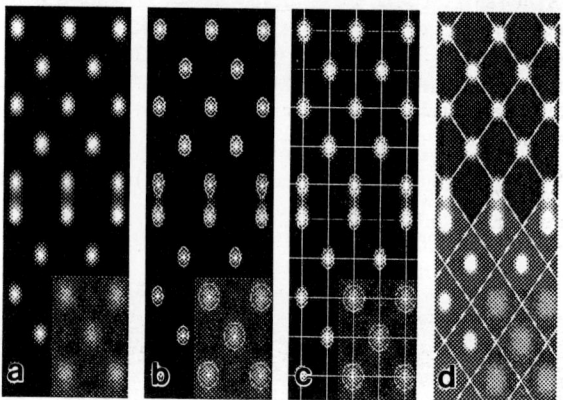

Fig. 3 : Image processing steps for compositional mapping for the example of a sharp Si/Ge interface: (a) original [110] image, (b) determination of geometrical dot centers, (c) image resampling and (d) resulting grey-level map of local composition (light grey: $x < 0.05$, dark grey: $x > 0.95$) Inset shows for comparison the influence of a high noise level on the image processing.

geometrically analysed concerning size and position of "local" projected unit cells. For this purpose the image is processed by a Laplacian-based edge-detection operator which determines the geometrical centers of bright contrast dots above atomic column or tunnel positions in the image (b). This procedure is very stable against random image noise, as shown by the insets representing image regions of high noise level, and is also unaffected by slowly varying modulations of the image background intensity. In the second step, neighbouring contrast dot centers are connected by lines which result in a mosaic-like mesh of image cells (c). These lines are chosen as the boundaries of locally projected unit cells and the image intensity within each projected cell is resampled onto a 16×16 pixels grid using a cubic interpolation routine. The result is a geometrically corrected image (d) in which all projected unit cells have the same pixel size and are exactly matched to the underlying lattice periodicity. In the third step, the Fourier coefficient $J(g_1)$ is obtained for each unit cell of the resampled image by a discrete Fourier transform. Local composition values $x_{det.}$ are finally derived for each cell from the measured value $J(\underline{g}_1)$ and using the relation $x_{\text{det.}} = (J(\underline{g}_1) - J_{\text{Ge}})/(J_{\text{Si}} - J_{\text{Ge}})$, in which J_{Si} and J_{Ge} are values of $J(\underline{g}_1)$ for reference layers of pure Si and Ge, respectively. The result is a two-dimensional compositional map in which different local compositions, which represent average values over the projected sample thickness, are displayed by different grey-levels (d).

The statistical error of the method is determined by fluctuations of $J(\underline{g}_1)$ for a fixed x, as described by the standard deviation σ of the distribution of $J(\underline{g}_1)$. J_{Ge} and J_{Si} can be considered to be noise-free since the reference values for $J(\underline{g}_1)$ in Si and Ge are obtained by averaging over many cells. For optimized imaging conditions we have $J_{\text{Ge}} = -J_{\text{Si}}$, and the statistical error is $\triangle x_{\text{det.}} = \sigma(J(\underline{g}_1))/(2J_{\text{Ge}})$. For our experimental images we obtain $\sigma = J_{\text{Ge}}/10$, resulting in a statistical error $\triangle x_{\text{det.}} = \pm0.05$ which also takes into account composition fluctuations of the random alloy on an atomic scale. By taking also systematic errors into account, for instance small variations of the imaging conditions within the sample area analysed, the total error in the local composition determination by our method amounts to $\triangle x_{\text{det.}} \leq \pm0.1$.

4. INTERFACE CHARACTERIZATION OF Si/Si$_x$Ge$_{1-x}$ MULTILAYERS

As an example, we show an application of our method to the compositional and structural characterization of a Si$_9$Ge$_6$ superlattice structure which has been grown by molecular beam epitaxy at 500° C on a Si$_x$Ge$_{1-x}$ buffer layer. Fig. 4a shows a 7-beam [110] cross-section lattice image (JEOL 4000EX, 400 kV) taken under optimized imaging conditions (Tab. 1) at $t = 10$ nm (as determined from $J_{000}^{\text{Si}} = J_{000}^{\text{Ge}}$) and $\triangle f = 50$ nm (as determined from $J_{111}^{\text{Si}} = -J_{111}^{\text{Ge}}$). Differences of the local Si content are directly imaged by different contrast patters, in agreement with the theoretical predictions (Fig. 1). Fluctuations in the layer thickness and in the local composition within the layers are clearly visible on an atomic scale. Fig. 4b shows a corresponding grey-level representation of the local Ge content which was obtained from a quantitative analysis of Fig. 4a by applying our method. By using the first layers of nominally pure Si and pure Ge as reference, an average composition value of $x_{\text{det.}} = 0.51$ is determined for the Si$_x$Ge$_{1-x}$ buffer layer. This value for the buffer layer composition is in excellent agreement with average composition values which were determined by lattice parameter measurements using selected area electron diffraction and by energy-dispersive X-ray spectroscopy (Jäger et al 1992).

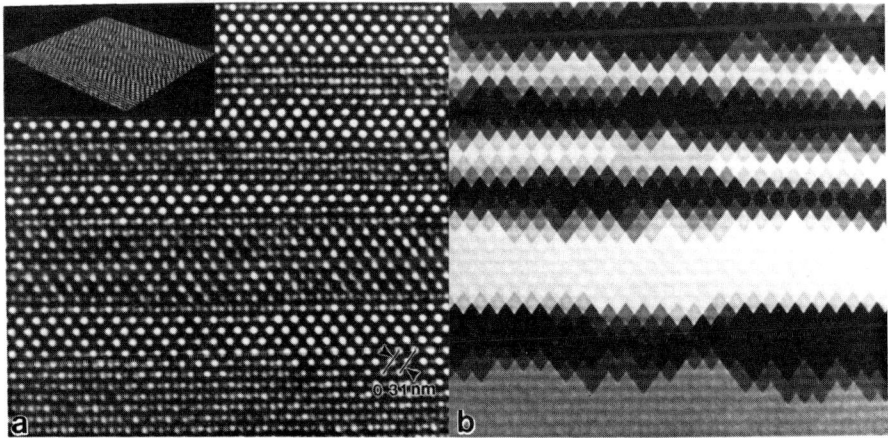

Fig. 4: (a) 7-beam [110] cross-section lattice image of a Si$_9$Ge$_6$ strained-layer superlattice on a Si$_x$Ge$_{1-x}$ buffer (bottom layer). (b) Quantitative grey-level representation of the local Si content as determined by our method. The areas of different grey levels represent regions with a Si content of less than 12.5% (light grey), between 12.5% and 37.5% (light/medium grey), between 37.5% and 62.5% (medium grey), between 62.5% and 87.5% (medium/dark grey) and more than 87.5% (dark grey), respectively.

5. SUMMARY AND CONCLUSIONS

We have described a method for structural and compositional analysis of coherent inter-faces of strained Si/Si_xGe_{1-x} alloy multilayers by lattice imaging in [110] and [100] crystal projections. Optimized imaging conditions for an electron energy of 400 keV have been de-rived from an analysis of linear and non-linear beam-interference contributions to the image intensity for which a continuous change of the image contrast occurs over the full range of x. For the quantitative composition determination, an image-processing algorithm has been developed which takes fully into account geometrical image distortions caused by the lattice mismatch between compositionally different layers but is applicable to lattice-matched and lattice-mismatched heterosystems in general. Composition values can be determined with an accuracy of $\triangle x \leq 0.1$ at close-to atomic resolution, and the geometric interface position can be determined with monolayer accuracy for sharp Si/Ge interfaces (Fig. 3).

"Real-space" pattern-recognition methods like, for instance, the cross-correlation method proposed by Ourmazd et al (1990), have been applied successfully to the compositional anal-ysis of $GaAs/Al_xGa_{1-x}As$ heterostructures. In principle, such methods are applicable also to lattice images of Si/Si_xGe_{1-x} heterosystems provided that the effects of strain on the lattice images are taken into account. Contrary to these "real-space" methods for which the result of pattern recognition depends on the entirety of image Fourier coefficients, our method uses only those image Fourier coefficients for composition determination which are strongly composition sensitive and which show a quasi-linear dependence on x. The application of our method relies on optimized imaging conditions which are chosen such that differences in the image contrast are maximum for different values of x. If the cross-correlation method is applied to lattice images taken under such conditions (Tab. 1), deviations from a linear relation between the re-sult of cross-correlation and the composition x occur due to the influence of additional image Fourier coefficients which themselves show a non-linear or even non-monotonous dependence on x (e.g. $J(g_0)$ and $J(2g_1)$). Also the statistical error in the determination of x is larger for this case compared to our method due to the contribution of fundamental image Fourier coefficients with a high noise level (Stenkamp and Jäger 1992b). Compared to a previous method for the quantitative composition analysis at Si/Si_xGe_{1-x} interfaces (Hull et al 1985), which relies on photodensitometrical measurements of the local image background intensity, our method yields a substantial reduction of the statistical error in x at an increased spatial resolution.

An exciting perspective would be the imaging of identical interface regions in the two dif-ferent crystal projections [110] and [100] under compositional-sensitive imaging conditions. By combining our method with three-dimensional reconstruction, an even more complete charac-terization of the interface in terms of position and chemical nature of atoms would be feasible.

REFERENCES

Hull R, Gibson J M and Bean J C 1985 Appl. Phys. Lett. **46** 179
Ishizuka K 1980 Ultramicroscopy **5** 55
Jäger W, Stenkamp D, Ehrhart P, Leifer K, Sybertz W, Kibbel H, Presting H and Kasper E
    1992 Thin Solid Films **222** 221
de Jong A F and Van Dyck D 1990 Ultramicroscopy **33** 269
Ourmazd A, Baumann F H, Bode M and Kim Y 1990 Ultramicroscopy **34** 237
Stadelmann P A 1987 Ultramicroscopy **21** 131
Stenkamp D and Jäger W 1992a Electron Microscopy **1** 545
Stenkamp D and Jäger W 1992b submitted to Ultramicroscopy
Thoma S and Cerva H 1991 Ultramicroscopy **38** 265

Inst. Phys. Conf. Ser. No 134: Section 1
Paper presented at Microsc. Semicond. Mater. Conf., Oxford, 5–8 April 1993

21

Characterization of $Al_xIn_{1-x}As$/InP heterointerfaces by transmission electron microscopy

E Carlino, M Catalano[1], C Giannini, L Tapfer, E Tournié[2], Y H Zhang[2] and K H Ploog[2]

Centro Nazionale Ricerca e Sviluppo Materiali (CNRSM), S.S. 7 per Mesagne, I-72100 Brindisi, Italy
[1]Università di Lecce, Dip. Scienza dei Materiali, Via Arnesano, I-73100 Lecce, Italy
[2]Max-Planck-Institut für Festkörperforschung, Heisenbergstr 1, D-7000 Stuttgart 80, Germany

ABSTRACT: Lattice matched $Al_xIn_{1-x}As$ epitaxial layers grown by molecular beam epitaxy on InP (100) substrates are investigated by Transmission Electron Microscopy and Double Crystal X-ray Diffractometry. Four different regions close to the $Al_xIn_{1-x}As$/InP heterointerfaces can be detected. The first region, located at the heterointerface is caused by an intermixing of P and As atoms, while the other regions are caused by a small gradient of In content in the epilayer.

1. INTRODUCTION

Semiconductor quantum-well heterostructures based on $Al_xIn_{1-x}As$ materials lattice matched to InP (100) are of great interest for high-speed and low-power dissipation electronic devices as well as for optoelectronic devices. The electronic and optical properties of these materials are strongly correlated to their structural features, namely heterointerface quality, homogeneity of the chemical composition and structural defects in the layers (Tournié et al 1991). Recently, a detailed strain analysis by High Resolution X-Ray Diffraction (HRXRD) of $Al_xIn_{1-x}As$/InP heterointerface reported the existence of distinct interface regions with different chemical compositions(Giannini et al 1993). In this work, a deeper investigation of the same heterointerface is performed by Transmission Electron Microscopy (TEM) and HRXRD.

2. EXPERIMENTAL

The $Al_xIn_{1-x}As$ layers were grown by solid-source Molecular Beam Epitaxy (MBE) on Fe-doped [100]-oriented InP substrates. Prior to growth, the InP substrates were heated up to 560°C under an As_4 pressure of $\approx 10^{-5}$ Torr and then cooled down to the growth temperature (520 °C). MBE growth was performed under constant As_4 flux ($\approx 10^{-5}$ Torr) and at a rate of 500nm/h. The samples were doped with Be ($p \approx 9 - 10 \times 10^{18}$ at/cm³) and capped by a 2 nm thick $Ga_{0.47}In_{0.53}As$ layer. The growth was monitored in situ by Reflection High Energy Electron Diffraction (RHEED). The estimated total thickness of the epilayer is 240 nm.

For the HRXRD experiments a 12 kW generator and a copper rotating anode as x-ray source were used. The x-ray beam was monochromatised by an asymmetrically cut crystal of Ge [001]-oriented and was detected by a NaI scintillator. Sample rotation was performed by a computer-controlled stepping motor with a resolution of 7.6 μrad. The simulations of the experimental spectra were performed by using the recursive formalism of the dynamical theory (Vardanyan et al 1985) for distorted crystals. The specimens for TEM analysis were prepared in the cross-section geometry. The TEM sample preparation of materials containing P and In is a very critical point particularly during the ion milling (Chew & Cullis 1987). Good results have been obtained according to the procedure shortly described below: the samples were glued with the film faced to a Si buffer 0.3 mm thick. Two other Si buffers 1.5

mm thick were glued to get the desired thickness. The wafer so obtained was sliced by a diamond saw and from each slide a 3mm disc about 1 mm thick was obtained. The disc was then mechanically thinned down to 20 μm. The ion mill thinning was performed by using argon ions accelerated by a voltage of 4 KV and impinging the wafer surface with a tilt of 15 degrees. A single ion gun was used with no rotation of the specimen holder. The specimen was oriented with its interface perpendicular to the ion beam. A liquid nitrogen cooled stage was used in order to minimise the generation of artefacts in the specimen. The quality of the final result strongly depends on the temperature of the samples during the ion mill process. The samples obtained according to this preparation procedure show   very low structural damage. If present, the damaged regions are localised in the InP without affecting the interface or the epilayer. In order to investigate the chemical and structural perfection of the $Al_xIn_{1-x}As$ film with good lateral resolution, High Resolution Electron Microscopy (HREM), Convergent Beam Electron Diffraction (CBED) and Electron Energy Loss Spectroscopy (EELS) were used. The TEM experiments were performed using a Philips CM 30 TEM/STEM microscope. The defect analysis by HREM was performed in the [110] zone axis (Mallard 1989). The analysis of HOLZ lines was performed scanning a small probe in different areas across the interface. Image and diffraction pattern simulations were done by the EMS program. In the simulation of the HOLZ lines the pattern obtained from the InP was used as reference. The patterns were taken in [212] zone axis at an accelerating voltage of 150 kV. A Gatan parallel electron energy loss spectrometer was used to collect EELS spectra. A probe of about 8 nm in diameter was scanned across the interface between the substrate and the epilayer.

3. RESULTS AND DISCUSSION

The RHEED pattern showed different features as a function of the growth thickness. In fact, during the growth of the first 20 nm the pattern was spotty, then the pattern gradually recovered and showed the typical streaks as expected for a smooth surface. The low magnification HREM image in fig.1a shows a variation of the contrast in the epilayer as a function of the distance from the interface. Four different zones can be distinguished.

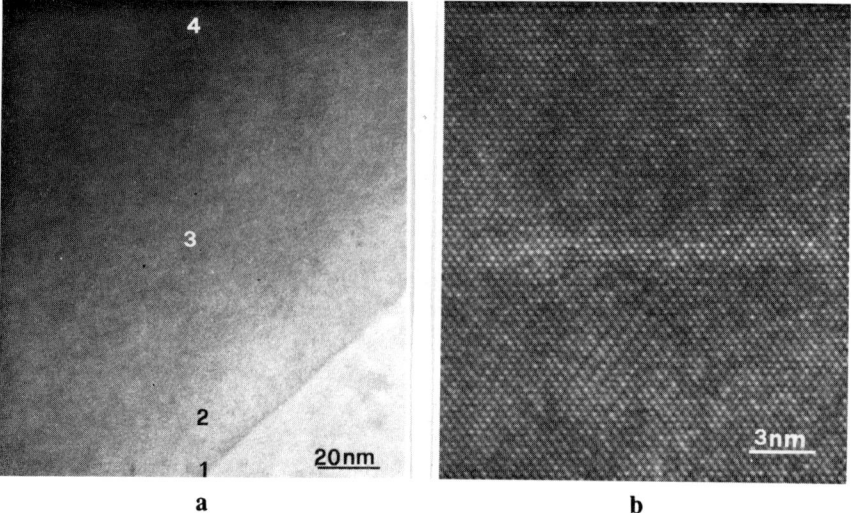

a   b

Fig.1   a) Low magnification HREM image in [011] zone axis of a cross-section of $Al_xIn_{1-x}As$ film. b) HREM image in [011] zone axis of zone 1.

Zone 1 is at the interface between substrate and epilayer. It is few monolayers thick (one monolayer is defined as half of the lattice parameter in the growth direction) and shows a separation surface relatively abrupt and flat. In materials with a common group V element the same definition of sharp interfaces is troublesome (Tapfer et al. 1989). Furthermore the measurement of the thickness of this first layer is not straightforward due to the problems in locating exactly the position of the interfaces in III-V compound HREM images (Suzuki & Okamoto 1985, Ourmazd et al. 1990). As far as regions 2, 3 and 4 are concerned, the separation surfaces are corrugated with an irregularity of the order of a few nanometers. This behaviour was observed in several HREM images even if the contrast among the layers shows different features as a function of the thickness of the observed area. This behaviour could be attributed to a compositional variation in the epilayer. An evaluation of the average width of the different zones was performed by using several HREM images. These values were used as input parameters for the simulation of the experimental x-ray patterns.

The structure resulting from the x-ray simulation is composed of 4 epitaxial layers. At the interface with the substrate there is an $InAs_zP_{1-z}$ layer 1.2 nm thick followed by 3 $Al_xIn_{1-x}As$ layers with different chemical composition and thickness (Table 1). According to these results, the variation of the contrast in the HREM images was attributed, respectively, in the zone 1 to the presence of an $InAs_zP_{1-z}$ layer, and, in the other zones, to a gradient of the In content in the $Al_xIn_{1-x}As$.

ZONE	THICKNESS (nm)	COMPOSITION
1	1.2±0.3	$In As_{0.5860} P_{0.4140}$
2	23±1	$Al_{0.4710} In_{0.5290} As$
3	70±2	$Al_{0.4750} In_{0.5250} As$
4	160±2	$Al_{0.4770} In_{0.5230} As$

Tab.1   Calculated thickness and composition for the epilayer (the error in the composition is $\Delta = 5*10^{-4}$).

The formation of the $InAs_zP_{1-z}$ layer is due to the pre-growth treatment of the InP substrate. In fact, when the temperature of the substrate exceeds the temperature of desorption of the P (360 °C), the vacancies on the surface are replaced by As atoms (Moison et al 1986, Hollinger et al 1990, Averbeck et al 1991). The lattice parameter of this compound is a = 0.59812±0.00002 nm consequently a large strain is localised at this interface due to the mismatch between substrate (a = 0.58684±0.00002 nm) and epilayer (a = 0.58728±0.00002 nm). A further variation of the strain localised in few areas of zone 1 is the origin of the V-shaped dislocation shown in fig.2. The increase of the strain in these areas may be due to local increasing of the thickness and/or As content in the $InAs_zP_{1-z}$ layer.

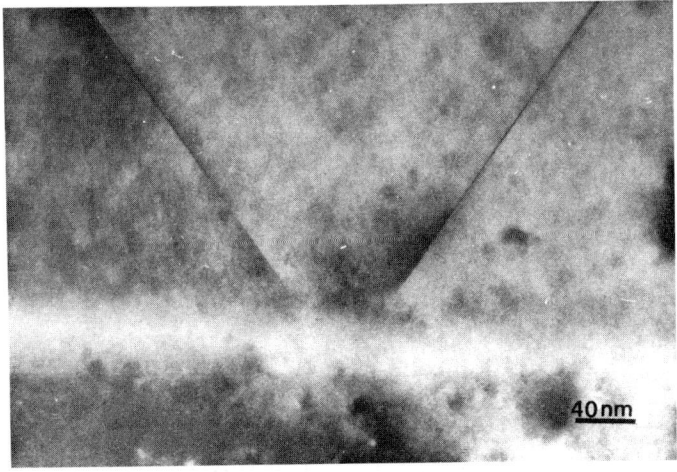

Fig.2   V-shape dislocation at the interface between substrate and epilayer

Several authors report a defect analysis in this kind of material (Chu et al 1985, Peirò et al 1992). The samples analysed in this work show a low defect density. We observed that defects localised deep inside the InP substrate propagate through the epilayer. However, in this work our attention was principally focussed on defects located at the substrate-epilayer interface and in the epitaxial layer. A few V-shaped dislocations (Nakahara et al 1986) were observed originating at the heterointerface.

HOLZ line pattern analysis was also performed, but no valuable difference between different film areas was observed, due to the limited sensitivity of this technique in the particular case of a relatively small lattice parameter variation (Steeds 1985).

No valuable differences were observed in the EELS spectra collected in the different zones of the epilayer from the analysis of both the low loss and the core loss region.

The experimental results obtained from the TEM and x-ray investigations can be summarised as follows. The $InAs_zP_{1-z}$ interface layer, a few monolayers thick (Zone 1), is formed by exposing the InP surface to the As atmosphere during the pre-growth substrate preparation. In the subsequent growth of the $Al_xIn_{1-x}As$ a decrease of the InAs mole fraction (In content) is observed. The relatively high In content in zone 2 to 4 can be explained by the "overshoot" effect, which may occur at the opening of the shutters of the effusion cells and may cause during this first stage of growth a 3-D nucleation, also observed in the RHEED pattern. It should also be noted that, after opening the shutters, the InP substrate is directly exposed to the effusion cells which may cause a small increase in the substrate temperature. Turco et al (1988) found that an increase in the growth temperature may produce a decrease in the In content. The formation of the V-shaped dislocations is due to the local variation and/or As content in the $InAs_zP_{1-z}$. The pre-growth treatment of the substrate and the "overshoot" effect and/or the increase in the substrate temperature may explain the experimental results obtained combining x-ray diffraction measurements and TEM observations.

ACKNOWLEDGEMENT

The authors are grateful to E. Pesce for the skilful sample preparation. We would like to thank P G Merli and the CNR-LAMEL team for their support. We would also like to thank R W Carpenter for a helpful discussion.

REFERENCES

Averbeck R, Riechert H, Schlötter H and Weimann G (1991) VI European Conference on Molecular Beam Epitaxy and Related Growth Methods, Finland
Chew N G and Cullis A G (1987) Ultramic. 23 175-198
Chu N G, Macrander A T, Strege K E and Johnston W D Jr. (1985) J. Appl. Phys. 57 249
Giannini C, Tapfer L, Tournié E, Zhang Y H and Ploog H K (1993) Appl. Phys. Lett. 62 149
Hollinger G, Gallet D, Gendry M, Santinelli G and Viktorovitch (1990) J. Vac. Sci. Technol. B 8 832
Mallard R E (1989) Can. J. Phys. 67 262-267
Moison J M, Bensoussan M and Houzay F Phys. Rev. B 34 2018
Nakahara S, Chu N G and Stall R A (1986) Phil. Mag. A 53 (3) 403-414
Ourmazd A, Baumann F H, Bode M, Kim Y (1990) Ultramic. 34 237-255
Peirò F, Cornet A, Herms A, and Morante J R, Georgakilas A, Halkias G (1992) J. Vac. Sci. Technol. B 10 (5) 2148-2152
Steeds J W (1985) "Convergent Beam Electron Diffraction" in Microscopia elettronica in trasmissione e tecniche di analisi di superfici nella scienza dei materiali. Parte b. Serie simposi ENEA 219- 280
Suzuki Y and Okamoto H (1985) J. Appl. Phys. 58 3456-3462
Tapfer L, Stolz W and Ploog K (1989) J. Appl. Phys. 66 3217-3219
Tournié E, Zhang Y H, Pulsford N J and Ploog K (1991) J. Appl. Phys. 70 7362-7369
Turco F, Guillaume J C and Massies J (1988) J. Cryst. Growth 88 282-290
Vardanyan D M, Manoukyan H M and Petrosyan H M (1985) Acta Cryst. A 41 212

Inst. Phys. Conf. Ser. No 134: Section 1
Paper presented at Microsc. Semicond. Mater. Conf., Oxford, 5–8 April 1993

25

Ga$_2$Te$_3$: a new interfacial phase in ZnTe/(100)GaSb

C T Chou[1], J L Hutchison[2], M-J Casanove[2], D Cherns[1], J W Steeds[1], B Lunn[3] and D A Ashenford[3]

[1] H H Wills Physics Laboratory, University of Bristol, Tyndall Avenue, Bristol BS8 1TL UK
[2] Department of Materials, University of Oxford, Parks Road, Oxford OX1 3PH UK
[3] Department of Engineering Design and Manufacture, University of Hull, Hull HU6 7RX UK

ABSTRACT: The phase Ga$_2$Te$_3$ has been identified as an interfacial layer between ZnTe and a (100)GaSb substrate. We describe a novel approach to HREM imaging of its superlattice, based on an analysis of CTF behaviour over a wide defocus range. The superlattice is thus imaged with enhanced contrast, enabling a likely structure to be derived and used for image simulations.

1. INTRODUCTION

Many wide band-gap II-VI semiconductor thin layers are now grown on lattice matching III-V substrates (Gunshor et al 1990). At the II-VI/III-V interfaces, formation of thin interfacial layers consisting mainly of III-VI compounds has been reported (Zahn et al 1987, Wilke et al 1990; Halsall et al 1992), based on indirect experimental methods such as Raman scattering, etc. Interest in these III-VI phases is increasing but despite efforts to study their structures using high resolution electron microscopy (Wright and Williams 1991, Wright et al 1992) little significant progress has been made in imaging them. In this paper we show that under unusual imaging conditions HREM can be used to detect and analyse a new Ga$_2$Te$_3$ phase in the ZnTe/GaSb interface.

MBE growth details for the ZnTe layers are given in Duddle et al (1992). $<110>$ cross-section samples have been prepared by reactive I$^+$ ion-milling at liquid nitrogen temperature. Samples were then examined at 400 kV in a JEOL 4000EX II electron microscope (Cs = 0.9mm).

2. EXPERIMENTAL HREM APPROACH AND CONTRAST INTERPRETATION

Under the Scherzer defocus condition, typical $<110>$ structure images of sphalerite structures were observed on both sides of the interface with no sign of any Ga$_2$Te$_3$ superlattice. However under certain unusual defocus conditions, areas displaying 2x2 superlattice contrast were detected on the ZnTe side of the sample. An example is shown in fig. 1(a-c). Assuming that an m-times superlattice (with m probably 2) based on the sphalerite structure was the basis of a possible atomic structure of Ga$_2$Te$_3$ and the fact that four {111}-type reflections are the dominant reflections in a [110] diffraction pattern of the sphalerite structure, we calculated for our particular electron microscope the conditions for which the {111} reflections were close to the positions of contrast minima of the contrast transfer functions (CTFs). The conditions for extinguishing the CTF at a particular spatial frequency $\mathbf{u}$ is given by the expression:

$$\Delta f = n/\lambda u^2 - C_s \lambda^2 u^2/2, \quad \text{or} \quad \Delta f = nd^2/\lambda - C_s \lambda^2/2d^2$$

where $\mathbf{n}$ is an integer. We substitute the relevant values appropriate for our system with $d_{111} = 0.3404$ nm ($a_o = 0.5896$ nm). This gives $\Delta f = (70.5n-10.5)$nm. CTF plots for Scherzer defocus -47.1 nm and overfocus values 60 and 130.5 nm for n = 1 and 2 respectively are shown in fig. 1(d-f).

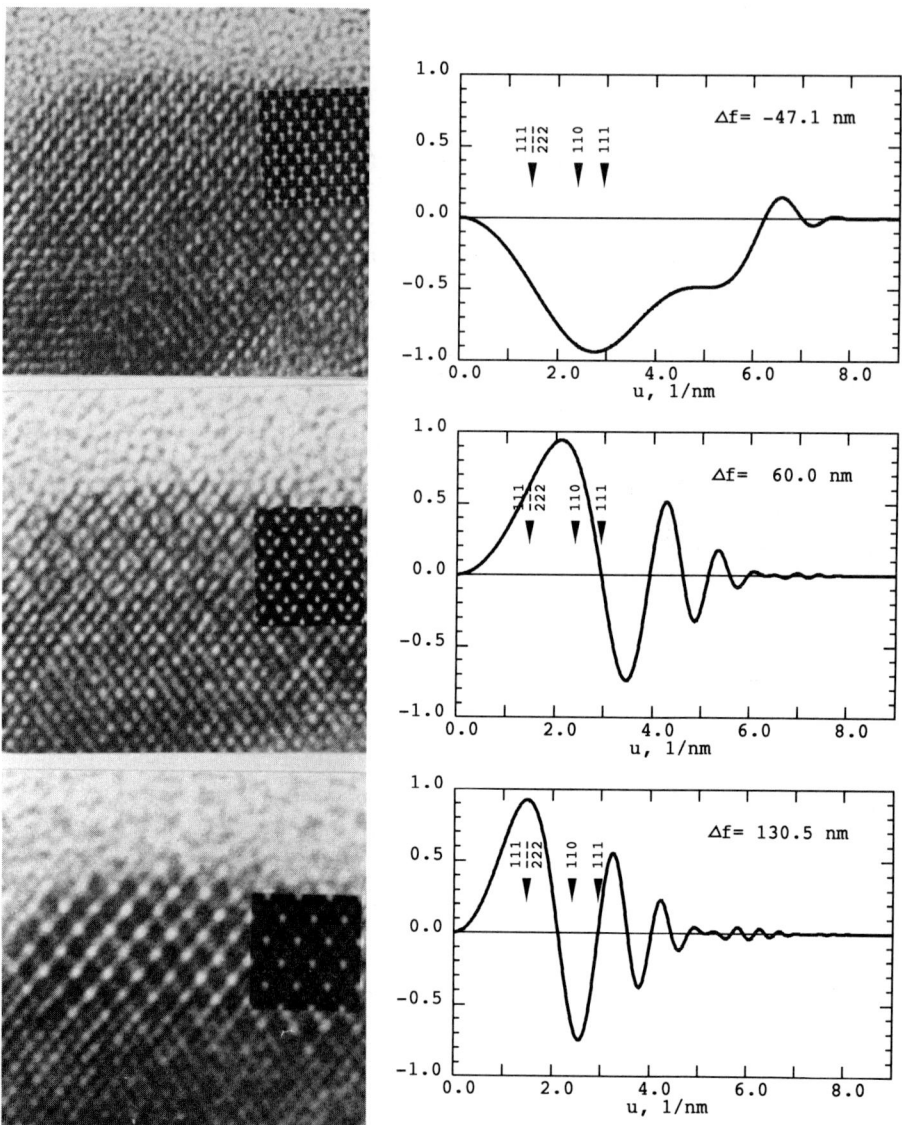

Fig. 1 [110] lattice images of Ga$_2$Te$_3$, and corresponding CTF plots.

From these it is obvious that Scherzer defocus would *not* be useful according to our criterion for imaging any superlattice, and the $\Delta f = -10.5$ nm (n=0) case would be expected to produce very low contrast in thin crystals. On the other hand the defoci at -150, -80, +60, and +130 nm could in principle be used to enhance superlattice contrast. For the last two conditions the very rapid oscillations in the CTF, and very strong chromatic damping effects would exclude them from "normal" HREM imaging. The superlattice areas appear in domains of 5 - 10 nm in size, in the ZnTe side of, and usually 2 - 4 nm away from the interface (see fig. 2a). An electron diffraction pattern of the interface region (Fig. 2b) contains superlattice spots (arrowed) showing significant deviation from the nearby fundamental spots, demonstrating that the lattice parameter of the superlattice *is shorter along the growth direction.*

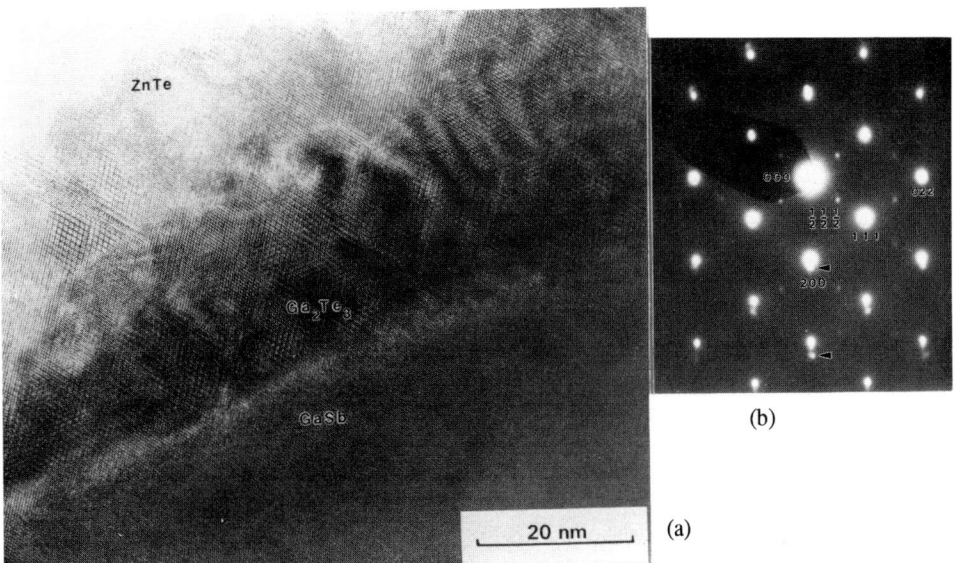

Fig 2(a). Low magnification [110] image of ZnTe/GaSb interface showing the distribution of Ga_2Te_3.
(b). Corresponding SAED pattern showing superlattice reflections.

It is reasonable to assume that the regular distribution of strong and weak spots in fig.1c arises from different occupancies of Ga atoms in the cation sublattice. We therefore concluded that the strong spots correspond to the cation sublattice sites with high Ga occupancy, whereas the weak spots are related to sites of low Ga occupancy. If the high occupancy sites are fully occupied by Ga atoms, then the low occupancy sites must have 5/9 Ga occupancy. Based on this ordered distribution of Ga on the cation sublattice sites in the [110] projection, shown in fig.3a, we can then derive all of the different <110> projections. The relation between the superlattice unit cell and sphalerite is as follows (subscript "s" indicates the sphalerite unit cell, with lattice parameter a_o):

$$a \parallel [110]_s, \ b \parallel [110]_s, \ c \parallel [001]_s, \ \text{with } |a| = (\sqrt{2}/2)a_o, \ |b| = \sqrt{2}a_o, \ |c| = 2a_o$$

The c-axis projection of the new cell is shown in fig.3b. In the attached table the Z coordinates of atoms, atom species and the occupancies are given. Related symmetry elements can be derived from this unit cell giving the space group Amm2. From the electron diffraction pattern we deduce that its c-axis is only 1.117 nm, less than $2a_o = 1.179$ nm. Multislice image simulations based on this model have been performed and confirm the structure, see insets in fig. 1a-c.

3. CONCLUSION

The study revealed a domain structure from which we derived a plausible structure model for the Ga_2Te_3 in ZnTe/GaSb. The model suggested that the vacancies in the structure are highly mobile in the cation sublattice resulting in lattices with an effective 5/9 Ga occupancy whereas the fully occupied Ga sites form continuous and regularly arranged chains along the $[110]_s$ direction. Our model disagrees with previous X-ray and electron diffraction studies of bulk III/VI compounds which suggest superlattice structures based on either fully occupied or completely vacant cation sites in the sphalerite cell (Lubbers and Leute 1982).

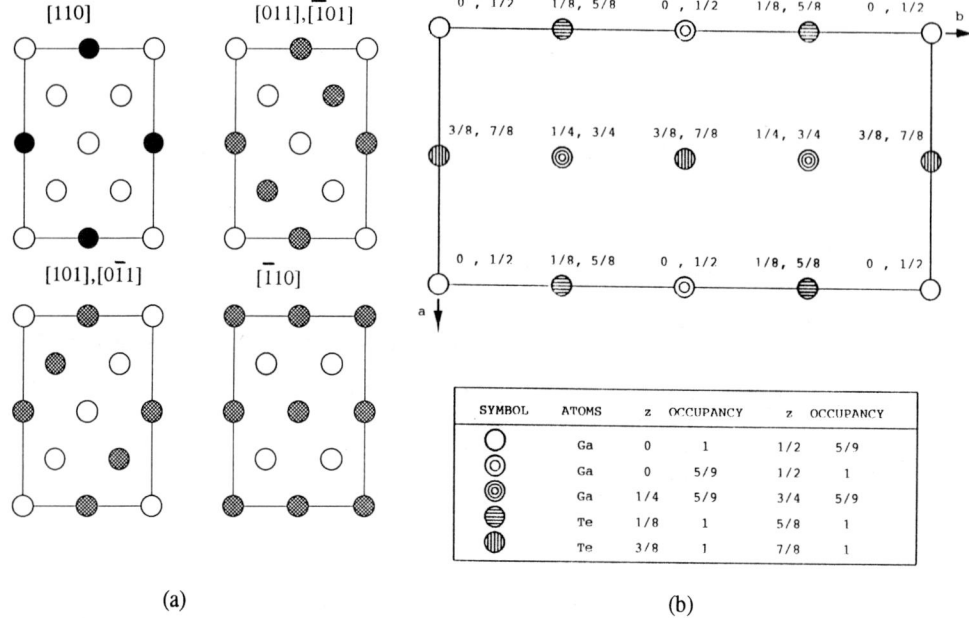

(a)   (b)

Fig 3 (a). Cation sublattice projections along different $<110>_s$. Solid circles represent fully occupied Ga sites, empty circles 5/9 occupied sites. Hatched circles indicate the alternate arrangement of these two positions. (b) C-axis projection of the unit cell. Atom types and coordinates are given in the table.

ACKNOWLEDGEMENTS

We are grateful to the Science and Engineering Research Council for their support of this work. One of us (M.-J.C.) thanks the Royal Society and CNRS for a Visiting Fellowship. We also thank the Materials Modelling Laboratory at the University of Oxford for provision of computing facilities.

REFERENCES

Duddles N J, Nicholls J E, Gregory T J, Hagston W E, Lunn B, and Ashenford D A 1992
  J. Vac. Sci. Technol. B, 10 912
Gunshor R L, Nurmikko A V, Kolodziejski L A, Kobayashi M, and Otsuka N, 1990
  J. Crystal Growth 101 14
Guymont M, Tomas A, and Guittard M 1992 Phil. Mag. A66 133
Halsall M P, Wolverson D, Davies J J, Lunn B, and Ashenford D A 1992 Appl. Phys. Lett. 60 2129
Lubbers D and Leute V 1982 J. Solid State Chem. 43 339
Wilke W G, Seedorf R, and Horn K 1990 J. Crystal Growth 101 620
Wright A C and Williams J O 1991 J. Crystal Growth 114 99
Wright A C, Williams J O, Krost A, Richter W and Zahn D R T 1992
  J. Crystal Growth 121 111
Zahn D R T, Mackey K J, Williams R H, Münder H, Geurts J and Richter W 1987
  Appl. Phys. Lett. 50 742

Inst. Phys. Conf. Ser. No 134: Section 1
Paper presented at Microsc. Semicond. Mater. Conf., Oxford, 5–8 April 1993

The characterization of delta doping by fresnel contrast analysis

R E Dunin-Borkowski*, W M Stobbs* and D D Perovic†

* Department of Materials Science and Metallurgy, Cambridge University, Cambridge CB2 3QZ
† Department of Metallurgy and Materials Science, University of Toronto, Ontario, Canada M5S 1A4

ABSTRACT: Quantitative information about a dopant profile is notoriously difficult to obtain for nanometre thick layers using any technique currently available. Here Fresnel Contrast Analysis is applied to the characterisation of a boron delta doped layer in silicon. It is demonstrated that the approach can be applied accurately despite the presence of absorption. For a well localised layer less than 1at% boron in silicon can be profiled using this technique.

1. INTRODUCTION

Layers of dopant atoms in a semiconducting material can be incorporated during growth as two dimensional sheets, nominally only a single atomic layer thick. Such 'delta doped' layers remain important as quasi 2D electron gas systems, both for device applications and for fundamental studies. The accurate modelling of electronic behaviour in such systems relies on a knowledge of layer compositions to monolayer accuracy but this is currently impossible using any of the established techniques. Compositional data obtained using SIMS and X-ray diffraction are usually, for example, either too qualitative or too indirectly related to the actual layer. Sub-nm layers are usually only visible using standard high-resolution and CTEM techniques where clustering of dopant atoms has occured or high concentrations of dopant have spread over many unit cells. However, it should be possible to obtain quantitative information about the dopant profile of layers containing concentrations of less than a tenth of an atomic layer (10^{13} atoms cm^{-2}) using Fresnel Contrast Analysis in the TEM. The absence of visible Fresnel contrast at a given imaging condition is also useful in the provision of a lower limit for the diffuseness of such a doped layer. As a demonstration of how the approach can be applied we assess here the diffuseness of a nominally delta doped boron layer in silicon. The lower limit of the concentrations in such layers which can be detected using the technique is also investigated.

2. RESULTS OF FRESNEL METHOD APPRAISAL OF BORON δ-DOPING

The delta doped layer examined here was grown on {001} silicon with a nominal boron sheet density of 1.6×10^{14} cm^{-2} (24 at%), as measured from the area under a SIMS profile. This is more than two orders of magnitude higher than the solubility limit at the growth temperature and corresponds to a change in the scattering potential relative to Si of $\Delta V = 1.7V$, assuming no change in the silicon lattice parameter. Boron atoms are expected to be fully substitutional for the low growth temperature of 480°C used. This rather high concentration should enable successful profiling even if the layer has diffused during growth over several unit cells. The experimental methods used and the general approach needed for the appraisal of Fresnel contrast have been described elsewhere (e.g. Ross and Stobbs 1991a). The Fresnel series examined comprised 13 plates and was taken using a JEOL 2000FX, a LaB$_6$ filament and a fairly large objective aperture (Airy disc radius 0.18nm). The beam convergence was 0.3mrad and the crystal was tilted to a systematic row 5.7° from {110}. Accurate defoci were fitted to the images using power spectra from the contamination at the specimen edge, and the defocus step size between images was calculated to be 208nm. Particular care was taken experimentally both to ensure that the edge-on specimens examined had the layering vertical and to maintain a low contamination level (which causes background phase contrast noise). Accordingly, the specimen surface was cleaned using 2kV Ar ions, and the Fresnel series was taken at an accelerating voltage of 100kV (below the threshold voltage of 145kV for knock-on damage in silicon). Full atomic multislice simulations were computed to simulate the experimental image series using consistent parameters with $C_s=2.3$mm, a focal spread of 30nm (the experimental energy spread is uncertain but makes little difference to the Fresnel contrast) and a sampling of 0.012nm/pixel.

Examples of images from the series are shown in Fig.1. The crystal thickness at the position of the boron layer was measured to be 100nm using weak-beam thickness fringes. Every second plate was digitised and the images were scaled to provide absolute electron intensities. The low signal to noise ratio necessitated careful projection of each image over a length of 50nm to minimise contributions from random noise. From the measured incident electron intensity, it is disturbing to note that 50% of electrons are lost in the image at the specimen edge despite the considerable care taken with specimen preparation and the minimisation of radiation damage in the microscope. The can only be attributed to ion damage and contamination caused during specimen preparation, but fortunately we can compensate for absorption caused by events outside the crystal by dividing the intensity profile near the interlayer by a linear background. Fig.2a shows the original profiles, divided by a linear background and projected over 50nm of the layering. Absorption due to the dopant is visible at zero defocus, indicating that contrast caused by electron diffraction out of the aperture and phase contrast effects are of comparable importance. Given the low atomic number of B relative to that of Si the greater absorption of B must be related to the diffuse scattering associated with local site irregularities and this is consistent with the way such layers exhibit bright contrast in High Angle Dark Field (HADF) images (e.g. Perovic at al 1991). Similar relative differences in absorption effects were seen for Fresnel contrast image series of GaAs / AlAs multilayers by Ross and Stobbs (1991b) who demonstrated that they could be dealt with successfully by dividing the entire series by the in-focus profile. A similar approach was used here, and the resultant profiles are shown in Fig. 2b. The more familiar form of Fresnel contrast is now evident. For a weakly diffracting orientation, as used here, we can ignore any asymmetry in the profiles when measuring an average profile and the symmetrised versions of 2a and 2b shown in Figs 2c and 2d were in the main used in the assessment of the doping profile. Small contrast features in these profiles are artefacts of the image processing, and we concentrate on comparing the more gross experimental and theoretical fringe contrast and spacing as defined in Fig.2d. Graphs of the experimental fringe contrast and spacing are shown in Fig.3 for the original profiles after they had been divided by the in-focus image. This division, taking account of differential absorption, leads to a shift of the minimum contrast back to zero defocus and it then remains to note the low experimental contrast values for which the approach can then be applied.

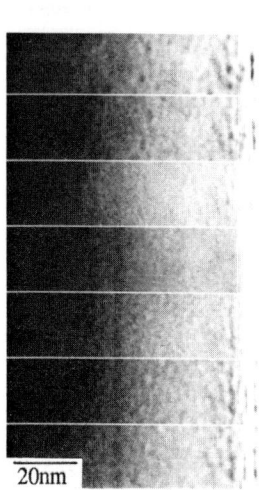

Fig.1 Images from a through-focal series of a boron delta doped layer in silicon at the defoci shown (nm).

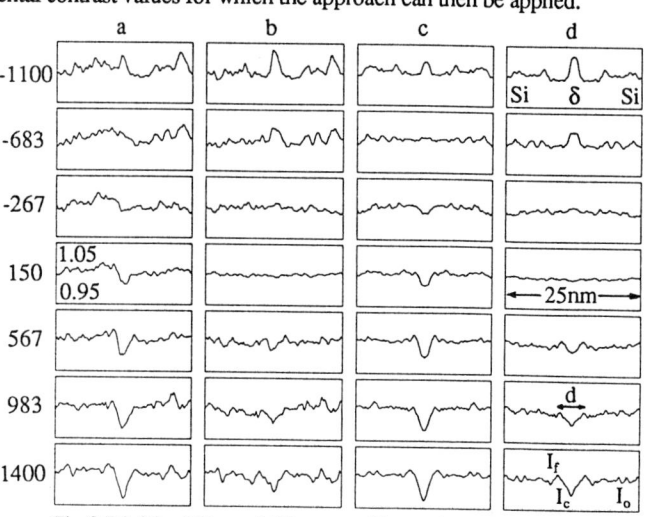

Fig.2 Digitised Fresnel fringe profiles at the defoci shown (nm) a) averaged over 50nm parallel to the layer; b) now divided by the in-focus image; c) and d) are symmetrised versions of a) and b). The symbols I_c, I_f, I_o and d are defined in d).

It is useful to remember that when matching simulations with experimental data it is a profile for the mean forward scattering potential that is being fitted, and this can be affected both by the compositional profile and by a change in the lattice parameter of the interlayer. Let us look firstly at the effect of changing the dopant profile (assuming it to be completely substitutional), retaining throughout the lattice parameter of the 'matrix' silicon. Figs.4a, b and c respectively show profiles for 100, 24 and 2.4 at% B confined to a single atomic sheet. The contrast values at 2.4 at% are considerably greater than that exhibited by our experimental data. This demonstrates that, if there is no change in the silicon lattice parameter associated with the presence of the dopant, then a true delta layer containing less than 2 at% boron could easily be profiled. If we now assume the integrated sheet density of 24 at% boron as measured using SIMS to be correct (though further confirmation using high-angle contrast would seem advisable), the effect of spreading the boron *evenly* over 8 and 20 atomic layers can be seen in Figs. 4d and 4e. The corresponding graphs of fringe contrast and spacing are plotted together with the experimental graphs on the same scale in Fig. 5. It is clear from the width of the best-fitting potential profile that the boron must have diffused over at least 2nm. The simulated contrast, however, is obviously too high, especially close to zero defocus. Neither interstitial atoms nor carbon or oxygen contamination can be responsible, as these would both increase the contrast.

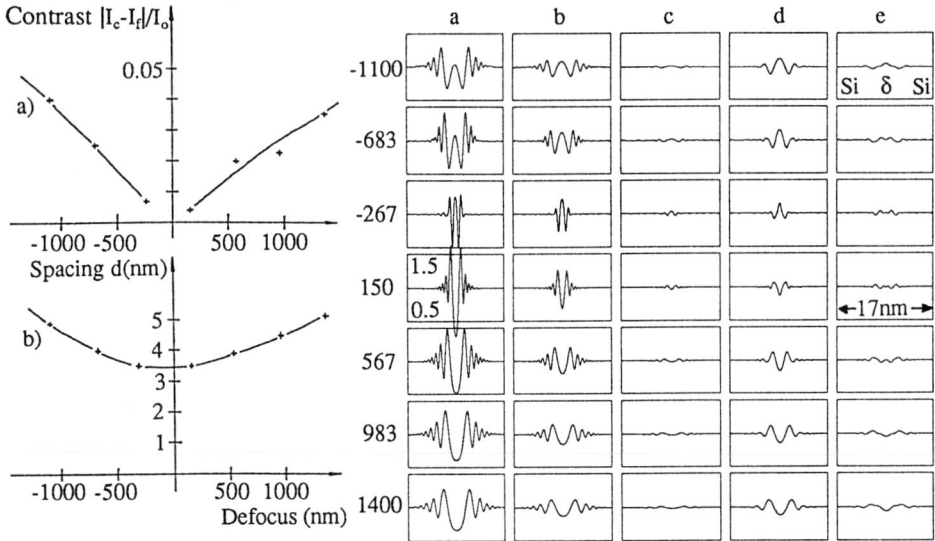

Fig.3 Experiment: a) first fringe contrast; b) first fringe spacing vs defocus for the profiles of Fig.2 d).

Fig.4 Simulations at the defoci shown (nm) and a crystal thickness of 100nm; a)100, b)24 and c)2.4at% boron in a single atomic layer; d) and e) show 24at% boron spread evenly from a sheet over 8 and 20 layers.

Retaining still the bulk silicon lattice parameter, we can next investigate the effect of 'diffuseness' on the contrast, incorporated here by spreading the boron in the form of a 'triangular' compositional profile. The profiles in Fig.6 show the effect of such triangular profiles with mean spreads of 20 and 16 atomic layers. It is clear that the latter profile is a better fit to the fringe spacing, and the contrast has the correct form as it approaches zero defocus even if it is still too high. Consequently we finally investigate the effect of changing the lattice parameter of the doped interlayer. Holloway and McCarthy (1993) have measured an experimental lattice contraction parameter for boron dopant in bulk silicon for lower boron contents, which we apply here by extrapolating their lattice parameter changes to the relevant higher B contents and by constraining the interlayer to match the lattice parameter of the silicon parallel to the interfaces, while retaining the predicted bulk volume. Examples of the contrast profiles then obtained are shown for an even compositional spread over 1 and 8 atomic layers in Figs. 7a and b respectively. Fascinatingly the predicted contrast has reversed for this full accommodation, and so no longer matches the sense of the experimental contrast. The explanation for this is clear when it is realised that the mean potential of boron in silicon when contracted to its experimental bulk volume is higher than that of silicon, and not lower as for any amount of boron on a substitutional lattice with silicon's planar spacing. Whatever the concentration of boron in our layer, it therefore cannot have retained the volume it would have as a bulk dopant. The effect of changing the lattice parameter by a smaller amount is also shown in Fig.6. A change in the lattice parameter of the interlayer by only 0.2% for a 24at% sheet spread evenly over 20 atomic layers is dramatic - the contrast has gone down by 30%. Although the contrast could be matched with any profile by changing the lattice parameter of the interlayer appropriately, we must be restricted here to match the fringe spacing to a profile that is spread over at least 2nm and this is not possible using the extrapolated lattice parameter changes based on the data used above. Consequently it would appear that either the lattice parameter reductions associated with the incorporation of B at the levels present is lower than expected or that the B concentration present is less than the SIMS data suggest. The former possibility is being investigated using convergent beam measurements of the tetragonality and the latter by the use of HADF imaging.

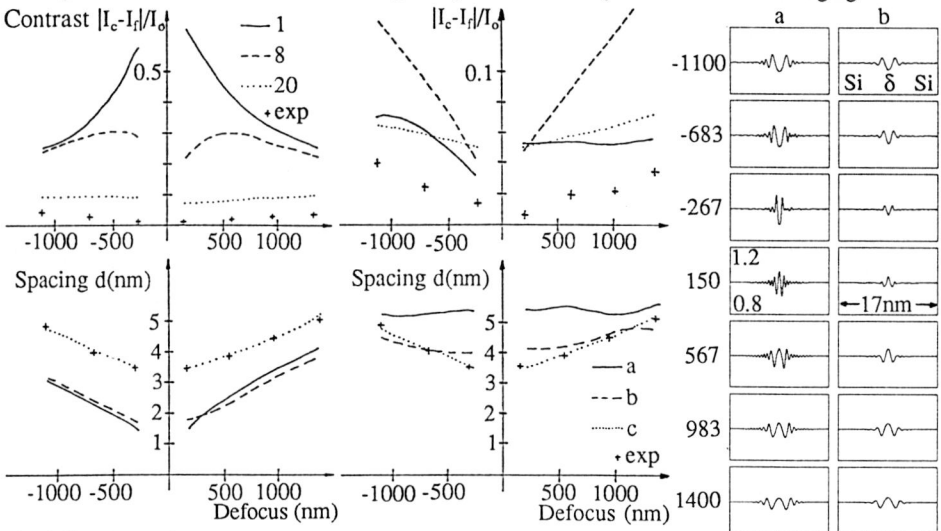

Fig.5 Simulated fringe contrast and spacing for 24at% boron (sheet density) spread evenly over 1, 8 and 20 atomic layers.

Fig.6 Diffuse profiles with mean spreads of a) 20 and b) 16 layers; c) interlayer plane spacing 0.2% ↓ for an even spread over 20 layers.

Fig.7 Even spread of B over a) 1 and b) 8 layers, retaining the predicted bulk volume.

In conclusion:

1. For this specimen, whatever the absolute boron concentration, it has diffused on average over about 2nm and has the form of a diffuse profile. If we believe that we started with a sheet density of 24at%, then the lattice must have contracted, but by much less than for the equivalent concentration in the bulk. In any case we here demonstrate that profiling concentrations of the order of 1at% boron or less is possible using the method. Significantly, the technique still works in the presence of absorption, which has not precluded an accurate characterisation.

2. In general, the Fresnel Method should be applicable to true, sub-nm delta doped layers with concentrations of 1at% or less.

3. In principle, the technique could therefore be used for measuring the diffusion coefficient of MBE-deposited boron in defect free material; measurements are normally made on implanted layers for which the diffusion is affected by implantation damage.

We thank Prof CJ Humphreys for provision of laboratory facilities, the SERC and GEC for financial support, Dr CB Boothroyd and Dr PD Brown for discussion, and Dr R A Kubiak and Prof EHC Parker at Warwick University for the Si:B sample examined here.

REFERENCES

Holloway H and McCarthy S L 1993 J. Appl. Phys. 73 103
Perovic D D, Weatherly G C, Egerton R F, Houghton D C and Jackman T E 1991 Phil. Mag. A63 757
Ross F M and Stobbs W M 1991a Phil. Mag. A63 37
Ross F M and Stobbs W M 1991b Ultramicroscopy 36 331

Inst. Phys. Conf. Ser. No 134: Section 1
Paper presented at Microsc. Semicond. Mater. Conf., Oxford, 5–8 April 1993

Structure refinement of the atomic model of the {113} defect in Si

S Takeda and M Kohyama*

Department of Physics, College of General Education, Osaka University, Toyonaka, Osaka 560, Japan
*Glass and Ceramic Material Department, Government Industrial Research Institute, Osaka, 1-8-31 Midorigaoka, Ikeda, Osaka 563, Japan

ABSTRACT: Structure refinement of the atomic model of the {113} defects which was recently proposed from HRTEM observations has been carried out. Electron diffraction intensity and HRTEM images have been simulated based on a *relaxed* atomic model. The *relaxed* coordinates of atoms have been determined with the energy minimization procedure by utilizing the Stillinger-Weber potential. Simulated electron diffraction intensities agreed with observations and simulated HRTEM images have shown improved matching with the observed ones.

1. INTRODUCTION

The {113} defect is a well-known secondary defect in Si and Ge, which is an aggregate of supersaturated self-interstitials (Ferreira Lima and Howie 1976, Salisbury and Loretto 1979). The defects are introduced by ion-implantation, electron irradiation or thermal annealing (for example, Bourret 1987). The atomic model of the defects which was recently proposed from HRTEM study showed that the lattice is reconstructed in the interior of a Si or Ge crystal involving supersaturated interstitials (Takeda 1991). The model has been evaluated by the energy calculation utilizing the Stillinger and Weber type potential (Kohyama and Takeda 1992). The result has indicated that the estimated increase of energy per interstitial atom in the model is less than 0.9eV. This value is apparently smaller than that estimated for an isolated interstitial and, hence, it is considered that the stability of the model is well understood theoretically.

In the course of the energy calculation, the atomic positions were slightly shifted in order to attain the energy minimum, with the atomic coordination in the original model unchanged. This lattice relaxation procedure gave the energetically favorable atomic positions in the framework of the empirical potential. With the results, the electron diffraction intensity from the defect is simulated and compared with experiment in this work. The HRTEM images are also re-simulated based on the relaxed model, and show better agreement with the observed images.

2. ATOMIC MODEL

The atomic model was already described in the previous paper (Takeda 1991). Here, let us describe the model briefly. It is constructed with two kinds of structural units. Both units have the periodicity in the <110> direction, which is parallel to the anisotropic direction of the defect of peculiar rod-like morphology. One unit is incorporated with chains of additional atoms (aggregated self-interstitials) in the <110> direction, belonging to the 5-, 6- and 7-membered rings. The unit will be called the I unit. The 6 membered-ring in the unit forms a different atomic arrangement from the diamond structure, but is a part of the Wurtzite lattice. The other unit called the O unit consists of an unusual 8-membered ring. The O unit is obtained by removing the chains of additional atoms from the I unit and by subsequent small displacement of atoms.

3. REFINEMENT

3.1 Lattice Relaxation Procedure

Detailed results of the energy calculation can be seen in the previous paper (Kohyama et al. 1992). Stability of the unusual 8-membered ring was quantitatively explained, and several interesting phenomena concerning the defect were considered theoretically.

Fig. 1(a) shows a part of the original model. An unrelaxed structure as the initial structure before lattice relaxation treatment is depicted in Fig. 1(b). Notice the symmetry of the initial structure is different from the original model in Fig. 1(a). This will be discussed in the next section. Fig. 1(c) shows the relaxed configuration after the lattice relaxation treatment, in which the symmetry of the original model is well recovered, and the large distortion of length and angle of the covalent bonds particularly at the 8-membered ring in Fig.1 (a) are removed.

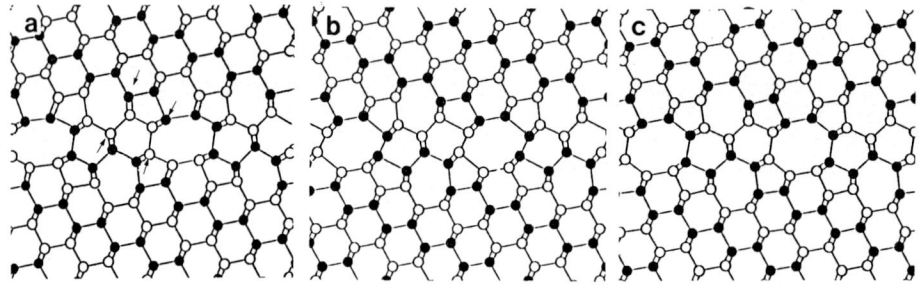

Fig. 1 (a) A part of the original atomic model . (b) The initial structure before lattice relaxation treatment. (c) The energetically favorable structure after the treatment.

3.2 Electron Diffraction Intensity from the Defect

Fig. 2(a) and (b) show a set of weak beam images of the same region with different operating g-vectors. The defects were induced by 2MeV electron irradiation for the experimental conditions in the previous study (Takeda 1991). The specimen was thinned after electron irradiation as in the previous paper (Takeda, Muto and Hirata 1990). Only the defect indicated by the arrow could be located in a selected-area aperture and exhibited electron diffraction. The defect became invisible near the plan view geometry as in Fig. 2(b), but it was possible to set the aperture by inspecting the neighboring defects. The plan view pattern from this isolated (113) defect is shown in Fig. 2(c). It is very obvious that the pattern exhibits several faint extra spots in addition to the intense Bragg spots resulting from a perfect crystal. The location of the extra spots in this particular diffraction pattern can be interpreted as the intersections of the Ewald shpere and the reciprocal rods originating from the two-dimensional lattice assumed on a (113) plane. It should be stressed that the diffraction pattern from the {113} defects usually exhibits diffuse scattering in the <332> direction, which results from the non-periodic arrangement of the two kinds of structural units as observed in the HRTEM images. The patterns in Fig. 2 were, however, carefully analyzed because it is rather simple to simulate electron diffraction intensity of a periodic object based on an atomic model. The translational vectors on (113) in order to explain the pattern are expressed as,

$$A=a/2-b/2 \quad \text{and} \quad B=-2a-5/2b+3/2c,$$

in which a, b and c are the translation vectors of a perfect Si crystal. The translational vectors is the same as those of the periodic I IIOI structure (Kohyama et al. 1992) , and the sequence of the

IIOIIO.... is one of the dominant sequence as observed in the HRTEM images. Electron diffraction patterns was simulated based on the *relaxed* atomic positions of the I IIO I periodic structure. The simulated pattern is shown in Fig 2(e). With the aid of the arrows indicating the corresponding positions between the computed and observed pattern, it is confirmed that the computed intensity distribution reproduces the experiment well.

The pattern taken with the [001] incidence of the same defect is shown in Fig. 2(d), in which the interesting extinction rule was found. This can be explained if a pair of atoms is located at the

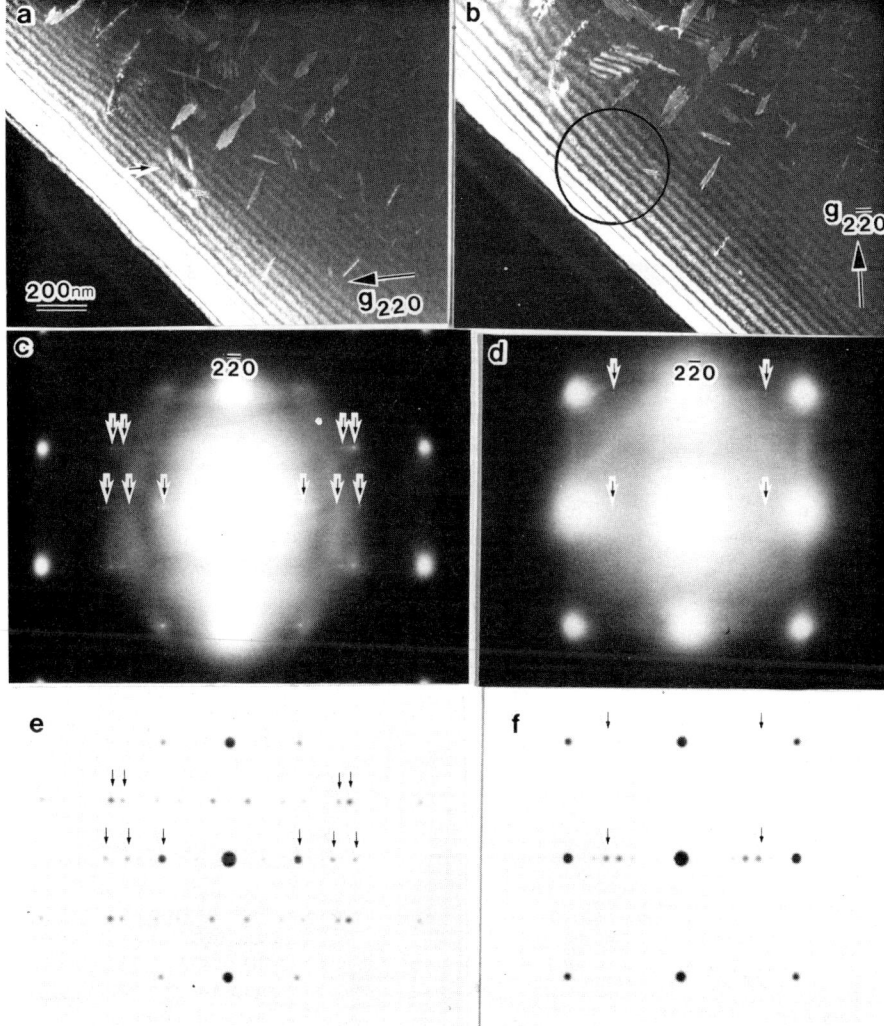

Fig. 2 (a), (b) Weak beam images of the same area with diffrent operating *g*-vectors. The image in (b) was taken near the plan-view geometry. The circle in (b) indicates the selected area aperture. (c) Planview electron diffraction from the defect indicated by the arrow in (a). (d) The [001] pattern from the same defect. Notice the systematic absence of the rows of diffraction spots. (e) The simulated planview pattern based on the relaxed configuration of the model. The arrows in the patterns indicate the corresponding positions. (f) The simulated [001] pattern.

positions

$$xA+yB+zc \quad \text{and} \quad (x+1/2)A+ yB+ z'c$$

Pairs of atoms satisfying the relationship above are indicated by the arrows in Fig. 1(a).   Note the relation is violated in Fig. 1(b), but recovers after lattice relaxation in Fig. 1(c). Fig. 2(f) shows the simulated pattern exhibiting the extinction.

3.3 HRTEM Images

It is interesting how HRTEM images are affected by lattice relaxation treatment. HRTEM images were re-simulated based on several hypothetical periodic structures after the treatment. It is assumed that the model is periodic in the <332> direction as well as in the <110> direction, and is sandwiched by vacuum layers in the <113> direction.   The parameters for image simulation were the same as the previous ones (Takeda 1991). Fig. 3 depicts the simulated images with the atomic positions before and after lattice relaxation (shown in Fig. 1(a) and (c) , respectively).   The basic features of the image are not affected, but the image matching seems better particularly at the large white patch in the image resulting from the eight-membered ring.

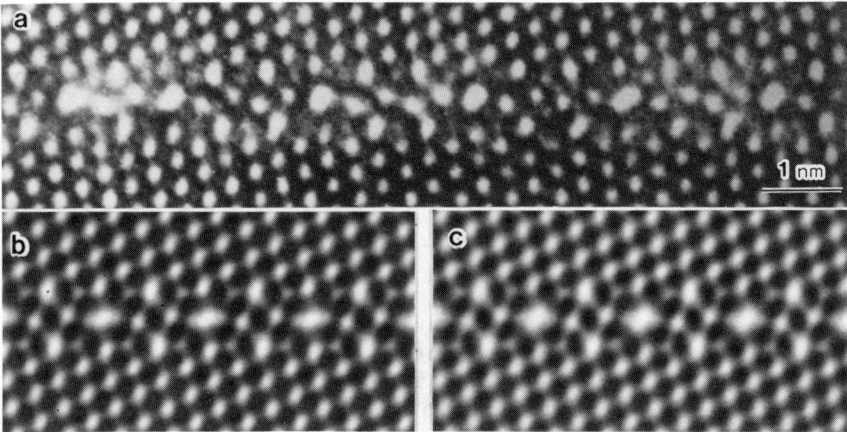

Fig. 3   Comparison of the observed and simulated HRTEM images before and after lattice relaxation treatment. (a) Observed.   Simulated image of the original model in (b) and after the lattice relaxation treatment in (c).

4. CONCLUSION

A refined atomic model of the {113} defects has been obtained.   Intensity distribution of electron diffraction has been simulated based on the refined atomic positions, and has agreed well with the observation from an isolated {113} defect in Si. The HRTEM images which have been re-simulated with the refined model have shown better matching with observations.

REFERENCES

Bourret A 1987 Inst. Conf. Ser. **87** 39
Ferreira Lima C A and Howie A 1976 Philos. Mag. **34** 1057
Kohyama M and Takeda S 1992 Phys. Rev. **B46** 12305
Salisbury I G and Loretto M H 1979 Philos. Mag. **A39** 317
Takeda S 1991 Jpn. J. Appl. Phys. **30** L639
Takeda S, Muto S and Hirata M 1990 Jpn. J. Appl. Phys. **29** L1698

Inst. Phys. Conf. Ser. No 134: Section 1
Paper presented at Microsc. Semicond. Mater. Conf., Oxford, 5–8 April 1993

Early stages of rod-like defect formation in silicon

P Werner, M Reiche and J Heydenreich

Max-Planck-Institut für Mikrostrukturphysik, Weinberg 2, 0 - 4050 Halle/S, F R Germany

ABSTRACT : The early stages of rod-like defect (RLD) formation in FZ silicon caused by 400kV electron irradiation is investigated by high-resolution electron microscopy (HREM). Previously, the crystals were ion implanted (B^+,Sb^+,Si^+) to study the reaction with different point defects. Starting their growth on {113} planes different agglomerates of ≤3nm in size were identified. The RLDs are mainly a form of precipitate of self-interstitials; however, their early stages and further growth are influenced by the existing point defect complexes in the Si bulk.

1. INTRODUCTION

Rod-like defects (RLDs) are a typical feature of the agglomeration of interstitial atoms in silicon. Their morphology is characterized by a plate-like shape on {113} planes (zig-zag shapes as well as simple plates may occur) and by an elongation in <110> direction. RLDs have been intensively investigated by different analytical techniques such as TEM during the last decade. While at first structures of RLDs were interpreted as precipitates of oxygen, nitrogen or metallic impurities, nowadays strong arguments confirm the agglomeration of self-interstitials. Different structure models of the RLDs are being discussed in the literature. Salisbury and Loretto (1979) discussed the segregation of di-self-interstitials with dangling bonds; Tan et al.(1981) proposed an atomic model of fully bonded self-interstitial atoms. Essential efforts in characterizing the atomic structure of RLDs have been made by interpretation of HREM images. Corresponding investigations by Bourret (1987), and Bender and Vanhellemont (1988) confirmed the possibility for self-interstitials to precipitate in the form of hexagonal silicon on {113} planes. Recently, Takeda et al (1992) proposed a defect structure also including 8-membered rings.

However, numerous investigations have also shown that the formation of RLDs, their density and morphology are strongly influenced by parameters of the silicon bulk, e.g. by the kind and content of impurities and dopants, as described by Aseev et al.(1992), and by the thermal treatment of the crystal. It can be supposed that this different behaviour is influenced by the initial stage of agglomeration. The electrical and crystallographic structures of such complexes influence their further growth leading to RLDs or types of dislocation dipoles. A large variety of complexes are identified: self-interstitial/impurity atom (C, B, Al) complexes, vacancy- impurity complexes, di-vacancies. Their supposed structure is characterized by different crystallographic symmetries (e.g. C_{2v}, C_{2h}), which influence further agglomeration.

Electron irradiation (E> 200keV) at room temperature generates intrinsic defects (self-interstitial/vacancy complexes), which are mostly trapped by impurities as reported by Newman (1973). The further growth of such metastable complexes is influenced by parameters such as irradiation temperature and dose, their different electrical states in p-doped and n-doped material, respectively. If the defects reach a larger size($\emptyset \geq 3$nm) owing to a further interstitial agglomeration the defects can be analysed by TEM imaging techniques. Their crystal structure can be investigated by HREM along the <110> directions.

It is the aim of this paper to investigate the early stages of the agglomeration of self-interstitials in p- and n-doped material during electron irradiation (400keV) by HREM. At room temperature this irradiation creates interstitials and vacancies, which can be trapped by impurities or dopants. To avoid the complex situation of the RLD formation in connection with the oxygen precipitation in thermally annealed Czochralski grown silicon (see e.g. Bourret (1986), the experiments were carried out on FZ crystals.

2. EXPERIMENTAL

The following FZ silicon crystals have been investigated: Sb-doped and B-doped [001] material (electrical resistivity 20-30Ωcm).Furthermore, dopant profiles were created in some crystals by means of ion implantation (B$^+$, Si$^+$, E= 100keV, doses 5x10^{13} and 5x10^{14}cm^{-2}, respectively) to study the dependence of the RLD structure on the dopant content . After implantation the specimens were thermally annealed at temperatures of 450, 550 and 800°C, for 30 min and 1 h., respectively, to ensure the formation of small self-interstitial aggregates of high density. Electron irradiation (intensity $\approx$5x10^{20} e$^-$cm^{-2}s^{-1}) has been carried out in a JEM 4000 EX microscope at room temperature.

For the TEM investigations plan-view as well as cross-sectional samples were prepared using the conventional techniques. For the final thinning the specimens were etched by ion (Ar$^+$)-milling and furthermore chemically polished. TEM studies of the specimens were carried out in a JEM100C microscope (100kV, resolution ~3Å) and afterwards in a JEM 4000EX (400kV, resolution ~1.9Å). Image processing techniques were applied to analyse the HREM micrographs.

3. RESULTS AND DISCUSSION

In the B-doped as well as Sb-doped crystals RLDs were generated after a special incubation time depending on the irradiation density and dopant concentration. The small defects grew to a width of about 20nm. In the higher-doped p-material a defect concentration of about 10^{14}cm^{-3} was estimated, whereas in the lower n-doped crystals the defect concentration generated was one order of magnitude lower.

In specimens of p-doped crystals, which were Si$^+$ implanted at room temperature and thermally annealed RLDs were generated immediately after a short electron irradiation within a surface layer of about 1 μm. Fig.1 shows a cross-section image of such a region. The defects showed a high concentration, which was estimated to about 10^{15}cm^{-2}. They had been growing relatively quickly on {113} planes reaching a width of about 10 to 15 nm.

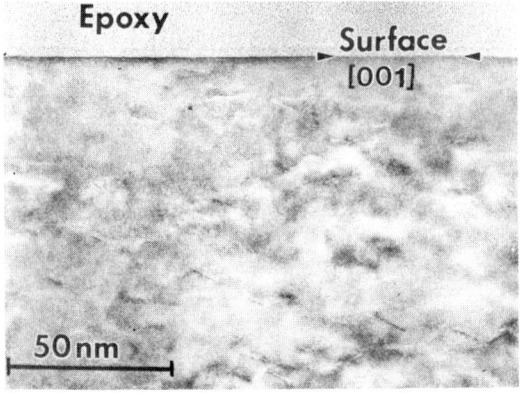

Fig.1 Cross-section micrograph of the surface region in a Si$^+$ implanted [001] Si crystal. The rod-like defects, which are visible in their projection along <110>, are generated during irradiation with 400kV electrons in TEM at room temperature.

In its initial stage the RLDs observable by HREM, had a width of about 1nm along the {113} planes. Comparing their image contrast with numerical contrast simulations revealed that in <110> directions they were more than 10nm in length. Fig.2 presents some kinds of early stages of RLDs. The contrast feature in a) can be interpreted by the agglomeration of two chains of di-interstitials. As proposed by Salisbury and Loretto (1979) at free tetrahedral sites these dimers have dangling bonds, and the Si lattice is not reconstructed. Such a lattice structure has often been observed in the Sb-doped crystals. The further growth during electron irradiation creates defect structures as demonstrated in Fig.3 a) and b). The inserted simulated image c) supports the assumption of an agglomeration of di-interstitials without the reconstruction of the crystal lattice. The defect in Fig.2b shows two symmetrical segments consisting of 5 and 7-membered rings marked by arrows. This interpretation can also be derived from contrast simulations. Groups of small agglomerates were also found. Their further growth, however, has seldom been observed.

In Sb-doped crystals V-shaped defects often grow, which are characterized by a mirror symmetry on a {100} plane and a small Burgers vector in <100> direction. The contrast analysis suggests a reconstructed fault plane of 5-7-membered Si rings.

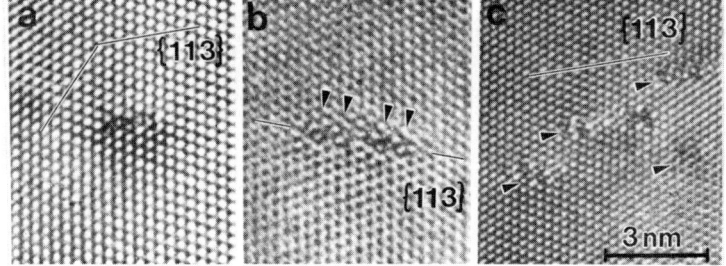

Fig.2 HREM micrographs presenting early stages of defect formation on {113} planes generated during electron irradiation. Viewing direction: [110]. a) Agglomerate of di-interstitials in Sb-doped Si, b) initzial growth along {223} direction, Sb-doped material, specimen thickness ≈30nm, c) group of agglomerates in B-doped Si

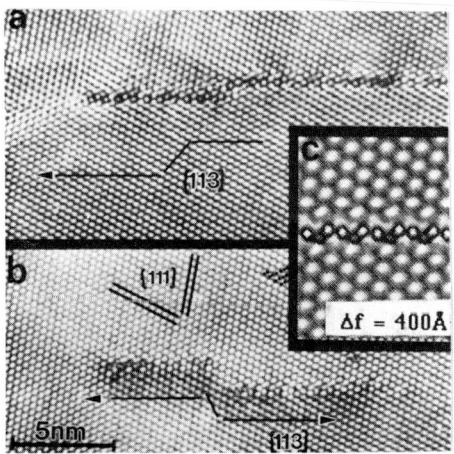

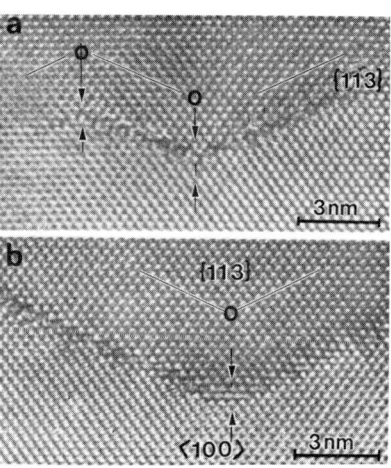

Fig.3 a) and b) RDLs in Sb-doped Si, having grown after e⁻-irradiation. c) Computer-simulated image of a RLD of non-reconstructed di-interstitials.

Fig.4 V-shaped RLDs started to grow from the central position marked by O. Viewing direction: <110>. Mirror plane {Ī10} = <100> direction.

This kind of RLD seems to start to grow in the position of the mirror plane. In Fig.4a the central structure is estimated as a single complex of a 5-7 ring with a mirror symmetry.

Assuming that the {113} growth of the RLD's starts at small agglomerates their structures and symmetries should be influenced by them. Therefore we investigated the structure contrast of some point defect agglomerates and the possibility of detecting them by HREM. Different defect types proposed in the literature, e.g. by Tan, Föll , Mader and Krakow (1981), were tested by computer simulation. Some examples of calculated crystal lattice images are presented in Fig.5. a) and b) showing images of two types of an extrinsic 90° edge dislocation dipole. An agglomeration of vacancies is given in c) displaying an intrinsic edge dipole. Defects in a) and c) are characterized by a (110) mirror plane symmetry (C2v), whereas the defect in b) has an inversion axis. Point defects , such as in a) and c), might initiate the growth of V-shaped RLDs. While HREM images suggest the existence of cases a) and b), there is no experimental proof for case c), up to now.

4. CONCLUSION

The above examples of the early stages of RLD formation should support the assumption that the growth of these defects, their morphology as well as their density is strongly influenced by existing point defects, e.g. dopants. The electron irradiation of Si at room temperature creates self-interstitial -vacancy complexes, which may be trapped by impurities or small agglomerates (Ø<1nm) of them. The electrical and crystallographic properties of them strongly influence the further kind of growth mostly on {113) planes.

The RDLs themselves represent a specific form of precipitation of self-interstitials (dimers, 5 and 7-membered rings or hexagonal structure). HREM is helpful in analysing such defects. In future, investigations have to be directly combined with DLTS and EPR measurements.

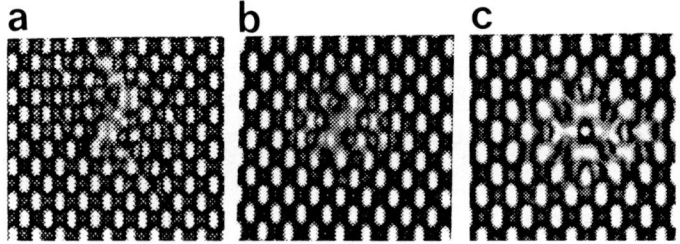

Fig. 5 Computer simulated HREM images of point defect agglomerates seen in <110> orientation. Crystal thickness: 25nm, objective defocus: 20 nm.

REFERENCES

Aseev A L .et al. 1989 Gettering and Defect Eng. in Semicond. Mat., ed. M Kittler, (Vaduz:Sci. Tech. Publ.) p. 235
Bender H. Vanhellemont J 1988, phys.stat.sol. a107, 455
Bourret A 1986 Inst. Phys. Conf.Ser. No. 59 p.129
Bourret A 1987 Inst.Phys.Conf.Ser. No. 87 p.39
Newman R C 1973 Infra-Red Studies of Crystal Defects,(London: Taylor & Francis Ltd) p.88
Salisbury I G. Loretto M H 1979 Phil. Mag. A39, 317
Takeda S. Muto S. Hirata M 1992, Mat. Sc. Forum 83-87, 309
Tan T Y. Föll H. Mader S. Krakow W 1981, Defects in Semiconductors, eds J Narayan and T Y Tan (New York: Noth Holland Publ.) p.179

Inst. Phys. Conf. Ser. No 134: Section 1
Paper presented at Microsc. Semicond. Mater. Conf., Oxford, 5–8 April 1993

41

{113} Defects in He$^+$ implanted germanium

J L Hutchison, A L Aseev* and L I Fedina*

Department of Materials, University of Oxford, Parks Road, Oxford OX1 3PH, UK
*Institute of Semiconductor Physics, Russian Academy of Sciences, Siberian Branch,
630090 Novosibirsk, Russia

ABSTRACT: HREM of Ge crystals implanted with He$^+$ ions at 1.23 Me is used to investigate the implantation damage structures. They are shown to consist of "voids" - probably He bubbles - and {113} defects. A model is postulated for the {311} defects which incorporates hexagonal Ge and 8-membered rings. A growth mechanism for the formation of {113} defects is proposed and compared with earlier data for Si and Ge.

1. INTRODUCTION

Various atomic models for {113} defects in Ge and Si have been postulated, (Salisbury and Lorretto 1979, Tan 1981, Bourret et al 1984, Bender 1984). Most recent investigations of "rod-like defects" (RLDs) involving zig-zag, narrow {113} defects in ion implanted (Bourret 1987; Hutchison 1990) and annealed silicon (Bender and Vanhelemont 1988) suggest a hexagonal arrangement of self-interstitials within these defects. Takeda (1991) and Takeda et al (1991) reported a hexagonal structure of more extended {113} defects in electron-irradiated silicon and germanium. In these cases the main problem is the origin of the {113} habit plane, since self-interstitials in a hexagonal arrangement are equivalent to a stacking fault with {111} habit plane (Hornstra 1958).

We report here an investigation of {113} and other defects in He$^+$ implanted Ge crystals (Aseev 1983). Since high energy He$^+$ ions form a damage layer some distance from the entrance surface of an irradiated crystal there is less influence of impurity atoms diffusing inwards from the surface during ion implantation, or during *in-situ*, high-voltage electron irradiation of thin crystals (Matthews and Ashby 1973; Oshima and Fujita 1986).

2. EXPERIMENTAL

n- and p-type Ge crystals with resistivity of 1 and 5 Ωcm were irradiated at room temperature by 1.2 Me He ions (current density 1.3 μAcm^{-2} and dose 10^{16} cm^{-2}). Plan-view and cross-sectional TEM and HREM specimens were prepared so that the implanted layer was revealed. Diffraction contrast and phase contrast of defects was carried out at 100 keV; HREM imaging was performed at 400 keV using a JEOL 4000EX(II).

3. RESULTS

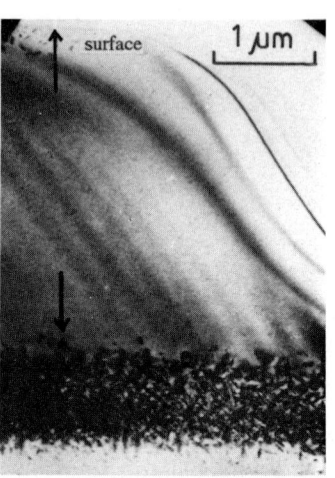

Fig. 1. TEM image of cross-sectioned specimen of helium implanted germanium crystal. Energy of ions: 1.2 Me; irradiation dose: 10^{16} cm^{-2}. Implanted surface and maximum of $F_d(x)$ (depth distribution of helium ions) are indicated by arrows.

Fig. 1 shows an example of a TEM image of a cross-sectional Ge specimen following He+ implantation. Arrows show narrow, rod-like defects (RLDs) and small, planar defects at a relatively low density, in the region of the implanted surface, and at a much higher density at depths corresponding approximately to that for maximum ion energy loss due to elastic collisions $F_d(x)$ and the mean projective length for helium ions R. At the same time an extensive region between this depth and the surface is seen to contain little or no damage. Out-of-focus phase contrast TEM revealed another type of defect in the area containing the high density of RLDs and other planar defects. Their diameters vary between 10 and 15 nm and they usually coincide with the positions of the RLDs and other defects. The spatial distribution of the voids, RLDs and planar defects is shown in Fig. 2, along with the depth distribution of $F_d(x)$ and the concentration of helium atoms n(x), (Burenkov et al., 1985).

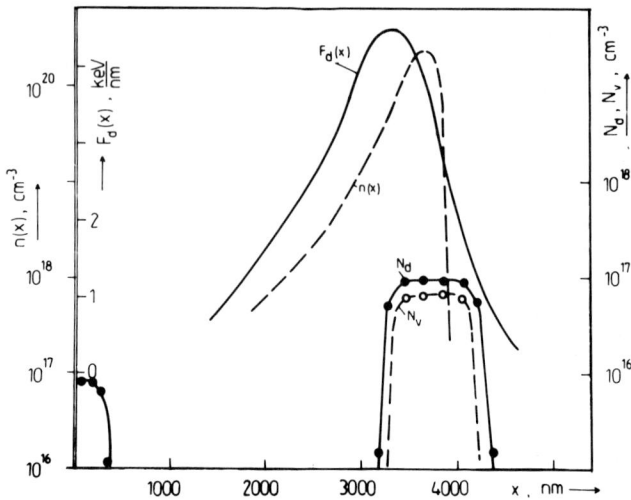

Fig. 2. Depth distribution of concentration of helium ions n(x) and elastic collision energy losses $F_d(x)$ for helium ions with 1.2 Me energy and dose of 10^{16}cm^{-2}. Results of measurement of density of "voids" N_v and interstitial clusters N_d are shown.

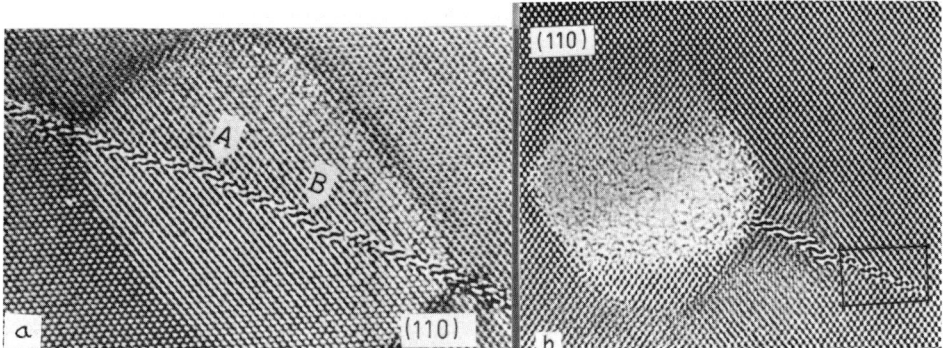

Fig. 3a. HREM image of a planar defect and a "void" in a {110} oriented germanium
crystal. The changing of the habit plane in areas A and B is indicated.
  3b. {113} defect and "void" in helium implanted germanium. Foil thickness for crystal
area in outlined area is less than 10 nm.

Typical HREM images of both types of damage structures are illustrated in Figs. 3 (a
and b). In Fig. 3a an extended defect with {113} habit plane and also other habit planes is
clearly seen. From the displacement of the lattice on either side of the defects, (i.e.
approximately $\frac{1}{4}a_o$, 0.14 nm) it is concluded that it is an interstitial-type {113} defect. We
note that many of the defects observed appear to be in contact with voids. In Fig. 3b the
projected shape of the "void" is clearly seen to correspond to a truncated octahedron, and the
lighter overall contrast combined with the perforation of the specimen are both evidence that
these objects indeed correspond to voids.

The HREM image of a {113} defect in contact with a void, shown in Fig. 3b, was
obtained at the thinnest part of the specimen, where the specimen may be regarded
approximately as a weak phase object; for images obtained at Scherzer defocus dark spots
in the lattice image may be interpreted in terms of projected pairs of closely-spaced atomic
columns. This allows an intuitive interpretation of the experimental image of the {113}
defect in Fig. 3b. The image contrast for the defect in the immediate vicinity of the void is
disrupted, due to slight lattice distortion. The postulated atomic structure of the area outlined
in Fig. 3b containing a {113} defect is presented in Fig. 4. It is possible to recognise
segments of hexagonal ring structure labelled "H" in the defect region, along with five- and
seven-membered rings connecting these hexagonal fragments to the host lattice. Hexagonal
rings are separated aperiodically by eight-membered rings.

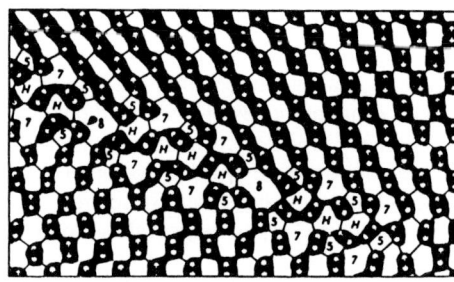

Fig. 4. Postulated atomic structure of the
damaged area imaged in Fig. 3b. The
positions of the atomic columns and inter
atomic bonds are shown. Five-, seven- and
eight-membered rings as well as hexagonal
rings (H) are indicated. This is the first
observation of hexagonal structure in a {113}
defect in an ion-implanted germanium crystal.
The arrangement of the contrast in images of
these defects in thicker parts of the specimen
is also consistent with hexagonal structure.

4. DISCUSSION

4.1 Damage in He+ implanted germanium

From TEM images of cross-sectional specimens, a non-uniform distribution of damage is clearly seen (Fig. 1) The density of voids and interstitial clusters in the form of rod-like and planar defects is maximum near the depths of maximum ion energy loss due to elastic collisions and the maximum of the helium ions distribution n(k). This fact, together with the very strong dependence of the defect density on depth (Fig. 2) indicates the existence of many-atom interactions during defect formation. The generation of a large number of atom displacements occurs at the depth of the maximum energy losses of the helium ions due to elastic collisions (for a discussion of this problem see e.g. Mayer, Ericsson and Davis (1979). In this way the appearance of RLDs and planar defects close to the entrance surface for the He$^+$ ions (Fig. 1) is not associated with any peculiarities in the spatial distributions of n(x) and F_d(x). According to the predictions of Matthews and Ashby (1973) and Hasebe et al. (1986) this phenomenon is due to the effect of impurity atoms in the native oxide layer being forced into the lattice by the helium ions. The correlation between n(x) and N_v(x) gives rise to the conclusion that the "voids" probably contain He atoms; i.e. the "voids" are in fact He bubbles.

The other feature of helium implanted Ge is associated with the large sizes of voids, and the interstitial clusters incorporating significant numbers of point defects, as many as 10^3, at the relatively low irradiation temperature not exceeding 40°C. We assume this to be due to the effective separation of vacancies and interstitials which takes place because of the energy barriers for mutual annihilation of point defects. The existence of these barriers for Si crystals has been proposed previously in papers by Antoniadis and Moskowitz (1982) and Gosele, Frank and Seeger (1983). The formation of the observed defects at the above-mentioned temperature confirms also the conclusion about the high mobility of point defects in irradiated silicon as well as for germanium, reported previously by Watkins (1975).

4.2 Atomic structure of {113} defects

The hexagonal structure of {113} defects in He$^+$ implanted germanium is consistent with experimental HREM data for implanted and annealed silicon, described in papers by Bourret (1987), Bender and Vanhellemont (1988) and Hutchison (1990). The structure shown in Figures 6 and 7, with fragments of hexagonal structure separated by eight-membered rings is very similar to the one proposed by Takeda (1991) for electron-irradiated silicon and Takeda et al. (1991) for electron-irradiated germanium. The presence of five- and seven-membered rings in addition to six-membered atomic rings provides the {113} habit plane, and differentiates the {113} defect from both intrinsic and extrinsic {111} stacking faults with regular arrangements of six-membered rings. According to the results of Bourret (1987) and Bender and Vanhellemont (1988) the formation of a hexagonal structure within {113} defects occurs due to a decrease in the energy associated with dangling bonds or distorted bonds in the defect region, and probably the model for formation of {113} defects includes clustering of split-interstitials in the manner proposed by Tan (1981) followed by further evolution with relaxation of deformed atomic bonds during reconstruction to the hexagonal structure.

The extending of a strictly planar hexagonal structure on {113}, as suggested by Bourret (1987), leads to an increase in strain in the defect area. From this point of view the formation of eight-membered atom rings decreases the strain and provides a planar {113} habit plane. The irregular disposition of the eight-membered rings in the defect plane, as observed experimentally in the present work, demonstrates the step-wise relaxation of elastic strain during the growth of a {113} defect in the <332> direction in the habit plane; this is accompanied by the transformation of metastable configurations of interstitial to the hexagonal structure and confirms the evolutionary model of formation of {113} defects. From this point of view the more regular arrangement of eight-membered rings observed by Takeda (1991) and Takeda et al. (1991) in electron irradiated silicon and germanium is probably due to the higher irradiation temperatures (400 and 300°C respectively), compared to the implantation temperature used in the current work. The formation of eight-membered rings leads to a decrease in the density of interstitials compared with a fully occupied {113} atomic plane, as has been pointed out previously by Takeda (1991).

To confirm the evolutionary model for the transformation of hexagonal structure we have compared the energy of {113} defects with various atomic structures. For example the energy of the defect with atomic structure corresponding to the model of Salisbury and Lorretto (1979) is approximately 6 J.m^{-2}; an estimate of the energy of {113} defects in Ge with the structure corresponding to the Tan model (1981) leads to the values 4.7 - 6.6 Jm^{-2} (see paper by Masuda and Kojima 1983).

We have estimated the energy of a {113} defect with hexagonal structure (Fig. 4) using the equation taken from the papers of Baraff, Kane and Schluter (1980) and Nandedkar and Narajan (1987):

$$E = \frac{3\alpha}{16a_o^2} \sum (r_i.r_j - a_o^2)^2 + \frac{3\beta}{8a_o^2} \sum (r_i.r_j + \frac{a_o^2}{3})^2 \qquad (1)$$

Here a_o is the interatomic bond length in the perfect lattice; r_i, r_j the interatomic distances in the distorted region, the values of force constants for Ge, ie. $\alpha = 38$ Nm^{-1}, $\beta = 12$Nm^{-1} are taken from the paper of Keating (1966). From equation (1) we obtain an energy of 1.3 Jm^{-2} for the "hexagonal" structure for the {113} defect in Ge. Since this is much smaller than the energy obtained for the Tan structure, we conclude that the formation of the hexagonal structure decreases the defect energy associated with the deformation of atomic bonds in the defect region.

The formation of five- and seven-membered atomic rings separated by a hexagonal arrangement of atomic rings provides a key to understanding the origin of the {113} habit plane, along with the origin of the {113} habit plane of the Tan model.

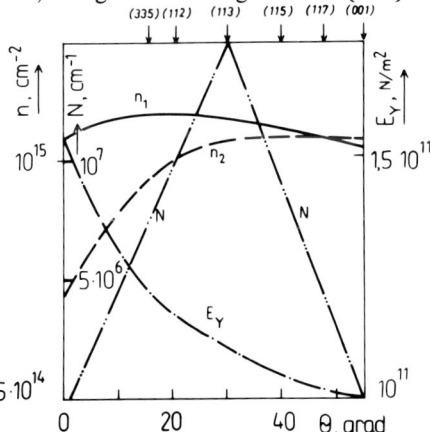

Fig. 5 shows several parameters for a set of atomic planes with a common <110> direction in the case of a Ge crystal (see Aseev 1987). On this figure it is evident that the {113} plane is different from the others in having the highest density of atomic steps. i.e. *the plane {113} is atomically rough*. This fact can explain the nature of the {113} habit plane: it is favourable for the complete closing of interatomic bonds between interstitial atoms in five- and seven-membered atomic rings separated by hexagonal fragments on the one side and perfect crystal lattice on the other side.

Fig. 5. The dependence of Young's modulus (E_y), density of tetrahedral interstitial sites (n_1), density of dangling bonds (n_2) and density of monatomic steps (N) on the orientation of atomic planes with a common <110> direction.

5. CONCLUSION

The structure of helium implanted Ge crystals has been investigated by means of TEM and HREM. The clustering of vacancies and interstitials leads to the formation of He bubbles and {113} defects, respectively. The hexagonal structure of the {113} defects has been found to be similar to those in electron-irradiated silicon and germanium crystals, as well as in ion-implanted and annealed silicon. The evolutionary model for the formation of {113} defects has been confirmed.

ACKNOWLEDGEMENTS

We thank Dr V P Popov (Institute of Semiconductor Physics, Novosibirsk) for the implantation of the germanium crystals. Part of the work was carried out during the stay of one of the authors (A.L.A.) at the University of Oxford, and the support of the Department of Materials is gratefully acknowledged.

REFERENCES

Antoniadis D A and Moskowitz I 1982 J.Appl.Phys., **53**, 6788.
Aseev A L, Ivakhnishin, V M Stas V F and Smirnov L S 1983 Sov.Phys.Sol.St., **25**, 3097.
Aseev A L 1987 in "Defects in Crystals" ed.Mizera (Singapore: World Sci.Publ.) p.161.
Baraff G A, Kane E O and Saluter M 1980 Phys. Rev. A. **21**, 5662.
Bender H 1984 Phys. Stat. Sol.(a). **86**, 245.
Bender H and Vanhellemont J 1988 Phys. Stat. Sol.(a), **107**, 455.
Bourret A, Thibault-Desseaux J and Seidman D N 1984 J. Appl. Phys. **55**, 825.
Bourret A 1987 in "Microscopy of Semiconducting Materials" eds. A G Cullis and
    D B Holt (Institute of Physics, London, Bristol), p.39.
Burenkov A F, Komarov F F, Kumakov M A and Temkin M M 1985 "Spatial distributions
    of energy released at atomic collisions in solid states" (Atomic Energy Pub. Moscow)
Gosele U, Frank W and Seeger A 1983 Solid. State Commun. **45**, 31
Hasebe M, Oshima R and Fujita F E 1986 Jap.J.Appl.Phys. **25**, 159.
Hashimoto H 1985 Ultramicroscopy, **18**, 19.
Hornstra J 1958 J.Phys.Chem.Solids. **5**, 129.
Hutchison J L 1990 in "Growth and Characterization of Semiconductors" eds.
    R.A.Stradling and P.C.Klipstein (Adam Hilger, Bristol) p.225.
Kaletta D 1976 J.Nucl.Mater. **63**, 347.
Keating P N 1966 Phys. Rev. A., **145**, 637.
Masuda K and Kojima K 1983 J.Phys.Soc.Japan, **52**, 10.
Matthews M D and Ashby S J 1973 Philos. Mag., **27**, 1313.
Mayer J W, Eriksson L and Davis J A 1970 "Ion Implantation in
    Semiconductors", (Academic Press, London and New York).
Nandedkar A S and Narayan J 1987 Philos. Mag. A, **56**, 625.
Salisbury I G and Lorretto M H 1979 Philos. Mag. A, **39**, 317.
Takeda S 1991 Jap.J.Appl.Phys., **30**, L639.
Takeda S, Hirata M, Muto S, Hua G -C, Hiraga K and Kiritani M 1991,
    Ultramicroscopy, **39**, 180.
Tan T Y 1981 Philos. Mag. A, **44**, 101.
Watkins G D 1975 "Lattice Defects in Semiconductors", (Institute of Physics,
    London and Bristol) p.1.

Inst. Phys. Conf. Ser. No 134: Section 1
Paper presented at Microsc. Semicond. Mater. Conf., Oxford, 5–8 April 1993

47

HREM characterization of small CdTe particles

R Hillebrand, H Hofmeister, K Scheerschmidt and J Heydenreich

Max-Planck-Institut für Mikrostrukturphysik, D O-4050 Halle, Weinberg 2, Germany

ABSTRACT: Small particles of the II-VI compound CdTe were studied by 400kV HREM. The crystallographic nature of the particles was determined by computer-simulated contrast tables for image matching, supported by image processing and optical diffraction. The contrast features (black-white reversal, lateral distribution) may be matched utilizing different defocus values. Reflection intensity / thickness profiles calculated for various crystallographic models exclude those others than the --A-B-- stacking to be significant in the CdTe microcrystals.

1. INTRODUCTION

The II-VI semiconductors possess band gaps ranging from the near UV to the far IR, which make these materials interesting for opto-electronics because of their non-linear optical and piezoelectric coefficients. The growth of high-quality layers of CdTe, ZnTe, or CdHgTe is being pursued for solar energy conversion, infrared detection and opto-electronic applications; see Panish (1987). As Feuillet (1989) reported, the epitaxial growth of CdTe films is affected by structural imperfections such as dislocations and planar defects. The occurrence of stacking faults and of twinning during the growth of CdTe, irrespective of the growth technique, is a general problem, described e.g. by Vere et al. (1983). Studies of the growth of CdTe thin films and the formation of defects in such films by Bykov et al. (1990) have included the study of crystalline island films of CdTe by high resolution electron microscopy (HREM). The present paper gives a crystallographic characterization of crystallites grown with their close-packed planes parallel to the substrate.

2. METHOD

The CdTe island films were prepared by vapour deposition on NaCl cleavage faces (substrate temperature 200 $^{\circ}$C) under a vacuum of about 1×10^{-4} Pa. After CdTe deposition the surface was coated with a carbon evaporation layer as a support to the small CdTe crystallites. The preparation procedure is described in more detail by Bykov et al. (1990). The HREM experiments were carried out in a JEM 4000 EX operating at 400 kV.

Computer simulations of the HREM image contrast - cf. Stadelmann (1987) and Hillebrand et al. (1992) - image processing and optical diffraction were applied to study the microstructure of the CdTe particles. Image matching procedures were used to correlate structure models and imaging parameters with experimental results.

3. RESULTS

In the CdTe island films, as presented in the survey micrograph of Fig. 1(a), a few of the crystallites are grown with their close-packed planes parallel to the substrate. Fig. 1(b) shows a 400kV HREM micrograph of such a crystallite at low magnification. Obviously, it consists of two regions, the contrast difference of which may result from a planar defect; see also Glaisher et al. (1987). In the following, the discussion will be focussed on subregions of the crystallites shown in Fig. 1(c), which appear darker and exhibit more pronounced contrast features of the crystal lattice structure.

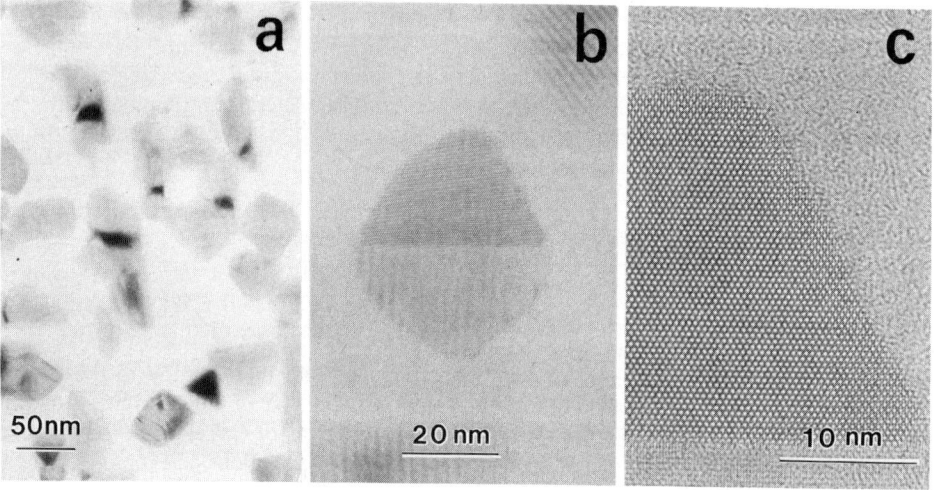

Fig. 1. (a) Survey of a CdTe island film grown by vapour deposition on NaCl at 200°C, (b) HREM image of a crystallite with its close-packed plane parallel to the substrate, (c) enlarged detail of the CdTe particle.

In accordance with the local crystallite thickness, the image contrast changes inside the crystallite. This is demonstrated for two different defocus values, recorded at over-focus (Fig. 2a) and at under-focus (Fig. 2b), both being proved by optical diffraction in the amorphous regions surrounding the crystallite. Applying Fourier analysis to the images reveals a sixfold (threefold resp.) symmetry of reflections arranged in two ring systems typical of the crystallite subregion examined.

Even for thin crystallites, dynamical electron diffraction causes contrast features, which cannot be interpreted in terms of the crystal projection. Image processing, including local FFT, was used to systematically study the lateral contrast features of Fig. 2.

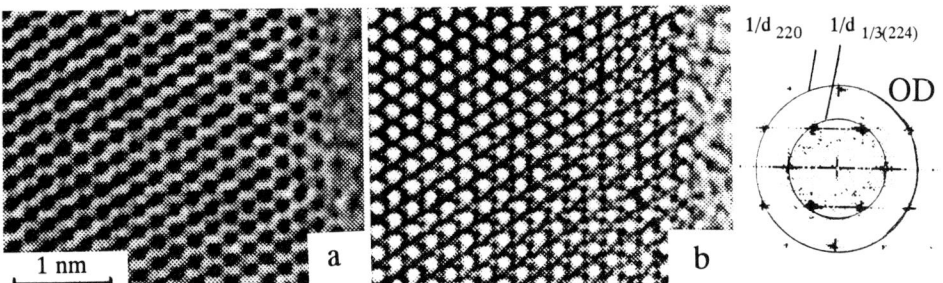

Fig. 2. Experimental micrographs of CdTe, V=400kV; (a) Δ=-50nm, (b) Δ=50nm (cf. diffractogram)

4. DISCUSSION

Assuming the cubic form of CdTe (space group $F\bar{4}3m$, a=0.648nm, sphalerite structure), complete cells of the (111) Laue zone axis provide only the (220) fringe system (1/d=4.36 nm^{-1}) in the range of resolution considered. Hillebrand et al. (1993) discuss the corresponding table of simulated HREM patterns, with the contrast phenomena exhibiting vertical (defocus) and horizontal (thickness) black-white periods. The fine structure seen in the experimental micrographs (cf. e.g. Fig. 2b) could not be fitted.

The alternative crystallographic description of CdTe (hexagonal, space group P6₃mc, a=0.457nm, c=0.748nm, wurtzite structure) yielded the simulated contrast table (thickness / defocus maps), shown in Fig. 3, that allows one to interpret the experimental results unambiguously. For an increasing specimen thickness, the contrast features are mainly controlled by 1/3(224) with 1/d= 2.52 nm^{-1}. The interference of two different 1/d-systems modifies the simple extinction behaviour occurring in the cubic case.

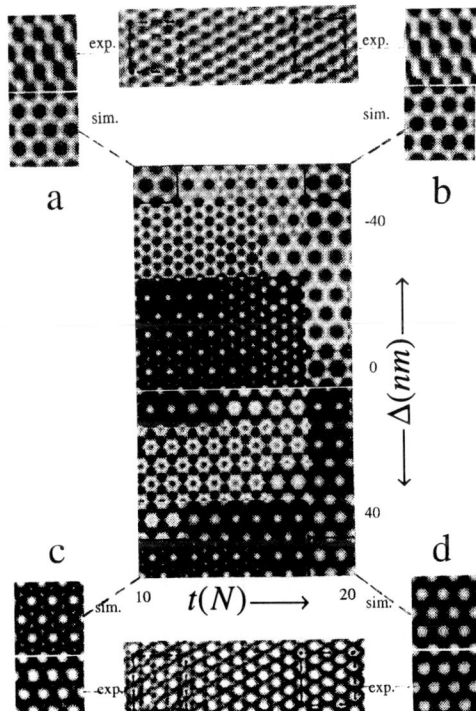

Fig. 3. Image matching CdTe(00.1), V=400kV, C$_s$=1mm (t=NΔz, Δz=0.374nm).
(a) Δ=-50nm, t=4 nm, (b) Δ=-50nm, t=8 nm,
(c) Δ= 50nm, t=4 nm, (d) Δ= 50nm, t=8 nm.

The contrast features (black-white reversal, lateral distribution) may be matched fairly well with the simulated contrast table. This finding is confirmed by a 2(t) x 2(Δ) consistency of experimental and simulated patterns. A quantitative comparison of experiment and simulation yields exactly 4 points in the thickness / defocus plane. The defocus values are additionally confirmed by optical diffractograms in the amorphous range of the micrographs and their correlation with the CTF.

Theoretical assumptions of upper layer effects, the consideration of higher-order Laue zones as well as different models of stacking faults and twins could not explain the strong influence of 1/3(224) on the image contrast. This is demonstrated by reflection intensity / thickness profiles calculated in a paper of Hillebrand et al. (1993). The different stacking irregularities assumed there were modelled using subsliced cubic [111] supercells having a thickness of 1/3 of that of the cell diagonal. The exact ordering of stacking in the crystals is described by the slice sequence in the multi-slice algorithm.

If the CdTe film is prepared at lower substrate temperatures crystallites, grown in (110), are detected additionally exhibiting stacking sequences of wurtzite structure in a faulted

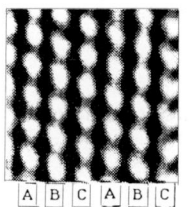
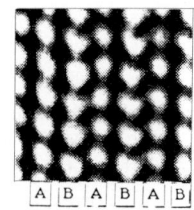

sphalerite lattice. A relevant example is shown in Fig. 4 directly indicating the structural deviations. The image clearly reveals that within the crystallite there is a high density of planar defects, exhibiting   "cubic"   as   well   as "hexagonal" subregions.

Fig. 4. Stacking disorder in a CdTe crystallite grown at 85°C in (110) orientation. The stacking sequence along <111> is denoted by letters. ("cubic - ABC / hexagonal - AB")

5. CONCLUSIONS

HREM contrast simulations for interpreting EM micrographs proved the existence of small CdTe crystallites with a hexagonal close-packed lattice having a wedge angle of steep descent. Calculated contrast tables and the corresponding reflection intensity / thickness profiles of alternative crystallographic models exclude others than the --AB-- stacking to be significant for these crystallites. Hence, HREM is a useful tool for directly studying crystallographic modifications of compound semiconductor microcrystals.

REFERENCES

Bykov V N, Nepijko S A, Hofmeister H and Junghanns T 1990 Proc. 11th Allunion Conf.
    Electron Microsc., ed V I Petrov (Moscow: IKAN) p 72
Feuillet G 1989 in: Evaluation of Advanced Semiconductor Materials, ed D Cherns (New-York: Plenum)
    p 33
Glaisher R W, Kuwabara M, Spence J C M and McKelvy M J 1987 in:
    Microscopy of Semiconducting Materials, eds A G Cullis and P D Augustus (Bristol: JOP) p 349
Hillebrand R and Scheerschmidt K 1992 Proc. 32. Course of the Int. Centre of Electron Microsc.,
    eds. J Heydenreich and W Neumann (Halle: Elbedruckerei) pp 85-94.
Hillebrand R, Hofmeister H, Scheerschmidt K and Heydenreich J 1993 Ultramicr. 49 252
Panish M B 1987 in: Materials for Infrared Detectors and Sources, eds: R F C Farrow, J F Shetzina
    and J T Cheung (Pittsburgh: MRS) pp 3
Stadelmann P 1987 Ultramicr. 21 131
Vere A W, Cole S and Williams D J 1983 J. Electronic Mater. 12 551

Acknowledgements:
    We are grateful to the Volkswagenstiftung for financial support. Thanks are due to Dr. V. N. Bykov for preparing the CdTe specimens.

Inst. Phys. Conf. Ser. No 134: Section 1
Paper presented at Microsc. Semicond. Mater. Conf., Oxford, 5–8 April 1993

Two beam high resolution dark field imaging: applications in epitaxial semiconductor materials

A C Wright and J O Williams

Advanced Materials Research Laboratory, NEWI Deeside, Connah's Quay, Clywd, CH5 4BR

ABSTRACT: We have applied two beam dark field lattice imaging for the characterisation of semiconductor materials. Only the (002) and (004) beams are used to form the image and under these conditions the image consists of a one dimensional set of fringes. In multilayer structures such as AlGaAs/GaAs, very strong contrast between the individual layers is obtained, making the measurement of layer widths easy. In addition, composition gradients can be imaged. Applications of the technique are shown in this paper.

1. INTRODUCTION

Modern advanced semiconductor devices often consist of very thin layers of semiconductor material with varying composition. The characteristics of these devices is a sensitive function of layer width and their composition. In addition, the structural perfection of the interfaces (roughness, chemical abruptness) between the various layers is also a major factor influencing device performance. Of all the techniques available for the characterisation of these parameters, transmission electron microscopy (TEM) is possibly the most direct. Clearly, the phase contrast (high resolution) imaging mode of TEM is best able to perform direct measurement of layer widths as the lattice spacings imaged as interference fringes have known spacings. It is now well known that with certain projections of the sphalerite structure (eg the <100> direction) the differences in composition between one layer and the next can be imaged as shown by Hetherington et al. (1987) and Wright and Williams (1991). Nevertheless, the interpretation of these complex many-beam images in terms of the position of the interface and determination of the absolute composition remains difficult; extensive computer based methods have been developed for this purpose (Ourmazd et al. (1989) and Thoma and Serva (1991)). Additionally, the contrast between layers in images produced by this technique are often low. By comparison, single beam imaging (in dark field) using the (002) reflection produces images with a much simpler (and stronger) contrast but lacks the inherent spatial calibration of the high resolution approach. However, the strong compositional sensitivity of the (002) beam has lead to the development of methods by which (for the $Al_xGa_{1-x}As$ alloy system at least) the composition may be determined (Bithell and Stobbs (1989)).

It would seem logical therefore, to try to combine the best characteristics of both the multi and single beam methods with (hopefully) none of their drawbacks. The simplest way of producing a phase contrast image (for spacial measurement) is to use just two beams to form the lattice image. Next, we have to select two beams which are strong functions of composition. Clearly the (002) beam is an obvious candidate. We could also use the zero-order (transmitted) beam but this has two drawbacks: first is that the image signal will be largely dominated by the very strong (000) beam, making for weak contrast and second is the large inelastic component close to the (000) position, which may complicate the overall relation between image intensity and composition. As most substrates tend to be (100) oriented, the next most obvious choice is the (004) beam. Both the intensities of the (002) and (004) beams are a function of composition for the sphalerite structure; for the (002) the intensity is proportional to the square of the difference in the scattering factors between the metal and non-metal sublattice of the sphalerite structure while for the (004) beam it is proportional to the sum squared. However, a simple calculation shows that when these effects are summed (coherently), there is no variation of intensity with composition under these kinematical conditions. Nevertheless, an experimental trial shows that a strong composition dependence of the overall image intensity does arise under dynamical conditions as shown by Williams et al. (1991) when used to image the thin layers of the compound Ga_2Se_3 found to occur at ZnSe/GaAs interfaces.

2. METHOD

The technique clearly involves high resolution imaging in dark field. The aperture size used needs to be large enough to include the two beams without their touching the edge of the aperture. To ensure that both beams suffer the same level of phase shift via the contrast transfer function, the optic axis is placed midway between the (002) and (004) positions. Thus the effective resolution needed is only one half of that expected from the difference in **g** value between the (002) and (004) beams. The crystal has to be tilted such that the (002) beam is at the Bragg position. Under these conditions the (004) beam is still strong as the Ewald sphere is fairly flat (at 300kV as used in all our experiments). We have performed all our imaging with specimens cut for the <100> orientation but tilted so as to create two beam conditions and to avoid strong HOLZ reflections. The specimens used were grown using AlGaAs alloys by both MBE and MOVPE. To characterise the composition sensitivity, a sample was prepared in which the AlGaAs layers were stepped up from 15 atomic % to 50 atomic % in 5% steps. Between each AlGaAs layer was GaAs. All layers in this sample were 10nm wide. The technique was also found useful for particle size measurement in Erbium doped GaAs. In this system, small particles of rocksalt ErAs occur for Er concentrations above $10^{17}/cm^3$ (Poole et al. 1992).

3. RESULTS

Figure 1 shows a conventional dark field (002) image of a GaAs/Al$_x$Ga$_{1-x}$As device structure (x=0.4) and reveals the expected strong contrast between the bright Al-rich barrier layers and the dark GaAs matrix. By comparison, the bright field many-beam phase contrast image of figure 2 shows very poor contrast between the individual layers. In addition, the exact positions of the interfaces is not easy to define and considerable changes of contrast can be seen to occur, particularly at interfaces. Figure 3 shows the same specimen imaged using the two beam approach. The contrast between layers is now similar to that obtained with the single (002) beam but with the added advantage of direct spatial calibration. Note that there are no variations in layer to layer contrast over the specimen and no anomalous contrast at interfaces.

Figure 1. Standard (002) beam image of AlGaAs/GaAs device structure. AlGaAs layers are bright.

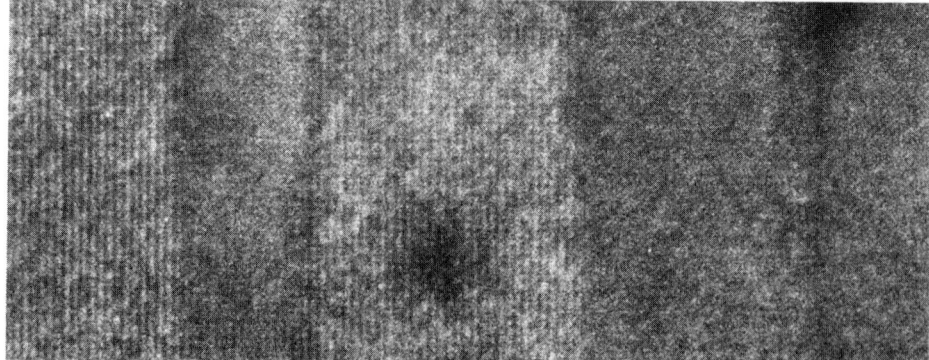

Figure 2. Bright field <001> zone axis image of same AlGaAs/GaAs device. Note poor contrast between layers.

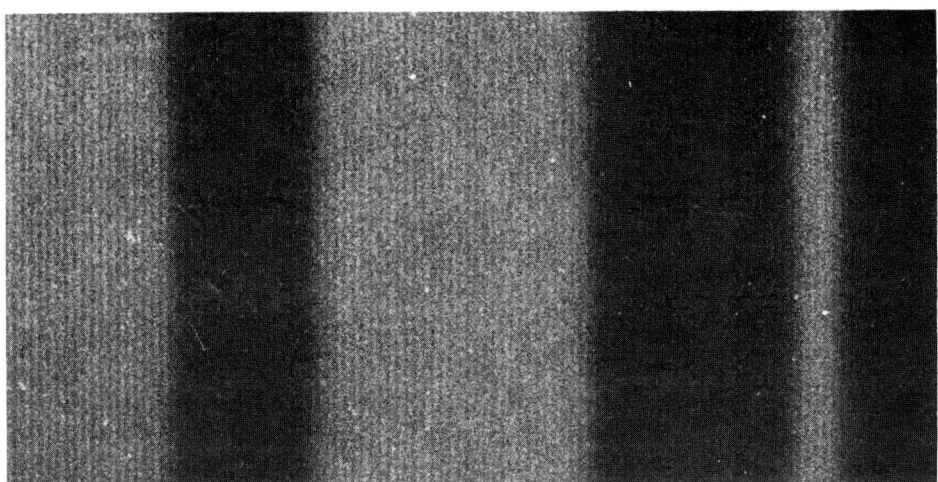

Figure 3. Dark field two beam (002)-(004) image of same AlGaAs/GaAs device. Note good contrast between layers with direct spatial calibration. Fringe width is half unit cell spacing (0.28nm).

Indeed, it is possible from Fig. 3 to infer a compositional grading from the fall off of fringe intensity across the interface. Using a digitised image it is possible to obtain an noise free profile across the fringe pattern by projecting (and averaging) the pixel intensities along the fringes, providing these are colinear with the plane of the layers. It then may be possible to apply a similar ratio of the signal for the Al-rich layers to that of the GaAs background as used by Bithell and Stobbs (1989) so as to obtain information on the Al content. Figure 4(a) shows a low magnification image of the stepped Al concentration structure with intensity ratio plotted as figure 4(b). The accuracy of such an approach will depend on how well the background fit is performed, but the profile obtained from this sample as seen here is fairly linear. Assuming that the actual aluminium concentration in this sample is linearly stepped then this may infer that the variation of image intensity with aluminium composition is also linear (approximately) for the $Al_xGa_{1-x}As$ system (at least under the diffraction conditions used here).

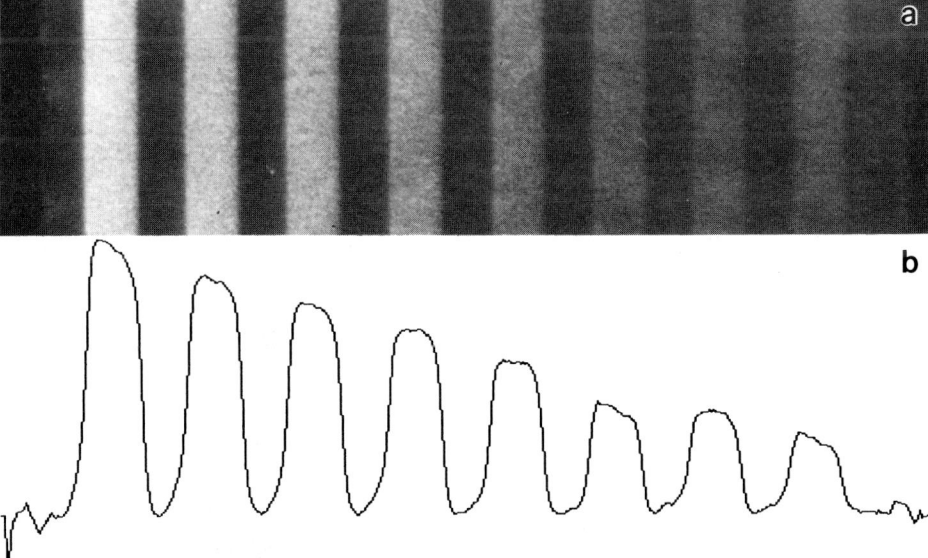

Figure 4. (a) Low magnification dark field (002)-(004) beam image of $Al_xGa_{1-x}As$/GaAs structure with x=0.5 at left and x=0.15 at right. (b) Ratio of AlGaAs layer peak intensity to GaAs background intensity.

The method works equally well when applied to particles of ErAs in a GaAs matrix, figure 5(a). The structure factor for the (002) reflection for the rocksalt structure is high while that for GaAs is low and so the particles are imaged as white on the dark GaAs background. The average size seen here is around 2nm. Again, the (002) single beam image shows very good contrast but without the advantage of spatial calibration, figure 5(b).

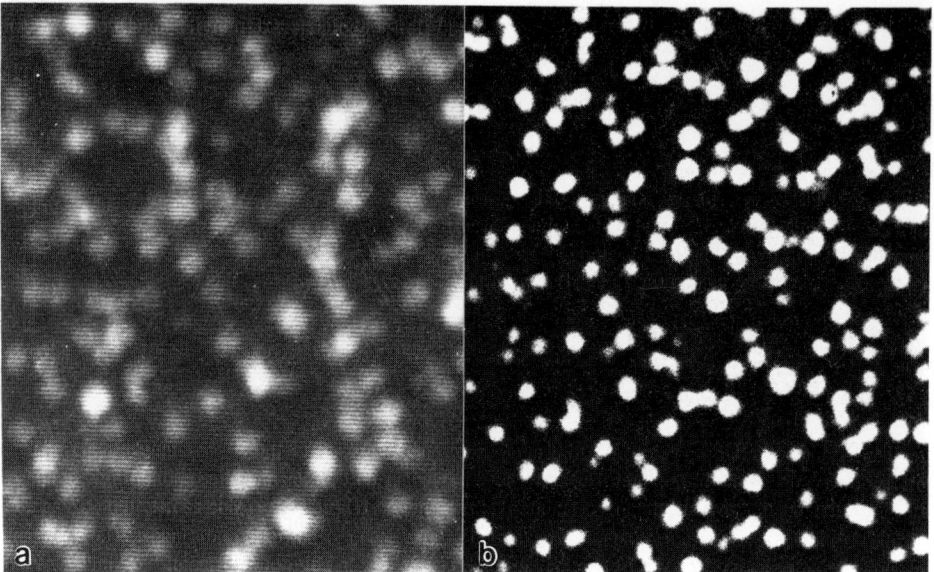

Figure 5. (a) (002-004) beam image of ErAs particles in GaAs. (b) (002) beam only.

4. CONCLUSION

The two beam method clearly has advantages over the conventional methods of imaging semiconductor device structures. The image is restricted to a simple set of one dimensional fringes by the use of only two beams, thus avoiding many of the complex image features seen for bright field many beam using both the <110> and <100> projections. Despite the use of only diffracted beams to form the image, the exposure times are not long in practice, all images recorded here were taken using 0.5 seconds (LaB$_6$ source). The only drawback is that the interfaces are not imaged edge on as with the conventional high resolution approach. The need to tilt the crystal by 0.199 degrees (at 300kV) leads to overlaps of interfaces by 0.174 nm for a 50nm thick section. This error increases as the accelerating voltage is reduced where better two-beam conditions can be obtained. Even this effect can be (in principle) be removed by deconvolution methods as the blurring function is fixed by the projection geometry. Such work is now in progress.

ACKNOWLEDGEMENT

ACW would like to thank Messrs I Brough ,G Cliff and P Kenway for valuable technical assistance during the many years at UMIST.

REFERENCES

Bithell E G and Stobbs W M (1989) Phil. Mag. A60 No 1 39-62
Hetherington C J D, Eaglesham D J, Humphries C J and Tatlock G J (1987) IOP conf. ser. 87 665-674
Poole I, Singer K E, Peaker A R and Wright A C (1992) J. Crystal Growth 121 121
Ourmazd A, Taylor D W and Cunningham T, Phys. Rev. Lett. (1989) 62 933
Thoma S and Serva H Ultramicroscopy (1991) 38 265-289
Wright A C and J O Williams (1991) J. Crystal Growth 114 99-106
Williams J O, Wright A C and Yates H M (1992) J. Crystal Growth 117 441-453

Inst. Phys. Conf. Ser. No 134: Section 2
Paper presented at Microsc. Semicond. Mater. Conf., Oxford, 5–8 April 1993

55

Dynamic characteristics of dislocations in semiconductors

Koji Sumino

Institute for Materials Research, Tohoku University, Sendai 980, Japan

ABSTRACT: Three important issues concerning dynamic activity of dislocations, namely generation, motion and immobilization, in Si and some III - V compound semiconductors are given in connection with experimental techniques to investigate them. *In situ* techniques of electron microscopy and X-ray topography as well as intermittent techniques have been adopted for the investigation. Each of the techniques has its advantage and disadvantage. The overall knowledge obtained by the different approaches is shown to be effective in establishing comprehensively the dynamic characteristics of dislocations in semiconductors.

1. INTRODUCTION

The dynamic behaviour of dislocations, such as generation, motion and immobilization, in semiconducting materials is attracting a great deal of attention in both the practical field of electronic device technology and the basic field of the physics of lattice defects.

It is clear that the most reliable way to investigate the dynamic behaviour of dislocations in a crystal is to carry out real time observation of dislocation processes which proceed under stress. Dislocations in a semiconductor crystal are mobile only at elevated temperatures. Hence, real time observation is possible in semiconductor crystals only with a transmission electron microscope or an X-ray topographic system installed with a high temperature stressing stage. The applicability of *in situ* techniques is limited for several reasons. Thus, the intermittent technique has widely been adopted especially in the measurements of dislocation mobility. In this technique the positions of dislocations are determined at room temperature by etching or X-ray topography, but dislocations are moved at elevated temperature by the application of stress. Naturally, it contains many sources of error as discussed by Sumino (1987). However, the technique is simple and applicable to a variety of materials.

Each of the experimental techniques has some limitation or drawback together with advantages. This paper aims to show how a comprehensive picture is established of the dynamic characteristics of dislocations in various kinds of semiconductor by combining information obtained with different techniques. Recent reviews of the dynamic characteristics of dislocations in a variety of semiconducting crystals are given by the present author (Sumino 1989a, 1993).

2. EXPERIMENTAL PROBLEMS

In situ transmission electron microscopy reveals the details of dislocation processes occurring on a microscopic scale of order of 1 μ m. Such a high resolution is the advantage of this experimental technique over all of other techniques. The disadvantage lies in the inaccuracies in determination of temperature and stress in the specimen originating from the small heat capacity of the stage or non-uniformity of the specimen thickness. Hence, any information obtained is qualitative.

Accurate determination of temperature and stress is possible in *in situ* X-ray topography (Sumino and Harada 1981) since a bulk crystal is subject to observation. Dislocation processes can be followed as a function of stress or temperature.   One of the disadvantages of X-ray topography lies in its low resolution and low magnification.   It requires highly perfect specimens with no or very low densities of dislocations, which are not common for most compound semiconductors at present.   Another disadvantage of X-ray topography is that it is applicable only to the material such as Si which has a low absorption coefficient for X-rays.   X-ray topographic observation of dislocations is possible in III - V compounds with high absorption coefficients by using the anomalous transmission of X-rays at relatively low temperatures, but not at high temperatures.   This makes *in situ* observation of dislocation processes difficult in the temperature range of interest.

The intermittent technique is usually used in the measurement of the dislocation velocity.   It assumes that dislocations have been moving continuously at constant velocity during the application of stress.   It also assumes that the positions and state of dislocations do not change during the heating and cooling of the specimen.   When the specimen contains certain kinds of impurity, dislocations getter impurity atoms from their vicinity during such a thermal cycle and, as a result, are immobilized.   This makes the measurements of the velocities of isolated dislocations unreliable in impure crystals.

3. DISLOCATION GENERATION

When a Si crystal initially free from dislocations is stressed at a high temperature, dislocations are observed to be generated heterogeneously under a stress lower than the theoretical stress for dislocation generation by orders of magnitude.   Usually, they are generated at the surface of the crystal and propagate into the bulk crystal if the crystal contains no structural irregularities inside which facilitate dislocation generation.   Such observation leads to the idea that any crystal has some irregularities on the surface that act as preferential generation centers of dislocations under stress.   The irregularities are thought to be microscopic regions with strongly disturbed atomic structure which may be formed by some energetic stimulation such as mechanical shock or chemical reaction at the surface.

Recent transmission electron microscopic observations have revealed that an extremely tiny region of amorphous Si accompanying a dislocated crystalline region is induced by scratching, abrasion or indentation of a Si surface at room temperature (Clarke et al 1988, Minowa and Sumino 1992).   A transmission electron micrograph of an amorphized region underneath a light scratch is shown in Fig. 1 (Minowa and Sumino 1992).   Such an amorphized region is observed to recrystallize into a dislocated microregion when the crystal is brought to a temperature higher than about 600 °C .   A large strain field around the flaw is released on recrystallization.   If the crystal is under stress, some of dislocations which have the maximum Schmid factor come out of the microregion and penetrate into the matrix crystal and expand on a macroscopic scale.   This process has been traced by means of *in situ* X-ray topographic observations (Sumino and Harada 1981).

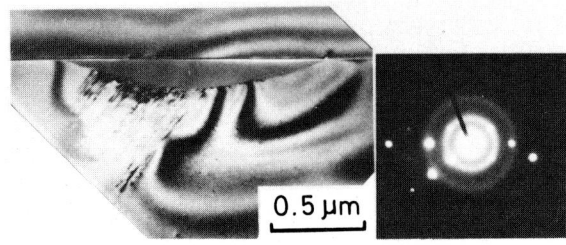

0.5 µm

Fig. 1 Cross-sectional transmission electron micrograph of a region around a scratch made on a Si surface at room temperature, with diffraction pattern.   A region having uniform dark contrast is the amorphized region and dark lines are images of dislocations.

Dislocations are observed to be easily generated from the surface flaw even under a stress lower than 1 MPa at high temperatures in a highly pure Si crystal grown by the floating-zone technique. However, in a Czochralski-grown Si crystal there is a critical stress below which no dislocation generation takes place even if the amorphized region transforms to a dislocated microregion. The critical stress increases with an increase in the temperature. Since the mobility of a dislocation in a Si crystal is not affected by O impurity dissolved in the crystal and increases with an increase in the temperature, the critical stress for dislocation generation is not related to the resistance to dislocation motion due to O atoms dissolved in the crystal. The critical stress is related to the release stress of dislocations in the microregion around the flaw that are immobilized by the gettering of O impurity during heating of the crystal. Immobilization of dislocations due to impurity gettering will be treated in section **5**.

As-grown crystals of compound semiconductors available now are usually dislocated. When such a crystal is stressed, dislocations are generated heterogeneously in the surface region under a stress which is lower than that needed to start grown-in dislocations moving. This is in contrast to the case of Si. Such observation suggests that there are some preferential generation centers for dislocations in the surface region of the compound semiconductor crystal which are activated under a rather low stress and that grown-in dislocations are firmly immobilized by impurities or for some other reason.

Transmission electron microscopic observations have revealed that mechanical treatment, such as indentation, abrasion or scratching, of the surface of GaAs induces a dislocated microregion with no amorphized region. The dislocation generation centers in compound semiconductor crystals are thought to be such dislocated microregions introduced accidentally by some energetic stimulation during surface preparation processes. The surfaces of GaAs or InP crystals prepared by chemical polishing also contain preferential nucleation sites for dislocations. It is thought that the process of chemical polishing includes some process to form some structural irregularities that act as preferential generation centers for dislocations.

As in the case of Si, impurities affect the generation of dislocations in compound semiconductors. Any dislocation loop in III - V materials with the sphalerite structure generated under stress consists of three basically different types of dislocation, termed α, β and screw dislocations, which are different from each other in the core structure.

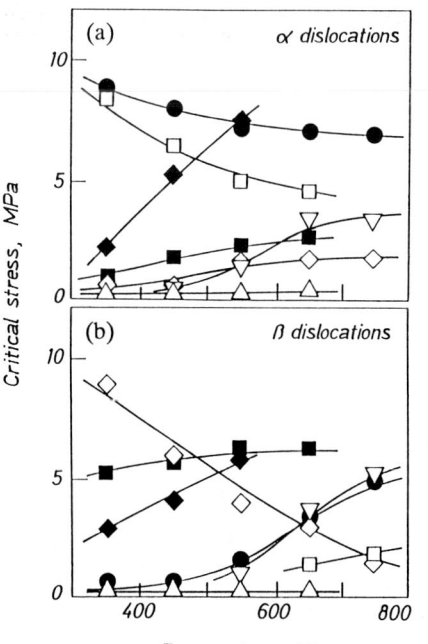

(a) α dislocations

(b) β dislocations

Critical stress, MPa

Temperature, °C

$\triangle$ undoped
$\blacksquare$ Zn: 2×10^{19} cm^{-3}
$\triangledown$ Si: 4×10^{18}
$\square$ Te: 6×10^{18}
$\blacklozenge$ S: 1×10^{17}
$\lozenge$ Al: 3×10^{18}
$\bullet$ In: 2×10^{20}

Fig. 2 Critical stresses for generation of (a) α dislocations and (b) β dislocations from surface scratches in non-doped and doped GaAs plotted against temperature. Species and concentrations of impurity are shown above.

On the basis of the knowledge obtained for Si by means of *in situ* X-ray topographic observation, we can use the etch pit technique very effectively to investigate the impurity effect on dislocation generation in Ⅲ - Ⅴ compound semiconductors.

Figure 2 shows the critical stresses for generation of α and β dislocations from surface scratches in GaAs crystals doped with different kinds of impurity measured against the stressing temperature (Yonenaga and Sumino 1989). The scratches were drawn with a diamond stylus under an identical load at room temperature and all the crystals were brought to the test temperature directly after scratching. Any type of dislocations in undoped GaAs are generated under a very low stress over the whole temperature range. There are two kinds of trend in the dependence of the critical stress on the temperature in crystals doped with impurities. The first is that the critical stress is high at low temperatures and decreases with increasing temperature. This trend is observed for α and screw dislocations in GaAs doped with a high concentration of In, for α dislocations in Te-doped GaAs, for β dislocations in Al-doped GaAs, and for screw dislocations in Zn-doped GaAs. The second kind of trend is that the critical stress is low at low temperatures and increases gradually as the temperature increases, becoming appreciable at temperatures higher than about 500 °C . Such trend is observed for all the types of dislocations in Si-doped GaAs, for α dislocations in Al- or Zn-doped GaAs, and for β dislocations in In- or Te-doped GaAs. An intermediate trend is observed for β dislocations in Zn-doped GaAs.

The first kind of trend in the temperature dependence of the critical stress is observed also in InP crystals co-doped with Ga and As for α dislocations while the second kind of trend is observed for all the types of dislocations in InP doped with S or Zn (Yonenaga and Sumino 1993a).

The dependencies of the critical stress for dislocation generation on the type of dislocation with respect to any given impurity are in accordance with those of the release stress for dislocations immobilized by gettering of the impurity illustrated in section **5**. Thus, as in Si, the suppression of dislocation generation in impurity-doped compound semiconductor crystals is also concluded to be caused by the immobilization of dislocations due to impurity gettering.

4. DISLOCATIONS IN MOTION

A number of reports have so far been published on dislocation velocities in various kinds of semiconducting material. The velocity v of a dislocation in a semiconductor crystal has been found to be well expressed by the following equation as a function of the shear stress τ and the temperature T over a certain temperature range where the crystals are ductile and not very close to the melting point :

$$v = v_0 (\tau / \tau_0)^m \exp (- Q / k_B T), \qquad (1)$$

where v_0, m and Q are material constants, k_B the Boltzmann constant, and $\tau_0 = 1$ MPa. The magnitudes of m and Q have been measured to be $1 - 2$ and $1.5 - 2.5$ eV, respectively, for a variety of semiconductor materials.

Figure 3 compares the velocities of various types of dislocations in a series of semiconducting materials under an applied stress of 20 MPa plotted against the reciprocal temperature $1/T$ (Si: Imai and Sumino 1983, GaAs: Yonenaga and Sumino 1989, GaP: Yonenaga and Sumino 1993b, InP: Yonenaga and Sumino 1991, GaSb and InAs: Choi et al 1978, InSb: Mihara and Ninomiya 1975).

In situ transmission electron microscopy revealed that dislocations in a Si crystal move in a viscous manner at elevated temperatures on the microscopic scale of 1 μ m (Sato and Sumino 1977, Sumino and Sato 1979, Louchet 1981). The dislocation velocity measured on such a scale is approximately equal to the macroscopic velocity measured by *in situ* X-ray topography (Imai and Sumino 1983). The magnitude of m in high purity Si determined by means of *in situ* X-ray topographic observation is unity. A dislocation consists of straight segments parallel to the $\langle 110 \rangle$ directions.

Light element impurities dissolved in Si that are electrically inactive do not affect the

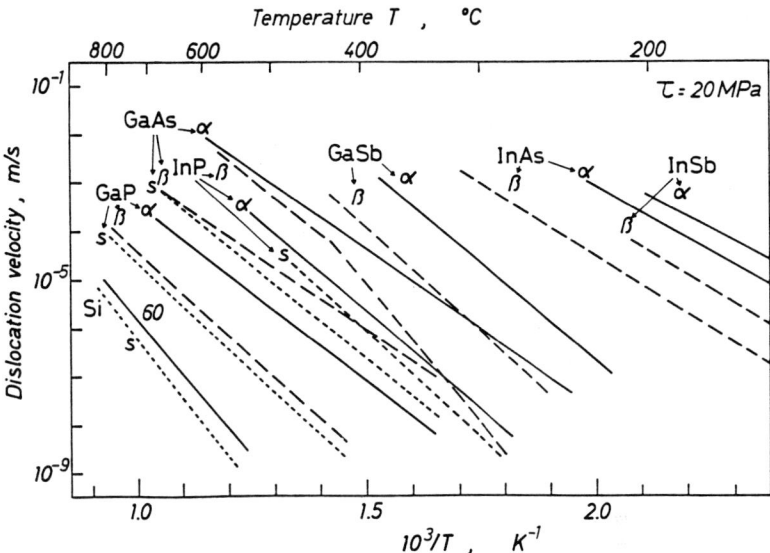

Fig. 3 Velocities of various types of dislocation under a stress of 20 MPa in undoped Si and a series of Ⅲ - V compounds plotted against the reciprocal temperature $1/T$.

velocity of dislocations when the dislocations move under high stresses. However, they affect the dynamic behavior of dislocations under low stresses depending on their concentrations. *In situ* X-ray topography revealed that dislocations originally in motion under a high stress cease to move when the stress is reduced lower than certain critical values. The velocity of dislocations in Si doped with O impurity decreases more rapidly than in high purity Si as the stress is decreased toward the critical stress for the cessation of dislocation motion. The shape of the segments of a moving dislocation in such a case is perturbed from the $\langle 110 \rangle$ straight lines and becomes irregular. The magnitude of m is determined to be apparently larger than unity in such a low stress range (Imai and Sumino 1983). The critical stress for the disturbance in shape and that for the cessation of dislocation motion both depend on the species and concentration of impurity and also on temperature and correlate well with each other.

There is agreement among several groups on the observations that donor impurities such as P, As and Sb enhance the velocity of both 60° and screw dislocations (Erofeev et al 1969, Erofeev and Nikitenko 1971 a,b, Patel et al 1976, Kulkarni and Williams 1976, George and Champier 1979, Imai and Sumino 1983). The increase in the dislocation velocity compared to that in high purity Si due to doping with donor impurities under any stress at any temperature is determined by only the concentration and is not influenced by the species of impurity. Acceptor impurities such as B in Si affect the dislocation velocity very little.

The different types of dislocation in any given Ⅲ - V material have different mobilities and show different characteristics in the interaction with impurities. Measurements of the dislocation velocities in these materials are usually conducted by means of the intermittent technique, watching the leading dislocation in an array of dislocations which are generated from a surface scratch under stress applied by means of three point bending. This method is advantageous to apply to the materials such as Ⅲ - V compounds, of which availability is commercially limited, since it allows us to obtain a rather large number of data points with one specimen. However, the velocity measured by this method is expected to be somewhat higher than that of an isolated dislocation due to the effect of the repulsive forces from successive dislocations within the array. Nevertheless, it serves as a good quantity in representing the dislocation mobility, especially when we are concerned with the mobility dependence on the type of dislocation as well as on the species and concentration of impurities in the crystals.

The dislocation mobilities are approximately the same in GaAs and InP. However, there is an essential difference in the dynamic characteristics of dislocations between GaAs and InP. Namely, the mobility of α dislocations is higher than that of β dislocations by about two orders of magnitude in GaAs over the whole temperature range. On the other hand, β dislocations have a higher mobility than α dislocations in InP in a temperature range higher than 350 °C, while the situation is reversed in the lower temperature range. In both materials screw dislocations are the slowest.

Impurity atoms dispersed within III - V materials give rise to a rich variety of effects on the velocities of dislocations in motion depending on the electrical activity of the impurity atoms and the types of dislocation with which they interact. The characteristics of such effects are different from material to material.

As has been mentioned in section **3**, no generation of dislocations takes place under low stresses in GaAs or InP doped with certain kinds of impurity. The velocity of dislocations is measured to be zero in a low stress range in such materials and to increase rapidly with stress once the stress exceeds the critical stress for generation by means of the intermittent technique. In such a case, the velocity versus stress relation usually shows a break at some high stress; from there the velocity increases rather slowly with increasing stress at approximately the same rate as that in undoped materials (Yonenaga and Sumino 1989, 1993a). Since the absence of dislocation generation in impurity-doped materials is interpreted to be caused by the immobilization of dislocations due to segregation of the impurities on them, the dislocation velocities under high stresses after the break reflect the effects of impurities dispersed within the crystal on the velocities of dislocations in motion.

Impurity effects on the mobilities of various types of dislocation are shown in a

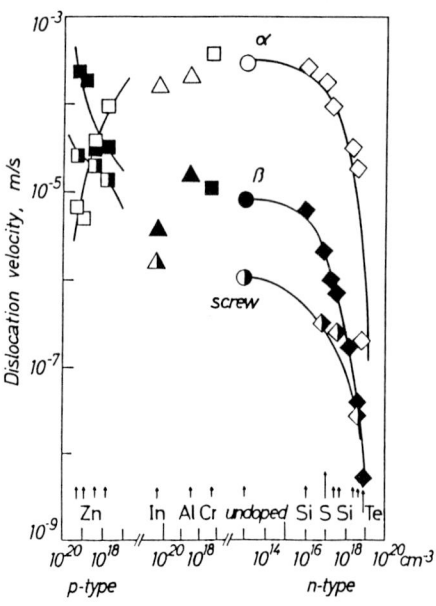

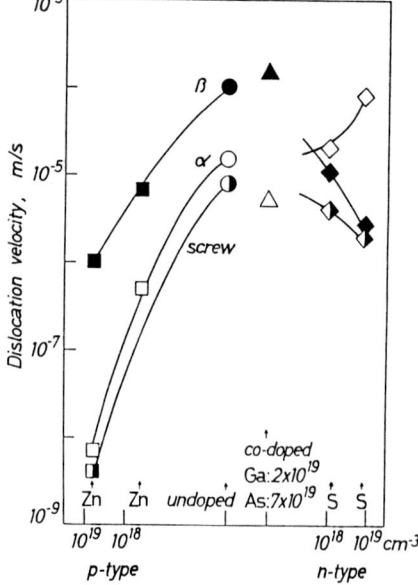

Fig. 4   Impurity effect of the velocities of various types of dislocation under a shear stress of 20 MPa at 450 °C in GaAs.

Fig. 5   Impurity effect on the velocities of various types of dislocation under a shear stress of 20 MPa at 450 °C in InP.

semi-quantitative way in Figs. 4 and 5 for GaAs and InP, respectively, in which the velocities of various types of dislocation under a stress of 20 MPa at 450 °C are compared for crystals doped with various species and concentrations of impurity. It is to be noted that isovalent impurities in Ⅲ - Ⅴ compounds give no appreciable influence on the mobility of any type of dislocation.

The characteristics of the morphology of a dislocation in motion in a Ⅲ - Ⅴ compound are seen from the topographic images of dislocations in a specimen which was stressed at elevated temperature to move dislocations and then cooled to room temperature without removing stress. When the mobility of a α or β dislocation is remarkably different from that of a screw dislocation, a moving dislocation loop usually does not assume the shape of a hexagon of which segments are parallel to the $\langle 110 \rangle$ directions, but is smoothly curved. On the other hand, a dislocation loop in motion assumes a regular hexagonal shape as in Si when the velocities of all types of dislocation are almost the same, as in GaAs doped with Zn at a certain concentration (Yonenaga and Sumino 1993c).

5. IMMOBILIZATION OF DISLOCATIONS BY REACTION WITH IMPURITIES

A fresh dislocation which was mobile even under a low stress becomes immobile when the crystal contains certain kinds of impurity and is aged at an elevated temperature while the dislocation is at rest. Such a process was first investigated successfully by means of *in situ* X-ray topography in Si doped with O, N or P impurity (Sumino and Imai 1983). The process of dislocation immobilization in Si clarified by the *in situ* technique opened the way to investigation of the immobilization processes in Ⅲ - Ⅴ compounds, by means of the intermittent technique using the etching method.

The immobilization of a dislocation takes place because impurity atoms segregate on the dislocation. The segregation of impurity atoms on a dislocation in a semiconductor crystal is not related to the development of the so-called Cottrell atmosphere, but is due to some special reaction occurring at the dislocation core which incorporates the impurities. A detailed discussion of this matter has been given by Sumino (1989b).

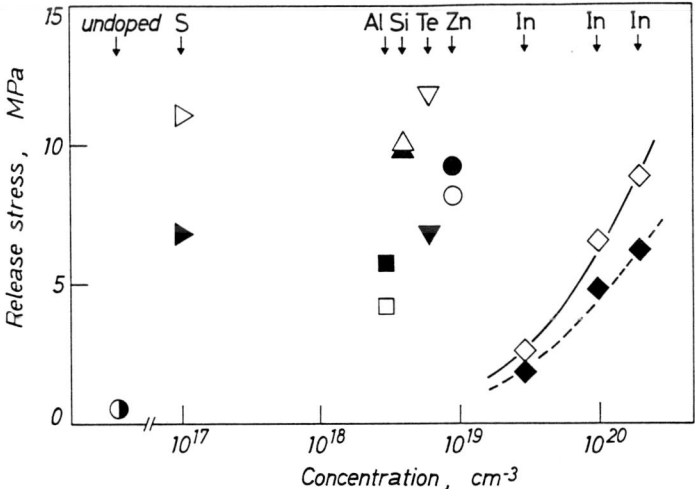

Fig. 6 Release stresses for α dislocations (open marks) and β dislocations (filled marks) at 500 °C in GaAs doped with various kinds of impurity after ageing at 550 °C for 315 min. Species of impurity are shown at the top of the figure and their concentration along the abscissa.

The release stress for a dislocation locked by impurities is determined by the state of gettered impurity atoms and the temperature at which the dislocation is released if the state of gettered impurity atoms is given.   The state of gettered impurity atoms depends on the type of the dislocation, the species and concentration of the impurity, and the thermal history of the crystal during which impurity atoms are gettered.

A 60 ° dislocation and a screw dislocation in Si show no essential difference in the features of locking by impurities.   On the other hand, a quite rich variety is found in locking of dislocations in III - V compounds. α dislocations in GaAs are selectively immobilized by In or Te impurity while β dislocations are immobilized by Al or Zn impurity.   Si impurity immobilizes both types of dislocation in a similar way (Yonenaga and Sumino 1989).

Figure 6 compares the release stresses for α and β dislocations at 500 °C in GaAs doped with various kinds of impurity after ageing at 550°C for 315 min.   The abscissa shows the concentration of impurity.   The dependence of the release stress on the type of dislocation related to any kind of impurity is in accordance with that of the critical stress of dislocation generation from a scratch seen in Fig. 2.   Thus, we may conclude that the suppression of dislocation generation in impurity-doped GaAs is caused by the immobilization of dislocations due to impurity gettering.   Dislocations in a microregion acting as a generation center mentioned in section 3 are immobilized before they come out under stress. A stress exceeding the release stress must be applied to start the dislocations moving.   Thus, the release stress is directly related to the critical stress for dislocation generation.

REFERENCES

Choi S K, Mihara M and Ninomiya T 1978 Jpn. J. Appl. Phys. 17 329
Clarke D R, Kroll, M C, Kirchner P D, Cook R F and Hockey B J 1988 Phys. Rev. Lett. 60 2156
Erofeev V N, Nikitenko V I and Osvenskii V B 1969 Phys. Stat. Sol. 35 79
Erofeev V N and Nikitenko V I 1971a Soviet Phys. JETP 33 963
Erofeev V N and Nikitenko V I 1971b Soviet Phys.-Solid State 13 116
George A and Champier G 1979 Phys. Stat. Sol. (b) 53, 529
Imai M and Sumino K 1983 Phil. Mag. A 47 599
Kulkarni S B and Williams W S 1976 J. Appl. Phys. 47 4318
Louchet F 1981 Philos. Mag. A 43 1289
Mihara M and Ninomiya T 1975 Phys Stat Sol (a) 32 43
Minowa K and Sumino K 1992 Phys. Rev. Lett. 69 320
Patel J R, Testardi L R and Freeland P E 1976 Phys. Rev. B 13 3548
Sato M and Sumino K 1977 Proc. 5th Int. Conf. on High Voltage Electron Microscopy, eds T Imura and H Hashimoto (Tokyo: Japanese Society of Electron Microscopy) pp 459-462
Sumino K 1987 Proc. 7th Int. School on Defects in Crystals, ed E Mizera (Singapore: World Scientific) pp 495-513
Sumino K 1989a Inst. Phys. Conf. Ser. No. 104 pp 245-256
Sumino K 1989b Point and Extended Defects in Semiconductors, eds G Benedek, A Cavallini and W Schröter (New York/London: Plenum) pp 77-94
Sumino K 1993 Handbook on Semiconductors, ed S Mahajan (Amsterdam: North-Holland) in press
Sumino K and Harada H 1981 Phil. Mag. A 44 1319
Sumino K and Sato M 1979 Kristall und Technik 14 1343
Yonenaga I and Sumino K 1989 J. Appl. Phys. 65 85
Yonenaga I and Sumino K 1991 Appl. Phys. Lett. 58 48
Yonenaga I and Sumino K 1993a J. Appl. Phys. in press.
Yonenaga I and Sumino K 1993b J. Appl. Phys. 73 1681
Yonenaga I and Sumino K 1993c J. Cryst. Growth 126 19

Inst. Phys. Conf. Ser. No 134: Section 2
Paper presented at Microsc. Semicond. Mater. Conf., Oxford, 5–8 April 1993

Structural properties of dislocations in plastically deformed InP

M Luysberg, D Gerthsen and K Urban

Institut für Festkörperforschung, Forschungszentrum Jülich GmbH, Postfach 1913, D-W5170 Jülich, Germany

ABSTRACT: The dislocation structure induced by the uniaxial compression along the <123> direction was studied by high-resolution and conventional transmission electron microscopy (TEM). The evolution of the dislocation structure was investigated for samples deformed to the easy-glide region (stage I), the first work hardening stage (stage II) and the first recovery stage (stage III). The measurements of the dissociation widths yielded a stacking fault energy of $18.7 \pm 1.3 \text{mJ/m}^2$. Lattice images of edge dislocations were compared with simulated images.

1. INTRODUCTION

The structural properties of dislocations in metals, elemental semiconductors and GaAs induced by plastic deformation have been extensively examined in the past (e.g. Steeds 1966, Alexander 1968, Boivin et al 1990). However, only few TEM studies of deformation-induced dislocations in InP exist. Gall et al (1987) observed dislocations along the <112>-directions after the deformation at 750°C. Brasen and Bronner (1983) found a cell structure of dislocations aligned along the <110>-directions induced by a deformation at lower temperatures. In samples deformed at 600°C Gottschalk et al (1978) found dislocation dipoles to be the dominating dislocation type. Apart from the stress-dependent dislocation morphology the atomic structure of the dislocation cores is another structural property of interest which has - to our knowledge - not been studied in InP.

Our goal was the investigation of the dislocation properties on different scales. The evolution of the dislocation arrangement with increasing strain and the stacking fault energy γ of undoped and Zn-doped InP crystals were investigated by conventional TEM. High-resolution transmission electron microscopy (HRTEM) was used to study the properties dislocations on an atomic scale.

2. EXPERIMENTAL TECHNIQUES

Undoped and Zn-doped (up to $2 \times 10^{18} \text{cm}^{-3}$) InP single crystals were deformed in compression at 500°C and 600°C with constant strain rates around 10^{-4} s^{-1} along the $[3\bar{2}\bar{1}]$ axis.

To study the dislocation morphology foils parallel to the primary glide plane $(1\bar{1}1)$ were prepared. For the HRTEM samples, foils with normals parallel to the dominant dislocation line directions ([121]- and <110>) were chosen. The conventional characterization of the dislocations included the determination of the line directions, Burgers vector analyses with at least two extinction conditions and the evaluation of the dislocation density N_{dis}. The dissociation widths were examined by the weak-beam technique. The atomic structure of the dislocations was studied with a JEOL 4000EX electron microscope with a spherical aberration constant of 1 mm, spread of focus 10 nm and an angle of beam convergence of 1.8 mrad.

3. EXPERIMENTAL RESULTS AND DISCUSSION

As for other semiconductors several stages in the stress-strain curves $\tau(\varepsilon)$ can be distinguished, which are characterized by different work-hardening coefficients $\Theta = d\tau/d\varepsilon$ (Siethoff et al 1988). At the lower yield point in the easy-glide region, a high dislocation density $(5 \times 10^8 \text{cm}^{-2})$ of the primary slip system is observed. The dominating dislocation type is the edge-dislocation dipole with a [121] line direction (Luysberg et al 1992), which most probably is formed by double cross slip of screw dislocations.

Θ increases in the first work hardening stage due to the activation of a secondary slip system at higher stresses. The interaction of these dislocations with the primary dislocations leads to the formation of a nearly rectangular dislocation network, which is extended parallel to the primary slip plane (Fig. 1). Lomer-Cottrell dislocations are formed as a result of these interactions. In contrast to stage I, only few dislocation dipoles and dislocation loops can be observed (arrows in Fig. 1).

On the further increase of the strain a recovery stage is reached which is characterized by a reduction of Θ. The recovery can be ascribed to the formation of subgrains (Fig. 2) with cell interiors almost free of dislocations surrounded by cell walls with $N_{dis} > 10 \times 10^{10} \text{cm}^{-2}$. Besides a two-dimensional network consisting mainly of screw dislocations, the interaction of dislocations of the primary and several secondary slip systems leads to a three-dimensional network in the subgrain boundaries. More detailed analyses indicate that the dominant recovery mechanism could be the cross slip of screw dislocations.

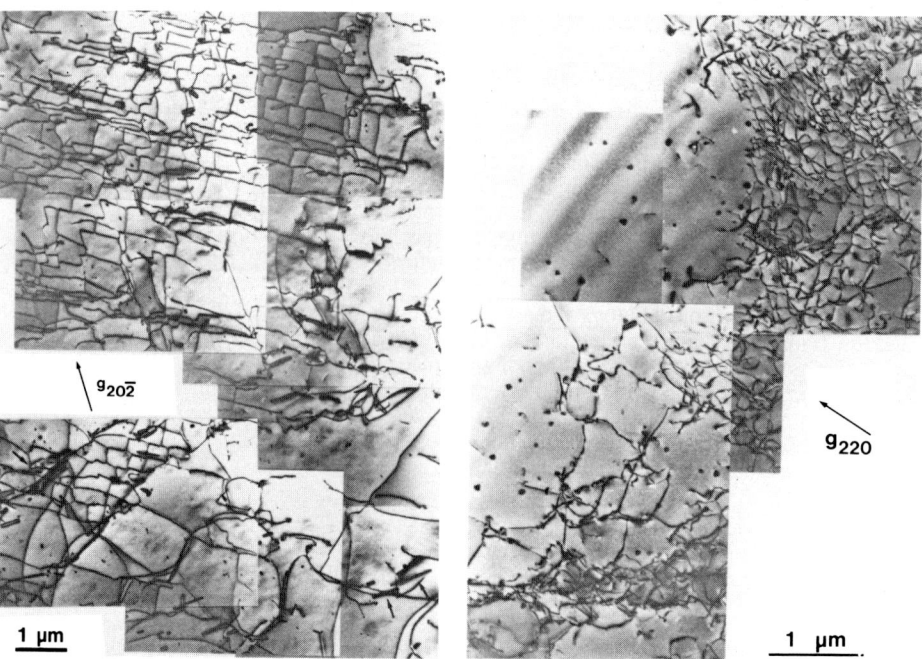

$g_{20\bar{2}}$

g_{220}

1 µm

1 µm

Fig.1: Dislocation structure in stage II (imaging vector $g = [20\bar{2}]$): The interaction of primary and secondary dislocations leads to a nearly rectangular network structure in the main slip plane.

Fig.2: Dislocation structure in stage III: Subgrain with a cell interior almost free of dislocations and cell walls with $N_{dis} > 10^{10} \text{cm}^{-2}$. The black dots result from surface contamination.

Weak-beam dark field images show that almost all dislocations are dissociated into Shockley partial dislocations. The measured dissociation widths Δ are plotted into Fig. 3 for three different samples deformed to the lower yield point (Zn-doped InP: crosses, undoped InP: circles and triangles). Significant differences cannot be observed indicating that the Zn-doping does not influence Δ within the margin of error. Using anisotropic elasticity theory and a least square fit, γ was determined to be $18.7 \pm 1.3 \mathrm{mJ/m^2}$ in agreement with the value of Gottschalk et al. (1978).

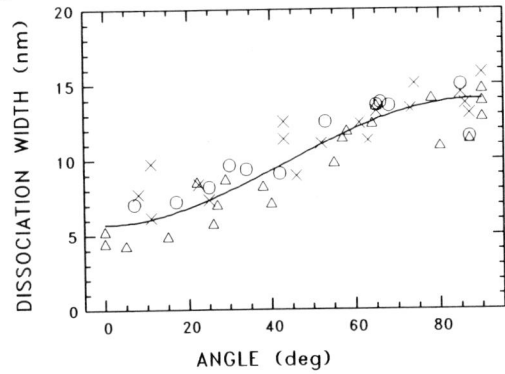

Fig. 3: Plot of Δ versus the angle between Burgers vector and dislocation line for two undoped samples (triangles and circles) and a Zn-doped ($10^{18} \mathrm{cm^{-3}}$) sample (crosses). The error bars of about $\pm 1 \mathrm{nm}$ for the Δ values are omitted for clarity. The solid line was calculated for a stacking fault energy of $18.7 \mathrm{mJ/m^2}$.

Several types of dislocations (screw-, 60°-dislocations, Z-dipoles, and edge-dipoles) have been analysed by HRTEM. As an example, an end-on image of a 60° partial dislocation of a dissociated edge dislocation along the [121]-projection is shown in Fig.4. The Burgers circuits around the core indicate one inserted $(20\bar{2})$-lattice plane. The dislocation is shown at two different underfocus values $\Delta f = 25 \mathrm{nm}$ and $\Delta f = 5 \mathrm{nm}$ at a thickness $t = 22 \mathrm{nm}$. Δf and t were determined by the comparison of the experimental images in the vicinity of the dislocation with simulated images of the perfect InP structure. The polarity, i.e. the position of the In and P $(1\bar{1}1)$ planes, was obtained from a third image of the same series at $\Delta f = 100 \mathrm{nm}$. The dislocation was therefore determined to be a β-dislocation, and the possible core configurations could be narrowed down to the P shuffle set or In glide set. The models for the multi-slice image simulations were obtained in a first approximation by applying isotropic elasticity theory (Hirth and Lothe 1982). The simulated images yielded a stronger bending of the $(1\bar{1}1)$ planes than the experimental images. Straight $(1\bar{1}1)$ planes were therefore assumed in the two atomic models in Fig. 4. The simulated images of both models agree well with the experimental images - in particular the behaviour of the differences in the dilatational and compressional regions around the dislocation cores. However, no significant difference is observed at the dislocation core. Thus at these Δf the identification of the 'true' model of the dislocation core is not possible.

ACKNOWLEDGMENTS

The authors are grateful to J. Völkl (Siemens AG, Erlangen) for supplying InP samples and H. G. Brion (Institut für Metallphysik, Göttingen) for carrying out the deformations.

REFERENCES

Alexander H 1968 phys. stat. sol. <u>27</u> 391 and 725
Boivin P, Rabier J and Garem H 1990 Phil. Mag. A <u>61</u> 619
Brasen D and Bronner W A 1983 Mat. Sci. Eng. <u>61</u> 167
Gall P, Peyrade P, Coquille R, Reynaud F, Gabillet S and Albacette A 1987 Acta Met. <u>35</u> 143
Gottschalk H, Patzer G and Alexander H 1978 phys. stat. sol. (a) <u>45</u> 207
Hirth J P and Lothe J 1982 Theory of Dislocations (NewYork: Wiley) chap. 5

Luysberg M, Gerthsen D and Urban K 1992 Phil. Mag. Lett. <u>65</u> 121
Siethoff H, Ahlborn K, Brion H G and Völkl J 1988 Phil. Mag. A <u>57</u> 235
Steeds J W 1966 Proc. Roy Soc. A <u>292</u> 343

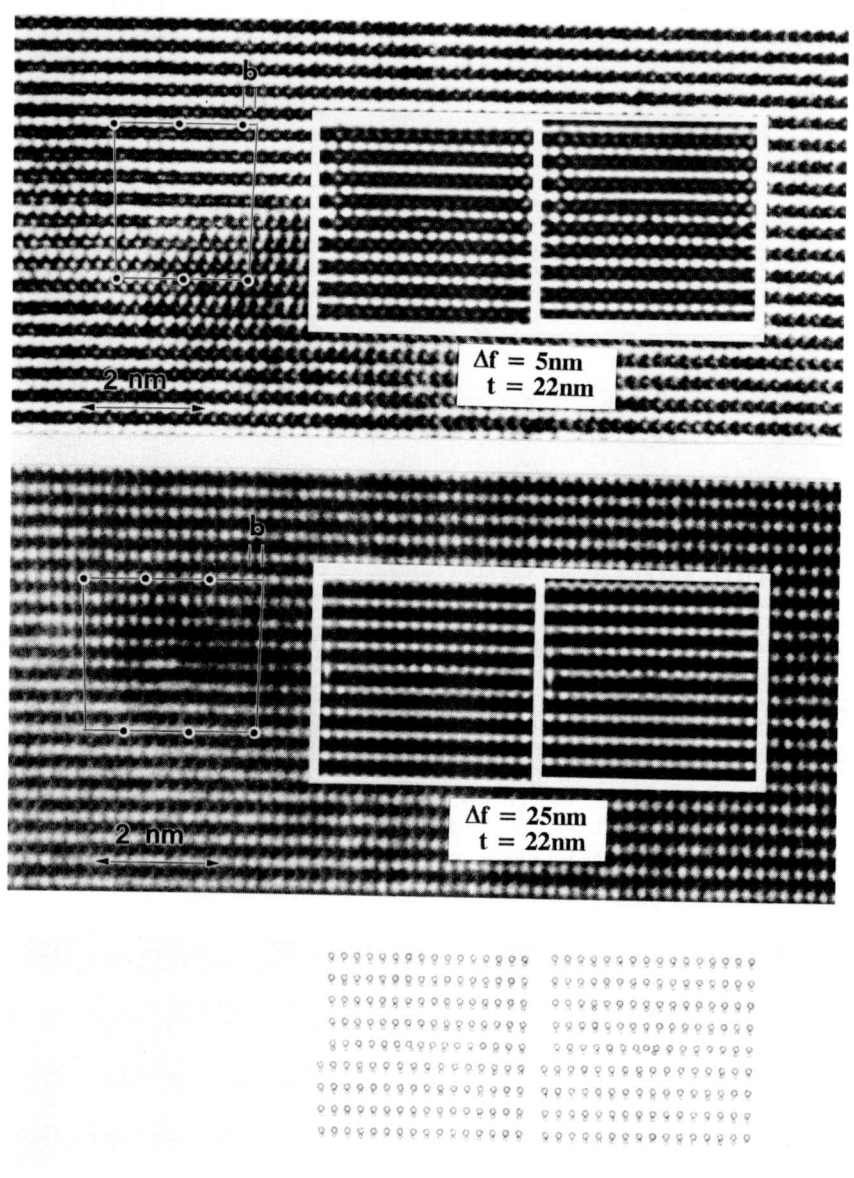

P SHUFFLE-SET   In GLIDE-SET

Fig. 4: HREM images af a 60° partial dislocation at two different underfocus values. The insets show simulations obtained for the two dislocation models at the bottom.

Inst. Phys. Conf. Ser. No 134: Section 2
Paper presented at Microsc. Semicond. Mater. Conf., Oxford, 5–8 April 1993

Investigation of dislocations in plastically deformed InP

E Le Bourhis, A Zozime, A Rivière and J-P Rivière

Laboratoire de Physique des Matériaux, CNRS-Bellevue, F-92195 Meudon cédex, France

ABSTRACT: Dislocations generated by microhardness indentation at 400°C on the (001) surface of InP:S are investigated by transmission electron microscopy and by scanning electron microscopy in the cathodoluminescence mode.

1. INTRODUCTION

It has been shown that the electrical properties of dislocations in InP are correlated with their type (Zozime and Schröter 1989, 1990). However, there is still a lack of information about the structure of dislocations generated by microhardness indentation. We investigated, in the present work, the structure of α and β dislocations by transmission electron microscopy (TEM) and scanning electron microscopy (SEM) in the cathodoluminescence (CL) mode.

2. SAMPLE PREPARATION

We used sulphur doped InP crystal grown by the liquid encapsulated Czochralski (LEC) technique ($n = 10^{18}$ cm^{-3}). The <100>-oriented samples were chemically mechanically polished with 0.5% bromine-methanol solution. They were then deformed by microhardness indentation performed with a Vickers indentor in atmospheric air at 400°C using a load of 0.25 to 1N for 30s.

3. MICROSTRUCTURE OBSERVATION IN SEM

Secondary electron images show steps corresponding to the emergence of glide planes in the <110> directions and cracks whose lengths increase with the load (see Fig. 1).

The indentation rosettes observed in CL show three zones (see Fig. 2): i) a central zone in the vicinity of the indentation characterised by strong recombination at the defects, ii) a zone where glide planes cross and iii) the ends of the rosette arms composed of black dots.

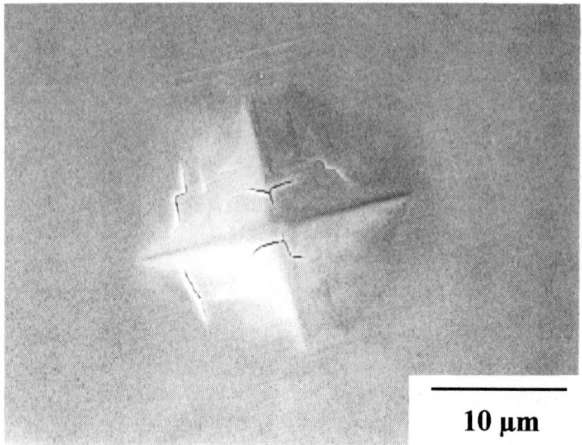

Fig. 1.   Secondary electron image of a Vickers indentation (load 0.25N).

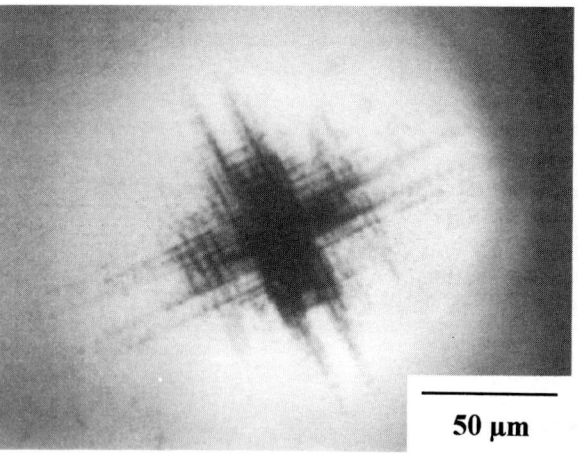

Fig. 2.   Cathodoluminescence image of an indentation rosette (load 0.25N).

4. MICROSTRUCTURE OBSERVED IN TEM

The rosette arms are composed of partial dislocations separated by stacking faults with a width of between 100 and 500nm (see Fig. 3). Let us note that such a stacking fault width is more than one order of magnitude larger than the dissociation width of a dislocation. The mechanism by which these partial dislocations are generated is not well understood. The formation of groups of partial dislocations can be observed (see Fig. 4).

Fig. 3. Transmission electron image of the end of a rosette arm (**g** 220-type).

Fig. 4. Transmission electron image showing a group of partial dislocations.

5. CONCLUSION

We have obtained results similar to those found by Haswell et al (1991) for GaAlAs and GaInAs deformed by microhardness indentation at room temperature. Both rosette arms showed the same microstructure, contrary to the observations of Lefebvre et al (1987) and Höche and Schreiber (1984) in GaAs deformed at room temperature and annealed at 350°C for 1h, where one arm was composed of perfect α dislocations and the other one of partial β dislocations.

Our investigations have shown that each black dot observed in SEM/CL at the end of rosette arms in InP:S deformed at 400°C corresponds to the electrical activity of either one or several partial dislocations.

ACKNOWLEDGEMENT

We are grateful to G Jacob (Crismatec Inpact, Plombière, France) for providing the InP samples.

REFERENCES

Haswell R, Bangert U and Charsley P 1991 Phil. Mag. Lett. **63**, 67
Höche H R and Schreiber J 1984 Phys. Stat. Sol. **a86**, 229
Lefebvre A, Androussi Y and Vanderschaeve G 1987 Phys. Stat. Sol. **a99**, 405
Zozime A and Schröter W 1989 Phil. Mag. **B60**, 565
Zozime A and Schröter W 1990 Appl. Phys. Lett. **57**, 1326

Inst. Phys. Conf. Ser. No 134: Section 2
Paper presented at Microsc. Semicond. Mater. Conf., Oxford, 5–8 April 1993

An EBIC investigation of alpha, beta and screw dislocations in gallium arsenide

S A Galloway, P R Wilshaw and T S Fell

Department of Materials, University of Oxford, Parks Road, Oxford, OX1 3PH, U.K.

ABSTRACT: Well characterised alpha, beta and screw dislocations have been created by micro-indentation and subsequent bending of GaAs specimens and studied, using the EBIC technique, in terms of their behaviour as recombination centres. It is found that all dislocations in LEC and LPE material behave in approximately the same way with their recombination strength increasing with decreasing temperature. In "ultra-pure" MOCVD material the dislocation recombination efficiency is different to that of LEC and LPE material and it is concluded that different recombination centres are active in this case. In MOCVD material large differences in behaviour are found depending on the way the dislocations are produced. This may indicate a dependence of recombination on different core configurations or point defect decoration of the dislocations.

1. INTRODUCTION

In group III-V semiconductors the cores of α and β dislocations are associated with a local loss of stoichiometry arising from the "extra half planes" which terminate respectively in a row of either all group V or all group III atoms, if the glide configuration for the core is assumed. Since III-V semiconductors are partially ionic, the "extra half plane" represents a line of charge of atomic dimensions which although rapidly eletrostatically screened may possess interesting electrical properties. This ionic charge of the dislocation is additional to that due to any deep levels which may be introduced by the disrupted atomic bonds at the dislocation. One method of studying the electrical activity of such dislocations is Electron Beam Induced Current (EBIC) microscopy. In this case experiments can be performed to measure the minority carrier recombination efficiency of the defects as a function of temperature, doping concentration, and minority carrier concentration which, provided a model exists, allows the data to be interpreted in terms of the electronic structure of the defects. Considerable progress has been made in this way for silicon (Wilshaw et al 1989) whilst the situation for GaAs and other III-Vs is generally less well advanced. This may in part be due to difficulties in obtaining straight, well characterised dislocations in high quality material. In the present work we have attempted to overcome these difficulties so that the recombination activity of individual α and β dislocations can be ascertained in "ultra-pure" Metal Organic Chemical Vapour Deposition (MOCVD) epitaxial GaAs and also "normal" Liquid Encapsulated Czochralski (LEC) and Liquid Phase Epitaxy (LPE) GaAs.

2. EXPERIMENTAL

Specimens with well characterised dislocations were produced by indentation of bars of GaAs on an (001) surface using loads of 0.2 to 2g and a diamond indentor. These bars were then heated to 400°C in a molybdenum 4-point bending apparatus and a load was applied to produce a tensile stress (<10MPa) along the long axis of the bar which was cut to be either [110] or [1$\bar{1}$0]. This process can be thought of as producing two sets of dislocations. The first, which are punched out by the residual stress of the indent to form a rosette pattern are generated independently of the application of a tensile stress and move up to 40μm from the indent. Their Burgers vector is parallel to the surface and the surface threading segments are α or β (60°) dislocations depending on whether they are punched out along [110] or [1$\bar{1}$0] directions (Qin Cai-Dong 1989).

The second set of dislocations experience a resolved shear stress due to the applied tensile bending stress which causes them to expand as half loops, the leading segment moving perpendicular to the applied tensile stress. The dislocations tend to move as long as the load is applied, and in the specimens studied in this work these dislocations were made to travel up to 200μm from the indent. In the following the first set of dislocations will be referred to as "punch-out" dislocations whilst the other will be called "bend" dislocations.

The bend process produces dislocation half loops with Burgers vector inclined to the surface and the surface parallel segment is either α or β depending on the tensile direction. The line direction of the dislocation segment which threads the surface can be determined by repeated etching using the Dilute Sirtl with Light technique (DSL) (Weyher and Van De Ven 1986), and optical microscopy. Following the work of Di Persio and Abbas (1989) and Yonenaga and Sumino (1993) who performed X-ray topography studies of dislocations in specimens produced in a similar way, it is thought that a threading bend dislocation segment is 60° when the angle between it and the surface parallel segment is 120°, and it is a screw segment if its line direction "flips back" so that it makes an angle of 60° with the surface parallel segment. Repeated etching of our specimens showed that β threading dislocations could be produced in this way. It was however found difficult to generate threading α bend segments since their high mobility compared to screw dislocations always led to the threading segment flipping back thus producing a screw segment. TEM investigation of some of the dislocations produced showed that punch-out dislocations tended to be curved whilst β bend dislocations were nearly straight lying parallel to <110> directions. Thus punch-out dislocations, although referred to in the rest of the text as α or β, were not pure 60° in character.

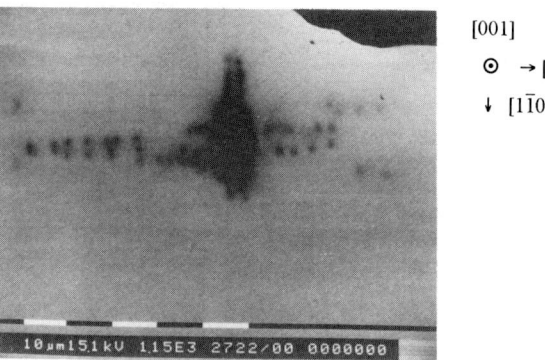

[001]

⊙ → [110]

↓ [1$\bar{1}$0]

Figure 1. EBIC micrograph of a rosette of punch-out dislocations and also some bend dislocations produced by indentation and bending of an MOCVD specimen. The bend dislocations show less contrast and have moved furthest from the indent along the [110] direction.

In summary for a (001) specimen and by choice of either [110] or [1$\bar{1}$0] directions the indentation and bending process produced the following dislocations threading the surface; curved α and β punch-out dislocations, screw bend dislocations and straight β bend dislocations. Figure 1 shows an EBIC micrograph of an indent with both the punch-out dislocations and some of the bend dislocations visible.

In addition a single (111) LPE wafer was indented and this also produced α and β dislocations in the form of rosettes with Burgers vectors parallel to the surface. Bending with a [21$\bar{3}$] tensile direction caused the more mobile α threading segments to travel further away from the indent than the β threading segments and the two dislocation types were thus easily distinguished under EBIC observation.

Au Schottky barriers and In/Au ohmic contacts were fabricated for the EBIC investigations and only results obtained from surface threading dislocation segments are reported here. The materials examined were "ultra-pure" 4×10^{15}cm^{-3} and 10^{16}cm^{-3}, n-type, 10μm thick MOCVD epitaxial layers grown on conducting GaAs substrates by Epitaxial Products International Ltd, 10^{16}cm^{-3}, n-type LPE and also 10^{17}cm^{-3}, n-type LEC GaAs. Minority carrier diffusion length L_D, measurements performed using the method of Wu and Wittry (1978) on deformed samples in regions far from any dislocations showed that L_D was always ≥15μm in the MOCVD specimens, whilst in the LEC and LPE material respectively, L_D decreased from 1.2 to 0.25μm and 1.6 to 0.4μm over the temperature range 350 to 100K. L_D was also found to decrease significantly in the LEC and LPE material when the electron beam current I_b was reduced from 10^{-9} to 10^{-12}A. These trends are in agreement with the results of Alexander et al (1990).

3. SIMULATION OF EBIC CONTRAST

In order that the EBIC technique can be used to obtain electrical information concerning dislocations, it is necessary to be able to relate the measured contrast to the actual recombination efficiency γ, of the defect. In addition, for measurements of dislocations threading the surface Schottky barrier, part of their length is in a depleted region where the electrostatic field will alter the recombination process compared to that in the bulk. Thus, a numerical simulation was developed based on Donolato's model for the EBIC contrast of point defects (Donolato 1978), so that experimental results could be related to the actual dislocation recombination efficiencies. In this simulation the dislocation was approximated by 2000 identical point defects in a row inclined to the surface at an angle appropriate for the dislocation being studied. Calculations, which will be reported in more detail elsewhere, were then made of the contrast variation as a function of minority carrier diffusion length, accelerating voltage, and depletion region width. Good agreement was found between this model and experiment for the dependence of contrast on accelerating voltage for a specimen with a measured diffusion length of 15μm, thus indicating the accuracy of the model. When the model was used to simulate the experimental conditions used to obtain quantitative data, it was found that in no case did more than 15% of the measured contrast originate from the segments of dislocation in the depletion region. It is thus a good approximation that the experimental results represent the recombination behaviour of dislocations in the bulk material, ie. not in the depletion region.

Further, it was also found that for the case of MOCVD material where the diffusion length was always ≥15μm, the measured EBIC contrast was directly proportional to γ and hence any measured changes in the EBIC contrast directly reflect changes in γ. However, for the LEC and LPE materials for which L_D was ≤1.6μm and changed as a function of other experimental parameters, the EBIC contrast varied strongly with L_D. Thus to obtain the variation in γ from measured changes in the EBIC contrast in these materials, the variation in L_D had to be taken into account using the simulation.

4. RESULTS

Measurements were made of the temperature dependence of the EBIC contrast of α, β and screw dislocations in the LPE and LEC material. In each case the contrast was found to decrease as the temperature was varied between 350 and 100K. Figure 2 shows data obtained from a β bend dislocation in 10^{17}cm^{-3} doped LEC material measured with I_b equal to 10^{-10}A at 15kV. The contrast decreases from 15 to 8%. Also plotted is the true recombination efficiency of the dislocation which has been calculated from the contrast data using the experimentally measured diffusion length values shown in figure 3 and the simulation described in Section 3. The recombination efficiency is found to increase with decreasing temperature which is the opposite of what might have been expected from the contrast data if the diffusion length variation had not been taken into account. Very similar shaped curves for recombination efficiency were found for all α, β and screw dislocations measured in LPE and LEC material and demonstrate that the trends found by Eckstein et al (1990), Eckstein and Habermeier (1991) are true for all dislocation types studied in LPE and LEC material.

Figure 2. EBIC contrast and recombination efficiency (AU) for a β bend dislocation in 10^{17}cm^{-3} n-type LEC GaAs as a function of temperature with $I_b=10^{-10}$A at 15kV.

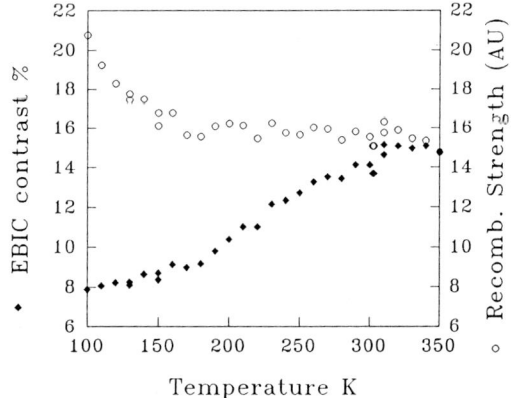

Temperature K

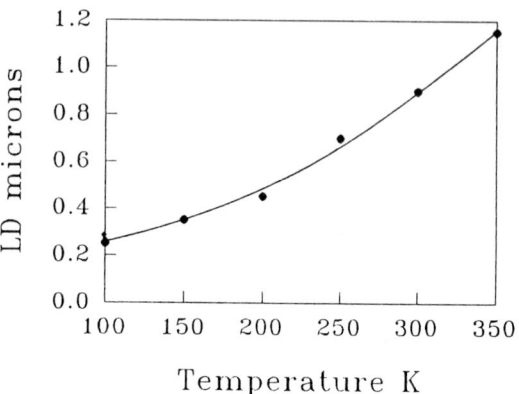

Figure 3. The temperature dependence of the minority carrier diffusion length L_D measured using the method of Wu and Wittry (1978) using a beam current of 10^{-10}A. Data obtained from the LEC GaAs specimen of figure 2.

Measurements were also made on dislocations in the "ultra-pure" MOCVD material for which any differences in contrast as a function of temperature or between specimens are directly proportional to the true recombination strength of the dislocations studied (see Section 3.). Very different behaviour was found for dislocations in MOCVD material to that of the dislocations in LPE and LEC material. The contrast and hence recombination efficiency of the bend dislocations in the MOCVD material was found to be only about half that of the punch-out dislocations and the temperature dependence was different. In all cases in MOCVD material, α and β punch-out dislocations were found to display contrast largely independent of temperature but with a slight decrease in contrast at lower temperatures whilst β bend dislocations, with lower contrast, varied more strongly with a maximum at 200-250K. Typical behaviour is shown in figure 4.

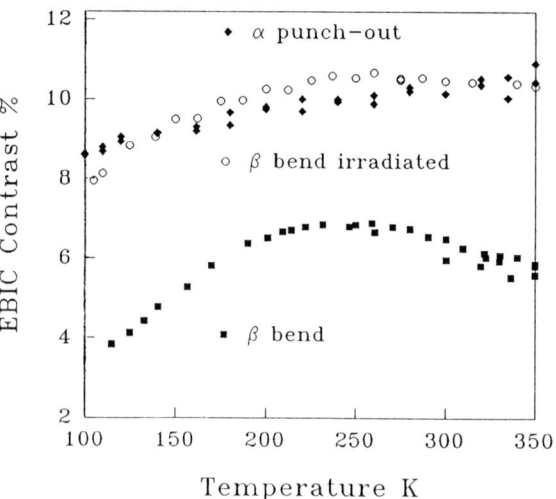

Figure 4. The temperature dependence of three dislocations in 4×10^{15}cm^{-3} n-type MOCVD GaAs with $I_b=10^{-11}$A, at 30kV.

Another effect was found whereby the EBIC contrast of β and screw bend dislocations increased as a function of the time during which the dislocation and the surrounding area was irradiated. For example, using a beam current of 10^{-10}A at 30kV and 300K the contrast of a screw bend dislocation was found to increase from ~5% to ~10% and then remain approximately constant under further irradiation, figure 5. Figure 4 shows the temperature dependence of an irradiated β bend dislocation.

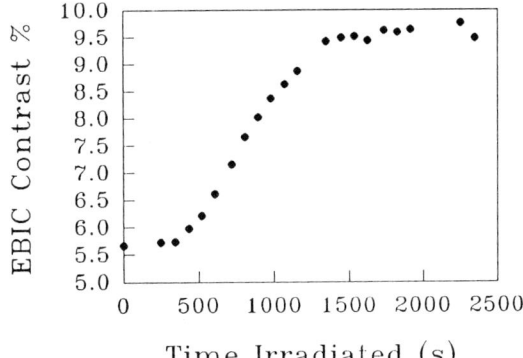

Figure 5. EBIC contrast of a screw bend dislocation in $4 \times 10^{15} cm^{-3}$ n-type MOCVD GaAs as a function of time whilst irradiated with $I_b = 10^{-10} A$ at 30kV and 300K.

The following table summarises EBIC contrast and its dependence on electron beam irradiation for the various dislocations and materials studied in this work.

Dislocation type	Material	Doping (cm^{-3})	Typical [1] contrast (%)	Dependent on electron irradiation	Typical [1] contrast after irradiation (%)
β bend	MOCVD	4×10^{15}	4.5	YES	8.5
0° bend	MOCVD	4×10^{15}	4.5	YES	8.5
β bend	MOCVD	1×10^{16}	6.5	YES	11
β punch-out	MOCVD	4×10^{15}	9	NO	
β punch-out	MOCVD	1×10^{16}	13.5	NO	
β punch-out [2]	LPE	$\sim 10^{16}$	15	NO	
β bend	LEC	$\sim 10^{17}$	12.5	?	?
0° bend	LEC	$\sim 10^{17}$	12.5	?	?
α punch-out	MOCVD	4×10^{15}	9	NO	
α punch-out	MOCVD	1×10^{16}	13.5	NO	
α bend [2]	LPE	$\sim 10^{16}$	~ 15	NO	

1. Typical contrast using 20kV accelerating voltage and $I_b = 1 \times 10^{-11} A$ at 300K.
2. Produced in a (111) wafer with $\langle 21\bar{3} \rangle$ tensile axis.

5. CONCLUSIONS AND DISCUSSION

The following conclusions are made:

(i) Dislocations in LEC and LPE material behave similarly to each other but very differently to those in "ultra-pure" MOCVD material.
(ii) In LEC and LPE material recombination efficiency increases with decreasing temperature, particularly at low temperature.
(iii) Bend dislocations show less contrast than punch-out dislocations in MOCVD material.
(iv) Dislocation contrast and recombination efficiency increase with doping concentration in MOCVD material.
(v) β and screw bend dislocations in MOCVD material are sensitive to electron irradiation and when irradiated show similar contrast values to punch-out dislocations.
(vi) In MOCVD material the dislocation's contrast and hence recombination efficiency is largely independent of (for punch-out), or decreases with decreasing temperature at low temperatures (for bend dislocations).
(vii) Only small differences were found between α, β and screw dislocations.

We conclude from the different temperature dependence of the recombination behaviour in LEC and LPE compared with the MOCVD material, that different sites at the dislocation are active in each case. It seems likely that in the case of LEC and LPE material which contain high concentrations of point defects that the recombination behaviour may be controlled by the segregation of such defects to the dislocation and that recombination at dislocations in the "ultra-pure" MOCVD material is probably more representative of the dislocations themselves. It is possible however, that smaller quantities of the same or different point defects are present even at dislocations in the "ultra-pure" material.

There are differences in recombination behaviour between punch-out dislocations and bend dislocations in MOCVD material. The punch-out dislocations move in response to the residual stress field around the indent, they are curved and rapidly come to rest during the specimen's high temperature treatment. The bend dislocations on the other hand are straight, move continuously in response to the externally applied stress during the high temperature bend and then are rapidly quenched to room temperature thereby leaving very little time for the dislocations to be stationary at a high temperature. It is thus likely that the differences in electrical behaviour are due to either a different core configuration caused by the different stress regimes affecting the bend and punch-out dislocations, or different point defect concentrations which agglomerate to the stationary and moving dislocations. We postulate that the effect of the electron beam on β and screw bend dislocations is either to allow recombination enhanced dislocation motion to enable the dislocations to take up different core configurations, or to stimulate recombination enhanced diffusion of point defects which then getter to the dislocations. It is noted that although substantial differences in the electronic properties of α and β dislocations have not been found, this may yet be due to effects obscuring their true behaviour in a manner similar to those which produce different activity at bend and punch-out dislocations.

ACKNOWLEDGEMENTS

The authors are very grateful to Epitaxial Products International Ltd. for kindly providing the MOCVD wafers, to Dr R E Mallard for TEM investigations, to Dr S G Roberts and Dr P D Warren for helpful discussions concerning dislocation geometries, and to The Royal Society and SERC for financial support.

REFERENCES

Alexander H, Dietrich S, Huhne M, Koble M and Weber G 1990 Phys. Stat. Sol.(a) 117 417
Di Persio J, Abbas M 1989 Inst. Phys. Conf. Ser. No. 100 391
Donolato C 1978 Optik 52 19
Eckstein M, Habermeier H-U 1991 Journal De Physique IV Coll. C6 23
Eckstein M, Jakubowicz A, Bode M, and Habermeier H-U 1990 SPIE 1284 236
Qin Cai-Dong 1989 D. Phil Thesis, Department of Materials, University of Oxford, UK
Weyher J L, Van De Ven J 1986 J. Crystal Growth 78 191
Wilshaw P R, Fell T S, and Booker GR 1989 "Point and Extended Defects in Semiconductors", ed.
    Benedek G, Cavallini A, Schroter W Plenum Publishing Corp. 243
Wu C J, Wittry D B 1978 J. Appl. Phys. 49 5 2827
Yonenaga I, Sumino K 1993 J. Crystal Growth 126 19

Structure of dislocations formed in Ge(Si) by glide on secondary glide planes

M Albrecht[1], D Stenkamp[2], H P Strunk[1], W Jäger[2], P O Hansson[3] and E Bauser[3]

[1]Universität Erlangen-Nürnberg, Institut für Werkstoffwissenschaftwen VII, Cauerstr. 6, D-8520 Erlangen, F.R.G.
[2]Forschungszentrum Jülich GmbH, Institut für Festkörperforschung, Postfach 1913, D-5170 Jülich, F.R.G.
[3]Max-Planck-Institut für Festkörperforschung, Heisenbergstr.1, D-7000 Stuttgart 80, F.R.G.

ABSTRACT: We analyse by high resolution transmission electron microscopy the structure of perfect dislocations ($b=a/2<110>$) glissile in the {311} and {110} secondary glide planes of Ge grown onto Si (001) and Si(110). Dislocations in the Ge/Si(011) interface have $<112>$ line directions and show a constricted core structure. Dislocations at the Ge/Si(001) interface have a core that is dissociated by climb into two partials ($b=a/4<110>$). The stacking fault formed by climb has an energy of $\gamma=120$ mJm^{-2}. A possible process that leads to this configuration is discussed in terms of a short circuit diffusion process.

1. INTRODUCTION

In semiconductors with diamond lattice structure the $a/2<110>${111}glide system has been observed generally (e.g. Alexander 1961). This observation is explained in terms of a high Peierls stress, which is a consequence of the covalent binding. The results are obtained by investigating dislocations in deformed single crystals. It should be noted that these experiments are limited to resolved shear stresses $\tau<300$ MPa.

Recently it has been demonstrated that secondary glide systems ($a/2<110>${110}, $a/2<110>${113}) can become active (Bonar et al 1992, Albrecht et al. 1992) when resolved shear stresses induced by the lattice mismatch in heteroepitaxial layers reach values comparable to the theoretical shear strength of G/10 (G is the shear modulus). The particular conditions are essentially ruled by the Schmid factor. The glide plane activation can be explained by a mechanical equilibrium analysis, which takes into account a lattice frictional force due to a static Peierls barrier (Chidambarrao et al 1990, Albrecht et al 1993). This paper summarizes results concerning structure and dissociation of dislocations in these secondary glide systems as evidenced by high resolution transmission electron microscopy (HRTEM).

2. EXPERIMENTAL

Ge$_x$Si$_{1-x}$ (x=0.85 or 0.5) layers are grown by liquid phase epitaxy from Bi solution at 810°C on both (001) and (110)-orientated Si substrates (P.O.Hansson et al.1990). These layers have a lattice mismatch of 3.6% and 2.1% for Ge concentrations of x=0.85 and x=0.5 respectively. The layers have a thickness of 300 nm. The growth rate is around 2 nm/s. Specimens are prepared for both plan view and cross sectional transmission electron microscopy (TEM) by mechanical grinding and polishing to about 10 μm followed by ion milling to electron transparency. The dislocation structure is analysed in plan view using the conventional weak-beam dark-field technique in a Philips EM 400 operating at 120 kV. High resolution micrographs are obtained in a JEOL 4000 EX operating at 400 kV. Projections in <010> and <112> line directions of dislocations in the secondary slip systems are utilized.

3. RESULTS AND DISCUSSION

3.1 a/2<110> dislocations with <100> line direction in the {110} glide plane

Figure 1a shows a square grid of misfit dislocations at a Ge$_{0.85}$Si$_{0.15}$/Si(001) interface with a misfit f=3.6%. The two perpendicular sets of misfit dislocations lie along [100] and [010] line directions and have a dislocation interdistance of 8 nm. In Figure 1b only the set of dislocations is visible, which has [010] line direction. Contrast analysis shows the dislocations to have Burgers vectors **b**=a/2[101] and **b**=a/2[011]. Thus the dislocations are of 90° type with the Burgers vector inclined by 45° to the interface. The efficiency of these dislocations to compensate the lattice mismatch, characterised by the edge component **b**$_\perp$, which lies perpendicular to the dislocation line within the interface, is higher than for the generally observed 60° dislocations (**b**$_\perp$=a/2 instead of a/√8 in the case of 60° dislocations). An end-on view by HRTEM of this dislocation type along the [010] line direction is shown in Fig. 2. The closing vector of the Burgers circuit corresponds to a perfect Burgers vector, which lies 90° with respect to the dislocation line, i.e. within the image plane. Two inserted {220} lattice planes (marked by arrows) can be revealed. These inserted lattice planes correspond to partial dislocations which have Burgers vectors **b**$_p$ =a/4[101]. Both partials are separated by a stacking fault which extends over 10 {220} lattice planes perpendicular to the glide plane of the dislocation (contrary to the dissociation stacking faults parallel to the glide plane, as observed in the usual a/2<110>{111} glide system (Ray and Cockayne 1970)). A stick and ball model of this configuration as obtained from the HRTEM micrograph shown in Figure 2b.

According to Frank's criterion (Frank 1949) the observed dissociation a/2[101]→a/4[101]+a/4[101] is energetically favoured and thus expected. The energy of the stacking fault can be assessed from the observed dissociation width of the partial dislocations. An equlibrium is obtained if the repulsive force F between the two straight edge partials perpendicular to their glide plane (i.e. the elastic climb force (Hirth and Lothe 1982)) balances the attractive force, which rises from the stacking fault energy γ, i.e. γ=F:

$$\gamma = \frac{Gb^2}{2\pi(1-v)d} \qquad (1)$$

If we substitute the observed dissociation width d=2nm of Fig 2a, the shear modulus

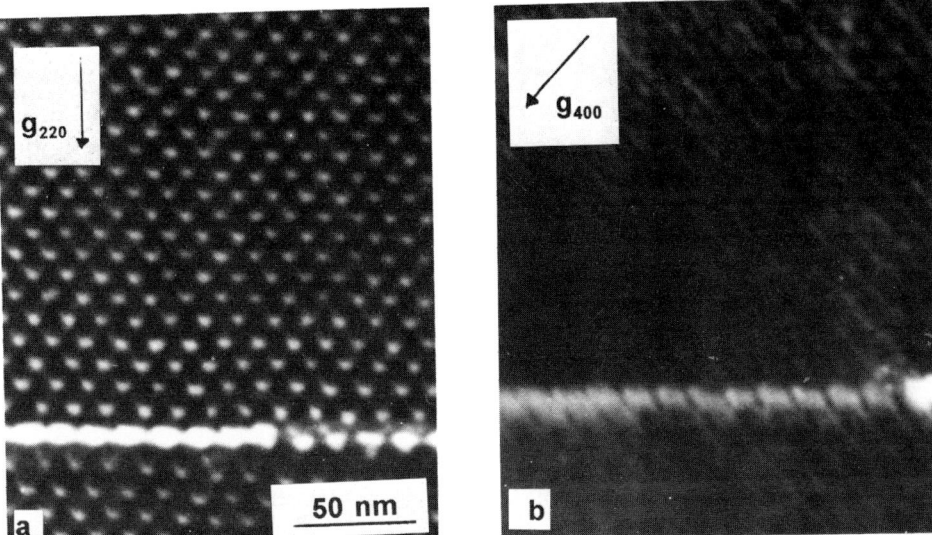

Figure1. Misfit dislocation network at the $Ge_{0.85}Si_{0.15}/Si(001)$ interface. Weak-beam dark-field micrograph in plane view. (a) A perfect network, consisting of two sets of perfect dislocations, which have [100] and [010] line directions is visible (Burgers vectors b=a/2[011] and b=a/2[101] respectively). (b) One set of misfit dislocations is visible (Burgers vector b= a/2[101].

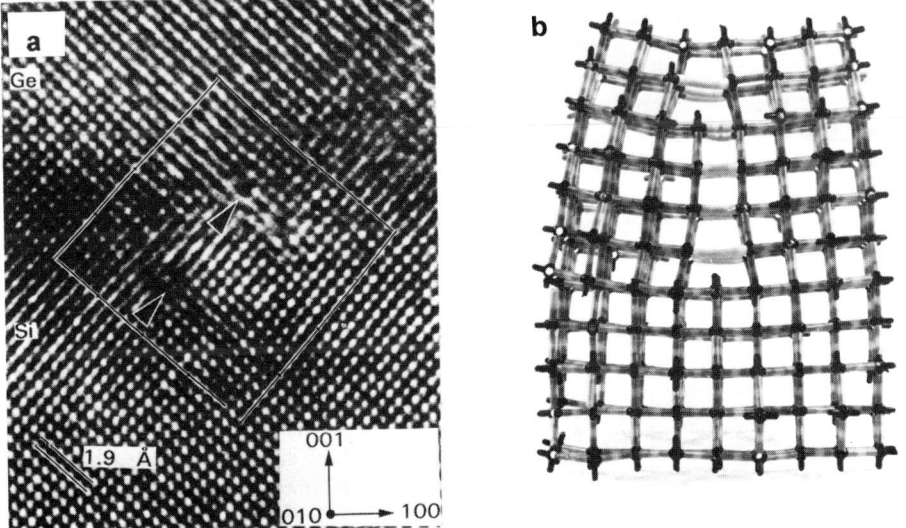

Figure 2. Dissociated 90° dislocation with {110} glide plane in the $Ge_{0.85}Si_{0.15}/Si(001)$ interface. (a) Cross sectional high resolution micrograph in [010] projection. Two additional {220}lattice planes are marked by arrows. The closing vector of the Burgers circuit corresponds to the Burgers vector of a perfect dislocation (b=a/2[10-1]). (b) Stick and ball model of the atomic configuration obtained from Fig 2a.

G=41.1GPa and Poissons ratio ν=0.28 (Landolt-Börnstein 1982) we obtain a stacking fault energy of γ=180 mJm^{-2}. This energy is about three times the value obtained for stacking faults in {111} planes (Ray and Cockayne 1973). It is, however, an upper limit for the stacking fault energy as this dissociation requires climb and the true equilibrium width may not be reached during the growth process of the samples. In a few cases a dissociation width of d=3 nm, corresponding to γ=120 mJm^{-2} is found. A possible process that leads to the observed configuration based on self diffusion is illustrated in Fig 3a-c. Starting from the gissile configuration in Fig 3 a, one row of core atoms diffuses into the position of Fig. 3b. This diffusion process can be described by formation and lateral drift of positive partial jog pair in one {220} lattice plane and a negative partial jog pair on the adjacent lattice plane. The thereby formed stacking fault may favour diffusion within the plane of the stacking fault of the second row of core atoms into the position of Fig. 3 c. As the growth temperature of T=810°C is sufficiently high to support self diffusion an external source of point defects is not required and this short circuit diffusion may repeat itself until the energetic equilibrium is reached.

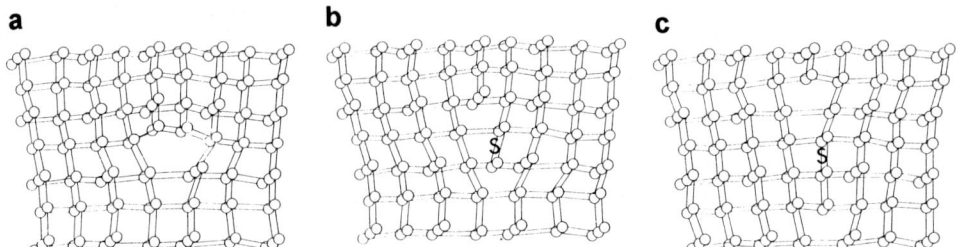

a   **b**   **c**

Figure 3. Mechanism of climb dissociation based on a short circuit diffusion process. (a) unreconstructed core of a 90° dislocation, which is glissile in {110}. (b) Diffusion of one row of core atoms leads to formation of a stacking fault S. (c) Diffusion of the next row of core atoms along the stacking fault.

3.2 a/2<110> dislocations with <112> line directions in the {110} glide plane

In Ge$_{0.5}$Si$_{0.5}$ layers grown on (110) orientated Si substrates, networks of perfect dislocations (**b**=a/2[1-10]) compensate the misfit of 2.1%. These networks consist of two sets of dislocations having <112> type line directions, which have 2 nm dislocation interdistance. These dislocations form by glide in the (110) interface. Dislocations formed by glide in {111} lattice planes would have [1-10] line directions and are not observed in our samples. Fig. 4a shows a high resolution micrograph of this type of dislocation in [1-1-2] projection along the Ge(Si)/Si interface. The inserted {111} lattice plane is clearly visible. The closing vector of the Burgers circuit in this case corresponds to the projection of a perfect Burgers vector orientated at an angle of 54° to the dislocation line. The dislocation core is constricted. A dissociation of the dislocation into partials could not be found at all. A stick and ball model of this dislocation with an unreconstructed dislocation core is shown in Fig. 4b. The number of dangling bonds is 1.63 per lattice vector. Reconstruction of the core can eliminate dangling bonds (see Fig 4c). The reconstructed core is then characterized by a five fold and seven fold ring. Our preliminary HRTEM investigations do not permit to distinguish which is the appropriate structure of this dislocation.

Figure 4. Perfect 54°-dislocation with {110} glide plane in the $Ge_{0.85}Si_{0.15}/Si(110)$ interface. (a) Cross sectional high resolution micrograph in [112] projection. The additional {111}lattice plane is marked by an arrow. The closing vector of the Burgers circuit corresponds to the Burgers vector of a perfect dislocation (**b**=a/2[1-10], which lies in the Ge/Si interface. (b) Stick and ball model of the dislocation with an unreconstructed dislocation core. (c) Stick and ball model of a reconstucted dislocation core which is characterized by five and seven fold rings.

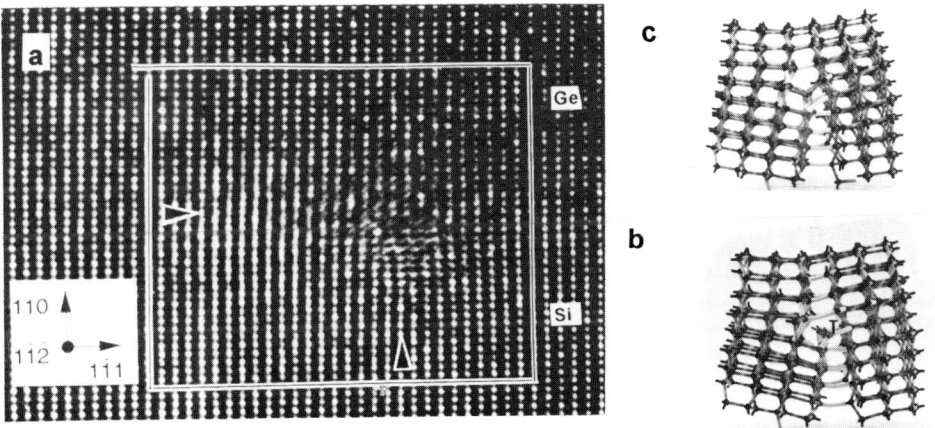

Figure 5. Perfect 73°-dislocation with {113} glide plane in the $Ge_{0.85}Si_{0.15}/Si(110)$ interface. (a) Cross sectional high resolution micrograph in [112] projection. Two additional lattice planes (one (220) and one (1-11)) are marked by an arrows. The closing vector of the Burgers circuit corresponds to the Burgers vector of a perfect dislocation (**b**=a/2[011]). (b) Stick and ball model of the dislocation with an unreconstructed dislocation core. The row of core atoms which is marked with T exhibits two dangling bonds. (c) Reconstructed dislocation core. The number of dangling bonds is reduced to 0.82 per lattice vector instead of 2.45 in Fig 5b.

3.3 a/2<110> dislocations with <112> line directions in the {113} glide planes

A misfit dislocation network, which is found at Ge(Si)/Si(110) interfaces of layers having a misfit f > 3.2%, consists of two sets of misfit dislocations which have <112> type line directions. In contrast to the dislocations discussed in 3.2 these dislocations are formed by glide in {311} lattice planes as determined by geometrical arguments (Burgers vector and line direction). The perfect Burgers vector $b=a/2<011>$ is established by appropriate contrast experiments and is orientated at 73° to their line direction. Figure 5a shows a high resolution micrograph of this dislocation along the [1-1-2] line direction in the Ge/Si(110) interface. Two inserted lattice planes can be revealed (one {111} and one {220} lattice plane) and are marked by arrows.The dislocation core, whose position cannot be determined on an atomic level shows a blurred contrast. A possible structure of this dislocation with an unreconstructed core as proposed by Hornstra (1958) is shown in the stick and ball model in Fig. 5c. This structure will probably be unstable as one of the core atoms (marked with T) has two broken bonds and may easily diffuse away as an interstitial. The reconstruction in the dislocation core proposed in Fig. 5c reduces the density of dangling bonds to 0.82 per lattice vector instead of 2.45 and thus is energetically favoured.

4. CONCLUSION

Dislocations in heteroepitaxial Ge/Si layers with misfits of a few percent are observed to glide in secondary {113} and {110} glide planes. Schmid factor and lattice mismatch determines the conditions for glide. This result is of general importance for other heteroepitaxial semiconductor systems (Bonar et al. 1992). Investigations by HRTEM show dislocations in Ge grown onto Si(011) to be constricted. In Ge on Si(001) the dislocations dissociate by climb into two partials $b_p=a/4<110>$. This dissociation is explained by a short circuit diffusion process. This investigation analyses the plastic behaviour of materials under resolved shear stresses that are two orders of magnitude larger than those accessible in conventional single crystal deformation experiments.

REFERENCES

Albrecht M, Strunk H P, Hansson P O and Bauser E 1992 Proc.Mater. Res.Soc. **238** 79
Albrecht M, Strunk H P, Hull R and Bonar J M 1993 Appl.Phys.Lett. **62**
Alexander H 1961 Z.Metallk. .**52** 344
Bonar J M , Hull R, Malik R J and Walker 1992 Appl.Phys.Lett. **60** 1327
Chidambarrao D, Srinivasan G R, Cunningham B and Murthy C S 1990,Appl.Phys.Lett. **57** 1001
Frank F C 1949 Physica **15** 131
Hansson P O, Werner J H, Tapfer L, Tilly L P and Bauser E 1990 J.Appl.Phys.**68** 2158
Hirth J P and Lothe J 1982 Theory of Dislocations (New York: Wiley)
Hornstra J 1958 J.P.Phys.Chem.Solids **5** 129
Landolt-Börnstein 1982 Data and Functionasl relationships in Science and Technology, New Series Vol. 17 a, Semiconductors:"Physics of Group IV Elements and III-V Compunds" (Berlin: Springer)
Ray I L F and Cockayne D J H 1973 J.Microsc. **98** 170

Inst. Phys. Conf. Ser. No 134: Section 2
Paper presented at Microsc. Semicond. Mater. Conf., Oxford, 5–8 April 1993

83

HREM and DLTS of $\Sigma = 37(610)[001]$ tilt grain boundary in Ge

N I Bochkareva, S S Ruvimov, R Scholz*, K Scheerschmidt* and L M Sorokin

Ioffe Physical-Technical Institute, Russian Academy of Science, 194021 St.-Petersburg, Russia
*Max-Planck-Institut für Mikrostrukturphysik, O-4050 Halle/S, Germany

ABSTRACT: DLTS and HREM studies have been performed on a $\Sigma=37(610)/[001]$ tilt boundary in Ge. A new approach is proposed to analyse DLTS spectra of the grain boundary (GB). HREM reveals the atomic structure of primary and secondary GB dislocations as well as microsteps and facets on the GB plane. Fluctuations in the potential barrier are due to the formation of vacancy-type oxygen complexes of donor like states at E_c-0.21 eV.

1. INTRODUCTION

A number of DLTS studies have been made of the GB electronic states (Broniatowski 1985). The DLTS spectra are usually analysed under the assumption that the transient capacitance of the GB is due to the capture and emission of charge carriers by the GB electronic states. However, there is a large spread of the ionization energies of the GB states revealed by DLTS and anomalously large carrier capture cross sections, which justifies the search for new approaches of analysing DLTS spectra at structurally characterized boundaries. Preliminary studies revealed similar electrical characteristics and DLTS spectra for Ge bicrystals of different misorientation angles ($1° < \theta < 20°$). The present paper investigates in detail the $\Sigma=37(610)/[001]$ ($\theta =18.92°$) tilt grain boundary.

2. SAMPLES AND EXPERIMENTAL TECHNIQUES

The samples were Czochralski-grown on a double seed (Sb-doped: $3 \cdot 10^{13} cm^{-3}$, dislocation density in the bulk: $10^4 cm^{-3}$). Samples for DLTS measurements ($3\times3\times9$ mm^3) were cut out of the bicrystal with the boundary parallel to 3×9 mm^2 faces, on which ohmic contacts were prepared. The structure of the $\Sigma=37(610)/[001]$ GB was determined using JEOL JEM-7A, JEM-100C (100 kV) and JEM-4000 EX (400 kV) electron microscopes.

3. STRUCTURE OF $\Sigma =37(610)/[001]$ TILT GRAIN BOUNDARY

The Ge bicrystals studied differ in the degree of the faceting of the GB planes. The surface of a strongly faceted boundary is corrugated, consisting of a set of plane regions, 20 to 500 nm long, both symmetrical and asymmetrical. The latter are deflected from the symmetrical position by an angle of up to 20°. Some of the GBs consist of alternately asymmetric regions with median planes (160) and ($\bar{1}$60). Hence, on the average the boundary remains symmetrically oriented (Scholz et al. 1990). The weakly faceted GB, which is the main subject of this study, contains a small number of facets. However, in its plane there are steps associated with the absorption by the boundary of lattice dislocations (Fig.1). GBs contain both primary and intrinsic secondary GB dislocations (GBDs) in the form of rows parallel to the tilt axis. On the average the distances between the secondary GBDs are 15-

25 nm. A strongly faceted GB also contains a dislocation network caused by the twist component of the grain misorientation.

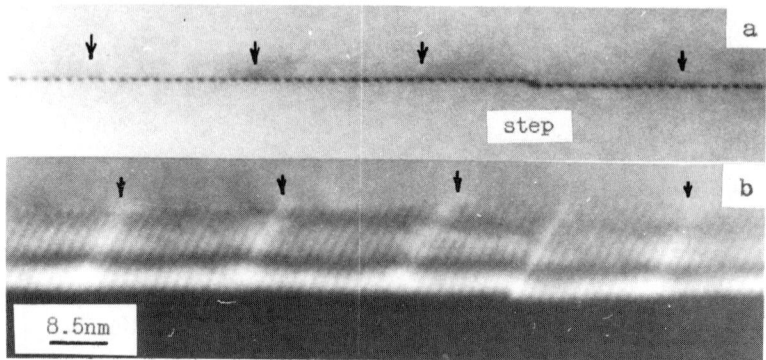

Fig.1. Conventional TEM images (g=220) of GB Σ=37(610)/[001] at different inclinations relative to the electron beam: normal (a) and inclined (b) to [001]. Arrows indicate secondary GBDs.

The structure of the symmetrical boundary region can be described in terms of Hornstra's (1960) model or a structural unit one (Bourret & Rouviere 1989, Levi, Smith & Wetzel 1991). An analysis of HREM images (Fig.2) suggests that the boundary contains a sequence of 45° dislocation pairs. The multislice image simulations (see inserts of Fig.2) were applied to relaxed boundary model (CERIUS, Molecular Simulation Corp.). While the general contrast features are rather well fitted the further model refinements are necessary to fit a core structure.

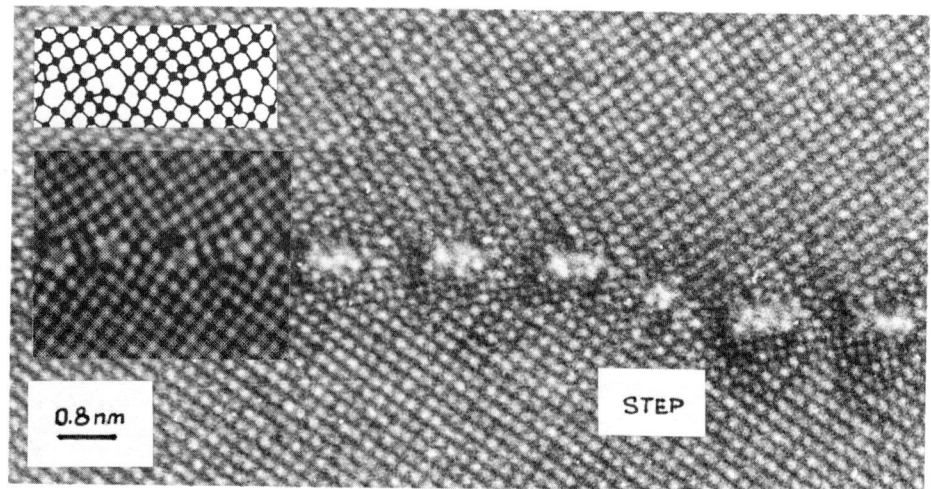

Fig.2. 400 kV HREM image (five beams) of GB. The inserts are computer simulated image (thickness: 25.5 nm, defocus value: -70 nm, atoms are white) and the boundary model.

The splitting of primary GBDs becomes obvious on the image (Fig. 3). The core of the secondary dislocation (Fig. 2) is localized inside the primary period, thus reducing it from 1.7 to 1.5 nm, which corresponds to the primary period of the Σ=13(510)/[001] GB. These conclusions agree with the study of the atomic structure of $\Sigma=25$ and $41/$ [001] tilt boundaries in Ge bicrystals (D'Anterroches and Bourret 1984)

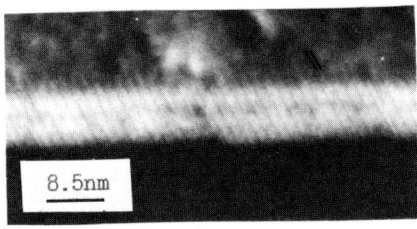

Fig.3.   TEM image revealing the dissociation of primary   GBDs (g=220).

4. ELECTRONIC PROPERTIES OF A $\Sigma=37(610)/[001]$ BOUNDARY

4.1. I-V characteristics

We selected samples having a weakly faceted GB. The I-V characteristics of the samples studied close to room temperature exibit a current saturation for both current directions. If the temperature is reduced, at low bias voltages (U) a sublinear current growth is observed changing to superlinear with increasing bias. Typical temperature dependences of the zero-bias resistance R' and resistance R measured across the GB, are shown in Fig. 4a. The activation energy derived from the Arrhenius plots in the temperature range of 230-300 K is 0.74 ± 0.05 eV for different samples. For T< 230 K, the activation energy decreases approaching 0.39 eV.

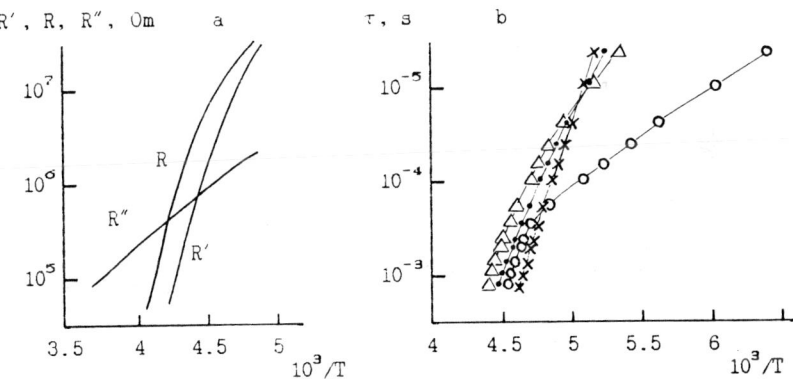

Fig.4. Temperature dependences of the zero-bias resistance R' (U=5mV), resistance R of U=0.5 V, differential resistance R" of U=1V(a), and capacitance recovery time constant (b) as derived from DLTS spectra in Fig.5. Pulse amplitude U'(V): 0.05 ($\cdot$); 0.3 (x) ; 0.6 (Δ); 1 (o) .

4.2. DLTS

Fig.5 presents typical DLTS spectra. The spectra measured for a pulse amplitude of U'>0.3 V are asymmetrically   shaped peaks with a characteristically   low-temperature tail. The temperature dependences of the capacitance recovery time constant derived from DLTS spectra are shown in Fig.4b. The relaxation is exponential only for a small pulse amplitude (U'=0.05 V), with the corresponding activation energy being 0.74 ± 0.04 eV for different samples. There is a correlation between the slopes of the temperature dependences log R', R (1/T) and logτ (1/T). At low U', the values of τ and R'C' are comparable, with C' = 3 10⁻¹⁰ pF being derived from the temperature dependence of the capacitance. There is also a correlation between the temperature shifts of these dependences as U and U' are increased. An increase of U' to 1 V results in a decrease down to 0.21 eV of the slope of the log τ (1/T) dependence in the low-temperature region , which corresponds to the slope of the temperature dependence of the differential resistance logR"(1/T) measured at U=1 V.

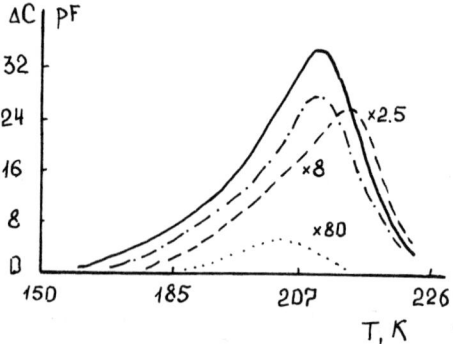

Fig.5. DLTS spectra of the Σ=37(610)/[001] GB. Frequency - 150 kHz, the rate window - 5.16 ms, duration of pulse - 1ms, bias voltage U=0 V,pulse amplitude U'(V): 0.05 (···); 0.3 (--); 0.6 (---); 1 (_).

The equal values of τ and R'C' cause one to suggest that the observed DLTS spectra are not associated with the thermal emission of charge carriers from the GB states to the band, but rather reflect the decay of the excess charge at the GB owing to the conduction current across the barriers on both sides of the GB when the voltage dropped to zero. The observed activation energies of 0.74±0.04 and 0.39eV, which are close to the width of the forbidden gap E_g and E_g/2, are related to the thermal generation of electron-hole pairs in the quasi-neutral and the depleted regions of the bicrystal, whereas the activation energy of 0.21 eV is related with the "leakage" conductance across the boundary. In connection with DLTS measurements the equivalent circuit of the bicrystal can be represented in the form of two oppositely connected n-p-junctions shunted by a "leakage" conductance. For small U' values, the transition processes are determined by the value of R'C', and for large ones, they are determined by the dependence on the voltage of each current component. One of the possible explantations of this are the GB barrier fluctuations owing to the spatial distribution of the point defects forming the regions with a high recombination activity. The activation-assisted conductance at the activation energy of 0.21 eV is associated with the donor-like centres at E_c-0.21 eV belonging to vacancy-type oxygen complexes, their concentration correlating with the dislocation density (Bochkareva 1991). This permits one to propose the following model of the potential barrier fluctuations arising at the boundary. It is well-known that the position of the Fermi level at the boundary determining the barrier height is close to the top of the valence band. We believe that the donor states of the oxygen complexes in the GBs can compensate, as happened for the boundary acceptor states, thus pinning the Fermi level near E_c-E_g/2-0.21/2, or E_c-0.21 eV.

REFERENCES

D'Anterroches C and Bourret A 1984 Phil. Mag .A **49** 783
Bochkareva N I 1991 Sov.Phys.Sem. **25** 323
Bourret A & Rouviere J L 1989 Polycrystalline Semiconductors, eds J H Werner , H J Möller & H P Strunk (Berlin: Springer Verlag) p.8
Broniatowski A 1985 Polycrystalline Semiconductors, ed G Harbeke (Berlin:Springer) pp 95-117
Hornstra J 1960 Physica **26** 198
Levi A A, Smith D A & Wetzel J T 1991 J.Appl.Phys. **69** 2048
Scholz R, Lubalin M D, Mosina G N, Ruvimov S S and Sorokin L M 1990 Col.Abstr.14th Sov.Conf. on Electron Microscopy (Moskow:Suzdal), pp 127-8

Twist Boundaries in silicon: a model system

R Gafiteanu, S Chevacharoenkul[1], U M Gösele and T Y Tan

School of Engineering, Duke University, Durham, North Carolina 27708-0302, USA
[1]Center for Microelectronic System Technologies, MCNC, PO Box 12889, North Carolina 27709-2889, USA

ABSTRACT: A combination of layer transfer and hydrophobic silicon wafer bonding was used to obtain a structure consisting of a single crystalline Si layer ($\cong 2\mu m$) rotationally misoriented on top of a (100) Si wafer. Because of the close proximity of the boundary to the surface and non-damaging method of preparation, this structure constitutes an adequate model system for studying the impact of processing on twist boundaries in silicon. Low-angle <001>Σ=1 twist boundaries before and after high concentration phosphorus diffusion have been studied using bright- and dark-field imaging conditions, electron diffraction and lattice-fringe imaging. Further possible applications are discussed.

1. INTRODUCTION

The potential use of multicrystalline silicon in solar cells constitutes a reason for studying silicon grain boundaries (GB's). Grain boundary dislocations (GBD's) drastically affect the efficiency of solar cells, especially when point defects and impurities are associated with the GBD's. The GBD's are not necessarily perfect edge or screw in character, but may have dissociated into various configurations. The change in GBD configuration under point defect injection conditions, e.g., due to the Si self-interstitial supersaturation induced by high concentration P diffusion during solar cell fabrication, may be beneficial or harmful to solar cells, and may influence the hydrogen passivation efficiency of the GBD's.

(001) silicon wafers can be used to produce twist GB's consisting of a square array of screw GBD's with a/2<110> Burgers vectors. For point defect injection and hydrogen passivation a controllable and reproducible structure with access within a couple of micrometers to the twist boundary is desirable. Any other defects located in the structure, with a potential of influencing results, should be avoided. To produce such GB's, Föll and Ast (1979) and Gleichmann et al (1991) have used compression bonding at 1200°C, while Perreault et al (1991) used minimum force bonding in the same temperature range. Compression bonding provides only small samples which are plastically deformed, and also with a high risk of being contaminated. The method of minimum force bonding overcomes some of these drawbacks, but primarily lacks reproducibility.

In this paper we report the results of using layer transfer and hydrophobic bonding of Si wafers to produce such GB's. The specimen size is the Si wafer size, and it is highly reproducible.

2. EXPERIMENTAL METHOD

In the layer-transfer/wafer-bonding method two 4" (001) Si wafers were used, one being a Si on insulator (SOI) wafer and the other a normal Si wafer. The SOI wafer has a ~2µm single crystalline Si layer on top of a SiO_2 insulator layer, with the latter being used in our experiments as a convenient etch-stop layer. The direct wafer bonding technique was used to produce the GB between the SOI layer and the front surface of the regular wafer. Figure 1 shows the wafer configurations as well as the bonding process. Prior to bonding, the native oxides on the wafer surfaces were removed by dipping the wafers in HF and then rinsing them in deionized water. Bonding was performed in a 'microcleanroom' setup developed at Duke University (Stengl et al 1988). This is a device which creates a particle-free space between two wafer surfaces, thus allowing high quality wafer bonding to be carried out in a non-cleanroom laboratory environment. On contact at room temperature, the two (hydrophobic) wafers bond to each other by Van der Waals forces between hydrogen, fluorine and possibly some water molecules adsorbed on the contact surfaces. Upon annealing at a high temperature, covalent bonding between the wafer surface Si atoms will develop. Our wafers were annealed at 1100°C for 3 h in a wet oxidation ambient, which additionally resulted in the growth of a 1,500Å thick oxide layer on the backsides of each wafer, capable of withstanding 15 h of KOH-etching (15% solution) at 70°C (Fig. 1b). The backside oxide on the SOI wafer was removed by HF prior to the KOH-etching (Fig. 1c). The bulk Si of the SOI wafer was completely removed after 13 hours of KOH-etching. Subsequently, a buffered oxide etch was used to remove the 1μm thick SOI oxide, resulting in the desired GB wafer shown in Figure 1d.

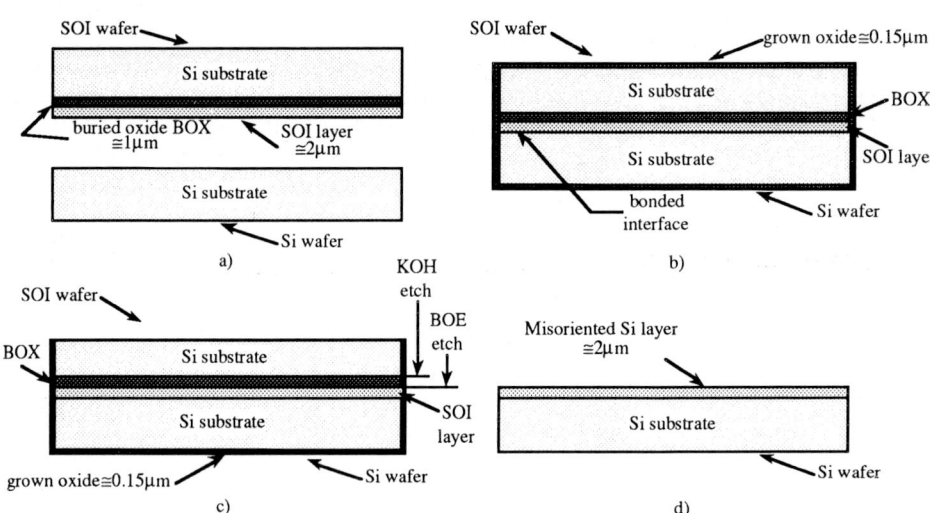

Fig. 1 Schematic of twist boundary formation by layer transfer technique using hydrophobic wafer bonding.

Several GB wafers, with a series of rotational misorientations of ~1-3°, were prepared. Samples of each type were diffused with P at 900°C using a spin-on-glass P source material.

Plan view and cross-sectional samples, from both as fabricated and P diffused GB wafers, were prepared and examined using TEM.

3. EXPERIMENTAL RESULTS

TEM examinations revealed that the present method of producing GB's is highly reproducible. The GB's contain not only a twist component, but also a tilt component. The presence of the latter obviously results from the original Si wafers being tilted away from the (001) orientations, which, in our case, is up to ~4°. In plan-view observations at low magnifications, a network of pure edge dislocations with irregular periodicity is observed, Fig. 2a, which accounts for part of the wafer tilt. At higher magnifications, the array of screw dislocations becomes observable, Fig. 2b and 2c. In Fig. 2b and 2c the dislocation spacing is ~9 nm, corresponding to a twist angle of ~2.5°, which was confirmed from SAD patterns. The screw dislocations are seen displaced along almost straight lines running at an angle to the <110> directions. A similar phenomenon has been observed by Föll and Ast (1979) and by Carter et al (1981). This is believed to be caused by dislocations with an edge component, resulting from wafer tilt, and are known as 'extraneous' or 'extrinsic' dislocations.

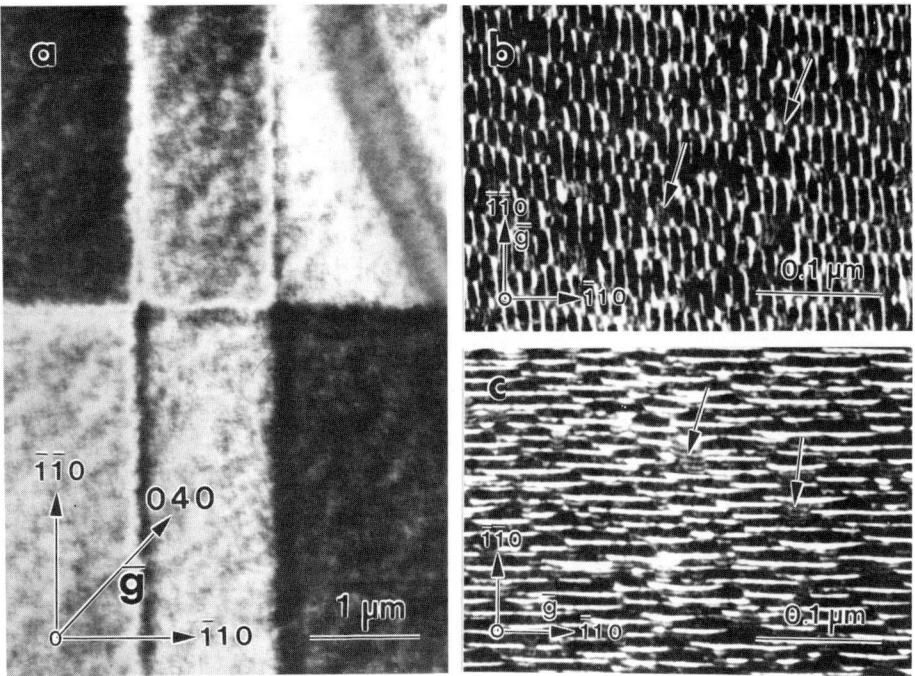

Fig. 2 Plan-view images of dislocations in the bonded interface: a) array of orthogonal edge dislocations; b) and c) weak-beam images of the screw dislocation network.

Cross-sectional views offered additional confirmation of the boundary structures. Figure 3a is an image of a dislocation network in a tilted sample. In Fig. 3b, the end view of one set of the screw dislocations is in contrast. In Fig. 3c, the features indicated by arrows are oxide precipitates and correspond to the dark patches in Fig. 2b and 2c. Lattice fringe images obtained in the HRTEM mode confirmed the amorphous nature of the oxide precipitates.

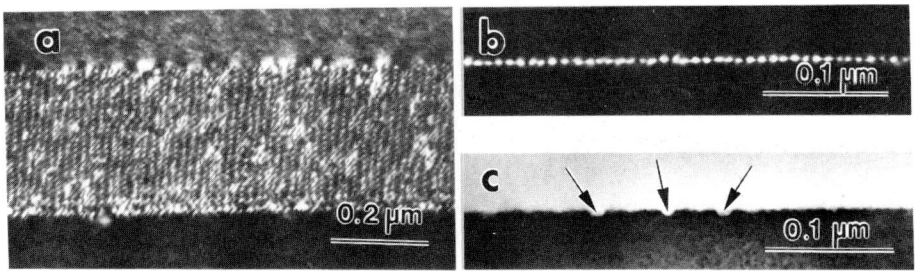

Fig. 3 Cross-sectional images of the boundary structure: a) weak-beam image of the GBD's, $(g,3g)_{\bar{2}20}$; b) weak-beam image of one set of end-on screw dislocations, $(g,3g)_{\bar{2}20}$; c) bright-field image of the interface showing the presence of oxide precipitates.

4. DISCUSSION

The layer-transfer/wafer-bonding method has been found to be a reliable way of producing large twist grain boundaries, which can obviously be extended to obtain GB's with other GBD structures. The transferred/bonded layer, using commercially available SOI wafers, is of only ~2μm, which is very convenient for further studies, e.g., TEM or EBIC studies of the GB structures and P in-diffusion effects on the GB structures. The GB's are, however, found to have structures more complicated than expected. This is partly due to the non ideal (001) wafer orientation, and partly due to the existence of oxide particles at the bonding interface. Overall, the quality of the interface as judged from high resolution studies, shows that 'microcleanroom' wafer bonding appears to be competitive with the UHV wafer bonding.

It is well-known that under the influence of a supersaturation of silicon self-interstitials, screw dislocations climb into a helix structure. No difference has been found in the GBD structures before and after P in-diffusion treatment generating a self-interstitial supersaturation, most likely because the screw dislocation array is too closely spaced. Assuming that P in-diffusion at 900°C induces a Si self-interstitial supersaturation $C_I/C_{I0} \cong 100$, and that the number of turns per unit length (n_L) of a climbed screw dislocation is given by (Weertman 1957)

$$n_L = kT/(2\pi Gb^4) \ln (C_I/C_{I0}),$$

we get $n_L \cong 10.7$ turns/μm ('G' is the shear modulus, 'b' is the Burgers vector, 'k' is Boltzmann's constant and 'T' is the absolute temperature). The maximum misorientation ($\Delta\theta$) which will allow screw dislocations to form helices under these circumstances, can be obtained by using the well-known relationship $d=b/(2\sin(\Delta\theta/2))$. Assuming that the dislocations are pinned at the network nodes, and therefore $d=1/n_L$ holds, we calculate that the misorientation has to be below $0°15'$ in order to observe helix formation.

On the other hand, closely spaced screw dislocations allow the study of the change of the electrical activity of screw dislocations due to the in-diffusion of phosphorus, e.g., by using EBIC, without the development of an edge component. We believe that the presence of oxide particles at the GB can be avoided by eliminating the water rinse step in preparing the hydrophobic Si wafer surfaces and by adjusting the annealing process.

ACKNOWLEDGMENTS

Helpful discussions with Dr. M. Reiche and his support in HRTEM investigations are gratefully acknowledged. Thanks are due to Dr. Q.-Y. Tong for assisting in sample preparation. This work was financially supported by the National Renewable Energy Laboratory under subcontract No. XD-2-11004-1.

REFERENCES

Carter C B, Föll H, Ast D G and Sass S L 1981 Phil. Mag. 43(2) 441
Föll H 1979 Phil. Mag. 40(5) 589
Gleichmann R, Blumtritt H, Höpner A and Sullivan T D 1991 Proc. 2nd Int. Conf. on Polycrystalline Semiconductors, eds J H Werner and H P Strunk (Berlin: Springer-Verlag) pp 103-108
Perreault G C, Hyland S L and Ast D G 1991 Phil. Mag. 64(3) 587
Stengl R, Ahn K-Y and Gösele U 1988 Jpn. J. Appl. Phys. 27 L2364
Weertman J 1957 Phys. Rev. 107 1259

Inst. Phys. Conf. Ser. No 134: Section 3
Paper presented at Microsc. Semicond. Mater. Conf., Oxford, 5–8 April 1993

91

Defects and silicon integrated circuit

J M Poate, D J Eaglesham, H J Gossmann, G S Higashi and G Pietsch

AT&T Bell Laboratories, 600 Mountain Avenue, Murray Hill, NJ 07974, USA

ABSTRACT: This review will detail three areas of Si processing research where the understanding of defect production and structure is crucial to successful device fabrication; (i) The effect of wet chemical cleaning on surface roughness at the atomic level and correlation with gate oxide integrity. (ii) The enhancement of B diffusion by Si self-interstitials. (iii) Er light emission in the presence of O defects.

1. INTRODUCTION

The Oxford Conferences on the Microscopy of Semiconducting Materials have successfully demonstrated over the years the increasing sophistication of analytic techniques which has resulted in the better understanding of semiconductor structures and defects. It could now be realistically asked what are the future directions of research. There are several answers to this question which we will try to detail using examples from our own work. Ultimately the studies of semiconducting materials revolve around the desire to produce better electronic and photonic devices. The electronics field is completely dominated by Si technology and several key defect issues are emerging with the shrinking dimensions of devices.

The shrinking of devices demands ever more processing steps and we will firstly discuss a processing area that is becoming more important; the cleaning and etching of Si surfaces to maintain surface smoothness at the atomic level. The shrinking of dimensions requires shallower ion implantation depth distributions and for dopants such as B the depth distribution can be greatly perturbed by interactions with Si self-interstitials. We are measuring these interactions using low temperature crystal growth techniques to produce unique B diffusion markers. Both these defect studies represent areas which are important in Si processing technologies. Our final example involves research into possible light emission from Er in Si and the role that impurity interactions play. In all these three examples, understanding of Si defects at the atomic level is required.

2. STRUCTURE AND CHEMISTRY OF TECHNOLOGICAL Si SURFACES

The complexity and sophistication of present day processing equipment such as ion implantation or thin film deposition systems tend to leave one with the impression that Si processing takes place in a vacuum. In fact, a large number of processing steps involve wet chemical cleaning whose use is based on the established result that such cleans give the best morphological and impurity control of Si surfaces. The importance of wet chemistry in Si manufacturing practice (Szkudlapski 1973) has long been recognized. Here we will describe one specific but crucial step in Si processing -- the cleaning of the surface prior to the growth of the gate oxide. Gate oxides are now approaching 100Å in thickness in production and are projected to shrink to $40-50$Å for future technologies. The electrical integrity of these oxides and channel mobilities depend critically on such factors as transition metal contamination and interface roughness. For example, dielectric breakdown will occur at unacceptable yield levels for these thinnest oxides if Fe contamination in the oxide or at the interface exceeds 10^9

atoms/cm^2. In this section we will focus on the issue of interface roughness.

The correlation of surface roughness with electronic device properties has been established by Hahn and Henzler (1984), Heyns et al. (1989) and Ohmi et al. (1992). The work by Ohmi and colleagues beautifully demonstrates not only the advances in the experimental techniques but also that surface roughness can be a real show stopper in Si processing technology. Fig. 1 shows their data of channel electron mobilities for 100 Å thick SiO_2 gate oxides in MOS devices as a function of the Si-SiO_2 interface roughness. The surface or interface roughnesses were varied by controllably etching the surfaces prior to oxide growth. The gate oxides were etched away following the electrical measurements and surface roughness measured by atomic force microscopy. These data show that RMS roughnesses of only $5-10$ Å can substantially degrade device performance.

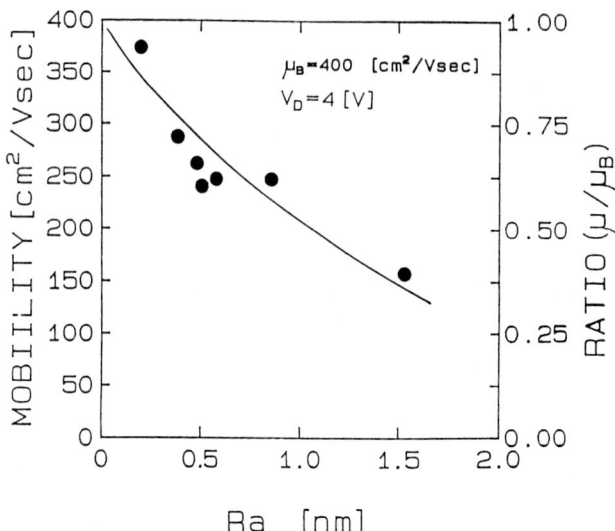

Fig. 1   Correlation of channel electron mobility (μ) to Si$-$SiO$_2$ interface roughness (R_a); the channel mobility, normalized to the bulk mobility (μ_B) is also plotted. From Ohmi et al. (1992).

There are approximately 20 processing steps before the gate oxides are grown by a dry thermal oxidation on Si and half of them involve wet chemical cleans. The goal of these cleans is to remove contaminants without uncontrollable etching of the surface. There are two fundamental types of cleans: hydrophillic and hydrophobic. These cleans have been in use for over 20 years but it is only recently that a detailed understanding of their physical and chemical workings has been gained. This

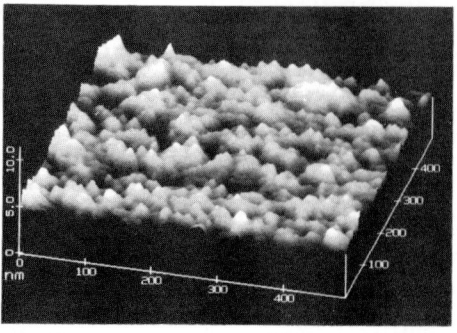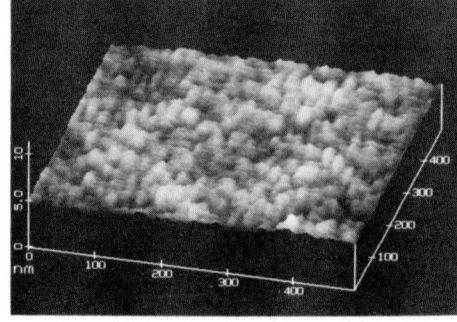

Fig. 2   Atom force micrograph of two (100) Si surfaces etched by HF. The left is a high pH clean resulting in a surface roughness of 5Å RMS and right is a low pH clean with 2Å RMS roughness.

field has recently been reviewed by Higashi and Chabal (1993). The standard hydrophillic clean is the RCA Standard Clean which is a sequential two-step clean where wafers are immersed first in a 1:1:5 solution of $NH_4OH:H_2O_2:H_2O$ at 80°C and then a 1:1:5 solution of $HCL:H_2O_2:H_2O$ at 80°C. The other popular hydrophillic clean is the Piranha etch which consists of an immersion in 4:1 $H_2SO_4:H_2O_2$ at a temperature somewhat in excess of 100°C. These cleans, which leave the surface passivated by 12-15Å of hydroxylated oxide, are very efficient at removing transition metals or particles. The data shown in Fig. 1 were generated using a hydrophillic clean.

The hydrophobic cleans developed at AT&T Bell Laboratories over 30 years ago (Buck and McKim 1958) are based on the use of dilute or buffered HF solutions. It was thought until quite recently that passivation occurred through fluorine termination of the surface dangling bonds. Recent work (Higashi and Chabal 1993) has demonstrated conclusively that passivation occurs by hydrogen termination. This passivation is noteworthy in that the surface can be remarkably stable. For example, etching (111) Si surfaces in buffered (pH ~8) HF solutions will lead to a preferential etching and flattening of the H-terminated surfaces. Indeed it is possible to form atomically flat and stable, unreconstructed (1x1) surfaces in this fashion (Higashi et al. 1991). This result is of fundamental importance but devices are constructed on (100) Si where the use of buffered HF will increase roughness due to the exposure of (111) facets. Fig. 2 shows on the left hand side an atomic force micrograph of (100) surface etched by buffered (pH ~8) HF solution. The roughness of this surface is 5Å. Etching (100) Si in dilute (pH ~2) HF will result in a roughness of only 2Å as shown on the right. Electrical breakdown measurements of 100Å oxides grown on these surfaces demonstrate superior oxides on the smoother surface.

Silicon Surface Cleaning

	hydrophilic			hydrophobic		
cleaning by ...	ultrathin oxide growth			surface removal		
method	RCA SC1	SC2	Piranha	HF conc	dilute	HF/NH$_4$F (buffered HF)
bulk Si attack	yes	no	no	no	slightly	yes
roughness	>	≅	≅	≅	≳	(111): ≤ (100): >
removal of ...						
particles	+	+/–	+/–	–	–	–
metals	–	+	+	+/–	+/–	+/–
hydrocarbons	+	–	+	–	–	–
applicable to patterned surf.	yes	yes	yes	no	yes	no [(111): restricted]

Fig. 3   Table of the various wet cleaning protocols and beneficial properties.

The recent application of the scanning probe technologies, in particular the atomic force microscope, to technological Si surfaces has established the correlation of atomic roughness and oxide electrical properties. Of course this is only one facet of the cleaning problems. We have not discussed contamination issues (e.g. hydrocarbons, particles or metals) or indeed the preferred way to grow oxides (e.g. wet or dry furnace oxides or oxides grown by rapid thermal annealing). The choice of cleans or growth technologies depends on many factors. In Fig. 3 we summarize some of the positive and negative aspects of the various cleans.

3. OXIDATION ENHANCED B DIFFUSION AND Si SELF-INTERSTITIAL DIFFUSIVITIES

A perennial challenge in Si processing is detecting and elucidating point defect generation and diffusion. For example, the diffusion of B, which occurs by a hybrid interstitialcy mechanism, can be dominated by the interaction with Si self-interstitials. The magnitude of this effect is demonstrated by the data of Fig. 4 where B diffusion is completely dominated by the release of Si self-interstitials from Si implantation. We are studying these phenomena using sharp B doping spikes in Si. These samples are made possible by the use of low temperature MBE techniques where dopant spikes can be incorporated without dopant segregation (Gossmann and Eaglesham 1992). Typical growth temperatures are in the region of 200-300°C where defect free layers of (100) Si can be realized with thicknesses of hundreds to thousands of Angstroms. Fig. 5 shows a B-doping superlattice which we have utilized for these studies (Gossmann et al. 1993). The doping superlattice was made out of six doping spikes each of width 100Å and sheet concentration 3.7×10^{13} cm^{-2} spaced on 1000Å centers. The depth profiles of the B were obtained by secondary ion mass spectrometry (SIMS).

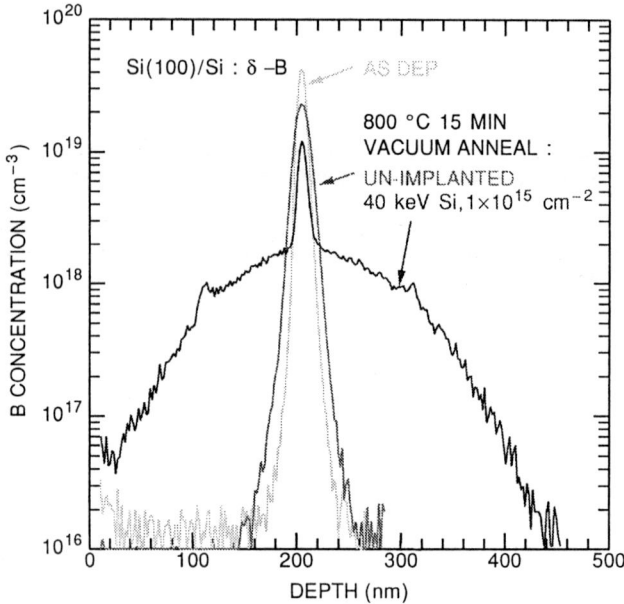

Fig. 4   Secondary ion mass spectrometry depth profiles of a B-dopant spike buried under 200 nm of Si, as deposited, after furnace annealing, and after ion implantation and annealing. In both cases annealing took place in vacuum at 800°C for 15 min. Implantation was done with 40 keV Si to a dose of 1×10^{15} cm^{-2}. The large enhancement of diffusion due to the point-defects introduced by the implantation step is clearly seen. This also leads to precipitation of a significant fraction of the B atoms, as evidenced by the central spike that does not appreciably widen during annealing. The smaller spikes at ≈100 nm and ≈310 nm are due to trapping of diffusing B in the region of the end-of-range-damage and in the residual defects located at the film-substrate interface, respectively.

We have injected interstitials by annealing in dry oxygen where the growth of SiO$_2$ at the surface leads to an essentially infinite source of interstitials. The diffusion of the B can then be used to map the Si interstitial concentration as a function of depth. Heating at 800°C for 15 minutes produces the oxidation enhanced diffusion (OED) profiles shown by the dotted lines. The B diffusion coefficients can be extracted directly from this data and they are plotted in Fig. 6. These diffusivities map the Si self-interstitial concentrations as a function of depth and the line represents the fit which uniquely gives the Si self-interstitial diffusivity at 800°C. In a similar fashion experiments have been carried out in the temperature range of 750-900°C and Fig. 7 shows an Arrhenius plot of the interstitial diffusivities. An exponential fit (solid line) yields an activation energy of 3.1 ± 0.4 eV.

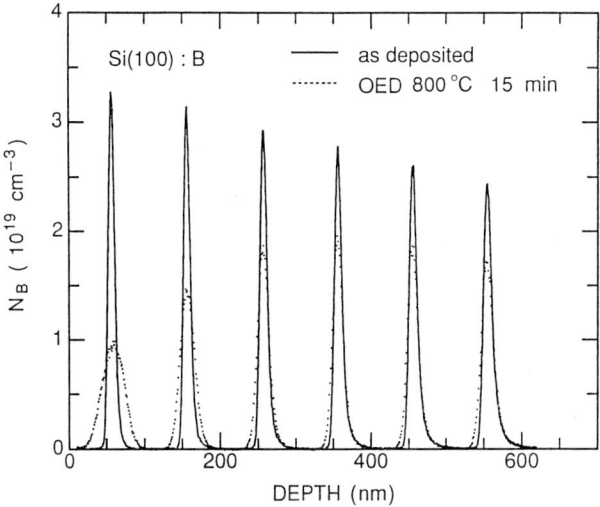

Fig. 5   Depth profiles of a B-doping superlattice before (solid line) and after (dotted line) oxidation in dry oxygen at 800°C for 15 min. Oxidation of Si generates Si self-interstitials at the oxide/Si interface which diffuse into the bulk of the sample. The enhancement of diffusion in the surface region due to the injection of Si self-interstitials is clearly seen. This enhancement decays with depth as the concentration of interstitials diminishes. From Gossmann et al. (1993).

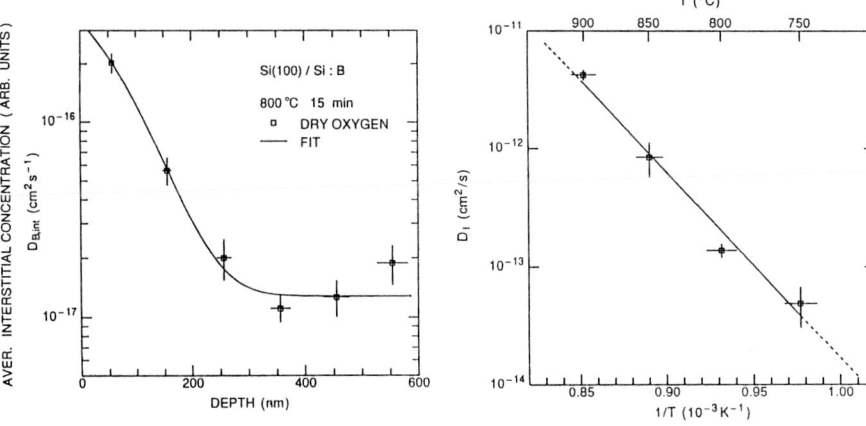

Fig. 6   B diffusivities, extracted from a comparison of the profiles in Fig. 7, as a function of depth; alternatively, average Si self-interstitial concentration as a function of depth. The solid line is the result of a fit to this average interstitial concentration profile that yields, assuming a constant surface concentration, an interstitial diffusivity of 1.4×10^{-13} cm^2/s at 800°C. From Gossmann et al. (1993).

Fig. 7   Diffusivities of Si self-interstitials as a function of inverse temperature. The solid line is an exponential fit to the data, yielding an activation energy of 3.1 ± 0.4 eV; the dashed lines are extrapolations of this fit. From Gossmann et al. (1993).

These results are noteworthy as we believe we have a system to directly monitor the diffusion of Si self-interstitials produced by oxidation. We can then apply such ideas to unraveling the much more complex case of interstitials arising from ion implantation damage. The diffusivities we have presented here are intriguing as they are much smaller than those obtained from kick-out studies using hybrid diffusers such as Au or Pt.

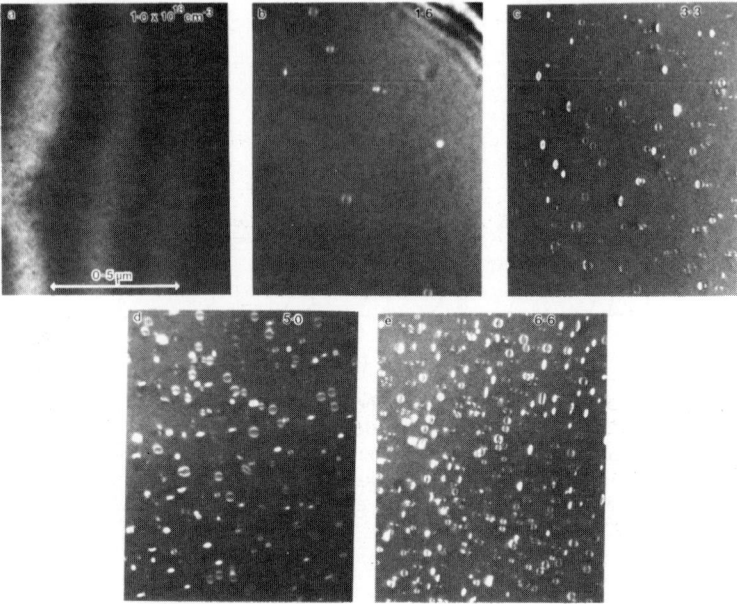

Fig. 8   Series of plan-view images showing the precipitate density following a 30 minute anneal at 900°C for Er implant doses of 1.0, 1.65, 3.3, 4.9 and $6.6 \times 10^{18} \, \text{cm}^{-3}$, respectively. The (220) weak-beam dark-field images show loop-like contrast. More detailed experiments show that the defects are platelet in character, possibly $ErSi_2$. From Eaglesham et al. (1991).

4. LIGHT EMISSION FROM Er IN Si

The previous section detailed defect studies which are intimately involved with Si integrated circuit fabrication. We now discuss some of our studies from a very different arena. The attainment of efficient light emission from Si for optoelectronic integration is one of the holy grails (Tennyson 1869) of Si technology. We have demonstrated (Benton et al. 1991 and Michel et al. 1991) that implantation of Er of MeV energies into Si leads to efficient photoluminescence and electroluminescence at low temperatures. Our interest in Er comes from the fact that the emission occurs at a wavelength of 1.54 µm which matches the absorption minimum in silica-based optical fibers and is the wavelength used for long distance optical communication.

A key to fabricating efficient Si based devices is the understanding of the local defect environment and solubility of the Er in Si. The luminescence arises from $^4I_{13/2} \rightarrow {}^4I_{15/2}$ transitions involving nonbonding 4f core electrons so the transitions are relatively insensitive to the nature of the host material. Nevertheless, we observe that the strength of the transition is strongly dependent on the O content of the Si. Implanting Er in Czochralski (CZ) Si can increase the emission strength by over two orders of magnitude as compared to implantation in float-zone (FZ) Si. Using O-rich Si we have observed Er emission at room temperature. Here we will describe transmission electron microscope (TEM) studies to determine Er solubility in Si and extended x-ray absorption fine-structure (EXAFS) measurements of the atomic environment.

Erbium incorporation in Si is limited by precipitation. Figure 8 shows TEM plan-view electron micrographs of Si samples implanted to different concentrations of Er at an energy of 500 keV. The sequence clearly shows a sharp onset for the production of defects at $\approx 1 \times 10^{18}$ cm^{-3}. Analysis of the contrast from these defects shows that they are not simply dislocation loops, in that they appear in all g-vectors in both 2-beam and weak-beam conditions; cross section HREM also suggests that a second phase may be present within these defects (Eaglesham 1991). Annealing shows that, unlike normal implantation-damage, these defects will not grow with annealing. The defects are probably platelet precipitates of ErSi$_2$; from the link between the onset of formation of these defects and the onset of damage formation from implantation they may well be caused by gettering of Er to the normal dislocation loops formed by the implant damage.

These experiments were carried out in CZ Si but very similar results for the onset of precipitation in terms of identical concentrations and temperatures were observed for FZ Si. The onset of precipitation in this environment cannot be identified unambiguously with bulk equilibrium solubilities. Nevertheless this threshold number ($\sim 1 \times 10^{18}$ cm^{-3}) is probably a more useful working definition and illustrates the limitation in incorporating high concentrations of Er in Si for useful devices.

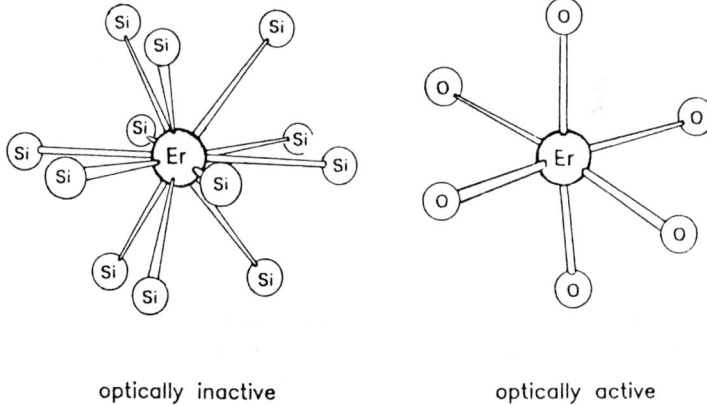

optically inactive   optically active

Fig. 9   Schematic picture of the first coordination shell surrounding Er in FZ (left) and CZ-Si (right). From Adler et al. (1992).

It has been a real analytical challenge to understand the structural role that O plays in the enhanced light emission. This defect structure has been solved by Adler et al. (1992) from EXAFS measurements on the AT&T Bell Laboratories beamline at the Brookhaven synchrotron. The CZ Si samples revealed a local six-fold coordination around the Er of an O cage. The O atoms were at an average distance of 2.25Å. In contrast, similar concentrations of Er in FZ Si, showed the Er to be coordinated to 12 Si atoms at a mean distance of 3.00Å. These results, which are shown schematically in Fig. 9, demonstrate an unusual defect structure in Si. The challenge is to elucidate what role this defect cage plays in enhancing the light emission.

ACKNOWLEDGEMENTS

We are indebted to our colleagues at AT&T Bell Laboratories for their enthusiastic collaboration in these studies; in particular David Adler, Janet Benton, Tom Boone, Paul Citrin, Michael Green, Karrie Hanson, Dale Jacobson and Conor Rafferty.

REFERENCES

Adler D L, Jacobson D C, Eaglesham D J, Marcus M A, Benton J L, Poate J M and Citrin P H (1992) Appl. Phys. Lett. 61 2181

Benton J L, Michel J, Kimerling L C, Jacobson D C, Xie Y H, Eaglesham D J, Fitzgerald E A and Poate J M (1991) J. Appl. Phys. 70 2667

Buck T M and McKim F S (1958) J. Electrochem. Soc. 105 709

Eaglesham D J, Michel J, Fitzgerald E A, Jacobson D C, Poate J M, Benton J L, Polman A, Xie Y H and Kimerling L C (1991) Phys. Lett. 58 2797

Gossmann H J and Eaglesham D J (1992) in Semiconductor Interfaces, Microstructures, and Devices: Properties and Applications, ed Z C Feng (Inst. Phys., Bristol, UK).

Gossmann H J, Rafferty, C S, Luftman H S, Unterwald F C, Boone T and Poate J M (1993) submitted to Appl. Phys. Lett.

Hahn P O and Henzler M (1984) J. Vac. Sci. Technol. A2 574

Heyns M, Hasenack C, DeKeersmaecker R and Falster R (1990) Proc. of the 1st Int. Symp. on Cleaning Technology in Semiconductor Device Manufacturing, Fall ECS 1989, eds by J Ruzyllo and R. E. Novak, PV 90-9 293 (Electrochemical Society, Pennington, NJ)

Higashi G S, Becker R S, Chabal Y J and Becker A J (1991) Appl. Phys. Lett. 58 1656

Higashi G S and Chabal I (1993) in press in Handbook of Semiconductor Wafer Cleaning Technologies, ed W Kern

Michel J, Benton J L, Ferrante R F, Jacobson DC, Fitzgerald E A, Xie Y H, Poate J M and Kimerling L C (1991) J. Appl. Phys. 70 2672

Ohmi T, Mujashita M, Itano M, Imaoka T, Kawanabe I (1992) IEE Transactions on Electron Devices 39 537

Szkudlapski A H (1973) Chapter 7 Surface Cleaning Practice in Environmental Control in Electronic Manufacturing ed P W Morrison (New York-Van Nostrand) pp 146-206

Tennyson A (1864) The Holy Grail and Other Poems (R R Clarke, London)

Inst. Phys. Conf. Ser. No 134: Section 3
Paper presented at Microsc. Semicond. Mater. Conf., Oxford, 5–8 April 1993

99

Structural transformations of As-doped polycrystalline silicon layers used as emitter contacts

C Spinella, A Cacciato, E Rimini, G Fallico[1], G Ferla[1] and P Ward[1]

IMETEM-CNR, Corso Italia 57, I-95129 Catania, Italy
[1]SGS-Thomson, Stradale Primosole 50, I-95100 Catania, Italy

ABSTRACT: The epitaxial realignment of As-doped polycrystalline layers has been investigated in structured substrates. Rectangular strips of width in the 0.2-100 μm range were used. The realignment starts by nucleation of epitaxial columns at grain boundaries in contact with the poly-Si/substrate interface. In the 1-100 μm strips the realignment proceeds by two-dimensional growth, while in the 0.2-0.3 μm strips by one-dimensional growth. The growth kinetics are slowed down in the small width geometry and the thermal budget needed to realign the films increases.

1. INTRODUCTION

In the fabrication of bipolar transistors heavily doped polycrystalline silicon films (poly-Si) deposited onto silicon single crystal substrates (c-Si) are routinely used as diffusion sources for the emitter region and as emitter contacts. The structural changes the poly-Si layer experiences when subjected to high temperature treatments are of relevance to the electrical behaviour of the device. Under the effect of thermal treatment, the native oxide at the poly-Si/c-Si interface agglomerates into small beads to reduce its surface energy (Ajuira et al 1991), the grains of the polycrystalline layer grow, and the epitaxial realignment with the underlying substrate can start (Wolstenholme et al 1987, Delfino et al 1989). The studies carried out up to now have been performed on infinite plane films. In order to investigate the structural transformation under conditions similar to those encountered in actual devices poly-Si films were deposited onto patterned wafers. We demonstrate that the kinetics of epitaxial realignment are strongly modified as the lateral extent of the contact area with the c-Si substrate is reduced to a width comparable with the grain size of the poly-Si film.

2. EXPERIMENTAL

The XTEM micrograph of Fig. 1a and the schematic of Fig. 2b show the typical configuration of an emitter-base structure in a bipolar transistor. The first polycrystalline silicon film (poly-Si 1) serves as contact to the base and collector regions, whilst

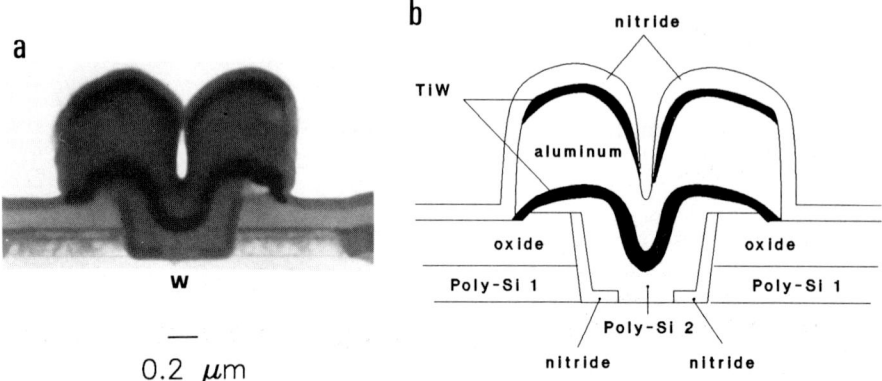

Fig. 1 (a) XTEM micrograph shown an emitter-base structure realized by two self-aligned polycrystalline silicon films. (b) Labels of the single layers.

the oxide layer and the nitride spacers act as isolating dielectrics. The polycrystalline silicon film (poly-Si 2) is in contact with the c-Si substrate through the window of width W. This film is heavily doped with As and it is used as diffusion source for the emitter and as contact. We investigated the epitaxial realignment of the poly-Si 2 film as a function of the window size W. Patterns consisting of 3 mm long strips and of different width (0.2 - 100 μm) were engraved on the same (001) c-Si wafer. The poly-Si 2 film was deposited by low pressure chemical vapor deposition (LPCVD) at 620 °C. The morphology of the samples is similar to that shown in Figs. 1a,1b with exception of the metalization layers which were missing. The samples were implanted with 60 keV of As ions to a dose of 1×10^{16}cm^{-2} and subjected to rapid thermal anneal (RTA) at 975 °C for several times. The fraction of the realigned material in the poly-Si layer versus the annealing time was measured by TEM plan view analysis. Samples were thinned by a backside final ion milling with 3 - 4 keV Ar ions at an incident angle of 12 degrees.

3. RESULTS AND DISCUSSIONS

At the early stages of the structural transformation induced by thermal treatment the native oxide present at the poly-Si/c-Si interface agglomerates into nanometer beads and the epitaxial realignment starts in the poly-Si film where a grain boundary termination is in contact with the interface (Fig. 2). By increasing the annealing time these epitaxial sites quickly reach the surface and, in the case of the poly-Si film deposited onto the larger strip ($W = 100\mu$m), they grow laterally at the expense of the randomly oriented crystal grains (Benyaich et al 1992). The plan view TEM micrograph of Fig. 3 shows the structure of 1 μm (Fig. 3a), and 0.25 μm (Fig. 3b) wide strips, from a wafer annealed at 975 °C for 85 s. The micrographs are taken in dark field conditions by using the (220) reflection coming from regions with the same substrate orientation. Thus the bright regions represent portions of the poly-Si film

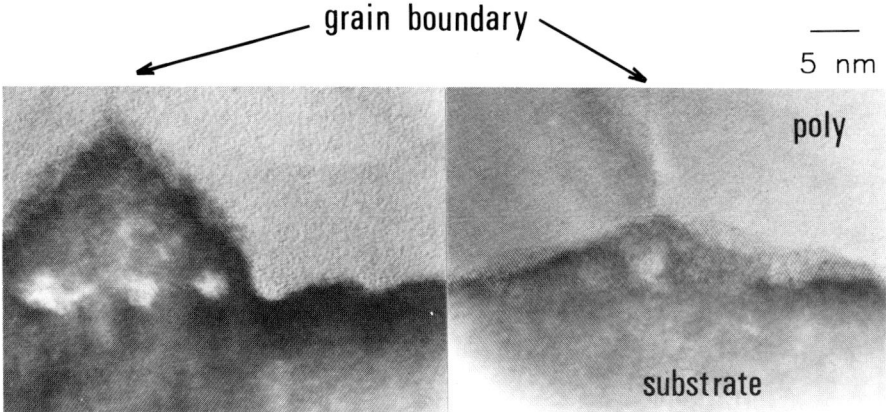

Fig. 2 High resolution TEM micrograph showing the microcrystalline structure of the poly-Si/c-Si interface after annealing at 975 °C for 50 s. Bubbles of oxide are revealed as brighter contrast at the interface. Single crystal protuberances appear where a grain boundary is in contact with the interface.

realigned to the substrate. In the large contact window (Fig. 3a) the film is completely realigned, whilst the narrowest strip (Fig. 3b) is still partially polycrystalline.

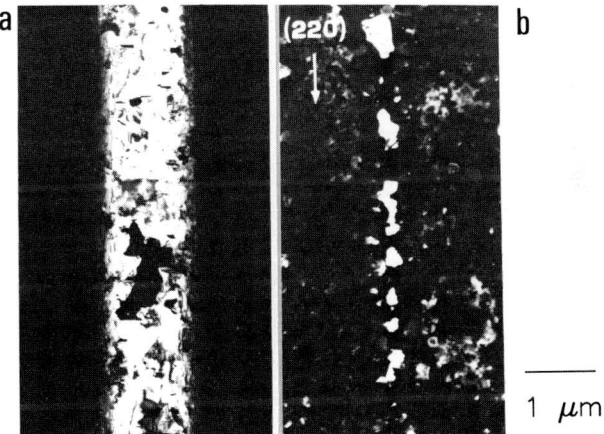

Fig. 3 Plan view TEM images in dark field ($\vec{g} = (220)$) of As-doped poly-Si films deposited onto 1 μm (a) and 0.25 μm (b) wide strip, after anneal at 975 °C for 85 s.

Fig. 4 reports the variations of the realignment fraction f versus annealing time t for a 1 μm (■) and for a 0.25 μm wide strip (○). A longer annealing time ($\sim$ 200s) is required to completely realign the poly-Si film in the narrowest strip. The continuous lines in the plot of Fig. 4 are best fits to the experimental data obtained by using the classical formalism for nucleation and growth of a new phase inside a film (Avrami 1939, Thompson et al 1939). The realigned fraction will be given by:

$$f(t) \quad = \quad 1 - \exp\left[-\left(\frac{t-\tau_0}{\tau_c}\right)^n\right] \tag{1}$$

where τ_0 is the incubation time for the interfacial oxide rupture and for the starting of epitaxy at the poly-Si/c-Si interface, τ_c a characteristic time related to the density of realigned columns and to their growth rate. In the absence of the nucleation phenomenon, the value of n is equal to 3, 2, or 1 for respectively tri-, bi-, and one-dimensional growth. From the fitting procedure   results, $n = 2$ in the 1 μm wide strip and $n = 1$ in the 0.25 μm wide strip. This demonstrates that the in-plane isotropy loss, induced by the reduction of the lateral extent of the poly-Si film, reduces the dimensionality of the growth. This change slows down the realignment kinetics and increases the thermal budget needed to complete the total realignment.

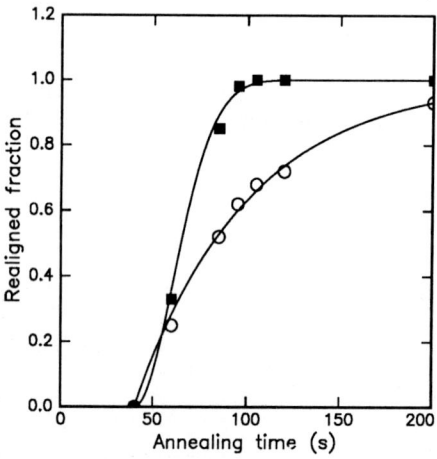

Fig. 4 Areal realigned fraction versus annealing time at 975 °C for As-doped poly-Si films deposited onto 1 μm (■) and onto 0.25 μm (O) wide strip.

REFERENCES

Ajuira S and Reif R, 1991 J. Appl. Phys. **69**, 662

Avrami M, 1939 J. Chem. Phys. **7**, 1103

Benyaich F, Priolo F, Rimini E, Spinella C, and Ward P, 1992 J. Appl. Phys. **71**, 638

Delfino M, Groot J L, Ritz K N, and Maillot P, 1989 J. Electrochem. Soc. **136**,215

Thompson W A and Mehl R F, 1939 Trans. AIME **136**, 416

Wolstenholme G R, Jorgensen N, Ashburn P, and Booker R, 1987 J. Appl. Phys. **62**, 225

Grown-in microprecipitates in CZ–Si

Katsuhiko Nakai, Tsuneo Nakashizu and Hiroyo Haga

Electronics Research Laboratories, Nippon Steel Corporation,
3434 Shimata Hikari, Yamaguchi 743, Japan

ABSTRACT: Micro-precipitates in as-grown Czochralski (CZ) silicon crystals have been studied by transmission electron microscopy (TEM) and high resolution electron microscopy (HREM). We have observed grown-in micro-precipitates with a number density of 10^{10} - 10^{11} /cm^3. The HREM observation of the micro-precipitates has revealed that the size is about 10 nm and that the structure is amorphous. Annealing treatments at 700 °C and 1100 °C have shown that they are stable at 700 °C and grow into large ones at 1100 °C. According to their morphology and annealing behavior, it has been concluded that they were formed at high temperatures (~1100 °C) during the crystal growth.

1. INTRODUCTION

The defects in the CZ-Si affect the electrical properties and gettering efficiency in Si devices (Tan et al 1977, Lin 1990). The defects, such as oxygen precipitates or oxidation induced stacking faults (OSF), are formed during the device fabrication process. Since the density of these defects after wafer annealing is greatly influenced by the condition of crystal growth, it has been considered that the nuclei are formed during crystal growth. These nuclei, which will be referred to as "grown-in defects", become the oxygen precipitates or OSF.

These grown-in defects have been previously observed by some workers. Oxygen precipitates of about 400 nm in size were observed by Wright etching and TEM (Wada et al 1982) in wafers of 3 inches in diameter. Their density was about 10^6 /cm^3. These large precipitates have never been found in the wafers of larger diameter. Interstitial type dislocation clusters of about 300 nm were found by Secco etching and also observed by TEM (Abe et al 1993). Their number density is in the range of 10^3 - 10^5 /cm^3. However, oxygen precipitates are usually formed in the range of 10^6 - 10^{12} /cm^3 depending on the annealing temperature. The number density of the precipitates is always larger than that of dislocation clusters, so these dislocation clusters cannot be the dominant nuclei of oxygen precipitates. Therefore, it should be considered that other defects which are undetectable by etching exist in as-grown CZ-Si wafers.

In order to find such defects, we have studied as-grown CZ-Si wafers using TEM and HREM.

2. EXPERIMENTAL

CZ-Si (100) wafers were used in this experiment. The wafers were n-type (phosphorus doped, 4×10^{14} /cm^3) and the oxygen concentration was 9.5×10^{17} atoms/cm^3, measured by Fourier transform infrared absorption spectroscopy (the conversion factor: 3.03×10^{17} /cm^2).

Sample preparation for TEM was carried out as follows. Discs of 3 mm in diameter were mechanically cut out from the centre parts of the wafers by an ultrasonic cutter and then chemically polished to have a small hole in the center of discs. Cross sectional samples of <110> orientation were also obtained by an argon ion milling method. The TEM used in this experiment was a JEOL JEM 2010.

In order to investigate the formation kinetics and annealing behavior of the micro-precipitates, isochronal annealings at 700 °C or 1100 °C for 128 hours in a nitrogen ambient gas were performed.

3. RESULTS

3.1 Observation of grown-in micro-precipitates

Figure 1 shows TEM photographs of micro-precipitates in an as-grown CZ-Si wafer. The contrast of the micro-precipitates is opposite to the fringe contrast. No strain field is observed around the micro-precipitates. They could not be detected as etch pits. Figure 2 shows the lattice image of a grown-in precipitate . The structure is amorphous and the size is about 10 nm. Although the morphology is not very clear, facets along {111} planes can be recognized. Figure 3 shows the size distribution of twenty micro-precipitates. It is notable that precipitate size is confined to a very narrow range. The precipitate density was estimated to be in the range of $10^{10} - 10^{11}$ /cm^3.

3.2 Densities of defects formed after wafer annealing

The as-grown wafers were annealed at 700 °C or 1100 °C for 128 hours in a nitrogen ambient gas to investigate the behavior of oxygen precipitates. The results are summarized in Table 1, which shows changes in number den-

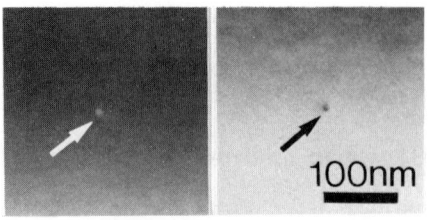

Fig. 1   TEM Photographs of grown-in micro-precipitates taken from <100> orientation.

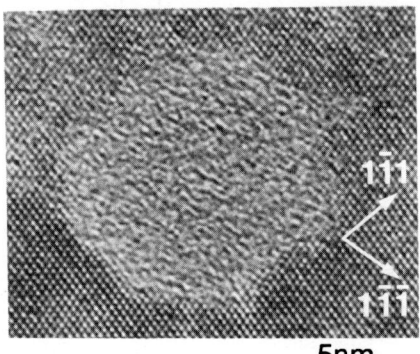

Fig. 2   Lattice image of grown-in micro-precipitates taken from <110> orientation.

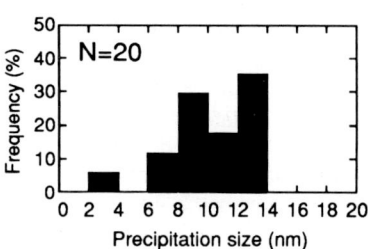

Fig. 3 Size distribution of grown-in micro-precipitates

Table. 1   Differences in number density, morphology and size of the precipitates in as-grown and annealed wafers.

	Number density (cm^{-3})	Morphology	Size (nm)	
as grown	$10^{10} \sim 10^{11}$	octahedra	~10	
after annealing				
700°C, 128h	1.9×10^{12}	platelets	~50 (diagonal size) ~4 (thickness)	
	$10^{10} \sim 10^{11}$	octahedra	~10	
1100°C, 128h	3.4×10^{6}	octahedra	~500	
	$10^{10} \sim 10^{11}$	octahedra	~10	

sity, morphology, and size for as-grown wafers and after annealing at 700 °C or 1100 °C. Figure 4 shows a TEM micrograph taken after annealing at 700 °C. Two kinds of precipitates were observed. The grown-in precipitates (A) retain both their original size and density. And another kind of precipitate (B,C), so called "platelets", were formed with a number density of $1.9 \times 10^{12}/cm^3$. After annealing at 1100 °C, large "octahedra" were formed with a lower number density of $3.4 \times 10^6/cm^3$ and also the grown-in micro-precipitates were observed with a similar size and density to those observed before the annealing.

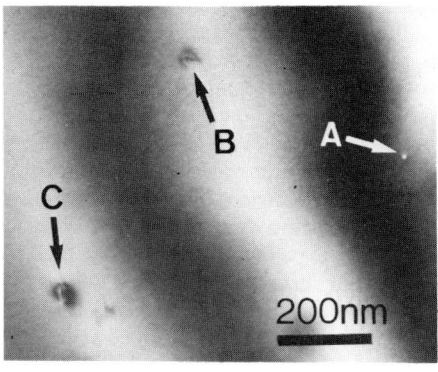

Fig. 4 Oxygen precipitates formed after annealing at 700 °C for 128 hours in nitrogen ambient gas. A is a grown-in micro-precipitate, and B,C are platelets.

4. DISCUSSION

4.1 The comparison of the morphology between the grown-in micro-precipitates and the oxygen precipitates grown in wafer annealing

The morphology of oxygen precipitates depends on their growth temperature (Bourret 1986, Ponce and Hahn 1989).

We have also confirmed that {100} amorphous "platelets" are formed with large strain field in the temperature range of 700-1000 °C and that amorphous "octahedra" with facets along {111} and {100} are observed with no strain field around them. It is considered that the morphology of the precipitates is determined by strain energy and interface energy. At low temperatures strain energy cannot be easily relaxed, so platelet precipitates form parallel to the {100} Si plane to minimize the strain energy (Hasebe et al 1991). At high temperature (~1100 °C) the strain can relax almost entirely so precipitates are able to grow into a more spherical shape. But to minimize the interfacial energy, their shape becomes an octahedron surrounded by {111} and {100} planes. As shown in Figure 2, the morphology of the grown-in micro-precipitates can be regarded as octahedral. This suggests that they are nucleated above 1100 °C at which the octahedral precipitates are formed.

4.2 Defects after wafer annealing

The result of annealing at 700 °C indicates that the grown-in micro-precipitates cannot be the nuclei of the platelets and that the platelets grow from other nuclei with the density of $10^{12}/cm^3$. These nuclei might be too small to be detected by TEM. At low temperatures such as 700 °C, the strain energy cannot be easily relaxed. Besides, the strain-energy increase caused by attaching an oxygen atom to the precipitates is larger in an octahedron than in a platelet. Thus only the platelets can grow. On the other hand, the grown-in micro-precipitates remain stable and keep their original state at 700 °C.

During the annealing at 1100 °C, the octahedral precipitates can grow because the strain field can be relaxed easily. Some of the grown-in micro-precipitates can grow and become large precipitates after the annealing.

4.3 Formation mechanism of grown-in micro-precipitates

We consider the formation of the grown-in micro-precipitates during the crystal growth.

Figure 5 shows the temperature change of the crystal in the pulling furnace as a function of time.   Assuming a temperature at which the grown-in micro-precipitates start growing, we can estimate the size of precipitates by application of the oxygen-diffusion controlled growth model (Zener and Wert 1950, Ham 1958).

The calculation results show that precipitates of over 10 nm can be formed during the crystal cooling.   If precipitates begin to grow at 1100 °C during the crystal cooling, the expected size of grown-in precipitates is about 60 nm from the calculation.   This value is larger than the actual size (~10 nm).   However if we postulate that octahedral precipitates are formed at higher temperatures and stopped their growth at 1100 °C, the size of the precipitates that grow at 1100 - 1120 °C becomes about 10 nm in the calculation.   This narrow range of the growth temperature explains well the narrow size distribution range of the precipitates.

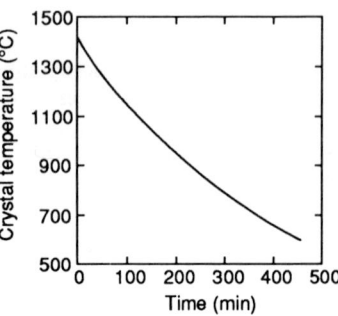

Fig. 5   Temperature change of the crystal in the growth process.

5. CONCLUSION

Grown-in oxygen micro-precipitates in as-grown CZ-Si crystals are studied by TEM and HREM.   These micro-precipitates are 2 - 14 nm in size and have an amorphous structure.   The grown-in precipitates are octahedral shapes with no strain field and they are similar to those observed after wafer annealing in the temperature range above 1100 °C.   They are considered to be formed above 1100 °C during the crystal growth in the crystal pullers.

ACKNOWLEDGEMENTS

The authors would like to thank K. Kawakami and A. Ikari for calculations, and M. Hasebe for enlightening discussions.

REFERENCES

Abe T 1993 Materials Science Forum 17 117
Bourret A 1986 Mat. Res Soc. Symp. Proc. 59 223
Ham F S 1958 J. Chem. Phys. of Solids 6 335
Hasebe M, Corbbett J W and Kawakami K 1991 Defects in Semiconductors 16, eds
    G. Davies, G. G. Deleo and M. Stavola (Pennsylvania: Trans Tech Publications) 1475
Inoue N, Osaka J and Wada K 1982 J. Electrochem. Soc. 129 2780
Lin W 1990 Semiconductor Silicon, eds R. Huff, K. G. Barraclough and J. Chikawa
    (Montreal: The Electrochemical Society) 569
Ponce F A and Hahn S 1989 Materials Science and Engineering B4 11
Tan T Y, Gardner E E and Tice W K 1977 Appl. Phys. Lett. 30 175
Wada K, Nakatani H, Takaoka H and Inoue N 1982 J. Crystal Growth 57 535
Zener C and Wert C 1950 J. Appl. Phys. 21 5

HRTEM investigation of microdefects in FZ-silicon crystal grown at high rate

L M Sorokin, J L Hutchison*, N B Ponomoriova and E S Falkevich

Ioffe Physical-Technical Institute of Russian Academy of Sciences, St. Petersburg 194021, Russia
*Department of Materials, University of Oxford, Parks Road, Oxford OX1 3PH, UK

ABSTRACT: (111) wafers of FZ silicon obtained at high growth rate have been studied by HREM. Crystalline microdefects thus revealed were identified as γ-Cu_5Si. Disk-like areas of both amorphous and crystalline structure were also found. Near them extra lattice planes and striations along the traces of $\{113\}$ lattice planes were also observed. The possible nature of these irregularities, and the mechanism of formation of the disk-like defects are discussed.

1. INTRODUCTION

Microdefects (MD) of various types, denoted by A, B and D (Veselovskaya et al. 1977) are known to occur in FZ-Si crystals grown at different rates. They differ from each other in size, structure and spatial distribution. It was shown that in crystals grown at a rate of 5 -8 mm/min the D-type MDs, mainly of interstitial type, were dominant (Sitnikova et al. 1984). Previous investigations had led to the assumption that these MDs can have both crystalline and amorphous structure (Sitnikova et al. 1988). In order to gain an understanding of the MD formation it was necessary to examine the structure of the crystals grown at the highest practical growth rates (9-10 mm/min). Such crystals had not previously been examined by HREM.

2. EXPERIMENTAL

A FZ-Si crystal of p-type (resistivity approximately 1 KΩ.cm) of diameter 30mm was grown in vacuum at varying rate (5 - 10 mm/min). Slices cut perpendicular to the [111] growth axis were investigated. X-ray topographs were obtained by the Lang method ($\mu t < 1$). The samples corresponding to the maximum growth rate (9 - 10 mm/min) have been studied by HREM at 400 kV accelerating voltage. (111) and (112) lattice images were obtained; the latter were obtained by tilting the (111) specimens by 19° around the [$\bar{1}$10] axis. Images were recorded close to the optimum defocus ($\Delta f = -48$nm). Optical diffractometry was used to assist in image interpretation. Double crystal X-ray diffractometry was also used to estimate overall crystal quality.

3. RESULTS AND DISCUSSION

The X-ray topography indicated that the crystal grown at v=9-10 mm/min had a high (10^6 cm^{-2}) density of evenly distributed dislocations. The half-widths of the rocking curves were almost twice the calculated values. Such broadening results from the block misorientation due to   randomly distributed dislocations. The diffuse X-ray scattering may arise from the MDs, caused by the high growth rate followed by quenching. A MD showing crystalline structure was revealed by HREM (Fig. 1a). The optical diffraction pattern (ODP) corresponding to this area shows one of the sets of matrix {220} type reflections to have significantly higher intensity than the other {220} set. This is due to the superposition of MD reflections with those of the matrix. This is shown in Fig.1 b,c.

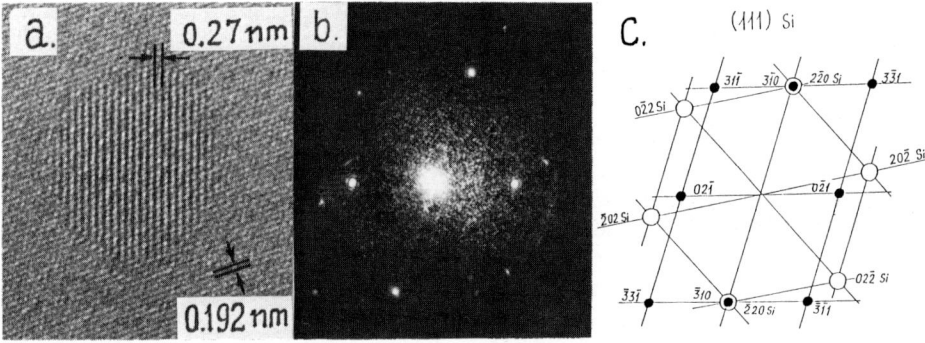

Fig.1 (a) (111) lattice image of microdefect with crystalline structure; (b) ODP of area (a); (c) schematic of (b), with reflections identified.

To identify this crystalline particle it was necessary to consider the range of phases which might be expected in FZ-Si crystals. Pirouz (1990) and Cerva (1991) have reported the existence of non-cubic Si phases. As found by Sadamitsu et al.(1988, 1991) particles containing Cu and Fe may also be present. The lattice spacings of the particle observed and of possible phases are presented in Table 1. Although the measured lattice spacings seem to correspond to those for several possible phases (γ-Cu$_5$Si, ξ-CuSi, α-FeSi$_2$), comparison of the {hkl} inter-planar angles on the ODP (Fig.1c) with calculated data for the above compounds indicates a best fit with the γ-Cu$_5$Si phase studied by Fageberg (1935).

TABLE 1. Comparison of experimental lattice spacings with those of known phases

d (Å) measured data	γ-Cu$_5$Si	ξ-CuSi	FeSi	α-FeSi$_2$	Fe$_3$Si	Si
	cubic	cubic	cubic	tetragonal	cubic	hexagonal
2.69	2.78	2.68	2.59	2.69	2.83	2.93
1.92	1.97	1.97	2.01	1.90	1.99	1.92
1.87	1.88	1.89	1.83	1.86	1.70	1.84
1.38	1.43	1.40	1.35	1.35	1.41	1.41

Apart from the particle described above there are also disk-like defects (DLD) with both amorphous and crystalline structure, about 10 - 40 nm in size. Some of these are presented in Fig. 2a.

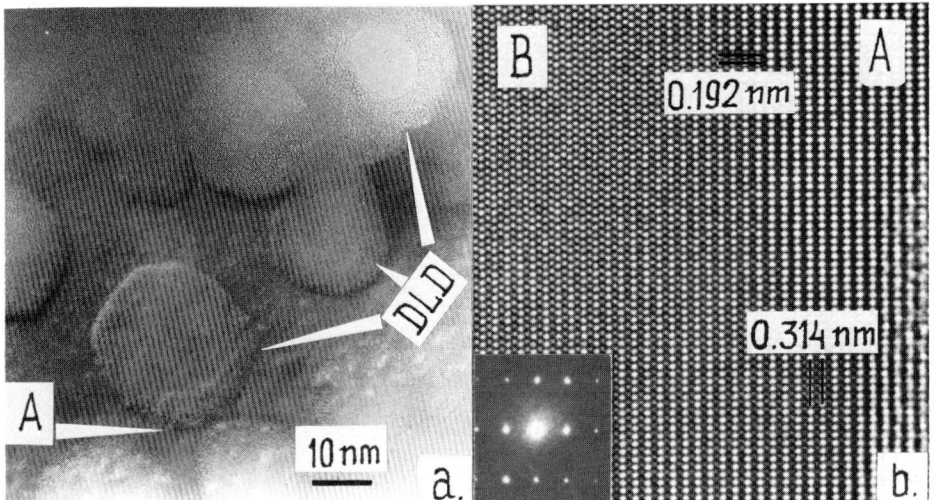

Fig.2. (112)Si: (a) experimental image of two types of DLD; (b) lattice image of area close to amorphous DLD with ODP inset. Note the extra planes in area B.

Narrow amorphous areas are commonly found at the edges of thin Si crystals when studied by HREM, due to ion-milling damage or the presence of thin oxide surface layers. In areas such as shown in Fig. 2a the amorphous region is much larger; we consider therefore such areas to be amorphous DLDs within the bulk of the crystal. Near the DLDs of both amorphous and crystalline type, as well as at the disk-matrix boundaries, there appears to be a specific structure which is different from the arrangement of dark and light spot contrast which is characteristic of the (112) image of perfect crystal. One such image is presented in Fig.2b. It shows contrast corresponding to both perfect (112) Si (A) and unusual structure (B). As indicated by Ourmazd (1985) contrast in the perfect region may in favourable conditions be interpreted as an approximation of the projected atomic structure. Close to the interface between the DLD image and the matrix the contrast contains an additional array of bright spots, parallel to the (220) lattice planes, and nonsymmetric about them. It can also be seen that the extra spots are aligned almost parallel to the traces of {113} matrix lattice planes, but slightly displaced from them. As a consequence, the {113} lattice planes in the image reveal a "corrugated appearance".

Near the DLD areas "cross-hatching" formed by two sets of fringes which are the traces of {113} planes are clearly seen (Fig.3a). This marked change in contrast could be due to high strain between the DLD and the matrix. At the high growth rate it is likely that the crystallization front captures liquid microdrops, which subsequently solidify forming both amorphous and crystalline DLDs. The thermal stress close to the droplets can cause the collective shift of the atoms of one "sublattice" with respect to another one along 1/6[121] in the (311) lattice plane (Fig. 3c). Another possible source of extra atomic planes is associated with the agglomeration of Si interstitial atoms in the {113} planes by the filling of tetrahedral positions. As shown by Salisbury (1980) this is the more energetically favourable case.

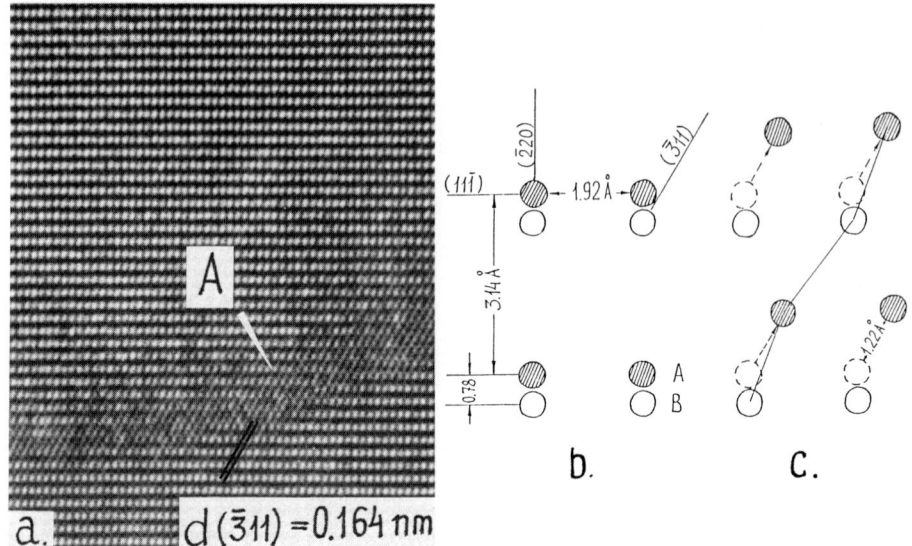

Fig.3. (112) lattice image of area A of Fig. 2a; (b) projected perfect structure;
(c) schematic diagram explaining appearance of extra planes in Fig.2b and striations
in Fig.3a (shear of A atoms with respect to B atoms)

4. CONCLUSION

In FZ-Si grown at high rates a crystalline microdefect was identified as cubic γ-Cu_5Si.
Specific structures of disk-like defects (both crystalline and amorphous) were observed,
which may be due to the peculiarities of the rapid crystallization process. Near the DLDs
extra lattice fringes and striations in $<112>$ images were observed. These could not be
interpreted in terms of the known structure of silicon. We recognise that the interpretation
of these novel features is tentative, and needs to be supported by image simulations; these
are planned.

REFERENCES

Cerva H 1991 J.Mat.Res. 6 2324
Fageberg S and Westgren A W 1935 Metallwirtsch. 14 265
Ourmazd A, Ahlborn K, Ibe K and Honda T 1985 Appl.Phys.Lett. 47 685
Pirouz P, Chaim R, Dahmen and Westmacott K H 1990 Acta Metall.Mater.38 313,323,329
Sadamitsu S, Sasaki A, Hourai M, Sumita S and Fujino N 1991 Jap.J.Appl.Phys. 30 1591
Sadamitsu S, Sumita S, Fujino N and Shiraiwa T 1988 Jap.J.Appl.Phys. 27 L1819
Salisbury I G 1980 J.Microscopy 118 75
Sitnikova A A, Sorokin L M, Talanin I E, Sheikhet E G and Falkevich E S 1984
    Phys.stat.sol (a) 81 433
Sitnikova A A, Sorokin L M and Sheikhet E G 1988 Sov.Phys.Sol.State (Translation)29 1513
Veselovskaya N V, Sheikhet E G, Neimark K N and Falkevich E S 1977 in "Growth and
    doping of semiconductor crystals and films" publ. by Academy of Sciences, Moscow.

Formation and dissolution of Si–O–C amorphous precipitates in Czochralski silicon heavily doped with carbon

S Chevacharoenkul and W Wijaranakula*

Centre for Microelectronic Systems Technologies, MCNC, P.O. 12889, Research Triangle Park, NC 22709-2889, USA
*Research and Development Department, SEH America Inc., 4111 Northeast 112th Avenue, Vancouver, WA 98682-6776, USA.

ABSTRACT: The nature of defects formed in annealed Czochralski silicon doped with carbon to the concentration of 2×10^{17} atoms.cm^{-3} was examined using transmission electron microscopy. After a 100 hour anneal at 850°C, a low density of square, thin amorphous precipitates with {100}-type habit planes and sides predominantly parallel to <110> crystallographic directions are detected. Punched-out dislocation loops at the precipitate platelets are absent implying that lattice strain associated with precipitate formation in this material is small. Further annealing of the silicon at 1050°C for 50 hours causes the precipitates to dissolve and dislocation loops to form on the same plane previously occupied by each precipitate. Some amorphous residues resulting from the decomposition are left inside the loops. In addition, dislocation loops are seen decorated, possibly with SiC and/or SiO$_x$. The high-temperature instability and small lattice strain associated with the precipitation suggest that the precipitates formed at 850° are possibly a Si-O-C glass.

1. INTRODUCTION

The roles of carbon and oxygen precipitation in Czochralski (Cz) silicon have been reported by several researchers. Pinizzotto and Schaake (1981) studied both low ($<2 \times 10^{16}$ cm^{-3}) and high (1.3×10^{17} cm^{-3}) carbon concentrations in 5-10 ohm-cm boron doped (100) Si and concluded that carbon causes heterogeneous nucleation of oxygen precipitation in Cz silicon by lowering the interfacial energy of the precipitates. Gupta et al (1992) studied Cz silicon having $9.4 \pm 0.3 \times 10^{17}$ cm^{-3} oxygen and $5.1 \pm 0.2 \times 10^{17}$ cm^{-3} carbon using the small-angle neutron scattering technique. They found that, for annealing at 750°C, precipitation kinetics in heavily carbon-doped silicon were rapid and the process was substantially over in less than 24 h (more than 100 h was required in low-carbon material). Furthermore, a larger number of smaller precipitates were formed and they were either separate precipitates of amorphous SiO$_2$ and SiC or co-precipitates containing both SiO$_2$ and SiC. A HRTEM micrograph of co-precipitation of SiC and SiO$_x$ was shown by Bender and Vanhellemont (1992) but experimental details were not given. While those studies indicated that carbon promotes oxygen precipitation, Bourret (1986) found carbon to have a retardation effect. Finally, using the static SIMS technique, Shimura et al (1985) showed that co-precipitation of C and O occurred in their study but the technique did not provide phase information, i.e. whether single or multiple phases were involved.

In this study, the Cz Si was p-type (8×10^{14} cm^{-3} boron). The initial C content was 2×10^{17} cm^{-3} determined using the FTIR technique while that of O was 8.5×10^{17} cm^{-3}. Anneals at both 850°C and 1050°C were done in an N$_2$ atmosphere. The annealed samples were prepared into [011] X-TEM samples in the normal way. A Philips 430 TEM, operated at 300 keV, was used for diffraction contrast imaging and a JEOL 200CX, having a top-entry specimen holder and a high resolution pole piece, was used for phase contrast imaging.

2. RESULTS

2.1   850°C, 100 h Anneal

After this low temperature anneal, the O and C contents were reduced to 5.2 x 10^{17} and 1.62 x 10^{17} cm^{-3}, respectively.   Precipitates were square thin platelets on (001) planes as shown in Fig. 1a.   This is the same morphology as those formed in low carbon Cz Si. However, the Si/precipitate interface appears rougher.   These precipitates were often intersecting as shown in Fig. 1b.   The nature of the platelets on (001) planes is also clearly seen in this figure.   Figure 2a shows that the precipitate density was low and that no other defects such as elongated prismatic loops, punched-out loops or stacking faults were present. The precipitate density was determined to be 1±0.4 x 10^{11} cm^{-3}, which is about two orders of magnitude lower than that reported by Tsai (1985) for a similar carbon content but with two step annealing.

The high resolution transmission electron micrograph of a precipitate shown in Fig. 3 reveals that it is amorphous.   Attempts were made to determine the chemical composition of such precipitates by energy dispersive X-ray analysis using an ultra-thin window detector. However, the results were not meaningful because of a carbon build-up at sample surfaces.

2.2 850°C, 100 h + 1050°C, 50 h Anneal

The defect distribution in a sample which received further annealing at 1050°C for 50 h is shown in Fig. 2b.   The defect density in this sample was found to be 4±0.5 x 10^{11} cm^{-3} which is slightly higher than that in the low-temperature annealed sample.   Other defects such as prismatic loops and stacking faults were present in this sample.   The O and C contents after the two step anneal were 2.94 x 10^{17} and 1.78 x 10^{17} cm^{-3}, respectively.

Detailed examinations of the defects reveal that the precipitates previously formed during the 850°C anneal had decomposed and were replaced by dislocation loops occupying the same habit planes.   Inside the loops, there were amorphous droplets which are believed to be a product of the precipitate decomposition.   The loops were also square having sides along [011] directions.   However, loops which formed around two intersecting precipitates had a more complex shape.   On the dislocation line, there were small precipitates decorating the loops.   Figures 4a and b show the two loops described above.

3. DISCUSSION

The annealing behaviour of the precipitates in high C Cz Si described above indicate that they are different from those formed in low C Cz Si.   The precipitates which form during an 850°C anneal are likely to be composed of Si, O and C forming a Si-O-C glass.   A formation mechanism may involve diffusion of interstitial C and O atoms to Si-C nuclei followed by growth.   Si and C are known to form a strong bond as manifested in the high strength of SiC, thus Si-C centres should be stable nuclei.   Since C causes contraction and O causes dilatation of Si lattice, it is likely that their co-precipitation can result in a lattice strain minimization associated with the defect formation.   This is supported by the fact that precipitation-induced loop-punching does not occur in this sample.   A positive proof of Si-O-C glass formation can be obtained using an STEM with an X-ray mapping capability.

The defect density was found to increase after the two-step anneal and there was a further reduction of O.   Both findings can be interpreted as carbon retarding oxygen precipitation. This is in agreement with the work reported by Bourret (1986).

The fact that these precipitates are unstable for 1050°C annealing further supports the proposal that they are Si-O-C glass because earlier work by Tan and Tice (1976) and Bender et al (1992) on SiO_x precipitation indicate that SiO_x precipitates are stable and grow in size upon annealing at 1050°C.

Lastly, dislocation loops in Si normally are formed on (011) planes.   For a loop to be stable on (001), as found in the two-step annealed sample, there must be some factors stabilizing it.   These factors are believed to be the inability of the dislocations to climb and pinning by amorphous residues inside the loop and by materials decorating the loop.

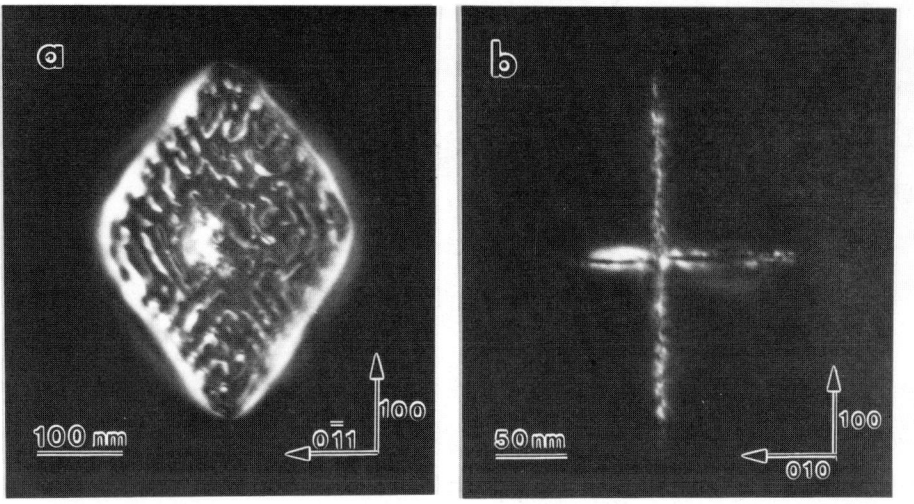

Fig.1 a) (011) projection of a precipitate formed at 850°C, 100 h anneal showing a sqaure morphlogy with irregular Si/precipitate interface. b) (001) projection of two intersecting precipitates showing platelet nature and their (100)-type habit planes.

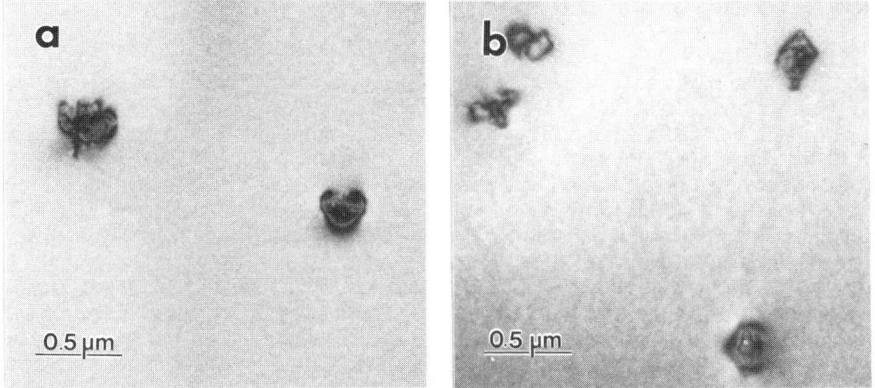

Fig. 2 Low magnification micrographs showing defect distribution in a) 850°C, 100 h anneal and b) 850°C, 100 h + 1050°C, 50 h anneal. Defect density increases after two-steps annealing.

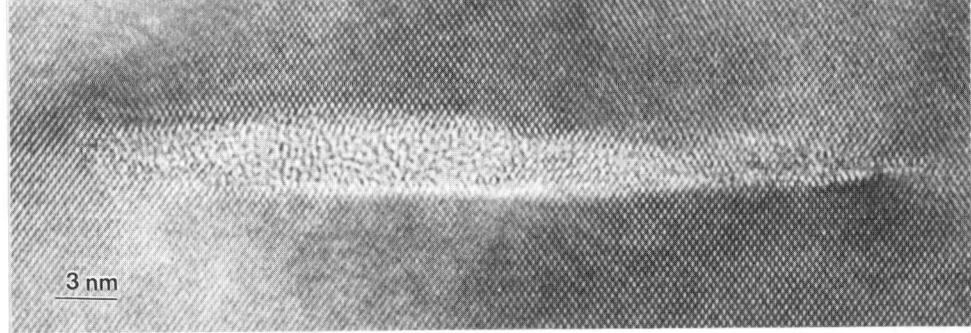

Fig. 3 HRTEM micrograph of a precipitate formed at 850°C, 100 h anneal. The precipitate is amorphous and contains a single phase.

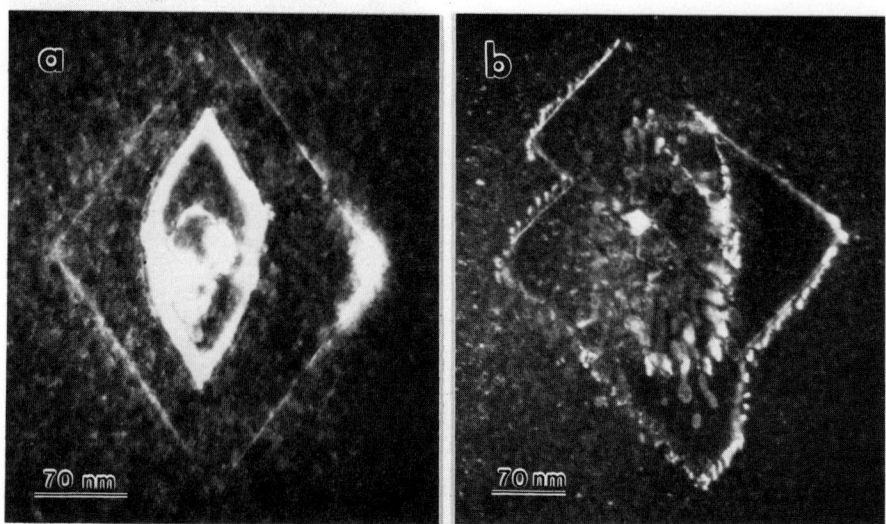

Fig. 4. TEM micrographs of defects after a 850°C, 100 h + 1050°C, 50 h anneal showing a) a square dislocation loop surrounding a partially decomposed precipitate and b) the loop is not square when formed at intersecting precipitates. Also shown in b) are amorphous residues inside the loop and material decorating the loop.

4. CONCLUSIONS

a) High C concentration does not alter the morphology of precipitates from those which form in low carbon Cz Si for anneals at 850°C. Furthermore, in our study, C seems to retard the rate of oxygen precipitation.

b) Co-precipitation of O and C forming Si-O-C glass upon 850°C annealing is probable because the volume occupied by Si-O-C glass should be smaller than that by SiO_x and SiC combined. These precipitates are the most likely sink of C interstitials which disappear from the matrix after annealing.

c) The precipitate instability for 1050°C anneals indicates that they are not the same SiO_x precipitates formed in the low carbon material. Those precipitates grow in size or assume a new morphology at this annealing temperature and produce punched-out dislocation loops. This action was not seen in high carbon Cz silicon.

d) The decomposition products as deduced from TEM micrographs consist of: i) glassy droplets of possibly higher C content, ii) regrown Si replacing the volume left behind by the precipitates iii) O atoms diffusing away or reforming SiO_x at other locations, iv) dislocation loops surrounding the glassy droplets and v) material seen decorating the loops can be small SiC precipitates and/or SiO_x.

REFERENCES

Bender H and Vanhellemont J, 1992 Proc. Mat. Res. Soc. Symp. Vol. 262 eds S Ashok et al (Pittsburgh: Mat. Res. Soc.) pp 15-29

Bourret A, 1986 Proc. Mat. Res. Soc. Symp. Vol. 59, eds J C Mikkelsen, Jr. et al (Pittsburgh: Mat. Res. Soc.) pp 223-236

Gupta S, Messoloras S, Schneider J R, Stewart R J and Zulehner W, 1992 Semicond. Sci. Tech. 7 (1) 8

Pinizzotto R F and Schaake H F, 1981 Defect in Semiconductors, eds J Narayan and T Y Tan (Pittsburgh: Mat. Res. Soc.) pp 387-391

Shimura F, Hocket R S, Reed D A and Wayner D H, 1985 Appl. Phys. Lett. 47 (8) 794

Tan T Y and Tice W K, 1976 Phil. Mag 34 (4) 615

Tsai H L, 1985 J. Appl Phys. 58 (10) 3775

Inst. Phys. Conf. Ser. No 134: Section 3
Paper presented at Microsc. Semicond. Mater. Conf., Oxford, 5–8 April 1993

Recombination at dislocations associated with oxygen precipitation

A Cavallini*, M Vandini*, F Corticelli**, A Parisini** and A Armigliato**

*University of Bologna, Physics Dept. Via Irnerio 46, I-40126 Bologna, Italy
**CNR-LAMEL, Via Castagnoli 1, I-40126 Bologna, Italy

ABSTRACT: The effects of process-induced crystal defects in silicon power devices have been investigated by comparing their electrical activity and structural properties. Fz P-doped silicon has been subjected to boron diffusion and subsequent oxidation during prolonged heat treatments. The induced defects have been analyzed before and after the oxidation process. Their electrical activity has been explored by electron beam induced current (EBIC) method, and related to the structural properties as observed by transmission electron microscopy (TEM). TEM analysis has shown the presence of extended dislocation networks in both sets of samples as well as of polyhedral shaped precipitates, whose shape and distribution depend upon the thermal treatment. The nature of the precipitates has been investigated by electron energy loss spectroscopy (EELS) which revealed the presence of oxygen. The comparison between the EBIC and TEM results indicates that the observed electrical activity of the dislocations is significantly influenced by the oxygen precipitates.

1. INTRODUCTION

High resistivity float zone silicon has found use as a substrate material for the manufacture of high-voltage high-current power rectifiers. Fz-Si, due to the absence of a crucible in the growth technique, has a very low oxygen and carbon content and lower impurity concentration as compared to Cz-Si crystals. Nevertheless, the thermal treatments to which the material of the final device has been subjected can induce defects and introduce impurities which lead to a degradation of its performance.

The purpose of this contribution is to investigate the defect state induced by the heavy doping of the p⁺-end region and a subsequent oxidation, representing two typical steps of power rectifier technology. The occurrence of defects is dramatically effected since the above mentioned thermal treatments are performed at high temperature for very prolonged time.

Although the initial oxygen content is low, its role is however important because of the harsh processing conditions. The presence of oxygen precipitates and their gettering efficiency will be examined in relation to the electrical activity of the revealed extended defects.

2. EXPERIMENTAL

The starting material was float zone (111) silicon, dislocation free, phosphorous doped with carrier concentration $(N_D-N_A) = 8*10^{13}$ cm^{-3}. The initial content of oxygen was [O] $\leq$ 10^{16} cm^{-3} and of carbon [C] $\leq 5*10^{16}$ cm^{-3}. Boron doping was performed by lateral diffusion for 18 hours at 1250 °C, with a pre-annealing step at 600°C for 30 minutes. The dopant concentration of the p$^+$-end region was $(N_A-N_D) = 1*10^{20}$ cm^{-3}. Dry-wet-dry oxidation at 1050°C for 7 hours was subsequently carried out.

The analyses have been performed in the p$^+$-zone, from the sample edge to the depletion region of the p-n junction, after diffusion (group B samples) as well as after oxidation (group BO samples). Both sets of samples have been examined by the Electron Beam Induced Current (EBIC) method. The changes of the minority carrier diffusion length L and of the contrast C at the defects allowed us to evaluate the recombination activity evolution at different processing steps.

Transmission Electron Microscopy (TEM) studies have been carried out in order to relate the structural features of the defects to their electrical properties. Since some precipitates associated with the presence of extended defects were observed, Electron Energy Loss Spectroscopy (EELS) analysis has been used to make possible the identification of the precipitate composition.

3. RESULTS

The EBIC investigation of the B group samples revealed the presence of evenly distributed dislocation networks (Fig. 1a), in which the contrast at the nodes was found to be significantly higher than at the dislocation lines.

On the contrary, in the BO group samples no more networks have been observed.

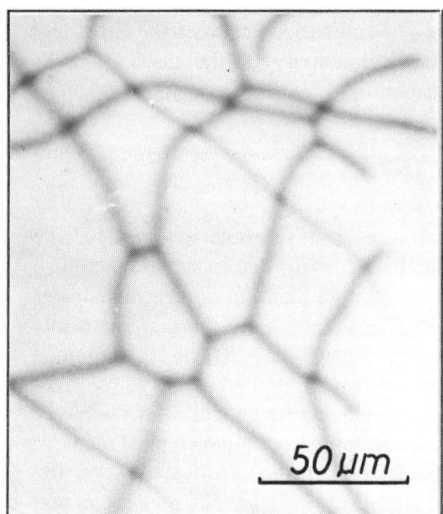 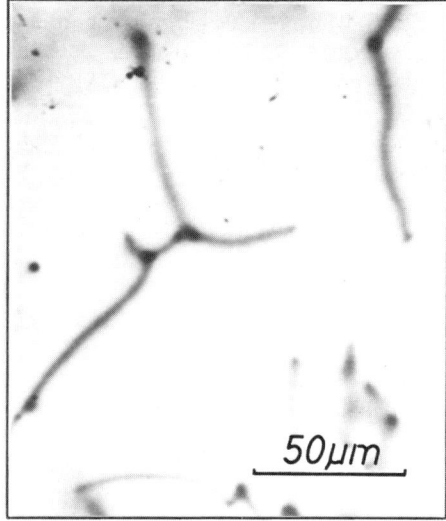

Figure 1a: EBIC micrograph of the p$^+$-side of a B sample at beam accelerating voltage equal to 15 kV.

Figure 1b: EBIC micrograph of the p$^+$-side of a BO sample at beam accelerating voltage equal to 15 kV.

Only short and non-homogeneously distributed dislocation segments are visible, with stronger recombining centres mainly occurring at their intersections (Fig. 1b).

In a more detailed examination of the as-diffused material, the dislocation networks appear to be electrically active along their whole length, showing stronger recombination centres at the nodes as well as along the dislocation lines (Fig. 2).

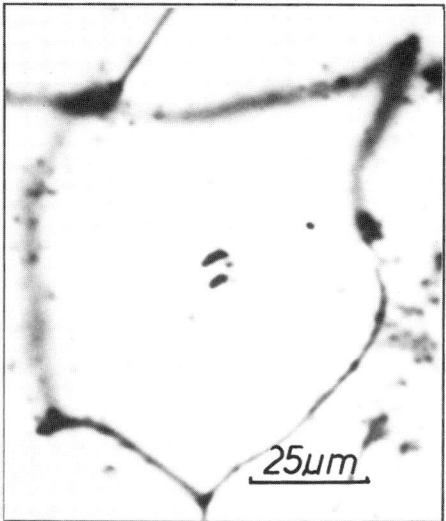

Figure 2: EBIC micrograph of a B sample in extreme contrast conditions (beam accelerating voltage equal to 10 kV).

Quantitative EBIC analyses showed an increase in both diffusion length L and contrast C at the defects (Table 1). After the oxidation process the contrast increase is especially marked at the nodes; in addition, the current profiles have given evidence of a bright contrast zone around the dislocations.

The values of contrast reported have been obtained by averaging the measurements performed on more than ten defects found in different samples. Furthermore the defects for contrast analysis have been selected in such a way that their geometry does not change significantly: only dislocation networks parallel to the sample surface were examined and the values of C reported in Table 1 represent an average over figures obtained at about the same beam energies.

TEM investigations have shown the presence of precipitates at dislocations in both sets of samples. However, their size and distribution were found to vary significantly before and after the oxidation process. In the as-diffused material (samples B) polyhedral precipitates of 50-150 nm in size are uniformly distributed along the dislocations as well as at the nodes (Fig. 3a); whereas, after oxidation (samples BO) the precipitation is mainly present at the nodes (Fig. 3b), and occasionally also along the dislocation lines. In addition, in the oxidized samples the precipitates showed a significant shrinkage with an average size of 20-50 nm. It is worthwhile noting that both before and after oxidation, the precipitates are three-dimensional, as observed in cross-section TEM. Furthermore, they have not been observed in the silicon matrix.

The nature of the precipitates has been investigated by EELS. A well defined k-edge was detected in the core loss region of the spectra taken inside the precipitates whereas no oxygen was revealed in the surrounding matrix. This suggests that the precipitates correspond to a SiO_x phase, which has been already observed in Cz-Si (Carpenter et al 1983, Yang et al 1978).

Table 1: Summary of quantitative EBIC results

THERMAL TREATMENT	L_{diff} (µm)	C % at dislocations	C % at nodes
Diffusion	5	10	12
Diffusion + Oxidation	10	16	30

It is important to stress that a parallel between EBIC and TEM analyses is not

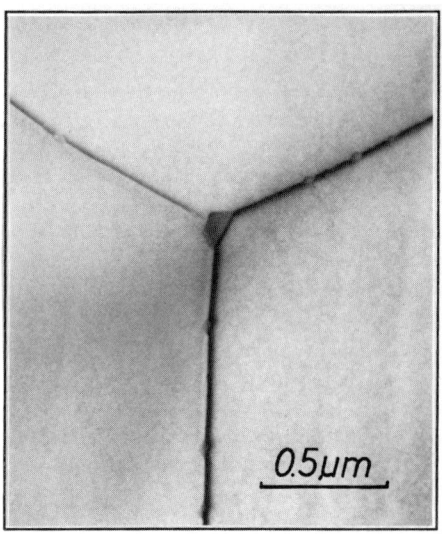

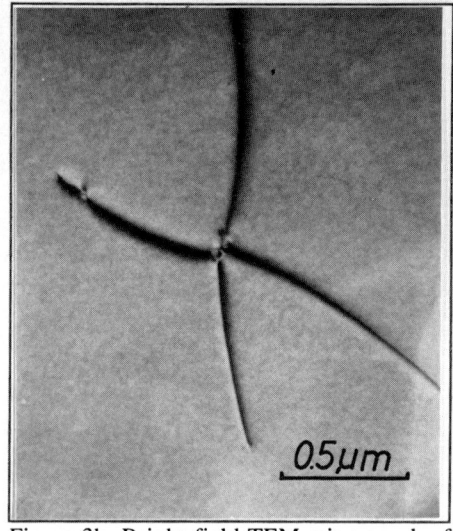

Figure 3a: Bright-field TEM micrograph of a type B sample at beam accelerating voltage equal to 300 kV.

Figure 3b: Bright-field TEM micrograph of a type BO sample at beam accelerating voltage equal to 300 kV.

straightforward. The different spatial resolutions of these two techniques prevent from directly comparing the structural and the electrical properties of the precipitates whose size, as revealed by TEM, is one order of magnitude below the EBIC limit of spatial resolution. Since the comparison cannot be made on the same scale, there is not a one to one correspondence between the precipitates observed in TEM (Fig. 3a) and the stronger recombination centres along the dislocations shown in EBIC (Fig. 2). However, the electrical activity induced by the presence of oxygen precipitates can be revealed in EBIC by the dotted appearance of the contrast at dislocations in the samples where, by TEM observations, precipitation was found along the networks.

4. DISCUSSION AND CONCLUSIONS

From the findings reported above on the defect state evolution it is evident that several different, and sometimes competitive, effects have to be considered to explain the processing outcome. The main factors playing a key role in both generation and interaction of dislocations are: presence of intrinsic point defects, oxygen introduction, stress due to the heavy doping, high temperature treatments, as well as contamination by impurities (Castaldini et al. 1992). The concurrence of all these factors justifies the oxygen precipitation in the otherwise high purity float zone silicon (Kolbesen et al. 1982).

However, the precipitation does not occur uniformly in the bulk, as usually happens in Cz-silicon. As has been well established (Radzmiski et al 1992, Sato et al 1985), extended defects act as preferential sites for precipitation, which is confirmed also in this case by the presence of oxide clusters at the dislocations only.

EBIC and TEM gave evidence that the precipitates, occurring mainly at the dislocation intersections, act as pinning sites affecting the dislocation motion. Furthermore, TEM has made it possible to observe that smaller precipitates (of about 50 nm in size) also occur evenly spaced along the dislocations, suggesting that they probably lie at kinks and/or jogs (Sato et al. 1985). Their polyhedral shape could be explained on the basis of thermodynamic considerations (Tiller et al. 1986). Moreover, their morphology and size change depending on the precipitate position, being larger if located at the mesh nodes than along the extended defects.

Other factors play a role during the oxidation process, strongly influencing both precipitate dimension and distribution. As already mentioned, a shrinkage of the precipitates at the dislocation nodes and the disappearance of most of them along the majority of the lines is clearly detected after the oxidation. This phenomenon could be explained by a supersaturation of silicon self-interstitials occurring during the oxidation process (Yamanaka et al. 1990), which is also enhanced by the interstitial dominated boron diffusion, re-activated by high thermal treatments (Gösele et al 1991).

The reaction controlling the growth and shrinkage of oxide precipitates is represented by the law-of-mass action:

$$xaO_i + bSi \rightleftharpoons aSiO_x + (b-a)I_{Si} \qquad (1)$$

where O_i denotes the interstitial oxygen atoms and I_{Si} the silicon self-interstitials, the latter generated by the volume expansion (of about a factor of 2) associated with oxygen precipitation.

During thermal oxidation silicon self-interstitials are generated at the sample surface. Some of these excess self-interstitials diffuse from the surface into the bulk during the prolonged heat treatment, that also re-activates the self-interstitial dominated boron diffusion. From equation (1) it can be deduced that a supersaturation of I_{Si} promotes the backward reaction, leading to the shrinkage of oxide precipitates. As a matter of fact, a competitive mechanism between the growth and the shrinkage takes place, and the critical radius above which precipitates grow is a function of both oxygen and silicon interstitial concentrations.

Although the model is not entirely satisfactory in the present case and also other mechanisms must be taken into consideration, our TEM observations of the oxidized material seem to confirm that the backward reaction of the law-of-mass action (1) is facilitated.

The structural and analytical results previously reported provide a powerful tool to explain the changes of the electrical activity of the defects observed by EBIC. TEM observations have revealed that in both groups of samples dislocations arranged in networks occur, whereas the precipitation features differ significantly. The results obtained by EBIC analyses have shown that the differences between the B and BO samples do not concern the morphology of the extended defects, but their electrically active portions and their recombination efficiency. By comparing these findings it can be deduced that the electrical activity of the defects is related to the presence of the oxide phase and to the effects that this produces on the impurity content of the surrounding area. As a matter of fact, the phenomenon of gettering of impurities by the precipitates has been widely utilized to explain the electrical activity of the dislocation lines (Falster et al 1990). Likewise the higher contrast at the nodes corresponds to a localized larger concentration of the SiO_x phase.

After the oxidation, parts of the networks become invisible in EBIC, that is the contrast is below the detection limit (about 2% in these samples). At the same time the recombination strength of the electrically active dislocations and precipitates is greatly enhanced, resulting in an increase of contrast at defects and minority carrier diffusion length (see table 1). The simultaneous occurrence of bright contrast in the close vicinity of the dislocations confirms the increase of the gettering efficiency of the extended defects (Blumtritt et al. 1989). It can be proposed that the reaction in the law-of-mass action (1) proceeds backwards, thus, together with silicon atoms that will reach an equilibrium position in the matrix or as point defects, oxygen interstitials are formed. These can either remain in solid solution, or react with metal impurities. Therefore the segregation of impurities is re-activated and enhanced by the oxidation process, thus explaining the observed effects as a reduction of the level of contamination in the bulk.

REFERENCES

Blumtritt H, Kittler M and Seifert W 1989 Inst. Phys. Conf. Ser. No 104, eds Roberts S G Holt D B and Wilshaw P R (Bristol: IOP) pp 233-38
Carpenter R W, Chan I, Tsai H L, Varker C and Demer L J 1983 Mat. Res. Soc. Symp. Proc. v 14, eds Mahajan S and Corbett J W, (New York: Elsevier) pp 195-99
Castaldini A, Cavallini A, Fraboni B and Giannotte E 1992 J. Appl. Phys.72 1
Falster R and Bergholz W 1990 J. Electrochem Soc. 137 1548
Gösele U M and Tan T Y 1991 Material Science & Technology v 4, ed Schröter W (Weinheim: VCH) pp 197-247
Kolbesen B O and Mühlbauer 1982 Sol. St. Electronics 25 759
Radzmiski Z J, Zhou T Q, Buczkowski A and Rozgonyi G A 1992 Appl. Phys. Lett. 60 1096
Sato M and Sumino K 1985 Proc. 9th Yamada Conf. on Dislocations in Solids, eds Suzuki H, Ninomiya T, Sumino K and Takeuchi S (Tokyo: University of Tokyo Press)pp 391-94
Tiller W A, Hahn S and Ponce F A 1986 J. Appl. Phys. 59 3255
Yamanaka H, Aoki Y and Samizo T 1990 Jap. J. Appl. Phys. 29 2450
Yang K H, Anderson R and Kappert H F 1978 Appl. Phys. Lett. 33 225

Inst. Phys. Conf. Ser. No 134: Section 3
Paper presented at Microsc. Semicond. Mater. Conf., Oxford, 5–8 April 1993

Microstructure and electronic behaviour of decorated stacking faults in silicon

Y Qian, J H Evans and A R Peaker

Centre for Electronic Materials and Dept. Electrical Engineering and Electronics,
University of Manchester Institute of Science and Technology, Sackville St,
P O Box 88,
Manchester, M60 1QD, U K

ABSTRACT: Gold and copper decorated oxidation-induced stacking faults in silicon have been studied by TEM, DLTS and EBIC. DLTS profiling indicates that the spatial extent of the electrical activity correlates with the location of the Frank partials bounding the stacking faults. We find that the detailed electrical properties of the deep levels are independent of the species of decorating impurity.

1. INTRODUCTION

The presence of oxidation induced stacking faults (OSF) in active device regions is quite a common problem in semiconductor processing. These defects are associated with the long thermal processing required for power devices but stacking faults also occur in processes such as pre-amorphisation by ion-implantation and SIMOX. Recently, the electrical effects of decoration of OSFs has been reported (Peaker et al, 1989, Berg et al, 1992); the authors show that, below the impurity concentration at which precipitates are visible in HRTEM, decoration of a stacking fault moves the associated deep levels in the bandgap towards mid-gap, irrespective of the species of transition metal used for decoration.

These studies were carried out on OSFs generated by polishing the surface prior to oxidation (Berg et al 1992 and references therein). As surface damage always results in near surface dislocations and OSFs, we report here a study of stacking faults generated by oxidation alone in order to avoid the creation of dislocations other than the Frank partials bounding the OSFs. We have characterised the OSF before and after decoration using TEM and EBIC. Depth profiles of the electrical activity of OSF-related deep defect states have been measured by Deep Level Transient Spectroscopy (DLTS), and the results are compared to TEM images.

2. FABRICATION OF OSF

The OSF were generated in (100) phosphorus-doped silicon grown by conventional vapour phase epitaxy on Czochralski substrates. The carrier concentration in the epitaxial Si was 2×10^{15} cm^{-3} as determined by capacitance-voltage measurements. A dry-wet-dry multi-oxidation process was carried out without prior surface damage. Dry oxidation was in 100% oxygen at atmospheric pressure at 1100°C for 10 minutes, and wet oxidation was in steam oxygen at 1100°C for 80 minutes. The density of stacking faults was in the range from 2×10^4 to 2×10^5 cm^{-2} as monitored by Nomarski optical micrography after a preferential Y3 etch (Yang 1984). Au or Cu diffusions were then carried out on the samples. For the Au diffusion, samples were first immersed in an aqueous solution containing 1% of chloroauric acid. For Cu decoration, the samples were lightly touched on the back with clean Cu wire. Diffusions were carried out at 750°C in a 100% nitrogen atmosphere for 30 minutes. The stacking faults were examined by plan view TEM and EBIC before and after the diffusions.

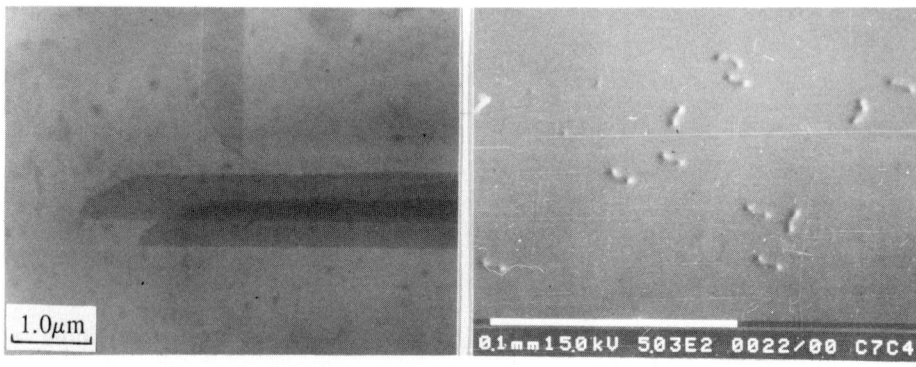

Figure 1 . Plan view TEM of typical nominally   Figure 2 . EBIC of OSF after Cu decoration
clean OSFs

Fig. 1 shows plan view TEM of typical nominally clean, undecorated OSF; the faults were found to occur at a range of depths below the surface, up to a maximum of 1.75 μm, and no dislocations other than Frank partials were observed. Fig. 2 shows an EBIC image of the OSF after Cu-decoration; it is clear that no other types of dislocations have been generated as a result of the diffusion.

3. ELECTRICAL ACTIVITY OF OSF

We carried out EBIC and DLTS measurements on the same diodes fabricated by evaporating semi-transparent Au Schottky barriers; aluminium was evaporated to form the back face ohmic contact. We found that the OSF densities were not uniform across the samples, and it was not possible to calculate the effect of decoration upon the trap concentrations.

Fig. 3 illustrates the clear correlation between the depth profile of the electrical activity and the expected distribution of Frank partials bounding the OSF. In the nominally clean sample at depths greater than 1.75μm, the trap concentration falls rapidly to the background level. As the Frank partials become decorated with Au, Fig. 3(b) or Cu, Fig.3(c), the integrated trap concentration does not alter within experimental error, but the depth distribution of the trap no longer follows the nominally clean 'ideal' stacking fault shape. This has been observed previously for Ag-decorated stacking faults, (Berg et al 1992) and is similar to the effect observed in computer simulations of depth profiles of electron trap occupation when the trap energy is a function of the occupation probability, (Shröter et al 1989). As an electrostatic potential builds up around the defect, the capture rate is modified and depth profiles of equilibrium occupation broaden.

Despite the changed depth distribution of the defect states after decoration, we do not believe that we have introduced any new type of capture centre as a result of our decoration process, for the following reasons. Figure 4 illustrates DLTS spectra of the OSF-related defect states before and after decoration. Fig. 4(a) shows the DLTS spectrum of the OSF trap before intentional decoration; two emission peaks are present. Decorating the stacking faults with Cu, Fig. 4(b) or Au, Fig. 4(c), leaves one remaining peak which appears at a higher temperature. The shift is probably different between the two samples because of different levels of decoration of Au and Cu (Lahiji et al, 1989).

Preliminary measurements of the capture cross section, σ_n, of both states in Fig. 4(a) indicate that it is the lower temperature trap that remains after decoration, and that the value of σ_n suggests that the deep state is negatively charged, as discussed later.

The linewidth of the DLTS peak is an indication of the strength of interaction between the measured activation energy, E_T, and the charge build-up in the vicinity of the OSF (Shröter et al 1989). The linewidths in Figure 4 are very similar in spectra 4(a), 4(b), and 4(c). This suggests that the detailed electrical properties of the capture centre have not been significantly modified by the decoration process. Work is in currently progress to model the interaction strength between E_T and the occupation factor of the OSF-related deep states.

Within the Shockley-Hall-Read (SHR) model for capture at isolated point defects, a small value of σ_n would normally be associated with a negatively charged deep level in n-type material (Lax 1960); however, it is anticipated that an SHR model is inappropriate for these extended multi-electron defects. It is likely that a potential barrier will build up around the defect during the DLTS experiment, modifying

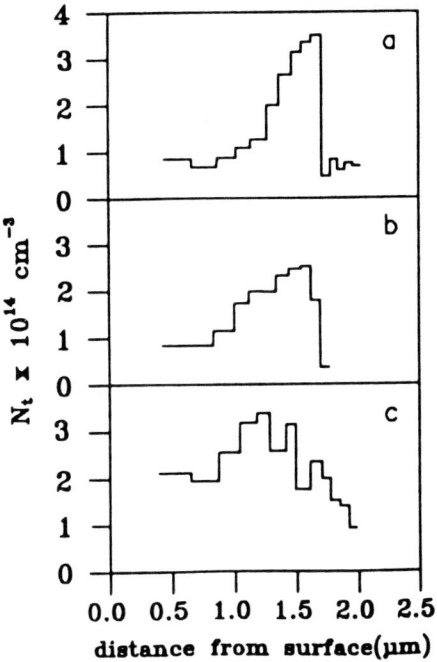

Figure 3. Deep trap profiles measured by DLTS (a) nominally clean OSF (b) Au decorated OSF (c) Cu decorated OSF

the capture rate. The rate of capture into the defects will probably reduce as charge builds up in the vicinity of the OSF, and preliminary investigations on these samples indicate that capture will be logarithmic.

Charge build up along dislocations has also been observed by EBIC (Wilshaw and Fell 1989), and it is thought that the amount of band bending increases with increasing dislocation charge. We have simulated this effect in the DLTS measurement by modulating the amount of captured charge in the vicinity of the OSF by changing the fill pulse length. We found that relatively long fill pulse times (15ms) were required in order to saturate the capture process, which is consistent with a small value of σ_n (and the doping level in the layers).

Fill pulse length (ms)	15.00	5.00	0.50	0.05	0.01
E_T (meV) (Cu-dec.)	480	505	538	564	570
E_T (meV) (Au-dec.)	465	482	500	549	473

TABLE 1: Measured deep level activation energy E_T as a function of fill pulse length for Au and Cu decorated OSF. The reverse bias was -2.5 V and the fill voltage was -2V.

Table 1 illustrates the measured values of E_T as a function of fill pulse length. Filling and emptying voltages were chosen to ensure the narrowest DLTS linewidths (Kaniewski et al 1992). The emission energy, E_T, clearly depends upon the fill pulse length, and therefore upon the stored charge at or near the Frank partials. It is also possible to show by EBIC measurements that the capture energy is sensitive to the amount of stored charge on the dislocation (Wilshaw and Fell 1989). E_T as a function of fill pulse length for the Cu decorated OSF, shown in Table 1, correlates well with the theoretical prediction of Shröter et al (1989) which states that any defect whose activation energy depends upon the defect occupation probability f_T will exhibit a lower temperature DLTS peak at longer fill pulse times (Shröter et al 1989). Therefore, a DLTS characterisation of E_T of OSF-related deep states yields similar conclusions to those drawn from EBIC measurements regarding the effect of stored charge.

4. CONCLUSION

We have demonstrated that the electrical activity of nominally clean oxidation induced stacking faults, as measured by DLTS profiling, correlates with the spatial location of the Frank partials bounding the OSFs, as determined by TEM. DLTS linewidths and preliminary electron capture cross-sections measurements before and after decoration suggest that the detailed electrical properties of the defect responsible for electron capture is independent of the decorating species and the amount of the impurity (below the concentration at which precipitates are visible in HRTEM). We also demonstrate that the phenomenon of charge build-up must be taken into consideration when attributing an activation energy to the trap.

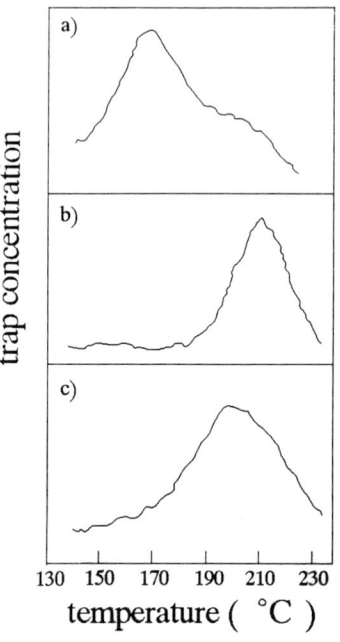

Figure 4 . DLTS spectra for a) nominally clean OSF b) Cu-decorated OSF c) Au-decorated OSF. The rate window was 200 s^{-1}.

We would like to acknowledge the financial support of the SERC for this work. One of us (YQ) would like to acknowledge support from the British Council. We would like to thank G R Booker and P R Wilshaw of Oxford University for useful discussions.

REFERENCES

Berg A, Brough I, Evans J H, Lorimer G and Peaker A R 1992 Semicond. Sci. and Technol. 7 A263
Kaniewski J, Kaniewska M and Peaker A R 1992 Appl. Phys. Lett. 1992, 60, 359
Lahiji G R, Peaker A R and Hamilton B, 1989, Inst. Phys. Conf. Ser., 104 239
Lax M 1960 Phys. Rev. 119 1502
Peaker A R, Hamilton B, Lahiji G R , Ture I E and Lorimer G 1989 Mat. Sci. & Eng. B 4 123
Shröter W, Queisser I and Kronewitz J, 1989 Inst. Phys. Conf. Ser. 104 75
Wilshaw P R and Fell T S 1989 Inst. Phys. Conf. Ser. 104 85
Yang K H 1984 J. Electrochem. Soc. 131 1140

Inst. Phys. Conf. Ser. No 134: Section 3
Paper presented at Microsc. Semicond. Mater. Conf., Oxford, 5–8 April 1993

125

Role of oxygen in copper precipitation at oxidation stacking faults in silicon

A Correia, D Ballutaud and J-L Maurice*

Laboratoire de Physique des Solides de Bellevue and *Laboratoire de Physique des Matériaux; CNRS-Bellevue, F-92195 Meudon cédex, France.

ABSTRACT: Thermal oxidation of Czochralski (Cz) and floating zone (FZ) silicon was carried out in a copper contaminated environment. In the Cz case, large copper precipitate colonies appeared in {110} (mainly) and {100} planes, nucleated at oxygen-containing clouds decorating the Frank partial dislocations bounding oxidation induced stacking faults (OSF). In the FZ samples, copper precipitated directly on these dislocations, which were in contrast free of oxygen, creating small colonies in the OSF {111} planes.

1. INTRODUCTION

The study of copper in silicon has been the object of extensive work over the last 40 years (e.g. Dash 1956, Thomas 1963, Fiermans and Vennik 1967, Nes and Washburn 1971, Nes 1974, Seibt and Graff 1988, Schröter et al 1991, Higgs et al 1992, Correia et al 1993). After contamination during a high temperature treatment, copper generally forms colonies upon cooling, either in {110} or {100} planes, composed of approximately spherical Cu_3Si particles with diameter rarely bigger than 30 nm. Formation of colonies occurs through the climb of edge dislocations, which is in turn fed by silicon self interstitials produced by the precipitation of Cu_3Si on these defects.

When oxidation induced stacking faults are present, Cu_3Si is likely to precipitate on the Frank partial dislocation bounding the fault (Schröter et al 1991), but this seems to only occur in floating zone materials (Higgs et al 1992), while the partial dislocation serves to nucleate colonies in {110} planes in Czochralski silicon (Thomas 1963). The aim of this paper is to give clear evidence of the respective roles of oxidation induced stacking faults on the one hand and of oxide precipitates on the other hand in the nucleation of copper colonies.

Electron beam induced current (EBIC) in the scanning electron microscope (SEM) was used to image recombinant defects. The EBIC collection efficiency vs. beam voltage curves were employed to measure minority carrier diffusion length and surface recombination velocity (Maurice 1993), before and after the different treatments. Analytical transmission electron microscopy (AEM) was then used to characterize the precipitates.

2. EXPERIMENTAL DETAILS AND RESULTS

The Czochralski (Cz) and floating zone (FZ) silicon were boron doped to 3×10^{16} cm^{-3} and 10^{16} cm^{-3}, respectively. Samples were mechanically polished with a 1/4 μm diamond paste. The heat treatments were carried out in sealed ampoules, under one atmosphere of dry oxygen, at 1000°C for 1 hour, the cooling rate being approximately 5°C s^{-1} (quenching of the ampoules in air). The conditions used (ampoule sealing and furnace) were known to copper-contaminate the samples (Delidais et al 1991). EBIC was performed after evaporation of an aluminium Schottky contact on the front surface. EBIC images were correlated with electron channelling patterns for defect orientation. Both plan-view and cross-sectional specimens were prepared for transmission electron microscopy (TEM). Plan-view samples were chemically thinned and cross sections were ion-milled. The microscope (Jeol 2000FX) allowed lattice imaging in the <110> directions and was equipped with an energy dispersive X-ray (EDX) spectrometer (Link) allowing light-element detection.

Thermal oxidation and contamination of the Cz material led to the formation of large recombinant defects, most of them extended along <110> directions (fig. 1), on which the EBIC contrast was between 80 and 90 % at 40 keV.

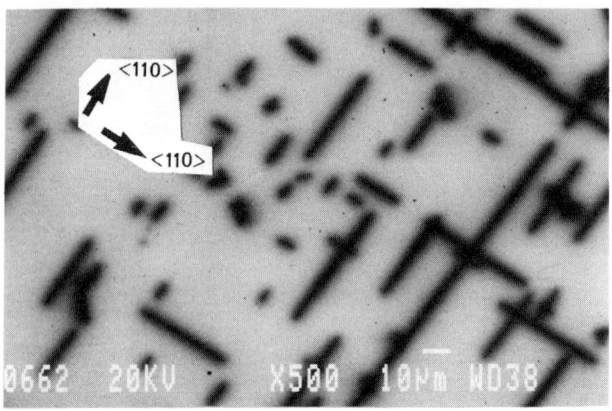

Fig. 1. EBIC image of the Czochralski sample after oxidation

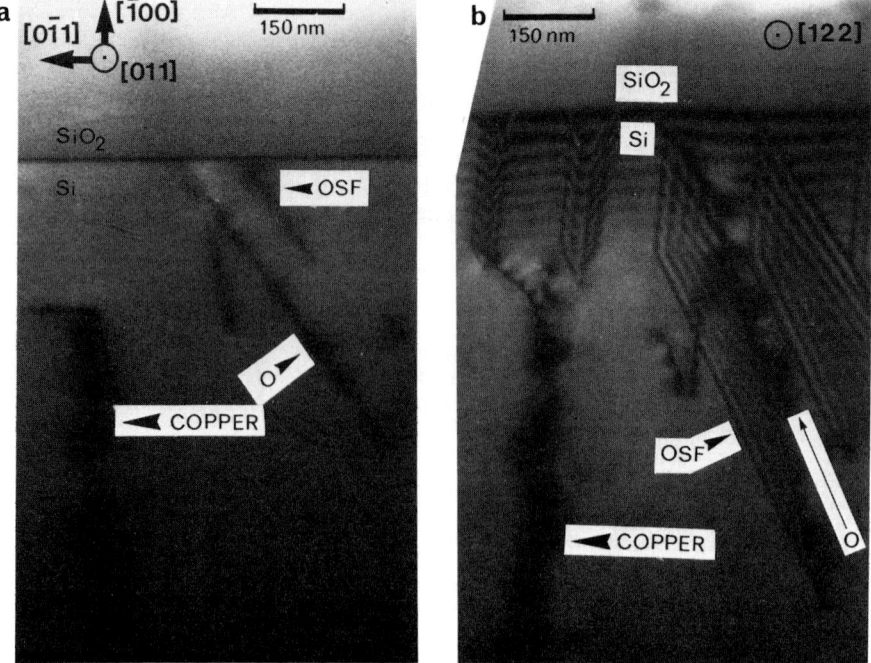

Fig. 2. Cross-sectional TEM view of the Cz sample after oxidation, (a) in [011] zone axis, (b) after tilt about [0$\bar{1}$1].

The minority carrier diffusion length was 4 μm in "defect-free" regions and 0.5 μm on the extended defects, compared to 100 μm in untreated references. The surface recombination velocity was everywhere below the measurement sensitivity (10^4 cms^{-1}). TEM plan-views exhibited a density of 10^8 cm^{-2} oxidation induced stacking faults (SF), which had no correspondence in EBIC, and large colonies of Cu_3Si precipitates, lying mainly in {110} planes perpendicular to the (100) surface, i.e. with traces along <110> directions on the surface, which were at the origin of the dark contrasts in EBIC. Cross-sectional views (fig. 2) showed the existence of a contact between stacking fault and colony, indicating a role of the fault partial dislocation in the nucleation of colonies. EDX analysis (fig. 3) showed the presence of oxygen in the clouds visible along this dislocation (fig. 2).

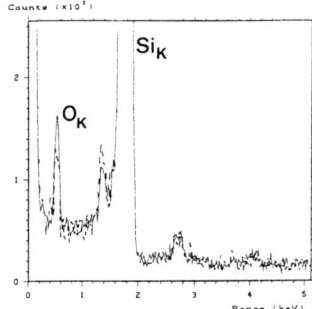

Fig. 3. EDX microanalysis of the decorated partial (full line) and of the nearby matrix (dotted line). Note the increase of oxygen K-line on partial.

The floating zone specimens exhibited a completely different aspect on EBIC images, as the large dark lines were absent and, in contrast, small dotted defects following traces of scratches were visible.

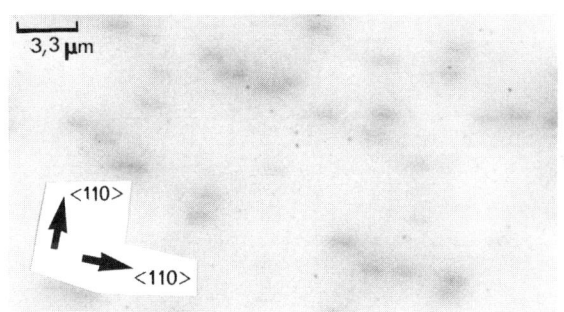

Fig. 4. EBIC image of the floating zone sample after oxidation

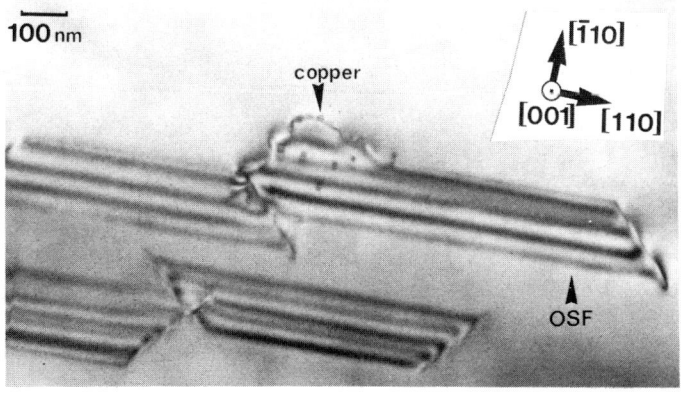

Fig. 5. TEM plan-view of the oxided floating-zone sample in [001] zone axis.

The diffusion length was 16 μm compared to 200 μm before treatment. However this value was unchanged in regions of higher concentration of the small defects, where the surface recombination velocity was in contrast increased to 3×10^5 cms^{-1} (compared to 1.4×10^5 in regions of lower density), indicating that these defects were close to the surface. These dots were elongated in a <110> direction (fig. 4), their length was about 1 μm. TEM plan-views exhibited a density of 6×10^7 OSF, about 15 % of which were decorated with Cu-based precipitates. In this case, the Frank partial was seen to have climbed in the {111} plane, enlarging the fault (fig. 5). These decorated stacking faults can unambiguously be identified as the electrically active <110> dot-like defects (same dimension and concentration). Their extension in the bulk, smaller than 0.3 μm, explains their influence on the surface recombination velocity rather than on the diffusion length.

3. CONCLUSION

As the difference between the two types of OSF (viz. in the Cz and FZ samples) was the presence of oxygen-containing clouds in the Cz, these observation are an indication that oxide precipitates are necessary to nucleate large Cu$_3$Si colonies in {110} planes (fig. 6a). In absence of oxygen, copper precipitation occurs with climb of the Frank partial dislocation: the colonies created are then in {111} planes (fig. 6b) and of smaller size. In other respects, the fact that copper-decorated OSF were visible in EBIC, while oxygen-decorated ones were not, gives qualitative evidence of the great difference in recombination power of these two types of precipitates.

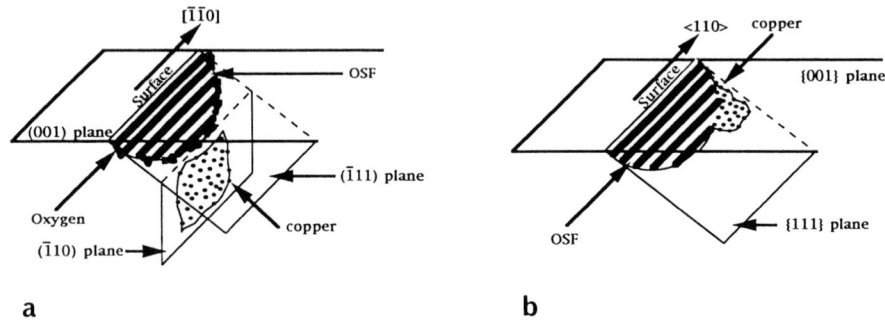

a   b

Fig. 6. Schematic of copper precipitation at OSF in (a) Cz and (b) FZ silicon

REFERENCES

Correia A, Ballutaud D, Maurice J-L and Cornier J-P 1993 Mater. Sci. Eng. B18 269
Dash W C 1956 J. Appl. Phys. 27 1193
Delidais I, Ballutaud D, Boutry-Forveille A, Maurice J-L, Zozime A and Aucouturier M 1991 Springer Proc. Phys. 54 217
Fiermans L and Vennik J 1967 Phys. Stat. Sol. 22 463
Higgs V, Goulding M, Brinklow A and Kightley P 1992 Appl. Phys. Lett. 60 1369
Maurice J-L 1993 J. Phys III France 3 603
Nes E 1974 Acta Met. 22 81
Nes E and Washburn J 1971 J. Appl. Phys. 42 3562
Schröter W, Seibt M and Gilles D 1991 Materials Science and Technology vol. 4, eds R W Cahn, P Haasen and J E Kramer, volume ed W Schröter (Weinheim: VCH) pp 539-89
Seibt M and Graff K 1988 J. Appl. Phys. 63 4444
Thomas D J D 1963 Phys. Stat. Sol. 3 2261

Inst. Phys. Conf. Ser. No 134: Section 3
Paper presented at Microsc. Semicond. Mater. Conf., Oxford, 5–8 April 1993

129

TEM and HREM studies of As-implanted high-dose oxygen implanted silicon at doses and energies suitable for thin-film silicon-on-insulator substrates

C D Marsh, Y Li*, A Nejim+, J A Kilner*, P L F Hemment+ and G R Booker.

Department of Materials, University of Oxford, Parks Rd, Oxford OX1 3PH.
*Department of Materials, Imperial College, London SW7 2BP.
+Department of Electronic and Electrical Engineering, University of Surrey, Guildford GU2 5XH.

ABSTRACT: The as-implanted microstructure of high-dose oxygen implanted silicon at doses and energies suitable for the fabrication of thin-film SIMOX structures has been investigated by TEM, HREM and SIMS. Several different morphologies of oxide precipitates have been identified and the defects present at the wafer surface are shown to be non-uniform across the implanted area, include intrinsic stacking faults, and the fraction of the implanted area that contains the defects is shown to depend on both the dose and beam current density.

1. INTRODUCTION

The separation by the implantation of high-doses ($>3 \times 10^{17}O^+/cm^2$) of oxygen (SIMOX) into silicon at elevated temperatures, followed by a high temperature anneal, is the leading technique for the fabrication of thin-film silicon-on-insulator wafers. SIMOX wafers consisting of a thin silicon film ($\approx 100nm$) above a buried oxide layer are potentially important for the fabrication of high-speed fully depleted CMOS devices and thin ($\approx 100nm$) buried oxide layers have the advantages, over the usual thickness of $\approx 400nm$, of reducing the implantation time, cost and circuit self heating, for both thin-silicon-film and standard (silicon $\leq 400nm$) SIMOX substrates.

The defect density after annealing depends on the as-implanted microstructure. Our previous work has shown that thin silicon layers above thin buried oxide layers can be fabricated by implanting at low energies (50-90keV) followed by an anneal (Li et al 1991, Marsh et al 1992 and Nejim et al 1993). In this paper we report the as-implanted microstructure, in particular the morphology of the oxide precipitates, over a wide dose range ($0.1-1.3 \times 10^{18}O^+/cm^2$) implanted at these energies.

Our previous work has also shown that thin buried oxide layers in standard SIMOX substrates (energy 200keV) can be fabricated by implanting doses of $\approx 0.7 \times 10^{18}O^+/cm^2$ (Marsh et al 1993). This work showed that for this dose the predominant crystallographic damage in the surface silicon layer are defects at the wafer surface which are non-uniform across the wafer surface. In the present paper we characterize the defects and report the distribution across the wafer surface as a function of dose and beam current density.

2. EXPERIMENTAL

Areas of 8 cm² in the centre of Fz silicon [100] wafers were implanted with single doses of O^+ ions in the range $0.1 \times 10^{18}O^+/cm^2$ to $1.3 \times 10^{18}O^+/cm^2$. The ions were implanted at energies of 50keV/O^+, 70keV/O^+, 90keV/O^+ and 200keV/O^+ using time averaged beam current densities in the range 2.5μA/cm² to 6.5μA/cm². Implantation was performed at 680°C with halogen lamp background and pre-heating (680°C for 10 minutes). The composition and microstructure were investigated by cross-section, planar and high resolution TEM and secondary ion mass spectroscopy (SIMS).

3. RESULTS AND DISCUSSION

3.1 Morphology of the Oxide Precipitates after Low Energy (50keV, 70keV and 90keV) Implants

For these energies, over the dose range investigated, five morphologies of oxide precipitates have been observed. These are spherical precipitates in cubic and random distributions, irregular precipitates, [100] columnar precipitates and precipitates elongated parallel to the surface (fig 1). The local oxygen concentrations have been calculated, assuming the oxygen concentration in the buried silicon dioxide layer to be 4.48×10^{22}/cm^3, from SIMS profiles using the results of Griffin (1989).

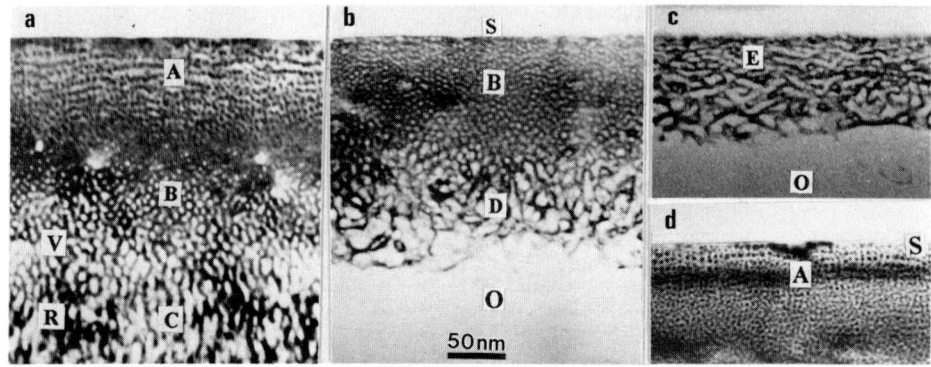

Fig 1 : Samples implanted at 90keV, doses 0.43×10^{18}O$^+$/cm^2 (a) & (d) and 1×10^{18}O$^+$/cm^2 (b), and at 50keV, dose 0.7×10^{18}O$^+$/cm^2 (c), showing the five types of precipitate morphology observed after low energy (50kev-90keV) implants. A = spherical precipitates in a cubic array, B = random spherical precipitates, C = [100] columnar precipitates, D = irregular precipitates, E = irregular precipitates elongated parallel to the surface, O = buried oxide layer, V = peak of vacancy production profile (90keV), R = peak of oxygen profile (90keV) and s = surface. Fig 1d <100> cross-section showing the precipitates near the surface in (a) are in a simple cubic array.

For doses $\leq$ 0.5, 0.7 and 1×10^{18}O$^+$/cm^2 at 50keV, 70keV and 90keV respectively, the morphology of the precipitates in the surface silicon layer depends primarily on the local oxygen concentration. For oxygen concentrations $<2.2 \times 10^{22}$/cm^3, the precipitates are spherical, suggesting that the shape is determined by the minimisation of surface energy and hence minimal strain is present. The precipitates are randomly distributed except close to the surface, for oxygen concentrations of 2-4×10^{21}/cm^3, where they are present in a cubic array (fig 1d). For concentrations $\geq 2.2 \times 10^{22}$/cm^3, which occurred either close to the buried oxide or at the peak of the oxygen distribution, larger irregular precipitates are observed (fig 1b), probably formed by the coalescence of adjacent precipitates.

For high doses, ie 0.7, 1 & 1.3 $\times 10^{18}$O$^+$/cm^2 at 50keV, 70keV and 90keV respectively, the precipitates in the surface silicon layer were irregular and elongated parallel to the wafer surface (fig 1c). The oxygen concentrations were in the range $1.3\text{-}4.4 \times 10^{22}$/cm^3. For lower dose implants spherical precipitates were present at these concentrations. The elongation of the precipitates parallel to the wafer surface suggests, because growth in the (100) plane generates the minimum lattice strain, that strain is determining the precipitate shape. For high doses a high interstitial concentration would be expected as the large volume of oxide precipitates present in the surface silicon would decrease the diffusion (diffusivity in SiO$_2$ << than in Si) of the interstitials generated by the internal oxidation, to the surface. As an accommodation volume is required for strain free precipitation, the presence of strain is probably due to the interstitial concentration being sufficiently large to annihilate the vacancies, generated during implantation, before they could provide the accommodation volume.

At the peak of the oxygen profiles (from SIMS), for concentrations $<1.3 \times 10^{22}$/cm^3, spherical precipitates were observed, while for oxygen concentrations $>1.3 \times 10^{22}$/cm^3, [100] columnar precipitates were observed (fig 1a). The origin of the spherical shape is discussed above. The minimum depth at which the columnar precipitates were observed coincides with the peak of the vacancy production profile from TRIM90 (fig 1a). This, the [100] shape and the presence at the peak of the oxygen profile suggests that a combination of the beam direction and high vacancy and oxygen concentrations are enabling precipitate growth in the [100] direction to take place.

3.2 Defects and Oxide Precipitates Near the Wafer Surface after 200keV Implants

Fig 2(a) shows a cross-section after implanting a dose of 0.6×10^{18}O$^+$/cm^2, at 200keV, using a beam current density of 6.25μA/cm^2. A large amount of damage exists at the peak of the oxygen distribution but the predominant damage in the surface silicon layer are defects at the wafer surface. These defects consists of dislocations and small {111} stacking faults (fig 2b). The nature of the stacking faults observed was difficult to determine but HREM images suggest they are intrinsic (fig 2c). Similar {111} defects have been observed by De Veirman (1991) but they were not characterised.

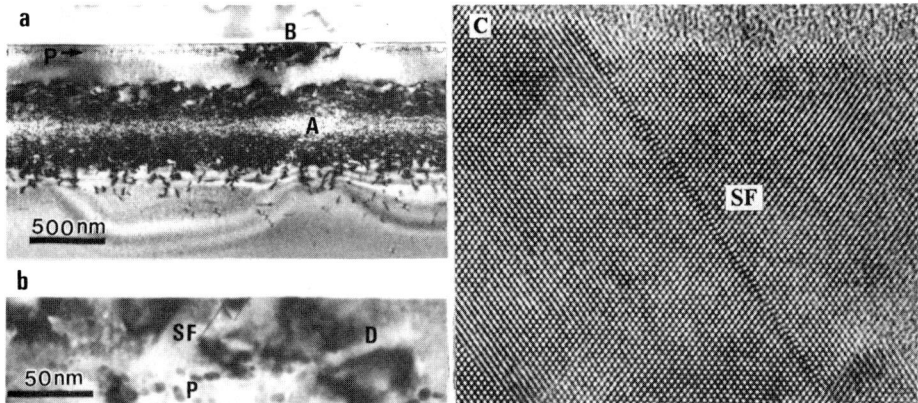

Fig 2 : Sample implanted at 200keV with a dose of $0.6 \times 10^{18} O^+/cm^2$ showing in a) damage at the peak of the oxygen distribution (A), the presence of regions of defects at the wafer surface (B), and a band (P) of spherical and oval oxide precipitates (see fig 4), b) the defects consist of stacking faults (SF) and dislocations (D). The precipitates arrowed in a) are indicated (P) and c) HREM image showing the stacking faults (SF) are intrinsic.

The distribution of these defects is not uniform across the implanted area. In plan-view the regions of silicon containing these defects are rectangular in shape with the edges approximately following the <100> crystallographic directions (fig 3). These "rectangular" regions of defects are randomly distributed across the implanted area. The edges of the defected regions are not abrupt, in plan-view, as the dislocations themselves do not follow the <100> crystallographic directions (fig 3).

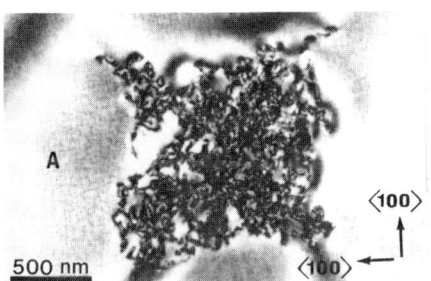

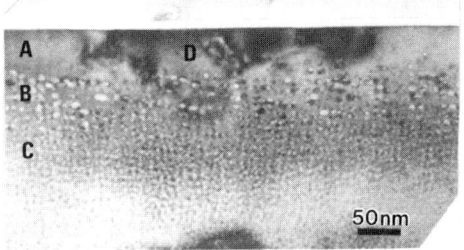

Fig 3 : [100] plan-view of a region of defects at the wafer surface (200keV, dose $0.6 \times 10^{18} O^+/cm^2$). Note the edges approximately following the <100> directions. The "tweed-like" background contrast (A) shows that the oxide precipitates (see fig 4) are also aligned along <100> directions.

Fig 4 : <100> cross section showing the precipitate denuded zone (A), a region of spherical and oval precipitates, (B) and smaller spherical precipitates (C). The precipitates are aligned along the <100> directions and the majority of the dislocations (D) are present only in the precipitate denuded zone.

The size of the "rectangular" regions of defected silicon (fig 3) and hence the percentage of the implanted area that contains these surface defects, decreased with both decreasing dose (table 1) and increasing time-averaged beam current density (table 2). When a large fraction of the surface area contained defects adjacent "rectangular" defect regions merged together.

	Dose ($\times 10^{18} O^+/cm^2$)				
	0.5	0.55	0.6	0.7*	1.2
% of area	3%	2%	25%	33%	100%

Table 1 : Percentage of implanted area containing surface defects as a function of dose (beam current density $6.25 \mu A/cm^2$, except * = $5 \mu A/cm^2$)

	Current density ($\mu A/cm^2$)		
	2.5	3.75	5
% of area	80 %	70%	33%

Table 2 : Percentage of implanted area containing surface defects as a function of beam current density (dose $0.7 \times 10^{18} O^+/cm^2$)

Fig 4 shows the oxide precipitates present in the surface silicon layer. The precipitate distribution consists of a denuded zone at the wafer surface, to a depth of 35nm, spherical and oval precipitates, diameters up to 7nm, between depths of 35nm and 95nm and a high density of small spherical precipitates, typical diameter 4nm, at depths deeper than 95nm. Plan-views show the precipitates are aligned along the <100> directions. During implantation the low oxygen concentration at the surface, resulting from the implantation profile and the out diffusion of oxygen, increases the critical radius for stability, inducing any precipitates to dissolve. The oxygen then either out diffuses or is gettered by the nearest stable precipitates. This gettering of the oxygen from the surface region is thought to explain why the precipitates at the edge of the denuded zone are larger than those further from the surface.

The majority of the dislocations were present in the precipitate denuded zone (fig 4). Where dislocations were observed deeper, dislocations were also present at a higher density in the denuded zone, and the larger precipitates observed at depths 35-95nm were present at lower densities. These suggest that the precipitates are modifying the dislocation distribution and the dislocations are influencing the precipitate distribution.

The random distribution across the implanted area of the regions of silicon containing defects suggests that the defects nucleate randomly. Therefore the number of defects nucleated would be expected to increase with implant time and hence the percentage of the area containing defects to increase with dose. The dislocations probably originate as dislocation loops, nucleated at the surface and formed from the excess of interstitials, generated by the internal oxidation during implantation. The presence of the oxide precipitates will restrict the movement of the dislocations and hence once nucleated the dislocations grow until they meet the precipitates which slow further growth. Therefore the dislocations tend to be present in the precipitate denuded zone at the surface and because the precipitates are aligned along the <100> directions this gives the defect regions <100> edges. The decrease in the fraction of the implanted area containing defects with increasing beam current may be the result of an increase in local annealing due to greater local beam heating with higher beam currents.

The interstitials, generated by the internal oxidation, that diffuse to the surface can adhere onto the surface and hence the surface can epitaxially grow during implantation. For a dose of $0.7 \times 10^{18} O^+/cm^2$ the surface can grow up to 66nm. The stacking faults were observed only in the top 35nm, ie. only at depths shallower than the position of the original surface. As the surface grows, if a defect (eg. contamination) generates a regrowth stacking fault this would grow with the surface. Hence the observations of stacking faults only at depths less than the original surface and their intrinsic nature, where, because of the excess of interstitials, extrinsic faults might be expected, suggests that these stacking faults are regrowth defects and not due to the coalescence of point defects.

4. CONCLUSIONS

The oxide precipitate morphology depends on the oxygen concentration, the vacancy production profile and the silicon interstitial concentration. For doses $\leq 1.2 \times 10^{18} O^+/cm^2$ implanted at 200keV, the predominant defects in the surface silicon layer are dislocations and stacking faults at the wafer surface. For doses $\leq 0.7 \times 10^{18} O^+/cm^2$ the defects are present in localised areas, ie not uniform across the implanted area. The fraction of the implanted area which contains these defect increases with increasing dose and decreasing beam current density. The stacking faults characterised were intrinsic. This and their presence only at the surface where the silicon has epitaxialy grown during implantation suggests that they are due to defective epitaxial growth rather than the coalescence of point defects.

5. ACKNOWLEDGEMENTS

This work as part of project IED1777 was funded by the Science & Engineering Research Council.

REFERENCES

De Veirman A, Van Landuyt J, Vanhellemont J, Maes H E and Yallup K 191 Vacuum 42 367
Griffin C, 1989 Ph D Thesis University of London
Li Y, Kilner J A, Robinson A K, Hemment P L F and Marsh C D 1991 J. Appl.Phys. 70 3605
Marsh C D, Booker G R, Nejim A, Giles L, Hemment P L F, Li Y, Chater R J, Kilner J A,
    Wainwright S and Hall S 1992 Proc. 1992 IEEE Int. SOI Conf. Ed Jerry Brandewie pp 8-9
Marsh C D, Nejim A, Li Y, Booker G R, Hemment P L F, Chater R J and Kilner J A 1993
    Nucl. Inst. & Meths B in press.
Nejim A, Li Y, Marsh C D, Hemment P L F, Chater R J, Kilner J A and Booker G R 1993 ibid.

Inst. Phys. Conf. Ser. No 134: Section 3
Paper presented at Microsc. Semicond. Mater. Conf., Oxford, 5–8 April 1993

Depth of amorphized ion implanted silicon zones predicted from point defect density calculations and TEM cross sections

H Cerva and G Hobler*

Siemens AG, Research Laboratories, Otto Hahn Ring 6, D-W8000 München 83, Germany,
*Institut für Allgemeine Elektrotechnik und Elektronik, University of Technology Vienna, Gusshausstraße 27, A-1040 Wien, Austria

ABSTRACT: For several technologically important high dose implantations into (100) single-, and polycrystalline silicon the width of the amorphous zone was determined by TEM cross sections and compared to point defect density calculations. In case of P^+, and As^+ implantations very good agreement is found for a critical point defect density of $1.15 \cdot 10^{22}$ cm^{-3} in one-, and two-dimensional cases. Point defect density calculations carried out for As^+ may also predict very reliably the widths of amorphous zones produced by Ge^+ implantations.

1. INTRODUCTION

Today ion implantation is a standard technique during the processing of modern microelectronic silicon devices. When using high dose implantations into the Si-substrate the surface is rendered amorphous which may have both advantageous and disadvantageous effects on a specific process or on the device performance. In most cases the width of the amorphous Si-zone which extends from the Si-surface or the interface of the scattering oxide with the Si-crystal to the amorphous/crystalline (a/c) interface in the Si-substrate, is of decisive importance. Moreover, the position of the a/c interface yields important information with respect to the dopant distribution in the as-implanted state (Cerva (1992a)). Hence, process optimization requires a large number of TEM investigations to obtain information on the amorphous zones. The depth X of the a/c interface measured from the sample surface is usually estimated from the projected range R_p and the range straggling ΔR_p of the dopant depth profile by the simple formula $X = R_p + n \cdot \Delta R_p$ (Prussin et al (1985)). The parameter n has to be adjusted for every specific ion and dose the latter because R_p and ΔR_p depend only on the primary ion energy. As an alternative, calculated spatial distributions of displaced target atoms ("point defects", Hobler (1988a)) or of the energy transferred to target atoms ("damage energy", Claverie et al (1988)) may be used. The results of the TEM cross sections may then be used to define a critical point defect density (PDD) or a critical damage energy density which marks the a/c transition.

2. POINT DEFECT CALCULATIONS

Investigations are restricted to high-dose implants where the amorphous zone extends to the Si-surface, and to implantation energies < 200 keV. The thickness of a possible thermal scattering oxide layer on top of the Si-crystal may safely be added to the thickness of the amorphous Si-zone or even be treated as amorphous Si since the stopping powers of ions in silicon and silicon dioxide are very similar (Hobler (1988a)). Damage profiles for various ions implanted into Si were calculated by Hobler (1988a) with Monte Carlo simulations which assume a displacement energy $E_d = 15$ eV. For 2-D cases the point defect concentration distribution at arbitrarily formed mask

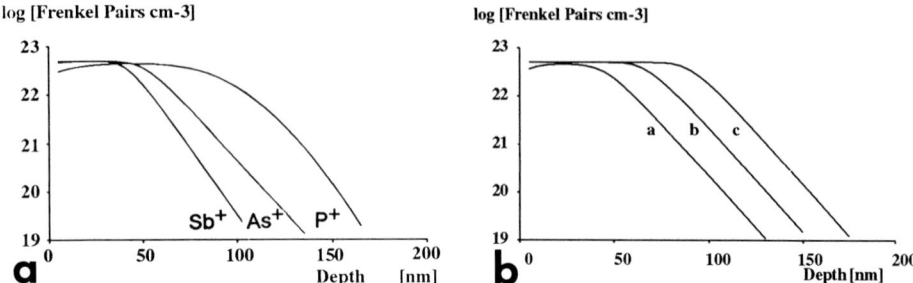

Fig. 1 Calculated point defect density (PDD) distributions for implantations into Si (100) (a) of P+, As+, and Sb+ ions with 50 keV, $1 \cdot 10^{15}$ cm-2, and (b) for As+ with 50 keV and a: $5 \cdot 10^{14}$ cm-2, b: $5 \cdot 10^{15}$ cm-2, $5 \cdot 10^{16}$ cm-2.

edges is calculated by the superposition of point responses according to Runge (1977). Hobler and Selberherr (1988b) have derived analytical functions for the 1-D point defect distribution and the 2-D point response of B+, P+, As+, and Sb+ implantations into Si. These analytical functions can rapidly be calculated and are used in this work. The damage profiles of P+, As+, and Sb+ in Fig. 1a simulated for 50 keV and $1 \cdot 10^{15}$ cm-2 demonstrate clearly that the damage zone becomes wider with decreasing atomic number of the implanted ion. Keeping the implantation energy constant and increasing only the dose leads to broader amorphous zones as becomes obvious from the PDD distributions in Fig. 1b. Thus, the thickness of a damage zone can be accurately tuned by varying atomic number, dose, and energy.

3. COMPARISON WITH TEM CROSS SECTIONS

Implantations were carried out with a tilt of 7° off the [100] Si-substrate surface normal around a <110> axis at room temperature.

3.1 One-Dimensional Damage Distributions

The width of the amorphous zones produced by four As+ implantations with three different

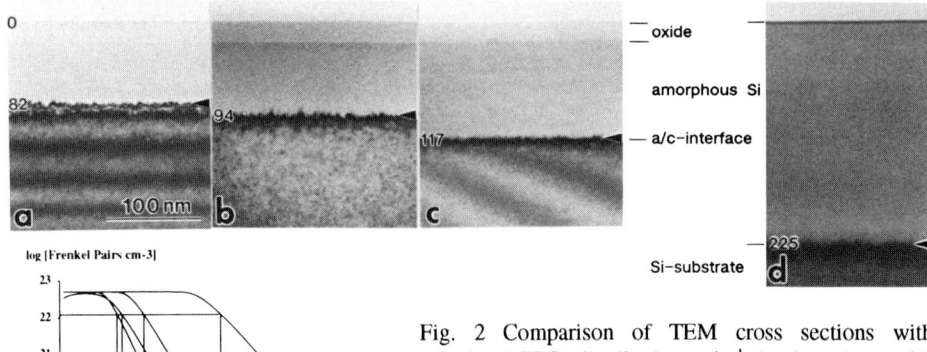

Fig. 2 Comparison of TEM cross sections with calculated PDD distributions. As+ implantations with (a) 50 keV, $5 \cdot 10^{15}$ cm-2, (b) 80 keV, $5 \cdot 10^{14}$ cm-2, (c) 80 keV, $5 \cdot 10^{15}$ cm-2, (d) 160 keV, $1 \cdot 10^{16}$ cm-2. (e) Calculated PDD profiles.

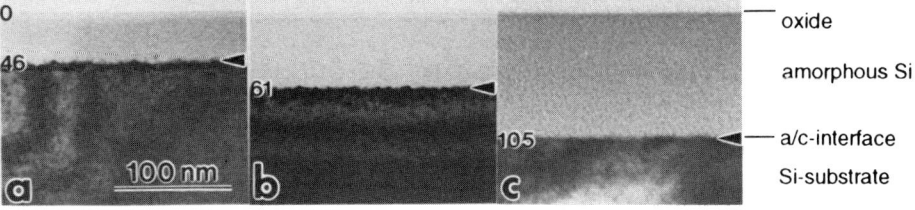

log [Frenkel Pairs cm-3]

Fig. 3 Comparison of TEM cross sections of Ge⁺ implanted sample: (a) 40 keV, $2\cdot10^{14}$cm⁻², (b) 40 keV, $2\cdot10^{15}$ cm⁻², (c) 80 keV, $1\cdot10^{15}$ cm⁻², with (d) calculated PDD distributions for As⁺.

energies and doses which are typical for device processing are indicated in the TEM cross sections of Figs. 2a-d. From the depths X in Figs. 2a,c a critical PDD of $1.15\cdot10^{22}$ cm⁻³ can be deduced. Using this value for the other two implant conditions leads to an error of about 5% in X. This error is much smaller than that using the formula $X = R_p + n\cdot\Delta R_p$ proposed by Prussin et al (1985) even when the data of R_p and ΔR_p are taken from recent calculations (Hobler (1988a)).

Ge⁺ implantations are used to preamorphize the Si-substrate to circumvent ion channeling during the subsequent B⁺ implantation. This allows the fabrication of very shallow p⁺ regions. In Fig. 3 TEM cross sections of three different implantation conditions are compared with PDD profiles calculated for As⁺. In two cases (Figs. 3a,c) the calculated depths X are slightly larger which may be due to the smaller atomic number of Ge. The accuracy is nevertheless <10%.

Polycrystalline silicon (poly-Si) for interconnects such as that shown in Fig. 4a on top of a thick oxide film may be doped by ion implantation. The intention was to amorphize the crystallinely deposited poly-Si film completely during doping with a high dose P⁺ implantation. As can be seen in Fig. 4a this did not happen and the residual grains at the bottom of the layer may act as growth nuclei during successive annealing and, therefore, lead to small grains and high sheet resistances. The calculated PDD profile a in Fig. 4b reveals that the position of the a/c interface is very well predicted (<5%) with a critical PDD of $1.15\cdot10^{22}$ cm⁻³ also for the P⁺ implantation. The profile b in Fig. 4b was calculated for a 50 keV, $5\cdot10^{15}$ cm⁻² P⁺ implantation resulting in an approximately

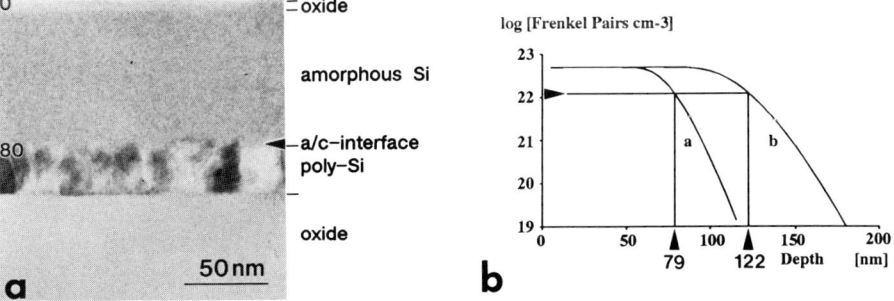

Fig. 4 P⁺ implantation (30 keV, $5\cdot10^{15}$ cm⁻²) into a 100 nm thick poly-Si layer through a 10 nm thick scattering oxide film. Comparison of TEM cross section (a) with the PDD calculation in (b).

122 nm thick amorphous zone which would have completely amorphized the 110 nm thick layer (scattering oxide plus poly-Si layer).

3.2 Two-Dimensional Damage Distributions

The effects of mask edge shape and lateral scattering under the mask edge become very important for source/drain implantations of metal oxide semiconductor transistors because they influence in particular the effective channel length of the gate and hence the transistor performance. The <110> aligned TEM cross section in Fig. 5a shows the amorphous zone at an oxide mask edge after source drain implantation (As^+, 50 keV, $5 \cdot 10^{15}$ cm^{-2}) carried out with a 7° tilt. In Fig. 5b the shaded area marks the extension of the amorphous zone including the scattering oxide as determined from the TEM image whereas the lower full line represents a calculated 2D-PDD contour for a critical PDD of $1.15 \cdot 10^{22}$ cm^{-3} yielding perfect agreement with the experimental result.

4. CONCLUSION

TEM cross sections of several high dose P^+, As^+, and Ge^+ implanted samples were compared with PDD calculations. A critical PDD of $1.15 \cdot 10^{22}$ cm^{-3} predicts the position of the a/c interface with an accuracy <5% for P^+, and As^+ implantations in 1-D and 2-D cases. For Ge^+ implantations the accuracy is still <10%. Though ion channeling was not included in the calculations the same results were found for P^+ and As^+ implantations into single-, and polycrystalline silicon. Moreover, excellent results were already reported for Si^+ implantations by Cerva and Hobler (1992b). Simulations for B^+ yielded no satisfactory results. However, for the process-relevant implantations described above, rapid and reliable predictions may now be obtained by PDD calculations.

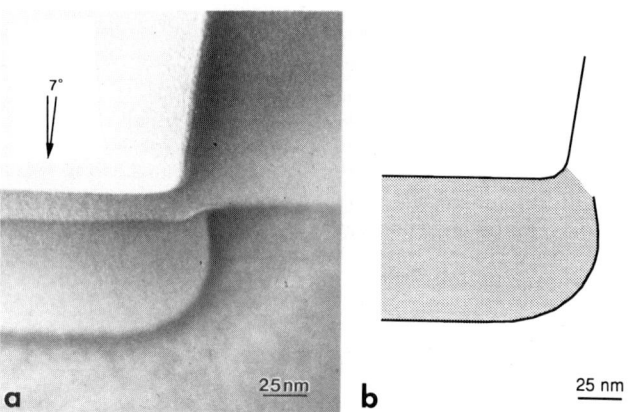

Fig. 5 As^+ implantation (50 keV, $5 \cdot 10^{15}$ cm^{-2}) at a mask edge. (a) <110> aligned TEM cross section. (b) 2D-PDD simulation of the a/c interface (full line), shaded area represents amorphous zone obtained from the TEM image.

REFERENCES

Cerva H 1992a J.Vac.Sci.Technol. B10 491
Cerva H and Hobler G 1992b J.Electrochem.Soc. 139 3631
Claverie A, Vieu C, Fauré J and Beauvillain S 1988 J.Appl.Phys. 64 4415
Hobler G 1988a Ph.D. Thesis, Technical University of Vienna, Austria
Hobler G and Selberherr S 1988b IEEE Trans. Computer Assisted Design CAD-7 174
Prussin S, Margolese D I and Tauber R N 1985 J.Appl.Phys. 57 180
Runge H 1977 Phys.Stat.Sol.A39 595

Inst. Phys. Conf. Ser. No 134: Section 3
Paper presented at Microsc. Semicond. Mater. Conf., Oxford, 5–8 April 1993

137

The crystalline-to-amorphous transition in Ge-implanted silicon and its role in the formation of end-of-range defects

M Seibt, J Imschweiler*and H-A Hefner*

IV.Physikalisches Institut der Universität Göttingen and Sonderforschungsbereich 345, Bunsenstr.13-15, W-3400 Göttingen, Federal Republic of Germany
*TELEFUNKEN electronic GmbH, Theresienstr.2, W-7100 Heilbronn, Federal Republic of Germany

ABSTRACT: We have used high-resolution transmission electron microscopy to study the morphology of the amorphous/crystalline interfacial region in Ge implanted silicon. By estimating the statistical properties of the amorphous/crystalline interfaces we observe a crystalline-to- amorphous transition accompanied by a decrease of interfacial roughness as a result of low temperature annealing. The interfacial properties are related to the type of end-of-range defects formed during solid-phase epitaxial regrowth.

1. INTRODUCTION

Doping by ion implantation is one of the most important processes in the fabrication of silicon integrated circuits. One major drawback of this technique is the necessity of post implantation heat treatments to remove damage and electrically activate dopant implants. Most of the defects can be prevented by choosing suitable implantation and/or annealing conditions. Extended end-of-range (EOR) defects, however, inevitably occur in the region below the amorphous/crystalline (a/c) interface after solid-phase epitaxial regrowth (SPEG) of the amorphous layer. These defects are of extrinsic type, i.e. they are agglomerates of silicon self- interstitials (Si_I) in form of {111} faulted or perfect loops (Jones et al 1988) or {113} - stacking faults (Lambert and Dobson 1981, Cerofolini et al 1986). In a recent paper, we have demonstrated that the type and location depth of EOR defects strongly depends on structural properties of the a/c- interface prior to SPEG. Especially, interfacial roughness and/or strain at the interface determine the nucleation conditions for EOR defects (Seibt et al 1992).

In this paper, we focus on the changes of a/c- interfacial roughness brought about by annealing at 450°C . We use HRTEM to image the interfaces and estimate statistical properties of the a/c interfacial region. We observe a decrease of interfacial roughness by a factor of about 3 and the formation of extended end-of-range defects which are attributed to {113} - stacking faults in their early stage of growth. Furthermore, our results imply an increase of the amorphous layer width due to the annealing.

2. EXPERIMENTAL

Starting material was B-doped FZ silicon ((001) oriented, 5-10Ωcm). ^{74}Ge or

[70]Ge was implanted at room temperature at energies between 60 and 150keV and doses of 2 - 9×10^{14}cm^{-2}resulting in amorphous layers of 50 - 150nm width extending to the surface for energies exceeding 70keV. A subsequent BF$_2^+$- implantation (40keV, 8×10^{13}cm^{-2}) was applied to some of the samples without noticeable effects on the results presented here. In order to study the effect of low- temperature annealing samples were cut from adjacent parts of the center of the wafers and either annealed at 450°C for 40 min in Ar or N$_2$ or left in the as-implanted state. Cross-sections in [110]- orientation were prepared by standard techniques involving ion beam thinning with the samples kept at liquid N$_2$ temperature. Seven-beam lattice images were taken at 120kV in a Philips 420ST using axial illumination.

To estimate the statistical properties of the a/c- interfacial roughness we use a procedure previously applied to Si/SiO$_2$- interfaces (Goodnick et al 1985). For this analysis the interface boundary is digitized by choosing the last visible lattice fringe of the Si lattice as the location of the interface. This one-dimensional approach is clearly not appropriate in cases where detached microcrystallites in the amorphous layer and amorphous pockets in the crystalline material are present. Hence, we restrict this analysis to samples implanted with 3×10^{14}cm^{-2}at 70 keV, where these features were not observed.

3. RESULTS

3.1 Nucleation of Extended End-of-Range Defects

Fig.1 shows lattice images of an as- implanted sample (Fig.1a) and a sample annealed at 450°C (Fig. 1b). The micrographs are aligned at the wafer surfaces (S). Besides considerable changes in the a/c interfacial region, which will be analyzed in Section 3.2, small dot-like regions of contrasts are visible in the annealed sample (arrows in Fig.1b). They are located in a zone starting about 10nm below the a/c interface, which appears to be defect- free in the as- implanted state (Fig.1a). We have shown previously, that {113} stacking faults are observed after solid- phase epitaxial regrowth at 550°C in the region of these contrast dots, which are ascribed to {113} stacking faults *in statu nascendi* (Seibt et al (1992)).

3.2 Analysis of Interfacial Roughness

In this section, we will estimate changes of the a/c- interfacial roughness brought about by annealing at 450°C . First, the position $\langle z \rangle$ of the interface as measured from the wafer surface is determined by averaging over a region of 100nm parallel to the interface. In this way, we obtain $\langle z \rangle = 76.5$nm and 80nm for as- implanted and annealed samples, respectively. This implies an *increase* of the amorphous layer width, which is a surprising result.

Changes in the roughness spectrum can be revealed by calculating the autocorrelation of the interface boundary after subtracting the dc values $\langle z \rangle$. Fig.2a shows the autocorrelation for as- implanted (dotted line) and annealed (solid line) samples demonstrating the overall decrease in interfacial roughness.

Due to the uncertainties involved in the determination of the a/c- interface position especially on the atomic scale, we have restricted the spectral analysis to wavelengths larger than 3nm. The power spectrum (normalized to the maximum amplitude occuring in the as- implanted state) obtained from the Fourier transform of the autocorrelation (Fig.2b) shows that except for the component corresponding to a wavelength of 6nm (arrow) all components are drastically reduced. If we estimate the interfacial width Δz from the total power P of the spectrum via $\Delta z = 2 \times P^{1/2}$ we obtain 2.38nm and 0.88nm prior to and after annealing, respectively.

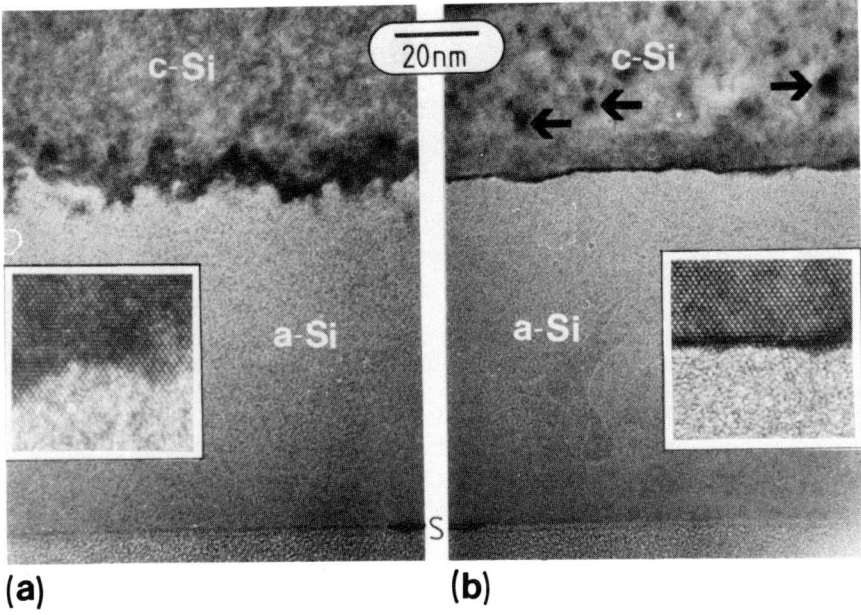

(a)   (b)

Fig.1:   Lattice images of Ge- implanted silicon; (a) as- implanted; (b) annealed
at 450°C ; the arrows indicate {113} - stacking faults *in statu nascendi*;
note the alignment of the micrographs at the wafer surfaces (S). The insets
show enlarged parts of the a/c- interfaces.

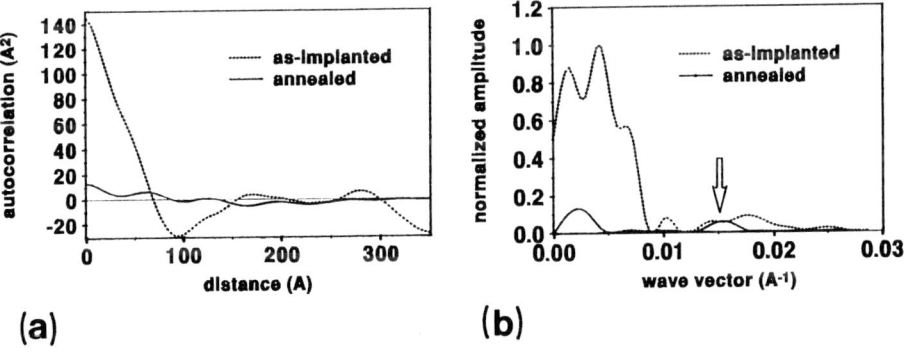

(a)   (b)

Fig.2:   Statistical properties of a/c- interface of as-implanted (dotted lines) and
annealed (solid line), respectively; (a) autocorrelation, (b) power spectrum
obtained from Fourier transformation of the autocorrelation.

4. DISCUSSION

We have analyzed the changes of a/c interfacial roughness due to annealing at 450°C of Ge- implanted silicon. We observe an *increase* of the amorphous layer width by about 3.4nm accompanied by a decrease of interfacial roughness by a factor of 2.7. A decrease of interfacial roughness as a result of heat treatments around 400°C has been reported previously (e.g. Maher et al 1986, Rozgonyi et al 1986). Concerning the increase of the amorphous layer width our results are in agreement with studies of Ge- implanted silicon (Rozgonyi et al 1986), but are at variance with the enhanced solid-phase regrowth *("incubation growth")* observed after Si^+- implantation at liquid N_2 temperature (Maher et al 1986).

The observation of the crystalline- to- amorphous transition relies on the assumption that the wafer surface can be used as a reference to measure the position of the a/c interface and that the as- implanted wafers are homogeneous in the region originally containing the annealed samples. The latter was checked experimentally by comparison of as- implanted samples cut from the extreme parts of this region. The former assumption might be affected by oxidation during the 450°C annealing although care was taken to avoid oxygen in the annealing system.

For a crystalline- to- amorphous transition to occur the crystalline part of the a/c interfacial region has to be unstable compared to amorphous silicon. For a rough estimate consider the size of the critical nucleus of crystalline silicon in an amorphous silicon matrix, which contains 110 atoms (Tu 1991). This value corresponds to a critical radius of about 1.3nm. Hence, crystalline parts of the a/c interface with a radius of curvature smaller than this value might be unstable. However, a verification and detailed modelling of our observation has to await future studies involving a more quantitative determination of the a/c interfacial roughness especially on the atomic scale, i.e. for wavelengths below 3nm.

In a previous paper we have shown that type and location depth of extended end-of-range defects are determined by the a/c- interfacial roughness prior to regrowth: direct solid-phase regrowth of as- implanted samples leads to {111} - faulted loop formation, whereas {113} stacking faults form in cases where the samples have been pre- annealed at 450°C (Seibt at al 1992). Additional studies show that the latter results in a considerable decrease of EOR defect density found after high- temperature RTA treatments (Seibt et al 1993). This implies that structural properties of the a/c- interface play a crucial role in the formation of extended end-of-range defects.

REFERENCES

Cerofolini G F, Meda L, Polignano M,Ottaviani M L, Bender G, Clays C, Armagliato A and Solmi S (1986) in: Semiconductor Silicon 1986, (The Electrochemical Society, Pennington) p 706

Goodnick S M, Ferry D K, Wilmsen C W, Lilienthal Z, Fathy D and Krivanek O L (1985), Phys Rev B 32 8171

Jones K S, Prussin S and Weber E R (1988) Appl Phys A 45 1

Maher D M, Seidel T E, Williams J S, Elliman R G, Knoell R V, Ellington M B, Hull R and Jacobson D C (1986) in: Semicondutctor Silicon 1986 (The Electrochemical Society, Pennington) p.678

Rozgonyi G A, Myers E, Sadana D K (1986) in: Semicondutctor Silicon 1986 (The Electrochemical Society, Pennington) p.696

Seibt M, Imschweiler J and Hefner H.-A. (1992) Mat.Res.Soc.Proc. Vol.262 1103

Seibt M, Imschweiler J and Hefner H.-A. (1993) to be published

Tu K N (1991) Appl Phys A53 32

Inst. Phys. Conf. Ser. No 134: Section 3
Paper presented at Microsc. Semicond. Mater. Conf., Oxford, 5–8 April 1993

TEM investigation of the influence of an overlayer on the structure of implanted p-n junctions in silicon

J Kątcki and A Bąkowski

Institute of Electron Technology, Al. Lotników 32/46, 02-668 Warsaw, Poland

ABSTRACT: The influence of an overlayer on the structure of implanted p-n junctions in silicon has been investigated. Silicon wafers were implanted with phosphorus, boron or arsenic. Each wafer was divided into three parts. The first part was covered with amorphous silicon (a-Si). An a-Si/Si_3N_4/SiO_2 structure was formed on the second one. The third part was left uncovered. Wafers were thermally annealed at the temperature of 920°C for 6 hours in a nitrogen ambient. TEM cross-sections were prepared from each part of the wafers. Comparison of micrographs of cross-sectional TEM specimens for P, B and As has proved that the layer deposited onto the implanted Si wafers changes the junction depth and influences the structure of silicon. Depth distributions of each dopant under different overlayers were measured using sheet resistance and spreading resistance methods. The possible mechanisms of defect formation in implanted p-n junctions are discussed.

1. INTRODUCTION

A very important technological problem in manufacturing CMOS integrated circuits is leakage current. In order to control the leakage current the correlation between lattice structure and electrical characteristics of p-n junction should be established. The aim of this work is to determine how an amorphous silicon (a-Si) layer deposited directly on the silicon wafer and/or an a-Si/Si_3N_4/SiO_2 multilayer deposited on the wafer can influence the structure of the junction during standard CMOS processing.

2. EXPERIMENTAL

The (100) oriented Czochralski grown silicon wafers were implanted with phosphorus, boron or arsenic with a dose of $1*10^{16}$ ions/cm^2. Implantation of boron and phosphorus was performed at an energy of 60 keV. Arsenic ions were implanted at an energy of 100 keV. Implanted wafers were oxidized at the temperature of 1000°C for 43 minutes in an O_2+2% HCl ambient, resulting in the formation of a 900Å thick SiO_2 film. After oxidation a 1000Å thick Si_3N_4 film was deposited on the wafers. Using a conventional photo-lithographic procedure Si_3N_4 and SiO_2 films were removed from the substrate leaving one of three parts of the wafers uncovered (part **C**). After cleaning a 2500Å thick amorphous silicon (a-Si) film was deposited on the wafers. Then by means of plasma etching the a-Si film was removed from part **A** of the wafers. This step was followed by wet etching of Si_3N_4 and SiO_2, resulting in the bare silicon surface in part **A** of the wafers. Last processing step was annealing at the temperature of 920°C for 6 hours in a nitrogen ambient. A schematic drawing of the structure prepared for the study is shown in Fig. 1.

TEM cross-sections were prepared from each part of the wafers. Specimens were observed in the

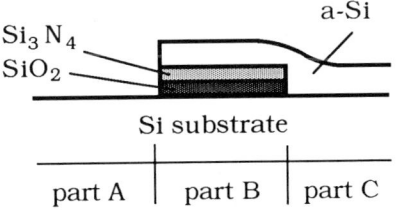

Fig. 1. Structure prepared for the investigation. Thicknesses of layers are: Si_3N_4 - 100 nm, SiO_2 - 100 nm, amorphous Si (a-Si) - 250 nm

Tesla BS540 TEM (120 kV). In order to reveal diffused areas the specimens were etched in the HF-HNO₃ solution for 5 sec. (Roberts *et al* 1985). Depth distributions of each dopant under different overlayers were measured using sheet resistance and spreading resistance methods.

3. RESULTS AND DISCUSSION

Measurements of the junction depth were performed both on cross-sectional TEM micrographs and with the spreading-resistance method. Results of the TEM measurements are summarized in Table 1. Our measurements have shown that the presence of an overlayer influences the junction depth. In the case of phosphorus and arsenic diffusion the a-Si overlayer decreases the junction depth. The same overlayer increases the junction depth for boron diffusion. The other overlayer, a-Si/Si₃N₄/SiO₂, decreases the junction depth for phosphorus and arsenic diffusion while it does not influence the junction depth for boron diffusion.

Fig. 2 shows cross-sectional TEM micrographs for phosphorus implanted specimens. Three kinds of dislocations can be distinguished. The first one (denoted by

Table 1. Depths of diffusion region

Implanted with	No overlayer	a-Si overlayer	a-Si/Si₃N₄/SiO₂ overlayer
phosphorus	1.72 μm	1.43 μm	1.22 μm
boron	1.22 μm	1.38 μm	1.22 μm
arsenic	0.73 μm	0.68 μm	0.54 μm

SD in Fig 2b) are dislocations exhibiting weak contrast. Dislocations of this type are visible as gray lines parallel to two crystallographic directions. They are observed in all samples. However, they have the highest density in samples covered with a-Si. In our samples they extend down to the depth of 0.7-0.85 μm. The second kind of dislocations are dislocation half-loops pinned at the surface or the interface (**HL** in Fig 2b). Those dislocations can be found in all phosphorus implanted samples. Their deepest edge is at the depth of three quarters of the junction depth in samples with the a-Si overlayer and of a half of the junction depth for the other samples. Since in the transparent area thickness of a thin foil does not exceed 2500Å in some cases only very small parts of dislocation half-loops are visible (**PL** in Fig. 2b). The third kind of dislocation are helical dislocations being present at the depth of two thirds of the junction depth (**HD** in Fig. 2b). The helical dislocations are observed in all

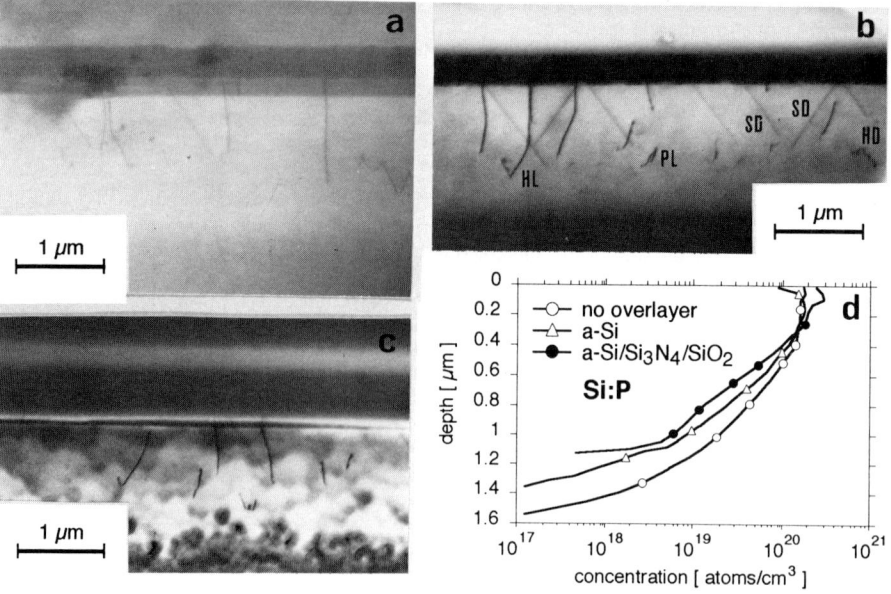

Fig. 2. Cross-sectional TEM micrographs of phosphorus implanted silicon a) without an overlayer, b) covered with a-Si, c) covered with a-Si/Si₃N₄/SiO₂; and d) Depth distribution of phosphorus for samples shown in fig. a) - c)

phosphorus implanted samples. The dopant depth distribution profiles for phosphorus implanted samples are drawn schematically in Fig. 2d.

Fig. 3 a) - c) presents cross-sectional TEM micrographs of boron implanted specimens. The major defects in these samples are dislocation half-loops which form a dense network of dislocations in the diffused area in samples without an overlayer and those with the a-Si overlayer. The dislocation network extends down to a half of the junction depth. In the samples with no overlayer at the depth of about a half of the junction depth dislocations parallel to the junction and dislocation loops are also observed. In samples with the a-Si/Si$_3$N$_4$/SiO$_2$ overlayer the density of dislocations is much lower than in the others. Characteristic defects for those samples are Y-shaped dislocations, denoted with **Y** in Fig. 3c. The Y-shaped dislocations are the simplest form of a dislocation network. The mechanism of formation of those defects is as follows. During the diffusion process two dislocation half-loops formed at the surface move towards the interior of the sample. At the certain depth they meet and coalesce to form a new dislocation lying parallel to the surface. In our samples that kind of a dislocation network extends to a depth of a half of the junction depth. The dopant depth distribution profiles for boron implanted samples are shown in Fig. 3d.

TEM cross-sections of arsenic implanted specimens are shown in Fig. 4 a) - 4 c). In samples with no overlayer the major defects are dislocation half-loops (Fig. 4a). The dislocations extend to the depth of four fifths of the junction depth and in spite of their high density do not form dislocation networks. A characteristic feature of the specimens with the a-Si overlayer is a very low density of dislocations (Fig. 4b). In specimens with the a-Si/Si$_3$N$_4$/SiO$_2$ overlayer (Fig. 4c), as with the boron implanted specimens, Y-shaped dislocations are observed, indicating dislocation network formation. The dislocation network forms at a depth of half of the junction depth. The dopant depth distribution profiles for arsenic implanted samples are drawn schematically in Fig. 4d.

The dislocation half-loops observed in the samples are typical diffusion-induced defects. It is known that, during the diffusion process stress is induced in the lattice (Lawrence 1966). This stress is governed by the difference in size between covalent radius of the diffused impurity and that of the host atom. This is why during phosphorus and boron diffusion in Si a tensile stress arises near the surface. The stress distribution due to P diffusion was given by Prussin (1961). Diffusion-induced stress causes the formation of dislocation half-loops pinned at the surface and their expansion towards the interior. The motion of the dislocation is restricted by the opposite stress present at a certain depth in the wafer (Itoh 1983). During processing the deepest parts of dislocation half-loops become parallel or almost parallel to their Burgers vector. Phosphorus diffusion is known to induce a large

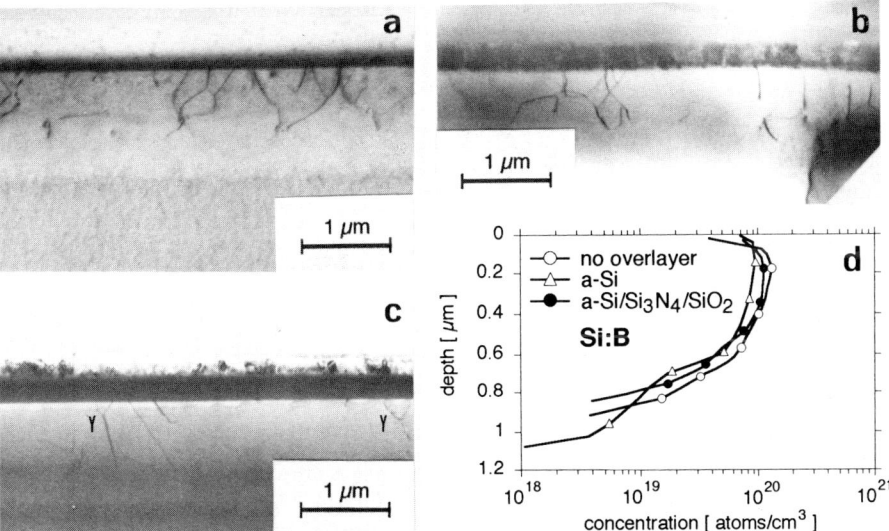

Fig. 3. Cross-sectional TEM micrographs of boron implanted silicon a) without an overlayer, b) covered with a-Si, c) covered with a-Si$_3$N$_4$/SiO$_2$; and d) Depth distribution of boron for samples shown in Fig. a) - c)

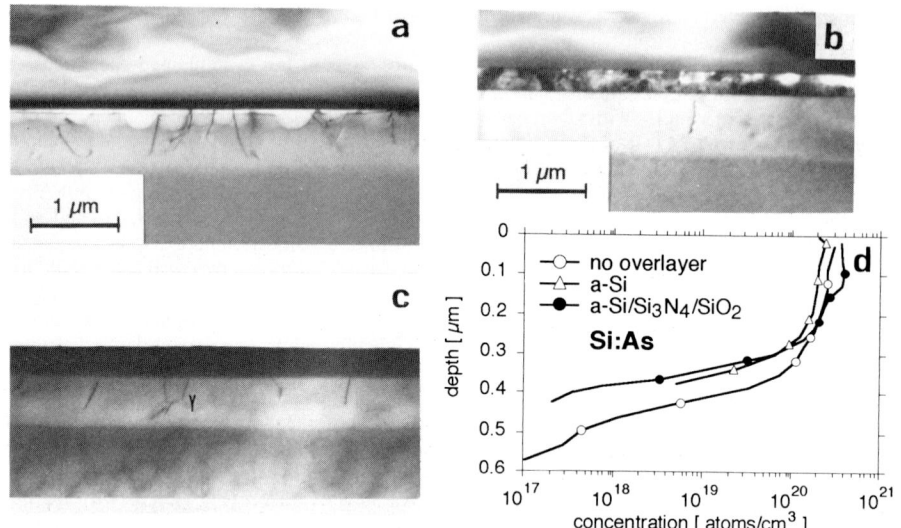

Fig. 4. Cross-sectional TEM micrographs of arsenic implanted silicon a) without an overlayer, b) covered with a-Si, c) covered with a-Si/Si$_3$N$_4$/SiO$_2$; and d) Depth distribution of arsenic for samples shown in Fig. a) - c)

number of self-interstitials in the bulk causing an effect such as "emitter-push-effect". The interstitials being saturated under the P diffused region are looking for a sink. A good sink for interstitials is a screw segment of a dislocation. Screw parts of dislocations absorbing Si self-interstitials climb and form helices.

4. CONCLUSIONS

In order to study the influence of an overlayer on the structure of implanted p-n junctions a-Si and a-Si/Si$_3$N$_4$/SiO$_2$ overlayers have been chosen. Our research proved that an overlayer being placed at the surface of silicon influences the junction depth. Both overlayers decrease the junction depth for phosphorus and arsenic implanted specimens. In case of boron implanted specimens covering the wafer with the overlayer of a-Si increases the junction depth while the a-Si/Si$_3$N$_4$/SiO$_2$ overlayer does not influence it.

The major defects in the diffused areas are dislocation half-loops and in the case of phosphorus implanted samples helical dislocations. Mechanisms of the formation of those defects were discussed. In order to combine diffusion mechanisms with processes of the defect formation further TEM analysis is necessary.

ACKNOWLEDGMENT

This publication is based on work sponsored by the Polish Government under the project # 80116.9101. The authors are very much indebted to Ms D Szczepańska for assistance in specimen preparation and Ms J Wiącek for careful preparation of micrographs.

REFERENCES

Itoh N, Nakau T 1982 Jap. J. Appl. Phys. <u>21</u> pp 817-821
Lawrence J E 1966 J. Appl. Phys. <u>37</u> pp 4106-4112
Prussin S 1961 J. Appl. Phys. <u>32</u> pp 1876-1881
Roberts M C, Yallup K J, Booker G R 1985 Microscopy of Semiconducting Materials 1985 (IOP Conf Ser No 76) p 483

Inst. Phys. Conf. Ser. No 134: Section 3
Paper presented at Microsc. Semicond. Mater. Conf., Oxford, 5–8 April 1993

145

TEM study of silicon after irradiation by 600 MeV protons

F Paschoud and M Victoria

Paul Scherrer Institute, CH-5232 Villigen-Psi, Switzerland

ABSTRACT: Silicon irradiated by 600 MeV protons has been observed with a transmission electron microscope. The microstructure after irradiation consists of a high density of small defects (diameter < 3nm). The defects show structure contrast, thereby they are considered to be amorphous zones. High resolution image simulation has been performed for a 1.5nm amorphous zone in perfect silicon crystals of different thicknesses. The contrast of the defect decreases rapidly with thickness and vanishes around 7nm.

1.0 INTRODUCTION

When a 600 MeV proton enters a silicon target, it looses its energy by ionization, on one hand, and by nuclear reactions with the silicon atoms on the other hand. The nuclear reaction, described in detail by Green (1984), produces a recoiling atom with a high recoil energy (hundreds of keV to some MeV). This energy is partly dissipated by displacement collisions in silicon lattice. This process is called a displacement cascade and produces structural defects in the material. As the range of 600 MeV protons in Si is about 0.8 m, the damage is produced homogeneously in the volume.

This study is motivated by the recent interest in radiation effects produced by energetics protons in electronic devices of satellites, since these components suffer irradiation by protons with a maximum energy of around 300 MeV. Even if the protons available have an energy twice higher, the recoil energy spectra are similar.

The correlation between the recoil spectra and the defects observed with the transmission electron microscope (TEM) have already been discussed (Alurralde 1993). In the present paper, the visibility of small amorphous zone in high resolution electron microscopy (HREM) will be discussed.

2.0 EXPERIMENTAL TECHNIQUES

2.1 Irradiation

Si single crystals, n-type, p-doped with [111] wafer normal and resistivities between 5 and 25 Ωcm were irradiated with 600 MeV protons in the PIREX II facility (Proton irradiation EXperiment) (Marmy 1989). The irradiation was performed on a 3mm wide strip under He gas cooling at 310K. The dose was determined to be 1.6×10^{23} p/m^2 by comparison with dosimetry experiments reported by Gavillet (1988 and 1993). This dose is below the saturation dose, at which the displacement cascades resulting from the energetic recoils are expected to start overlapping.

2.2 TEM Observations

For the TEM observations, 3mm discs were cut by ultrasound. They were then mechanically thinned to 100 μm and the final specimens were produced on one hand by thinning down in a HNO₃/HF solution at 300K, on the other hand by cleaving a 60° wedge as described by P.A. Buffat et al. (1989). This last method has the advantage of producing a clean surface and a well defined thickness profile of the specimen. As the interest is the proton irradiation defects, the ion milling procedure has to be avoided because of the production of undesirable defects (Chew 1987).

The TEM observations were performed with a Philips CM20 microscope (at the EPFL) at 120 kV. The wedge specimens were mounted with the [110] direction parallel to the electron beam. Preliminary high resolution observations were performed with a Philips EM430 at 300 kV.

2.3 Image Simulation

The simulations of high resolution images were performed with the EMS software developed by P. Stadelmann (1987), using the multislice formalism. For the multislice calculations, a spherical amorphous zone of 1.5nm in diameter has been assumed. The position of the atoms in the amorphous region comes from a molecular study of liquid silicon and was given by Ray (1993). No strain at the interface between the Si crystal and the amorphous region has been taken into account. The size of the supercell, including the 1.5nm damage zone, was 4x3nm for a <111> orientation.

For the computed images, the parameters for the Philips EM430 have been used (spherical aberration of 1.2mm, a convergent semi-angle of 0.8 mrad and a defocus spread of 8nm) with an aperture of 20nm⁻¹. Images with different thicknesses of the specimen were calculated by adding slices of perfect silicon on the top and the bottom of the supercell containing the amorphous zone.

3.0 RESULTS AND DISCUSSION

3.1 TEM Observations

The TEM observations were performed at 120 kV to avoid irradiation defects from the electron beam. By observing a perfect Si specimen, at the same conditions, it has been checked that no defect is produced during the observations.

Figure 1 shows the defects in a wedge specimen observed on a [110] zone axis, after proton irradiation. There is a high density of small defects with a maximum size of around 3nm. They are assumed to be small amorphous zones since they show structure contrast (contrast coming from the difference in the extinsion distance between the matrix and the damage zone). No black-white contrast is observed as would be the case for defects having an appreciable strain field. These observations are in agreement with other TEM, X-ray and channeling observations (Howe et al. 1981, Ruault et al. 1984 and G. Bai et al. 1991).

The formation of amorphous zones and layers by ion implantation has extensively been studied in Si (Veirman 1991 and Holland 1990). The annealing behavior and the recrystallization of these defects are the main interest. Some high resolution observations of amorphous zones have been reported (Herring 1987). They are usually bigger (around 10nm) than the specimen thickness and are clearly visible. In our experiment, the small defect size makes it necessary to know the visibility conditions. Image simulations have been performed and are discussed in the next paragraphs.

Fig.1: TEM micrographs of a 600 MeV proton – irradiated Si specimen showing the amorphous zones. Observation on a [110] zone axis.

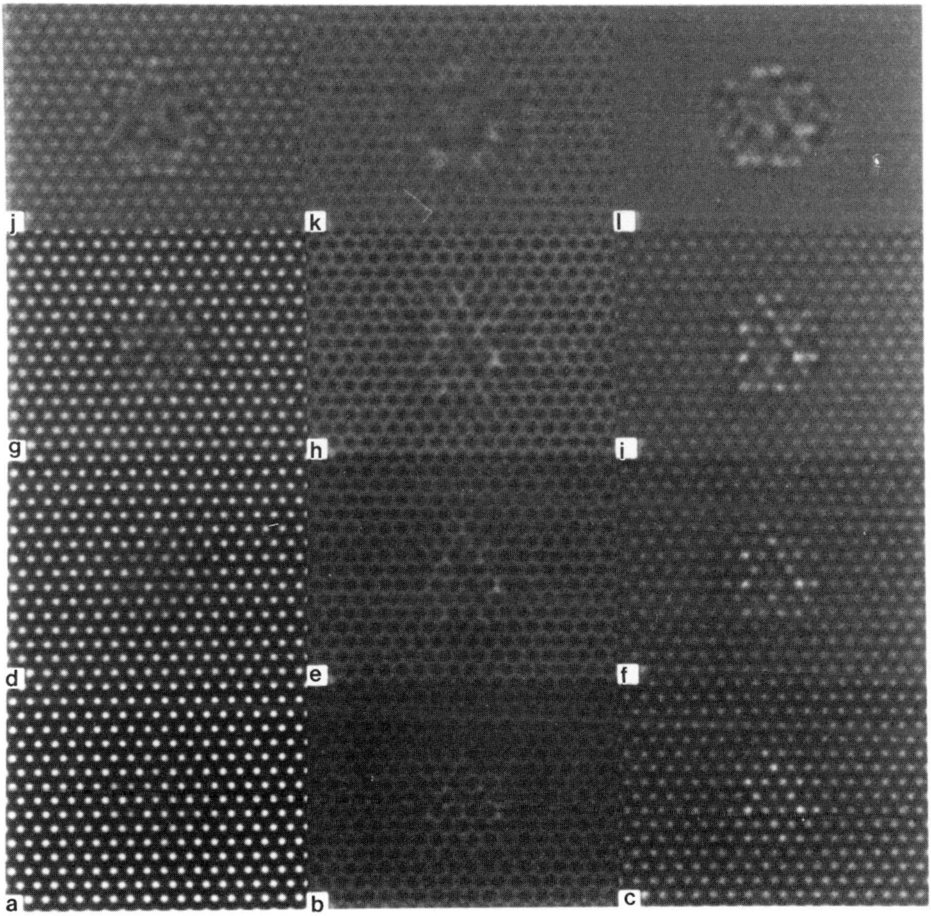

j	k	l
g	h	i
d	e	f
a	b	c

defocus	-60 nm	-74 nm	-88 nm

Fig.2: Computed HREM images of a 1.5nm amorphous zone in a Si of thickness: a-c 8.5nm, d-f 4.7nm, g-i 2.8nm and j-l 0.9nm.

3.2 Image Simulations

Figure 2. shows the image simulation for different thicknesses of the supercell, 0.9nm, 2.8nm, 4.7nm and 8.5nm. the complete calculation was done with defocusing from 60nm to 100nm in 5nm steps. Only three defocus settings have been kept in figure 2 for each thickness. In Fig 2j)-l), the defect zone is bigger than the thickness of the specimen and the regular structure is lost in the defect region. If the defect is embedded in a thicker Si crystal, the contrast decreases rapidly with thickness and the 1.5nm diameter defect becomes invisible for a value of around 7-8nm.The same kind of simulation has been performed by positioning the defect closer to the surface, but no important change in the contrast has been obtained.

Our results are comparable with the visibility study of a SiAs precipitate in Si (Armigliatto 1987) and they confirm that the amorphous zones are clearly visible when the specimen is thinner than the defect and become rapidly invisible with increasing thickness. Thus the difficulty of observing the very small amorphous zone by high resolution microscopy is shown.

4.0 CONCLUSIONS

The high resolution image simulations of a small amorphous zone (1.5nm in diameter) in a perfect silicon matrix reveal that the visibility of the defect decreases rapidly with the thickness of the specimen and vanishes for a thickness of around 7-8nm. When the specimen is thinner than the defect, it can be seen by the lost of atomic structure of perfect Si. Moreover, to be able to draw conclusions from a HREM observation of the small amorphous zones, great care in preparing the specimen is needed in order to avoid artifacts. The better way to prepare Si specimens is probably by cleaving.

REFERENCES

Alurralde M, Paschoud F, Victoria M and Gavillet D 1993 Proc. 8th Int. Conf. on Ion beam Modification of Materials, IBMM92, to be published in Nucl. Instr. Meth. in Phys. Res. B
Armigliato A, Bourret A, Frabboni S and Parisini A 1987 Inst. Phys. Conf. Ser. No. 87 pp 55
Bai G and Nicolet M A 1991J. Appl. Phys. 70 649
Buffat P A, Ganière J D and Stadelmann P 1989 in Evaluation of advanced semiconductor materials by electron microscopy, eds Cherns D, Plenum Press, Nato ASI Series B: Physics 203 319
Chew N G and Cullis A G 1987 Ultramicroscopy 23 175
De Veirman A and Van Landuyt J 1991 Phil. Mag. A64 513
Gavillet D 1988 PSI internal report - PIREX sample activity and dose rate TM-22-88-03
Gavillet D 1993 to be published
Green S L 1984 J. Nucl. Mat. 126 30
Herring R A, Venables J D and Fiore E M 1987 Inst. Phys. Conf. Ser. No. 87 pp 511-516
Holland O W, White C W, El-Ghor M K and Budai 1990 J. Appl. Phys. 68 2081
Howe L M and Rainville M H 1981 Nucl. Instr. and Meth. 182/183 143
Marmy P, Daum M, Gavillet D, Green S L, Hegedues F, Proennecke S, Rohrer U, Stieffel U and Victoria M 1989 Nucl. Instr. & Meth. in Phys. Res. B47 37
Ray R 1993 Private communication
Ruault M O, Chaumont J, Penisson J M and Bourret A 1984 Phil. Mag. A50 667
Stadelmann P 1987 Ultramicroscopy 21 131

Inst. Phys. Conf. Ser. No 134: Section 3
Paper presented at Microsc. Semicond. Mater. Conf., Oxford, 5–8 April 1993

149

In situ HVEM studies of electron irradiation induced {113}-defect generation in silicon

Jan Vanhellemont and Albert Romano-Rodríguez*

IMEC, Kapeldreef 75, B-3001 Leuven, Belgium
*Permanent address: LCMM, Dept. Física Aplicada i Electrònica, Universitat de Barcelona, Diagonal 645-647, E-08028 Barcelona, Spain

ABSTRACT: Results are presented of an in-situ study of dopant and strain dependence of {113}-defect generation in silicon by high flux 1-MeV electron irradiation. The enhanced {113}-defect nucleation close to compressively strained interfaces is explained by an increased capture of vacancies by the interface. The dopant type and profile dependence of {113}-defect nucleation is explained by its impact on the density of bulk sinks for intrinsic point defects.

1. INTRODUCTION

The high voltage electron microscope (HVEM) is a versatile tool to perform in-situ studies of electron irradiation induced effects in materials. The instrument allows to vary the electron flux over about nine decades between 10^{12} and 10^{21} cm^{-2}s^{-1}. Using the appropriate specimen holders, the irradiation temperature can easily be chosen between 77 and 1200K. Due to the high acceleration voltage silicon specimens can be used with thicknesses up to 5 μm, which are more representative of the bulk behaviour of the material. The defect nucleation or the transformation of the irradiated area can be followed and recorded in real time.

It is now well established that irradiation of silicon with high flux, high energy electrons leads to the formation of so called {113}-defects by the condensation of self-interstitials in agglomerates with {113} habit planes. In the present paper results are presented of an in-situ study of 1-MeV electron irradiation induced {113}-defect generation in silicon with dopant profiles and locally strained areas. Previous results have been published elsewhere (Romano-Rodríguez and Vanhellemont 1992a,b). Results on uniformly doped substrates have been published by Fedina and Aseev 1990 and by Aseev et al 1991, and are correlated with those obtained in the present work.

2. EXPERIMENTAL

Cross-section specimens are prepared of ion implanted and annealed (001) silicon wafers. Some irradiations are also performed on a local isolation structure to study the influence of interfaces and localised strained areas. The irradiations are performed in the HVEM of the University of Antwerp (RUCA) at temperatures between 300K and 640K, with a 1-MeV electron flux between 10^{19} to 2×10^{20} cm^{-2}s^{-1} and for times up to 80 min. Some specimens receive a second irradiation at another temperature.

3. OBSERVATIONS AND DISCUSSION

During irradiation at room temperature with electrons with energies larger than a few hundred keV, silicon atoms are displaced by knock-on collisions with electrons. A small fraction of the Frenkel pairs thus formed will separate and form two more or less independent populations of self-interstitials and vacancies. In Czochralski-grown silicon dominant sinks for intrinsic point defects are the interstitial oxygen, substitutional carbon and the surfaces of the sample. When the dopant concentration, which is for conventional substrates in the range of 10^{14} to 10^{15} cm^{-3}, increases to values that are comparable to that of the oxygen and carbon content, it can be expected that the self-interstitial concentration and thus also the {113}-defect density will be influenced. In the next paragraphs the influence of interfaces, strain and dopant concentration and type will be discussed.

3.1 Influence of Strain and Interfaces

Fig. 1 shows the result of a 10 min irradiation, at 640K with an electron flux of 2×10^{19} cm^{-2} s^{-1}, of a cross-section specimen of a LOCOS structure. During the irradiation, {113}-defects nucleate first near the tip of the bird's beak (area 2) where the compressive stresses are concentrated. Below the nitride mask, the pad oxide and also the silicon substrate are under compressive stress and also there the defect density is higher than under the field oxide (area 3) where the substrate is under tensile stress.

It was shown elsewhere (Vanhellemont and Romano-Rodríguez 1993b) that the position z_m of the {113}-defect-rich layer below the interface can be estimated from

$$z_m = \frac{1}{\sqrt{\kappa_I} - \sqrt{\kappa_V}} \ln \frac{\sqrt{\kappa_V}\,\sigma_I\,(\sigma_V + \sqrt{\kappa_V})}{\sqrt{\kappa_I}\,\sigma_V\,(\sigma_I + \sqrt{\kappa_I})}$$

with κ_I and κ_V, σ_I and σ_V the strengths of the substrate and interface sinks for self-interstitials and vacancies, respectively.

From the comparison with the observations it is concluded that the interface is a stronger sink for self-interstitials than for vacancies and that the change of position on the {113}-defect layer in area 1-3 is mainly determined by the change of σ_V. Compressive straining of the silicon substrate leads to an increase of σ_V and thus to an increased {113}-defect formation.

3.2 Influence of Dopants

A clear dependence of the {113}-defect nucleation on dopant concentration and type, on the local strain distribution and on the presence of interfaces is observed. Defects form preferentially in areas with well defined dopant concentration.

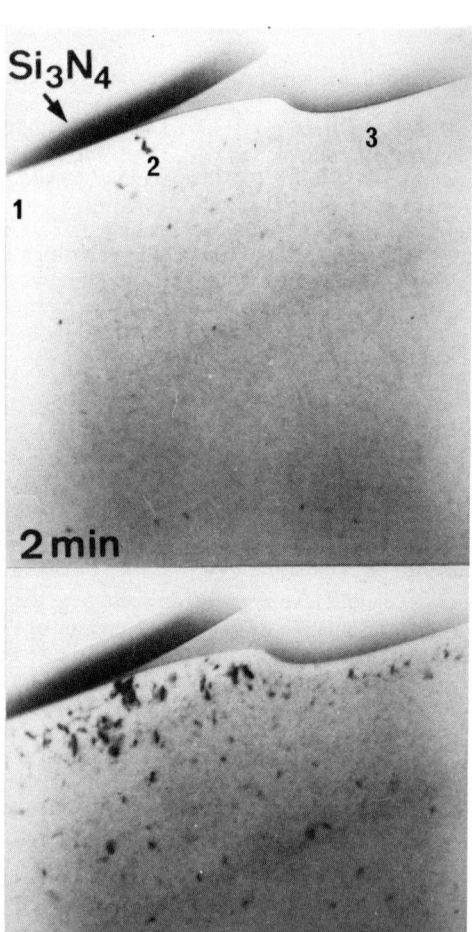

Fig. 1: LOCOS structure irradiated at 640K with an electron flux Φ of 2×10^{19} cm^{-2}s^{-1}.

3.2.1 Boron

In boron doped silicon extended defects are observed dominantly in areas with a dopant concentration C_B between 5×10^{17} and 5×10^{18} cm^{-3} as illustrated in Fig. 2. Varying the temperature has little or no influence on the position of the {113}-defect-rich layer but only on the nucleation and growth kinetics. The observed defect generation is not dependent on the ion implantation process as a similar behaviour was observed in epitaxial boron doped layers grown by chemical vapour deposition (Vanhellemont and Romano-Rodríguez 1993a). It is well known that boron atoms can easily be kicked-out from their substitutional position by a mobile self-interstitial. An increasing boron concentration will thus present an increasing sink for self-interstitials explaining the lack of {113}-defects in the areas with a high boron concentration.

3.2.2 Phosphorus and Arsenic

In phosphorus doped samples a strong dependence on the irradiation temperature is observed. As illustrated in Fig.3, below 410K defects nucleate mainly in areas with C_P > 10^{19} cm^{-3} while above this temperature, the defects nucleate in areas with C_P < 5×10^{17} cm^{-3}. It is well-known that phosphorus atoms are a sink for vacancies with which they form the so called E-centre (vacancy/phosphorus pair). From DLTS studies it is known that the E-centre is stable up to a temperature of 400K. The dissolution of the E-centre above this temperature will release a large number of vacancies which are then available for recombination with self-interstitials thus leading to a suppression of {113}-defect formation. Also for As doped silicon, two temperature regimes exist, i.e. above and below 450K corresponding also with the stability of the As/V pair.

3.3.3 Influence of Carbon Contamination

To evaluate the possible influence of cracking of hydrocarbon molecules in the microscope, one side of a thinned sample similar to the one in Fig. 2, was evaporated with carbon. The result of the subsequent irradiation is shown in Fig. 4. Already after two minutes irradiation a large number of {113}-defects is nucleating in the lowly doped substrate. After longer irradiations, however, these defects disappear again and a similar result is obtained to that without carbon contamination.

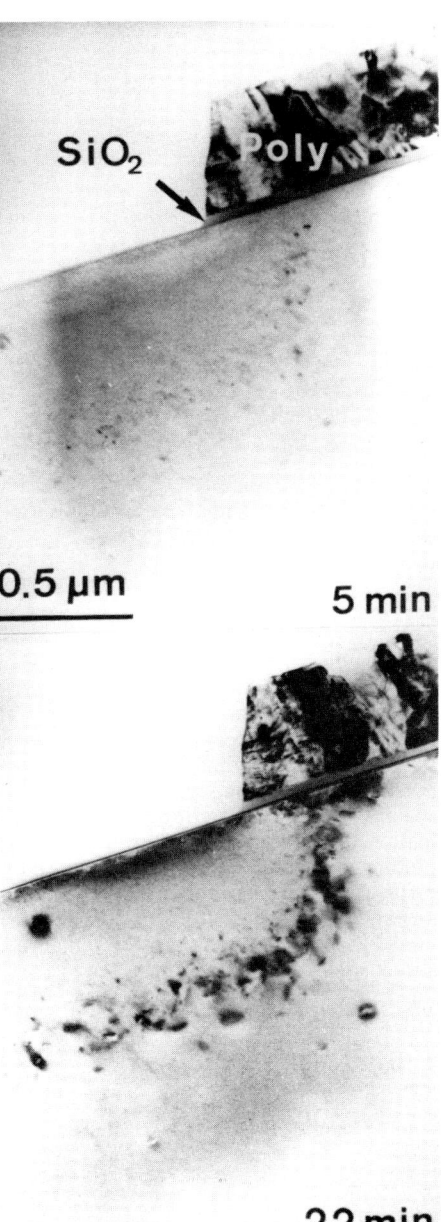

Fig. 2: Boron implanted sample irradiated at 640 K ($\Phi = 2 \times 10^{19}$ ecm^{-2}s^{-1}).

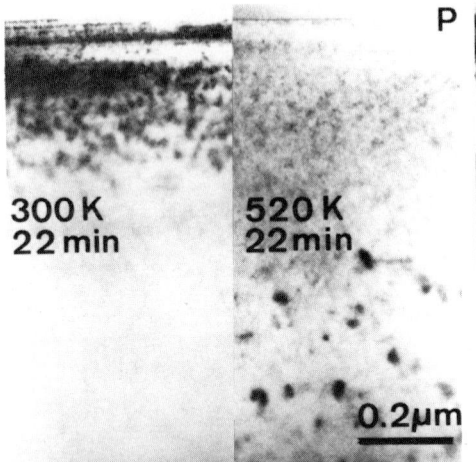

Fig. 3: Phosphorus implanted sample irradiated at two different temperatures.

This behaviour corresponds well with the observations of Hasebe et al (1986) obtained on uniformly doped specimens contaminated with carbon. They attributed the observed behaviour to an accumulation of the slower moving species, i.e. the vacancy. The theoretical results reported in paragraph 3.1 suggest that an alternative explanation is the presence of a strong vacancy sink at the specimen's surface leading to the rapid formation of {113}-defects close to it. During prolonged irradiation at 640K the carbon will diffuse into the specimen. As carbon is known to be a strong sink for self-interstitials, the {113}-defects in the lowly doped bulk shrink again.

*Part of this work was performed with financial support of the Spanish Ministry of Science and Education (MEC).

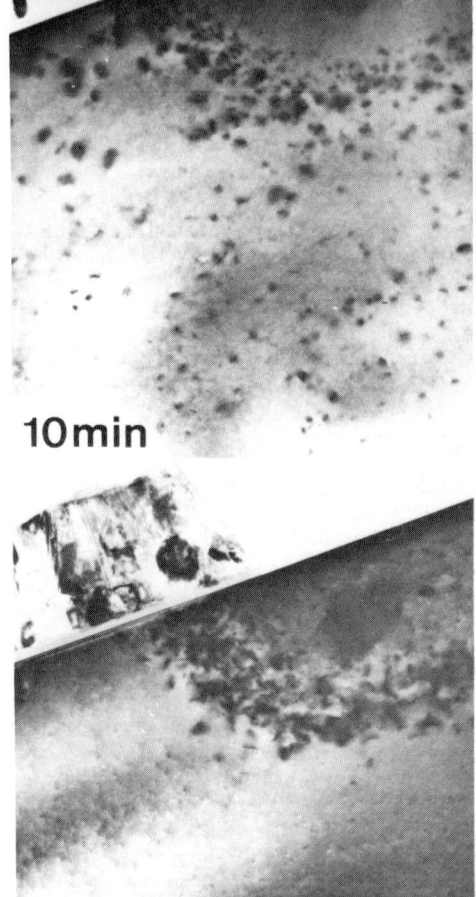

Fig. 4: Carbon contaminated sample of Fig. 2 irradiated at 640K, with the same flux.

REFERENCES

Aseev A L, Denisenko S G and Fedina L I
    (1991) Sov. Phys. Semicond. 25 352
Fedina L I and Aseev A L (1990) Sov. Phys.
    Solid State 32 33
Hasebe M, Oshima R and Fujita F E (1986)
    Jap. J. Appl. Phys. 25 159
Romano-Rodríguez A and Vanhellemont J (1992a) Materials Science Forum Vol. 83-87 303
Romano-Rodríguez A and Vanhellemont J (1992b) Mat. Res. Soc. Symp. Proc. 262 1091
Vanhellemont J and Romano-Rodríguez A (1993a) presented at MRS Fall '92, to be published
    in Mat. Res. Soc. Symp. Proc.
Vanhellemont J and Romano-Rodríguez A (1993b) Appl. Phys. Lett., submitted

Inst. Phys. Conf. Ser. No 134: Section 3
Paper presented at Microsc. Semicond. Mater. Conf., Oxford, 5–8 April 1993

Imaging of semiconductor defects using ion channelling

P J C King[1], M B H Breese[2], P R Wilshaw[1], G R Booker[1], G W Grime[2], F Watt[2] and M J Goringe[1]

[1]Department of Materials, University of Oxford, Parks Road, Oxford, OX1 3PH.
[2]Oxford University Scanning Proton Microprobe Unit, Nuclear Physics Laboratory, Keble Road, Oxford, OX1 3RH.

ABSTRACT: Using a Scanning Proton Microprobe, the techniques of Scanning Trasnmission Ion Microscopy and ion channelling have been combined to produce images of defects in semiconducting crystals. The technique is able to image individual misfit dislocations and bunches of dislocations in $Si_{1-x}Ge_x/Si$ crystals, and information on the Burgers vector of the dislocations can be obtained. Oxidation induced stacking faults in a silicon crystal have also been imaged. This method of defect imaging is able to use samples that are over an order of magnitude thicker than those required for TEM analysis.

1. INTRODUCTION TO THE TECHNIQUE OF SCANNING TRANSMISSION ION MICROSCOPY(STIM).

A focused ion beam can be used to provide a number of analytical techniques which give structural and compositional information on the sample being studied. One such technique is called Scanning Transmission Ion Microscopy (STIM) (Sealock et al. 1987). STIM gives information on a sample's density and thickness by measuring the energy loss of ions transmitted through thin specimens. This paper demonstrates how STIM, combined with ion channelling, can be used to image dislocations and stacking faults in semiconducting crystals, with samples that are over an order of magnitude thicker than those required for transmission electron microscopy analysis.

A Scanning Proton Microprobe (SPM) (Watt and Grime 1987) has been used to produce these results. The SPM provides a beam of MeV protons (typically 3MeV) which is focused down to a spot size of ~0.3μm for this work. In STIM, the protons are detected after being transmitted through the sample under investigation. A proton energy spectrum is built up, and from this an image can be produced which shows the average energy loss of the protons at each pixel. The experimental set-up is as shown in fig. 1.

2. COMBINING STIM WITH CHANNELLING.

The technique of ion channelling can be used to investigate samples that are crystalline. Channelling (Feldman 1982) occurs when the ions are incident along a major crystallographic plane or axis. The regular rows of lattice atoms steer the ions through the crystal, and the ions are shielded from making close collisions with the lattice atoms. One of the effects of channelling is to reduce the rate of energy loss of the ions to ~0.5 times that of non-channelled ions (Appleton et al. 1967), and this energy loss reduction can be seen in the transmitted ion energy spectrum.

A disruption in the regular arrangement of lattice atoms by a crystal defect such as a stacking fault or dislocation can cause channelled ions to be 'dechannelled'. This means

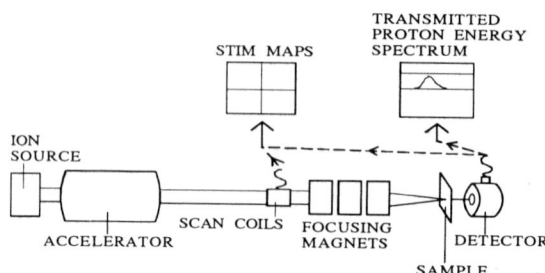

Fig. 1 Schematic diagram of STIM on the Oxford Scanning Proton Microprobe. A proton beam is produced by the ion source, accelerated to MeV energies, and focused onto the sample. Protons transmitted through the sample are detected, an energy spectrum is built up, and images based on the proton energy loss can be produced as the beam is scanned over the sample.

that they are no longer steered by the rows of lattice atoms, and they continue their journey through the crystal with their normal energy loss. Thus, ions which are initially channelled will be transmitted through a region of a crystal containing a defect with a higher energy loss than those transmitted through a region of good crystal because of the dechannelling the defect causes. This energy loss difference can be seen on a STIM average energy loss image, and provides the means for detecting the defects that cause dechannelling. The technique of using the local variations in channelled ion energy loss to produce images of regions of crystal where channelling is poor is called Channelling STIM, or CSTIM (Cholewa et al., 1990).

3. SAMPLE PREPARATION.

3MeV protons have a range of ~90μm in Si, so this was the upper limit on sample thickness for analysis. The samples described here were all 20-40μm thick. They were thinned with wet and dry paper from the back surface, and then polished to remove large scratches. The preparation process took about an hour for each sample. The thinned samples were then mounted on a four-axis goniometer for analysis by the Oxford SPM.

4. DISLOCATION IMAGES FROM A $Si_{0.95}Ge_{0.05}$/Si SAMPLE.

A sample consisting of a 1.8μm thick epitaxial layer of $Si_{0.95}Ge_{0.05}$ on an (001) Si substrate has been studied by the CSTIM technique. Misfit dislocations were present in localised regions at the layer-substrate interface of this sample running along [110] and [1$\bar{1}$0] directions; these could be imaged with CSTIM, and also made visible to optical microscopy by chemical etching (Gibbings et al. 1989). The etching process revealed the misfit line, and also produced an etch pit at each end of the line where a threading dislocation from the substrate met the sample surface; the etch pits enabled the number of dislocations that had been imaged with CSTIM to be determined.

Fig. 2a shows an etched sample taken with Nomarski phase contrast optical microscopy. It shows the slip trace of a single misfit dislocation lying along a [1$\bar{1}$0] direction (A) and the slip trace and threading dislocation etch pits of a group of 13 dislocations, (B). Fig. 2b shows a CSTIM image of the same region with the sample thinned to ~24μm, taken with the proton beam aligned with the (110) planes of the sample. The single dislocation can be clearly seen, with the bunch of dislocations giving stronger contrast next to it. The image took 2.5 hours to collect at a count rate of ~2000 protons per second. The CSTIM images shown in this paper are grey scale maps with light greys corresponding to lower energy loss and dark greys to higher energy loss.

As well as imaging the dislocations, CSTIM can provide information on the dislocation Burgers vector due to contrast changes that occur when the sample is tilted just off the channelling direction. Figs. 3a and b show two CSTIM images of a bunch of 8-9 dislocations; they were taken with the beam -0.2° and +0.2° respectively from the channelling direction. The vertical line due to the bunch of dislocations has changed in contrast between the two images and this asymmetrical contrast change with angle from the channelling direction is consistent with the dislocatons being of the 60° type as opposed to pure edge or pure screw (Breese et al. 1993).

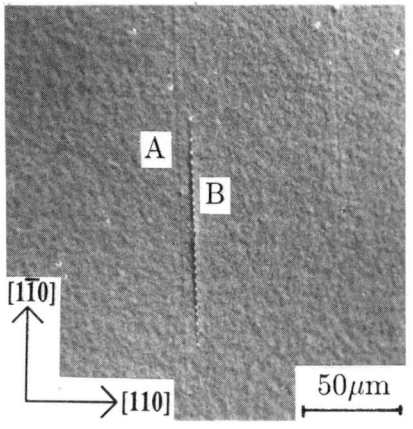 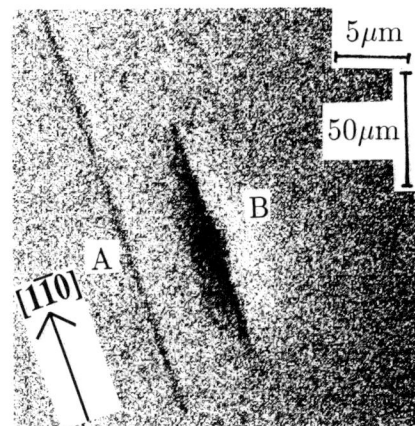

Fig. 2. a) Normarski phase optical microscope picture of an etched $Si_{0.95}Ge_{0.05}$/Si sample. A - single dislocation; B - bunch of 13 dislocations. b) A CSTIM energy loss image of the same region as in a). Darker is higher energy loss. Please note the different scales along the horizontal and vertical directions.

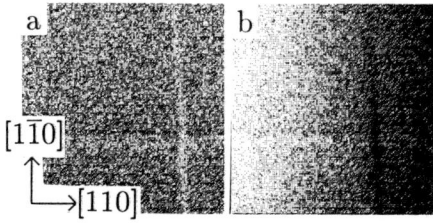

Fig. 3. Change in energy loss contrast of group of 8-9 dislocations (vertical line) when imaged with the beam $-0.2°$ (a) and $+0.2°$ (b) from the channelling direction. The asymmetrical contrast change with angle from channelling orientation gives information on the way the lattice planes are bent and hence on the dislocation Burgers vector. The images are 60 μm wide.

5. DISLOCATION IMAGES FROM A $Si_{0.85}Ge_{0.15}$/Si SAMPLE WITH MESAS.

A second SiGe/Si sample studied with CSTIM consisted of a 1μm thick $Si_{0.85}Ge_{0.15}$ layer deposited on to an (001) Si substrate which had 3μm high raised areas (mesas) etched on to it. Analysis by TEM of the non-mesa regions showed that bunches of misfit dislocations were present at the layer-substrate interface. The mesa regions were too thick for easy TEM analysis. Fig. 4. shows STIM energy loss images of an 80μm wide mesa, a) taken with the beam not channelled and b) taken with the beam channelled along the [001] axis of the specimen. The mesa is dark in the images as it was 3μm thicker than the non-mesa regions causing the protons to lose more energy in passing through it. Bands can be seen in the mesa and non-mesa regions in the channelled image which are not visible in the non-channelled image and are therefore due to the bunches of interface dislocations. The bands seen in CSTIM images of this sample showed similar asymmetrical contrast changes with tilt angle from the channelling direction to those exhibited by the bunches of dislocations in the $Si_{0.95}Ge_{0.05}$ sample. This is shown in fig. 5 and is again consistent with the dislocations being of the 60° type.

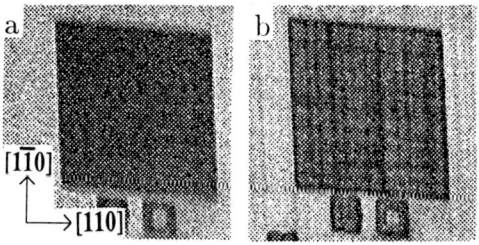

Fig. 4. CSTIM energy loss images of an 80μm wide mesa (dark) on the substrate of an $Si_{0.85}Ge_{0.15}$/Si sample. a) Beam not channelled. b) Beam channelled along the [001] axis; bands can be seen in the mesa and non-mesa regions running along [110] and [1$\bar{1}$0] directions due to bunches of dislocations at the layer-substrate interface. The images are 120μm wide.

6. STACKING FAULT IMAGES.

As well as being able to image dislocations, the CSTIM technique has been used to produce images of oxidation induced stacking faults in an (001) Si wafer. The faults were produced by damaging the wafer suface and then subjecting the wafer to oxidation treatments. Fig.6 shows CSTIM energy loss images of a ~37μm thick sample of the wafer with the beam aligned with the [001] axis of the crystal. The stacking faults were visible as 'D' shaped dark patches; the shape was due to the geometry of the faults, which were lying on {111} planes inclined at 54.7° to the sample surface.

7. CONCLUSIONS

The CSTIM technique has been able to image bunches of misfit dislocations at the layer-substrate interface of a $Si_{0.95}Ge_{0.05}/Si$ sample, and a single dislocation has been imaged on an etched sample. Contrast changes on tilting the sample just off the channelling direction give information on the Burgers vector. CSTIM has been used to image bunches of misfit dislocations in raised portions of an $Si_{0.85}Ge_{0.15}/Si$ sample that were too thick for easy TEM analysis, and images of oxidation induced stacking faults in an Si crystal have also been produced. The samples studied were over an order of magnitude thicker than those required for TEM analysis, greatly reducing sample preparation time and the chance of introducing artifacts during the preparation process.

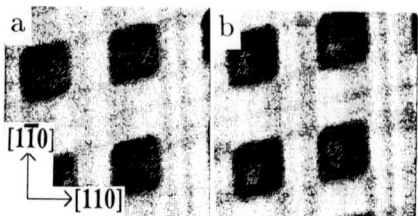

Fig. 5. Two 40μm wide CSTIM images of a group of four 10μm wide mesas. a) and b) were taken with the beam -0.2° and +0.2° from the channelling direction respectively. The vertical dark bands between the mesas have completely reversed in contrast between the two images, giving information on the dislocation Burgers vector.

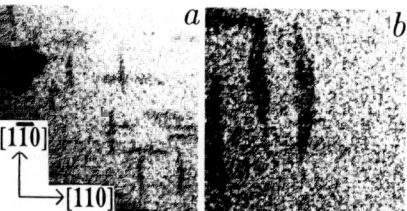

Fig. 6. CSTIM images of oxidation induced stacking faults in an (001) Si crystal. a) 100μm wide image, b) 35μm wide image. The images have been smoothed to show the fault shape more clearly.

8. ACKNOWLEDGEMENTS.

P.J.C. King would like to thank SERC for a studentship, and M.B.H. Breese would like to thank the Royal Commission of 1851 for support. Thanks are due to Dr. C. Tuppen and Dr. C. Gibbings of British Telecom, Professor E. Parker (Warwick University) and Mr. A. M. Gundlach (Edinburgh University), and Dr. R. Falster (MEMC) for providing respectively the $Si_{0.95}Ge_{0.05}/Si$, $Si_{0.85}Ge_{0.15}/Si$ and stacking fault samples. The help of M. deCouteau in the stacking fault work and T. Fell who etched the $Si_{0.95}Ge_{0.05}$ sample is gratefully acknowledged.

REFERENCES.

Appleton B R, Erginsoy C and Gibson W M 1967 Phys. Rev. 161 330

Breese M B H, King P J C, Whitehurst J, Booker G R, Grime G W, Watt F, Romano L T and Parker E H C 1993 J. App. Phys., in press

Cholewa M, Bench G, Legge G J F and Saint A 1990 App. Phys. Lett. 56(13) 1236

Feldman L C, Mayer J W and Picraux S T 1982 Materials Analysis by Ion Channeling (New York: Academic Press)

Gibbings C J, Tuppen C G, and Hockly M 1989 App. Phys. Lett. 54(2) 148

Sealock R M, Jamieson D N and Legge G J F 1987 Nucl. Inst. and Meth. B29 557

Watt F and Grime G (eds.) 1987 Principles and Applications of High Energy Ion Microbeams (Bristol: Adam Hilger)

Inst. Phys. Conf. Ser. No 134: Section 3
Paper presented at Microsc. Semicond. Mater. Conf., Oxford, 5–8 April 1993

157

Defects in polycrystalline CVD diamond layers analysed with the TEM

P Wurzinger, M Joksch and P Pongratz

Institut für Angewandte und Technische Physik, TU Wien, Wiedner Hauptstrasse 8-10, A-1040 Vienna, Austria

ABSTRACT: Dislocations and areas filled with microtwins in low pressure CVD diamond layers are analyzed using TEM diffraction contrast analysis and HREM. Dislocation glide is found to occur both on {111} and {100} glide planes. Dissociation of the dislocations glissile on {111} is observed and the partial separation is comparable to natural diamond. No evidence for dislocation climb is found. Primary twin lamellae parallel to all four {111} planes of the matrix and a small volume fraction of differently oriented crystals, together with virtually amorphous inclusions, are observed in the areas filled with microtwins. Very few partial dislocations are found at the twin lamellae and within the twin intersections.

1. INTRODUCTION

Because of the unique properties of diamond, low pressure chemical vapour deposition (CVD) of polycrystalline diamond layers on different substrates is gaining more and more technological interest. Scanning electron microscopy (SEM) and Raman spectroscopy are widely used to characterize the layers. However, TEM studies have shown that a large number of defects which can not be assessed by these techniques is present in the layers (Williams and Glass 1989, Zhu *et al.* 1989, Kaae *et al.* 1990, Narayan 1990, Suzuki *et al.* 1992). These authors reported the observation of areas which contain many thin twin lamellae and, apart from these regions, a high density of dislocations. The present paper gives an account on the results of an extensive study of these defects.

2. EXPERIMENTAL

Polycrystalline diamond layers were grown on SiAlON from CH_4/H_2 gas using microwave assisted CVD (growth temperature 700°C, total gas pressure 100 mbar, microwave power and CH_4 content: 1500 W/2.1 % and 1300 W/1.4 % respectively). SEM showed {111} facetting of the layers. Mechanical separation from the substrate, grinding on a fast rotating cast iron disc, and Ar^+ ion thinning were used for the TEM specimen preparation. The TEM investigations were performed on a JEOL 200 CX TEM operated at 200 kV. The defect analysis consisted of the stereoscopic determination of the dislocation geometry and diffraction contrast analysis and computer simulation of strong- and weak-beam images. A JEOL 4000 EX TEM operated at 400 kV was used for HREM investigations.

3. RESULTS

3.1 Dislocations

Apart from regions containing a large number of microtwins dislocations were observed in virtually every grain. Bright field contrast analysis (for which a dissociation of the dislocations was not considered) revealed $<110>/2$ Burgers vectors for all of them. The average dislocation density was about $5\text{-}10 \cdot 10^8/\text{cm}^2$. However, large deviations from this value were detected for single grains.

Most of the observed dislocations were glissile on $\{111\}$ planes and were curved in these planes. Figure 1 shows an interesting example of such dislocations: All of the marked dislocations lie within the accuracy of stereoscopic evaluation parallel to the (111) plane and are glissile. The change of curvature of dislocations A_1 and A_4 occurs within this glide plane.

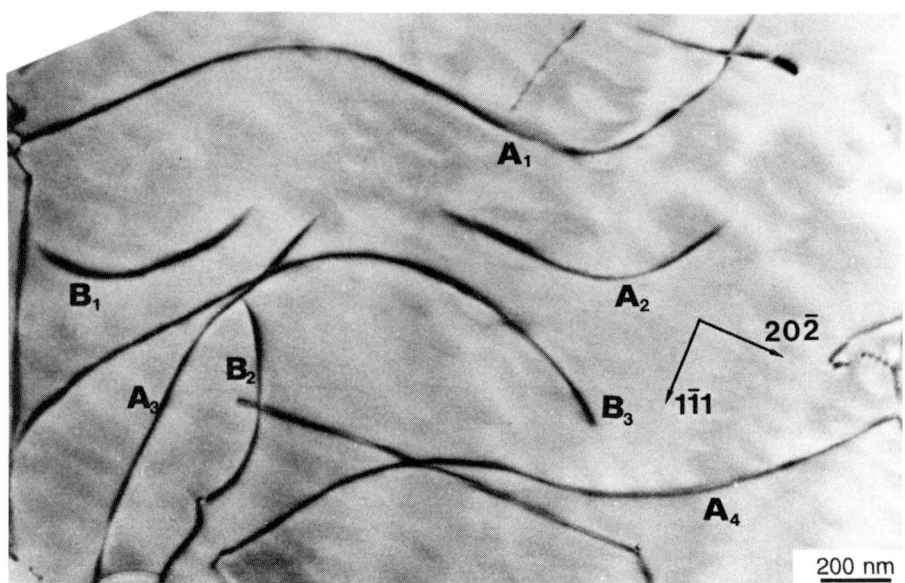

<u>Figure 1</u>: *Curved dislocations with $\{111\}$ glide plane; bright field image in the [121] pole; the marked dislocations are entirely glissile on the (111) plane; Burgers vectors: A: [011]/2, B: [110]/2*

Weak beam analysis revealed a dissociation into partial dislocations for several of the dislocations which were glissile in $\{111\}$ (fig. 2). The partial separation was found to be about 5 nm. However, an error limit of about 2 nm has to be assumed for our measurements.

About one third of all analyzed dislocations did not belong to the $<110>/\{111\}$ slip systems. A few of these dislocations were of edge type with $<100>$ line directions but most of them had a $\{100\}$ glide plane. This group consisted of Lomer dislocations (edge type dislocations with $<110>$ line direction) and of dislocations with the line direction near to $<100>$ (called 45°-dislocations further on). Some of the 45°-dislocations were curved (fig. 3). This is believed to be indicative of glide on the $\{100\}$ planes.

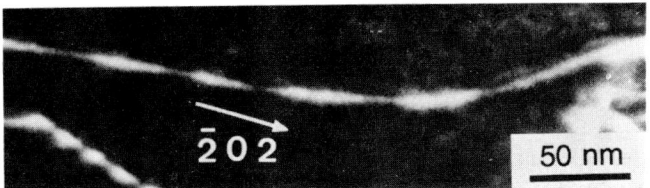

<u>Figure 2:</u> *Weak beam image of a dissociated dislocation on the {111} glide plane*

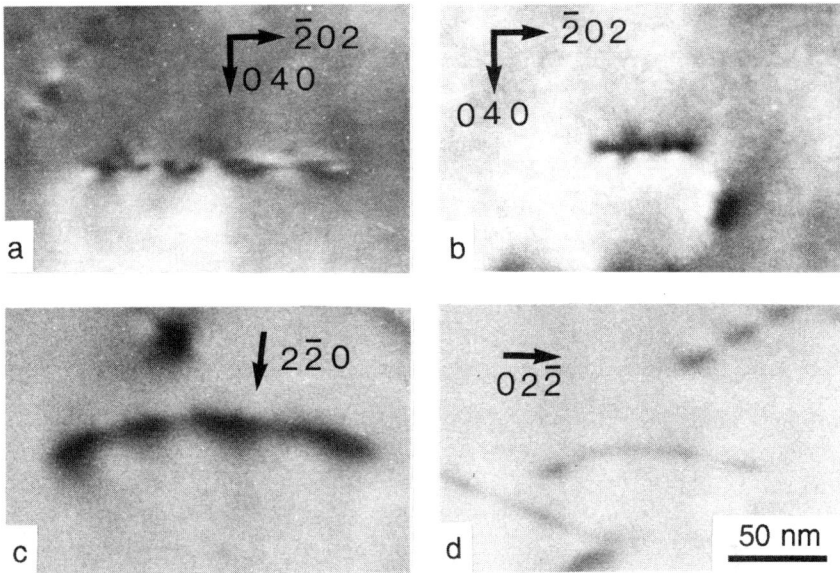

<u>Figure 3:</u> *Curved dislocations with {100} glide plane; a,b: Bright field images in the [101] pole; the dislocations, lying in the (010) plane (Burgers vector ±[10$\bar{1}$]/2), appear straight; c,d: Bright field images of the dislocations (a) and (b) respectively near the [111] zone axis*

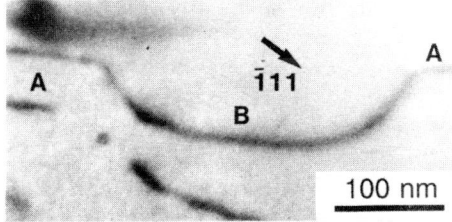

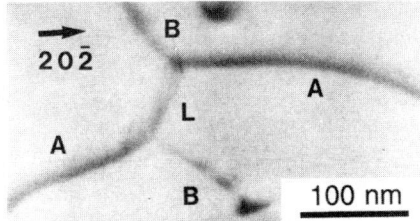

<u>Figure 4:</u> *Cross-slip from the [10$\bar{1}$]/(010) onto the [10$\bar{1}$]/(1$\bar{1}$1) slip system; bright field image near [101]; A: Dislocation segments glissile in the (010) plane; line direction near to the screw orientation; B: Bulge formed by cross slip in the (1$\bar{1}$1) plane*

<u>Figure 5:</u> *Dislocation reaction forming a Lomer lock; A: Glide plane (111), Burgers vector [10$\bar{1}$]/2; B: Glide plane (11$\bar{1}$), Burgers vector [1$\bar{1}$0]/2; L: Lomer dislocation, glide plane (100), Burgers vector [01$\bar{1}$]/2*

Some dislocation configurations proved to have been formed by cross-slip. Cross-slip was observed both between two < 110 >/{111} slip systems and from a < 110 >/{100} slip system onto a < 110 >/{111} slip system. An example for the latter is given in fig. 4, where a bulge in the (111) plane has developed from a screw segment of a dislocation which is glissile in the (010) plane.

Because of the high dislocation density the dislocations form networks, and dislocation reactions were observed frequently. Dislocation reactions within one {111} plane and several reactions forming a Lomer lock (fig. 5) could be analyzed uniquely. However, more complicated reactions for which the contrast analysis techniques failed to give unique results, because of the high dislocation density, were observed as well.

3.2 Areas Filled with Microtwins

Areas filled with microtwins consist of intersecting lamellae of first order twins parallel to all four {111} planes of the matrix. This can be seen e.g. from fig. 6: Twin lamellae parallel to the [011] zone axis of the matrix are visible both in (a) and (c). Furthermore, the fringes (I) in (c) indicate twin lamellae inclined to the [011] zone axis. Diffraction patterns (e.g. fig. 6b) show strong reflections of the matrix, weak spots corresponding to first order twins and double diffraction, and streaking because of stacking faults and thin twin lamellae (Williams and Glass 1989). There are no other spots visible which would indicate crystalline phases other than diamond or orientations other than first order twins.

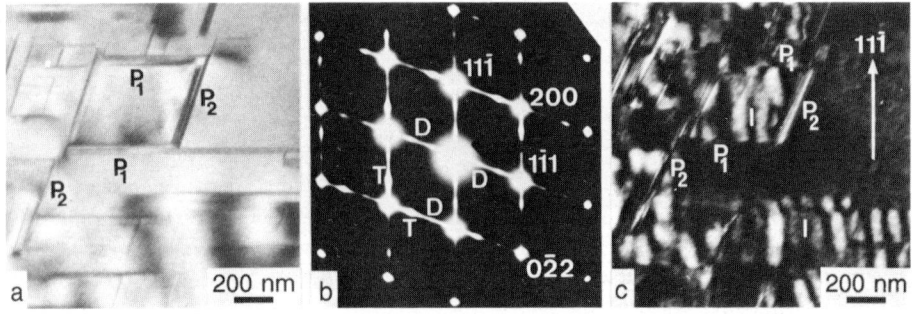

Figure 6: *Diffraction contrast analysis of the areas filled with microtwins*
a: Bright field image near the [011] zone axis; b: Diffraction pattern corresponding to (a);
c: Weak beam image near [011] with the (111) matrix reflection;
P_1 and P_2: Twin lamellae parallel to [011]; I: Twin lamella inclined to [011]; T: Twin reflections; D: Double diffraction spots; matrix reflections are indexed

Some twin intersections in the areas filled with microtwins were studied with HREM. Figure 7 shows the intersection of a 6 layers thick lamella (T_1) parallel to (111) with a 3 layers thick lamella (T_2) parallel to (111). Both twin lamellae end at the intersection. Although the atomic details of the intersection cannot be recognized from fig. 7, it seems that neither of the twins terminates the other. Parts of both twins end at a boundary to the matrix rather than to each other. Evaluation of the Burgers circuit around the intersection gives no resulting closure failure. Therefore, no resulting dislocation (other than a projecting screw dislocation which cannot be detected by the method) is present within the intersection. This is typical for most of the analyzed twin intersections. Resulting dislocations were only found at intersections where twins of a thickness other than 3n atomic layers met. However, even for intersections of rather thick twins the resulting Burgers vectors corresponded to single Shockley partial dislocations or stair-rod dislocations.

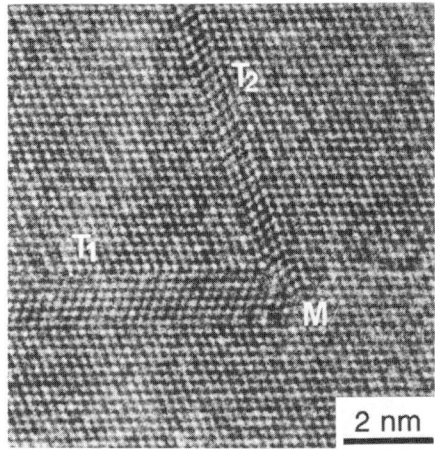

Figure 7: *HREM image of intersection of twins; [011] zone axis, pairs of atoms are black; T_1: 6 layers thick lamella parallel to (111); T_2: 3 layers thick lamella parallel to (111)*

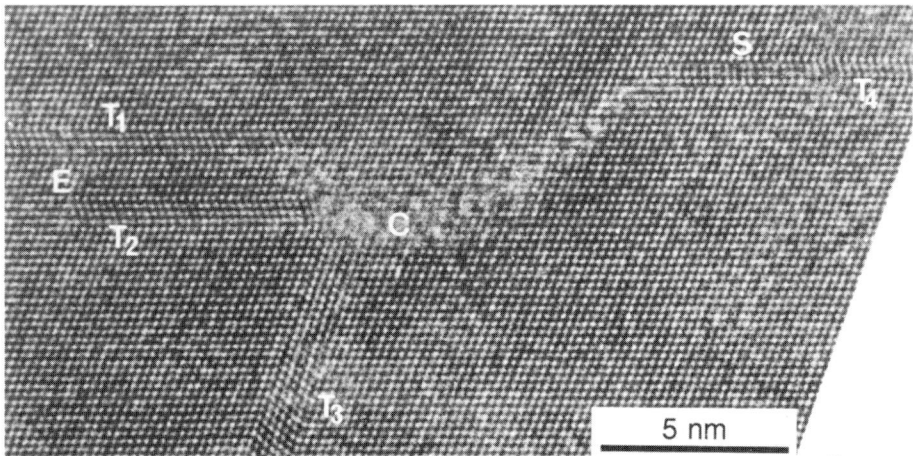

Figure 8: *HREM image of twins ending at a virtually amorphous inclusion; [1$\bar{1}$0] zone axis, pairs of atoms are black; T_n: Twin lamellae; C: inclusion; E: Abrupt ending of twin T_2; S: Lattice image confusion*

Twin lamellae were found to end not only at other twins but also abruptly in the matrix or at virtually amorphous inclusions. This can be seen from fig. 8 in which four twins (T_1-T_4) end in a curved inclusion (C) which seems to be amorphous. Twin T_2 extends only over a few nm and ends abruptly in the matrix (E). The 3-fold periodicity of twin T_3 near to the inclusion indicates vertical ending of the lamella within the specimen foil and overlapping with the matrix. The lattice image of the extrinsic stacking fault T_4 is confused near to the inclusion. This might be caused by a step in the stacking fault plane lying inclined to the specimen surface. Evaluation of the Burgers circuit around the inclusion gives a closure failure of [1 1 2]/12 which is the edge component of a 30° Shockley partial dislocation.

In contrast to expectations from the diffraction patterns, multiple twinning could be observed in the HREM images at a few twin intersections. However, the volume fraction of secondary or higher order twins was very small so that the absence of corresponding spots in the diffraction patterns is reasonable. An example of five fold twinning within the areas filled with microtwins is given in fig. 9. Matrix orientation (M), first (T_1) and second (T_2) order twins are marked. Slight misorientation and a rather thick specimen at this point are responsible for the contrast differences between the twins. The curvature of the $\Sigma 9$ boundary marked C might indicate an overgrowth of the spherical particle by the matrix.

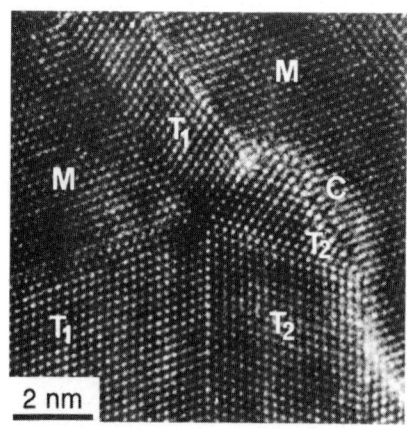

Figure 9: *5-fold twinning at an intersection of twin lamellae; M: Matrix orientation; T_1: First order twins; T_2: second order twins; C: Curved $\Sigma 9$ boundary*

4. DISCUSSION

The observed dislocation configurations show that dislocations in diamond are glissile both on the {100} planes and on the {111} planes even at a temperature of 700 °C or less. The number of observed dislocations in the two slip systems indicates that the predominant slip planes are {111}. The curvature of glissile dislocation segments is known to be a consequence of internal stress and can, therefore, be used for the stress determination. The equilibrium stress acting on the dislocations shown in fig. 1 was estimated using anisotropic elasticity theory (Scattergood and Bacon 1975). These calculations gave a stress of the order of 0.1 GPa which is considerably less than values (some GPa) reported from X-ray diffraction data (Baglio *et al.* 1992). This might be due to the interaction with the other dislocations (Alexander 1986) which also might be responsible for the rather inhomogeneous stress which is indicated by the change of curvature.

Pirouz *et al.* (1983) reported separations of 2.5 up to 4.3 nm for dissociated <110>/{111} dislocations in natural diamond. Our measurements agree with this within the error limit and, therefore, we conclude that the investigated samples have a stacking fault energy similar to that of natural diamond.

There are very few observations of glissile dislocations on {100} planes in the diamond cubic lattice (Korner *et al.* 1987, Rajan 1991). Karnthaler (1978) showed that <110>/{100} slip starting from constricted nodes of Lomer dislocations is possible and can be observed in fcc metals deformed at room temperature. The observation of Lomer dislocations in our specimens suggests that this mechanism might also be working in diamond. However, so far we do not know the exact source of the dislocations with {100} glide plane.

We did not observe any dislocation configuration which was obviously a result of dislocation climb. All analyzed configurations could be explained by glide and cross slip. Therefore, we assume that dislocation climb does not play an important role in the investigated samples. This is in marked contrast to observations by Suzuki *et al.* (1992). However, Suzuki investigated samples grown with a d.c. plasma jet, which might not be comparable to the samples investigated in this study.

In comparison to the crystal parts where dislocations have been analyzed the areas filled with microtwins contain a very high concentration of crystal imperfections. First order twinning is obvious and additionally there are crystalline and virtually amorphous inclusions observed by HREM. However, the matrix orientation is unique and has been maintained during crystal growth although parts of the matrix seem to be completely enclosed in boxes formed by microtwin lamellae.

It has been stated recently (Shechtman *et al.* 1993) that non-equilibrium configurations formed during the growth of the diamond layers can not transform to their equilibrium form because of the low growth temperature of the CVD process. According to this we assume that the inclusions which show in most cases curved, nonfacetted boundaries have been incorporated during the growth of the layers. Overgrowth of such particles by the matrix is likely to cause the development of strain or the formation of growth twins because of lattice mismatch.

Twinning can occur not only during the growth but also subsequently by glide of Shockley partial dislocations (e.g. Narayan 1990). A mechanism for the twin formation in Si because of elastic strain has been given by Twigg *et al.* (1990) and it is well known that large strain develops during the growth of CVD diamond layers. The observed inclusions could both cause elastic strain and operate as dislocation sources. Moreover, twin formation via gliding dislocations would contain the maintenance of the matrix orientation implicitly. However, neither our results nor the accounts of other groups (Narayan 1990, Luyten *et al.* 1992) indicate the presence of a large number of partial dislocations at the twin lamellae or within twin intersections. Therefore, as we do not have an explanation for the removal of the twinning dislocations, we have to conclude that the twin formation does not occur via the gliding of partial dislocations.

A model of the enhancement of crystal growth by re-entrant grooves formed at the intersections of first order twin lamellae with the surface was presented recently by Angus *et al.* (1992) for growth conditions leading to {111} facetting. According to this model a three atom nucleation step is needed to start growth of a new layer on a perfect {111} surface while only two atoms are needed at the re-entrant groove. These nucleation steps imply the maintenance of the matrix-twin configuration at the re-entrant groove while the layer nucleated at the perfect {111} surface can be oriented either as the matrix or as a first order twin but not differently. When the growth conditions are not perfect secondary nucleation or deposition of small particles of amorphous carbon may occur. However, the growth conditions for these deposits are not as favourable as for the matrix-first order twin system. Thus, the particles should be soon overgrown in the course of which process new nuclei of first order twin lamellae can be formed. The growth driven by the deposition of new material at re-entrant grooves can be continued at these twins. The matrix orientation would be maintained during all these processes and, therefore, the process considered here would explain the observed configurations. However, further investigations are needed to prove whether this is valid or not.

5. CONCLUSION

In conclusion, we have found that dislocations can glide in CVD diamond layers at 700 °C or less both on {111} and {100} planes. Lomer dislocations were detected as a possible source of dislocations with {100} glide plane. The curvature of dislocations with a {111} glide plane was used to determine the order of magnitude (0.1 GPa) of the stress acting on them. A change of curvature showed inhomogeneities of this internal stress. Dissociation of the dislocations on {111} planes was observed and the separation indicated the same value for the stacking fault energy as in natural diamond.

Intersecting primary twin lamellae parallel to all four {111} planes of the matrix and a small volume fraction of differently oriented crystallites and virtually amorphous inclusions were detected in the areas filled with microtwins. There were only very few dislocations observed at the twin lamellae and within the twin intersections. These observations were discussed in the light of a recently presented growth model.

ACKNOWLEDGEMENTS

The authors want to thank Prof. B. Lux and Dr. R. Haubner for the growth of the investigated samples. We are indebted as well to the SIEMENS AG in Munich and especially to Dr. H. Cerva for giving the opportunity to use their JEOL 4000 EX TEM. The authors gratefully acknowledge financial support of the present work by the Austrian Science Foundation (FWF) (Project No. S5903-TEC) carried out under the auspices of the trinational "D-A-CH" cooperation of Germany, Austria and Switzerland on the "Synthesis of Superhard Materials".

REFERENCES

Alexander H 1986 in *Dislocations in Solids*, Ed. Nabarro F R N, 7 113
Angus J C, Sunkara M, Sahaida S R and Glass J T 1992 J. Mat. Res. 7 3001
Baglio J A, Farnsworth B C, Hankin S, Hamill G and O'Neil D 1992 Thin Solid Films 212 180
Kaae J L, Gantzel P K, Chin J and West W P 1990 J. Mat. Res. 5 1480
Karnthaler H P 1978 Phil. Mag. 38 141
Korner A, Martinez-Hernandez M, George A and Kirchner H O K 1987 Phil. Mag. Lett. 55 105
Luyten W, Van Tendeloo G, Amelinckx S and Collins J L 1992 Phil. Mag. A 65
Narayan J 1990 J. Mater. Res. 5 2414
Pirouz P, Cockayne D J H, Sumida N, Hirsch P and Lang A R 1983 Proc. Roy. Soc. London A386 241
Rajan K 1991 Appl. Phys. Lett. 59 2564
Scattergood R O and Bacon D J 1975 Phil. Mag. 31 179
Shechtman D, Feldman A, Vaudin M D and Hutchison J L 1993 Appl. Phys. Lett. 62 487
Suzuki K, Ichihara M, Takeuchi S, Ohtake N, Yoshikawa M, Hirabayashi K and Kurihara N 1992 Phil Mag. 65 657
Twigg M E, Richmond E D and Pellegrino J G 1990 J. Appl. Phys. 67 3706
Williams B E and Glass J T 1989 J. Mater. Res. 4 373
Zhu W, Badzian A R and Messier R 1989 J. Mater. Res. 4 659

Inst. Phys. Conf. Ser. No 134: Section 4
Paper presented at Microsc. Semicond. Mater. Conf., Oxford, 5–8 April 1993

165

In situ TEM studies of the crystallization of amorphous silicon: the role of silicides

J L Batstone and C Hayzelden*

IBM Corporation, T J Watson Research Center, P O Box 704, Yorktown Heights, NY 10598 USA
*Division of Applied Sciences, Harvard University, Cambridge, MA 02138 USA

ABSTRACT: *In situ* transmission electron microscopy is a powerful tool for the dynamic study of phase transformations. The amorphous to crystalline phase transformation has been studied for amorphous silicon and Ni-implanted amorphous silicon. The mechanisms of interfacial propagation are discussed.

1. INTRODUCTION

The amorphous to crystalline phase transformation in silicon is of interest to the semiconductor device community. As device dimensions shrink in size, formation of shallow source and drain regions in metal oxide field effect transistors often employs a pre-amorphization stage followed by dopant implantation, diffusion and annealing. The subsequent silicon regrowth must result in high quality, defect-free single crystal silicon. In addition, heavily doped polycrystalline silicon (poly-Si) is commonly employed as a gate electrode. Transistor performance can be dominated by transport effects across grain boundaries and large grained poly-Si is generally required.

In this paper, we will describe recent *in situ* transmission electron microscopy (TEM) experiments designed to aid our understanding of poly-Si grain growth and morphology. Dynamic annealing experiments within the TEM offer valuable insight into the nucleation and growth processes. In addition, the atomistics of interfacial motion can be studied at the advancing amorphous-crystal (a-c) interface. Two different experiments are described. The first experiment illustrates the growth of poly-Si from a thin amorphous silicon (a-Si) film formed by electron beam deposition on an amorphous Si_3N_4 substrate. Crystallisation of a-Si typically occurs at temperatures of 650-750°C. This work has recently been published in Batstone (1993) and here we will describe the propagation of the a-c interface. The second set of experiments were designed to investigate the effects of silicides on the a-Si crystallisation process. Metal-induced crystallisation has been reported for a number of different metal-Si systems, and these have been reviewed recently by Spaepen et al (1992). We have previously reported that implantation of Ni into a-Si, followed by thermal annealing at 350-400°C, results in the formation of octahedral $NiSi_2$ precipitates buried in a layer of a-Si. Further annealing at 450-550°C leads to low-temperature crystallisation of the a-Si with the $NiSi_2$ precipitates acting as sites for Si nucleation. The crystallisation of a-Si proceeds via the migration of $NiSi_2$ precipitates until the entire film is transformed (Hayzelden et al 1992, Hayzelden and Batstone 1993). *In situ* TEM experiments allowed us to observe the initial stages of Si nucleation and the $NiSi_2$ precipitate migration parallel to <111> directions. This is usually the slowest growth direction during solid phase epitaxy (SPE) (Cseprегi et al 1976). However, for the silicide-mediated growth, the (111) growth rate was at least 30 times faster than for metal-free Si SPE. The crystallisation process was found to be diffusion-limited and effective diffusivities for transport across the silicide layer were obtained.

Data from recent *in situ* high resolution TEM experiments will be presented which show the atomistic nature of the interfacial propagation.

2. EXPERIMENTAL

Two sets of samples were prepared. The first set consisted of thin 40nm films of a-Si which were prepared by room temperature electron beam deposition of elemental Si onto a-Si$_3$N$_4$ coated Si substrates in which "windows" of Si$_3$N$_4$ were exposed by anisotropic etching of Si. The effect of silicides on the a-c phase transformation was studied by implanting 55keV Ni ions at doses of 5x10^{14}, 1x10^{15} and 5x10^{15}ions/cm^2 into a 95nm thick low pressure chemical vapour deposited a-Si film on an a-SiO$_2$/Si substrate. An approximately Gaussian Ni profile was formed with a Ni concentration peak located at a depth of ~50nm.   Samples were prepared for the TEM by selective removal of the Si substrate in a dilute HF:HNO$_3$ etch. For *in situ* high resolution TEM, additional ion-milling was required to remove the amorphous SiO$_2$ substrate. Further details of the thin film preparation procedures can be found in Hayzelden et al (1992) and Batstone (1993).

All samples were examined in a plan view orientation. *In situ* annealing experiments were performed   at 300keV and 150keV in a Philips EM430 high resolution electron microscope equipped with a Gatan single tilt water-cooled heating stage which was sufficiently stable to permit high resolution images to be obtained at temperatures in excess of 700°C. The sample temperature was estimated to be accurate to within ±25°C of the thermocouple reading (Batstone 1993). The samples were cooled to room temperature before the amorphous to crystalline phase transformations were complete and transferred to a JEOL 4000EX for additional HREM analysis at 400keV to study the a-c interfaces.

3. RESULTS

3.1 Amorphous Silicon

The a-Si films crystallised at ~700°C with the initial appearance of small crystallites, elongated along <112> directions with extremely dendritic a-c interfaces.   An example of several crystallites after annealing at 700°C for 10 mins is shown in Figure 1a.   The crystals were commonly found to be <110> oriented although large flake-like <111> oriented crystals were also observed.   The interface morphology is seen more clearly in Figure 1b.   As growth progressed, large crystals were found which had impinged upon neighbouring crystallites, with extremely rough, irregular grain boundaries separating them.   Device quality poly-Si is generally formed via LPCVD deposition of Si from silane followed by thermal annealing to promote secondary grain growth.   Twin bands are commonly observed.   Anderson (1973) has described poly-Si morphologies under varying deposition and annealing conditions.   For these electron beam deposited films, all crystals were heavily twinned with long parallel twin bands often visible which ran parallel to the long axis of the dendritic arms.   The twin bands can be seen clearly in Figure 2a.   Figure 2b shows the selected area diffraction pattern from these grains.   Twinning on the {111} planes gives rise to additional twin-related diffraction spots as well as extensive double diffraction.   The presence of twin bands and the tendency for growth to be fastest along <112> directions is intriguing.   The reasons for this become clear as the a-c growth front is examined in the high resolution TEM.   Figure 3 shows a <110> projection of the a-c interface at the leading edge of a crystallite which is growing in a {112} direction.   The parallel twin bands protrude at the a-c interface providing possibly favourable sites for atom addition.   The twins are formed during the initial stages of nucleation and growth and once present, propagate through the crystals. *In situ* high resolution TEM (Batstone 1993) showed that the fastest growth occurred where multiple twin bands emerged from the growth front.   Measurements of growth velocities at different temperatures allowed an activation energy of 3.36eV to be determined for growth.

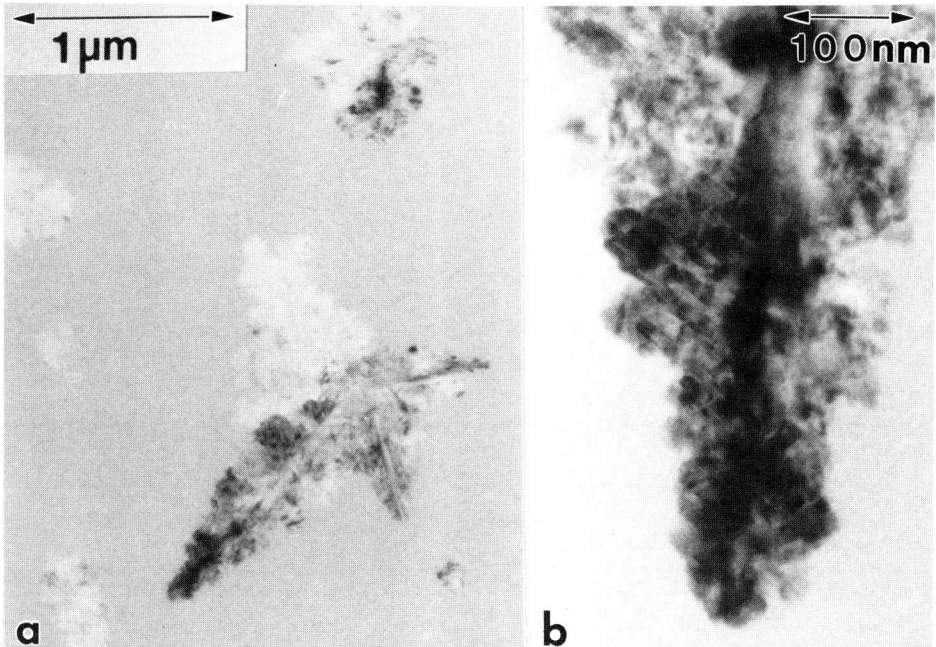

Figure 1. (a) Bright field image of Si crystallites forming in an a-Si matrix during *in situ* annealing at 700°C and (b) a higher magnification view showing the dendritic a-c interface.

Figure 2. (a) Bright field image of several impinged crystals with twin bands clearly visible and (b) selected area diffraction pattern showing twinning and double diffraction spots from the highly defective crystal.

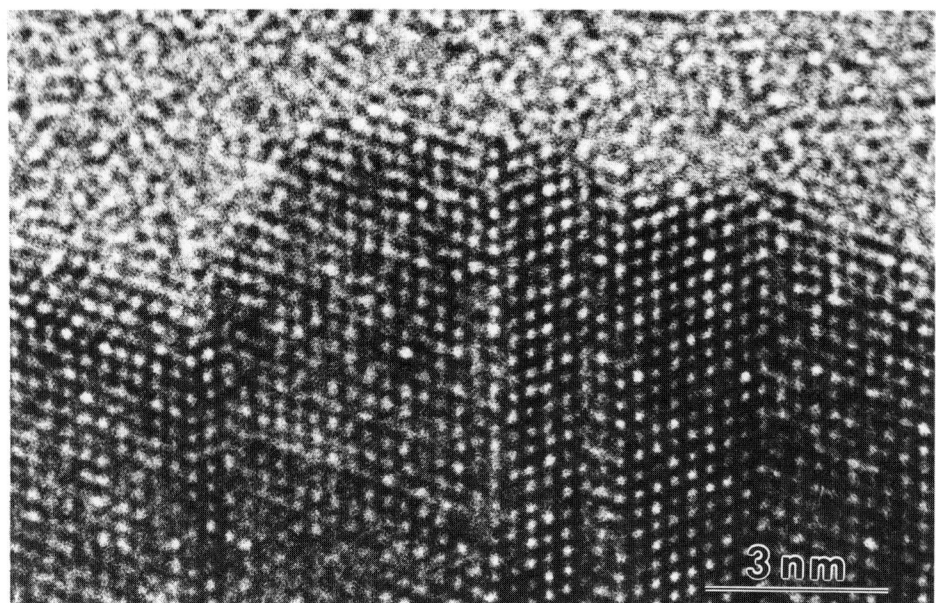

Figure 3. High resolution TEM image in a <110> projection of the a-c Si interface with twin bands emerging at the growth front.

Figure 4. (a) Bright field image of a partially recrystallized 1×10^{15}Ni ions/cm^2 implanted a-Si film and (b) high magnification image of a Si needle with a NiSi$_2$ precipitate during *in situ* annealing at 457°C.

3.2 Ni-implanted Amorphous Silicon

Annealing of Ni-implanted a-Si results in the formation of $NiSi_2$ precipitates followed by a-Si crystallisation via the migration of $NiSi_2$ precipitates which trail *needles* of epitaxial crystalline Si (c-Si) behind them. Details of this process can be found in Hayzelden and Batstone (1993). The migration of silicide precipitates was found to be diffusion-controlled and a dissociative diffusion mechanism was proposed. Figure 4a shows a bright field low magnification view of a partially crystallised film where needles of c-Si are all terminated with a narrow $NiSi_2$ precipitate. Figure 4b shows a high magnification view of one needle imaged in a <110> projection during *in situ* annealing at 457°C. The growth direction for all needles was parallel to <111>. The *in situ* experiments revealed that growth progressed via a ledge mechanism, with steps clearly visible at both the a-Si/$NiSi_2$ interface and the $NiSi_2$/c-Si interface. The steps ran along the interfaces, allowing the $NiSi_2$ precipitate to maintain an approximately constant thickness of ~5nm. Figure 5 shows examples of the presence of steps at the two interfaces during these dynamic experiments. The images shown in Figure 5 are ~20s apart and show the passage of steps or ledges. The $NiSi_2$/c-Si interface is epitaxial with a Type A silicide orientation and no evidence for misfit dislocations, (misfit dislocations would not be expected for a 5nm $NiSi_2$ thickness and a misfit of 0.4%). Figure 6 shows examples of steps at the Si/$NiSi_2$ interfaces. The steps were often found to vary in height with steps of d(111) and 3d(111) most commonly observed.

4. DISCUSSION

4.1. Mechanisms for growth

A mechanism for the motion of the a-c Si interface was proposed by Spaepen (1978) which involved the nucleation and propagation of ledges across (111) terraces along <110> directions. Growth proceeded via atom attachment which required the completion of six-membered rings in diamond cubic silicon. Refinements to this idea have been proposed and a number of them are reviewed by Williams (1983). Growth on Si(111) is slow because ledge nucleation must occur in order for the ledge to move across the terrace. Since the Si(100) surface always has Si(111) ledges protruding from it, growth on Si(100) is fast. Drosd and Washburn (1982) showed that on Si {111}, an atom cluster of 3 atoms is needed for crystallisation to proceed and this is illustrated schematically in Figure 7a. The atom cluster can attach in either the 'perfect' orientation to generate single crystal Si or in a rotated 'twin' orientation. Since the twinned orientation does not involve any nearest-neighbour stacking errors or bond distortions, formation of twins at an advancing Si{111} interface is statistically likely. Once twins have formed within a small Si crystallite, they provide favourable sites for atom attachment where the twin emerges at a growth front, as illustrated in Figure 7b. At the exit point of the twin plane, attachment of a 2 atom cluster results in a step which can then traverse the (111) plane for rapid growth.

Silicide-mediated crystallisation is remarkable since it occurs at extremely low temperatures and involves rapid growth of high quality twin-free Si in <111> directions. The *in situ* data shown in Figure 5 for growth at 457°C showed an increase in growth rate of three orders of magnitude compared with metal-free Si SPE. The high resolution images showed that growth occurred via a ledge mechanism with steps propagating along the (111) interfaces. Steps were observed at both the a-Si/$NiSi_2$ interface and the $NiSi_2$/c-Si interface. However, the rate-limiting step for precipitate migration was found to be diffusion-limited and not interface-controlled. A possible explanation for this is illustrated in Figure 7c which shows a ball and stick diagram of a $NiSi_2$/Si(111) interface containing a step of height 3d(111). In order for the steps to move to the right in this figure, bonds must be broken. However at the step itself, there are Si atoms with dangling bonds. In a dissociative-diffusion process, the diffusion occurs in a coordinated manner involving both Si and

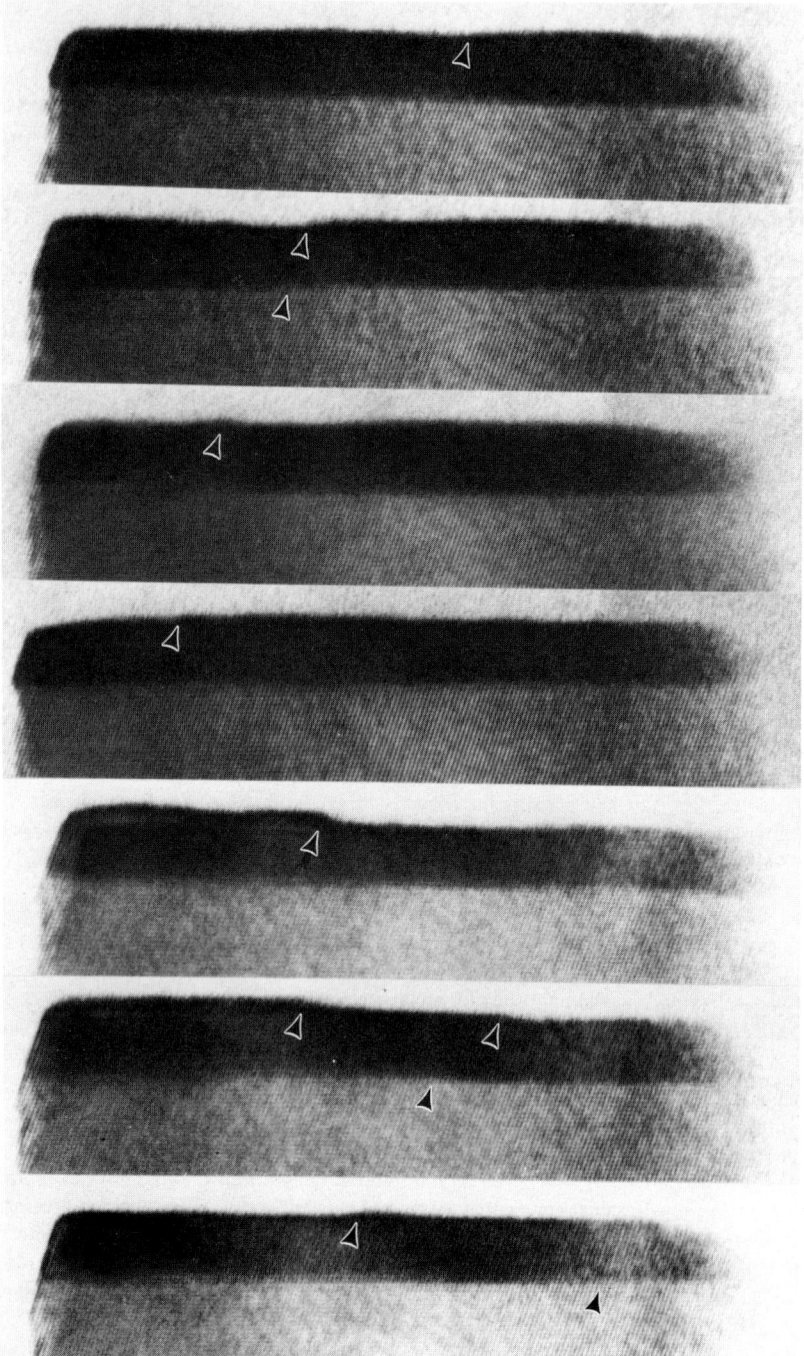

Figure 5. Sequential images of the NiSi₂ precipitate at the leading edge of a Si needle. The images are ~20s apart and steps can be seen moving across both interfaces.

Ni atom species. Diffusion of Ni towards the a-Si (i.e. in the growth direction) can occur by breaking the bonds around the Ni atoms at the step interface thus enabling Si atoms to hop into the vacated Ni lattice sites. This allows the step to advance and the $NiSi_2$ precipitate to migrate through the a-Si trailing behind high quality epitaxial c-Si.

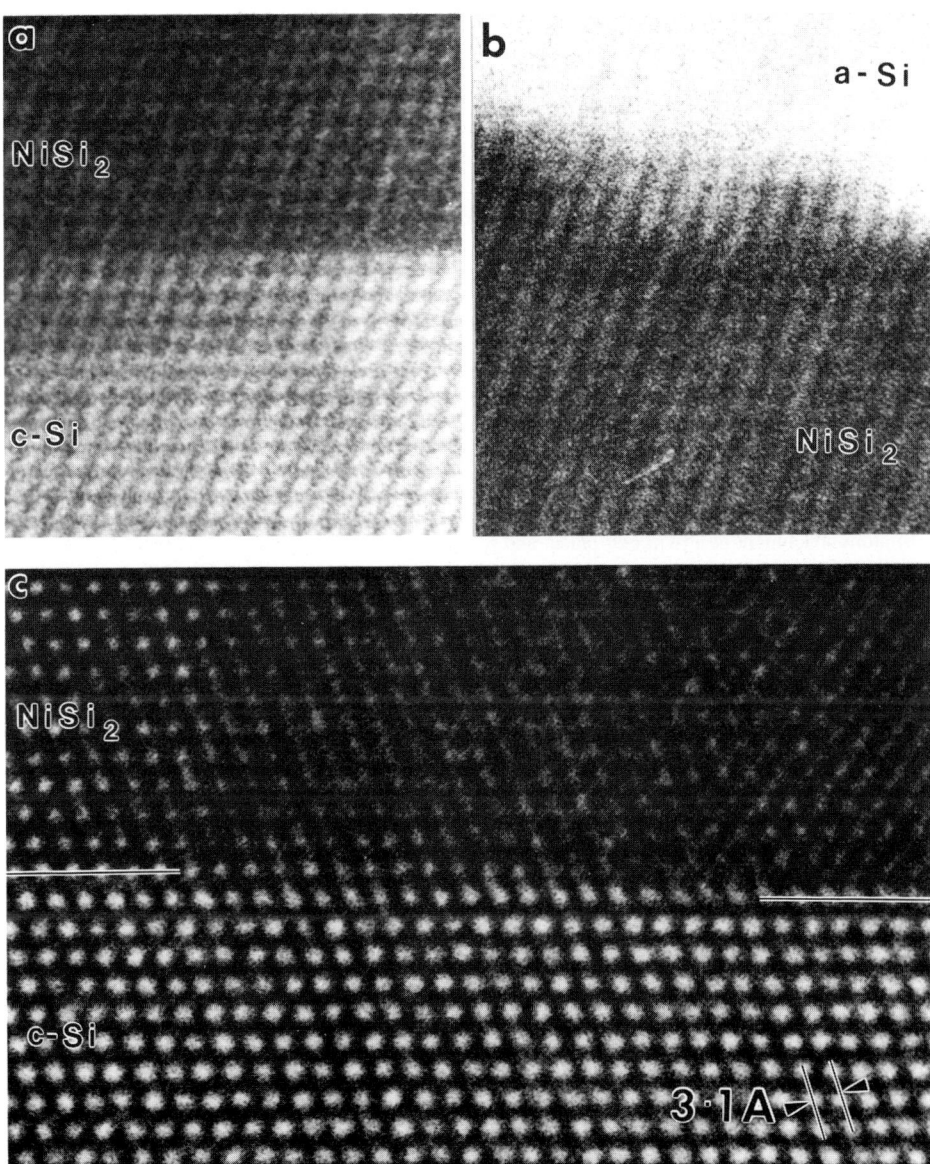

Figure 6. High resolution images of steps at interfaces during *in situ* annealing. (a) 3d(111) step at $NiSi_2$/c-Si interface, (b) 3d(111) step at a-Si/$NiSi_2$ interface and (c) d(111) step at $NiSi_2$/c-Si interface after sample was cooled to room temperature.

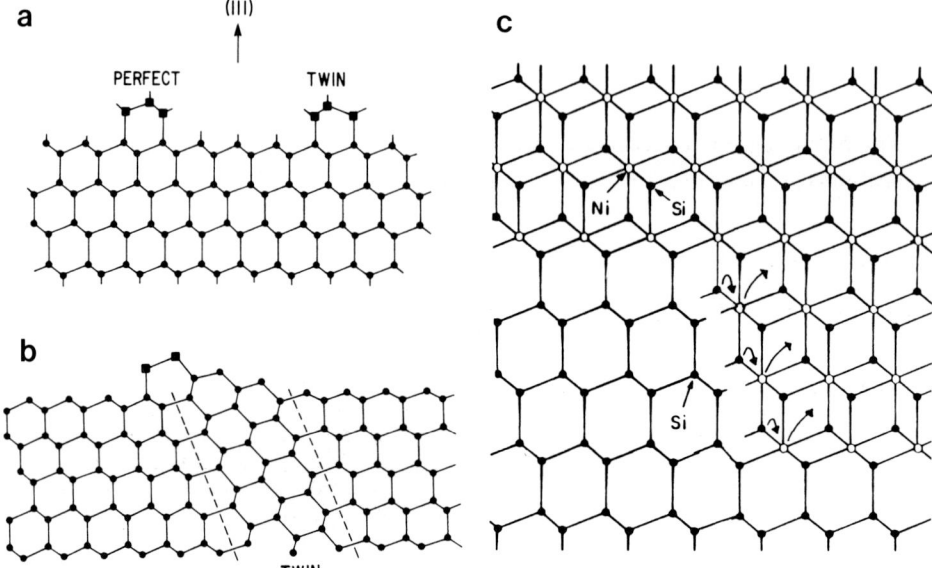

Figure 7. Ball and stick diagrams showing (a) atom attachment in a perfect or twin orientation, (b) atom attachment at a twin exit plane, and (c) the NiSi$_2$/Si(111) interface containing a 3d(111) step. Step motion can occur in conjunction with dissociative diffusion of Ni and Si atoms.

5. CONCLUSIONS

In conclusion, we have demonstrated the power of *in situ* TEM annealing experiments to study the atomistics of a-c interfacial propagation in Si. For both a-Si and Ni-implanted Si, atom attachment at an advancing interface appears to proceed via a ledge mechanism, either via step nucleation at emerging twin boundaries or at an epitaxial interface. Addition of Ni resulted in greatly increased growth rates for Si crystallisation.

REFERENCES

Anderson R 1973 J Electrochemical Society 120 1540
Batstone J L 1993 Philosophical Magazine A 67 51
Csepregi L, Mayer J W and Sigmon T W 1976 Applied Physics Letters 29 92
Drosd R and Washburn J 1982 J Applied Physics 53 397
Hayzelden C, Batstone J L and Cammarata R C 1992 Applied Physics Letters 60 225
Hayzelden C and Batstone J L 1993 J Applied Physics 73 8279
Spaepen F 1978 Acta Metallurgica 26 1167
Spaepen F, Nygren E and Wagner A V 1992 NATO Advanced Study Institute Series E : Applied Sciences 222 483 *Crucial Issues in Semiconductor Materials and Processing Technologies,* (eds Coffa A, Priolo F, Rimini E and Poate J M) Kluwer Academic, Dortrecht
Williams J S 1983 *Surface Modification and Alloying,* (eds Poate J M, Foti G, and Jacobson, D C) Plenum NY (and references therein).

Inst. Phys. Conf. Ser. No 134: Section 4
Paper presented at Microsc. Semicond. Mater. Conf., Oxford, 5–8 April 1993

173

In situ high resolution electron microscopy of metal-contact induced crystallization of amorphous semiconductors

Toyohiko J Konno and Robert Sinclair

Department of Materials Science and Engineering, Stanford University, Stanford, CA 94305, U.S.A.

ABSTRACT : We studied crystallization of amorphous semiconductors induced by metals in Si-Al and Ge-Ag layered systems. In situ transmission electron microscopy revealed that both Al and Ag grains migrate into the amorphous phase leaving the crystalline semiconductor phase behind. From this observation, we propose a model whereby the growth of the crystalline semiconductor phase is mediated by the metal phase, which provides the fastest reaction path for the semiconductor elements.

1. INTRODUCTION

Amorphous semiconductors must have thermal stability in order to be reliably used in applications. The introduction of amorphous Si (a-Si) as a photoreceptor to replace the thermally less stable amorphous selenium, for instance, has led to improvements in electrophotography. An important issue in the thermal stability of amorphous semiconductors is the crystallization of these materials when they are in contact with metals with which they form a simple eutectic system (e.g., Herd et al. 1972) or silicides (e.g., Batstone 1993).

This paper describes crystallization behavior of Si and Ge in a-Si/Al and amorphous Ge (a-Ge)/Ag layered systems. We employed in situ transmission electron microscopy (TEM) in order to observe the reaction in real time. In particular, we show the use of high-resolution TEM to elucidate the interfacial reaction at the atomic level.

2. EXPERIMENTAL PROCEDURE

We made a-Si/Al, a-Ge/Ag and a-Si/Ag layered films by rf/dc magnetron sputtering in a Ti-gettered Ar atmosphere on the following substrates: (100) Si wafers and slide-glasses coated with photoresist. The former is used for cross-section TEM. The latter is made into free-standing films after the photoresist is dissolved in acetone for differential scanning calorimetry (DSC).

We made cross-section TEM specimens following a standard procedure (Bravman and Sinclair (1984)). The ion-milling of the TEM specimen was carried out at 3kVx0.4mA at liquid nitrogen temperature in order to avoid any reaction during the sample preparation. We used a Philips EM430ST microscope operated at 300kV with a Philips single tilt heating holder for the in situ annealing TEM study. Ordinary precautions regarding the in situ experiments are fully taken into account (Sinclair et al. 1988). We used a Perkin-Elmer DSC7 for the calorimetric study.

3. RESULTS

3.1 a-Si/Al layered system

Fig.1 is a bright field (BF) image of an as-deposited Si(60Å)/Al(60Å) multilayer. The strongly-diffracting layers are crystalline Al (c-Al) layers, while homogeneous weakly-diffracting layers are a-Si layers. Inset diffraction patterns are low-angle and normal

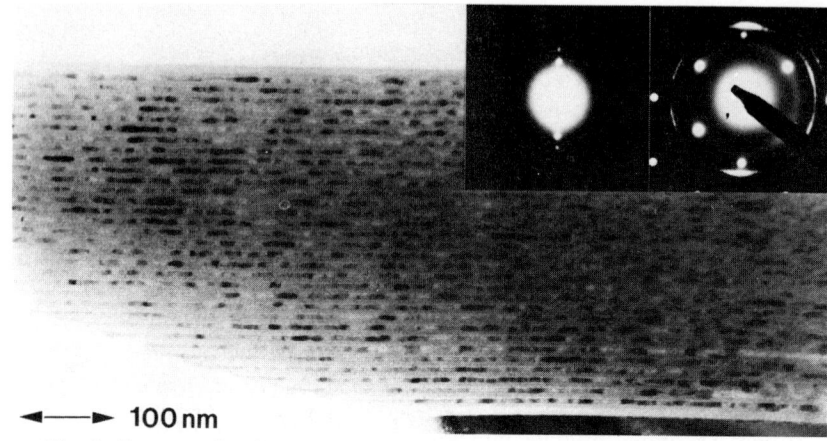

◄——► **100 nm**

Fig. 1 Cross-section TEM micrograph of as-deposited Al/Si multilayer. BF
image and normal (right) and low-angle (left) diffraction patterns

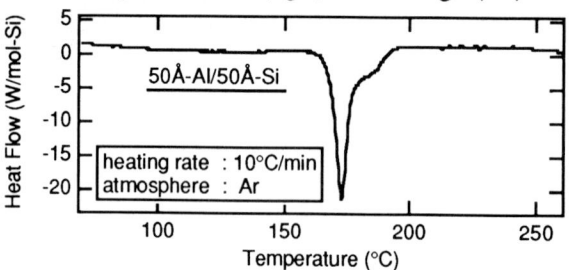

Fig. 2
DSC of Al/Si
multilayer film.

diffraction patterns. The former shows low-angle spots arising from the layered structure. The latter shows a halo at the Si (111) reciprocal distance, indicating that the Si is amorphous, while the diffraction ring at Al (111) indicates that c-Al grains possess a strong (111) texture. (Spots are due to the Si substrate.) Fig.2 is a result of the DSC scan taken for a Si(50Å)/Al(50Å) multilayer film. The peak temperature and the heat of the exothermic reaction are 175°C and 11±2kJ/mol-Si, respectively. These values can be compared with 600°C (peak temperature) and 11.6kJ/mol-Si for the crystallization of ion-implanted a-Si studied by Donovan et al. (1985).

Microscopy and diffraction show that the predominant reaction is the crystallization of the amorphous silicon. Fig.3 is a sequence of the unprocessed cross-section BF video images of a Si(200Å)/Al(30Å) multilayer film annealed at the nominal temperature of 200°C (time interval: about 1 min). As shown in the figure, the Si crystallizes and grows into the Al layer. On many occasions, it was observed that a thin Al grain was split into at least two parts, which were then moved aside as the c-Si penetrates the original Al layers. No amorphous phase intermixing nor (metastable) silicide formation was observed, by both in situ and ex situ TEM experiments, contrasting with the behavior of silicide forming metals (e.g., Holloway et al. (1989)).

3.2 a-Ge/Ag layered system

Fig.4 is a BF image of as-deposited a-Ge/Ag/a-Ge trilayer film. The inset diffraction pattern shows a halo due to the a-Ge and a strongly textured Ag (111) ring.

Fig.5 (a) is a BF image of a-Ge/Ag/a-Ge trilayer annealed in situ at 260°C. The left-hand side of the image shows that considerable mixing has occurred. As the picture shows, the Ag grain at the reaction front is "squeezed-out" from the original layer and migrates toward the amorphous Ge region.(This is most clearly seen, of course, in the in situ video-sequence.)

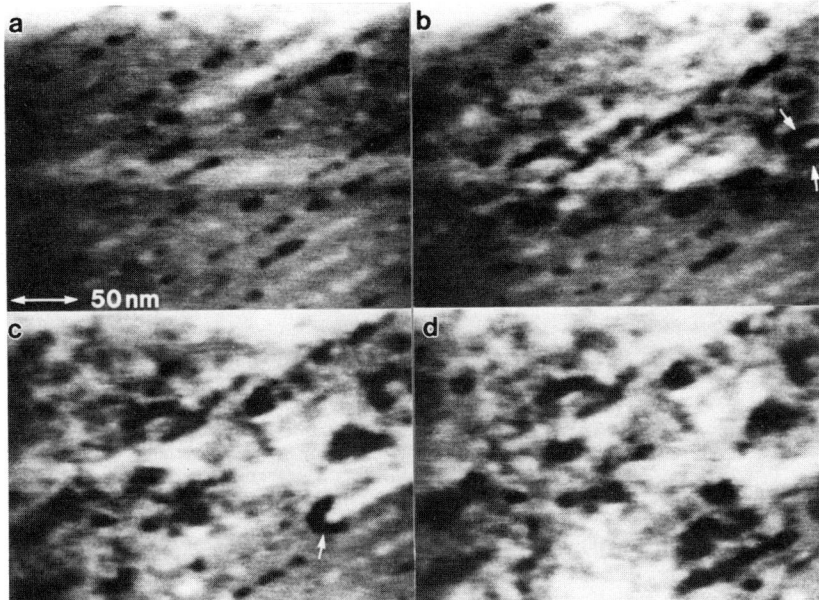

Fig. 3 A sequence of video images showing the crystallization of a-Si and intermixing of Al and Si grains (time interval, about 1min).

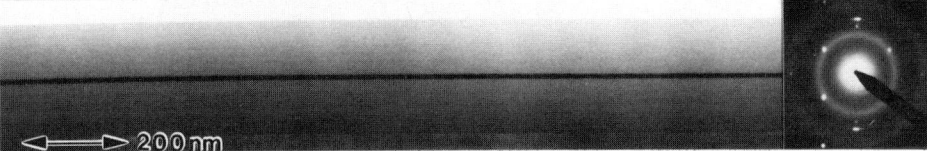

Fig. 4 Cross-section TEM micrograph of Ge/Ag/Ge trilayer.

Fig.5 (b) is a high-resolution image of the Ag grain at the reaction front. The Ag phase and c-Ge phase can clearly be identified by their {111} lattice spacings of 2.36Å and 3.27Å, respectively. Thus, the mixed region found in the BF image is due to the crystallization of a-Ge and the migration of Ag grains. Note that the "squeezed-out" Ag grain still possesses (111) planes parallel to those in the intact grains in the Ag layer on the right-hand side of the picture, indicating that orientation of the Ag grain does not change during its migration. In the in situ high resolution TEM annealing experiments, we observed in real time that the Ag grain migrates toward the a-Ge region, leaving the c-Ge phase behind. Especially in the Ge/Ag/Ge trilayer film, where the thickness of the a-Ge layer is nearly 100nm, the migration of the thin Ag layer (often only several nanometers thick) continues until the crystallization of the whole layer is complete. In order to explain this migration of the Ag grains and the creation of the c-Ge phase behind them, we can propose from the in situ observations that Ge diffuses from a-Ge to c-Ge through the Ag grains, with the driving force for the diffusion being a chemical potential difference between the a-Ge and c-Ge. We made essentially the same observation on a-Si/Ag layered system, in which the a-Si crystallizes at about 410°C.

4. SUMMARY AND DISCUSSION

We investigated metal-contact-induced crystallization of amorphous semiconductors in the a-Si/Al and a-Ge/Ag layered films. The in situ TEM studies revealed that the metal grains are ejected from the original layer as the crystalline semiconductor grains grow into the metal layer. This reaction was most strongly demonstrated in a Ge/Ag/Ge trilayer system, where the

Fig. 5 Cross-section TEM micrograph of Ge/Ag/Ge film annealed at 260°C. (a) BF image
and (b) high-resolution image of the reaction front shown in (a).

ejected grains migrated more than 100nm until the crystallization of all the a-Ge is complete.
Based on this observation, we propose that semiconductor atoms diffuse from amorphous to
crystalline phase through the metal phase because of the chemical potential difference between
the amorphous and crystalline semiconductor phases. The kinetic rate of the reaction is
consistent with solid-state diffusion through the metal (Konno and Sinclair (1992)). We also
suggest that this mechanism of crystallization provides the fastest reaction path for the system
to reduce its excess free energy.

This work is supported by the U.S. National Science Foundation (Grant No. DMR 8902232).

REFERENCES

Batstone J L and Hayzelden C 1993, Inst. Phys. Conf. Ser., these proceedings.

Bravman J C and Sinclair R 1984 J. Electron. Microsc. Technique $\underline{1}$ 53

Donovan E P, Spaepen F, Turnbull D, Poate J M and Jacobson D C 1985 J. Appl. Phys. $\underline{57}$
    1795

Herd S R, Chaudhari P and Brodsky M H 1972 J. Non-Crystalline Solids $\underline{7}$ 309

Holloway K, Sinclair R and Nathan M 1989 J. Vac. Sci. Technol. $\underline{A7}$ 1479

Konno T J and Sinclair R 1992 Phil. Mag. B $\underline{66}$ 749

Sinclair R, Yamashita T, Parker M A, Kim K B, Holloway K and Schwartzman A F 1988
    Acta Crystallogr. A $\underline{44}$ 965

Inst. Phys. Conf. Ser. No 134: Section 4
Paper presented at Microsc. Semicond. Mater. Conf., Oxford, 5–8 April 1993

177

Contrast arising from $\frac{1}{2}\langle 101 \rangle$ dislocations associated with steps in thin-film type-B $CoSi_2$/Si(111) interfaces

A C Daykin[1], C J Kiely and R C Pond

The Department of Materials Science and Engineering, The University of Liverpool, P.O.Box 147, Liverpool, L69 3BX, England.
[1] The Materials Research Laboratory, The University of Illinois at Urbana-Champaign, Urbana, Illinois, IL 61801, USA.

ABSTRACT : Ultra-thin films of B-type $CoSi_2$ grown on Si(111) substrates have been examined by transmission electron microscopy. In addition to the usual 1/6<112> misfit dislocations observed at this interface, we have found evidence for the existence of 1/2<101> defects which are associated with interfacial steps. These new defects have a 1/3[111] component of Burgers vector perpendicular to the interfacial plane which can lead to significant stress relaxation at the surface of ultra-thin films. Diffraction contrast experiments demonstrate that these defects have a short-range displacement field which resembles that of a disclination dipole.

1. INTRODUCTION

The accommodation of atomic scale steps at heterointerfaces has received relatively little attention in comparison with many other aspects of interface structure. The two basic ways that interface steps arise are (i) that they persist from steps already present on the substrate surface or (ii) they develop by some diffusion/reaction process occuring at the interface during fabrication. A study of the defect structure at the B-type $CoSi_2$/Si(111) interface has allowed us to gain some understanding of how steps arising from both these routes are accommodated.

It has been shown elsewhere (Daykin et al, 1992a,b) using the topological theory of interfacial defects, that the two smallest possible Burgers vectors for dislocations associated with interfacial steps in this system are 1/6<112> and 1/2<101>. The 1/6<112> dislocations have been observed in many previous studies (eg. Foll, 1982 and Gibson, 1982). They form a disordered hexagonal array, and since their Burgers vectors lie almost entirely in the interfacial plane, they are efficient at relieving the 1.2% misfit strain in this system. By comparison, the 1/2<101> dislocations, which have a 1/3[111] component of Burgers vector perpendicular to the interfacial plane, have not been reported previously. It is the characterisation of the distinctive contrast from these latter defects and their relationship with interfacial steps which are the topic of this paper.

2. EXPERIMENTAL

Two thicknesses of $CoSi_2$ (2nm and 5nm) were fabricated by depositing Co under UHV conditions onto clean (111)Si wafers held at room temperature. When less than 1nm of Co was deposited, islands of B-type $CoSi_2$ formed immediately (Tung et al, 1989). For thicker Co deposits a 1hr anneal in UHV at 700°C was required to transform the layer, via CoSi, to the desired $CoSi_2$ phase. Electron transparent specimens were prepared by backthinning plan-view samples with an HF/HNO₃ etch. Diffraction contrast experiments were carried out on a JEOL 2000FX transmission electron microscope.

3. RESULTS

3.1 Observations on 2nm thick $CoSi_2$ layers

In these samples the $CoSi_2$ phase forms directly at room temperature. Initial substrate surface features such as atomic scale depressions or mesas seem to persist as steps in the interface even

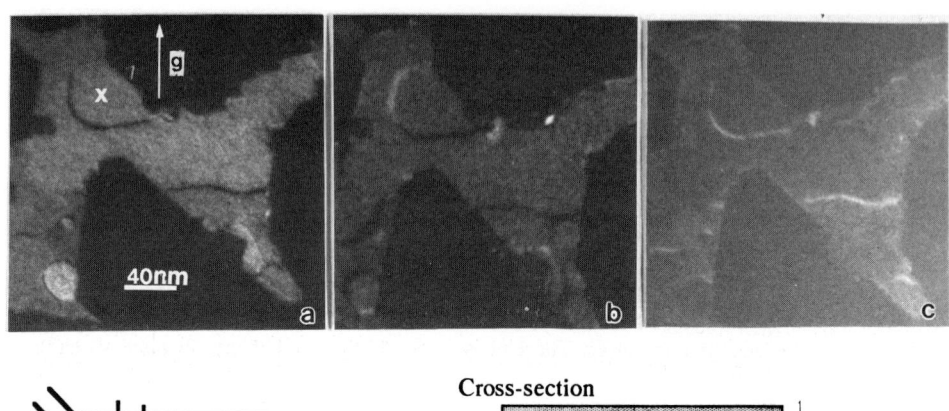

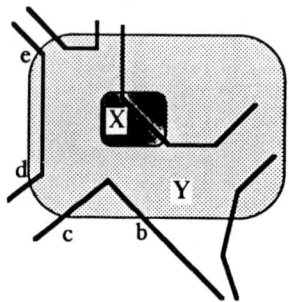

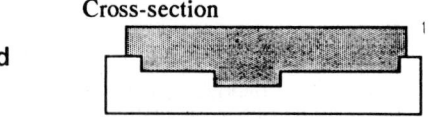

Cross-section

Fig.1 Dark field images of a 2nm thick CoSi$_2$ island showing 1/2<101> dislocations. These were taken in a **g** = 400 reflection with (a) **s**=0, (b) **s** positive and (c) **s** negative. The schematic diagram (d) illustrates the step configuration in this area.

though some of the surface Si atoms are consumed in forming the silicide layer. Figs 1(a), (b) and (c) are a series of dark field images of a B-type CoSi$_2$ island which has overgrown two concentric substrate surface steps, as illustrated in fig.1(d). Interfacial defects, which we propose in section 4 to be 1/2<101> dislocations, appear as broad dark lines in strong beam images (fig.1(a)). When imaged under conditions where the deviation parameter **s** is non-zero, each dislocation loop appears as a pair of arcs, one dark and one bright, which touch on a line approximately perpendicular to **g**. Furthermore comparison of figs. 1(b) and (c) clearly demonstrates that the bright/dark contrast reverses when the sign of **s** is reversed.

3.2 Observations on 5nm thick CoSi$_2$ layers.

In these thicker samples the CoSi$_2$ layer forms by a complicated solid phase epitaxy process, via a number of metal rich-silicide phases (Catana et al, 1992). Hence, the origin of the interfacial steps is less clear cut in his case. Weak-beam imaging experiments have revealed that most of the steps present in this interface are associated with the usual 1/6<112> misfit relieving dislocations (Daykin et al, 1992a). However, occasional isolated 1/2<101> defects, which appear as characteristic broad dark lines in certain strong-beam images, are also present (see Fig.2(a)). These defects have been studied under various **g**$^{\text{CoSi2}}$ darkfield imaging conditions and we find them to be strongly visible in **g**=400, but almost invisible in *both* the **g**=$11\bar{1}$ and **g**=$11\bar{1}$ reflections. Since *all* three of these reflections lie in the 011 reciprocal lattice plane, it is clear that the conventional invisibility condition, **g.b**$_\wedge$**u**=0, cannot be used to characterise the defect.

Fig. 2 shows a sequence of **g**=400 images of the same defect taken with differing values of **s**. The black/white assymmetry in defect contrast is once again apparent if one compares images taken with different signs of **s** (eg. Figs 2(c) and (e)). Another clear feature in this set of images is that at small values of **s** the defect appears as a double line image (fig.2(b)). As the magnitude of **s** is increased, these lines are observed to move continously together until they eventually merge to form a single line image (fig2(c)). Increasing **s** even further only serves to make the image rapidly fade away.

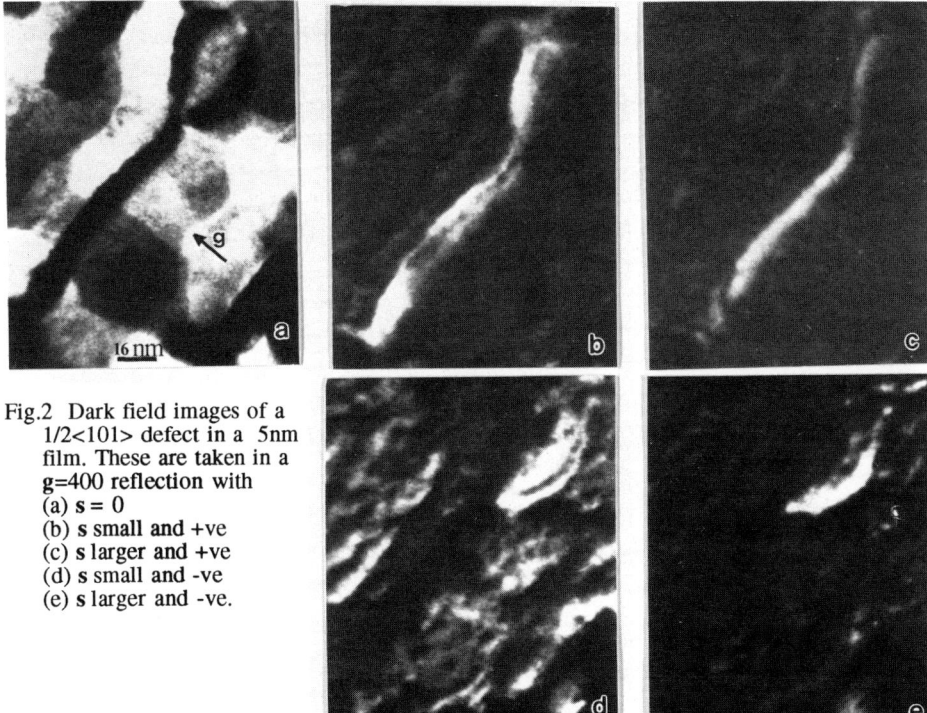

Fig.2 Dark field images of a
1/2<101> defect in a 5nm
film. These are taken in a
g=400 reflection with
(a) **s** = 0
(b) **s** small and +ve
(c) **s** larger and +ve
(d) **s** small and -ve
(e) **s** larger and -ve.

4. DISCUSSION

The defects exhibiting this unusual diffraction contrast are 1/2<101> dislocations associated with interfacial steps. Such dislocations will have a 1/3[111] component of **b** perpendicular to the surface and a misfit relieving in-plane component of 1/6<112>. The expected relaxation of the perpendicular component of **b**, by the splaying of atomic planes in the silicide film, is illustrated schematically in Fig.3(a). The very asymmetric displacement field of this dislocation can be equivalently described as a disclination dipole if one employs the ideas first proposed by Eshelby (1956). The vertical planes (B) in the silicide to the left of the defect are seen to rotate gradually more and more clockwise on crossing the defect, until they eventually reach a maximum rotation, ω_{max}, at the point labelled E. Beyond this, the planes gradually rotate back in an anticlockwise fashion until to the right of the defect (C) until they become vertical again. The local rotation ω of the planes as a function of distance x across the defect is shown schematically in fig.3(b).

Based on the nature of the displacement field shown, we are able to qualitatively explain all the contrast behaviour exhibited by these defects. The asymmetry in the the images on changing the sign of **s** occurs as a consequence of the displacement field effectively being only on one side of the defect. When **s**=0 (strong beam conditions) all the planes across the core of the defect are locally rotated from the Bragg condition, and so the defect appears as a dark line against a strongly diffracting background. When **s** is small, the spatially separated points D and F (in Fig 3(a)) are brought to the bragg condition and the image appears as two separated bright lines against a weakly diffracting background. As **s** is made progressively larger the two positions across the defect which satisfy the Bragg condition move closer together and so the bright line images get closer together. Eventually a value of **s** will be reached where only the position E is at the Bragg condition and a single bright line image against a dark background is observed. It should be noted that the image splitting is only expected to be resolvable in the thicker layers where the displacement field has developed a significant lateral spread. This is experimentally borne out by the fact that we have not observed double images in the 2nm thick films. The relative visibility of the defects imaged in different **g** vectors can be explained by considering simple geometrical Ewald sphere constructions (Daykin 1993). In fact the intensities should be strongly dependent on the relative magnitudes of the deviation parameters, s^{400}:$s^{\bar{1}1\bar{1}}$:s^{111} (which in this case is 4:1:1).

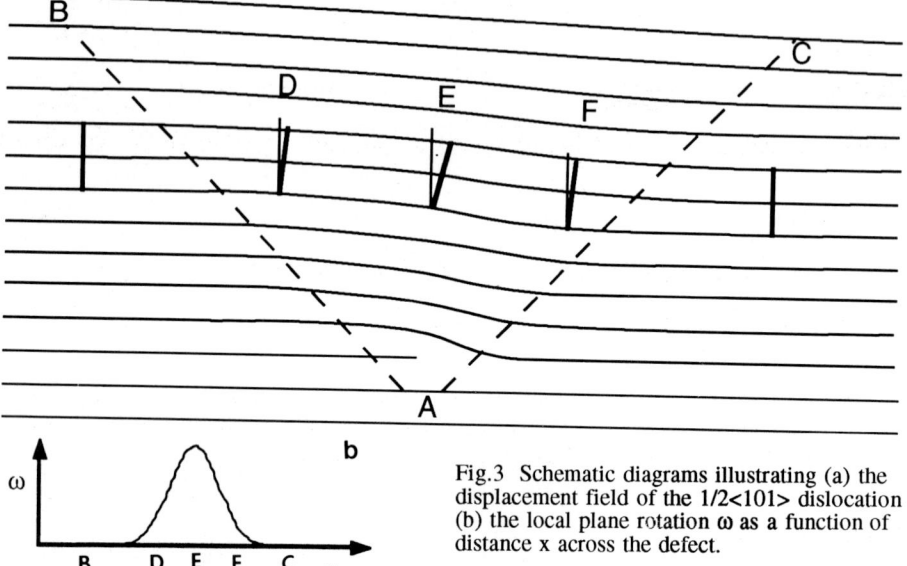

Fig.3 Schematic diagrams illustrating (a) the displacement field of the 1/2<101> dislocation (b) the local plane rotation ω as a function of distance x across the defect.

Several possible explanations for the existence of these defects can be postulated. Even though the 2nm thick films are below the critical thickness for misfit strain relief, any atomic scale interfacial step must have a dislocation associated with it. The fact that 1/2<101> defects are observed could suggest that the core energies of the 1/2<110> dislocations are lower than the core energies of 1/6<112> dislocations. This is plausible because in very thin films, where the elastic strain energy is comparatively small, the core energy may dominate the total energy of the defect. Another factor which may need to be taken into account is the exact sense of the step as discussed by Daykin et al (1992a, 1993). On a substrate containing a random distribution of steps, there will always be some steps, which if associated with a 1/6<112> dislocation would lead to an increase in the mismatch between epilayer and substrate. Formation of the appropriate 1/2<101> dislocation on such a step can lead to a configuration where the in-plane component of its Burgers vector is misfit relieving. This particular point is discussed in more detail in Daykin and Kiely (1993).

5. CONCLUSIONS

Two distinct types of interfacial dislocation, both associated with interfacial steps, have been found in the B-type CoSi$_2$/Si(111) bicrystal. The newly identified 1/2<101> dislocation has been shown to exhibit diffraction contrast behaviour which is consistent with them having a short-range displacement field which resembles that of a disclination dipole. Possible reasons for the existence of these 1/2<101> dislocations have been proposed.

The observations presented here are thought to be of considerable significance to the general area of strain relief in ultra-thin epitaxial films. Evidence has been presented showing that the displacement field of an interfacial dislocation having a significant component of **b** perpendicular to the film can be quite distinct from those of dislocations with in-plane Burgers vectors.

6. REFERENCES

Catana A, Schmid P E, Lu P and Smith D J, (1992), Phil.Mag.A, **66**, 933.
Daykin A C and Kiely C J, (1993), submitted to Phil. Mag A
Daykin A C, Kiely C J and Pond R C, (1992a), Acta. Metall.Mater., **40**, S195-S205.
Daykin A C, Kiely C J and Pond R C, (1992b), Proc.Mat.Res.Soc.Symp., **238**, 3-16.
Eshelby J D, (1956), Solid State Phys., **3**, 79.
Foll H, (1982), Phys.Stat.Sol.(a), **69**, 779.
Gibson J M, Bean J C, Poate J M and Tung R T (1982), Appl.Phys.Lett., **41**, 818.
Tung R T, Levi A F J, Schrey F and Anzlowar M, (1989), NATO ASI Series B: Physics **203**, 187.

Inst. Phys. Conf. Ser. No 134: Section 4
Paper presented at Microsc. Semicond. Mater. Conf., Oxford, 5–8 April 1993

181

TEM investigation of ion beam synthesized buried FeSi₂ in (001) silicon

J Tavares[1,2], H Bender[1], A Lauwers[1], K Maex[1] and M Van Rossum[1]

[1]IMEC, Kapeldreef 75, B-3001 Leuven, Belgium
[2]LSI, Universidade de São Paulo, Brasil

ABSTRACT: A continuous buried β-FeSi₂ layer is prepared by Ion Beam Synthesis in (100) silicon. The different epitaxial orientation relationships between the silicide and the silicon matrix are studied by HREM and electron diffraction.

1. INTRODUCTION

Iron disilicide is reported to occur in three phases : the tetragonal metallic α-FeSi₂ (a=b= 0.2695 nm, c=0.539 nm), the orthorhombic semiconducting β-FeSi₂ (a=0.9863 nm, b=0.7791 nm, c=0.7833 nm, Dusausoy et al 1971) and the metastable cubic γ-FeSi₂ (a=0.5431 nm). The orthorhombic phase is stable at temperatures below 960°C and can occur in different epitaxial orientation relationships with respect to the silicon lattice. In spite of the mismatch with the silicon, it has been shown that surface β-FeSi₂-layers can be grown epitaxially on both Si(111) (Cheng et al 1984) and Si(100) (Cherief et al 1989) by molecular beam epitaxy or chemical vapour deposition.

There is a special interest in β-FeSi₂ because of its semiconducting properties, having a direct bandgap of 0.87 eV. Thus it should be possible to produce opto-electronic devices in the infrared region. A prerequisite is that epitaxial and continuous β-FeSi₂ layers can be grown. The preparation of buried FeSi₂ layers by ion beam synthesis (IBS) has recently been proposed (Oostra et al 1991, Radermacher et al 1991, Hunt et al 1992, Bulle-Lieuwma et al 1992, Maex et al 1992).

In this paper the results will be discussed of a TEM investigation of buried FeSi₂ layers prepared by IBS in (100) silicon.

2. EXPERIMENTAL DETAILS

Iron ions are implanted with an energy of 200 keV to a dose of 3×10^{17} Fe⁺/cm² into (100) silicon. To study the phase formation as a function of the implantation temperature, three Si substrate temperatures are used : 350, 500 and 600°C. The samples are subsequently furnace annealed in a N₂ atmosphere at 850°C for 15 h.

Information on the phase formation and orientation relationships of the implanted layer with respect to the silicon substrate is obtained by combination of conventional and high resolution electron microscopy with electron diffraction for both the as-implanted and annealed samples. Because of the small difference in length between the b- and c-axis of the β-FeSi₂ (hkl)$_\beta$ planes cannot be distinguished from (hlk)$_\beta$ planes by electron diffraction or HREM. Therefore an ambiguity exists in the last indices of all β-planes.

The [011] cross-section and the plan view specimens are prepared by grinding and dimpling, and subsequently ion milling as outlined by Romano et al (1990). The TEM examinations are performed by HVEM with a JEM1250 (operated at 1000kV) and HREM images are obtained on a JEM 4000EX (400kV).

3. RESULTS AND DISCUSSION

A previous investigation (Maex et al 1992) by X-ray diffraction showed the presence of $(202)_\beta$ planes parallel with the silicon surface for the sample as-implanted at 350°C while in the sample as-implanted at 550°C the $(204)_\beta$ diffraction line and some α-lines occur. After annealing only β planes are observed in the XRD spectra. Again for the lowest implantation temperature the $(202)_\beta$ is present, while for the higher implantation temperature only $(204)_\beta$ is observed.

3.1 Morphology of the Buried Layer

Plan view and cross-section TEM reveals the presence of a continuous buried silicide layer consisting of grains with different preferential orientations with respect to the silicon. Plan view observation shows irregularly shaped grains with rounded edges after the implantation, whereas the grains have a polygonal shape in the annealed samples. In cross-section the grains appear irregularly shaped so that the interface between the silicide and top or bulk silicon is non-planar. Occasionally the grains reach the surface. The top silicon has a thickness of 50 to 100 nm. The silicide layer has a thickness of 80 to 120 nm. The planar grain size increases from 120-150 nm after implantation to 200-400 nm after annealing.

In the as-implanted material Moiré fringes in the top layer and below the buried layer indicate the presence of silicide precipitates embedded in the silicon. A further detailed analysis of the nature of these Moiré fringes is going on. Also the presence of silicon twins is observed in the as-implanted samples.

Beneath the buried layer residual radiation damage is present consisting of dislocation loops. The highest density of defects occurs upto $\sim$200 nm below the silicide, while a decreasing defect density is present up to 400 nm. In all cases the top layer is highly dislocated as well.

Table 1 : Relative orientations between FeSi$_2$ and Si as observed in $[011]_{Si}$ cross-section samples

	$\|[011]_{Si}$	$\|(200)_{Si}$	$\| (11\bar{1})_{Si}$	$\|(02\bar{2})_{Si}$
I	$[0\bar{2}1]_\alpha$	$(100)_\alpha$	$(112)_\alpha$	$(012)_\alpha$
II	$[0\bar{2}1]_\alpha$	$(124)_\alpha$	$(\bar{1}12)_\alpha$	
III	$[010]_\beta$	$\sim(20\bar{4})_\beta$	$\sim(\bar{2}0\bar{2})_\beta$	
IV	$[111]_\beta$	$(\bar{1}0\,82)_\beta$	$(\bar{2}02)_\beta$	
V	$[011]_\beta$	$(53\bar{3})_\beta$		$(4\bar{2}2)_\beta$
VI	undefined	$\sim(202)_\beta$	-	-

Fig. 1 : Diffraction pattern of $[010]_\beta$ $\| [011]_{Si}$, the angle between $(20\bar{4})_\beta$ and $(200)_{Si}$ is 6° (implantation at 500°C + anneal 850°C 15h).

3.2 Crystallographic Relationships

Previous HREM studies (Cherief et al 1989, Bulle-Lieuwma 1992) showed that independent of the substrate orientation the (202) or (220) planes of β-FeSi$_2$ are parallel to one of the $(111)_{Si}$ planes.

A detailed structural examination of our samples shows the occurrence of several epitaxial relationships of β-FeSi$_2$-grains with the substrate. In two cases (III and IV) the $(202)_\beta$ planes are nearly parallel with a $(111)_{Si}$ plane. The observed orientation relationships for the $[011]_{Si}$ cross-section samples are summarized in table 1.

The α-FeSi$_2$-grains with orientations I and II are only observed in the as-implanted samples. Orientation II is twinned with respect to the I orientation.

The $[010]_\beta$ orientation (III) is observed for all treatments (Fig. 1, 2). For the samples prepared with the 350°C implantation temperature, the $(20\overline{4})_\beta$ makes an angle of $\sim 12°$ with the $(200)_{Si}$ while the $(202)_\beta$ deviates $\sim 7°$ from the $(\overline{1}\overline{1}1)_{Si}$. This is in agreement with the absence of the $(204)_\beta$ XRD diffraction line for these samples. For the material prepared with a higher implantation temperature, the $(20\overline{4})_\beta$ is better aligned with the $(200)_{Si}$ surface (0 to 5° deviation), while the angle between $(202)_\beta$ and $(\overline{1}\overline{1}1)_\beta$ is then approximately -5 to 0° (Fig. 2).

Fig. 2 : HREM image of a silicide grain with $[010]_\beta \parallel [011]_{Si}$. A stepped interface $(200)_{Si}/(\overline{2}04)_\beta$ is present. The $(202)_\beta$ deviates by -5° from the $(11\overline{1})_{Si}$ (implantation at 600°C + anneal 850°C 15 h).

The second most observed orientation is $[111]_\beta \parallel [011]_{Si}$ (IV). For this orientation the $(\overline{2}02)_\beta$ is well aligned with the $(11\overline{1})_{Si}$ planes. In this case the surface plane is parallel with $(\overline{1}0\,82)_\beta$ (Fig. 3).

A few grains are detected with orientation V with the $(4\overline{2}2)_\beta$ plane aligned with the $(02\overline{2})_{Si}$ orthogonal to the wafer surface and a $(53\overline{3})_\beta$ plane parallel to the wafer surface.

Although XRD shows mainly the presence of $(202)_\beta$ planes parallel to the wafer surface for the samples with the lowest implantation temperature, only few grains could be detected which show $(202)_\beta$ planes nearly parallel with the $(200)_{Si}$. In these cases a misorientation of approximately 5° is observed. Probably grains with the $(202)_\beta$ parallel to the wafer surface give no HREM-resolvable orientations in the $[011]_{Si}$ oriented cross-section specimens.

3.3 Interfaces

Although the shape of the grains is irregular, a few types of planar interfaces can be distinguished. For orientation III these are:

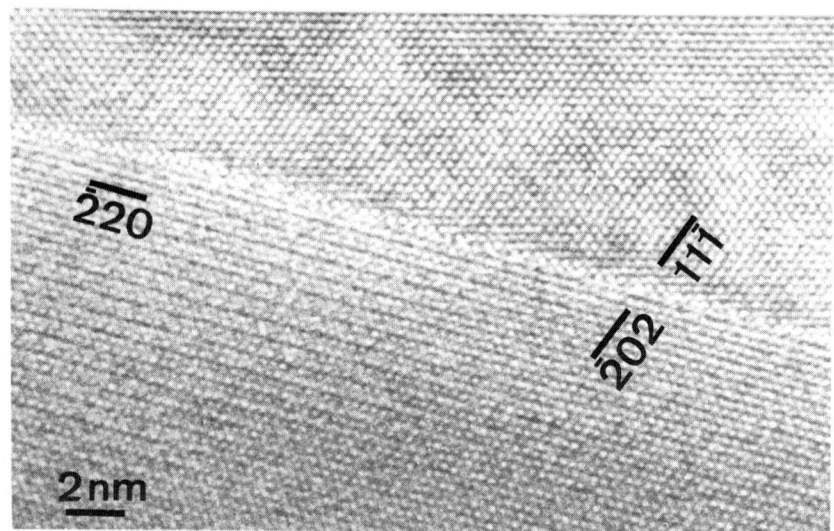

Fig. 3 : HREM image of $[111]_\beta \parallel [011]_{Si}$. The $(11\bar{1})_{Si}$ plane is aligned with the $(\bar{2}02)_\beta$. The interface planes is $(\bar{2}20)_\beta \parallel (5\bar{1}1)_{Si}$ (implantation at 600°C + anneal 850°C 15 h).

- $(20\bar{4})_\beta \parallel (200)_{Si}$ with a quite large misfit along $[01\bar{1}]_{Si}$ of -7.8 % ($[201]_\beta$ vs $3[01\bar{1}]_{Si}$). These interfaces are always stepped and are not very extended (Fig. 2)
- $(202)_\beta \parallel (3\bar{1}1)_{Si}$ with a misfit of -1.1 % along $[23\bar{3}]_{Si}$ ($2[10\bar{1}]_\beta$ vs $[23\bar{3}]_{Si}$)
- $(002)_\beta \parallel (3\bar{1}1)_{Si}$ with a misfit along $[23\bar{3}]_{Si}$ of -3.1% ($5[100]_\beta$ vs $2[23\bar{3}]_{Si}$)
The misfit along the $[010]_\beta$ direction is 1.4%.

For the orientation IV large flat interfaces are found with $(\bar{2}20)_\beta \parallel (5\bar{1}1)_{Si}$ (Fig. 3). The misfit along $[111]_\beta$ is in this case -4.1%. For orientation V the interfaces are irregular with a tendency to form $(11\bar{1})_\beta$ stepped interfaces. The misfit along $[011]_\beta$ is -3.6%.

The TEM observations are performed with the EM at the University of Antwerpen (RUCA).

REFERENCES

Bulle-Lieuwma C W T, Oostra D J and Vandenhoudt D E W 1992 Mat. Res. Soc. Symp. Proc. 279 in press

Cheng H C, Chen L J and Your T R 1984 Mat. Res. Soc. Symp. Proc. 25 441

Cherief N, D'Anterroches C, Cinti R C, Nguyen Tan T A and Derrien J 1989 Appl. Phys. Lett. 55 1671

Dusausoy Y, Protas J, Wandji R and Roques B 1971 Acta Cryst. B 27 1209

Hunt T D, Reeson K J, Gwilliam R M, Homewood K P, Wilson R J, Spraggs R S, Sealy B J, Meekison C D, Booker G R and Oberschachtsiek P 1992 Mat. Res. Soc. Symp. Proc. 260 239

Maex K, Lauwers A, Van Hove M, Vandervorst W and Van Rossum M 1992 Mat. Res. Soc. Symp. Proc. 279 in press

Oostra D J, Vandenhoudt D E W, Bulle-Lieuwma C W T and Naburgh E P 1991 Appl. Phys. Lett. 59 1737

Radermacher K, Mantl S, Dieker Ch and Lüth H 1991 Appl. Phys. Lett. 59 2145

Romano A, Vanhellemont J and Bender H 1990 Mat. Res. Soc. Symp. Proc. 199 167

Inst. Phys. Conf. Ser. No 134: Section 5
Paper presented at Microsc. Semicond. Mater. Conf., Oxford, 5–8 April 1993

Microscopy of ULSI devices

J M Brown, E G Boden and C Hoener

SEMATECH, 2706 Montopolis Drive, Austin, Texas 78746, U.S.A.

ABSTRACT: ULSI devices that will be manufactured in the 1990's have gate line widths below 0.35 μm and oxide thicknesses less than 90 angstroms. Their performance is closely linked with the ability to control and monitor the component architecture of the silicon chip. The requirement for high yields necessitates closely controlled processes and high precision metrology to characterize the materials.

1. INTRODUCTION

The silicon semiconductor industry has traditionally progressed to smaller device dimensions on a generation cycle of approximately 3 years. Technology presently in manufacture has gate lengths of 0.8μm with 0.5μm coming into pilot line production at the leading edge companies. Several research & development organizations are working on successive generations down to 0.1μm minimum lateral feature size. To obtain the required performance characteristics, device designers and process architecture specialists are continuously modifying the physical chip layout and materials requirements. Electrical test performance can often be used to infer the physical state of the device, however it is extremely important to understand the physical layout both horizontally and vertically to develop robust processes and to understand failure mechanisms. Materials and physical analysis are increasingly essential ingredients for successful research, development and manufacturing of ULSI devices.

This paper presents an overview of the applications of different forms of microscopy to the manufacture of ULSI devices. Particular emphasis is placed on Transmission Electron Microscopy (TEM), but the applicability of other techniques such as Scanning Electron Microscopy (SEM), Atomic Force Microscopy (AFM), Scanning Auger Microscopy (SAM) and Focused Ion Beam systems (FIB) are also discussed.

2. DEVICE ARCHITECTURE

Competition in semiconductor manufacturing has several effects on silicon device design & process architecture. Demands for high reliability, performance memory and

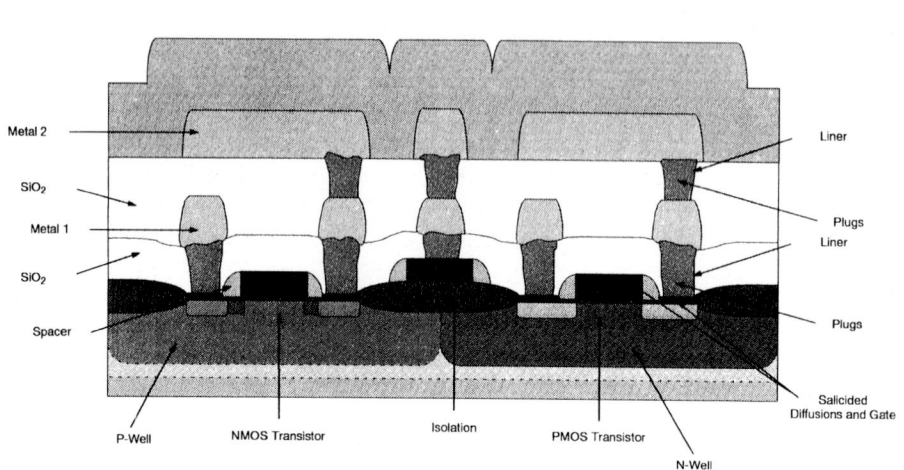

Fig. 1 Submicron Device Structure

logic chips at increasingly lower cost result in continual shrinking of device dimensions to improve performance and increase packing density. At the same time silicon wafer size increases, 200mm now being in production with moves to 300mm on the horizon; higher numbers of die per wafer reduces unit area cost of fully processed silicon. A schematic of a generic ULSI CMOS device is shown in figure 1.

Presently, 0.35μm gate length transistors are in pilot line production on 200mm diameter wafers, with development and research extending out to 0.25μm, 0.18μm and beyond. Some device design rules for the different technology nodes, 0.5μm, 0.35μm, 0.25μm are shown in table 1.

Table 1 - Scaling of MOS Dimension

Gate length	0.35μm	0.25μm	0.18μm
Contact	0.45μm	0.30μm	0.22μm
Gate Oxide	9nm	6.5nm	4.5nm
Junction Depths	0.15μm	0.09μm	0.05μm
Interconnect	1.0μm	0.60μm	0.40μm

As device dimensions become smaller, new processing methods and materials are introduced. Physical analysis techniques such as TEM, SEM & AES are increasingly essential to understand the mechanisms involved and problems that arise in the development and manufacturing stages.

3. APPLICATIONS

In order to ensure quality processing and manufacture, regular inspection points are inserted in the process flow. However, both the vertical and horizontal dimensions of the devices now in manufacturing have already decreased or are rapidly decreasing below the resolution limit of the generally accepted method of routine characterization and inspection in the factory, namely optical microscopy.

Over the last several years, electron microscopy has been used in conjunction with other materials characterization techniques to aid development engineers and research scientists to understand the mechanisms at work in the processing of experimental lots. It is becoming obvious that the analytical techniques that were once only found in research labs need to be applied on a regular basis to the solution of process development and manufacturing problems. A clear link needs to be established for example, between the defect detection tools and the SEM, SAM, SIMS & other materials analysis equipment in order to identify the chemistry of particulate contamination to better identify and remove the source.

The applications of microscopy and other analytical techniques to semiconductor development and manufacturing are many and varied:

- silicon substrate evaluation (dislocations, gettering state etc.)
- overall device architecture
- layer thickness
- interlayer positioning/overlay
- layer integrity & composition
- defect identification
- failure mode analysis

This paper will address the application of microscopical techniques to three different aspects of ULSI devices, namely materials characterization; materials development; critical dimension measurements; device architecture & failure mode analysis.

3.1 Materials Characterization

ULSI device manufacture demands the use of several different materials in addition to high quality silicon substrates. Several new materials have been introduced to address specific issues associated with the decrease in device size. In some cases the application of a new material necessitates increasing the complexity of the device. This is the case in interconnect technology. Smaller devices require smaller contacts and interconnect lines, and as a result new materials with lower resistivity are required. However, evidence has been found that some of these materials such as tungsten, can migrate from the contact through the underlying material to short the electrical junction (Georgiou et al 1987). Bravman and Saraswat (1985) identified the particles as being WSI_2. Tungsten metallization has been shown to be effective in many ULSI applications (Riley et al 1990) both as a 'plug' material and as a blanket conducting film.

Tungsten used in silicon processing is an extremely challenging material to evaluate by TEM. Sample preparation is very difficult due to the differential milling between it and silicon. Advances in sample preparation techniques by using Focused Ion Beams (Noshio et al 1990) and other techniques (Klepeis et al 1988) have made the study of this material system more commonplace in many company laboratories. It is still however, a very difficult area.

Titanium nitride is being routinely used as a barrier material at the base of both contacts and vias to prevent such metal penetration into the underlying silicon. It also has the ability to act as an adhesion layer in larger area films. The physical and crystallographic state of the TiN has to be closely controlled to ensure that there is no stress built up at the interface and that the grain structure provides a continuous film. TiN can be deposited in several different ways; sputtering of titanium metal in a nitrogen atmosphere; sputter deposition of titanium metal followed by nitridation at high temperature. The conditions for deposition have a marked effect on the stress, stoichiometry, grain structure and barrier characteristics of the films as has been shown by Mandl et al (1990).

A major problem associated with small contacts (<0.6µm), is the difficulty of ensuring deposition of material at the base. These contact and vias have high aspect ratios; a factor of 3 for a 0.35µm contact through a 0.8µm dielectric layer. Titanium is first sputtered into the base of the contact to reduce any SiO_2 present followed by TiN deposition. During conventional sputtering, the top of the contact hole will mask both the sides and the base resulting in very little coverage in these areas, see fig. 2. In the most severe case the top will be closed before the required thickness is deposited at the base. TEM is normally an excellent tool for examining these layer dimensions; 30-300A. However, alternative methods of evaluation such as the Atomic Force Microscope (Fig. 3) are being pursued to minimise errors introduced by the high curvature of small contact sidewalls. If the 3-dimensional nature of a 0.35µm diameter contact is considered, a typical TEM sample of 2000A in thickness would encompass a curvature at the sidewall of 300A, resulting in a possible error of 100% in a 150A sidewall layer. An excellent description of the applications of Atomic Force Microscopy to semiconductor processing has been given by Neubauer et al (1992).

Fig. 2 0.3µm convention TiN contact

Fig. 3 AFM image of collimated TiN

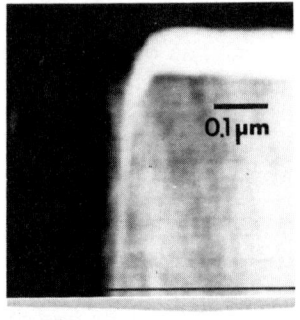

AES is an important part of the study of materials issues in small contacts. Depth profiles have been used to compare sputtered Ti/TiN deposited with and without collimation. TEM studies have shown the effectiveness of collimation in ensuring the successful deposition of the bi-layer at the base of the contact as compared to the single TiN layer observed using conventional deposition. TEM imaging and diffraction patterns clearly indicate the presence of both Ti and TiN at the base of the contact using collimated deposition (Fig. 4). In the un-collimated case no indication could be found for the Ti layer either from the image or by electron diffraction. The AES depth profile from this contact confirmed these results (Fig. 5).

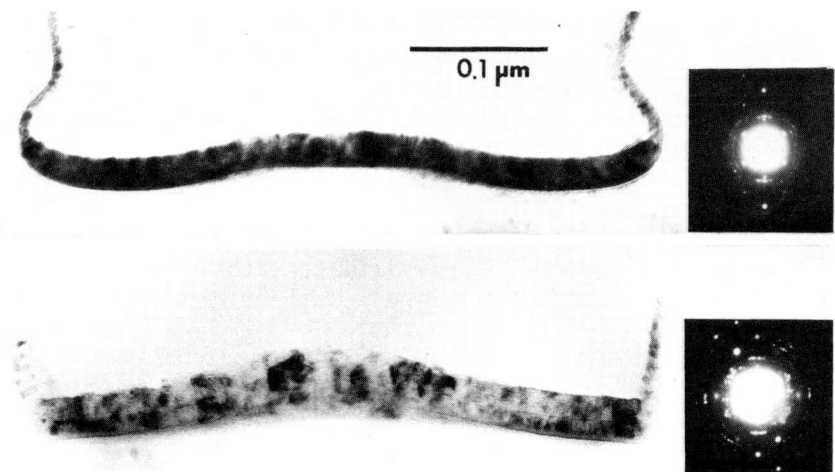

Fig. 4 TEM micrograph of TiN contact: conventional (top) collimated (bottom)

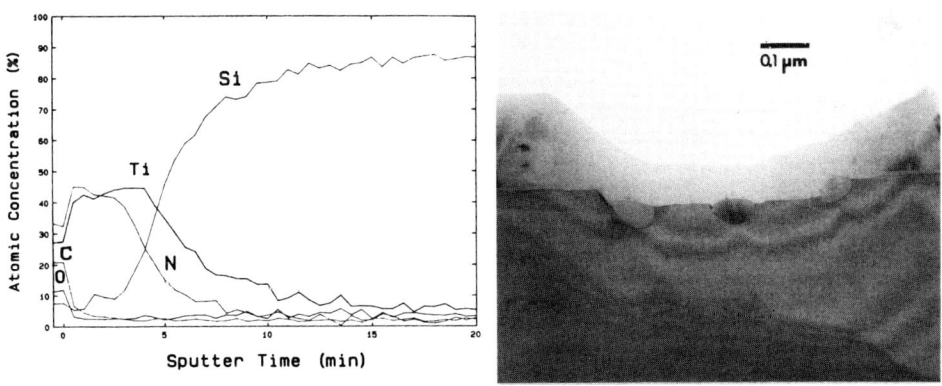

Fig. 5 AES depth profile - uncollimated   Fig. 6 $TiSi_2$ formed by RTP

Metallic silicides are used to provide low resistance contacts to both the source and drain regions of silicon substrate and to the polysilicon gate. $TiSi_2$, WSi_2, $MoSi_2$ and $CoSi_2$ all have applications in the industry and many papers have been published in this area (Murarka 1983). The reduced thermal budget of sub-half micron devices has led to the advent of rapid thermal processing (RTP) of a thin layer of metal to produce the silicide. RTP is a method by which high power lamps are used to elevate the temperature of a silicon wafer to high temperature (>750C) rapidly, maintain the temperature for a few seconds and rapidly cool back to room temperature. This provides for the minimum diffusion of any dopants already present in the silicon. Difficulties with this technique are reproducibility of the temperature, temperature uniformity across a wafer & temperature measurement. If the metal layer is too thin or the temperature varies across the wafer, discontinuous silicide will be formed as is illustrated in Fig. 6.

3.2 Critical Dimension (CD) Measurements

ULSI technology nodes are generally designated by the gate length, however the dimensions of other features scale accordingly. Conventional optical microscopy can no longer be employed to measure lateral dimensions less than $0.5 \mu m$ while also determining process variability. The Scanning Electron Microscope (SEM) is an integral part of any leading edge manufacturing line and can be found both in-line and in the support laboratories. SEM's can now accommodate 200mm wafers and have computer controlled stage drives with accuracy and repeatability of the order of $1 \mu m$. In-line SEMs are used routinely to monitor the CD control of the lithography tools and incorporate complex algorithms to determine both the size and shape of a lithographic feature.

During process development, such as the characterization of a new photo-resist, more accurate determinations of the line profiles are required and off-line SEMs are commonly used to examine cross-sections taken through specific areas. Modern high resolution field emission systems can be used to reveal both simple CD measurements of resist profiles and also to examine fully processed devices. For example, the electrical gate length of a device is determined by a combination of processing factors and knowledge of its physical structure is essential for comparison with models developed by technology computer aided design tools (TCAD) that are used to design the gate oxide process architecture. SEM is often used in conjunction with Focused Ion Beam Milling to section through specific devices that have been characterized electrically to compare the physical structure with that 'predicted' . Fig. 7 shows a cross-section through a $0.2 \mu m$ NMOS transistor which has been shown to exhibit good electrical performance. It is important to characterize several of the features of the device from a process development and device design standpoint. The thickness of the layers ranging from the metallization silicide to the sidewall and spacer oxides are essential to understand the device performance characteristics. The shape of the sidewall oxide is revealed by careful etching or 'staining' of the different layers of oxide. The shape and thickness of this layer will determine the dopant distribution in the source and drain regions. The micrograph also shows discontinuous $TiSi_2$ formation to the right of the gate.

A measurement that cannot be made in existing SEMs is that of the gate oxide thickness. The 65A oxide of a $0.25 \mu m$ device can only be accurately measured by a TEM high resolution cross-section. The micrograph shown in fig. 8 is taken from a TEM test structure that is repeated at different positions on the wafer in an attempt to identify any trends associated with temperature variations during oxide formation. The thickness, Si/SiO_2 interface quality and the broadening of the oxide layer at the edge of the gate (the 'bird's beak') can all be determined using TEM.

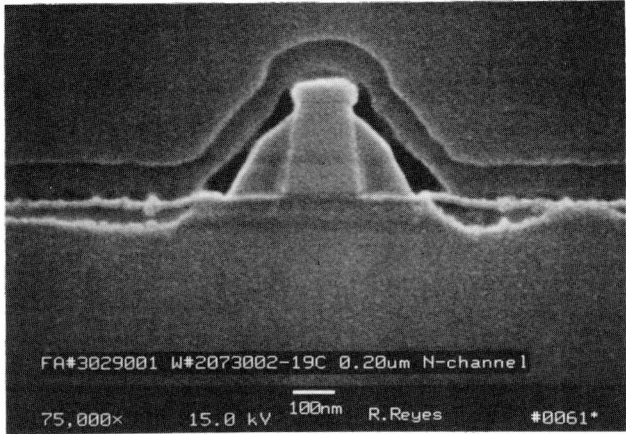

Fig. 7 SEM cross-section of a 0.20μm n-channel transistor

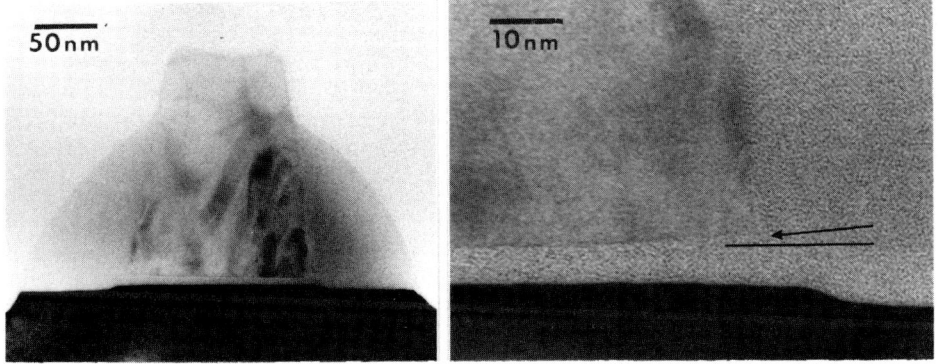

Fig. 8 TEM micrograph showing 0.18μm Gate

3.3 Failure Mode Analysis and Process Development

Visual inspection of silicon devices is a major part of failure mode analysis in the industry. Once again, the need for higher resolution has driven the evaluation to be made in the SEM and in some cases the features of a TEM are needed to fully understand the nature of a problem. In the past there has been a great deal of difficulty in ensuring that the device that has been identified to have failed is the exact device that is used to determine the failure mechanism. The Focused Ion Beam (FIB) system has made it possible to prepare samples for SEM analysis at the exact defect location site. The micrograph shown in Fig. 9 illustrates the use of the combination of the FIB and SEM in failure analysis. The image shows a change in the surface morphology of a metalized device. The cross-section produced by the FIB was examined in the SEM using secondary imaging and energy dispersive X-ray analysis. The problem can be clearly identified as an adhesion problem between the tungsten metal layer and the underlying titanium nitride.

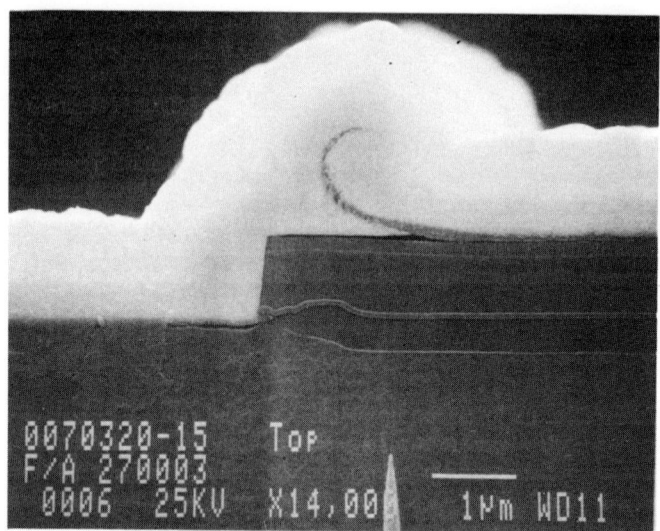

Fig. 9 Focused Ion Beam section through defect

4. SUMMARY AND CONCLUSIONS

The applications of microscopy to the research, development and manufacturing of ULSI devices has been described. The advent of increasingly smaller dimensions will put further constraints on the ability of existing techniques to support the analysis requirements. An increased focus on the development of equipment that is capable of handing the entire wafer is being driven by the need to have rapid feedback on development issues and the requirement for in-line inspection tools. Electron microscopy is one of the only techniques that can image the next generation device dimensions and provide measurement with the required accuracy. There is however an increasing concern over the statistical validity of gate oxide thickness data from a 500A slice taken from a 200mm wafer; correlation with electrical measurements is essential.

REFERENCES

Bravman J and Saraswat K 1985 Proc. Tungsten Workshop p 125
Georgiou G E, Brown J M, Green M L, Lui R, Williams D S & Blewer R S 1987 Proc. Tungsten Workshop p 227
Klepeis S J, Benedict J P and Anderson R M 1988 Proc. Mat. Res. Soc Vol 115 p 179
Mandl M, Hoffman H and Kucher P 1990 J. Applied Physics 68 (5)p 2127
Murarka S P 1983 Silicides for VLSI (London: Academic Press)
Neubauer G, Dass M L A and Johnson T J 1992 Proc. Reliability Physics p 299
Nishio N, Shiotani K, Kitakata M and Mikoshiba M 1990 Abst. Conference on Solid State Devices and Materials Sendai p 1169
Riley P E, Clark T E, Gleason E F and Garver M M 1990 IEEE Trans on Semiconductor Manufacturing vol 3 no 4 p 150

Inst. Phys. Conf. Ser. No 134: Section 5
Paper presented at Microsc. Semicond. Mater. Conf., Oxford, 5–8 April 1993

215

Measurement of sub-nanometre film thicknesses in silicon VLSI

P D Augustus, D K Skinner, *+M R Goulding, *S Nigrin and **R Beanland.

GEC-Marconi Materials Technology, Caswell, Towcester, NN12 8EQ.
*GEC Plessey Semiconductors, Cheney Manor, Swindon, SN2 2QW.
**Department of Materials Science and Engineering, The University of Liverpool, P.O. Box 147, Liverpool, L69 3BX.
+Present address; Seagate Microelectronics, Macintosh Rd., Livingstone, EH54 7BU.

ABSTRACT: Sub-nanometer silicon dioxide films sandwiched between silicon layers in VLSI device structures are studied by cross-sectional transmission electron microscopy and Auger bevel profiling. The oxide film roughens and balls up during annealing and whilst the electron microscopy is very important in showing the oxide morphology, Auger bevel profile analysis is shown to be an accurate way of quantifying the oxide present in the layer. Oxide equivalent to layer thicknesses of down to 0.3 nm have been measured by this technique.

1. INTRODUCTION

Modern developments in silicon integrated circuit technology have an emphasis on finer geometries in vertical as well as lateral dimensions. There is, therefore, an increasing emphasis in the precise control of processing conditions for the production of thin layers and an increasing demand for accurate layer measurement techniques. In modern bipolar processes the same polysilicon layer serves as both the diffusion source for the emitter and as the emitter contact. The crystallographic nature of the polysilicon and the way in which it acts, is dependent upon the thickness and integrity of the interfacial oxide between the polysilicon and the underlying single crystal silicon. To manufacture devices with consistent transistor parameters the control of thickness and uniformity of this oxide must be well understood; in addition, the behaviour of the oxide during annealing of the polysilicon and drive in of the dopant must be predictable and quantifiable.

The thickness of the initial interfacial oxide is a function of the treatment of the silicon wafer prior to polysilicon deposition. During rapid thermal annealing the interfacial oxide starts to roughen; it can then either remain continuous or break up into islands or balls of oxide. In previous transmission electron microscopy (TEM) (Wilson et al 1982 and Jorgensen et al 1985) it was shown how the oxide breaks up to form an array of oxide balls at the position of the original interface with epitaxial regrowth of the polysilicon occurring. The kinetics of the regrowth of the polysilicon were shown by Benyaich et el (1992) to be intimately related to its microcrystalline structure, to the morphology of the oxide film at the interface, and to the location of As dopant atoms. Cross-section TEM (XTEM) can easily detect the break up of the oxide and shows the extent of epitaxial re-alignment, but its use in giving a precise measurement of the thickness of thin uneven oxides is limited.

This paper shows how the measurement of total oxygen content at the interface by Auger bevel analysis can be translated into an equivalent thickness for a continuous oxide layer. It also shows how using information from plan view TEM as well as XTEM, an equivalent thickness for a continuous oxide layer can be determined which is in good agreement with the Auger results.

2. TRANSMISSION ELECTRON MICROSCOPY

Specimens were prepared for TEM analysis either by lapping and polishing followed by Ar ion beam milling for cross-section analysis or by jet thinning from the back using HF/HNO3 for plan view specimens. The precise imaging conditions used in TEM have a very important effect on the resultant micrograph and its interpretation. If a (110) cross-section is being examined, it is important to align the specimen close to the (110) pole so that the beam direction is parallel to the silicon to oxide interfaces. If this is not done double images will result due to the finite thickness of the specimen foil. Precise alignment can only be achieved where the interfaces lie precisely on the (001) orientation and are set parallel to the electron beam direction. As the interfaces roughen it becomes more difficult to measure the oxide thickness accurately. In bright field TEM images, minimum Fresnel fringe thickness can be achieved by choosing a thin region of the foil and a weakly diffracting condition. Even under these conditions the width of the fringes set the lower limit for the measurement of oxide thickness at about 20Å. Figure 1 shows a through focal series from a thin interfacial oxide in the as-prepared state, that is after polysilicon deposition, but before any ion implantation and rapid thermal annealing (RTA). The "at focus" image is taken in a minimum contrast condition where the oxide cannot be seen, slight under-focus is the normal imaging condition where a sharp Fresnel fringe is observed and over-focus produces a reversal of the fringes. It will be shown later that the actual oxide thickness in figure 1 is of the order of 5Å. A broadening of the "black line" appears with thicker oxide layers such that 30Å layers can be measured with a reasonable degree of certainty and 50Å measurements have shown excellent correlation with TEM lattice image measurements of oxide thickness. Studies of the Fresnel fringes have been made by Ross and Stobbs (1991), who by taking a through focal series and comparing these electron micrographs with computer simulations achieved precise thickness measurements of uniform oxide layers. During rapid thermal annealing the oxide roughens and finally breaks up into oxide balls at higher annealing temperatures; this is illustrated in the cross-section TEM micrographs of figure 2, where it can be seen that for a 30 sec. RTA anneal the oxide layer is substantially roughened at 1000 °C and at 1100 °C it has completely broken up with complete epitaxial realignment of the polysilicon. Lattice imaging, although offering higher magnification through improved spatial resolution, has its own problems. It still demands a TEM cross-section sample with a thickness of approximately 30 atomic layers and the interfaces must be parallel with the electron beam direction and with the crystal lattice. In addition, the polysilicon grain adjacent to the oxide must be aligned such that a lattice image from this grain is visible at the same time as that from the substrate. Where a crystalline lattice and an amorphous material are superimposed the image from the lattice predominates. This is a useful feature when TEM samples are covered in a thin oxide or an ion beam created amorphous film resulting from the specimen preparation, but where we are trying to pick out the amorphous filling in a crystalline sandwich a difficulty is presented in that the crystalline silicon is seen but not the oxide. This is illustrated in figure 3 for the case of an oxide film broken up into balls. Because the balls are sandwiched in a crystalline lattice it is the lattice image that predominates. In this electron micrograph a contrast difference resulting from the density difference of the oxide is visible such that the extent of the oxide ball can be seen.

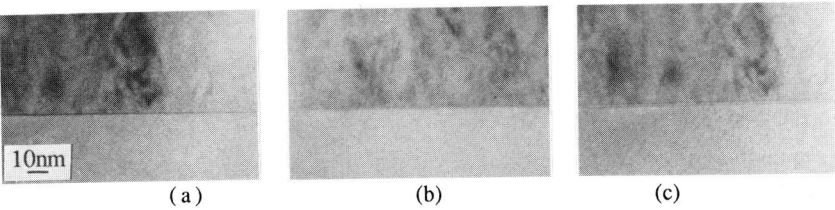

Figure 1: Bright Field transmission electron micrographs of a cross-section of polysilicon emitter contact showing the interfacial oxide (a) underfocus, (b) at focus, (c) over focus.

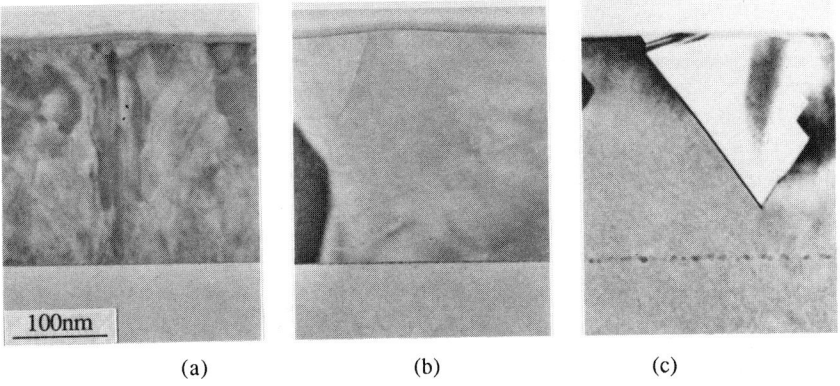

Figure 2. Bright Field XTEM micrographs of a polysilicon emitter implanted with P and annealed at (a) 900°C, (b) 1000°C and (c) 1100°C for 30 sec

This contrast difference is not seen in the case of a thin continuous interfacial layer and extensive examination of lattice images failed to produce a convincing image of a 5Å oxide layer. Jorgensen et al (1985) showed that the microscopic details of the initial recrystallisation can be examined using lattice imaging. In Figures 4 and 5 it is shown that for a phosphorus implanted sample definite evidence of oxide breakup is evident after 30 seconds at 1000 C, with this assertion being based upon the uneven progress of the substrate lattice into the polysilicon layer rather than the direct observation of 5Å oxide in the lattice image. In figure 2(b) the same sample imaged in conventional bright field showed a continuous Fresnel fringe but with some roughening of the layer being apparent. In neither figures 4 or 5 can a distinct oxide layer be seen but some small patches of oxide can be resolved. The most obvious signs of oxide breakdown were apparent in isolated regions such as that shown in figure 5 where more extensive recrystallisation has occurred at a grain boundary in the polysilicon. An alternative approach for the determination of the thickness of the initial oxide layer is to examine the images of the balled up layers created by a high temperature anneal. These oxide balls are quite stable and do not dissolve in the matrix silicon even after prolonged heating at high temperatures. Measurements from the TEM cross-section image were made to give the average depth of the spheres, measurements were then taken from plan view TEM images to give the areal fraction covered by the balls.

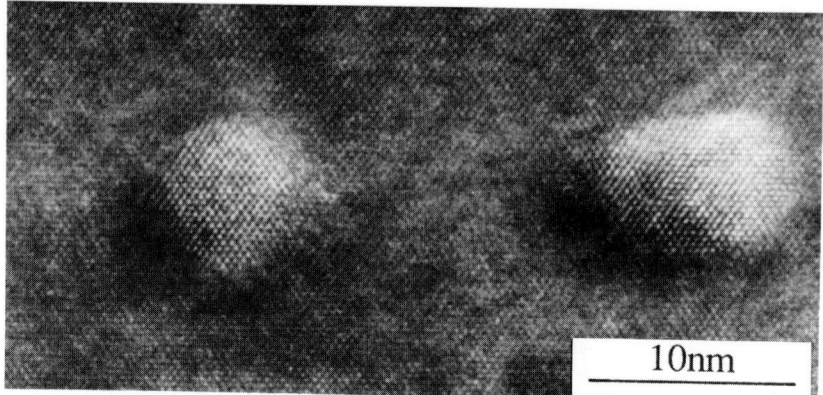

Figure 3   XTEM lattice image of oxide balls in silicon.

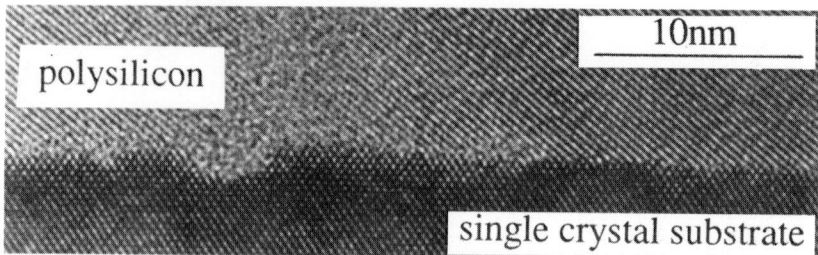

Figure 4   XTEM Lattice image of oxide interface heated to 1000°C for 30 sec.

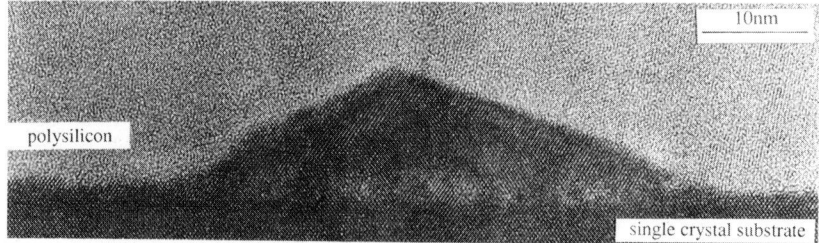

Figure 5   XTEM Lattice image of oxide interface heated to 1000°C for 30 sec, the recrystallisation is adjacent to a grain boundary in the polysilicon.

From these two measurements the volume of oxide was determined and the corresponding thickness of a continuous interfacial oxide calculated. If we consider an interfacial layer 5Å thick and this balls up to approximate to a 50Å cube the surface of the cube will occupy 10% of the surface area of the interface. A 50Å cube has the same volume as a sphere of 63Å diameter. The plan view TEM image from a layer heated to 1100°C is shown in figure 6, this image was taken in a weakly diffracting condition to minimise the contrast from crystallographic defects. The areal coverage of oxide balls was measured using a Quantimet 570 image analyser to give an average coverage of 12% consistent with 65Å diameter oxide balls and a continuous layer thickness of 5Å.

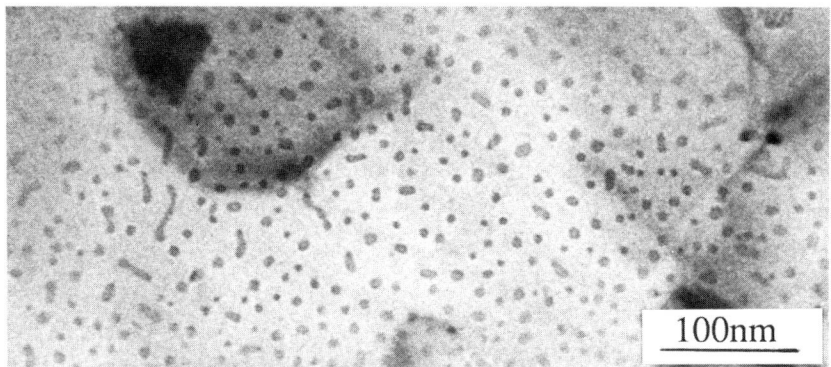

Figure 6 Plan view TEM image through a recrystallised polysilicon layer showing oxide balls covering 12% of the surface.

3. AUGER BEVEL MEASUREMENTS

The method developed for measuring thin oxide layers down to sub-nanometer thicknesses is based on the 'ion beam bevel section', Skinner (1989). In this technique, shown schematically in figure 7 (a), a finely focussed ion beam was electronically rastered in such a way as to produce a bevelled section that presents a magnified projection of buried layer structures. Auger line scans were obtained by monitoring the Auger electron intensity for oxygen whilst scanning the incident electron beam along the surface of the bevel.

The method of calibration adopted was to decide at the outset on a bevel depth that would be sufficient for the range of thicknesses of polysilicon overlayer to be studied. A reference sample containing a buried oxide of known thickness was then beveled and an Auger line scan for oxygen recorded. Oxide layers chosen for calibration purposes were 57 Å and 83 Å as measured by XTEM, using both conventional bright field and lattice imaging. The TEM image was also used to show that the calibration samples had uniform parallel oxide to silicon interfaces. The thickness of an unknown oxide was determined by;
(a) recording an Auger line scan for oxygen,
(b) integrating the area under the oxygen line scan,
(c) using the integrated area of the calibration sample to calculate the thickness of the unknown oxide.

Figure 7 (b) shows the oxygen line scan recorded from a 5Å oxide. Such a layer is broadened to around 50Å by surface roughening and atomic mixing. With a layer magnification of approximately 5000 times the width of the oxide in the bevel was increased to 25μm, well within the resolving power of the electron gun used for AES. The integration method of calculation, however, ensures that any further broadening resulting from the electron beam spot size did not compromise the accuracy of the thickness measurement.

The sources of error in this type of technique are mainly concerned with preferential sputtering effects which can distort the flatness of the bevel slope. For the measurement of very thin oxides we have shown that this is not normally a problem, due to the layer broadening introduced by atomic mixing. This broadening means that the oxide, exposed by the bevel, is in effect a dilute mixture of oxygen in silicon. The thinnest oxide we measured by this technique had a thickness of 3.5Å which coincides with the thickness expected from one monolayer of silicon dioxide. One monolayer is the most likely expected minimum interfacial oxide thickness and this gives additional credence to the measurement.

The advantages of bevel profiling over conventional Auger profiling are;
(a) a high data density can be accumulated from the regions of specific interest,
(b) any number of elements can be profiled under optimised conditions,
(c) unsuspected contaminants can be found and identified,
(d) the sample remains available for re-evaluation using the same bevel,
(e) there is a faster sample turn-round than for conventional profiling.

4. SUMMARY

It has been demonstrated that TEM has important limitations when it comes to measuring the thickness of very thin interfacial oxide layers which are roughened by RTA annealing. Where the oxide balls up, a simple plan view TEM measurement can be used to assess the thickness of the original continuous oxide layer.   Alternatively, it is shown that Auger bevel profiling can be used in the accurate measurement of the oxygen content of thin oxide layers.   This technique is equally valuable in the measurement of other thin interfacial layers in planar structures.

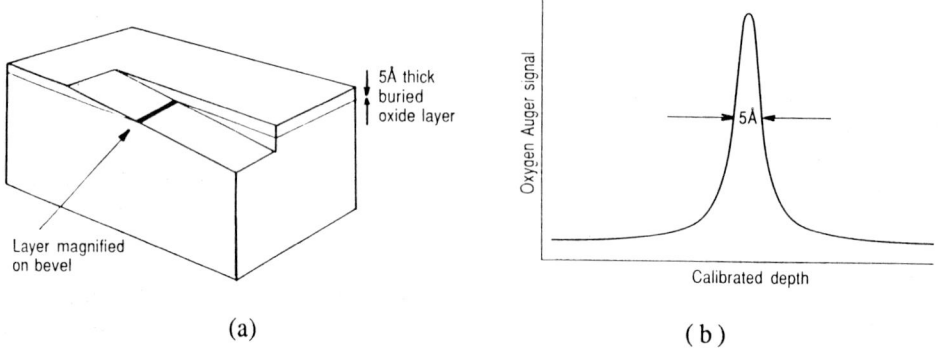

(a)   (b)

Figure 7(a) Schematic diagram of the shape of the ion beam machined bevel for AES measurement.   **(b)** Line scan for oxygen equivalent to a 5 Å thick interfacial layer.

ACKNOWLEDGEMENTS

We thank G B Davies for TEM specimen preparation and photographic assistance. This work was partly supported by the Department of Trade and Industry under programmes IED1540 and IED1594.

REFERENCES
Benyaich F, Priolo F, Rimini E, Spinella C and Ward P 1992 J Appl Phys 71 p638-647.
Ross F M and Stobbs W M 1991 Phil. Mag. A 63 No 1 1-36.
Skinner D K 1989 Surface and Interface Analysis 14   567-571.
Wilson M C, Ashburn P, Soerwirdjo B, Booker G R and Ward P 1982 J de Physique 43
     Suppl. No.10 C1-253.
Jorgensen N, Barry J C, Booker G R, Ashburn P, Wolstenholme G R, Wilson M C and
     Hunt P C. 1985 Inst Phys Conf Ser 76: Microscopy of Semiconducting Materials,
     Edited by A G Cullis and D B Holt, 471.

Inst. Phys. Conf. Ser. No 134: Section 5
Paper presented at Microsc. Semicond. Mater. Conf., Oxford, 5–8 April 1993

TEM observation of thin oxide-nitride-oxide layers

R M Anderson, J Benedict, P Flaitz and S Klepeis

IBM East Fishkill Facility, Surface/Materials Analysis, ZIP E-40, Hopewell Junction, NY 12533-0999, U.S.A.

ABSTRACT: The nature and origin of TEM contrast loss when observing thin oxide-nitride-oxide ONO layers is explained to be the result of radiation damage occuring in either the specimen preparation ion mill or via fast electron damage in the TEM.

1. INTRODUCTION

Thin oxide-nitride-oxide (ONO) layers are important structural components of high density VLSI and ULSI chips. ONO layers are relied upon to provide electrical insulation between various component devices and to serve as a dielectric in trench capacitor DRAM applications. Ideally, the surfaces or trench walls on which the ONO are formed are smooth and featureless, yielding ONO layers of uniform thickness and electrical properties. As this is often not the case, microscopic analysis is required to investigate the smoothness of the Si underlying the ONO and to determine the integrity and thickness of the ONO layers grown on Si asperities and depressions. As the growth rates of oxide on Si (100) and (110) surfaces are different, it is expected that the thickness of ONO layers grown on Si (100) trench walls will be different than the thickness of layers grown on (110) trench bottoms. Dielectric failure and electrical leakage via thin or defective ONO layers on trench bottoms and elsewhere, therefore, are major reliability concerns.

2. DAMAGE DURING SPECIMEN PREPARATION

ONO layers are too thin to be examined with light-optical microscopes. SEM examination is difficult again due to the extremely small width of the layers, less than 10nm, and the low contrast between Si-oxide and Si-nitride in an undecorated, flat SEM specimen. Specimen contrast enhancement of the ONO layers via chemical etching or high-angle ion milling produces topographic ridges that, in turn, produce SEM edge-effect contrast, which is frequently larger in apparent dimension than the actual ONO layer thickness. Utilizing fractured specimens to make SEM examination of ONO layers is likely to produce a number of mechanical and imaging artifacts. TEM analysis provides ample resolution of the thin layers but, because the ONO layers are amorphous, interlayer visualization is via weak Si-oxide/nitride mass contrast.

The method used to prepare ONO layer TEM specimens influences the visibility of the layers during examination. ONO layer contrast will be extra-weak, or non-existent, when specimen preparation consists of polishing or dimpling followed by several hours of ion milling despite the production of optimum thickness specimens. It can be shown that this contrast disappearance phenomena is due mainly to radiation damage occurring in the ion mill. Ion mill radiation damage takes the form of displacement cascades that intermix the oxide and nitride layers analogous to the production of thin artifactual amorphous layers on the surfaces of ion milled specimens.

Three-dimensional network solids like crystalline SiO_2 and Si_3N_4 are characterized by a high degree of directed bonding. The coordination is low and the number of bonding constraints is equal to or less than the degrees of freedom. Such solids can undergo polymorphic phase transitions under ion or electron irradiation as a consequence of induced structural disorder. Included here, especially for Si and SiO_2, is the *Metamict* transition from a crystalline to an amorphous state. For the specific case of ion milling induced radiation damage, one must appreciate the large amount of energy (5keV) that is deposited into a very small volume (about $10^{-25}m^3$) per incident ion, an event accompanied by a cascade of displaced atoms followed by a thermal spike. As shown by Sigmund (Sigmund, 1973), the damage region is pear-shaped with major axis along the direction of ion incidence. The region consists of extreme local disorder and many point defects plus a material-dependent degree of amorphisation of crystalline specimens. The amount of damage produced tracks the sputtering yield as a function of ion-beam angle of incidence. The 15° angle, relative to the surface, commonly used for ion milling TEM specimens maximizes sputtering yield, leading to rapid thinning, but also maximizes the displacement cascades into the specimen. For the case of ONO layers, the structure is already amorphous and all that is required is a radiation-induced degree of mixing of the constituent atoms to homogenize the structure and cause the visible nitride layer structure to disappear.

3. DAMAGE DURING OBSERVATION

Regardless of how TEM specimens are prepared, a similar radiation induced effect is produced during TEM observation (Cerva, *et al.* 1987). The transfer of kinetic energy from the incident electron to the atom nuclei constitutes a direct transfer of momentum to the atoms in the solid and should the energy transferred exceed the solid's bonding energy, it's possible that an atom will be displaced to a new site leaving a vacancy behind. The vacancy-displaced atom pair, so produced, constitutes the basic unit of radiation damage in this direct-momentum-transfer, or "knock-on" process. The critical transferred kinetic energy is called the displacement energy and is related to and usually greater than the bonding in the solid, since several bonds must be broken and others perturbed to displace an atom. For amorphous Si-oxide and Si-nitride, the covalent bonds are stronger than those in metals and the direct nature of the bonding results in an open structure. Therefore, the displacement energies are closer to the bonding energies and displacement mixing in the amorphous solid readily occurs. Atoms at or near surfaces are more easily displaced than interior atoms and this leads to faster mixing of species for the thinnest TEM specimens.

Radiolysis is another means whereby fast electron irradiation can cause atom displacements and possibly mass loss. SiO_2 is particularly susceptible to radiolysis, mainly via elevation of valence electrons to electron-hole pair states (excitons). When the potential energy of the excitation is greater than the displacement energy for the atom in the excited state, the potential energy of the excitation is converted into momentum of a departing nucleus sufficient to displace it through a saddle-point configuration into a new location.

The result of atomic displacement, whether by knock-on or by radiolysis, is that point defects are produced in crystalline solids. For the amorphous case considered here an analogous process to crystal point defect production is the mixing of species across a previously well-defined boundary. The stochastic nature of the local structural rearrangement during

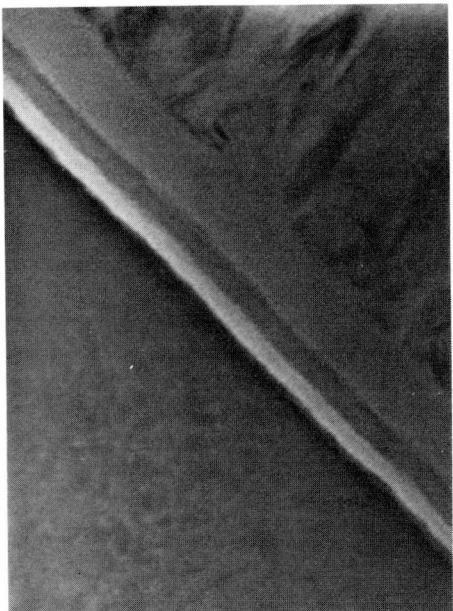

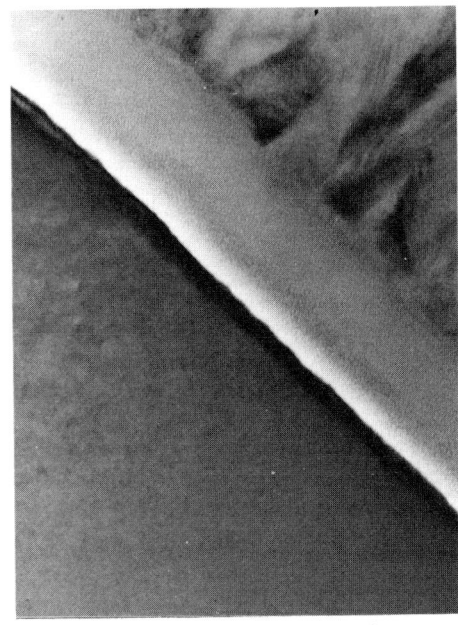

Figure 1a: As-polished specimen. Small aperture TEM photo.

Figure 1b: Ion milled for one minute at 12°, one gun, 0.5 ma.

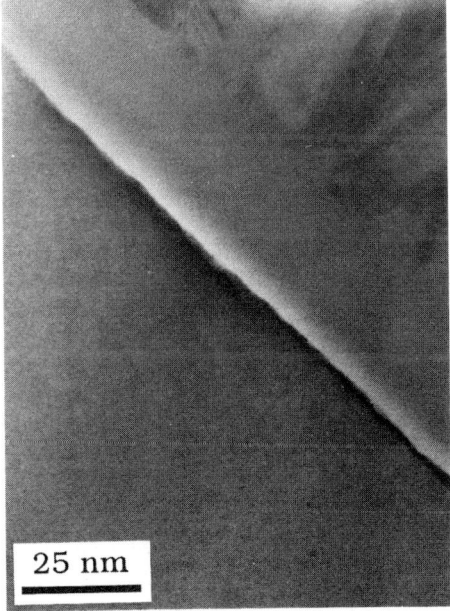

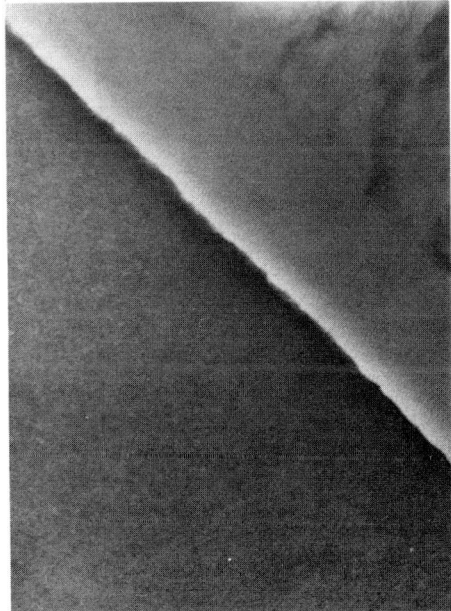

25 nm

Figure 1c: Ion milled for 3 minutes.

Figure 1d: Ion milled for 5 minutes.

electron and ion irradiation is essentially irreversible and can lead to a complete homogenization of the ONO layers into an amorphous Si-oxynitride with no discernible structure. One additional complication is that for TEM specimens ion milled for long periods of time (over 10 minutes), the very large number of defects existing in the specimens prior to TEM examination, and fast electron radiation damage will result in enhanced defect and atom diffusion during TEM analysis accelerating the oxide-nitride mixing process.

4. EXPERIMENTAL

To test these hypotheses, TEM specimens were prepared of ONO layers grown on DRAM capacitor trench walls without utilizing ion milling. The preparation method used (Klepeis, *et al*, 1987), consists of marking an appropriate row of trenches with either a pair of laser ablation dots or FIB square craters and then using a tripod polisher tool to polish into one side of the trench structures. The polishing medium is mylar diamond lapping film followed by 50nm colloidal silica abrasive. The specimen is then inverted and parallel polished from the second side until the total specimen thickness is about 300nm. At this point, an intentional wedge shape is polished into the specimen until the specimen attains zero thickness at the top surface of the specimen. Continued polishing is carried on until the edge of the sharpened wedge moves down into the trench area. This will yield a specimen well-suited for high resolution TEM imaging without ion milling. The specimen is finished by demounting it from the tripod tool and attaching a grid. Specimen preparation spatial resolutions of less than 0.5 microns are easily obtained in times of two to four hours for the complete specimen preparation procedure. Specimens were examined as-polished and after one, three, and five minutes of single-sided ion milling in a GATAN Model 600 Duo Ion Mill. Conventional TEM imaging was undertaken at 200keV in a Philips CM-20-UT. Lattice resolution imaging was performed in the CM-20-UT. Figure 1a, b, c, and d show small objective aperture TEM micrographs of an ONO structure as-polished, and ion milled one, three, and five minutes. The pictures were taken as rapidly as possible to minimize electron beam damage. The degradation of the nitride layer is seen easily.

REFERENCES

Cerva H, Hillmer T, Oppolzer H, and v Criegern R, (1987), Inst. Phys. Conf. Ser. **87**, 445

Klepeis (1987) and co-workers have evolved the specimen preparation method presented here over a number of publications, the first and last are referenced here: Klepeis S, Benedict J, and Anderson R, in "Specimen Preparation for Transmission Electron Microscopy of Materials," ed. Bravman, *et al.*, Mater. Res. Soc. Proc. **115**, Pittsburgh, PA USA p. 179 (1987). Anderson, R. in "Specimen Preparation for Transmission Electron Microscopy of Materials-III," ed. Anderson, *et al.*, Mater. Res. Soc. Proc. **254**, Pittsburgh, PA USA p. 141 (1992)

Sigmund, P, (1973), *J. Mater. Res.* **8**, 1543

Inst. Phys. Conf. Ser. No 134: Section 5
Paper presented at Microsc. Semicond. Mater. Conf., Oxford, 5–8 April 1993

225

On the quantitative characterization of strain distributions at thin film edges

K G F Janssens[1], J Vanhellemont[2], H E Maes[2] and O Van der Biest[1]

[1]Department of Metallurgy and Materials Engineering, K.U.Leuven, de Croylaan 2, B-3001 Leuven (Belgium) [2]IMEC, Kapeldreef 75, B-3001 Leuven (Belgium)

ABSTRACT: The characterization of submicrometer local stress fields is becoming of increasing importance in integrated circuit technologies. The only techniques which have the capabilities to study stress fields with micrometer to nanometer resolution are electron diffraction contrast imaging (EDCI) and convergent beam electron diffraction. In the present paper some calculated EDCImages of stress concentrations at thin film edges are shown to explore the possibilities of the technique.

1. INTRODUCTION

The impact of submicrometer local stress fields on device properties is becoming of increasing importance in integrated circuit technologies (Hu 1991). State of the art integrated circuit processing includes embedding and overlaying structural elements of materials of different elastic and thermal properties. As device processing requires several high temperature treatments, considerable stress develops during thermal cycling. In addition several structural elements exhibit "intrinsic" stresses due to their formation processes. Mechanical stress has an important influence on many physical processes. Recently it has become clear that mechanical stress can influence the reliability as well as the electrical characteristics of the circuits. In order to investigate the stress fields the technique used must have submicrometer to nanometer resolution.

2. TEM STRESS MEASUREMENT TECHNIQUES

One of the few techniques capable of studying stress distributions with sufficient spatial resolution is transmission electron microscopy (TEM). In TEM, convergent beam electron diffraction (CBED) is a rather well established technique for stress measurement but has the drawback of being a "point to point" technique. With electron diffraction contrast imaging (EDCI) one obtains a bright or dark field contrast image of a strained area using one or more diffracted electron beams. Then, from a mathematical model for the displacement field, the contrast image is simulated under the same diffraction conditions as determined in the experiment. The simulation is based on the dynamic theory of electron diffraction, taking into account normal as well as anomalous absorption effects. Depending on the recording method used in the experimental stage, the simulation can also account for the image transformation process that occurs when recording the image intensity distribution onto a micrograph. In a next step the experimental image is compared to the simulated image. Matching the images using visual criteria leads to a first qualitative correlation between the mathematical model and the experimental object. A more quantitative approach requires digitalization of the experimental contrast image. The acquired data can then be compared with the simulated contrast image, absolute intensity values can be compared for each pixel of the image. However, the practical details of the photographic recording process remain difficult to control in a quantitative manner, and are at the origin of a relevant increase in the overall error. We conjecture that the recently emerging slow scan CCD cameras may reduce this problem in the near future. From the above it is clear that although EDCI gives a global image of the strain field rather than a point to point value, the global procedure is more complex and the interpretation less straightforward.

3. DYNAMIC DIFFRACTION THEORY

The process of diffraction which occurs when an incident coherent electron beam interacts with a thin crystalline specimen of material is described by the dynamic diffraction theory (Hirsch et al 1965). This theory accounts for two or multi beam diffraction conditions and includes normal and anomalous absorption effects. The most stringent approximation used is the column approximation. It follows indirectly from this approximation that the lateral resolution in a simulated electron diffraction contrast image (EDCImage) can be of the order of 1 nanometer at the best.

In this paper only two beam diffraction conditions will be considered. When only considering the transmitted beam and one diffracted beam the theory is simplified to the well known Howie and Whelan equations (Howie and Whelan 1961).

4. EDCI SIMULATION: SIMCON

The wide variety of defect types and geometries necessitates a modular approach to the implementation of the diffraction theory. A modular software package "SIMCON" is being developed to calculate EDCImages of *arbitrary* strain fields (Janssens et al 1992). The complete image simulation is divided into three modules: the *defect module*, the *integration module* and the *display module*. The *defect module* computes the displacement field **R** resulting from the mechanical stress field. The *integration module* uses the stored values of **R** and integrates the Howie and Whelan equations. Finally, the *display module* converts the transmitted and diffracted amplitudes into a pixel image of intensities or, using a standard model for photographic processes (Dainty and Shaw 1974), into a more realistic photographic density image.

The crucial point of any kind of image simulation is of course the final comparison between experimental and simulated images. For some defect simulations it is sufficient to obtain an agreement which is only qualitative. The nonlinear character of photographic emulsions restricts further refinement of the image simulation and hence a visual comparison of photographic prints and a computed image will suffice. If a CCD camera is available (or any other image recording device with a linear response over a wide dynamic range of intensities), the agreement between computed and experimental images can be expressed in a more quantitative way: pixel-to-pixel comparison of two images then becomes feasible and a least squares analysis could be used to refine the extraction of the parameters of the displacement field. However, one should always be aware of the errors involved in the determination of the experimental conditions under which the image was obtained. A parameter analysis should therefore involve all parameters which have a significant influence on the final image.

5. EXPERIMENT

It should be clear from the preceding paragraphs that in order to be able to quantify the information in the EDCImage one has to take great care when performing the TEM experiments. In addition to recording good EDCImages one also has to perform the necessary operations in order to obtain all the parameters necessary for the simulation afterwards. Some of these operations are fairly standard, like the determination of the diffraction vector **g** and the Bragg deviation parameter s. It should be noted that the Bragg deviation parameter s has an important influence on the intensity distribution in the EDCImage (see for instance Janssens et al (1992), Fig.s 9 and 10).

Another parameter, which will strongly influence the EDCImage (see below), is the thickness or thickness distribution of the TEM specimen. The thickness distribution is of importance in three different ways: the *mean thickness* of the sample (even a plan parallel one) is a critical parameter, if there is a thickness *distribution* this can "superimpose" thickness fringes on the strain contrast, and in some cases the thickness distribution can also modify the displacement field substantially in comparison to the bulk field. In order to control the thickness distribution of the specimen one can use advanced preparation techniques (Benedict et al 1989) and in addition use CBED to measure the thickness at a well defined number of points.

A third set of parameters, like the absorption parameters, is less standard to measure. Most of the time they are derived from theoretical calculations and assumed constant for a particular material.

6. PRELIMINARY RESULTS

The figures below show simulated EDCImages of the stress field at the edge of a silicon local isolation structure. The experimental observation used for comparison can be found elsewhere in this volume (Armigliato et al 1993). It is assumed that the force field exerted by the film edge can be replaced by a single force in the plane of the figure (Fig. 1)

Fig. 1 shows the difference between two calculated EDCImages of the same stress field under the same diffraction conditions. The only parameter that is varied is the specimen thickness. The EDCImages on the left have been "gray-scaled": the minimum intensity is set to black while the maximum is set to white. From this it can be concluded that EDCImages from thicker specimens show more distinct features. In the images on the right zero intensity is set to black and full intensity (as from the incident electron beam) is set to white. It should be clear that, although thicker specimens theoretically show more distinct features, it will be nearly impossible to capture them with conventional TEM photography, due to the fact that the difference between minimum and maximum intensity becomes very small. We conjecture that Slow Scan CCD cameras may enhance the capacity of modern microscopes to record this kind of image , since their dynamic range can be fitted to the range of the EDCImage *on line* at the microscope.

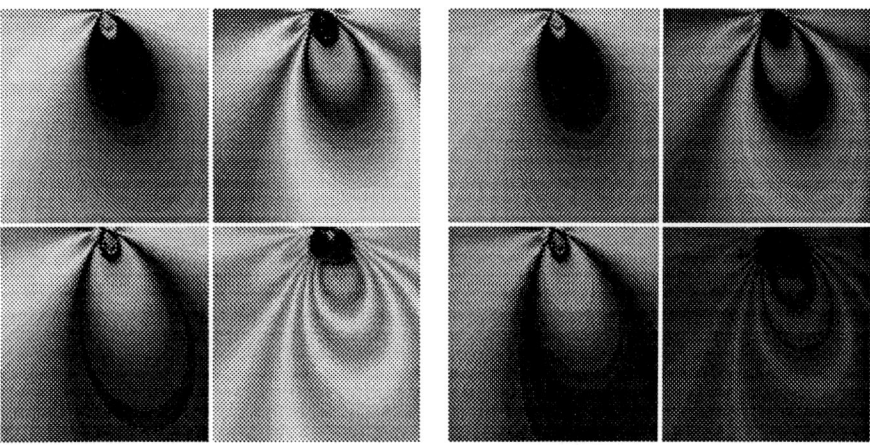

Fig. 1: Gray-scaled (left) and non-scaled (right) EDCImage simulations, specimen thickness 200, 400, 800 and 2000 nm.

The simulations in Fig. 2 show the influence of the parameters of the displacement model f and α at different specimen thickness (the specimen is assumed to be planparallel). It is also clear that the thickness of the specimen is an essential parameter in the EDCImage.

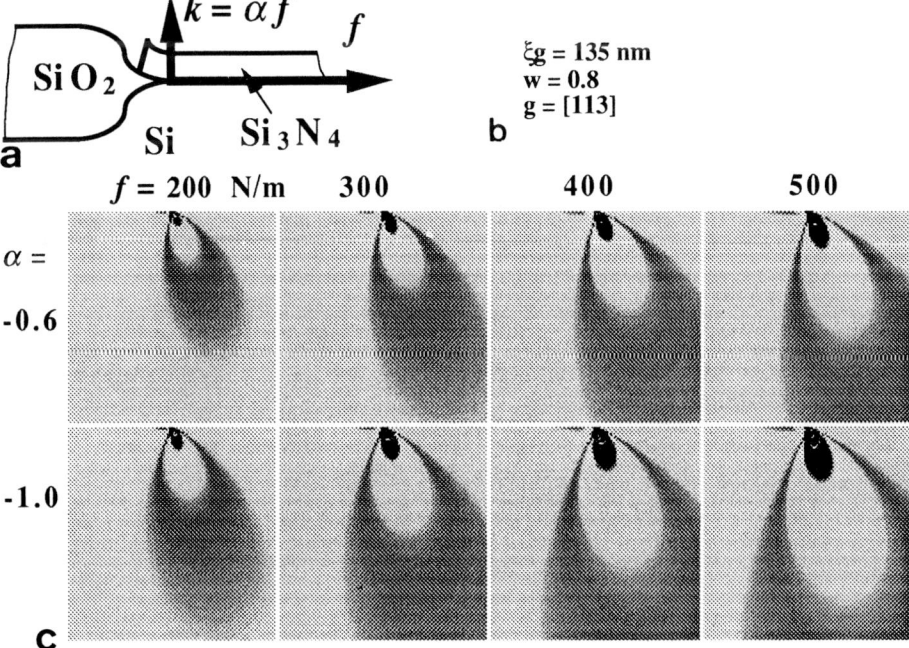

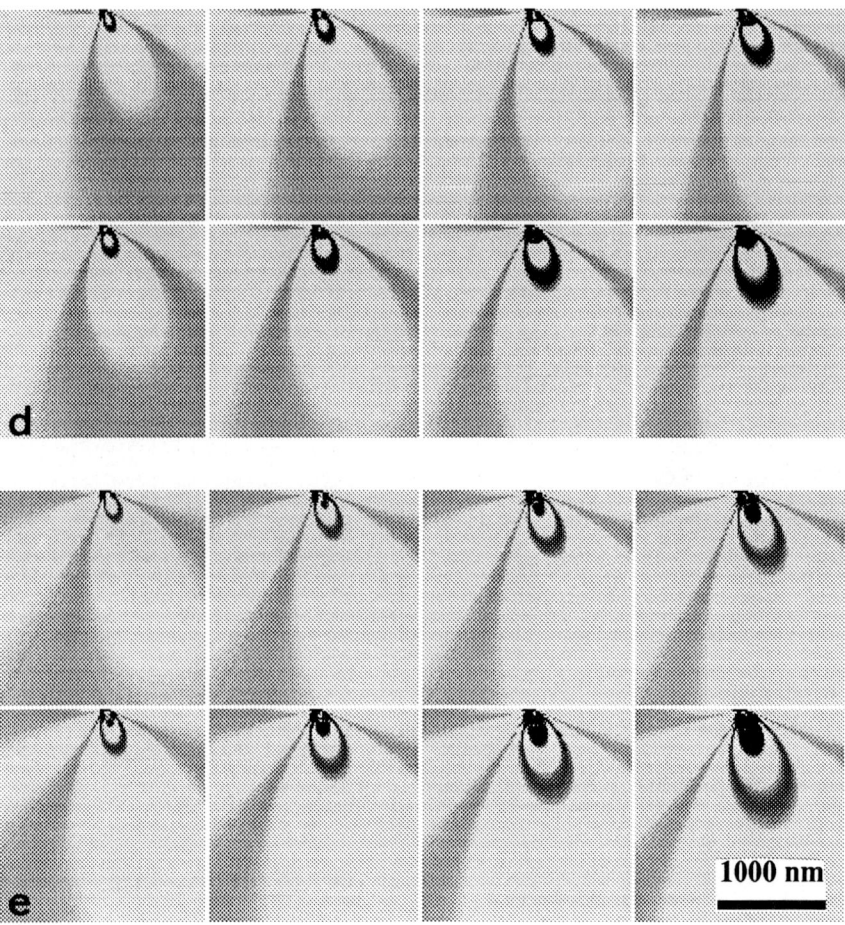

Fig. 2: EDCImage simulations transformed according to a standard photographic model.   a) The stress field model. b) The stress field parameters used for the simulations in c and d. c) EDCImages at specimen thickness 600 nm. d) The same EDCImages at specimen thickness 800 nm. e) At 1000 nm.

REFERENCES

Armigliato A, Balboni R, De Wolf I, Frabboni S, Janssens K G F and Vanhellemont J 1993 This Conference Proceedings

Benedict J P , Klepeis S J, Vandygrift W G and Anderson R 1989 EMSA Bulletin 19 74

Dainty J C and Shaw R 1974 Image Science: Principles Analysis and Evaluation of Photographic-type Imaging Techniques (Academic Press)

Hirsch P B, Howie A, Nicholson R B, Pashley D W and Whelan M J 1965 Electron Microscopy of Thin Crystals (London Butterworths)

Howie A and Whelan M J 1961 Proc. Roy. Soc. London A 217

Hu S M 1991 J. Appl. Phys. 70 R53

Janssens K G F, Vanhellemont J, De Graef M and Van der Biest O 1992 Ultramic. 45 323

Koenraad G. F. Janssens is indebted to the Belgian IWONL foundation for his fellowship.

Inst. Phys. Conf. Ser. No 134: Section 5
Paper presented at Microsc. Semicond. Mater. Conf., Oxford, 5–8 April 1993

Determination of lattice strain in local isolation structures by electron diffraction techniques and micro-Raman spectroscopy

A Armigliato*, R Balboni*, I De Wolf[+], S Frabboni*, K G F Janssens° and J Vanhellemont[+]

CNR-Istituto LAMEL, Via Castagnoli, 1, 40126 Bologna, Italy; * Dipartimento di Fisica, Università di Modena, Via Campi 213/A, 41100 Modena, Italy; [+] IMEC, Kapeldreef 75, 3001 Leuven, Belgium; ° Department of Metallurgy and Materials Science, University of Leuven, de Croylaan 2, 3001 Leuven, Belgium

ABSTRACT: To investigate the stress fields in local isolation structures, convergent beam electron diffraction (CBED), electron diffraction contrast imaging (EDCI) and micro-Raman spectroscopy (μRS) have proved to be powerful, complementary techniques. CBED and EDCI have a higher spatial resolution with respect to μRS but need elaborate sample preparation procedures to obtain thin cross sections of the structure. On the other hand μRS can measure stresses in the silicon substrate provided the overlayers are transparent; moreover, the recorded Raman shift is a convolution of shifts due to different stress components. The results of preliminary experiments performed on LOPOS structures having linewidths in the range 0.5-5 μm are reported and critically compared.

1. INTRODUCTION

The influence of the edges of the thin film structures on the device characteristics in integrated circuits is of increasing importance, due to their ever decreasing dimensions. The difference between the structural and mechanical properties of the films and those of the semiconductor substrate induces localized stress fields which, in turn, affect the electrical parameters (e.g. interface states and band gap narrowing) and the reliability of the devices.

In this paper convergent beam electron diffraction (CBED), electron diffraction contrast imaging (EDCI) and micro-Raman spectroscopy (μRS) have been applied to investigate the stress fields in local isolation structures.

2. EXPERIMENTAL

2.1 Sample Preparation

Different local isolation structures are fabricated on (001) silicon substrates. After growing a 10 or 18 nm pad oxide, a 50 nm polycrystalline silicon film is deposited in order to form local oxidation of polysilicon over silicon (LOPOS) structures. Afterwards, a 150 nm thick Si_3N_4 layer is deposited by low-pressure chemical vapour deposition (LPCVD). Using conventional photolithography, an oxidation mask consisting of parallel lines with different pitch is defined by wet oxidation at 950°C. TEM cross sections are prepared using the conventional procedures of mechanical polishing, dimpling and ion beam milling.

2.2 Equipment Employed

CBED experiments have been performed with a Philips CM30 TEM/STEM, operating at 100 kV, on cross-sectioned specimens kept at 100 K; the local strain is obtained by minimising the difference between the experimental patterns and the ones calculated by the EMS software package by Stadelmann (1987). EDCI images are taken with a Jeol 200CX TEM of the University of RUCA (Antwerp) at an acceleration voltage of 200 kV; the calculations of the diffraction contrast are performed by SIMCON, a recently developed software package, based on the dynamical diffraction theory (Janssens et al 1992).

The Raman spectra are recorded in back-scattering configuration using the 457.9 nm argon laser line. The scattered light is not analyzed. The laser spot is focused to a diameter smaller than 1 μm and moved across the sample perpendicularly to the direction of the oxidation mask lines with a minimum step size of 0.1 μm (De Wolf et al 1992).

3. A MODEL FOR THE STRAIN FIELD AT THIN FILM EDGES

The sketch of a (110) cross-section image of typical LOPOS structures studied in this work is reported in Fig.1. A simple model (Vanhellemont et al 1987a, De Wolf et al 1992) assumes that the film edge exerts a force F per unit length on the underlying substrate, which lies in the image plane and is inclined with respect to the surface. This force can be decomposed in a horizontal component f and a vertical component $k = \alpha f$. In this figure the x axis is oriented along the [110], the y axis along the [$\bar{1}$10] and the z axis along the [001] direction. For the thin TEM samples, a planar stress can be assumed, so a relaxation in the y-direction (if any, see §4.2.1) can occur, then in case of a single line of width w one obtains (Vanhellemont et al 1987a):

$$\sigma_{ij} = ijT(x) + (w - i)(w - j)T(w - x) \text{ where } T(x) = -2f\frac{x + \alpha z}{\pi(x^2 + z^2)^2}, \ i, j = x, z \quad (1)$$

From the expression for $T(x)$ the displacement field $\mathbf{R}$ near the film edge, which is needed for the EDCI simulations, can be calculated (Vanhellemont and van den Hove 1988):

$$\mathbf{R} = \mathbf{R}(x, z, f, \alpha, E, \nu) \quad (2)$$

where E and ν are the Young modulus and the Poisson ratio, respectively.

The components of the strain tensor can be expressed in the crystal axes system X, Y, Z as:

$$
\begin{aligned}
\epsilon_{XX} &= \epsilon_{YY} = (S_{11} + S_{12})\sigma_{xx}/2 + S_{12}\sigma_{zz}, \\
\epsilon_{ZZ} &= S_{12}\sigma_{xx} + S_{11}\sigma_{zz} \\
\epsilon_{XY} &= (S_{44}/4)\sigma_{xx} \\
\epsilon_{YZ} &= \epsilon_{XZ} = (S_{44}/2\sqrt{2})\tau_{xz}
\end{aligned}
\quad (3)
$$

where $S_{11} = 7.68 \cdot 10^{-12}\text{Pa}^{-1}$, $S_{12} = -2.14 \cdot 10^{-12}\text{Pa}^{-1}$ and $S_{44} = 12.7 \cdot 10^{-12}\text{Pa}^{-1}$.

Crystalline silicon has three optical Raman modes, from which only the longitudinal mode can be observed for backscattering from the (001) surface. In the presence of strain, the frequency of the Raman peak changes. For planar strain, a complicated relation is obtained between the Raman frequency and the three strain components (De Wolf et al, 1992) and it is not possible to calculate directly the magnitude of each of these components from the measured Raman frequency shift.

It will be shown in the next section that all the three techniques employed in this work can yield values for f and α and, at the same time, give partial and complementary tests of the model. In fact, μRS can determine these two parameters through eq.(1) by measuring σ_{xx} at the center of a LOPOS line; CBED measures ϵ_{ij} (i,j=X,Z), with an accuracy of $2 \cdot 10^{-4}$, wherefrom f and α can be found by combining eqs.(1) and (3); EDCI determines R and hence f and α through eq.(2).

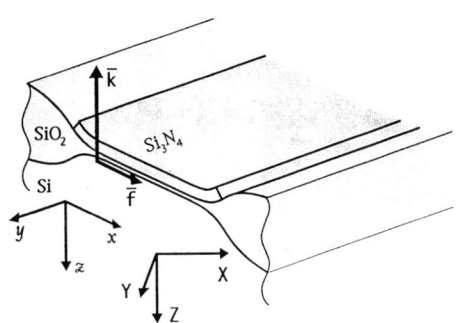

Fig.1 Sketch of the simple model assuming planar stress and the forces f and k exerted by the structure on the silicon substrate.

Fig.2 Raman frequency shift obtained during a scan across a 5 μm and a 2 μm line of a LOPOS sample.

4. RESULTS AND DISCUSSION

4.1 Micro-Raman Measurements

Fig. 2 shows the frequency of the silicon Raman peak (LO-peak) as a function of the position on a 5 μm and a 2 μm wide LOPOS sample. The vertical lines denote the border of the nitride film. Outside these lines, the Raman frequency is equal to the stress free value, ω_0, about 520 Rcm^{-1} for both widths. This indicates that the stress under the field oxide is very small. Under the nitride line, at about the position of the tip of the bird's beak, a large downward shift is detected, indicating tensile stress. This shift depends on the magnitude of both σ_{xx}, σ_{zz} and σ_{xz}, and it is not possible to give an estimation of the value of each of these stress components. It can be seen on the figure that this tensile stress is independent of the width of the lines, as the downward shift is the same for the 2 μm and the 5 μm samples. At the center of the 5 μm sample, the Raman frequency is larger than the stress free value, indicating compressive stress. For these broad lines, uniaxial stress can be assumed in the center of the line, so only σ_{xx} is non-zero. In this case the relation between Raman frequency, ω, and σ_{xx} is a linear function :

$$\omega - \omega_0 = -1.93 \cdot 10^{-9} \sigma_{xx} \tag{4}$$

and σ_{xx} is calculated to be $0.350 \cdot 10^9$ Pa. From this value and the relation between σ_{xx} and f at the center of the lines, the edge force f can be calculated to be 680 N/m (De Wolf et al, 1992).

For the 2 μm sample, the upward shift at the center is much smaller. For these widths, the influence of the vertical force k on the stress at the center cannot be neglected anymore, and also the Raman shift is influenced. As a first order approximation, it can be assumed that, also for these small lines, the effect of k on the stress at the center of the lines can be neglected. This would imply a value for σ_{xx} of $0.021 \cdot 10^9$ Pa and for f of 16 N/m. This contrasts with our simple model, which assumes that the film force f is independent of the width of the line, therefore for small lines the vertical force k also acts at the center and influences the stress. From eq.(1) and the assumption of a constant value of 680 N/m, k is calculated to be about -1700 N/m.

4.2 TEM Measurements

4.2.1 Convergent Beam Electron Diffraction (CBED)

In Fig. 3a is shown a cross-section TEM image of a 5 μm LOPOS line, where three points, among the ones selected for the CBED experiments, are indicated. Two of them (A and B) are close to the bird's beak, whereas the third one (C) corresponds to unde-formed silicon. The corresponding CBED patterns (both experimental and computed) are reported in Fig. 3b. The acceleration voltage is 100 kV and the crystallographic projection is < 130 >, where the central disc contains lines very sensitive to small strains and free from dynamical interactions (Balboni and Frabboni 1991). The effective voltage was determined by taking a CBED pattern on a region of the substrate, like the one labelled C in Fig. 3b, and assuming a silicon lattice parameter $a_0 = 0.54289$ nm, which is the one corresponding to a temperature of 100 K. The overall quality of these patterns is quite good, as can be judged from both the sharpness of the HOLZ lines belonging to the second Laue zone and the visibility of the fourth one.

From the comparison between the experimental and computed CBED patterns it comes out that in the probed points the cell is either monoclinic (point B) or triclinic (point A). From the corresponding lattice parameters, it is possible to deduce the ϵ_{ij} values (i,j =X,Y,Z), which allow in turn the stress components σ_{xx}, σ_{zz} and τ_{xz} to be determined. The values of the strain components are reported in Table I.

In Fig.4 are shown the cross section TEM images performed on a 2 μm and 1μm line of a LOPOS sample. The values of the ϵ_{ij} parameters deduced from the CBED patterns taken in the points indicated in this figure are reported in Table I. At the bird's beak regions of the 5 μm and 2 μm lines the strain components ϵ_{XX} and ϵ_{ZZ} are either tensile or compressive, which gives a more detailed description of the strain field with respect to the μRS analyses.

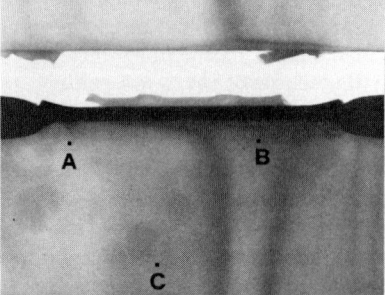
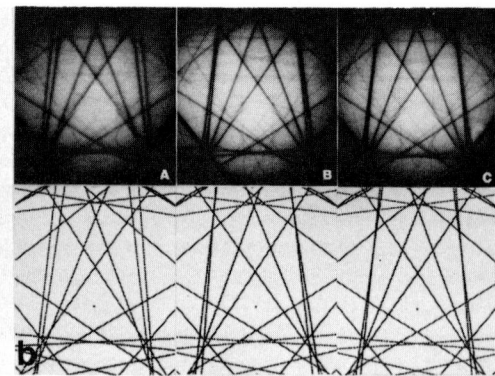

Fig. 3. (a) Cross section TEM image of a 5 μm line of a LOPOS structure. (b) Experimental (top) and computed (bottom) CBED patterns taken in the points labelled in (a). The point C has been taken in an undeformed silicon region.

The f values deduced from Table I are 250 N/m for the 5 μm line and 60 N/m for the 2 and 1 μm lines ($\alpha = -0.4$ for the three cases). While the former value is not too different from the one yielded by μRS (f=680 N/m), the latter f value is not in agreement with the hypothesis of a constant film force. This suggests that either f is not constant or the stress is not planar in the TEM sample.

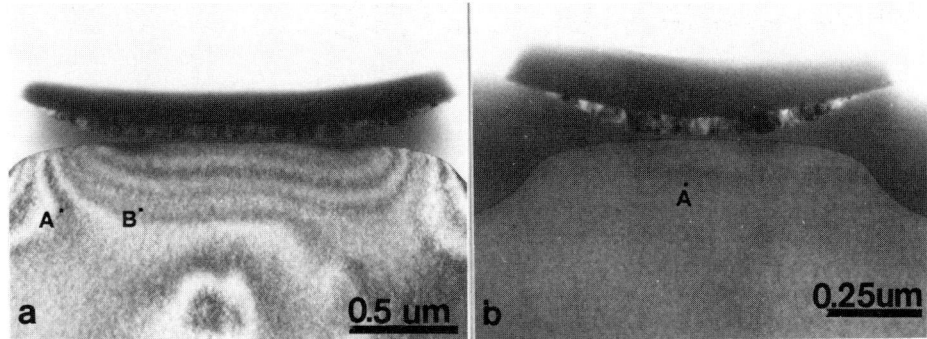

Fig. 4. Cross section TEM images of a 2 μm (a) and a 1 μm (b) line of a LOPOS structure. The values of the parameters reported in Table I have been obtained from CBED patterns taken in the labelled points.

Table 1: Values of the components of strain tensor in eq.(3), as deduced from CBED patterns taken in the points indicated in Figs. 3 and 4.

Linewidth (μm)	Point	Depth (μm)	Lattice Strain (x10^{-4})			
			ϵ_{XX}	ϵ_{ZZ}	$2 \cdot \epsilon_{XY}$	$2 \cdot \epsilon_{YZ}$
5	A	0.3	-7	+24	> -2	3
	B	0.3	0	-7	-7	0
2	A	0.3	0	-4	0	0
	B	0.3	> -2	+6	0	0
1	A	0.12	-4	+9	-3	> -2

4.2.2 Electron Diffraction Contrast Imaging (EDCI)

Electron Diffraction Contrast Imaging is a well established technique of which the principles have already been defined more than 30 years ago. In EDCI one obtains a bright or dark field contrast image of a strained area using one or more diffracted electron beams. Then, from a mathematical model for the displacement field, the contrast image is simulated under the same diffraction conditions as determined in the experiment. Recently work started to further develop EDCI taking full advantage of the progress which has been made in computing and image recording hardware (Janssens et al 1992 and 1993). An EDCI result obtained so far is reported in Figure 5, which shows a two beam electron diffraction image of the strain field in a 0.5 micron LOPOS line. The "best fit" calculated image is obtained by visual comparison between calculated and experimental images. The α and f values, -1.0 and 300 N/m, respectively, correspond well with those obtained with μRS and from the calculation of equilibrium dislocation distributions at local isolation structure edges (Vanhellemont et al 1987b). It is also in agreement with the value $f=250$ N/m obtained by CBED on a 5 μm line. A more quantitative comparison requires digitized images and a regression procedure to minimize the difference between experimental and calculated images.

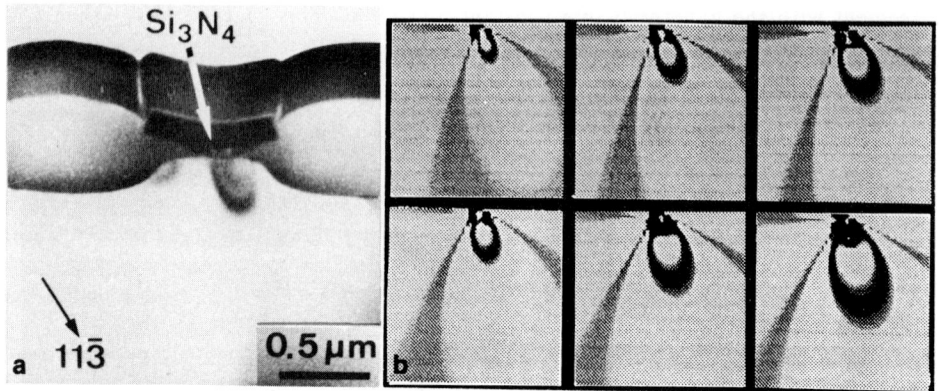

Fig. 5. a) Experimental two beam electron diffraction image of the strain field at the edge of a 0.5 μm LOPOS line (courtesy A.Romano-Rodríguez). b) Simulated EDCI images with α = -0.6,-1.0 (rows from top to bottom) and f=200, 300, 400 N/m (columns from left to right) and assuming a specimen thickness of 800 nm, g =[11$\bar{3}$] and w = 0.8.

5.CONCLUSIONS

The employed TEM techniques (CBED and EDCI) have a high spatial resolution (< 10 nm), but need the elaborate   sample preparation techniques, typical of TEM cross sections. On the other hand, μRS is non-destructive, but its spatial resolution is limited to about 1 μm. Nevertheless, useful complementary information has   been obtained by these two techniques in the 5 and 2 μm wide lines. Presently, the f values extracted with the three techniques are not yet in full quantitative agreement; probably the proposed model needs to be improved to better describe the case of TEM cross sections.

In any case, structures having a 1 μm or smaller size can only be measured by TEM. It is therefore expected that, proceeding towards deep submicron structures, TEM techniques like CBED and EDCI will play a dominant role in the field of the characterization of silicon lattice strains.

ACKNOWLEDGMENTS

The authors are indebted to H Norström for supplying the structures studied in this work and to F Corticelli and D Govoni for technical assistance. K G F Janssens is indebted to the Belgian Science Foundation (IWONL) for his fellowship.
This work was partially supported by CNR-Progetto Strategico MADESS II.

REFERENCES

Balboni R and Frabboni S 1991 Microsc.Microanal.Microstruct. 2 617
De Wolf I, Vanhellemont J, Romano-Rodríguez A, Norström H and Maes H E 1992 J. Appl. Phys. 71 898
Janssens K G F, Vanhellemont J, De Graef M and Van der Biest O 1992 Ultramicroscopy 45 323
Janssens K G F, Vanhellemont J, Maes H E and Van der Biest O 1993, this Conference
Stadelmann P 1987 Ultramicroscopy 21 131
Vanhellemont J, Amelinckx S and Claeys C 1987a J.Appl.Phys. 61 2170
Vanhellemont J, Amelinckx S and Claeys C 1987b J.Appl.Phys. 61 2176
Vanhellemont J and van den Hove L 1988 Inst.Phys.Conf.Ser. 93 423

Inst. Phys. Conf. Ser. No 134: Section 5
Paper presented at Microsc. Semicond. Mater. Conf., Oxford, 5–8 April 1993

235

On the formation of trench-induced dislocations in dynamic random access memories (DRAMs)

M Dellith, G R Booker, B O Kolbesen[*], W Bergholz[*] and F Gelsdorf[*]

Department of Materials, University of Oxford, Parks Road, Oxford OX1 3PH, England
[*]Siemens AG, Abteilung HL TB, Otto-Hahn-Ring 6, D-8000, Munich, Germany

ABSTRACT: The presence of dislocations near the edge of trench capacitor memory cells in 4M DRAMs leads to refresh failure of these cells. From a systematic TEM investigation of the development of these trench-induced dislocations (TIDs) from their initial stages the following is concluded: small dislocation half loops nucleate on damage clusters introduced by dry etching and stress from oxide layers then triggers the formation of large half loops and TIDs. The number of TIDs depends markedly on the types and thicknesses of the dielectric layers in the memory cells. This behaviour is attributed to differences in the local stresses created at the edges of the memory cells and/or dislocation pinning effects by nitrogen when Si_3N_4 layers are used.

1. INTRODUCTION

In early releases of 4M DRAMs with trench design (Fig. 1) electrical failure of a small number of trench capacitor memory cells reduced the performance of the devices. In these chips, dislocation densities of up to ~ 10^5 cm^{-2} occurred as determined by Secco-etching and optical microscopy, which means that only about one in a hundred trenches was affected. Higher dislocation densities correlated strongly with increased charge leakage and the need for shorter refresh times. These defects were termed trench-induced dislocations (TIDs) because their formation was dependent on the presence of the trenches. Several factors were found to influence the TID density, most importantly the trench dielectric materials and layer thicknesses. Later in the development of the 4M DRAMs the formation of TIDs close to the trench cells was suppressed (Kolbesen et al 1989, 1991a and b, 1992, Dellith et al 1991). The present work represents the first reported systematic analysis carried out by transmission electron microscopy (TEM) of pre-release DRAMs containing TIDs. The development of defects from the initial to the final stages is followed, the important stages in the processing giving rise to the defects are identified and defect formation mechanisms are discussed.

2. EXPERIMENTAL

TEM investigations were performed on the following wafers. A first series (group 1) corresponded to processing up to five different stages using trench capacitors with a thin SiO_2 layer. The stages were termed POLY 1 (trench partly filled with polysilicon), IOX (trench oxide sealing), POX 1 (transistor oxide), $p^+ - I^2$ (boron diffusion into periphery) and ZOX 1 (isolation oxide before bit line deposition). A second series (group 2) corresponded to processing up to the ZOX 1 stage but using trench capacitors with different dielectric materials and layer thicknesses. For the TEM studies the surface dielectric and metal layers were initially removed from the Si wafers by immersing them in concentrated (49%) HF for 24 hr. TEM plan-view specimens were prepared by standard procedures with thinning of the wafer from the back side only. The TEM examinations were performed using a Philips CM20 and an Akashi 200CX both with a beam energy of 200 keV. This procedure enabled defects present in those portions of the wafer lying between the surface and a depth of ~ 0.5 µm to be observed.

3. RESULTS

The TEM investigation of wafers taken out of the process at different stages (group 1) showed a remarkably consistent sequence in which small defects introduced during the initial processing subsequently developed to give the TIDs. After the trenches were formed by dry etching but before filling, the Si close to the edges of some of the trenches showed the presence of fine damage. After the POLY 1 stage, some of the trench edges showed groups of dislocation half-loops, each loop being typically ~ 100 nm long and ~ 5 nm wide with its two ends located at the trench edge (Fig 2). After the IOX stage, such regions contained either 1, 2 or 3 half-loops typically ~ 100 nm long and ~ 50 nm wide with their two ends again located at the trench edge (Figs 3, 4 and 5). After the POX 1 stage, these half-loops had occasionally developed into pairs of long dislocations but had mostly developed into single long dislocations, of which only portions were present in the thin foil specimens (Fig 6). In general each of these latter dislocations was curved and connected the trench edge, sometimes at a point deeper in the wafer than could be revealed by TEM of the thin foil specimen, to the Si wafer surface between this trench and an adjacent trench. After the $p^+ - I^2$ stage, and also after the ZOX 1 stage, there was no significant change, the long dislocations still being present. It was these dislocations (TIDs) which were present in the later and final stages of the fabrication that were revealed by etching and optical microscopy. Neither the fine damage nor the half-loops could be identified by the etching and optical microscopy.

The above TEM results were obtained for wafers with a thin SiO_2 trench dielectric layer. Subsequent TEM comparisons of wafers taken out of the processing after the ZOX 1 stage, but with trench capacitors containing different thicknesses and types of dielectric layer (group 2), showed the following. For a thin SiO_2 layer, the main defects were TIDs (as described above) with density ~ 10^5 cm^{-2}, although half-loops were occasionally present. For a thick SiO_2 layer, there were TIDs of density ~ 10^4 cm^{-3} and many more half-loops. For an SiO_2 layer of medium thickness, the distribution between TIDs and half-loops was intermediate. For either a $SiO_2/Si_3N_4/SiO_2$ triple layer or a SiO_2/Si_3N_4 double layer, etching revealed TIDs of density ~ 10^3 cm^{-2}. TEM showed some half-loops and occasionally a TID.

The geometries and Burgers vectors of typical examples of all of the dislocations observed were determined. This showed that the half-loops and TIDs mainly occurred on the inclined {111} planes and possessed a/2 <110> Burgers vectors lying within the particular {111} plane, i.e. the dislocations were glissile. For the individual TIDs, a significant length of the dislocation was often close to a 60° type orientation.

4. DISCUSSION AND CONCLUSIONS

The TEM examinations at progressive stages of fabrication of the DRAMs with the thin SiO_2 layers in the trenches reveals how the TIDs are nucleated and progressively develop. The dry etching procedure used to make the trench holes produces fine damage at the edges of the trenches. When the trenches are partly filled with polysilicon, the fine damage acts as nucleation sites for the formation of groups of small dislocation half-loops at the edges of some of the trenches. When the trench oxide sealing is performed, many of these half-loops are eliminated, some increase in size to give larger half-loops and a few develop further to give TIDs. When the transistor oxide is deposited, more of the half-loops develop to give TIDs. During the subsequent processing, only small changes in the dislocations arise. The initial development of the half-loops probably occurs by a combination of climb and glide processes, whereas the subsequent development of the TIDs from the half-loops occurs by glide only.

The point defects, e.g. Si interstitials, that cause the climb arise from thermal generation at the processing temperatures and injection at the trench and wafer surfaces caused by, for example, the growth of the SiO_2 layers. The stresses that cause the glide arise from the oxidation processes close to the trenches, the dielectric layers and dopant concentration gradients. The magnitudes of both the point defect concentrations and the stresses are different in the various local regions of the specimen, are different during the heating-up, processing and cooling-down periods for the individual processing stages, and are different for the various processing stages. They are complex and difficult to calculate. Nevertheless, such calculations have been performed (Fahey et al 1992, Hu 1988) for trench structures in Si wafers with surface dielectric layers based on simple models, and these have indicated that maximum stress often occurs at the trench edge near the wafer surface, i.e. where in the present DRAMs the initial dislocation activity takes place. Such stresses can, for example, be large during the IOX stage because an oxide wedge forms near the wafer surface between the trench polysilicon 1 layer

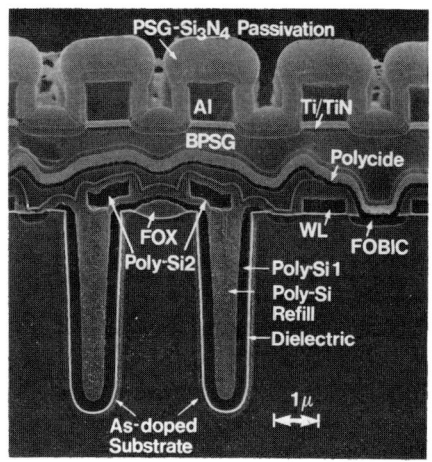

Fig. 1: Cross-section SEM of trench memory cells and bitline contact (FOBIC)

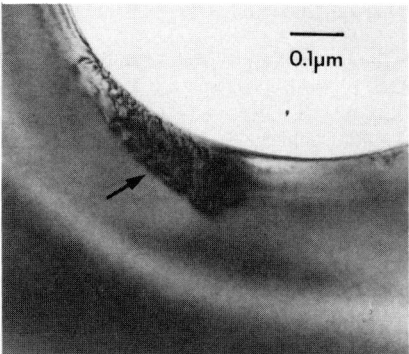

Fig. 2: Plan-view TEM of group of small half loops at trench edge

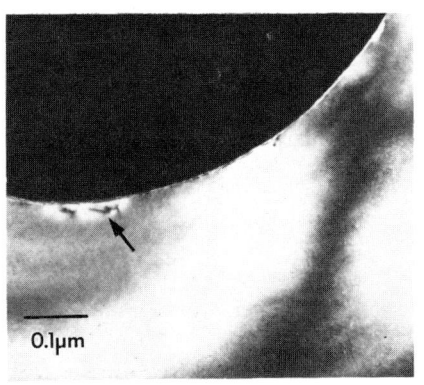

Fig. 3: Plan-view TEM of large half loops at trench edge

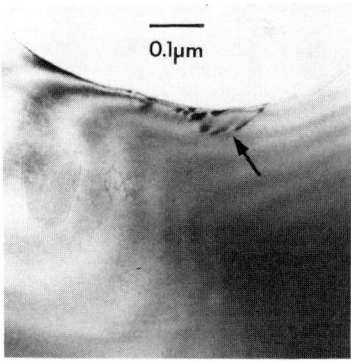

Fig. 4: Plan-view TEM of large half loops at trench edge

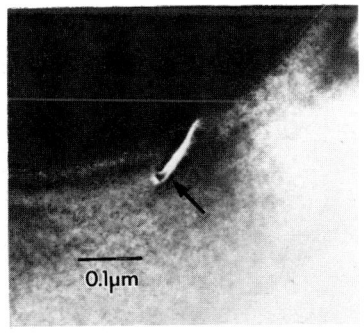

Fig. 5: Plan-view TEM of large half loop at trench edge

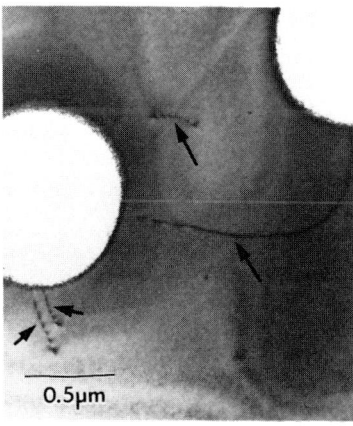

Fig. 6: Plan-view TEM of four trench-induced dislocations (TIDs) arising from loops

and the polysilicon filling (Fig 1), and this could trigger the major change from half-loops to TIDs that then occurs.

The present results showed that after the ZOX 1 stage, for trenches with a thin SiO_2 layer there were ~ 10^5 cm^{-2} TIDs and few large half-loops, while for trenches with a thick SiO_2 layer there were ~ 10^4 cm^{-2} TIDs and many large half-loops. This suggests that immediately prior to this stage there was approximately the same number of large half-loops for both types of trench, and that during this stage the local stresses were greater for the trenches with the thin SiO_2 layers and so more of the large half-loops were triggered to give TIDs in this case. However, initial considerations suggest that a thin SiO_2 layer is likely to produce smaller stresses in the adjacent Si than a thick SiO_2 layer. A possible explanation is that compensating stresses may be present at this particular processing stage. For example, the major stresses could be compressive and arise from the trench filling, while the minor stresses could be tensile and arise from the trench SiO_2 layer. The net stress at the trench edge would then be greater with the thin trench SiO_2 layer than the thick trench SiO_2 layer.

The present results also showed that after the ZOX 1 stage, for trenches with either a $SiO_2/Si_3N_4/SiO_2$ triple layer or a SiO_2/Si_3N_4 double layer, there were ~ 10^3 cm^{-2} TIDs, i.e. significantly less than when a SiO_2 layer only was used. Possible reasons for this are either a decrease in the number of injected Si interstitials resulting in decreased dislocation climb or the diffusion of nitrogen from the Si_3N_4 layers into the Si resulting in pinning of the dislocations and hence decreased glide (Sumino 1989). On the evidence presently available it is considered that the nitrogen pinning is probably the main mechanism operating. A more detailed account of these findings will be published elsewhere.

The work has shown how the TIDs are nucleated and develop and how their development can be either retarded or suppressed by modifying the trench dielectric layers. The detailed behaviours that occur have been elucidated by performing comprehensive TEM examinations and this would not have been possible by using only etching and optical microscope studies. With these particular 4M DRAMs, optimisation of the design and processing conditions enabled the TID density to be reduced to sufficiently low values that acceptable device yields were obtained. For future DRAMs with increased numbers of memory cells, decreased feature sizes and increased complexities, the TID density may again increase due to, for example, increased local stresses. The results of the present work will be helpful in understanding the dislocation behaviour which then occurs and possibly in indicating how the TID number density could be reduced.

REFERENCES

Dellith M, Gelsdorf F, Bergholz W, Booker G R and Kolbesen B O 1991 Inst.Phys. Conf. Ser. No. 117 169

Fahey P M, Mader S R, Stiffler S R, Mohler R L, Mis J D and Slinkman, J.A. 1992 IBM Journ. Res. Devel. 36 158.

Hu S M 1988 J. Appl. Phys 64 323.

Kolbesen B O, Bergholz W and Wendt H 1989 Materials Science Forum 38-41 1

Kolbesen B O, Bergholz W, Cerva H, Gelsdorf F, Wendt H and Zoth G 1991a Inst. Phys. Conf. Ser. No. 104, 421

Kolbesen B O, Gelsdorf F, Cerva H, 1991b Symposium on Advanced Science and Technology of Silicon Materials, Japan Society for the Promotion of Science, p. 75

Kolbesen B O, 1992 NATO ASI Ser. 'Crucial Issues in Semiconductor Materials and Processing Technology' (Kluwer Academic Publs, Netherlands) p. 3

Sumino K, 1989 Inst. Phys. Conf. Ser. No. 104 245.

Inst. Phys. Conf. Ser. No 134: Section 5
Paper presented at Microsc. Semicond. Mater. Conf., Oxford, 5–8 April 1993

239

A cross-section TEM investigation of SOI transistor structures

M C L Ward, A G Cullis, K M Brunson, P W Smith and A M Hodge

DRA Malvern, St Andrews Road, Malvern, Worcs WR14 3PS

ABSTRACT: Cross-sectional TEM has been used to perform a detailed study of FIPOS (Full Isolation by Porous Oxidised Silicon) and SIMOX (Separation by Implantation Of Oxygen) SOI (Silicon On Insulator) MOS transistors. FIPOS technology has potential advantages over SIMOX technology since the material quality should be superior, the buried oxide thickness can be controlled and the quality of the buried oxide should be similar to that of a thermal oxide. We have fabricated CMOS devices using the two technologies and compared the electrical performance of the two device types. These results have then been correlated with the results of a TEM study. We have identified different types of defect structure present in the two types of isolation oxide and differences in the nature of the silicon/ isolation oxide interface which we have correlated with the electrical properties of the devices.

1. INTRODUCTION

Silicon On Insulator (SOI) technologies offer many advantages over conventional bulk technologies for harsh environment applications. The presence of the isolation oxide between active devices and the substrate has several beneficial effects. The device isolation completely removes the cause of thyristor latch and so allows devices to be packed much closer together for a higher packing density. The reduction in device substrate capacitance allows faster operation due to the reduced substrate coupling. Finally, the junction areas and volumes are reduced in the SOI technology thus reducing leakage currents at elevated temperatures and providing a greater degree of radiation tolerance. There is therefore considerable interest in exploiting SOI technology for military and future civil applications.

There are several different types of SOI technology, SIMOX (Belz et al 1990), FIPOS (Brumhead et al 1987), SOS (Mayer et al 1984) and BESOI (Mumola et al 1992). SOS material is available commercially but the silicon is generally of a poor quality due to poor epitaxy and thermal mismatch. BESOI while offering the potential for very high quality material is not yet readily available. SIMOX is now available from a variety of sources and the quality of this material is improving rapidly. Even though FIPOS is not a commercial technology it can be produced under laboratory conditions using relatively simple equipment and should provide material of the highest quality. FIPOS and SIMOX material are therefore potential competitors to replace the already established SOS technology for the next generation of SOI devices that will require the higher quality material offered by the FIPOS and SIMOX technologies. In order to confirm the relative merits of these two technologies we have conducted a comparative study between MOS devices fabricated in FIPOS material produced in our laboratory and SIMOX material obtained from Ibis.

The FIPOS and SIMOX material was selected to have similar silicon and buried oxide thicknesses($0.2\mu m$ and $0.5\mu m$ respectively) and carrier concentrations (N-type $1E16cm^{-3}$). Devices were then fabricated in the two material types using the same process line and conditions. To minimise differences between the processing the two technologies were run as one batch. The devices were then characterised electrically and structurally using cross-sectional TEM.

2. SAMPLE PREPARATION

2.1 FIPOS

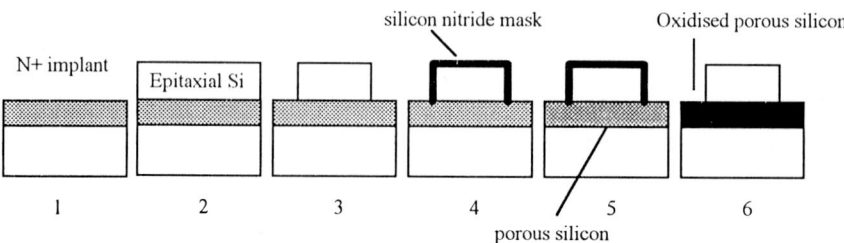

Fig. 1. Process sequence for the formation of FIPOS SOI

The process sequence for the formation of FIPOS material is shown in Fig. 1. A bulk wafer ((100), As doped $10^{15}cm^{-3}$) is implanted with Sb to form an N+ surface layer (1). It is this surface layer that will be anodised to porous silicon and finally converted to oxide. After an activation anneal, 0.26μm of silicon is epitaxially grown on to the surface (2). The required active areas of the layer are protected with a silicon nitride mask and the remaining epitaxial silicon removed by dry etching (3),(4). The wafer is then anodised in an HF/ethanol mixture (5). The N+ regions are selectively anodised, with pores penetrating beneath the masked island. The structure is then oxidised: the porous silicon is converted to a thermal oxide. The mask is finally removed leaving an isolated silicon island (6). The process was so arranged as to produce a final silicon island thickness of 0.2μm and a buried oxide thickness of 0.5μm.

2.2 SIMOX

The SIMOX material was obtained from Ibis. The characteristics of the supplied material were: silicon substrate orientation < 100 > , doping As $10^{16}cm^{-3}$, silicon over-layer thickness 0.2μm, buried oxide thickness 0.4μm. These were confirmed by SEM observations and optical characterisation. The required active areas were protected by a photo-resist mask and the remaining epitaxial silicon layer removed by plasma etching.

2.3 Device Fabrication

With both material technologies at the same stage, active area silicon defined, the two sets of material were processed through a common CMOS device process sequence. In order to limit any removal of the buried oxide during sacrificial gate oxidation, no sacrifical gate oxide was used. The gate oxide was therefore subjected to back and front gate implants. For this initial study no attempts were made to include advanced device structures such as LDD or corner/edge controls.

3. ELECTRICAL CHARACTERISATION

The wafers were extensively characterised for mobility, threshold voltage and subthreshold slope. The results are summarised below in Table 1. Despite the common process sequence the FIPOS and SIMOX devices displayed different threshold voltages. The FIPOS devices were closer to the designed threshold of 1.0V. Both also displayed good threshold slopes of less than 100mVdec^{-1}. Typical input characteristics of the devices are shown in Fig. 2.

Technology	Threshold Voltage (V)	Mobility ($cm^2V^{-1}s^{-1}$)	Subthreshold Slope(mV/dec)
SIMOX	1.54	570	103
FIPOS	0.85	600	79

Table 1. Summary of the electrical properties of the FIPOS and SIMOX devices.

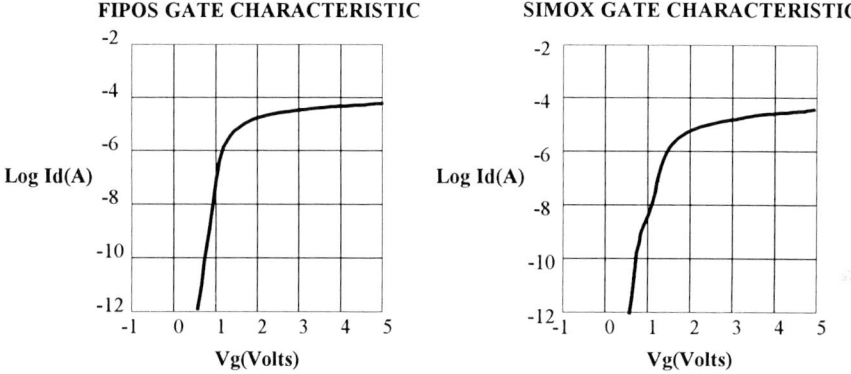

Fig. 2. Typical input characteristics of the FIPOS and SIMOX devices.

The SIMOX device with the 'kinked' characteristic displays clear evidence of 'side wall' transistor action. The electrical data suggest that the channel mobility of the FIPOS devices is higher than for the SIMOX devices. However the effect is small and further work is needed to confirm it.

In order to examine the quality of the back of the device and the buried oxide we have measured the breakdown on the back gate and the generation rate associated with the back interface. No breakdown could be found on the SIMOX wafers with up to 100V, while many of the FIPOS devices displayed clear breakdown at about 80V under similar conditions. Given the thicker oxide present in the FIPOS sample we are forced to conclude that the FIPOS oxide is of a lower quality than the SIMOX oxide.

We have also used a simplified lifetime measurement (Kimpton et al 1992) to compare the generation rate of minority carriers in the device. The experimental configuration is shown in Fig.3.

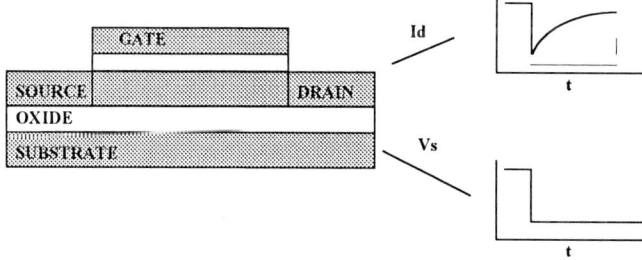

Fig. 3. Experimental configuration used for lifetime measurement.

A small drain bias (100mV) is applied to the SOI device. The substrate potential is then pulsed from ground to -100V. This pulse forces the back interface into accumulation, however the majority carriers required for accumulation are not immediately available and must be generated within the device. This leads to the expulsion of minority carriers from the front channel with a resultant drop in drain current. The generation of the majority carriers required for back interface accumulation can then be monitored by the rise in drain current. The results presented in Fig. 4,

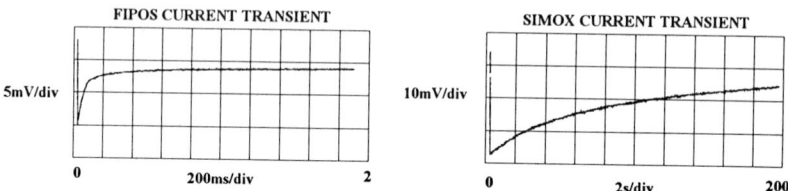

Fig. 4. Current transients of the devices.

clearly show that the recovery time in the FIPOS device is nearly an order lower than that of the SIMOX device. Since the two devices were fabricated under identical conditions and therefore share a similar structure, we are able to conclude that the greater generation rate of the FIPOS is due to a higher generation rate of the bulk depleted material or of the back interface.

4. TEM CHARACTERISATION

4.1 FIPOS

In order to determine details of the device structure and of the quality of the buried oxide and interface we studied the devices using cross-sectional TEM. A cross-section of the edge of a FIPOS island is shown in Fig. 5. The image shows that a thickening of the oxide at the edge of the island

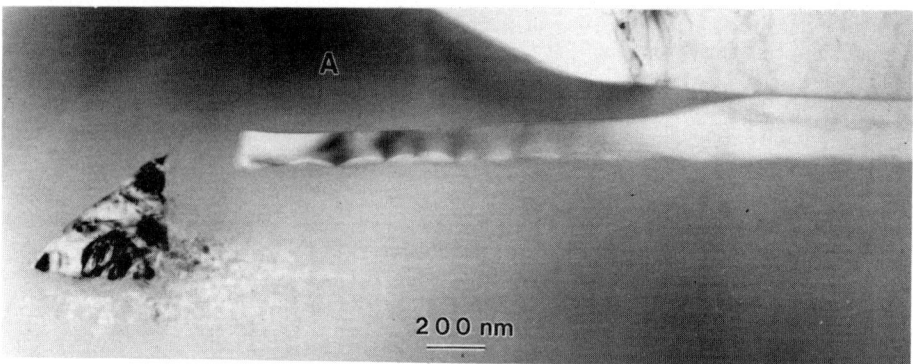

Fig. 5. TEM cross-section of the edge of a FIPOS transistor.

(A) has occurred. This thickening took place during oxidation of the porous silicon, indicated by the depth of the thickening and the position of the fillet of polysilicon gate material. This thickening lifts the polysilicon gate off the edge of the silicon island. Normally the gate electric field is increased at this point due to geometrical effects, locally reducing the threshold voltage of the device, leading to 'side wall' transistor action. However the increased separation of the gate now reduces the electric field intensity there, reducing the edge effect transistor action. The lateral extent of the oxidation at the edge (1.0μm) does limit the fabrication of narrow transistors, so from that point of view it detracts from the technology. The TEM cross-section of the center of a FIPOS device is shown in Fig. 6.

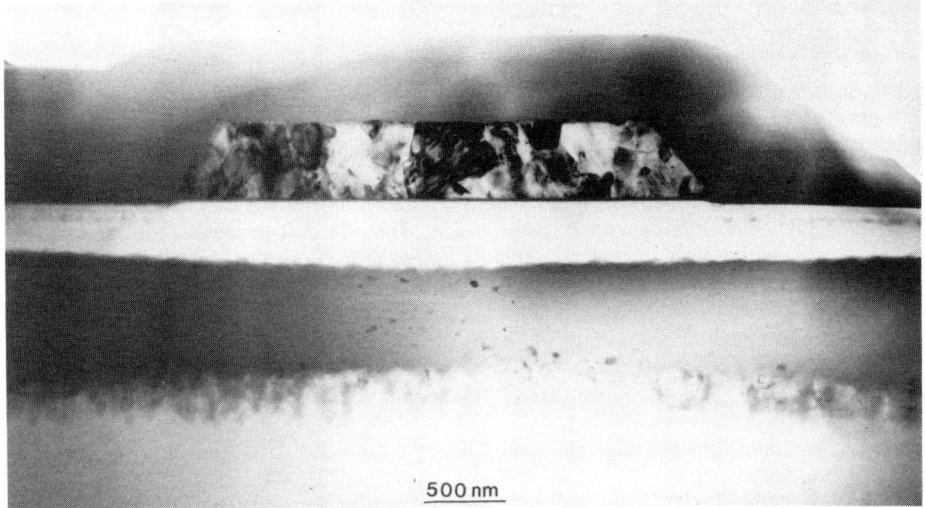

Fig. 6. TEM cross-section of a FIPOS transistor.

The buried oxide can be seen to contain some small regions of un-oxidised silicon. This is probably the cause of the poor back oxide breakdown characteristics of the buried oxide. The silicon island also shows considerable variation in thickness. This is caused by the limited range of anodisation in material of this carrier concentration. Thus at the middle of the island the extent of anodisation is reduced leading to the formation of thinner oxide. The interfaces between the oxide and silicon layers are complex displaying a great deal of structure.

4.2 SIMOX

Fig. 7. TEM cross-section of a SIMOX transistor island.

The SIMOX cross-section shown in Fig. 7 above is of a typical silicon island (A) after processing. From the electrical characteristics we do not expect to see any substantial thickening of the island edge. The SIMOX buried oxide (B) also shows the presence of silicon inclusions (C) within the oxide at the substrate interface. However these are far removed from the active device layer and so should have very little effect upon device performance. Comparison of the back interface of the FIPOS and SIMOX devices display a clear contrast. The SIMOX displays a relatively sharp, though not completely planar, well defined interface while the FIPOS interface shows more undulations and irregularities. The latter is caused by anodisation and subsequent oxidation.

5. CONCLUSION

We have correlated the back interface quality with the 'minority carrier' life time of the device, identified small remnants of silicon in the FIPOS oxide that may be associated with the breakdown of the FIPOS oxide and have concluded that the edge transistor action in FIPOS is controlled by the thickening of the oxide on the island edge that takes place during the oxidation of the porous silicon. We conclude that FIPOS can provide devices fully compatible with SIMOX quality, but that further development is required to make it technologically exploitable. The SIMOX devices were simple to fabricate and of good quality. The side wall transistor problem has been easily addressed with a suitable implant. The poor quality of the buried oxide and interface can be addressed by optimising the epitaxial growth of the silicon over layer.

REFERENCES

Belz J, Burbach G, Vogt H and Pieczynski J 1990 Proc. Fourth International Symposium on Silicon-on-Insulator Technology and Devices, ECS, **90-6**, p 518

Brumhead D, Castledean J G, Keen J M, Cole J M, Earwaker L G, Farr J P G, L'Ecuyer J, Loretto M H and Sturland I M 1987 Proc. ESSDERC'87 pp 391-394

Godfrey D J 1987 Proc. IEE Colloquium on Semiconductor Interfaces, Digest No.72

Kimpton D and Kerr J 1992 IEEE Trans. Nuc. Sci. **39** 6

Mayer D C, Vasudev P K, Lee J Y, Allen Y K and Henderson R C 1984 IEEE Electron. Lett., **EDL-5**, pp 156-158

Mumola P B, Gardopee G J, Clapis P J, Zarowin D B, Bollinger L D and Ledger A M 1992 Proc. IEEE International SOI Conference, Saw Grass, p 152

Changes in electrical device characteristics during the formation of dislocations *in situ* in the TEM

F M Ross*, R Hull, D Bahnck, J C Bean, L J Peticolas, C A King and R R Kola

AT&T Bell Laboratories, 600 Mountain Avenue, Murray Hill, NJ 07974, USA
*Lawrence Berkeley Laboratory, 1 Cyclotron Road, Berkeley, CA 94720, USA

ABSTRACT: By adding electrical connections to a specimen heating holder for a transmission electron microscope, we have measured the characteristics of electronic devices such as diodes while they remain under observation in the microscope. We have made electron-transparent specimens from metastable GeSi/Si p-n junction diodes and introduced dislocations by heating *in situ*. The combination of electrical measurement and real-time observation of dislocation formation allows us to examine the electrical properties of dislocations in individual devices and the influence of defects on device performance.

1. INTRODUCTION

The presence of structural defects in a semiconductor is known to be associated with poor performance in electronic devices fabricated from the material (Sze 1985). Although the relationship between material perfection and device performance is usually only considered for the case of defects in single crystal substrates, it also applies, of course, to any defects which may be introduced during the growth, patterning and operation of devices. We are particularly interested in the effects of misfit dislocations on the properties of devices made using strained layer epitaxy, and in this paper we describe experiments designed to probe the relationship between the formation of misfit dislocations and the performance of such devices.

The equilibrium critical thickness theory of Matthews and Blakeslee (1974) predicts that if a strained film is grown epitaxially onto a mismatched substrate, at a critical film thickness, h_c, it becomes energetically favourable for the epilayer to relax by the formation of misfit dislocations at the substrate-epilayer interface. For the layer thicknesses and lattice mismatch often necessary in semiconductor devices, this can result in the introduction of dislocations into active regions of structures and poor performance in, for example, III-V optoelectronic devices. Rather than considering complex structures, we have investigated the effect of misfit dislocations on simple devices fabricated in a model system. We have made p-n junction diodes from metastable GeSi/Si heterostructures, with the GeSi epilayers chosen to be above h_c so that dislocation formation is energetically favoured and will occur upon heating the material. By forming electron-transparent regions in the devices, we can observe the density of dislocations during heating *in situ* in a TEM. These *in situ* heating experiments are extremely interesting in themselves, as dynamical dislocation processes (such as movement, nucleation, and interaction) are visible (Hull et al 1988). However, to investigate the electrical properties of the dislocations we also need a method of recording the electrical characteristics of the specimen. We have done this by adding electrical feedthroughs to the specimen holder so that a curve tracer can be connected across the device without taking it out of the microscope. We can therefore measure the progressive change in parameters such as the reverse leakage current as dislocations are introduced.

We have used the results of these experiments to investigate the electronic properties of the dislocations, as well as to study the kinetics of the relaxation process in lithographically patterned and unpatterned structures. We find that a generation-recombination process (Sze 1985) occurring at the dislocation cores does not adequately explain the changes we observe in device characteristics, and we suggest that electrical changes due to the introduction of dislocations are related to the creation of point defects or the diffusion of impurities such as metals during dislocation formation. We also describe the kinetics of the relaxation process and discuss the strong dependence it has on the processing steps used to fabricate a device.

2. ELECTRICAL MEASUREMENTS IN THE TEM

Our experiments were carried out in a JEOL 2000FX electron microscope equipped with a Gatan image intensifier and a video recording system. The specimen holder we used is shown in figure 1. We modified a Gatan single-tilt heating holder by adding a top surface electrical contact via a Ta wire ring and a lower contact through the furnace itself (Ross et al 1993). The temperature was measured using a Pt thermocouple which has been calibrated to within 30°C (Hull et al 1991). The design of this holder is ideal for measuring the electrical properties of specimens fabricated in the plan view geometry. A suitable device consists of a p-type GeSi layer on an n-type Si substrate; some diodes also had a p-type Si capping layer. All the structures were grown by molecular beam epitaxy at a substrate temperature of 550°C and a rate of about 1 monolayer per second, using 3 keV ion beam doping. Doping levels were chosen so that the depletion region at zero bias extended about half way through the GeSi layer; the GeSi layer thicknesses chosen were above h_c, but not so much greater that significant relaxation occurred during growth. Upon heating the specimens, the movement of dislocations was clearly visible in electron-transparent regions formed by back etching with HF/HNO$_3$. These dislocations propagate in ⟨110⟩ directions in the plane of the GeSi/Si interface (Hirth and Lothe 1968), forming an orthogonal array, and the interfacial misfit segments are connected to the surface (or to another dislocation) by a threading arm.

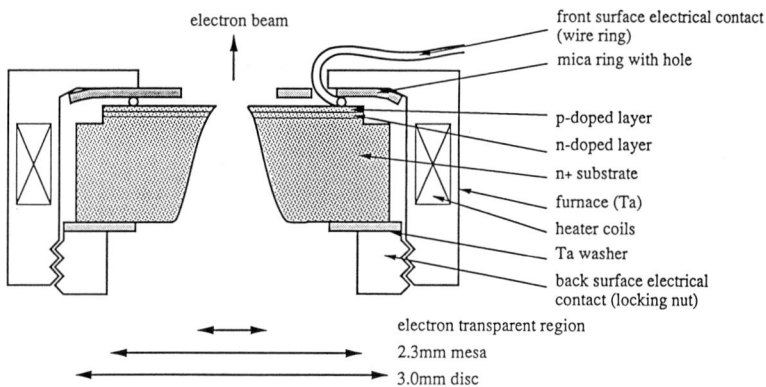

Figure 1. Schematic diagram of a p-n diode specimen in the specimen holder. Vertical dimensions are exaggerated for clarity.

3. ELECTRICAL PROPERTIES OF DISLOCATIONS

We find that the growth of dislocations during heating *in situ* is correlated with a degradation in the electrical characteristics of these diodes. Figure 2 shows that the reverse leakage current increases with the defect density, and that control structures with no Ge do not show this effect, demonstrating that these electrical changes are associated with the formation of misfit dislocations.

The effects we observe are not due to the threading arms acting as "short circuits" through the depletion region, because this would not lead to proportionality between leakage current and dislocation length. We have therefore modelled the misfit segments as generation sources for carriers in the depletion region (Oldham and Milnes 1964). Our experiments show that each meter of misfit dislocation generates 10^{-5} - 10^{-4} Amps, in other words 10^4 - 10^5 electron-hole pairs are emitted per Å per second (this number is given at -5V but is somewhat dependent on reverse bias voltage). Application of the Shockley-Read-Hall theory (see for example Sze 1985) then enables us to calculate the number of sources of electron-hole pairs (traps) which must be present to generate this leakage current, given values for the capture cross section and energy of the traps, and thus in principle address the issue of whether special sites at or around the dislocation lines, such as kink sites, are responsible for the generation of electron-hole pairs.

Both cross section and trap energy will be affected by the degree of dissociation and the possibility of decoration by metal impurities. However, reasonable values in silicon (Patel and Kimmerling 1981; Sze 1985) lead to surprisingly high values of N_t, typically 10^5 traps per nanometer of misfit line length. This is greater than the total number of potential sites along the misfit segments by a factor of 10^4 to 10^5, and it is therefore clear that a model in which the only sources are special sites along the dislocation cores can not explain these data. This large number of traps is therefore distributed over an extended volume around the cores and may thus be associated with point defects around the dislocations. Alternatively they may be associated with point defects remaining after the passage of the threading arms through the material; EPR studies suggest that high point defect densities are in fact present following the motion of dislocations (Kisielowski-Kemmerich and Alexander 1986). Even in this case, however, given the depletion layer widths in our materials there is still a shortfall of potential sites unless the disturbed region is very thick.

However, the presence of metal impurities is likely to be of even greater importance in these experiments. Even low levels of metal impurities are known to alter the electrical activity of dislocations significantly, as observed for example by cathodoluminescence (Higgs et al 1992) and electron beam induced current (Radzimski et al 1992). Thus diffusing metal species may decorate the planes of the threading arms or the dislocation cores, giving rise to traps with different cross sections and energy levels. We are presently attempting to determine the energy levels of traps associated with the dislocations in these GeSi/Si diodes, and we are also carrying out experiments in which the GeSi/Si interface is placed at different distances from the depletion region, so that the effects of the threading arms and the misfit segments can be examined separately.

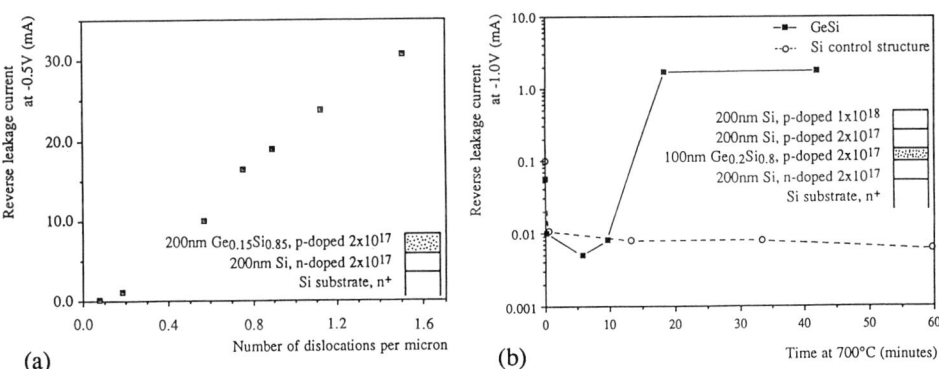

(a)

(b)

Figure 2 (a) Dislocation density against reverse leakage current for a 3mm diameter ultrasonically cut diode. The number of dislocations cutting a line is an average over several fields of view, and the current was measured at a fixed reverse voltage through the whole device. A dislocation density of $16\mu m^{-1}$ would correspond to complete relaxation: even prolonged heating does not lead to relaxation of more than 10-20% of the strain.
(b) The increase in reverse leakage current upon heating *in situ*, compared for a diode of the structure shown and an unstrained control structure. The initial improvement of the diode trace by a short anneal may be due to the removal of point defects produced during ion beam doping.

4. DEVICE PROCESSING AND DISLOCATION FORMATION

The relationship between dislocation formation and device performance is significant not just in the GeSi/Si system we have discussed here but also for III-V devices, for example in the formation of dark line defects in the active regions of laser structures (e.g. Grovenor 1989). It is clear that the kinetics of dislocation nucleation and growth, particularly their dependence on the patterning process which a strained heterostructure undergoes as it is fabricated into a working device, are extremely important.

We find that for these devices the nucleation of dislocations dominates the relaxation kinetics. Dislocations appear to nucleate from edges and surface irregularities in most of the materials we have examined, and the presence of a capping layer increases the stability. We also find that patterning has a very strong influence on stability. Etched, metallised diodes show a surprising stability to dislocation formation upon heating (Ross et al 1993). As well as the areal reduction in the nucleation opportunities for dislocations which occurs when a mesa is formed (Fitzgerald et al 1988, 1990), we also believe that the stress in the metallisation layer, particularly at the mesa edges, may diminish or remove the driving force for the formation or motion of dislocations. We have compared the relaxation kinetics of GeSi/Si heterostructures on which compressive and tensile W layers have been sputtered, and find that the sign of the stress in the metal overlayer can alter both the onset of relaxation and the ultimate dislocation density (Ross et al 1993). We are at present carrying out finite element calculations to model this effect, and are very interested in the possibility that the stability of strained materials may be improved by an appropriate choice of processing conditions.

5. CONCLUSIONS

We have measured diode characteristics and dislocation behaviour simultaneously during the relaxation of strained layer p-n junction diodes in the TEM and found a strong correlation between electrical degradation and the formation of misfit dislocations. We suggest that in our system electrically active defects are distributed widely around dislocation cores, or that the motion of threading arms is a significant feature of the degradation process, in terms of the creation of defects or the diffusion of metal impurities into the depletion region. It is clear that the opportunities for dislocation nucleation have a dominant role in determining the overall relaxation kinetics of metastable structures. The way in which a strained layer material is fashioned into a device is significant in determining the density of dislocations, and therefore the electronic quality of the finished result.

We would like to acknowledge valuable discussions with Janet Benton and Simon L. King. This work was supported by the Director, Office of Energy Research, Office of Basic Energy Sciences, Materials Science Division, U. S. Department of Energy under contract No. DE AC-03-76SF00098.

E. A. Fitzgerald, P. D. Kirchner, R. Proano, G. D. Pettit, J. M. Woodall and D. G. Ast, Appl. Phys. Lett. **52**, 1496 (1988)

E. A. Fitzgerald, Y.-H. Xie, D. Brasen, M. L. Green, J. Michel, P. E. Freeland and B. E. Weir, J. Electronic Materials **19**, 949 (1990)

C. R. M. Grovenor, *Microelectronic Materials*, Adam Hilger, Bristol (1989)

V. Higgs, M. R. Goulding, A. Brinklow and P. Kightley, Appl. Phys. Lett. **60**, 1369 (1992)

J. P. Hirth and J. Lothe, *Theory of Dislocations*, McGraw-Hill, N. Y. (1968)

R. Hull, J. C. Bean, D. J. Werder and R. E. Leibenguth, Appl. Phys. Lett. **52**, 1605 (1988)

R. Hull, J. C. Bean, D. Bahnck, L. J. Peticolas, K. T. Short and F. C. Unterwald, J. Appl. Phys. **70**, 2052 (1991)

C. Kisielowski-Kemmerich and H. Alexander, in *Proc. V International Symposium on Structure and Properties of Dislocations in Semiconductors*, Moscow (1986)

J. W. Matthews and A. E. Blakeslee, J. Cryst. Growth **27**, 118 (1974); **32**, 265 (1974)

W. Oldham and A. Milnes, Solid State Electronics **7**, 153 (1964)

J. R. Patel and L. C. Kimmerling, Crystal Research and Tech. **16**, 187 (1981)

Z. J. Radzimski, T. Q. Zhou, A. Buczkowski, G. A. Rozgonyi, D. Finn, L. G. Hellwig and J. A. Ross, Appl. Phys. Lett. **60**, 1096 (1992)

F. M. Ross, R. Hull, D. Bahnck, J. C. Bean, L. J. Peticolas and C. A. King, Appl. Phys. Lett. **62**, 1426 (1993)

F. M. Ross, R. Hull, D. Bahnck, J. C. Bean, L. J. Peticolas, R. R. Kola and C. A. King, to appear in Mat. Res. Soc. Proc. **280** (1993)

S. M. Sze, *Semiconductor Devices: Physics and Technology*, Wiley, N.Y. (1985)

Inst. Phys. Conf. Ser. No 134: Section 5
Paper presented at Microsc. Semicond. Mater. Conf., Oxford, 5–8 April 1993

High resolution scanning electron microscopy in mesoscopic physics

David A Williams, Eirik K Pettersen*, Douglas J Paul* and Haroon Ahmed*

Hitachi Cambridge Laboratory, Hitachi Europe Ltd., Madingley Road,
Cambridge CB3 0HE   U.K.
*Microelectronics Research Centre, Cavendish Laboratory, University of Cambridge,
Madingley Road, Cambridge CB3 0HE U.K.

ABSTRACT: Microfabricated semiconductor structures used for mesoscopic physics experiments usually have structure pertinent to the experiments on length scales from nanometres to millimetres. High resolution scanning electron microscopy is therefore extremely useful as a rapid and versatile characterisation technique, both during and after fabrication. Surface and cross-sectional views of mesoscopic devices are presented, using a range of accelerating voltages from 1 to 30 keV. Sidewall features are seen after reactive ion etching of trenches in silicon and silicon-germanium, which are not observable by any other means. The morphology of ohmic contacts after electron beam rapid thermal annealing is also shown.

1. INTRODUCTION

The mesoscopic regime covers the phenomena observed in solid state electron transport, between atomic scale localisation and bulk three dimensional behaviour (Altshuler et al. 1991). Usually, this involves confinement of electron and/or phonon transport to within about a wavelength in one or more directions, with a combination of material growth by molecular beam epitaxy in the vertical direction and lateral patterning using electron beam lithography. The fabrication of such devices requires many complex and critical processes, and analysis of the structures during and after fabrication is essential to ensure that the device finally measured has the structure that was intended.

This report presents observations made with a Hitachi S-900 scanning electron microscope, which uses a cold field emission source and the specimen inserted in the objective lens to give a resolution of 7Å at 20-30 keV and 35Å at 1 keV accelerating voltage. Structure is seen which was not previously observable with conventional field emission scanning electron microscopy. The devices were patterned using electron beam lithography and contacts were annealed by rapid thermal annealing in an electron beam annealer.

2. OBSERVATIONS

2.1 Quantum Wires in Silicon-Germanium

Figure 1 shows a quantum wire test structure fabricated from a silicon-germanium wafer by electron beam lithography and reactive ion etching. As shown, the quantum wire is central, running vertically in the micrograph, with source and drain regions to the top and bottom. The side arms are gates which can be used to constrict electron transport in the wire. Figure 2 shows an enlarged view of a similar device, taken with an accelerating voltage of 2 keV, and showing a source/drain region at the base and two gate arms on either side. The structure was fabricated to test the electron beam lithography and reactive ion etching, and demonstrates several features which are of interest in this type of observation. Small pillars can be seen which are the result of contamination in the reactive ion etching. These may charge up and so affect the characteristics of the device, or may interfere with subsequent lithographic stages such as the lift -off of a patterned metallic layer. Structure can be seen on the sidewalls of the gates. This is produced during etching, and appears to be topographic. It cannot be resolved at higher accelerating voltages due to the larger secondary electron volume, and is difficult to coat by sputtering or evaporation due to the geometry of the device.

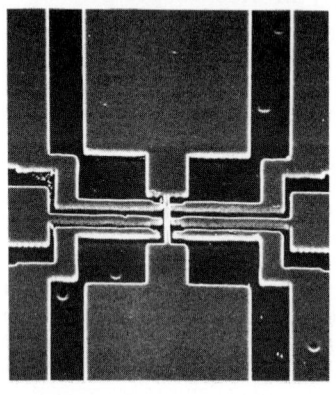

2 µm ⌐_____⌐   3 kV

Figure 1: A silicon-germanium test structure with three pairs of side gates.

300nm ⌐_____⌐   2 kV

Figure 2: A similar structure to Figure 1 showing a source/drain region, the start of the quantum wire, and two pairs of gates.

It should also be noted that this geometry, of steep walls and narrow trenches, precludes the use of scanning tunnelling microscopy to observe the sidewall structure. One pair of gates is seen to be closer to the wire than the other, due to different exposures being used in the electron beam lithography test.

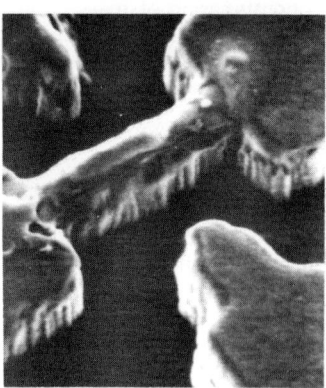

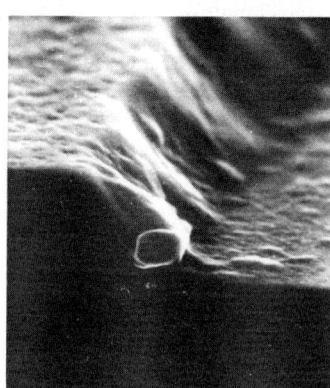

300 nm ⌐_____⌐   1 kV

Figure 3: Silicon-germanium gated wire.

150 nm ⌐_____⌐   30 kV

Figure 4: Wet-etched mesa edge in a GaAs:AlGaAs device.

Figure 3 shows a shorter quantum wire device in silicon-germanium with a nichrome surface layer, taken with an accelerating voltage of 1 kV. Again, surface structure can be seen, as can a layer contrast which is due to a delta doping layer of boron in the silicon-germanium. This is attributed to charging contrast between the more conductive doped layer and the rest of the semiconductor

The presence of a metallic layer on the surface of all these silicon-germanium samples poses great difficulty in imaging at high (>5 kV) accelerating voltages, as the greater secondary electron yield from the metal drowns out contrast in the uncoated semiconductor. Details of the etching damage, and the dependence of the surface contrast on beam acceleration voltage in similar silicon-germanium devices will be reported elsewhere.

2.2 Edge Profiles

Figure 4 shows a cross section of a gallium arsenide / aluminium gallium arsenide device from which a region has been isolated by wet etching. The profile here is very much shallower than that in the previous devices. Also, some surface structure can be observed which is different in character between the etched and protected regions. This structure is not visible in plan view or in conventional low resolution scanning electron microscopy, and it can have a profound effect on device performance if the conduction channel is near the surface. It is not generally necessary to use low accelerating voltages when observing uncoated GaAs, and it has been shown by Castell et al. (1993) that good contrast can be obtained between different materials in heterostructure and superlattice cross sections without coating.

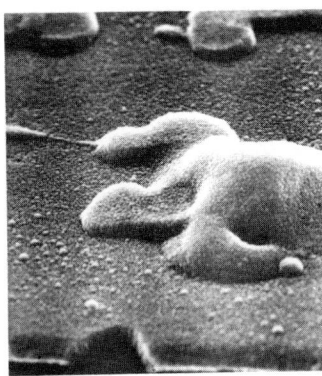
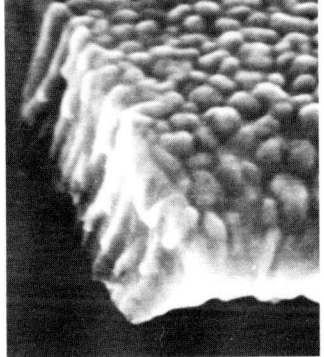

500 nm   30 kV   100 nm   30 kV

Figure 5: Surface of a rapid thermally annealed gold-germanium-nickel ohmic contact.

Figure 6: Enlargement of the edge of the contact in Figure 5.

2.3 Ohmic Contacts in GaAs:AlGaAs Devices

To make electrical measurements of these devices, ohmic contacts have to be made between an external bond wire and the conduction region of the device. This is usually done by sintering a metal film on the semiconductor surface to form an ohmic alloy. In one device structure of interest, (Williams et al. 1992), the conduction region is at a GaAs:AlGaAs interface buried 150 nm below the surface of the semiconductor, and there is another conduction region 100 nm below that. It is therefore necessary to anneal the metal so as to form an alloy at least 150 nm deep but not the extra 100 nm. This is particularly difficult with the most widely used alloy - gold/germanium/nickel - as it tends to alloy as spikes into the semiconductor. To control the sintering process with the required accuracy, rapid thermal annealing with a scanned electron beam is used.

Figure 5 shows a large scale view of a gold/germanium/nickel alloy on a gallium arsenide surface after a 4 second anneal. The surface has been covered with a nichrome - gold metallic bilayer for wire bonding, and the structure has been cleaved but not further coated. The morphology is seen to be very varied, with many surface irregularities. The edge region is enlarged in Figure 6 to show the change in grain structure of the metal overlayer and the smoothness of a region in which there is no adhesion to the substrate. Figures 7 and 8 are cross sections of a region containing rounded humps, voids and flat-topped mesa-like regions. The mesa regions are where the alloying is most uniform, giving good adhesion and electrical characteristics.   Figure 9 shows a void area, where the film has bubbled during annealing, giving no adhesion.

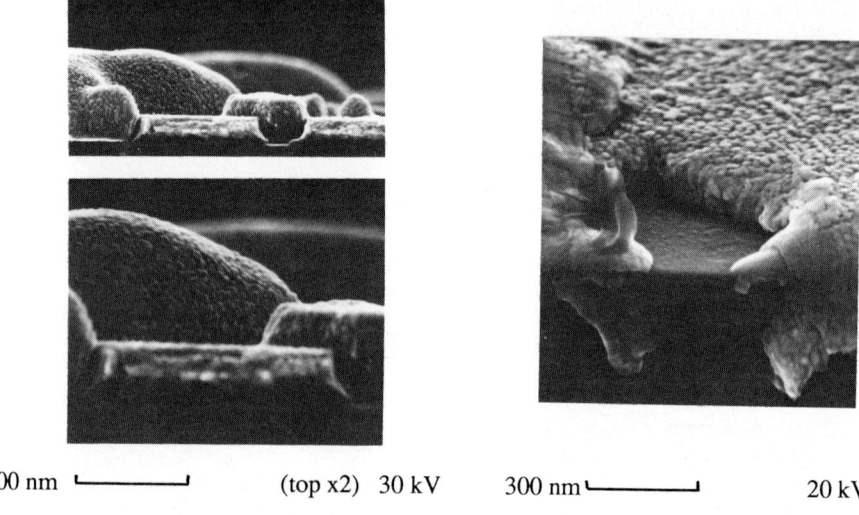

300 nm ⊢————⊣   (top x2)   30 kV   300 nm ⊢————⊣   20 kV

Figures 7/8: Cross sectional views of an ohmic contact   *Figure 9: The remains of a void.*

3. CONCLUSIONS

High resolution scanning electron microscopy from 1 -30 keV acceleration voltage is an invaluable tool in the fabrication of experimental devices for mesoscopic physics. In particular, surface structure can be observed which is not measurable by any other technique. The detailed dependence of the observed surface structural contrast with microscope accelerating voltage will be presented elsewhere.

The S-900 scanning electron microscope was donated to the Cavendish Laboratory by Hitachi Ltd.

REFERENCES

Altshuler B.L., Lee P.A. and Webb R.A. 1991 Mesoscopic Phenomena in Solids (Amsterdam: North-Holland)
Castell M.R., Howie A., Perovic D.D., Ritchie D.A., Churchill A.C. and Jones G.A.C. 1993 Phil.Mag.Lett. **67**(2) 89
Williams D.A., Mueller H.H., Chen W., White J.D., Allam J., Nakazato K. and Cleaver J.R.A.C. 1992 International Symposium on the Foundations of Quantum Mechanics (Tokyo) To be published in the Japanese Journal of Applied Physics.

Inst. Phys. Conf. Ser. No 134: Section 5
Paper presented at Microsc. Semicond. Mater. Conf., Oxford, 5–8 April 1993

Microstructure of GaAs/AlAs and InGaAs/AlAs resonant tunnelling diodes and its correlation with the electrical properties

Ch Dieker, D Gerthsen*, A Förster, J Lange and H Lüth

Institut für Schicht- und Ionentechnik, *Institut für Festkörperforschung, Forschungszentrum Jülich GmbH, Postfach 1913, D-W5170 Jülich, Germany

ABSTRACT: The influence of the growth temperature and the effect of the indium concentration on the interface microstructure of GaAs/AlAs and InGaAs/AlAs resonant tunneling diodes (RTDs) have been systematically studied by high-resolution transmission electron microscopy (HRTEM) using chemically sensitive imaging conditions. Both the growth temperature and the In content affect the interface roughness and the interdiffusion across the interfaces. The microstructure of the interface has a significant influence on the electrical properties which were determined by current/voltage (I-V) measurements.

1. INTRODUCTION

GaAs/AlAs RTDs were first grown by molecular beam epitaxy (MBE) by L. L. Chang et al. (1974). A RTD consists of two potential barriers separated by a potential well. RTD structures are promising candidates for the application in oscillator circuits in which frequencies up to 700GHz have already been achieved by Brown et al. (1991). Essential for the device performance is a high current density and an extended range of negative differential resistance in the I-V curves which is described by the current peak-to-valley ratio (PVR). To maximize the PVRs, a thorough understanding of the growth and structure dependent properties of the RTDs is necessary.

In this study, GaAs/AlAs and pseudomorphic InGaAs/AlAs diode structures were grown by MBE. The effect of the growth temperature and an increase of the In content on the microstructure of the interfaces was investigated by HRTEM using chemically sensitive imaging conditions as introduced by Ourmazd et al. (1990). The microstructure of the interfaces is characterized by the width of the transition between the GaAs (InGaAs) and the AlAs, i.e. the interdiffusion between the components, and the roughness laterally along the interface which is assessed by the distance between steps. The results of the I-V measurements, the determination and the simulation of the PVRs and the influence of other properties (conduction band offsets and effective masses) are given by Förster et al. (1993a,b). The microstructural properties of the interfaces on which this paper will focus can be well correlated with the PVRs.

2. EXPERIMENTAL TECHNIQUES

The RTD structures were grown in a conventional Varian MBE ModGenII System on highly n-doped GaAs(100) 2" wafers. On top of a GaAs buffer layer (Si doped, $n=2\times10^{18}cm^{-3}$) the AlAs double barrier region consists of two 6 monolayers (MLs) thick AlAs layers (corresponding to 1.7nm) which are separated by a 5 nm GaAs potential well. The growth temperatures were varied between 480°C and 680°C. The pseudomorphic AlAs/InGaAs structures consist of 4nm $In_xGa_{1-x}As$/1.7nm AlAs/5nm $In_xGa_{1-x}As$/1.7nm AlAs/4nm $In_xGa_{1-x}As$ ($0\leq x\leq0.35$) embedded in GaAs which were grown at a substrate temperature of 500°C.

[100] cross-section samples were examined which were prepared by dimpling and Ar^+ ion milling on a liquid-nitrogen cold stage. The TEM was carried out on a JEOL 4000 EX electron microscope characterized by a spherical aberration constant of 1 mm. High-resolution images were simulated with the EMS program package (Stadelmann 1987) with the following imaging parameters: radius of the objective aperture $15nm^{-1}$, semiangle of beam divergence 0.9mrad and defocus spread 10nm. Chemically sensitive imaging conditions were obtained by adjusting the objective lens defocus Δf between -10 nm and -35nm (underfocus) with specimen thicknesses between 8nm and 18nm. Under these conditions, the contrast of the AlAs is dominated by bright spots only at the position of the columns of the Al atoms. The contrast of the GaAs (and the InGaAs with low In content) is characterized by bright spots of equal or almost equal intensity which are located at the positions of the columns of both the group III and the group V elements. Intermediate compositions of the group III elements are identified by different intensities of the bright spots at the positions of the group III and V elements.

3. EXPERIMENTAL RESULTS

Fig. 1 shows an overview image of an GaAs/AlAs RTD grown at 580°C. The AlAs layers are completely straight and homogeneous in thickness over the whole observable section of a cross-section sample. Significant structural differences between the structures grown at different temperatures have not been observed by conventional TEM.

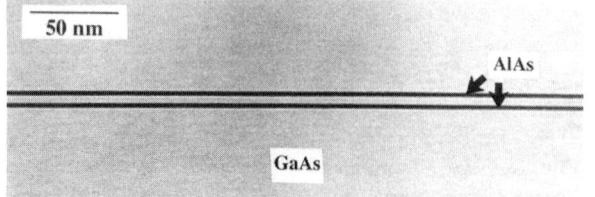

Fig.1: Bright-field image of an AlAs/GaAs RTD structure grown at 580°C taken with **g**=[040]

Nevertheless, the PVRs are strongly dependent on the growth temperature as shown in Fig.2. The maximum PVR is obtained at 580°C.

Fig. 3 shows [100] high-resolution images of the AlAs/GaAs RTDs grown at 480°C (Fig.3 a), 580°C (Fig.3 b) and 680°C (Fig.3 c) where distinct differences between the interface microstructures are observed. Fig.3 a shows a gradual transition between the image patterns of pure AlAs and GaAs which extends over 2 to 3 MLs for both the AlAs on GaAs (normal) interface and the GaAs on AlAs (inverted) interface. In contrast, the transitions between the image patterns in Fig. 3b occur fairly abruptly within 1 to 2 MLs. Occasionally, monolayer steps (as indicated by an arrow) are observed. The step distances range between 10 nm and 20 nm for the normal interface and slightly less for the inverted interface. The lower AlAs barrier is on the average 1 ML thinner than the upper barrier. In Fig.3c, a significant broadening of the AlAs barriers is found. The transition between AlAs and GaAs for the lower normal interface extends over 2 to 3 MLs. On a larger scale (approximately 20nm), the bottom interface of the lower barrier looks flat. The lower inverted interface appears slightly broader with fluctuations of the width on a 5nm to 10nm scale. Steps cannot be clearly localized. An increased broadening of the upper AlAs barrier compared to the lower barrier is observed.

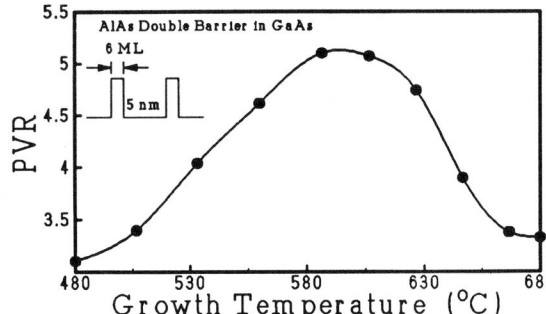

Fig. 2: Current PVRs for AlAs/GaAs RTDs as a function of the growth temperature.

Fig.3: [100] high-resolution cross-section images of the AlAs/GaAs RTD structures grown at 480°C (Fig.3 a), 580°C (Fig.3 b) and 680°C (Fig.3 c).

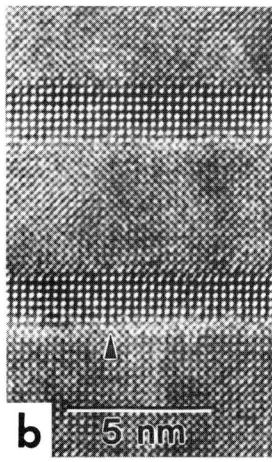

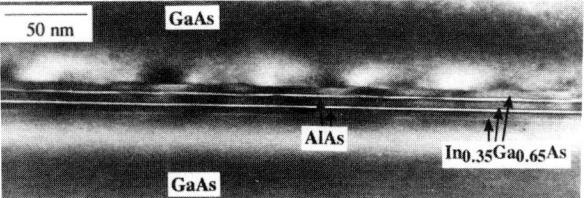

Fig.4: Bright-field image of the $In_{0.35}Ga_{0.65}As$/AlAs RTD structure taken with $\mathbf{g} = [040]$

For the $In_xGa_{1-x}As/AlAs$ RTDs, the maximum PVR of 6.35 was measured at $x=0.17$ with a sharp decrease of the PVRs towards higher indium concentrations (Förster et al. (1993b)). Fig. 4 shows an overview of the structure containing 35 % indium with a lattice-parameter mismatch between $In_{0.35}Ga_{0.65}As$ and GaAs of 2.4%. The thickness fluctuations increase with the overall thickness of the structure itself which is visualized by the straight lower and the undulated upper AlAs barrier. Contrast fluctuations in the barrier region resulting from the strain induced by the lattice-parameter mismatch are also observed. Misfit dislocations were not found.

Fig.5 displays a [100] high-resolution image of the structure containing 35% In. Both AlAs barriers are still between 6 and 8 MLs thick. The abrupt change between the image patterns at both interfaces of the lower AlAs barrier indicate sharp chemical transitions. Steps (indicated by the arrows) at the lower normal interface are clearly observed. Thickness fluctuations up to 6 MLs are found at the top of the InGaAs well.

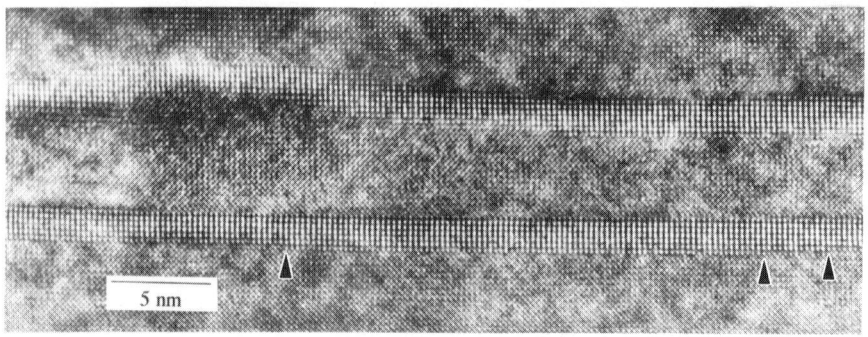

Fig. 5: [100] high-resolution image of the $In_{0.35}Ga_{0.65}As/AlAs$ RTD structure.

Significant thickness fluctuations were not observed for the structure containing 17% In. The HRTEM image (Fig.6 a) shows that the normal interfaces of the $In_{0.17}Ga_{0.85}As/AlAs$ RTD with the highest PVR are characterized by an abrupt transition between the two image patterns (1 to 2 MLs). Steps can be occasionally localized. The chemical transition at the inverted interfaces is considerably more extended. The region of the upper barrier is magnified in Fig.6 b. The image was digitized by a CCD camera and carefully filtered in the Fourier space. Small spatial frequencies up to $1nm^{-1}$ were masked to remove large scale contrast variations. The imaging conditions of Fig.6 b (specimen thickness 9nm and $\Delta f = -10nm$) can be accurately determined because the combination of contrast patterns of GaAs, $In_{0.17}Ga_{0.83}As$ and AlAs are only observed at the above conditions. The GaAs contrast is characterized by slightly brighter spots at the positions of the Ga atoms. The inserted image simulations agree reasonably with the experimental image if the intensities of the bright spots at the positions of the group V and III atoms are compared. The model is based on different structures for the normal and the inverted interface. At the normal interface, one intermediate ML containing 50% Al, 42% Ga and 8% In is assumed. For the inverted interface, the compositional transition is assumed to extend over 3 MLs with the compositions indicated in Fig.6 b. Further simulations have shown that the contrast is mainly determined by the Al content. Varying the In content between 0% and 15% at the expense of Ga does yield a significant contrast change. Concentrations of less than 20% Al or more than 80% Al are difficult to distinguish by eye from the pure components.

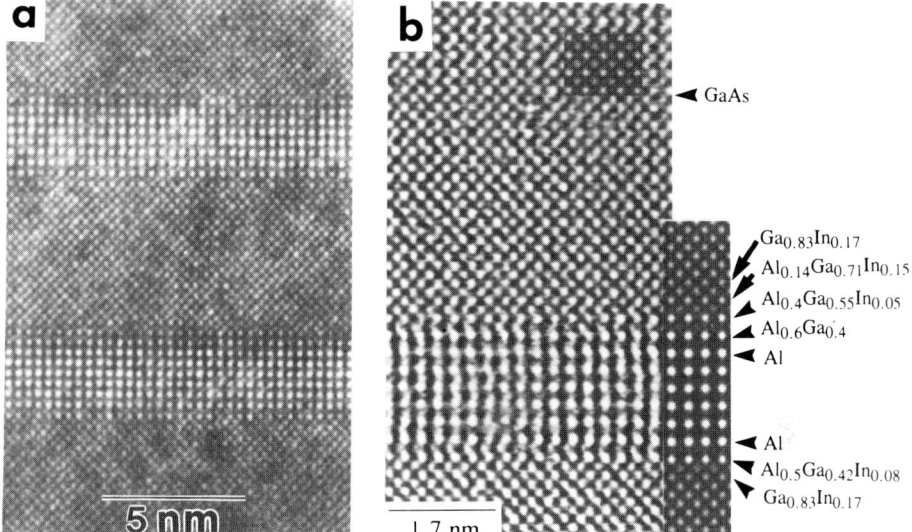

Fig.6: [100] high-resolution image of the $In_{0.17}Ga_{0.85}As/AlAs$ RTD structure (Fig.6 a). Magnification of the upper barrier (Fig.6 b) with image simulations of the GaAs and the InGaAs/AlAs barrier region at $\Delta f = -10nm$ and $t = 9nm$ inserted. The compositions of the group III element monolayers are indicated.

4. DISCUSSION

The goal of the HRTEM study was the investigation of the structural features of the barrier region which significantly influence the PVRs. An important criterion for the discussion of the PVRs is the density of interface steps which act as scattering centers for the charge carriers. Generally, steps can only be clearly localized in high-resolution cross-section images if the steps run approximately parallel to the electron beam. A sharp chemical transition is another requirement because steps will not be visible if the contrast gradually changes due to a gradual compositional transition. Another requirement for the visibility of steps are step distances larger than the sample thickness.

A broadening of the interfaces is observed for the 680°C GaAs/AlAs RTD which is nonuniform laterally along the interfaces. Mainly interdiffusion can be expected to lead to this effect. The thickness fluctuations cause nonuniform potential barriers and wells resulting in a significant decrease of the PVR. It can be assumed due to the broadening of the interfaces, that the barriers consist of AlGaAs rather than pure AlAs and this also contributes to the reduction of the PVR.

The structure grown at 480°C also exhibits nonabrupt transitions between the image patterns. However, interdiffusion is unlikely at this low temperature. The most likely explanation could be small step distances compared to the sample thickness which should exhibit an appearance similar to that of an interface where interdiffusion has occured. The reduction of the PVR can thus be explained by a high density of interface steps which act as scattering centers for the charge carriers if the step distances are smaller than the coherence length of the tunneling electrons.

The highest PVR of the GaAs/AlAs RTDs was observed at 580°C. HRTEM shows layers of homogeneous thickness with sharp compositional transitions (1 to 2 MLs) and monolayer steps with distances on the order of 20 nm. Therefore the main structural requirements for obtaining high PVRs are homogeneous layer thicknesses and a low density of interface steps.

The addition of In up 17% leads to an increase of the PVRs at a growth temperature of 500°C. Comparing the RTD structures grown without In (Fig.3 a) and with 17% In (Fig.6 a), the chemical transition at the normal interfaces of the RTD with indium has become more abrupt. Due to a possibly higher diffusivity of the In atoms at the growth surface, a significant increase of the step distances is obtained for the normal interfaces. The roughness of the inverted interfaces of the $In_{0.17}Ga_{0.83}As/AlAs$ RTD is comparable to the GaAs/AlAs RTD which is expected to be due to the low diffusivity of the Al atoms at low temperatures. The assumption of a high In diffusivity at the growth surface is confirmed by looking at the normal interface of the bottom barrier of the $In_{0.35}Ga_{0.65}As/AlAs$ RTD where the clear visibility of steps indicates a very sharp chemical transition.

The lattice-parameter mismatch increases up to 2.4% for 35% indium. At a similar mismatch, the transition between two-dimensional (layer) growth mode and three-dimensional (Stranski-Krastanov) growth mode was shown to occur by Lentzen et al. (1992). The inhomogeneous thickness especially of the InGaAs potential well, and - possibly - strain fluctuations, lead to significant local variations of the tunneling behaviour of the charge carriers and therefore to a dramatic reduction of the PVR.

V. CONCLUSIONS

Comparing the electrical and structural properties of GaAs/AlAs RTDs grown at different temperatures, the highest PVR is obtained at 580°C. At this temperature, little interdiffusion occurs leading to sharp chemical transitions. Sufficiently large step distances reduce the scattering of charge carriers compared to the structures grown at lower temperatures. The addition of indium at a growth temperature of 500°C yields a smoothening of the normal interface and a maximum of the PVR at 17% In. The transition towards the Stranski-Krastanov growth mode at higher In contents leads to locally varying tunneling properties and consequently a rapid reduction of the PVR.

REFERENCES

Brown E R, Söderström J R, Parker C D, Mahoney L J, Molvar K M and Mc Gill T C 1991 Appl. Phys. Lett. 58 2291
Chang L L, Esaki and Tsu R 1974 Appl. Phys. Lett. 24 593
Förster A, Lange J, Gerthsen D, Dieker Ch and Lüth H 1993a Appl. Phys. Lett. in press
Förster A, Lange J, Gerthsen D, Dieker Ch and Lüth H 1993b J. Vac. Sci. Technol. B in press
Lentzen M, Gerthsen D, Förster A and Urban K 1992 Appl. Phys. Lett. 60 74
Ourmazd A, Baumann F H, Bode M and Kim Y 1990 Ultramicorscopy 34 237
Stadelmann P A 1987 Ultramicroscopy 21 131

Inst. Phys. Conf. Ser. No 134: Section 5
Paper presented at Microsc. Semicond. Mater. Conf., Oxford, 5–8 April 1993

New techniques for the study of degradation modes in semiconductor lasers using a combination of transmission electron microscopy, focused ion beam sputtering and electroluminescence imaging

R Hull, D Bahnck, FA Stevie*, L R Harriott, L Koszi, S N G Chu and C Snyder.

AT&T Bell Laboratories, 600 Mountain Avenue, Murray Hill, NJ 07974, USA
*AT&T Bell Laboratories, 555 Union Blvd., Allentown, PA 18103, USA

ABSTRACT: We describe new opportunities for the study of defect and stress distributions in semiconductor laser diodes, using a combination of Transmission Electron Microscopy, Focussed Ion Beam Sputtering and Electroluminescence Imaging. Using these techniques, areas selected for cross-sectional imaging can be located to sub-micron accuracy with reference to the original laser diode structure, and correlated with optically degraded regions of the device.

1. INTRODUCTION

Characterization, understanding and avoidance of degradation modes in semiconductor diode lasers are of crucial importance to developing technological applications. Transmission Electron Microscopy (TEM) has played a central role in elucidating defect structures and generation mechanisms during device operation e.g. Ueda (1988), Chu et al. (1988), Szot et al. (1987), Romano et al (1989), Bangert et al. (1991), Fried et al. (1991), but many of the fundamental mechanisms of structural degradation remain poorly understood. Application of TEM to microstructural studies of degradation mechanisms has been hampered by the difficulty in efficiently obtaining reproducible, controllable and representative thin foil samples using conventional preparation techniques. In this paper we describe new techniques for the microstructural studies of degradation modes in semiconductor laser diodes We believe that these techniques substantially enhance the power of TEM analysis in this field.

The essential difficulties in obtaining suitable TEM samples using conventional preparation techniques arise from two primary sources: (i) The primary area of interest for microstructural analysis is the active laser stripe, which in most contemporary device implementations corresponds effectively to a wire of material, with typical dimensions of the order 0.1 μm high by 1 μm wide by 250 μm long, running the length of the laser chip. This is illustrated by a schematic illustration of the standard Channelled-Buried Mesa Heterostructure (CMBH) geometry in Figure 1(a). Successful isolation of this region of interest within the thin area of a TEM sample is thus a very difficult challenge using conventional preparation techniques. (ii) The different vertically-stacked layers in laser diode structures consist of very dissimilar materials (semiconductors, dielectrics, metals etc.). These materials thus typically have very different etching or sputtering rates using conventional ion thinning or chemical etching techniques. It is therefore extremely difficult to prepare laser device cross-sections in which all materials are simultaneously thin using conventional techniques.

Cross-sectional TEM sample preparation of laser diode devices to date has largely been restricted to standard ion-beam thinning and chemical polishing techniques, often with the device metallization layers pre-stripped and frequently with irreproducible sample quality and control. In this manuscript, we describe use of a Ga$^+$ focussed ion beam (FIB) for fabrication of thin membranes for cross-sectional TEM imaging of laser diode device structures. In conjunction with electroluminescence (EL) imaging, this technique also allows direct observation of optically degraded regions of the active laser stripe.

2. DESCRIPTION OF THE FIB/TEM/EL TECHNIQUE

FIB has previously been applied to fabrication of TEM samples, e.g. Kirk et al. (1989), Young et al. (1990), and has been applied to GaInP/AlGaInP laser diodes by Szot et al. (1992). A schematic illustration of the present FIB sample preparation technique is shown in Figure 1(b), by reference to a CMBH structure. An unbonded laser chip, of typical dimension 250 µm x 250 µm, is mechanically thinned (using fine grit SiC paper) to a dimension ~50µm measured along the buried laser stripe. A thin membrane is then prepared in the region of interest (typically containing the active laser stripe) by FIB removal of a trench of material either side of the remaining membrane. The sample is then mounted on edge in the TEM such that the electron beam runs along the trench and penetrates through the membrane for imaging.

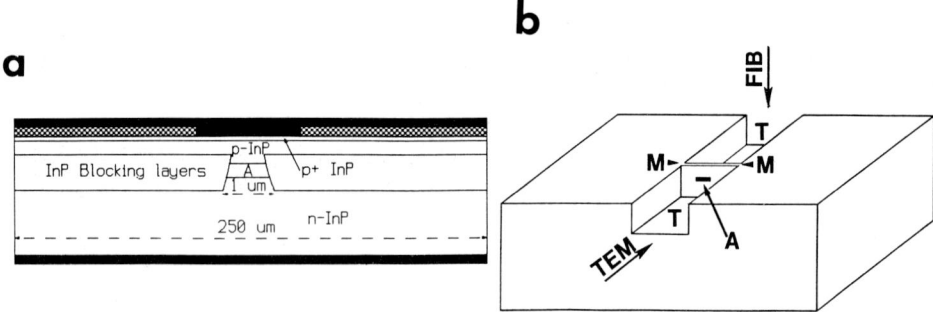

Figure 1: (a) Schematic cross-section illustration of a Channelled-Buried Mesa Heterostructure (CMBH) semiconductor laser diode structure. Metallization layers are shown with solid shading, and dielectric layers with cross-hatched shading. A is the active stripe. The laser chip thickness in the direction perpendicular to the Figure is of the order 250 µm. (b) Schematic illustration of the FIB technique used to prepare cross-sectional TEM samples. The directions of the FIB sputtering beam and TEM imaging beam are shown. The FIB is used to prepare a thin (~ 200 - 300 nm) membrane, M...M, by removal of a trench, T...T, on either side of the membrane. The trench is positioned so that it lies along the laser active stripe, such that a small portion of the active stripe, A, is included within the membrane.

The FIB used for these studies is a commercially-available FEI 611 with a gallium liquid metal ion source, operated at 25 kV. The area to be sputtered is software-defined on a secondary electron image generated within the FIB, and the ion beam is then automatically rastered over this area. Initial material removal is performed with the highest possible beam current (and hence material removal rate) of approximately 4 nA. At this beam current, the beam spot size is approximately 270 nm. As the final membrane thickness is approached, successively smaller beam currents and spot sizes are used (as many as nine can be selected), down to a minimum of about 64 pA into a 60 nm spot. The final membrane thickness which is routinely achievable in these structures is of the order of 200-300 nm (as subsequently checked by conventional scanning electron microscopy), which is highly suitable for diffraction contrast analysis using 200 kV electrons.

The technique described thus far is essentially similar to that reported previously by Szot et al. (1992), with the following differences: (i) Szot et al. emphasize the need to remove metallization layers before FIB thinning to eliminate membrane roughening due to differential sputtering rates from different metal grains (see next section). We find this to be a relatively unimportant phenomenon and thus retain the metallization through FIB thinning as it provides a protective surface to prevent erosion of the high sputtering rate InP as the membrane thickness is reduced. (ii) Szot et al. describe an experimental geometry where two membranes for TEM imaging are produced at either end of the FIB trench. Although appealing because of the ability to produce two membranes per sample, this geometry does not allow TEM examination in the ideal geometry with the electron beam parallel to the various interfaces in the laser diode structure. We therefore generally prefer single membrane samples.

Electron microscopy is performed in a JEOL 2000FX TEM operated at 200 kV. Cross-sectional images, formed using different diffracted beams near the <011> pole, of a CMBH laser structure grown by Organo-Metallic Vapor Phase Epitaxy (OMVPE) are shown in Figure 2. It is apparent that all different regions of the laser structure may be simultaneously imaged, from substrate to surface. A wealth of detail is clearly visible in this image. Of particular interest are the apparent lobes of stress at the edges of the active stripe. One powerful potential application we envisage for these FIB samples is the prospect of deconvoluting micro-stress fields by careful analysis of diffraction contrast images, given the excellent thickness uniformity of the FIB membrane.

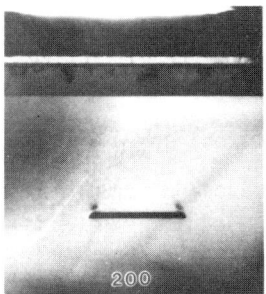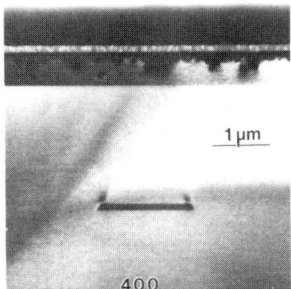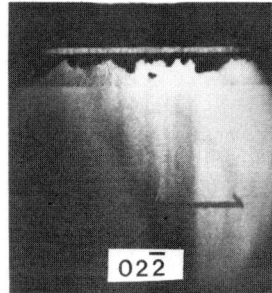

Figure 2: Cross-sectional TEM images of a FIB specimen from a CMBH structure using diffraction vectors near the <011> pole parallel to the (100) interfaces.

By combining FIB, TEM and electroluminescence (EL) imaging, a powerful technique for failure mode analysis (FMA) of degraded lasers can be developed. This arises from the ability to locate accurately the FIB membrane at a given position along the laser stripe. Optically degraded regions of an aged laser diode can then be located by EL imaging, and an FIB membrane positioned exactly within the degraded region. Experiments exploiting this ability are illustrated by Figures 3 and 4. Figure 3 shows a plan view EL image of this structure following aging for 24 hours at 100°C and a current of 200 mA. A dark region (arrowed) is clearly visible in the EL, corresponding to a highly degraded region in the laser stripe. Using FIB, we can locate the imaged membrane within this dark region by measuring from the facets at the ends of the stripe. A corresponding TEM image of this region is shown in Figure 4, which shows a relatively low magnification view of the region surrounding the active stripe, which includes all epitaxially grown and metallization layers. The following features are observed: (i) Contrast apparently associated with the morphology of the etched mesa (labeled M in the Figure); i.e. arising from the interface between the mesa sides and the epitaxial growth overlying these regions. (ii) Substantial stress fields around the active and cladding layer regions (labeled S in the Figure) (iii) Planar defects (labeled D in the Figure) running between the mesa and the quaternary capping layer. These defects appear as if they may have emanated from the top of the original etched mesa, and are thus likely to be associated with contamination or growth problems at the interface between the top of the etched mesa and the epitaxial growth over it. (iv) The InP/quaternary/metal interfaces above the active region are non-planar with substantial surface facets (labeled F in the Figure) of the order of tens of nm high. (v) An incipient dislocation loop at the active stripe sidewall is shown in a $g = [02\bar{2}]$ inset image.

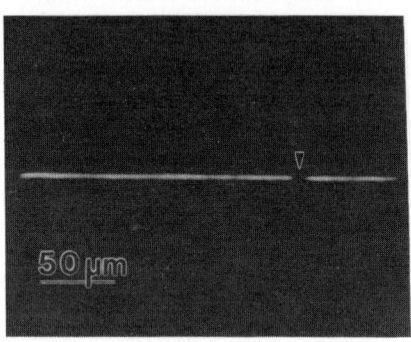 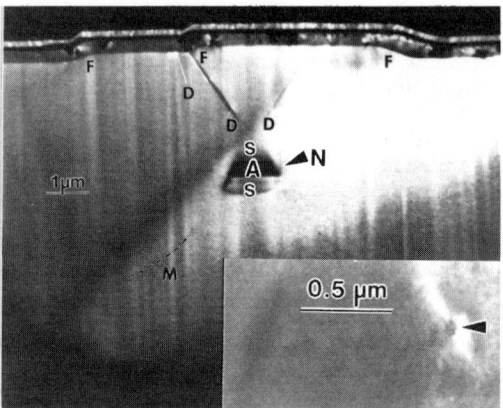

Figure 3: Electroluminescent image of a purged (24 hr at 200 mA, 100°C) CMBH structure, showing optically dead region (arrowed).

Figure 4: Cross-section TEM image (recorded close to (200) dark field diffraction conditions) of an FIB specimen from the optically dead region arrowed in Figure 3. The active region is marked A. For explanation of other labels, see text. An incipient dislocation loop (arrowed) at the active stripe sidewall is shown in a $g = [02\bar{2}]$ inset image.

3. A CRITICAL ASSESSMENT OF THE FIB/TEM/EL TECHNIQUE

To summarize the advantages of the FIB/TEM/EL techniques described above:

(1) The ability to select reproducibly the desired region of structure (e.g. the active stripe) in a cross-sectional membrane thin enough for TEM imaging.

(2) The ability to simultaneously image all vertical regions of the structure, which typically include materials incompatible for conventional TEM sample preparation, i.e. metal, dielectric and semiconductor layers.

(3) The ability to perform FMA, by locating the membrane for imaging at any cross-sectional area of the semiconductor diode, e.g. at optically dead regions of the active stripe in degraded structures, as determined from electroluminescence images.

(4) The potential ability to deconvolute micro-stress fields by diffraction contrast electron microscopy. This possibility arises from the known geometry of the FIB membrane. In conjunction with finite element analysis modeling of stress distributions within the membrane, we hope that modeling and interpretation of electron diffraction contrast will be possible.

The potential (or actual) disadvantages of the FIB technique are:
(a) Membrane specimen surface damage. In principle, the high very intensity Ga$^+$ ion beam could be expected to create a relatively large amount of surface damage. Inspection of the

TEM images in Figures 2 and 4, however, suggest that this surface damage is minor, and in our experience less than obtained from conventional Ar$^+$ ion milling techniques. The reason for this relatively low damage is presumably that the angle of the Ga$^+$ ion beam with respect to the membrane surface is essentially zero degrees, i.e. almost perfect grazing incidence. This ensures that the total ion momentum perpendicular to the membrane surface is very low. A quantitative test of the relative surface perfection obtained by FIB sputtering is illustrated in Figure 5, where we show Light-Current (L-I) output curves from a CMBH laser where one cleaved facet has been etched back using the FIB (Harriott et al., 1987). Both the threshold current and quantum efficiency of the laser with the FIB sputtered facet have degraded only by a few percent with respect to the same structure operated with the original cleaved facets. This comparison demonstrates the very high quality of the FIB-sputtered facet.

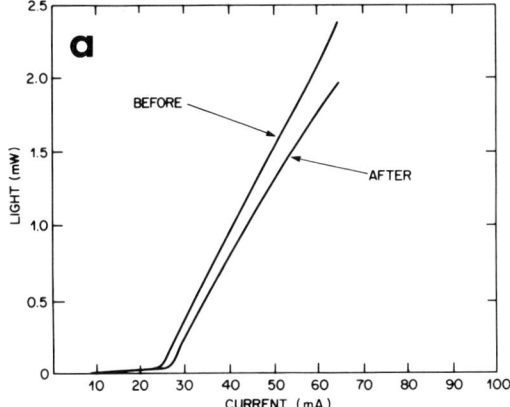 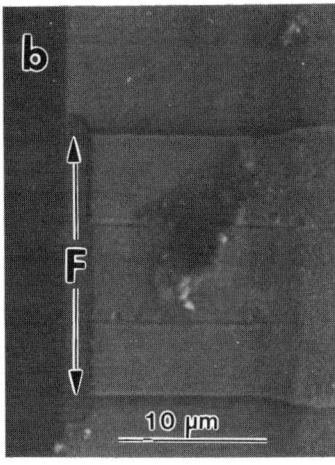

Figure 5: (a) Light output vs. current output before and after FIB milling of one facet in a CMBH laser diode structure. (b) SEM image of the FIB milled facet.

(b) Although differential sputtering yields between different planar layers of materials encountered sequentially along the ion beam direction are not a problem in FIB membrane preparation, differential sputtering yields in directions transverse to the ion beam direction can cause artifacts in the final sample. In practice, this problem arises in laser diode structures from the polycrystalline metallization layers, particularly the relatively thick (~1 μm) final gold layer common to most structures. Orientationally-dependent sputtering yields within the Au film can then produce texture within the final membrane sample. This is evident in some TEM images as surface morphology in the Au metallization layer. This effect can also produce a slight surface morphology within the semiconductor layers beneath the metallization. This is most evident as faint lines of contrast running perpendicular to the layer interfaces when imaged with a diffraction vector parallel to the interfaces (i.e. for the present geometry with (100) interfaces viewed along an <011> projection, g = <022>). Redeposition of sputtered metal can also occur, but typically this can be controlled by appropriate specification of the FIB scan geometry (Szot et al. 1992). In our experience the small amount of metal re-deposition onto the membrane is readily identifiable, and does not generally interfere with interpretation of the information in the image.

(c) Spatial definition in placing the membrane and correlating with optically degraded regions in EL images. In principle the FIB membrane can be located to sub-micron precision on the laser chip. In practice, the accuracy of the correlation between the membrane position and the optically degraded region in the active stripe is limited by the spatial resolution of the EL signal and imaging system. The effective EL resolution is of the order of a few μm, and thus the FIB membrane is much thinner than this signal width. The optically dead region in the EL image may, for example, correspond to a carrier recombination volume around a single line defect. In this case the best estimate for the FIB membrane location would generally be at the center of the optically dead region, but this does not guarantee that the non-radiative recombination center would be located within the membrane. We are attempting to overcome this sampling problem by fabricating membranes lying longitudinally along the active stripe (or at a slight acute angle to it), as opposed to the transverse membrane geometry of Figure 1(b), so that larger lengths of the stripe are imaged. (d) Sample preparation using this technique requires considerable investment in time, both of personnel and expensive equipment (the FIB). The material removal rate of InP at the highest beam currents in the FIB is of the order $1 \mu m^3 s^{-1} nA^{-1}$, requiring of the order 3-4 hours of FIB exposure per sample (much of which time comes from the final membrane polishing with successively smaller beam currents). We are currently developing lithographic techniques (see above) which should substantially reduce the FIB milling time, as well as offering other advantages. Weighed against this time investment are the quality and reproducibility of the samples obtained. Our success rate in obtaining high quality samples using our simplest pre-FIB mechanical thinning technique is of the order 80%, and the amount of information from each sample is very high. We therefore believe that the return on effort using this technique is very favorable.

4. CONCLUSIONS

We have demonstrated how a combination of FIB, TEM and EL may be used to reproducibly obtain high quality images from regions of semiconductor laser diode active stripe, which can be identically correlated to optically degraded regions. In addition, cross-sectional TEM samples from the very dissimilar layered materials in the laser diode structure may readily be obtained. The FIB membrane produces a very well-defined sample geometry which should allow deconvolution of micro-stress fields from diffraction contrast image intensities. Development of pre-FIB lithographic definition allows other sample geometries to be considered, including membranes parallel to the active stripe.

We would like to acknowledge FIB operation by T. Shane and C. Ringel, and useful discussions and encouragement from J. C. Bean, D. Ekholm, J.M. Geary and A.S. Jordan.

REFERENCES

Bangert U, Briggs A T R, Goodwin A R and Charsley P 1991 Inst. Phys. Ser. Conf. 117, 581 (IOP, Bristol)
Chu S N G, Nakahara S, Twigg M E, Koszi L A, Flyn E J, Chinn A K, Segner B P and Johnston W D 1988 J. Appl. Phys. 63 611
Fried A, Jakubowicz A, Newcomb S B and Stobbs W M 1991 Inst. Phys. Ser. Conf. 117, 585 (IOP, Bristol)
Harriott L R, Scotti R E, Cummings K D and Ambrose A F 1987 J.Vac. Sci. Technol. B5 207
Kirk E C G, Williams D A and Ahmed H 1989 Inst. Phys. Ser. Conf. 100, 501 (IOP, Bristol)
Romano A, Vanhellemont J, Bender H and Morante J R 1989 Ultramicroscopy 31 183
Szot J, Young D, Bourdillon A and Easterling K E, 1987 Phil. Mag. Lett. 55 109
Szot R, Hornsey R, Ohnishi T and Minigawa S 1992 J. Vac. Sci. Technol. B10 575
Ueda O 1981 J. Electrochem. Soc. 135 11C and references therein.
Young R, Kirk E C G, Williams D A and Ahmed H 1990 Microelectronic Engin. 11 409

Inst. Phys. Conf. Ser. No 134: Section 5
Paper presented at Microsc. Semicond. Mater. Conf., Oxford, 5–8 April 1993

Electron beam charging thermography—a novel technique to study thermal effects at mirrors of laser diodes

A Jakubowicz

IBM Research Division, Zurich Research Laboratory, 8803 Rüschlikon, Switzerland

ABSTRACT: The charging of insulating films in a scanning electron microscope is shown to be a highly useful thermographic technique to reveal hot regions in microelectronic devices with a spatial resolution in the submicron range. This technique has been applied to investigate mirrors of GaAs/AlGaAs graded index separate confinement single quantum well laser diodes. Thermographic images of these mirrors have been obtained with a spatial resolution of 0.25 μm. The proposed technique makes it possible to image fast thermal phenomena.

1. INTRODUCTION

Generation of heat in microelectronic devices is a problem that has been addressed for many years and that is becoming increasingly crucial as we endeavor to reduce the dimensions of microelectronic devices. This is also true for optoelectronic devices. Heat effects, for example, determine the operation stability of semiconductor laser diodes, heat production limits the laser's operation power, and thermal effects are responsible for catastrophic failure of these devices. GaAs/AlGaAs laser diodes are known to fail owing to mirror degradation. Intensive studies of this failure mode have stimulated the development of techniques to measure mirror temperatures and to image temperature distributions during laser operation (Brugger and Epperlein 1990, Epperlein 1990). The lateral resolution of thermographic methods is usually no better than 1 μm. As laser mirrors have lateral dimensions in the range of a few microns, and the entire thickness of the active region — consisting of various materials — is in the submicron range, there is a clear need for techniques offering better resolution. Only scanning probe microscopy-related thermographic techniques provide a lateral resolution in the submicron range (Weaver et al 1989, Williams and Wickramasinghe 1986, Williams and Wickramasinghe 1990). These techniques, however, are slow and not routinely applicable. Regarding the thermography of laser mirrors, a serious drawback is their near-field character; the mechanical thermometer will always disturb normal laser operation.

In this paper it is shown that electron beam charging of insulating coatings can provide unique submicron scale information on thermal effects at mirrors of laser diodes, and can thus serve as a basis for scanning electron microscopy (SEM) thermography of laser mirrors. It is demonstrated that the proposed technique, hereafter called electron beam charging thermography (EBCT), can compete with other thermographies in terms of spatial resolution and speed. Hot regions at mirrors of GaAs/AlGaAs graded index separate confinement single quantum well laser diodes have been revealed with a lateral resolution of 0.25 μm. The dynamics of thermal drift at the mirror of a laser operating at high power has been investigated.

2. METHOD

The proposed method entails charging a thin insulating film deposited on the laser mirror by scanning the film with a low-energy electron beam in an SEM. The low energy ensures shallow beam penetration, such that shorting the insulating layer by the beam is prevented. This is important not only for the charging process, but also for the elimination of standard voltage contrasts due to different local potentials at the device, underneath the insulating film. Under electron irradiation, when the beam energy, beam current and scanning rate are appropriately selected, a stable charge state is obtained in the insulating film. This film serves as thermographic medium. The appearance of a hot region on the device surface locally modifies the charge in the insulator. This modified charge distribution, induced by device operation, changes the local SE signal accordingly and thus produces a thermographic image. The beam parameters are selected so as to maximize the thermal contrast.

Figure 1 demonstrates the effect of temperature on charging an insulating film. In Fig. 1a, a thin Pt resistor is deposited on an amorphous silicon nitride layer. One of the resistor contacts is grounded. Under electron irradiation the insulating film becomes negatively charged, as deduced from the SE signal measured at various operation conditions of the SEM. Under current flow, the resistor heats up the entire structure. A thermocouple is used to measure the temperature. In Figs. 1b and 1c the SE signals are evaluated along line L, between points L_1 and L_2, when no current is flowing through the resistor (i.e. at room temperature) and under current flow, respectively. The curve in Fig. 1c was recorded, as the thermocouple indicated a temperature increase of 35 °C above room temperature. Note that the SE signal from the metal is nearly the same in both cases, while the signal coming from the insulator, when the sample is warmer, is significantly lower. This indicates that heating the insulator results in less effective electron beam charging.

The formation of EBCT contrast is demonstrated in Fig. 2. A thin-film Pt resistor structure, electrically isolated from a conductive substrate, was coated with a 0.3 μm thick layer of amorphous aluminum oxide (Fig. 2a). The resistor contacts were left uncoated. Figure 2b shows a selected part of this sample without current flow through the resistor. In this case the entire sample has the same temperature. As expected, no contrast is seen within the area covered by the insulating film (upper part of the image) (the residual, very weak contrast is due to surface topography). Clearly visible are only those features that are outside the insulating film (lower part of the image). Current flow (Fig. 2c) generates heat in the resistor. This heat modifies the charge accumulated in those regions of the electron irradiated insulating film that are in contact with the resistor. Thus a contrast is generated in the SEM's SE mode. This contrast is the

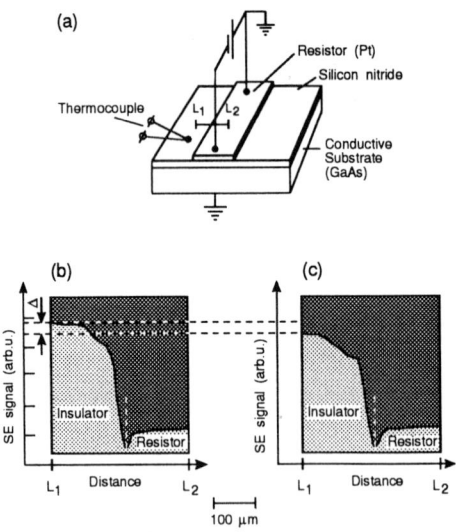

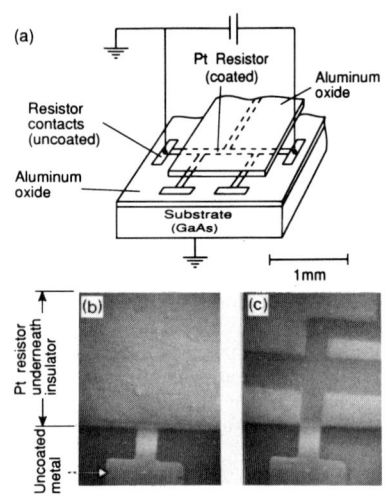

FIG. 1. The effect of temperature on electron beam charging an insulating film; (a) experimental structure; (b) and (c) secondary electron signals evaluated along line L, between points L_1 and L_2, without and under dc current flow, respectively; Δ is the thermally induced SE signal variation. [From Jakubowicz 1992].

FIG. 2. Formation of EBCT contrast; (a) test sample, (b) secondary electron image of a selected part of the sample at room temperature, without current flow through the resistor; (c) same part as in (b), taken at 200 mA. [From Jakubowicz 1992].

thermographic image of the scanned part of the sample. In Fig. 2c, the thermal contrast is caused by a temperature increase of the resistor by 35 °C, as estimated from the change of the metal resistance.

3. THERMOGRAPHY OF LASER MIRRORS

Ridge-type lasers with cleaved mirrors were investigated. A detailed description of these devices is given in Jaeckel et al (1991). Amorphous aluminum oxide and silicon nitride were employed as thermographic media. The thicknesses of the deposited films varied between 0.2 and 0.3 μm. The SEM was operated at 2-4 kV and beam currents between 10^{-10} and 10^{-8} A. All images were recorded by a computerized video acquisition system.

Figure 3 shows SEM images of the same laser mirror when the laser is off (a), and in operation (b)-(f). For electron irradiation of the non-operating laser, the entire scanned field becomes negatively charged. Operating the laser results in a reduction of this charge at the active region (arrow in Fig. 3b). This dark contrast represents the thermographic image of the mirror. It disappears after switching off the laser.

When operated in the SEM's TV mode, EBCT offers the unique ability to observe/record complete thermographic images — in practical terms — continuously in time. Furthermore, since EBCT operates in combination with standard SEM one can localize the hot regions very quickly and with high precision. Figure 3 gives an example of how this can be beneficial for studies of dynamic thermal effects. The series of images (b)-(f) was taken at a high laser current (close to 160 mA), with a time interval between each consecutive image of about 10 seconds. Though

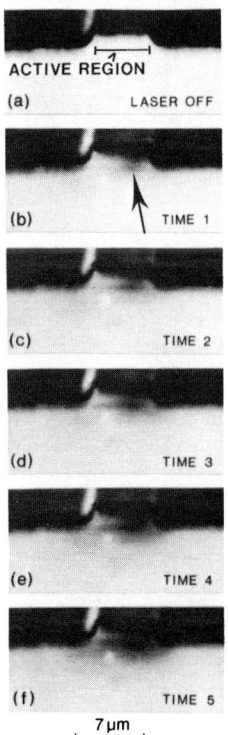

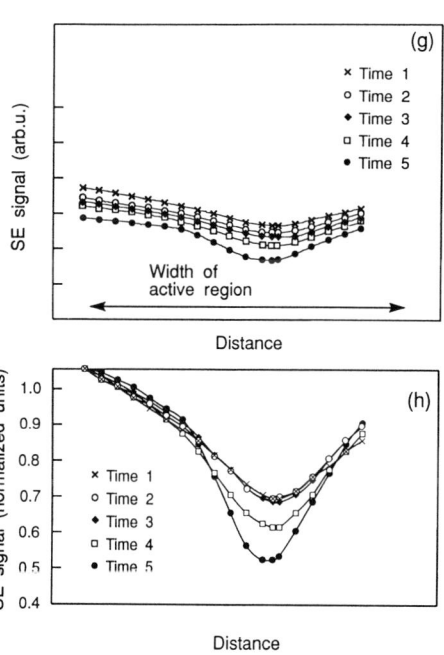

Fig. 3. Observation of the dynamics of thermal drift for a laser operating at a high constant current; (a) secondary electron image of the mirror when the laser is off; (b)-(f) secondary electron images of the same mirror as in (a), taken during laser operation; the time interval between each consecutive image is 10 seconds; (g) secondary electron signal versus distance along the laser's active region, evaluated from images (b)-(f); (h) same curves as in (g), in normalized units. [From Jakubowicz 1992].

the current was kept constant, a fast increase of the charge contrast (EBCT contrast) in time was observed. The observed behavior has been attributed to thermal drift at the laser mirror (increase of temperature at constant laser driving current), which initializes the thermal runaway (Ueda 1988). Figure 3g shows the respective SE signal profiles evaluated along the active region. The same profiles, but in normalized units, are shown in Fig. 3h. In the experiment discussed here, EBCT revealed first a large diffuse warm area with a maximum at the active region. In time, the contrast not only increased as a whole, it also became sharper, i.e. increasingly concentrated on a small spot located in the active region. This behavior is likely to have been caused by thermal focusing (Thompson 1988).

Figure 4 demonstrates the potential of EBCT in terms of spatial resolution. It shows a thermographic image of a laser mirror when the laser operates in a pulsed mode (50 ns pulses with a repetition period of 2 μs). Two hot spots are clearly revealed (Fig. 4a). They are attributed to the two maxima of the laser's near-field emission. The curve in Fig. 4b represents the contrast at the center of the hot spots, along the laser's active region. From the width of the two peaks measured at 75% of their height, assuming both peaks to be of triangular shape and using the definition of point resolution (Holt 1989), the latter is concluded to be 0.25 μm. It was possible to obtain this superior resolution since the pulses were shorter than the time needed to obtain thermal equilibrium in the semiconductor.

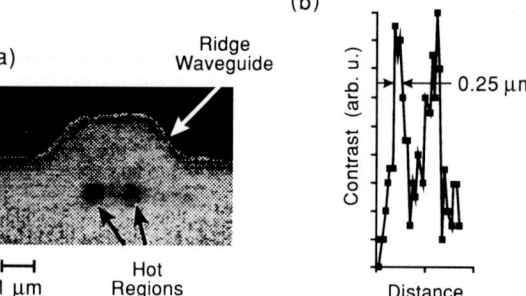

FIG. 4. (a) Thermographic image of a laser mirror; the laser is operating in a pulsed mode (pulse width 50 ns, repetition period 2 μs); (b) contrast at the center of the hot spots along the laser's active region. [From Jakubowicz 1992].

4. DISCUSSION

From earlier investigations of insulators it is known that charge build-up under electron irradiation depends on temperature (Vigouroux et al 1985). Annealing of SiO_2, after being negatively charged by electron irradiation, resulted in a reduction of the charge in time, the relaxation time decreasing with increasing temperature (Vigouroux et al 1985). The present results are consistent with this behavior. In terms of microscopy, details concerning the physical mechanisms involved in the formation of EBCT contrast must be clarified. In this respect, not only questions related to the local thermal transport and thermally induced charge exchange between the semiconductor and the insulating film have to be addressed. Open also are questions related to the role of defects, recombination and trapping processes in the insulating film. Though these and many similar problems have been addressed in the literature on dielectrics, there is no consistent theory that might serve as a theoretical base for EBCT. All this means that mainly qualitative investigations are possible at present. Quantitative measurements are feasible, but this requires performing appropriate calibration procedures.

Brugger H and Epperlein P W 1990 Appl. Phys. Lett. **56** 1049
Epperlein P W 1990 Inst. Phys. Conf. Ser. **112** 633
Holt D B 1989 in SEM Microcharacterization of Semiconductors, eds D B Holt and D C Joy (Academic Press: London) pp 10-15
Jaeckel H et al 1991 IEEE J. Quantum Electron. **27** 1560
Jakubowicz A 1992 submitted to J. Appl. Phys.
Thompson G H B 1988 in Physics of Semiconductor Laser Devices (Wiley: Chichester) p 382
Ueda O 1988 J. Electrochem. Soc. **135** 11C
Vigouroux J P et al 1985 J. Appl. Phys. **57** 5139
Weaver J M R et al 1989 Nature **342** 783
Williams C C and Wickramasinghe H K 1986 Appl. Phys. Lett. **53** 1587
Williams C C and Wickramasinghe H K 1990 Nature **344** 317

Inst. Phys. Conf. Ser. No 134: Section 6
Paper presented at Microsc. Semicond. Mater. Conf., Oxford, 5–8 April 1993

269

Composition variations, clustering and composition fluctuations in III–V alloys

F Glas

France Telecom, Centre National d'Etudes des Télécommunications, Paris B, Laboratoire de Bagneux, 196 avenue Henri Ravéra, BP 107, 92225 BAGNEUX, France

ABSTRACT: In the TEM images of the III-V alloys, several types of contrast have been interpreted as arising from composition variations. We review some cases where this is justified; in particular, the 'coarse quasiperiodic contrast' has been proven by X-ray microanalysis to arise from composition modulations. On the contrary, there is so far no proof that the ubiquitous 'fine contrast' is also due to proper composition variations. Moreover, the effects of the statistical fluctuations of the atomic distributions (inevitable in these alloys) and of the induced static atomic displacements have up to now been neglected. We give a comprehensive calculation of these effects on both diffraction patterns and images, and this makes most of the hitherto unexplained or wrongly explained observations understandable. We demonstrate in particular that even homogeneous alloys must display a considerable fine scale contrast.

1. INTRODUCTION

The formation of III-V alloys involves the substitution of different atoms on one or both of the sublattices (termed 'mixed') of the sphalerite structure. The III/V stoichiometry remains nearly perfect and only the atomic distributions on the mixed sublattice(s) can vary substantially. To deal with spatial non-uniformities in these materials, the Virtual Crystal Approximation (VCA) is a convenient starting point; it describes the alloy (1) as a perfect sphalerite crystal, (2) whose lattice parameter variation with composition follows Vegard's law, and (3) whose sites are occupied by 'average' atoms, one for each sublattice (Phillips 1973).

Although only assumption (2) is valid (and only for the average lattice, accessible through the usual diffraction methods), the VCA is widely used in TEM, if only implicitly. But obviously average atoms do not exist and, indeed, an alloy without full atomic ordering on the mixed sublattice(s) cannot be a perfect periodic crystal. Moreover, except for $Ga_xAl_{1-x}As$, not considered here, the atoms sharing the mixed sublattice(s) have different covalent radii. Thus, any 'chemical' deviation from the VCA (i.e. affecting the atomic distributions) induces some strain with respect to the VCA lattice, because the actual atomic radii are either smaller or larger than the appropriate (III or V) average radii, which determine the average lattice parameter.

Now, chemical deviations from the VCA may be of two types. We shall speak of genuine inhomogeneities (or composition **variations**) if there exist regions of the material where the distributions of some atomic species deviate in a statistically significant fashion from those expected in a random alloy. The associated strains can usually be described by linear elasticity. However, in a real alloy, even ideally disordered, any site is occupied by a given atom. This induces a microscopic inhomogeneity, which is nothing else than the statistical **fluctuations** in the distributions of the various atomic species sharing the mixed sublattice(s). This kind of inhomogeneity is more fundamental, since it is inevitable (except for perfect ordering, not considered here), and it also induces strains with respect to the VCA lattice. This is so because, as was revealed by EXAFS

(Mikkelsen and Boyce 1983), the distance between two nearest neighbour (NN) atoms in any alloy differs little from the sum of their actual covalent radii (indeed, only this 'bond length conservation' allows us to speak of radii), whilst the VCA NN distance is the sum of the two average radii. These strains, the static atomic displacements (SAD) from the VCA sites, must be described by microscopic elasticity models. It is surprising that, although some experimental consequences of these composition fluctuations have been recognized long ago, their effect on diffraction patterns and TEM images has been considered only recently.

We shall review in section 2 the composition variations on a scale large enough to have undoubtedly been proven genuine by TEM. In section 3, we shall ask ourselves if TEM would not demonstrate, as is very often assumed, the existence of genuine composition variations at a smaller scale (these might be termed clustering). A critical examination of the results available will show that it is not so. In section 4, we shall thus investigate the possible consequences of the inevitable composition fluctuations on the diffraction patterns and TEM images, and discover that most of the 'evidence' supposed to support the existence of small scale proper composition variations is merely the effect of these fluctuations, and in particular of the SAD field they induce.

2. GENUINE COMPOSITION VARIATIONS

2.1 A short review

Genuine composition variations may be either intentional or spontaneous. We shall not discuss here the intentional variations, the most common of which are the artificial superlattices. Spontaneous modulations along the growth direction have been reported for alloys containing P (Maksimov and Nagdaev 1979) or Sb (see the review by Norman et al 1993). The former have been described (Maksimov 1991) as a metastable state arising during growth due to kinetic phenomena, namely an autocatalytic process consisting of the preferential incorporation of like atoms at the surface steps. The latter are usually accompanied by extended defects and are thought to result from the formation of large islands during growth. Small coherent inclusions differing markedly from the surrounding matrix have, to our knowledge, only been observed once, in $In_xGa_{1-x}P$ (Glas 1986).

Spontaneous lateral modulations (i.e. with wavevectors normal to the growth direction) have also been observed. Sometimes, they occur in the individual layers of very short period strained layer superlattices. Such lateral modulations were first observed as a columnar growth by Goldstein et al (1985), and then with a much shorter and better defined wavelength by Hsieh at al (1990). To interpret both results, our elasticity calculations (Glas 1987) demonstrating that a strain-inducing coherent modulation parallel to the substrate costs less energy than the same modulation occurring along the growth direction, have been invoked (Glas 1987, Cheng et al 1992).

2.2 The quasiperiodic composition modulations

We now discuss in more detail the first spontaneous lateral composition modulations to have been observed (Hénoc et al 1982). Unlike those listed above, they occur in nearly perfectly lattice-matched epitaxial layers, and consist of a 'quasiperiodic' variation of the alloy composition in directions parallel to the substrate. These coarse modulations have often been observed in layers of the In-Ga-As-P system (Glas et al 1982, Launois et al 1983, Mahajan et al 1984, Ueda et al 1984, Treacy et al 1985, Norman and Booker 1985, Glas 1989a, Bons et al 1989, Peiró et al 1991) and also in the In-Ga-As-Sb (Argunova et al 1989) and In-Al-As (Peiró et al 1992) systems. They still give rise to some misunderstandings which we want to correct here, restricting ourselves to the In-Ga-As-P system.

The one manifestation of these modulations that all authors report is a TEM contrast in the plan-view images of some layers (to be listed later) epitaxially grown on (001) substrates (Fig. 1). The contrast consists of dark and bright bands oriented along the two [100] and [010] directions parallel to the substrate and spaced by about 100-200 nm. Although the absence of extended defects suggests that the contrast is due to composition variations, this could only be proved directly by local chemical

TEM X-ray microanalysis experiments (Hénoc et al 1982, Cherns et al 1987). These showed that the composition of the alloy is significantly modulated (in the same directions and with the same spatial quasiperiodicity as the contrast) between extreme compositions corresponding respectively to small and large intrinsic lattices; the Ga/As concentration ratio deviates typically between ±1 and ±10 % from the specimen average (Fig. 2). Particular care was taken (Glas et al 1982, Glas 1986) to ensure that these experiments were free of any possible artefact, such as thickness variation or primary beam channelling.

However, the relation between composition modulations and contrast had to be clarified. The latter is not the direct effect of the composition modulations on the structure factors of the beams contributing to the image (Treacy et al 1985, Glas 1989a). On the contrary, its behaviour upon changing diffraction conditions (only the bands not parallel to the operating reflexion **g** are visible) shows that it is a strain contrast (Hénoc et al 1982, Glas 1989a), whose origin Treacy et al (1985) explained very convincingly: it is due to the bending of the reflecting lattice planes caused by coherent elastic relaxation near the top and bottom free surfaces of the thin specimen. According to both experiment and this model, composition modulations and contrast have the same wavelength. The agreement between the experimental and calculated contrasts, the microanalytical results and tilting experiments (Glas 1986, Cherns et al 1987) strongly suggest that, at least for typical specimen thicknesses (up to 100-200 nm), the composition does not vary much along the growth direction. Nevertheless, some authors (Mahajan et al 1984, McDevitt et al 1992) maintain that the contrast is merely an artefact due to the thinning of the specimen which would be modulated in composition, but only at a much scaller scale. In addition to X-ray microanalysis, several facts contradict this hypothesis: the contrast exists whatever the specimen preparation method (Glas 1989a); in a given specimen, its quasiperiod is independent of specimen thickness (Treacy et al 1985); finally, the fine scale contrast (see section 3), taken by Mahajan et al (1984) as a proof of the smaller scale composition modulation, exists irrespective of the presence of the quasiperiodic contrast, and the surprising suggestion that these hypothetical fine modulations would induce a contrast at a scale 10 or 20 times larger than their own remains totally unsubstantiated. In layers lattice-matched on average, the quasiperiod is around 100-200 nm, but larger periods may be observed in mismatched layers, together with a decrease of the quasiperiod with layer thickness (Peiró et al 1991). Only Charsley and Deol (1986) have reported similar modulations along a [110] direction, with a quasiperiod varying in the depth of the layer.

These coarse modulations were observed in layers grown mainly by Liquid Phase Epitaxy but also by Molecular Beam Epitaxy (Norman et al 1985, Peiró et al 1991, 1992) and by Vapour Phase Epitaxy (Chu et al 1985, Charsley and Deol 1986). For a given average alloy composition, the modulations only appear below a certain critical growth temperature; the detail of their domain of existence is given in Fig. 3.

Both the simple regular solution model (de Crémoux et al 1982) and sophisticated 'first principles' calculations (Wei et al 1990) predict that these alloys should be unstable at low temperature with respect to incoherent decomposition. However, when the elastic energy accompanying coherent decomposition is taken into account (Stringfellow 1982), the homogeneous alloys become stable in bulk form. We showed (Glas 1987) that this 'strain stabilization' was reduced when the elastic energy was correctly calculated for an epitaxial modulated layer, because of strain relaxation at its free surface and at the substrate/layer interface; the two elastically soft <100> directions parallel to the substrate are then indeed the preferred decomposition directions. However, this reduction is insufficient to account for the high critical temperatures observed for the appearance of the composition modulations. This implies that no coherently decomposed state is the equilibrium state of the layers. Most authors now reckon that the modulations appear during growth and are subsequently frozen, without solid state autodiffusion, very slow in these alloys, playing any part during growth or cooling. Annealing experiments will be discussed in section 3. It has been suggested that a (coherent) surface phase diagram, possibly exhibiting phase separation, should thus be considered, but this has not been substantiated further. At present, the origin of the coarse composition modulations remains unclear.

3. DOES TEM BRING ANY EVIDENCE OF SMALL SCALE COMPOSITION VARIATIONS ?

The occurrence of all the genuine composition variations discussed in section 2 depends strongly on the particular III-V alloy considered and on the technique and conditions used to grow the layers. On the other hand, some features are observed by TEM in all III-V alloys except $Ga_xAl_{1-x}As$. They are often believed to result from proper composition variations. We show in this section that there is no evidence to support this assumption.

3.1 The fine scale contrast in TEM plan view images

One universal feature of the TEM plan view micrographs of III-V alloys is the 'fine scale contrast' (Roberts et al 1981, Hénoc et al 1982, Glas et al 1982, Launois et al 1983, Gowers 1983, Mahajan et al 1984, Ueda et al 1984, Glas et al 1985, Treacy et al 1985, Chu et al 1985, Norman and Booker 1985, Charsley and Deol 1986, Glas 1986, Glas 1989a, Bons et al 1989, Peiró et al 1991). It is observed in all the plan view micrographs of all alloys, except $Ga_xAl_{1-x}As$ (the only alloy with no size effect, since Ga and Al have the same covalent radius); it is also absent from the III-V binaries (Roberts et al 1981, Glas 1989a). This contrast is best characterized by its behaviour in images of layers grown on (001) substrates taken close to a 2-beam condition for a reflexion g (Glas et al 1985):
(i) Its amplitude $(I_M-I_m)/(I_M+I_m)$, where I_M and I_m are the maximum and minimum image intensities across the contrast elements, varies with specimen thickness and diffraction conditions; it may be more than 0.5 and remains visible at high thicknesses (several extinction distances).
(ii) If g is of the 400 type, it consists in elements elongated normally to g, whose most prominent repeat distance along g is about 10 nm for $w_g=0$, although close examination reveals a large range of spatial frequencies. If g is of the 220 type, it is more isotropic. Thus there is no correlation between the images of the same area for various g (apart from the inversion when g is changed in -g).
(iii) Its size diminishes when the modulus of the deviation parameter w_g increases (Fig. 4). Consequently, there is little or no correlation between images taken with different $|w_g|$ (Glas 1989a). These earlier data have been complemented recently:
(iv) There is no correlation between images taken (for $w_g=0$) at different primary beam accelerating voltages E_0 (Fig. 5).
(v) For layers grown on (110) and (111)$_A$ substrates, the micrographs and associated optical diffractograms of McDevitt et al (1992) show unambiguously that the fine scale contrast is again nearly always elongated normally to g.
These results amply show that the fine scale contrast is also a strain contrast. What we see is not the direct image of domains whose structure factors would differ from that of their environment, and indeed no feature remains unaltered by the change of diffraction conditions (ii to iv), as opposed to what happens for the coarse scale contrast. But the presence of a strain field does not in itself prove the existence of proper composition variations. And no TEM microanalytical experiment has ever revealed small scale composition variations in these specimens. We return to this question in section 4 to show that, surprisingly, fine scale contrast and homogeneity are compatible.

3.2 The contrast in TEM cross sectional images

In cross sectional specimens, several authors (Norman and Booker 1985, Chu et al 1985, Bons et al 1989, McDevitt et al 1992) observe a strong contrast with elements elongated normally to g for g parallel to the substrate, and of a much weaker contrast for g normal to the substrate. Somewhat surprisingly, most authors associate the strong contrast with the fine contrast visible in plan view images. This remains an assumption. Firstly, the former has components of much larger wavelengths than the latter. Secondly, the stresses exerted on the thin area by the unthinned parts of the substrate have never been taken into account. Finally, most observations are made (a) on (110) sections (b) of specimens with genuine quasiperiodic composition modulations (section 2.2). This obscures the

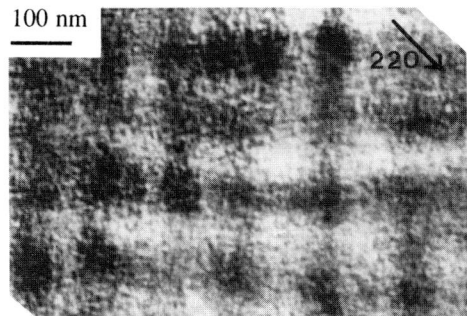

Fig. 1: 220 dark field image of the quasiperiodic contrast induced by the composition modulations. The fine scale contrast is also clearly visible.

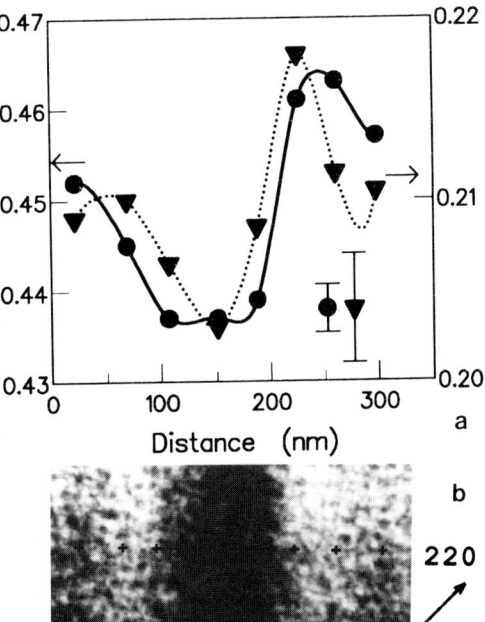

Distance (nm)

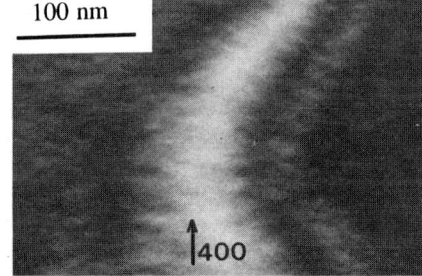

Fig. 2: Results of X-ray microanalysis (a) in an $In_xGa_{1-x}As_yP_{1-y}$ epitaxial layer at points marked + in (b). Disks, left scale: Ga/As (atomic concentrations); triangles, right scale: P/As (arbitrary units).

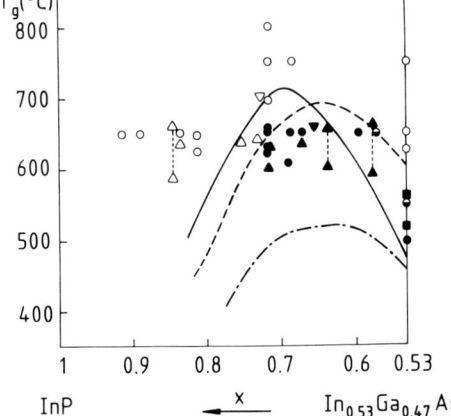

Fig. 3: Presence (full symbols) and absence (empty) symbols) of the coarse quasiperiodic contrast in $In_xGa_{1-x}As_yP_{1-y}$ layers grown on (001) InP (1-x ~ 0.47y), as a function of their average x and growth temperature T_g. o,•: our LPE results; Δ,▲: other LPE results (Gowers 1983, Mahajan et al 1984, Norman and Booker 1985, Cherns et al 1987, Bons et al 1989, McDevitt et al 1991); ■: MBE (Norman and Booker 1985, Peiró et al 1991); ∇,▼: VPE (Chu et al 1985, Charsley and Deol 1986). Lines show several calculated spinodal and miscibility gaps (Glas 1986).

Fig. 4: 400 bend contours in $In_xGa_{1-x}As_yP_{1-y}$. The fine contrast gets finer as the deviation parameter w_{400} increases, ie moving away from the bright central fringe (w_{400}=0).

Fig. 5: Same area of an $In_xGa_{1-x}As$ specimen for primary beam energies of (a) 120 and (b) 200 keV.

argument, since then the [100] and [010] directions of modulation are both seen at an angle of 45° and the beam travels through regions of varying composition. What is then seen is probably the superposition of a contrast due to the quasiperiodic modulation (itself not unlike the superposition of broad features and fringes observed in largely tilted (001) plan views by Glas (1986, 1989a) and Cherns et al (1987)) and of a fine scale contrast. This is confirmed by what is seen when constraint (a) or (b) is eliminated: the images of (010) sections of modulated specimens (Norman and Booker 1985) show much more clearly this superposition (since now the quasiperiodic modulations produce only broad features) and the images of (110) sections of non-modulated specimens (Bons et al 1989) show nothing more than the fine scale contrast.

3.3 Diffraction, EXAFS and PLAP

Another ubiquitous observation is that of very characteristic diffuse patterns by X-ray or electron diffraction (Glas et al 1985, Charsley and Deol 1986, Glas et al 1990). They consist of portions of broad lines parallel to all the <110> reciprocal directions and passing slightly away from the main diffraction spots (Fig. 6a). Tilting reveals that these lines are the sections of diffuse planes parallel to all the {110} directions. Where these lines intersect are diffuse maxima. These maxima have been interpreted as diffuse satellites by Mahajan et al (1984), whose patterns however show clearly the diffuse lines, and McDevitt et al (1992). Whereas this might be only a matter of terminology, concluding that these features are caused by clustering is totally unjustified, since composition variations are of course not the only possible source of diffuse scattering: SAD in a fully disordered homogeneous alloy also produce diffuse scattering (Warren et al 1951). Here again, as for the fine scale contrast, what is disclosed is mainly, if not only, a strain field. Section 4 will show that these strains are compatible with homogeneity.

EXAFS shows the existence of SAD (section 1). In a disordered alloy whose atoms have different covalent radii, these are *a priori* compatible with homogeneity (Mikkelsen and Boyce 1983, Podgórny et al 1985, Glas and Hénoc 1987). When used to investigate homogeneity, EXAFS detects no clustering in bulk alloys (Mikkelsen and Boyce 1983) or in thin layers (Takeda et al 1990).

A few other experiments indicate that **in some specimens** small scale composition variations exist. Some Transmission Electron Diffraction (TED) patterns show sharp satellites to the main diffraction spots, which are hardly compatible with full alloy disorder; however, these have been reported by only two groups (Norman and Booker 1985, McDevitt et al 1992). Usually no such satellites are observed (Mahajan et al 1984, Glas et al 1985, Charsley and Deol 1986, Bons et al 1989), and the fine scale contrast appears to be the same with or without satellites. The Pulse Laser Atom Probe (PLAP) experiments of Mackenzie et al (1991) detected clustering in some epitaxial layers and none in others; again the fine scale contrast was the same in both cases. These are indeed strong indications that the fine scale contrast is intrinsic and **not** due to composition variations.

3.4 Annealing

This is further supported by the behaviour of the contrasts upon annealing. Early experiments carried out at 650°C (Launois et al 1983) had shown no change either in the coarse or the fine contrasts. Recent experiments performed at much higher temperatures (McDevitt et al 1991) on samples exhibiting as grown quasiperiodic composition modulations show that the latter disappear whereas a strong fine scale contrast remains visible. The authors however report a decrease of its scale. Very surprisingly, they interpret these experiments as confirming that the fine scale contrast is due to composition variations whereas the coarse one is an artefact. For us, the annealing procedure has here been sufficient to make the coarse composition variations vanish by solid state diffusion. If any fine scale contrast were also due to composition variations, it would disappear **before** the coarse one. That it does not strengthens the conclusion of the previous paragraph. However, the possible existence of two kinds of fine contrast, a very fine intrinsic one and a coarser one, the latter observed in all as grown specimens with or without coarse quasiperiodic composition modulations, the former intrinsic and becoming visible after annealing, should be considered further.

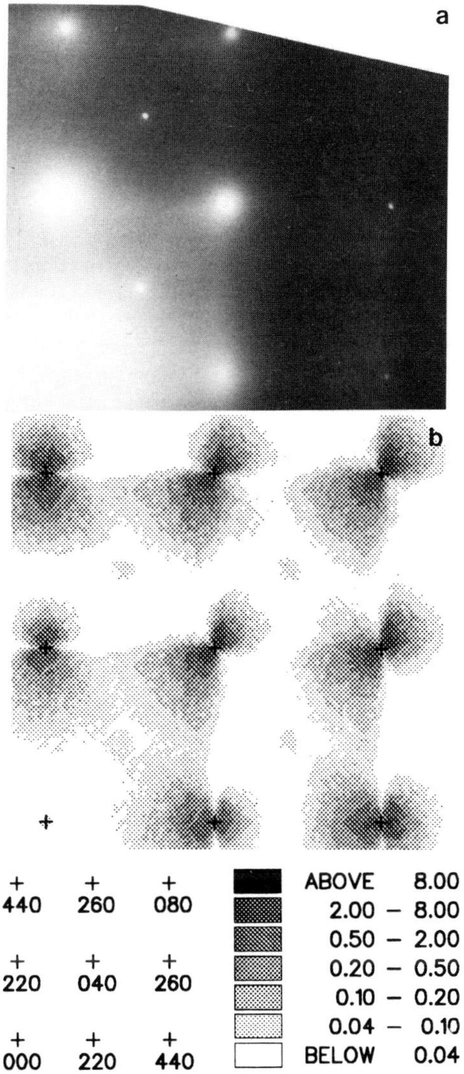

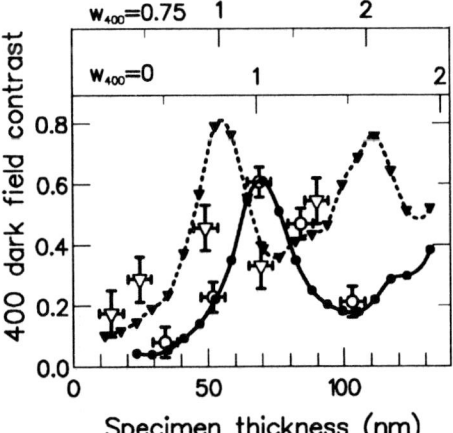

+ 440	+ 260	+ 080
+ 220	+ 040	+ 260
+ 000	+ 220	+ 440

![black]	ABOVE 8.00
	2.00 – 8.00
	0.50 – 2.00
	0.20 – 0.50
	0.10 – 0.20
	0.04 – 0.10
	BELOW 0.04

Fig. 6: Experimental (a) and calculated (b) diffuse scattering patterns for a [001] reciprocal plane of $In_{0.53}Ga_{0.47}As$. Crosses indicate the positions of the diffraction spots.

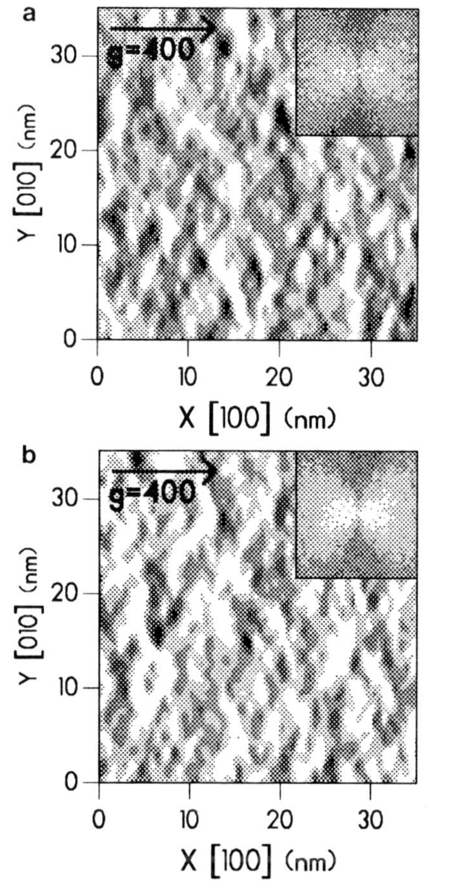

Fig. 8: Variation with specimen thickness of the amplitude of the fine scale TEM contrast calculated (full symbols) and measured (empty symbols) in 400 dark field images of $In_{0.5}Ga_{0.5}As$ for $w_{400}=0$ (circles, full lines) and $w_{400}=0.75$ (triangles, dashed lines). Upper horizontal scales give the thickness normalized to the respective effective extinction distances. $E_0=100$ keV.

Fig. 7: Calculated 400 dark field images of an homogeneous random $In_{0.5}Ga_{0.5}As$ specimen 132 nm thick, for (a) $w_{400}=0$ (contrast=0.38) and (b) $w_{400}=1$ (contrast=0.7). Inserts: image power spectra (axes: -0.85 to 0.85 nm^{-1}). $E_0=100$ keV.

From this survey, we conclude that, contrary to common belief and apart from the rare observation of satellites, TEM and TED have so far given no evidence of the existence of proper small scale composition variations in the epitaxial layers of III-V alloys.

4. THE EFFECT OF THE STATIC ATOMIC DISPLACEMENTS ON THE DIFFRACTION PATTERNS AND TEM IMAGES OF III-V ALLOYS

4.1 Introduction

Deviations from the VCA are inevitable in all non-perfectly ordered III-V alloys (section 1). They assume two related aspects: the occupation of any site by a given atom, not the average atom, and the SAD from the VCA sites. The realistic situation which would most closely approach the VCA is ideal alloy disorder, the absence of any correlation between the occupations by the atomic species of the sites of the mixed sublattice(s). Such an alloy could be considered as truly homogeneous, but the SAD would remain, because of the tendancy to bond length conservation revealed by EXAFS. The existence of SAD in disordered alloys, or Atomic Size Effect (ASE), has been known for long and Warren et al (1951) calculated their effect on diffraction patterns, without considering their possible anisotropy. What was only recently investigated is the consequence of ASE on the diffraction patterns (Glas 1989b, Glas et al 1990) and TEM images (Glas 1993) when, even in presence of perfect chemical disorder, the SAD are strongly anisotropic.

The Valence Force Field (VFF) model of Keating (1966) and Martin (1970) extended from the III-V binaries to the alloys by Podgórny et al (1985) yields the SAD for given atomic occupations of all sites, by minimizing a microscopic elastic energy depending on the differences between actual and ideal NN distances and bond angles. The diffraction patterns and TEM images of any given specimen can then be calculated by using known methods.

To isolate the effect of the SAD from that of any compositional inhomogeneity, we first performed these calculations for ideally disordered alloys. What matters is that, although the correlation length of the composition fluctuations is then zero, that of the SAD is finite. It is well known that a localized deviation, such as a defect or an inclusion, creates an elastic strain field extending far beyond itself. VFF calculations show that it is also true for a single impurity in a III-V binary (Glas et al 1990). Moreover, the SAD are strongly anisotropic: they are much larger along the <110> rows passing through the impurity than along the <100> rows. In the alloy, any site of the mixed sublattice(s) is occupied by an 'impurity', *ie* not the average VCA atom; the total strain is the net effect of all of them. Nevertheless, its wavelength spectrum is not white noise, as is that of the composition fluctuations: not only are some wavelengths favoured, but it is also strongly anisotropic. We now show that these features of the strain spectrum, together with the ability offered by TEM to probe only part of it by setting particular diffraction conditions, produces the strongly anisotropic diffuse patterns and images which are too often taken as indicating true inhomogeneity.

4.2 Calculation of the diffuse scattering patterns of disordered III-V alloys

Large homogeneous disordered $In_xGa_{1-x}As_yP_{1-y}$ crystals were simulated by drawing the atom present at any site according only to the respective average concentrations. The equilibrium SAD were found by minimizing the VFF energy without allowing the atoms to exchange sites. The diffuse scattering patterns were calculated using standard methods (see details in Glas et al 1990). In Fig. 6, the calculations are compared with experiment for a [001] zone axis diffuse pattern of $In_xGa_{1-x}As$. They reproduce the broad <110>-oriented segments, passing slightly away from the main diffraction spots, and the pseudo-satellites produced by their intersection near *e.g.* the 220 spots (and thus appearing along the <100> directions). These calculations do show that perfect disorder is compatible with the strongly anisotropic experimental diffuse patterns (section 3.3). However, calculations for alloys with significant clustering show very similar diffuse patterns (Glas et al 1990). These patterns are thus a universal feature of non-ordered III-V alloys but bring little information about possible inhomogeneities. In no way should they be taken as indicative of such inhomogeneities.

4.3 The TEM contrast due to the atomic size effect in homogeneous III-V alloys

The same full description of a simulated III-V alloy thin crystal in terms of site occupation and of SAD calculated from the VFF model allows us to calculate the TEM images for any diffraction conditions. Up to now we restricted ourselves to conventional TEM images. Since, in a disordered crystal, the SAD may vary rapidly from site to site, the usual column approximation cannot be used *a priori*; instead the equations of Howie and Basinski (1968) must be used. Since however the SAD strain field has long range components, very large crystals have to be simulated. The figures present results for crystals of surface 60x60 cubic unit cells and up to 225 cells thick (132 nm). The beam direction is close to the [001] layer normal; deviation from the Bragg condition for reflection **g** is as usual measured by the deviation parameter w_g. The main results (Glas 1993) are the following:
- Both bright and dark field images display a large contrast (Figs. 7,8).
- If **g**=400, the contrasts elements are elongated normally to **g** (Fig. 7).
- The contrast pattern is inverted if **g** is changed to **-g**.
- The contrast pattern of any area changes rapidly with w_g. It gets finer if $|w_g|$ increases (Fig. 7).
- The contrast possesses a large range of wavelengths along **g** (Fig. 7). For **g**=400, the image power spectra display a first peak around 10-15 nm if w_g=0; when $|w_g|$ increases, the higher frequencies are reinforced (Glas 1993).
- The contrast amplitude oscillates with specimen thickness (Fig. 8).
- The contrast pattern changes also rapidly with primary beam energy E_0.
- Changing the absorption parameters or taking into account the finite resolution of the microscope modifies slightly the contrast without altering the previous conclusions (Glas 1993).

The calculated contrast thus behaves in the same way as the fine scale contrast (section 3.1) observed in the III-V alloys with ASE. The 10-15 nm wavelengths, prominent in the experiments, might not show up as much in the present simulations, which do however reproduce them, since the lateral size of our specimens is still only about twice as large. All our other computational results have been observed experimentally. What remains open to discussion is the possible existence of two types of fine scale contrast in these alloys. What our simulations prove is that considerable fine scale contrast, produced by the SAD, exists in the III-V alloys with ASE, **even homogeneous ones**.

5. SUMMARY

We reviewed cases where TEM proves the existence of genuine spontaneous composition variations in III-V alloys. On the other hand, we showed that, even in homogeneous alloys, the inevitable static atomic displacements from the sites of the average lattice produce a considerable fine scale contrast which should in no way be taken as indicative of composition variations.

This paper is dedicated to the memory of Pierre Hénoc (1933-1993).

REFERENCES

Argunova T S, Baranov A N, Ruvimov S S, Sorokin L M and Sherstnev V V 1989 Sov. Phys. Solid State <u>31</u> 1365
Bons A J, Oei Y S and Schapink F W 1989 Microscopy of Semiconducting Materials 1989, Inst. Phys. Conf. Ser. No 100, eds A G Cullis and P D Augustus (Bristol: Institute of Physics) pp 161-166
Charsley P and Deol R S 1986 J. Cryst. Growth <u>74</u> 663
Cheng K Y, Hsieh K C and Baillargeon J N 1992 Appl. Phys. Lett. <u>60</u> 2892
Cherns D, Greene P D, Hainsworth A and Preston A R 1987 Microscopy of Semiconducting Materials 1987, Inst. Phys. Conf. Ser. No 87, eds A G Cullis and P D Augustus (Bristol: Institute of Physics) pp 83-87

Chu S N G, Nakahara S, Strege K E and Johnston D 1985 J. Appl. Phys. 57 4610
de Crémoux B 1982 J. Physique (Paris) 43 C5-19
Glas F 1986 Thesis, Université Paris XI
Glas F 1987 J. Appl. Phys. 62 3201
Glas F 1989a Evaluation of Advanced Semiconductor Materials by Electron Microscopy, NATO Advanced Science Institutes Ser. B 203, ed D Cherns (New York: Plenum) pp 217-232
Glas F 1989b Microscopy of Semiconducting Materials 1989, Inst. Phys. Conf. Ser. No 100, eds A G Cullis and P D Augustus (Bristol: Institute of Physics) pp 167-172
Glas F 1993 to be published
Glas F, Gors C and Hénoc P 1990 Philos. Mag. B 62 373
Glas F and Hénoc P 1987 Philos. Mag. A 56 311
Glas F, Hénoc P and Launois H 1985 Microscopy of Semiconducting Materials 1985, Inst. Phys. Conf. Ser. No 87, eds A G Cullis and P D Augustus (Bristol: Institute of Physics) pp 251-256
Glas F, Treacy M M J, Quillec M and Launois H 1982 J. Physique (Paris) 43 C5-11
Goldstein L, Glas F, Marzin J Y, Charasse M N and Le Roux G 1985 Appl. Phys. Lett. 47 1099
Gowers J P 1983 Appl. Phys. A 31 23
Hénoc P, Izrael A, Quillec M and Launois H 1982 Appl. Phys. Lett. 40 963
Howie A and Basinski Z S 1968 Philos. Mag. 17 1039
Hsieh K C, Baillargeon J N and Cheng K Y 1990 Appl. Phys. Lett. 57 2244
Keating P N 1966 Phys. Rev. 145 637
Launois H, Quillec M, Glas F and Treacy M M J 1983 GaAs and Related Compounds 1982, Inst. Phys. Conf. Ser. No 65, ed G E Stillman (Bristol: Institute of Physics) pp 537-544
Mackenzie R A D, Liddle J A and Grovenor C R M 1991 J. Appl. Phys. 69 250
Mahajan S, Dutt B, Temkin H, Cava R J and Bonner W A 1984 J. Cryst. Growth 68 589
Maksimov S K 1991 Microscopy of Semiconducting Materials 1991, Inst. Phys. Conf. Ser. No 117, eds A G Cullis and N J Long (Bristol: Institute of Physics) pp 491-496
Maksimov S K and Nagdaev E N 1979 Sov. Phys. Dokl. 24 297
Martin R M 1970 Phys. Rev. B 1 4005
McDevitt T L, Mahajan S, Laughlin D E, Bonner W A and Keramidas V G 1991 Microscopy of Semiconducting Materials 1991, Inst. Phys. Conf. Ser. No 117, eds A G Cullis and N J Long (Bristol: Institute of Physics) pp 477-483
McDevitt T L, Mahajan S, Laughlin D E, Bonner W A and Keramidas V G 1992 Phys. Rev. B 45 6614
Mikkelsen J C and Boyce J B 1983 Phys. Rev. B 30 6217
Norman A G and Booker G R 1985 Microscopy of Semiconducting Materials 1985, Inst. Phys. Conf. Ser. No 87, eds A G Cullis and P D Augustus (Bristol: Institute of Physics) pp 77-82
Norman A G, Seong T-Y, Ferguson I T, Booker G R and Joyce B A 1993 Semic. Sci. Technol. 8 S9
Peiró F, Vilà A, Cornet A, Herms A, Morante J R, Clark S and Williams R H 1991 Microscopy of Semiconducting Materials 1991, Inst. Phys. Conf. Ser. No 117, eds A G Cullis and N J Long (Bristol: Institute of Physics) pp 519-522
Peiró F, Cornet A, Morante J R, Clark S and Williams R H 1992 Appl. Phys. Lett. 59 1957
Phillips J C 1973 Bonds and Bands in Semiconductors (New York: Academic Press) pp 13, 20-23, 212-214, 221
Podgórny M, Czyzyk M T, Balzarotti A, Letardi P, Kisiel A and Zimnal-Starnawska M 1985 Solid St. Commun. 55 413
Roberts J S, Scott G B and Gowers J P 1981 J. Appl. Phys. 52 4018
Stringfellow G B 1982 J. Electron. Mat. 11 903
Takeda Y, Oyanagi H and Sasaki A 1990 J. Appl. Phys. 68 4513
Treacy M M J, Gibson J M and Howie A 1985 Philos. Mag. A 51 389
Ueda O, Isozumi S and Komiya S 1984 Jap. J. Appl. Phys. 23 L241
Warren B E, Averbach B L and Roberts B W 1951 J. Appl. Phys. 22 1493
Wei S-H, Ferreira L G and Zunger A 1990 Phys. Rev. B 41 8240

Inst. Phys. Conf. Ser. No 134: Section 6
Paper presented at Microsc. Semicond. Mater. Conf., Oxford, 5–8 April 1993

279

Nature and origin of atomic ordering in III–V semiconductor alloys

A G Norman, T-Y Seong[1], B A Philips[2], G R Booker[1] and S Mahajan[2]

Interdisciplinary Research Centre for Semiconductor Materials, The Blackett Laboratory, Imperial College, Prince Consort Road, London SW7 2BZ, UK
[1]Department of Materials, University of Oxford, Parks Road, Oxford OX1 3PH, UK
[2]Department of Materials Science, Carnegie Mellon University, Pittsburgh PA 15213, USA

ABSTRACT: Epitaxial layers of ternary and quaternary III-V alloys grown by a variety of techniques have been found to exhibit several types of atomic ordering which can have marked effects on the electrical and optical properties of the layers. In this paper we summarise the main observations of ordering behaviour to date and discuss the possible origins of the ordering. We concentrate in particular on the most commonly observed type, CuPt, which is surface induced and whose origin is thought to be linked to the occurrence of surface reconstruction during growth.

1. INTRODUCTION

Ternary and quaternary III-V alloys possess a wide range of optical and electrical properties and are thus of interest for a variety of devices. The majority of these alloys are predicted to be thermodynamically unstable in the bulk and epitaxial forms, exhibit miscibility gaps and show a tendency towards clustering and phase separation at low temperatures (e.g. Onabe 1982). Miscibility gaps have been experimentally observed for a number of bulk alloys and electron microscopy studies have revealed evidence of composition modulations and clustering resulting from phase separation in epitaxial layers of many III-V alloys (e.g. Mahajan et al 1989, Norman et al 1993 and Glas 1993). Due to this tendency for clustering and phase separation at low temperatures it was not expected that these alloys would exhibit atomic ordering. Srivastava et al (1985), however, using first-principles local-density total-energy minimisation calculations performed on both ordered and disordered models of bulk $Ga_xIn_{1-x}P$ predicted that certain ordered intermediate phases could be thermodynamically stable at low temperatures. These ordered phases were concluded to be stable because they are strain reducing, since they can simultaneously accommodate the different GaP and InP bond lengths in the alloy in a coherent fashion, introducing less strain than would arise in a random alloy. Subsequently several types of atomic ordering have been observed in a wide range of III-V epitaxial layers, often occurring at the same time as phase separation. The commonly observed ordered structures, however, are not those predicted to be the most stable in the bulk alloys since they are surface induced during epitaxial growth and are subsequently frozen in to the bulk of the layers. The occurrence of atomic ordering is often associated with significant changes in the electrical and optical properties of the layers. An understanding of the nature, origin and effects of this atomic ordering is thus crucial to the optimisation of the performance of devices fabricated from these materials. In addition, if the atomic ordering can be controlled, the different properties of ordered layers may be used in novel device structures. In this paper we first describe some of the possible ordered structures for III-V alloys and then summarise the main experimental observations to date. We then discuss the current understanding of the origins of the different types of atomic ordering concentrating in particular on the most commonly observed type, CuPt.

2. NATURE OF ORDERING

2.1 Possible Ordered Structures

Unordered ternary and quaternary III-V alloys have the zinc-blende crystal structure which consists of two interpenetrating face-centred cubic (FCC) sublattices, one composed of group III atoms and the other of group V atoms, displaced from each other by a/4[111], see Fig. 1, with the atoms arranged randomly on the sites of their respective sublattices. When atomic ordering occurs in a ternary alloy, e.g. GaInAs, the Ga and In atoms are arranged in an ordered fashion on the group III FCC sublattice, leading to an increase in periodicity along certain crystal directions and the formation of a superlattice (SL) structure. Fig. 1 shows three of the possible SL structures for which perfect ordering would occur for $A_{0.5}B_{0.5}C$ alloys. The increase in lattice periodicity in these SL structures means that normally forbidden SL reflections become allowed in x-ray and electron diffraction and the characteristic arrays of SL reflections for each SL structure, Fig. 1, enables them to be distinguished. For the CuAu I, CuPt and chalcopyrite structures the ordering occurs on {110} and {100} planes, {111} planes, and {210} planes respectively. Other SL structures are also possible such as famatinite (Cu_3Au-type) ordering on {100}, {110} and {210} planes and luzonite (Al_3Ti-type) ordering on {100} and {110} planes for which full ordering would occur at compositions of $A_{0.25}B_{0.75}C$ or $A_{0.75}B_{0.25}C$. Perfect ordering is not usually observed since only partial ordering of the atoms occurs to the correct sites, enabling the different SL structures to occur over a wide range of compositions. Antiphase boundaries (APBs) are also normally present.

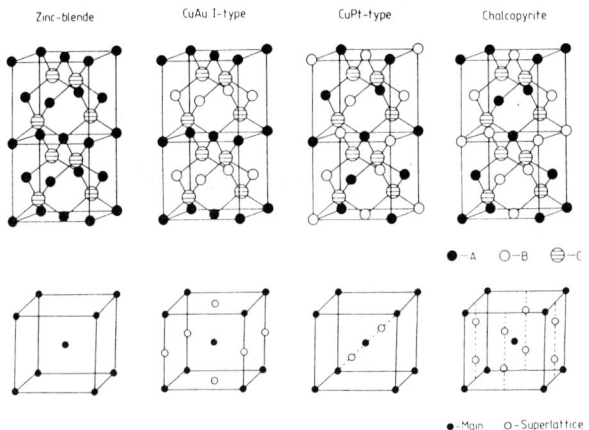

Fig. 1. Atomic models and corresponding sections of reciprocal space for random zinc-blende structure and CuAu I, CuPt and chalcopyrite superlattice (SL) structures for $A_{0.5}B_{0.5}C$ alloy, e.g. GaInAs.

2.2 Experimental Observations

Kuan et al (1985) reported the first experimental evidence of atomic ordering in III-V alloys. They observed CuAu I-type ordering using transmission electron diffraction (TED) in $Al_xGa_{1-x}As$ layers grown by molecular beam epitaxy (MBE) and metal organic chemical vapour deposition (MOCVD) on (110) GaAs substrates. Identical ordering was later found in MBE $Ga_xIn_{1-x}As$ layers grown on (110) InP substrates (Kuan et al 1987) with a high density of APBs present in the ordered regions. More recently Ueda et al (1991a, b) have investigated this ordering in (110) MBE GaInAs in more detail. Transmission electron microscopy (TEM) and TED studies revealed that growth on vicinal substrates oriented a few degrees towards the <001> directions from (110) resulted in stronger ordering. Plate-like microdomains tilted slightly from the (110) plane and a large variation in strength of ordering with growth temperature were also observed. Chin et al (1991) and Nakata et al (1991) reported enhanced two-dimensional electron gas mobilities in modulation doped GaInAs/AlInAs heterostructures grown on vicinal (110) InP substrates with partially CuAu I-type ordered GaInAs layers. Jen et al (1986) observed the simultaneous presence of both CuAu I and chalcopyrite type ordering in MOCVD $GaAs_{0.5}Sb_{0.5}$ layers grown on (001) InP. Growth temperature, growth rate and substrate orientation were found to have a significant effect on this ordering (Jen et al 1987a, b). Similar TED patterns to Jen et al (1986) were reported by Nakayama and Fujita (1986) and Norman (1989) from liquid phase epitaxy (LPE) GaInAs and MOCVD AlInAs and GaAlInAs grown on (001) InP substrates. They concluded from TED and high resolution electron microscopy results that famatinite ordering was present in their samples.

Murgatroyd et al (1986a, b) reported the first CuPt-type ordering in III-V alloys when it was observed in MBE $GaAs_ySb_{1-y}$ layers grown on (001) GaAs and InP substrates. A modulation in the [110] direction of period $4d_{110}$ was also present in layers with $y > 0.5$. Since then CuPt-type ordering has been found in a wide range of ternary and quaternary III-V (001) layers with either mixed group III or mixed group V atom sublattices, grown by MOCVD, MBE and other vapour phase epitaxy (VPE) techniques and is by far the most often reported type of ordering. It has been found in GaAsSb (Murgatroyd et al 1986a, b, 1990, Ihm et al 1987), InAsSb (Jen et al 1989a, Seong et al 1990, Kurtz et al 1992), AlInAs (Norman et al 1987, Hull et al 1987, Ueda et al 1989, Baxter et al 1993), GaInAs (Shahid et al 1987, Norman 1987, 1989, Seong et al 1991), GaAsP (Plano et al 1988, Jen et al 1989b), InAsP (Jaw et al 1991), InPSb and GaPSb (Stringfellow 1989, Stringfellow and Chen 1991), AlInP (Yasuami 1988, Nozaki et al 1988) and GaInAsP (Shahid et al 1987, 1988, Plano et al 1988, Chu et al 1992). Most reports of CuPt-type ordering behaviour, however, have been for the GaInP and GaAlInP systems grown by MOCVD with early work being performed by Gomyo et al (1987, 1988), Ueda et al (1987, 1988), Bellon et al (1988), McKernan et al (1988), Goral et al (1988), Dabkowski et al (1988), Kondow et al (1988a, b, c,), Yasuami et al (1988), Morita et al (1988), Nozaki et al (1988) and Gavrilovic et al (1988). Significantly CuPt-type ordering has never been observed in LPE layers or in AlGaAs layers where the Al and Ga atoms are almost identical in size.

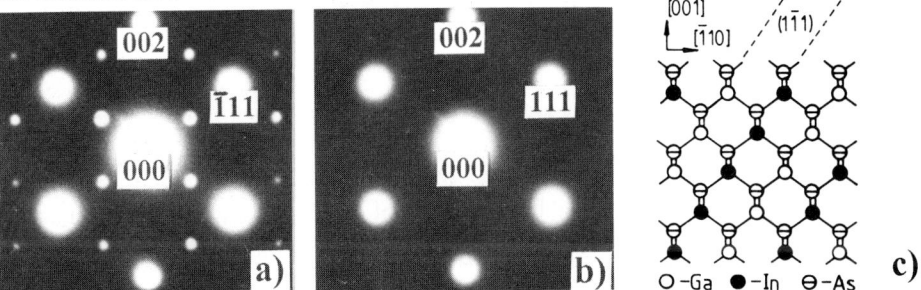

Fig. 2. (a) [110] and (b) [$\bar{1}$10] cross-section TED patterns of MOCVD $Ga_{0.47}In_{0.53}As$ layer grown on (001) GaAs at 550°C and a rate of 0.2 nm/s. (c) [110] projection of $Ga_{0.5}In_{0.5}As$ crystal perfectly ordered on ($1\bar{1}1$) planes.

CuPt-type ordering is only observed in layers grown on substrates whose orientation is close to (001) suggesting that its origin is closely linked to the structure of the (001) surface. For both mixed group III and mixed group V alloy layers only two, the ($\bar{1}11$) and ($1\bar{1}1$) or $CuPt_B$ variants, of the four possible variants are observed (Murgatroyd et al 1990, Chen et al 1990) and this has been associated by many authors to the anisotropy of bonding at the (001) surface. Fig. 2 shows this behaviour for a MOCVD GaInAs layer grown on (001) GaAs. Superlattice spots at 1/2{111} positions are visible in the [110] cross-section TED pattern corresponding to ordering on the ($\bar{1}11$) and ($1\bar{1}1$) planes, the $CuPt_B$ variants. No 1/2{111} superlattice spots are present in the [$\bar{1}$10] TED pattern indicating no ordering occurring on the (111) and (11$\bar{1}$) planes, the $CuPt_A$ variants. The orthogonal <110> cross-sections were distinguished using convergent beam electron diffraction and anisotropic chemical etching techniques performed on the GaAs substrates (Bellon et al 1989, Murgatroyd et al 1990). A [110] projection of a $Ga_{0.5}In_{0.5}As$ crystal perfectly ordered on ($1\bar{1}1$) planes is shown in Fig. 2(c). Fig. 3 shows similar behaviour in a MBE GaAsSb layer grown on (001) GaAs with 1/2 {111} superlattice spots only visible in the [110] pattern. Weak [001] streaks of diffuse intensity are, however, faintly visible in the [$\bar{1}$10] cross-section of this MBE GaAsSb sample at ±n/8 g$\bar{2}$20 positions from the main diffraction spots corresponding to a [110] modulation in the crystal of periodicity $4d_{110}$ (Murgatroyd et al 1986b, 1990) which breaks down in the [001] growth direction. Similar [110] modulations but of period $3d_{110}$ were observed by Jen et al (1987b) in (001) MOCVD GaAsSb layers. Growth on vicinal (001) substrates offcut a few degrees towards either the [$\bar{1}$10] or [1$\bar{1}$0] directions leads to the preferential formation of the ($\bar{1}$11) and (1$\bar{1}$1) variant respectively (Suzuki et al 1988, Bellon et al 1989, Chen and Stringfellow 1991) indicating that the presence and motion of [110] surface steps plays an important role in the ordering process.

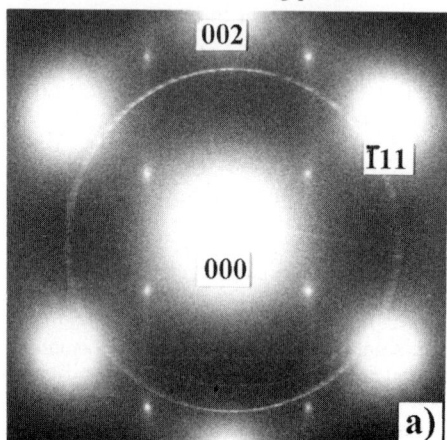 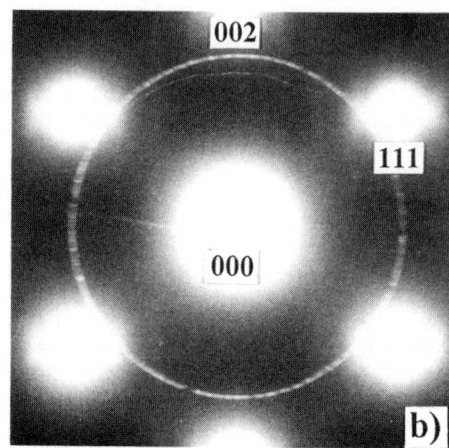

Fig. 3. (a) [110] and (b) [$\bar{1}$10] cross-section TED patterns of MBE GaAs$_{0.55}$Sb$_{0.45}$ layer grown on (001) GaAs at 525°C.

The ordering behaviour described above strongly suggests that the CuPt-type ordering observed in the layers develops at the surface during growth and remains frozen into the bulk of the layers during subsequent growth. Total energy calculations (e.g. Bernard et al 1990) and post growth annealing experiments (Plano et al 1988) support this view since they predict that the CuPt-type ordering is unstable in the bulk of the layers and reveal that it can be destroyed by bulk diffusion with the rate of disordering enhanced by the presence of impurities. The observation of strong ordering in layers grown under certain conditions indicates that disordering in the bulk can be kinetically limited during growth.

The degree of ordering, morphology and size of domains and density of APBs are very sensitive to growth conditions such as temperature, rate and V/III ratio. The degree of ordering in layers is observed to be low in samples grown at low temperatures, passes through a maximum at intermediate temperatures and then to fall to zero at high temperatures with the growth temperature for maximum ordering a function of the growth rate. It has been proposed that this complex growth condition dependence of ordering observed for both GaInP and GaInAs layers is a result of the competing processes of surface induced ordering and bulk disordering which occur during growth (Kurtz et al 1990, Seong et al 1991). In general larger regions of the two variants containing a lower density of APBs are observed in samples grown at low rates and intermediate temperatures (Morita et al 1988, Mahajan et al 1989, Seong et al 1991, Baxter et al 1991a). Samples grown at high rates and low temperatures tend to contain smaller ordered regions and a higher density of APBs resulting in smaller domains. Often arrays of APBs are present tilted from the (001) plane leading to thin plate-like domains of the two variants (Morita et al 1988, Cao et al 1991, Seong et al 1991, Baxter et al 1991a, b). The arrays of thin plate-like domains are tilted in opposite senses in the two variants and this manifests itself in TED patterns as an elongation of the superlattice spots arising from the two variants along directions rotated in opposite senses from [001]. Baxter et al (1991a, b) have reported a much more complex microstructure in CuPt-type ordered MOCVD GaInP grown on close to exact (001) orientation GaAs substrates. In these samples alternating (001) thin laminae of the two variants were present with the laminae of each variant grouped together into coarser plate-like domains inclined in opposite senses to the (001) plane.

The occurrence of CuPt-type ordering is found to have significant effects on the optical and electrical properties of the layers. For example Gomyo et al (1987, 1988) reported a 50-100 meV reduction in the band gap of CuPt-type ordered GaInP layers in comparison to random layers of the same composition which affected the performance of visible light lasers. Similar band gap narrowing has also been observed recently for CuPt-type ordered InAsSb layers (Kurtz et al 1992). Wei and Zunger (1990, 1991) have theoretically predicted significant band gap narrowing for a wide range of perfectly CuPt-type ordered III-V alloys including InAsSb and GaInSb of interest for long wavelength detectors. The band gap difference between CuPt-type ordered and random III-V alloys has already been used to fabricate novel laser and light-emitting diode devices (Ueno et al 1990, Lee et al 1992). Friedman et al (1991) and Lee et al (1991) have reported reduced carrier mobilities in partially CuPt-type ordered GaInP layers in comparison to random alloys of the same composition and attributed this to scattering introduced by the presence of the ordered domains and APBs. Perfectly ordered layers, if they can be grown, are expected to possess higher mobilities than random layers due to the elimination of alloy scattering and hence could be of interest for high speed devices.

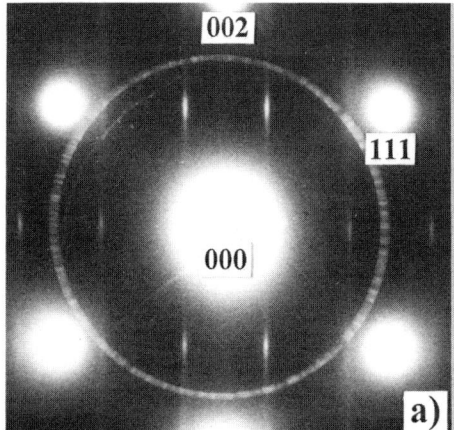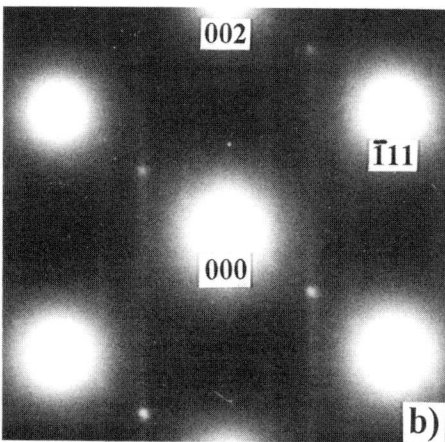

Fig. 4. (a) $[\bar{1}10]$ cross-section TED pattern of MBE $GaAs_{0.83}Sb_{0.17}$ grown at 625°C and (b) [110] cross-section TED pattern of MBE $Ga_{0.5}In_{0.5}As$ grown at 500 °C onto (001) GaAs substrates showing novel forms of ordering present.

We have recently observed a new type of ordering in $GaAs_ySb_{1-y}$ alloys (y >0.75) grown at temperatures higher than 600°C. The [110] TED patterns of these layers reveal only very weak CuPt-type ordering present. The $[\bar{1}10]$ TED patterns obtained from these layers, Fig. 4(a), however, contain relatively strong [001] rods of diffracted intensity at 1/8 $\mathbf{g}220$ positions either side of the main spots with quite strong maxima in intensity at {1/4, 1/4, 1}, {3/4, 3/4, 0} and {5/4, 5/4, 0} etc. positions indicating the existence of a new SL structure. Baxter et al (1993) have recently reported a novel form of ordering in (001) MBE AlInAs and GaAlInAs layers. Conventional CuPt$_B$ ordering was present in the layers with the 1/2{111} SL spots in [110] TED patterns connected together by sharp rods of diffracted intensity along [001] indicating abrupt changes occurring in the ordering along this direction. However, sharp [001] rods of diffracted intensity were also present in the same [110] TED pattern passing through 1/8 $\mathbf{g}220$ positions indicating a modulation present in the crystal along $[\bar{1}10]$ of period $4d_{110}$. We have observed similar ordering in (001) MBE GaInAs layers as shown in Fig. 4(b).

3. ORIGINS OF ORDERING

3.1 CuAu I in (110) and Chalcopyrite and CuAu I in (001) Layers

The exact mechanism for CuAu I-type ordering in (110) layers seems to be unclear at present and needs further investigation. The experimental evidence suggests it is surface induced and that atomic size differences in the alloys do not appear to be important since it is observed in both AlGaAs and GaInAs. Thus bond energy or electronic effects probably play a dominant role in its formation. Wang et al (1985) suggested a charge transfer dipole mechanism to explain the long range ordering observed in (110) AlGaAs layers but this was later shown to be unlikely by the total energy calculations of Wei and Zunger (1988) for bulk $AlGaAs_2$ which predicted phase separation followed by CuPt-type ordering to be the most stable configurations and suggested that the ordering was most likely due to a surface or kinetic effect. Van Vechten (1985) proposed that the ordering observed in (110) AlGaAs layers was a result of a kinetic segregation associated with a saturation of kink sites at growth steps on the surface arising from the different energies released by the Ga and Al atoms as they attach to the kink sites. Exchange reactions occurring during growth, as suggested by Petroff et al (1982) to explain coarser scale composition modulations found in (110) MBE AlGaAs layers, are also a possible source of this ordering. Chalcopyrite ordering, e.g. as observed in (001) MOCVD GaAsSb layers by Jen et al (1986), is predicted to be thermodynamically stable in coherent epitaxial layers of size-mismatched alloys (Bernard et al 1990) and is the expected form of ordering if surface effects, e.g. reconstruction, are ignored. The simultaneous occurrence of CuAu I-type ordering in these layers is unexpected since it is calculated to be unstable in coherent epitaxial layers of GaAsSb (Bernard et al 1990). Experimental results suggest both types of ordering occur at the growing surface (Jen et al 1987a,b). Further work is required to determine the origin of this ordering.

3.2   CuPt-type Ordering in (001) Layers

The widespread occurrence of this form of ordering was surprising since it was predicted to be the highest in energy of all the structures for size-mismatched alloys in both the bulk and coherent epitaxial forms (Bernard et al 1990). However these calculations did not include surface effects and as we have seen experimental results strongly suggest it is surface induced during growth. An (001) monolayer of the observed two $CuPt_B$ variants consists of alternating [110] rows of the different types of atoms on the mixed atom sublattice as can be seen in Fig. 2(c). To form 3 dimensional regions of either of the two variants successive ordered (001) monolayers of this type need to be stacked in phase with each other to avoid the formation of APBs or order twin boundaries between variants. Models proposed for this ordering therefore need a mechanism for creating ordered (001) monolayers of the correct type and also a phase locking mechanism for the lateral alignment of subsequent ordered monolayers to produce extended regions of the two observed variants. Early models to explain the occurrence of this ordering in mixed group III alloys suggested that it arose from a surface ordering of group III atoms which occurred to minimise the strain energy associated with incorporating the different sized atoms (with their different bond lengths) on the growing surface (Norman 1987, Suzuki et al 1988, Bellon et al 1989). A surface arrangement of group III atoms corresponding to an (001) monolayer of the two observed variants of CuPt-type ordering was concluded to be the lowest in energy. Phase locking of successive ordered monolayers to give extended regions of CuPt-type ordering was associated with the incorporation of atoms at surface micro-facets and atomic steps (Suzuki et al 1988, Bellon et al 1989). These models, however, wrongly predicted that the $CuPt_A$ variants should occur in mixed group V alloys which is contrary to the observation of only the $CuPt_B$ variants in these alloys by Murgatroyd et al (1990) and Chen et al (1990).

It is now generally believed that the origin of the observed $CuPt_B$ atomic ordering is related to the occurrence of surface reconstruction during growth. On an unreconstructed group V terminated (001) surface of a III-V semiconductor e.g. GaAs each surface group V atom has two dangling bonds aligned along the $[\bar{1}10]$ direction. The surface can lower its energy by reconstructing with the group V atoms forming $[\bar{1}10]$ oriented dimers and so reducing the number of dangling bonds. Similar dimerisation of group III atoms is thought to occur on group III atom terminated surfaces. In Fig. 5(a) is shown the missing dimer model (Chadi 1987, Pashley et al 1988) for the (2x4) reconstructed (001) surface of GaAs which is commonly observed by reflection high energy diffraction (RHEED) during the MBE growth under ultra high vacuum of III-V alloys using group V rich surface conditions. Recently Kamiya et al (1992a,b) have found using reflectance difference spectroscopy that almost identical surface reconstructions occur during MOCVD growth at atmospheric pressure. The (2x4) reconstructed surface consists of blocks of 3 group V atom dimers aligned along the [110] direction separated by [$\bar{1}$10] rows of missing dimers. The 2x periodicity of the surface along [$\bar{1}$10] comes from the dimerisation of the group V atoms whilst the 4x periodicity along [110] comes from the regular spacing of the rows of missing dimers. The blocks of dimers can either lie in phase across the missing rows corresponding to (2x4) surface reconstruction or lie out of phase corresponding to c(2x8) reconstruction. In Fig. 5(b) is shown a [110] projection through a 2x reconstructed As-terminated (001) surface of GaAs. The formation of the [$\bar{1}$10] oriented surface As dimers is expected to induce stresses beneath the crystal surface similar to those reported for (2x1) reconstructed (001) Si surfaces (Appelbaum and Hamann 1978). The atomic sites in the third and fourth atomic layers beneath the surface are placed alternately under compression directly below the dimers and tension directly below the gaps between the dimers along the [$\bar{1}$10] direction as shown in Fig. 5(b).

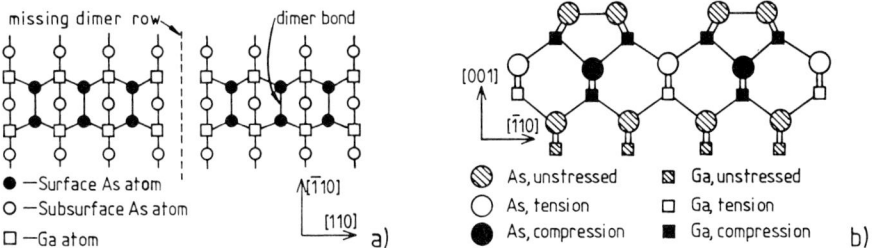

Fig. 5. (a) [001] projection of missing dimer model for (2x4) reconstructed (001) GaAs surface. (b) [110] projection of 2x reconstructed As-terminated (001) GaAs surface showing the nature of dimer-induced subsurface stresses.

Murgatroyd et al (1990) linked the origin of CuPt-type ordering and [110] modulations found in (001) MBE GaAsSb layers to the occurrence of (2x4) surface reconstruction during growth. It was suggested that As-Sb dimers would preferentially form at the growing reconstructed surface due to chemical interactions and that these dimers would be all oriented in the same sense forming the alternating [110] rows of As and Sb atoms at the surface required for an (001) monolayer of CuPt$_B$ ordering. No mechanisms for this alignment of dimers or for the phase locking of subsequent ordered (001) monolayers to give extended regions of the two variants were given. This model was also unable to explain the CuPt$_B$ ordering observed in mixed group III alloys. The origin of the $4d_{110}$ modulations observed in these layers was linked to the rows of missing dimers present on the (2x4) reconstructed surface. Chen et al (1991) extended these ideas to try and explain the CuPt$_B$-type ordering observed in MOCVD GaAsP and other mixed group V alloy layers. They proposed that a selectivity of incorporation of the group V atoms would occur during growth as [110] step edges move across a 2x reconstructed surface due to a difference in bonding introduced by the dimerisation. This proposed mechanism results in the formation of As-P dimers on the surface with the As and P atoms aligned in [110] rows as required for an (001) monolayer of CuPt$_B$ type ordering. For mixed group III alloys the mechanism was less clear but it was suggested that a similar selectivity of incorporation of the different group III atoms was introduced by the motion of kinks along the [110] step edges on a 2x reconstructed surface leading to the required atomic arrangement for an (001) monolayer of the observed CuPt$_B$-type ordering. No calculations were performed to test these ideas. A simple feedback mechanism where the group V atom dimer rows in the surface layer form directly above the [110] rows of the smaller group III or group V atoms in the underlying ordered (001) monolayer so as to minimise the dimer-induced subsurface stresses was proposed to lead to the phase locking required to produce large domains of the observed two ordered variants.

Suzuki et al (1991, 1992) reexamined their earlier mechanism proposed for the formation of CuPt$_B$ ordering in mixed group III alloys taking MOCVD GaInP on GaAs substrates as an example. In their new model it was assumed that 2x surface reconstruction occurs during MOCVD growth leading to the formation of [$\bar{1}$10] oriented group V atom surface dimers. The concept of step-terrace-reconstruction (STR) was then proposed in which it was suggested that on a non planar 2x reconstructed surface containing [110] steps the terrace widths between step edges would be such that all the group V atoms on the terraces could form dimers thus minimising the surface energy. This is shown in Fig. 6(a) for a 2x reconstructed vicinal (001) GaAs surface. It was then shown that growth on such a reconstructed stepped surface would lead to the formation of large domains of the two observed CuPt$_B$ variants if preferential incorporation of one of the group III atoms occurred at the [110] step edges, e.g. Ga due to bond energy effects or In due to steric effects, followed by the incorporation alternately of [110] rows of Ga and In atoms due to a combination of bond energy, steric and stress minimisation effects. Provided that in local areas the direction of motion of [110] step edges was the same, phase locking of successive ordered (001) monolayers was shown to automatically occur for growth on a step-terrace-reconstructed surface. To explain the CuPt$_B$-type ordering in mixed group V alloys a selectivity of incorporation of the group V atoms at [110] step edges on the reconstructed surface was tentatively suggested (Suzuki et al 1992) similar to the model proposed by Chen et al (1991). Again calculations were not performed to check these proposals.

Following the work of Appelbaum and Hamman (1978) and Kelires and Tersoff (1989), LeGoues et al (1990) proposed that the CuPt-type ordering observed in (001) MBE SiGe alloy layers was a result of a lateral ordering of Si and Ge atoms which occurred in the third and fourth atomic layers beneath the (2x1) reconstructed surface driven by the local stress field induced by the surface dimerisation. The smaller Si atoms were predicted to segregate to the atomic sites directly under the dimers which are under compression whilst the larger Ge atoms were predicted to segregate to the atomic sites underneath the gaps between dimers which are under tension. To explain the large domains of all four variants of CuPt-type ordering observed in the layers further assumptions were needed. Growth was assumed to proceed by the motion of double height steps. The rate of diffusion in the third and fourth atomic layers beneath the surface was assumed to be rapid enough to enable the lateral ordering to occur but diffusion deeper in the crystal insufficient to subsequently destroy the ordering.

We have extended the ideas of LeGoues et al (1990) to produce an improved model to explain the origin of the CuPt$_B$ ordering observed in (001) ternary and quaternary III-V alloy layers (Mahajan 1991, Mahajan and Philips 1991, Norman et al 1991, Norman et al 1993, Philips et al 1993). We first assume that the layers are grown under group V rich surface conditions and that the surface group V atoms reconstruct to form dimers as shown in Fig. 5. The dimerisation of the surface As atoms induces subsurface strains in the GaAs crystal similar to those observed for 2x reconstructed Si (Appelbaum and Hamman 1978). Atomic sites in the third and fourth atomic layers beneath the surface are placed alternately under compression directly below the dimers and under tension below the gaps between dimers along the [$\bar{1}$10] direction, Fig. 5(b). During growth of a mixed group III alloy e.g. GaInAs on such a reconstructed surface it is proposed that the larger In atoms will segregate to the [110] rows of atomic sites under tension in the fourth atomic layer beneath the surface. The smaller Ga atoms will segregate to the [110] rows of sites under compression. This segregation minimises the strain energy associated with the dimerisation and incorporating the different sized atoms in the crystal and results in the formation of the alternating [110] rows of Ga and In atoms required for an (001) monolayer of the observed CuPt$_B$ ordering. It is assumed that disordering of the ordered structure in the bulk deeper below the surface during further growth is kinetically limited. Similarly for the growth of a mixed group V alloy, e.g. GaAsSb, on such a reconstructed surface it is proposed that the larger Sb atoms will segregate to the [110] rows of atomic sites under tension and the smaller As atoms to the [110] rows of sites under compression in the third atomic layer beneath the surface again leading to the required atomic arrangement for an (001) monolayer of the observed CuPt$_B$ ordering. Thus the model easily explains why the same two variants of CuPt-type ordering are observed in both mixed group III and mixed group V alloys and also why CuPt-type ordering has not been observed in AlGaAs layers where the atomic sizes of Al and Ga are almost identical resulting in little driving force for the subsurface lateral segregation. These proposals are supported by the results of valence force field (VFF) calculations (Philips et al 1993) which indicate for the growth of GaInAs on (2x4) reconstructed (001) GaAs that the preferential occupation of the dilated [110] rows of atomic sites in the fourth atomic layer beneath the surface by the larger In atoms reduces the strain energy by ~ 100 meV/dimer site over the less favourable arrangements. The simulation of growth of GaAsSb and GaAsP epitaxial layers on identically reconstructed (001) GaAs gave similar results with the difference in strain energies between the favourable configuration, i.e. the larger atoms in dilated sites and smaller atoms in compressed sites on the third atomic layer beneath layer surface, and unfavourable configurations calculated to be ~ 120 meV/dimer site.

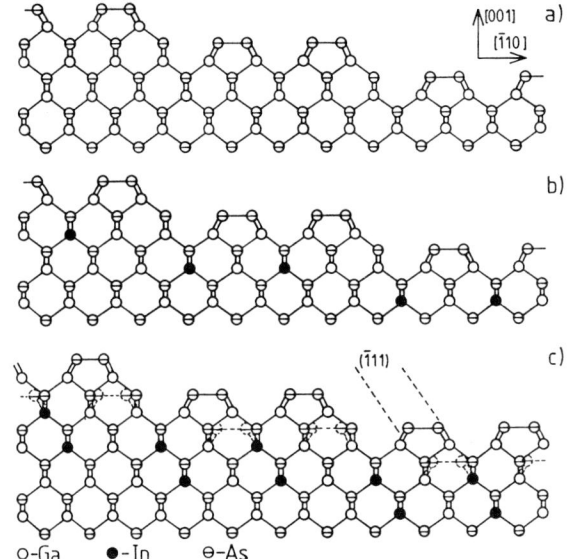

o–Ga   •–In   ⊖–As

Fig. 6. (a) Step-terrace-reconstruction (STR) model of 2x reconstructed vicinal (001) GaAs surface.
(b) STR model of 2x reconstructed vicinal (001) GaInAs surface showing dimer-induced subsurface segregation of Ga and In atoms.
(c) Situation after growth of next monolayer of GaInAs showing evolution of CuPt-type ordering on ($\bar{1}$11) planes due to STR-induced phase locking of (001) ordered monolayers.

The second part of our model concerns the effect of the presence and motion of [110] surface steps which have been found to play an important role in the ordering process leading to the preferential formation of the ($\bar{1}$11) or (1$\bar{1}$1) variants on vicinal (001) surfaces offcut a few degrees towards the [$\bar{1}$10] or [1$\bar{1}$0] direction respectively (Suzuki et al 1988, Bellon et al 1989, Chen and Stringfellow 1991). We first assume that the [110] steps on 2x reconstructed vicinal surfaces preferentially occur with the unbonded riser configuration shown in Fig. 6(a), as also suggested by Suzuki et al (1991, 1992), so that terraces between steps contain an even number of surface group V atoms. This configuration is thought to be the lowest in energy since all the group V atoms on the terraces between steps can form dimers. Consider now the growth of GaInAs on such a surface. The ordering mechanism described above will lead to the segregation of the smaller Ga atoms to the compressed atomic sites directly under the dimers and the larger In atoms to the dilated atomic sites underneath the gaps between dimers as shown in Fig. 6(b). It is assumed that growth proceeds by the incorporation of atoms at the step edges which continuously move towards the [$\bar{1}$10] direction with each newly attached pair of surface As atoms reconstructing to form dimers and hence causing further subsurface segregation of the different sized group III atoms to occur. The situation after the deposition of the next monolayer of crystal is shown in Fig. 6(c). It can be seen that the newly deposited As dimers are shifted by $a/2\sqrt{2}$ in the [1$\bar{1}$0] direction from those that were in the underlying layer. Consequently the [110] rows of dilated and compressed atomic sites beneath the newly dimerised surface are also displaced similarly with respect to those in the lower terrace leading to the generation of CuPt-type ordering of the Ga and In atoms on the ($\bar{1}$11) planes as observed experimentally for this type of vicinal surface. If a terrace of width corresponding to an odd number of group V atoms was present on the reconstructed surface then it can easily be seen that this would lead to the formation of an APB in the ordered structure for our mechanism. The observation of fairly large domains containing few APBs in samples suggests that the majority of terraces are an even number of group V atoms wide.

Zunger and coworkers (Froyen and Zunger 1991, Bernard et al 1991, Osório et al 1992) have performed first-principles total-energy calculations investigating the origin of spontaneous surface induced CuPt-type ordering in (001) GaInP layers. They showed that an electronically driven surface reconstruction (dimerisation, buckling and tilting) of a cation (Ga, In) terminated (001) surface would act to stabilise the CuPt$_B$ surface topology over other structures including phase separation by about 84 meV per surface atom. However as growth is normally performed under group V (anion) rich surface conditions it is unlikely that such cation atom surface reconstructions would occur. For an anion (P) terminated surface, expected under normal growth conditions, little driving force was found for ordering in the cation atom layer immediately beneath the surface. VFF calculations were performed to investigate the lowest energy topologies of atoms in deeper subsurface layers. For cation terminated surfaces a weak preference for the topology of one of the CuPt$_A$ variants was indicated for the first subsurface cation layer whilst for the second subsurface cation layer the topology required to produce true 3d CuPt$_B$ type ordering was found to be the lowest in strain energy. For anion terminated surfaces a large energy preference, ~90meV per surface atom assuming unbuckled P dimers, was predicted for the CuPt$_B$ topology in the second cation layer beneath the surface. This is induced by strains associated with the dimerisation of the surface P atoms and is identical to the mechanism proposed by LeGoues et al (1990) to explain CuPt-type ordering in (001) MBE SiGe layers and by us above to explain the CuPt$_B$ ordering in (001) III-V alloy layers. Similar first-principles total-energy calculations are required for mixed group V alloys e.g. GaAsSb.

4. SUMMARY

We have seen that several types of atomic ordering can arise during the epitaxial growth of III-V semiconductor alloys. This ordering may have significant effects on the electrical and optical properties of the layers. The most often observed form of ordering in (001) layers, CuPt, is surface induced and its origin is believed to be associated with the occurrence of surface reconstruction during growth. We have presented an improved model which can explain the main features observed for the CuPt ordering. In this model the CuPt ordering is proposed to arise as a result of a subsurface lateral segregation of atoms which occurs to minimise the strain energy associated with dimerisation of group V atoms on the reconstructed surface and the incorporation of the different sized atoms on the mixed sublattice in the crystal. The nature of [110] surface atomic steps is found to play an important role in the formation of extended regions of the two observed variants.

ACKNOWLEDGEMENTS

The work at Imperial College and the University of Oxford was funded by the Science and Engineering Research Council and TYS would also like to thank the British Council for financial support. The work at Carnegie Mellon was carried out under the auspices of the Office of Naval Research and Department of Energy and BAP and SM gratefully acknowledge the awards of grants N00014-92-J-1124 and DE-FG02-87ER45329.

REFERENCES

Appelbaum J A and Hamann D R 1978 Surface Science 74 21
Baxter C S, Stobbs W M and Wilkie J H 1991a Microscopy of Semiconducting Materials 1991, Inst. Phys. Conf. Ser. No. 117, eds A G Cullis and N J Long (Bristol: Inst. Phys.) pp 469-472
Baxter C S, Stobbs W M and Wilkie J H 1991b J. Cryst. Growth 112 373
Baxter C S, Stobbs W M, Broom R F and Reithmaier J P 1993 Submitted to J. Cryst. Growth
Bellon P, Chevalier J P, Martin G P, Dupont-Nivet E, Thiebaut C and André J P 1988 Appl. Phys. Lett. 52 567
Bellon P, Chevalier J-P, Augarde E, André J-P and Martin G P 1989 J. Appl. Phys. 66 2388
Bernard J E, Dandrea R G, Ferreira L G, Froyen S, Wei S-H and Zunger A 1990 Appl. Phys. Lett. 56 731
Bernard J E, Froyen S and Zunger A 1991 Phys. Rev. B 44 11178
Cao D S, Reihlen E H, Chen G S, Kimball A W and Stringfellow G B 1991 J. Cryst. Growth 109 279

Chadi D J 1987 J. Vac. Sci. Technol. A5 834

Chen G S, Jaw D H and Stringfellow G B 1990 Appl. Phys. Lett. 57 2475

Chen G S, Jaw D H and Stringfellow G B 1991 J. Appl. Phys. 69 4263

Chen G S and Stringfellow G B 1991 Appl. Phys. Lett. 59 324

Chin A, Chang T Y, Ourmazd A and Monberg E M 1991 Appl. Phys. Lett. 58 968

Chu S N G, Logan R A and Tanbun-Ek T 1992 J. Appl. Phys. 72 4118

Dabkowski F P, Gavrilovic K, Meehan K, Stutius W, Williams J E, Shahid M A and Mahajan S 1988 Appl. Phys. Lett. 52 2142

Friedman D J, Kibbler A E and Olson J M 1991 Appl Phys. Lett. 59 2998

Froyen S and Zunger A 1991 Phys. Rev. Lett. 66 2132

Gavrilovic P, Dabkowski F P, Meehan K, Williams J E, Stutius W, Hsieh K C, Holonyak N Jr., Shahid M A and Mahajan S 1988 J. Cryst. Growth 93 426

Glas F 1993 This proceedings

Gomyo A, Suzuki T, Kobayashi K, Kawata S, Hino I and Yuasa T 1987 Appl. Phys. Lett. 50 673

Gomyo A, Suzuki T and Iijima S 1988 Phys. Rev. Lett. 60 2645

Goral J P, Al-Jassim M M, Olson J M and Kibbler A 1988 Mater. Res. Soc. Symp. Proc. Vol. 102 583

Hull R, Carey K W, Fouquet J E, Reid G A, Rosner S J, Bimberg D and Oertel D 1987 GaAs and Related Compounds 1986, Inst. Phys. Conf. Ser. No. 83, ed W T Lindley (Bristol: Inst. Phys.) pp 209-214

Ihm Y-E, Otsuka N, Klem J and Morkoç H 1987 Appl. Phys. Lett. 51 2013

Jaw D H, Chen G S and Stringfellow G B 1991 Appl. Phys. Lett. 59 114

Jen H R, Cherng M J and Stringfellow G B 1986 Appl. Phys. Lett. 48 1603

Jen H R, Cherng M J, Jou M J and Stringfellow G B 1987a GaAs and Related Compounds 1986 Inst. Phys. Conf. Ser. No. 83, ed W T Lindley (Bristol: Inst. Phys.) pp 159-164

Jen H R, Jou M J, Cherng Y T and Stringfellow G B 1987b J. Cryst. Growth 85 175

Jen H R, Ma K Y and Stringfellow G B 1989a Appl. Phys. Lett. 54 1154

Jen H R, Cao D S and Stringfellow G B 1989b Appl. Phys. Lett. 54 1890

Kamiya I, Tanaka H, Aspnes D E, Florez L T, Colas E, Harbison J P and Bhat R 1992a Appl. Phys. Lett. 60 1238

Kamiya I, Aspnes D E, Tanaka H, Florez L T, Harbison J P and Bhat R 1992b Phys. Rev. Lett. 68 627

Kelires P C and Tersoff J 1989 Phys. Rev. Lett. 63 1164

Kondow M, Kakibayashi H and Minagawa S 1988a J. Cryst. Growth 88 291

Kondow M, Kakibayashi H, Minagawa S, Inoue Y, Nishino T and Hamakawa Y 1988b J. Cryst. Growth 93 412

Kondow M, Kakibayashi H, Minagawa S, Inoue Y, Nishino T and Hamakawa Y 1988c Appl. Phys. Lett. 53 2053

Kuan T S, Kuech T F, Wang W I and Wilkie E L 1985 Phys. Rev. Lett. 54 201

Kuan T S, Wang W I and Wilkie E L 1987 Appl. Phys. Lett. 51 51

Kurtz S R, Olson J M and Kibbler A 1990 Appl. Phys. Lett. 57 1922

Kurtz S R, Dawson L R, Biefeld R M, Follstaedt D M and Doyle B L 1992 Phys. Rev. B 46 1909

Lee M K, Horng R H and Haung L C 1991 Appl. Phys. Lett. 59 3261

Lee M K, Horng R II and Haung L C 1992 J. Appl. Phys. 72 5420

LeGoues F K, Kesan V P, Iyer S S, Tersoff J and Tromp R 1990 Phys. Rev. Lett. 64 2038

Mahajan S, Shahid M A and Laughlin D E 1989 Microscopy of Semiconducting Materials 1989, Inst. Phys. Conf. Ser. No. 100, eds A G Cullis and J L Hutchison (Bristol: Inst. Phys.) pp 143-153

Mahajan S 1991 Proc. of the 5th Brazilian School on Semiconductor Physics, ed J R Leite (Singapore: World Science Publishers)

Mahajan S and Philips B A 1991 Proc. of NATO Meeting on Intermetallics, Irsee, Germany

McKernan S, De Cooman B C, Carter C B, Bour D P and Shealy J R 1988 J. Mater. Res. 3 406

Morita E, Ikeda M, Kumagai O and Kaneko K 1988 Appl. Phys. Lett. 53 2164

Murgatroyd I J, Norman A G and Booker G R 1986a MRS Spring Meeting 1986 paper B5.2

Murgatroyd I J, Norman A G, Booker G R and Kerr T M 1986b Proc. XIth Int. Cong. on Electron Microscopy 1986, J. Electron Microsc. 35 Supplement, eds T Imura, S Maruse and T Suzuki (Tokyo: Jpn. Soc. Electron. Microsc.) pp 1497-1498

Murgatroyd I J, Norman A G and Booker G R 1990 J. Appl. Phys. 67 2310

Nakata Y, Ueda O and Fujii T 1991 Jpn. J. Appl. Phys. 30 L249

Nakayama H and Fujita H 1986 GaAs and Related Compounds 1985, Inst. Phys. Conf. Ser. No. 79 , ed M Fujimoto (Bristol: Inst. Phys.) pp 289-294

Norman A G, Mallard R E, Murgatroyd I J, Booker G R, Moore A H and Scott M D 1987 Microscopy of Semiconducting Materials 1987, Inst. Phys. Conf. Ser. No. 87, eds A G Cullis and P D Augustus (Bristol: Inst. Phys.) pp 77-82

Norman A G 1987 D. Phil. Thesis, University of Oxford

Norman A G 1989 Evaluation of Advanced Semiconductors by Electron Microscopy 1988, NATO Adv. Sci. Inst. Ser. B 203, ed D Cherns (New York: Plenum) pp 233-253

Norman A G, Seong T-Y and Booker G R 1991 Unpublished work

Norman A G, Seong T-Y, Ferguson I T, Booker G R and Joyce B A 1993 Proc. Int. Conf. Narrow Gap Semiconductors 1992, Semicond. Sci. Technol. 8, eds R A Stradling and J B Mullin (Bristol: Inst. Phys.) pp S9-15

Nozaki C, Ohba Y, Sugawara H, Yasuami S and Nakanisi T 1988 J. Cryst. Growth 93 406

Onabe K 1982 Jpn. J. Appl. Phys. 21 L323

Osório R, Bernard J E, Froyen S and Zunger A 1992 Phys. Rev. B 45 11173

Pashley M D, Haberern K W, Friday W, Woodall J M and Kirchner P D 1988 Phys. Rev. Lett. 60 2176

Petroff P M, Cho A Y, Reinhart F K, Gossard A C and Wiegmann W 1982 Phys. Rev. Lett. 48 170

Philips B A, Norman A G, Seong T-Y, Mahajan S, Booker G R, Skowronski M, Harbison J P and Keramidas V G 1993 Submitted to J. Cryst. Growth

Plano W E, Nam D W, Major J S Jr., Hsieh K C and Holonyak N Jr. 1988 Appl. Phys. Lett. 53 2537

Seong T-Y, Norman A G, Booker G R, Droopad R, Williams R L, Parker S D, Wang P D and Stradling R A 1990 Mater. Res. Soc. Symp. Proc. Vol. 163 907

Seong T-Y, Norman A G, Hutchison J L, Booker G R, Cullis A G, Bass S J and Taylor L L 1991 Microscopy of Semiconducting Materials 1991, Inst. Phys. Conf. Ser. No. 117, eds A G Cullis and N J Long (Bristol: Inst. Phys.) pp 463-468

Shahid M A, Mahajan S, Laughlin D E and Cox H M 1987 Phys. Rev. Lett. 24 2567

Shahid M A and Mahajan S 1988 Phys. Rev. B 38 1344

Srivastava G P, Martins J L and Zunger A 1985 Phys. Rev. B 31 2561

Stringfellow G B 1989 J. Cryst. Growth 98 108

Stringfellow G B and Chen G S 1991 J. Vac. Sci. Technol. B 9 2182

Suzuki T, Gomyo A and Iijima S 1988 J. Cryst. Growth 93 396

Suzuki T and Gomyo A 1991 J. Cryst. Growth 111 353

Suzuki T, Gomyo A and Iijima S 1992 Ordering at Surfaces and Interfaces, eds A Yoshimori, T Shinjo and H Watanabe (Berlin: Springer-Verlag) pp 363-375

Ueda O, Takikawa M, Komeno J and Umebu I 1987 Jpn. J. Appl. Phys. 26 L1824

Ueda O, Takikawa M, Takechi M, Komeno J and Umebu I 1988 J. Cryst. Growth 93 418

Ueda O, Fujii T, Nakada Y, Yamada H and Umebu I 1989 J. Cryst. Growth 95 38

Ueda O, Nakata Y and Fujii T 1991a Appl. Phys. Lett. 58 705

Ueda O, Nakata Y Nakamura T and Fujii T 1991b J. Cryst. Growth 115 375

Ueno Y, Fujii H, Kobayashi K, Endo K, Gomyo A, Hara K, Kawata S, Yuasa T and Suzuki T 1990 Jpn. J. Appl. Phys. 29 L1666

Van Vechten J A 1985 J. Cryst. Growth 71 326

Wang W I 1985 J. Appl. Phys. 58 3244

Wei S-H and Zunger A 1988 Phys. Rev. Lett. 61 1505

Wei S-H and Zunger A 1990 Appl. Phys. Lett. 56 662

Wei S-H and Zunger A 1991 Appl. Phys. Lett. 58 2684

Yasuami S, Nozaki C and Ohba Y 1988 Appl. Phys. Lett. 52 2031

Inst. Phys. Conf. Ser. No 134: Section 6
Paper presented at Microsc. Semicond. Mater. Conf., Oxford, 5–8 April 1993

291

On the occurrence of phase separation in InGaAs/InP systems

F Peiró[1,2], A Cornet[1] and J R Morante[1]

[1]LCMM Dept. Física Aplicada i Electrònica. Universitat de Barcelona. Diagonal 645. 08028 Barcelona, Spain.
[2]Serveis Científico-Tècnics. Univ. Barcelona. Lluis Solé i Sabarís 1-3, 08028 Barcelona, Spain.

ABSTRACT: This work presents the evolution of the fine contrast modulation observed in the $In_xGa_{1-x}As$/InP system, with the layer mismatch, layer thickness and growth temperature. The results show that the fine structure appears during the growth and has a dynamic behaviour as growth proceeds. Some results about the coarse modulation are also commented on.

1. INTRODUCTION

Electron microscopy characterizations of most of the III-V compounds grown on InP have shown the appearance of two quasi-periodic contrast modulations along the $<010>$ directions, whose wavelengths are hundred of nm (coarse modulation) and tens of nm (fine modulation), with similar features even in alloys grown by different epitaxial techniques as LPE, VPE and MBE. Although it is extensively assumed that phase separation induced by an immiscibility gap (Stringfellow 1983) is the origin of the fine speckle modulation, two opposite models based on bulk (Norman 1985) and surface atomic diffusion phenomena (McDevitt 1992), have been proposed. A greater controversy remains about the origin of the coarse structure. Some authors (Norman 1985) relate it to composition modulations due to surface diffusion during growth and recently, McDevitt (1992) claimed that the coarse structure is an artifact of TEM thin foils as a result of the accommodation of stresses due to the presence of the fine modulation.

Our aim is to study the effect of the main technological parameters, namely layer thickness (t), layer mismatch (x) and growth temperature (T_g), on the final configuration of layers with contrast modulation, to provide new experimental results for the understanding of such structures.

2. EXPERIMENTAL DETAILS

We have studied InGaAs layers grown on (100) InP substrates in a VG Semicon V8OH MBE system: Group A consists of five $In_xGa_{1-x}As$ layers with thickness (t) 0.3, 0.5 ,0.7, 1, and $2\mu m$, grown with fixed composition $x=54.3\% \pm 0.2\%$ at $T_g=515°C$. Group B is formed by samples changing in composition, and hence in mismatch, with x values of 54.1%, 54.6%, 59%, and 62.5%, with the fixed values $t=0.5\mu m$ and $T_g=515°C$. Samples of group C have $x=54\%$ and $t=1.8\mu m$, but have been grown at $T_g=450°C$, 475°C, 500°C, 525°C and 550°C.

TEM specimens have been obtained by mechanical polishing and Ar^+ bombardment, in a liquid nitrogen cooled stage. For plan view observations, an accurate control of the layer surface milling time, allows us to have TEM foils with the same thickness, and to locate the thinner regions at different distances from the interface. So, we are able to study the microstructure of the whole epilayer, being sure that their features are not due to thickness differences of the regions examined.

3. RESULTS AND DISCUSSION

The TEM images of the (100) zone axis present similar results for all the epilayers: a fine modulation, with dark bands along the $<010>$ directions, in strong contrast for $g = <022>$ bright field two beam condition (Fig. 1a) and with extinction of the dark lines parallel to g when $g = <040>$ beams are used (Fig. 1b and 1c). The diffraction patterns of the layers do not show satellite spots, but an elongation of the 040 and 004 reflections towards the [010] and [001] directions is clear (inset of Fig. 1a). This means that the periodicity of the fine modulation is not small enough for its corresponding spot to be resolved from the main 040 type reflections. Moreover, in group A samples, the long wavelength (Λ) modulation has been also observed, showing the same behaviour as the fine modulation, when the imaging conditions are changed. The details of this coarse structure have been published elsewhere (Peiró 1991).

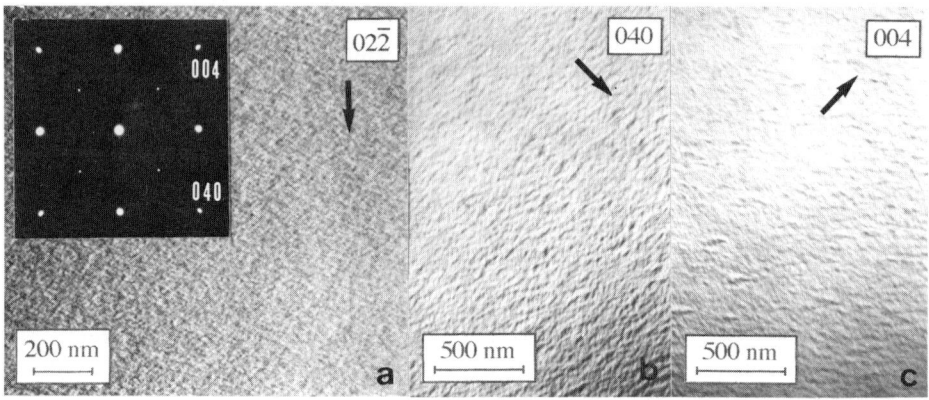

FIG: 1: This image corresponds to a InGaAs layer (group B), with $x = 54.1\%$, $t = 0.5\mu m$ and $T_g = 515°C$. a) $g = 022$, the fine contrast modulation along [010] and [001] is observed. The inset shows the diffraction pattern, with the $<040>$ spots elongated towards the $<010>$ directions. b) $g = 040$, only the lines along [001] are visible. c) $g = 004$, only the lines orthogonal to g remain in contrast.

The value of the wavelength λ of the fine speckle is approximately constant all over the layer. Nevertheless λ changes when the growth parameters are modified. The evolution of λ with the layer composition, growth temperature and layer thickness, is shown in Fig. 2a, b and c respectively. The general tendency of λ is to move towards lower values as t, x and T_g rise.

The fine speckle wavelength λ seems to evolve with layer thickness in the same way as the wavelength Λ of the coarse modulation (Peiró 1991). This fact suggests a dynamic nature of both structures, which may modify their λ and Λ values as growth proceeds. We have also studied the evolution of the coarse contrast upward from the interface, in the thicker samples of group A. Fig. 3a corresponds to the sample with $t = 2\mu m$, in the region near the layer-substrate interface. There, the coarse and fine modulations are both present, whereas, imaging the top of the epilayer (Fig. 3b), only the fine structure is visible. XTEM examinations confirm the results. In Fig. 3c, besides the fine modulation, there are some dark bands along the [100] direction reaching $0.5\mu m$ above the interface, whereas in the upper regions, only the fine speckle remains. Fig. 3d corresponds to the $0.3\mu m$ thick InGaAs layer, for which both the fine and coarse modulations are present in the whole film. Therefore, our results show that it is possible to have fine modulation and not the coarse structure in specimens with the underlying substrate totally removed; this proves that a buckling of the thin film due to stresses associated with the fine modulation (McDevitt 1992) is not the cause of the coarse structure. Moreover, the fine speckle in [0$\overline{1}$1] XTEM images is only observable under g=022, appearing then as a columnar structure along [100], and contrast changes orthogonal to the growth direction (Fig. 3d). Under g=200, the layer does not show any type of contrast variations, indicating that the fine modulation is a twodimensional structure developed on the surface during growth, in agreement with the results of Mc.Devitt (1992).

As far as the x-dependence of λ is concerned, Mahajan (1989) proposed that, if the growth conditions were identical, the wavelength of the fine speckle would be the same whatever the composition of the quaternary InGaAsP alloy was. The apparent contradiction with our results may be explained taking into account that, for $In_xGa_{1-x}As_yP_{1-y}$, a wide range of x and y values are possible, always assuring lattice matching to the InP substrate. Nevertheless, the wavelength of the fine modulation is influenced not by the composition variations themself, but by the stresses induced by lattice mismatch. Indeed, the most evident drop of λ takes place for parameters directly related to the increment of the elastic energy of the system, i.e. layer thickness and intrinsic mismatch, the latter driven by the changes in In molar fraction.

Conversely, the growth temperature has less influence on λ values, although it seems that λ decreases for higher T_g, opposite to the results shown by McDevitt (1992) in InGaAsP layers. If further experiments in a wider range of T_g values confirmed the decrease of λ as T_g increases, it would be in clear contradiction with atomic diffusion phenomena, which would be favoured at higher T_g.

The fine speckle can be explained on the basis of a non-random location of III-V type atoms (Glas 1989), due to the dissimilar bond lengths of the constituents (In-As and Ga-As, 7% different). The state of bond strain in

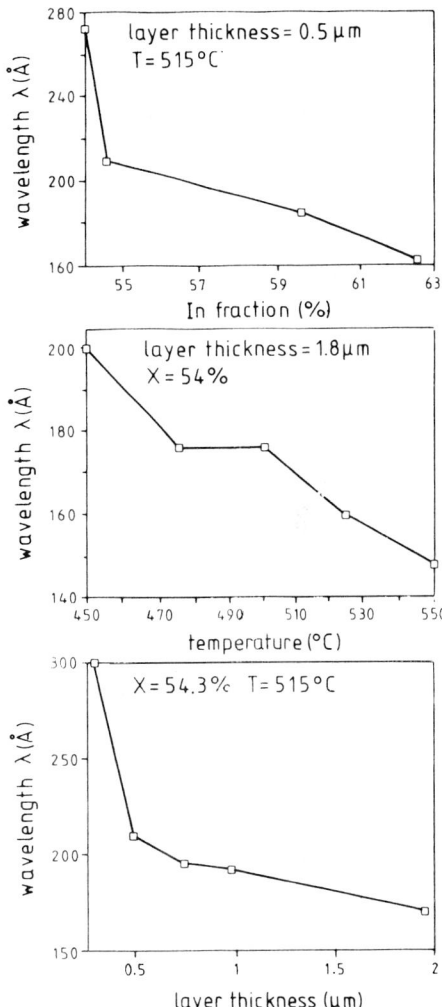

FIG. 2: Evolution of the fine modulation wavelength λ with a) layer-substrate mismatch, b) growth temperature, and c) layer thickness.

ternary or quaternary compounds, with respect to their original binaries, is the main term of the mixing enthalpy and dominates even over the entropy term (Ichimura 1989). So, we suggest that a non-randomn distribution of preferential sites for atom incorporation could be present, giving rise to a short-order clustering configuration, i.e. In rich or Ga rich zones and local lattice parameter variations responsible for the fine contrast modulation, even in matched layers. For coherent epitaxial growth, the increment of the elastic energy of the system, as layer thickness increases, induces a redistribution in order to minimize the Gibbs free energy, conferring to the system a dynamic behaviour. It is well known that the elastic strain induces a III-V alloy stabilization inside the immiscibility gap (Stringfellow 1983), nevertheless, Kazakov (1992) has shown that the stabilization criteria must be modified to account for the anisotropy of these compounds, and that, even in this situation of short-order clustering, the alloy may fall into an unstable state due to the complexity of the Gibbs free energy curves.

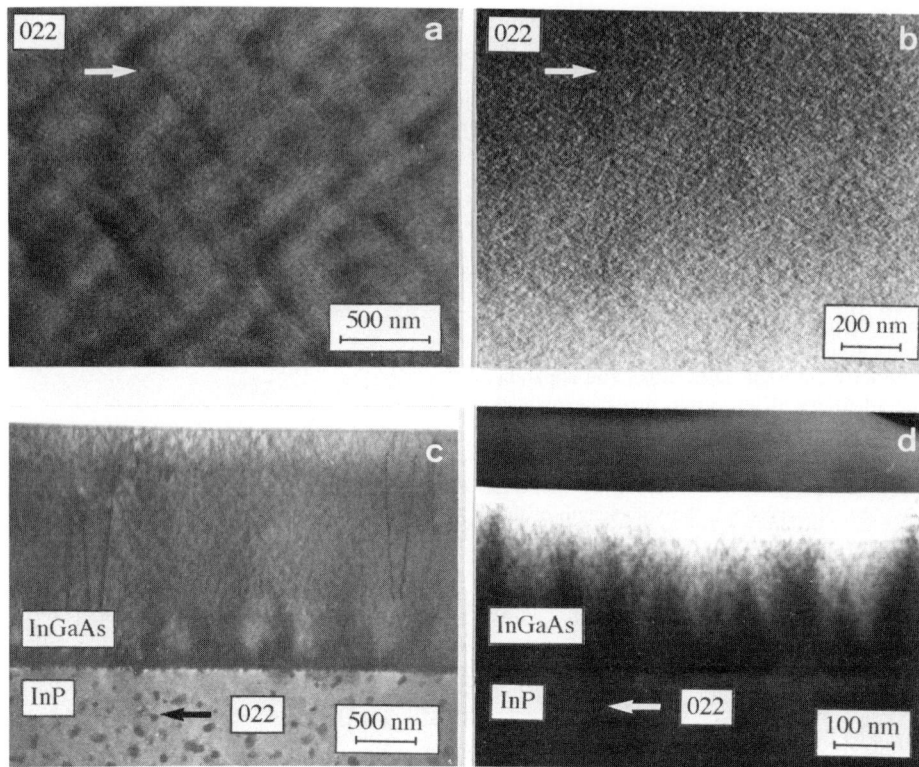

FIG. 3: *Samples of group A. a) $t=2\mu m$. (100) zone axis. Both the coarse and fine modulations are observed near the interface. b) Top of the layer, only the fine modulation remains in contrast. c) BF [0̄11] XTEM of the same sample. The dark coarse contrast near the interface reaches only 0.5μm above InP. c) DF image of the sample with $t=0.3\mu m$. The fine and coarse contrasts cover all the layer.*

4. CONCLUSIONS

We have found that the fine modulation observed in InGaAs layers has lower wavelength values as layer thickness, layer mismatch and growth temperature increase. The hypothesis of element distribution according to differences in bond lengths accounts for the evolution of λ with x, t. Our results have demonstrated that the buckling of the thin TEM foils due to stresses associated with the fine modulation is not the cause of the coarse pattern.

REFERENCES

Glas F (1989) NATO ASI Series <u>B203</u> (New York: Plenum Press) pp 217-16
Ichimura M and Sasaki A (1989) J. Cryst. Growth <u>98</u> 18
Kazakov A I and Kishmar I N (1992) J. Cryst. Growth <u>125</u> 509
Mahajan S, Sahid M A, Laughlin D E (1990) Inst. Phys. Conf. Ser. <u>100</u> 143
McDevitt T L, Mahajan S and Laughlin D E (1992) Phys. Rev. <u>45</u> 6614
Norman A G and Booker G R (1985) J. Appl. Phys. <u>57</u> 4716
Peiró F, Cornet A, Morante J R, Clark S and Williams R H (1991) Appl. Phys. Lett. <u>59</u> 1958
Stringfellow G B (1983) J. Cryst. Growth <u>65</u> 454

Inst. Phys. Conf. Ser. No 134: Section 6
Paper presented at Microsc. Semicond. Mater. Conf., Oxford, 5–8 April 1993

The study of MBE grown (001) $In_xGa_{1-x}P$/GaAs strained layer heterostructures by TEM and CL

Jiannong Wang, John W Steeds and Mark Hopkinson[*]

H.H.Wills Physics Laboratory, University of Bristol, Tyndall Avenue,
Bristol BS8 1TL, U.K.
*Dept. of Electronic and Electrical Eng., University of Sheffield,
Sheffield S1 3JD, U.K.

ABSTRACT: $In_xGa_{1-x}P(x=0.436-0.546)$/GaAs heterostructures grown on (001) GaAs substrates by molecular beam epitaxy have been studied using transmission electron microscopy (TEM) and cathodoluminescence (CL) techniques. TEM study of plan-view and cross-sectional samples of the epitaxial layers revealed no evidence of CuPt-type atomic ordering in the InGaP layers, but two types of composition modulation in the InGaP layers were observed. Different strain relaxation mechanisms for the layers in tension and compression were also observed. CL analysis was carried out for both strained and relaxed InGaP layers. The dependence of CL emission energy and of peak width on the misfit strain is discussed and correlated with the TEM study. Biaxial compressive strain increases the CL emission energy and reduces the emission peak width. Biaxial compressive strain relaxation near to the microcracks was also studied by CL.

1. INTRODUCTION

The semiconductor alloy $In_xGa_{1-x}P$, which matches the lattice constant of GaAs at x=0.486, attracts attention mainly for its applications as a visible light source and as a potential alternative to AlGaAs in GaAs based heterostructures and devices. InGaP grown on GaAs by molecular beam epitaxy (MBE) was employed as an active layer in double-heterojunction lasers with emission in the range 660-690nm (Asahi et al, 1983), as a wide band gap emitter in a GaAs heterostructure bipolar transistor (Razeghi et al 1990a) and as a heterostructure insulated gate field effect transistor (Razeghi et al 1990 b). High quality InGaP/GaAs single and multiple quantum well structures have also been reported (Garcia et al, 1991). In this work, we investigated a series of $In_xGa_{1-x}P$ epilayers with x in the range 0.436 to 0.546, grown by MBE on (001) GaAs substrate. Transmission electron microscopy (TEM) has been used to examine the crystallographic and defect properties of the layers in plan-view and cross-sectional specimens. Although no evidence was found of atomic ordering in these materials, growth rotation and elastic strain related composition modulations were observed. Different lattice relaxation mechanisms were found for the layers in tensile and compressive stress. The scanning transmission electron microscopy (STEM) cathodoluminescence (CL) technique was applied to measure luminescence peak energies and the full width at half maximum (FWHM) of the peaks from both strained and relaxed InGaP layers and to study strain field around microcracks in an InGaP layer in tension.

2.   EXPERIMENTAL DETAILS

The InGaP layers studied in this work were grown by MBE on (001) semi-insulating GaAs substrates using conventional metal cells and a cracked elemental phosphorus source. The In composition of the layers was determined by double crystal x-ray diffractometry. A series of $In_xGa_{1-x}P$ layers with In composition from x=0.436 to 0.546 and layer thickness varying in the range 1.6 to 6.5μm were examined. The relative lattice mismatch (f) between the GaAs substrate and the InGaP epilayers was varied from 0.45% to -0.39%, where f is defined as $(a_{InGaP} - a_{GaAs}) / a_{GaAs}$. When f is negative the InGaP layer is in tension and when positive in compression.

The studies of the microstructure of the InGaP layers were carried out on a Philips EM430 transmission electron microscope at an operating voltage of 250kV. The layers were examined in directions both parallel (plan-view specimens) and perpendicular (cross-sectional specimens) to the growth direction. Low temperature CL experiments were performed on an extensively modified Philips EM400 STEM. To determine the strain effect on the luminescence properties of the materials, CL spectra were acquired from both bulk specimens, where GaAs substrate had not been removed, and from thin plan view specimens, where the GaAs substrate had been removed from the central region by selective chemical etching leaving a large thin window of InGaP about 1mm in diameter.

3.   MICROSTRUCTURE OF InGaP LAYERS

Extensive TEM examination of the plan-view and cross-sectional specimens of the InGaP layers revealed two forms of ordered microstructures. In some of the samples, with smaller misfit strain(-0.08%<f<0.18%), a regular periodical structure was observed along growth direction in <110> cross-sectional specimens, as illustrated in Figure 1(a). The periodical contrast only appeared in 002 two-beam TEM images and, as this reflection is most sensitive to structure factor differences, it suggests that the contrast originated from composition modulation along the growth direction. We believe that this composition modulation was caused by the rotation of the samples during MBE growth. The period of the structure was measured as about 21Å a value consistent with the growth rate of 5Å/Sec. and the rotation rate of 15RPM used in this work.

We also observed another type of quasiperiodic contrast perpendicular to the growth direction in cross-sectional specimens, that is illustrated in Figure 1(b) and (c). This form of contrast was observed in all of the samples studied but stronger contrast (see figure 1(b)) was found in those samples with larger residual strains and much weaker contrast (see figure 1(c)) was found in the samples with smaller residual strains.

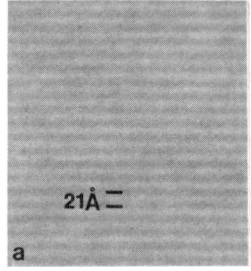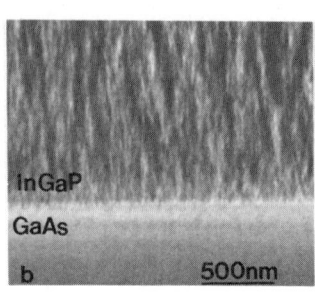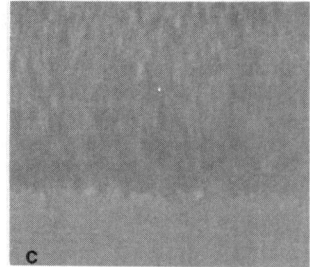

**Figure 1. Cross-sectional TEM images of InGaP/GaAs
with (a) f=-0.08% (b) f=0.39% (c) f=0.18%**

The contrast did not show up when the reflection used for imaging was parallel to the growth direction, i.e. in the case of 002 and 004 reflections, but was greatest for reflections perpendicular to the growth direction. TEM observations of plan-view specimens of the layers showed no detectable contrast. According to the analysis of Treacy et. al.(Treacy et. al. 1985,1986) and Glas(Glas 1989) we interpret the origin of the quasiperiodic structure contrast observed in cross-sectional TEM examinations as a result of the surface relaxation of strain, related to composition-modulation in the layers. The fact that this quasiperiodic contrast was observed in cross-sectional specimens along the two perpendicular <110> directions with similar periodicities suggests that the composition modulation takes a columnar form with the column axis parallel to growth direction. In the case of the bulk sample, each column would be under two-dimensional compressive or tensile elastic strain relative to its neighbours. In the direction perpendicular to the column axes in the case of cross-sectional specimens, for columns near to the specimen surface half of its surrounding columns are removed so that the strain field within the column is significantly altered. This surface relaxation effect is the principal reason for the contrast which we observed. On the other hand, in the case of plan-view specimens, the specimen surfaces intersect the columns in such a way as to cause little change in the strain field within the columns, since each column is still surrounded by the same neighbours as in the bulk material. An absence of the contrast in plan-view samples is therefore expected. However, convergent beam electron diffraction experiments on the plan-view specimens revealed the asymmetric smearing out of fine higher order Laue zone lines characteristic of the microstrains associated with the compositional modulations.

Under the optical microscope, the surfaces of these layers were smooth and contrastless. However, for layers with larger compression and tension strain cross-hatchings were observed. There was an asymmetric distribution of hatchings along the two surface <110> directions. Further studies by TEM revealed that these cross-hatchings corresponded to misfit dislocations in the case of layers in compression and to microcracks in the case of layers in tension. Figures 2(a) and (b) are cross-sectional TEM images that show two types of micro-cracks which were observed in a sample with a tensile relative mismatch f=-0.39% and a layer thickness of 1.6mm. As indicated by the arrow in Figure 2(a), we can see that a misfit dislocation was first generated at the interface and, as layer thickness increased, a micro-crack formed that cut through the dislocation. This particular micro-crack was subsequently healed by later growth of a highly defective layer on top. On the other hand, the micro-crack shown in Figure 2(b) was introduced at late stage of growth, possibly during the cooling process from the growth temperature to room temperature. In both cases the microcracks extended into the GaAs substrate but along different planes, {110} for the crack shown in Figure 2(a) and {111} for the crack shown in Figure 2(b). At the crack tips, the high stress concentration can sometimes cause propagation of dislocations into the substrate as is illustrated in Figure 2(c).

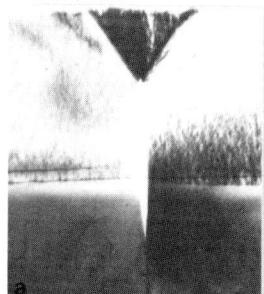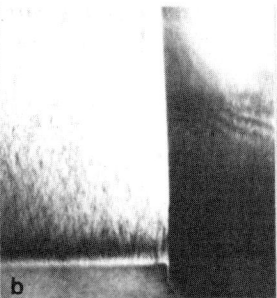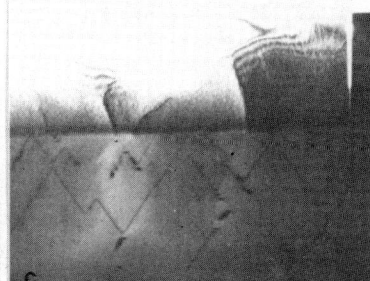

Figure 2 TEM cross-sectional images of InGaP/GaAs with f=-0.39%

4.   CATHODOLUMINESCENCE OF InGaP LAYERS

The interpretation of spectral emission from InGaP layers is incomplete at present. Most of our results consisted of a single strong emission peak with a weak and broad band centred about 30meV to the low energy side of the peak. To estimate the strain effect on luminescence properties, CL has been applied to both plan-view bulk specimens and plan-view thin specimens of InGaP layers free of interface defects. For bulk specimens the InGaP layers were under biaxial stress arising from the lattice mismatch with the GaAs substrate. For thin specimens of layers in compression, strain relaxation of the InGaP layers occurred following the removal of the GaAs substrate. This relaxation resulted in the buckling of the thin film in the centre of the specimens. However, in the case of thin InGaP films in tension, strain relaxation was prevented by the surrounding ring of thick material. This produced a perfect flat thin film in the centre of the specimens in sharp contrast with the thin films produced from the samples in compression which were buckled. As a result, thin film interference caused modulation of the spectra from the thin films in tension but not in the case of films in compression. The microcracks which occurred in some layers, with larger mismatch, produced local regions where stress relaxation was possible even without the removal of the substrate.

CL spectra taken from an InGaP layer with x=0.524 (f=0.28%) are shown in Figure 3. Spectrum (a) was acquired from the thin InGaP film, where substrate removal and strain relaxation had occurred. Spectrum (b) was acquired from the InGaP layer in a region where the substrate remained in place so that the layer was in a state of tetragonal strain, caused by a relative mismatch f=0.28%. Comparing the two spectra we can see that there is a noticeable difference between the CL emission peak energies. For the relaxed layer the CL emission energy was shifted towards the low energy side by 14meV. Another interesting feature which may be noted in Figure 3 is that the FWHM of the peak is increased as the strain is relaxed.

CL spectra from the thin and bulk regions of an InGaP layer in tension are shown in Figure 4. This specimen had an In content of 48.1%, corresponding to a relative mismatch of 0.04%. As we can see, CL emission from the thin InGaP film (see spectrum (a)) is modulated in comparison with the bulk (see spectrum (b)). An obvious explanation is that thin film interference of photons is modulating the spectrum. The modulation induced by the photon interference in the thin tensile InGaP films complicates the study of strain effects. However, CL spectra and monochromatic image techniques have been applied to the study of strain relaxed regions around orthogonal microcracks in the case of bulk InGaP layers. The CL spectra taken from a region containing orthogonal cracks of an InGaP layer in tension, with In composition 43.6% and relative mismatch 0.39%, showed an variation of 15meV in emission energies between the position away from the both cracks (Figure 5 spectrum (a)) and in the vicinity of the crack intersection (Figure 5 spectrum (b)). This shift corresponds to a significant stress relaxation of the InGaP layer at the crack intersection. Images (a) and (b) in Figure 6 were obtained by using the emission energies at 2.024eV and 2.039eV respectively. The spatial distribution of the regions corresponding to different states of strain relaxation are clearly shown.

ACKNOWLEDGMENTS

Discussions with Dr.Y.Rebane were very much appreciated. J.Wang would like to thank the SERC for financial support.

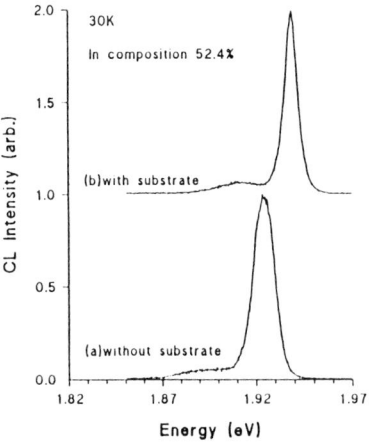

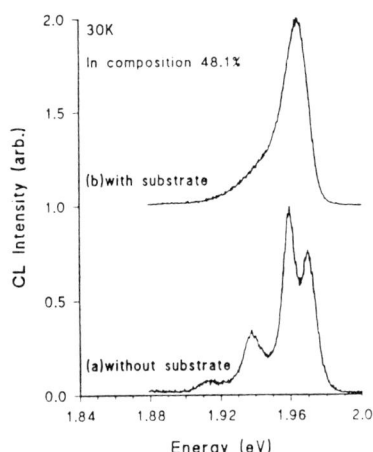

Figure 3 CL spectra from InGaP layer in compression.

Figure 4 CL spectra from InGaP layer in tension.

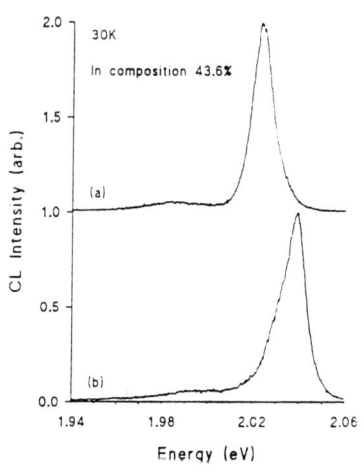

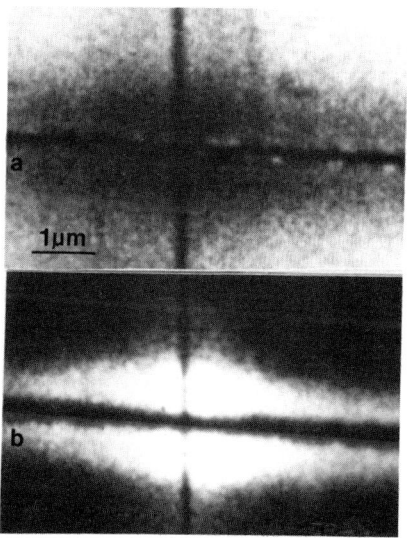

Figure 5 CL spectra taken away from(a) and in the vicinity (b) of orthogonal cracks

Figure 6 CL monochromatic images taken with energy (a) 2.024eV, (b) 2.039eV

REFERENCES

Asahi H, Kawamura Y and Nagai H, 1983 J Appl Phys **54** 6958
Garcia J C, Maurel P, Bove P and Hirth J P, 1991 Japanese J Appl Phys **30** 1186
Glas F 1989 NATO ASI Series B Physics **203** 217 (edited by D Cherns)
Razeghi M, Omnes F, Defour M, Maurel Ph, Hu J, Wolk E and Pavildis D, 1990
    Semicond Sci Technol **5** 278
Razeghi M, Omnes F, Defour M, Maurel Ph, Chan Y J and Pavildis D, 1990 Semicond
    Sci Technol **5** 274
Treacy M M, Gibson J M and Howie A, 1985 Phil Mag A **51** 389
Treacy M M and Gibson J M, 1986 J Vac Sci Technol B **4** 1458

TEM and TED studies of fine scale contrasts in InGaAs layers grown by metalorganic chemical vapour deposition

T-Y Seong, G R Booker and A G Norman[+]

Department of Materials, University of Oxford, Parks Road, Oxford OX1 3PH, UK
[+]IRC for Semiconductor Materials, Imperial College, London SW7 2BZ, UK

ABSTRACT: TEM and TED were used to investigate the fine scale contrasts of a MOCVD InGaAs layer grown on a (001) InP substrate at 717°C and 0.95nm/s, both before and after annealing at 600°C for 744hr. An ~15nm modulated contrast oriented along the [100] and [010] directions and an unoriented ~5nm speckled contrast were simultaneously present in the as-grown layer, but only the speckled contrast occurred in the annealed layer. The modulated contrast is attributed to spinodal decomposition and the speckled contrast to random local displacements of the In and Ga atoms on the Group III sublattice.

1. INTRODUCTION

Group III-V ternary and quaternary epitaxial layers often show a fine scale contrast when examined by TEM (Gowers 1983, Mahajan et al. 1984, Norman and Booker 1985a and b, Chu et al. 1985, Glas et al. 1985, 1990). Norman and Booker (1985a and b) investigated LPE InGaAsP layers grown at 600°C on (001) InP substrates using (001) plan-view and (010) cross-section specimens. For an InGaAsP layer for which bulk theory predicted that spinodal decomposition would occur, TEM showed pronounced modulated contrast along both the [100] and [010] directions on a scale of ~15nm. TED showed pairs of satellite spots associated with each of the (200) and (020) spots, whose directions and spacing corresponded with the 15nm modulated contrast. There was no modulated contrast along the [001] direction and there were no satellite spots associated with the (002) spot. This was interpreted in terms of composition variations along [100] and [010] arising from spinodal decomposition, but no composition variations along [001]. For an InGaAsP layer for which bulk theory predicted that spinodal decomposition would not occur, TEM showed weak modulated contrast along [100] and [010] and TED did not show satellite spots. Norman and Booker (1985b) also observed fine scale contrast and satellite spots for MBE InGaAs layers grown at 560°C and VPE InGaP layers grown at 750°C.

Glas et al (1985, 1990) investigated InGaAsP and InGaAs layers grown on (001) InP substrates using plan-view specimens. For all the layers, TEM showed a fine scale contrast and TED patterns showed lines of diffuse intensity along the [110] and [1$\bar{1}$0] directions close to the main spots, but no satellite spots. These lines were interpreted as arising from static atomic displacements associated with a totally random distribution of the In and Ga atoms on the Group III sublattice, based on comparisons between experimental and computer simulated TED patterns. It was considered that this effect gave rise to the observed TEM fine scale contrast.

In the present work two types of fine scale contrast are distinguished and these are interpreted as arising from spinodal decomposition and random static atomic displacements (Seong 1991).

2. EXPERIMENTAL

An atmospheric pressure MOCVD $In_{0.53}Ga_{0.47}As$ layer 2μm thick was grown at 717°C and

0.95nm/s on a (001) InP substrate. A piece of the specimen was subsequently annealed at 600°C for 744hr in a sealed quartz ampoule containing crushed InP and GaAs powders and evacuated to 10^{-6} torr. TEM plan-view specimens were prepared by chemical thinning from the substrate side only and cross-section specimens were prepared by ion-beam thinning with the specimen at 77K.

3. RESULTS

For the as-grown layer, TEM examination of (001) plan-view specimens showed that a fine scale modulated contrast and a fine scale speckled contrast were simultaneously present. Two-beam **g(220)** micrographs (Fig.1(a)) showed that the modulated contrast consisted of dark and bright blobs elongated approximately along either [100] or [010], corresponding to [010] and [100] modulations

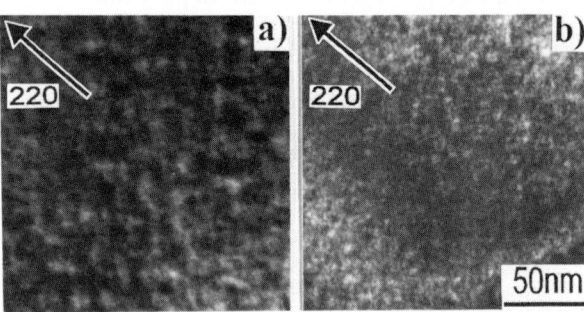

Fig.1. InGaAs layer, TEM (001) plan-views, two-beam **g(220)**, a) as-grown, b) annealed.

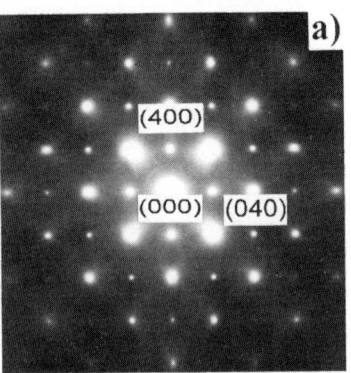

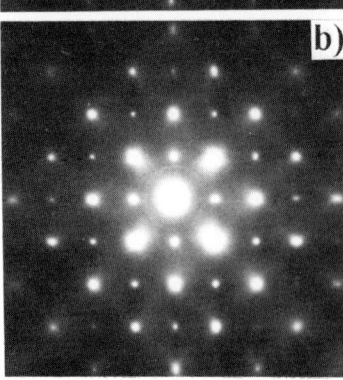

Fig.2. [001] TED patterns from specimens of Fig.1, a) as-grown, b) annealed.

respectively, with the blobs of width of ~15nm. The speckled contrast consisted of dark and bright blobs without any specific orientation and of size ~5nm. For the annealed layer, a similar examination (Fig.1(b)) showed only an ~5nm speckled contrast.

For the as-grown layer, TED patterns from the (001) plan-view specimens (Fig.2(a)) showed the standard zinc-blende spots. However, there were diffuse lines of intensity along both [110] and [1$\bar{1}$0], these lines running close to the (2m 2n 0) spots, where m and n are integers taking +, − and 0 values. Furthermore the (800) and ($\bar{8}$00) spots were elongated along [100] and the (080) and (0$\bar{8}$0) spots were elongated along [010]. Separate satellite spots were not observed. For the annealed layer, the corresponding TED patterns (Fig.2(b)) were the same.

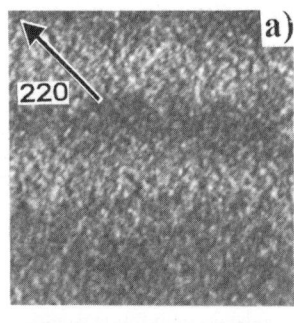

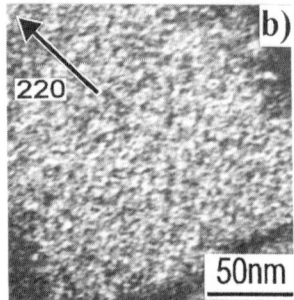

Fig.3. TEM (001) plan-views, weak-beam **g(220)3g**, a) as-grown, b) annealed.

For the as-grown layer, weak-beam **g(220)3g** micrographs from (001) plan-view specimens (Fig.3(a)) showed only a 5nm speckled contrast. For the annealed layer, a similar examination (Fig.3(b)) also showed only a 5nm speckled contrast.

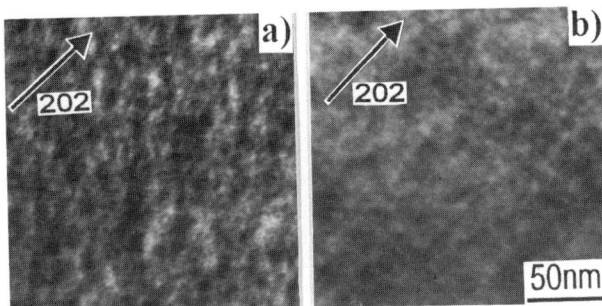

Fig.4. TEM (010) cross-sections, two-beam **g(220)**, a) as-grown, b) annealed.

For the as-grown layer, two-beam **g(220)** micrographs from the (010) cross-section specimens (Fig.4(a)) showed the simultaneous presence of a 15nm modulated contrast and a 5nm speckled contrast. The modulated contrast consisted of dark and bright blobs elongated along [001], corresponding to [100] modulation. There was no blob elongation along [100], i.e. no significant [001] modulation. For the annealed layer, a similar examination (Fig.4(b)) showed only a 5nm speckled contrast.

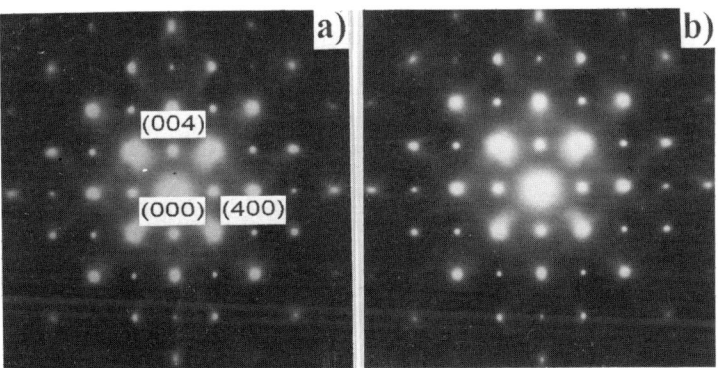

Fig.5. [010] TED patterns from specimens of Fig.4, a) as-grown, b) annealed.

For the as-grown layers, TED patterns from the (010) cross-section specimens (Fig.5(a)) showed the standard spots together with <110> diffuse lines of intensity and {800} elongated spots. For the annealed layer, the corresponding TED patterns (Fig.5(b)) were the same.

4. DISCUSSION

The TEM results reveal two types of fine scale contrast. The speckled contrast is observed for plan-view and cross-section specimens and for both as-grown and annealed layers. It has a size of 5nm, shows no orientation effects and appears similar in all cases. The TED patterns show <110> diffuse lines of intensity for plan-view and cross-section specimens and for both as-grown and annealed specimens and the lines appear similar in all cases. The speckled contrast and the diffuse lines can both be considered as specimen non-orientation-dependent and 3D-distributed. We correlate the speckled contrast with the diffuse lines of intensity. Glas et al (1985, 1990) previously correlated the diffuse lines of intensity with static atomic displacements for randomly distributed In and Ga atoms. We now correlate the speckled contrast with such static atomic displacements. The modulated contrast is observed for the as-grown layer but not for the annealed layer. It is modulated along [100] and [010] with a width of 15nm, but is not modulated along [001]. The modulated contrast can be considered as specimen orientation-dependent and 2D-distributed. There are no features in the present TED patterns that correlate with the modulated contrast.

The questions arise as to why in the TED patterns the speckled contrast gives in addition elongated {800} spots, and the modulated contrast does not give (200) and (020) satellite spots. For the elongated {800} spots the probable reason is that which was proposed previously by Glas et al (1985). Thus, for an [001] pattern and an (800) spot, the [110] and [1$\bar{1}$0] lines of diffuse intensity intersect one another at two points on either side and close to the (800) spot. These two broad intersections are joined to the main spot and have the effect of elongating it along the [100] direction.

With regard to the modulated contrast and satellite spots, the most pronounced example of this was observed by Norman and Booker (1985a and b) for an LPE InGaAsP layer. Their micrographs of the modulated contrast, especially those for cross-section specimens, showed well-defined elongated blobs. For the InGaAs layer of the present work, for which bulk theory predicts that spinodal decomposition would not occur, the micrographs of the modulated contrast (Fig.1(a) and 4(a)) show less well-defined elongated blobs. These results suggest that composition variations due to spinodal decomposition often occur in Group III-V ternary and quaternary layers but to widely varying degrees, depending on alloy type and composition and growth temperature. These composition variations readily give rise to modulated contrast in the micrographs, but it is only when the variations are pronounced and well defined that they give detectable satellite spots in the TED patterns.

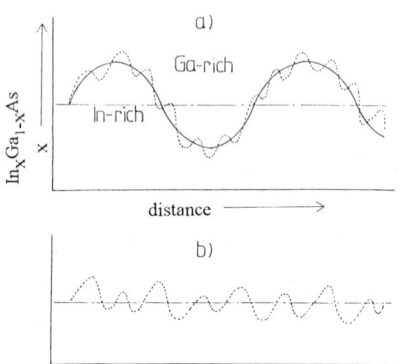

Fig.6. Diagrammatic model for cause of modulated contrast and speckled contrast, a) as-grown, b) annealed.

The modulated contrast arises mainly from lattice parameter variations caused by the composition variations and is a diffraction contrast effect. This view is supported by weak-beam micrographs of the plan-view specimens for the as-grown InGaAs layer (Fig.3(a)), which showed no modulated contrast, whereas modulated contrast was observed for this layer in similar two-beam micrographs (Fig.1(a)). The larger tilt angle required for the weak-beam micrographs has eliminated the modulated contrast.

A suggested qualitative model for the occurrence of the two types of contrast is as follows. For the as-grown layer (Fig.6(a)), the modulated contrast corresponds to a sinusoidal composition modulation of In- and Ga-rich regions (continuous line). The speckled contrast arises by the mechanism proposed by Glas et al (1990) which is equivalent to an analogous composition variation but with smaller amplitude and smaller wavelength. The combination of these mechanism gives the dotted line. For the annealed layer (Fig.6(b)), only the speckled contrast composition variations are present, the modulated composition variations having been eliminated during the anneal by atomic diffusion. The bulk theory may give an indication of whether spinodal decomposition will occur in the as-grown layer but may not be precise because the decomposition and hence modulated contrast may grow in at the layer surface.

ACKNOWLEDGEMENTS: The authors wish to thank the British Council, DTI and SERC for financial support and S.J. Bass and L.L. Taylor for providing the epitaxially grown layer.

REFERENCES

Chu S N G, Nakahara S, Strege K E and Johston WD 1985 J. Appl. Phys. <u>57</u> 4610.
Glas F, Hénoc P and Launois H 1985 Inst. Phys. Conf. Ser. <u>76</u> 251.
Glas F, Gors C and Hénoc P 1990 Phil. Mag. B<u>62</u> 373.
Gowers J P 1983 Appl. Phys. A<u>31</u> 23.
Mahajan S, Dutt B.V, Temkin H, Cava R J and Bonner W A 1984 J. Cryst. Growth <u>68</u> 589.
Norman A G and Booker G R 1985a J. Appl. Phys. <u>57</u> 4715.
Norman A G and Booker G R 1985b Inst. Phys. Conf. Ser. <u>76</u> 257.
Seong T-Y 1991 D.Phil. thesis, Oxford University.

Inst. Phys. Conf. Ser. No 134: Section 6
Paper presented at Microsc. Semicond. Mater. Conf., Oxford, 5–8 April 1993

Influence of strain and composition on TEM speckle contrast and the morphology of quaternary layers

U Bangert and A J Harvey[*]

Department of Pure and Applied Physics, UMIST, Manchester M60 1QD
[*]Physics Department, University of Surrey, Guildford GU2 5XH

ABSTRACT: Contrast fluctuations (fine scale and very large scale) in $Ga_xIn_{1-x}As_yP_{1-y}$ multilayer systems, with a wide range of values x an y are analyzed and discussed with respect to a possible relationship between composition fluctuations and strain induced non-planar growth modes.

1. INTRODUCTION

The growth of quaternary and ternary multilayer systems of $Ga_xIn_{1-x}As_yP_{1-y}$ of any combination x and y is of great technological interest for electronics and optoelectronics. A current problem in the MOCVD growth of strained quaternary multiple quantum well (MQW) structures is the breakdown of good epitaxial growth after a certain number of layers in a MQW stack. The cause for this phenomenon is not yet clear; it has, however, to do with non-planar growth modes, which eventually evolve into wavy layer growth. There is no evidence of misfit dislocations, but very large scale contrast modulations, progressing in columns in the growth direction through the MQW structure, can be observed in 90^0-wedge specimens in the TEM. The behaviour of these MQWs might be affected by the miscibility gap (MG) (de Cremoux 1981, Stringfellow 1982), which is predicted to exist in quaternary alloys. This encloses compositions, which can undergo spinodal decomposition when the system is in thermodynamic equilibrium. This issue has been controversially commented on ever since it was raised, because epitaxial growth is a non-equilibrium process. Considerations of composition fluctuations as a stable growth mode in thin foils have been brought forward by Glas (1987) and implications of composition fluctuations on 3-dimensional growth and strain relaxation have been investigated by Peiro (1992). Based on a study of many samples with a wide range of quaternary (and ternary) compositions we attempt in this paper to correlate composition and fine speckle contrast, and to explain the very large scale contrast modulations and the wavy layer growth in MQWs in terms of strain induced phenomena.

2. EXPERIMENTAL

We invstigated over 40 samples with different compositions, strain values, number and thickness of their epitaxial layers, but we only discuss a representative cross section in this paper. A more detailed report will be published elswhere (Bangert et al 1993). The samples were all MOCVD grown structures on (100)-InP. Growth of all structures took place at 650^0C. Fig.1 yields information about the sample compositions and their position with respect to the MG. The dashed line represents compositions, which are lattice matched with InP, the dotted lines to the left and the right represent compositions, which have 1% compressive and tensile strain with respect to InP. The general sample geometry of the MQWs consists of an InP substrate, a quaternary/quaternary or quaternary/ternary strained well/barrier MQW structure, followed by an InP cap. Sometimes the MQW stack was additionally embedded in quaternary lattice matched waveguide material. Some structures were single layers grown on InP (for sample details see table 1). Samples were prepared for TEM as 011-cross sections or as 90⁰ cleaved wedges. TEM was undertaken at 200 kV in a Jeol 2000 FX instrument.

3. RESULTS

In the cross sections in fig.2 the development of the fine speckle contrast for layers grown at different points in the composition diagram is shown. Fig.2a shows sample 1, a thick single quaternary layer, with a composition outside the MG. No speckle contrast can be observed. Fig.2b is of sample 2, a MQW with layers and barriers both grown outside the MG. The sample shows no pronounced fine speckle contrast, though fine scale mottling with occasional clustering can be seen. Fig.2c shows sample 3, a thick quaternary layer with a composition inside the MG. The speckle contrast is very pronounced here. Fig.2d shows sample 4, containing a MQW with wells close to, and barriers inside the MG. The fine speckle contrast is very pronounced with occasional larger scale clusters. Thus the development of the fine speckle contrast appears to depend on the distance of the composition from the MG and also seems to be influenced by the strain state of the structure.

In the following we concern ourselves with very large scale contrast fluctuations. For this purpose 90⁰ wedges were made out of strained MQW structures. Cleaved wedges are extremely suitable for bulk strain relaxation studies, due to their absolutely planar sample surfaces, the precisely known sample geometry and their bulk-like behaviour, in which the usual thin foil effects due not occur. The electron diffraction contrast can be predicted and therefore any observed deviation from these predictions is a direct reflection of the bulk strain distribution. Fig.3a shows sample 4, described above. Only the top 13 wells are shown here. No large scale contrast fluctuations have developed. The interfaces appear as sharp fringed triangles due to the sample tilt. A 32-well structure of the same compositions showed break-down of planar growth and wavy layers after some 20 wells (not shown here). Fig.3b shows sample 5, a 9-well MQW structure with wells grown outside the MG. Very large scale contrast streaks, starting at the bottom of the MQW system and running in the growth direction through it, can be seen. Fig.3c is of sample 6, a 6-well MQW structure, where wells and barriers were both grown outside, but on opposite sides, of the MG. The very large contrast modulations dominate the structure and the upper three wells were observed to have grown in a wavy fashion. Considering the results of some 40 samples with different compositions, strains and well numbers, of which the samples shown here are a representative cross section, we observed the following: a) In strained systems with at least the wells grown outside the MG the very large scale contrast variations occur early on in epitaxial growth (after only 3 wells in sample 6), when the connecting lines between well and barrier composition cross the MG (see fig.1). b) In strained systems with wells and barriers grown inside the MG the fine speckle contrast seems to undergo occasional clustering, but on the whole the very large scale modulations appear to be inhibited until the layer system has reached a considerable thickness. Then these contrast modulations are on a smaller scale and lead eventually to wavy layer growth. c) In strained MQW samples with well and barrier compositions outside, but on the same side of the MG, the very large scale contrast modulations were not observed (sample 2, wedge results not shown here). The above phenomena were also not observed in ternary MQW structures of GaInAs at strain values as high as 2%, and neither did they occur in unstrained quaternary MQW systems.

4. DISCUSSION

The critical condition, which promotes the large scale contrast variations and the wavy layer growth observed in wedges, appears to be a succession of barriers and wells, in which the barriers can, and the wells cannot decompose; it also seems to happen when barriers and wells both cannot decompose, but when their interfacial compositions (i.e. the average composition of well and barrier) can. We propose that lateral variations in the composition at the growth surface have somehow occurred during quaternary growth. These will create fluctuations in the strain. This strain pattern, added to the misfit strain, will be carried through in a subsequently grown, strained layer, which cannot decompose (and hence has no possibility to form counteracting composition fluctuations, which produce the fine speckle contrast in layers grown inside the MG). These lateral strain fluctuations might induce faster growth of the next layer, grown with a relative strain, in places were the strain is smallest. If this next layer can decompose, local changes in the Ga/In and As/P ratio might additionally occur to minimize the strain. We have frequently observed fluctuations in the well thickness of the first few layers in a stack. This island-like growth mode might also be the initial stage of wavy growth. Misfit strains appear to crucially influence this process. Dislocation mobilities are small in quaternary compared to binary material (Bangert, unpublished results) and hence island growth might be the preferred route of strain relief. In fact no misfit dislocations have been observed in any of the MQW structures. This process of strain related growth rate and composition variations should in principle be possible in all system, in which fluctuations in the composition can occur (i.e. in alloys in general). It will also depend on other parameters like growth temperature and point defect concentrations etc. and has indeed been observed in the Si/SiGe system (Vescan et al 1992). It is anticipated, however, that it would be facilitated in spinodally decomposing systems, because here composition fluctuation can occur more easily.

ACKNOWLEDGEMENTS

The work was carried out under the LINK project IED2/440/30/005, for which BNRE and BT supplied the samples.

REFERENCES

Bangert U, Harvey A J, Wilkinson V A, Dieker C, Jowett J M, Smith A D, Perrin S D and Gibbins C J 1993, to be published in J Cryst. Growth.
De Cremoux B, Hirtz P and Ricciardi J 1981 Inst. Phys. Conf. Ser. 56 115
Glas F 1987 J. Appl. Phys. 62 3201.
Peiro F, Cornet A, Herms A, Morante J R, Clark S A and Williams R H 1992 Electron Microscopy 2 EUREM Granada, 159.
Stringfellow G B 1982 J. Cryst. Growth 58 194
Vescan L, Jaeger W, Dieker C, Schmidt K, Hartmann A and Lueth H 1992 Proc. MRS Spring Meeting, San Fran.

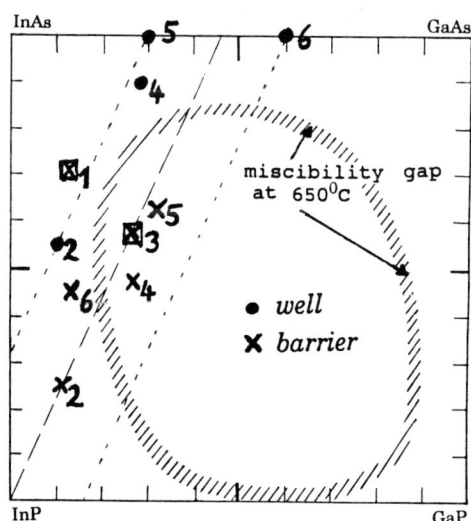

Figure 1.   Position of sample compostions.

Sample No	1	2	3	4	5	6
well/Barrier	Q/-	Q/Q	Q/-	Q/Q	T/Q	T/Q
Well No	Single layer	8	Single layer	16	9	6
Well thickness	2000	25	1000	40	22	80
Barrier strain	-	us	-	zns	zns	zns
Well Strain	+1.15	+1	0	+0.8	+1	-1

Table 1: Q: quaternary; T:ternary; us: stack has unstrained barriers; zns: stack has counterstrained barriers to achieve zero net strain; well thickness are given in Angstrom, strain values are given in %.

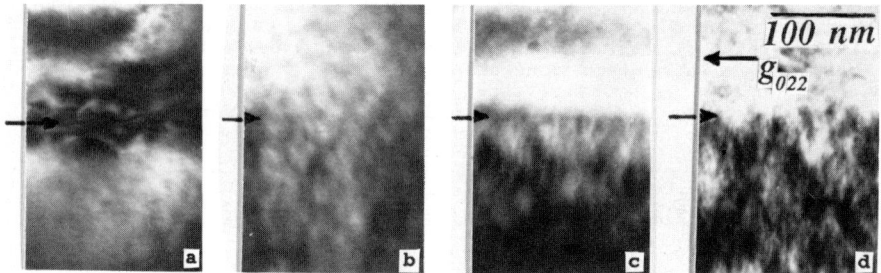

Figure 2.   Change of fine speckle contrast with composition. The arrows assign the interface between InP (top) and quaternary (bottom).

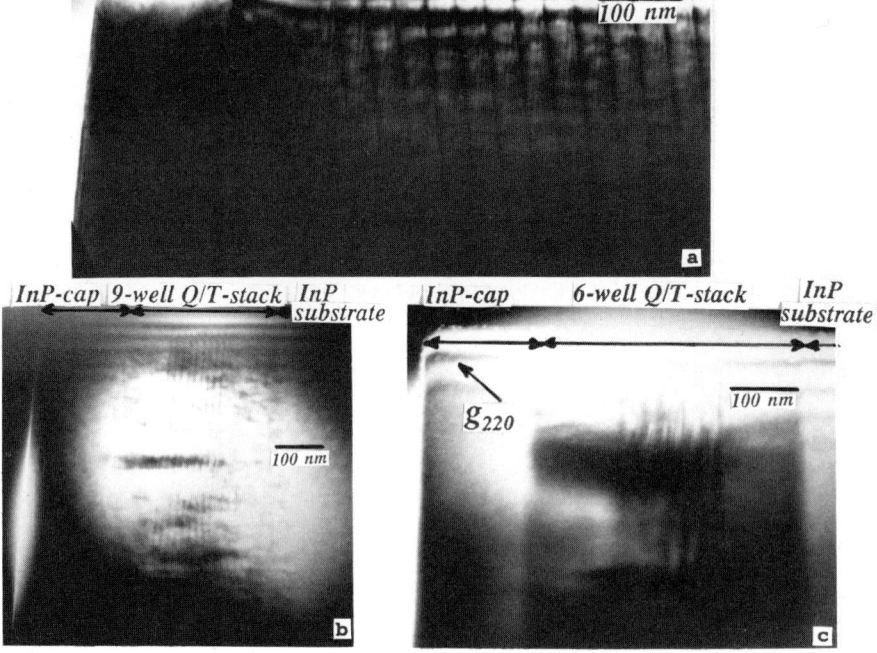

Figure 3.   Very large contrast fluctuations in wedges b) and c).

Elastic strain-induced surface roughening and the nucleation of misfit dislocations in heteroepitaxial thin films

D D Perovic, D C Houghton[†], J P Noël[†] and N L Rowell[†]

Department of Metallurgy and Materials Science, University of Toronto, 184 College Street, Toronto M5S 1A4 Canada
† Inst. for Microstructural Sciences, National Research Council, Ottawa, K1A 0R6 Canada

ABSTRACT: The transition from 2-D layer-by-layer growth to the initial stages of a 3-D growth morphology is considered as a precursor to misfit dislocation injection in Ge_xSi_{1-x}/Si heterostructures with misfit strains as low as 0.6 %. The *'Double Half-Loop'* misfit dislocation source is believed to originate from elastic strain-induced atomic-scale (< 1.5 nm) interfacial perturbations which can exist in areal densities as high as ~10^9 cm^{-2}. Changes in photoluminescence behaviour with increasing strained layer thickness have been used to study strained layer growth morphology at low misfits.

1. INTRODUCTION

Relatively few workers have experimentally explored the detailed micro-mechanisms of interfacial dislocation nucleation during the *initial* stages of strained layer coherency breakdown in low misfit systems; the majority of experimental work has been carried out on the Ge_xSi_{1-x}/Si and In_xGa_{1-x}As/GaAs strained layer systems (Fitzgerald 1991). Of the possible misfit dislocation sources that have been suggested and/or observed, three mechanisms have become popular to date: (i) reorientation of pre-existing threading dislocations originating from the substrate, (ii) surface injection of dislocation half-loops and (iii) precipitate-induced threading dislocation nucleation at the substrate/epilayer interface followed by reorientation into misfit stress relieving segments. Although these mechanisms have been shown to play an important role in certain specific cases, they generally cannot explain existing strain relaxation data since: (i) the density of pre-existing dislocations in present-day substrates is very low (10-10^4 cm^{-2}), (ii) the injection of surface half loops cannot account for complete loops localized within multilayers at specific interfaces and (iii) microscopic precipitates present at the buffer-substrate interface do not generate a sufficiently high density of threading dislocations (< 10^4 cm^{-2}).

In this study we present further TEM evidence on the origin of a novel misfit dislocation nucleation source at Ge_xSi_{1-x}/Si ($x<~ 0.25$) interfaces, the *'Double Half-Loop'* (Perovic and Houghton 1991). In parallel, we employ photoluminescence (PL) spectroscopy to carefully study elastic strain-induced roughening as a function of strained layer thickness.

2. EXPERIMENTAL

Several <100>-oriented Ge_xSi_{1-x}/Si heterostructures were grown by solid source MBE at temperatures between 450-600 °C using procedures described elsewhere (Houghton 1991). The composition (x) and strained layer thicknesses (t) were chosen to provide a wide range of effective stress (0 - 1 GPa) to drive misfit dislocation formation. Some as-grown specimens were rapid thermal annealed in a forming gas ambient for times and temperatures ranging from 5-200 sec and 450-1000 °C respectively. Electron diffraction contrast was used to study the dislocation nucleation process in detail. PL spectra were recorded using a Fourier transform infrared spectrometer at liquid He temperatures. The excitation wavelength was 458 nm at ~1 W/cm^2 and the luminescence was detected by a Ge crystal at 77K.

3. ELECTRON MICROSCOPY

Fig. 1 is a plan-view image of an as-grown, three-period Ge$_{0.18}$Si$_{0.82}$/Si multilayer showing a typical *'Double Half-Loop'* misfit dislocation source originating at a strained layer interface. The concentric strained Ge$_x$Si$_{1-x}$/Si layers are revealed due to surface lattice plane bending effects following specimen thinning (Perovic and Weatherly 1991). The two half-loop segments lie on different {111} plane variants whose lines of intersection with the [100] interface are orthogonally opposed; the Burgers vectors of each half-loop are most often of different $a/2<110>$ 60°-types.

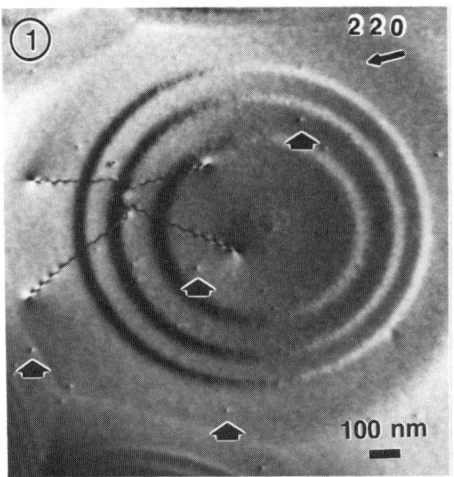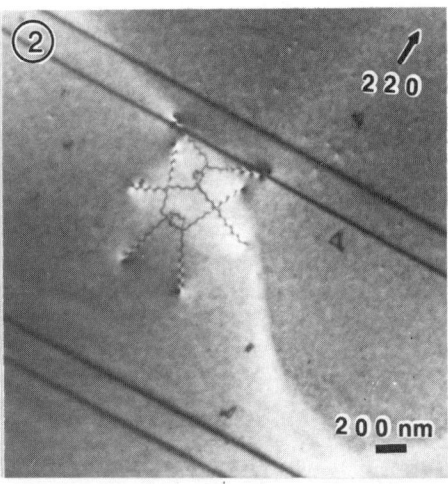

Fig. 1: Three period Ge$_{0.18}$Si$_{0.82}$/Si superlattice showing a *'Double Half-Loop'* misfit dislocation source. The layer thicknesses are: t_{GeSi}= 7.5 nm, t_{Si}= 30 nm. Arrows delineate anomalous strain field contrast associated with interstitially strained perturbations.
Fig. 2: Single Ge$_{0.15}$Si$_{0.85}$ strained layer (t_{GeSi}= 60 nm) following thermal activation.

Following post-growth annealing the half-loops can glide such that each *'double half-loop'* generates an orthogonal 60° array directly at the misfitting interface. Although this event is difficult to observe in light of statistical limitations, some evidence has been obtained thus far (Perovic and Houghton 1991). Fig. 2 is a plan-view image from a single Ge$_{0.15}$Si$_{0.85}$ strained layer following post growth annealing at 700 °C for 2 min. Two *'Double Half-Loop'* sources are present which have begun to extend as misfit dislocations following thermal activation. Furthermore, several incipient dislocation loops have expanded to an observable size.

In addition to the study of metastable heterostructures in the as-grown and post-growth annealed conditions, heterostructures which are unstable with respect to misfit dislocation generation at growth temperatures have also been studied. Fig. 3a is a plan-view image of an as-grown Ge$_{0.25}$Si$_{0.75}$ strained epitaxial layer possessing a very high density of 60° misfit dislocations piled up along different {111} plane variants. Cross-sectional analysis, as shown in Fig. 3b, demonstrates that the misfit segments nucleate at the strained interface and subsequently force several misfit segments deep into the buffer layer analogous to previous observations on graded layer GeSi/Si heterostructures (LeGoues *et al.* 1992).

Under both kinematic (weak-beam) or dynamic diffraction conditions, there is no detectable heterogeneity observed at the dislocation sites within the resolution limit of diffraction contrast imaging (~1.5 nm). However, several perturbations possessing interstitial strains and plate-shaped symmetry can be revealed within the various Ge$_x$Si$_{1-x}$/Si structures that have been studied (cf. Figs.1-3). The interstitial platelets were never observed in the

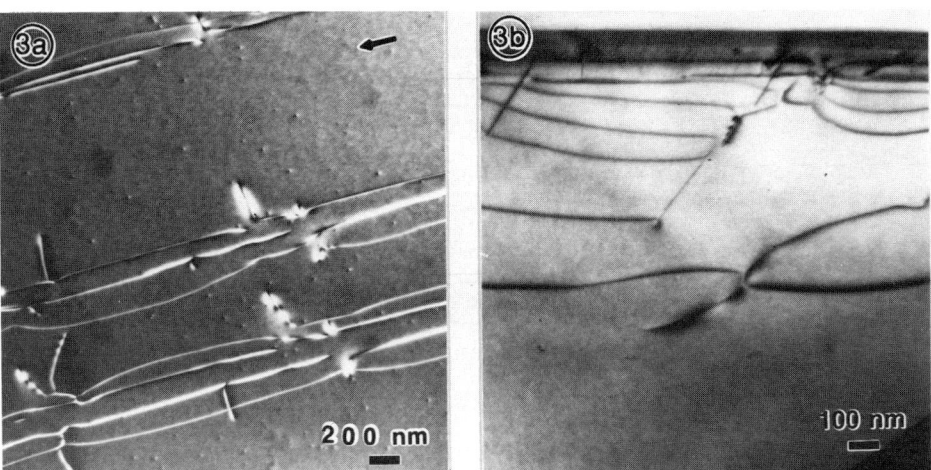

Fig. 3: Ge$_{0.25}$Si$_{0.75}$ strained layer (t_{GeSi}= 100 nm) grown at 580 °C which is unstable with respect to misfit dislocation generation: (a) plan-view, (b) cross-sectional view.

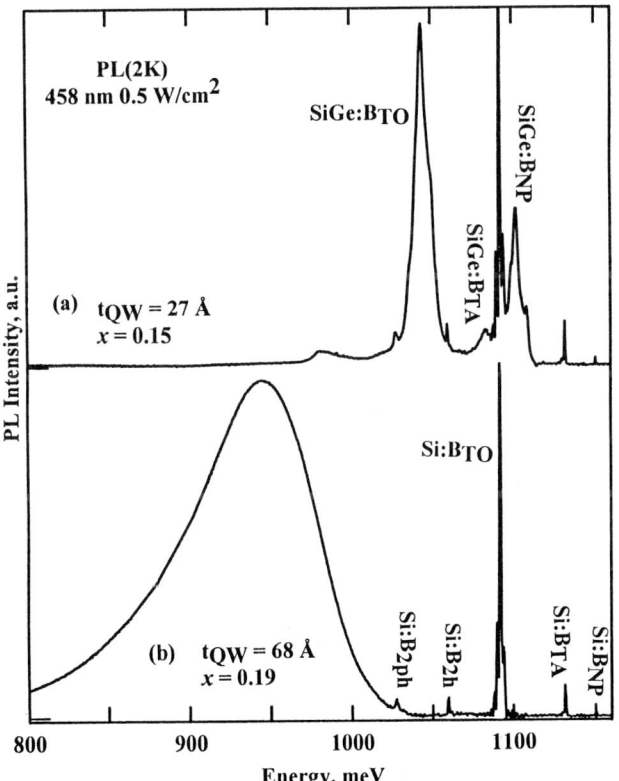

Fig. 4: 2K PL spectra of two 20-period Ge$_x$Si$_{1-x}$/Si heterostructures with t_{Si}= 20 nm and: (a) x= 0.15, t_{GeSi}= 2.7 nm, (b) x= 0.19, t_{GeSi}= 5.2 nm; (see text for details).

thinnest foil regions away from the heterostructure (ie. in the Si cap) which indicates that they are intrinsic to the strained layer structure and cannot be ascribed to specimen damage-related effects. The density of these sources can be as high as $\sim 10^9$ cm^{-2} depending on the strained layer composition and thickness. Further information on the origin of these sources has been obtained from PL spectroscopy as discussed below.

4. PHOTOLUMINESCENCE SPECTROSCOPY

Several heterostructures were studied in parallel using photoluminescence spectroscopy (PL) to elucidate the dependence of optical properties on MBE growth parameters (Noël *et al.* 1992). The PL spectrum in Fig. 4a is from a dislocation-free GeSi/Si multilayer possessing a relatively low density (2×10^6 cm^{-2}) of interstitially strained perturbations. Fig. 4a shows phonon-resolved, near band-edge luminescence similar to bulk, unstrained Ge$_x$Si$_{1-x}$ material with the exceptions that: (i) the transition energies are shifted to lower energies due to coherency strains and (ii) quantum confinement effects shifted the transitions to higher energies for very thin layers. The two no-phonon (NP) peaks (B$_{NP}$) indicate that compositional variations (x) between strained layers were ~ 0.01. The NP peaks and their phonon replicas (TA= transverse acoustic, TO= transverse optical) originate from excitons bound to shallow B-atoms in the 2.7 nm thick Ge$_{0.15}$Si$_{0.85}$ layers.

With a twofold increase in the heterostructure alloy layer thickness under otherwise identical conditions, a concomitant increase in the interstitially strained perturbation density (7×10^8 cm^{-2}) was observed. Fig. 4b indicates that phonon-resolved PL is no longer present. However, intense, broad-band PL is now observed with its high energy edge near the B$_{NP}$ energy. This change in PL behaviour has been attributed to the large increase in interstitially-strained platelet density. The occurrence of the broad PL band, which is shifted lower in energy than the phonon-resolved PL, is attributed to exciton localization in lower band-gap Ge-rich platelets (Noël *et al*). Since the platelets vary in size and Ge level, a broad PL band is thus expected reflecting the distribution of Ge-profiles in the platelet regions.

5. SUMMARY

From our combined study using TEM and PL spectroscopic analyses, it is believed that the interstitially strained incipient sources observed are associated with surface step kink regions which act as preferential sites for the accommodation of misfitting atoms during heteroepitaxial growth by way of lateral segregation across the terraces to the step edges (Jesson *et. al.* 1992). This is to be expected since Ge and Si adatom mobilities and surface binding energies are not likely to be identical which would thus favour Ge-rich, quasi 3-D island formation. It will be shown elsewhere (Perovic and Houghton 1993) that the shear stress at an interstitial platelet is sufficiently high such that dislocation shear loops can nucleate spontaneously in order to relieve the misfit stress. Accordingly, it is believed that elastic strain-induced surface roughening acts as a precursor to misfit dislocation injection through the operation of a novel misfit dislocation source, the '*Double Half-Loop*'.

REFERENCES

Fitzgerald E A, 1991, *Mater. Sci. Rep.*, 7 87.
Houghton D C, 1991, *J. Appl. Phys.*, 70 2136.
Jesson D E, Pennycook S J, Baribeau, J-M and Houghton D C, 1992, *Phys. Rev. Lett.*, 68 2062.
LeGoues F K, Meyerson B S and Morar J M, 1991, *Phys. Rev. Lett.*, 66 2903.
Noël J-P, Rowell N L, Houghton D C, Wang A and Perovic D D, 1992, *Appl. Phys. Lett.*, 61 690.
Perovic D D and Houghton D C, 1991, *Inst. Phys. Conf. Ser.*, 117 641.
Perovic D D and Houghton D C, 1993, (to be published).
Perovic D D and Weatherly G C, 1991, *Ultramicroscopy*, 35 271.

Inst. Phys. Conf. Ser. No 134: Section 6
Paper presented at Microsc. Semicond. Mater. Conf., Oxford, 5–8 April 1993

313

Generation of misfit dislocations in ultra-thin semiconductor films

F A Ponce*, L Gonzalez°, A Mazuelas° and F Briones°

(*) Palo Alto Reseach Center, Xerox Corporation, Palo Alto, CA 94304, USA
(°) Centro Nacional de Microelectrónica, Serrano 144, E-28006 Madrid, SPAIN

ABSTRACT: The lattice structure of thin films of GaP on GaAs has been studied for thicknesses around the critical value for the generation of misfit dislocations. Atomic layer molecular beam epitaxy was used for growth at low temperatures, on flat surfaces, and with accurate in-situ measurement of thickness. TEM images show surprisingly uniform distributions of misfit dislocations for thickness above 6 monolayers of GaP. A diffusionless, long-range cooperative mechanism is proposed for the relaxation process.

1. INTRODUCTION

Lattice mismatch in heteroepitaxy can be satisfied in various ways. Under some circumstances it can be relieved solely by the chemical bonding configuration at the interface between the thin film and the substrate (Ponce 1981) leading to an incoherent interface. When there is a tendency to maintain local coherency, the introduction of misfit dislocations is necessary to compensate for the existence of lattice mismatch (Frank and van der Merwe 1949, Ponce et al 1986). Misfit dislocations have been shown to exist after a critical film thickness is exceeded (Matthews and Blakeslee 1974). The understanding and control of misfit dislocations is important in most applications of heteroepitaxial semiconductor films. A perfectly periodic array of misfit dislocations should be in principle a desirable feature of the interface structure, and should not have an impact on the electrical and optical properties of the film. In practice, however, the interface structure is not always easy to control, and misfit dislocations are often seen to emerge into the film in the form of threading dislocations. In this work, we have attempted to understand the generation of misfit dislocations once the critical thickness is reached. Using atomic layer molecular beam epitaxy (ALMBE, Briones and Ruiz 1991) we have achieved three important objectives for the material in this experiment: (a) low temperature growth to prevent the thermally induced rearrangement of the dislocation structure, (b) atomically flat surfaces for the growth of uniformly thick layers, and (c) accurate in-situ monitoring of the thickness of the layer. The final thickness of the film was measured by x-ray interference which corroborated the in-situ thickness measurement. Transmission electron microscopy has been used to observe the atomic structure of the interfaces.

2. EXPERIMENTAL DETAILS

The GaP films were grown on GaAs and they were subsequently capped with a 200nm thick GaAs film, in a sandwich-like configuration. The ALMBE growth

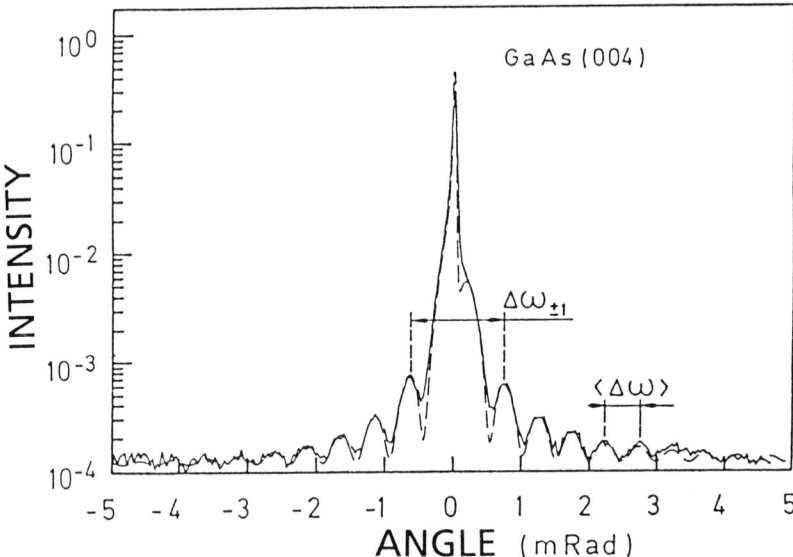

Figure 1. Rocking curve x-ray diffractogram (continuous line) of the 6 monolayer GaP sample and simulated diffractogram (broken line) corresponding to 5.6 monolayer of GaP.

was performed at 350C, with alternating Ga, As, and P pulses at a rate of about one monolayer per second. Normally MBE growth of GaP and GaAs are done at much higher temperatures. Using lower temperatures increasesthe opportunity to observe the structure before any thermal rearrangements occur. The Ga flux is

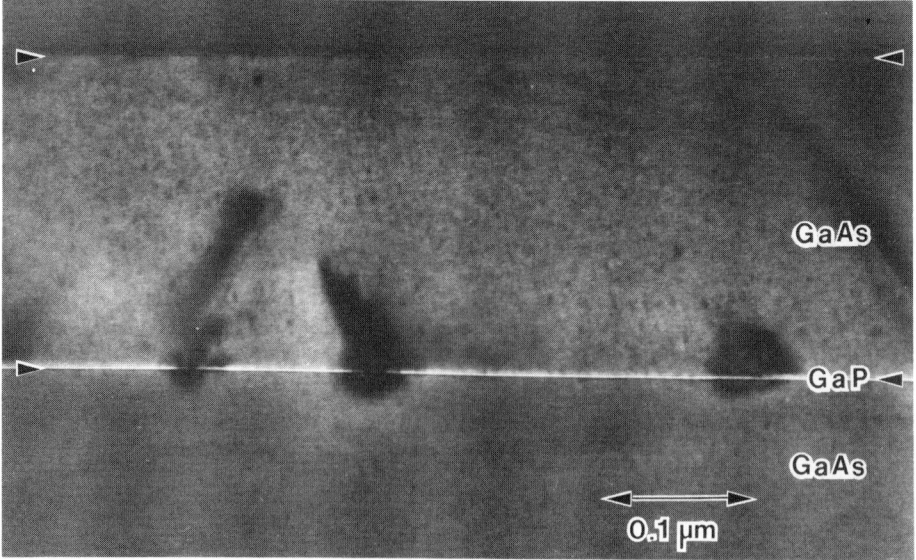

Figure 2. Transmission electron micrograph taken with the $<200>$ strong diffracting conditions. Note defects at average spacings of 300nm. The GaP film appears as a bright line.

carefully adjusted by measuring the growth rate using the well known RHEED oscillations technique during conventional MBE growth. A series of samples were produced with an integral number n of monolayers (n = 2, 3, 4, 5, 6, 10, 250). The GaP film thickness was subsequently checked with the Pendellösung x-ray interference technique (Mazuelas et al 1992, Recio et al 1990), which is very sensitive for the measurement of thickness in this type of geometry. The structure of the films was studied with transmission electron microscopy.

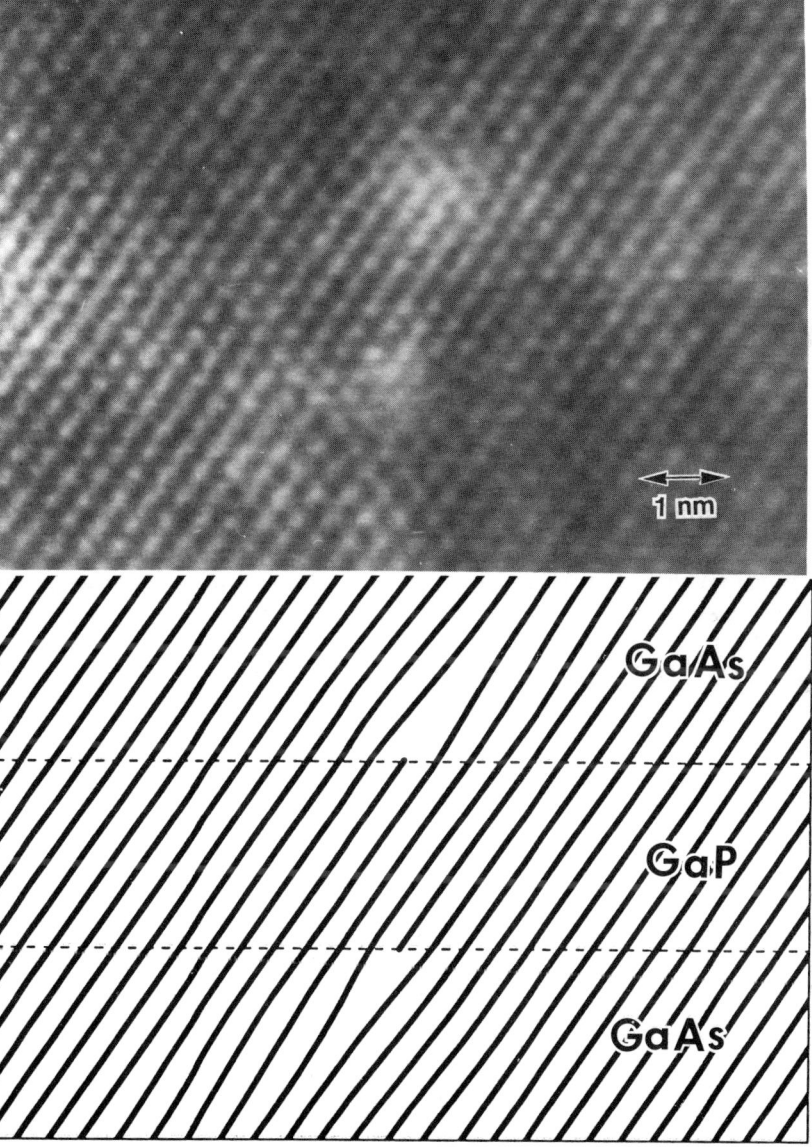

Figure 3. (a) High resolution lattice image of the 6 monolayer GaP film, showing dislocation dipole. (b) Diagram illustrating dislocation dipole structure in lattice image in (a).

RESULTS AND DISCUSSION

TEM images showed no dislocations present for films with thickness less than 6 nominal monolayers, in good agreement with the Matthews and Blakeslee model (1974). It should be noted that during growth the Ga pulse duration corresponded to that needed to produce a full monolayer in the GaAs film. Thus, it also corresponded to a full GaP strained monolayer, with the same surface density as the GaAs substrate. Figure 1 shows the x-ray interference measurement for the six monolayer thick GaP film. The calculated curve that matches the experimental curve corresponds to a GaP thickness of 15.3Å, equivalent to 5.6 monolayers of relaxed GaP film. The difference with the nominal thickness (6monolayers) may be due to growth rate indetermination. Figure 2 shows a two-beam transmission electron microscope image of a cross section of the sample. The GaP layer appears bright under $<200>$ imaging conditions. Defects are observed at an average spacing of .3µm. Misfit dislocations are not readily observable at this magnification. Closer observation using lattice imaging conditions shows that the dislocations are arranged as dipoles, with opposite 60° dislocations on either side of the film located on the GaAs side of the film (See Figure 3). The strain field relatively far from the dipoles is null, thus the misfit dislocation contrast observed under diffraction conditions is very weak. The dislocations observed in Figure 2, correspond to un-matched single dislocations. Some of these dislocations seem to disociate and form stacking faults. The observation of such close periodicity in these samples is rather surprising. For temperatures below 350C (the growth temperature), we expect very little if no mobility of dislocations. Our data indicates that the dislocation arrangement occurs even at low temperatures with a high degree of order. If the introduction of dislocations were by glide from the top surface of the GaP film, a highly cooperative and interactive process would have to take place in order to locate the dislocations in the proper positions. Given the low mobility of dislocations at these temperatures, that mechanism is unlikely to exist. We propose a cooperative mechanism for the relaxation process. This mechanism is similar to a martensitic transformation, and does not require slip nor atomic diffusion. It basically consists of a phase transformation from a highly deformed state (substrate-like) to a relaxed configuration with the bulk lattice parameter and atomic bonding configuration of the epitaxial film (in this case GaP). The misfit dislocations appear as a result of the change in the atomic bonding configuration, and not as a result of plastic deformation.

REFERENCES

Briones F and Ruiz A 1991 J. of Crystal Growth 111 pp 194-199.

Frank F C and van der Merwe F H 1949 Proc. R. Soc. London, Ser. A 198 (1949) 205.

Lowe W, MacHarrie R A, Bean J C, Peticolas L, Clarke Dos Passos R W, Brizard C and Rodricks B 1991 Phys. Rev. Lett. 67 pp 2513-2516.

Matthews J W and Blakeslee A E. 1974 J. Cryst. Growth 27 pp 118-125.

Mazuelas A, Tapfer L, Ruiz A, Briones F, and Ploog K 1992 Appl. Phys. A 55 pp 582-585.

Ponce F A 1982 Appl. Phys. Lett. 41 pp 371-373.

Ponce F A, Anderson G B and Ballingall J M 1986 Surf. Science 1986 pp 564-570.

Recio M, Armelles G, Ruiz A, Mazuelas A and Briones F 1990 Surf. Sci. 228 139.

Inst. Phys. Conf. Ser. No 134: Section 6
Paper presented at Microsc. Semicond. Mater. Conf., Oxford, 5–8 April 1993

317

A comparison of the growth mode and strain relief behaviour of $In_xGa_{1-x}As$ layers on GaAs(001) and (110)

X Zhang, D W Pashley*, P N Fawcett, J H Neave and B A Joyce

The Interdisciplinary Research Centre for Semiconductor Materials, Imperial College, London SW7 2AZ.
* The Department of Materials, Imperial College, London SW7 2BP.

ABSTRACT: Layers of $In_xGa_{1-x}As$ ($0<x\leq1$) grown by MBE on both GaAs(001) and GaAs(110) were studied using RHEED and TEM. In the case of InAs/GaAs both the growth mode and the strain relief behaviour were very different on the two surfaces. With reduced strain, $In_{0.25}Ga_{0.75}As/GaAs(110)$ showed different strain relief mechanisms compared to the higher strained InAs/GaAs (110). These changes in strain relief behaviour are explained by the geometry of the {111} slip planes in a (110) epilayer.

1. INTRODUCTION

The growth of $In_xGa_{1-x}As$ on (001) GaAs has been widely studied. The major reported studies on (110) substrates are the RHEED and X-ray measurement by Munekata et at (1987a, b), who observed that the InAs grew in 2D fashion on GaAs (110) and that the residual strains in the [1$\bar{1}$0] direction varied with layer thickness. No account of surface geometry was given. In the present study $In_xGa_{1-x}As$ ($0<x\leq1$) layers were grown on both GaAs (001) and GaAs (110) by MBE. The growth mode and the strain relief mechanisms on the two types of substrates are compared.

2. EXPERIMENTAL PROCEDURE

The MBE growth was carried out using a specially designed system for reflection high energy electron diffraction (RHEED) studies. For each growth run a (001) and a (110) GaAs substrate were loaded side by side, which eliminates the uncertainty in growth conditions when observations on the two substrates are compared. The study can be categorised into three sections: (1) comparison of growth mode and morphology of InAs on GaAs (001) and (110); (2) the strain relief of InAs on GaAs (110) with different layer thicknesses and relaxation mechanisms; (3) the effect of degree of misfit on the strain relief mechanisms using $In_xGa_{1-x}As$ ($0<x<1$) on GaAs (110). The growth mode was studied by monitoring RHEED patterns during the $In_xGa_{1-x}As$ deposition. TEM studies were carried out in both plan and cross-sectional views. The plan view samples were prepared using chemical thinning ($HCl/H_2O_2/H_2O$) and the cross-sections using standard Ar^+ milling. Two JEOL microscopes, 2000-FX and JEM-2010, both operated at 200kV, were used in the TEM study.

3. RESULTS AND DISCUSSION

3.1 Growth mode and morphology of InAs on GaAs (001) and (110) under identical growth conditions

Fig.1 shows typical RHEED patterns after the initial 3MLs of InAs deposition on GaAs (a) (001) and (b) (110) respectively. On the (001) surface, the spotty reflections indicate that the 3MLs of InAs is 3D in nature. On the (110) surface, the same amount of InAs is in a 2D form as indicated by the continuous RHEED streaks. These patterns remained more or less unchanged (except the local intensity variation) with further deposition up to at least the thickness of 60Å. TEM plan view images of the two samples are shown in fig.1(c) and (d), and it is clear that InAs shows a network with exposed GaAs on the (001) surface and a continuous film on the (110) surface. The above observations clearly demonstrate that the growth mode of InAs on GaAs has changed from 2D(not shown, see Zhang et al 1993a) $\rightarrow$3D on the (001) substrate to 2D only on the (110). It is not yet clear why such a change in growth mode should take place. However, it may be associated with the fact that the (110) surface has lower energy and its non-polar surface has a more open structure than the (001), hence allowing easier surface migration.

3.2. The strain relief mechanisms for InAs on GaAs(110)

Three InAs layers with thicknesses of 30, 60 and 400Å on GaAs(110) were studied. The residual strains in the InAs films along the two orthogonal directions, [001] and [1$\bar{1}$0], were measured from the (110) electron diffraction patterns. Fig. 2(a) is an example taken from the 60Å layer. The two orthogonal reflections are enlarged in the inserts, where the arrows indicate the positions where fully relaxed InAs are expected. It can be seen that the 2$\bar{2}$0 reflection is much nearer to the arrowed position than the 004, indicating that the strain relief is more complete along the [1$\bar{1}$0] direction than along the [001] (Zhang et al 1993b). The residual strains are calculated in both directions from the diffraction patterns and the results, together with the measurements from the 30Å and 400Å films, are shown in Table.1. The strain relief in the [001] direction increases significantly with increasing layer thickness, whereas in the [1$\bar{1}$0] direction it remains more or less independent of layer thickness. As a result the degree of strain relief is different along the two orthogonal directions during the initial stage of growth when the InAs layer is thin. The study of the misfit dislocations shows the following:

Lomer type: lying along [001] with $\underline{b}$=a/2[1$\bar{1}$0];
60° types: lying along [1$\bar{1}$0] with $\underline{b}$=a/2<101>.

The above asymmetric strain relief can be explained by the orientation of the {111} slip planes with respect to (110) surface. On this surface, only two out of the four {111} slip planes, (111) and (11$\bar{1}$), are inclined to (110) and both intersect (110) along the [1$\bar{1}$0] direction. During the strain relaxation process, Lomer dislocations which are most efficient in relieving strain can be formed easily along the [001] direction, with $\underline{b}$=a/2[1$\bar{1}$0], from the beginning of growth and before completion of a continuous film. Hence it appears that the strain relief in [1$\bar{1}$0] is more or less complete at the early stages of growth. In the orthogonal direction the formation of Lomer type dislocations with Burgers vector of 1/2<110> type is not possible. However because of the presence of the two inclined {111} planes, 60° type dislocations with $\underline{b}$=a/2<101> may form subsequently, and slip along these planes to reach the interface. The strain along [001] is relieved via the component of $\underline{b}$ along [001], similar to the critical

Alloy type	film thickness, Å	$\varepsilon_{1\bar{1}0}$ %	ε_{001} %
InAs	30	0.9	4.8
	60	0.7	2.7
	400	0.4	0.7
$In_{0.25}Ga_{0.75}As$	350	1.0	> 0.1

Table.1 Residual strains in InAs and $In_{0.25}Ga_{0.75}As$ films on GaAs(110)

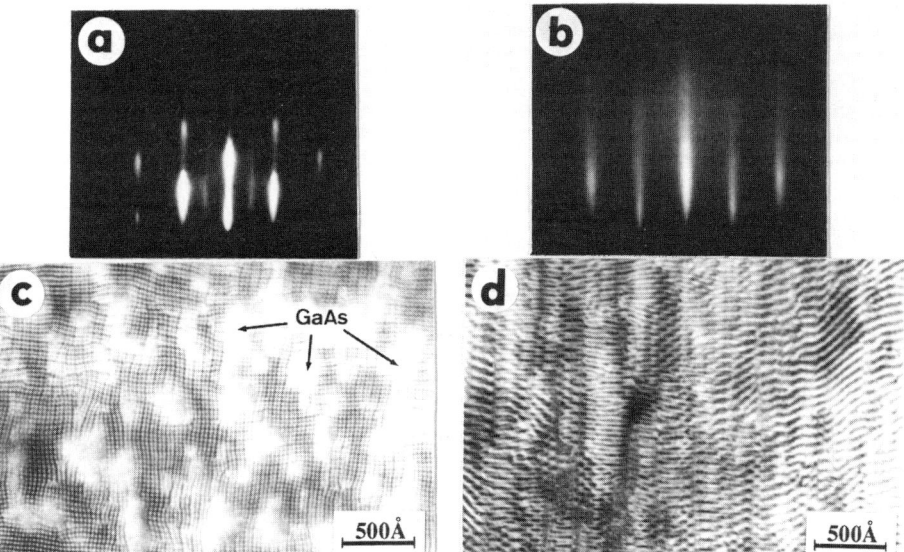

Fig.1 RHEED patterns of 3MLs of InAs on GaAs (a) (001) and (b) (110) substrates. (c) and (d) are the plan view TEM micrographs of 60Å thick InAs on GaAs (001) and (110) respectively.

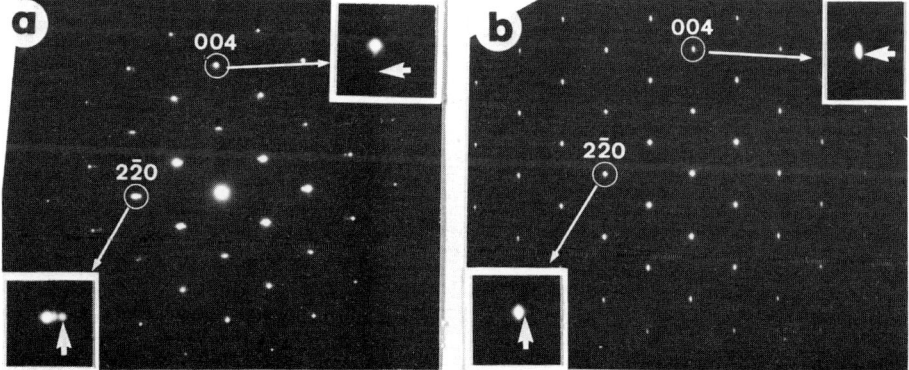

Fig.2 Plan view electron diffraction patterns of (a) 60Å InAs and (b) 350Å $In_{0.25}Ga_{0.75}As$ on GaAs(110). The two orthogonal reflections are enlarged in the inserts. The arrows mark the positions where reflections from the fully relaxed alloys are expected.

thickness model in the small misfit epitaxial systems on (001) substrates.

3.3. The effect of degree of misfit on the strain relief mechanisms of (110) growth

InAs/GaAs(110) has a misfit strain of $\approx 7.2\%$. To see the effect of reduction of strain on the strain relief mechanism, $In_xGa_{1-x}As$ alloys with x=0.75, 0.5 and 0.25 on GaAs (110), corresponding to misfit strains of 5.4%, 3.6% and 1.8% respectively, were examined. Among the three compositions, the former two showed a behaviour of strain relief similar to InAs/GaAs as discussed in section 3.2. With $In_{0.25}Ga_{0.75}As$, however, a major change of strain relief mode was observed. Fig.2(b) is a (110) electron diffraction pattern of a 350Å thick $In_{0.25}Ga_{0.75}As$ film on GaAs(110). Again, the arrows in the inserts of $2\bar{2}0$ and 004 reflections mark the expected positions of a fully relaxed alloy. In this case the 004 reflection is nearer to the arrowed position than the $2\bar{2}0$, implying that the relaxation in the [001] direction is more complete than in the $[1\bar{1}0]$ (see Table.1 for the value of residual strain). This is effectively opposite to the observation of InAs/GaAs(110) shown in fig.2(a). Study of the misfit dislocations shows that along the $[1\bar{1}0]$ direction there exists a high density of 60° type dislocations, stacking faults and microtwins. There are many less dislocations present along the orthogonal [001] directions and they are in the form of short segments. At this stage, the exact Burgers vectors of these segmented dislocations have not been determined; they appear in strong contrast under $2\bar{2}0$ two-beam conditions and disappear under 004 two-beam conditions. Hull et al (1991) have made similar observations for the growth of Si_xGe_{1-x} on Si(110) oriented substrates.

The change of strain relief mechanism observed in $In_{0.25}Ga_{0.75}As/GaAs$ (110) (as opposed to InAs/GaAs(110)) lies in its reduced strain. In this case, the alloy grows psuedomorphically during the initial stage of growth. At the point when the introduction of misfit dislocations becomes favourable, the 60° dislocations can be formed, in a manner similar to that discussed previously, to relieve the strain along the [001] direction. No normal slip systems exist for the relief of strain along the orthogonal $[1\bar{1}0]$ direction. Consequently an abnormal strain relief process must be involved. Several possible mechanisms have been considered by Pashley (1993). Further investigation is being undertaken to clarify the point.

4. SUMMARY

Layers of $In_xGa_{1-x}As$ on GaAs (110) grown by MBE were studied. The strains in the films in the two orthogonal directions are relieved via different mechanisms. This can be explained as due to the orientation of the {111} slip planes with respect to the (110) surface.

REFERENCES

Hull R, Bean J C, Paticolas L and Bahnck D 1991 Appl, Phys. Lett. <u>59</u> 964
Munekata H, Chang L L, Woronick S C and Cao Y H 1987a J. Cryst. Growth <u>81</u> 237
Munekata H, Segmuller A and Chang L L, 1987b Appl. Phys. Lett. <u>51</u> 587
Pashley D W 1993 Phil. Mag., in press
Zhang X, Pashley D W, Neave J H, Fawcett P N, Zhang J and Joyce B A 1993a J. Crystal Growth, in press.
Zhang X, Pashley D W, Fawcett P N, Neave J H and Joyce B A 1993b J. Crystal Growth, in press.

Inst. Phys. Conf. Ser. No 134: Section 6
Paper presented at Microsc. Semicond. Mater. Conf., Oxford, 5–8 April 1993

321

Relaxed epitaxial layers: the effect of an added interface

P Kidd[1], D J Dunstan[2], R Grey[3], J David[3], P F Fewster[4], N L Andrew[4], S I Molina[5,6] and C J Kiely[5]

[1]Department of Materials Science and Engineering, [2]Department of Physics, University of Surrey, Guildford GU2 5XH, UK.
[3]University of Sheffield SERC III-V Semiconductor Facility, Department of Electronic and Electrical Engineering, P.O. Box 600, Mappin St, Sheffield S1 4DU, UK.
[4]Philips Research Laboratories, Cross Oak Lane, Redhill, U.K.
[5]Department of Materials Science and Engineering, University of Liverpool, Liverpool L69 3BX, UK.
[6]Permanent Address: Departamento de Química Inorgánica, Universidad de Cádiz, Apdo. 40 11510-Puerto Real Cádiz, Spain.

ABSTRACT: The residual strain and dislocation distribution in a relaxed two layer sample comprising $In_xGa_{(1-x)}As$ on (001) GaAs has been investigated using High Resolution X-ray Diffraction (HRXD), Transmission Electron Microscopy (TEM) and Convergent Beam Electron Diffraction (CBED). It is shown that the average residual strain of the relaxed composite structure taken as a whole obeys the relaxation law previously observed for single layers of the same compositions. However, the distribution of strain between the layers is dependent upon dislocation interactions at the two interfaces.

1. INTRODUCTION

In order to be able to design buffer layer structures which can provide an in-plane lattice parameter different from that of the substrate, it is necessary to study the relaxation of strained layer structures and to ascertain the criteria necessary to predict and measure relaxation behaviour. As part of this programme we have examined firstly, single surface layers of $In_{.1}Ga_{.9}As$ and $In_{.2}Ga_{.8}As$ on (001) GaAs with a variety of layer thicknesses above the critical thickness for relaxation. The results of these investigations have been presented elsewhere (Dunstan et al 1991, Kidd et al 1993a,b). We have observed that the residual strain, e, in these layers, as measured by x-ray diffraction, can be predicted from the layer thickness, d, alone provided that a relaxation critical thickness has been exceeded (see Kidd 1993b) and that the layer growth is of good quality, i.e. 2D layer by layer growth (Dunstan et al 1991). This has provided us with a semi-empirical relaxation curve e = k/d (k ~ 0.8) depicted in Fig 1, which we use as a starting point in the design of two-layer step graded multilayer structures. The design premise is that, provided the second interface is coherent, the strains from the two layers can be superimposed and thus a two layer structure with strains e_1 and e_2 and thicknesses d_1 and d_2, can be treated as a single layer with an average strain given by: $e_{av} = (e_1d_1 + e_2d_2)/(d_1 + d_2)$. This average strain in the multilayer stack can thus be predicted from the relaxation curve and is a function of the total thickness of the stack provided that the stack has exceeded the relaxation critical thickness i.e. $(d_1 + d_2) > k/e_{av}$. In this paper we verify that this is indeed what happens, and investigate the dislocation distributions in the two interfaces which give rise to relaxation in the two constituent layers.

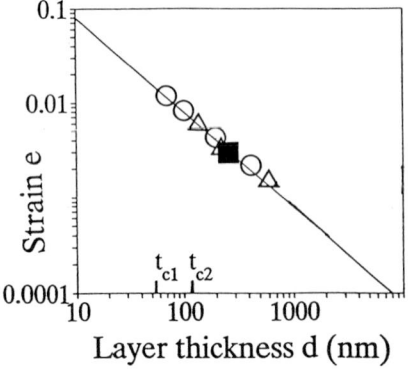

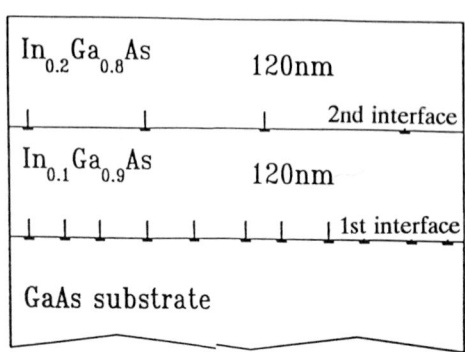

Fig 1 Relaxation curve for residual strain versus layer thickness for $In_xGa_{(1-x)}As$ layers on GaAs, as measured by x-ray diffraction. Open triangles x=0.09, open circles x=0.2, filled square multilayer (see Fig 2).

Fig 2 Schematic representation of the multilayer stucture studied

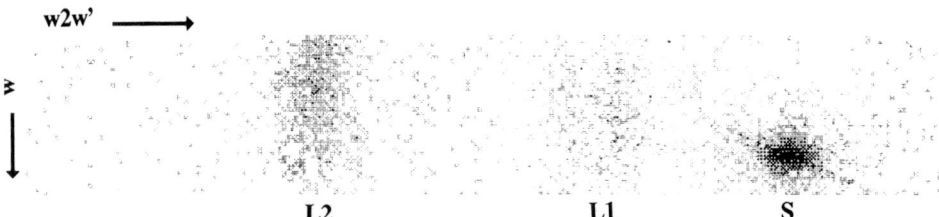

Fig 3 004 Bragg reflection reciprocal space map of the multilayer structure depicted in Fig 2. showing the diffraction peaks from the substrate, S, the first layer L1 and the second layer L2. The reciprocal space map separates lattice strain measurement (w2w' axis) from lattice tilt (w). the diffracted intensity is plotted in logarithmic grey scale contrast.

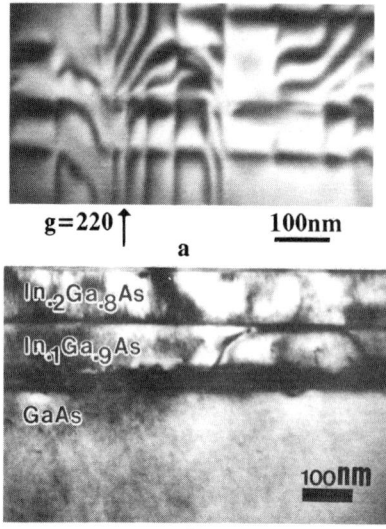

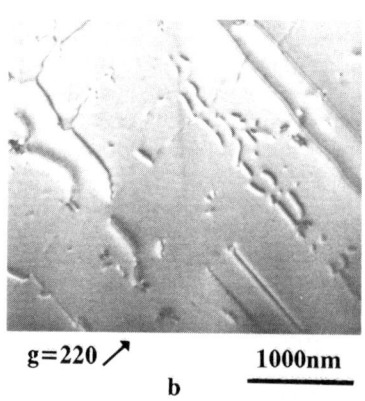

Figs4a-c TEM micrographs of the multilayer sample illustrated in fig 2. Figs 4a and 4b are plan views of the first and second interfaces respectively. Fig 4c is a {110} cross-section showing both interfaces.

2. EXPERIMENTAL

The sample was grown by molecular beam epitaxy (MBE) at 510°C, the details of which have been presented elsewhere (Dunstan et al 1991). The double layer sample consists of a layer of In_1Ga_9As of thickness 120nm followed by the same thickness of In_2Ga_8As as shown schematically in Fig 2. The thickness of each layer was chosen so that the 10% In layer would be just at critical thickness as defined by the relaxation curve, and would therefore start to relax plastically as soon as the 20% In alloy was grown on top. The compositions and residual strains in the constituent layers were measured by HRXD using reciprocal space maps which provide measurement of the lattice strain and tilts for the in-plane (a,b) and perpendicular (c) lattice unit cell parameters (Fewster 1991) averaged over a layer volume of approximately $10^8 nm^3$. High Order Laue Zone (HOLZ) line positions in the transmitted disks of convergent beam patterns (CBPs) were used to measure the relative strains (Randle et al 1989) for a, b and c lattice parameters for a probing volume of $\sim 4 \times 10^3$ nm^3. Comparison of the two measurements thus gives an indication of the inhomogeneous distribution of lattice relaxation on a microscopic scale. The layer thicknesses were measured by cross-section TEM and the dislocation spacings and distributions from plan-view and cross section TEM micrographs.

3. RESULTS AND DISCUSSION

3.1 Residual Strain

TABLE 1		MEASURED LATTICE PARAMETER			
	relaxed (x)	(a)	(b)	(c)	scatter
LAYER 2					
HRXD	5.7381 (.21)	5.697	5.709	5.771	(0.002)
CBED		5.718	5.720	5.760	(0.003)
LAYER 1					
HRXD	5.6887 (.09)	5.674	5.682	5.699	(0.002)
CBED		5.677	5.679	5.703	(0.005)

Table 1 summarises the relative lattice parameter measurements for the two layers, where the value used for the GaAs substrate lattice parameter is 5.6535Å. Scatter in the HRXD measurement arises from the diffuseness of the peaks in the reciprocal space maps. Fig 3 shows an example of one of the 004 maps used in the measurement. The layer peaks are diffuse in both the w2w' and w reciprocal lattice directions, as a result of inhomogeneous distribution of microscopic strains and tilts. The integrated intensities of the peaks for each layer are the same, since the layers have the same thicknesses. However the peak from the first layer is 50% broader than the second along w2w' indicative of a greater degree of lattice disruption from dislocation strainfields. Measurement of the unit cell distortion shows that during relaxation the tetragonal symmetry is lost in the strained layer; this occurs as a result of an asymmetry in burgers vector distribution amongst the interfacial dislocations and from the microscopic tilts around these dislocations. Firstly it can be seen that there is very good agreement between the CBED and HRXD lattice parameter values for the first layer, which has the highest relaxation, whereas the agreement between measured values for the top layer cannot be made within the probed volume scatter. As shown in the next section, the dislocation spacing at the second interface is around 500nm, the strainfields of the dislocations will be confined laterally to a distance of the order of the layer thickness (120nm) (Kidd 1993b) thus the strain in the layer above the interface will vary over a long range comparable to the CBED probe with diameter $\sim$ 20nm. In contrast, the dislocation spacings in the first interface are of the order 100nm, it is more likely that their strainfields overlap and that the lateral variation of the strain is less than the CBED probe volume. The residual strain in the layers is calculated by comparing the difference between the relaxed cubic lattice parameter and the average of the measured a and b parameters values with the substrate lattice parameter namely: $e = (a_r -(a + b)/2)/a_s$. We find, using HRXD values, that $e_2 = 0.0062$ and $e_1 = 0.0019$, giving an average residual strain for the whole structure $e_{av} = 0.004$, this is in good agreement with the value $e = 0.0035$ predicted from our relaxation curve, and is shown plotted on Fig 1. Theory would predict that a perfect structure would relax to $e = 0$ at the first layer, leaving the second layer at $e = 0.007$. However, we find empirically, that the first layer has some residual strain and that the second layer has relaxed more in compensation. This corresponds to the presence of some extra dislocations in the second interface.

3.2 Dislocation Distribution

The dislocation distributions are shown in Figs 4a-c. Fig 4a is a plan view of the first interface; Fig 4b is a plan view of the second interface. Fig 4c is a cross-section micrograph showing the two layers and a segment of dislocation line threading between the layers. The mean dislocation spacings in the two orthogonal directions are 500nm and >1000nm for the second interface and 100nm and 65nm for the first interface. The dislocations at the first interface show a semi-regular array typical of interfacial dislocation arrays, whereas the dislocations in the second interface are irregular in spacing and bow in and out of the interface region. From the cross-section micrograph it can be seen that the dislocations are generally located in or close to the interfaces except when they are threading between the two. The irregular nature of the dislocations in the second interface would imply that they have been pinned at this interface. It is not clear whether the pinning is due to the compositional change at the interface or rather due to a momentary growth interrupt. Work is in process to examine this further. It is however interesting to note that even when dislocations are pinned and the structure fails to reach its lowest energy configuration, the relaxation law still applies.

4. CONCLUSIONS

We have shown that a composite structure taken as a whole, follows the same relaxation law as observed for single layers. This process allows us to relax layers in a multilayer stack via a controlled and predictable method provided layer composition and thickness are chosen carefully. However, subtle departures of the individual strains of the component layers from the ideal case show that disloations are easily pinned at the interface and that this may prove a limiting problem in the creation of large multilayer stacks.

ACKNOWLEDGEMENTS

We are grateful to the Royal Society and the Science and Engineering Research Council (UK) and to the Commission of the European Communities for financial support for this work. One of us (SIM) would like to thank the Spanish Ministry of Science and Education for financial support and Prof. P J Goodhew for the use of laboratory facilities at Liverpool.

REFERENCES

Dunstan D J, Kidd P, Howard L K, Dixon R H 1991 Appl. Phys. Lett. **59** 3390-3392
Fewster P F 1991 Appl. Surf. Sci. **50** 9-18
Kidd P, Dunstan D J, Grey R, David J P R, 1993a Submitted to Appl. Phys. Lett.
Kidd P, Fewster P F, Andrew N L, Dunstan D J 1993b, presented at this conference
Randle V, Barker I and Ralph B, 1989 J. Electr. Micr. Tech. **13** 51-65

Inst. Phys. Conf. Ser. No 134: Section 6
Paper presented at Microsc. Semicond. Mater. Conf., Oxford, 5–8 April 1993

325

TEM and XRD study of strain release in GaAs/InP and InP/GaAs heterostructures grown by MOVPE

L Lazzarini, G Attolini, D Bertone*, P Franzosi, C Pelosi and G Salviati

Maspec-CNR Institute, via Chiavari 18/A, 43100-I Parma
*CSELT, via G Reiss Romoli 274, 10148-I Torino

ABSTRACT: The strain release has been studied by high resolution X-ray diffraction in highly mismatched Atomic Layer MOVPE InP/GaAs and conventional MOVPE GaAs/InP grown heterostructures. The InP/GaAs structures follow the predictions of the equilibrium theory, while the strain release in the GaAs/InP structures occurs at slightly larger thickness values. ALE achieves 2-Dimensional growth of InP from thickness of 5 nm while growth islands are still present in 20 nm thick GaAs layers. The crystal defects have been investigated by transmission electron microscopy and their characteristics have been correlated with the strain release. No differences in the defect density and structure have been observed between the two systems.

1. INTRODUCTION

The GaAs/InP and InP/GaAs heteroepitaxial systems are very attractive for application purposes in view of the integration of GaAs high speed electronics with InP based optical devices.

The layers are affected by a large lattice mismatch of about 3.8%; the InP epilayers are under compressive strain conditions, whereas the GaAs ones are under tensile strain conditions. Due to the large lattice mismatch, many crystal defects are generated in the epitaxial layers even at low thickness. The defects strongly degrade device performance, except when they can be confined far from the active regions. The difference in thermal expansion coefficients of the two materials can provide an additional source of defects during cooling after growth. Furthermore, the generation of some defects is related to the growth mechanisms.

In this paper we report on structural analyses performed by High Resolution X-ray Diffractometry (HRXRD) and by Transmission Electron Microscopy (TEM) on InP/GaAs and GaAs/InP heterostructures grown by Atomic Layer (AL) Low Pressure Metal Organic Vapour Phase Epitaxy (MOVPE) and conventional Atmospheric Pressure (AP) MOVPE respectively.

The study of the influence of the strain sign, the growth temperature and the nucleation mechanisms on the strain release and the crystal defects was the main aim of the present work.

2. EXPERIMENTAL

InP layers with thicknesses ranging between 5 and 150 nm have been grown entirely by AL epitaxy at a growth temperature of 340° C on good quality (001) oriented GaAs substrates. GaAs epitaxial layers with thicknesses ranging between 15 nm and 7 μm have been obtained by AP-MOVPE at 600° C on (001) InP substrates 2° off towards (110). The epilayers were directly grown on the substrates without any buffer layer in both the heterostructures. The growth procedures have been described elsewhere (Bertone 1992 and Attolini et al 1993, respectively).

The elastic lattice strain was investigated by HRXRD, using a Philips high resolution diffractometer equipped with a four crystal monochromator (Ge, 220 reflection). The Cu $K\alpha_1$ radiation and the 004 symmetric reflections were used; under these conditions, the Bragg peak of a high quality GaAs crystal exhibits a full width at half maximum of about 15 arcsec.

The TEM investigations were performed in a JEOL 2000FX TEM, with a resolution of 0.31 nm. (001) plan and (110) cross sections for the TEM observations were prepared by a conventional mechano-chemical procedure followed by room temperature argon and iodine ion milling (Cullis et al 1988). The observations in the high resolution mode (HREM) have been carried out in the bright field (110) zone axis condition.

3. RESULTS AND DISCUSSION

The values of the epilayer lattice parameter a perpendicular to the interface were obtained by HRXRD. The determination of the Bragg peak centre was more difficult in thinner layers; the peaks became broader and weaker as the layer thickness decreased. For example, a 25 nm thick layer exhibited a peak width of about 1500 arcsec and a diffracted intensity maximum of about 8 counts/sec to be compared with the background level slightly smaller than 1 count/sec and with the substrate peak maximum of about 5×10^4 counts/sec. In spite of these difficulties, reliable a data were obtained down to a thickness of 5 nm. Assuming that continuum elasticity theory holds, the values of the epilayer lattice parameter a and the residual strain ε parallel to the interface were calculated. The results are presented in Fig. 1, that shows the absolute value of the parallel strain ε as a function of the layer thickness t for both InP/GaAs and GaAs/InP heterostructures. The same figure also shows the predictions of the equilibrium theory of Matthews and Blakeslee (1974) (solid curve). It is apparent that there is quite a satisfactory agreement between the InP/GaAs data and the theoretical predictions; in contrast, the GaAs/InP data show the same slope as the theoretical curve, but they are shifted towards larger t values. Moreover, the residual strain tends to become constant for $t > 1 \, \mu m$; the constant residual strain exhibited by the thicker layers is due to the thermal mismatch, i.e. the difference in the linear thermal expansion coefficients. In the GaAs/InP structures grown at 600° C, the thermally induced strain is expected to be about 1×10^{-3} (Olego et al 1992a); in the InP/GaAs structures grown at 340° C, a value of about 6×10^{-4} can be roughly evaluated.

Olego et al (1992b) found the opposite behaviour of the two systems with respect to elastic strain relaxation, thus leading us to rule out the influence of the strain sign in the process. The role of the temperature also seems not to be critical because we found a larger relaxation in InP/GaAs despite the low growth temperature.

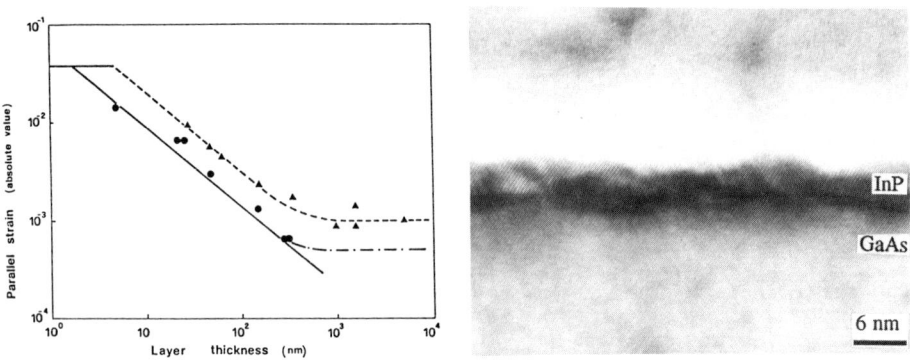

Fig. 1   Fig. 2

Fig. 1: Absolute value of the residual parallel strain as a function of the layer thickness. Dots: InP/GaAs; triangles: GaAs/InP; solid curve: theoretical prediction of the equilibrium theory; dashed and dash-dotted curves are guides for the eye.

Fig. 2: (110) Bright Field(BF)-Zone Axis HREM micrograph of an InP/GaAs 5 nm thick epilayer. The growth was already following the 2-Dimensional mechanism.

The TEM investigations of the early stages of growth revealed that even the thinnest InP sample is a continuous 5 nm thick layer as shown in Fig. 2. The corrugated epilayer surface is a consequence of the rough surface of the substrate, due to the absence of a buffer layer. The thinnest GaAs epilayer, obtained with a growth time of 1.5 sec, presented growth islands on average 15 nm thick, bounded by {111} planes (Fig. 3a). With a growth time of 5 secs, islands of the same thickness were still present but showed larger lateral dimensions. The islands coalesced with a 10 sec growth time, giving a continuous 30 nm thick layer shown in Fig. 3b. Therefore, the ALE method achieves 2-Dimensional growth process at layer thickness considerably lower than the conventional MOVPE (Olego et al 1992a) and MBE (see for instance Biegelsen et al 1987 and Lee et al 1991) when employed for growing heterostructures with similar mismatch. In agreement with Olego et al (1992a), we find that an earlier island coalescence results in a larger strain relaxation.

The cross sectional TEM observations gave the common result (see for instance Chu et al 1989 and Olego et al 1992a) that the distribution of the crystal defects across the thickness of the epilayer is not uniform. A very large density of misfit dislocations (MDs) is observed at the interface. Planar defects (PDs) such as stacking faults and microtwins, originating at the interface, propagate along the {111} planes and affect all the investigated layers. In addition, threading dislocations (TDs) have been revealed in the specimens thicker than 70 nm. The PDs and TDs density decreases from the interface region to the top of the layers due to annihilation

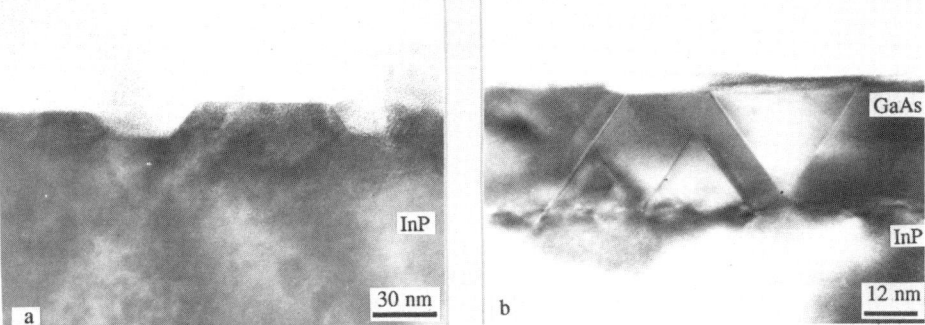

Fig. 3: a) CTEM picture of a 15 nm GaAs/InP thick layer. GaAs growth islands are clearly visible. b) (110) CTEM cross section of a 30 nm thick layer: the GaAs islands coalesced resulting in a flat epilayer surface.

reactions. The great majority of TDs, according to the standard contrast extinction rules, have been found to be 60° mixed type, gliding on {111} planes. Pure edge dislocations have only occasionally been observed. The TDs are randomly distributed in the interface region and the pyramid dislocation tangles structures described by Chu et al (1989) have not been observed.

The TEM plan view investigations showed that the MDs at the interface are not arranged in regular arrays in both the systems. As an example, a top view of a GaAs/InP 0.1 μm thick layer is shown in Fig. 4. MDs are present in the left hand side of the micrograph, where the sample still contains the interface region. The nature of the MDs was determined by constructing Burgers circuits on HREM micrographs. The MDs were mostly 60° type in the thickest specimens, but some 90° type MDs were also observed at the interface as shown in Fig. 5.

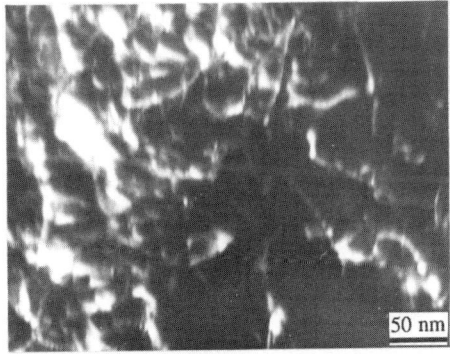 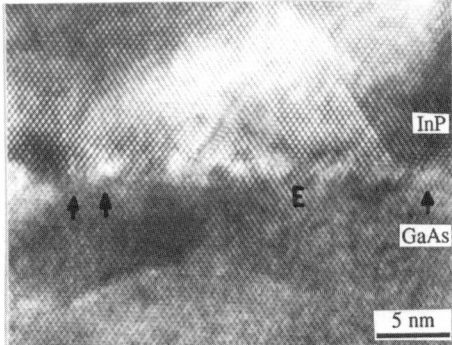

Fig. 4   Fig. 5

Fig. 4: DF top view of the interfacial region of a GaAs/InP heterostructure.
Fig. 5: (110) BF-Zone axis HREM image of the interface region of a 150 nm InP thick layer. Arrows and E indicate 60° and edge dislocations respectively.

There are no apparent differences between the two systems from the point of view of the extended defect density and structure. This result is not in agreement with the previous findings of Chu et al (1989) and Olego et al (1992a, 1992b). An evaluation of the MD linear density at the heterointerface, performed in the HREM mode, gave a value of about 8×10^5 cm^{-1} in the thinnest layers and of about $2x10^6$ cm^{-1} in the thickest ones where the strain was almost completely released. According to the HRXRD measurements, the same amount of strain is released at slightly larger thickness in the GaAs/InP heterostructure than in the InP/GaAs one. The difference in the MD linear density in the two systems, calculated by the X-ray measurements assuming that all the strain is released by 60° type MDs, should be about 25 % for layer thickness t= 20 nm and reduces to 5% for t=100 nm. Such differences can hardly be resolved in the HREM mode. However, a satisfactory agreement between XRD and TEM results was found.

In the region close to the interface the PDs density remains practically the same in all the specimens investigated. For this reason and because the measured MD density corresponds to the density required for the measured strain release in all the samples investigated, it is possible to deduce that the planar defects play a minor role in the strain release process. It seems that they are growth-related rather than misfit strain-related, because they have been found to be present in both high (see for instance Gerthsen et al 1990, Schwartzmann and Sinclair 1991) and low (Ernst and Pirouz 1988, Franzosi et al 1991) mismatch epitaxial heterostructures. The origin of these defects is still unclear.

4. CONCLUSIONS

Highly mismatched InP/GaAs and GaAs/InP heterostructures have been grown by Atomic Layer MOVPE at 340° C and by conventional MOVPE at 600° C respectively.

The TEM investigations of the initial growth stages show that AL epitaxy produces 2-Dimensional growth from 5 nm InP layer thickness, despite the large lattice strain. Conversely, growth islands are still present in GaAs layers 15-20 nm thick; the layers exhibited flat growth surfaces for thickness ≥ 28 nm.

HRXRD analyses showed that the same amount of strain is released at slightly larger thickness in the GaAs/InP heterostructure than the InP/GaAs one, which was found to follow the theoretical predictions of the equilibrium theory. Thus, an earlier island coalescence results in a larger strain release. It was also possible to deduce that the strain sign as well as the temperature seem not to substantially influence the strain relaxation process.

Large densities of MDs, PDs and TDs have been observed in all the samples investigated. The TEM measured linear MD densities are in agreement with those obtained from the X-Ray Diffraction measurements of the parallel lattice mismatch. No meaningful differences between the two systems have been observed from the point of view of the crystal defect density and structure.

ACKNOWLEDGMENTS

The authors are indebted to Dr M Urchulutegui for her kind cooperation and to Mr M Scaffardi for his technical assistance.

REFERENCES

Attolini G, Franzosi P, Pelosi C, Lazzarini L and Salviati G to be published in J. Electronic Mat.
Bertone D 1992 J. Electronic Mat. 21 265
Biegelsen D K, Ponce F A, Smith A J and Tramontana J C 1987 J. Appl. Phys. 61 1856
Chu S N G, Tsang T H, Chiu T H and Macrander A T 1989 J. Appl. Phys. 66 (2) 520
Cullis A G and Chew N G 1988 MRS Symp. Proc. 115 3
Gerthsen D, Biegelsen D K, Ponce F A and Tramontana J C 1990 J. Cryst. Growth 106 157
Ernst F and Pirouz P 1987 J. Appl. Phys. 64 4526
Franzosi P, Lazzarini L, Salviati G, Scaffardi M and Timò G 1991 Inst. Phys. Conf. Ser., 117, 399
Lee H P, Liu X, Malloy K, Wang S, George T, Weber E R and Liliental-Weber Z 1991 J. Electronic Mat. 20 179
Matthews J W and Blakeslee A E 1974 J. Cryst. Growth 27 118
Olego D J, Okuno Y, Kawano T, Tamura M 1992 a J. Appl. Phys. 71 (9) 4492
Olego D J, Okuno Y, Kawano T, Tamura M 1992 b J. Appl. Phys. 71 (9) 4502
Schwartzman A F and Sinclair R 1991 J. Electronic Mat. 20 805

Inst. Phys. Conf. Ser. No 134: Section 6
Paper presented at Microsc. Semicond. Mater. Conf., Oxford, 5–8 April 1993

329

HREM strain measurement of ultra thin ZnTe and MnTe layers grown in CdTe

P H Jouneau, A Tardot, G Feuillet, H Mariette[*] and J Cibert[*]

Equipe CNRS-CEA "Microstructures de semiconducteurs II-VI"
CEA - Département de Recherche Fondamentale sur la Matière Condensée, BP 85X, 38041 Grenoble , France
* CNRS - Laboratoire de Spectrométrie Physique, Université Joseph Fourier, Grenoble, France

ABSTRACT: High Resolution Electron Microscopy is used to investigate the morphology of ultrathin pseudomorphic (001) ZnTe and MnTe strained layers grown in CdTe. Local distortions of the crystal lattice are measured directly on HREM images by use of image processing software. In the case of ZnTe/CdTe superlattices, the method yields the location and the total amount of Zn per period. For MnTe layers embedded in CdTe, one can deduce the critical thickness and the atomic morphology of the interfaces.

1. INTRODUCTION

II-VI semiconductor strained heterostructures are promising candidates for optoelectronic devices, such as lasers (Cibert et al 1993), operating from the mid-infrared to the blue region. The optical and electronic properties of such heterostructures could be strongly dependent on the morphology of the interfaces (abruptness and roughness) between the different layers, and on the possible presence of misfit dislocations. For accurate characterization of these interfaces, High Resolution Electron Microscopy (HREM) is a tool of choice, as it provides images of each interface with atomic resolution. However, the image contrast between different materials of the same crystalline structure is often weak, especially in the <110> projection which is the most commonly used direction for imaging sphalerite semiconductors. Therefore, the methods proposed by Ourmazd et al (1990) or Thoma et al (1991) to extract a "chemical lattice image" are very efficient but difficult to implement routinely. An alternative way to extract information from HREM images in the case of strained multilayers has been proposed recently by Bierwolf et al (1993). It consists in measuring local distortions within the crystal lattice directly on HREM images. These distortions are due both to the lattice parameter difference between the materials and to the elastic deformations of the strained layers. Assuming the elasticity theory is valid, quantitative information on the chemical profile at the interfaces can be deduced.

We report here on the application of such a method to evaluate ultrathin ZnTe and MnTe layers grown in CdTe by Molecular Beam Epitaxy (MBE). These three materials, when grown by MBE, exhibit the same sphalerite structure, with lattice parameters of 6.481 Å for CdTe, 6.103 Å for ZnTe and 6.340 Å for MnTe. Therefore the lattice mismatch is 6% between ZnTe and CdTe, 2.2% between MnTe and CdTe. The paper is organised as follows: section 2 presents the basic principles of our distortion analysis method, then the results on ZnTe/CdTe and MnTe/CdTe are presented on section 3 and 4 respectively. For all samples, HREM analysis was performed at 400 keV using a JEOL 4000EX microscope (with a point to point resolution of 1.7 Å). The specimens were prepared in <110> cross-section by mechanical polishing and subsequent Ar[+] ion milling.

2. DESCRIPTION OF THE DISTORTION ANALYSIS METHOD

The method used to measure local displacements on HREM lattice images is similar to that developed by Bierwolf et al. Its successive stages are summed up in Fig. 1.
i) First, the HREM image (Fig. 1.a, here 2 monolayers (ML) of ZnTe embedded in CdTe) is digitized by a CCD camera with a resolution of 768 by 512 pixels. On this image, which is obtained in the <110> orientation, each dot represents an atom-pair, and therefore no information is available on the Cd-Te (or Zn-Te) bonds.
ii) Secondly, the positions of all the atom-pairs are determined very accurately. This is a key point of the process. As the displacements typically correspond to a few tenths of a pixel, we need a precision on the position better than one pixel. To achieve that, we do not use a Fourier filtering technique and a contrast enhancement algorithm as done by Bierwolf et al, but we determine the position of each dot representing an

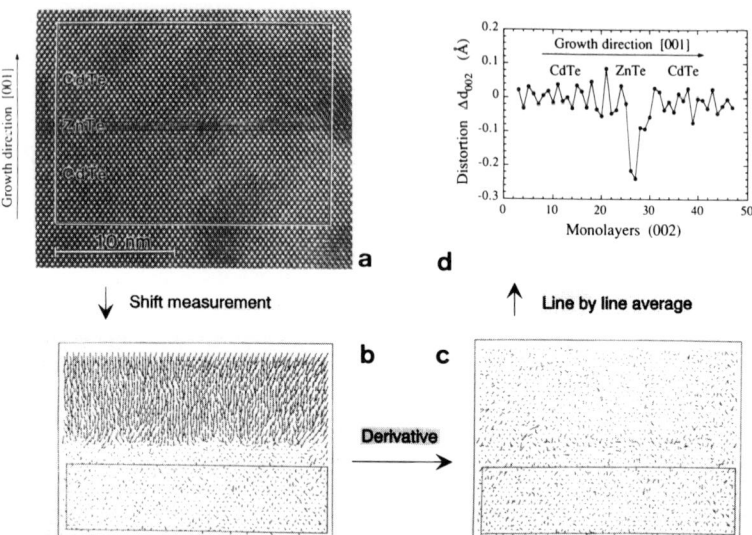

Fig. 1 The four stages of the distortion analysis method. See text for explanation.

atom-pair from a centre of mass calculation on the intensity profile. This calculation is done many times recursively to achieve a better resolution.

iii) A reference area is chosen in the undisturbed CdTe region, marked by a rectangle in Fig. 1.b. From this area, we compute an ideal lattice (that is to say 6 parameters: an origin and two vectors) thanks to a least square method on the position of each dot representing an atom-pair. This ideal lattice is extrapolated on the whole image.

iv) Then we measure the displacements between each point of the ideal lattice and each experimental atom-pair position. These displacements are represented by arrows, with a magnification of 10, on Fig. 1.b. On such a map, some noise is observed in the reference area, giving an indication on the accuracy of our measurements. On the rest of the image, the 2 ML ZnTe strained layer induce a rigid body translation of the lattice, which reflects the total amount of Zn and the strain state.

v) To obtain information on the local distortions, we calculate the "derivative" of the measured displacements: for each dot, we determine the difference between the displacements at that point and the mean displacement at two neighbouring points in the [001] direction (Fig. 1.c). To improve the signal/noise ratio, the result is plotted on Fig. 1.d after averaging along each line of the image. It still can be improved by repeating the analysis on several images of the same layer.

To sum up, the analysis method leads to a measurement of local distortions between two (002) atomic planes directly from HREM images. In order to check the validity of the method, we have tested it on calculated HREM images (using the EMS software of Stadelmann (1987)), and then applied it to study experimental HREM images of ultrathin ZnTe and MnTe layers.

3. STUDY OF ULTRATHIN ZnTe LAYERS EMBEDDED IN CdTe

The growth of 2D pseudomorphic planes of ZnTe inserted in CdTe/CdZnTe superlattices was recently shown to be a very powerful tool to probe the optical properties of such heterostructures (Pelekanos et al 1993). This justifies our study by HREM and HRXRD (High Resolution X-ray Diffraction) on the structural quality of CdTe/ZnTe superlattices with ultrathin ZnTe layers.

Three samples have been grown by MBE on $Cd_xZn_{1-x}Te$ substrates (with x=2.7%), at a temperature of 280°C. They are respectively referred to as sample 27/1, 57/2 and 88/3, the two numbers corresponding to the nominal number of CdTe monolayers and of ZnTe monolayers per period. To ensure pseudomorphic growth (i.e. without strain relaxation and misfit dislocation), two criteria must be fulfilled: i) Each individual layer must be thinner than its own critical thickness; for ZnTe/CdTe, this thickness is only 5 monolayers, for $CdTe/Cd_{0.97}Zn_{0.03}Te$, it is about 1200 monolayers (Jouneau et al 1991). ii) The total superlattice thickness must be lower than the critical thickness of a layer with the same average lattice parameter. To fulfil this second condition, we have chosen the layer thicknesses in order to obtain a strain symmetrization, i.e. the average lattice parameter of the superlattice along the growth axis is the same as the substrate lattice parameter.

In this way, the average strain of the superlattice is close to zero and the pseudomorphic growth is maintained whatever the total thickness of the superlattice may be (Ponchet at al 1990). Then the CdTe layers are in weak compression, and the ZnTe ones in strong dilatation.

The dark-field image of the sample 57/2 is presented in Fig. 2. The ZnTe layers look quite uniform, without any discontinuity and without interface dislocations, attesting to the pseudomorphic growth. Fig. 3 shows the result of our distortion analysis on the three samples. Even though the distortions are mainly localised on one, two or three monolayers, there is still some residual distortion extending on a few planes on either side of the inserted layers. Furthermore, the maximum distortion observed is lower than the expected one. This undoubtedly indicates the presence of non-abrupt interfaces, resulting from the interfacial roughness and also from some Zn segregation during the growth.

Moreover, if we assume the validity of the elasticity theory for such thin layers, we can quantify from these data the amount of Zn in each (002) plane, and then the total amount of Zn per period in the superlattice. These values, reported in the following table, are compared with the ones given by HRXRD (the complete results will be published in a future paper). It is indeed well established that the zero order diffraction gives a precise value of the average lattice parameter of a superlattice, and then allows a determination of the total amount of Zn per period.

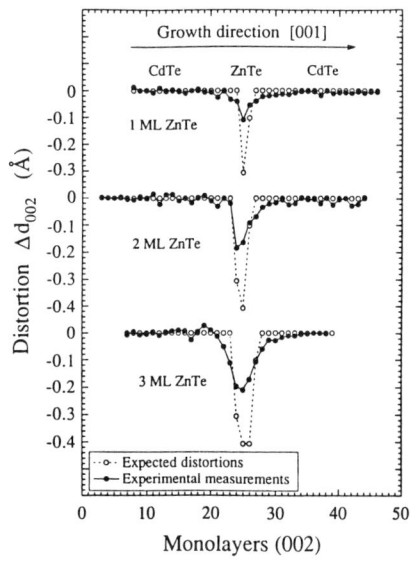

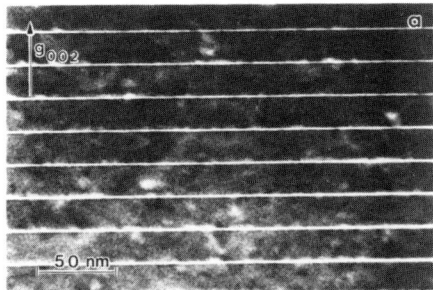

Fig. 2 Dark-field micrograph (g=002) of the 57/2 sample. ZnTe layers appear in white, CdTe in dark.

Fig. 3 Distortion analysis results on HREM images of samples 27/1, 57/2 and 88/3.

Sample n CdTe/m ZnTe	measured period (HREM)	total amount of Zn per period (HREM)	total amount of Zn per period (HRXRD)
27/1	27.5 ± 0.5 ML	0.86 ± 0.05 ML	0.8 ± 0.1 ML
57/2	59 ± 0.5 ML	1.38 ± 0.05 ML	
88/3		2.32 ± 0.05 ML	2.2 ± 0.1 ML

A very good agreement is obtained between the two sets of values, confirming the validity of our method. In all cases, the total amount of Zn inserted during the growth has been found lower than the one expected from the growth rate calibrations. We would like to point out that a unique advantage of such HREM study is to yield the **location** of Zn atoms within a period.

4. STUDY OF ULTRATHIN MnTe LAYERS EMBEDDED IN CdTe

As the lattice mismatch between cubic MnTe and CdTe is smaller than between ZnTe and CdTe, the critical thickness is expected to be thicker, but, to our knowledge, no experimental value can be found in the literature. To determine this value, multiple MnTe layers with progressive thicknesses (from 2 ML to 100 ML) enclosed between thick CdTe layers on a relaxed CdTe buffer layer were grown by MBE. Fig. 4 shows a dark-field TEM micrograph of this sample. Clearly many defects originate inside the 100 ML thick layer, and a few can be detected inside the 50 ML thick layer. None of them are seen in the thinner layers: this gives evidence for a critical thickness for MnTe/CdTe between 40 ML and 50 ML. Plastic relaxation occurs via the formation of stacking faults inside the layers, bounded by partial dislocations lying at the interfaces.

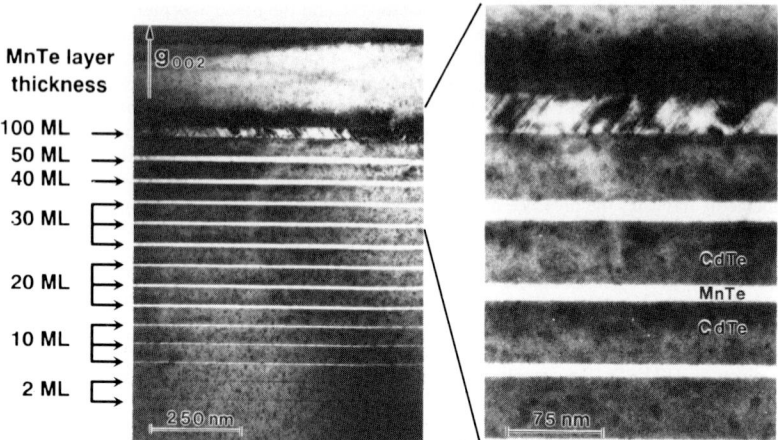

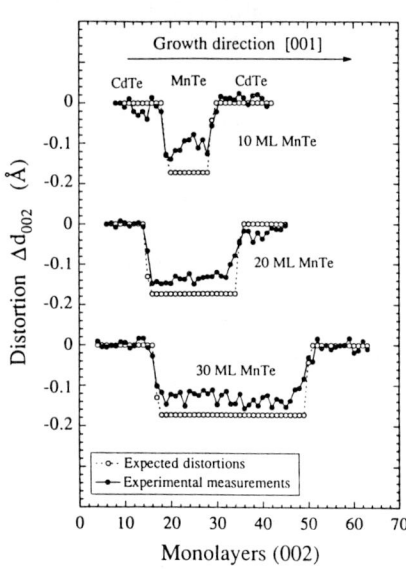

Fig. 4 Dark-field image of a structure with MnTe layers of progressive width (from 2 ML to above the critical thickness) embedded in thick CdTe layers.

Fig. 5 shows the result of the distortion analysis on HREM images on three MnTe layers of this sample, with increasing thicknesses (10, 20 and 30 ML). We notice that the MnTe→CdTe interface is much broader than the CdTe→MnTe one. Moreover, its abruptness decreases with the thickness of the MnTe layer. This is ascribed to a surface diffusion lower for MnTe than for CdTe at the growth temperature (280°C), and we are trying at present to improve the interface quality by controlling growth temperatures and flux ratios. Another important piece of information supplied by the distortion analysis is the average distortion in the MnTe layers which provides a direct measurement of the MnTe strain state. A value lower than the one expected from the elasticity theory is systematically found. It may be due to erroneous values of the elastic coefficients used for cubic MnTe when extrapolated from values for the $Cd_{0.5}Mn_{0.5}Te$ alloy (Maheswaranathan 1985), or to a departure from linearity in the elasticity theory for such important deformations.

Fig. 5 Distortion analysis results on HREM images of pseudomorphic MnTe layers with increasing thicknesses (10, 20 and 30 ML) embedded in CdTe.

REFERENCES

Bierwolf R, Hohenstein M, Phillipp F, Brandt O, Crook G E and Ploog K 1993 Ultramicroscopy 49 273

Cibert J, Bodin D, Le Si Dang, Feuillet G, Jouneau P H, Molva E, Accomo R and Labrunie G 1993 Materials Science and Engineering B16 279

Jouneau P H, Cibert J, Feuillet G, Mallard R E, Saminadayar K, Tatarenko S and Le Si Dang 1991 Inst.Phys.Conf.Ser. No117 623

Maheswaranathan P, Sladek R J and Debska U 1985 Phys.Rev. B31 5212

Ourmazd A, Baumann F H, Bode M and Kim Y 1990 Ultramicroscopy 34 237

Pelekanos N T, Peyla P, Mariette H, Jouneau P H, Tardot A, Magnea N 1993 Accepted in Phys.Rev. B

Ponchet A, Lentz G, Tuffigo H, Magnea N, Mariette H and Gentile P 1990 J.Appl.Phys. 68 6229

Stadelmann P A 1987 Ultramicroscopy 21 131

Thoma S and Cerva H 1991 Ultramicroscopy 38 265

Relaxation phenomena in strained $Si_{1-x}Ge_x$ layers on planar and patterned Si substrates

E Bugiel, P Zaumseil, B Dietrich and H J Osten
Institute of Semiconductor Physics, Walter-Korsing-Straße 2, D-O-1200 Frankfurt (Oder), Germany

ABSTRACT: The electronic properties of $Si_{1-x}Ge_x$ strained layers grown on (001)Si substrates strongly depend on Ge content and strain. The concentration of Ge and the relaxation of partly pseudomorphically grown thin SiGe layers on Si can be found independently by a combination of standard X-ray double-crystal diffractometry (DCD) and transmission electron microscopy (TEM). DCD and TEM determine the lattice constant variations of the lattice planes parallel and perpendicular to the surface. Epitaxial $Si_{1-x}Ge_x$ layers have been also grown on mesa-like structures. The film growth, the crystallographic perfection, the relaxation mechanism and the control of dislocations on top of the mesas are discussed.

1. INTRODUCTION

The determination of the Ge fraction x in the highly mismatched SiGe/Si system is more difficult than in systems with small lattice mismatch , e.g., $Ga_{1-x}Al_xAs$ on GaAs, where strain and x value are proportional to each other and the X-ray diffraction (XRD) method developed by Hornstra and Bartels (1978) can be used. Especially for SiGe layers with a thickness above the "critical thickness", predicted by the model of Matthews and Blakeslee (1974), the generation of misfit dislocations during the epitaxial growth process has to be taken into acount leading to an only partly pseudomorphically grown SiGe layer. Thus, the simultaneous determination of the composition requires the measurement of the lattice relaxation caused by the dislocations, and consequently the measurement of both lattice parameters of the epilayer parallel $a_\parallel$ and perpendicular $a_\perp$ to the substrate surface (see Fig.1). XRD in the symmetrical Bragg geometry directly provides the lattice constant of the SiGe layer $a_\perp$ perpendicular to the surface. TEM generally allows the determination of the in-plane strain represented by $a_\parallel$ in two different ways, namely from the distance of misfit dislocations or of moire fringes. This combination of XRD and TEM has some advantages in comparison to a characterisation by X-ray diffraction only. The determination of $a_\parallel$ by TEM is independent of the layer thickness, while with XRD the width of the SiGe peak increases with decreasing thickness and increasing dislocation density. Moreover, taking into account that an asymmetric X-ray reflection (Fatemi and Stahlbush 1991) detects only the projection of $a_\parallel$ in the direction of the reflection vector, TEM seems to be more accurate. Furthermore, TEM offers more direct information about the structural defects present in the layer.

The influences of structured Si substrates or structured films on Si on the growth, strain and defect distribution of such layers have been investigated by Fitzgerald et al (1990), Prokes et al (1992), and Hull et al (1992). A lot of results have been published for III-V compound semiconductor growth on patterned substrates (Grundmann et al 1992, Colas et al 1992, and Mirin et al 1992). If the strained epitaxial layer is grown on large-size patterned substrates, then the movement of threading dislocations stops at the edges of the epitaxial layer. Therefore it is possible to reduce the dislocation density drastically by using patterned substrates compared with the dislocation density in a layer on an unpatterned substrate with the same density of seeds for dislocation generation. This geometrical effect was described for example by Fitzgerald et al (1990), and Hull et al (1992). From in situ TEM of SiGe films on planar substrates we expect (for a 100 nm thick film containing 20% Ge and annealed at 600 °C) an average dislocation spacing of the order of 100 nm, in aggreement with other authors (Hull et al 1989). Therefore it would be interesting to look for relaxation on narrow mesa stripes with widths in the range of several hundred nanometers to a few micrometers. A pseudomorphic SiGe film on top of such a narrow mesa stripe is not only compressed perpendicular to the stripe, but also the top of the Si mesa is dilated. Therefore, there should be a mesa with such a width, that only one dislocation can move on the mesa stripe parallel to the longer edges. A second dislocation is not able to move, because the residual strain is too small.

2. EXPERIMENTAL

The samples 1,2 and 3 with low, medium and high Ge content were prepared using a three chamber MBE equipment, without sample rotation. After in situ cleaning of the Si(001) substrate Si was deposited from an electron beam evaporator with a growth rate of 3 nm/min and Ge from an effusion cell. We deposited Si first, followed by $Si_{1-x}Ge_x$ (calibrated for the (001) surface).

X-ray measurements were carried out with a double crystal diffractometer in symmetrical Bragg case geometry using CuK_α radiation and 004 reflection.

The structure and defect distribution of the films were investigated by TEM and HREM at 300 kV on plan-view as well as cross-sectional samples. A single tilt heating holder was used for in situ annealing experiments in order to observe relaxation processes.

The substrates were patterned by anisotropic wet chemical etching in KOH (see Hashimoto et al 1989) and by a short isotropic etching. We obtained 1.4 μm high mesa stripes with widths at the top of 0.2μm up to several microns (see also Bugiel et al 1993).

3. RESULTS AND DISCUSSION

3.1. Growth on planar substrates

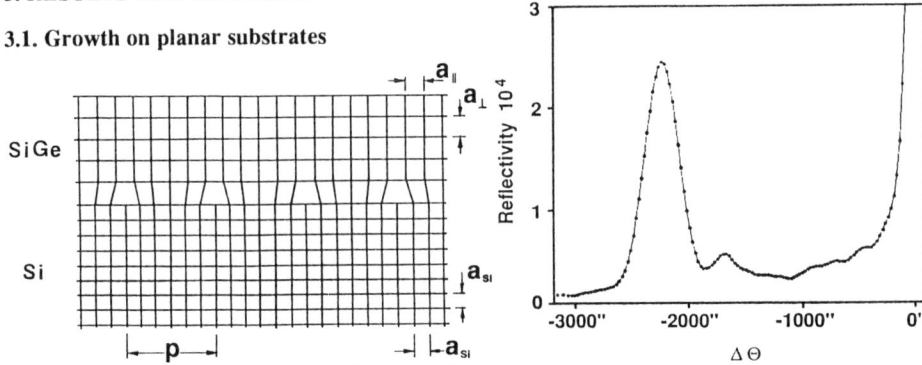

Fig.1: Scheme of a partly relaxed SiGe layer. The lattice constants and the average distance of misfit dislocations are defined.

Fig.2: Double-crystal diffractometer of sample 1, CuK_α radiation, 004 reflection

Fig. 2 shows the rocking curve of sample 1. It is directly related to the lattice constant perpendicular to the surface $a_\perp$ by

$$\Delta\Theta_{SiGe} = -\tan\Theta_B \frac{a_\perp - a_{Si}}{a_{Si}} , \quad (1)$$

were Θ_B is the Bragg angle of the used reflection. The lattice constant $a_\perp$ is related to the lattice of the Si substrate in the two limiting cases of an alloy like (cubic, relaxation factor R=1) and a pseudomorphic (tetragonal distorted, R=0) lattice, respectively, according to the linear interpolation approximation as follows
$a_\perp = a_{Si} \cdot (1 + 0.0418 \cdot x)$ (R=1) and $a_\perp = a_{Si} \cdot (1 + k\, 0.0418 \cdot x)$ (R=0), where 0.0418 is the misfit between the pure cubic Ge and Si lattice ($0.0418 = (a_{Ge} - a_{Si})/a_{Si}$),and k describes the tetragonal distortion of the biaxial strained SiGe layer, by which $k = 1 + 2c_{12}/c_{11}$ for 001 surface orientation, c_{ij} are elastic constants. Since k is nearly equal for Si and Ge the interpolated value of k = 1.72 can be assumed for any alloy. The relaxation factor for a partly relaxed SiGe layer with a given Ge content x can be derived as

$$R = \frac{1}{k-1} \cdot \left(k - \frac{\Delta a_\perp}{a_{Si}\, 0.0418 \cdot x} \right) \quad (2)$$

with $\Delta a_\perp = a_\perp - a_{Si}$.

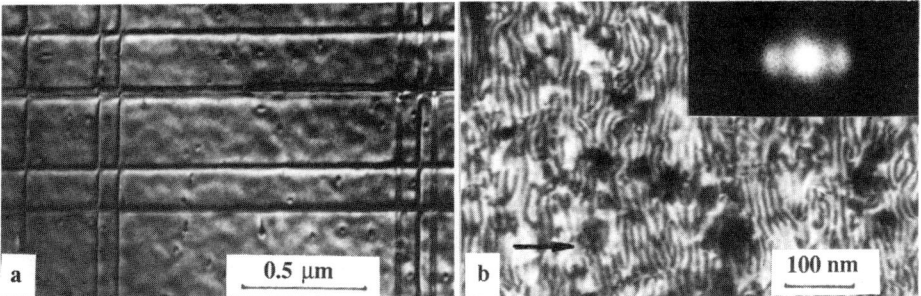

Fig.3: Plan-view TEM micrographs of two SiGe layers with a) low (sample 1) and b) higher Ge content (sample 2), resp.

Fig. 3 presents TEM micrographs of two samples in plan-view. Sample 1 with low Ge content shows individual dislocations (Fig. 3a). At higher x values and obviosly higher relaxation factors moire fringes can be observed (Fig. 3b).In the first case (low x), the SiGe lattice constant parallel to the surface $a_\parallel$ can be expressed by the average distance p between the misfit dislocations measured in the TEM image :

$$a_\parallel = a_{Si} (1 + b_{eff}/p) \qquad (3)$$

b_{eff} is the absolute value of the component of the Burgers vector acting as relaxation. Most of the misfit dislocations with $\vec{b} = a(110)/2$ are of the 60°-type: $b_{eff} = 2a_{Si}/4$ (see e.g. Rajan et al 1987) In the case of SiGe films with higher Ge content the average distance D_{hkl} of moire fringes (see LeGoues et al 1991) was measured by optical diffraction (insert in Fig. 3b). D_{hkl} is given by the corresponding netplane distances of the substrate and the layer :

$$D_{hkl} = \frac{1}{1/d_{Si} - 1/d_\parallel} \quad \text{with} \quad d = \frac{a}{h^2 + k^2 + l^2}$$

The lattice constant a can be calculated by

$$a_\parallel = \frac{a_{Si} D_{hkl} \sqrt{h^2 + k^2 + l^2}}{D_{hkl} \sqrt{h^2 + k^2 + l^2} - a_{Si}} \qquad (4)$$

Similar to the lattice parameter perpendicular to the surface two limiting cases can be discussed : $a_\parallel = a_{Si} \cdot (1 + 0.0418 \cdot x)$ (R = 1)
$$a_\parallel = a_{Si} \qquad (R = 0)$$
In the second case no dislocations $(p \rightarrow \infty)$ and no moire fringes $(D \rightarrow \infty)$ can be observed by TEM. For a partly relaxed SiGe layer the relaxation factor is given by

$$R = \frac{\Delta a_\parallel}{a_{Si} \cdot 0.0418 \cdot x} \qquad (5)$$

with $\Delta a_\parallel = a_\parallel - a_{Si}$.

The combination of equations (2) and (5) allows the determination of x and R :

$$x = ((k-1)\Delta a_\parallel + \Delta a_\perp)/(k \, a_{Si} \, 0.0418) \qquad (6a)$$

$$R = k \, \Delta a_\parallel /((k-1) \, \Delta a_\parallel + \Delta a_\perp) \qquad (6b)$$

Fig. 4 illustrates the curves (Eq. 2 and 5) for fixed values of $\Delta a_\parallel$ and $\Delta a_\perp$ in a x-R-matrix. x and R can be determined from the cross points.

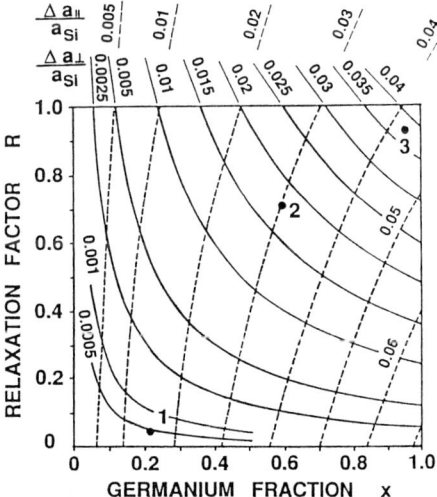

Fig. 4:Curves of fixed values of relative lattice costant variations $\Delta a_\perp$ (Eq.2) and $\Delta a_\parallel$ (Eq.5). The results of sample 1,2 and 3 are given.

3.2 Growth on patterned substrate

Figure 5 shows TEM cross-sectional (Fig.5 a) and plan-view (Fig. 5b) micrographs of a sample grown on a patterned substrate (see Bugiel et al 1993). The mesa stripe shown in Fig. 5 has a width of about 1 μm. Only one dislocation parallel to the edges could be found. This situation does not change during a heat treatment up to 900 °C. For stripes with a width smaller than 0.7 μm no parallel dislocations could be found. A heat treatment at 900 °C only increases the number of dislocation perpendicular to the stripe direction but does not create any parallel dislocation. Wider mesa structures show a complete network of dislocations. The degree of relaxation on extended (001) areas increases from R=0.12 up to R=0.18 during heat treatment.

After heat treatment we observe relaxation effects. A part of the misfit dislocations between the mesas is generated from sources on top of the mesas. One source on top of a mesa punched out dislocation half loops moving down on one or two different {111}glide plane.

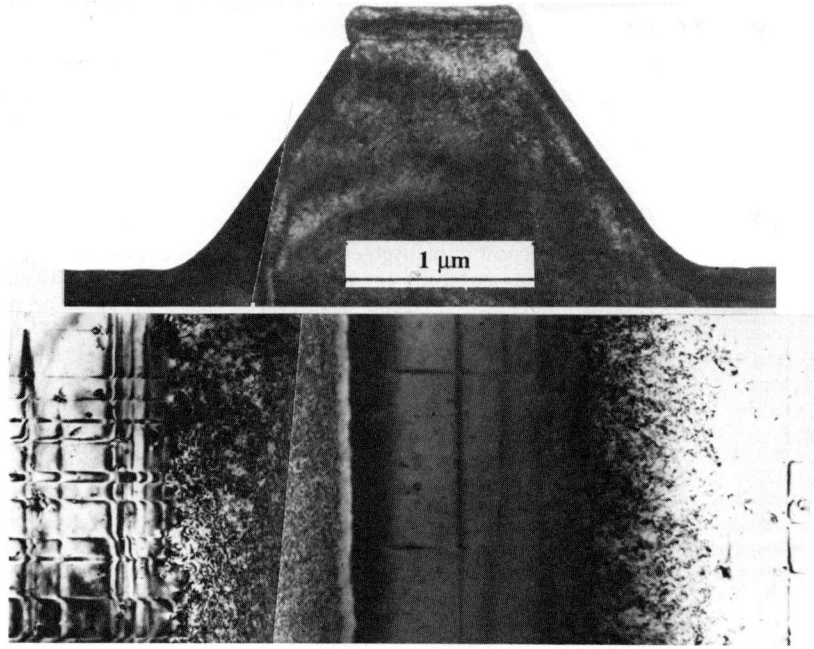

Fig.5: Cross-sectional and plan-view TEM micrographs of a mesa stripe with a width of 1 μm. Only one dislocation parallel to the edge was generated on top of the mesa.

REFERENCES

Bugiel E, Dietrich B and Osten H J 1993 J. Cryst. Growth in press
Colas E, Nihous G C and Hwang D H 1992 J. Vac. Sci. Techn. A 10 691
Fatemi Mand Stahlbush R E 1974 Appl. Phys. Lett. 58 825
Fitzgerald E A, Xie Y-H, Basen D, Green M L, Michel J, Freeland P E and Weir B E 1990 J. Electron. Mater. 19 949
Grundmann M, Krost A, Bimberg D, Ehrmann O and Cerva H 1992 Appl. Phys. Lett. 60 3292
Hornstra J and Bartels W J 1978 J. Cryst. Growth 44 513
Hull R, Bean J C, Eaglesham D J, Bonar J M and Buescher C 1989 Thin Solid Films 183 117
Hull R, Bean J C, Higashi G S, Green M L, Peticolas L, Bahnck D and Brasen D 1992 Appl. Phys. Lett. 60 1468
LeGoues F K, Horn-Von Hoegen M, Copel M and Tromp R M 1991 Phys. Rev. B44 12894
Matthews J W, and A.E. Blakeslee A E 1974 J. Cryst. Growth 27 118
Mirin R P, Tan I-H, Weman H, Leonard M, Yasuda T, Bowers J E and Hu E L 1992 J. Vac. Sci. Techn. A 10 697
Prokes S M and Rai A K 1992 Appl. Phys. Lett. 60 568
Rajan K and Dendhoff M 1987 J. Appl. Phys. 62 1710

Thermal stress-induced defects in GeSi/Si heterostructures

B Roos and F Ernst

Max-Planck-Institut für Metallforschung, Institut für Werkstoffwissenschaft,
Seestraße 92, 7000 Stuttgart 1, Germany

ABSTRACT: Defects generated by plastic deformation under thermal stresses are studied in GeSi layers grown epitaxially on {111}-Si substrates. TEM observations after different annealing treatments lead to a model for the formation of thermal stress-induced dislocation networks.

1. INTRODUCTION

Thick epitaxial GeSi layers on Si substrates are of growing interest for opto-electronic devices. Besides the lattice misfit between GeSi and Si that needs to be accommodated during layer growth at elevated temperatures, GeSi/Si heterostructures also suffer from a "thermal mismatch". Without plastic deformation the difference in thermal contraction during cooling to room temperature would result in biaxial stresses of hundreds of MPa. Thus, plastic deformation is expected to relax the thermal stresses, thereby generating crystal defects.

So far, little systematic work has been dedicated to clarify the nature of defects originating from thermal stresses in GeSi/Si heterostructures. In the present paper we have systematically studied the formation of such defects in GeSi/Si layers under different annealing treatments.

$Ge_{0.9}Si_{0.1}$ layers with a thickness of 1.2 μm were grown on {111}-Si substrates by liquid phase epitaxy (LPE) at 1093 K. The deposition of GeSi from a eutectic Bi-Ge-Si-solution was initiated by a small supercooling of 0.1 K. Subsequently, layer growth was conducted at a constant cooling rate of less than 2 K/h. After growth the layers were cooled to room temperature. The cooling rate was varied between ≈ 100 K/min and 0.25 K/min. Some samples were exposed to thermal cyclic annealing (TCA) between room temperature and 1073 K.

2. PLASTIC DEFORMATION UNDER THERMAL STRESS

The biaxial thermal stresses introduced during cooling are tensile in the GeSi layer and compressive in the Si substrate. These stresses cause a tetragonal elastic distortion of the GeSi layer (Hinckley and Singh 1990). The elastic distortion $\varepsilon^{\parallel}$ of the GeSi layer parallel to the GeSi/Si interface was measured in a double-crystal X-ray diffractometer. The method is described in the literature (Krishnamoorthy et al 1992).

For all samples grown at 1093 K we measured $\varepsilon^{\parallel} = (0.0018 \pm 0.0001)$. Within the experimental error $\varepsilon^{\parallel}$ is independent of the cooling rate and the number of annealing cycles. Without plastic deformation the thermal elastic distortion would amount to

$$\Delta = \int_{RT}^{1093\,K} [\, \alpha_{GeSi}(T) - \alpha_{Si}(T) \,]\; dT = 0.0024, \tag{1}$$

where α_{GeSi} and α_{Si} denote the thermal expansion coefficients of $Ge_{0.9}Si_{0.1}$ and Si, respectively (Touloukian et al 1975). After crystal growth the epitaxial layer is left with a residual compressive elastic strain of $\eta^{\parallel} = 0.0003$ (Roos and Ernst). Taking the latter into account, the difference of $\Delta - \varepsilon^{\parallel} - \eta^{\parallel} = (0.0003 \pm 0.0001)$ between the theoretical and the measured strain suggests that about 1/8 of the thermal stress is relaxed by plastic deformation.

3. TEM STUDY

3.1 Dislocation networks accommodating misfit during layer growth

Weak beam dark field images revealed that two different dislocation networks form in the {111}-GeSi/Si interface to accommodate the lattice mismatch of 3.7 % at the growth temperature. Both these networks consist of edge-type misfit *partial* dislocations with <110> line directions and $a/6$<211> Burgers vectors. In one network the partials are straight, while in the other they form a honeycomb pattern. Details of these coexisting primary misfit dislocation networks are described elsewhere (Ernst et al 1992).

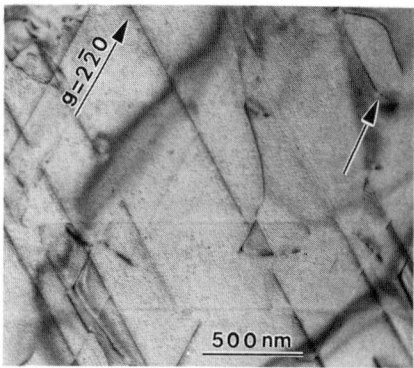

Fig. 1: Trigonal pseudo network

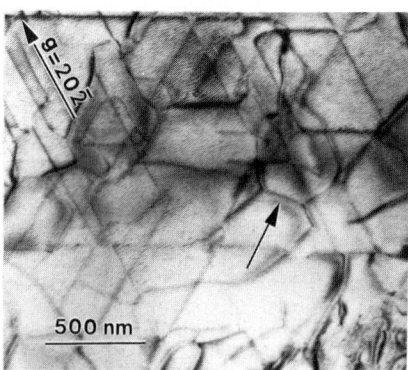

Fig. 2: Hexagonal network (fragments)

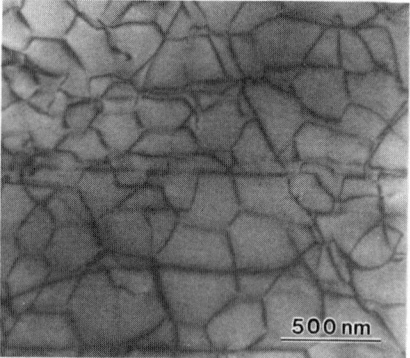

Fig. 3: Hexagonal network (developed)

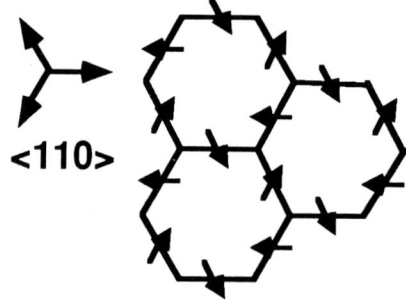

Fig. 4: Hexagonal network (idealized)

3.2 Thermal stress-induced dislocations

3.2.1 As-grown samples

Fig. 1 is a bright-field image of a plan-view sample of an as-grown layer (cooled once), recorded under a two-beam condition with $\mathbf{g}$ = (2̄20). The fine periodic pattern results from the superposition of a moiré contrast with the contrast of the primary misfit dislocation network. According to the lattice mismatch of 3.7 %, both the moiré pattern and the primary dislocation network have a spacing of ≈6 nm along <110> directions. Besides some irregular dislocations fig. 1 shows a coarse-meshed trigonal configuration of straight dislocations along <110> directions. The horizontal set of dislocation lines merely exhibits residual contrast because $\mathbf{g} \cdot \mathbf{b} = 0$, but $\mathbf{g} \cdot \mathbf{b} \times \mathbf{u} \neq 0$, where $\mathbf{u}$ denotes the line direction. TEM of cross-sectional foils revealed that these dislocations lie about 10 nm above the GeSi/{111}Si interface plane. Fig. 1 suggests that the dislocations intersect in nodes. Stereo microscopy, however, revealed that they usually do *not* intersect with each other but lie in different {111} planes parallel to the interface. We will refer to this configuration as a "pseudo"-network in the following.

Stereo micrographs also proved that the straight dislocation lines actually belong to dislocation half-loops. The marker in fig. 1 points at one of the threading dislocation segments connecting the extended straight segments with the surface of the GeSi layer. The character of the dislocations in fig. 1 was determined by $\mathbf{g} \cdot \mathbf{b}$ analysis of two-beam images obtained with <220> and <224> diffraction vectors parallel to the {111} GeSi/Si interface. The dislocations possess $a/2$ <110> Burgers vectors that are *inclined* against the GeSi/Si interface and make an angle of 60° with the dislocation line. Thus, these dislocations are glissile on {111} planes inclined against the {111} plane of the GeSi/Si interface. Micrographs of cross-sectional specimens cut parallel to one of the {111} glide planes inclined against the GeSi/Si interface show that the threading arms also lie in these {111} glide planes. Stereo images of plan-view specimens show entire dislocation half-loops at the surface of the GeSi layer.

The parallel segments in the coarse-meshed trigonal pseudo-network possess an average spacing of $S = (500\pm100)$ nm. Correspondingly, the edge components of their Burgers vectors parallel to the interface relax a lateral elastic distortion of $\delta^{\parallel} = (0.00035 \pm 0.00008)$, provided that the sign of the Burgers vectors is always such that the dislocations increase the lateral dilatation of the GeSi layer. The sign was determined from the contrast behaviour of the dislocation image across an extinction contour (Howie and Whelan 1962). In three of three spot checks the Burgers vectors proved to possess the correct sign. The figure for $\delta^{\parallel}$ corresponds to 1/7 of the plastic deformation required for complete relaxation of the thermal stresses. This result agrees well with the X-ray data and implies that 1/7 of the thermal stresses is relaxed, while 6/7 are frozen-in. Varying the cooling rate did not have a pronounced effect on the dislocation density.

Besides the trigonal pseudo-network, the as-grown layer occasionally exhibits fragments of a coarse-meshed hexagonal dislocation network, which consists of dislocations in <110> directions. Fig. 2 includes fragments of this configuration (arrowed), along with dislocations of the trigonal pseudo-network. Stereo microscopy indicated that the configuration in fig. 2 is a *real* network in the sense that the dislocations intersect in nodes.

According to diffraction contrast experiments the hexagonal network consists of complete 60° dislocations with $a/2$<110> Burgers vectors *parallel* to the {111}-GeSi/Si interface. The segments of the network are mostly parallel to the interface. Some segments, however, have also been observed on inclined {111} planes, yielding a wavy topology of the network. Fig. 4 presents the idealized geometry of the hexagonal network, with arrows indicating the Burgers vectors. The spacing of parallel segments in the hexagonal network is approximately equal to that in the trigonal network, i.e. about 500 nm. A quantitative estimate yields that about 2/7 of the thermal stress is relaxed in those regions covered by the hexagonal dislocation network.

3.2.2 Layers exposed to post-growth thermal cyclic annealing

Fig. 3 shows a plan-view TEM image of a GeSi layer after three cooling cycles. The hexagonal network already encountered in fig. 2 is fully developed here. Compared to fig. 2 the average spacing of the parallel segments is reduced by a factor of $\approx 2/3$, implying that 3/7 of the thermal stresses are relaxed. The hexagonal network, however, only covers less than 1/3 of the total interfacial area. In these regions it almost completely replaced the trigonal pseudo-network. The absence of a moiré pattern from fig. 3 indicates that the region contained in the TEM foil as well as the hexagonal dislocation network lie notably above the GeSi/Si interface. Cross-sectional specimens revealed a distance of 50 - 100 nm between the network and the GeSi/Si interface.

4. DISCUSSION

The experimental observations lead to the following conclusions. The thermal stresses developing on cooling after growth relax by the formation of dislocation half-loops at the surface of the GeSi layer. These half-loops glide down to the GeSi/Si interface, where they form the trigonal pseudo-network of 60° dislocations with $a/2<110>$ Burgers vectors inclined against the interface. On the average, only 1/7 of the thermal stresses are relaxed in this way. Relaxation only starts when the thermal stresses exceed a critical value, which is reached about 200 K below the growth temperature. From the measured plastic deformation it can be calculated that dislocation activity freezes at a temperature in the range of the brittle-ductile transition temperature (Roos and Ernst), which lies at ≈ 823 K in $Ge_{0.9}Si_{0.1}$ (Alexander 1961). Further relaxation does not occur, albeit this would be energetically favourable. At a given elastic strain ε, stress relaxation by plastic deformation is favourable if the GeSi layer exceeds a critical thickness $h_c(\varepsilon)$. Applying the theory of Matthews et al (1976) for the residual elastic distortion of $\varepsilon^{\parallel} = 0.0018$ we obtain $h_c(\varepsilon^{\parallel}) < 0.2$ µm, while the actual layer thickness is much larger (1.2 µm). Therefore, the incomplete stress relaxation must be explained by kinetics rather than energetics.

The observed dislocation density fully accounts for the plastic deformation expected from the X-ray diffractometry. Microtwin lamellae parallel to the {111}-GeSi/Si interface , which have been proposed to form under thermal stresses (Albrecht and Strunk 1991) have not been observed here.

Thermal cycling promotes the transformation of the trigonal pseudo-network into a hexagonal network. The 60° dislocations of the latter have their $a/2<110>$ Burgers vectors parallel to the GeSi/Si interface and thus are more efficient in relaxing thermal stresses. When relaxing a given thermal stress the total line length and thus the total line energy of the hexagonal network is three times lower than for the trigonal pseudo-network. Since the dislocations of the hexagonal network can only glide *parallel* to the interface they cannot form directly by half-loop nucleation at the layer surface but must result from reactions between the dislocations of the trigonal pseudo-network. A possible reaction, using Thompson's tetrahedron with the plane ABC parallel to the {111} GeSi/Si interface, is:

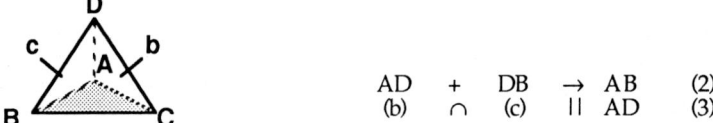

$$AD \quad + \quad DB \quad \rightarrow \quad AB \qquad (2)$$
$$(b) \quad \cap \quad (c) \quad \parallel \quad AD \qquad (3)$$

According to Hirth (1961) this reaction is energetically favourable. The planes specified in (3) are the slip planes of the reacting dislocations. While (2) yields the correct Burgers vector the line direction of the product dislocation is inclined to the interface. As mentioned above, such inclined dislocation segments have been experimentally observed. In order to form the more frequently observed segments parallel to the {111}-GeSi/Si interface the inclined segments may re-orient themselves by cross slip on a {111} plane parallel to the interface, or by climb.

ACKNOWLEDGEMENT

The authors would like to thank E.Bauser for her assistance in growing the GeSi layers,R.Dieter for the X-ray diffractometry, and R.Hull and K.Rajan for helpful discussions.

REFERENCES

Albrecht M and Strunk H P 1991 Polycryst. Semiconductors II (Berlin: Springer) pp 503
Alexander H 1961 Z. Metallk. 52 344
Ernst F, Pirouz P and Bauser E 1992 phys. stat. sol. (a) 131 651
Hinckley J M and Singh J 1990 Phys. Rev. B 42 3546
Hirth J P 1961 J. Appl. Phys. 32 700
Howie A and Whelan M J 1962 Proc. Roy. Soc. A267 206
Krishnamoorthy V, Lin Y W, Calhoun L, Liu H L , Park R M 1992 Appl.Phys.Lett. 61 2680
Matthews J W, Blakeslee A E and Mader S 1976 Thin Solid Films 33 253
Roos B and Ernst F, to be published
Touloukian Y S, Kirby R K, Taylor R E and Lee T Y R 1975 Thermal expansion,
    nonmetallic solids (New York - Washington: IFI/Plenum)

Electron microscopy of epitaxial structures

N A Kiselev, A L Vasiliev, O I Lebedev, E I Givargizov, A N Stepanova, A N Kiselev and J L Hutchison[1]

Institute of Crystallography, Russian Academy of Sciences, 59 Leninsky pr, Moscow 117333, Russia
[1]Department of Materials, University of Oxford, Parks Road, Oxford OX1 3PH

ABSTRACT: This paper describes recent applications of SEM and HREM to the study of epitaxial structures, carried out mainly at the Institute of Crystallography, Russian Academy of Sciences. Areas of importance are: semiconductor epitaxial layers and interfaces, high-T_c superconductor epitaxial layers on different substrates, edge-type Josephson junctions with PrBaCuO barrier layers, and nanometric tips prepared from epitaxially-grown silicon whiskers using the one-stage and two-stage techniques.

1. INTRODUCTION

The effect of epitaxy, when oriented over-growth of one crystal on another occurs, is widely used in modern technology. High resolution electron microscopy (HREM) of epitaxial layers and interfaces is closely related to the development of semiconducting and superconducting materials physics. Fine silicon tips are used for field-emission studies and are a new tool for nanostructural physics and technology. SEM and HREM are playing an important part in these developments.

2. EPITAXIAL LAYERS AND INTERFACES

The following epitaxial structures are being investigated at the Institute of Crystallography in collaboration with the Institute of Semiconductor Physics (Novosibirsk) and the Physical Technological Institute (Moscow) of the RAS: InAs/(001)GaAs (Karasev et al 1991), CaF2/(001)Si (A Kiselev et al 1993), TiSi2/(111)Si and (100)Si (Vasiliev et al 1991) and some others. Studies were carried out on a Philips EM 430 ST at 200 kV and 300 kV.

InAs/(001)GaAs is an example of an interface with similar structure and large lattice mismatch (7%). Two types of MBE grown InAs/(001)GaAs specimens were investigated: strained layer superlattices (SLS) and specimens with varying InAs layer thickness (0.6-0.9 nm for the first layer; 1.2 nm for the second and the third one; 1.8 nm for the fourth and the fifth layers and 10 nm for the sixth one).

Cross-sectional (110) images of SLS and the first InAs layer of the heterosystem show that the interface is coherent, i.e. does not contain MDs. Thus, pseudomorphic layers are observed. The lattice mismatch is compensated by distortions of the layers lattice: it is extended in the growth direction and compressed in the (001) plane, resulting in a relative rotation of the InAs and GaAs {111} plane by a kink angle $\psi k\{111\}=3°$.

There are two ways to minimize the system energy, the strain energy included. The first is the elastic strain relaxation with creation of MDs, keeping the surface energy constant. The second is collapse of the film into islands in order to minimise the surface energy. These two mechanisms may operate simultaneously.

At h>0.9 nm (second layer) the heteroepitaxial strain is relaxed by MDs generation and the initially continuous layer is transformed into an island one. Small islands are visible at the interface.   One of the islands (Fig. 1a) is free from dislocations.   The lattice mismatch is compensated by very small distortion of the island lattice.   On the other island (Fig. 1b) extra half-planes in the (111) and (001) families are arrowed.   Most probably this is a 60° dislocation with the Burgers vector $b_0 = 1/2\ a[110]$.   In the left part of the island {111} atomic plane kink is observed.   So in this island the InAs/GaAs lattice mismatch is accommodated partly by lattice deformations and partly by single dislocation.   The fifth layer of the specimen demonstrates an increase of the island dimensions as well as the number of 60° dislocations on the upper and lower interfaces.

At h=10 nm (sixth layer) the islands coalesce, the layer is continuous and contains a quasi-periodic MD network at the interface.   Two dislocation types are observed: Lomer dislocations and two closely situated 60° dislocations.   The presence of two dislocation types gives evidence that Lomer dislocations in the InAs/GaAs system are formed as the result of two 60° dislocation interaction.

Titanium disilicide films are used in electron device technology.   The $TiSi_2$ film has an orthorhombic lattice and may give large mismatch with the Si substrate. $TiSi_2$/Si is a reactive solid/solid interface.   The $TiSi_2$/Si system was investigated by HREM (see for example, Catana et al 1990).   It was shown that (101) $TiSi_2$ is a preferential plane for epitaxial growth on Si (111).

Vasiliev et al (1991) used electron-beam evaporation from Ti and Si targets onto a heated substrate.   Polycrystalline film was obtained with the following epitaxial relations $(202)TiSi_2//(111)Si$, $[010]TiSi_2//[110]Si$, which are the same as described by Catana et al (1990).   In cross-sectional micrographs of $TiSi_2$/Si(111) contrast variations with a mean spacing of 2.8-3 nm caused by elastic strain were observed along the interface.

Superposition of "surface lattices" of (111)Si and (101)$TiSi_2$ $([010]TiSi_2//[110]Si)$ shows that in (110) cross-sections the mismatch along the interface is 11.9%.   In cross-sectional micrographs (Fig. 2) it is possible to observe regions, where a few (202)-$TiSi_2$ and (111)Si planes fit rather well.   These regions are separated by 9 or 10 (202)$TiSi_2$ planes and by wedge-shaped defects formed by extra planes in $TiSi_2$.   Near the wedge-shaped defects the lattice in $TiSi_2$ is distorted.   This distortion most probably is the nature of periodic strain observed along the interface.   Sometimes inclined lattice planes are visible.   All this gives evidence that the misfit compensation at this interface is complex.

3.   MICROSTRUCTURE OF EDGE-TYPE JOSEPHSON JUNCTIONS (EJJ)

In collaboration with the Institute of Applied Physics, RAS YBCO films, obtained by laser ablation on different substrates were investigated (Vasiliev et al 1990).   It was shown that films on $SrTiO_3$ (100) are single crystals.   For such films, the c-axis is oriented normal to the interface and the critical current density (Jc) reaches 7.5 x 10^6 A/cm^2.

The lattice mismatch for $YBa_2Cu_3O_{7-x}$ (orthorhombic phase)/$SrTiO_3$ at room temperature in different directions is: fa=2.07%; fb=0.33% and fc/3=0.07%.   The system could be related to materials with different structure and small mismatch.   At the deposition temperature fc/3>fab.   This is one of the possible explanations, why the c-axis is oriented normal to the interface.

An EJJ with $PrBa_2Cu_3O_{7-x}$ barrier layer is schematically shown in Fig. 3a (Lebedev et al 1992): Y1 is the YBCO superconducting layer, PI is the $PrBa_2Cu_3O_{7-x}$ insulating layer, PB is the $PrBa_2Cu_3O_{7-x}$ barrier layer and Y2 is the YBCO superconducting layer.   All layers were obtained by laser ablation on a $SrTiO_3$ substrate.   So the EJJ is a multilayered system of epitaxially grown superconducting and non-superconducting, i.e. insulating, films, properties of which are strongly dependent on stoichiometry, deposition conditions and state of the underlying film.

The structure was formed in two stages.   In the first stage Y1 and PI layers were deposited in one vacuum cycle.   Either a $PrBa_2Cu_3O_{7-x}$ or a non-superconducting $YBa_2Cu_3O_{7-x}$ layer, deposited at 500°C temperature served as an insulator.   The edges were formed by ion sputtering (Ar$^+$, 2 KeV) using a photoresist mask.   After removal of the photoresist the specimens were ultraviolet treated for restoring the edge surface properties.   After that, the 20-25 nm thick barrier layer and the Y2 were deposited.   EJJ demonstrates Josephson conductivity at Tc=77K, giving Jc-10^4 A/cm^2 at characteristic voltage Uc~50μV.

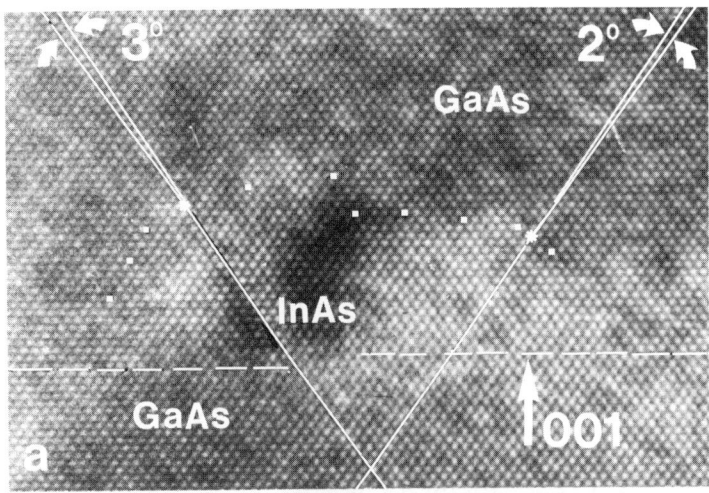

Fig. 1 Cross sectional (110) HREM images (300 kV) of an InAs/(001) GaAs interface. In island (a) the lattice mismatch is compensated by InAs lattice distortion, in (b) MD halfplanes are arrowed.

Fig. 2 Cross sectional (110) digitally filtered image of TiSi$_2$/Si(111) interface. Extra planes in TiSi$_2$ and distorted layer are arrowed.

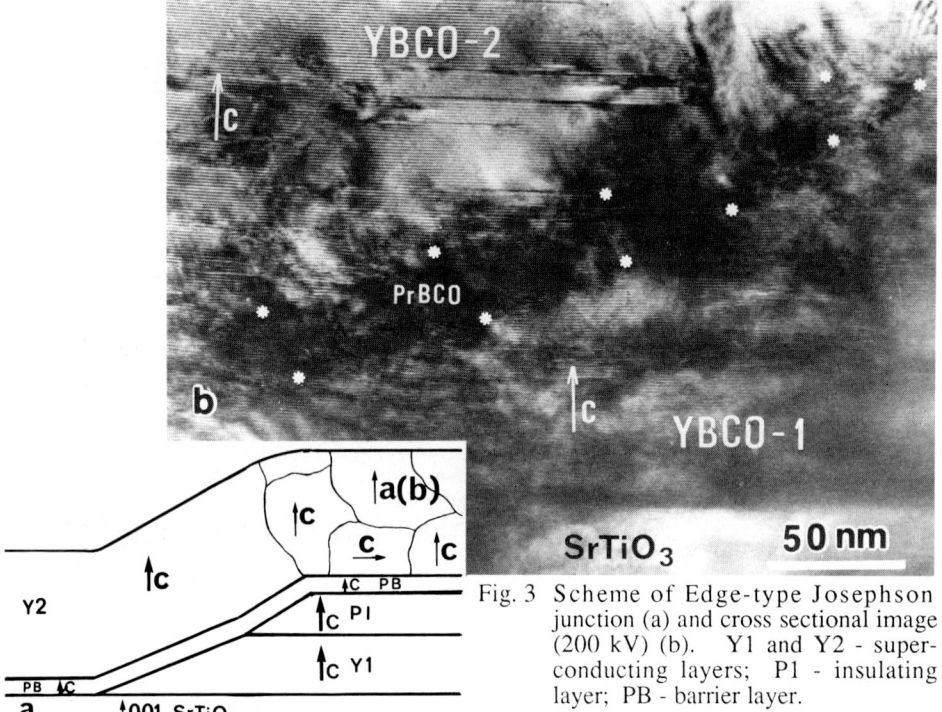

Fig. 3 Scheme of Edge-type Josephson junction (a) and cross sectional image (200 kV) (b).   Y1 and Y2 - superconducting layers; P1 - insulating layer; PB - barrier layer.

Cross-sectional low magnification image of the EJJ is shown in Fig. 3b.   Y1, PI and PB films are single crystalline with the c-axis normal to the substrate surface.   In the right hand side region the terminating Y2 layer is polycrystalline (the grain c-axis may be normal or parallel to the interface).   The reason for this may be an insufficiently flat PB surface or its contamination after the lithography process.   The PB/Y1 interface in the region of the edge contact is inclined by 20-35° to the substrate.   The PB layer thickness in this region is 20-25 nm.   Lattice resolution images reveal epitaxial growth of both the PB and Y2 films.   On the Y1/PB interface the APBs-like boundaries are observed.   The Y2 film in the edge contact region is characterised by a high APB concentration and by CuO double layers.

4.   NANOMETRIC Si TIPS

The most important part of the Scanning Tunnelling Microscope (STM) and all Scanning Probe Devices (SPDs) is an ultra sharp tip.   For SPDs not only the very end, but also the general shape of the tip is very important.   Thus, for profilometry a specially shaped tip with relatively thick body (necessary for higher vibration stability) and a very thin end with a small curvature radius is extremely desirable.   Sharp silicon microtips have potential application as field emitters and as electrical or mechanical microsensors.

The whiskers were prepared by the vapor-liquid-solid (VLS) technique (Wagner and Ellis 1964).   Small islands of gold were deposited onto a (111) Si wafer as a substrate. According to the phase diagram of these two elements, they can form a low-melting point (about 360°C) eutectic.   At the temperature of 800-900°C, a droplet of the Si-Au alloy is formed on the substrate.   In the presence of a gaseous mixture $SiCl4 + 2H2 \rightarrow Si + 4HCl$ above the substrate silicon is absorbed by the droplet, the Si-Au solution becomes supersaturated in respect to Si, and the excess Si is deposited on the solid-liquid interface. As a result, the liquid droplet is pushed out from the original substrate, remaining on the tip of the growing crystal.   Gold droplets may be obtained by heating gold films evaporated onto the substrate, by deposition through a mask, or with the help of photolitigraphy.

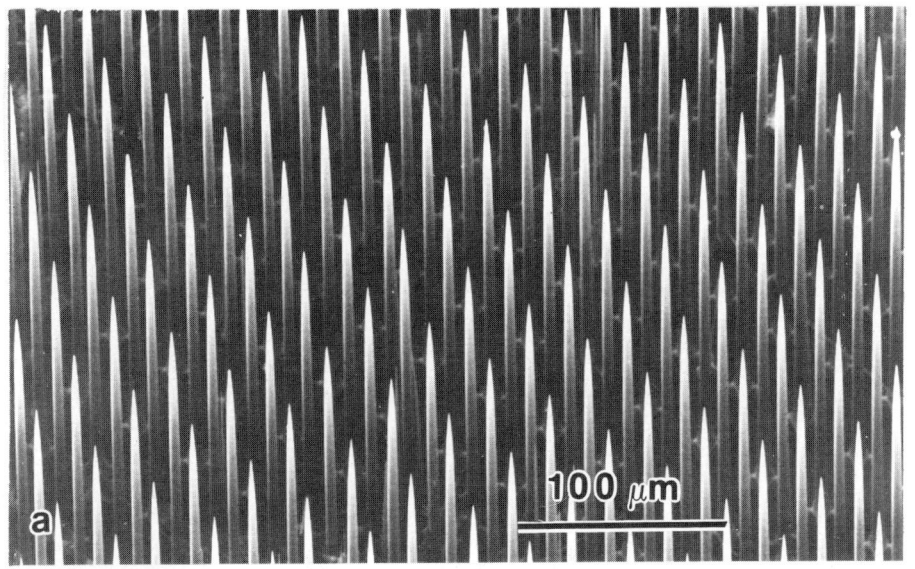

Fig. 4 SEM images of the Si tips obtained by one-stage (a) and by two-stage (b) growing processes.

Fig. 5 Schematic representation of the tips geometry: (a) One-stage process; (b) Two-stage process.

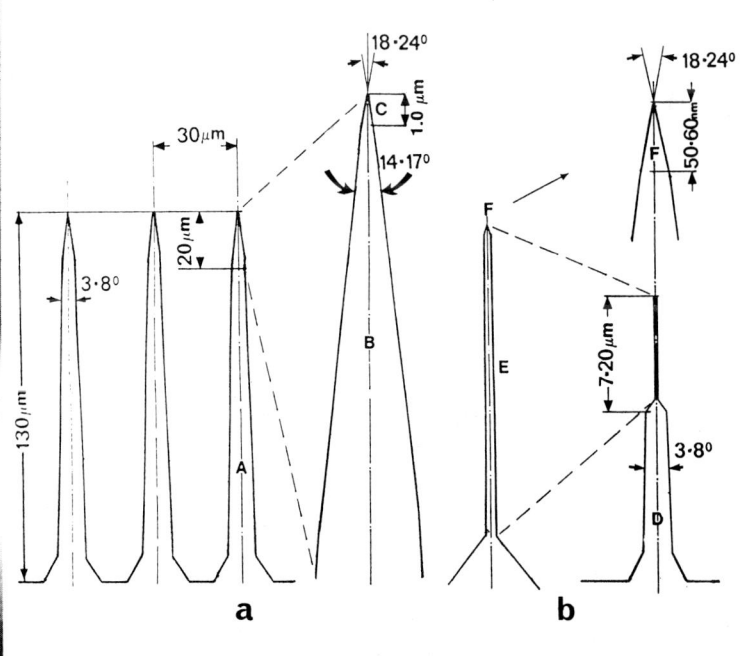

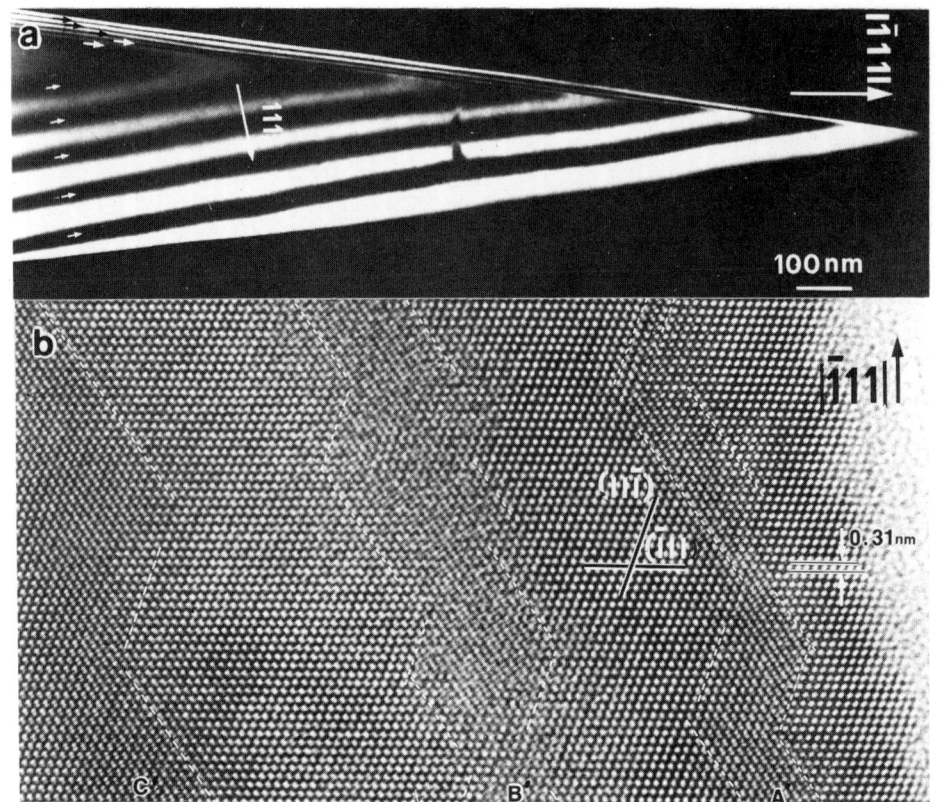

Fig. 6 Si tips (one stage process): (a) Two-beam diffraction contrast image.   Equal
thickness fringes are arrowed; (b) HREM image (390 kV) of tip's edge.   Areas of
half-spacing contrast are marked A', B', C'.

Whiskers were treated in a solution of HF:HNO3:H2O (1:40:1), which selectively
etched the interface between the Si whiskers and the Si-Au "cap" (Thomas et al 1972, 1974).
After that the tips were additionally sharpened by oxidation with subsequent etching (Marcus
et al 1990).

Narrow tips are desirable for investigating trenches with perpendicular walls.   The
tip length should be at least 2 µm and the diameter as small as 100 nm.   When a normal
etching procedure is used the probes thus obtained are very sensitive to vibration.   The two-
stage VLS whisker growing process solves this problem to some extent.   First, a relatively
thick basis is grown and then, by lowering the growth temperature, the thin upper part is
formed (Givargizov et al 1993).

All stages of whiskers growth were studied in JEOL JSM-840 instrument in the
secondary electron mode at 20 kV.   Fig. 4b is an SEM image of the Si-tips forming a square
lattice with a spacing between tips of 30 µm.   Tips are very uniform in height (130 µm).
Such a lattice of tips is covering an area 15 mm in diameter.   This kind of specimen was
prepared for field-emission studies.

Specially prepared plate-like substrates 0.23 x 2 x 10 mm in size were used to
investigate the tips in orientation optimal for HREM.   The substrates were oriented using X-
ray diffraction techniques.   "Wide" substrate sides corresponded to (110)-Si, and the "butt-
ends" to ($\bar{1}$11)Si.   After the whisker growth and the necessary sharpening treatment, the
specimens were glued onto supporting rings with the "butt-end" approximately in the center
of the ring.

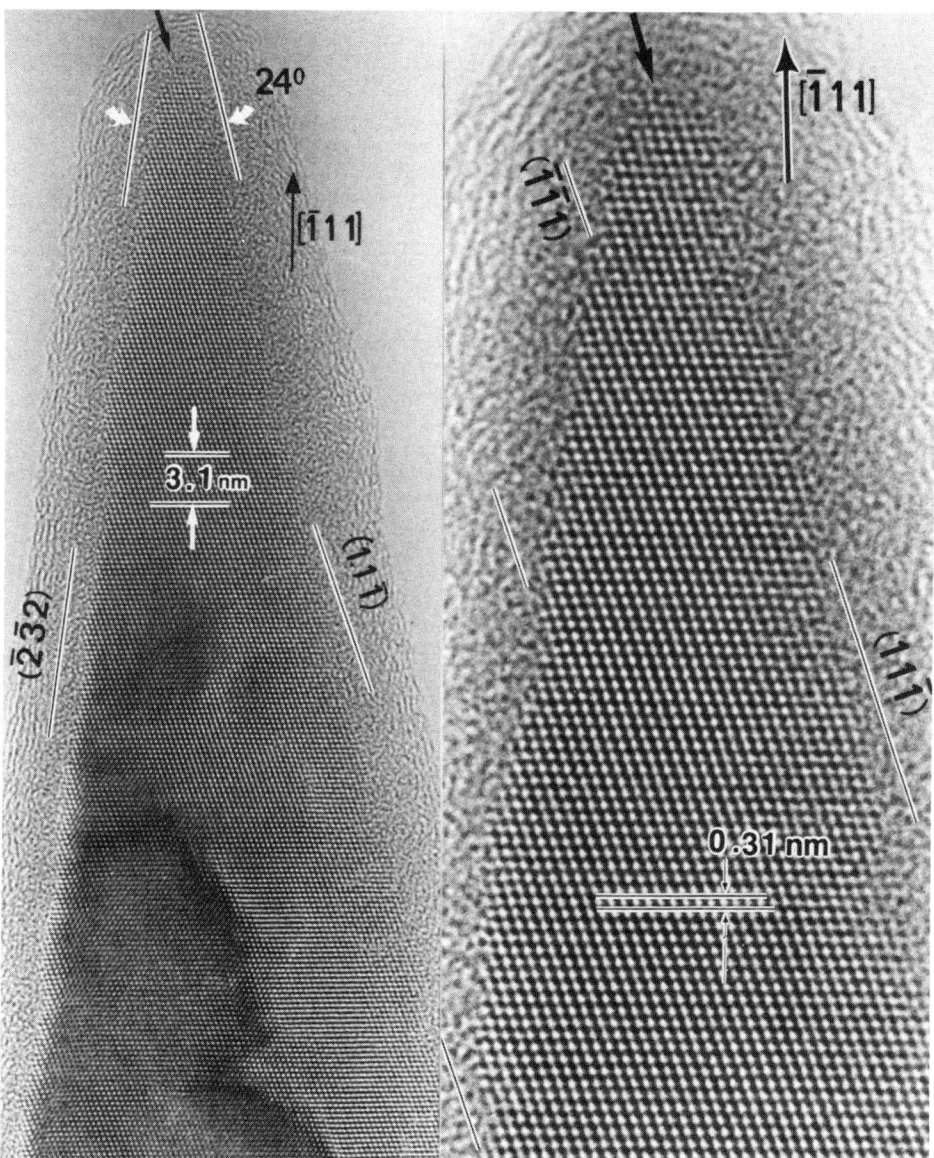

Fig. 7 HREM image (390 kV) of the Si tip (one stage process).

Specimens were examined in a Philips EM-430 ST microscope at 200 kV, and a JEOL JEM 4000 EX at 390 kV. The lattice image in JEM 4000 EX was obtained, using the central beam, and all diffracted beams out to 0.15 nm⁻¹. Two-beam diffraction contrast images (g=111) were also obtained.

The geometry of the tips is schematically shown in Fig. 5. SEM can give information concerning region (A) and partially (B). TEM and HREM are most effective for investigations of the upper part of (B) and the end of the tips (C).

Two-beam diffraction contrast imaging of region C (Fig. 6a) usually revealed a striped structure, formed from alternation dark and bright bands, parallel to the tip sides.

These bands represent equal thickness fringes, caused by dynamic diffraction effects (revealing electron extinction distances). The orientation of these bands indicates that the tips must be triangular in cross-section, with truncated corners. In the ideal case, when one side of the triangle is precisely parallel to the electron beam, no fringes must be seen at one of the sides. In reality, despite precise [110] orientation, one can see fringes on the left side of the tip as well. This phenomenon can be explained by step-like surfaces of the tip. The truncated triangle, enveloping the surface of the tip, is slightly rotated.

HREM lattice image of the tip (C region) is shown in Fig. 7. In etched silicon six {111} planes are the most stable ones. Judging from HREM data in the case of tips there are two combinations of three {111} planes (two types of etching modes): (i) main stable grains are (111), ($1\bar{1}1$) and ($11\bar{1}$) and (ii) ($\bar{1}\bar{1}1$), ($\bar{1}11$) and ($1\bar{1}\bar{1}$). The tips cone angles are 14-17° (B region) and 18-24° (C region). The extreme end of the tip (Fig. 7) is usually formed by a few atomic columns, viewed along <110>; the lattice image in this direction is an image of pairs of atomic columns. The very end of the tip can be thus considered as "atomically sharp".

The triangular cross-section of the tips is also confirmed by HREM data (Fig. 6b). The image character changes with growing distance from the tip's edge. Some areas demonstrate a "half-spacing contrast", appearing in regions where the sample thickness is a multiple of the extinction distance of the (000) beam. Areas A', B', C', etc. are typical for wedge-shaped crystals, i.e. triangular in cross-section.

Typical SEM image of nanometric tip, obtained using the two-stage growing process with subsequent sharpening, is shown in Fig. 4b. The diameter of the thin part (E, Fig. 5) of most tips is 10-25 times smaller than that of the base (D). The length of the tips' thin part (E) is 7-20 µm. On TEM micrographs it is possible to see that part E of the tips is becoming narrower towards the end of the tip. The angle which forms the edge profile of the tip with <111> growing direction is 2-3°. The diameter of the tips in this region is 50-100 nm. From HREM images a conclusion can be made that faceting of the tips is along (111) and slightly pronounced. Thus, at low magnification the tips look very smooth.

Sharp end (F, Fig. 5) cone angles are 18-24°, which means they are the same as for the tips grown by the one-stage process (B and C, Fig. 5). But the length of F region is only 50-60 nm, while the whole thin part (E) with small (2-3°) cone angle is usually one or two orders larger. The end of the tip is usually atomically sharp.

ACKNOWLEDGEMENTS

The authors express their thanks to V Yu Karasev for micrographs of InAs islands and to L.N. Obolenskaya for tips preparation.

REFERENCES

Catana A, Schmid P E, Heintze M, Stadelman P, Bonnet R 1990 J. Appl. Phys. 67 1820-5
Givargizov E I, Kiselev A N, Obolenskaya L N, Stepanova A N 1993 Appl. Surf. Sci. accepted for publication
Karasev V Ju, Kiselev N A, Orlova E V, Gribelyuk M A, Gutakovsky A K, Kanter Yu O, Pintus S M, Rubanov S V, Stenin S I, Fedorov A A 1991 Ultramicroscopy 35 11
Kiselev A N, Velichko A A, Okomelchenko I A 1993 J. Cryst. Growth 129 163
Lebedev O I, Vasiliev A L, Kiselev N A, Maso L A, Gaponov S V, Paveliev D G and Strikovsky M D 1992 Physica C. 198 278
Marcus R B, Ravsi T S, Gmittes T, Chin K, Liu D, Orvis W J, Ciasio D R, Hunt C E, Trujillo J 1990 Appl. Phys. Lett. 56 236
Thomas R N and Nathanson H C 1972 Appl. Phys. Lett. 21 384
Thomas R N, Wickstron R A, Shroder D K and Nathanson H C 1974 Appl. Phys. Lett. 17 155
Vasiliev A L, Kiselev N A, Dovidenco A M, Gaponov S V, Kalyagin M A 1990 Superconductivity 3 N4 557 (in Russian)
Vasiliev A L, Kiselev N A, Lebedev O I, Orlova E V, Orlikovsky A A 1991 Inst. Phys. Conf. Ser. No 117 297
Wagner R S, Ellis W C 1964 Appl. Phys. Lett. 4 89

Networks of growth steps on as-grown GaAs(001) vicinal surfaces

T Marek, HP Strunk, E Bauser° and YC Lu°,

Universität Erlangen, Institut für Werkstoffwissenschaften VII, Cauerstr. 6, 8520 Erlangen, Germany;
°Max-Planck-Institut für Festkörperforschung, Heisenbergstr. 1, 7000 Stuttgart, Germany.

ABSTRACT: Reflection electron microscopy, capable of imaging surfaces in high resolution, reveals that in a range of growth temperatures between 650°C and 760°C the treads and risers of terrace growth surfaces are characterized by a network of growth steps. Two sets of growth steps can be discriminated, which we term orientation steps (height around 10 monolayers) and exchange steps (height up to a few monolayers). Details of this network will be discussed.

1. INTRODUCTION

The exact knowledge of the as-grown surface microtopology is a precondition for the understanding of the growth-mechanisms on a molecular scale. For studying the as-grown vicinal surfaces we investigated GaAs-epilayers grown onto GaAs(001) substrates by liquid phase epitaxy.

In the optical microscope, these growth surfaces exhibit the well-known stepped topology that is commonly described in terms of treads and risers (Bauser and Strunk 1984).

Reflection electron microscopy (REM) reveals that on a more microscopic scale these treads and risers are formed by a network of microsteps. This network consists of two sets of microsteps, that essentially differ in height (Marek et al 1993). We call these microsteps orientation and exchange steps. The different local slopes in the treads and risers are a function of changes only in the dimensions of the microsteps in this network.

2. EXPERIMENTAL TECHNIQUES

GaAs epitaxial layers are grown from Ga-solution onto spherically shaped GaAs substrates. The spherical shape yields substrate surfaces the deviations of which lie between 0° and 2° off (001). Growth is performed in a slider boat system (Benz et al 1980). First the Ga solution is saturated with As at the growth temperature. Then, growth is initiated with a supersaturation (undercooling) of 0.2 K using a cooling rate of 0.2 K·min⁻¹. Growth is terminated by withdrawing the Ga solution. Finally the furnace is cooled to room temperature.

The topologies of the as-grown vicinal surfaces are inspected and preselected by optical microscopy in Nomarski differential interference contrast (NDIC) and afterwards investigated in a reflection electron microscope (REM).

For REM investigations the samples are mounted into a special specimen holder and inspected in a Philips EM 400 electron microscope. For imaging the preselected areas, a higher order Bragg reflection of the surface-parallel (001) planes is used. Since the Bragg angle is of the order of a few degrees the obtained micrographs represent a glancing view of the microtopology. In the micrographs an additional perpendicular scalemarker accounts for the corresponding foreshortening. For details about REM see Yagi (1989).

It should be noted that our approach permits the as-grown surface to be imaged without any prior preparation.

3. RESULTS

Fig. 1 shows a typical example of a network on a solution-grown GaAs(001) vicinal surface. The two sets of growth steps (orientation and exchange steps) are marked by O and E. The treads of these microsteps represent atomically smooth surfaces.

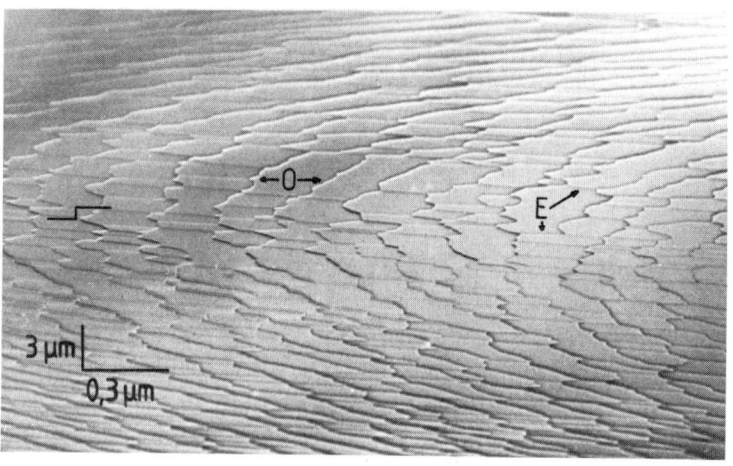

Fig. 1 A network of microsteps on an as-grown vicinal surface. Imaged in a reflection electron microscope (REM). The step sketched into the figure indicates the direction of the overall inclination of the surface. O = orientation step, E = exchange step.

The orientation steps account for the average local slope in the treads and risers. If the average distances between these microsteps increase (decrease) the local slope decreases (increases). If the average heights of these steps increase (decrease) the local slope increases (decreases) too. Fig. 2 shows examples for the average heights of the orientation steps as a function of the growth temperature.

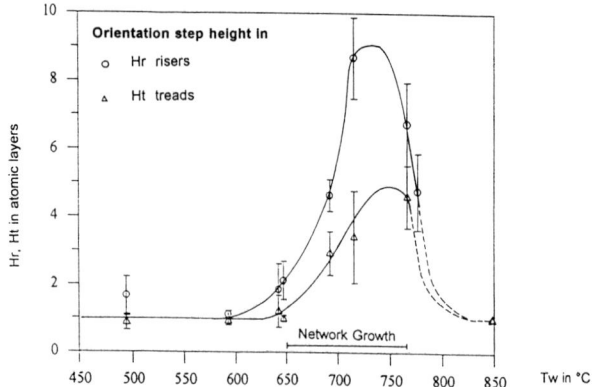

Fig. 2 The average heights of the orientation steps in treads and risers as a function of the growth temperature (Tw).

The average heights of the orientation steps were determined by measuring the average local slopes with a Tencor profiler and calculating the average height using the number density of the orientation steps in the electron micrographs of the same area.

The exchange steps are confined to the microtreads and connect in many cases two of these successive orientation steps. In Fig. 3, which shows a magnified section of Fig. 1, some of these steps are marked by E. The estimated height of these steps ranges between one and a few monolayers. It is interesting to note that many segments form a kind of half loop (e.g. EL in Fig. 3) that lies at the base of an orientation step.

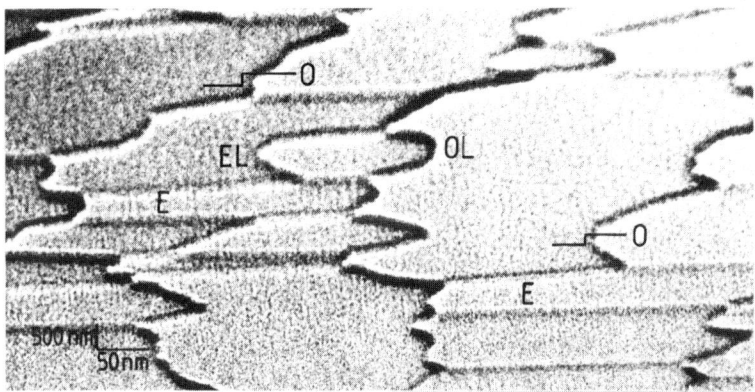

Fig. 3 Details of the network, enlarged area from Fig. 1. Note the correspondence of exchange steps (some marked EL) with a segment in the corresponding orientation step (marked OL) that lags behind its neighbouring segments. E = exchange step, O = orientation step.

We interpret this configuration as an early stage immediately after nucleation and regard it to be characteristic of the terrace growth mode. A further typical structural element associated with the exchange steps is that they cause the corresponding orientation step segment to lag behind its neighbouring segments; one of these lagging segments is labelled OL in Fig. 3.

As indicated in Fig. 2, the results for the 'network growth' hold for a range of growth temperatures from ca. 650°C to ca. 760°C.

4. DISCUSSION

In discussing details of the network we assume that an exchange step nucleates at the base of the orientation step, very probably at a site preformed in the structure of the microriser. This nucleated exchange step, under the action of local supersaturation, then sweeps across the tread surface ahead of it, say at EL in Fig.3. Its velocity is due to its rather small height much higher than that of the original comparably high orientation step. The nucleated step, once having caught up and recombined with the preceding orientation step, forms two links between orientation steps (Fig. 3). Such links may further move parallel to the orientation steps until they encounter another exchange step or a junction made by two orientation steps. Corresponding situations are sketched in Fig. 4; this sketch also symbolizes the fact that subsequent orientation steps know of each other's presence since this information is mediated or exchanged by the exchange steps.

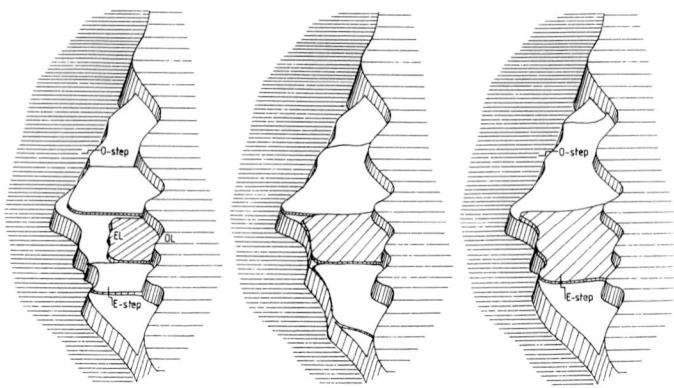

Fig. 4 Sketch summarizing the aspects associated with exchange steps in the network.
O = orientation step, E = exchange step.

Within the outlined notion of the role of exchange steps we may understand how the lagging segments in orientation steps (OL in Fig. 3) may be caused by the nucleated exchange steps (EL in Fig. 3). Following the model of Burton, Cabrera and Frank (1951) we can assume two paths the growth species may select to take to the growth step: direct diffusion from the solution and diffusion of adatoms along the liquid/solid interface. This interface diffusion is of interest here.

Once an exchange step forms at the re-entrant corner of an orientation step, it competes for the adatoms with its pertaining orientation step segments and thus reduces the flux of adatoms to this orientation step segment. This segment thus reduces its lateral velocity and lags behind its neighbouring segments, which are not "shielded" by an exchange step ahead of them (EL, OL in Fig. 3).

ACKNOWLEDGMENT

The financial support of the Deutsche Forschungsgemeinschaft (Str 277/2) is gratefully acknowledged.

REFERENCES

Bauser E, Strunk H P (1984) J. Crystal Growth 69 561
Benz K W, Bauser E, 1980 'Crystals', Vol. 3: 'III-V Semiconductors',
    ed. H.C.Freyhardt, (Heidelberg: Springer) pp.21-48
Burton W K, Cabrera N, Frank F C, (1951) Phil. Trans. Roy. Soc. 243 299
Marek T, Strunk H P, Weishart H, Bauser E, submitted to J. Cryst. Growth
Yagi K, (1989) Advances in Optical and Electron Microscopy 11 57

Inst. Phys. Conf. Ser. No 134: Section 6
Paper presented at Microsc. Semicond. Mater. Conf., Oxford, 5–8 April 1993

TEM analysis of the initial stage of GaAs ALMBE layers on Si

A Vilà[1], A Cornet[1], J R Morante[1] and P Ruterana[2,+]

[1] LCMM, Dept. de Física Aplicada i Electrònica, Universitat de Barcelona, Diagonal 645-647, 08028-BARCELONA (Spain)
[2] I²M, Ecole Polytéchnique Fédérale de Lausanne, 1015-LAUSANNE (Switzerland)

ABSTRACT: The initial stages of GaAs growth on (001) Si by conventional MBE and ALMBE have been investigated by using HREM. Although the growth mode is three-dimensional for the two techniques, the thickness at which coalescence occurs is smaller in ALMBE. Evidence for a critical thickness at which planar defects are nucleated is also found, and, in thicker islands, strain relaxation gives rise to interface dislocations and facetting. A study of the transition between coherent and dislocated 3D growth modes is also presented.

1. INTRODUCTION

The epitaxial growth of GaAs on Si (001) substrates has been a subject of considerable interest since it offers an opportunity to combine the best properties of the two materials. However, defect free GaAs epitaxial layers are difficult to grow due to lattice mismatch, different expansion coefficients and polarity. A main goal is therefore to find experimental conditions improving the quality of epitaxial layers.

Several methods have been tested in order to decrease defect density. Although these methods are effective and a reduction of the defects threading up to the surface of the epilayer is obtained, earlier experimental work (such as that of Fang et al. 1990) shows that the best crystalline quality would be obtained if the growth takes place in a layer-by-layer way. However, the initial stages of epitaxial growth of GaAs on Si (100) is found to be clearly three-dimensional, 3D, at the temperatures currently used (Hull et al. 1987, Tsai et al. 1989). This 3D nucleation favours the presence of defects at the island edges, the production of threading dislocations and planar defects after coalescence. In consequence, an improvement of the two-dimensional growth mode at the first stages will be more effective for producing layers of better crystalline quality. Several attempts, as the use of an initial Ga, As or Ge prelayer or a modification of some of the standard conditions in the MBE technique, have been developed in this way. Among these possibilities, González et al (1992) have developed the Atomic Layer MBE, ALMBE, technique as an alternative epitaxial growth method based on a continuous supply of group III flux and periodic short pulses of group V flux.

The aim of this work is to study the first stages of the nucleation provided by these techniques. Two fundamental aspects are analyzed in order to understand how the epitaxy proceeds. Firstly the growth mode, and secondly the evolution of defects relieving stresses.

[+] Present address: LERMAT, Université de Caen, 6 BLD Maréchal Juin, 14050-CAEN (France)

2. EXPERIMENTAL

Two series of samples were studied: one grown by conventional MBE and the other by the ALMBE technique (at $T_s = 300\,°C$) with thicknesses up to ~20 nm on top of (001) Si substrates. The thin GaAs layers were capped with a thick amorphous As layer for protection. Details about the procedure are given by González et al. (1992). Standard methods were used for sample preparation for microscopy, using mechanical polishing and final thinning by Ar+ milling. Eventually, low current and temperature conditions were used to avoid annealing effects. The characterisation was made by High Resolution Transmission Electron Microscopy (HRTEM) on a Philips EM 430ST microscope operated at 300 keV.

3. RESULTS AND DISCUSSION

Although growth is basically three-dimensional in both cases, a difference between the growth mode obtained by ALMBE and conventional MBE techniques is observed. The first stages are shown in figure 1, where nominally ~1.8 nm thick layers grown by MBE and ALMBE are presented (fig. 1.a and 1.b respectively).

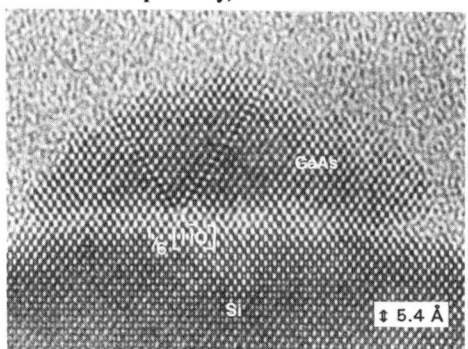

Figure 1.a. ~1.8 nm thick GaAs MBE layer. Some of the islands are relaxed by stair rod dislocations or stacking faults.

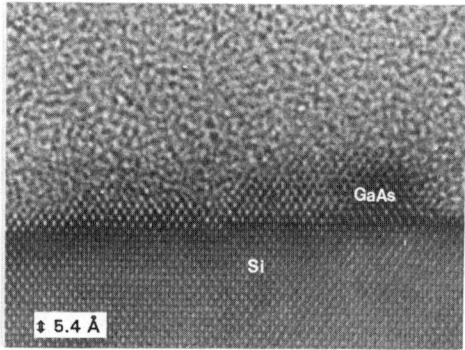

Figure 1.b. Nominal 1.8 nm thick ALMBE layer. Islands are small and very close.

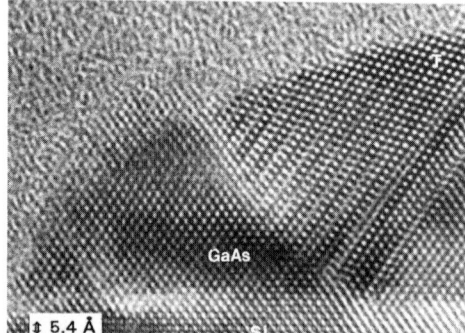

Figure 2.a. ABCABC stacking transformed into an ACBACB twin (T) in the nominal 10 nm MBE layer.

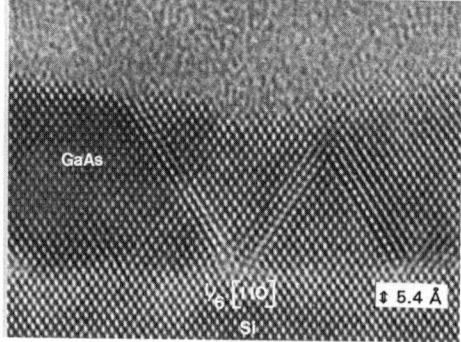

Figure 2.b. Continuous GaAs layer obtained after 10 nm ALMBE deposition. Most defects are dissociated dislocations.

For MBE, widely spaced islands (by several tens of nm) are found. When their dimensions are large enough to relax, stair rod dislocations and planar defects are present. In contrast, ALMBE samples show small islands without defects. The distance between them is smaller (10 nm or less) giving a better surface coverage.

From these observations, it appears that epitaxy progresses by growing previously existent islands rather than nucleating new ones. Figures 2a and 2b show cross-sectional electron micrographs of the MBE and ALMBE layers when a nominal layer of 10 nm has been grown. Individual islands are significantly larger than in the previous case, better covering the surface. However, in the MBE layer, they are still separated while in the ALMBE neighbouring islands have coalesced. Later, a 2D growth mode starts, as predicted by the Stranski-Krastanov theory. In figure 2, it is clear that this transition takes place in an earlier stage of growth for ALMBE.

Consequently, a 3D nucleation in the first stages of both MBE and ALMBE, followed by a Stranski-Krastanov transition after coalescence, gives rise to continuous GaAs layers above a critical thickness. The influence of this growth on the final layer quality will critically depend on the size and defective state of the coalescing islands. So, it is fundamental to analyze the evolution of defects in order to understand epitaxial mechanisms.

Islands start to grow coherently on the substrate, as shown by the absence of defects in the smallest ones (figure 1.b), in which stress is distributed in a fringe near the interface. As the epitaxy continues, a larger amount of stress is concentrated in the deformed area. When the island dimensions are large enough, the total energy collected by coherent growth becomes large enough for dislocations to appear. It is energetically convenient to relax in a way determined by the activation energy for each type of defect. Our results show that relaxation does not depend only on the island height but also on its length, in agreement with the theory of Hull and Fisher-Colbrie (1987). This is shown in figure 3, where the relaxation is represented by island height versus ratio of height to length. We also report in this figure the theoretical predictions of the 3D coherently strained-dislocated transition following the analysis of Luryi and Suhir (1986), using critical thickness models by People and Bean (1986) and Matthews and Blakeslee (1974).

It is well known that defects best relieving interfacial stress in FCC materials are dislocations with a Burgers vector of $\mathbf{b} = a/2 < 110 >$: mainly the Lomer type, having $\mathbf{b}$ parallel to the (001) interface, followed by 60° type, with $\mathbf{b}$ at an angle of 45° from the interface and an efficiency which is half of that of the Lomer type. However, as shown in figure 1.a, the first stage of relaxation is reached by stair rod dislocations and planar defects. These results are in agreement with the work of Lee et al. (1991) which suggests that although these types of defects relieve less stress, they are more convenient for starting relaxation because their activation requires smaller energy. As the deposition continues, more defects are nucleated at the island edge, giving series of stacking faults located in {111} planes parallel to limiting facets. They give rise to disordered areas which do not reorganise due to the many planes involved. This is the case in micrograph 2.a, where a large defect giving an overlapped twin is visible.

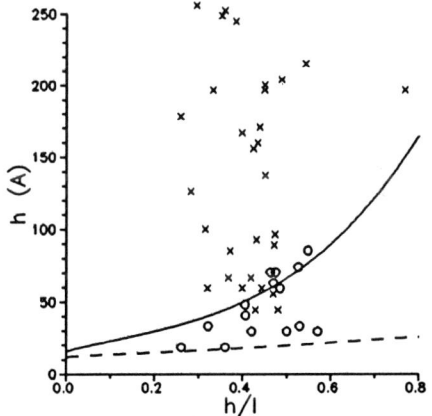

Figure 3. Height vs. height/length ratio for relaxed (x) and non-relaxed (o) islands. Full and dashed lines are obtained using People & Bean or Matthews & Blakeslee models.

The stress accommodation also influences the island shape. Figure 4 shows the two most characteristic types of facetted islands observed in thickest nominal layers. In fig. 4.a. islands are slightly {111}-facetted. There, main defects are stacking faults and microtwins. In contrast, other islands can be larger, and clearly {113}-shaped before coalescence. In this case, we find very few extended planar defects inside, as shown in fig. 4.b.

At the first stages of deposition, islands grow coherently to the substrate. When they reach dimensions large enough they relax stress by nucleating partial dislocations. Next, planar defects parallel to the limiting facets of the islands appear at their edges. As deposition continues an accumulation of them at these limiting facets is observed. If coalescence is attained when only a few planes are involved in the stacking defects (as seen in ALMBE), reorganisation can take place and the thick layer will contain fewer defects. On the contrary, if coalescence is reached after the formation of larger and more defective islands (as in MBE case) a large density of defects will propagate inside the epitaxial layer.

 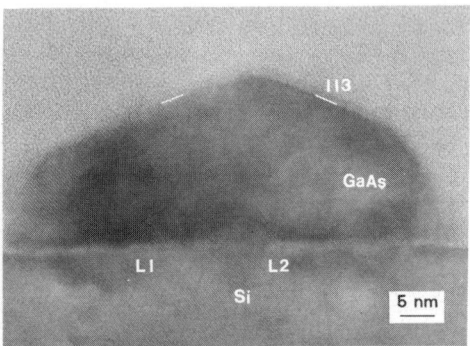

Figure 4.a. Highly defective island in ~ 20 nm layer grown by MBE containing intrinsic and extrinsic stacking faults and microtwins.

Figure 4.b. Large {113}-shaped island containing only interface dislocations of Lomer type more or less extended (L_1 ~ 1 nm, L_2 ~ 6 nm).

4. CONCLUSIONS

A three-dimensional nucleation in the first stages of both MBE and ALMBE gives rise to continuous GaAs layers above critical thicknesses notably different in both cases by means of a Stranski-Krastanov transition. However, the main epitaxial mechanisms relieving stress are similar. Initially, epitaxial growth is coherent to substrate. Next, stair rod dislocations and stacking faults are generated. When their area is too large to be stable, partial dislocations limiting them recombine giving misfit dislocations. Defects remaining finally in the layer depend strongly on the size and state of each island when coalescence takes place.

REFERENCES

Fang S F, Adomi K, Iyer S, Morkoç H and Zabel H 1990 J. Appl. Phys. **68** R31
González Y, González L, Briones F, Vilà A, Cornet A and Morante J R 1992 J. Crys. Growth **123** 385
Hull R and Fischer-Colbrie A 1987 Appl. Phys. Lett. **50** 851
Lee H P, Liu X, Malloy K, Wang S, George T, Weber E and Liliental-Weber Z 1991 J.Elec. Mat. **20** 179
Luryi S abd Suhir E 1986 Appl. Phys. Lett. **49** 140
Matthews J W and Blakeslee A E 1974 J. Cryst. Growth **27** 118
People R and Bean J C 1986 Appl. Phys. Lett. **49** 229
Tsai H L and Matyi R 1989 Appl. Phys. Lett. **55** 266

TEM study of the GaAs/Si system grown by two-step MBE/SPE method

S Nahm, M C Paek, K I Cho, O J Kwon and W K Choo

Electronics and Telecommunication Research Institute, Daedog Science Town, P.O. Box 8, Daejeon, South Korea 305-606

ABSTRACT: In the initial stage of SPE growth of GaAs films on Si, the size of most GaAs islands was observed as ~ 10 nm but large islands (~ 40 nm) were also seen. Misfit dislocations and stacking faults were already formed at this early stage of growth. Stacking faults and misfit dislocations consisting of Lomer and 60° dislocations were observed in the GaAs films grown at 580 °C by a modified two-step MBE method. However, after RTA treatment at 900 °C for 10 seconds, only Lomer dislocations with a spacing of ~ 10 nm were found.

1. INTRODUCTION

The epitaxial growth of GaAs films on Si substrates has been a subject of great interest because of the desire to combine the best of two semiconductors' properties. However, the high density of defects formed due to the differences in lattice parameters and thermal expansion coefficients limits the quality of GaAs films. The formation of crystal defects depends on the details of growth conditions. In this work, GaAs films were grown on the vicinal Si (001) substrate by a modified two-step molecular beam epitaxy (MBE) method, and the interfacial defect structure at each growth stage was investigated using transmission electron microscopy (TEM).

2. EXPERIMENTAL DETAILS

In order to study the initial stage of growth, an amorphous GaAs film deposited on a Si substrate at 80 °C in an MBE system was furnace-annealed (solid phase epitaxy: SPE) at 300 °C for 10 minutes in a N_2 atmosphere (sample 1). For the complete growth of the GaAs film, the substrate temperature was slowly increased to 580 °C and the main GaAs film of 0.5 µm was grown after crystallization of the amorphous GaAs buffer layer in the MBE system (sample 2). Finally, to decrease the density of defects, rapid thermal annealing (RTA) treatment was carried out on sample 2 at 900 °C for 10 seconds (sample 3). The Si substrate used in this work was tilted by 4° from [001] to [110] direction. The details of substrate cleaning and growth procedures have been discussed elsewhere (Cho 1991). The cross-sectional and plan-view TEM samples were prepared by mechanical grinding and subsequent ion milling at liquid nitrogen temperature. A Philips CM20 T/STEM microscope was used to observe the samples.

3. RESULTS AND DISCUSSION

Figure 1 shows a bright field plan-view image of sample 1. GaAs islands identified by moiré pattern were observed. Moreover, they can be divided into two groups according to size and the spacing of moiré pattern. For most GaAs islands, the size is ~ 10 nm and the spacing of the moiré pattern is about 5 nm(see the islands marked as A). The spacing of a parallel moiré pattern due to 220 beams is calculated as 4.8 nm, which agrees with our experimental value. For the samples 2 and 3 where GaAs films were completely grown on Si substrates tilted by 4°, the spacing of moiré patterns due to 220 beams is about 5 nm. Koch et al. (1987) also found moiré patterns with 5 nm spacing for a sample grown by MBE. However, for some islands marked as B in Fig. 1, their size is large (~ 40 nm) and the spacing of moiré patterns is ~ 2.5 nm. More study is needed to explain the difference in the spacing of moiré patterns between the small and large islands.

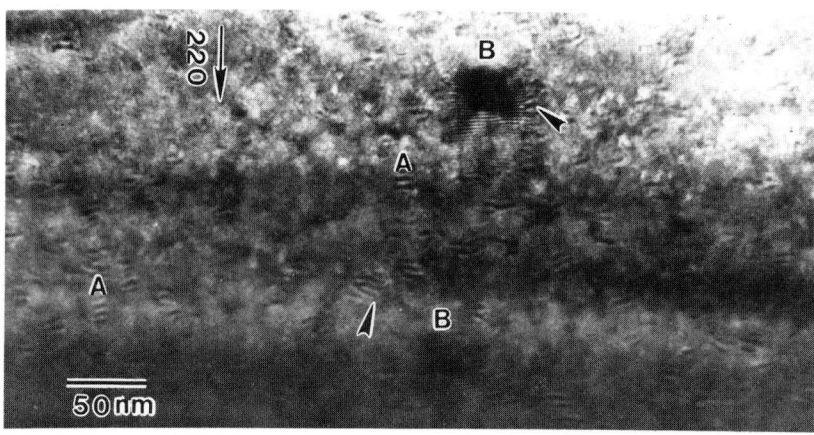

Figure 1. Bright field image of GaAs film deposited at 80 °C and annealed at 300 °C for 10 minutes.

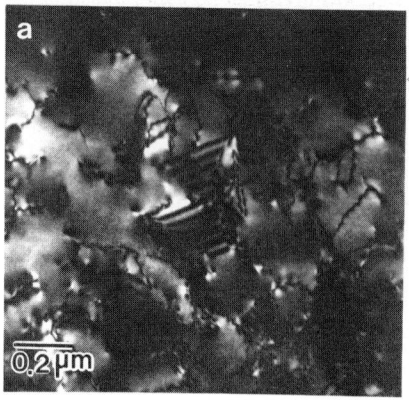 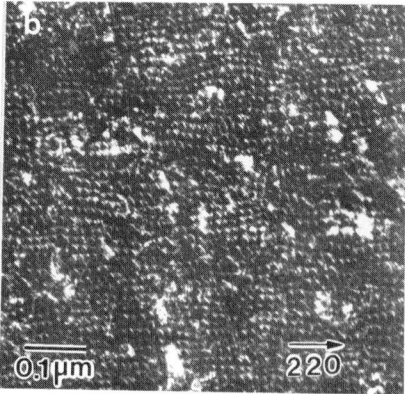

Figure 2. (a) 220 dark field image of the area away from the interface, and (b) $g_{(400)}$ and 3g weak beam image at the interface of GaAs/Si system grown at 580 °C.

Misfit dislocations can be observed in both small and large GaAs islands as indicated by the arrow heads in Fig. 1. Even if stacking faults were not found in Fig. 1, they were observed in cross-sectional high resolution TEM images (Choo 1992). Therefore, the lattice mismatch between GaAs films and Si substrates is accommodated by misfit dislocations and stacking faults in this early stage of growth.

Figure 2(a) is a 220 dark field image of sample 2 taken in the area away from the interface between the GaAs film and Si substrate. Stacking faults and a high density of threading dislocations can be observed in Fig. 2(a). In order to study the defect structure at the GaAs/Si interface, ion milling was carried out for 15 minutes more on the GaAs surface. Figure 2(b) shows a 400 weak beam image taken at the interface. Grid-like dislocations running along the [110] and [$\bar{1}$10] directions are shown in Fig. 2(b). According to the previous study, both Lomer and 60° dislocations are present in this sample (Choo 1991). The above results indicate that for sample 2, the misfit between the GaAs film and Si substrate is accommodated by both stacking faults and misfit dislocations.

In order to decrease the defects in the system, one of samples 2 was annealed at 900 °C for 10 seconds. A high resolution lattice image with a [110] beam direction is shown in Fig. 3. Lomer dislocations distributed with a spacing of ~ 10 nm are observed (see the arrow heads in Fig. 3) but there is no evidence of formation of 60° dislocations and stacking faults (or micro-twins). A plan-view sample was also investigated and 400 weak beam and bright field images are shown in Figs. 4(a) and 4(b), respectively. The distance between the grid-like dislocations in Fig. 4 is about 10 nm which is the same as that between Lomer dislocations in Fig. 3. Therefore, all the dislocations present in this sample are considered to be Lomer dislocations. The above results indicate that rapid thermal annealing at 900 °C drastically reduces the density of 60° dislocations and stacking faults and that the misfit between the GaAs film and Si substrate is relieved by Lomer dislocations.

Figure 3. High resolution lattice image of GaAs/Si sample with RTA treatment at 900 °C for 10 seconds.

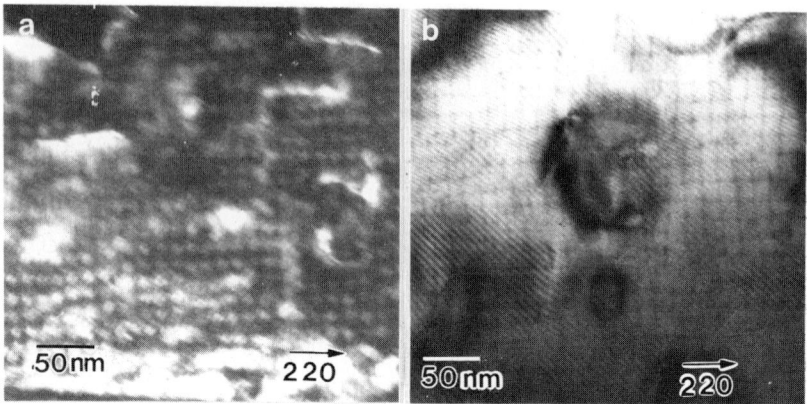

Figure 4. (a) $g_{(400)}$ and 3g weak beam image, and (b) bright field image of GaAs/Si sample with RTA treatment at 900 °C for 10 seconds.

4. CONCLUSION

For GaAs films deposited at 80 °C and annealed at 300 °C, the size of most islands was observed to be ~ 10 nm but large islands (~ 40 nm) were also observed. Misfit dislocations were found in both small and large islands at this early stage of growth and stacking faults are expected to exist. Grid-like dislocations consisting of Lomer and 60° dislocations were observed at the interface of the GaAs/Si system grown at 580 °C. Stacking faults were also observed in the area away from the inetrface. However, after RTA treatment at 900 °C for 10 seconds, only Lomer dislocations were observed. The distance between Lomer dislocations was measured as ~ 10 nm.

REFERENCES

Cho K I, Choo W K, Lee J Y, Park S C, and Nishinaga T 1991 J. Appl. Phys. <u>69</u> 237.

Choo W K, Cho K I, Lee J Y, Park S C, and Kwon O J 1991 Mat. Res. Soc. Symp. Proc. vol. <u>202</u> 573.

Choo W K, Kim I, Lee J Y, Cho K I, Lee J L, and Kim Y J 1992 Mat. Res. Soc. Symp. Proc. vol. <u>263</u> 131.

Koch S M, Rosner S J, Hull R, Yoffe G W, and Harris J S Jr 1987 J. Crystal Growth <u>81</u> 205.

Inst. Phys. Conf. Ser. No 134: Section 6
Paper presented at Microsc. Semicond. Mater. Conf., Oxford, 5–8 April 1993

361

TEM characterization of GaAs on Si layers grown by low pressure OMCVD

J Meddeb, M Ambri, M Pitaval, N Draidia[*], R J Dieter[*] and F Scholz[*]

Département de Physique des Matériaux Bat. 203 ; Université Claude Bernard Lyon 1 . 43 Bd du 11 Novembre 1918- 69622 Villeurbanne cedex France.
[*] 4.Physikalisches Institut, Universität Stuttgart, Pfaffenwaldring 57, D-7000 Stuttgart 80, Germany.

ABSTRACT: The effect of Thermal Cyclic Growth (TCG) combined with a Thermal Annealing Cycle (TAC) to reduce dislocation density in GaAs/Si layers grown by low pressure OMCVD and the first step growth have been studied by TEM and HRTEM. A structure of one symmetric grain boundary has been examined by HRTEM and image simulation.

1. INTRODUCTION

Growing GaAs on Si substrates is an interesting topic for possible optoelectronic applications. This heteroepitaxy technology combines the well developed GaAs electronics with Si mature technology. However two major problems exist in the growth of GaAs on Si. One problem is the formation of Antiphase Boundaries (APB) due to the growth of a polar semiconductor on a non polar substrate. This problem is severe on (001) Si which is the most desirable orientation for devices and processing. The use of substrates cut a few degrees off <001> inhibits the formation of these APB.

The other problem in the growth of GaAs on Si is the 4.1% lattice mismatch and the difference between thermal expansion coefficients which develop thermoelastic strain during post-growth cooling.This results in the presence of almost 10^{12} cm^{-2} misfit dislocations lying in the heterointerface. Dislocations are frequently observed to thread from this heterointerface to the epilayer and affect the electronic and optical properties of GaAs.

Much effort has been made to reduce this threading dislocation density using different methods : Adequate post growth annealing, introduced Strained Layers Superlattices, Buffer layers, localized epitaxy...(see Fang 1990 for example).

In this context, we used Conventional TEM and HRTEM to evaluate the efficiency of introduced TCG combined with one TAC in reducing threading dislocation density.

2. EXPERIMENTAL CONDITIONS

The observations were performed using a high resolution electron microscope (JEOL 200 CX) with a point to point resolution of 2.3 Å. [110] thin cross-sectional foils were prepared by usual thinning techniques. For plan-views, specimens were thinned mechanically from the back side and then finished with a HF/HNO$_3$ chemical etch.

GaAs layers were grown in Aixtron kit Aix-200 with horizontal reactor operating at reduced pressure of 20 hPa on (001) Si substrates, 2° misoriented towards <110> direction. TCG process consists of : 15 nm nucleation step at 400°C and conventional growth of 1.2 μm at 700°C, the sample was cooled down to 400°C and a 50 nm GaAs layer was deposited . The temperature was raised to 900°C, then after cooling to the normal growth temperature of 700°C, 50nm GaAs was grown. This cycle was repeated three times. After that a TAC step consisting of cooling the sample down to 200°C, heating to 900°C and holding the temperature for 10min, again cooling to 200°C and raising the temperature to 700°C to achieve the growth (see Dieter et al (1992) for more details).

3. RESULTS AND DISCUSSION

On the [110] cross section of Fig. 1, we can see a drastic decrease in defect density away from the interface. The GaAs layer exhibits bent   dislocations at different depths corresponding to the TCG levels. The thermal gradient created during this process leads threading dislocations to extend in the plane of growth and annihilate showing the efficiency of this process to reduce their density. Dislocation loops, planar defects with stacking faults and some microtwins have been also detected but were at the limit of the TEM detection threshold ($<10^6$ cm^{-2}). The dislocation density has been estimated at 10^7cm^{-2} on TEM plan views.

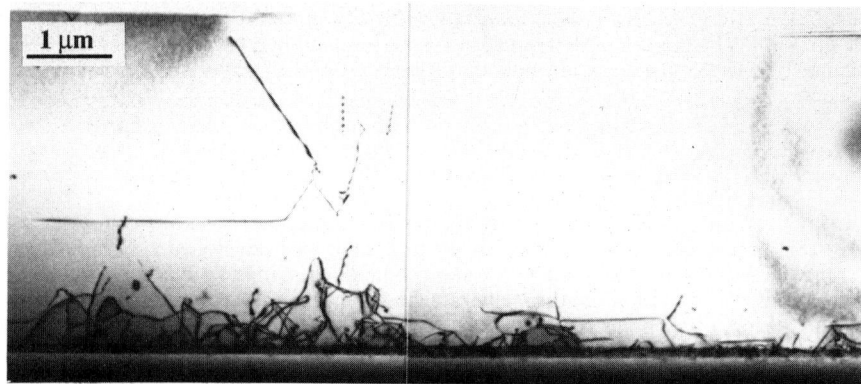

Fig. 1. [110] Cross section of GaAs layer with a Thermal Cycle Growth.

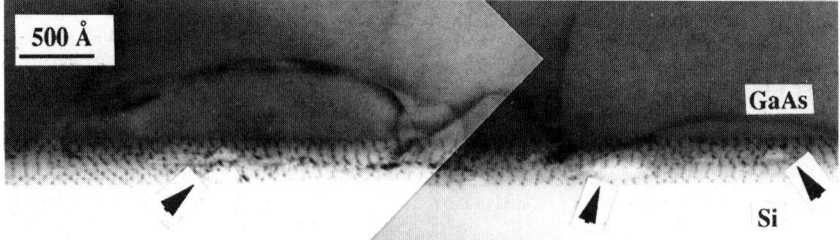

Fig. 2. Arrows point to small dislocation half loops lying on the GaAs/Si interface.

In the same way the GaAs/Si interface has been investigated under non conventional conditions from cross-sections. The (440) planes were under Bragg reflection, and the transmitted beam was selected. We observed both [110] and [1$\bar{1}$0] directions. On Fig. 2 we can identify [110] parallel Lomer dislocations with a mean spacing of 9 nm whereas it is 9.8 nm for a fully relaxed layer with only 90° dislocations showing a residual strain persisting in the layer. 60° dislocations are easily recognizable by a shift of half mean spacing.

On the same Fig. 2 small dislocation half loops with a 20-50 nm mean length and 7-15 nm height have been observed lying on the GaAs/Si interface. They are parallel and present in both [110]and [1̄10] directions with sharp contrast. These defects are supposed to be originated from the high silicon diffusion through the interface induced by the different high temperature annealing steps. SIMS profiles show a 10^{17} cm^{-3} Si concentration in the GaAs layer with peaks corresponding to the TCG temperature steps.

These defects led us to investigate the first steps of the growth. Two 15 nm thick GaAs prelayers have been studied. The first one was grown at 400°C without annealing and the second one was annealed at the second step temperature growth 700°C for 5 min. [110] cross section of Fig. 3a with the electron beam perpendicular to step ledges shows a complete substrate surface coverage but a rough surface: it can be supposedthat a three-dimensional growth occurred at the early nucleation stage. This prelayer is highly defected and contains many twins and stacking faults. Solely raising the temperature to 700°C eliminated the major part of these planar defects as shown on Fig. 3c.

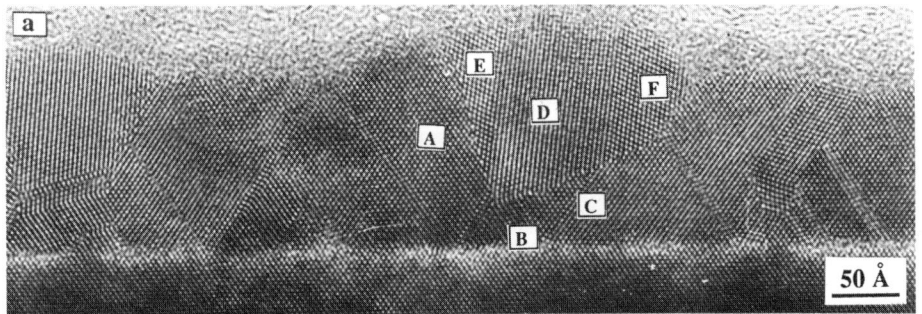

b	Rotation angle θ around [110] common axis [Grimmer 1959]	
Σ 3	70.°53	A/B B/C B/D D/F
Σ 9	38.°86	D/C
Σ 27	31.°59	F/C
Σ 19	26.°52	A/E

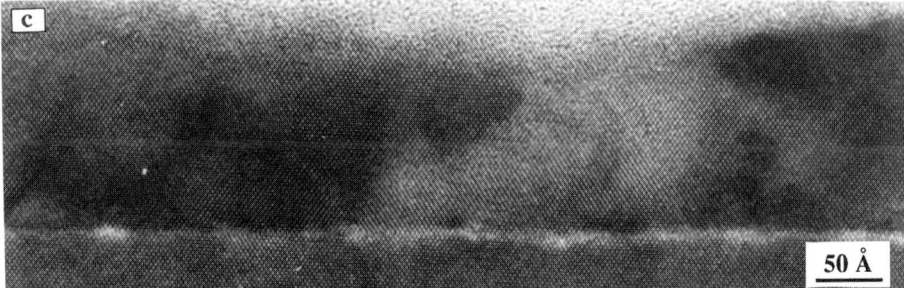

Fig. 3 (a). [110] High Resolution cross section of a prelayer deposited at 400°C with many planar defects and twin boundaries.(b) Various grain boundaries of Fig. 3a.with their rotation angles. (c) Drastic reduction of planar defects after annealing.

Frequent grain boundaries (GB) formed by successive twinning have been identified (Fig.3b) such as symmetric Σ19 grain boundary with (133)A/(133)E twinned planes with rotation angle of θ= 26°30' around [110]. Enlarged image Fig. 3a of this grain boundary has been compared to computer simulated images. The atomic arrangement of this boundary is suggested in Fig. 4 a. The image of this GB was located in the first extinction distance of 000 beam. An optical difractogram obtained from a region containing amorphous GaAs confirmed that the image was recorded near the Scherzer defocus ($\Delta f=-480\text{Å}$). On Fig. 4b we can see the inset,corresponding to a 30Å thickness,matching well with the experimental image.

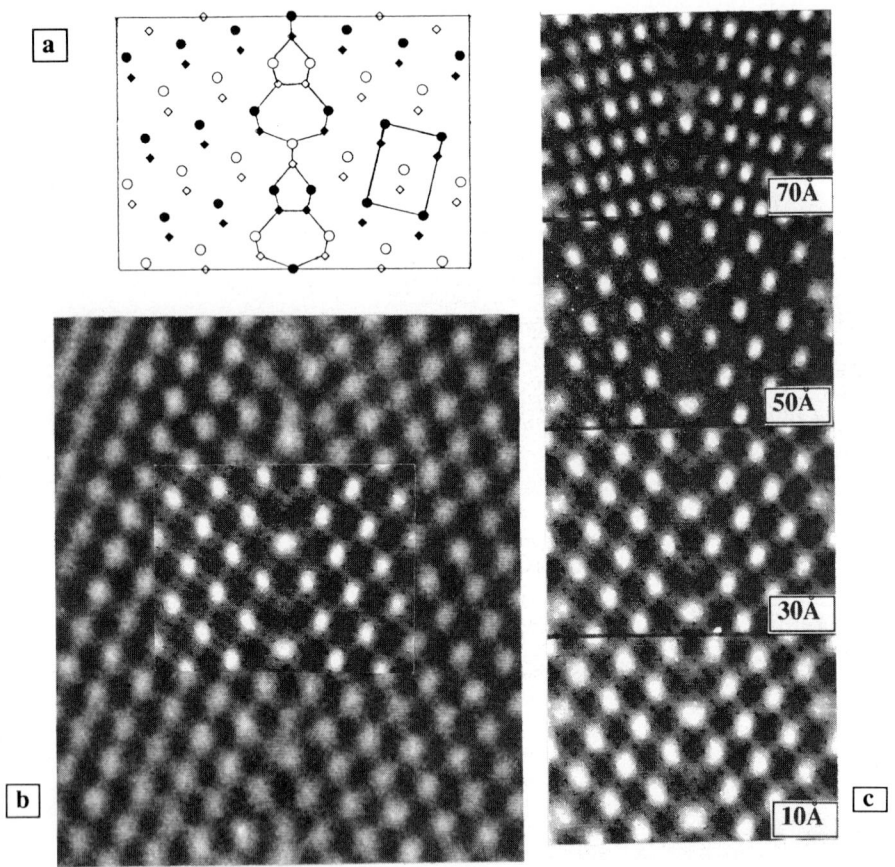

Fig. 4 **(a)** A model of the atomic configuration for a symmetrical (133)A/(133)E twin boundary. **(b)** Enlargement of high magnification micrograph of Σ 19 grain boundary, inset is the simulated image (crystal thickness t=30 Å, $\Delta f=-480$ Å). **(c)** computer simulated images at increasing foil thickness t=10, 30, 50 and 70 Å near Scherzer defocus $\Delta f=-480$ Å.

REFERENCES

Dieter R J, Scholtz F, Martin W, Hangleiter A, Dôrnen A, Michler P, Kürner W, Lu B, Frese V, Hilgarth J and Rash KD, Microelectronic Engineering **18**, 189-205.

Fang S F, Adomi K, Iyer S, Morkoç H and Zabel H 1990 J. Appl. Phys., **68**, R31.

Grimmer H, Bollmann W and Warrington D H 1974 Acta Cryst. **A30** 197.

Inst. Phys. Conf. Ser. No 134: Section 6
Paper presented at Microsc. Semicond. Mater. Conf., Oxford, 5–8 April 1993

365

HREM study of GaAs layers grown on Ge by the close-spaced vapour transport technique

N Guelton, G Feuillet[*], RG Saint-Jacques and JP Dodelet

INRS Energie et Matériaux CP 1020 Varennes Québec Canada, J3X 1S2.
[*]Centre d'Etudes Nucléaires de Grenoble DRF/SPh/PSC 85X, 38041 Grenoble CEDEX, France.

ABSTRACT: The effects of operating conditions on the microstructure of GaAs grown on Ge substrates by close-spaced vapor transport have been investigated by conventional and high resolution electron microscopy. High quality single domain layers were obtained by a pre-growth thermal treatment of Ge vicinal substrates in H_2 prior to growth.

1. INTRODUCTION

GaAs epitaxial layers have been grown on Ge by close-spaced vapor transport (CSVT). This technique is similar to CVD. The source (a commercial GaAs wafer) and the Ge substrate are in close proximity (<1 mm) in the reactor and are heated at 800°C and 750°C, respectively. Water vapor, transported in a H_2 stream, reacts with the source to yield volatile Ga and As compounds, according to the following reaction:

$$2GaAs + H_2O \rightleftarrows Ga_2O + As_2 \text{ (or } \tfrac{1}{2}As_4) + H_2.$$

The inverse reaction takes place at the substrate. The driving force of the deposition is the 50°C temperature gradient between the source and the substrate. Most of the work previously done on GaAs grown by CSVT deals with the modelling of the growth rate (Chavez et al 1983; Côté et al 1986) and the control of the electrical properties of the layers (LeBel et al 1993). In this paper, the film structure is examined using transmission electron microscopy. It reports that a low temperature baking combined with a high temperature pre-growth treatment of a vicinal substrate leads to a defect free interface.

2. EXPERIMENTAL

Standard degreasing and etching procedure are used for the chemical preparation of the substrate. The source and the substrate are loaded into the reactor and baked at 200°C for 10 min in order to remove the adsorbed water. This is a crucial step toward the control of the nucleation and growth of GaAs because water is the transport agent of the reaction. An oxide desorption step is also necessary. It consists in heating the source and the substrate for 30 min at 750 and 800°C, respectively. This pre-growth thermal treatment leads also to step doubling such that the terraces at the surface belong to the same sublattice. The deoxidised Ge surface and the source are then cooled down to 400°C and finally heated back to 750°C and 800°C respectively. The deposition begins as soon as water is injected. Exactly oriented (100) substrates and vicinal substrates, whose surface was tilted away from the (100) plane by 3° towards the [011] axis, were used. Typical GaAs layers were 2 to 3µm thick. The {011} cross sections were prepared using standard mechanical polishing and argon ion milling procedures. Through focus HREM images were taken around the Scherzer defocus with the beam direction aligned to the [011] or [01$\bar{1}$] thin foil axis.

3. RESULTS AND DISCUSSION

Fig.1 shows the {011} cross sectional micrographs of the three investigated samples. The GaAs layer shown on fig.1a has been grown on an exactly oriented (100) substrate with no water desorption baking or pre-growth thermal treatment. The layer of fig.1b has been deposited on an exactly oriented (100) substrate which has only been baked. Fig.1c has been taken using a g_{200} reflection in dark field. It shows a typical layer grown on a baked and thermally etched vicinal substrate. It is clear that baking at 200°C and pre-growth heating at 800°C of a misoriented substrate lead to a significant improvement of the quality of the epilayer: baking eliminates microtwins and threading dislocations and a pre-growth treatment allows an effective self-annihilation of antiphase boundaries (APB) within a very short distance from the interface.

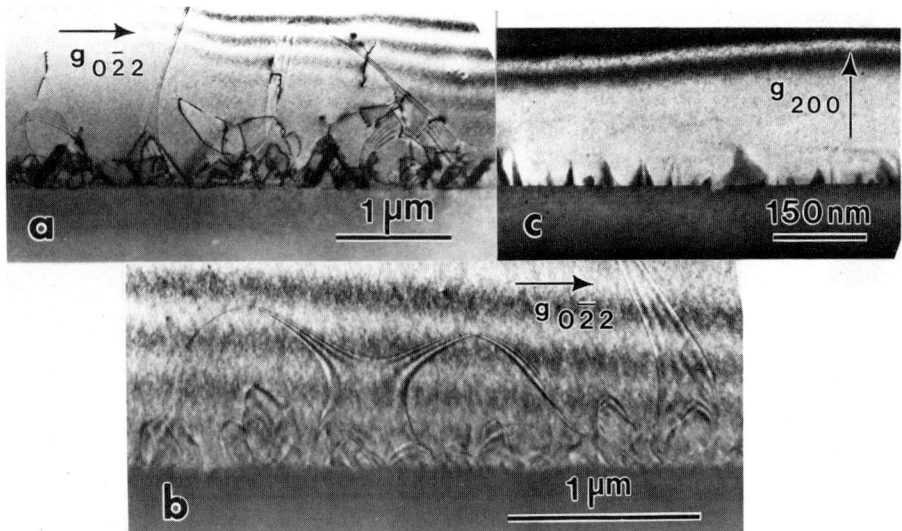

Fig.1: a) GaAs on (100)Ge with no baking or pregrowth treatment. b) GaAs on a baked (100)Ge substrate. c) [200] dark field image of GaAs on a baked and thermally etched (100)Ge vicinal substrate.

Fig.2 is the electron diffraction pattern (EDP) taken from the layer of fig.1a. It shows that, along the two <111>* directions of the reciprocal lattice, two extra spots come in between two typical [011]* azimuth diffraction spots related to the bulk lattice. One of them, the strongest one marked by t, is a twin spot while the other one, marked by d, is attributed to the double diffraction. Thus the presence of these extra spots makesthe periodicity three times smaller along the two <111>* directions. The corresponding high resolution lattice image of the EDP is shown in fig.3a: the region where the periodicity is no longer equal to the spacing d_{111} of the {111} planes but equal to $3d_{111}$ can be explained by the overlapping of the twin and its matrix related to each other by a 180° rotation around the [111] or [1$\bar{1}\bar{1}$] twin axis (Wu et al 1991). Thus, twins can stop naturally and disappear from the growing layers. A second example, a Σ=3 {211} incoherent twin boundary viewed edge-

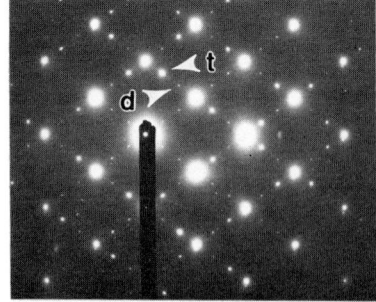

Fig.2: [011] EDP of the heterostructure shown in fig.1a. Twin reflection spots are marked by t and double diffraction spots by d.

on, appears in fig.3b. But twins can also propagate until they meet a terminating obstacle such as another twin (fig.1a) or an APB, as shown in fig. 3c: the APB is the diffuse bright band marked by arrows. The stopping boundary, the (11$\bar{1}$) plane viewed edge-on, gives the orientation of the antiphase facet which interacts with the twin.

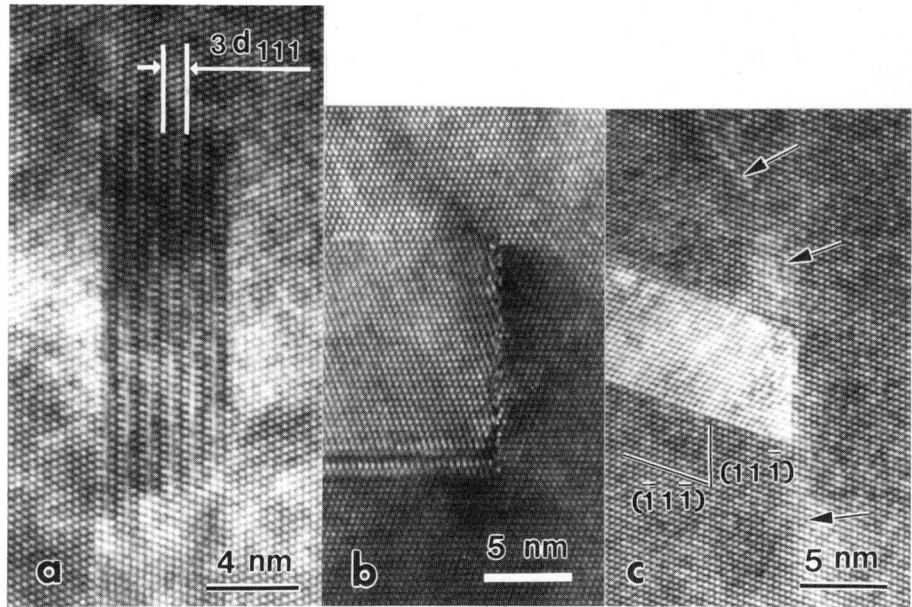

Fig.3: a) (111) or (1$\bar{1}\bar{1}$) boundary between a twin and its matrix. In the region where they overlap, the periodicity is multiplied by 3. b) Σ=3 {211} incoherent twin boundary. c) Interaction between a twin and an APB (marked by arrows): the twin stops at the (11$\bar{1}$) plane.

The fundamental difficulty involved with the growth of a polar compound such as GaAs on a non polar substrate such as Ge is the antiphase disorder. A uniform As or Ga prelayer on a (100) Ge surface that is either atomically smooth or has exclusively double-atom height steps will result in single-domain GaAs. Kroemer (1987) suggested that high-temperature anneal of off-axis (100) crystals leads to surfaces with only double steps of the same sublattice. The dark field imaging with a g_{200} superlattice reflection, which is an imaging condition sensitive to antiphase domains (APD) as reported by Posthill et al (1988), revealed that small APBs remain, in spite of the pre-growth treatment. This result indicates that step doubling has not been

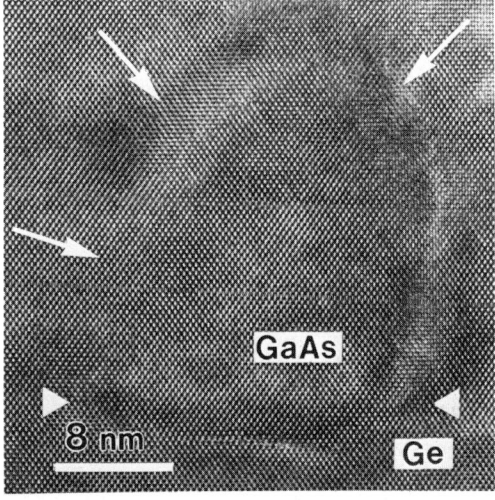

Fig.4: HREM picture of one of the APDs of fig.1c.

completed or that Ga and As are simultaneously transported in the beginning of the deposition. However, APDs are observed to annihilate within about 50 nm from the interface and this result is similar to what is obtained by Adomi et al (1991). Fig.4 shows a picture of one of these APDs: its height does not exceed 28 nm.

The size of the APDs corresponds to the size of the initial terraces at the surface of the substrate, as suggested by the comparison between fig.1b and 1c. Close to the interface, the APDs have inclined boundaries which make their annihilation easier. Beyond 1 μm from the interface, the APBs are rather vertical (fig.1b), their separating distance increases and their elimination becomes less probable. Thus, if a substrate of higher misorientation had been used, the height of remaining APBs would have probably been shorter.

Misfit dislocations have also been investigated by HREM although the good match between GaAs and Ge (Δa/a=0.07%) makes their observation difficult. However one dislocation has been found in the sample of fig.1c and is viewed in fig.5: it is a 60° dissociated dislocation. The 90° partial is located in the interface and the 30° partial is in the film so that the extra half-plane belongs to the film in tension. For a low mismatch heterostructure it is easier to observe the misfit dislocations that run parallel to the foil because they propagate long lengths before threading in the layer or interacting with another misfit dislocation. Examination of this sample under weak beam dark field imaging conditions revealed that all the misfit dislocations were dissociated. This result confirms that the defect of fig.5 is a misfit dislocation.

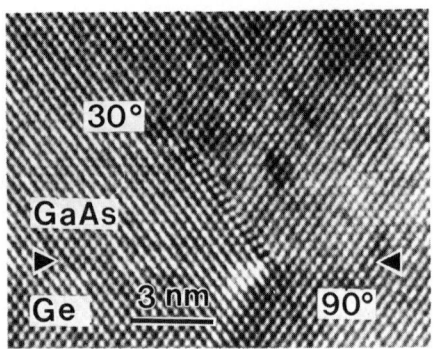

Fig.5: a 60° dissociated misfit dislocation.

The nucleation mechanism of the misfit dislocations of this heterostructure has been investigated in detail by Burle et al (1993) using transmission X ray topography. From various layers whose thicknesses ranged from 0.3 to 4 μm, a critical thickness equal to 1.3 μm has been found.

4. CONCLUSION

High quality GaAs epitaxial layers have been grown by CSVT on vicinal (100) Ge substrates that have been successively baked at 200°C and thermally etched at 800°C prior to growth. Microtwins and threading dislocations have been entirely eliminated. APBs still occur but they annihilate each other within 50 nm from the interface. HREM has been a useful technique to show different twin boundaries and the dissociated misfit dislocations.

REFERENCES

Adomi K, Strite S, Morkoç H, Nakamura Y and Otsuka N 1991 J. Appl. Phys. 69 (1) 220
Burle N, Guelton N, Pichaud B and Saint-Jacques RG 1993 This Conference
Chavez F, Mimila-Arroyo J, Bailly F and Bourgoin J C 1983 J. Appl. Phys. 54 6646
Côté D, Dodelet J P, Lombos B A, Dickson J I 1986 J. Electrochem. Soc. 133 1925
Kroemer H 1987 J. Vac. Sci. Technol. B 5 1150
LeBel C, Cossement D, Dodelet J P, Leonelli R, DePuydt Y and Bertrand P 1993 J. Appl. Phys. 73 (3) 1288.
Posthill J B, Tarn J C L, Das K, Humphreys T P and Parikh N R 1988 Appl. Phys. Lett. 53 1207
Wu X J, Li F H and Hashimoto H 1991 Phil. Mag. B63 931

Inst. Phys. Conf. Ser. No 134: Section 6
Paper presented at Microsc. Semicond. Mater. Conf., Oxford, 5–8 April 1993

369

Defect study of InAs/GaAs and GaP/GaAs double heterostructure

G Aragón, S I Molina, Y González*, L González*, F Briones* and R García

Departamento de Química Inorgánica, Universidad de Cádiz, Apdo. 40, Puerto Real, 11510 Cádiz. Spain.
*Centro Nacional de Microelectrónica, CSIC, Serrano 144, 28006 Madrid. Spain.

ABSTRACT: A defect characterization of two highly lattice-mismatched semiconductor heterostructures (InAs/GaAs and GaP/GaAs) grown by Atomic Layer Molecular Beam Epitaxy has been performed by Transmission Electron Microscopy. The experimental results show that layers grown under compression exhibit a lower planar defect density than those grown under tension. This behaviour can be explained by the dissociation state of dislocations which depends on the sign of the stress in each layer.

1.INTRODUCTION

In the last few years a big effort has been made to study the influence of the stress sign on the defect structure of lattice-mismatched epitaxial layers. It has been established that layers grown under compression contain fewer stacking faults than those grown under tension (Petruzzello and Leys 1988). Based on the nucleation mechanisms of partial (30° and 90°) and perfect (60°) dislocations, two models have been developed to explain the defect structure of epitaxial layers (Mareé et al 1987, Park et al 1992). Both models conclude that the nucleation of 90° partial dislocations is predominant in tensile layers while the nucleation of 60° perfect dislocations constitutes the main relaxation mechanism in compressive layers. Therefore, dislocations tend to be dissociated in tensile layers but perfect in compressive layers.

However, the experimental data which support these models can be misleading as the layers undergoing stress of different sign have different chemical compositions, and therefore exhibit different physical properties which may control the relaxation mechanisms.

In order to clearly determine the dependence of the defect structure on the stress sign, layers of the same material system (GaAs in our experiments) suffering tensile or compressive stress were studied. With this aim we have grown GaAs layers sandwiched between InAs for the tensile case and GaP for the compressive case. The density and type of defects contained in GaAs layers are clearly different for both cases. A relation between the planar defect density in GaAs layers and the stress sign (tensile and compressive) for each layer will be shown.

2.EXPERIMENTAL

Two double heterostructures have been grown by Atomic Layer Molecular Beam Epitaxy (ALMBE) (Briones et al 1989) at 350°C on an (001) oriented GaAs substrate. The structures consist of a GaAs buffer layer followed by two layers (InAs and GaP in samples A and B, respectively) sandwiching a GaAs layer of 200 nm width (Fig 1a and 1b).

Cross sectional samples were prepared along the two <110> directions which are contained in the (001) growth plane. A TEM study has been performed using a JEOL 1200

EX transmission electron microscope operating at 120 kV and a JEOL 2000 EX microscope operating at 200 kV.

300nm	InAs(2)	ALMBE
200nm	GaAs(2)	ALMBE
300nm	InAs(1)	ALMBE
300nm	GaAs(1)	-buffer-

GaAs (001)

Fig 1a. Sample A: InAs/GaAs double heterostructure

200nm	GaP(2)	ALMBE
200nm	GaAs(2)	ALMBE
200nm	GaP(1)	ALMBE
600nm	GaAs(1)	-buffer-

GaAs (001)

Fig 1b. Sample B: GaP/GaAs double heterostructure

3.RESULTS

3.1 Sample A: InAs/GaAs double heterostructure

Fig 2 shows the defect distribution in the different layers of sample A. At first sight it is clearly observed that a high density of planar defects appears in the GaAs(2) layer. The measured values of the dislocation (N_{TD}) and planar defect (N_{MT}) densities are presented in Table 1. It is worth noting that the planar defect density is higher than the dislocation density in the GaAs(2) layer (tensile).

sample A	N_{TD}	N_{MT}
InAs(2)	$12. \, 10^{10}$ cm^{-2}	$1.5 \, 10^{10}$ cm^{-2}
GaAs(2)	$6.9 \, 10^{10}$ cm^{-2}	$7.4 \, 10^{10}$ cm^{-2}
InAs(1)	$9.0 \, 10^{10}$ cm^{-2}	$0.3 \, 10^{10}$ cm^{-2}

Table 1. Dislocation (N_{TD}) and planar defect (N_{MT}) densities of different layers of sample A.

3.2 Sample B: GaP/GaAs double heterostructure

Fig 3 shows a cross section TEM image of sample B. Defect density measurements are presented in Table 2. The GaAs(2) layer grown under compressive stress shows a density of planar defects which is lower than the dislocation density and this behaviour is opposite to that observed for the GaAs(2) layer in sample A.

sample B	N_{TD}	N_{MT}
GaP(2)	$3.8 \, 10^{10}$ cm^{-2}	$6.7 \, 10^{10}$ cm^{-2}
GaAs(2)	$3.6 \, 10^{10}$ cm^{-2}	$2.9 \, 10^{10}$ cm^{-2}
GaP(1)	$3.1 \, 10^{10}$ cm^{-2}	$3.7 \, 10^{10}$ cm^{-2}

Table 2. Dislocation (N_{TD}) and planar defect (N_{MT}) densities of different layers of sample B.

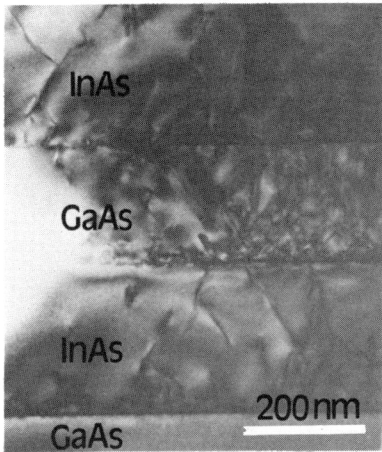
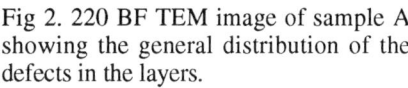
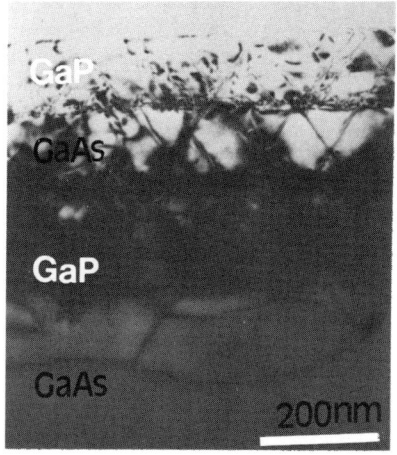

Fig 2. 220 BF TEM image of sample A showing the general distribution of the defects in the layers.

Fig 3. 220 BF TEM image of sample B with its defect distribution. Note the existence of large dislocation loops in the GaAs buffer layer.

4. DISCUSSION

Data from Tables 1 and 2 show that the values of dislocation densities in GaAs(2) and sandwiching layers are different for both samples. Differences in the values for the different layers are qualitatively explained by the lattice misfit which is markedly different for samples A (7.2%) and B (3.6%). However, arguments based only on the different magnitude of the lattice mismatch can not explain why the majority defects appearing in a GaAs(2) layer grown under tensile stress (sample A) are planar defects, while the predominant defects are perfect dislocations in a GaAs(2) layer grown under compressive stress (sample B). This difference in the ratio between planar defect and dislocation density can be explained by a correlation of "stress sign-planar defect density" (Marée et al 1987 and Park at al 1992). When a GaAs layer is grown under tension, dislocations tend to be dissociated and stacking faults appear. This tendency is expected to extend to the majority of dislocations and, therefore, the planar defect density becomes larger than the dislocation density.

The same trend is observed in the sandwiching layers of the two samples. The InAs layers in sample A are grown under compression and have a planar defect density smaller than their dislocation density, whereas the GaP layers in sample B are grown under tension and show a reverse relation between the two types of defects. However, in spite of the importance of the stress sign, other parameters (e.g. the dislocation mobility and the growth mechanism) may influence the defect structure of materials systems.

5. CONCLUSION

Experimental data on the influence of the stress sign on the type of defects in strained layers have been presented. The results are in good agreement with theoretical predictions. The stress sign is an important parameter to be taken into account for the control of the planar defect density in III-V/III-V heteroepitaxy.

ACKNOWLEDGEMENT

The present work has received financial support from the Spanish Interministerial Commission of Science and Technology, CICYT, Project MAT 1205/91 and from The Junta de Andalucía under the group 6020. This work has been carried out at The Electron Microscopy Facilities of The University of Cádiz.

REFERENCES

Petruzzello J and Leys M R 1988 Appl. Phys. Lett. <u>58</u> 2414
Marée P M J, Barbour J C, Van der Veen J F, Kavanagh K L, Bulle-Lieuwman C W T and Viegers M A P 1987 J. Appl. Phys. <u>62</u> 4413
Park H H, Lee J K, Lee E H, Lee J Y and Hong S K 1992 Mat. Res. Soc. Symp. Proc. <u>263</u> 479
Briones F, González L and Ruiz A 1989 Appl. Phys. A<u>49</u> 729

Inst. Phys. Conf. Ser. No 134: Section 6
Paper presented at Microsc. Semicond. Mater. Conf., Oxford, 5–8 April 1993

373

The effect of growth interrupts on CBE grown InP

PD Brown, EG Bithell[†], CJ Humphreys, PJ Skevington,* PJ Cannard* and GJ Davies*

Department of Materials Science and Metallurgy, University of Cambridge, Pembroke Street, Cambridge, CB2 3QZ.
[†]Now at Department of Materials, University of Oxford, Parks Road, Oxford OX1 3PH.
*B.T. Laboratories, Martlesham Heath, Ipswich, IP5 7RE.

ABSTRACT: InP layers were grown by chemical beam epitaxy (CBE) and paused during growth. Surfaces stabilised under P_2, As_2 and no group V gas for variable interrupt times were examined using the combined techniques of TEM, SIMS and XRD. TEM contrast was only obtained for samples paused under arsine, confirming that As substitution for P occurs. SIMS and XRD data indicate that substitution of underlying layers continues right up to the longest interrupt time rather than stabilising at one monolayer As coverage. Strain features associated with longer interrupt times are possibly indicative of the early stages of precipitation.

1. INTRODUCTION

During InP//InGaAs growth by CBE a growth interrupt is used to stabilise the InP surface prior to InGaAs deposition. It is necessary to optimise this switching sequence from InP to InGaAs as there is a critical balance between the pumpdown time for P_2 and the time for stabilisation under As_2, during which time site substitution might occur [Foxon and Joyce 1978, Davies *et al* 1980]. The group V gases are precracked in CBE and so surfaces are stabilised under P_2 and As_2 molecules rather than PH_3 and AsH_3. These interrupts may in turn have consequences for compositional grading across the interface which can significantly alter the shape of a quantum well, and hence, the electrical and optical properties of a structure [Jusserand *et al* 1985]. The additional introduction of regions of high strain may have adverse effects on device lifetimes and stability. Hence, there is a need to characterise this stage of the growth process.

2. EXPERIMENTAL

A series of test samples was grown in a modified VG Semicon V80H system [Davies *et al* 1991] at 500°C to investigate the effect of pausing InP growth under P_2, no group V and As_2. The gas switching sequence employed for samples interrupted under As_2 is shown in Fig. 1. A 0.5μm InP buffer layer was grown and initially paused under P_2 to allow the trimethylindium (TMIn) to fully switch, leaving a P-terminated surface. The P_2 was then pumped out and the surface exposed to As_2 for a variable time Δt. The As_2 was then switched out and the P_2 flux re-established before introducing TMIn and resuming InP growth. 200Å InP spacer layers separated successive growth interrupts and the sequence of 1,2,4,8,16 and 32 seconds was repeated 4 times. A further 0.3μm InP capping layer contained an InAs marker layer. The same sequence was used when pausing samples under P_2, in which case TMIn was switched out for a variable time; and again for the case of no group V, whereby all

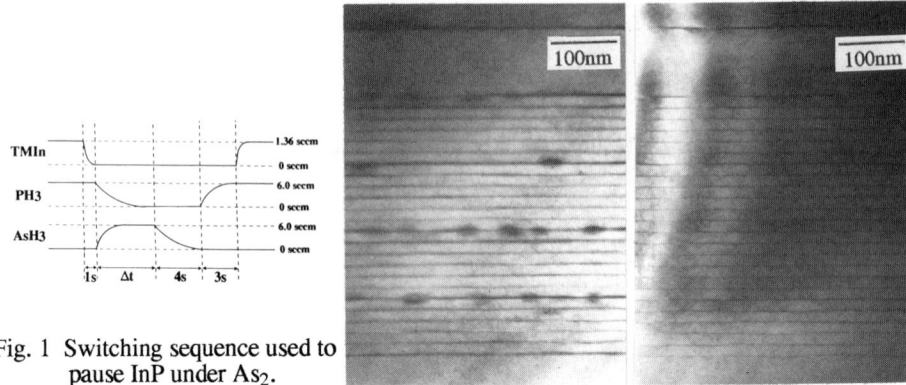

Fig. 1 Switching sequence used to
       pause InP under As$_2$.

Figs. 2a) [110] and b) [1$\bar{1}$0] CTEM images of arsine interrupted InP (g=002 bright field and
       dark field respectively).

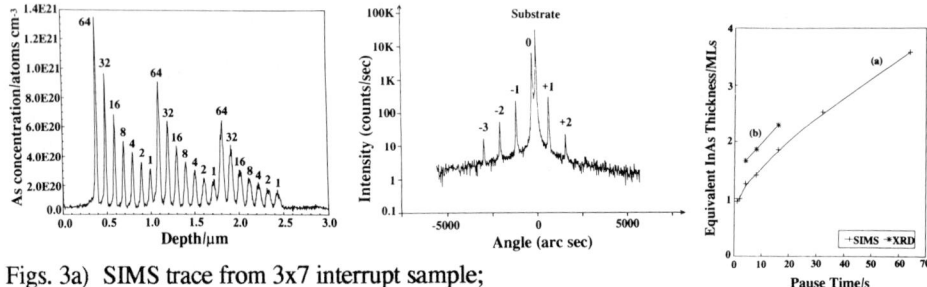

Figs. 3a) SIMS trace from 3x7 interrupt sample;
b) X-ray trace from 25x8 second interrupt sample and
c) plot of equivalent InAs MLs introduced by As$_2$ interrupts from SIMS and X-ray data.

the gases were pumped out. TEM observations were performed using JEOL 2000FX and
4000FX instruments. SIMS analysis was carried out in a Cameca IMS 3F using a low energy
(2.5keV) O$_2^+$ primary ion beam and detecting positive secondary ions. X-ray rocking curves
were measured using the 004 symmetrical reflection on a Philips high resolution
diffractometer.

3. RESULTS AND DISCUSSION

Only the samples paused under arsine exhibited any contrast using conventional TEM
imaging techniques and the 4x6 interrupt layers are shown in Figs. 2a and 2b, imaged along
orthogonal [110] and [1$\bar{1}$0] projections respectively. Absolute polarity of these samples was
determined using CBED [Spellward and James 1991; Burgess et al 1993]. The sixth layer of
each set corresponding to the longest (32 second) growth interrupt shows strongest contrast.
Steps typically 25Å in size are apparent along the first few interrupts for the [1$\bar{1}$0] sample
projections (Fig. 2b). It is interesting to note that these features, which we believe to be due to
bunching of monolayer surface steps, appear immediately once growth interruption occurs
($\Delta t=1$ sec in this instance) and, in the absence of any gross substrate misorientation, suggests
that step bunching occurs extremely quickly or that large steps already exist on the advancing
InP growth surface. Undulations were apparent along the lower interrupt interfaces imaged in
the [010] projection, which is consistent with a 3D growth mode superimposed on some slight
sample tilt. The orthogonal [110] projection (Fig. 2a) exhibited additional strain features on the
longest interrupt layers, and hence anisotropy also plays a role in the formation of these

Fig. 4 HREM images of arsine growth interrupts for a) [1 1̄ 0] and b) [010] projections.

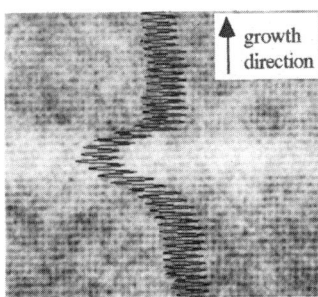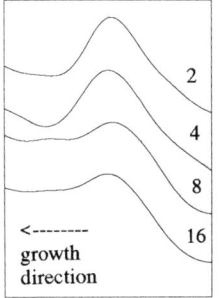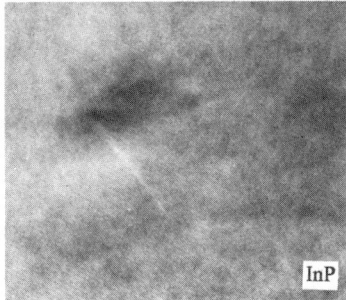

Fig. 5a projected line trace across lower interrupt of 4b.

Fig. 5b smoothed line profiles across 2, 4, 8 and 16 second interrupts.

Fig. 6 HREM image of strain feature at 32 second growth interrupt.

interfacial (presumably InAsP) layers. The sample was highly susceptible to damage by the (200keV) electron beam with the development of damage (background speckle) after a few minutes observation. The sample examined by SIMS (Fig. 3a) contained a sequence of 3x7 interrupts. Resolution of the profile decreases with depth due to roughening at the crater bottom. Concentrations were quantified to an accuracy of about ±20% by comparison with an ion implanted calibration specimen of As in InP. The ion implanted specimen had been calibrated independently by neutron activation analysis. The quantity of As incorporated at each interrupt was calculated by integrating areas under the peaks and averaging the three values obtained for each Δt. A plot of equivalent InAs monolayers introduced with time is shown in Fig. 3c. This confirmed that As substitution of underlying layers continues with increasing interrupt time rather than saturating at one monolayer coverage. Samples examined by XRD contained a sequence of 25 identical interrupts thereby generating a superlattice (SL) structure. Three samples were grown with Δt values of 4, 8 and 16 seconds. Satellite peaks characteristic of strongly periodic structures were clearly visible in the rocking curves (*e.g.* Fig. 3b from the 25x8 second interrupt sample). The mean mismatch of each superlattice was derived from the separation of the substrate and zero order peaks, whilst superlattice periods were derived from satellite peak spacings. The equivalent amount of InAs introduced is approximated by (mean SL mismatch x SL period)/mismatch of InAs, and this data is also plotted in Fig. 3c, and while indicating a slightly higher proportion of As incorporation as compared with the SIMS data, the same trend of increasing substitution with interrupt time is evident.

Figs. 4a and 4b show the line of dark contrast associated with these arsine interrupts in high resolution, imaged along the [1$\bar{1}$0] and [010] projections respectively. A line of sharp contrast showing undulations to within 1 unit cell is present at the top surface of each growth interrupt, while a more gradual change in contrast is associated with the region below each growth interrupt, and this is more clearly illustrated in the projected line scan shown in Fig. 5a. This effect is again consistent with As swapping for P during the growth pause, but not for P swapping with As after InP growth has been resumed. Smoothed projected line profiles from interrupts corresponding to 2,4,8 and 16 seconds are shown in Fig. 5b. It is apparent that there is little difference in the form of each curve for the range of interrupt times.

HREM simulations of [010] projected InP/InAsP/InP wedges using Stadelmann's EMS software allowed the quick assessment of a wide range of thickness and defocus conditions, and was used to gain some appreciation of the level of detectability of As within InP by HREM. Consideration of just variation in interrupt composition demonstrated that even a monolayer of $InAs_{0.1}P_{0.9}$ should be detectable within $\geq$ 300Å thick foils in HREM, with a monolayer of InAs being detected within a 200Å foil. However, HREM micrographs showing consecutive interrupts, *i.e.* incorporating the range of growth interrupt times from 1 to 32 seconds showed no great difference in the contrast of each interrupt. The general lack of variation in visible contrast across successive interrupts suggests that either interfacial strain is more likely to be dominant than compositional effects, or else that substitution occurs at a fairly low rate for this range of interrupt times, in accordance with SIMS and X-ray data.

Counting 002 fringes between the interfaces shown indicated a spacing of 200±5Å for both the [1$\bar{1}$0] and [010] oriented sample foils, which is demonstrative of the excellent lateral uniformity and control afforded by the CBE technique. The orthogonal [110] projection also revealed contrast around the strain related features shown in the CTEM micrograph. The dark/bright lobe contrast around this feature (Fig. 6) suggests that it is due to the initial stages of precipitate formation. Again, these interfaces were strongly influenced by the imaging (400keV) electron beam and became washed out after a few minutes observation.

The lack of visible contrast from P_2 and no group V interrupts indicate that such procedures are likely to cause no problem during CBE of InP//InGaAs. Both SIMS and X-ray data suggest that a finite coverage of InAs (0.9 - 1.4MLs) will be obtained for zero interrupt time under As_2. This is believed to reflect the contribution from arsenic pumpdown, and subtracting this from the plotted data indicates that less than 1ML equivalent of InAs will be formed in the 15 - 18 seconds used to stabilise the As_2 flux in the standard InGaAs/InP switching sequence, and hence that substitution by As is not expected to have a major detrimental effect on the quality of InP/InGaAs interfaces.

ACKNOWLEDGEMENTS

This work was carried out under the LINK/ASM/MOMBE initiative. PDB would like to thank Chris Boothroyd, Rafal Dunin-Borkowski, Caroline Baxter, Angus Kirkland and George Burgess.

REFERENCES

Burgess W G, Saunders M, Bird D and Humphreys C J 1993 Microbeam Analysis '93
Davies G J, Skevington P J, Scott E G, French C L and Foord J S 1991 J. Crystal Growth 107 999
Davies G J, Heckingbottom R, Ohno H, Wood C E C and Calawa A R 1980 Appl. Phys. Lett. 37 290
Foxon C T and Joyce B A 1978 J. Crystal Growth 44 75
Jusserand B, Alexandre J, Paquet D and Le Roux G 1985 Appl. Phys. Lett. 47 301
Spellward P and James D 1991 Inst. Phys. Conf. Ser. No. 119 375

Defect characterization in MOCVD InP/GaInAs/InP layers

N Y Jin, G R Booker and R Blunt*

Department of Materials, University of Oxford, Parks Road, Oxford OX1 3PH
*Epitaxial Products International Ltd, Cypress Drive, St. Mellons, Cardiff, CF3 0EG, UK

ABSTRACT: This paper investigates room-temperature (RT) lattice-matched InP/GaInAs/InP layer structures grown on (001) InP substrates by MOCVD. For some specimens the final as-grown layers exhibited surface features not seen on control specimens included in the same growth run. Correlations between extended crystallographic defects revealed by TEM and such surface features have been established. A possible nucleation mechanism for interface dislocations that occurred is proposed based on the TEM results.

1. INTRODUCTION

Hetero-epitaxial layers of III-V compound semiconductors are widely used for optoelectronic and high speed microwave devices. For good quality devices, epitaxial layers free from misfit interface dislocations are required. The conditions for the growth of thick GaInAs layers on InP without interface dislocations were studied by Nakajima et al (1980). The nucleation and propagation of interface dislocations in heteroepitaxial layers have been widely discussed, e.g. by Tan (1981), Eaglesham et al (1989) and Maree (1987). In addition to interface dislocations, both stacking faults and microtwins have been observed in some III-V epitaxial layers to accommodate the mismatch strains. This paper investigates nominally lattice-matched InP/GaInAs/InP layer structures which may be used for PIN infra-red detectors. Correlations between extended crystallographic defects revealed by TEM and surface features are established. A possible nucleation mechanism for the interface dislocations is proposed based on the TEM results.

2. EXPERIMENTAL

The layers grown by MOCVD at 650°C were (starting from the surface) 0.5μm InP(capping)/4μm $Ga_{.47}In_{.53}As$/0.7μm InP(buffer) on (001) InP Fe-doped LEC substrates and for specimens of surface type D only 2μm InP(capping)/4μm $Ga_{.47}In_{.53}As$/2μm InP(buffer) on (001) InP S-doped VGF-grown substrates. The RT lattice mismatch for all these specimens was measured by X-ray diffraction to be less than 2×10^{-4}. The final surface of the as-grown layers was examined by optical microscopy (OM) with Nomarski contrast. Two types of thin foils were prepared for TEM observations. Plan-view (P-) specimens were obtained by mechanical polishing and chemical thinning from the substrate side with a Cl_2-methanol solution until penetration. A thin layer from the top surface was removed by a short time exposure in the solution. Cross-sectional (X-) specimens were made from two strips cleaved perpendicularly from the wafer and glued face to face, then mechanically polished and finally ion-milled with Ar^+ at liquid-N_2 temperature on both sides.

3. RESULTS

3.1 Surface Morphology
Although the growth of the epitaxial layers was under the same control, some specimens of the final as-grown layers exhibited surface features not seen on control specimens included in the same growth run when examined by OM. Four types of wafer (Table) were distinguished. For type A, no features occurred. For type B, surface decorations, i.e. shallow surface pits, occurred of density ~ 2×10^4cm^{-2} (Fig. 1a) . For type C, surface decorations and cross-hatch occurred, the latter consisting of orthogonal lines along <110> directions (Fig.1b). For type D, heavy cross-hatch occurred (Fig.1c).

Table Correlation between layer surface and microdefects

Type	Surface by Nomarski	Defects by TEM
A	good	-
B	surface decorations	SFP, TD*
C	cross-hatch and surface decorations	ID T/B, Cor. ID, TD
D	heavy cross-hatch	ID C/T, ID T/B

ID C/T: interface dislocation at InP-capping/GaInAs interface.
ID T/B: interface dislocation at GaInAs/ InP-buffer interface.
Cor.ID: correlated interface dislocations at GaInAs/buffer and buffer/substrate interfaces with a 55° configuration.
TD: threading dislocation.
SFP: stacking-fault pyramid.
*Threading dislocation starting at buffer/substrate interface.

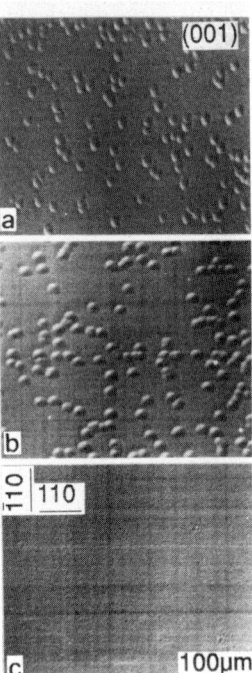

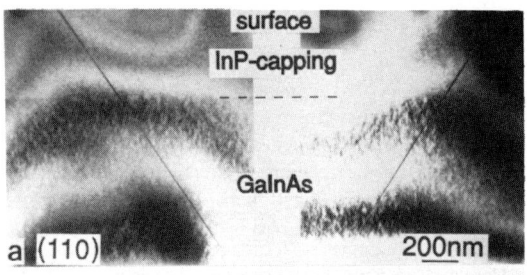

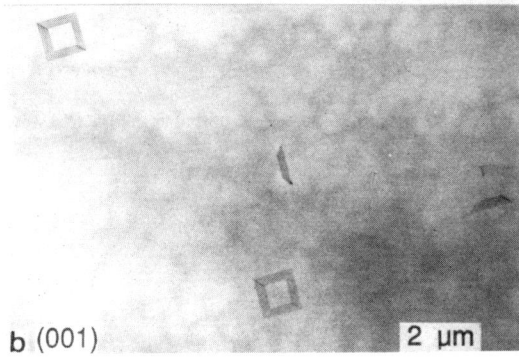

Fig. 2 SFPs observed by TEM in a wafer of type B: (a) SFP on {111} planes in a X-specimen penetrating GaInAs and capping layers; (b) truncated SFP in plan-view specimen.

Fig. 1 Surface morphology observed by OM: (a) surface decorations; (b) cross-hatch and surface decorations; (c) heavy cross-hatch.

3.2 TEM Observations
No extended crystallographic defects were observed by TEM in wafers of type A. The general features observed in wafers of type B viewed in X-specimens were stacking-faults (SFs) on the four {111} planes (Fig.2a) penetrating the GaInAs and InP-capping layers. The distance between the intersections of the two SFs in Fig.2a with the top surface of the InP-capping layer was ~8μm. In P-specimens, truncated stacking-fault-pyramids (SFPs) and their residuals were observed (Fig.2b). The mean density of these SFPs measured from such P-specimens was ~5×10^4cm^{-2}. The

due to chemical thinning was not known. Fig.3 provides direct evidence from a X-specimen that the initiation of the SFPs was at the substrate/buffer interface. The depth of the apex of a SFP calculated from the distance of ~8μm between the intersections of the two SFs in Fig.2 gave the same result. Analysis of the contrast of a truncated SFP for various diffraction conditions indicated that all the SFs of the pyramid were extrinsic. Defects running along the epitaxial growth direction [001], possibly dislocation dipoles, were occasionally observed and these also started at the buffer/substrate interface. No interface dislocations were observed in this wafer.

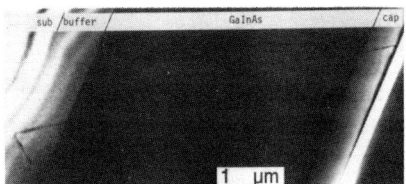

Fig. 3 Initiation of a SFP at the substrate/buffer interface.

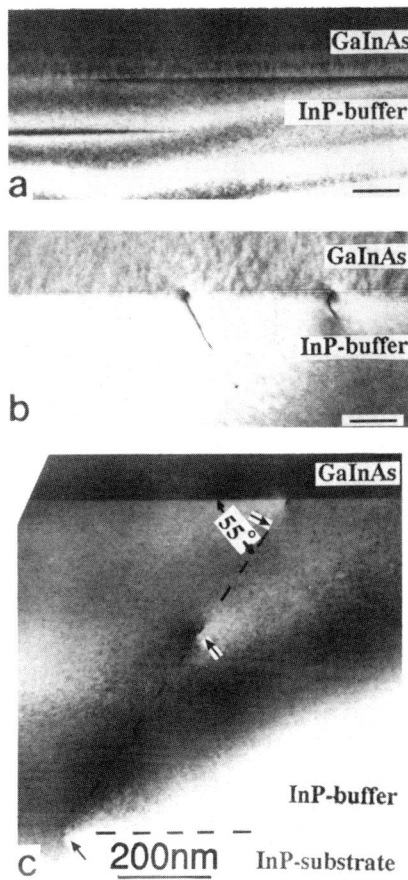

Fig. 4 Dislocations at the GaInAs /buffer and buffer/substrate interfaces in a wafer of type C: (a) in-plane dislocations at the GaInAs/buffer interface; (b) two end-on dislocations at the GaInAs/buffer interface; (c) correlating interface dislocations.

In X-specimens of the wafers of type C, dislocations were frequently observed at the GaInAs/InP-buffer interface along the two <110> directions (Fig.4). When these dislocations were in the foil plane (Fig.4a) they were typically some tens of microns long, while when end-on (Fig.4b) their observed lengths were limited by the foil thickness. The distances between these interface dislocations varied, and were occasionally as small as 0.6μm (Fig.4b). Threading dislocation segments, connecting with the end-on interfacial dislocations, were on the InP-buffer layer side (Fig.4b). Often the end-on interface dislocations at the GaInAs/buffer and buffer/substrate interfaces were correlated, i.e. the line drawn between them made an angle of ~55° with the interface (Fig.4c), indicating that the two end-on interfacial dislocations were lying on the same inclined {111} plane. Some threading dislocations which penetrated the substrate/buffer/GaInAs layers were also observed.

In the wafers of type D, dislocations were observed in X-specimens at both the InP capping/GaInAs and GaInAs/InP buffer interfaces and along the two <110> directions. The distance between these dislocations was ~0.6μm at the former interface and ~1.2μm at the latter interface. Examples of the dislocations at the GaInAs/buffer interfaces are shown in Fig.5. The density of interface dislocations was generally higher than for wafers of type C. All the threading dislocation segments were located within the InP-buffer layer.

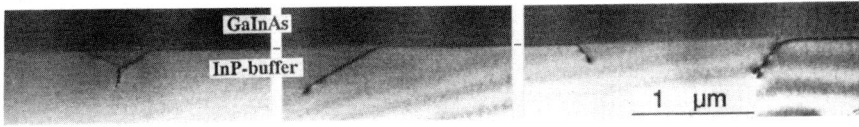

Fig. 5 Dislocations in a wafer of type D at GaInAs/buffer interface.

4. SUMMARY AND DISCUSSION

A summary of the main defects observed by OM and TEM in these wafers is given in the Table. The surface decoration correlates with threading defects including SFPs and threading dislocations, and the cross-hatch correlates with dislocation networks at the interfaces. The latter interface dislocations were mainly observed at the substrate/buffer and the GaInAs/buffer interfaces except in the case of the heavy

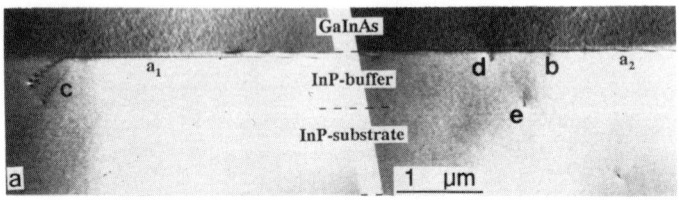

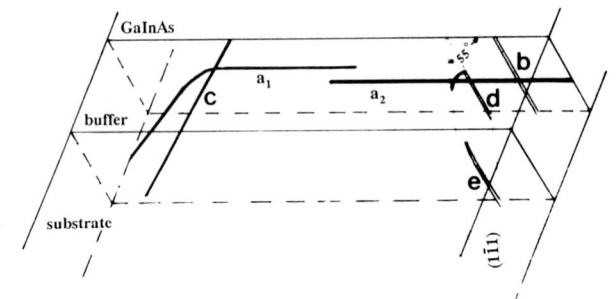

Fig. 6 Dislocation configuration in the GaInAs/InP buffer/InP-substrate layers for a wafer of type C: (a) TEM; (b) schematic diagram of the geometry of the dislocation configuration shown in (a).

cross-hatch, where they were also observed at the capping/GaInAs interface. Threading segments of these dislocations were always present in the buffer layer. These results suggest that the nucleation of the interface dislocations occurs in the buffer layer. A typical dislocation configuration in the region of the GaInAs/InP buffer/InP substrate observed in wafers of type C is shown in Fig.6a. Stereomicrographs provided a three dimensional view of the dislocations, which is schematically illustrated in Fig.6b. Burgers vector analyses indicated that both in-plane and end-on interface dislocations, e.g. a and b, were of 60°-type, and the threading dislocation c was not of edge type. However, the Burgers vectors of a_1 and a_2 were different, lying on $(\bar{1}11)$ and $(1\bar{1}1)$ planes respectively. The two correlated end-on interfacial dislocations on the GaInAs/buffer and buffer/substrate interfaces respectively, d and e, were lying and dissociated into partials on the same $(1\bar{1}1)$ plane, and were bent towards each other. These results, therefore, suggest that such 55° configurations of end-on interface dislocations result from dislocation loops initiated at the buffer/substrate interface during growth and subsequently expand along the {111} planes and meet the GaInAs/buffer interface.

ACKNOWLEDGEMENTS

The authors would like to thank DTI and SERC UK under a LINK initiative for their support and D. Baker for growing the specimens.

REFERENCES

Eaglesham D J, Kvan E P, Maher D M, Humphreys C J and Bean J C 1989 Phil. Mag. A 59 1059
Maree P M J, Barbour J C and van der Veen J F 1987 J. Appl. Phys. 62 4413
Nakajima K, Komiya S, Yamaoka T and Ryuzan O 1980 J. Electrochem. Soc. 127 1568
Tan T Y 1981 in 'Defects in Semiconductors', MRS Proc., Elsevier North-Holland, vol 2 p.163

Measurement of composition and thickness of thin semiconductor layers by EPMA

U Zeimer, F Bugge and S Gramlich

Ferdinand-Braun-Institut für Höchstfrequenztechnik, Rudower Chaussee 5, D-1199 Berlin, Fed. Rep. Germany

ABSTRACT: An EPMA-deconvolution technique, based on the method of Pouchou and Pichoir (1990) is applied to composition investigation of thin single- and multilayer structures in the InGaAs/GaAs- and AlGaAs/GaAs-system. Down to layer thicknesses of ≥ 10 nm the results are in good agreement with other analytical methods like Low Temperature Photoluminescence (LT-PL) and X-ray Diffractometry (XRD).
EPMA results do not depend on strain and quantum size effects. Therefore this method can also be used for the composition determination of partly relaxed InGaAs-layers on GaAs. For the investigation of multilayer structures we used samples bevelled at a very small angle ($\alpha \approx 0.1°$).

1. INTRODUCTION

Semiconductor devices like laser diodes or transistors often are based on heterostructures of thin layers. For optimisation and process control a precise evaluation of thickness and composition of the different layers is essential. Our special device of interest is a separated confinement InGaAs/GaAs/AlGaAs-QW-strained layer laser. We used EPMA for the investigation of such a device structure, because PL and XRD cannot give information about all layers within the structure.

The EPMA deconvolution technique was calibrated on perfect single AlGaAs- and InGaAs- layers on GaAs by the results of PL and XRD. Then it was applied to bi- and multilayer systems.

2. APPLIED METHODS

2.1 EPMA

Conventional quantitative $\Phi\rho z$-deconvolution (i.e. XPP-program by KEVEX) was used in a first step to determine the composition of thin ($\leq 0.5~\mu m$) ternary layers. Here the low energy L-lines of the characteristic X-ray emission of the heavy elements were analyzed.. In this case one can reduce the energy (and thus the penetration depth) of the exciting electron beam to achieve a better depth resolution. The thin layer then can be considered as "bulk" material. The results of this approach were compared to the thin film deconvolution procedure STRATA after Pouchou and Pichoir (1990).

The idea is to compare the intensity of a certain line of characteristic emission from a layered sample and from a bulk standard. The standard has to contain all the elements, which are found in the layer. For several different energies (and penetration depths) of the electron beam both intensities have to be measured. Then the intensities (k-values) are used for a deconvolution with the generation function in the bulk material and in the layered specimen. If the layer only consists of elements that are not present in the underlying substrate, the thin film program deduces both composition and layer thickness. Unfortunately, in our case (AlGaAs- or InGaAs-layer on GaAs-substrate) overlayer and substrate contain common elements. Here one of the parameters has to be known for the determination of the other. We measured the layer thickness by a different method to give a starting value for the program to find the composition. The final thickness was deduced from the best fit of the theoretical k-curves to the experimental data.

2.2 Low Temperature PL

To allow for an assessment by PL, the layers have to be of high optical quality. The samples were excited by an Ar^+-laser at 10 K sample temperature. The emission wavelength λ_{max} depends on the composition of the ternary layer. For the lattice matched AlGaAs-system the AlAs atom fraction of thick layers, where quantum size effects must not be taken into account, can be determined by the formula:

$$x = (E_g - 1.516)/1.4 \quad , \quad E_g = 1.239/\lambda_{max} \quad (1).$$

For the highly mismatched InGaAs/GaAs-system the situation is more complicated. The emission wavelength depends not only on the composition, but also on the degree of lattice relaxation due to misfit dislocations. In the case of fully relaxed layers the formula for InGaAs/InP, found by Goetz et al (1983) can be applied to the InGaAs/GaAs-system, Nickel et al (1990). If the layer is fully strained (without relaxation) the formula of Andersson et al (1988) can be used:

$$E = 1.515 - 1.144x + 0.255x^2 \quad (2).$$

For partly relaxed layers it is difficult to determine the composition from the emission wavelength. Furthermore a high doping level, often present in devices, shifts λ_{max} and makes an analysis difficult. For very thin layers (QW s) λ_{max} depends very strongly on the layer thickness due to the quantum size effect, Kolbas et al (1988). So one of the parameters (d or x) has to be known for the interpretation of the PL results.

2.3 X-ray Diffractometry

The shape of the measured X-ray rocking curves (004-reflection) depends on misfit strain within the layered structure and the crystal perfection. If the layer is nearly perfect, the composition can be deduced from the peak separation and the thickness from the oscillations by comparison to rocking curves calculated from the Takagi-Taupin formula.

If the layers are partly relaxed, a broadening of the layer peak, a decrease of intensity and a peak shift toward the substrate peak occur. Assessment of such layers is possible by analysis of asymmetric reflections. However, this analysis is difficult and fails if the degree of relaxation varies with depth. In case of very thin single layers on a substrate usual XRD is not possible due to the low intensity from such small scattering volumes.

3. RESULTS AND DISCUSSION

3.1 Single Layers

In the lattice matched $Al_xGa_{1-x}As$/GaAs-system layers with different thickness but identical composition (x = 0.22) were grown by MOCVD. First conventional EPMA (ZAF-correction) at different e^--beam energies was carried out using the Al-K_α-line (1.487 keV) and the L_α-lines of Ga (1.096 keV) and As (1.282 keV). Due to the low line energies even layers as thin as 100 nm can be considered as "bulk" material. The results were compared to that of the thin film STRATA deconvolution, which was applied to all samples. Excellent agreement was found. In Tab. 1 the results of the STRATA procedure are compared to other methods.

Tab. 1: Comparison of measured AlGaAs composition

sample-Nr.	d [nm]	x-EPMA	x-PL	x-XRD
#1	50	0.229	-*	-**
#2	100	0.230	0.210	-**
#3	300	0.225	0.226	0.228

*   PL-intensity too low, **   no layer peak

For the AlGaAs single layers on GaAs the STRATA deconvolution was the only method to determine the composition of layer #1. The reproducibility of this method was about ± 0.5 At% and therefore the same as known for conventional EPMA.

For $In_xGa_{1-x}As$-layers on GaAs, grown by MOCVD, the situation is more complicated due to the large lattice mismatch. For x≈0.25 there is an onset of lattice relaxation by misfit dislocations for layer thicknesses ≥15 nm, Ballingall et al (1990). Therefore, no XRD investigations could be made due to the deterioration of the rocking curves. The EPMA results depend only on the composition of the ternary layer and are **not** influenced by strain. However, for the excitation of the In-L_α-line (3.287 keV) a high beam energy (> 7keV) is needed. This makes thin film deconvolution necessary for all layers with d < 0.5 μm. The results of the composition and thickness determination are shown in Tab.2.

Tab.2: Comparison of measured InGaAs thickness and composition

sample-Nr.	d [nm]	d (EPMA)	x_V***	x-EPMA	x-LT-PL	x-XRD
#1	11*	13	0.266	0.222+	0.200	-
#2	50**	58	0.300	0.222	0.223	-
#3	80**	80	0.300	0.300	0.227	-
#4	380**	380	0.250	0.208	0.206	-

* measured by ellipsometry, ** measured by SEM , *** In/Ga-concentration in the vapour phase
+ error in d-determination of 20% (2 nm) causes error of the x-value of 15 % (0.03) using the same k-values

Except for sample #3 a good agreement between PL and STRATA-results is obtained. The difference for this sample is probably due to concentration inhomogeneities within the sample and different diameter of the excited area (PL: 150 μm, EPMA: 10 μm).

Our results show that thin film deconvolution is a powerful method for composition investigation of single layers, especially for partly relaxed layers down to layer thicknesses of 10 nm.

3.2 Multilayers

First bilayers, InGaAs with a GaAs overlayer (270 nm) were grown on GaAs-substrate. The thickness of the $In_xGa_{1-x}As$ (x≈0.25) layer was varied (#1 =10 nm, #2 = 20 nm, #3 = 50 nm). The aim was to investigate the lower limit of InGaAs-thickness, measurable by STRATA. Measuring from the top at different energies only from sample #3 an information could be obtained due to the absorption of In-L_α-radiation by the GaAs overlayer. So we used ion beam etched samples bevelled at a very small angle ($\alpha \leq 0.1°$).

At equidistant points the line intensities were measured. The actual layer thickness was determined from geometry knowing α, measured by a tally-step. The results of the thin film deconvolution are indicated in Tab.3.

Tab.3: Comparison of measured thickness and composition of InGaAs in GaAs/InGaAs/GaAs-bilayer

sample-Nr.	d [nm]	d(EPMA)	x(top)+	x(α)+	x-PL	x-XRD
#1	10*	-	-	-	0.22	0.21
#2	20 *	18	-	0.25	0.24	-
#3	50**	50	0.27	0.26	-	-

* measured by XRD, ** measured by SEM, + STRATA-results, measured from the top and on the bevelled sample

For the thinnest layer (#1) x and d could be measured by XRD, because the InGaAs was fully strained. The thickness, obtained from the XRD-simulation, was used for estimation of composition from the PL-peak. EPMA could be applied to samples #2 and #3. For layer thicknesses < 20 nm the error of STRATA results increases due to errors in thickness estimation on the wedge and low In-peak intensity.

Finally, the thin film deconvolution was used for the investigation of a separated confinement $Al_xGa_{1-x}As$/GaAs/InGaAs-laser structure. This structure consists of a large number of layers with a wide range of layer thicknesses (1 μm to 10 nm) and different compositions, where the active region is buried under thick cladding layers. The EPMA-STRATA was carried out on a bevelled sample. The Al-concentration in the thick cladding (d = 1 μm) and thin waveguide layers (d < 100 nm) was determined successively starting the measurement from the substrate. Measured data of the underlying layers are used for the deconvolution of the k-values of the layer currently under study.
This way the Al-concentration of the thin waveguide layer can be determined, which is not accessable by XRD and PL.

4. CONCLUSIONS

1. Quantitative EPMA of thin layers is possible, if a deconvolution technique (i.e. STRATA) is used. If common elements are found in layers and substrate the layer thickness has to be measured by a different method.
2. For ternary single layers (AlGaAs and InGaAs) on GaAs a composition determination can be performed for thicknesses down to 10 nm. The results are in excellent agreement with PL and XRD on perfect layers.
3. The results are **not** affected by strain, dopant level or crystal perfection. That way partly relaxed layers are accessible for composition determination.
4. For the analysis of thin buried layers in thick multilayer samples (eg. DH lasers) bevelled samples can be used.

REFERENCES

Andersson T G, Chen Z G, Kulakovskii V D, Uddin A and Vallin J T 1988 Phys. Rev. B 37 4032
Ballingall J M, Ho P, Martin P A, Tessmer G J, Yu T H, Lewis N and Hall E L 1990 J Electr. Mat. 19 509-13
Goetz K H, Bimberg D, Jürgensen H, Selders J, Solomonov A V,Glinskii G F and Razeghi M 1983 J. Appl. Phys. 54 4534
Kolbas R M, Anderson N G, Laidig W D, Sin Y, Lo Y C, Hsieh K Y and Yang Y J 1988 IEEE J-Q 24 1605-13
Nickel H, Lösch R, Schlapp W, Leier H and Forchel A 1990 Surface Science 228 340-3
Pouchou J L and Pichour F 1990 Scanning 12 212-24

Localized thinning of semiconductor nanostructures for cross-sectional transmission electron microscopy

C Vieu, A Pepin, G Ben Assayag, J Gierak and F-R Ladan

L2M/CNRS, 196 avenue H. Ravera, BP 107 92225 Bagneux, France

ABSTRACT: Two new productive techniques, one combining electron lithography and reactive ion etching and the other based on focused ion beam micromachining to thin high quality cross-sectional TEM samples are proposed. Electron transparent areas are generated with a high degree of localization, within 0.1 µm, over distances of several mm in GaAs/GaAlAs heterostructures. TEM observations demonstrate that no artefacts are introduced during the preparation and different conditions of illumination are achievable.

1. INTRODUCTION

Recent development of semiconductor technology, as well as progress on quantum nanostructures, involves the fabrication of submicron devices. In many cases the reduction of dimensions prevents the use of transmission electron microscopy (TEM) as a systematic characterisation tool because of the difficulty in locating the region of electron transparency right on the area of interest. Indeed, with classical cross-sectional preparation of thin specimens involving mechanical polishing and ion beam milling, it is very difficult to pre-select the observable area with an accuracy better than 1 µm. Various attempts have been performed to overcome this problem by advanced microstructuring techniques (Kirk et al. 1989, Sweeney 1985). The purpose of this work is to present two complementary methods enabling a high degree of localization and good reproducibility. The first one uses high resolution electron beam lithography (EBL), lift-off and reactive ion etching (RIE). The second one combines electron or optical lithography with micromachining using a focused ion beam (FIB) column coupled to a scanning electron microscope (SEM).

In the first part of the paper the experimental procedures are described . The TEM observations are presented in a second section and finally, the respective advantages and limitations of the two techniques are discussed.

2. EXPERIMENTAL PROCEDURES

2.1 High resolution EBL and RIE

High resolution EBL is performed with a JEOL nanowriter at a voltage of 50 kV on a PMMA resist layer   150 nm thick. After exposure and development a 40 nm thick film of Ni is deposited on the sample and metallic lines are transferred on the surface by lift-off. The position of these lines can be easily aligned on selected regions with an accuracy better than 0.1 µm. The orientation of the lines can be chosen arbitrarily so that observations along different azimuths are possible. Then, by SiCl$_4$ RIE nearly vertical crystalline 2 μm high walls are fabricated. A very anisotropic low pressure process at 1W/cm^2 , 300 V for 20 min. is commonly used on a conventional RIE planar diode station. In comparison with previous studies (Sweeney 1985, Dobisz et al 1986), our conditions of etching enable us to define the transparent areas in a single step without any pedestal underneath the thin walls. Figure 1 presents SEM micrographs at different steps of the procedure. Figure 1a) exhibits the pattern of lines after the lift-off process. The small lines are attached to larger pads for a good mechanical stability of the features to avoid an eventual collapse of the walls during the glueing operations. The lines are segmented in small sections of different widths ranging from 50 nm to 300 nm. This makes it possible to obtain, after etching, observable areas of various thickness as can be seen in figure 1b). The etched surface of the sample is very clean, this being a crucial point to avoid any shading of the transparent areas during TEM observation. In some cases an uneven under-etching can occur as illustrated in figure 1c). Our special pattern of lines enables us to compensate this effect. Indeed, due to the different widths of the Ni etch-mask it is always possible to find a suitable area for TEM observation. In figure 1c) the smallest segments have been eroded under the Ni mask which stands suspended between the wider sections of the wall where an observation of the sample up to the surface remains possible.

After etching, the Ni mask can be removed in dilute HNO$_3$. Then, the sample is cleaved or microsawed in small pieces of around 300 μm which are glued on a grid rotated at 90° so that cross-sectional TEM observation through the thin walls becomes possible.

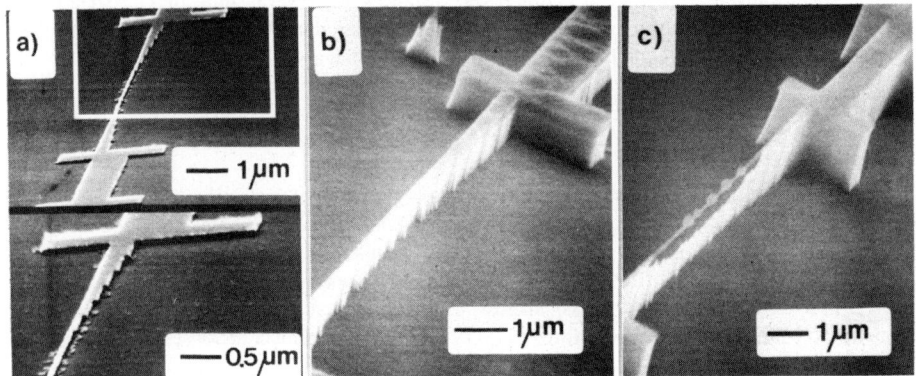

Figure 1: SEM micrographs showing the surface features at different steps of the process. Figure 1a) shows the Ni lines after lift-off, figure 1b) the thin walls of different widths after etching and figure 1c) a case of under-etching where very thin sections can be found for TEM observation.

2.2 FIB/SEM micromachining

The principle of this technique recently proposed by Young et al (1990) and Szot et al (1992) is similar to the previously described one. Thin walls are etched to allow electron transparency, but the etching is operated without mask, using a focused ion beam spot. Our FIB system developed by Ben Assayag et al (1990) in the laboratory, delivers a 30 keV Ga$^+$ spot of 30 nm FWHM diameter with a typical current of 60 pA. The originality of our approach relies on the coupling of our FIB apparatus with a SEM equipment enabling the in-situ control of the micromachining. The transparency of the wall can be visualized and the process is stopped when the thickness of the wall is optimum for a high quality TEM specimen preparation. In our conditions, the GaAs erosion speed is 10 μm^3/min, thereby most of the material must be removed by classical etching process prior to localized FIB thinning. Deep 5 μm wide mesas are realised by either optical or electron lithography following wet or dry etching. This operation roughly defines the area of observation and allows the FIB thinning to be performed in a reasonable time (around 5 min). Each mesa structure is then cleaved or sawed and mounted on a TEM grid for cross-sectional observation before FIB intervention, thus avoiding any damage on the thin areas to be caused during the manipulations. The final FIB thinning is then performed with the incident ion beam normal to the surface of the sample to obtain a mirror quality on the facets (Ben Assayag 1990). The front face is machined by scanning the ion beam from the edge of the mesa towards the inside, in order to avoid any redeposits of removed material. The back face is trimmed in the same way and the thinning is stopped when the remaining wall appears transparent to the SEM electrons. Of course, the thickness can be accurately controlled and reproduced by computer. The position of the observable area is adjusted within an accuracy of around 0.1 μm using the SEM image. Figure 2 presents a SEM micrograph of two walls of constant thickness. The top of the foil remains intact enabling TEM observation of the surface region while some details of the mesa behind the wall are already visible by transparency to electrons of 20 keV.

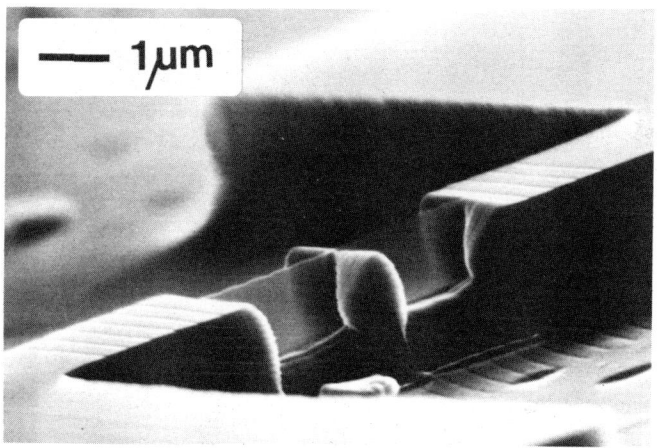

Figure 2: SEM micrograph of two transparent areas machined by FIB. The surface of the sample has been implanted with the FIB system prior to the thinning. A 0.8 μm period array of lines is clearly seen even in the thin windows

3. TEM OBSERVATIONS

Figure 3a) shows a 200 dark field image of four GaAs/Ga$_{0.67}$Al$_{33}$As quantum-wells (QWs) grown by molecular beam epitaxy (MBE). For this experiment, a wall of constant thickness of 40 nm was etched by RIE after EBL along a <110> direction. A 1 cm x 1 cm sample was prepared in various locations where photoluminescence (PL) characterization had been made. The total length of the walls which were suitable for TEM inspection was more than 1 cm and for this application we could relate the variations of the PL peak to the fluctuations of the size of the QWs from one side to the other of the starting sample. As seen on the figure no artifacts due to the preparation can be observed. Some darker regions are sometimes visible in the image. These weak contrasts arise from wall width fluctuations, as can be seen in figure 1b), presumably due to small variations of the size of the Ni line caused by the lift-off step. A lattice image, obtained in the seven-bright field mode at medium magnification is shown in figure 3b), the heterointerfaces of the second QW are clearly seen. On this special sample, the totality of the thin wall turned out to be suitable for high resolution imaging, revealing the homogeneity of the transparent area over long distances. Figure 3c) shows a striking example where a high degree of localization is required. Several 5 μm x 5 μm arrays of lines were implanted by FIB on a GaAs substrate. The sample is then prepared by EBL and RIE in order that each observable area coincides with the implanted features. Finally, on a single TEM specimen it was possible to observe the different arrays corresponding to many specific conditions of implantation. Figure 3c) shows one of these arrays for a Ga^+ 25 keV 4 x 10^{15} ions/cm^2 FIB implantation at a grating of 225 nm. The amorphous zones generated under each line overlap, leading to a wavy crystal-amorphous interface.

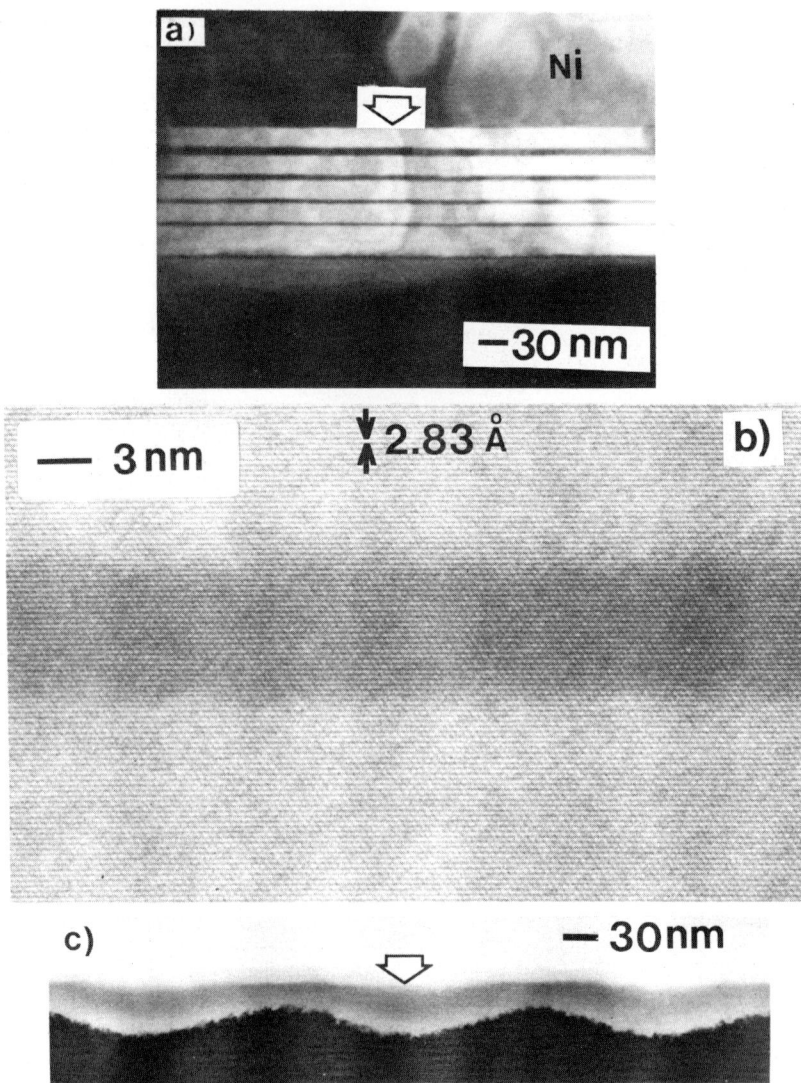

Figure 3: Examples of cross-sectional TEM observations along [110] for III-V specimens prepared by EBL and RIE. Figure 3a) is a 200 dark field image of four GaAs/Ga$_{0.67}$Al$_{33}$As QWs grown by MBE. Figure 3b) is a lattice image of the second QW (24 monolayers thick). Figure 3c) presents an array of FIB implanted lines on GaAs, where a wavy amorphous layer is formed due to the overlap of the damaged regions under each line. The surface of the sample is indicated by an arrow.

Figure 4 presents a GaAs/GaAlAs heterostructure grown by MBE after the FIB/SEM thinning procedure. It consists in a 6 nm GaAs/Ga$_{0.67}$Al$_{.33}$As QW inserted between two short period GaAs(1 nm)/AlAs(1 nm) superlattices. This sample was used for experiments on intermixing generated by SiO$_2$ encapsulation and annealing. The image was obtained after the deposition of a 100 nm thick SiO$_2$ film by rapid thermal chemical vapour deposition (RTCVD). A very good contrast is obtained under 200 dark-field illumination at the heterointerfaces, demonstrating that the heterostructure is not affected by the SiO$_2$ deposition except near the surface where small defects can be observed. As seen on the figure, the transparent area seems very uniform without artifacts of preparation. Moreover, no intermixing has been induced even for the short period superlattice which is very sensitive to the disordering. However, on several samples, when no attention is paid to the dwell-time, the presence of precipitates induced by the FIB thinning have been observed and attributed to Ga-oxided droplets. Reducing the time per pixel eliminates these artefacts. In addition, when a very thin wall (<80 nm) is machined, the amorphous layers generated on each side of the wall by the FIB irradiation begin to overlap, preventing the observation of the structures. This ultimate size should be reduced using a lower Ga$^+$ ion energy to thin the specimen. In every case, the transparent areas machined by FIB were always of uniform thickness, demonstrating the quality of the facets as can be seen on figure 2.

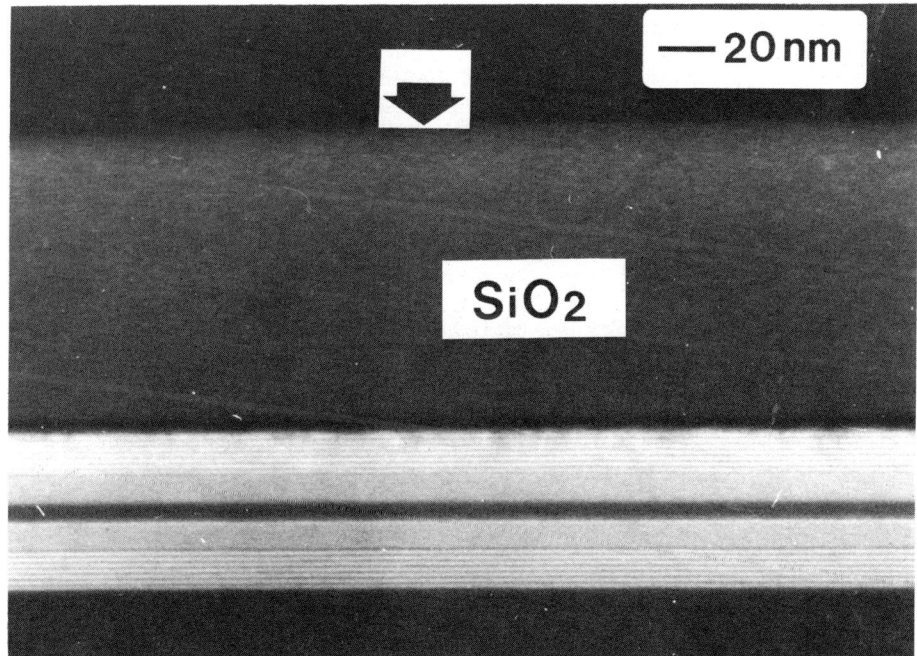

Figure 4: Dark-field 200 micrograph of a 6 nm GaAs/Ga$_{0.67}$Al$_{.33}$As QW inserted between two short period GaAs(1 nm)/AlAs(1 nm) superlattices after deposition of a SiO$_2$ layer by RTCVD. The arrow indicates the top of the SiO$_2$ film.

4. CONCLUSION

Two new techniques have been developed to produce cross-sectional TEM samples of good quality. Both of them are able to locate the region of transparency with an accuracy better than 0.1 μm. A large number of observable areas can be realised over a long distance, the only limitation being the size of the sample compatible with the specimen holder. Lateral nanostructures can be observed due to the high localization of the thin sections. Different examples of application have been presented which demonstrate that thin walls without artefacts can be obtained up to the surface. Compared to time-consuming conventional techniques, these preparations are rapid, with a high success rate. Despite the high cost of these procedures, it is interesting to point out that the instruments necessary to implement these techniques are commonly used by people involved in lateral nanostructure fabrication, who are the most interested users of highly localized preparation.

The procedure based on EBL and RIE can produce very thin sections suitable for high-resolution images but is only applicable to materials susceptible to be etched through choice of appropriate mask and reactive gases (Wetzel et al 1989). It can also suffer from slight changes in the RIE station leading to excessive under-etching. Due to small fluctuations in the size of the lines obtained by lift-off, some slight variations in the thickness of the transparent area are visible particularly for very thin walls.

In comparison, the method relying on FIB/SEM machining has a success yield close to 100%, thanks to the in situ SEM visualization of the etching that can be interrupted at any time. This procedure can be applied to any material since no chemistry is involved during etching. The facets of the walls processed by FIB are smoother than the features fabricated by RIE. However, artifacts can be generated for thin sections of less than 80 nm due to amorphous layers generated by the ion beam at an energy of 30 keV. In the future, thinning experiments will be carried out at a lower energy in order to evaluate the minimum thickness achievable without artefacts.

The two techniques thereby appear as very complementary, and are currently implemented in parallel to increase the chance of   efficient TEM characterization. Indeed, the large pads generated by EBL and RIE next to the thin walls (see figure 1) can be thinned by FIB/SEM machining. In a given area of the sample it is then possible to observe a transparent area generated by RIE and another one thinned by FIB, thus combining the respective advantages of the two methods.

REFERENCES

Ben Assayag G, Sudraud P, Gierak J, Remiens D, Menigaux L and Dugrand L 1990
    Microcircuit Engineering 11 413
Dobisz E A, Craighead H G, Beebe E D and Levkoff J 1986 J. Vac. Sci. Technol.
    B4 850
Kirk E C G, Williams D A and Ahmed H 1989 Microscopy of Semiconducting
    Materials Inst. Phys. Conf. Ser. 100 501
Sweeney J 1985 J. Vac. Sci. Technol. B3 918
Szot J, Hornsey R, Ohnishi T and Minagawa S 1992 J. Vac. Sci. Technol. B10 575
Wetzel J, Jost M, Rishton S A, Fryer P M, Kwietniak K T, Klaus D,
    Bucchignano J J, Hu C K and Brown T J 1989 Ultramicrosc. 29 110
Young R, Kirk E C G, Williams D A and Ahmed H 1990 Microcircuit Engineering 11 409

Inst. Phys. Conf. Ser. No 134: Section 6
Paper presented at Microsc. Semicond. Mater. Conf., Oxford, 5–8 April 1993

Recent advances in ion milling techniques and instrumentation for TEM specimen preparation of materials

R Alani and P R Swann

Gatan R&D, 6678 Owens Drive, Pleasanton, CA 94588, USA.

ABSTRACT: In recent years, TEM specimen preparation techniques and instrumentation have been significantly advanced. For example, a specimen post designed for low angle milling in conventional ion mills has considerably simplified the production of difficult specimens. The excessive milling time associated with low angle milling stimulated the development of a Penning ion gun, which was subsequently incorporated into a new ion milling system. The latest version of this system, equipped with an ion beam modulator and a CAIBE attachment, is described along with examples of high quality TEM specimens.

1. INTRODUCTION

The preparation of high quality TEM specimens has been significantly simplified by recent advancements in ion milling techniques and instrumentation. For example, the introduction of a single-side specimen post for low angle ion milling in conventional ion mills opened up new possibilities to alleviate the traditional ion milling problems such as differential thinning rates, surface roughness and limited electron transparent areas (Alani, Jones and Swann 1990a). The application of this technique was rapidly expanded to include cross sectional TEM (XTEM) specimens of multilayered semiconductors, metals, as well as powders and fibers (Alani 1990b). In conventional ion mills, this technique has now become a preferred method for the production of difficult TEM specimens (Humiston et al 1991, Anderson et al 1991 and Mitwalsky 1992).

The excessive milling times required to obtain the results of low angle ion milling in conventional ion mills stimulated the development of an improved ion mill using newly designed, high powered Penning ion guns. The design and construction of an earlier version of the system has been reported elsewhere along with first examples of high quality TEM specimens that were obtained of semiconductors, ceramics and metals (Alani and Swann 1991). This instrument has been continuously upgraded by new design elements and attachments to prepare higher quality TEM specimens in shorter times for a wider variety of materials.

In this article, we describe some features of the new ion mill and a technique for thinning XTEM specimens with high differential milling rates. This technique combines very low angle ion milling (<4°) capabilities of the ion mill with its ion beam modulation system. We also describe a new chemically assisted ion beam etching (CAIBE) attachment to the original instrument that facilitates the production of artifact free TEM specimens of certain compound semiconductors, e.g., InP, with enhanced milling rates.

2. SYSTEM DESCRIPTION

In the course of designing the new ion mill, the first objective was to provide much higher thinning rates. This was achieved by incorporating miniature rare earth magnets into a modified Penning ion gun. The second objective was to provide high precision gun alignment capability to aim the ion beams accurately at the center of the specimen. This feature is particularly important when milling at very low angles.

The standard unit contains two ion guns (120° apart) that mill the specimen at a fixed milling angle of 4°. The fixed angle was chosen to simplify the operation of the unit. The specimen holder used in the instrument is the same single post used in the conventional ion mills for low angle ion milling. The guns are operated at 1keV to 6keV. Careful design of the gun ion optics has virtually eliminated cathode aperture erosion. As a result, gun maintenance is rarely required and specimen contamination from the ion gun is minimal. Specimen contamination is also reduced by using an oil free vacuum system which is maintained by a diaphragm pump backing a molecular drag pump. Short specimen exchange times, excellent beam alignment, improved specimen mounting which minimizes beam heating effects and precise termination of thinning are among other features of the instrument.

Shown below are some typical milling rates at 4° obtained for various materials using one ion gun operating at 5keV with no specimen rotation:

Material	Milling rate(μm/hr/gun)
Copper	18
Silicon	15
Silicon Carbide	8
Stainless steel 316	7
Tantalum	4

With these high thinning rates it is quite practical to prepare specimens at 4° in reasonably short times, particularly with two guns in operation. On the other hand, it is seen that materials differ in milling rates even at 4°, which may lead to a serious differential thinning rate problem in composites. One solution for this problem would be to lower the milling angle. The design of the instrument allows the selection of milling angles less than 4° by using a spacer under the specimen post. However, in the case of XTEM specimens, even at very low milling angles, uneven milling can still occur, when the ion beam travels along the multilayer interface (epoxy line). To solve this problem an ion beam modulator was designed and incorporated into the unit.

3. ION BEAM MODULATION

The ion beam modulator combines fast electronic (on/off) switching of the ion beam with variable speed specimen rotation. It was possible to design such a system, because the new ion guns were able to deliver stable ion beams after passing through rapid on/off cycles. In this system, the specimen rotates at 1rpm to 6rpm (adjustable) for the slow milling sector and 12 rpm for the fast sector. The ion beam is off and the rotation is maximum when the ion gun is aligned along the interface (epoxy line). The angular range of the slow (beam on) sector perpendicular to the epoxy line is 60°. The system allows selection of a single or double slow sector mode.

The effectiveness of the ion beam modulator with a milling angle of 4° has been illustrated by the results of complex multilayered cross section of various materials (Alani and Swann 1993). However, it has been noticed that some special XTEM specimens required even lower milling angles than 4° for proper preparation. Fig. 1 shows a cross sectional image of such a specimen, taken in a TEM at the accelerating voltage of 120keV. It can be seen that all the layers including Al, W, TiN, Ti, SiO_2 and the Si substrate are electron transparent, despite their wide spread chemistry (atomic number) and high differential milling rates. The initial wafer was cross sectioned and dimpled to about 7μm using a technique described elsewhere (Alani et al 1990a). The dimpled specimen was milled at 5keV under the ion beam modulation system with a milling angle of 3° for the dimpled side and 2° for the flat side. The total milling time was about 3 hrs.

4. CAIBE ATTACHMENT

Another new attachment to the instrument is a CAIBE system. This required an upgrade in the vacuum system to handle the reactive gases. Initially, CAIBE was designed for conventional ion mills and become an established method for TEM specimen preparation of certain materials (Alani, Jones and Swann 1990a). In this method, the specimen is exposed to a reactive gas

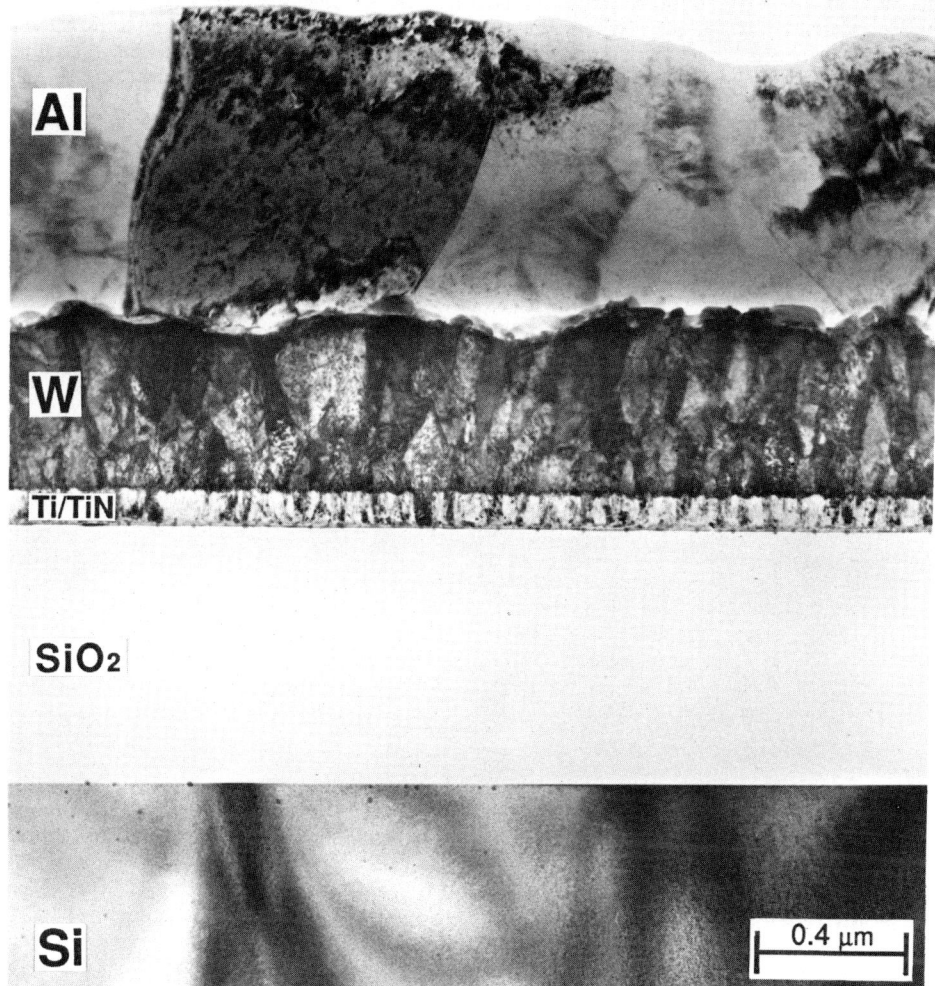

Fig. 1. XTEM image of a Al/ W/TiN/Ti/SiO2/Si specimen. Uniform thinning was achieved by ion milling at very low angles (<4°) in the new ion mill under a beam modulation system. (Bulk specimen courtesy of Dr. A. Mitwalsky, Siemens Research Labs, Munich, Germany)

through a jet assembly and an inert gas ion beam. Therefore, the thinning occurs by chemical reactions and physical sputtering enhancing the milling rate. The reactive gas can be generated from a solid source which sublimes e.g. iodine or it can be injected directly from a pressurized gas bottle e.g. nitrous oxide. Nitrous oxide in combination with a xenon ion beam has been used for cross sectioning TEM specimens of diamond films on silicon (Mardinly 1991).

A well known application of the iodine CAIBE method is the thinning of InP. In order to test the capability of the new iodine CAIBE attachment, a plan view specimen of InP was dimpled to a thickness of 10μm and milled at 4° at room temperature. Only one gun was used (5keV) to gain better control over the milling process. The iodine flow was adjusted until the specimen chamber pressure reached ≈ 8x10^{-5} Torr. Fig. 2 shows a typical TEM image of the InP specimen after the iodine CAIBE milling. It is observed that the specimen is free from indium island artifacts that are formed after conventional argon ion milling. The features seen on the micrograph are dislocations produced during mechanical prethinning of the specimen. The total milling time was about 10 minutes, corresponding to a thinning rate of 1μm/min/gun.

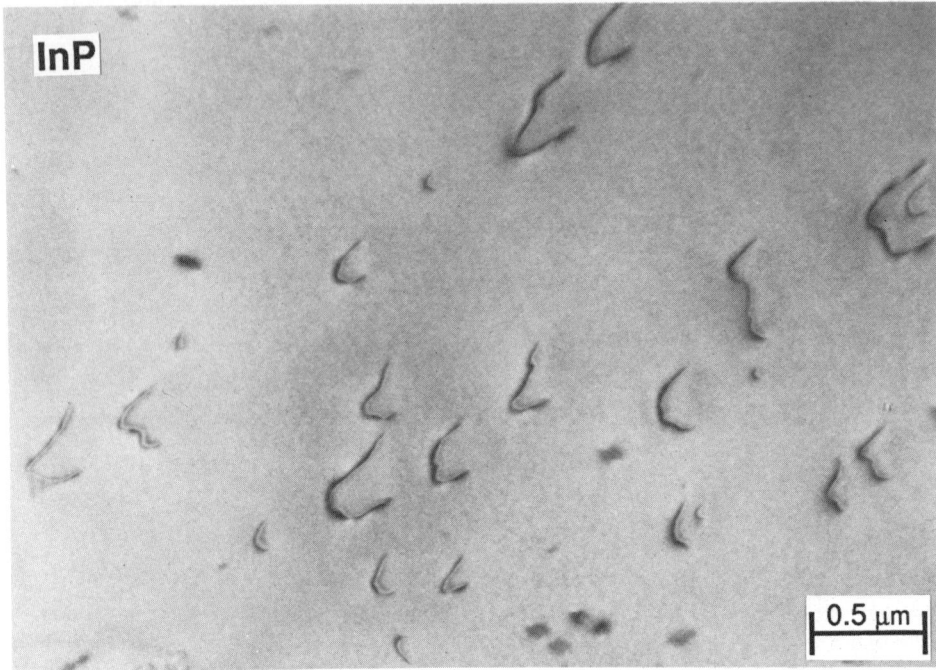

Fig. 2. Typical bright field TEM micrograph of InP after low angle (4°) iodine CAIBE milling at room temperature in the new ion mill. Note, the specimen is free from indium islands artifacts.

5. CONCLUSION

It is shown that the new low angle ion mill is capable of very high thinning rates, which facilitates a rapid production of high quality TEM specimens. With attachments of the ion beam modulator and the CAIBE system a wider variety of specimens/materials can now be readily prepared.

REFERENCES

Alani R, Jones J S and Swann P R 1990a Proc. MRS Symp. 119, Workshop on Spec. Prep. for TEM of Materials II, ed R M Anderson, pp 85-101
Alani R 1990b Proc. of Int. Symp. on Electron microscopy in China, eds K Kuo and J Yao (Hong Kong: World Scientific), pp 461-478
Alani R and Swann P R 1991, Proc. MRS Symp. 254 Workshop on Spec. Prep. for TEM of Materials III, eds R M Anderson, B M Tracy and J C Bravman, pp 43-63
Alani R and Swann P R to be published in Proc. of Microbeam Analysis Society 1993
Anderson R and Benedict J 1991 Proc. MRS Symp. 254, Workshop on Spec. Prep. for TEM of Materials III, eds R M Anderson, B M Tracy and J C Bravman, pp 141-148
Humiston H L, Tracy B M and Dass L 1991, Proc. MRS Symp. 254, Workshop on Spec. Prep. for TEM of Materials III, eds R M Anderson, B M Tracy and J C Bravman, pp 211-221
Mardinly A J 1990 private communication
Mitwalsky A 1992 private communication

Inst. Phys. Conf. Ser. No 134: Section 6
Paper presented at Microsc. Semicond. Mater. Conf., Oxford, 5–8 April 1993

395

Surfactant effect of Bi and Sb on Si/Ge/Si(001) heteroepitaxy

H Matsuhata, K Sakamoto, K Kyoya*, K Miki and T Sakamoto

Electrotechnical Laboratory, 1-1-4 Umezono, Tsukuba, Ibaraki, 305 Japan
*On leave from Meiji University

ABSTRACT: The surfactant effect of 1ML Bi or Sb on Si/Ge/Si heteroepitaxial growth was investigated using SIMS, HRTEM and RHEED analysis techniques. Improvement in the abruptness at the heterointerface of Si/Ge(4ML)/Si was observed using SIMS. While a small amount of residual Sb was detected in the grown Si layer, no significant Bi was detected in the specimen except at the top surface. In the cross-sectional HRTEM observations, though vast improvement was seen in the abruptness at the heterointerface, the existence of distinct atomic steps at the heterointerfaces indicates that the flatness at the heterointerfaces still needs improvement.

1. INTRODUCTION

Recent development of the molecular beam epitaxy (MBE) technique enables us to fabricate Si/Ge heterostructures with precise control of crystal growth at the atomic layer level. However, high-resolution transmission electron microscope (HRTEM) observations of such heterointerfaces have indicated that a Si/Ge heterointerface is neither significantly more abrupt nor flatter than a Ge/Si heterointerface. This unsatisfactory morphology at the Si/Ge heterointerface is partly due to the Stranski-Krastanov (S-K) mode growth of Ge on a Si surface, and the degraded abruptness at the heterointerface is due to the surface segregation of Ge during the growth of a Si layer on a Ge layer.

To overcome these problems, an application of a surfactant, which is always segregated on the growing surface and which also reduces the surface free energy, was proposed (Copel et al, 1989). Though the exact mechanism of a surfactant in the conversion from an S-K-type growth mode to Frank-Van der Merwe (F-VM) type in Ge/Si heteroepitaxy is still somewhat uncertain (Copel et al, 1990), such application has recently been attracting considerable attention. Already certain elements, eg Sb, As, Ga, have been investigated for such application in epitaxial growth (eg Copel et al, 1989, 1990). When surfactants are applied to materials used in electronic devices, they should be removed after fabrication so that they are not incorporated into the materials. Even if incorporated, however, they should not have harmful deep impurity levels. Since Bi has one of the smallest segregation coefficients and since it forms a shallow donor level, it is a preferable candidate for a surfactant. Reflection high-energy electron diffraction (RHEED) experiments confirmed that Bi behaves as a surfactant and can be easily removed by re-evaporation at relatively low temperatures (Sakamoto et al, 1993). Also, Sb is a known surfactant investigated using in situ RHEED in epitaxial growth experiments of Ge/Si (Sakamoto et al, 1992). In this report the effect of one monolayer (1ML) Sb or 1ML Bi layer on the heterointerface of a Si/Ge/Si(001), is investigated in more detail using secondary ion mass spectroscopy (SIMS) and HRTEM techniques to see how the morphology at the heterointerfaces is altered.

2. EXPERIMENTS

For SIMS analysis and HRTEM observations, specimens with a heterostructure illustrated in

Fig. 1 were fabricated using the MBE method at 673K. The sample consisted of four Ge layers, each having a 4ML thickness, separated by 30 nm Si layers. For saturation coverage (1ML), either Bi or Sb atoms were deposited on the top surface of the third Ge layer. Each MBE grown specimen was then cut into two pieces: one for SIMS analysis and one for TEM observations. SIMS measurements were made to detect ^{74}Ge and ^{209}Bi secondary ions with O_2^+ primary ions accelerated at 3keV, 80nA. Cross-sectional specimens for HRTEM observation were prepared using standard Ar-ion etching techniques. A 400keV HRTEM was operated at 200keV to reduce the interdiffusion induced by the radiation.

3. RESULTS

The results obtained using SIMS analysis are shown in Figs. 2a and 2b for 1ML Sb and 1ML Bi, respectively. In both figures, the Ge profile exhibits an abrupt change at the third Ge layer where the surfactants were deposited. The concentration gradients in the Ge profiles above the third Ge layer are steeper and the minimum Ge concentration in the Si spacer layers is lower than those without a surfactant. The decay length, in which the Ge content decays to e^{-1} determined from the slope of Ge in Fig. 2, improved from 1.3nm to 0.6nm after surfactant adsorption. The change in the Ge profile clearly shows that both elements suppress the surface segregation of Ge atoms during Si overgrowth and improve the abruptness of the Si/Ge interface. Even though Bi atoms were deposited on the third Ge layer, the Bi profile shows that most Bi atoms were segregated at the top surface. The counts of Bi secondary ions in the grown layer were at the noise level, indicating that the amount of Bi incorporated was below the detection limit of the SIMS instrument, *ie*, 5×10^{16} counts/cm^3. On the other hand, as much as 10^{19} counts/cm^3 corresponded to residual Sb incorporated in the Si layer. Since these layers were then heavily doped n-type, Fig. 2a shows that Sb is not a particularly suitable surfactant for electronics application.

Figure 3a shows a cross-sectional TEM image of the first 4ML Ge fabricated without surfactants. This layer is the first Ge layer in Fig. 1, and corresponds to the first Ge peak in Fig. 2b. In this Fig. 3a, particularly white and straight contrast designated by an arrow is observed at the Ge/Si heterointerface, showing that the interface is smooth and flat. Since the size of a white dot in the contrast in the high-resolution picture corresponds to the length of a dumbbell-shaped array of 2 atoms in the [110] orientation, the dark Ge layer seems thicker than 4ML. Indented interfaces of Si/Ge may be due to the S-K growth mode of Ge, since a 4ML Ge layer is already beyond the critical thickness for this mode. Also, the contrast of the Si/Ge heterointerface is not clear due to the surface segregation of Ge to the overlaid Si.

Figure 3b is the third layer of the specimen where Bi was deposited on Ge(4ML)/Si. This layer is the third layer in Fig. 1, and corresponds to the third Ge peak in Fig. 2b, showing that the abruptness at the Si/Ge heterointerface was improved.

Figure 3c shows the fourth layer of the specimen with Bi. This layer is the fourth Ge layer in Fig. 1, and corresponds to the fourth peak in Fig. 2b, showing that the abruptness at the Si/Ge heterointerface was greatly improved. Atomic steps at the heterointerface can clearly be observed. These steps at the heterointerface can be seen not only at the Si/Ge heterointerface, but also at the Ge/Si heterointerface due to the surfactant effect. The steps are not clearly visible at the Ge/Si heterointerface of the other layers, suggesting the smoothness at the Ge/Si has been degraded by the surfactant. The same heterointerface morphology changes were observed also in the specimen with Sb as a surfactant.

4. DISCUSSION AND CONCLUSIONS

In Fig. 3a, fine, dark contrast was observed inside the Si layer around the Ge/Si heterointerface. While it is difficult to conclude whether this contrast is mainly due to the interdiffusion of Ge into the Si layer underneath or due to the strain around the Ge/Si heterointerface, very little interdiffusion has been reported at the Ge/Si heterointerface (Marèe *et al*, 1987). Hence, contrast observed in the Si substrate side may be mainly due to the strain around the Ge/Si heterointerface. A similar contrast change was observed also in the specimen with Sb. In Fig. 3c, the contrast around the Ge/Si heterointerface has also been

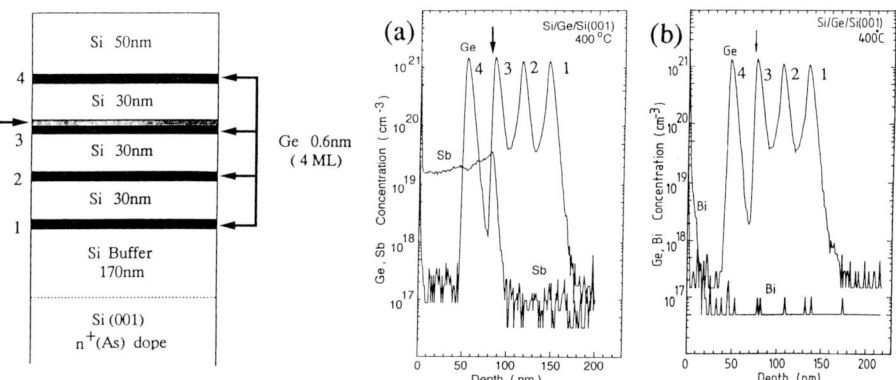

Fig. 1 Schematic of a heterostructure fabricated using MBE.
Fig. 2 SIMS results for the heterostructure (a) with a 1ML Sb, and (b) with 1ML Bi. The arrows indicate the positions where the surfactants were deposited.

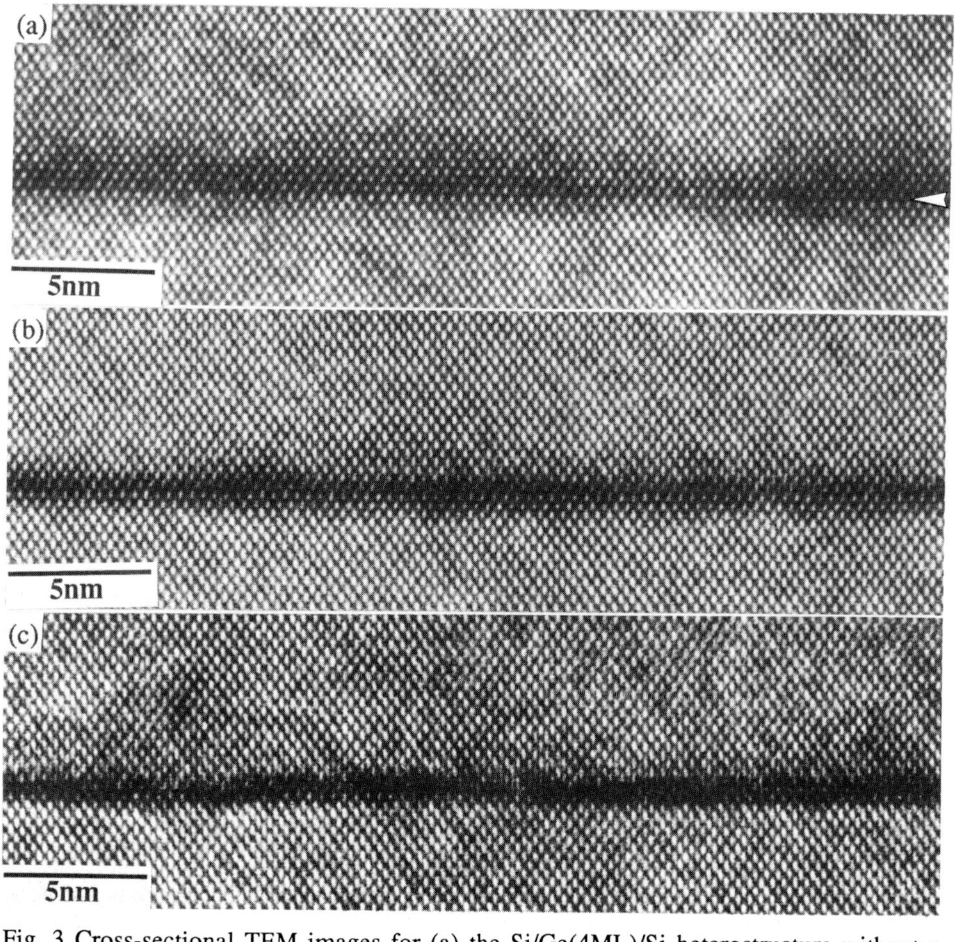

Fig. 3 Cross-sectional TEM images for (a) the Si/Ge(4ML)/Si heterostructure without a surfactant, (b) Si/Bi(1ML)/Ge(4ML)/Si heterostructure, (c) Si/Ge(4ML)/Si with 1ML Bi as a surfactant.

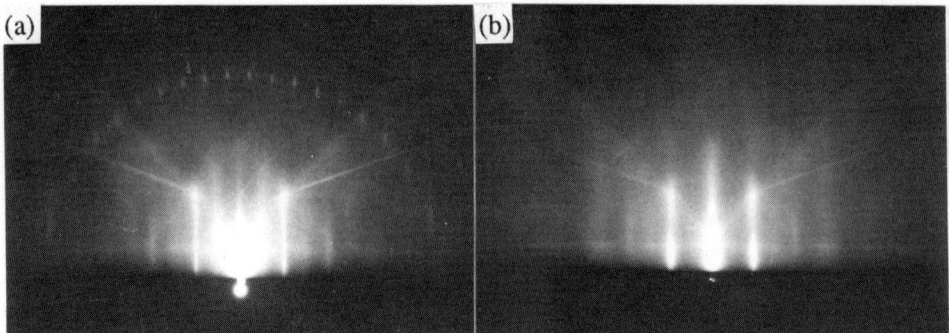

Fig. 4 (a) RHEED pattern taken after a 50ML deposition of Si on 4ML Ge/Si heterostructure, (b) and that of Si(50ML)/Sb(1ML)/Ge(4ML)/Si heterostructure.

altered: the Si side of the Ge/Si heterointerface is brighter than that at the first and the third Ge layers shown in Figs. 3a and 3b. This may indicate change in the strain around the Ge/Si heterointerface.

Figure 4 shows the RHEED pattern taken after a 50ML of Si was grown on a 4ML Ge/Si(001) without a surfactant Fig. 4a, and with a surfactant Fig. 4b. The RHEED pattern from the specimen with the surfactant shows a poor growing surface morphology, which is due to suppression of surface diffusion on Si caused by the coverage of 1ML Sb (Sakamoto *et al.*, 1992). The appearance of distinct atomic steps seen at the heterointerface of Si/Ge/Si in Figs. 3c may be due to the suppressed diffusion of atoms at the growing surface.

For fabrication of a Ge/Si heterostructure, improvement in the critical thickness limit of layer-by-layer growth of Ge and suppression of the tendency towards S-K growth by using a surfactant was reported for As (LeGoues *et al.*, 1990) and for Sb (Sakamoto *et al.*, 1992) using RHEED. These results are also considered to be due to the reduction of surface mobility of the growing species by surfactants, and consequently resulting in the reduction of island formation and agglomerated growth. However, the main role of the surfactants is to reduce the surface free energy (Copel *et al.*, 1989). Details of the mechanism which alters the growth mode from S-K to F-VM are still unknown.

Our TEM and SIMS results indicate both suppression of the segregation and improvement in the abruptness at the heterointerface with degraded growing surface and heterointerface morphology by surfactants. In the SIMS analysis, incorporation of residual Sb into the Si layer was detected, while Bi was detected only at the top surface. Since Bi can be easily evaporated by slight increase in the specimen temperature, Bi is a useful surfactant in the fabrication of electronic device materials.

ACKNOWLEDGMENT

High-resolution microscopy was carried out at Kyushu University. Prof. Tomokiyo and Mr. Manabe are gratefully acknowledged for allowing us the opportunity to use the electron microscope.

REFERENCES

Copel M, Reuter M C, Kaxiras E and Tromp R M (1989) Phys. Rev. Lett. **63** 632.
Copel M, Reuter M C, von Hoegen H and Tromp R M (1989) Phys. Rev. B**42** 11682.
Fujita K, Fukatsu S, Yamaguchi H, Igarashi T, Shiraki Y and Ito R (1990) Jpn J. Appl. Phys. 29 L1981.
Sakamoto K, Miki K, Sakamoto T, Yamaguchi H, Oyanagi H, Matsuhata H and Kyoya K, (1992) Thin solid films, **222** 112.
Sakamoto K, Kyoya K, Miki K, Matsuhata H and Sakamoto T (1993) Jpn. J. Appl. Phys. **32** L204.
LeGoues F K, Copel M and Tromp R M (1990) Phys. Rev. B**42** 11690.
Marèe M J , Nakagawa K , Mulders F M and van der Veen J F (1987) Surf. Sci. **191** 305.

Inst. Phys. Conf. Ser. No 134: Section 6
Paper presented at Microsc. Semicond. Mater. Conf., Oxford, 5–8 April 1993

399

The structure of $Si_{1-x}Ge_x$/Si heterostructure waveguides

Z Yang, G Shao[*] and B L Weiss

Department of Electronic and Electrical Engineering
[*]Department of Materials Science and Engineering
University of Surrey, Guildford, Surrey, GU2 5XH, UK

ABSTRACT Cross-sectional Transmission Electron Microscopy (XTEM), Convergent Beam Electron Diffraction (CBED) and convergent beam shadow imaging (CBSIM) techniques have been used to determine the microstructure of $Si_{1-x}Ge_x$/Si heterostructure waveguides. The strain is characterized using CBSIM and off-axis CBED techniques. The results show that for the strained waveguides the losses increase at a wavelength of 1.15μm and decease at 1.523μm, with increasing Ge concentration. A high density of threading dislocations and roughness of the interface can lead to high scattering loss in the waveguide.

1. INTRODUCTION

There is a growing interest in $Si_{1-x}Ge_x$/Si heterostructures for electronic and optoelectronic devices. Low loss single mode optical waveguides for operation around 1.523 μm are important components which have been fabricated in these structures (Soref et al 1990 and Pesarcik et al 1992), where light is confined to the SiGe layer with the upper and lower cladding layers being formed by air and the Si substrate, respectively. It has been shown (Weiss et al 1992a) that changes of the Si:Ge ratio result in changes in the strain and dislocation distribution in the heterostructure as well as the refractive index and absorption coefficient. Since both the defects and the strain affect the waveguide loss characteristics, it is very important to characterize these waveguide structures. This allows the waveguide guiding layer thickness and composition to be optimised for the particular application. In this paper, XTEM and CBED techniques have been used to study threading and misfit dislocations and the interfacial quality of the various $Si_{1-x}Ge_x$/Si samples. CBSIM and off-axis CBED have been used to characterise the strain at the interface.

2. RESULTS AND DISCUSSION

The SiGe/Si heterostructures used in these experiments were grown on an undoped Si substrate using atmospheric pressure MOCVD. XTEM samples were prepared with ion beam thinning and the TEM analysis was carried out using a JEOL JEM 2000-fx microscope, as described previously by Shao et al (1993). Waveguide samples were prepared by cleaving and polishing both input and output facets optically flat and parallel. The insertion loss was measured using HeNe lasers operating at wavelengths of 1.15μm and 1.523μm, using endfire coupling. The propagation loss was calculated by subtracting the sum of the Fresnel loss and the modal mismatch loss from the insertion loss and expressing the loss in dB/cm.

The XTEM image of 33.6% Ge sample, see Figure 1, shows a rough interface with a high density of dislocations at the interface, which are due to the strain relaxation since the epitaxial layer thickness is greater than the critical thickness. Also, a large number of threading dislocations are seen in the epitaxial layer, which are thought to originate from misfit dislocations, since no dislocations were found in the strained samples (i.e. those with 1.3% and 2.3% Ge); this also suggests that the Si substrate was initially free of threading dislocations. Banding contrast parallel to the interface has been found, see Figure 1. The splitting of spots in the selective area electron diffraction (SAD) pattern in Figure 1 is evidence for the difference of lattice parameters between these bands, indicating that this banding contrast is due to compositional fluctuations along the growth direction.

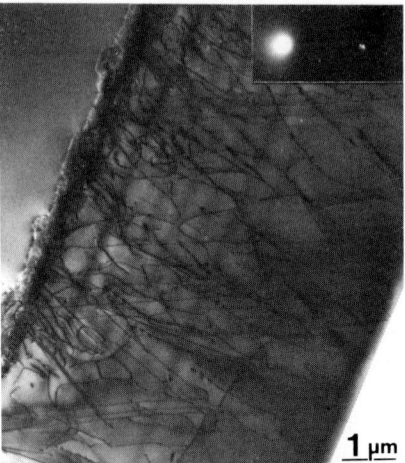

Figure 1:XTEM image of $Si_{0.664}Ge_{0.336}$, superimposed with a part of a SADP.

Off-axis CBSIM of low Ge concentration samples (1.3% and 2.3% Ge), see Figure 2, shows that there is no shift of the n(0 2 2) ZOLZ lines across the SiGe/Si interface. This shows that the epitaxial layer has the same orientation as the substrate, and the (0 2 $\bar{2}$) lattice spacing of the epitaxial layer maintains the same orientation as that of the substrate. Consequently, no strain relaxation occurs in 1.3% and 2.3% Ge heterostructures, and the strain in these materials has been determined by measuring the lattice spacings with (0 8 $\bar{8}$) and (6 0 0) ZOLZ line pairs in the double beam off-axis CBED patterns, see Table 1.

The layer thickness of the 10% Ge sample is 8 μm which is greater than the calculated critical thickness of 1 μm. Consequently a layer with a high density of dislocations is found at the interface, although

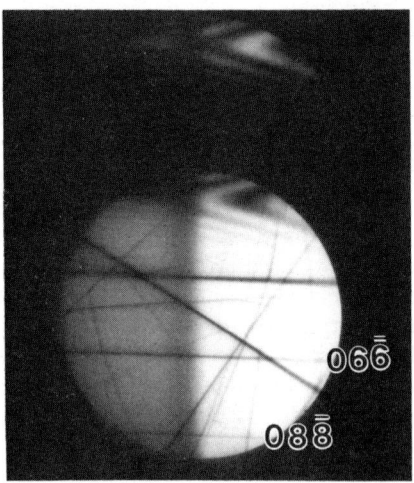

Figure 2: CBSIM of the $Si_{0.977}Ge_{0.023}$/Si strain layered structure.

only a limited number of threading dislocations are found in the epitaxial layer, see Figure 3. By examining the CBED patterns taken from both inside and outside the dislocation layer (Shao et al 1993), it is believed that the strain is gradually relaxed across the interface dislocation layer. The off-axis CBSIM images taken from the interface of 10% Ge sample, see Figure 4, shows the displacement of the ZOLZ lines across the interface, indicating a relative rotation φ between the $[0\ 2\ \bar{2}]_{SiGe}$ epitaxial layer and the $[0\ 2\ \bar{2}]_{Si}$

substrate around [2 0 0], which is given in Table 1.

It is interesting to note that in the $Si_{0.9}Ge_{0.1}$ sample, the interface dislocation band which is about 2500Å thick, is smaller than the critical thickness, as calculated by People et al (1985). Therefore, there is only a limited possibility for the interface dislocations to penetrate the layer and to act as sites for the generation of threading dislocation. However, for the $Si_{0.664}Ge_{0.336}$ sample, the interface dislocation band is greater than the critical thickness and results in a high density of threading dislocations.

The results show that for waveguides with a high Ge concentration, such as 33.6%, the propagation losses of TE and TM modes are ~20dB/cm at wavelengths of 1.15μm and 1.523μm. As the Ge concentration decreases down to 10%, the TE and TM mode loss reduce significantly, with the reduction being greater for a wavelength of 1.523 μm. When the Ge concentration is decreased further, the propagation loss is seen to reduce and increase for wavelengths of 1.15 μm and 1.523 μm, respectively. The measured waveguide characteristics (Weiss et al 1992b) show that a narrow band of interfacial dislocations in $Si_{0.9}Ge_{0.1}$/Si waveguide does not contribute significantly to the propagation loss since the majority of the epitaxial waveguide layer is free from dislocations. The waveguide loss of this sample at a wavelength of 1.15μm can be explained by the phonon-assisted bandedge absorption (Weiss et al 1992a), while much lower loss (~1dB/cm) at the wavelength of 1.523μm is mainly due to the lower radiation

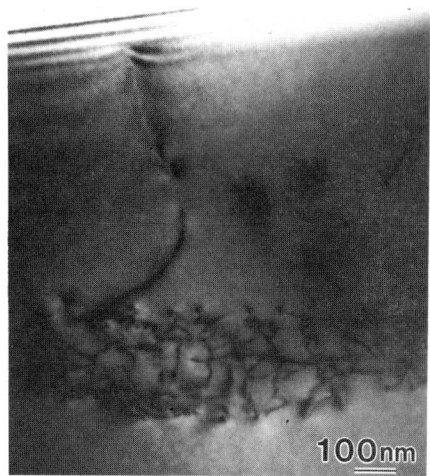

Figure 3: XTEM of $Si_{0.9}Ge_{0.1}$.

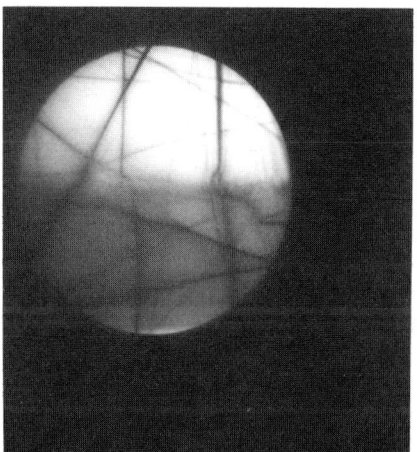

Figure 4: CBSIM of $Si_{0.9}Ge_{0.1}$.

loss resulting from the larger refractive index difference between the waveguiding layer and the substrate. For the waveguides with high Ge concentration and layer thicknesses much greater than the critical thickness, the rough interface and large number of threading dislocations lead to high scattering loss at the interface and within the waveguiding layer. For the waveguides in $Si_{0.987}Ge_{0.013}$ and $Si_{0.977}Ge_{0.023}$ strained structures, the propagation loss increases and decreases at wavelengths of 1.15μm and 1.523μm respectively, with increasing Ge concentration. Considering the effects of strain-induced absorption edge shift and strain-induced refractive index change, it has been found that the strain in our samples causes the bandedge to shift towards longer wavelength, contributing to higher absorption loss at 1.15μm. The absorption loss for both samples, however, approaches zero at 1.523μm, meaning that the strain effect on refractive index change is the dominant mechanism in this case, since the increase of the strain can lead to a larger difference of refractive index and hence better confinement and lower radiation loss in waveguides.

Table 1: Results of TEM characterization

Samples	Defects present	Strain	ϕ
$Si_{0.664}Ge_{0.336}$	A lot of dislocations; Rough interface.	Completely relaxed.	0.149°
$Si_{0.90}Ge_{0.10}$	Interface dislocation band ~2500Å; a few threading dislocations.	60% interface fractional relaxation.	0.043°
$Si_{0.977}Ge_{0.023}$	No dislocations; Smooth interface.	Strained. $\varepsilon_{(0\,2\,\bar{2})}=0.104\%$ $\varepsilon_{(2\,0\,0)}=0.21\%$	0
$Si_{0.987}Ge_{0.013}$	No dislocation; Smooth interface.	Strained. $\varepsilon_{(0\,2\,\bar{2})}=0.047\%$ $\varepsilon_{(2\,0\,0)}=0.14\%$	0

3. CONCLUSION

In summary, the propagation losses of SiGe/Si planar waveguides with different Ge concentrations are quite different due to the influence of structural defects and the strain distribution. Both epitaxial and misfit dislocations should be minimised in order to reduce the scattering loss. For the strained waveguides, the Ge concentration should be carefully chosen to obtain both low absorption and low radiation losses in terms of different wavelengths.

The authors wish to thank Dr F Namavar for the growth of the SiGe/Si structures and Prof A P Miodownik for stimulating discussions.

REFERENCES

Soref R A, Namavar F and Lorenzo J P 1990 Optics Lett. <u>15</u> 270
Pesarcik S F, Treyz G V, Iyer S S and Halbout J M 1992 Elect. Lett. <u>28</u> 159
Shao G, Yang Z and Weiss B L 1993 Jpn. J. Appl. Phys. <u>32</u>(1B) 404
People R and Bean J C 1985 Appl. Phys. Lett. <u>47</u> 322
Weiss B L, Yang Z, Shao G and Namavar F 1992a, Mat. Res. Soc. Fall 92 Meeting, Boston, USA. December, Paper D10.6
Weiss B L, Yang Z and Namavar F 1992b, Electron. Lett. <u>28</u>, 2218

Inst. Phys. Conf. Ser. No 134: Section 6
Paper presented at Microsc. Semicond. Mater. Conf., Oxford, 5–8 April 1993

The epitaxial quality of silicon grown by LPCVD as a function of *ex-situ* wafer preparation

J M Bonar and G J Parker

Department of Electronics and Computer Science, The University of Southampton, Highfield, Southampton, SO9 5NH, UK

ABSTRACT: The epitaxial quality of silicon layers grown by Low Pressure Chemical Vapour Deposition (LPCVD) has been examined as a function of ex-situ clean and growth temperature. The ex-situ cleaning techniques examined are the RCA clean followed by an in-situ high temperature bake and the hydrofluoric acid (HF) dip clean followed by an in-situ bake at the growth temperature. Minimizing the time a wafer is at elevated temperature is crucial for preservation of sharp dopant profiles in advanced device designs.

1. INTRODUCTION

The initial condition of a silicon wafer surface is crucial to the success of epitaxial growth, however the drive towards lower temperature processing to preserve sharp dopant profiles has made the high temperature bake which is usually part of in-situ cleaning a liability. In-situ silicon wafer preparation which thermally desorbs native oxide requires temperatures in excess of 1100°C, while chemical preparations such as the RCA (Kern et al, 1970) or Shiraki (1986), which replace the native oxide with a more easily desorbed oxide, still require a bake at temperatures near 950°C. The HF dip clean has become popular because it leaves the silicon surface terminated by hydrogen (Meyerson et al, 1991), which desorbs at temperatures as low as 500°C (Eaglesham et al, 1991).

Successful epitaxy using an HF dip as the only ex-situ clean and with no high temperature in-situ bake has been reported by Meyerson et al (1991). Increased carbon contamination at the substrate-epitaxy interface has been reported by Srinivasan et al (1990) for epitaxy performed without a high temperature in-situ bake. A certain amount of carbon can be tolerated at the substrate-epitaxy interface as it will incorporate into the lattice. However, if the active region of a device includes the substrate-epitaxy interface, the carbon contamination may degrade device performance. As the RCA clean requires a subsequent high temperature bake, carbon contamination at the interface is expected to be less of a problem.

In addition to the issue of carbon contamination, the level of cleanliness in terms of the presence of oxide on the silicon surface is crucial for successful epitaxy. A relationship exists between the levels of oxygen and water vapour present in the growth ambient and the wafer temperature, this relationship details the conditions necessary for the maintenance of a clean silicon surface and was elucidated by Smith and Ghidini (1982, 1994). At temperatures lower than a critical temperature for a given partial pressure of water and oxygen, oxide will form on the silicon surface. This seriously degrades the quality of the epitaxy as stacking faults and dislocations are expected to form. Therefore the amount of oxide present on a wafer surface at the initiation of epitaxy and the water and oxygen background are both crucial. An indirect method of determining the level of oxygen and water vapour cleanliness is to examine the level of stacking faults and dislocations formed in epitaxial growth as a function of both growth temperature and wafer cleaning method.

The RCA clean is well a characterized ex-situ cleaning technique, and epitaxy following the RCA clean and an in-situ high temperature bake would be expected to be of very high quality because the interfacial oxide formed by the clean will be reliably and thoroughly removed at the 950°C bake temperature used. The HF dip clean operates by a very different mechanism, by first removing the native oxide and then passivating the surface with hydrogen. Although the hydrogen surface is reported to be stable in air (Dumas et al, 1990), rinsing the surface of the wafer with water after the HF dip has been shown by Higashi et al (1990) to increase the level of oxide present with rinse time. As the in-situ bake following a HF dip is merely stabilization of temperature at the value to be used for growth, any oxide formed will not have been removed by a high temperature treatment. Thus the HF dip clean might be expected to have a higher level of "epitaxial mistakes" due to the possible higher level of oxide present.

2. EXPERIMENTAL

The samples examined in this study were grown by LPCVD epitaxy using SiH_4 as a source gas at temperatures varying from 960°C to 700°C. At each temperature, two layers with the same growth parameters (ie gas source, pressure, flow and composition) but different ex-situ cleaning techniques were compared. For layers grown following an RCA clean, the wafers received a 960°C bake before lowering the temperature (where necessary) to the growth temperature. Wafers receiving an HF dip were brought to the growth temperature before epitaxy was initiated, but no high temperature bake was employed.

The initial study involved examination of the 800°C, 850°C, 900°C and 960°C samples by plan-view Transmission Electron Microscopy (TEM). The plan-view samples were prepared by mechanical thinning followed by final etching using an $HF:HNO_3:CH_3OOH$ 3:5:3 solution, as is standard. The samples were examined in a JEOL 2000FX microscope using standard 2-beam diffraction conditions. Cross-section TEM observations were made of some samples in order to qualitatively examine the substrate-epitaxy interface. Samples were prepared by the standard method: wafer sections were epoxied face to face and mechanically thinned and polished, before final thinning by ion beam milling.

In addition to TEM techniques, all samples were prepared by etching for 30 seconds in a Sirtl (1961) etch and examined using Nomarski contrast at 200X magnification. Secondary Ion Mass Spectroscopy (SIMS) analysis was performed in order to quantify interfacial contaminants on a sample consisting of two consecutively grown epitaxial layers. Prior to the growth of the first layer the sample was cleaned by an HF dip, and an in-situ bake and growth at 960°C were performed; the sample was then removed from the chamber and given an RCA clean, with the second in-situ bake and epitaxial growth also performed at 960°C. The cross-section TEM and SIMS analysis samples were grown at a lower growth pressure than the other samples in the study, but the gas flow and composition were exactly the same. In general it is more difficult to achieve good quality epitaxy, as measured by surface quality, using low pressure growth conditions. These conditions are used to achieve selective growth, ie epitaxial growth in silicon windows opened in an oxide mask.

3. RESULTS AND DISCUSSION

Examination of the layers in plan-view TEM was inconclusive, as the dislocation density was less than 1 dislocation in the field of view at magnifications of around 5000X, which rendered comparisons ineffective. The results of the Nomarski study proved more conclusive, as the defect density varied with wafer temperature and ex-situ wafer clean, as shown in Figure 1. The types of defects observed included stacking faults, dislocation, and saucer etch pits. At the highest growth temperatures studied the RCA and HF samples had similar etch pit densities, which are typical of good quality epitaxial growth (Meyerson 1986, Galewski et al 1990). As the growth temperature is reduced below 900°C, the etch pit density increases for both types of cleans, but the increase is greater for HF cleaned than for RCA cleaned wafers. There appear to be more dislocation etch pits at reduced temperature in HF cleaned samples than in RCA cleaned samples. At 700°C growth for both types of clean, the contrast of the samples suggests seriously degraded epitaxy. The defect densities observed in these samples are misleading as this material would not be suitable for further device processing. In the 850°C, 800°C and 700°C HF cleaned samples, the defects occur in a spatially localized fashion, suggesting some part of the etching, rinsing or drying procedure is acting insufficiently, or material is being deposited on the wafer which degrades the epitaxy. The presence of large numbers of these streaks of defects accounts for the large increase in defect density at these temperatures. In the 700°C RCA cleaned sample, areas of relatively low defect density were found as well of areas of higher defect density, which is reflected in the very large error bar for this datum. (The error bars for the other data points are within the size of the symbols.) This suggests that the high temperature bake followed by growth at a low temperature does not reliably produce a surface sufficiently clear of oxide for high quality epitaxial growth.

Examination of the selectively grown layers provided clear evidence of interfacial contamination,

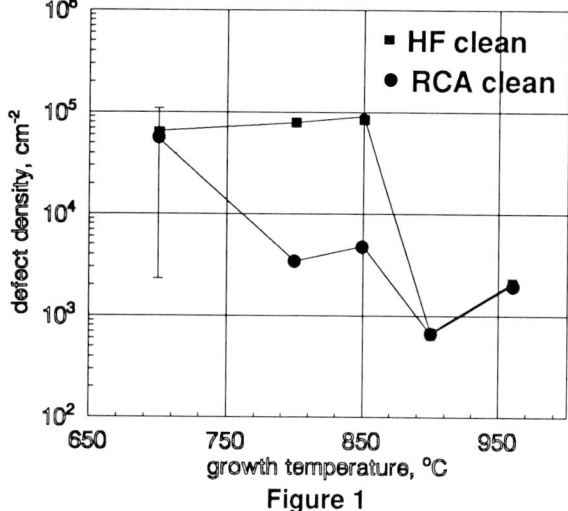

Figure 1

as can be seen in Figure 2. In this micrograph, the edge of the oxide mask feature is visible, with some adjacent highly defective material. The substrate-epitaxy interface is level with the bottom of the oxide bar, and is clearly demarcated. The SIMS analysis, performed on the double selective growth, provided evidence of carbon contamination at both interfaces with a carbon level in the epitaxy of $3 \times 10^{18} cm^{-3}$, comparable with reports of other material grown by this technique. At the HF dip interface an oxygen level of $1 \times 10^{20} cm^{-3}$ was measured while at the RCA interface any oxygen present was less than the background at $8.5 \times 10^{19} cm^{-3}$. This oxygen background is relatively high, so only an upper limit can be placed on the oxygen level in the epitaxy. The higher level of oxygen detected at the HF dip interface than the RCA interface suggests insufficient oxide removal may be a factor in the degraded epitaxial quality at this interface. The excellent electrical quality of material produced by this LPCVD kit has been confirmed by mobility measurements made on thin film samples grown with conditions similar to those in the main part of the study, which have the same mobility as bulk silicon.

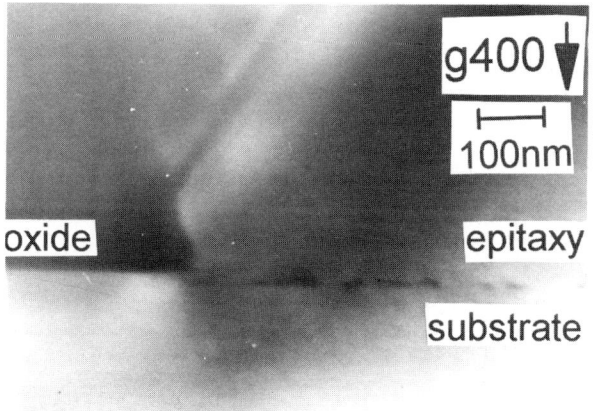

Figure 2

4. CONCLUSIONS

In general, the RCA clean followed by a high temperature bake provides higher quality epitaxy than an HF dip clean followed by a reduced temperature bake, especially at temperatures below 900°C. At the lowest temperature studied, 700°C, the RCA clean plus high temperature bake is not completely reliable, however additional factors such as the cleanliness of the H_2 gas used in the bake process should also be examined. The provision of a dedicated HF cleaning area (separate from the rest of the clean room operation) to be used only for LPCVD preparation will remove the possibility of cross contamination from the HF cleaning process, and this along with the move of the LPCVD kit and all substrate preparation operations to clean room conditions is expected to reduce the effects of dust on degraded epitaxial quality. Improvement in load lock purging, and the fitting of purifiers to all source, carrier and dopant gasses is expected to reduce the levels of oxygen present in the material. Use of wafer support platens not contaminated by silicon-carbide is expected to reduce the carbon levels. The effect of these improvements will be determined by monitoring the defect density using Nomarksi microscopy.

ACKNOWLEDGEMENTS

The authors would like to acknowledge A. Jury for technical assistance in Nomarski microscopy, and B.A. Cressey and T. Khan of the Electron Microscopy Centre at the University of Southampton for use of the facility.

REFERENCES

Dumas P, Chabal Y J and Higashi G S 1990 Physics Review Letters **65**(9) pp1124-1127
Eaglesham D J, Higashi G S and Cerullo M 1991 Appl. Phys. Lett. **59**(6) pp685-687
Galewski C, Lou J-C and Oldham W G 1990 IEEE Transactions on Semiconductor Manufacturing **3**(3) pp93-98
Higashi G S, Chabal Y J, Trucks G W and Raghavachari K 1990 Appl. Phys. Lett. **56**(7) pp656-658
Ishizaka I and Shiraki Y 1986 J. Electrochem. Soc.: Electrochem. Sci. and Tech. **133**(4) pp666-671
Kern W and Pustinen D A 1970 RCA Review **134** p188
Meyerson B S 1986 Appl. Phys. Lett. **48**(12) pp797-799
Meyerson B S, Himpsel F J and Uram K J 1991, Appl. Phys. Lett. **57**(10) pp1034-1036
Sirtl E and Adler A 1961 Z Metalk. **52** p529
Smith F W and Ghidini G 1982 J. Electrochem. Soc.: Solid State Science and Technology, **129**(6) pp1300-1306
Smith F W and Ghidini G 1984 J. Electrochem. Soc.: Solid State Science and Technology, **131**(12) pp2924-2928
Srinivasan G R and Meyerson B S 1990 J. Electrochem. Soc.: Solid State Science and Technology **137**(7) pp2323-2327

Inst. Phys. Conf. Ser. No 134: Section 6
Paper presented at Microsc. Semicond. Mater. Conf., Oxford, 5–8 April 1993

407

Epitaxy of CdTe on exotic (h11) GaAs surfaces

G Patriarche, A Tromson-Carli*, J-P Riviere, J-L Maurice, R Druilhe* and J Castaing

Laboratoire de Physique des Matériaux and *Laboratoire de Physique des Solides de Bellevue, CNRS-Bellevue, F-92195 Meudon cédex, France

ABSTRACT: Metal organic chemical vapour deposition (MOCVD) of CdTe on (h11) (h= 2, 3, 5) GaAs surfaces at 365°C is reported. The epilayers are misoriented about the $[01\bar{1}]$ axis in such a way that the traces of (111) planes perfectly coincide at the interface.

1. INTRODUCTION

Direct deposition of Cd(Hg)Te on GaAs is interesting for future integrated optoelectronics operating in the far infrared. But the large relative variation of lattice parameter (14.6% for pure CdTe), has the effect of introducing at the interface and in the epitaxial layers a high density of crystalline defects. In order to minimise defect generation, several authors have tried to tilt the substrate surface in various orientations, growth being carried out using either molecular beam epitaxy (MBE) (Feuillet et al 1989, Pain et al 1991, Nakamura et al 1992, Sasaki et al 1992) or metal-organic chemical vapour deposition (MOCVD) (Cinader and Raizman 1992). In this paper, we report our results of growth by MOCVD of CdTe on (211), (311) and (511) surfaces of GaAs at 365°C. The misorientation obtained between the substrate and the epilayers appears to obey a simple rule of continuity of the {111} planes that are at the smallest angle to the surface. However the epilayer may be twinned with respect to these planes, depending on the polarity, roughness and density of coincidence sites at the interface.

The stress is put here on the analysis of the general geometrical features characteristic of growth of Cd(Hg)Te on GaAs surfaces parallel to $[01\bar{1}]$ (i.e. of type (h11) or (100)), the importance of polarity on the twinning of the epilayers has been discussed in a previous paper (Patriarche et al 1993).

2. EXPERIMENTAL CONDITIONS AND RESULTS

The experimental techniques have already been described (Patriarche 1992, Tromson-Carli et al 1993). The substrate surface was chemically etched using a H_2O_2-NH_4 solution and the specimens were annealed for 30 min under hydrogen flow prior to MOCVD. Both A and B surface polarities were used for each (h11) orientation. Cross-sectional transmission electron microscopy (TEM) specimens were then prepared, using Ar ion milling under low-dose conditions (Patriarche 1992).

Growth of CdTe on GaAs at 365°C gave a misorientation to the epilayers in all the cases observed. For CdTe on GaAs (311) (fig. 1), two orientations appeared simultaneously, twin related to each other with respect to the (111) planes, which were in turn in a perfect continuity with the GaAs{111} planes at the smallest angle to the surface. Twinning was much more frequent in layers grown on the surface of polarity A (fig. 1b). The measured misorientation was about 5° and 114°, respectively, in the normal and twin cases; this is consistent with the standard twinning angle of 109.28° about a <110> axis. Consequently, the interface on the CdTe side was near (411) and (011) in the respective orientations. Dislocations every 2.6 nm appeared at the interface of the normal orientation, which were absent in the twin case (fig. 1). The observations were fairly similar in the case of GaAs (211), except that twinning was far denser on the GaAs surface of polarity B (Patriarche et al 1993). In this case, the misorientation was 3° and 112°, respectively, and the interface was close to (211) and (133) CdTe planes, for the normal and twin orientations. First observations of layers grown on GaAs (511) also indicated a continuity of (111) planes.

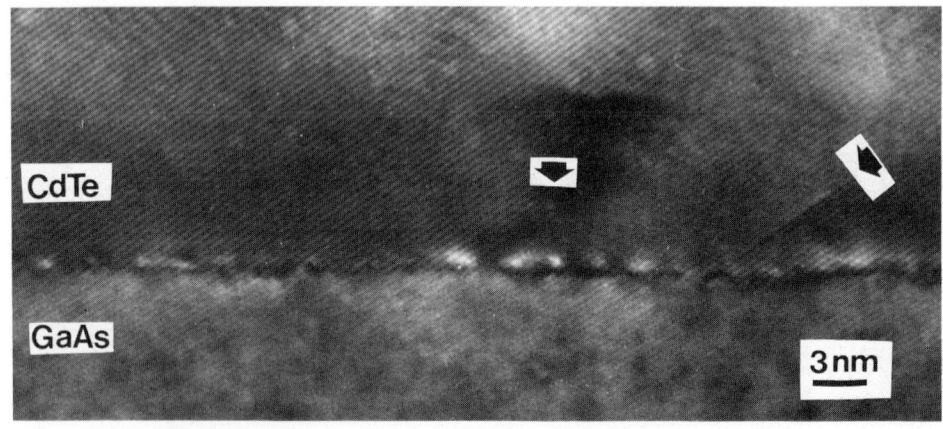

Fig. 1 (a) CdTe on GaAs (311) B in [01$\bar{1}$] zone axis. Note dislocations and stacking faults (arrows). (b) CdTe on GaAs (311) A in [01$\bar{1}$] zone axis. Dislocations (arrowed) are present when the orientation of CdTe is regular (tilt but no twin) and absent in the case of twinning

3. DISCUSSION

The continuity of (111) planes can also be described as perfect coincidence at the interface of the traces of these planes. This coincidence then naturally implies a misorientation (fig. 2), the angle Φ of which obeys a refraction-like rule, first given by Feuillet et al (1989):

$$d_{GaAs} \sin(\Theta + \Phi) = d_{CdTe} \sin\Theta,$$

where Θ is the angle between (111) GaAs planes and the interface, and d_{GaAs} and d_{CdTe} are the respective (111) interplanar distances.

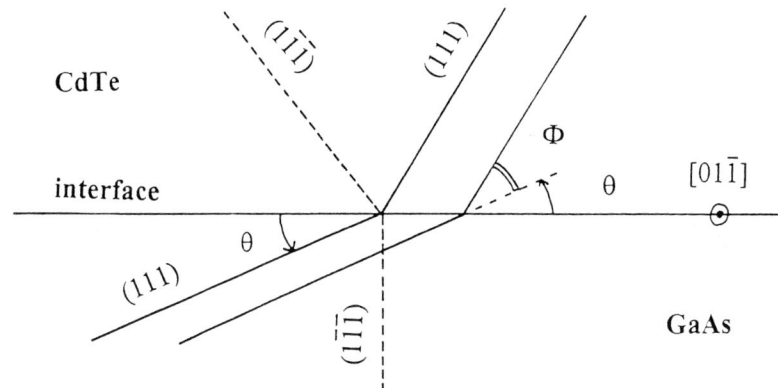

Fig. 2: Schematic of the continuity of (111) planes at the interface

The presence of dislocations in only one of the two orientations on GaAs (311) may be explained considering the relation through the interface of the other family of {111} planes parallel to [01$\bar{1}$]. The traces of these planes at the interface have a misfit of 15.6% in the case of the regular orientation, which supposes the presence of a dislocation every 2.3 nm, a value fairly close to the observed one (2.6 nm). In contrast, in the twin case, these planes are in continuity with the CdTe (200) ones, with which the mismatch is only -2.2%, which is consistent with the absence of dislocations in TEM micrographs.

The phenomenon of {111} plane coincidence is however not synonymous of coincidence of sites at the interface, in the way suggested for instance by Zur and McGill (1984) for CdTe (111) on GaAs (100). These authors proposed that heteroepitaxy should follow simple rules in which the planar coincidences of atomic sites play a major role. The fact that the [01$\bar{1}$] axis - where the mismatch is 14.6% - is always conserved in the misorientation of the epitaxial layers would indicate in our observations that such considerations are secondary. It is interesting to check, however, the coincidences in the direction lying in the interface perpendicular to [01$\bar{1}$]. In order to evaluate this, we looked for the indices of the CdTe planes the closest to the interface for (h11) and (100) GaAs, and calculated the smallest lattice translation perpendicular to [01$\bar{1}$] in it; the spatial period of the coincidence lattice could then be inferred by comparison with the equivalent lattice translation in GaAs (tab. 1). Precise orientations were calculated using the above formula.

Except in the untwinned (100) case, the periods obtained are smaller for misoriented layers than for corresponding aligned ones. They are close to the one in the [01$\bar{1}$] direction which is 3.2 nm (with a 0.3% mismatch). The twin orientations have smaller periods, and therefore appear favoured, compared to regular ones, for growth on (211), (311) and (100). In contrast, the case of twin on (511) is unfavourable with a period above 10nm: if this criterion of coincidence were effective, then growing CdTe on such a surface should give better epilayers with respect to the density of twins.

4. CONCLUSION

Our observations then lead to a straightforward conclusion: growth proceeds by joining atoms to the {111} substrate planes that make the largest terraces on the surface, with either a regular or reversed (twinned) pile-up, depending mainly on in-plane coincidences at the interface.

Table 1: In-plane unidirectional coincidences perpendicular to $[01\bar{1}]$.

GaAs surface	(211)	(311)	(511)	(100)
unit vector perpendicular to $[01\bar{1}]$	$a_{GaAs}[\bar{1}11]$	$\dfrac{a_{GaAs}}{2}[\bar{2}33]$	$\dfrac{a_{GaAs}}{2}[\bar{2}55]$	$\dfrac{a_{GaAs}}{2}[011]$
CdTe aligned: coincidence period (0.3% mismatch)	7.8nm	10.6nm	16.3nm	3.2nm
CdTe misoriented: regular: plane nearest to interface (and angle)	(944) (0.1°)	(411) (0.9°)	(911) (0.34°)	$(\bar{5}11)$ (1.11°)
unit vector perpendicular to $[01\bar{1}]$	$\dfrac{a_{CdTe}}{2}[\bar{8}99]$	$a_{CdTe}[\bar{1}22]$	$\dfrac{a_{CdTe}}{2}[\bar{2}99]$	$\dfrac{a_{CdTe}}{2}[255]$
coincidence period mismatch	4.9nm −0.5%	3.9nm −2.2%	4.2nm 0.5%	2.4nm −0.7%
CdTe misoriented: twin: plane nearest to interface (and angle)	(133) (0.3°)	(011) (1.06°)	$(\bar{1}44)$ (0.52°)	$(\bar{1}11)$ (1.21°)
unit vector perpendicular to $[01\bar{1}]$	$\dfrac{a_{CdTe}}{2}[\bar{6}11]$	$a_{CdTe}[100]$	$\dfrac{a_{CdTe}}{2}[811]$	$\dfrac{a_{CdTe}}{2}[211]$
coincidence period mismatch	2nm 2%	1.3nm −2.2%	10.4nm 1.4%	0.8nm −0.7%

ACKNOWLEDGEMENT: The authors want to thank R. Triboulet and Y Marfaing for valuable discussions.

REFERENCES

Cinader G and Raizman A 1992 J. Appl. Phys. 71 2202
Feuillet G, Cibert J, Gobil Y, Saminadayar K and Tatarenko S 1989 Mater. Res. Soc. Symp. Proc. 148 389
Nakamura Y, Otsuka N, Lange M D, Sporken R and Faurie J P 1992 Appl. Phys. Lett. 60 1372
Pain G N, Sandford C, Smith G K G, Stevenson A W, Gao D, Wielunski L S, Russo S P, Reeves G K and Elliman R 1991 J. Cryst. Growth 107 610
Patriarche G 1992 Doctoral Thesis, Université Paris 6
Patriarche G, Tromson-Carli A, Rivière J P, Triboulet R, Marfaing Y and Castaing J 1993 submitted for publication
Sasaki T, Tomono M and Oda N 1992 J. Vac. Sci. Technol. B 10 1399
Tromson-Carli A, Patriarche G, Druilhe R, Lusson A, Marfaing Y, Triboulet R, Brown P D and Brinkman A W 1993 Mater. Sci. Eng. B16 145
Zur A and McGill T C 1984 J. Appl. Phys. 55 378

Inst. Phys. Conf. Ser. No 134: Section 6
Paper presented at Microsc. Semicond. Mater. Conf., Oxford, 5–8 April 1993

411

The characterization of CdS/CdTe thin films for solar cell applications

M M Al-Jassim, F S Hasoon, K M Jones, B M Keyes, R J Matson and H R Moutinho

National Renewable Energy Laboratory, Golden, CO 80401, USA

ABSTRACT: The morphology, microstructure and the luminescent properties of CdS and CdTe films were studied. As-grown CdS films were polycrystalline with a grain size in the 200-600 Å range. Significant grain growth and reduction in defect density were observed in these films after heat treatment. The CdTe films exhibited a high density of grain boundaries and structural defects within the grains. Both grain boundaries and intragrain defects are very active recombination sites. Heat treatment effected a marked grain growth and reduction in the recombination efficiency of the grain boundaries and intragrain defects.

1. INTRODUCTION

Although conversion efficiencies exceeding 15% have been reported in CdS/CdTe solar cells (Ferekides 1993), the defect structure and the electrical and luminescent behaviour of the defects in these cells are not well understood. A better understanding of the nucleation, growth, and microstructure of these films will facilitate a better control of the structural and electrical properties, thereby improving the device characteristics. Furthermore, it has been reported that S and Te interdiffusion takes place at the CdS/CdTe interface (Birkmire et al 1992a). However, to date, no systematic study of this interface has been carried out. In this work, the morphological and structural properties of polycrystalline thin films of CdS and CdTe were investigated as a function of deposition method, deposition conditions and post-deposition heat treatment. The CdS/CdTe interface properties and the recombination behaviour of defects were studied in as-deposited and heat-treated films.

2. EXPERIMENTAL

In this study, the CdS films were grown on either SnO_2-coated glass or on SnO_2-coated (100) single crystal Si substrates. SnO_2 films in a solar cell structure form a transparent front Ohmic contact. The CdS growth was performed by the solution growth method, which uses a reaction among a Cd salt ($CdSO_4$), NH_4OH, and thiourea [$CS(NH_2)_2$] in an aqueous solution (Chu et al 1991). The film thickness used was in the 1000-Å to 1400-Å range. The CdTe films were deposited by either close space sublimation (CSS) (Chu et al 1992) or physical vapour deposition (PVD) (Birkmire et al 1992b) on either glass/SnO_2/CdS, or crystalline Si/SnO_2/CdS substrates. Both as-deposited and heat-treated samples were examined. The heat treatment was carried out for 20-30 minutes at 400-450 °C. A commonly used (McCandless and Birkmire 1991) fluxing agent, namely $CdCl_2$, was also used in conjunction with the heat treatment. This was applied by dipping the CdTe films in a saturated solution of $CdCl_2$ in methanol prior to heat treatment. The characterization of these films was carried out by SEM, atomic force microscopy (AFM), TEM, TED, and energy dispersive X-ray spectroscopy (EDS) in the TEM. The luminescent properties of the defects in these films were studied by cathodoluminescence (CL) and time-resolved photoluminescence (TRPL).

3. RESULTS

3.1 CdS

AFM examination showed that as-deposited CdS films are polycrystalline with fine grains (Fig. 1a). The grain size ranged from 200 Å to 600 Å. These films were subsequently annealed in N_2 at 350°C, 450°C and 550°C for 60 min. No grain growth was observed after the 350°C anneal, and only a small increase in grain size resulted after the 450°C, anneal (Fig. 1b). SEM examination of as-deposited and heat-treated CdS films clearly revealed the change in morphology. While a fine-grained material was observed in as-deposited films, heat treatment in N_2 at 550°C for 60 minutes resulted in a grain growth up to 1500 Å (Fig. 2a). The largest grains, however, were obtained by a combination of $CdCl_2$ and heat treatment. Figure 2b shows the surface morphology of a CdS film after being heat treated in N_2 at 550°C for 60 min in the presence of $CdCl_2$. Clearly, grains as large as 5000 Å are obtained. Another desirable effect of $CdCl_2$ treatment is the prevention of S loss from the film surface. X-ray photoemission spectroscopy (XPS) examination showed that films treated at 550°C without $CdCl_2$ exhibited severe S loss from the surface region, while no such a loss was detected upon using $CdCl_2$ (Niles and Hasoon 1993).

TEM and TED studies further corroborated the AFM and SEM results and revealed the microstructure of the films. As-deposited films exhibited a very fine-grain microstructure with some amorphous characteristics and a tendency to a preferred orientation. Furthermore, the grains are heavily faulted with a high density of stacking faults and microtwins. Heat-treated films at 550°C (without $CdCl_2$), on the other hand, exhibited no preferred orientation, considerably larger grains, and a similar defect density within the grains. Films that were heat treated in the presence of $CdCl_2$, however, exhibited a markedly different microstructure, as the grains contained considerably fewer planar defects. In some grains, the planar defect density approaches zero. This is a significant finding as our work on the generation and propagation of defects in CdTe films showed that planar defects in the underlying CdS tend to propagate across the CdS/CdTe interface. TEM studies also showed that CdS has a predominantly hexagonal lattice in both as-deposited and heat treated forms. The significance of this will be clear in the later part of this paper, as the CdTe has a cubic (zinc blend) lattice.

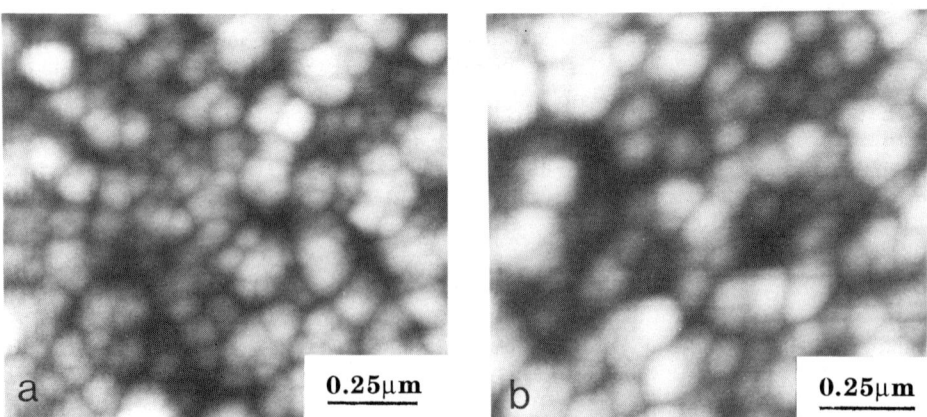

a   0.25μm   b   0.25μm

Fig. 1 AFM micrographs of CdS films; (a) as-deposited, (b) annealed at 450°C for 1h.

3.2 CdTe Grown by Close Space Sublimation

Cross-sectional as well as plan-view SEM investigations of as-deposited films showed that the grain size of CdTe grown on CdS scales directly with the film thickness. While grain sizes of 1.5-3 μm were observed in 4-μm thick films, the size increased up to 10 μm for 20-

Fig. 2 SEM micrographs of CdS films; (a) heat treated at 550°C for 1 h, (b) heat treated at 550°C for 1 h with CdCl$_2$.

25-µm thick films. The grain size, however, does not increase linearly with film thickness, as 40-µm thick films yielded grain sizes in the 10-20-µm range. The as-deposited film surface is usually rough, sometimes displaying facetted morphology. SEM cross-sectional examination revealed that the grains are smaller and more columnar in shape close to the CdS/CdTe interface. As the film thickness increases, the grains become more equiaxed.

Time-resolved photoluminescence (TRPL) measurements of as-deposited films showed that the minority carrier lifetime is very short (<100 ps) regardless of the grain size. On the other hand, TRPL measurements of heat-treated films yielded markedly longer lifetimes and showed that the latter depends strongly on the grain size. Films having a grain size ≤ 0.25 µm exhibited short minority-carrier lifetime of the order 200 ps, while large grain size (2-4 µm) films gave rise to lifetimes in the 2-3-ns range. However, we observed a considerable variation in lifetime at constant film thickness and grain size. PL does not have the spatial resolution to examine small areas or individual grains in a film to investigate such variations. To probe the recombination characteristics of individual CdTe grains, panchromatic CL experiments were performed on large (5-20-µm) grain material. Figures 3 and 4 show some of our findings. Figure 3 shows a typical spatially resolved luminescence image of a large-grain, as-deposited film. Clearly, the luminescence properties of the grains are very diverse. Some grains display a high luminescence yield, indicative of a low non-radiative recombination rate (a desirable quality for solar cell material) while other grains do not luminesce at all. These grains possess a high density of non-radiative recombination centers. In order to quantify these data and measure the recombination efficiency of various defects, CL linescan measurements were performed on as-deposited, heat-treated, and CdCl$_2$-treated samples from the same wafer (Fig. 4). In as-deposited samples (Fig. 4a), a grain boundary recombination

Fig. 3 Panchromatic CL micrograph of an as-deposited CdTe film.

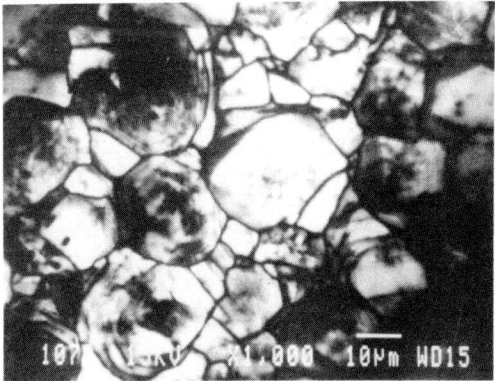

efficiency of 78% was obtained. The same measurement was repeated on the same sample after heat treatment in He at 450°C for 60 min. The grain boundary recombination efficiency decreased to 62%-66%. After heat treatment at 400°C for 20 min in the presence of CdCl₂, however, a dramatic decrease of the recombination efficiency to 38% was observed (Fig. 4b). A similar CL study was carried out on the defects within the CdTe grains. Recombination efficiencies in the 20%-40% range were measured in as-deposited films. This clearly demonstrates that although the defects within the grains are less efficient non-radiative recombination sites, they are a serious loss factor due to their high density.

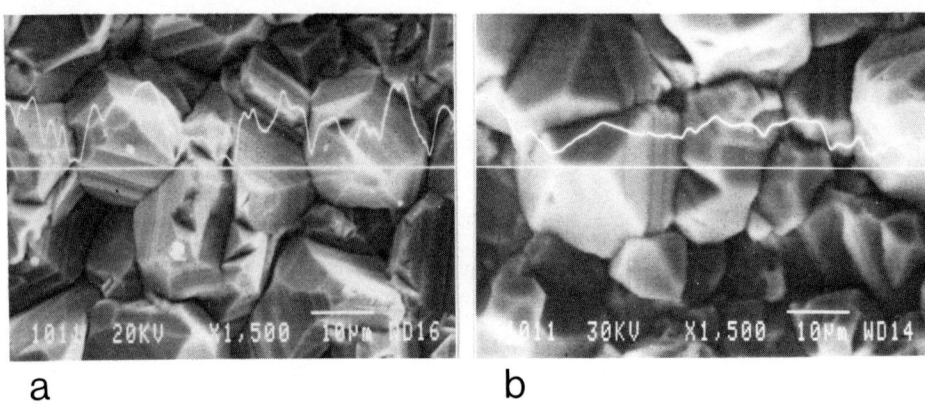

a   b

Fig. 4 CL linescan measurements of : (a) as-deposited CdTe film, and (b) after heat treatment at 400°C for 20 min with CdCl₂.

The work presented above unambiguously shows that structural defects in CdTe are active non-radiation recombination centers that could cause large carrier losses. To investigate the nature, density, and distribution of these structural defects, TEM examinations of plan-view specimens were carried out. This revealed a large variety of intragranular structures. While some grains are highly faulted and exhibit twins, threading dislocations, and stacking faults in densities up to 10^9 cm⁻², other grains are virtually defect free. Often these grains are adjacent to each other, as can be seen in Fig. 5. These results are in agreement with the panchromatic CL examination discussed earlier, and they explain the large variation in CL intensity across the film.

3.3 CdTe Grown by Physical Vapour Deposition

The cells used here have the following structure: Si substrate/SnO₂/CdS/CdTe/Cu-Au. Both CdS and CdTe films were grown by PVD. The thickness of these films is 0.5 μm for CdS and 2-3 μm for CdTe. In some cases, the CdS was heat treated at 400°C for 20-30 minutes prior to the deposition of CdTe to enhance the grain size. After the deposition of the latter, the samples were heat treated in the presence of CdCl₂ at 400°C for 20-30 min in air. TEM, STEM/EDS, and SIMS were used to characterize the CdTe and CdS films and the CdS/CdTe interface.

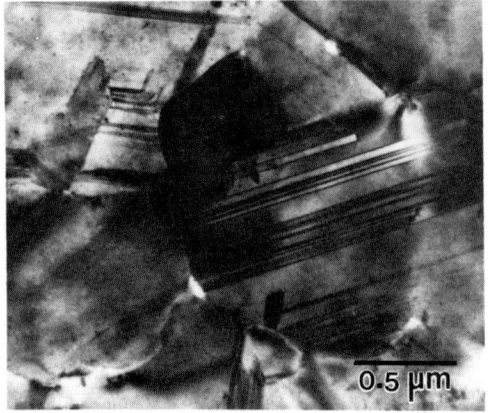

Fig. 5 TEM micrograph of a CdTe film

TEM cross-sectional examination of an as-deposited structure (Fig. 6) showed that the grain size in the CdTe films is similar to that in CdS. Detailed TEM study of the CdS/CdTe interface revealed that the majority of grain boundaries and planar defects propagate across the interface. Furthermore, high-resolution lattice imaging showed that a large percentage of the CdTe grains are epitaxially related to the underlying CdS grains. Selected area electron diffraction examination of the films showed that the CdS is constituted of a mixture of cubic and hexagonal phases, while the CdTe is predominantly cubic. Such a difference in the crystal structure is thought to complicate the interface structure considerably. STEM/EDS measurements were performed in several areas across the interface. No evidence of Te or S interdiffusion was observed in the as-deposited structures. These results were corroborated by SIMS measurements.

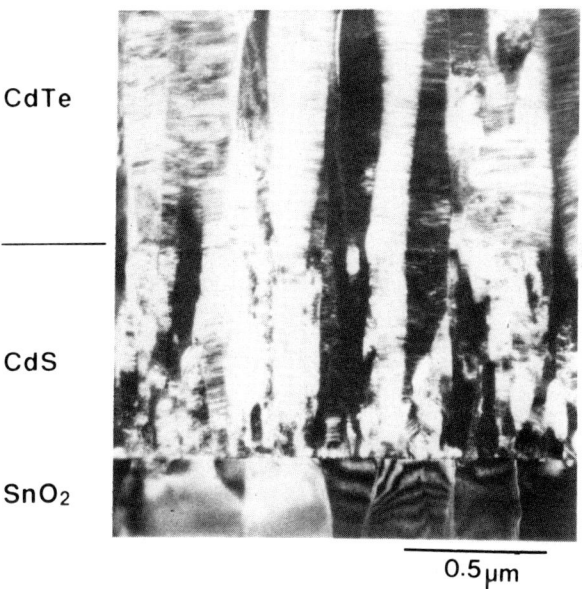

CdTe

CdS

SnO$_2$

0.5 μm

Fig. 6 TEM cross-section of a PVD-deposited CdS/CdTe structure before heat treatment

Heat-treated samples exhibited a markedly different microstructure and interface properties. Figure 7 is a TEM cross-section that demonstrates the considerable grain growth that was observed in both CdS and CdTe films. However, the grain growth was an order of magnitude higher in CdTe than in CdS. Additionally, high-resolution lattice imaging of the CdTe/CdS interface shows that the CdTe grains appear less oriented with respect to the underlying CdS grains. This is believed to be due to the fact that the CdS grains are larger in number and exhibit no preferred orientation. Hence, a CdTe grain can only be oriented epitaxially with one CdS grain. Unlike as-deposited structures, STEM/EDS analyses across the interface of heat treated samples detected sulphur diffusion into CdTe and lesser amount of tellurium diffusion into CdS. The concentration of sulphur varied considerably along the interface. In the CdTe grains, increased levels of sulphur were detected in regions with a high density of planar defects. The EDS results obtained so far, however, are qualitative. These findings were supported by SIMS analysis of the CdS/CdTe interface.

4. SUMMARY

This study showed that as-deposited, solution-grown CdS thin films are polycrystalline having a grain size in the 200-600Å range. Heat treatment in the presence of CdCl$_2$ at 500°C or above resulted in a grain growth up to 5000 Å. This will have a significant effect on the

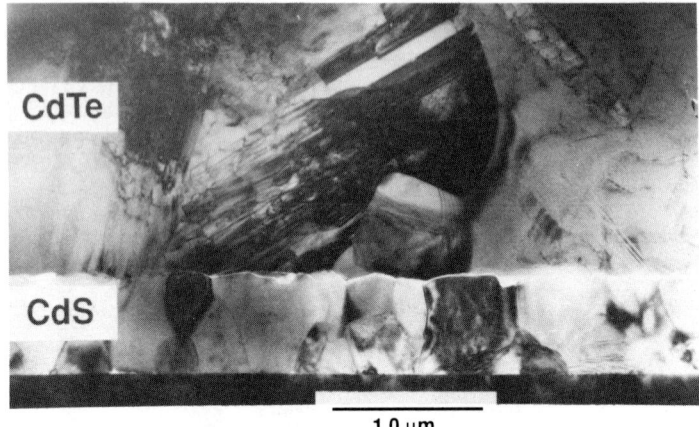

1.0 μm

Fig. 7 TEM cross-section of a PVD-deposited CdS/CdTe structure after heat treatment

nucleation of CdTe, as our work showed that the grain size in as-deposited CdTe films is dictated by the grain size of the underlying CdS. Additionally, $CdCl_2$ treatment lowered the structural defect density within the CdS grains. This means that fewer defects will propagate to the CdTe film. Structural investigation of CdTe films showed that the structure of the grains is very diverse. While some grains are virtually defect free, adjacent grains have defect densities up to the 10^9 cm^{-2} level. Similar conclusions can be drawn from CL measurements. While some grains do not show any luminescence yield, others have a high luminescence yield. This is due to a varying number of deep levels that act as nonradiative recombination sites for the excess charge carriers. Grain boundaries in CdTe also act as sites of enhanced nonradiative recombination, with a recombination efficiency more than twice that of defects within the grains. $CdCl_2$ treatment proved to lower that by a factor of two, indicative of its defect passivation capability.

ACKNOWLEDGMENTS

The authors wish to acknowledge T.L. Chu and S. Chu for providing the CSS-grown CdTe films, R. Noufi and D. Du for growing the CdS films by the solution-growth method, and R. Birkmire for providing the PVD-deposited CdTe films. The authors also wish to acknowledge the technical help of S. Asher of NREL for the SIMS measurements and D. Niles for the XPS measurements of the CdS.

REFERENCES

Birkmire R W, Hegedus S S, McCandless B E, Phillips J E, Russell T W F, Shafarman W N, Verma S and Yamanaka S 1992a AIP Conf. Proc. DOE/NREL PV AR&D, 268, pp 212
Birkmire R W, McCandless B E and Hegedus S S 1992b Int. J. of Solar Energy 12, No. 1/4, 145
Chu T L, Chu S S, Ferekides C, Wu C Q, Britt J and Wong C 1991 Proc. 22nd IEEE PVSC, 2, pp 952
Chu T L, Chu S S, Britt J, Chen G, Ferekides C, Schultz N, Wang C and Wu C Q 1992 Proc. PV AR&D, AIP Conf. Proc., 268, pp 88
Ferekides C Univ. of South Florida, private communication
McCandless B E and Birkmire R W 1991 Solar Cells 31 527
Niles D W and Hasoon F S 1993 Accepted for Publication, Progress in Photovoltaics

Inst. Phys. Conf. Ser. No 134: Section 6
Paper presented at Microsc. Semicond. Mater. Conf., Oxford, 5–8 April 1993

417

Anomalous "anomalous weak beam contrast" of stacking faults in ZnSe on (100) GaAs

C Y Chen and W M Stobbs

Dept. of Materials Science and Metallurgy, Pembroke St., Cambridge CB2 3QZ, UK

ABSTRACT: The five configurations of fault groupings that are observed to form in thin (< 100nm) layers of ZnSe grown on GaAs are described and of these the characterisation of the apparently isolated and paired formats is discussed. Conventional and weak beam TEM methods were used and while paired faults proved to be intrinsic the apparently isolated faults exhibited an anomalous form of "anomalous weak beam contrast" suggesting, contrary to the interpretation of their two beam contrast, that they are in fact thin twins.

1. INTRODUCTION

It was Cullis and Booker (1972) who first noted that intrinsic stacking faults in Si imaged under weak-beam (WB) conditions show a marked change in contrast if the **g** vector is reversed. The magnitude of the contrast change on reversing **g** for extrinsic faults is generally accepted to be larger than for intrinsic faults but over the last twenty years neither have all the observations described for different materials been mutually consistent nor has a generally agreed origin been found for all the contrast effects which have been seen. Current, if slightly differing, understandings of the "anomaly" can be found in the work of Cockayne et al. (1984) Wilson et al. 1985 and Bithell et al. (1989). It is the proximity of apparently different types of fault in a layer of constant thickness which makes the comparison of their contrast for a range of imaging conditions in an epitaxially grown film of ZnSe on (001) GaAs a worthwhile exercise, since the weak beam contrast of faults can often be sensitive to small changes in the foil thickness and deviation parameter.

2. OBSERVATIONS

The fault arrays which are described here were observed in thin (<100nm) epitaxial films of ZnSe grown on (001) oriented GaAs by molecular-beam epitaxy (MBE) and provided by Dr.K.A.Prior (Herriot Watt University). Both plan view and cross-sectional TEM samples were examined as prepared by ion milling using a JEOL2000FX at 200 kV. Standard bright field and centred dark field TEM techniques were used to determine the nature of the faults. Both extrinsic and intrinsic faults were then examined under a range of weak-beam imaging conditions and here we concentrate on the way some of the apparently extrinsic faults exhibited contrast inconsistent with their apparent nature.

The faults which form in the (001) ZnSe films lie on {111}planes and are of equilateral triangular shape with an apex at, or very close to, the interface with the GaAs substrate as may be seen from the 002 dark field image of the film, as viewed edge-on with a beam direction near to (110), in Fig.1(a). Their edges are delineated by <110> directions and the free surface of the epilayer. Five different configurations of groupings of such faults were observed and examples of all of these can be seen in plan view in the 040 weak-beam dark field image shown in Fig.1(b). The different types of grouping build up to varying degrees a pyramid with its apex at the substrate interface and the five configurations seen are shown schematically in Fig.2 as viewed from the substrate. Configuration "1" is simply an isolated triangular fault while in configuration "2" two such faults are paired on opposite {111} planes meeting at their apex at the substrate interface. Configurations "3" and "4" have three and four triangular sides respectively, meeting along <110> directions. The fifth configuration, like the second, has two sides but now the faults are on {111} planes meeting in a common <110> direction.

We will concentrate here on the contrast exhibited by the faults in configurations "1" and "2" examples of which are shown, as imaged under a range of dark field conditions in Fig.3. The plan view foil retains a uniform thickness of the ZnSe epilayer and the interface with the substrate, having been thinned from the GaAs side. Conventional methods were first used to determine their nature and, as may be seen from the strong dark field images in Figs.3(a-b), the faults in configuration "2" (marked I) were consistently found to be intrinsic. The apparently isolated faults (whether marked E_a or E_b), on the other hand exhibited contrast in this type of image which indicated that they are extrinsic. The experimental observation that reversing g can result in a change in the intensity of the fringes in a weak-beam image is associated with a change in the sign of α ($=2\pi g.R$). This can be changed either by reversing g and keeping R the same or by keeping g the same and changing the sense of R (the direction of R is dependent on the sense of inclination of the fault as is demonstrated for the opposite sides of the faults in configuration "2" in the strong-beam images in Fig.3(a-b)). Let us now consider the weak-beam contrast exhibited by the different type of fault. Figs.3(c-h) show 220 weak-beam images under $2g$, $3g$ and $4g$ diffraction conditions. In each case the images for these progressively increased deviation parameters are paired for opposite directions of g.

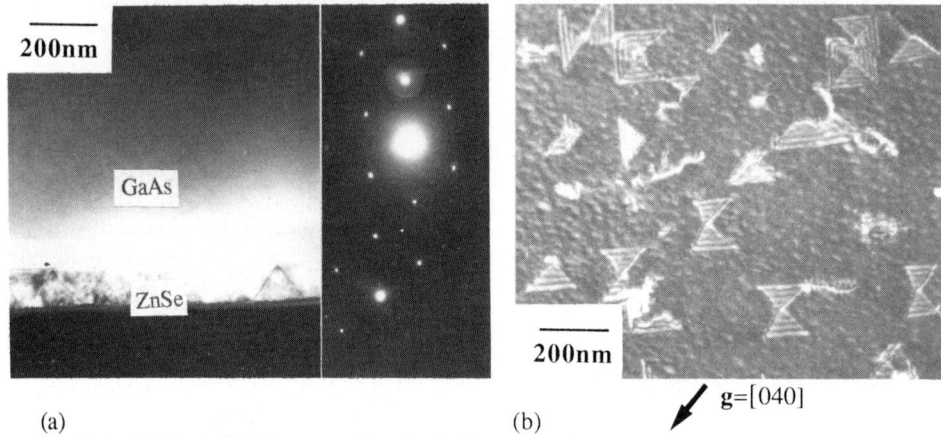

(a)   (b)

Fig.1(a) 002 dark field image of the ZnSe/GaAs interface viewed edge-on.
        (b) 040 weak beam dark field image of stacking faults in the ZnSe/GaAs in plan view.

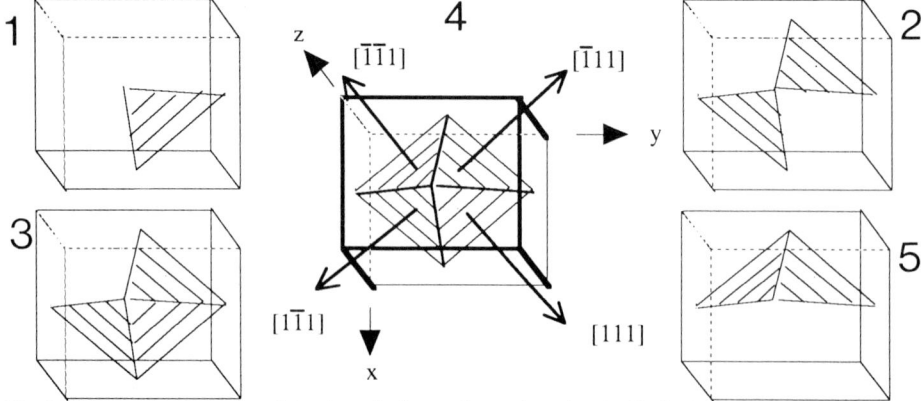

Fig.2 Schematic diagram of the five fault configurations in the ZnSe layers on (001) GaAs as viewed from the substrate. Fringes are shown for each of the faults of the form that would be seen under weak-beam conditions.

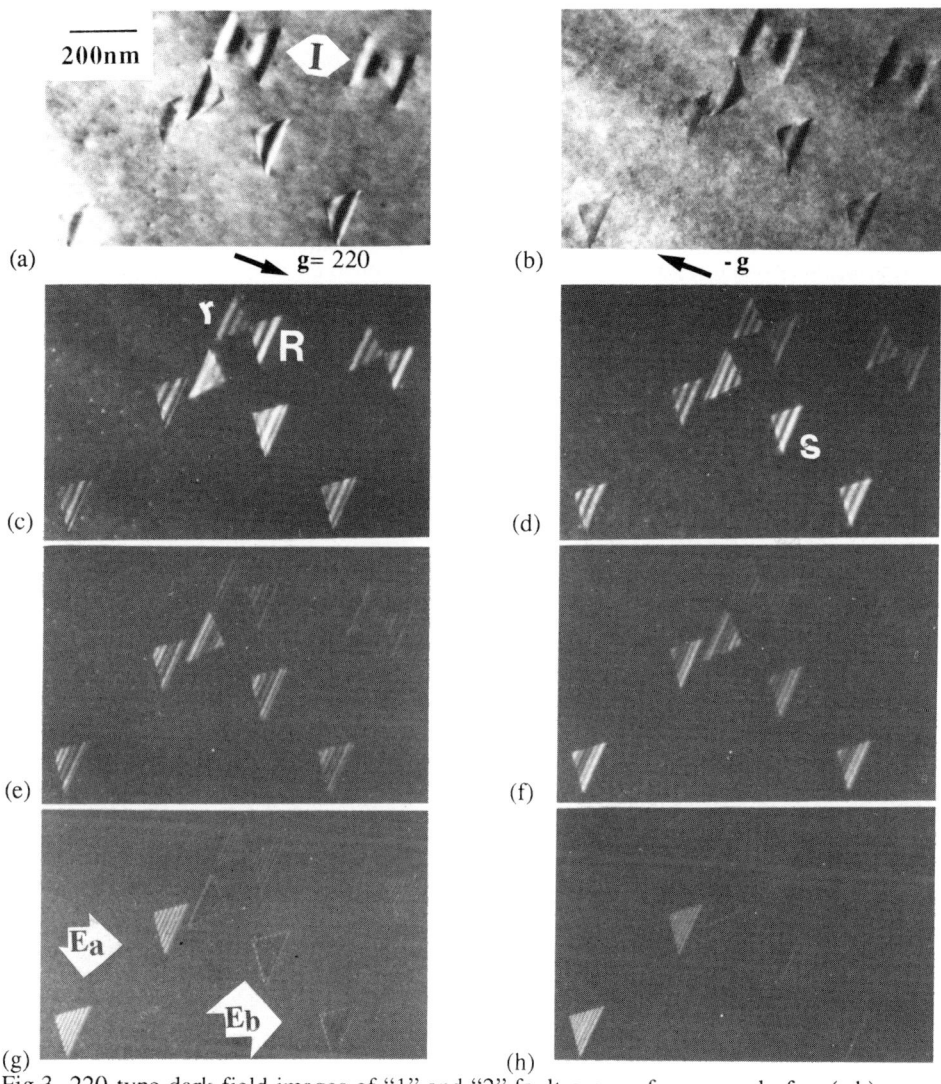

Fig.3 220-type dark field images of "1" and "2" fault groups for reversal of **g**: (a,b) strong dark field; (c) (**g**, 2**g**), (d) (-**g**, -2**g**), (e) (**g**, 3**g**), (f) (-**g**, -3**g**), (g).(**g**, 4**g**), and (h) (-**g**, -4**g**) diffraction conditions.

Considering firstly the intrinsic faults in configuration "2" we see that for a **g**,2**g** condition (figs.3(c-d)) the fringe contrast and average intensity are higher with α positive (as at **R**) than with α negative (as at **r**). For images at higher deviation parameter the effect becomes progressively less pronounced but in both contexts the contrast behaviour is as would be expected from previous work on the "anomalous weak beam contrast" of such faults. The behaviour of the apparently isolated and, on a conventional analysis, extrinsic faults for the different weak beam conditions is more complicated and requires us to consider them as two different types. Those marked E_a again exhibit the contrast behaviour that might be conventionally expected. For example, as the imaging conditions are made progressively more kinematic (as in Fig.3(g-h)) their visibility becomes increasingly greater than that of the intrinsic faults while their visibility is the grater when α is negative However those marked E_b exhibit the "anomalous" anomalous weak beam contrast of interest to us here. They exhibit negative contrast relative to the background under the **g,4g** conditions for which the E_a faults exhibited classical strong contrast. Even at a low deviation parameter (Fig.3(c-d)) their relative visibility for opposite inclinations (as at S) tends to be opposite to that normally expected. A possible explanation is that these apparently isolated E_b faults are actually microtwins and the presence of such defects would not be surprising on the basis of current analyses of the way coherency can be lost (Angelo et al., 1993). As we can see from Fig.4, and using a kinematic approach which should be viable for the high deviation parameter of interest for a **g,4g** condition, if we consider the separation of the centres of the two phase amplitude circles after the phase changes caused by traversing three sheared planes the average visibility of a microtwin could indeed be reduced for either sign of α as is found here (Figs.3(g-h)). Unfortunately however the classical strong-beam extrinsic contrast of these faults (Figs.3(a-b)) is then surprising and it is possible that varying segregation to the different faults might also be of relevance. It should be noted that the varying substrate thickness below the uniformly thick epilayer would not appear to provide an explanation for the anomaly we have described. Contrast simulations are now required for the various potential overlap configurations that might be possible and in this context it is also interesting that the changes in visibility for the E_a faults on reversing **g** was small for moderate deviation parameters and this too needs further investigation.

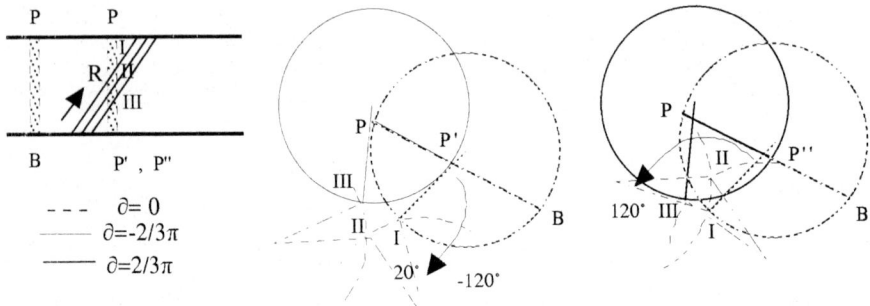

Fig. 4. Amplitude-phase diagram for the contrast after three successive phase shifts of $2/3\pi$ or $-2/3\pi$. P starting point; B end point for background contrast; I, II, III, points of first, second and third phase shift; P', P'' end points for fault with $\alpha=2/3\pi$ and $\alpha=-2/3\pi$, respectively.

3. CONCLUSIONS

1. Faults formed at the interface of (001) ZnSe layers grown on GaAs tend to group in the epilayer in five different configurations to form varyingly incomplete pyramids the apices of which are at the interface.
2. The paired faults prove to be intrinsic and to exhibit the classically expected anomalous weak beam contrast for faults of this nature.
3. The isolated faults exhibited uniformly consistent extrinsic contrast under strong-beam conditions. However they exhibited two different classes of behaviour under weak beam conditions. Some exhibited conventionally anomalous contrast behaviour under kinematic conditions. However, others showed contrast darker than the background under a (**g,4g**) diffraction condition, a possible explanation being that these are actually microtwins.

We are grateful to the INER for financial support and to Dr. K.A. Prior for the provision of the sample examined.

REFERENCES

Angelo, J. E., Britania, G., Sorba, L., Franciosi, A., Stobbs, W. M., Gerberich, W.W., 1993, *Phil. Mag.Let.*, in press.
Bithell, E. G., Donovan, P. E., and Stobbs, W. M., 1989, *Phil. Mag.*, 59, 63.
Cockayne, D. J. H., Pirouz, P., Liu, Z., Anstis, G. R., and Karanthaler, P., 1984, *Phys. Stat. Sol. (a)*, 82, 425.
Cullis, A. G., and Booker, G. R., 1972, *Proceedings of the Fifth European Conference on Electron Microscopy, Manchester*, p. 532.
Wilson, A. R., and Cockayne, D. J. H., 1985, *Phil. Mag.*, 51, 341.

Inst. Phys. Conf. Ser. No 134: Section 6
Paper presented at Microsc. Semicond. Mater. Conf., Oxford, 5–8 April 1993

421

XTEM observations of MOVPE grown $(Hg_xMn_{1-x})Te$ epilayers

H Tatsuoka, K Durose and M Funaki

Department of Physics, University of Durham, Durham, DH1 3LE, UK.

ABSTRACT: $(Hg_xMn_{1-x})Te$ (MMT) layers grown on GaAs(001) with CdTe(001) buffer layers have been examined by TEM. The alloy MMT has been grown by MOVPE using direct alloy growth (DAG) and inter-diffused multilayer (IMP) modes. There is a considerable reduction in the threading dislocation density at the MMT/CdTe interface. XTEM investigation of layers in the $[1\bar{1}0]$ and $[1\bar{1}0]$ orientations revealed anisotropy in the dislocation distribution. The IMP sample showed improved crystalline quality having a lower dislocation density, and no stacking faults or sub-grain boundaries.

1. INTRODUCTION

The dilute magnetic semiconductor alloy MMT is a material of relevance for infrared magneto-optics (Furdyna 1988 and Rogalski 1991). MMT was first produced in bulk single crystal by Delves and Lewis (1963). Epitaxial layers were first grown by Reno *et al.*(1985) by MBE. MOVPE grown MMT layers using DAG and IMP were reported by Al-Allak *et al.*(1991) and Funaki *et al.*(1993). The IMP-grown layers, which rely upon species interdiffusion, are of particular interest since MnTe has the NiAs lattice and HgTe that of sphalerite. On the other hand, for heteroepitaxial growth, the MMT layer is suitable for dislocation filtering on CdTe/GaAs, since the lattice parameters of HgTe (6.46 Å) and cubic MnTe(6.34 Å) are both smaller than that of CdTe(6.48 Å)(Durose *et al.*1993). The dislocation filtering by a DAG-grown MMT layer was first observed by Al-Allak *et al.*(1991). But the details of the dislocation nature were not described. In this paper, DAG and IMP-grown MMT layers are characterised by TEM, and the crystalline quality of each of these layers is compared.

2. EXPERIMENTS

The $(Hg_xMn_{1-x})Te$ x=0.9 layers on GaAs(001) with 1 μm CdTe(001) buffer layers have been grown by MOVPE using di-*iso*propyl telluride, elemental mercury and tricarbonyl methyl cyclopentadienyl manganese. Both DAG and IMP modes of growth were examined. The growth temperature was 380 °C. The IMP layer consisted of 100 HgTe-MnTe periods. The details of the growth procedure are described elsewhere (Funaki *et al.*1993). The layer thicknesses were 0.5 and 1.8 μm for DAG and IMP-grown layers, respectively.

The first process in the preparation of the specimens for cross-section TEM was to cleave the samples along orthogonal {110} planes lying mutually perpendicular to the surface. The slices were bonded face to face with epoxy resin. Then, they were prepared by mechanical polishing followed by Ar^+ ion milling. Typical operating conditions employed are an acceleration voltage of 4-6 kV, a beam current of less than 10 μA, and a beam incidence angle of 16°. Plan-view specimens were made by thinning from the substrate side only. Perforated samples were examined in a JEOL 100CX transmission electron microscope operated at 100 kV. Crystal polarity was determined using the technique of microdiffraction (Lu and Cockayne 1986, Brown *et al.*1990).

3. RESULTS AND DISCUSSION

3.1 DAG-grown layers

Fig.1(a) and (b) show typical bright-field, XTEM images of the layer, for [110] and [1$\bar{1}$0] orientations, respectively. The high threading dislocation content with the CdTe buffer layer is reduced near the MMT/CdTe interface. The threading dislocations in these images can be classified into two principal types: (i) irregularly shaped segments elongated nearly parallel to [1$\bar{1}$0] direction and (ii) long dislocations propagated to the surface on mainly {111} planes. The Burgers vectors of the former dislocations (type i) were determined by contrast experiments to have **b**=1/2a<011> lying in {111} planes. From these figures, it is considered that a three dimensional misfit dislocation array is created in the MMT layer near the interface, and the array is distributed up to a distance of about 0.15 μm from the interface. In addition, an anisotropic array of relatively straight dislocations elongated along [1$\bar{1}$0] is evident.

It has been suggested that the observed defect anisotropy is dependent on the differential motion of α and β dislocations in the sphalerite structure. In addition consideration must be given to the geometrical factors arising from the stress component in the epitaxial layer concerned. The MMT layer is under tensile stress on CdTe(001) buffer layer. It is suggested that the α dislocation is the more mobile dislocation in this MMT layer.

The three dimensional arrays were observed in graded heteroepitaxial (In,Ga)P layers (Abraham *et al.*1975). In this case, the interdiffusion of Cd and Hg or Mn takes place during the growth. The lattice parameter of the MMT layer near the interface is changed by the interdiffusion. Thus a graded layer region is created at the interface. The interdiffusion distance is roughly estimated as 0.15 μm using the data after Leute *et al.*(1981), which agrees with the width of the dislocation array along the growth direction. The three dimensional array of misfit dislocations is considered to be caused by climb of dislocations during the growth.

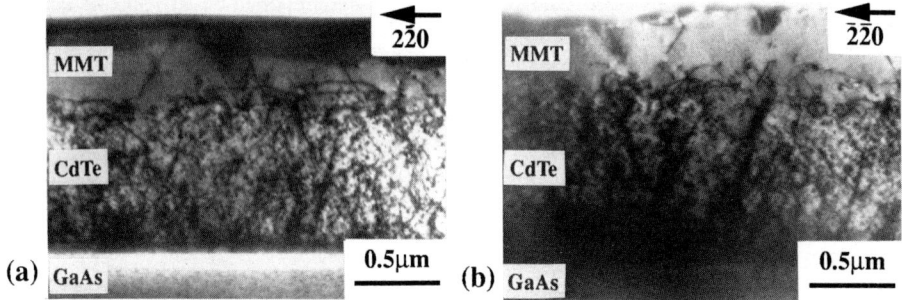

Fig.1 TEM micrographs of DAG-grown MMT/CdTe/GaAs viewed along (a)[110] and (b)[1$\bar{1}$0].

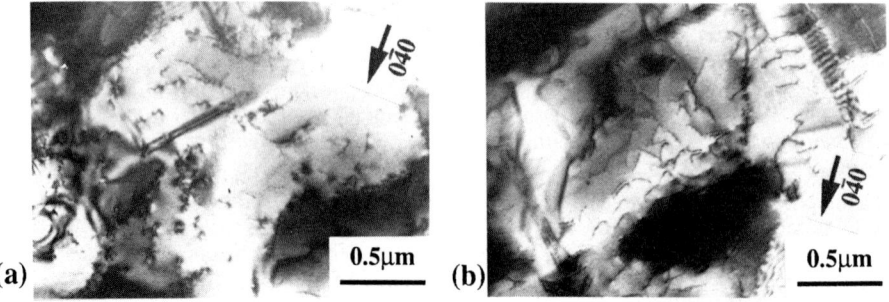

Fig.2   Plan-view TEM micrographs of DAG-grown MMT layer near surface.

Fig.2(a) and (b) show plan-view, bright field images of dislocations at the MMT surface. A large number of threading dislocations still exit near the surface, although many dislocations are bent near the MMT/CdTe interface. The dislocation density in the MMT surface is approximately 10^9 cm^{-2}. In these figures, the following characteristic features are observed.

In the Fig.2(a), four stacking faults on the adjacent {111}planes forming a "pyramid" are observed. The origin of the pyramid is considered to be a localised defect at the interface ; this corresponds with known thickness of the layer assuming that the stacking fault penetrates to the surface from the MMT/CdTe interface. In Fig.2(b), a large number of stacking faults, which are intrinsic type, bounded by partial dislocations are also observed. In addition to ordinary dislocations, several dislocation pairs were observed. It is determined from contrast experiments that the Burgers vector of each dislocation is 1/2a<110> type parallel to the layer surface. The spacing between them is about 250 Å, and is constant for all pairs observed. The generation mechanism of the pairs is not understood, and is under investigation.

It is observed in these figures that the layer consists of sub-grains. The dimension of each grain is about 1 μm. It is found from diffraction pattern analysis that the grains are tilted with respect to each other by 0.7°. It is considered that the generation of a sub-grain boundary is caused by three dimensional growth at the initial stage of the growth.

3.2 IMP-grown layers

Fig.3 shows a bright-field, XTEM image for the MMT layer grown by IMP. It is confirmed that the interdiffusion between MnTe and HgTe is complete through the layer. No dislocations are observed at the surface of the layer.

Fig.4(a) and (b) show typical images of the layer near the interface, for [110] and [1$\bar{1}$0] orientations, respectively. It is also observed that the high threading dislocation content with the CdTe buffer layer is drastically reduced near MMT/CdTe interface. In IMP-grown layer, only irregularly shaped dislocation segments nearly parallel to the interface are observed, and the density of dislocations reaching the surface is even lower than for the DAG sample examined. Indeed the dislocations propagated to the surface are eliminated. For all observable parts of the sample, no dislocations intersecting the surface were observed. The near interface array is anisotropic and dislocations tend to lie along [1$\bar{1}$0], as is the case for the DAG layer.

Fig.3   TEM micrograph of IMP-grown MMT viewed along [1$\bar{1}$0].

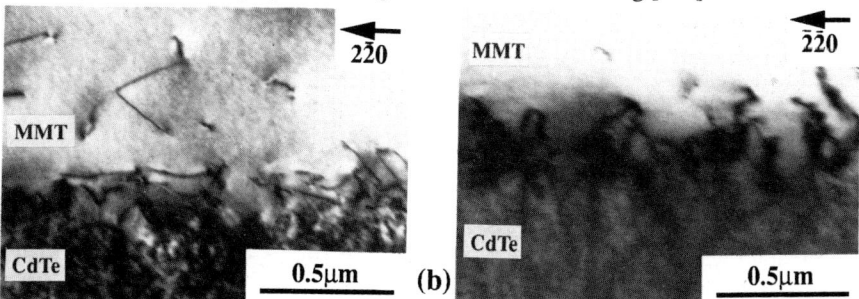

Fig.4   TEM micrographs of IMP-grown MMT/CdTe viewed along (a)[110] and (b)[1$\bar{1}$0].

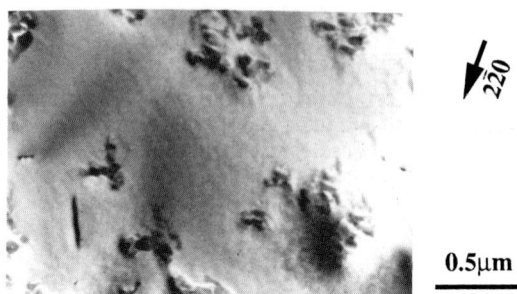

0.5μm

Fig.5   Plan view TEM micrograph of IMP-grown MMT layer near surface.

Fig.(5) shows a typical plan-view, bright field image of dislocations at the surface. The threading dislocation density is approximately 10^7 cm^{-2}. The lattice mismatches of MnTe/CdTe and that of HgTe/MnTe are 2.2×10^{-2} and 1.9×10^{-2}, respectively. These lattice mismatches and multilayer growth bend a lot of threading dislocations in IMP-grown layers near the interface, compared with DAG-grown layers. Also stacking faults and dislocation pairs are eliminated in the IMP-grown layer. In addition, sub-grain boundaries are not observed. IMP growth may promote two dimensional growth, and might act to eliminate sub-grains. The dislocation density of an IMP-grown layer can not be compared with that of DAG-grown layer directly, because the layer thicknesses are different. It is clear, however, that stacking faults, dislocation pairs and sub-grains are eliminated in IMP-grown layer. The results show that IMP growth improves the crystalline quality of the layer. The results is also confirmed by X-ray diffraction measurements (Tanner 1993).

4. CONCLUSION

MMT layers were grown on CdTe/GaAs(001) hybrid substrates using the MOVPE DAG and IMP processes and were characterised by TEM. In both layers the threading dislocations are blocked near MMT/CdTe interface. However, large number of dislocations, stacking faults, dislocation pairs, sub-grains still exist in DAG-grown layer. On the other hand, IMP-growth eliminates such defects, and improves the crystalline quality of the layer.

ACKNOWLEDGMENTS

The authors would like to thank Dr.A.W.Brinkman and Dr.Al-Allak for useful discussions.

Abrahams M S, Buiocchi C J and Olsen G H 1975 J.Appl.Phys. 46 4259
Al-Allak H M, Brinkman A W, Clifton P A and Brown P D 1991 Mat.Res.Soc.Symp.Proc. 216 35
Brown P D, Durose K, Russell G J and Woods 1990 J.Cryst.Growth 101 211
Delves R T and Lewis B 1963 J.Phys.Chem. Solids 24 549
Durose K, Turnbull A and Brown P 1993 Mater.Sci.Eng. B16 96
Funaki M, Lewis J E, Hallam T D, Li Chaorong and Tanner B K 1993 Semicond.Sci.Technol. 8 S200,   Funaki M, Brinkman A W, Hallam T D and Tanner B K 1993 Appl.Phys.Lett. in press
Furdyna J K 1988 J.Appl.Phys. 64 R29
Leute V, Schmidtke H M, Stratmann W and Winking W 1981 phys stat sol. (a)67 183
Lu G and Cockayne D J H 1985 Philos.Mag. A53 307
Reno J, Sou I K, Wijewarnasuirya P S and Faurie J P 1985 Appl. Phys.Lett. 47 1168
Rogalski A 1991 J Appl.Phys. 64 117
Tanner B K private communication

Inst. Phys. Conf. Ser. No 134: Section 6
Paper presented at Microsc. Semicond. Mater. Conf., Oxford, 5–8 April 1993

425

Structural quality of lattice matched GaAs/Sc$_{0.2}$Yb$_{0.8}$As/(001) GaAs structures

B Guenais, A Poudoulec, A Guivarc'h, Y Ballini, V Durel and C d'Anterroches[*]

France Telecom / CNET, LAB/OCM/MPA, BP 40, F22301, Lannion, France.
[*] France Telecom / CNET, CNS/DCF/CAP, BP 98, F38243, Meylan, France.

ABSTRACT : The influence of the growth parameters on the structural quality of lattice matched GaAs / Sc$_{0.2}$Yb$_{0.8}$As / (001) GaAs structures with different layer thicknesses was studied by Transmission Electron Microscopy and Rutherford Backscattering. Three main parameters were studied : the growth temperature of the GaAs overlayer, the Sc$_{0.2}$Yb$_{0.8}$As layer thickness, and the use of a vicinal surface. The Sc$_{0.2}$Yb$_{0.8}$As layer is matched to GaAs and grows in a two-dimensional mode, leading to a high structural quality. But the GaAs overlayer contains a high density of structural defects, due to three-dimensional growth. Its quality is much improved when growth occurs at 550°C and the best results are obtained for growth on a thin metallic layer on a vicinal surface. The nature of the structural defects were characterized by Transmission Electron Microscopy and conclusions are discussed with respect to the growth conditions.

1 INTRODUCTION

Epitaxial structures with a metallic layer embedded in a III-V semiconductor matrix are potentially rich in applications in the field of electronic and optoelectronic devices. However the growth of such structures is particularly difficult because of the difference in structure and chemistry of the metal and the semiconductor. On a GaAs substrate, among the possible metallic compounds, the rare-earth monoarsenides (REAs) with the NaCl structure offer the advantages of stability and of lattice matching with GaAs, by forming ternary alloys, for example Sc$_{0.2}$Yb$_{0.8}$As (Durel 1991) ; in the following, this particular composition will be indicated by the simple term "ScYbAs".

In spite of the high quality that can be achieved for the lattice matched metallic layer, the quality of the GaAs overlayer is limited by a high density of structural defects (Guivarc'h et al 1993). This can be related to a common feature of the GaAs / REAs / (001) GaAs structures : the growth of the metallic layer occurs in a two-dimensional mode whereas, due to a poor wetting of the (001) REAs surfaces, the GaAs overlayer growth is of island-type (Le Corre et al 1990, Zhu et al 1990).

Several structures having different thicknesses and substrate orientations have been characterized by Transmission Electron Microscopy (TEM) and Rutherford Backscattering (RBS). TEM results on some samples will be presented in this paper to illustrate the influence of three parameters on the structural quality : the overlayer growth temperature, the ScYbAs layer thickness, and the substrate misorientation.

2 EXPERIMENTAL

The samples were grown by Molecular Beam Epitaxy (MBE) (Durel et al 1991). First a 500 nm GaAs buffer layer was grown at 600°C, then the temperature was lowered to about 450°C in an As$_4$ flux, and the ScYbAs layer was grown at a deposition rate of 0.02 nm s^{-1}. Last a 200 nm thick GaAs overlayer was grown at a 0.5 μm h^{-1} ; the GaAs overlayer growth temperature will be noted " T$_{ov}$ ".

Three parameters were varied : first the overlayer growth temperature T$_{ov}$, second the thickness of the metallic layer, and third the orientation of the substrate. T$_{ov}$ was either " low ", about 400 °C, or " high ", about 550 °C. The thickness of the metallic layer was either " thin ", of the order of 2 nm, or " thick ", of the order of 40 nm. The substrate was either nominal or 4 ° misoriented about the [1-10] axis.

Combining these three parameters, eight structures were grown and characterized by RBS and TEM. TEM results on three of these structures will be presented :
- Sample 1 : (200 nm) GaAs / ScYbAs (40 nm) / GaAs ; T_{ov} = 400°C, nominal surface.
- Sample 2 : (200 nm) GaAs / ScYbAs (2 nm) / GaAs ; T_{ov} = 550 °C ; nominal surface.
- Sample 3 : (200 nm) GaAs / ScYbAs (2 nm) / GaAs ; T_{ov} = 550 °C ; misoriented surface.

The samples were characterized in-situ by reflection high energy electron diffraction (RHEED) and ex-situ by 1.8 MeV He$^+$ RBS. TEM observations were performed on (110) cross-sections at 120 kV and some observations at 400 kV for high resolution TEM (HRTEM).

3 RBS RESULTS

The overall structural quality of the GaAs overlayer was estimated from RBS aligned spectra : the χ_{min} (ratio of aligned to random backscattering yield) is measured at the overlayer surface and serves as a quality criterion, by reference to the GaAs substrate which χ_{min} is 3 %. The RBS results point out the dramatic effect of the overlayer growth temperature : for a "low" T_{ov} the χ_{min} ranges from 50 to 80 %, depending on the samples, whereas for a "high" T_{ov} the χ_{min} ranges from 5 to 30 %. The χ_{min} value is 50 % for sample 1, 10 % for sample 2 and 5 % for sample 3, which is the best one. For a "high" T_{ov} we note two favourable effects: the use of a misoriented surface (compare for example sample 2 and sample 3) and the growth on a "thin" ScYbAs layer. These effects are not appreciable for structures grown at "low" T_{ov}.

4 TEM RESULTS

4-1 GaAs overlayer grown at "low" temperature : sample 1.

Fig. 1 is a (002) dark field image of sample 1. The quality of the overlayer appears rather poor ; indeed the layer is polycrystalline. Only two orientations are found : the first one is the same as the substrate and the second one is $[111]_{overlayer}$ // $[001]_{substrate}$; two variants of this orientation coexist : $[1-10]_{ov}$ // $[110]_{sub}$ or $[1-10]_{ov}$ // $[1-10]_{sub}$. The size of the domains is about 50 nm. The same features are observed whatever the thickness of the metallic layer (either 2 or 40 nm) or the orientation of the substrate (either nominal or vicinal). These results are perfectly consistent with the RBS results reported before. Clearly a "low" temperature, in our growth conditions, must be avoided to obtain a monocrystalline GaAs overlayer.

Fig. 1 : (002) dark field cross-section TEM image of sample 1 : GaAs (200 nm) / ScYbAs (40 nm) / GaAs ; T_{ov} = 400 °C, nominal surface.

4-2 GaAs overlayer grown at "high" temperature : samples 2 and 3.

There is a fundamental difference with the preceeding observations : all the samples grown in such conditions were found to be monocrystalline, whatever the substrate orientation or the thickness of the metallic layer. This explains the quality difference pointed out by RBS results.

Fig. 2(a) and 2(b) are bright field images of sample 2 and 3. For the two samples, the dominant defects in the overlayer are microtwins and stacking faults which originate from the interface with the metallic layer. What is particularly salient on the image of sample 3 is that the density of the planar defects is reduced by suppression of the defects in one {111} plane. No structural defects were found in the metallic layer which is lattice matched to GaAs. But Fig. 3, which is a HRTEM image of the metallic layer in sample 3 taken in a very thin area, shows that the ScYbAs layer contains pinholes (1 to 10 nm) separating heavily facetted islands (1.5 to 30 nm). As RHEED results show a two-dimensional growth of the ScYbAs layer (Durel 1991), we think that pinholes occur when the temperature is raised to 550°C for the GaAs overlayer growth ; the metallic layer, when particularly thin, has a tendency to agglomerate. A plan-view of the sample (not shown here) shows that in spite of these pinholes, the metallic layer is continuous. Fig. 3 shows also a twin in the overlayer occuring in the vicinity of a pinhole ; twins are always leaning in the same sense with respect to the misoriented surface.

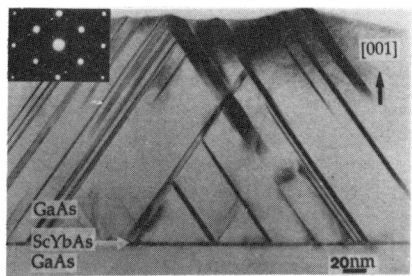 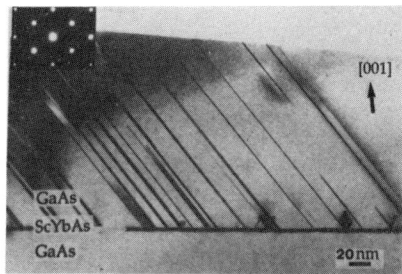

Fig 2 (a) : bright field cross-section TEM image of sample 2 : GaAs (200 nm) / ScYbAs (2 nm) / GaAs ; T_{ov} = 550 °C ; nominal surface. (b) : bright field cross-section TEM image of sample 3 : GaAs / ScYbAs (2 nm) / GaAs (200 nm) ; T_{ov} = 550° C ; misoriented surface.

Concerning the origin of these planar defects, we can notice that the lattice mismatch is not responsible for the occurence of this type of defects : this was previously underlined for other systems (Ernst et al 1989). Their origin was discussed elsewhere (Guenais et al 1992) : large GaAs islands are presumed to nucleate in the pinholes. These islands develop most probably {111} facets which move laterally and attach small misoriented islands which have nucleated on the metallic surface, leading to twin lamellae. Such "coalescence" microtwins are reported in other systems (Matthews and Allison 1963). The misoriented surface would result in GaAs islands having an anisotropic shape : one of the two facets parallel to the tilt axis will be stabilized to the prejudice of the other one. Fig. 4 shows the initial stage of the growth of a GaAs overlayer for sample 3 and confirms the proposed mechanism : a GaAs island is clearly shown nucleating in a ScYbAs pinhole and developing one {111} facet ; on this facet a twin lamella such as shown in Fig. 3 could occur.

Fig. 3 : HRTEM cross-section image (400 kV) of sample 3 : GaAs (200 nm) / ScYbAs (2 nm) / GaAs ; T_{ov} = 550° C ; misoriented surface.

Fig 4 : HRTEM cross-section image (120 kV) of the initial stage of the top GaAs layer growth ; T_{ov} = 550 °C, misoriented surface.

For a sample containing a "thick" metallic layer , the seeding of large well oriented GaAs islands through the pinholes does not occur, leading to a higher proportion of misoriented nuclei, and a higher density of planar defects (χ_{min} = 30 %). The use of a vicinal surface has a favourable effect (χ_{min} = 20 %), but this is less spectacular on a cross section than for sample 3. This could be related to the roughness of the top metallic layer which screens the vicinal nature of the substrate. A plan view analysis would be necessary to clarify the distribution of defects in the four {111} planes ; there is probably a decrease in the density rather than a supression of planar defects in one of the {111} planes.

5 CONCLUSION

The structural quality of the GaAs overlayer is closely related to the growth temperature T_{ov}. For a low T_{ov} (400°C) the overlayer is polycrystalline because of the occurence of misoriented domains having the [111] axis parallel to the growth axis. For a "high" T_{ov} (550°C), the top GaAs layer is monocrystalline ; the majority of structural defects are microtwins which result from the three-dimensional growth of GaAs. These defects develop on {111} facets of GaAs islands. Depending on the temperature, it could be more favourable for a misoriented (111) island to turn into a twin relationship when coalescence occurs with a (001) well oriented island, rather than keeping its own orientation.

Similar structures containing embedded lattice matched ScErAs layers into GaAs were studied by TEM (Zhu 1991). The results show also a strong dependence of the overlayer quality on T_{ov}. However their conclusions are opposite : for Tov < 450 °C the majority of the layer is (001) oriented but heavily twinned. For Tov > 450°C, (111) oriented domains are found. This points out that the tendency for island growth and wrong orientation on REAs' surfaces is closely related to the temperature but also to the chemical nature of the surface atoms (Er or Yb) ; note that it is totally independent of any mismatch effect.

On the other hand, GaAs / ErAs / (001) GaAs structures grown by migration enhanced epitaxy (MEE) at 320 °C (Yamaguchi et al 1992) have (001) oriented GaAs overlayers ; however their structural quality can be influenced by the very low thickness of the metallic layer which ranges from 1 to 3 monolayers and most probably contains pinholes.

The high quality obtained for a GaAs overlayer grown at high T_{ov} on a thin metallic layer and a vicinal surface is explained first by the existence of pinholes which induce a seeding of the GaAs top layer in the form of large islands, and second by the misorientation which, in this particular case leads to the suppression of microtwins in one of the {111} planes.

As a general conclusion the TEM results show that the quality of the GaAs overlayer on ScYbAs layers is mainly limited by the island growth of GaAs on the metallic surface. Another growth process has to be found to improve the structural quality of the top GaAs layer : the introduction of a "surfactant" allowing   two-dimensional growth could offer this advantage.

REFERENCES

Durel V 1991 Thesis, University of Rennes France n° 673 (Septembre 1991).
Durel, V, Caulet J, Ballini Y, Minier M, Guenais B, Dupas G, Guivarc'h A 1991 Proc. EUROMBE-91 Tampere Finland Fp 10.
Ernst F and Pirouz P 1989 J. Mater. Res. 4 834.
Guenais B, Poudoulec A, Guivarc'h A, Ballini Y, Durel V and d'Anterroches C 1992 Microsc. Microanal. Microstruct. 3 299.
Guivarc'h A, Guenais B, Ballini Y, Auvray A, Caulet J, Minier M, Dupas G, Ropars G and Regreny A 1993 J. Cryst. Growth, in press.
Le Corre A, Guenais B, Guivarc'h A, Lecrosnier D, Caulet J, Minier M and Ropars G 1990 J. Cryst. Growth 105 234.
Matthews J W and Allison D L 1963 Philos. Mag. 8 1283.
Yamaguchi H, Horikoshi Y 1992 Appl. Phys. Lett. 60 2341.
Zhu J G, Carter C B, Palmstrom C J, Mounier S 1990 Appl. Phys. Lett. 56 1323.
Zhu J G 1991 Thesis, Cornell University USA.

Microstructural characterization of the heteroepitaxy PbSe/BaF$_2$/CaF$_2$ on (111) Si

V Mathet, G Padeletti, J Olivier, P Galtier and F Nguyen-Van-Dau

THOMSON CSF-LCR, Domaine de Corbeville, F-91404 Orsay Cedex, France.

ABSTRACT: Narrow gap lead selenide layers grown on (111) Si substrates by molecular beam epitaxy (MBE) with the aid of an intermediate fluoride buffer have been studied by transmission electron microscopy (TEM). In the case of the CaF$_2$/ Si system, we measure the residual strain as a function of thickness. We also observe that the reduction of defects in the PbSe layer depends on crystalline quality and thickness of the fluoride layers. Results from TEM observations are completed and correlated with other techniques such as metallographic analysis and atomic force microscopy (AFM).

1. INTRODUCTION

Recently, there has been a growing interest in the heteroepitaxy of narrow gap lead salts (LSC) on (111) Si (Zogg et al 1991). Heteroepitaxy can be achieved using (Ba, Ca)F$_2$ buffer layers, in order to overcome the important difference between lattice parameters (typically 12-15%) and thermal expansion coefficients (8 times larger in LSC than in the Si). However, detailed structural features such as defects in the layers and interface microstructure remain to be studied.

This paper describes results obtained on PbSe/ BaF$_2$/ CaF$_2$ layers on (111) Si substrates. TEM, as well as AFM (Nanoscope III) and metallographic analysis are used to investigate the type and density of the defects present in the layers as well as their evolution with the growth parameters.

2. EXPERIMENTAL METHODS

The (111) orientation of the Si substrates is chosen because it leads to a preferential two dimentional growth mode for the fluorides (Blunier et al 1988). The growth is started by 20-200 Å of CaF$_2$, to prevent twinning in the subsequent 3000 Å thick BaF$_2$ layer (Zogg et al 1987). Typical growth rates are about 0.3 μm/h. The PbSe layer is then grown using PbSe and Se sources to obtain p-conducting layers as needed for device fabrication. Growth rates are 0.5-1 μm/h at a substrate temperature of 170-250°C.

Metallographic analysis has been performed on the fluoride layers. It gives a macroscopic view of the distribution of defects, such as dislocations (Huber et al 1983). The samples were also studied using plan-view and cross section transmission electron microscopy in an AKASHI 002B microscope operating at 200 keV. AFM is used to directly image the insulating fluoride surfaces.

3. CaF$_2$ ON (111) Si

The effect of the growth temperature on the CaF$_2$ surface morphology has been investigated. Fig. 1 presents AFM observations of the surface of two samples of 200 Å thick CaF$_2$. At 650°C (Fig. 1a), we observe small islands. They have a height of 200 Å and a lateral extension around 1500 Å. At 570°C (Fig. 1b), we only observe some spirals and a surface roughness of ≈15 Å. It is interesting to note that the small islands as well as the spirals exhibit a density of about 10^9 cm^{-2}. Thus, we think that these two features are closely related. This can be compared to TEM observations on a plan-view (111) image of the CaF$_2$ layer grown at 650°C (Fig. 2). Moiré fringes are the clearest features on the micrographs. We also observe contrasts with a circular shape (Fig. 2a). Their density (10^9 cm^{-2}) and their size (about 1500 Å) demonstrate that they correspond to the islands observed in AFM. A g=[2-20] bright field image of the same area (Fig. 2b) shows that the moiré fringes are not continuous across the islands. This is the signature of threading dislocations (Hirsch et al 1977). These two TEM observations allow us to conclude that each circular contrast is associated with a threading dislocation.

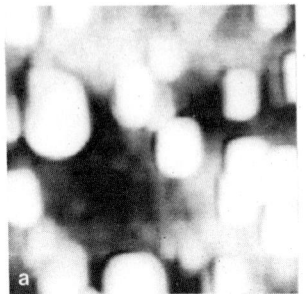 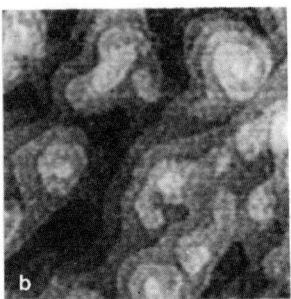

Fig. 1. AFM images of 200 Å thick CaF_2 layers on (111) Si. (a) at 650°C, (b) at 570°C. The observed area is 1 μm* 1 μm in both cases.

By correlating AFM and TEM experiments, we are able to associate the small islands as well as the spirals, with threading dislocations. It is well known (Frank 1949) that the presence of steps on a surface plays a great role in the growth mechanisms. Steps can be due to the emergence of a dislocation at the surface, with at least a screw component. The step height is equal to the normal component of the Burgers vector. During growth, such a step rolls up into a spiral, the growth rate being greater in the vicinity of the dislocation. The observation of such effects allows us to conclude that the growth is dominated by surface diffusion. If the growth temperature is too high, the growth of the spirals is favoured and they form the islands observed at 650°C. The growth of the CaF_2 layer is performed at 570°C in all the following experiments.

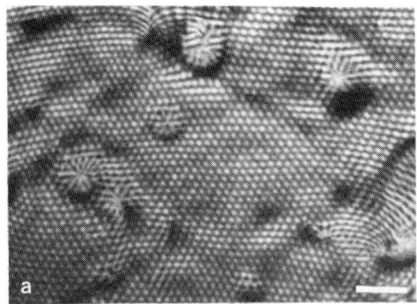 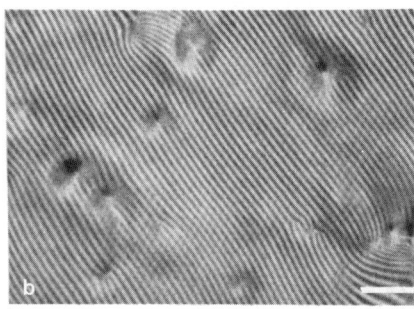

Fig. 2. Plan-view images of a 200 Å thick CaF_2 layer grown at 650°C. (a) observation in the zone axis (111), (b) same area with g=[2-20]. The bars correspond to 0.1 μm.

The Moiré patterns observed on the micrographs of fig. 2b are due to double diffraction (Hirsch et al 1977). The observed fringe spacing (178± 10 Å) has to be compared with the 330 Å expected from bulk parameters; the difference corresponds to a tensile residual strain, in the CaF_2 layer, parallel to the surface. Thus, moiré fringes can be used to study the variation of residual strain with the CaF_2 layer thickness. We do not observed moiré fringe for 20 Å thick CaF_2, which shows that the layer is totally strained, with the same lattice parameter as the Si substrate. Thus, the growth mode of CaF_2 on (111) Si appears to be pseudomorphic with a critical thickness (h_c) larger than 20 Å. This result is in good agreement with AFM experiments (Olivier et al 1993), and agrees with calculations (Mathet et al 1993) which consider the equilibrium theory using Matthews's expressions (Matthews 1975).

A detailed analysis of the moiré fringes also provides information concerning the Burgers vector of the dislocations (Hirsch et al 1977). Two kinds of **b** vectors are observed, those with b= a/2<1-10> (parallel to the interface plane) and those with b= a/2<110> (inclined to the interface plane).

4. BaF₂ ON CaF₂/ (111) Si

Detailed studies on plan-view samples were not possible on BaF_2 because of the radiolysis effect (Fathauer et al 1985). Many reliable results have been obtained with the use of the metallographic technique. It has allowed the optimisation of the growth temperature of the BaF_2 layer (Mathet et al 1993). Fixing the growth temperature at 340°C, we have grown samples with different thicknesses (t). Metallographic analysis is used to check the evolution of dislocation densities with t. We observe a t^{-1} law. Following Kroemer et al (1989), we believe that this decrease of the dislocation density corresponds to a binary recombination process of the dislocations. Thus, the BaF_2 layer thickness is chosen to be around 0.3 μm, in order to obtain a buffer layer as thin as possible which allows the epitaxy of the PbSe, with a reasonable dislocation density (typically in the low 10^8 cm^{-2} range).

5. PbSe LAYERS

The growth of the p-PbSe layer on fluoride buffer layers, which consist of 0.3 μm BaF_2 on 100 to 200 Å CaF_2, induces the formation of macroscopic defects (blisters) up to 10 μm in diameter. This phenomenom is not observed when growing n-type layers (evaporation of PbSe only) or p-type layers on bulk (111) BaF_2 (Zogg et al 1986). Moreover, this problem can be totally avoided by reducing the CaF_2 thickness to less than h_c. The formation of such blister shape defects seems to be correlated with the diffusion of Se as well as with the plasticity of the CaF_2 layer (Mathet et al 1993). The blister-free layers are then characterised using plan-view TEM. Near the interface we observe 5.10^9 dislocations cm^{-2}, and for a 2 μm thick p-PbSe layer the emergent dislocation density at the surface is below 10^7 cm^{-2}. Using the two-beam technique we have determined the Burgers vector of these dislocations to be either a/2<110> (inclined to the interface) or a/2 <1-10> (parallel to the interface plane).

6. OBSERVATIONS OF THE INTERFACES

HRTEM observations of the different interfaces present in the heteroepitaxy are shown in fig. 3 and fig. 4. It can be seen on the micrograph of fig. 3a that the CaF_2 is twinned compared to the silicon (type B) (Schowalter et al 1986). No dislocation can be seen at the interface, but the distance which could be observed in HRTEM experiments is limited to 500 Å. Blunier et al (1992) have studied the strain relief in this system. They observed that the lattice mismatch is mainly relieved by the glide of dislocations, which have been nucleated in half loops at the surface, in the {100} glide plane.

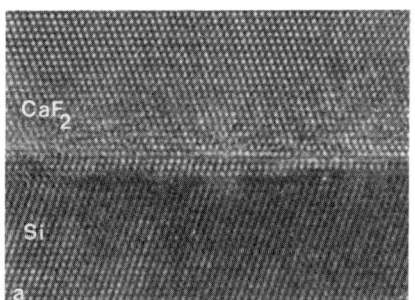 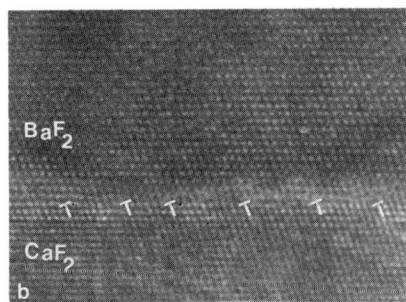

Fig. 3. HRTEM images of the CaF_2/ Si (a) and BaF_2/ CaF_2 (b) interfaces.

The BaF_2/ CaF_2 interface is shown in fig. 3b. The lattice mismatch at the growth temperature is 13.5%. The lattice orientations of BaF_2 and CaF_2 are the same. The electron beam is parallel to the <1-10> direction of the sample. Dislocations can be seen at the interface, whose lines are parallel to <1-10> directions. Burgers circuits reveal that the in-plane components of the Burgers vector (b) lies along the <11-2> direction, which means that b is parallel to the (111) interface. The average distance between dislocations is seven interatomic distances, which corresponds to the lattice mismatch being overcome. Thus, the lattice mismatch is accomodated by a grid of interfacial dislocations (Blunier et al 1992).

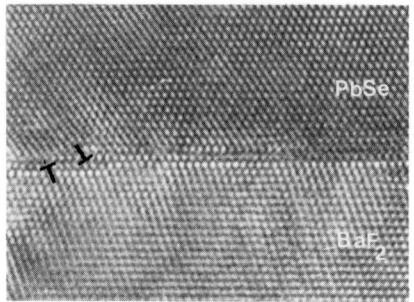

Fig. 4. HRTEM image of the PbSe/ BaF$_2$ interface.

Fig. 4 is a HRTEM observation of the PbSe/ BaF$_2$ interface. The PbSe layer is of type A with respect to Si. The Burgers circuits reveal that the in-plane component of the vector associated with the defects lie along the (11-2) direction. The vector is then parallel to the (111) interface. At present, we do not know if it corresponds to dislocations with 1/2<1-10> Burgers vector, because the contrast is extended and similar to those observed in the case of a monolayer-thick microtwin (Hwang et al 1991).

7. CONCLUSION

We have described microstructural characterisation of the heteroepitaxy PbSe/ BaF$_2$/ CaF$_2$ on (111) Si. Moiré fringe analysis has allowed a study of the residual strain in the CaF$_2$ layer as a function of thickness. The growth mode appears to be pseudomorphic with h$_c$ greater than 20 Å. This is confirmed by an AFM study, which gives a h$_c$ between 20 and 60 Å. This result has contributed to improving the crystalline quality of the PbSe layer. As a matter of fact, we obtained monocrystalline PbSe layers despite the large lattice mismatch and the difference of thermal expansion between PbSe and Si. HRTEM observations of the interfaces were also achieved. Such a study allows the optimisation of the growth process, and the fabrication of PbSe active layers with structural characteristics suitable for technological applications.

Acknowledgements: We wish to thank Dr. H Zogg for fruitful discussion. This work was supported by the French Ministry of Defense under contract DRET No. 91.34.483.

REFERENCES

Blunier S, Zogg H, Weibel H 1988 Mat. Res. Soc. Symp. Proc. 116 425.
Blunier S, Zogg H, Maissen C, Tiwari A N, Overney R M, Haefke H, Buffat P A, Kostorz G 1992 Phys. Rev. Lett. 68 3599.
Fathauer R W, Lewis N, Schowalter L J and Hall E L 1985 J. Vac. Sci. Technol. B3 736.
Frank F C 1949 Disc. Farad. Soc. 5 48.
Hirsch P, Howie A, Nicholson R, Pashley D W and Whelan M J 1977 Electron Microscopy of Thin Crystals ed Kreiger R E.
Huber A M, Laurencin G, Razeghi M 1983 J. Phys. C4 410.
Hwang D M, Schwarz S A, Ravi T S, Bhat R and Chen C Y 1991 Phys. Rev. Lett. 66 739.
Kroemer H, Lui T Y, Petroff P M, 1989 J. Cryst. Growth 95 96.
Mathet V, Galtier P, Nguyen-Van-Dau F, Padeletti G, Olivier J 1993 submitted
Matthews J W 1975 J. Vac. Sci. Technol. 12 126.
Olivier J, Padeletti G, Mathet V, Nguyen-Van-Dau F, Bisaro R 1993 to be published in Microscopy of Semiconducting Materials.
Schowalter L J, Fathauer R W, Ponce F A, Anderson G, Hashimoto S 1986 Mat. Res. Symp. Proc. 67 125.
Zogg H, Maeir P, Melchior H 1987 J. Cryst. Growth 80 408.
Zogg H, Maier P and Norton P 1986 Mat. Res. Soc. Symp. Proc. 56 235.
Zogg H, Maissen C, Masek J, Hoshino T, Blunier S and Tiwari A N 1991 Semicon. Sci. Tech. 6 C36.

Transmission electron microscopy in the study of the growth of Ga₂Se₃ thin films by the heterovalent exchange mechanism

A C Wright, J O Williams, M von der Emde[*], D R T Zahn[*], A Krost[*] and W Richter[*]

Advanced Materials Research Laboratory, NEWI Deeside, Connah's Quay, Clwyd, CH5 4BR
[*] Institut für Festkörperphysik, TU Berlin, Hardenbergstraße 36, D-1000 Berlin 12, Germany

ABSTRACT: Cubic Gallium Selenide can be prepared by the direct reaction of heated GaAs substrates with H_2/H_2Se gas. We present evidence that the thin films formed using this mode of growth (the heterovalent exchange mechanism) grow in two concurrent stages. The lower part of the film results from a direct replacement of GaAs with Ga_2Se_3. By contrast, the upper part of the layer forms from the reaction of excess gallium liberated from the substrate with the vapour at the free surface. We also show that films prepared under optimum conditions can be highly ordered.

1. INTRODUCTION

Cubic Gallium Selenide, Ga_2Se_3, has potential for the fabrication of short wavelength optoelectronic devices. Recent work by von der Emde at al. has shown that a direct energy gap exists at approximately 3eV making it a direct competitor to conventional II-VI materials such as ZnSe. The structure of Ga_2Se_3 is based on that of sphalerite but valence requirements lead to a metal sublattice which is 30% vacant. In the beta form, these vacancies are ordered giving a monoclinic unit cell (Lubbers and Leute 1982).

While both MBE and MOMBE have previously been applied for thin film growth, a simple route to the synthesis of epitaxial material has been developed by Kolodziejczyk et al. called the Heterovalent Exchange Mechanism. In this method, heated single crystal III-V substrates such as GaAs are exposed to an atmosphere of flowing H_2Se/H_2. The epitaxial film forms at the surface by gradual replacement of GaAs with subsequent loss of arsenic. Wright et al. used cross-sectional transmission electron microscopy to reveal the internal structure of these films and showed that an epitaxial bilayer structure had formed in all cases of growth.

A mechanism for film formation was proposed by Wright et al. in which the film grows in two concurrent processes. The lower layer forms as a result of direct replacement of GaAs by Ga_2Se_3 and the upper layer forms from excess gallium liberated from the formation of the lower layer. Thus the join between the two layers marks the position of the original substrate surface. The loss of arsenic and migration of excess gallium is facilitated by the large vacancy population on the metal sublattice.

In this paper we present additional evidence to support this mode of growth and also show that more than one vacancy ordering scheme can exist.

2. EXPERIMENTAL

Thin films were grown using a VPE system and a procedure as described by Kolodziejczyk et al. Using X-ray diffraction, a substrate temperature of 850K was found to give the best crystallinity. X-ray diffraction also showed that reflections from the beta form were evident. The variation in the X-ray diffraction data over a range of substrate temperatures and H_2Se flow rates has been described by Zahn et al. Transmission electron microscopy was performed at 300kV using a Philips EM430T fitted with EDX. Specimens for TEM were prepared in both <100> and <110> sections. Iodine ion milling on uncooled samples was used for final thinning (5keV and 30 microamps). To provide additional evidence for the growth mechanism, two experiments were performed. One involved the growth on a substrate which had an 0.5 micron AlGaAs layer (20% Al) on top. The other used a GaAs substrate which was partially masked with an evaporated SiO_2 layer to define the position of the surface.

3. RESULTS

A cross-section of a Ga_2Se_3 film grown under optimum conditions is shown in figure 1. It is clear that there are two main layers present and that their structure differs. The bottom layer is quite porous in nature while the upper layer is fully dense. The observed porosity concurs with the overall growth mechanism whereby arsenic is replaced by selenium in the lower layer giving rise to Kirkendall type voids when differences in the diffusion rates between counter diffusing species arise. Clearly no such problems arise in the case of the upper layer as new material is formed at the interface with the gas phase. Microanalysis of the junction between the two layers showed that this region was rich in arsenic.

Figure 1. Cross-section of bilayer. Note dense upper layer and porous lower layer. Insets are diffraction patterns from each layer.

Diffraction patterns (insets on figure 1) taken along <100> from the two layers differ quite markedly: the lower layer displaying only the sphalerite subcell reflections, while the upper layer clearly shows the additional spots expected from the monoclinic unit cell of the ordered beta phase.

Dark field images taken using a pair of superlattice spots show that quite large domains of order have formed in the upper layer, figure 2, and contrast with the very small domains observed in material not grown under optimum conditions as observed by Wright et al. High resolution imaging of the top layer along <100> revealed the superlattice associated with the monoclinic structure of the ordered beta phase, figure 3. The inset is a simulation of this structure and shows a clear match. When viewed along <110>, high resolution imaging also revealed regions of another type of superstructure as shown in figure 4. This is clearly very similar to that seen in some ordered III-V alloys and involves ordering on the (111) planes of the

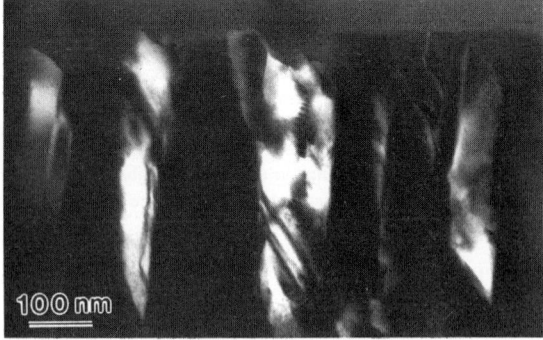

Figure 2. Dark field image of top layer taken using two superlattice spots showing domains of the monoclinic beta phase.

sphalerite sublattice. This ordering scheme has not previously been observed in Ga_2Se_3 and may be peculiar to this mode of growth where mass transport via the solid state may lead to non-equilibrium structures. Another possibility is that the presence of arsenic (in transit through the layer) may modify the crystal chemistry. Annealing treatments in the absence of H_2/H_2Se may remove excess arsenic from the layer and convert the CuPt structure type to the conventional monoclinic beta phase.

Figure 3. Phase contrast image of the ordered beta phase with simulation as inset. Brighter spots correspond to atomic planes containing vacant gallium sites.

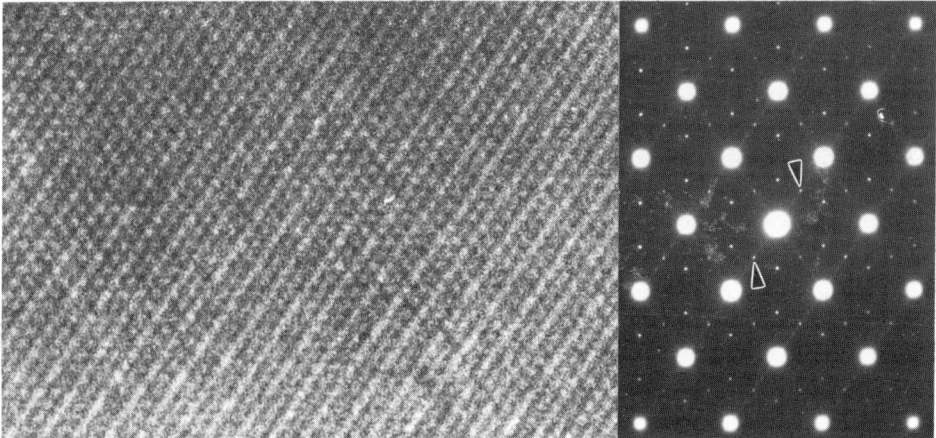

Figure 4. Phase contrast image of CuPt type order. Diffraction pattern shows extra spots (arrowed).

Material grown using the AlGaAs substrate is shown in figure 5. Diffraction patterns, taken from both layers differ in that only the lower layer is single crystal, the upper layer is now polycrystalline.

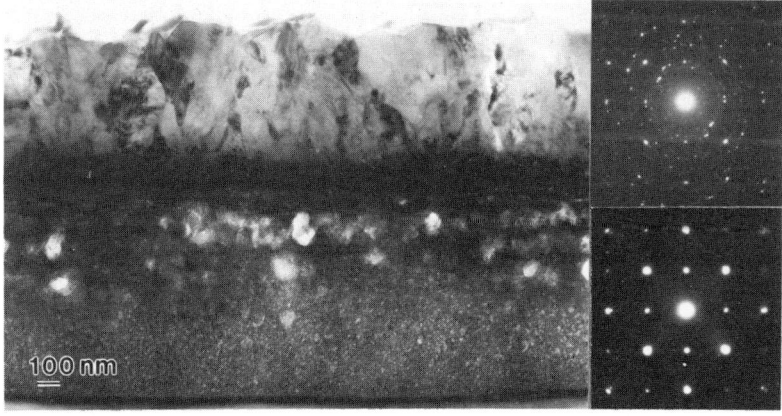

Figure 5. Cross-section of Ga_2Se_3 film grown on Al_2Ga_8As substrate. Diffraction patterns show that while the lower layer is epitaxial, the upper layer is polycrystalline.

A cross-section of the masked substrate clearly shows that the position of the surface is at the same level as the join between the two layers, figure 6. There appears to have been growth both under and over the edge of the mask.

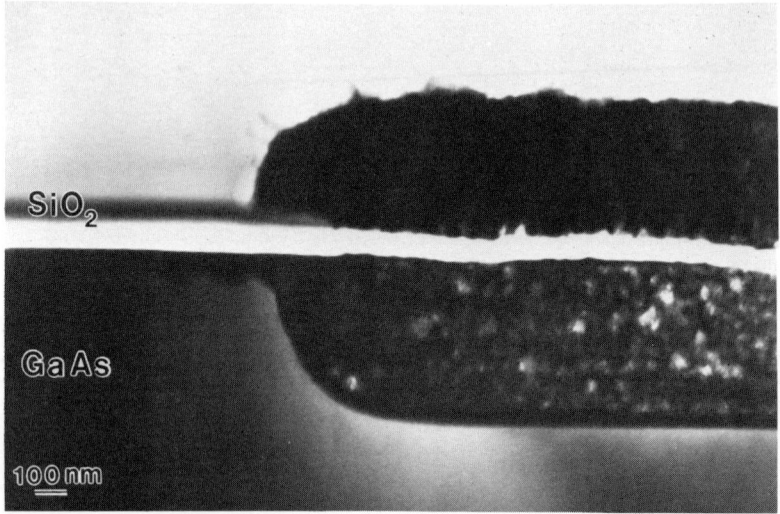

Figure 6. Cross-section of sample with an SiO mask (left). The position of the original interface is coincident with the boundary between the upper and lower Ga₂Se₃ layers.

4. DISCUSSION AND CONCLUSION

These results clearly show that the original growth mechanism proposed by Wright et al. is correct. This is bourne out by the masked specimen which unequivocally marks the position of the original surface. In addition, growth on AlGaAs substrates also supports the idea of transport of metal (Ga,Al) via the solid state as the presence of aluminium will reduce the atomic mobility thus resulting in the poorer crystallinity of the upper layer. The lower layer, which is formed by   the direct replacement of the substrate lattice by the Ga₂Se₃ lattice will be unaffected.

The structural quality of the material presented in this study is clearly superior to that shown by Wright et al. Nevertheless, this growth mode presents some difficulty for fabrication of optoelectronic devices as the lower layer, which is porous, is poorly attached to the substrate. In addition, heat conduction to the substrate will be poor. Films produced using this method will suffer from a residual arsenic level which will act as a dopant; this element clearly has some difficulty in diffusing out of the grown film as evidenced by the build up of arsenic at the junction between the two Ga₂Se₃ layers. There seems little scope for improvement in these respects when using this method of producing semiconductor films.

ACKNOWLEDGEMENT

We would like to thank the SERC for financial support of our programme on III-VI compound semiconductors.

REFERENCES

Kolodziejczyk M, Filz T H, Krost A, Richter W and Zahn, D R T. J. Crystal Growth 117 (1992) 549
Lubbers D and Leute J. J. Solid State Chemistry 43 (1982) 339
Wright A C, Williams J O, Krost A, Richter W and Zahn D R T. J. Crystal Growth 121 (1992) 111
von der Emde M, Zahn D R T, Rossow U, Kudlek G, Krost A, Richter W, Morley S, Wright A C and J O
    Williams. International Conference on the Formation of Semiconductor Interfaces (ICFSI-4). To be held in
    Julich, July 1993. Accepted for publication.
Zahn D R T, Krost A, Kolodziejczyk M and Richter W. J. Vac. Sci. Technol. B10 (1992) 2077

Inst. Phys. Conf. Ser. No 134: Section 6
Paper presented at Microsc. Semicond. Mater. Conf., Oxford, 5–8 April 1993

437

A study of tin doped indium oxide thin films grown by electron beam evaporation

A S A C Diniz, C J Kiely. I Elfallal[1], R D Pilkington[1] and A E Hill[1]

Department of Materials Science and Engineering, University of Liverpool, Liverpool, U.K.
[1] Department of Electronic and Electrical Engineering, University of Salford, Salford, U.K.

ABSTRACT : Thin films of tin-doped In_2O_3, (ITO), with excellent electrical and optical characteristics, have been produced by a new electron beam evaporation and post annealing route. The microstructure of the ITO layers has been characterised at each stage of the fabrication process using transmission electron microscopy (TEM) and X-ray diffraction (XRD). Correlations are drawn between the film microstructure and the measured electrical and optical properties at each step in the fabrication procedure.

1. INTRODUCTION

Thin films of tin doped indium oxide, ITO, are highly transparent in the visible range yet have good electrical conductivities. These properties combined with a high infrared reflectance and excellent adherence to many types of substrate ensure that ITO films have a wide range of applications. They are now finding uses as active layers or window coatings in solar cells. ITO thin films have been prepared by a number of different techniques such as conventional vacuum evaporation, sputtering from mixed oxide targets, reactive sputtering from metal alloy targets and chemical vapour deposition. The properties of ITO films, however, are found to be strongly dependent on particular growth technique used and the tin concentration of the film. Ideally ITO retains the cubic bixbyte structure of indium oxide until the solid-solubility limit of SnO_2 in In_2O_3 is reached (Frank et al, 1979). At low doping levels (<10at.% Sn) In^{3+} is substitutionally replaced on the cation sublattice by Sn^{4+} ions. Each dopant atom provides an additional electron for conduction and thus contributes to the free electron carrier density (Vossen, 1977, Frank et al, 1982). At higher doping levels, the conduction mechanism becomes more complicated due to the additional presence of oxygen vacancies, interstitial In^{3+} ions and a variety of tin-oxide complexes (Frank et al, 1982, Hamberg et al, 1986).

In spite of the technological importance of ITO thin films only a limited number of microstructural studies have been published. X-ray diffraction (XRD) (Frank et al, 1979, Parent et al, 1992) has been used for structure identification and the use of transmission electron microscopy (TEM) has been limited to the characterisation of features such as grain size distribution in fully oxidised films (Rauf et al, 1991). Elfallal et al (1992,a,b) have recently optimised an electron beam evaporation and post-annealing procedure for reproducibly producing ITO thin films with excellent electrical characteristics and optical transmission values in the visible region. In this study we have used TEM and XRD to characterise the microstructure of ITO films at each stage of this fabrication process and we correlate our observations with electrical and optical measurements on the films.

2. EXPERIMENTAL PROCEDURE

Details of the electron beam evaporation technique and post-annealing sequences have been presented elsewhere (Elfallal et al, 1992a,b) and will only be outlined here. ITO thin films were deposited onto (001)Si and glass substrates which were held at 20°C, using a mixed In_2O_3/SnO_2 pellet as the source material. A deposition rate of 10 Å/s and an oxygen partial pressure of 3×10^{-4} Torr was maintained during the evaporation. Film thicknesses of 0.4μm were typical. Two series of samples were prepared containing 4.7 and 8.1 at% Sn respectively. All as deposited ITO films had a brownish visual appearance and poor electrical characteristics. Sets of identical samples were annealed at 350 °C in air for times ranging between 15 minutes and 1 hour and were subsequently analysed by XRD and TEM. The one hour anneal was found to be sufficient to turn the films transparent. A further anneal in a reducing N_2/H_2 (4:1) atmosphere at 350°C for 1 hr was carried out on selected films in order to improve their electrical characteristics. Our samples were electrically characterised at each process stage using room temperature Van der Pauw measurements. Optical transmission measurements were obtained using a Cary 17D UV-VIS-IR spectrophotometer (Elfallal et al 1992,b). Electron transparent TEM samples were prepared in two ways. Cross-sectional ITO/Si specimens were made using conventional preparation techniques. ITO films deposited on glass substrates were simply delaminated and supported on a holey carbon film.

3. RESULTS

The variation of sheet resistance with post deposition annealing time for a series of 4.7 and 8.1at% Sn doped films are presented in Fig.1. It is clear that the sheet resistance decreases significantly during the first 30 minutes of annealing in air and then tends to level off. The films with the higher Sn content always have a lower sheet resistance value than their lower doped counterparts as expected. We consistently found that a significant improvement in the sheet resistance could be achieved by giving the layer a second anneal in a reducing N_2/H_2 atmosphere as illustrated by the dotted curve in Fig1. The electrical, optical and microstructural characteristics of all these films have been assessed. In this paper we only report our findings on three of these samples which are labelled A (unannealed), B (1hr anneal in air) and C (1hr anneal in air followed by 1hr anneal in N_2/H_2) in Fig1.

Fig.2 shows an electron micrograph and corresponding XRD spectrum from sample A. It is clear from Figs.2(a) and (b) that the as-deposited film was amorphous in nature. When observing such films in the TEM we noticed that they tended to crystallise under the influence of the electron beam forming a nanocrystalline structure as shown in Fig 2(c).

Fig.3 shows an electron micrograph and corresponding XRD spectrum from sample B. This film is clearly crystalline in nature with a grainsize ranging between 10 and 75nm. The strongest peaks in the XRD spectrum can be indexed to the cubic bixbyte structure with a lattice parameter of 1.0103nm. However, also present in electron diffraction ring patterns and XRD patterns are weak peaks which are inconsistent with the bixbyte structure. We can tentatively assign these to the presence of various tin oxide phases.

Fig.4 shows an electron micrograph and corresponding XRD spectrum from sample C. Again the film is random polycrystalline in nature, but with a slightly larger grainsize (in the range 20-95nm). The XRD spectrum fits well with the cubic bixbyte structure but with a slightly increased lattice parameter of 1.0152nm. Such an increase in lattice parameter is consistent with there being more substitutional Sn in the In_2O_3 lattice (Frank et al, 1982). This is borne out by the fact that the weaker (non-bixbyte structure) peaks decrease in intensity.

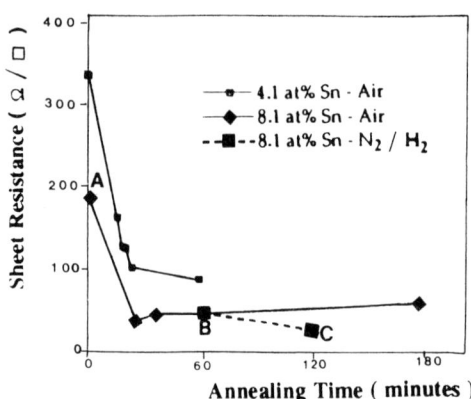

Fig 1. The variation in ITO sheet resistance with post-deposition annealing time

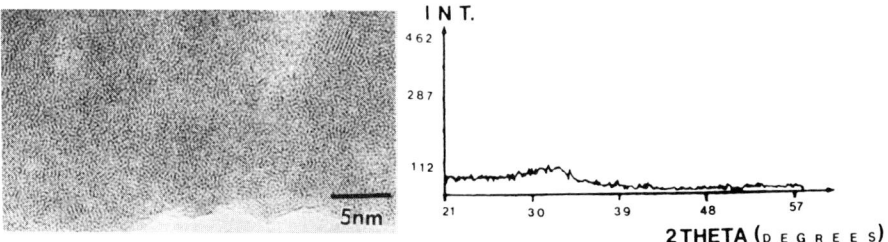

Fig. 2(a) - Electron micrograph of sample A   Fig. 2(b) - XRD spectrum from sample A

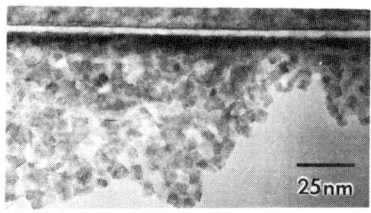

Fig. 2(c) - An HREM image showing the nanocrystalline ITO structure induced by electron beam
irradiation

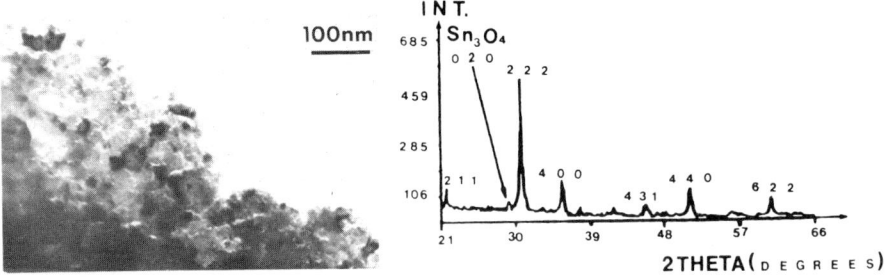

Fig. 3(a) - Electron micrograph of sample B   Fig. 3(b) - XRD spectrum from sample B

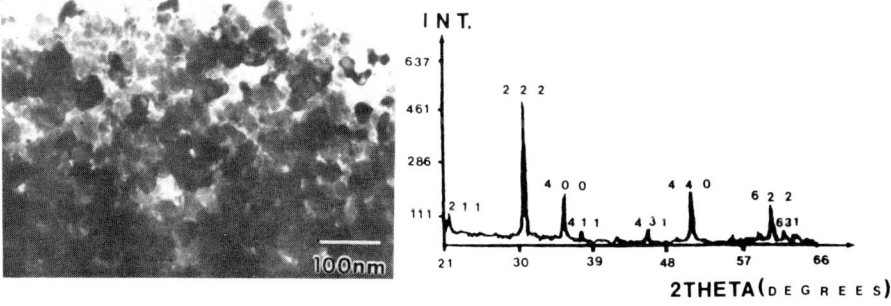

Fig. 4(a) - Electron micrograph of sample C   Fig. 4(b) - XRD spectrum from sample C

We have summarised the measured electrical characteristics (sheet resistance, carrier concentration and carrier mobility) of these three samples in Table 1. After annealing, all samples were n-type with a carrier concentration greater than 10^{21} cm^{-3}. No significant differences were observed in these electrical parameters for comparable films on Si or glass substrates. For completeness we have also included optical transmission and microstructural data in Table 1. A further point to note here is that all our films, whether amorphous or crystalline, show some degree of porosity. However, it is difficult to be quantitative about the fraction of porosity from our TEM micrographs.

Samples	Grain - size range (nm)	Lattice Parameter (nm)	Sheet Resistance ($\Omega/\square$)	Mobility (cm^2/Vs)	Carrier Concentration ($\times 10^{21}$cm^{-3})	Transmission at 1200 nm (%)
A	Amorphous	-	190	1	Indeterminate	<10
B	10 - 75	1.0103	44	10	3.2	90
C	20 - 95	1.0152	23	18	4.0	70

Table 1. Variation in microstructural, electrical and optical properties of samples A, B and C.

4. DISCUSSION

As deposited ITO films have poor visible light transmission and high sheet resistances (due to their low carrier concentrations and mobilities). These poor characteristics are a consequence of the film having an amorphous structure. The loss of long range order is likely to cause a smearing of the band edges and the introduction of mid-gap trapping states.

The sharp decrease in sheet resistance and improvement in optical transmission observed during the anneal in air can be correlated with the amorphous to crystalline transition. The ITO primarily adopts the bixbyte structure although some evidence has been observed for diffraction peaks from secondary phases. Annealing for times longer than 1 hr in air is seen to slightly degrade the electrical properties of the layer, although the films do generally become clearer with extended annealing.

A secondary annealing step in a N_2/H_2 atmosphere was seen to further improve the carrier concentration and mobility. It was also observed to cause several changes to the film microstructure. The fact that the ITO lattice parameter increases significantly leads us to deduce that more Sn^{4+} ions have been incorporated substitutionally into the In_2O_3 lattice, which will consequently increase the number of electrons available for conduction. This increased Sn^{4+} uptake into the ITO lattice is also reflected in the fact that the amount of Sn-containing secondary phase seems to decrease. The increase in carrier mobility may be associated with the increase in average grain size and hence decrease in grain boundary scattering in this material. The optical transmission of the films annealled in N_2/H_2 is seen to decrease at 1200 nm in accordance with classical Drude theory, which predicts that free carrier absorption dominates in the near infrared (Elfallal, 1993).

REFERENCES

Elfallal I, (1993) PhD thesis, University of Salford.
Elfallal I, Pilkington RD and Hill AE, (1992a), 2nd World Renewable Energy Congress, 1 530.
Elfallal I, Pilkington RD, Hill AE, Tomlinson RD, Diaz R, Leon M, Galan L and Rueda F (1992b), -11th European Photovoltaic Conference, 925.
Frank G and Kostlin H, (1982), J.Appl.Phys. A27 197
Frank G, Kostlin H and Rabenau A,(1979) Phys.Stat.Sol. (A) 52 231.
Hamberg I and Granqvist C, (1986), J.Appl.Phys. 60 (1986) R 126.
Machet J, Guille J, Saulnier P and Robert S, (1981) Thin Solid Films 80 149.
Parent P, Dexpert H and Tourillon G,(1992), J. Electrochem.Soc. 139 276.
Rauf I and Walls M, (1991), Ultramicroscopy 35 19.
Vossen J, (1977), Phys. of Thin Films 9 1.

Inst. Phys. Conf. Ser. No 134: Section 7
Paper presented at Microsc. Semicond. Mater. Conf., Oxford, 5–8 April 1993

441

Electron microscope investigations of heterostructures, nanostructures and misfit dislocations

C J Humphreys

Department of Materials Science and Metallurgy, University of Cambridge, Pembroke Street, Cambridge CB2 3QZ, UK

ABSTRACT: The optical moiré technique to aid the study of high resolution electron micrographs is described and illustrated. Reflection electron microscopy and high resolution electron microscopy have been performed on field-emitting silicon nanocones used in vacuum microelectronic devices and regions of hexagonal silicon have been found in the silicon nanocones. A novel fabrication method for producing silicon nanocolumns is described and energy filtered images of such structures are presented. The puzzle of the nucleation of misfit dislocations in strained epilayers on dislocation-free substrates is reviewed.

1. INTRODUCTION

This paper will review, and present the latest results on, some key problems in semiconductor materials, and the relevant EM techniques for solving these problems. First, we will describe a very useful technique, the optical moiré technique, for detecting and magnifying small distortions in high resolution electron micrographs. This technique aids the extraction of the maximum amount of information from high resolution images of a wide variety of crystal defects. Second, we will consider the application of reflection electron microscopy (REM) to studying silicon nanocones used for field emitters in vacuum microelectronic devices, and the unexpected discovery of hexagonal silicon in some of these nanocones. Third, we will describe the recent discovery of the fabrication of silicon nanocolumns in a controlled manner by the electron irradiation of silicon dioxide, and the imaging of such nanocolumns using energy filtered imaging. Lastly we will consider the problem of the introduction of misfit dislocations into strained epilayers when the substrate is dislocation free.

2. HREM AND THE OPTICAL MOIRÉ TECHNIQUE

Moiré fringes have long been observed in electron microscopy when two crystals with small differences in their lattice parameters or orientations overlap (see, for example, Hirsch et al, 1979). These patterns essentially magnify distortions produced by crystal defects and enable the distortions to be more easily detected and measured. Thus extra half-planes at dislocations were first revealed by parallel moiré patterns formed by overlapping Pd and Au layers (Pashley et al, 1957) long before electron microscopes had sufficient resolution to resolve directly lattice fringes in a lattice image.

Small distortions in HREM lattice images are often difficult to detect visually. Hetherington (1990) and Dahmen et al (1990) have proposed a very useful "optical moiré" technique to magnify such distortions and render them immediately visible. In this technique, moiré patterns are generated by superimposing a reference lattice upon the lattice image on the negative or print. The reference lattice could be a lattice image from a perfect region of the same specimen, but higher contrast results are obtained from a simple array of black dots on a clear background, with the array being of a spacing similar to that of the original lattice.

Using this technique, extra half planes at dislocations become easily visible and lattice rotations on images can be accurately measured. It must of course always be borne in mind that extra half planes or lattices rotations in an HREM image do not necessarily correspond to identical features in the specimen, and dynamical electron diffraction calculations must normally be performed to interpret the image contrast observed, and magnified using the optical moiré technique. An example of this technique is shown in Fig.1(a) and (b), due to Hetherington and Brown (private communication) Fig.1(a) shows an HREM image of dissociated dislocations in Si, and Fig.1(b)

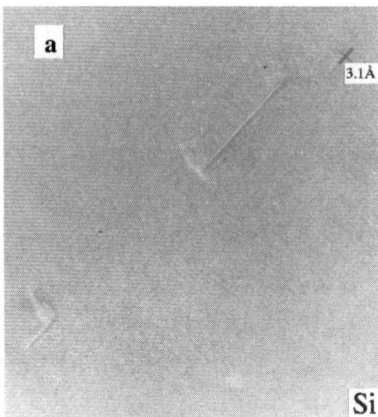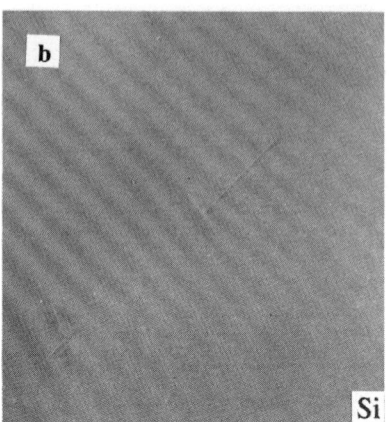

Fig.1   (a) HREM image of dissociated dislocation in Si and (b) optical moiré
image revealing extra half planes

shows the optical moiré image of the same area. Note how the extra half planes at the dislocations are immediately apparent, and the optical moiré image gives information about rotations and displacements over the whole of the negative simultaneously, avoiding the need to examine figures at a glancing angle. The technique is useful for the study of dislocations, twins and displacements at boundaries, including small subsidiary displacements at boundaries.

3.   SILICON NANOCONES AND HEXAGONAL SILICON

Until the end of the 1950's, most electronic circuits were based upon thermionic valves. The microelectronics revolution then occurred, transistors replaced thermionic valves in most applications, and silicon based integrated circuits now dominate the electrical device market. However integrated circuits are unsuitable for certain harsh environments (for example, high temperatures or high radiation levels) and there is considerable interest in the use of vacuum microelectronic valves for such environments.

The basic structure of these devices is a cathode which ejects electrons into a vacuum, a grid which can apply a modulating bias, and a collector anode. The devices can be made very small if the cathode is not a thermionic emitter, but a field emitter, and if the cathode tip is very sharp only a small voltage needs to be applied to obtain field emission. Fig.2 shows a reflection electron micrograph (taken in a Philips 400T) of a typical silicon cone. The width of the base of the cone is typically 1 μm and the cone tapers to a fine point.

REM is very useful for locating the cones on the specimen before performing TEM and HREM. REM causes a severe foreshortening of the image so three scale markers are shown on Fig.2: one for measurements parallel to the tilt axis (i.e. the width of the cones), one parallel to the axis of the cones, and one perpendicular to these two. The contrast of such REM images of cones is different from that normally observed in REM since there is significant transmitted intensity through the thin regions near the tip. Thus the contrast observed is a combination of specular reflection and Bragg reflection, with the regions around the tips being essentially dark-field transmitted images.

In practice, arrays of such cones are used and the main problems are non-uniformity of emission from the cones, and fracture of the tip of the cones during emission. In order to investigate the reasons for these problems an HREM study of such cones has been performed (Morgan, Preston and Humphreys, 1992). It was extremely surprising to find in one cone a region of hexagonal silicon near the tip (Fig.3). Hexagonal silicon is a rare phase which normally only forms under high pressure, for example under indentation of silicon (Pirouz et al, 1990). It is possible that the interface between hexagonal silicon and cubic silicon is relatively weak, and that the tip could fracture at this interface when subjected to the large stresses involved in field emission: this is not yet known.

The sharp silicon tips were produced by an oxidation sharpening method and it seems likely that the stresses generated by this technique have produced the region of hexagonal silicon in Fig.3. The power of HREM in studying problems with vacuum microelectronic devices is clear.

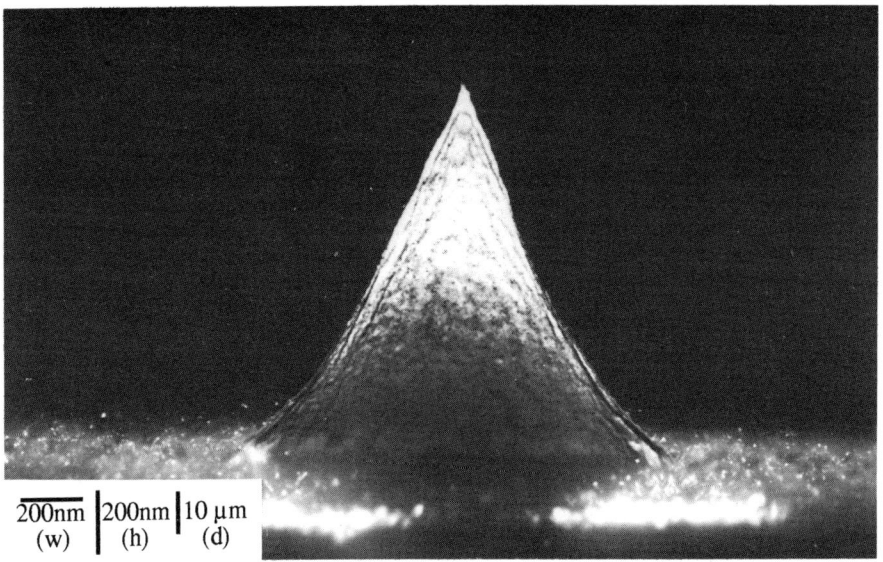

Fig.2   Reflection electron micrograph of silicon cone with scale markers for width, height and (foreshortened) depth

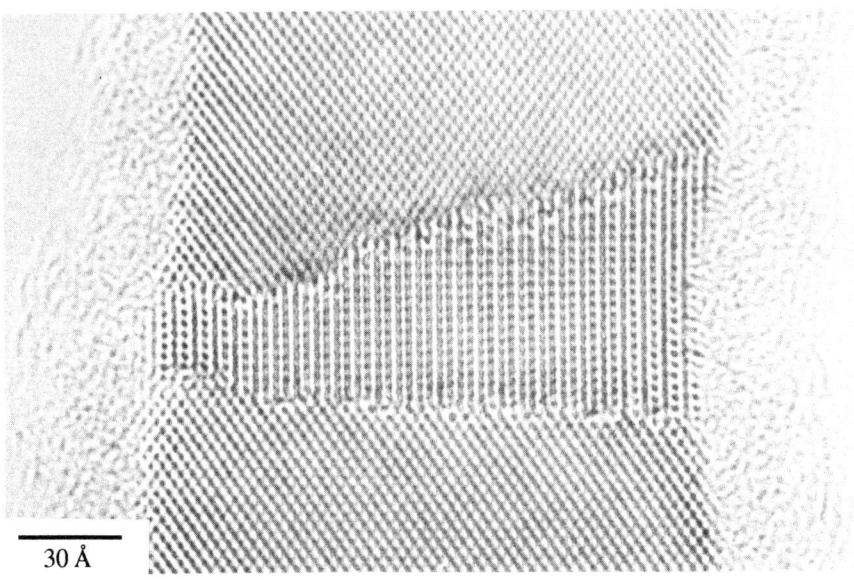

Fig.3   A region of hexagonal silicon formed near tip of nanocone and imaged on JEOL 4000EX

4.   EELS STUDY OF NANOMETRE DIAMETER SILICON COLUMNS IN A SiO_2 MATRIX

Chen et al (1993 a and b) have found that if a thin film of amorphous SiO_2 is irradiated with a high current density electron beam, the SiO_2 is reduced to Si.   Thus if the electron beam has diameter 2nm, a column of silicon of diameter 2nm is formed.   The less intense tails of the electron beam reduce the SiO_2 to SiO_{2-x}, leaving an annular cylinder of SiO_{2-x} around the Si column.   Thus the resulting heterostructure consists of an amorphous Si column surrounded by an amorphous SiO_{2-x} annulus embedded in a SiO_2 matrix.   An array of such structures can be formed by moving the electron beam and irradiating an array of points on the SiO_2 thin film.   When crystallised, such nanometre diameter Si columns may be of considerable potential interest for their possible light-emitting properties, and as model structures for light-emitting porous silicon (Canham 1990; Cullis and Canham, 1991).

In order to determine and demonstrate that these structures are indeed silicon nanocolumns surrounded by SiO_{2-x} and embedded in SiO_2, EELS and energy-filtered imaging are extremely useful techniques.   Fig.4 shows an array of irradiated areas in 15nm thick amorphous SiO_2 formed using a 4 second electron beam dwell time, equivalent to a dose of $5 \times 10^9 Cm^{-2}$ per irradiated spot.   The beam diameter (FWHM) was 2nm.   Fig.4 is an energy-filtered electron micrograph using electrons which have lost energies in the window 16 to 17eV, corresponding to the Si plasmon loss. Bright dots, about 2nm in diameter, appear in the centre of each irradiated area.   We identify these

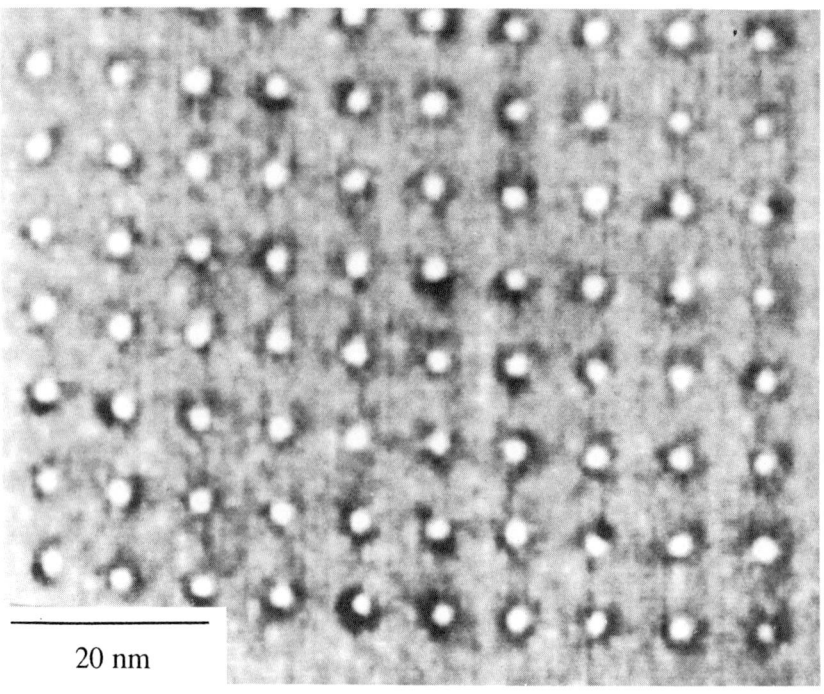

20 nm

Fig.4   Energy filtered micrograph, with amorphous Si columns imaged as bright dots surrounded by the darker SiO_{2-x} annuli

bright dots as Si since they are present in this Si plasmon energy-filtered image, and also present in bright-field and annular dark-field images, but absent from a SiO_2 plasmon energy-filtered image (taken using electrons which have lost energies between 22-24eV).   The SiO_2 matrix appears

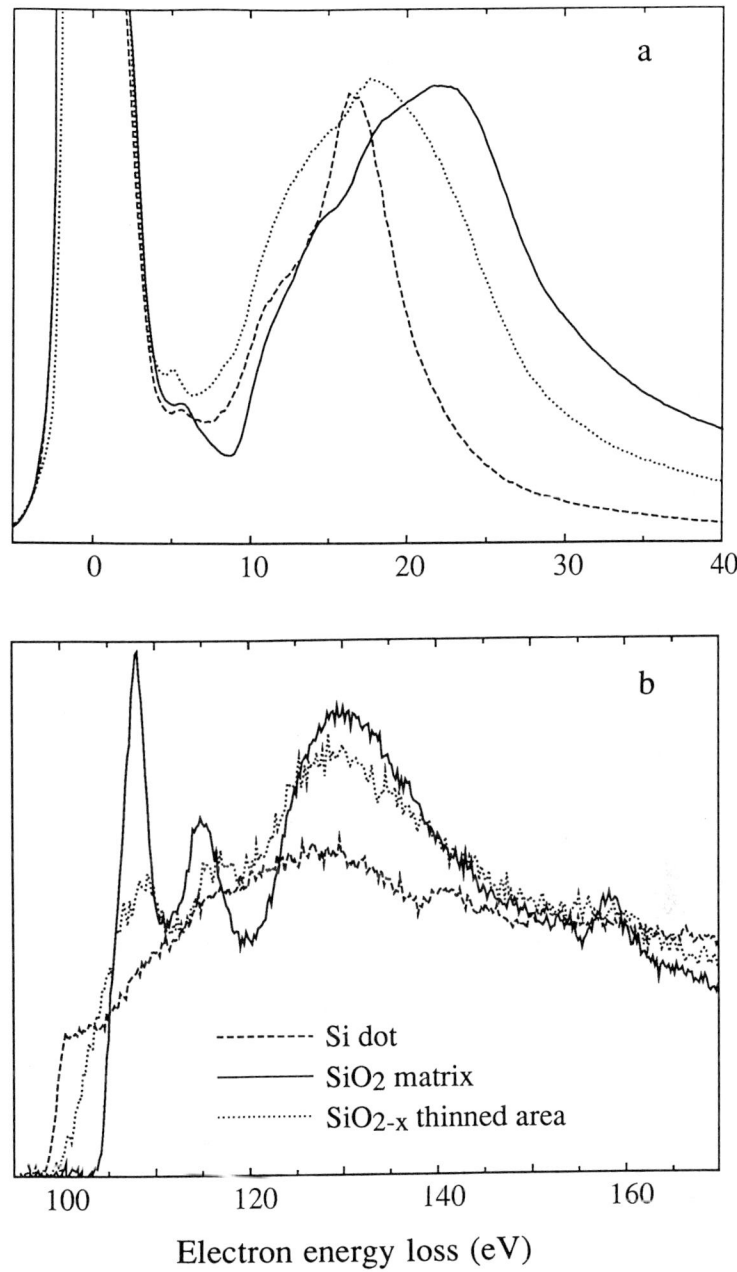

Fig.5   Electron energy loss spectra (a, low loss and b, higher loss spectra around
Si L edge) from Si dot, SiO$_2$ matrix and SiO$_{2-x}$ region around dot

reasonably bright in this Si plasmon energy-filtered image because the broad SiO_2 plasmon loss peak (centred at 23eV) overlaps the narrow Si plasmon loss peak.

Direct confirmation of the composition of the structures is provided from electron energy loss spectra, using an electron beam of about 1nm in diameter, taken from the Si dot, the SiO_2 matrix and the SiO_{2-x} region surrounding the dot. Fig.5(a) shows low loss spectra for energy losses of less than 40eV and Fig.5(b) shows higher loss spectra around the Si L edge. It is clear from these spectra that the bright dots in Fig.4 are indeed images of 2nm diameter columns of Si, since the plasmon loss and the L edge both correspond to silicon. Similarly, it is clear that the matrix is SiO_2: the plasmon loss is typical of SiO_2 and the Si L edge from SiO_2 is chemically shifted as expected. The annular region around the Si has spectra intermediate between Si and SiO_{2-x}, and we therefore identify this region as SiO_{2-x}. The power of energy-filtered imaging and of EELS in identifying and characterising these nanometre-scale Si/SiO_2 heterostructures is clear, as is the high spatial resolution of these techniques.

A preliminary attempt to crystallise the amorphous silicon nanocolumns has been made. Fig.6 shows an HREM image of a crystallised nanodomain in an amorphous SiO_2 matrix. Unfortunately analysis of the lattice fringes shows that the crystals are not of Si but are probably SiC, arising from contamination of the specimen. Further work is in progress to obtain crystalline silicon nanocolumns and to measure their optoelectronic properties.

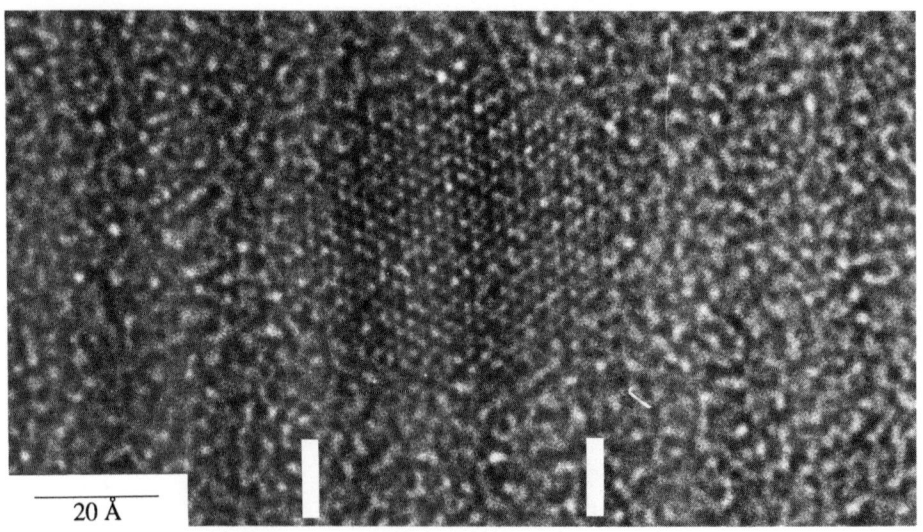

20 Å

Fig.6   HREM image of a single column after rapid annealing shows crystal of probably SiC (the white markers delimit the crystal width)

5.   MISFIT DISLOCATIONS IN STRAINED LAYERS

The relaxation of strain and the introduction of misfit dislocations in strained layers is a topic of considerable importance, however the nucleation of misfit dislocations in many semiconductor systems is still not well understood. We consider now the present state of our understanding.

In a classic paper Van der Merwe (1963) argued that as a thin epitaxial layer grew coherently upon a substrate with different lattice parameter a critical thickness, h_c, would be reached at which it was energetically favourable to accommodate the lattice misfit using an array of dislocations rather than by increasing the elastic strain in the epilayer. This is clearly an equilibrium argument which does not take into account either the dislocation introduction mechanisms or any energy barriers to nucleating the misfit dislocations.

Matthews (1975) and Matthews et al (1976) provided a model for the introduction of misfit dislocations, based on the existence of dislocations already in the substrate and threading up to the substrate/epilayer interface. According to their theory, the critical thickness occurs when the epilayer stress becomes sufficient to cause the existing threading dislocations in the substrate to bend over at the interface and form misfit dislocations. This model is also an equilibrium model and Willis et al (1990) have shown the exact equivalence of the equilibrium theories of Van der Merwe (1963) and Matthews (1975) and Matthews et al (1976). Jain et al (1992) have considered in detail interactions between the misfit dislocations and have shown that when these interactions are taken into account the critical thickness is always lower than the value of h_c derived from the theories of Van der Merwe (1963) or Matthews (1975). Willis et al (1991) have considered the driving force required for introducing the 'last' misfit dislocation to complete a periodic array, as distinct from considering the introduction of the 'first' misfit dislocation.

All of the above models (Van der Merwe 1963, Matthews 1975, and the improved model of Willis et al 1990) provide a good description of the behaviour of many systems (e.g. Kuk et al 1983). However it is evident that they cannot adequately explain h_c for epilayer growth on low dislocation density semiconductor substrates in which the observed critical thickness is far greater than that predicted by the equilibrium theories (Van der Merwe, 1963; Matthews, 1975). This implies that the kinetics of dislocation nucleation and propagation are central to our understanding of epitaxial semiconductor systems. However the apparent critical thickness depends strongly on the experimental technique which is used (Fritz, 1987), and Eaglesham et al (1988) have demonstrated that the critical thickness for GeSi/Si is far lower when measured using X-ray topography than when measured using electron microscopy, because of the lack of sensitivity of the latter technique in detecting dislocations when the density is lower than 10^5 cm^{-2}. In fact Eaglesham et al (1988) concluded that, at least in low mismatched systems, there was no sharply defined critical thickness.

For epilayer growth on a dislocation free substrate, what is the source of the first misfit dislocations? At least five sources for the nucleation of misfit dislocations have been found so far for the GeSi/Si system: (i) regions of crystalline damage at the edges of GeSi/Si wafers (Tuppen and Gibbings 1989); (ii) internal precipitates in the epilayers (Tuppen et al 1989a and b); (iii) precipitates at the Si substrate/buffer interface due to inadequate substrate cleaning (e.g. carbon remaining at the substrate surface results in the formation of SiC precipitates, Perovic et al 1989); (iv) precipitates at the epilayer surface due to impurity contamination after growth (Higgs et al 1991); (v) "diamond defects", which are internal stacking faults in the epilayer (Eaglesham et al 1989a and b, Humphreys et al 1989, 1991). These five sources of dislocations give rise, respectively, to misfit dislocations originating (i) near the wafer edge, (ii) internally in the epilayer, (iii) at the substrate/buffer or, in the absence of a buffer layer, substrate/epilayer interface, (iv) internally in the epilayer, (v) at the epilayer surface. At least some of these heterogeneous dislocation sources are expected to occur in a wide range of lattice mismatched systems.

However, if growth techniques improve so that all of the above dislocation sources are eliminated, the question remains as to how such dislocations can be nucleated in a system with a dislocation-free substrate. Frank (1950) and Hirth (1963) have shown that the lowest energy route is through the nucleation and propagation of a dislocation half-loop from the growth surface. Matthews et al (1976) have considered in detail dislocation half-loop nucleation and propagation in strained epilayers and have given expressions for the critical radius a dislocation half-loop must have for it to propagate and the corresponding activation energy required. It was concluded that the nucleation barrier could not be overcome (at typical growth temperatures) for misfits below about 2% for *any* epilayer thickness. Eaglesham et al (1989a) have re-examined these calculations using a higher value of the dislocation core parameter (probably appropriate for dislocations in semiconductors) and calculate that the nucleation energy for a critical-radius half-loop is significantly higher than that calculated by Matthews: the new value is about 100eV at 2% misfit, and it increases to about 1 000 eV as the misfit tends to zero.

Hence for low misfits and very careful growth it appears that very thick strained epilayers might be grown, totally dislocation free, because of the difficulty in nucleating dislocations. However, Perovic and Houghton (1992) have recently suggested that the nucleation of misfit dislocations can occur spontaneously during growth, although it is not clear what the driving force is in their model to cause the dislocation half loop spontaneously nucleated to reach a critical radius, and hence to grow rather than to shrink.

Thus the current theoretical situation for the introduction of misfit dislocations in high quality low misfit strained layers grown on dislocation-free substrates ranges from this being very easy (Perovic and Houghton, 1992) to virtually impossible (Matthews et al, 1976; Eaglesham et al, 1989a). The current experimental situation is that the introduction of misfit dislocations is easy if, because of poor growth techniques, various internal sources are present. However, if growth techniques improve, and internal sources are eliminated, whether or not misfit dislocations are easily introduced at the critical thickness remains an open question yet to be resolved.

ACKNOWLEDGEMENT

The author is grateful to C J D Hetherington for help in preparing this manuscript and to C B Boothroyd, P D Brown, G S Chen, C J D Hetherington and C J Morgan for supplying figures.

REFERENCES

Canham L T 1990 Appl. Phys. Lett. 57 1046
Chen G S, Boothroyd C B and Humphreys C J 1993b Appl. Phys. Lett. 62 1949
Chen G S, Hetherington C J D, Boothroyd C B and Humphreys C J 1993a These proceedings
Cullis A G and Canham L T 1991 Nature 353 335
Dahmen U, Hetherington C J D and Westmacott K H 1990 Proc. XIIth Int. Conf. on Electron
    Microsc. (San Fransisco Press) 338
Eaglesham D J, Kvam E P, Maher D M, Humphreys C J, Green G S, Tanner B K and Bean J C 1988
    Appl. Phys. Lett. 53 2083
Eaglesham D J, Kvam EP, Maher D M, Humphreys C J and Bean J C 1989a Philos Mag 59 1059
Eaglesham D J, Maher D M, Kvam E P, Bean J C and Humphreys C J 1989b Phys. Rev. Lett. 62
    187
Frank F C 1950 Symposium on Plastic Deformation of Crystalline Solids (Carnegie Inst. of
    Technology Pittsburgh)89
Fritz I J 1987 Appl. Phys. Lett. 51 1080
Gomez A, Cockayne D J H, Hirsch P B and Vitek V 1975 Phil. Mag. A 31 105
Hetherington C J D 1990 Mat. Res. Soc. Symp. Proc. 183 123
Higgs V, Knightley P, Goodhew P J and Augustus P D 1991 Appl. Phys. Lett. 59 829
Hirsch P B, Howie A, Nicholson R B, Pashley D W and Whelan M J 1979 Electron Microscopy of
    Thin Crystals (Krieger: New York) 169
Hirth J D 1963 In Relation between Structure and Strength in Metals and Alloys (HMSO: London)
    218
Humphreys C J, Maher D M, Eaglesham D J and Salisbury I G 1989 Inst. Phys. Conf. Ser. No. 100
    241
Humphreys C J, Maher D M, Eaglesham D J, Kvam E P and Salisbury I G 1991 J. Phys. III 1 1119
Jain S C, Gosling T J, Willis J R, Totterdell D H J and Bullough R 1992 Phil. Mag A65 1151
Kuk Y, Feldman L C and Silverman P J 1983 Phys. Rev. Lett 50 511
Matthews J W 1975 J. Vac. Sci. Technol 12 126
Matthews J W, Blakeslee A E and Mader S 1976 Thin Solid Films 33 253
Morgan C J, Preston A R and Humphreys C J Electron Microscopy Vol 2 EUREM 92 Granada
    Spain 1992 147
Pashley D W, Menter J W and Bassett G A 1957 Nature 179 752
People R and Bean J C 1985 Appl. Phys. Lett. 47 327
Perovic D D and Houghton D C 1992 Mat. Res. Soc. Symp. Proc. 263 391
Perovic D D, Weatherly G C, Baribeau J-M and Houghton D C 1989 Thin Solid Films 183 141
Pirouz P, Chaim R, Dahmen U and Westmacott K H 1990 Acta Metall. Mater. 38 313
Tuppen C G and Gibbings C J 1989 Thin Solid Films 183 133
Tuppen C G, Gibbings C J and Hockly M 1989a J. Cryst. Growth 94 392
Tuppen C G, Gibbings C J and Hockly M 1989b Mat. Res. Soc. Symp. Proc. 130 141
Van der Merwe J H 1963 J. Appl. Phys 34 123
Willis J R, Jain S C and Bullough R 1990 Phil. Mag. A62 115
Willis J R, Jain S C and Bullough R 1991 Appl. Phys. Lett. 59 920

Correlation between the structural and optical properties of AlAs/GaAs quantum well structures

T Walther[1], D Gerthsen[1], R Carius[2], A Förster[2] and K Urban[1]

[1]Institut für Festkörperforschung, [2]Institut für Schicht- und Ionentechnik, Forschungszentrum Jülich GmbH, Postfach 1913, D-W5170 Jülich, Germany

ABSTRACT: AlAs/GaAs quantum well structures were grown by molecular beam epitaxy on (001) GaAs substrates. The influence of growth interruptions on the interface characteristics has been studied by high-resolution transmission electron microscopy (HRTEM) and photoluminescence spectroscopy (PL). A pattern recognition procedure was applied to determine the composition in the interface region along the $<100>$ and $<110>$ projections on a near-atomic scale.

1. INTRODUCTION

The properties of AlAs/GaAs quantum well structures (QWSs) have been extensively studied in the recent past by different experimental techniques. HRTEM has been frequently applied, e.g. by Ikarashi et al (1990), to study the microstructure of the interfaces between GaAs and AlAs. Using chemically sensitive imaging conditions (Ourmazd et al 1990) the chemical composition in the interface region can be determined on a near-atomic scale.

The optical properties of the QWSs have been examined by PL measurements (e.g. Katzer et al 1990, Gammon et al 1990, Kopf et al 1991). The PL line shapes were successfully fitted by the model of Singh et al (1984, 1986) and the model of Christen and Bimberg (1990) which are based on the assumption of abrupt chemical transitions between the quantum wells and barriers. The half width (FWHM) of the PL lines is mainly determined by roughness of the interface, i.e. the extension and height of interface islands with respect to the exciton diameter (≈ 20nm). Narrow linewidths are interpreted in terms of atomically smooth interface regions which are large compared to the exciton diameter and separated by steps of one monolayer in height.

There has been a controversy about the interpretation of the results obtained by those two methods. HRTEM images reveal substantial interface roughness on a nanometer scale (Bode and Ourmazd 1992) although PL line splitting is observed. Also, chemically graded interfaces are not in contradiction to narrow PL lines (Bimberg et al 1992). As already pointed out by Warwick et al (1990) and Gammon et al (1991), HRTEM and PL probe the interface roughness in different ranges of spacial frequencies. A quantitative correlation of these two methods therefore is necessary to describe interface roughness properly.

2. EXPERIMENTAL TECHNIQUES

Quantum well structures with GaAs well thicknesses of 3nm, 5nm and 8nm were grown in a Varian MBE ModGenII System at 650°C on GaAs(001) substrates. The GaAs wells are separated by 25nm AlAs barriers. Two wafers were grown with a beam equivalent pressure (BEP) of 28 for As/Ga and of 230 for As/Al with and without growth interruption (denoted as wafer 1 and wafer 2). Another set of wafers was grown with reduced BEPs of 14 and 115. Only the results from wafer 1 and wafer 2 will be discussed in this report. However, the results obtained from wafers with the reduced BEPs are consistent with the results presented in the following.

$<100>$ and $<110>$ cross-section specimens were prepared by the standard "sandwich" technique (Bravman and Sinclair 1984). The final thinning was carried out with 2 keV Ar^+-ions under an angle of 10° to remove amorphous damage layers induced at higher acceleration voltages. A JEOL 4000 EX electron microscope with a spherical aberration constant of 1mm was used. High-resolution images were simulated with the EMS program package (Stadelmann 1987) with the following parameters: objective aperture diameter 15.2nm^{-1} ($<100>$ zone axis) and 10.4nm^{-1} ($<110>$ zone axis), semiangle of beam convergence 0.9mrad, defocus spread 10nm.

The PL measurements were performed at a temperature of 2 K with an excitation energy of 2.41 eV and a SPEX 1404 monochromator with an attached photon counting unit.

A pattern recognition procedure (PRP) was applied to determine the chemical composition in the interface region. The procedure - similar to that suggested by Thoma and Cerva (1990) - is based on the Fourier transformation of image unit cells of the size of 0.28nm^2 for the $<100>$ zone axis and 0.28x0.566nm^2 for the $<110>$ zone axis. Applying the PRP on simulated images, the following optimum imaging conditions for AlAs/GaAs interfaces were determined regarding the sample thicknesses t and the underfocus values Δf.

a) $<100>$ zone axis: 6.8nm $\leq$ t $\leq$ 13.6nm, 15nm $\leq \Delta f \leq$ 30nm
The Fourier amplitude of the {200}-image frequencies A_{200} depends linearly on the Al concentration x within a precision of 10 atomic percent (Fig. 1).
b) $<110>$ zone axis: t > 14.0 nm, 55nm $\leq \Delta f \leq$ 65nm
The amplitudes of all relevant image frequencies are not linearly correlated with the Al concentration - even in the above t and Δf range. Thus an exact determination of the imaging conditions is necessary for obtaining quantitative information from the experimental images.

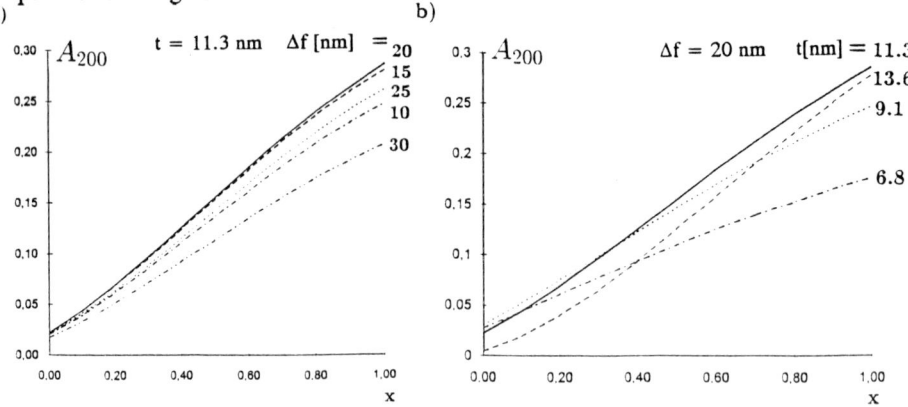

Fig.1: Amplitude of the {200} image frequency as a function of the Al concentration. (a) in the underfocus range 10nm$\leq\Delta f\leq$30nm at t=11.3nm, (b) in the thickness range 6.8nm$\leq$t$\leq$13.6nm at Δf=20nm

Sections from HRTEM micrographs are digitized with a 512x512 pixel CCD camera. The local image unit cells are determined and resampled onto a perfect grid with 16 pixels corresponding to 0.28nm. The Fourier amplitude of the relevant image frequency (i.e. A_{200} in the case of the $<100>$- orientation and A_{111} or A_{200} for the $<110>$ orientation) of each image unit cell is computed and related to the local chemical composition.

3. EXPERIMENTAL RESULTS

Conventional TEM images show GaAs wells of homogeneous thickness. A typical example of a processed $<100>$ image of the 5 nm GaAs quantum well of the wafer 2 is presented in Fig.2(a). In the underlying HRTEM image, bright spots in the AlAs are located at the positions of the Al atoms. The image pattern of GaAs is characterized by bright spots at the positions of both the Ga and As atoms. The six grey levels from dark grey to very light grey refer to Al concentrations of $<10\%$, 10%-30%, 30%-50%, 50%-70%, 70%-90% and $>90\%$. Depending on the noise on the micrograph, the standard deviation of the A_{200} Fourier amplitude in the binary materials far away from the interfaces corresponds to composition deviations between 11 and 16 atomic percent Al. Fig.2.(b) shows the averaged compositions of each monolayer parallel to the interface.

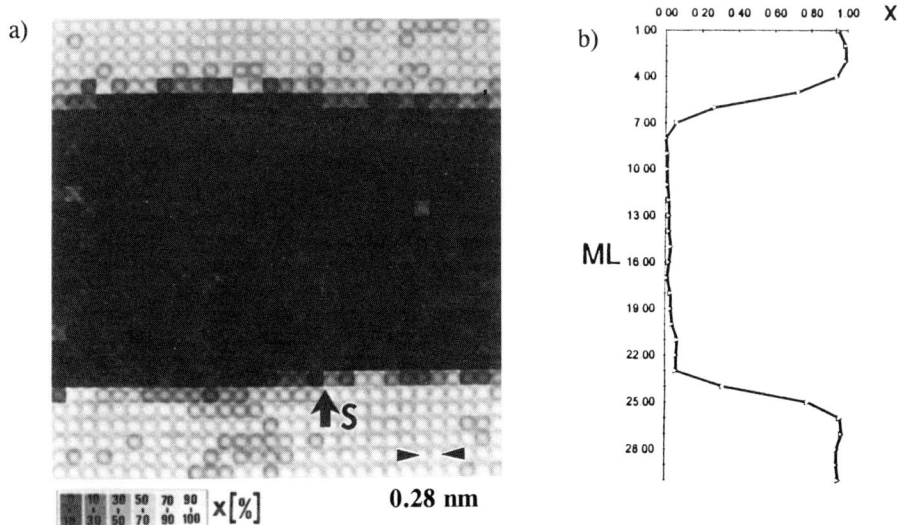

Fig.2: (a) $<100>$ chemical lattice image of the 5 nm quantum well of wafer 2, (b) averaged compositions of each monolayer parallel to the interface. S marks a step.

Wafers 1 exhibits considerable roughness on an atomic scale (<10nm) at both interfaces, which is significantly reduced for the GaAs on AlAs interface of wafer 2. This is illustrated by the average widths of the transition from $Al_{0.8}Ga_{0.2}As$ to $Al_{0.2}Ga_{0.8}As$ of the inverted (GaAs/AlAs) D_i and normal (AlAs/GaAs) interfaces D_n determined from the averaged compositions parallel to the interface. Table 1 summarizes the results for the 5nm quantum wells. The number of interface steps with distances <10nm (N_n and N_i for the normal and inverted interfaces) from comparable interface sections is also listed. Due to the bad statistics, the number of steps with distances >10nm does not yield significant information.

Table 1: average interface widths D_i and D_n and step densities (N_n, N_i) for both wafers

	D_n [MLs]	D_i [MLs]	N_n	N_i
wafer 1	2.3	3.3	12	18
wafer 2	1.9	1.9	10	10

Fig. 3 shows a processed <110> image of the 8nm quantum well of the wafer 2 (t=15nm and Δf=60nm). The nonlinear dependence of the A_{200} image amplitude results in a reduced chemical sensitivity displayed by only three grey levels corresponding to Al contents of <33% (light grey), 33%-67% and >67% (dark grey).

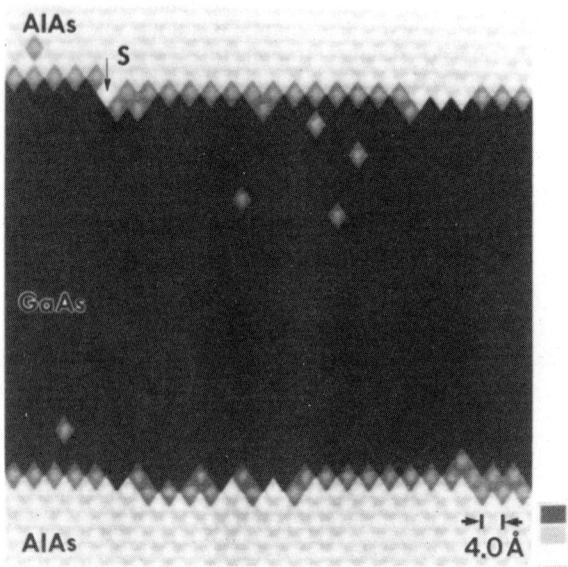

Fig.3: <110> processed lattice image of the 8nm quantum well of wafer 2 at t=15nm and Δf=60nm. S marks a step.

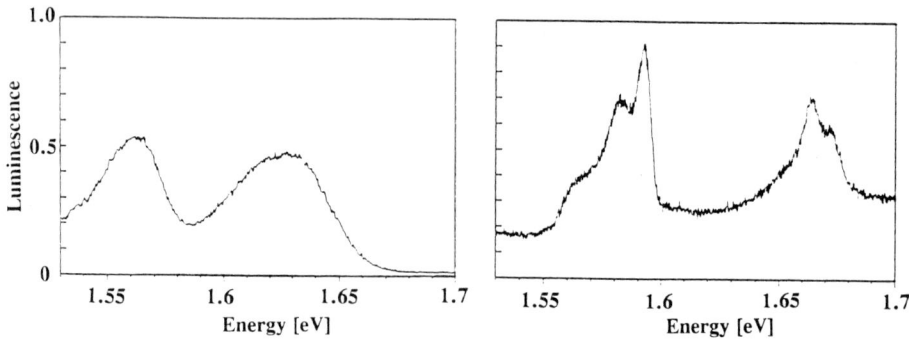

Fig.4: PL spectra of wafer 1 (left) and wafer 2 (right) taken at 2K.

The PL-spectra of wafer 1 and wafer 2 are given in Fig.4. For wafer 2, the PL lines of the 5nm and 8nm quantum wells are composed of several lines which are clearly separated at photon energies around 1.67eV and 1.58eV, respectively. The PL spectrum of wafer 1 is characterized by single broad lines. The further evaluation will focus on the 5nm quantum well because more HRTEM images are available in this case for a more significant comparison. The quantum well thickesses L_z of the 5nm quantum well were calculated from the peak energies E_{max} according to Christen (1990) in Table 2 with the following material properties: effective electron masses for GaAs $0.0665m_e$ and for AlAs

$0.124m_e$, effective heavy hole masses for GaAs $0.475m_e$ and for AlAs $0.5m_e$, conduction band offset 0.269eV, valence band offset 0.44eV. Due to the inaccuracies of the material properties, the calculated L_z may differ from the real values. However, the differences of 1 ML for wafer 2 are significant within the margin of error.

Table 2: peak energies E_{max} of the PL lines and calculated L_z for the 5nm quantum well for both wafers

	E_{max} [meV]	L_z [MLs]
wafer 1	1625 ± 3	21.9 ± 0.4
wafer 2	1674 ± 3	16.3 ± 0.3
	1664 ± 1	17.2 ± 0.1
	1654 ± 1	18.0 ± 0.1

4. DISCUSSION

Prior to comparing PL and HRTEM results, the interpretation of the HRTEM images requires a clarification of the significance of the data and the effects leading to nonabrupt interfaces.

1. Interdiffusion causes a finite compositional transition which cannot be completely excluded for a growth temperature of 650°C and the strong gradients of the chemical potential at the interface. Due to a constant growth temperature, the interdiffusion profile should be of comparable width for all wafers. Its upper extension limit of 2 MLs can be estimated from D_n of wafer 2 from interface sections without steps.

2. Steps on the growth surfaces induce an interface roughness which is interpreted with respect to the PL line widths. However, step distances can only be quantitatively measured from HRTEM images if the step distances are larger than the specimen thickness. Small step distances (< 10nm) cannot be quantitatively evaluated due to the projection through the sample thickness. The use of the < 110 > cross-section samples provides an advantage in this respect because steps run preferably along the < 110 > directions (Ikarashi et al 1992). The density of small step distances is, however, a qualitative indicator of the degree of the *atomic scale* roughness. A high density of steps obviously results in extended transition widths D_i and D_n (Table 1).

The HRTEM images are quantitatively evaluated on a scale smaller than the exciton diameter. A qualitative evaluation of high-resolution images by visual inspection is possible on a somewhat larger scale.

In agreement with a previous study (Bimberg et al 1992) growth interruptions yield a considerable reduction of the atomic scale roughness of the GaAs on AlAs interface because both the density of the steps N_i and the transition width D_i decrease. The effect is less pronouced for the AlAs on GaAs interface.

The changes of the interface microstructure as a consequence of the growth interruptions induce significant changes in the PL spectra. The property which mainly determines the PL line widths is the distribution of L_z values - on a lateral scale of the exciton diameter - over the laser spot diameter. In the case of wafer 2, the observation of different lines from one quantum well - despite the presence of roughness on an *atomic scale* (see Fig.2) - confirms that the energy of the excitons is not sensitive to these *atomic scale* fluctuations of L_z. The occurence of the *atomic scale* roughness can, however, lead to PL line energies corresponding to noninteger values of L_z (Table 2). This was previously shown by Warwick et al (1990) and Gammon et al (1991) by the measurement of PL spectra as a function of the position on the wafer. Although the quantum well thickness changes in wafer 2 by several monolayers over the laser spot diameter, the fluctuations of the average well thickness on the scale of the exciton diameter is smaller than one monolayer as indicated by the PL line width. Obviously, the atomic scale roughness is not large enough to generate a *continuous* spectrum of L_z values. In contrast, the higher degree of atomic scale roughness leads to a *continuous* spectrum of L_z in the case of wafer 1 that inhibits the observation of discrete lines. Therefore, truly smooth interfaces are not a necessity for PL line splitting. However, the atomic scale roughness may indirectly influence the line widths because a continuous distribution of well widths is produced, if a high degree of atomic roughness is present, which could be associated with the occurance of double layer steps (as observed for wafer 1).

5. CONCLUSIONS

HRTEM and PL measurements have been combined to study the interface properties of AlAs/GaAs quantum well structures grown with and without growth interruptions. A pattern recognition procedure was applied to HRTEM images along the <100> and <110> zone axes to quantitatively determine the composition at the interfaces on a near-atomic scale. The atomic scale roughness is significantly reduced - in particular for the GaAs on AlAs interface - if growth interruptions are applied.

The atomic scale roughness was found to have an indirect influence on the PL line widths. Fluctuations of L_z on a lateral atomic scale are not in contradiction with narrow PL lines. However, a high degree of atomic scale roughness leads to a continuous spectrum of L_z - on a lateral scale of the exciton diameter - and therefore to broad PL lines.

ACKNOWLEDGMENTS

The authors thank D Stenkamp who wrote the image processing and pattern recognition program and A Thust for fruitful discussions on the image simulations.

REFERENCES

Bravman J C and Sinclair R 1984 J. Electron. Microsc. Tech. 1 53
Bimberg D, Heinrichsdorff F and Bauer R K, Gerthsen D, Stenkamp D, Mars D E and Miller J N 1992 J. Vac. Sci. Technol. B 10 1793
Bode M and Ourmazd A 1992 J. Vac. Sci. Technol. B 10 1787
Christen J 1990 Festkörperprobleme 30 239
Christen J and Bimberg D 1990 Phys. Rev. B 42 7213
Gammon D, Shanabrook B V and Katzer D S 1990 Appl. Phys. Lett. 57 2710
Gammon D, Shanabrook B V and Katzer D S 1991 Phys. Rev. Lett. 67 1547
Ikarashi N, Tanaka M, Sakaki H and Ishida K 1992 Appl. Phys. Lett. 60 1360
Katzer D S, Gammon D, Shanabrook B V and Tadayon B 1990 Superlattices and
    Microstructures 8 19
Kopf R F, Schubert E F, Harris T D and Becker R S 1991 Appl. Phys. Lett. 58 631
Ourmazd A, Baumann F H, Bode M and Kim Y 1990 Ultramicroscopy 34 237
Singh J, Bajaj K K and Chaudhuri S 1984 Appl. Phys. Lett. 44 805
Singh J and Bajaj K K 1986 Appl. Phys. Lett. 48 1077
Stadelmann P A 1987 Ultramicroscopy 21 131
Thoma S and Cerva H 1990 Ultramicroscopy 35 77 and 265
Warwick C A, Jan W Y, Ourmazd A and Harris T D 1990 Appl. Phys. Lett. 56 2666

Dimensional reduction of GaAs/AlGaAs quantum well cut by dislocation slip observed by transmission electron microscopy

H Atmani, A Rocher, C Guasch[*], JP Peyrade[*], F Voillot[*], M Goiran[*] and E Bedel[+]

Centre d'Elaboration des Matériaux et d'Etudes Structurales, CNRS, F-31055 Toulouse
[*] Laboratoire de Physique des Solides, INSA, F-31077 Toulouse
[+] Laboratoire d'Automatique et d'Analyse des Systèmes, CNRS, F-31077 Toulouse, France

ABSTRACT: A new method of generating one dimensional semiconductor structures is proposed. This method is based on intrinsic dislocation slip properties. Dislocations emitted by one source work as an atomic saw to cut a GaAs quantum well with a sharp lateral confinement without loss of material.

1. INTRODUCTION

A lot of effort is devoted to both theoretical and experimental work in the optical and electronic properties of one dimensional (1D) and zero dimensional (0D) semiconductor structures. Whereas two dimensional (2D) structures are actually obtained with an accuracy of the order of one monolayer, experimental studies on 1D and 0D are still limited by technology.

Concerning the realization of 1D structures, i.e. quantum wires, one can arbitrarily distinguish two major ways : the direct growth of wires and the patterning of an existing 2D structure. Either vicinal surfaces (Tsuchiya et al 1989) or V-grooved surfaces (Kapon et al 1992) are used as substrates to achieve direct growth of the wires. The techniques to pattern a well defined 2D structure are, first, nanometer scale lithography combined with dry etching followed by an epitaxial overgrowth (Izrael et al 1991, Lehr et al 1992) and second selective ion implantation followed by controlled interdiffusion (Cibert et al 1986, Leier et al 1990, Vieu et al 1992). Intrinsically these techniques have to deal with some imperfections; the nano-meter scale lithography has to face the problem of the surface recombination; the selective ion implantation leads to a relatively smooth potential barrier. In both cases, the size dispersion is small but a lot of active material is lost during the processes and the wire density is limited.

Recently Peyrade et al (1992) proposed a new method based on the possibility for dislocations to generate directly 1D structures with a sharp lateral confinement. We present here the TEM observations on GaAs/GaAlAs quantum well (QW) structures deformed under two different experimental conditions.

2. BASIC PRINCIPLE OF CUTTING

The cutting principle can be described as follows: a GaAs crystal, subjected to a stress higher than its elastic limit, deforms plastically when the temperature is sufficient to allow the motion of the dislocations in their slip planes. A dislocation loop emitted by a source increases its size until it reaches the free surface of the crystal. At this point, one part of the crystal is shifted with respect to the other by an amount equal to the Burgers vector b of this dislocation and the crystal perfection is restored in the volume. In this condition, a GaAs QW layer of thickness e sandwiched between the two GaAlAs barriers will be cut when the source has emitted n dislocation loops, with n satisfying the condition $n.b_{eff} > e$, where b_{eff} is the component of the Burgers vector along the growth direction.

3. EXPERIMENTS

A (GaAs/AlGaAs) specimen was grown by Molecular Beam Epitaxy on a semi-insulating (001) GaAs substrate. The structure consisted of 18±1 monolayers of GaAs sandwiched between GaAlAs barriers of thickness d = 100 nm, with an aluminium concentration x = 0.38, and a GaAs cap layer of thickness c = 20 nm.

We choose to use a pure bending test which is a simple way to apply a uniaxial stress: no cutting stress is involved, only a bending moment exists. In addition, from the (001) face to the (00$\bar{1}$) face, the stress varies continuously from tension to compression. This situation will activate dislocation sources located close to the surface near the active region.

The bending test applied along [$\bar{1}$10] involves the activation of at least two slip planes and favours four different Burgers vectors: [10$\bar{1}$] and [0$\bar{1}$1] in (1$\bar{1}$1), [0$\bar{1}$1] and [101] in ($\bar{1}$11). The net result of this operation is a bending of the sample along [$\bar{1}$10], which is characterized by its radius of curvature. The samples have been also studied by photoluminescence (PL) before and after deformation.

TEM observations have been performed using cross section (XTEM) specimens thinned perpendicular to the deformation axis in order to demonstrate the feasibility of the method by direct observation of the dislocation cut of the QW. The specimens were prepared by conventional ion milling.

3.1 Specimen weakly deformed

Sample QW1 was plastically deformed in a four point bending setup at 673 K for 40 hours. The net result of this operation is a bending of the sample along the [$\bar{1}$10] direction. The curvature radius obtained is about 20 mm. This value is constant over 10 mm along the [$\bar{1}$10] direction. The average strain rate of 10^{-7} s^{-1} at the surface gave a deformation of approximatively 1.5% in the surface.

Figure 1 is a XTEM 002 dark field image of sample QW1 after deformation. Two shifts, A and B separated by 0.6 μm, are revealed in this image of the whole 2D structure. The shift A (48Å) is much sharper than the shift B (28Å).

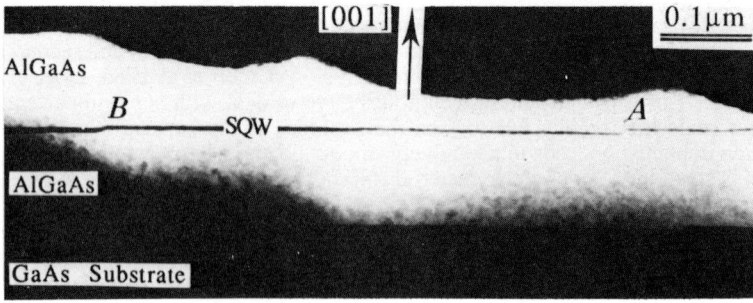

Fig.1 : XTEM : 002 dark field image of the SQW cut by dislocations: two shifts labelled A and B related to the deformation are revealed.

The shift A is enlarged in fig. 2 where a corresponding shift A' is observed at the AlGaAs barrier/GaAs buffer interface. The AA' direction is [$\bar{1}$12] which confirms that the slip plane is (1$\bar{1}$1). The sharpness of the cut indicates that only one dislocation source has been activated to produce this shift by the emission of 17 dislocations. The shift B seems to be due to dislocations emitted by at least three different sources working on parallel planes separated by few a nm.

Few dislocations are observed in the active region. This observation is in good agreement with the optical properties of the structure studied by Goiran et al (1993). PL measurements

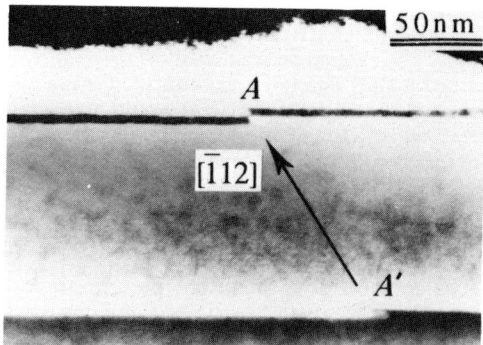

Fig.2 : Enlargement of shift A. Note the shift A' at the GaAlAs barrier/GaAs buffer interface. The AA' direction corresponds to the trace of (111) plane on the (110) image plane.

have been carried out at 4.2K on the SQW before and after deformation, exciting the sample with a 700nm laser line. After deformation, the exciton peak has decreased by a factor of approximately 2 and the FWHM has increased by a factor of about 2.5. The exciton peak intensity is not shifted by this process. This result is directly explained by the large wire width observed to be a minimum of 0.6 μm, which is not small enough to create quantum confinement.

3.2 Specimen strongly deformed

Sample QW2 has been plastically deformed using a three point setup at 673 K. The average strain rate was s = 10^{-7} s^{-1} at the surface. At the end of the deformation, the sample presents an heterogeneous bending along the [$\bar{1}$10] direction. The deformation radius obtained in the central part is about 4 mm. The maximum deformation is located in the central region and is 7.5% in the surface.

XTEM shows in fig.3 that the GaAs quantum well layer has been heterogeneously deformed: i) first in fig. 3a it appears to be wavy with a periodicity of about 50 nm, ii) cut with large transition areas observed on fig. 3b. This cut involves both cross slip and the two {111} active slip planes containing the bend axis. Large steps with a width > 1 μm and a height > 25 nm are also observed. They accomodate about 40% of the deformation.

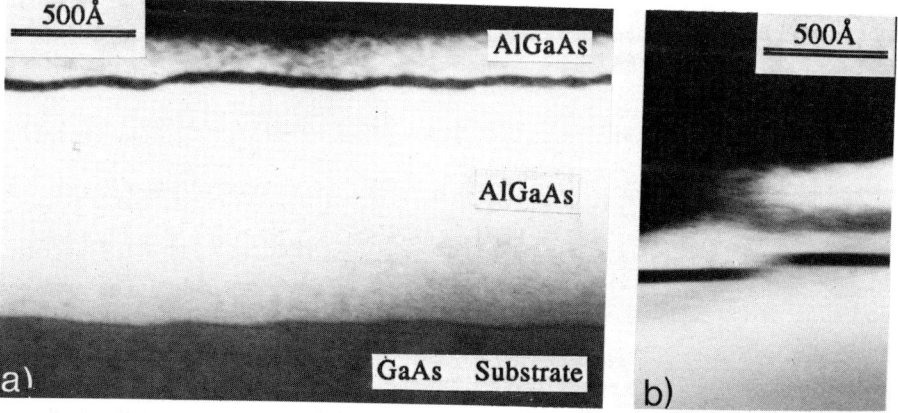

Figure 3 : XTEM dark field image of the SQW strongly deformed. Note a) the wavy shape of the SQW. b) the large transition area between GaAs layers.

A shift of the PL spectra has been observed in this sample. The shift of the exciton peak varies continuously from 5 meV at the maximum of deformation to zero in the undeformed regions. The exciton peak intensity is reduced by a factor of about 10 after deformation. If we assumed a shift due to lateral confinement, a calculation using the theoretical approach of Bockelmann and Bastard (1992) would give a wire width of about 30 nm. Nevertheless, the correlation between the shift measured by photoluminescence and the wavy shape of the GaAs layer observed by XTEM is in this case not so obvious.

4. CONCLUSION

The method described here presents several advantages in comparison with other techniques used to pattern a 2D structure: the existence of an homogeneous barrier around the wire and the conservation of the total active material. Moreover, due to the principle of the method, the range of the potential discontinuity is a distance of the order of the lattice parameter when only one slip plane is active. One of the major features of the PL results is the small loss in the exciton peak intensity with respect to the mechanical and thermal treatment endured by the samples. This point is related to the dislocation properties which conserve the crystal perfection after slipping.

The cutting process should be improved in at least one way: the distribution of wire widths should be adjusted by control of the location of the dislocation sources. The results presented in this paper have been obtained with natural sources using low dislocation density substrates. The adjustment of an artificially created source distribution will control the lateral size of the wires and limit the creation of defects in the QW during the deformation. Work is in progress in order to control the slip process by choosing more appropriate stress conditions, deformation temperature, dislocation source distribution, etc.

REFERENCES

Bockelmann U and Bastard G 1992 Phys. Rev. B **45** (4) 1688

Cibert J, Petroff P M, Dolan G J, Pearton S J, Gossart A C and English J H 1986 Appl Phys Lett **49** 1275

Goiran M, Guash C, Voillot F, Peyrade J P, Bedel E, Munoz-Yague A, Atmani H and Rocher A, submitted to Europhysics Letter 1993

Izrael A, Marzin J Y, Sermage B, Birotheau L, Robein D, Azoulay R, Benchimol J L, Henry L, Thierry-Mieg V, Ladan F R and Taylor L 1991 Jap J Appl Phys **30** (11B) 3256

Kapon E, Walther M, Christen J, Grundmann M, Caneau C, Hwang D M, Colas E, Bhat R, Song G H and Bimberg D 1992 Superlattices and Microstructures **12** (4) 491

Leier G, Bergmann R, Rudeloff R, Scholz F and Scheitzer H 1992 Appl Phys Lett **61** (5) 517

Peyrade J P, Voillot F, Goiran M, Atmani H, Rocher A and Bedel E 1992 Appl. Phys. Lett. **60**, 20, 2481

Tsuchiya M, Gaines J M, Yan R H, Simes R J, Holtz P O, Coldren L A and Petroff P M 1989 Phys Rev Lett **62** 466

Vieu C, Schneider M, Benassayag G, Planel R, Birotheau L, Marzin J Y and Descours B 1992 J Appl Phys **71** (10) 5012

Voillot F, Goiran M, Guasch C, Peyrade J P, Dinh L, Rocher A and Bedel E, to be published in Journal.de Physique III (1993).

Inst. Phys. Conf. Ser. No 134: Section 7
Paper presented at Microsc. Semicond. Mater. Conf., Oxford, 5–8 April 1993

459

Intermixing mechanisms in single GaAs/GaAlAs quantum well samples under SiO$_2$ capping

A Pepin, C Vieu, M Schneider, G Ben Assayag, F–R Ladan, R Planel, H Launois, Y Nissim* and M Juhel*

Laboratoire de Microstructures et de Microelectronique / CNRS, *Centre National d'Etudes des Telecommunications, 196 avenue Henri Ravera, BP 107, 92225 Bagneux, FRANCE

ABSTRACT : Intermixing of AlGaAs–based structures is known to be enhanced by capping samples with a layer of SiO$_2$. Experiments on undoped single quantum well structures capped with rapid thermal chemical vapour deposition SiO$_2$ performed by transmission electron microscopy in conjunction with low temperature photoluminescence indicated a strong, homogeneous interdiffusion without significant degradation of the optoelectronic properties. Secondary ion mass spectroscopy validated a Ga vacancy–induced interdiffusion mechanism. Patterning of the SiO$_2$ by high resolution electron beam lithography was also achieved and the stress dependence of the intermixing was confirmed.

1. INTRODUCTION

Built–in shape alteration of semiconductor quantum wells (QWs) in order to locally modify their effective bandgap energy and thereby the exciton transition energies has been extensively studied during the past few years. Novel optoelectronic devices should emerge from the ability to selectively control the lateral confinement of carriers in low dimensional quantum structures. Among the several existing intermixing techniques, SiO$_2$ capping–induced intermixing was selected for presumably being an impurity–free method based on the thermally activated diffusion of point defects. The most commonly proposed mechanism involves the excess vacancies created under the SiO$_2$/GaAs interface, by the preferential absorption of the Ga atoms by the SiO$_2$ layer during anneal, and their diffusion down to the QW heterointerfaces where they stimulate the exchange of the barrier and well atoms (Deppe et al 1986 , Koteles et al 1989). Cross–sectional transmission electron microscopy (TEM) and low temperature photoluminescence (PL) were chosen as probes to characterize the induced interdiffusion.

2. EXPERIMENTAL PROCEDURE

Two different undoped QW structures were grown by molecular beam epitaxy (MBE) for this study. The first sample (S1) contained a single 6 nm–thick QW embedded in a symmetrical set of upper and lower 10 nm–thick Al$_{0.33}$Ga$_{0.67}$As barriers followed on the top and bottom by two short period 14 nm–thick GaAs/AlAs superlattices (SLs). To evaluate the dependence of the intermixing upon well thickness and distance from the surface, a second sample (S2) was grown which consisted of three individual QWs –QW1, QW2 and QW3– of respective nominal widths 6.2, 3.1 and 1.7 nm separated by 20 nm–thick barriers, the thinnest well (QW3) located farthest from the surface. Both samples were capped with a thin GaAs cap layer to prevent direct contact between SiO$_2$ and the Al contained in the top SL (in the case of S1) or top barrier (S2).

Samples were coated with a thin layer (<600 nm) of rapid thermal chemical vapour deposited (RTCVD) SiO$_2$. This unique dielectric deposition technique involves the pyrolitic decomposition of SiH$_4$ and O$_2$ gases in an inert flowing N$_2$ atmosphere, accelerated by the rapid heating (700°C) of the sample holder by halogen lamps. The layer obtained presents physical and compositional properties very close to those of thermally grown SiO_2, as could be verified by infrared absorption spectroscopy.

Thermal anneals were performed in a classical furnace, for 10 min at 900°C or 3h at 850°C, using an enhanced As overpressure GaAs proximity cap technique.

Samples were characterized by PL before and after SiO_2 deposition, and after heat treatment (with or without SiO_2 capping) in conjunction with cross sectional TEM. Special TEM sample preparation was realized using reactive ion etching (RIE) (Dobisz et al 1986) and focused ion beam (FIB) (Szot et al 1992) micromachining. TEM observations were performed in the bright field, dark field and high resolution image configurations. PL was carried out at 2K using an argon laser. The position and width of the peak due to the n=1 electron to heavy hole transition is directly related to the magnitude of the intermixing. The shift towards higher energies, due to the modification of the potential from an abrupt shape (as–grown QW) into an error function–like shape, is treated in a variational calculation of the energy levels from which a value of the interdiffusion length Δ can be extracted (Schlesinger and Kuech 1986).

Secondary ion mass spectroscopy (SIMS) was also used as a complementary method to observe the expected Ga pumping by SiO_2 and the eventual diffusion of Si and/or O species inside the semiconductor material. In order to investigate the dependence of this intermixing technique upon stress, that is the stress essentially generated during heat treatment by the difference in thermal expansion coefficients between SiO_2 and GaAs, a sample (S1 type) containing both uniformly capped regions and patterned SiO_2–capped regions was also realized by high resolution electron beam lithography and characterized by PL after annealing.

3. RESULTS AND DISCUSSION

PL spectra presented on Fig. 1 illustrate the high amount and quality of the intermixing obtained with RTCVD SiO_2 capping and annealing.

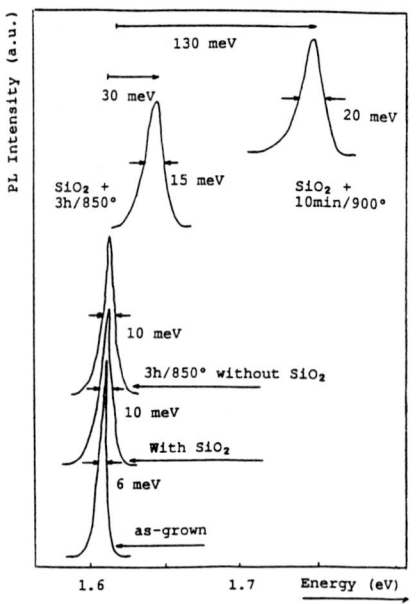
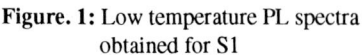

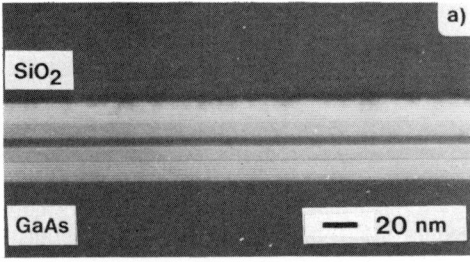

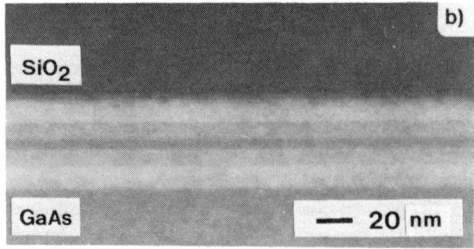

Figure. 1: Low temperature PL spectra obtained for S1

Figure. 2: (200) dark field cross–sectional TEM micrographs of SiO_2 capped S1, a) after capping, b) after capping and 3h/850°C anneal

While uncapped samples show very good stability against heat treatment (slight broadening equal to that induced by dielectric deposition, but no noticeable energy shift nor intensity loss), capped samples experience significant blue shift: 130 meV for the 10 min /900°C anneal case, which corresponds to an interdiffusion length $\Delta = 1.5$ nm. The smaller 30 meV blue shift obtained for the 3h/850°C case reveals the critical dependence of the mechanism upon annealing temperature. The strong disordering was confirmed by 200 dark field observations as is shown on Fig. 2. The clear compositional contrast observed between well and barrier material before anneal becomes blurry after heating, and the SL Al modulation is greatly affected, although the SL located closest to the SiO_2/GaAs interface seems to have experienced less interdiffusion than the deeper one. In accordance with the good conservation of both intensity and width of the PL emission after intermixing, TEM images show no extended defects. The homogeneous interdiffusion obtained by SiO_2 capping is in contrast with the results obtained with other sources of intermixing such as ion implantation where unavoidable residual defects are observed and considerably affect the electronic and optical properties.

SIMS measurements, showing evidence of the Ga–pumping inside SiO_2 but no significant diffusion of the Si and/or O species into GaAs, allowed us to validate the vacancy–induced intermixing hypothesis. A Ga concentration inside SiO_2, and thus a Ga vacancy concentration generated under the interface, of 5.10^{19} /cm^3 could be roughly estimated from analysis as opposed to the corresponding established equilibrium vacancy concentration in GaAs of 10^{17} /cm^3 (Chiang and Pearson 1975).

A series of experiments were carried out on S1 samples with different oxide thicknesses. The mixing rate was found to increase with the thickness of the SiO_2 layer, showing the sensitivity of the mechanism to the total number of vacancies available at the surface.

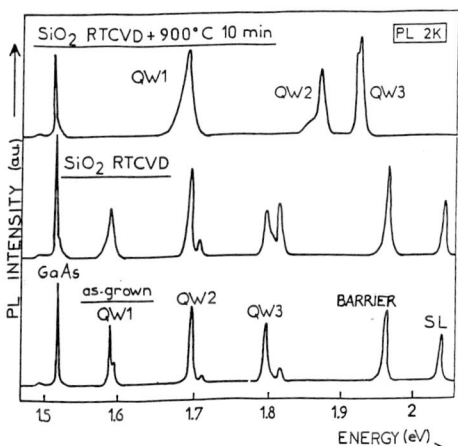

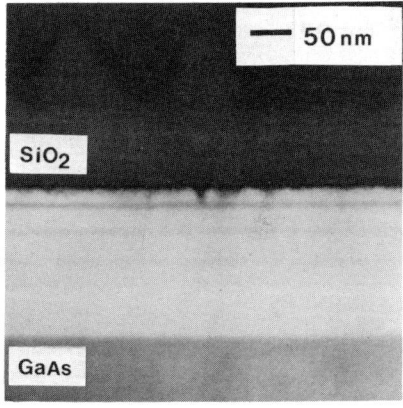

Figure. 3: Low temperature PL spectra obtained for S2

Figure. 4: 200 dark field cross–sectional TEM micrographs of SiO_2 capped S2, after capping and 10min/900°C anneal

The PL spectra obtained with S2, presented in Fig. 3, show very original features. The as–grown multiple line spectrum reveals the formation during epitaxial growth of large islands differing in thickness by one monolayer. The dielectric deposition step causes minute change to the spectra, within the uniformity of the sample, the relative intensity differences being attributable to a slight change in the location of the region probed in PL. After anneal, the QWs experience significant blue shifts (about 110 meV for QW1, 180 meV for QW2, 130 meV for QW3) but still exhibit one monolayer–related splittings. Interdiffusion lengths deduced from theoretical calculation give $\Delta = 1.9$ nm for QW1, $\Delta = 3$ nm for QW2 and $\Delta = 4$ nm for QW3, revealing a quasi complete mixing of QW3, yet narrowness and intensity conservation of the emission suggest a homogeneous mixing and a strong carrier transfer inside the altered well. QW1, situated closest to the surface shifts less than QW2 and 3, which seems to be in contradiction with the behaviour expected from an interdiffusion generated by vacancies coming from the surface and thus depending on the distance of a given QW from the surface. These results were corroborated by TEM observation, as shown on Fig. 4.

While the source of the interdiffusion has been identified as the Ga vacancies, the above behaviour suggests the existence of an accelerating force acting on the diffusion process, which we attribute to the stress field created during heat treatment in the semiconductor by the oxide layer. Clearly, such a driving force would induce an anisotropic vacancy diffusion, from the surface towards the depths of the structure. Due to the stress gradient, this anisotropy is higher in the near surface region than deeper in the sample, where a higher degree of disordering can be induced. The fabrication of one and zero dimensional structures of high lateral selectivity can therefore be expected from a suitable exploitation of the directionality of this process.

The patterning experiments carried out on an S1 type sample lead in this sense to most interesting consequences. Under uniformly capped 500 μm x 500 μm regions the large PL blue shifts previously reported were reproduced while no noticeable shift could be measured under 40 μm x 40 μm regions patterned with arrays of SiO$_2$ wires of different sizes and periods, ranging from 20 nm /period 100 nm to 300 nm / period 1.2 μm. The intermixing occurrence clearly appears to have been dramatically modified by stress redistribution subsequent to patterning.

4. CONCLUSION

We were able to achieve strong, homogeneous, reproducible, Ga vacancy–driven intermixing of GaAs QWs under RTCVD SiO$_2$ uniform capping, without inducing significant degradation of the electronic and optical properties. The combination of TEM and PL characterization proved efficient in the analysis of the mechanisms involved. QWs located closer to the surface were found to have experienced less interdiffusion than deeper wells. The effect of stress redistribution on the intermixing process was observed, and a simplistic model involving thermal stress is proposed to explain the results obtained.

Experiments on stress distribution control on patterned SiO$_2$ samples are currently being undertaken and should soon allow us to evaluate the critical parameters involved in this type of intermixing and further investigate the applicability of this method to the realization of low dimensional quantum structures of high lateral selectivity.

REFERENCES

Deppe D G, Guido L J, Holonyak N Jr, Hsieh K C, Burnham R D, Thorton R L and Paoli T L 1986 Appl. Phys. Lett. 49 510

Koteles E, Elman B, Holmstrom R P, Melman P, Chi J Y, Xin Wen, Powers J, Owens D, Charbonneau S and Thewalt M L W 1989 Superlattices and Microstructures 5 321

Dobisz E A, Craighead H G, Beebe E D and Levkoff J 1986 J. Vac. Sc. Technol. B4 850

Szot J, Hornsey R, Ohnishi T and Minagawa S 1992 J. Vac. Sc. Technol. B10 575

Schlesinger T E, Kuech T 1986 Apply. Phys. Lett. 49 519

Chiang S Y, Pearson G L 1975 46 2986

Inst. Phys. Conf. Ser. No 134: Section 7
Paper presented at Microsc. Semicond. Mater. Conf., Oxford, 5–8 April 1993

463

A low temperature EBIC study of quantum wires fabricated in GaAs/AlGaAs heterostructures using ion-implanted gates

R J Blaikie, B Fraboni*, J R A Cleaver† and H Ahmed†

Hitachi Cambridge Laboratory, Cavendish Laboratory, Madingley Road, Cambridge CB3 0HE
*Department of Physics, University of Bologna, 40126 Bologna, Italy
†Microelectronics Research Centre, Cavendish Laboratory, Madingley Road, Cambridge CB3 0HE

ABSTRACT: Quantum wires have been fabricated using p-type implanted gates to confine laterally the two dimensional electron gas (2DEG) of a GaAs/AlGaAs heterostructure. These have been imaged in the EBIC mode at T = 7 K and T = 300 K. Two types of EBIC signals were observed and a study of their temperature and implanted-gate bias dependencies revealed their origins. The effect of electron beam irradiation on the electrical characteristics of the wires was also investigated. Electron beam irradiation induces parallel conduction in the δ-doped donor layers so that it is the depletion regions in these layers as well as in the 2DEG which are imaged.

1. INTRODUCTION

Quantum effect devices offer a unique opportunity for investigating phenomena that only occur if the dimensionality of the electrical conduction is reduced and the device is operated at low temperature where scattering effects are minimal. Confinement to a two dimensional electron gas (2DEG) can be achieved by growing heterostructures, with stacked layers of materials of well controlled thicknesses, doping densities and band gaps. Device dimensionality can be further reduced by lateral confinement in the x or y directions (for examples see Reed and Kirk, 1989). With confinement techniques such as etching or Schottky gate depletion the devices have considerable surface topography and are readily imaged using secondary electron (SE) contrast in a scanning electron microscope (SEM). However other confinement techniques leave planar surfaces after fabrication and so different imaging techniques must be employed.

In this paper we present results of an Electron Beam Induced Current (EBIC) study of quantum wires fabricated using p-type implanted-gates to provide the lateral confinement. Strong EBIC signals are found more than 100 µm away from the collecting junction which arise because of the built-in potentials of the heterojunction and collection through a weakly p-type substrate. The microscope used for this study was an Hitachi S-800 SEM with an Oxford Instruments liquid Helium cold stage, giving a temperature range of 7 K – 300 K. Connections to the sample allow *in-situ* measurements of the devices' electrical characteristics. The electron beam voltages used ranged from 2 kV to 8 kV and the beam current from 200 pA to 60 nA.

2. SAMPLE FABRICATION

The technique for fabricating quantum wires using implanted p-type gates to provide the lateral confinement is shown schematically in Fig. 1; 20 keV Be$^+$ is implanted on either side of the wire, and rapid thermal annealing is used to activate the Be ions as well as to anneal contacts to these implanted-gates (Blaikie *et al.*, 1992a). Biasing these gates varies the widths of the depletion regions surrounding them, so varying the width of the quantum wire.

At low temperatures (< 100 K) all conduction in the quantum wire occurs through the confined 2DEG layer, since carrier freeze-out occurs in all other layers. Measurements

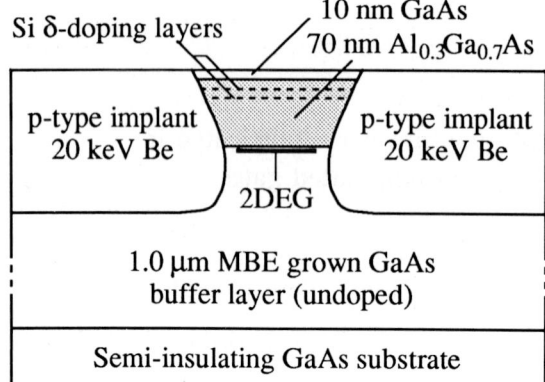

Fig. 1 Schematic diagram of the device structure.

performed on 10 µm long multi-probe wires show a number of anomalous magnetoresistance effects (Blaikie *et al.*, 1992b), and in order to quantitatively compare these results with the predictions of a classical billiard-ball model measurements are required for various geometrical parameters of the wires. Any imaging technique must therefore delineate the *p*-type implants with high resolution. Because the device surface is planar, SE imaging in a SEM provides little contrast between implanted and unimplanted regions. However the device consists of lateral *p-n* junctions, with independent electrical connections to each side, which makes EBIC the natural choice for obtaining high-contrast images of the implanted regions.

3. EBIC SIGNAL DEPENDENCE ON TEMPERATURE

EBIC micrographs of a 10 µm long, multi-probe wire are shown in Fig. 2, both at room temperature (a) and with the sample at 7 K (b). In both cases the *p*-type implanted regions are clearly delineated, as is the 0.6 µm wide, 10 µm long central wire. In this device the upper and lower channels were designed to be wider than the central wire, which can be seen in Fig. 2(b). The room temperature image is dominated by a strong background signal, labelled I in Fig. 2(a), which dies away only at distances greater than 100 µm from the *p*-type implants. When the sample is cooled these type I signals decay more quickly and another type of signal is observed at distances less than 1 µm away from the implants, a narrow bright band labelled II in Fig. 2(b). The intensity of the type II signals relative to that of the type I signals increases as the beam energy is reduced, indicating that they originate from closer to the sample surface.

Type I signals are strongest in the room temperature EBIC images, and have a decay length which varies with temperature. A decaying EBIC signal is usually associated with minority carrier diffusion; however, the decay of the type I signals will imply a room temperature minority carrier diffusion length >100 µm and this is much greater than values previously measured in *n*-type

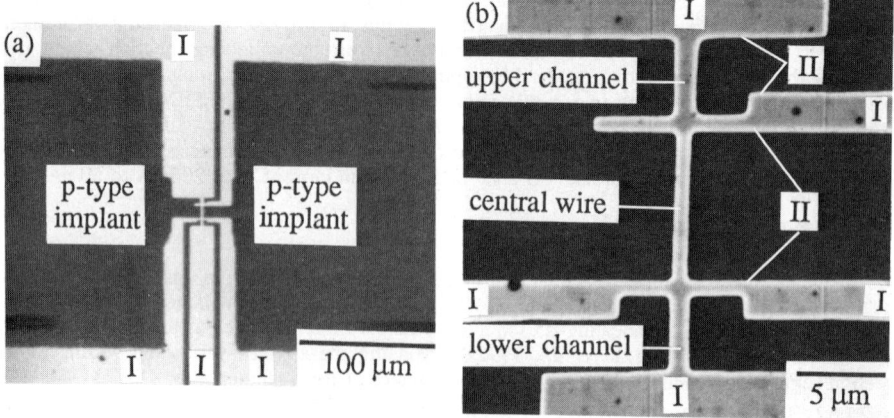

Fig. 2 (a) EBIC micrograph with the sample at 300 K, showing the type I signals which are observed far from the implants. (b) 7 K EBIC image of the central region of the device, showing the type II signals which are observed close to the implants. Beam voltage 3 kV.

GaAs (*e.g.* Wu and Wittry, 1978, Sieber and Carton, 1991). Long decay length EBIC signals have been observed in *n*-type GaAs, attributed to the self-detection of infra-red photons (Wittry and Kyser, 1965). These were two orders of magnitude weaker than the signal from the depletion region at the *p-n* junction. However in Fig. 2 the long decay length signals are of similar strength to the type II signals, associated with the depletion regions near the surface, and thus the type I signals are not attributed to infra-red photon absorption.

We propose that the type I signals are generated in the depleted buffer layer by the built-in fields associated with the GaAs/AlGaAs heterojunction, as shown schematically in Fig. 3. For this model to be correct the buffer layer would have to show weak *p*-type conductivity in order to complete the EBIC circuit. The decay of the signal away from the implants would then be caused by the series resistance of the buffer layer. As the sample is cooled the resistivity of the buffer layer will increase, which explains the shorter decay length which is observed in Fig. 1(b).

In order to confirm this theory for the origin of type I signals a sample was cleaved across a large area *p*-type implant, and this cleaved face was imaged by EBIC. Fig. 4 shows an EBIC image for the cleaved sample tilted at approximately 55°. The type I signals can be seen to be generated within the top 1 μm of the specimen, which is consistent with our model. Fig. 4 also shows that there is no depletion region present directly beneath the implant, so the implant must terminate in a *p*-type layer. The semi-insulating substrate was found to be *n*-type from Hall measurements which implies that the 1 μm thick buffer layer must be weakly *p*-type, as is often the case due to unintentional carbon impurity doping during growth (Goodridge, 1990).

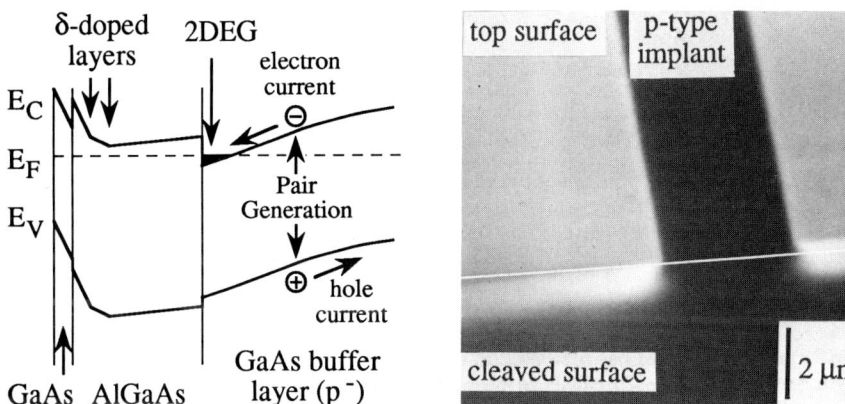

Fig. 3 Band-diagram showing the generation of an EBIC signal in the buffer layer.

Fig. 4 EBIC image at 3 kV of a device cleaved across a large-area p-type implant (T = 300 K).

4. IMPLANTED GATE BIAS VARIATION

At low temperatures type II signals become clear, so they have been investigated by EBIC microscopy with the sample at 7 K. Furthermore, the effects of applying a bias V_{ig} to the implants can also be studied since at low temperatures the implanted-gate leakage current is less than 1 nA. The type I and type II signals behave differently when the implants are biased; the intensity of the type I signals remains unchanged whereas the intensity of type II signals increases under reverse bias. Fig. 5(a) shows 7 K EBIC line scans taken across the upper channel of the device shown in Fig. 2, for various V_{ig}. An EBIC image showing where the line scans were performed is shown in Fig. 5(b) together with low temperature resistance versus V_{ig} characteristics of the upper channel. These were measured *in situ* both before and after imaging the sample.

The type II signals arise from the depletion regions in the surface layers which have high electron density: the two δ-doping layers and the 2DEG. If these signals originate solely from the 2DEG then the EBIC images would provide an observation of the lateral confinement in the quantum wire. However a comparison between the EBIC line scans and the device electrical characteristics shows this not to be the case. The EBIC line scans across the top channel show two depletion regions with an approximately constant separation of 0.7 μm, even for the highest reverse biases applied. In contrast the electrical characteristics of the upper channel taken before imaging show that conduction in the 2DEG is only expected for $V_{ig} > 0$. The electrical characteristic taken after imaging reveals the origin of this discrepancy. After electron beam irradiation the upper channel remains conducting for all V_{ig}. This is because the electron beam has induced additional electrons into the δ-doped donor layers, taking these layers out of the freeze-out regime. Conduction now occurs in these donor layers which makes biasing the implants less effective in changing the channel width. In the EBIC images we are therefore observing the lateral confinement in these donor layers, as well as in the 2DEG.

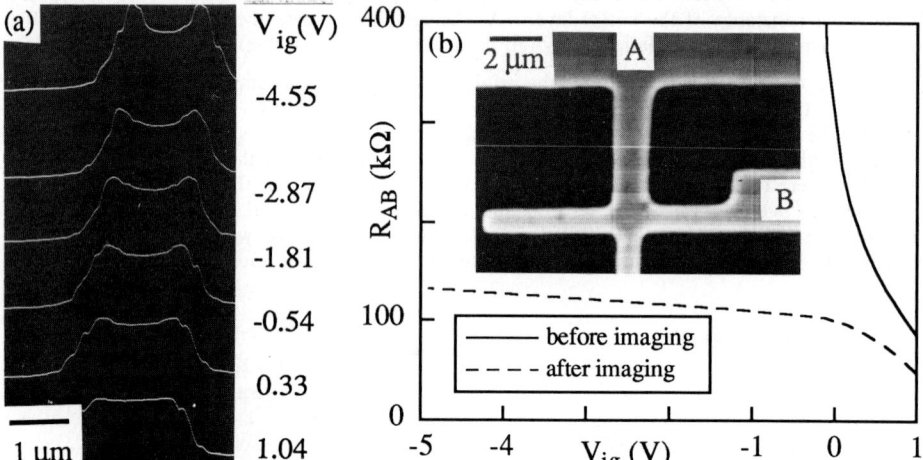

Fig. 5 (a) Line scans across the upper channel of the device shown in Fig. 2 at various V_{ig}.
(b) Resistance versus V_{ig} characteristics of the upper channel, measured immediately before (—)
and after (– – –) imaging the device. The inset shown the position where the line scans were
measured and the two areas (A and B) between which the resistance was measured.

5. SUMMARY

EBIC contrast in a SEM has been used to image quantum wires fabricated in a GaAs/AlGaAs heterostructure using *p*-type implants to provide the lateral confinement. Two types of EBIC signals are observed in these devices and their origins have been investigated by studying their temperature and V_{ig} dependencies. The type I signals originate from the built-in fields of the heterostructure and are collected because the MBE-grown buffer layer shows weak *p*-type conductivity. The type II signals originate from the depletion regions in the active layers near the surface. The electron beam causes significant changes to the electrical characteristics of the devices, so that direct observation of the lateral confinement in the 2DEG is not possible.

We thank T Tanoue of the Hitachi Central Research Laboratory, Tokyo for growing the heterostructure material used in this study. One of us (BF) is aided by a grant in aid of research from the National Academy of Science through Sigma Xi Scientific Research Society.

REFERENCES

Blaikie R J, Cleaver J R A, Ahmed H and Nakazato K 1992a Appl. Phys. Lett. <u>60</u> 1618
Blaikie R J, Nakazato K, Cleaver J R A and Ahmed H 1992b Phys. Rev. B <u>46</u> 9796
Goodridge I H 1990 Properties of Gallium Arsenide (London: INSPEC) pp 53-57
Reed M A and Kirk W P eds 1989 Nanostructure Physics and Fabrication (New York: Academic)
Sieber B and Carton P 1991 Phys. Stat. Sol. (A) <u>127</u> 423
Wittry D B and Kyser D F 1965 J. Appl. Phys. <u>36</u> 1387
Wu C J and Wittry D B 1978 J. Appl. Phys. <u>49</u> 2827

Indentation plasticity of single and multiple layer GaAs–AlAs heterostructures

MR Castell, A Howie, DD Perovic*, DA Ritchie, AC Churchill and GAC Jones

Cavendish Laboratory, Madingley Road, Cambridge CB3 0HE, UK.
*Department of Materials Science, University of Toronto, Toronto M5S 1A4, Canada.

ABSTRACT: We present a high resolution scanning electron microscope study of the mechanical behaviour of GaAs-AlAs superlattices and AlAs surface layers that have been deformed through microindentation. Transverse cleavage through the indentation sites reveals the pattern of the deformed material, showing in particular regions of high compressive strains, slip bands of the $\{111\}<011>$ type and no evidence of indenter penetration of the surface layers.

1. INTRODUCTION

The elastic/plastic response of materials to microindentation is not well understood either in an experimental or theoretical context, nor is there enough experimental information available to support proposed models (Yoffe 1982, Chiang et al 1982) or computer simulations (Laursen and Simo 1992). Advances in indentation testing (Pethica et al 1983) and surface microscopy mean that the deformation of very small volumes can be examined through techniques such as scanning tunnelling microscopy (Castell et al 1992, Walls et al 1992) and high resolution scanning electron microscopy (SEM). Indentation studies provide a convenient and well controlled method for testing the mechanical properties of surface layers, but little sub-surface information is gained directly unless additional sample preparation like thinning for transmission electron microscopy is undertaken (Page et al 1992). Other approaches are described by Castell et al (1993) and Chaudhri (1993).

To map sub-surface deformation at high spatial resolution we use single crystal superlattices that are made up of GaAs-AlAs layers and cleave through the indentation sites. The deformed layers are imaged with an immersion-lens SEM equipped with a field emission gun which is capable of spatial resolutions of less than 1nm. These superlattice structures which have periodicities that would normally be beyond the resolution of a conventional SEM can now be imaged directly using secondary electrons (SE) or backscattered electrons (Ogura 1991).

2. EXPERIMENTAL PROCEDURES

The materials used were fabricated by molecular beam epitaxy (MBE) in a VG Semicon V80H machine where the GaAs-AlAs material was grown on (001) GaAs substrates. The following work discusses results from a GaAs-AlAs superlattice with 160 nm periodicity and a 675nm AlAs surface layer on a GaAs substrate; it was necessary to grow an additional 10nm layer of GaAs on top of the AlAs layer to prevent it from oxidising.

Indentations in the load range of 15gf to 300gf were made using an Ernst Leitz 'Miniload' Hardness Tester which was fitted with a diamond Vickers indenter. Our experimental procedure involved aligning the diagonals of the indenter with the [110] sample direction and then making a line of indentations in the [110] direction. The sample was then cleaved along the indentations which produced a cross-sectional (1$\bar{1}$0) plane through the volume of the deformed material. Without additional preparation, samples were transferred to the ultra high resolution Hitachi S900 SEM which was typically operated at accelerating voltages of 20kV.

3. RESULTS AND DISCUSSION

The SE micrograph in Fig.1a shows the (1$\overline{1}$0) cleavage plane which lies at right angles to the surface and has produced a cross-section through a 15gf indentation made into a GaAs-AlAs superlattice of 160nm periodicity. The exposed material appears dark and light corresponding to the AlAs and GaAs layers respectively and the boundary of the deformed volume can be readily identified. A comparison of the layer positions and separations directly under the bottom of the indentation with those that are far away indicates a compressive strain

Fig.1

(a) A 15gf indentation where there is little c r a c k i n g   o r delamination. __ Two (111) and one ($\overline{1}$11) slip band are indicated in the micrograph.

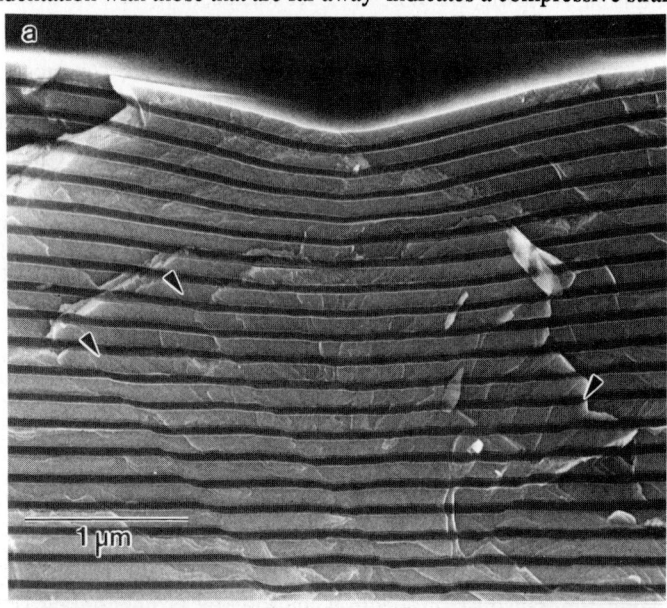

(b) A close-up of a (111) slip band where about 80 individual slip processes have occurred on approximately 70 planes.

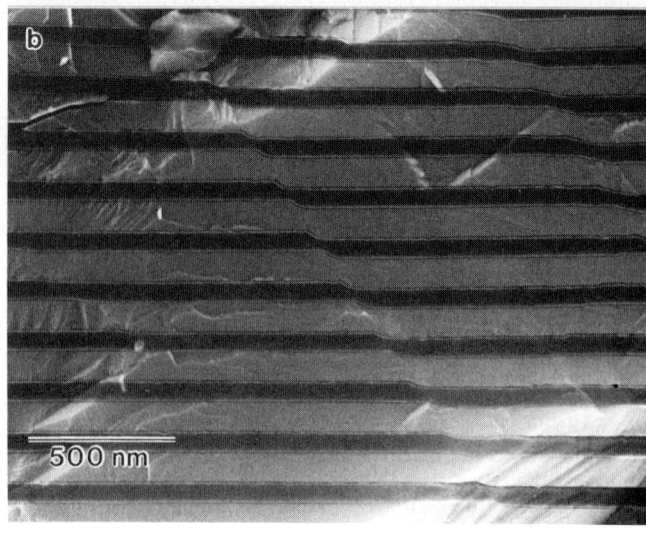

of about 15% which implies a large amount of plastic shear flow within the layers. The strain is higher for indentations made at greater loads which also exhibit layer delamination and cracking (Castell et al 1993). The region below the indentation shows clear evidence of slip bands which make an angle of approximately 56° with the layers and are indicated by arrows in Fig.1a. A (111) plane slipping in the [$10\bar{1}$] or [$01\bar{1}$] directions and a ($\bar{1}11$) plane slipping in the [$0\bar{1}1$] or [$10\bar{1}$] directions would both have an angle of 54.73° between the slip planes and the layers. Our experimental evidence certainly suggests that these slip systems have been activated in the deformed material seen in the micrographs. The off-set of the layers can be measured from the micrographs and thus an estimate can be made of how many slip processes have occurred in one region. For the slip band shown in Fig.1b, found under a 15gf indentation, the off-set perpendicular to the layers is approximately 23nm which corresponds to about 80 individual slip processes on the (111) planes slipping in the [$10\bar{1}$] or [$01\bar{1}$] directions. From the micrograph one can also measure that the number of planes that have slipped is of the order of 70, which might lead one to the conclusion that each (111) plane has slipped approximately once.

Fig. 2 shows a series of SE images of the deformation of a single 675nm AlAs layer on a GaAs substrate made at increasing loads; there are a number of features in the micrographs that are interesting from the deformation aspect. The first is that the AlAs layer is never penetrated although it is softer than the GaAs substrate. AlAs and GaAs have micro-hardness values of 5 GPa and 7.5 GPa respectively (Lide 1991) and some authors (Fabes et al 1992) believe that if the surface layer is softer than the substrate, the indenter will penetrate the layer. As can be seen from the micrographs, this is certainly not the case. The layer becomes thinner within the indentation, but this is out of necessity of having to cover a larger area compared with the layer in the unindented state.

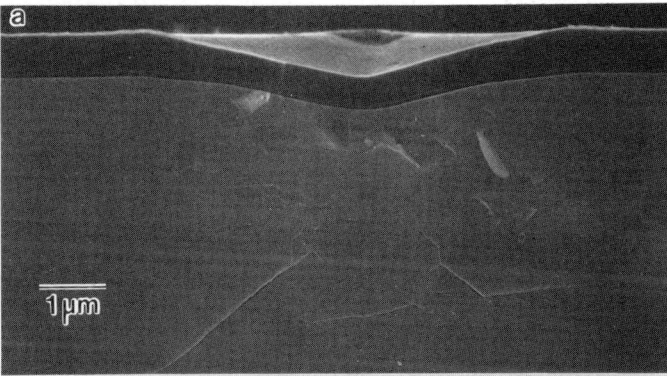

Fig. 2

Cross-sectional SE images of microindentations into a 675nm AlAs layer on a GaAs substrate. Indentation loads are 15gf, 50gf and 200gf for images (a), (b) and (c) respectively.

Fig.2 cont.

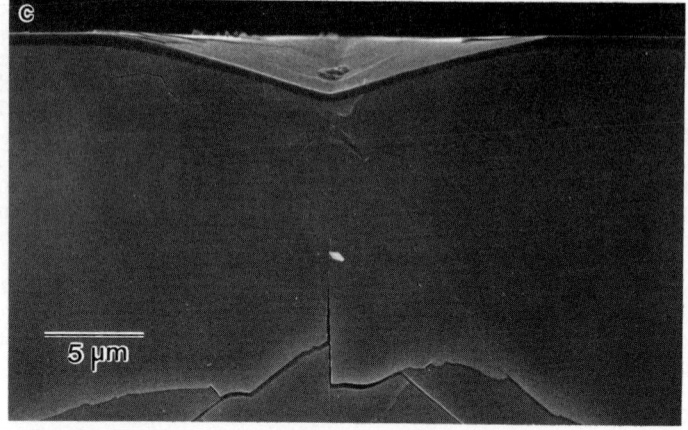

Another aspect of the deformation behaviour that is observed in the micrographs of Fig. 2 concerns subsurface cracking. In all the images there is an element of cracking that originates under the indentations. Rough inspection of the cracks shows that they always originate some distance below the indentation and that there is usually an uncracked zone, that has undoubtedly been plastically deformed, underneath the indentation. If subsurface cracking had set in during the loading stage, then the cracks would be closer to the bottom of the indentations as the indentations become larger. As this is not the case, it is almost certain that the cracking was caused during the unloading stage of the indentation experiment.

4. CONCLUSIONS

Our study has revealed the usefulness of layered structures in studying indentation plasticity and has highlighted some of the mechanisms that are active in the deformation of semiconductor superlattices and surface layers. Using information from Fig.1 and assuming isotropic flow and no densification of the material, one could calculate the precise flow pattern of the material under the indenter.

ACKNOWLEDGEMENTS

We are grateful to Dr. DA Williams of the Hitachi Cambridge Laboratory who rendered valuable assistance in the SEM operation. We would also like to thank the SERC for funding the Cambridge MBE programme and MRC's CASE studentship who is also supported by the National Physical Laboratory. We are grateful to Prof. LM Brown, Dr. MM Chaudhri and Dr. G Shafirstein for their advice and assistance.

REFERENCES

Castell MR, Walls MG and Howie A 1992 Ultramicroscopy 42-44 1490
Castell MR, Howie A, Perovic DD, Ritchie DA, Churchill AC and Jones GAC 1993 Phil. Mag. Lett. 67 89
Chaudhri MM 1993 Phil. Mag. Lett. 67 107
Chiang SS, Marshall DB and Evans AG 1982 J. Appl. Phys. 53 298
Fabes BD, Oliver WC, McKee RA and Walker FJ 1992 J. Mater. Res. 7 3056
Laursen TA and Simo JC 1992 J. Mater. Res. 7 618
Lide DR 1991 CRC Handbook of Chemistry and Physics 77 CRC Press
Ogura K 1991 JEOL News 29e 26
Page TF, Oliver WC and McHargue CJ 1992 J. Mater. Res. 7 450
Pethica JB, Hutchings R and Oliver WC 1983 Phil. Mag. A 48 593
Walls MG, Chaudhri MM and Tang TB 1992 J. Phys. D 25 500
Yoffe EH 1982 Phil. Mag. A 46 617

High resolution TEM and XRD study of interface composition inhomogeneities in CBE and MOCVD grown in (GaAs)P/InP multiple quantum wells

G Salviati, C Ferrari, L Lazzarini, F Genova*, C Rigo*, GM Schiavini* and F Taiariol*

CNR-MASPEC Institute, via Chiavari 18/A, 43100-I Parma
*CSELT, via G Reiss Romoli 274, 10148-I Torino

ABSTRACT: In(As)P/InP and InGaAs(P)/InGaAs "false" QWs have been investigated for studying interface compositional changes during the growth. An As and P incorporation depth of about 3-4 monolayers, quite independent of the growth conditions both for CBE and MOCVD samples, was found. The As and P concentration increased by increasing the growth interruption time, the dominant effect being the group V atom incorporation at the InGaAs/InP heterointerfaces. Further, contamination effects (carry over) of As atoms in the growth chamber were also observed in lattice matched MQW laser structures. Finally, HREM and high resolution x-ray diffraction investigations of the same structures revealed different amounts of elastic strain at the InGaAs/InP and InP/InGaAs interfaces.

1. INTRODUCTION

The problem of interface sharpness, both for composition and planarity, is of great importance for InGaAs(P)/InP multiple quantum well (MQW) based optoelectronic devices. Composition variations at the interfaces arise due to the need to prevent the surfaces from thermal degradation by a group V atom flux during growth interruptions . As a consequence, changes in the well energy profiles and in the position of the quantized levels as well as local strain at the heterointerfaces occur (see for instance Streubel et al 1992a). In the last few years many groups have been studying both the growth procedures for improving the structural and optical quality of the QWs and the compositional and structural properties of the interfaces by TEM, High resolution X-Ray diffraction (HRXRD), secondary ion mass spectroscopy and photoluminescence techniques (see for instance Ourmazd et al 1987, Carey et al 1987, Barnett et al 1988, Wang et al 1990, Long et al 1991, Meyer et al 1992). The majority of the groups suggest that As atoms are much more effective than P ones in incorporating in the QW layers, the As incorporation being explained in terms of As carry over into the InP barrier after growing InGaAs.

In this work, the incorporation due to As or P substitution inside the InP and InGaAs layers respectively during the GI will be distinguished from the carry over due to the contamination of the next layer to be grown. In order to study separately the P and the As effect on both InP/InGaAs and InGaAs/InP interfaces, "false" MQWs (FMQWs) have been intentionally grown. The FMQWs result from the incorporation of the protecting group V species different to that one present in the growing layer during the periods of growth interruptions (Genova et al 1992). The FQWs were grown at different growth temperatures and with different growth interruption (GI) times by Chemical Beam Epitaxy (CBE) and Metal Organic Chemical Vapour Deposition (MOCVD). Interface roughness and compositional changes in InP(As)/InP and InGaAs(P)/InGaAs FMQWs have been studied by comparing Conventional (CTEM) and High Resolution Transmission Electron Microscopy (HREM) and High Resolution X-ray Diffraction (HRXRD) results.

2. EXPERIMENTAL

In order to grow the FMQWs, the InP and the InGaAs growths were periodically interrupted and the specimen surfaces were exposed for different times to AsH_3 or to PH_3 flux respectively. The CBE and MOCVD systems have already been described extensively elsewhere (Mircea et al 1986, Antolini et al 1992).

The HRXRD measurements were performed with a Philips diffractometer in the $CuK\alpha_1$ 400 symmetrical reflection geometry. The four 220 diffractions in the Bartel monochromator reduced the beam divergency to 12 seconds of arc. The period, Δ, and the average composition $\langle x \rangle$ of the FMQWs were obtained from the MQW peak spacing $\delta\Theta$ and from the separation $\Delta\Theta$ of the zeroth order MQW peak from the substrate peak respectively (Holy et al 1990):

$$\Delta = 1 / (2\ \delta\Theta\ \cos\Theta_B) \qquad \text{and} \qquad \langle x \rangle = (d(x)-d(0))/(d(1)-d(0))$$

in which $d(x)$ is the lattice spacing corresponding to the alloy at composition x. The $d(x)$ value is calculated from the separation $\Delta\Theta$ and corresponds to the average MQW lattice parameter. The As or P distributions within the period were obtained by simulating the experimental rocking curves on the basis of the x-ray diffraction dynamical theory (Taupin 1964). The FMQW computer simulations are based on the assumption that the As or P atoms are uniformly distributed in a layer of thickness t. Since $\langle x \rangle$ and Δ can be directly deduced by the experimental rocking curves, only the parameter t had to be evaluated by the best fitting procedure, the incorporation x being given by:

$$(x \cdot t) / \Delta = \langle x \rangle \qquad\qquad 1)$$

In the frame of this simple model (Ferrari et al 1993) increasing MQW peak intensities are obtained when the incorporation depth decreases, as shown by the best fit of Fig. 1.

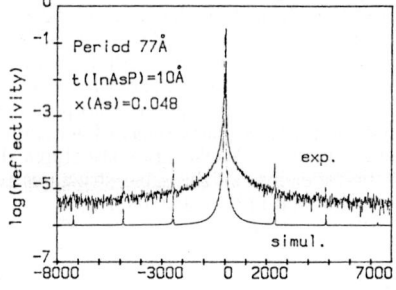

Fig. 1. Experimental and simulated HRXRD rocking curves for sample #32 (see Table A).

Table A: Summary of XRD and TEM results and of different growth conditions.

Sample	growth T (°C)	GI time (sec)	QW period (Δ) (Å)	$\langle x \rangle$	$\langle x \rangle \cdot \Delta$ (Å)	incorporation depth (Å) R x	TEM
MOCVD (AsH$_3$ flux)							
137	560	1	90	0.037	3.33	10	
114	585	0.5	61	0.025	1.50	10	9±3
156	585	0.5	101	0.022	2.21	15	
135	585	1	91	0.026	2.37		
157	585	1	93	0.0324	3.01	15	
141	585	2	91	0.065	5.91		10±3
32	610	1	77	0.0065	0.48	10	no QW contrast
MOCVD (PH$_3$ flux)							
138	585	0.5	130	0.006	0.80		10±3
158	585	0.5	154	0.003	0.45	15	no QW contrast
139	585	1	102	0.018	1.84		12±3
142	585	2	101	0.026	2.63	12	
159	585	3	104	0.046	4.78	19	15±3
29	610	1	68	0.033	2.21	10	9±2
CBE (AsH$_3$ flux)							
205	490	2	92	0.010	0.92	15	12±3
420	490	2	74	0.012	0.93	15	
209	490	10	84	0.032	2.69	15	12±3
CBE (PH$_3$ flux)							
419	490	2	80	0.0056	0.45	15	
206	490	5	90	0.020	1.80	12	
189	490	10	90	0.036	3.24	15	13±3

Samples for the transmission electron microscopy (TEM) analyses were thinned in the (001) and (110) cross section geometries by standard mechanochemical procedures followed by low temperature Ar ion milling. An ultimate low voltage Iodine ion milling was then carried out in order to minimize the presence of artefactual specimen structures (Cullis and Chew 1988). Thinned specimens were examined in a JEOL 2000FX TEM working at 200 KV

(Scherzer resolution ≈ 0.31 nm) using both conventional bright-field or dark-field, g=002 diffraction contrast imaging (Petroff 1977) and bright-field, axial illumination [001] and [110] lattice imaging. The results of TEM and HRXRD studies are summarized in Table A.

3. RESULTS AND DISCUSSION

3.1 MOCVD specimens

The results of Table A reveal a good agreement between XRD and TEM analyses. In particular, TEM investigations evidenced an average incorporation depth of about 3-4 monolayers (MLs) in all the samples. First the influence of the growth temperature was determined. Three rocking curves corresponding to In(As)P samples grown at 560, 585 and 610 °C respectively with a constant GI time of 1 sec are shown in Fig. 2. According to the equation 1, the quantity $<x> \Delta$ is unaffected by the error in the determination of the incorporation depth t, and by the different period Δ among the samples. The As and P values of $<x> \Delta$ in MOCVD grown samples are reported in Fig. 3a as a function of different growth temperatures. The lower As incorporation at higher temperatures is in agreement with the phase diagrams both for MOCVD (Mircea et al 1986) and MBE (Antolini et al 1992) specimens.

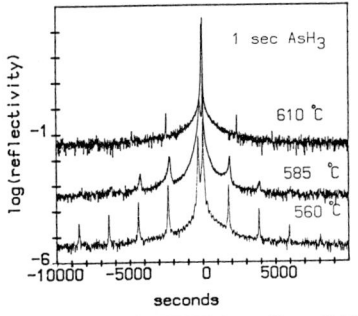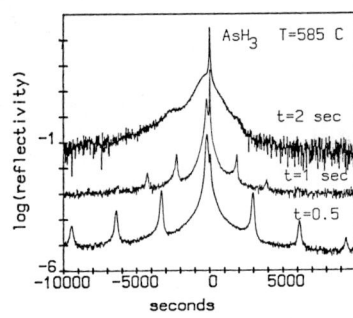

Fig. 2: Experimental HRXRD profiles of MOCVD samples grown at:
a) different temperatures and 1 sec of GI;   b) same temperature and different GI times.

Samples grown at a fixed temperature of 585 °C but with different GI times were then analysed. The data shown in Fig. 3b indicate a larger As and P incorporation at increasing GI times. This result is in agreement with previous results of Streubel et al (1992a,b).

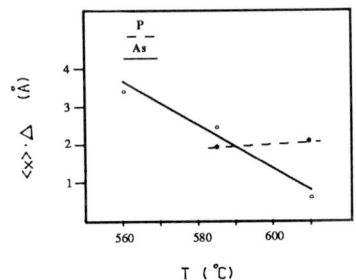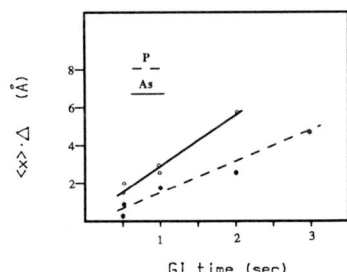

Fig. 3: a) As and P incorporation amount, $<x>\Delta$, for MOCVD samples grown at different temperatures with 1 sec of GI time. b) As and P $<x>\Delta$ values in samples grown at 585 °C with different GI times.

The HRXRD profiles showed an increase of the asymmetric tail of the MQW peaks for higher GI times (Fig. 2b). This has been interpreted in terms of a "memory effect" due to As deposition in the growth chamber (Streubel et al 1992b), which leads to an increase of the average As content during the growth. Such an effect explains the broadening of the MQW peaks for GI times of 2 sec. For longer GI times the sample surface appeared completely deteriorated. An example of incorporation depth in MOCVD grown FMQWs is shown in Fig. 4. Lattice resolution TEM micrographs of both [001] and [110] oriented cross sections of a sample grown with a GI of 1 sec reveal InAsP layers with an average thickness of about 3 MLs.

In order to show possible differences between the As and P incorporation behaviour, InGaAs layers were grown at a fixed growth temperature of 585°C. In this case the phosphine flux during the GI time leads to an InGaAsP additional layer. The results reported in Table A and in Fig. 3b indicate a lower P incorporation efficiency than As but, again, very similar incorporation depth as in the case of As. From the point of view of the HRXRD, a good planarity of the interfaces is achieved only at GI times of 2 sec. or more. FMQWs peak splitting observed in the X-ray rocking curves, for samples with GI times lower than 2 seconds, can be attributed to a period fluctuation along the growth axis. This result was confirmed by TEM pictures in which nearly 3 MLs thick InGaAs(P) layers were observed (Fig. 5).

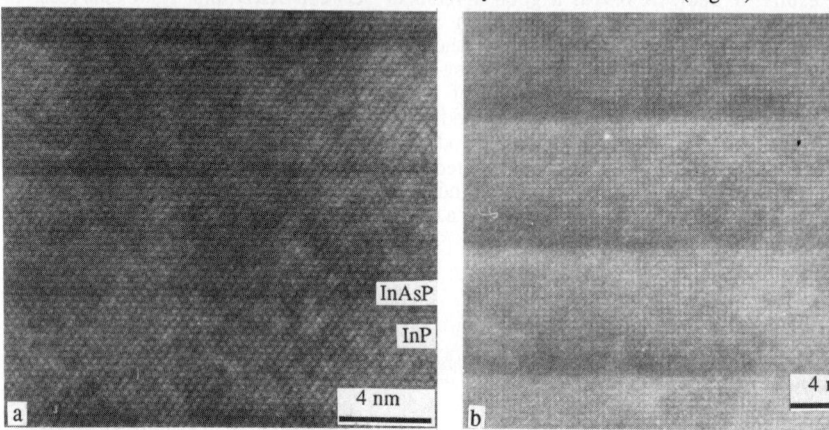

Fig. 4   Comparison between a) [110] and b) [001] oriented BF zone axis HREM micrographs of sample #114. InAsP layers with an average thickness of about 3 MLs are shown.

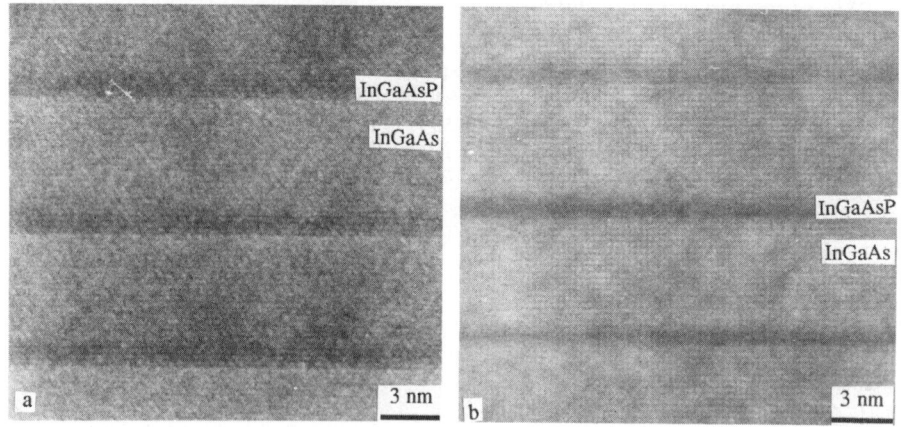

Fig. 5   Comparison between: a) [110] oriented three beam dark field, g=002, and   b) [001] oriented BF zone axis micrographs of sample #29. InGaAsP layers of about 3 MLs are shown in both the pictures.

3.2 CBE specimens

In the case of CBE grown FMQWs, the optimum growth temperature of 490 °C was chosen (Antolini et al 1992). Six FMQWs specimens were grown with with different GI times. The TEM estimation of the InAsP layer thickness gave a value of about 3-4 MLs for the samples with GI of 2 and 10 seconds respectively (Fig.6). As for MOCVD samples, HRXRD analysis showed MQW period fluctuations that were ascribed to the poor interface planarity as evidenced by TEM (Fig. 6a). Further, an increase of P incorporation in InGaAs(P) FMQWs grown at the same temperature by fluxing PH_3 during GI times of 5 and 10 sec has been also observed by HRXRD. A broadening of the high order MQW peaks was found in HRXRD profiles for the lower GI time sample, indicating again a poor QW planarity. It is worth noting that the As or P concentration at the interface is comparable to that found in MOCVD samples.

In order to compare these results with real structures, InGaAs/InP MQWs, presenting quite good structural and optical quality (Antolini et al 1992) have also been studied.

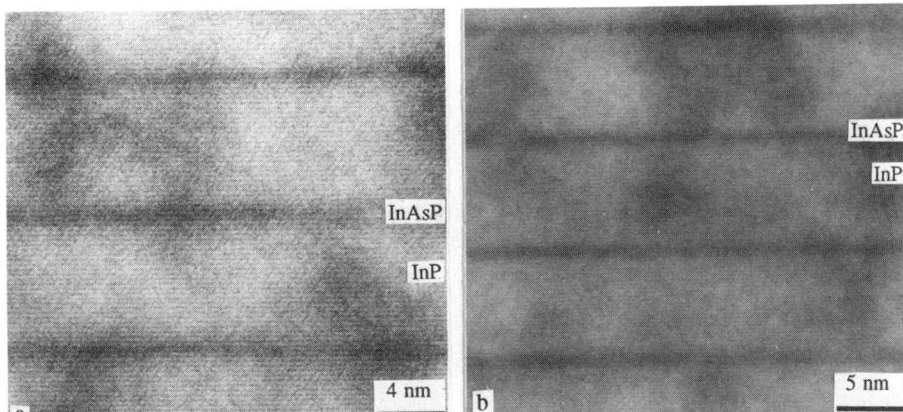

Fig. 6: DF, g=002 type, lattice resolution three beam (110) oriented TEM micrographs showing a comparison between CBE FMQWs grown with different GI times. a) GI=2 sec. b) GI=10 sec.

HRXRD data revealed that the lower (InGaAs/InP) and upper (InP/InGaAs) interfaces presented different amounts of elastic strain. As an example of the structures investigated, Fig. 7a shows HRXRD rocking curves of a 40x(73Å InGaAs + 73Å InP) MQW containing a 8600 Å In.$_{511}$Ga.$_{489}$As cap. The structure was grown at 490 °C by CBE with GI times of 1 sec at both the interfaces. An As incorporation of about 30% and a P incorporation of about 13% in 3 ML thick layers at the InGaAs/InP/InGaAs interfaces, giving rise to InAsP and InGaAsP extra layers have been found. Further, As carry over of about 3.5% through the whole InP barrier had to be introduced in the computer simulation in order to fit the experimental profiles. It is interesting to stress at this point that the HRXRD technique does not allow us to distinguish at which interface the As or P incorporation occurs. Fig. 7b shows an HREM micrograph of the same MQW. In order to avoid possible intermixing of elements at the heterointerfaces due to radiation damage effects (Humphreys 1989, Deveaud et al 1990), the HREM micrographs were taken first by defocusing the electron beam while adjusting the instrument and then by reducing the exposure time at the minimum values compatible with the best recording conditions. Despite the lower InGaAs/InP interface appears quite smooth, the {111} InP lattice planes in proximity of the heterointerface show a lattice distortion, consistent with an increase of the lattice parameter, along the growth direction which is not detectable at the upper InP/InGaAs interface. Assuming a tetragonal distortion of the lattice planes, a maximum mismatch of $2\pm1\times10^{-2}$ can be calculated from the approximated equation:
$$\Delta\theta = \Delta d/d \sin (2 \phi)/2 \qquad 2)$$
in which $\phi=54.7°$ is the angle between the {111} planes and the (001) growth plane, $\Delta\theta\leq1.5°$ is the estimated {111} lattice planes bending and $\Delta d/d$ is the lattice mismatch. The increase of the lattice parameter at the lower InGaAs/InP interface is consistent with an As incorporation in the InP barrier surface (Streubel et al. 1992a.). Finally, despite both carry over and incorporation of As are present through the MQW interfaces, the lattice strain is mainly due to the As incorporation induced by the switching of the complementary hydride into the reactor in order to prepare the growth of the next layer. Since no carry over has been revealed by HRXRD simulation in FMQWs, this seems to be related to the total As or P flux duration.

4. CONCLUSIONS

In(As)P/InP and InGaAs(P)/InGaAs FMQWs have been grown by CBE and MOVPE under different growth conditions as test structures for studying compositional changes in the growth of MQWs for optoelectronic applications. It has been shown that a significant As and P incorporation occurs in the FMQWs even at very short GI times. Further, the GI times should not be too short in order to allow the reconstruction of the growth surface and should not be too long in order to avoid the surface degradation. The results of HRXRD and TEM investigations can be summarized as follows:

i) An As and P incorporation depth of about 3-4 MLs was determined from TEM and HRXRD techniques. This value appeared almost independent on the growth conditions for both CBE and MOCVD samples.

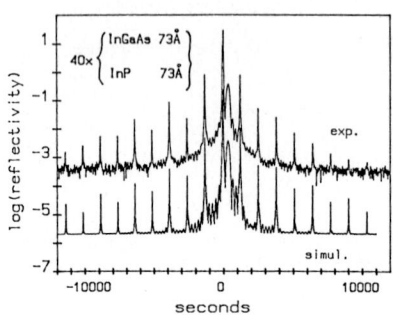 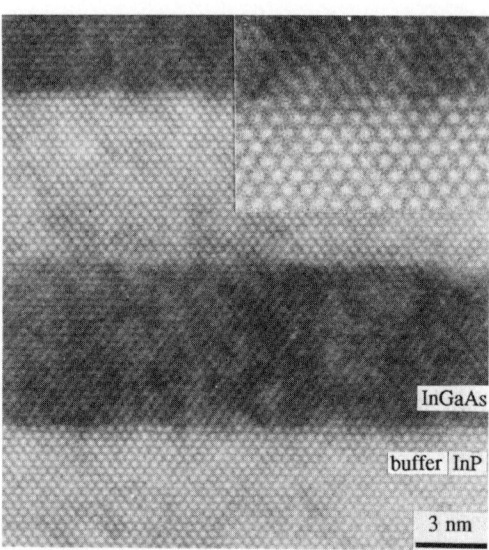

Fig. 7 a) Experimental and simulated HRXRD rocking curves of an InGaAs/InP CBE grown MQW. b) [110] oriented BF zone axis lattice image of the same MQW as in a). A lattice plane tetragonal distortion at the lower InGaAs/InP interface can be seen in the inset. Growth direction upside page. See text for details.

ii) a slightly larger As incorporation amount, with respect to the P one, was found.

iii) As and P incorporation increased in all the samples by increasing the GI time.

iv) the interface planarity is influenced by the GI times. Exposure times up to 1 sec of the InGaAs layer to P atoms did not result in a smooth surface; on the contrary, the InP surface degraded for As exposure times longer than 1 sec.

v) contamination of As atoms in the growth chamber was found.

vi) As and P incorporation during the GI time in the barrier and in the well respectively seem to be more effective than the carry over in real MQWs.

References

Antolini A, Bradley PJ, Cacciatore C, Campi D, Gastaldi L, Genova F, Iori M, Lamberti C, Morello C, Papuzza C and Rigo C 1992 IEEE J. Electron. Mater. EM-21 233

Barnett SJ, Brown GT, Courtney SJ, Bass SJ and TaylorLL 1988 64(3) 1185

Carey KW, Hull R, Fouquet JE, Kellert FG and Trott GR 1987 Appl. Phys. Lett 51 910

Chew NG and Cullis AG, 1984 Appl. Phys. Lett. 44 142

Deveaud B, Guennais B, Poudoulec A, Regreny A, d'Anterroches C 1990 Phys. Rev. Lett. 65 2317 and answer of Ourmazd A and Cunningham J 1990 Phys. Rev. Lett. 65 2318

Ferrari C, Lazzarini L, Salviati G, Gastaldi L, Taiariol F, Schiavini G, Rigo C 1993 IEEE Proc. of the Fifth International Conference on "InP and Related Materials", Paris,

Genova F, Antolini A, Francesio L, Gastaldi L, Lamberti C, Papuzza C and Rigo C 1992 J. Cryst. Growth 120 333

Holy V, Kubena Jand Ploog K 1990 Phys. Status Solidi (b) 162 347

Humphreys CJ 1989 Ultramicroscopy 28 357

Long NJ, Norman AG, Petford-Long AK, Butler BR, Cureton CG, Booker GR and Thrush EJ 1991 Inst. Phys. Conf. Ser. No 117 69

Meyer R, Hollfelder M, Hardtdegen H, Lengeler B and Luth H 1992 J. Cryst. Growth 124 583

Mircea A, Mellet R, B. Rose B, P. Dastè P, G.M. Schiavini GM, 1986 J. Cryst. Growth 77 340

Ourmazd A, Tsang WT, Rentschler JA and Taylor DW 1987 Appl. Phys. Lett. 50 1417

Petroff PM 1977 J. Vac. Sci. Technol. 14 973

Streubel K, Harle V, Scholz F, Bode M, Grundmann M, and references therein enclosed 1992a J. Appl. Phys. 71(7) 3300

Streubel K, Wallin J, Amiotti M and Landgren G 1992b J. of Cry. Growth 124 541

Taupin D 1964 Bull. Soc. Franc. Minér. Crist. 87, 496

Wang TY, Jen HR, Chen GS and Stringfellow GB 1990 J. Appl. Phys. 67(1) 563

Inst. Phys. Conf. Ser. No 134: Section 7
Paper presented at Microsc. Semicond. Mater. Conf., Oxford, 5–8 April 1993

477

Modification of layer interdiffusion in GaInAs/GaInAsP multiple quantum wells through the controlled introduction of dislocations

R E Mallard, E J Thrush[+], S A Galloway, E M Allen[+] and G R Booker

Department of Materials, University of Oxford, Parks Road, Oxford, OX1 3PH, UK
+ BNR Europe Limited, London Road, Harlow, Essex, CM17 9NA, UK

ABSTRACT: Interdiffusion between the well and barrier layers of GaInAs/GaInAsP multiple quantum wells (MQW) results in a modification of the luminescence wavelength of the material, and can be induced by thermal annealing at temperatures in excess of 650°C. We describe how the extent of interdiffusion in MQWs from this system, due to annealing at 700°C to 750°C, can be spatially varied through the controlled introduction of dislocations. This process, which induces a wavelength shift in excess of 100nm in dislocation free regions of the MQW, produces a negligible shift in regions within approximately 300μm of intentionally introduced dislocation tangles. The magnitude of the luminescence modification is spatially correlated to the extent of MQW interfacial broadening, as observed by TEM. We propose that dislocations in the material have the property of being able to getter point defects which participate in the diffusion mechanism, and thereby are able to locally suppress layer interdiffusion.

1. INTRODUCTION

State-of-the-art optoelectronic telecommunications systems presently rely on the use of InP-based materials in devices which exploit quantum confinement effects, such as multiple quantum well (MQW) lasers and modulators. An important issue in the production of these devices is the preservation of the structural integrity of the MQW layers throughout the processing and lifetime of the component. It has been shown that the emission wavelength of GaInAs/GaInAsP MQWs grown on InP can be modified (specifically, blue shifted) by annealing at temperatures as low as 650°C (Nakashima et al 1987, Razeghi et al 1987, Temkin et al 1987, Glew et al 1992, Mallard et al 1993). This is in the range of the metalorganic chemical vapour deposition (MOCVD) growth temperature for these materials, and as such, poses a problem in the development of a process for the fabrication of optoelectronic integrated circuits, which may involve several high temperature epitaxial growth stages.

It has recently become apparent, that the propensity for these materials to undergo a blue shift is especially pronounced when the layers are grown onto substrates which are characterised by a low etch pit density (Glew et al 1992). MQW samples grown on S doped InP substrates have been shown to be susceptible to large blue shifts upon annealing whereas equivalent MQWs grown on higher defect density Fe doped InP substrates are much less liable to undergo a wavelength modification. This effect is not due to the participation of the substrate doping via impurity enhanced layer interdiffusion (Mallard et al 1993). This surprising result suggests that the presence of threading dislocations in the MQW may actually retard layer interdiffusion. In this work we describe the first transmission electron microscope (TEM) study of such specimens which exhibit thermally induced blue shift in the vicinity of dislocations, and show that this is indeed the case.

2. EXPERIMENTAL

A layer structure was grown by low pressure MOCVD at a temperature of 650°C, consisting of a 1μm buffer layer of InP grown on a (100) S-doped InP substrate, followed by 0.2μm of $Ga_{0.21}In_{0.79}As_{0.45}P_{0.55}$ and a 6 period MQW region consisting of 11nm of $Ga_{.21}In_{.79}As_{.45}P_{.55}$ barriers and 7nm of $Ga_{.47}In_{.53}As$ wells and was capped with 0.2μm of $Ga_{0.21}In_{0.79}As_{0.45}P_{0.55}$ and 0.1μm of InP. The structure was undoped and nominally InP-lattice matched throughout. The S doped substrate was selected because of its characteristically low defect density.

In order to evaluate the thermal stability of the MQW in relation to defect density, dislocations were intentionally introduced into some areas of several samples. This was accomplished by making a line of microindentations in the wafer surface at 10μm intervals using a diamond stylus, and subsequently deforming the samples in a three point bending rig at 300°C in an argon ambient. The microindentations act as sources for dislocations which glide into the sample bulk in response to the bending stress. Slip lines extending approximately 100μm away from the microindentations, in the direction perpendicular to the line of indents, are observed when the deformed samples are examined in the optical microscope. Photoluminescence (PL) analysis shows that there is no significant change in the emission wavelength of the MQW following deformation and prior to annealing. although some quenching in the intensity is observed. Following deformation, the samples were overgrown with a further 1μm of InP at 650°C and then "*in situ*" annealed under an equilibrium atmosphere of PH_3 for 1h at 750°C. A "control" sample of the original undeformed wafer was annealed in the same growth run. For comparison, a second group of similarly processed samples was furnace annealed in a sealed quartz tube for 1h at 700°C (in place of the second stage MOCVD growth and anneal). The samples were subsequently examined by cross sectional TEM and PL. These examinations were performed on regions from the same specimen which were both adjacent to, and far removed from, the dislocations, as well as on the undeformed "control" samples. The PL analysis was performed at room temperature, with a spatial resolution defined by the size of the incident excitation probe, of 1mm.

3. RESULTS AND DISCUSSION

Fig 1a shows photoluminescence spectra from a control sample (undeformed), before and after annealing *in situ* for 1h at 750°C. The MQW luminescence blue shifts by 82nm in response to this heat treatment. TEM analysis of the sample, using the "chemically sensitive" (200) reflection, shows that the interfaces are chemically diffuse in the annealed sample, and have a width of approximately 3-4nm, as shown in fig 2d.

Conversely, the deformed sample from an adjacent portion of the wafer, annealed at the same time, exhibited a blue shift of approximately 20nm in regions where the dislocation density was high, at points I and II in fig 1b. In relatively undeformed regions of this sample, points III-V, the blue shift is comparable to that observed in the control sample. In addition, there is some evidence to suggest that the area within which there is a reduction in blue shift extends beyond the deformed regions of the sample: when the heavily deformed portion was "cleaved away", the edge of the sample which had been adjacent to the dislocations exhibited less than 30nm of blue shift. Qualitatively similar results were observed in both the furnace and *in situ* annealed samples.

TEM observations of the extent of the layer interdiffusion support these findings, and show that it is the reduction in interdiffusion which is responsible for the reduced blue shift. Fig 2a is a cross section TEM micrograph of the edge of one of the intentionally introduced dislocation tangles; it is apparent that the interfaces of the MQW are chemically abrupt. Furthermore, as suggested by the PL data, the interfaces remain abrupt up to several hundred microns away from these tangles, as shown in fig 2b. Only in regions of the deformed sample which were a long way away from the intentionally introduced dislocations (in the case shown in fig 2c, by approximately 1mm) was the interfacial abruptness, and indeed the extent of the blue shift, comparable to that observed in the control sample, shown in fig 2d.

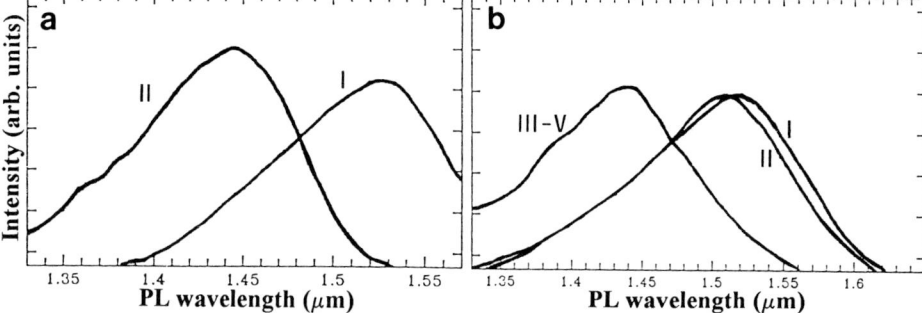

Fig 1: (a) PL spectra from as-grown "control" sample, I, and *in situ* annealed control sample, II. (b) spectra from various positions, designated I-V, on the deformed sample, annealed together with sample in (a). Points I and II lie within the heavily deformed region; III-V are from relatively defect free regions.

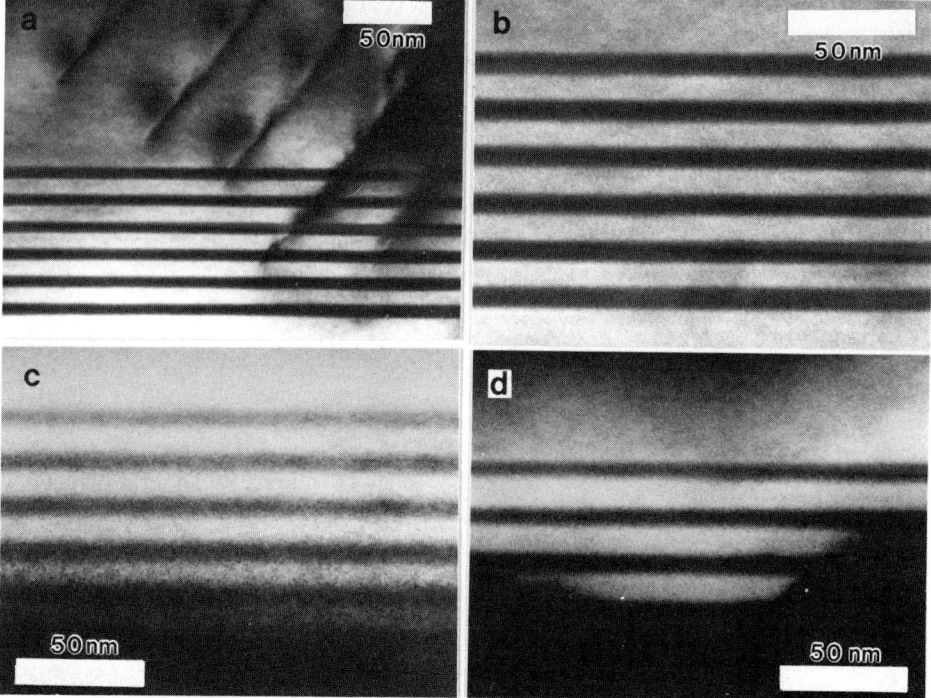

Fig 2: (200) DF TEM micrographs of furnace annealed MQW samples. (a) adjacent to the intentionally introduced dislocation clusters, (b) 300μm away from the dislocation cluster shown in (a), (c) 1mm away from the dislocation cluster shown in (a). Micrographs (a) to (c) were all taken from the same microscope specimen. (d) Undeformed "control" sample annealed under identical conditions.

The widths of the variations in contrast across the MQW interfaces shown in fig 2 were quantitatively measured by digitising the images and then projecting the image intensity onto a plane perpendicular to the interfacial direction. The interfacial width is taken as the distance across the interface for which 20% to 80% of the total contrast change between the layers occurs. This analysis reveals that the interfacial "diffuseness" in the region adjacent to the dislocation tangles (fig 2a) is approximately 12Å. This value increases to 16Å at a distance of $300\mu m$ from the dislocations (fig 2b), and is approximately 25-30Å both at a distance of 1mm from the dislocations (fig 2c) and in the control sample (fig 2d). We are thus able to identify a positive correlation between the extent of the spatially varying blue shift and the interfacial abruptness in the deformed sample. These results confirm that the presence of extended lattice defects such as threading dislocations can influence the thermal stability of GaInAs/GaInAsP multilayers by retarding layer interdiffusion. This effect is likely due to a phenomenon whereby a "participant" in the solid state interdiffusion mechanism, such as a vacancy or interstitial species, is gettered by the dislocations. Similar behaviour has been observed in Si, where dislocation tangles have been used to getter transition metal impurities over large distances (see for instance, references in Fair et al 1985). This could explain our observation that within a radius of $300\mu m$ from the intentionally introduced dislocation clusters, the blue shift and interdiffusion were minimal. Future work will apply the technique of spatially-resolved PL in order to better define the extent of the non-interdiffused regions.

In summary, we have presented the first direct evidence that the presence of dislocations in a GaInAs/GaInAsP MQW can suppress layer interdiffusion. This effect persists within some distance from the location of the dislocations themselves. It is evident that a better understanding of the mechanism of solid state diffusion in the system is required in order to fully understand the effect of the presence of dislocations on interdiffusion, and thereby to optimise the processing methods for the fabrication of optoelectronic devices.

4. ACKNOWLEDGEMENTS

This work was carried out as part of the DTI-SERC funded LINK ASM programme. We wish to acknowledge R W Glew and P R Wilshaw for valuable discussions.

REFERENCES

Fair R, Pearce C and Washburn, eds, 1985, MRS Symposium on Impurity Diffusion and Gettering in Silicon, <u>36</u>
Glew R W, Briggs A T R, Greene P D, and Allen E A, 1992, Proc 4th Int'l Conf on InP and Related Materials, Newport RI, (New York: IEEE), 234
Mallard R E, Thrush E J, Martin R W, Wong S L, Nicholas R J, Pritchard R E, Hamilton B, Long N J, Galloway S A, Chew A, Sykes D L, Thompson J, Scarrott K, Jowett J M, Satzke K, Norman A G, and Booker G R, 1993, Semicond Sci Tech, in press
Nakashima K, Kawaguchi Y, Kawamura Y, Asahi H, and Imamura Y, 1987, Jpn J Appl Phys, <u>26</u>, L1620
Razeghi M, Acher O and Launay F, 1987, Semicond Sci Tech, <u>2</u>, 793
Temkin H, Chu S N G, Panish M B, and Logan R A, 1987, Appl Phys Lett, <u>50</u>, 956

Inst. Phys. Conf. Ser. No 134: Section 7
Paper presented at Microsc. Semicond. Mater. Conf., Oxford, 5–8 April 1993

Microstructural characterization of heterointerfaces in the MOCVD-grown InGaAs/GaAs strained layer system using cross section transmission electron microscopy

M C Paek, H H Park and O J Kwon

Electronics and Telecommunicatins Research Institute, Daedog Science Town P.O. Box 8, Daejeon Korea 305-606

ABSTRACT : Microstructures of heterointerfaces in InGaAs/GaAs strained layer systems were characterized using cross section transmission electron microscopy. The epitaxial InGaAs layers were grown by MOCVD on (001) GaAs substrates. Both misfit and threading dislocations were observed at the interface of $In_{0.21}Ga_{0.79}As$/GaAs strained layer and a part of misfit dislocation showed a multiplication features lying parallel direction to the interface. In the $In_{0.1}Ga_{0.9}As$/GaAs system, only misfit dislocations with small segments were observed in the region near the interface. In quantum well structures of both compositions, threading dislocations were found to be extended through the wells to the surface.

1. INTRODUCTION

Lattice-mismatched heterostructures of semiconductor epitaxial layers have been actively investigated in both materials science and device-related research. Built-in strains at the interface make materials with a wide variety of electrical and optical properties. In this paper, we have considered the InGaAs/GaAs strained layer structure as an example of an important lattice-mismatched heterointerface. The strains at the interface depend on the composition and the thickness of the deposited layer. The misfit stresses accumulate at the interface elastically with increase of thickness, and at a certain critical thickness the strains are relaxed by generation of lattice defects. The critical layer thicknesses of lattice-mismatched heterostructures have been determined and calculated by many workers with measurement and modeling analysis [Matthews et. al (1970), Fritz et. al.(1987), Ladig et. al.(1984), Zou et. al.(1991)] The best device properties can be obtained within this critical thickness range, which varies with the layer composition, and is the major factor determining the band structure of the materials.

In this work, the defect structure of the interface in InGaAs/GaAs systems has been characterized for a single epi-layer of 2000 Å thickness and quantum well structures of various well thicknesses using cross section transmission electron microscopy. The shapes and distributions of the dislocations in the vicinity of the interfaces have been observed.

2. EXPERIMENTAL

GaAs substrates of (001) 2 inch wafers were precleaned before epitaxial growth. The metal organic chemical vapor deposition (MOCVD) technique was

employed to grow InGaAs epitaxial layers. The growth temperature was fixed at
600° C and the chamber pressure was 76 torr. Source gases for the In, Ga and
As were trymethylindium (TMIn), trymethylgallium (TMGa) and arsine (AsH$_3$),
respectively. Two different compositions of In$_x$Ga$_{1-x}$As layers (x=0.1 and x=0.21)
were grown as 2000 Å thick single epitaxial layers and multi-quatum well
structures with well thicknesses of 100, 150 and 200 Å. In the growth of the
quantum well structures, a gas interruption technique has been adopted to make
uniform and abrupt composition interfaces. To observe cross section electron
micrographs, the thin foil specimens were prepared by the sandwich method
attaching two wafers face to face with M-bond 610 glue. Mechanical
grinding/polishing followed by Ar/I$_2$ ion milling to a final thickness of electron
transparency has been carried out in a liquid nitrogen cryo chamber to minimize
milling damage.

The prepared specimens were observed and characterized by transmission
electron microscopy in a Philips CM20T using a LaB$_6$ filament. Dislocations
were observed by conventional bright field imaging and high resolution images
were taken using [110] projection.

3. RESULTS AND DISCUSSION

Cross section electron micrographs of the In$_x$Ga$_{1-x}$As single epi-layer system
reveal the generation of misfit dislocations at the interfaces as shown in fig.
1(a) and (b). In the case of x=0.21 in fig. 1(a), some of the dislocations
originate from the interface and develop vertically to the surface and some
dislocations bend down to the interface, and have a direction parallel to the
interface. Parallely bent dislocations show multiplication features, suggesting that
the strain relaxation continues during the epitaxial layer growth process.
However in fig. 1(b), x=0.1 case, the density of dislocations is much reduced
and no threading dislocations are observed in the layer.   At the interface, a
relatively small number of dislocation segments are observed within a limited
region.

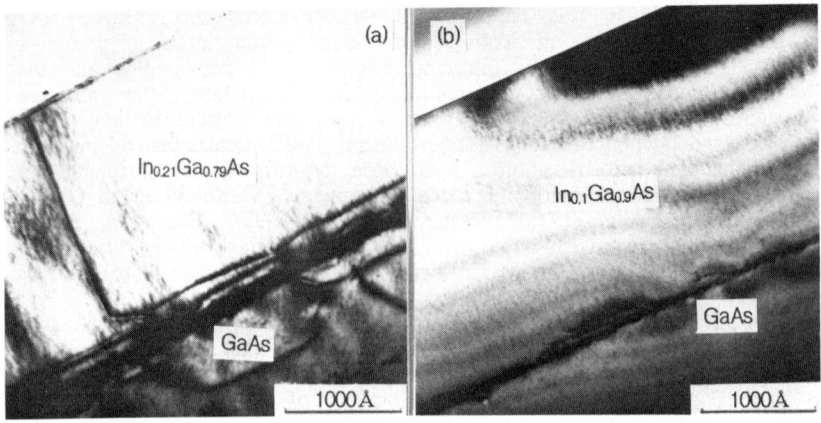

Fig.1 Cross section electron micrographs of (a) In$_{0.21}$Ga$_{0.79}$As/GaAs and
    (b) In$_{0.1}$Ga$_{0.9}$As/GaAs heterointerfaces

It is suggested that at low lattice-mismatch (x=0.1; 0.7%), the accumulated stresses during layer growth are relaxed to give perfect misfit dislocations lying along the interfacial directions of <110> type and showing chain-like undulations of small segments of bent dislocation. In this case, the development of threading dislocations is prevented by dislocation reactions. At the interface of relatively higher lattice-mismatch, threading and misfit dislocations are generated and developed. Especially the misfit dislocations lying parallel to the interface show double or triple lines, indicating that the dislocation multiplication process had occurred. Jesser et. al (1990) have reported threading dislocation pinning effects due to the segregation and/or impurities at the interface. In the present results, the multiplication of dislocations can be explained by a pinning effect and the large lattice-mismatch. Figure 2 is a cross section high resolution electron micrograph of the $In_{0.1}Ga_{0.9}As$/GaAs interface along the [110] direction, showing that {111} planes of both layers have lattice mismatch point at certain area where extra half planes of GaAs meet. At that point, we can find 2 sets of (111) extra half-planes making an angle of 70.7° and one (002) extra half-planes along the interface-parallel direction, revealing that the type of misfit dislocation is pure edge which can be formed by the reaction of two glide 60° mixed type dislocations. [Sharan et. al. (1991)] A pure edge 90° dislocation is normally sessile at the interface and prevent the development of threading dislocations. The 0.7% of mismatch is considered insufficient to generate threading dislocations.

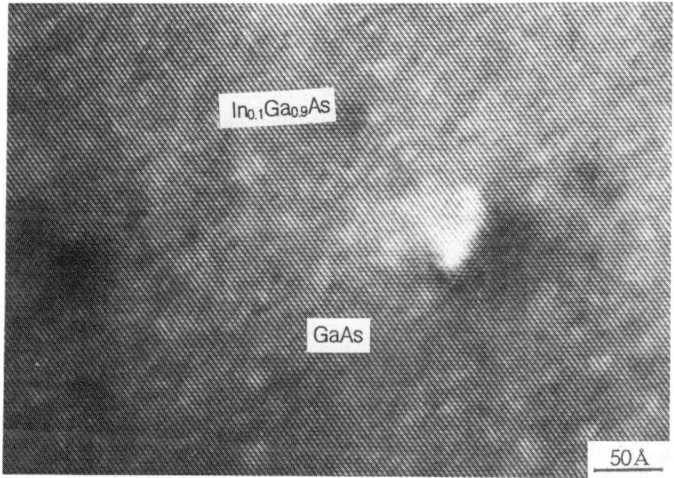

Fig. 2 High resolution cross section electron micrograph of the $In_{0.1}Ga_{0.9}As$/GaAs interface, showing the lattice-mismatched region.

To investigate the dependence of the dislocation behavior on the layer thicknesses, we have formed multi-quantum well structures of various well thicknesses. Figure 3(a), and (b) are cross section electron micrographs of the $In_xGa_{1-x}As$/GaAs (x=0.1, 0.21) quantum well structures. In both structures, threading dislocations and interface-parallel misfit dislocations are observed which originate from wells of different thicknesses. In fact, the trend of dislocation generation is not regular in this experiment and the critical layer thickness at which the defect generation begins can not be clearly identified.

(a)   (b)

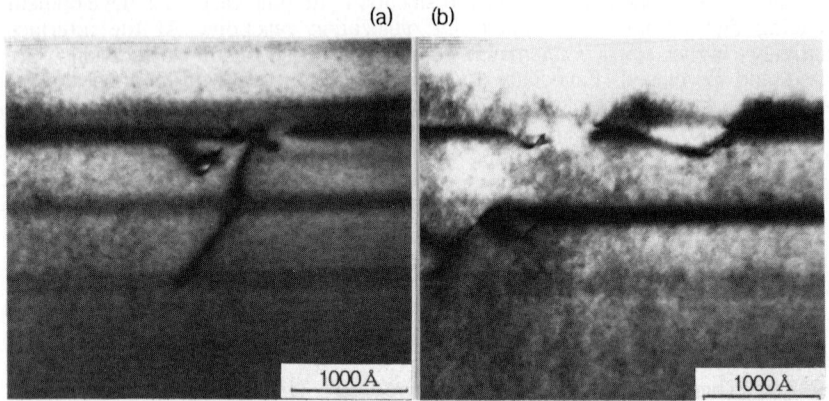

Fig. 3 Cross section electron micrographs of (a) $In_{0.1}Ga_{0.9}As/GaAs$ and (b) $In_{0.21}Ga_{0.79}As/GaAs$ multi-quantum well structures of various well thicknesses; 100, 150 and 200 Å.

4. CONCLUSIONS

The dislocation behaviour of the lattice-mismatched InGaAs/GaAs hetero-interface system has been studied by cross section transmission electron microscopy. In the highly mismatched interface of x=0.21 composition, misfit dislocations' multiplication features and many threading dislocations were found, and in the low mismatch interface of x=0.1, small segments of bent dislocations lie along directions parallel to the interface. High resolution micrographs revealed that there exist pure edge sessile misfit dislocations at the interface, which prevent the development of threading dislocations. In multi-quantum well structures of various well thicknesses, defect generation was observed to be irregular.

REFERENCES

Fritz I J, Gourley P L and Dawson L R 1987 Apply. Phys. Lett. 51 1004
Jesser W A and Fox B A 1990 J. Elec. Mat. 19 1289
Ladig W D, Peng C K and Lim Y F 1984 J. Vac. Sci. Tech. B2 181
Matthews J W, Mader S and Light T B 1970 J. Appl. Phys. 41 3800
Sharan S, Narayan J and Fan J C C 1991 J. Elec. Mat. 20 779
Zou J, Usher B F, Cockayne D J H and Glaisher R 1991 J. Elec. Mat. 20 855

Inst. Phys. Conf. Ser. No 134: Section 7
Paper presented at Microsc. Semicond. Mater. Conf., Oxford, 5–8 April 1993

485

TEM observation of modulations in strained GaInAsP multilayers grown by gas source molecular beam epitaxy

A Ponchet, A Rocher, J Y Emery[1], C Starck[1] and L Goldstein[1]

CEMES/LOE-CNRS, 29 rue Jeanne Marvig, BP 4347, 31055 Toulouse, FRANCE
[1] ALCATEL ALSTHOM RECHERCHE, Route de Nozay, 91460 Marcoussis, FRANCE

ABSTRACT: Alternating strained GaInAsP multilayer grown by Gas Source Molecular Beam Epitaxy which possessed altered structural and optical characteristics was examined. A strong lateral modulation of thickness, strain and probably chemical composition was observed by Transmission Electron Microscopy experiments. This modulation was anisotropic, with a periodicity of about 50 nm along the [110] direction. Although its origin is not fully accounted for yet, it seems to allow partial elastic relaxation of tensile layers.

1. INTRODUCTION

Gas Source Molecular Beam Epitaxy (GSMBE) has demonstrated a great potential for producing high quality GaInAsP/InP multiple quantum wells (MQW) designed for lasers emitting at 1,5 μm. Strained compressive MQW with lattice matched barriers have been developed with enhanced optical performance relative to lattice matched MQW (Starck et al 1992). However misfit dislocations may appear to relax the elastic energy if the layer thickness exceeds the critical thickness value. Use of tensile rather than lattice matched barriers is expected to allow an increase in the total MQW thickness by stabilising the whole structure. However, macroscopic characterisation has shown that the introduction of tensile barriers instead of lattice matched ones has caused the optical and structural quality to decrease (Emery et al 1993). The purpose of the structural study reported here is to understand why the alternate strain introduced may induce such degradation.

2. EXPERIMENTS

A zero-net strain $Ga_{0,18}In_{0,82}As_yP_{1-y}$ MQW has been grown on (001) InP by GSMBE at substrate temperature of 500°C. Lattice mismatches with InP and layer thicknesses are respectively -0,5% and 16 nm in tensile barriers (y=0,2) and +1% and 8 nm in compressive wells (y=0,7). Because these thicknesses are lower than the critical thickness, the MQW is expected to be tetragonally distorted. The [110] and [$\bar{1}$10] directions were revealed by chemical etching figures.

The Transmission Electron Microscopy (TEM) study was performed at 200 kV on a Phillips CM20 microscope using both cross-sectional and plan view observations. Thin cross-sectional specimens were obtained by mechanical polishing followed by Ar^+ beam milling with liquid nitrogen cooling to avoid Indium drop formation. Plan view specimens were prepared by ion milling using the same conditions as for cross-sectional specimens.

Dark field (DF) images were taken with g=002, which is highly sensitive to chemical composition variations, and with g=220 which is highly sensitive to strain fields.

3. RESULTS

3.1 Quasi-periodic thickness modulations

A quasi-periodic thickness undulation of the barrier / well interfaces is shown by g=002 DF observations on a ($\bar{1}$10) cross-sectional sample. Its amplitude at the second barrier is about 1 to 2 nm (Fig. 1). The amplitude of these interface undulations increases rapidly with each further deposited period. Compressive quantum well layers tend to smooth interfaces. However from the 5th well undulation begins to be visible at the well / barrier interface too.

Each layer exhibits lateral contrast modulation. The g=002 diffracted intensity is larger in the thickest part of the tensile layer than in the thinnest one. In the compressive layer the diffracted intensity is larger in the thinnest parts.

Two remarkable results should be pointed out : first the quasi-periodic modulation has a columnar structure throughout the whole multilayer with a lateral periodicity of about 50 nm. Secondly undulation of the well to barrier interface spreads out with opposite phase to the undulation of the barrier to well interface, as seen in Fig. 1.

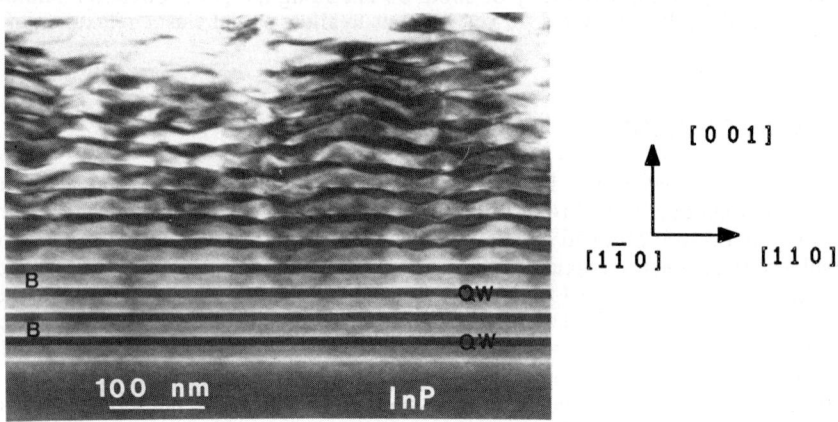

Fig. 1 : g=002 DF ($\bar{1}$10) cross-sectional micrograph showing the InP buffer layer , tensile barriers (B) and compressive quantum wells (QW). Interface undulations became visible after the 2nd barrier and the 5th or 6th well, with opposite phases.

3.2 Modulation anisotropy

Both cross-sectional and plan-view observations have demonstrated that the thickness modulation is strongly anisotropic. Using a double tilt goniometer allowing a +/- 45° rotation around the growth direction [001], a (100) cross-sectional sample was observed both in projection on ($\bar{1}$10) plane and in projection on (110) plane. Contrary to the first projection, the second one presents no modulation and increasingly blurred interfaces (Fig. 2). On the other hand, only modulations along the [110] direction are observed in plan views with g=200 and g=020 [Ponchet et al 1993] and g=220 (Fig. 3). The average period measured on plan views is of 48 nm and the coherence length in the [$\bar{1}$10] direction is greater than 200 nm.

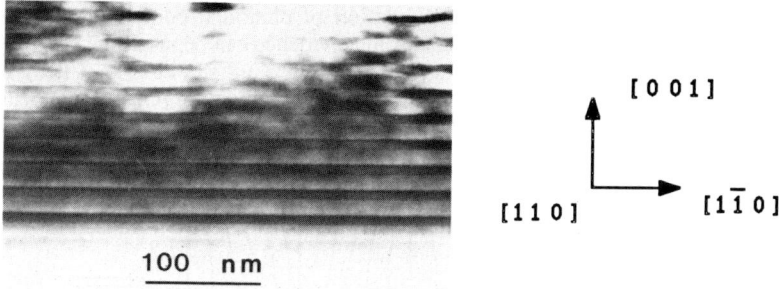

Fig. 2 : Projection in (110) plane. Thickness of blurred interfaces is of the same order of magnitude as the undulation amplitude seen in Fig. 1.

3.3 Displacive modulation

Observations made on ($\bar{1}$10) cross-section specimens with g=220 show systematically a columnar structure with a lateral periodicity well correlated to the thickness modulation described above. Moreover, while g=220 and g=440 micrographs (Fig. 3a) present periodic contrast consistent with the g=220 cross-sectional observation, the g=$\bar{2}$20 one shows no significant contrast (Fig. 3b). These results are consistent with a displacive modulation along the [110] direction but not along the [$\bar{1}$10] direction : (110) interplanar distances are modulated while ($\bar{1}$10) interplanar distances are not.

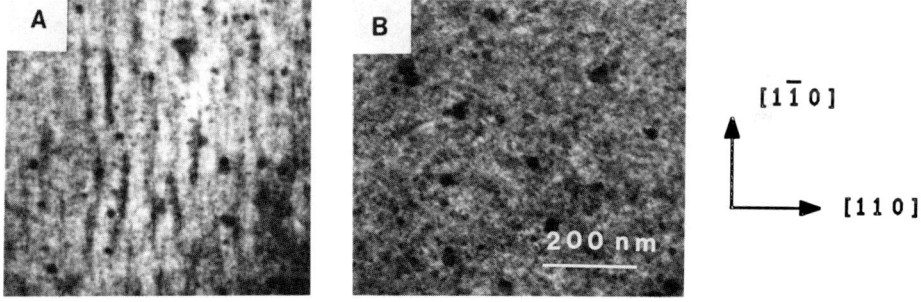

Fig. 3 : DF plan views with (a) g=220 (b) g=$\bar{2}$20. With g=220 the modulated contrast is periodic along the [110] direction, while with g=$\bar{2}$20 it completely disappears.

4. DISCUSSION

The morphology of modulation can be accounted for by assuming that the thickest zones are partially relaxed compared to the thinnest ones. Indeed, if the thickest zones of the tensile barrier have started to relax, the (110) plane spacing is smaller than that of the InP substrate, while in the thinnest zones the (110) plane spacing is larger, as suggested in Fig. 4. This is reversed in compressive layers whose relaxed parameter is larger than the InP one.

Phase opposition and amplification of modulation result from this mechanism. It should be pointed out that in this case, deformation is not tetragonal as expected, but orthorhombic. On the other hand, the contrast analysis of our g=002 cross-sectional micrographs is consistent with the existence of a lateral modulation of chemical composition, which could help relaxation of strained layers, but no absolute evidence can be produced yet.

Comparison with another strained system [Cullis et al 1992, Cheng et al 1992] and calculation of energy minimization due to lateral modulation [Glas 1987, Grilhe 1993] agree with this hypothesis. They are discussed elsewhere with details [Ponchet at al 1993].

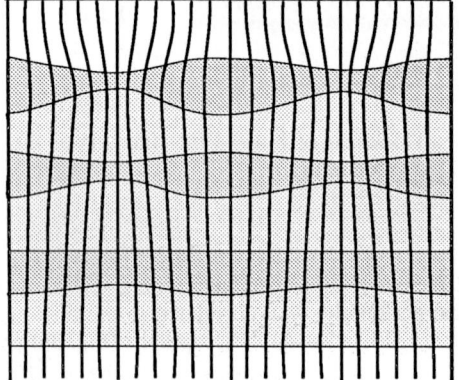

InP cap layer

compressive well

tensile barrier

compressive well

tensile barrier

compressive well

tensile barrier

InP buffer

Fig. 4 : Schematic description of (110) plane modulation and interfaces.

5. CONCLUSION

TEM experiments have proved that the optical and structural degradation of the alternating strained MQW is due to a lateral modulation of thickness, strain and probably chemical composition within the layers, which is periodic and strongly anisotropic : (110) plane spacings are modulated while ($1\overline{1}0$) ones are not.

Analysis of TEM experiments suggests that the amplification of modulation and the opposed behavior of tensile and compressive layers could be accounted for by a partial elastic relaxation of tensile layers. Comparison with another system agrees with this assumption. Based on this analysis, a schematic description of the distortion modulation is proposed.

REFERENCES

Cheng K Y, Hsieh K C and Baillargeon J N 1992, Appl. Phys. Lett. 60, 2892
Cullis A G, Robbins D J, Pidduck A J and Smith P W 1992, J. Cryst. Growth 123, 333
Emery J-Y, Starck C, Goldstein L, Ponchet A and Rocher A 1993, to be published in J. of Cryst. Growth
Glas F 1987, J. Appl. Phys. 62, 3201
Grilhé J 1992, to be published in Acta Met.
Ponchet A, Rocher A, Emery J-Y, Starck C and Goldstein L 1993, submitted to J. Appl. Phys.
Starck C, Emery J-Y,Simes R J, Goldstein L and Barrau J 1992, J. Cryst. Growth 120, 180

Inst. Phys. Conf. Ser. No 134: Section 7
Paper presented at Microsc. Semicond. Mater. Conf., Oxford, 5–8 April 1993

489

TEM studies of strain relaxation in $Ga_xIn_{1-x}As$ strained-layer multiple quantum well structures grown by MOCVD with InP and quaternary barriers

X Zhou, P Charsley and U Bangert[*]

Department of Physics, University of Surrey, Guildford, Surrey GU2 5XH, UK;
*Department of Pure and Applied Physics, UMIST, P O Box 88, Manchester M60 1QD, UK

ABSTRACT: Different behaviours of strain relaxation were observed in MOCVD grown $Ga_xIn_{1-x}As_yP_{1-y}$ lattice mismatched multiple quantum well structures in which the barrier material was either InP or quaternary lattice-matched to InP. Plastic relaxation occurred in samples with InP barriers but not in those with quaternary barriers. In the latter, however, both coarse and fine speckle composition modulation contrast was observed. It is suggested that the composition modulation may be a cause of strain relaxation.

1. INTRODUCTION

Strained-layer quantum well structures based on $Ga_xIn_{1-x}As_yP_{1-y}$ alloys on InP substrates have two major advantages compared with conventional lattice-matched ones, in terms of their application in semiconductor lasers used for optical-fibre telecommunications: first, a more flexible choice of wavelength range can be achieved, particularly for long-wavelength applications; second, the properties of the lasers are expected to be improved substantially due to the effect of strain on energy band structures (O'Reilly, 1989). To optimise device performance it is vital to understand and control the growth characteristics of strained epitaxial layers. It has been accepted that plastic relaxation will occur when the thickness of a lattice mismatched epilayer exceeds a critical value (t_c) (e. g. Matthews and Blakeslee, 1974). Also, there are extensive reports of composition modulation in lattice-matched ternary and quaternary III-V compound semiconductor heterostructures (mostly grown by LPE) or in strained ternary heterostructures (Mahajan et al 1984; Treacy et al 1985; Norman & Booker 1985; Glas 1993; Peiró et al 1992). However, little is known to date about these issues in the $Ga_xIn_{1-x}As_yP_{1-y}$ strained layer system. In this paper we present TEM observations of MOCVD grown strained layers with approximately constant values of compressive strain with a range of layer thicknesses. We compare two samples which differ only in the barrier material to determine the effects of the barrier on critical thickness. It is found that the microstructure for the specimen with quaternary barriers is significantly different from the specimen with InP barriers.

2. EXPERIMENTAL

Two multiple quantum well specimens used in this study were grown by the metallorganic chemical vapour deposition (MOCVD) technique on (001) InP substrates at 650 °C. The structures of the specimens are illustrated in Fig. 1. These two specimens had identical structures except that one contained InP barriers (referred to as Sample 1) and another quaternary barriers lattice-matched to InP (referred to as Sample 2). The $Ga_xIn_{1-x}As_yP_{1-y}$ wells had a fixed composition of x=0.64, y=0.95, which results in a compressive strain of 0.5%. The quaternary barrier material was $Ga_{0.63}In_{0.37}As_{0.79}P_{0.21}$. The barrier thickness was fixed at 200 nm and the wells had thicknesses 50 nm to 150 nm.

Cross-section TEM and 90°-wedge specimens were prepared by   standard techniques, cross-section samples initially thinned by Ar^+ ion beam at liquid nitrogen temperature and finally thinned by I^+ ion beam at room temperature. Plan-view TEM specimens were prepared by jet chemical thinning using Br-CH_3OH (1:10) solution. The TEM observations were carried out by using a JEOL 2000FX electron microscope operating at 200 keV.

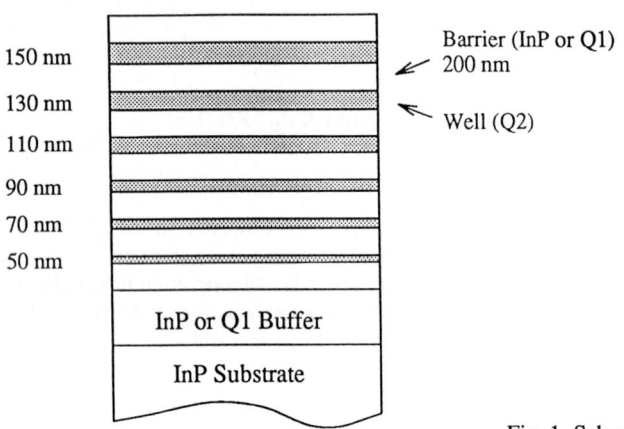

150 nm

130 nm

110 nm

90 nm

70 nm

50 nm

Barrier (InP or Q1)
200 nm

Well (Q2)

InP or Q1 Buffer

InP Substrate

Fig. 1 Schematic diagram of the structures of the specimens used in this study. Q1: $Ga_{0.64}In_{0.36}As_{0.95}P_{0.05}$, lattice-matched to InP; Q2: $Ga_{0.63}In_{0.37}As_{0.79}P_{0.21}$.

3. RESULTS

Cross-sectional or 90°-wedge TEM micrographs of Sample 1 and Sample 2 are shown in Fig. 2. To reveal dislocations Sample 1 was tilted away from the [110] pole. It can be seen from Fig 2(a) that dislocations occurred in Sample 1 when the well thickness exceeded 90 nm. The highest dislocation density was observed at the interface of the 110 nm well. Dislocations were at both interfaces between the top and bottom side of this well and the barriers. Dislocations can also be seen with a lower density at the interfaces between the 130 nm or the 150 nm well and the barriers. However, no dislocations were observed in the area where the well thicknesses were below 90 nm. Plan-view TEM observation (Fig. 3) showed that the dislocations were lying along two perpendicular <110> directions, as has been seen frequently in plastically relaxed strained layer systems.

In Sample 2, as shown in Fig. 2(b), no dislocations of any kind were observed, even for the thickest well. Instead, plan-view TEM revealed strong coarse contrast modulation, with a period of a few hundreds of nanometres (Fig. 4). The contrast consisted of two sets of bands which were roughly elongated along [100] and [010] directions. For two beam conditions with g=220 the two sets were in contrast, while when g=040 only the set of bands perpendicular to g remained visible. In addition, fine speckle contrast with a period of about 10 nanometres was also observed. These types of contrast in $Ga_xIn_{1-x}As_yP_{1-y}$ epilayers are commonly recognised as the results of composition modulation during the process of layer growth (Treacy et al 1985).

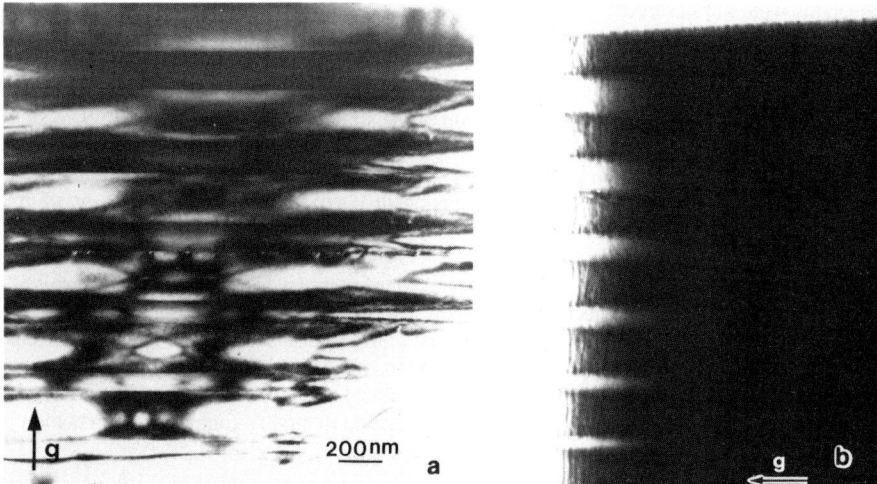

Fig. 2 (a) Bright-field cross-sectional image of Sample 1 (with InP as the barriers), $g=004$; (b) 90°-wedge image of Sample 2 (with $Ga_{0.63}In_{0.37}As_{0.79}P_{0.21}$ as the barriers), $g=040$.

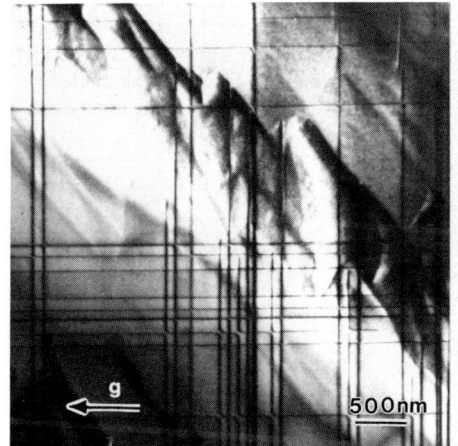

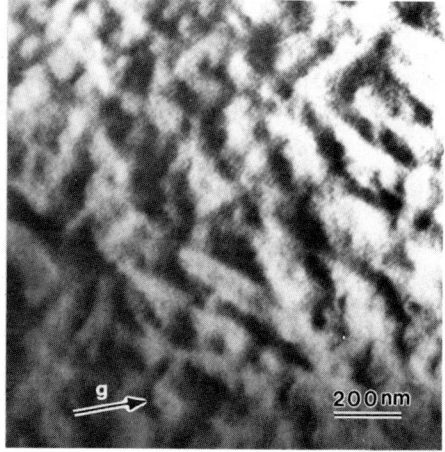

Fig. 3 Plan-view bright-field image of Sample 1, showing the misfit dislocation arrays. $g=220$

Fig. 4 Bright-field image of Sample 2, showing both coarse modulation and fine speckle contrast, $g=220$.

4 DISCUSSION

The above results indicate the significant difference in microstructure between two 0.5% strained multiple quantum well structures which consisted of different types of barrier but otherwise were identical in structure. The occurrence of plastic relaxation in Sample 1 indicates that either an individual well thickness or the overall well thicknesses had exceeded the critical layer thickness (t_c). It is not possible to determine the value of t_c in such a system because of the possible contribution of several layers to the strain at each interface, but it is likely that at least the thickness of the top well (150 nm) is greater than the critical thickness. In Sample 2, composition modulation was significant, but dislocations were not observed.

It is believed that the coarse contrast modulation pattern is due to the localized composition fluctuation (Treacy et al 1985). In Sample 2 no dislocations were observed for any layer thickness although the strain due to mismatch must be close to the values for Sample 1 before plastic relaxation occurs. There is no direct evidence for the magnitude of the strain retained in the layers in Sample 2. We suggest two possible mechanisms which might explain the microstructure in that specimen: (1) the observed composition modulation partially accommodates strain in the layers; (2) the fluctuations in composition inhibit the movement and multiplication of dislocations which, in the case of InP barriers, lead to significant plastic relaxation; this implies an increase in the critical thickness for plastic relaxation.

We are currently comparing the strain retained in single quaternary layers of different thickness, with quaternary barriers, using the wedge technique (Harvey et al 1993). By comparing such measurements with similar measurements using InP barriers it is hoped to distinguish between the two mechanisms. It should be noted that the two mechanisms are not strictly independent. Stress relaxation due to composition fluctuation would release the strain energy available for plastic relaxation and, conversely, if dislocation processes are inhibited the strain energy available to modify composition fluctuation is increased.

4. CONCLUSIONS

Different behaviours of strain relaxation were observed in two 0.5% compressively strained multiple quantum well specimens which had different types of barriers (InP or quaternary), but were otherwise identical in structure. Plastic relaxation occurred in the samples with InP barriers but not in those with quaternary barriers, in which, instead, composition modulation contrasts were observed. The results indicate that composition modulation in a strained layer system is probably a cause of strain relaxation and this will occur before the plastic relaxation, though the layer has exceeded the critical thickness.

ACKNOWLEDGEMENT

This work was carried out under the UK DTI/SERC LINK project. The authors would like to thank Mr. A. D. Smith of BNR Europe for providing the specimens.

REFERENCES

Glas F   1993   this proceeding volume
Harvey A J, Faux D A, Bangert U and Charsley P   1993   Phil. Mag. A $\underline{67}$ 433
Norman A G and Booker G R   1985   J. Appl. Phys. $\underline{57}$ 4715
Matthews J W and Blakeslee A D   1974   J. Cryst. Growth $\underline{27}$, 118
Mahajan S, Dutt B V, Temkin H, Cava R J & Bonner W A   1984   J. Cryst. Growth $\underline{68}$ 589
O'Reilly E P   1989   Semicond. Sci. Technol. $\underline{4}$ 121
Peiró F, Cornet A, Herms A, Morante J R, Clark S A and Williams R H   1992   Electron Microscopy
   vol. 2   Proc. EUREM 92, (Garanada, Spain: University of Garanada Publication) pp. 159-160
Treacy M, Gibson J M and Howie A   1985   Phil. Mag. A $\underline{51}$ 389

Inst. Phys. Conf. Ser. No 134: Section 7
Paper presented at Microsc. Semicond. Mater. Conf., Oxford, 5–8 April 1993

493

The effect of the imaging electron beam on InP/InGaAs MQW structures

PD Brown, EG Bithell[†], CJ Humphreys, PJ Skevington*, SD Perrin* and GJ Davies*

Department of Materials Science and Metallurgy, University of Cambridge, Pembroke Street, Cambridge, CB2 3QZ.
[†] Now at Department of Materials, University of Oxford, Parks Road, Oxford, OX1 3PH.
*B.T. Laboratories, Martlesham Heath, Ipswich, IP5 7RE.

ABSTRACT InP//InGaAs MQW structures irradiated with a 400keV electron beam damage preferentially at alternate interfaces. Which interface of a pair is damaged depends on sample <110> orientation; *i.e.* the InP/InGaAs interface becomes damaged for the [110] projection, while the InGaAs/InP interface becomes damaged for the orthogonal [1$\bar{1}$0] projection, and may be explained in terms of the combined effects of interfacial strain, electron beam channelling and crystal polarity.

1. INTRODUCTION

Considerable effort is being directed towards the extraction of quantitative chemical and compositional information from semiconductor heterostructure interfaces. The photoluminescence (PL) technique is unable to separate composition, thickness and abruptness effects, while X-ray rocking curves are unable to separate compositional grading from interfacial undulations (if they have the same averaged compositional profiles). Electron beam techniques of analysis facilitate investigation of these interfaces on the very fine scale and may usefully be used to complement X-ray and PL techniques. However, it is necessary to gain some appreciation of the extent of sample damage by the imaging electron beam. For an $In_{1-x}Ga_xAs$ epilayer on InP with x<<0.47 (strained in compression) interfacial dislocations are formed, while for x>>0.47 (in tensile strain) cracking within the epilayer occurs [Chew et al 1987]. When x=0.47, the lattice parameters of InGaAs and InP are matched, but strain is still present at these interfaces as shown by 002 dark field and HREM image contrast, for example. Simple stacking considerations indicate that there will be a monolayer of InGaP at the InGaAs/InP interfaces [Vandenberg et al 1988, Lyons et al 1990]. In addition, localised alloy clustering within the ternary alloy will be present, and interdiffusion possibly across both sublattices might be expected to occur. In this context, the effect of a 400keV electron beam on InP//InGaAs MQW structure is examined.

2. EXPERIMENTAL

Samples of MOVPE and CBE grown InP//InGaAs were cross-sectioned using conventional techniques, with iodine reactive ion sputtering being used for the final stage of foil preparation, and then immediately inserted into the microscope (JEOL 4000FX, column vacuum of $\approx 2x10^{-7}$torr). Orthogonal <110> sample projections were irradiated and examined. The MOVPE InP//InGaAs MQW structure examined was grown using trimethylindium, trimethlygallium:triethylphosphine adduct, arsine and phosphine at 650°C and a growth pressure of 824torr.

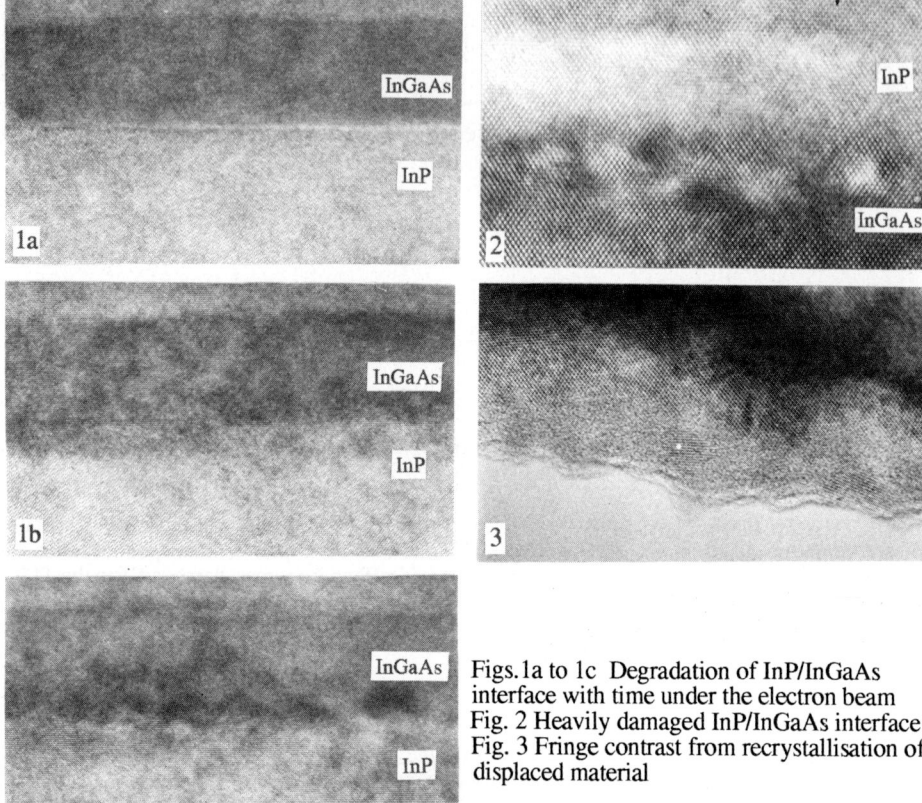

Figs.1a to 1c Degradation of InP/InGaAs interface with time under the electron beam
Fig. 2 Heavily damaged InP/InGaAs interface
Fig. 3 Fringe contrast from recrystallisation of displaced material

The CBE grown sample was grown using trimethylindium, triethylgallium, cracked arsine (As$_2$) and cracked phosphine (P$_2$) at 500°C and a growth pressure of 10^{-4}torr.

3. RESULTS AND DISCUSSION

Rapid preferential degradation of alternate InP//InGaAs interfaces was found to occur, on a time scale of a few minutes, during routine high resolution imaging (e.g. spot 2, 150pA/cm^2 screen beam current). This damage takes the form of a gradual roughening of the interface leading to the eventual formation of a broken band of bright contrast (Figs. 1a to 1c), and is probably related to interdiffusion, change in strain and material loss. The micrograph shown in Figure 1c is typical of a MOVPE grown InP//InGaAs sample which had been observed for a few minutes before recording the images. Alternate bright and dark layers can be seen delineating the interfaces, with alternate interfaces being damaged to different extents. Similar behaviour was found for the CBE grown InP//InGaAs sample. Both InP/InGaAs and InGaAs/InP interfaces initially showed a line of dark contrast. With time one interface became bright while the other faded and became washed out. This effect is attributed to damage induced by the imaging electron beam. This indicates the requirement to accurately align the microscope away from the region of interest and record images immediately if one is to be confident that they represent

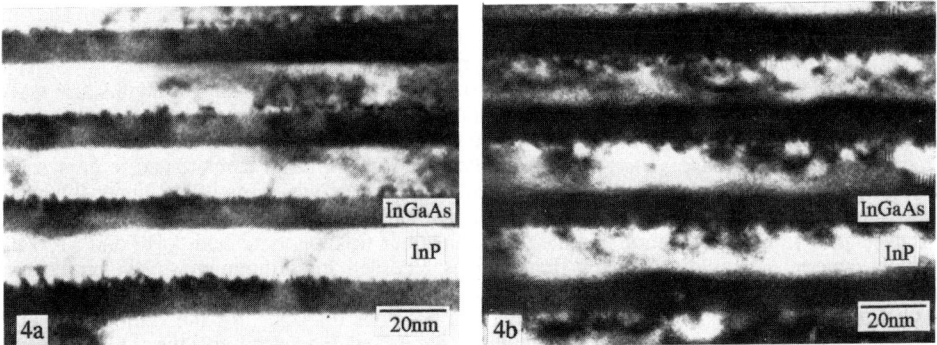

Fig 4   002-dark field images from a) [110] and [1$\bar{1}$0] sample projections showing damage at InP/InGaAs and InGaAs/InP interfaces respectively and triangular features extending into the InP.

material which is free of electron beam induced artefacts, *e.g.* Fig. 1a. (Alternatively, one may use a smaller spot size and condenser aperture to limit beam damage, but this may not be practicable depending on the TV detection system sensitivity.)   No interfacial defects are apparent and the interface appears to be abrupt to within 1 unit cell. InGaAs regions of material generally show slightly darker contrast in high resolution images and sharp interfacial contrast may generally be obtained when there is stronger relative 002 contribution to the InGaAs image[Mallard et al 1987].   The line of bright contrast along the InGaAs/InP interface of Fig. 1a is attributed to an 002 contrast effect associated with the inclination of the layer to the beam direction, rather than a compositional change [Williams et al 1991].

Fig. 2 shows a HREM image of heavily damaged CBE grown InP//InGaAs sample.   Some bending of fringes around the light contrast features and some 60° dislocations are apparent, and this is possibly explained by material collapsing around some vacancy platelets lying on {111} planes.   Displaced material is found to recrystallise and fringes obtained from such material showed spacings of 2.87±0.05Å (Fig. 3).   Candidates for this displaced material are therefore, In (d_{101}=2.72Å); In_2O_3 (d_{222}=2.92Å); Ga (d_{111}=2.95Å) and As (d_{102}=2.77Å). Given the volatility of the group V species, it would seem reasonable to suggest that the displaced material is some InGa alloy.   In addition, a <110> diffraction pattern recorded after damage showed the presence of faint polycrystalline rings, though whether or not this is related to i) damage at the interfaces, or ii) the crystallisation of some thin residual amorphous layer left on the sample from the milling process, or iii) is from material displaced by the imaging electron beam recrystallising on the sample surface, as indicated by the Moiré fringes in Figs 4a,b remains unclear.

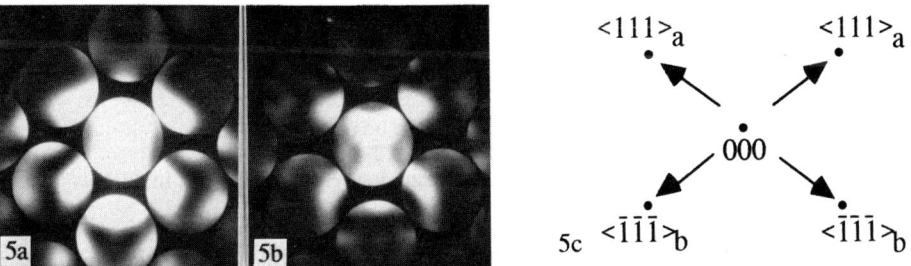

Fig. 5a,b) CBED patterns for <110> InP for two different sample thicknesses and c) indexed schematic after Burgess et al.

Preferential damage at InP//InGaAs interfaces must be related to interfacial strain and the localisation of weak bonds. Damage within electron beam irradiated semiconductors results from either sub-threshold or direct ionisation mechanisms depending on accelerating voltage and material ionicity. The effect of interfacial strain will be to modify bond strengths and make this region more susceptible to damage. Additionally, the localisation of point defects at interfaces (gettered by the strain field) will give rise to shared or dangling bonds which in turn facilitates ease of damage. In order to examine the extent of beam damage and to gain some insight into the nature of the mechanisms operative, samples were simply left under the beam for 15 to 20 minutes. Which InP//InGaAs interface is damaged by the electron beam is found to depend on sample orientation and the same behaviour was found for both CBE and MOVPE grown material. The two 002-dark field images recorded for orthogonal <110> projections (Figs. 4a, 4b) clearly demonstrate that the InP/InGaAs interface becomes damaged for the [110] projection while the InGaAs/InP interface becomes damaged for the orthogonal [1$\bar{1}$0] projection; the damage in crystallographic in nature as evidenced by the {111} faceted structures extending into the InP wells in each case, and this is consistent with material loss. The attack of different interfaces for orthogonal layer projections corresponds to the same sense of InP crystal polarity away from each of these interfaces. These observations suggest that the damage mechanism is associated with the combined influence of electron channelling and crystal polarity. The strain fields associated with the interfaces means that these regions of material become preferentially damaged, while crystal polarity, and hence sample orientation, determines which alternate interfaces are attacked. The large difference in the atomic scattering factors of In and P precludes the use of the microdiffraction technique to determine absolute sample polarity. However, this information may be obtained by use of CBED [Spellward and James 1991, Burgess et al 1993], and experimental patterns for <110> zone axes at two sample thicknesses are shown in Figs. 5a,5b and an indexed schematic is shown in Fig. 5c.

Simply irradiating the InP substrate, or single crystal InAs, with the electron beam (spot 1, 500pA/cm^2) for 30min caused the formation of an irregular hole around which an amorphous region extended for ≈10nm. Platelets showing fringe contrast again appeared on the edge of the sample foil. No crystallographic features were associated with this damaged region unlike the InP/InGaAs interfaces. However, a similar experiment on (Hg,Mn)Te gave rise to the formation of regular holes with triangular facets, the 001 edges of which showed reconstructed surfaces [Brown et al 1993]. It is considered that the plastic nature of this material led to recovery of the crystal structure following damage by the imaging electron beam and hence, some analogous behaviour may be present during damage at the InP/InGaAs interfaces.

This study demonstrates the need for caution when recording and interpreting high resolution images of InP/InGaAs MQWs since damage is easily introduced by the imaging electron beam. By gaining some appreciation of the nature and the extent of the damage it becomes possible to establish conditions such that more confidence can be given to the images obtained, particularly when trying to quantify measurements.

ACKNOWLEDGEMENT

This work was carried out under the LINK/ASM/MOMBE initiative.

REFERENCES

Brown P D, Kirkland A and Humphreys C J 1993, to be published
Burgess W G, Saunders M, Bird D and Humphreys C J 1993 Microbeam Analysis '93
Chew N G, Cullis A G, Bass S J, Taylot L L, Skolnick M S and Pitt A D, Inst. Phys. Conf. Ser. No. 87 (1987) 231
Lyons M H, Scott E G and Halliwell 1989 Inst. Phys. Conf. Ser. No. 100 473
Mallard R E, Waddington W G and Spurdens P C 1987 Inst. Phys. Conf. Ser No. 87 21
Spellward P and James D 1991 Inst. Phys. Conf. Ser. No. 119 375
Vandenberg J M, Panish M B, Temkin H and Hamm R A 1920 Appl. Phys. Lett 53 1920
Williams E, Bithell E and Stobbs 1991 Inst. Phys. Conf. Ser. No. 117 83

Inst. Phys. Conf. Ser. No 134: Section 7
Paper presented at Microsc. Semicond. Mater. Conf., Oxford, 5–8 April 1993

High resolution STEM Z-contrast imaging and XRD: two new approaches for the characterization of GaInP/GaAs heterostructures

H Lakner, B Bollig, P Volmich, [#] Q Liu, [#] F Scheffer, [#] A Lindner and [#] W Prost

Werkstoffe der Elektrotechnik, [#] Halbleitertechnik/Halbleitertechnologie, Sonderforschungsbereich 254, Universität Duisburg, D 47048 Duisburg, FR Germany

ABSTRACT: High-resolution Z-contrast imaging, performed in a field-emission STEM, is able to show both the exchange of As and P at interfaces and the ordering of GaInP-layers on $\{111\}$ planes directly. STEM bright-field images and, for the problem of As and P exchange, Z-contrast image simulations support these results. X-ray diffraction (XRD) was used for the first time to observe ordering in GaInP-layers. Samples with ordered structure show an additional $\{1/2, 1/2, 5/2\}$ reflection which is absent in disordered GaInP or in GaAs. This result is confirmed by transmission electron diffraction.

1. INTRODUCTION

The benefits of the $Ga_{0.51}In_{0.49}P$/GaAs material system (i.e. a large valence band discontinuity, significant decrease of oxidation compared to AlGaAs/GaAs, and a high doping capability without deep level effects) make it very promising for the fabrication of e.g. Heterojunction-Bipolar-Transistors and Heterostructure-Field-Effect-Transistors. But as well there exist some drawbacks. The abruptness of GaInP/GaAs interfaces can be deteriorated by an exchange of P and As across the interface depending on the growth temperature and the gas switching procedures used during MOVPE growth (Garcia et al 1991). Additionally, the growth conditions determine whether or not the GaInP-layers exhibit CuPt-type ordering (Baxter et al 1991) on $\{111\}$ planes which has an effect on the band gap energy of such layers. It is the intention of this paper to describe two new approaches in characterization techniques which allow one to investigate the effects mentioned above. High-resolution Z (atomic number) contrast imaging in the scanning transmission electron microscope (STEM) which is sensitive to the local chemical composition of specimens (Pennycook et al 1991, Lakner et al 1992) can be used for the analysis of interface quality and reveals an exchange of P and As across the interface directly. As well we will show that this technique is sensitive to ordering. Additionally, we will demonstrate that x-ray diffraction (XRD) allows the observation of ordering effects in GaInP. Transmission electron diffraction (TED) was used to check and to confirm the latter.

2. EXPERIMENTAL PROCEDURE

The STEM micrographs and diffraction patterns have been recorded in a field-emission STEM (VG Microscopes: HB 501) operated at 100 keV which is equipped with a high-resolution pole-piece (spherical aberration coefficient C_s = 1.3 mm). Z-contrast images are obtained by the detection of elastically and quasi-elastically scattered electrons using a high-angle annular dark field detector which collects all electrons with scattering angles from 58 to 180 mrad. Such high-angle scattering is very sensitive to the atomic number of the scattering atoms and is therefore sensitive to the chemical composition. Bright field images are recorded by means of an axial detector. The modulation of the angle of incidence of the electron probe relative to the specimen (angular scan) and synchronous detection of the transmitted intensity with the axial detector delivers TED patterns.

STEM Z-contrast image simulations are performed with simulation programs from the School of Applied and Engineering Physics (Prof. Silcox) at Cornell University which were developed by Earl J Kirkland (Kirkland et al 1987).

XRD rocking curves were obtained using a STOE STADI P triple-crystal diffractometer with a 2xGe monochromator in (220) setting and Cu-Kα_1 radiation (λ = 0.154 nm). {115} reflections were used to investigate ordering on the {111} planes in the column III sublattice. Ordered GaInP behaves like a GaP/InP superlattice in <115> directions with a period D which is approximately twice the period of d_{115} in GaAs. The corresponding angle of incidence for x-rays (see fig. 1) is equal to ω =4.94° and the corresponding Bragg-angle is equal to ϑ =20.733°.

A horizontal low pressure (20 mbar) MOVPE-reactor was used for the growth (growth rate: 1.4 μm/h) of the GaInP layers on exactly (001)-oriented s.i. GaAs substrates (substrate temperatures from 600° C to 730° C). Trimethylgallium and -indium, arsine and phosphine with a V/III ratio of 380 were used as precursors. Si doping was carried out with disilane diluted in hydrogen.

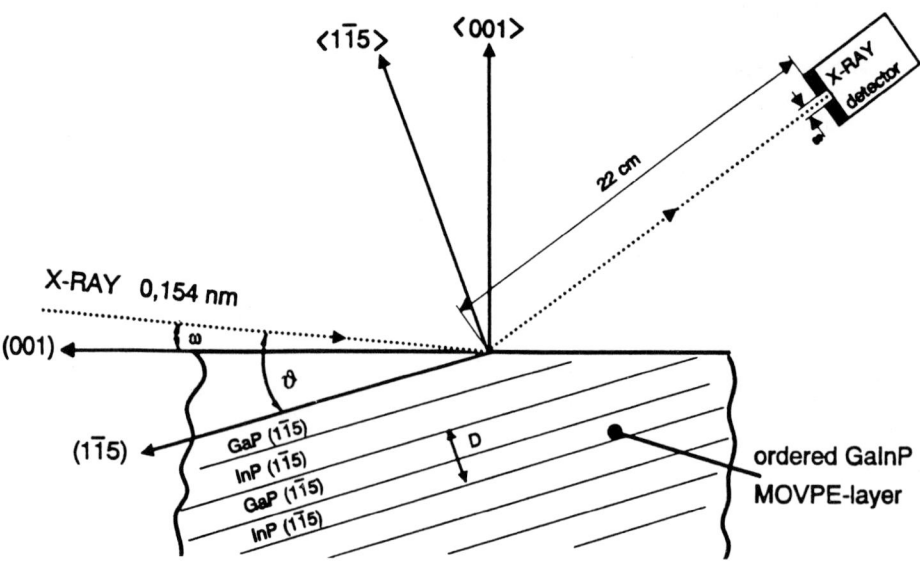

Fig. 1: XRD-set-up for the analysis of ordering in GaInP

3. RESULTS AND DISCUSSION

3.1 The GaAs/GaInP-interface

An STEM Z-contrast image of a GaAs/GaInP interface (grown at 650°C) which is exactly oriented along the <110> zone axis is shown in figure 2a as well as a line scan (which represents an average of 20 lines) across that interface. The exact position of the interface was determined by bright field imaging (compare fig. 3). The thickness of the specimen is decreasing from left to right as indicated by the shape of the line scan. Additionally, the line scan exhibits a relative minimum (maximum) close to the left (right) side of the interface which can not be attributed to thickness effects. However, a possible exchange of As ($Z = 33$) and P ($Z = 15$) across the interface as indicated in figure 2a would exactly explain the observed minimum and maximum because the image (and the line scan) intensity is strongly influenced by the atomic number Z of the scattering atoms. In order to support the interpretation of the observed contrast behaviour line scans across a perfect GaAs/GaInP-interface are simulated (see fig. 2b). The results show that no such minimum and maximum as observed in figure 2a can be expected for a perfect interface. The dashed curve represents the simulated intensity for the case that atomic columns are resolved. The simulated intensity in the GaInP-layer is slightly less compared to GaAs and this is in accordance with the Z-contrast model. Therefore we conclude that the observed contrast close to the interface in figure 2a is due to an exchange of As and P so that quaternary material with a thickness of a few nm exists at the interface. The theoretical consideration of an exchange of Ga and In would yield a contrast behaviour which is inverse to the observed one and therefore this assumption can be excluded.

3.2 Ordering in GaInP

Ordering in GaInP can be observed by STEM bright field imaging (which is similar to TEM bright field imaging) directly as demonstrated in figure 3a. The {111} planes in GaAs are resolved. Additionally, in GaInP lattice planes with twice the spacings of the {111} planes are visible indicating the presence of ordering. However, bright field imaging cannot answer the question which planes are e.g. In-rich and to which amount they are In-rich. STEM Z-contrast imaging can be used for a quantitative approach for the determination of ordering in GaInP. First examples for this are shown in figure 3b and 3c. In the GaAs-layer (see fig. 3b) the {111} planes are resolved but contrast is very weak. But now In-rich (Ga-rich) planes in GaInP can be identified directly because they appear bright (dark) in the Z-contrast image and their brightness increases with increasing In content. As well, boundaries where planes which are In-rich become Ga-rich can be recognized directly (see arrow in fig. 3c).

The results of the x-ray diffraction and the electron diffraction experiments are shown in figure 4. Ordered GaInP (grown at 650°C) exhibits a peak in the x-ray rocking curve (curve a) which was recorded according to figure 1. This peak is absent (curves b and c) for the case of GaAs and disordered GaInP (grown at 600°C). This is confirmed by TED-patterns where again ordered GaInP exhibits the typical additional ordering spots compared to the TED-patterns of GaAs and disordered GaInP where only the basic zinc-blende pattern for <110> orientation is visible. The peak in the rocking curve is caused by the {1/2,1/2,5/2} Bragg reflections of the GaP/InP ordered superlattice in GaInP. The broad FWHM of the

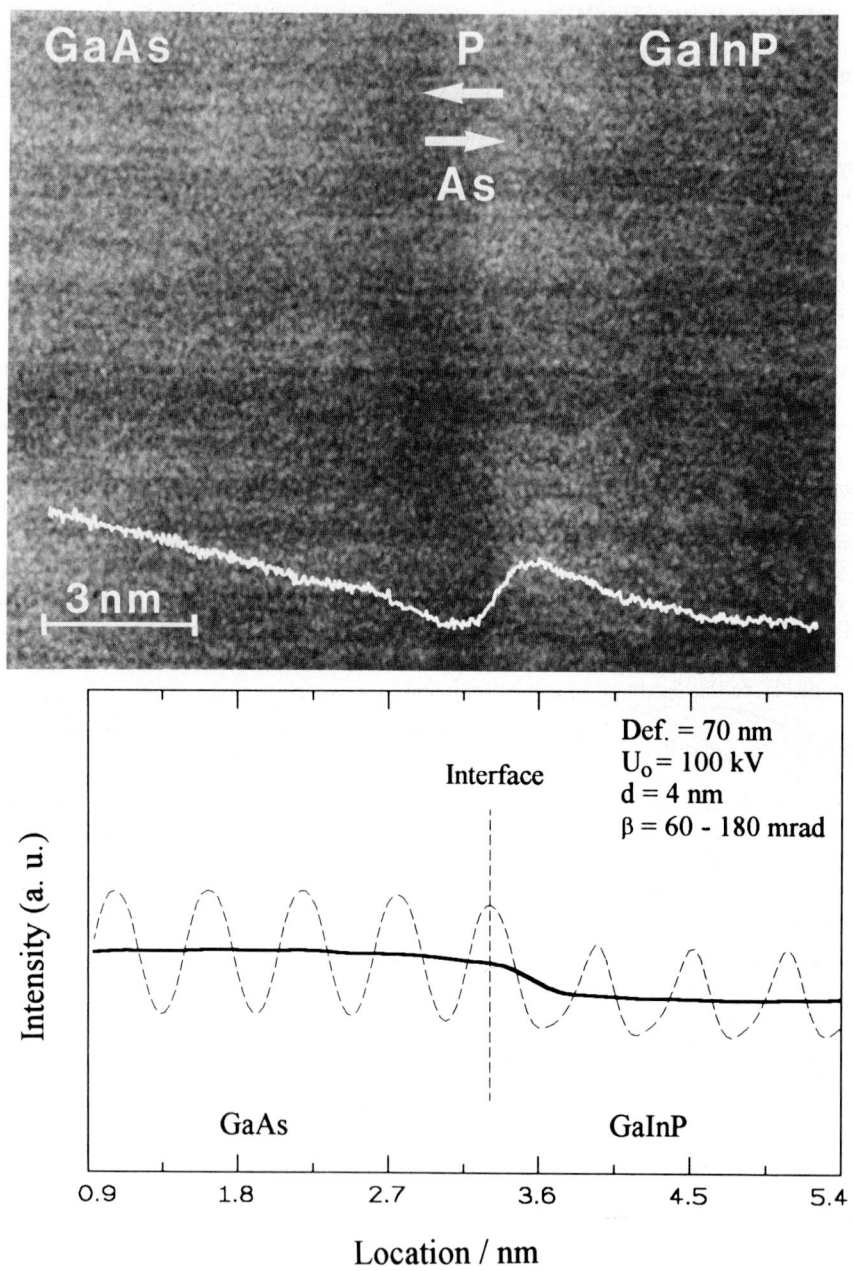

Fig. 2: a)   STEM Z-contrast image and line scan of an GaAs/GaInP-interface
       b)   Simulated STEM-Z-contrast line scan across a perfect GaAs/GaInP interface
              (Ordering in the GaInP-layer is not considered here)

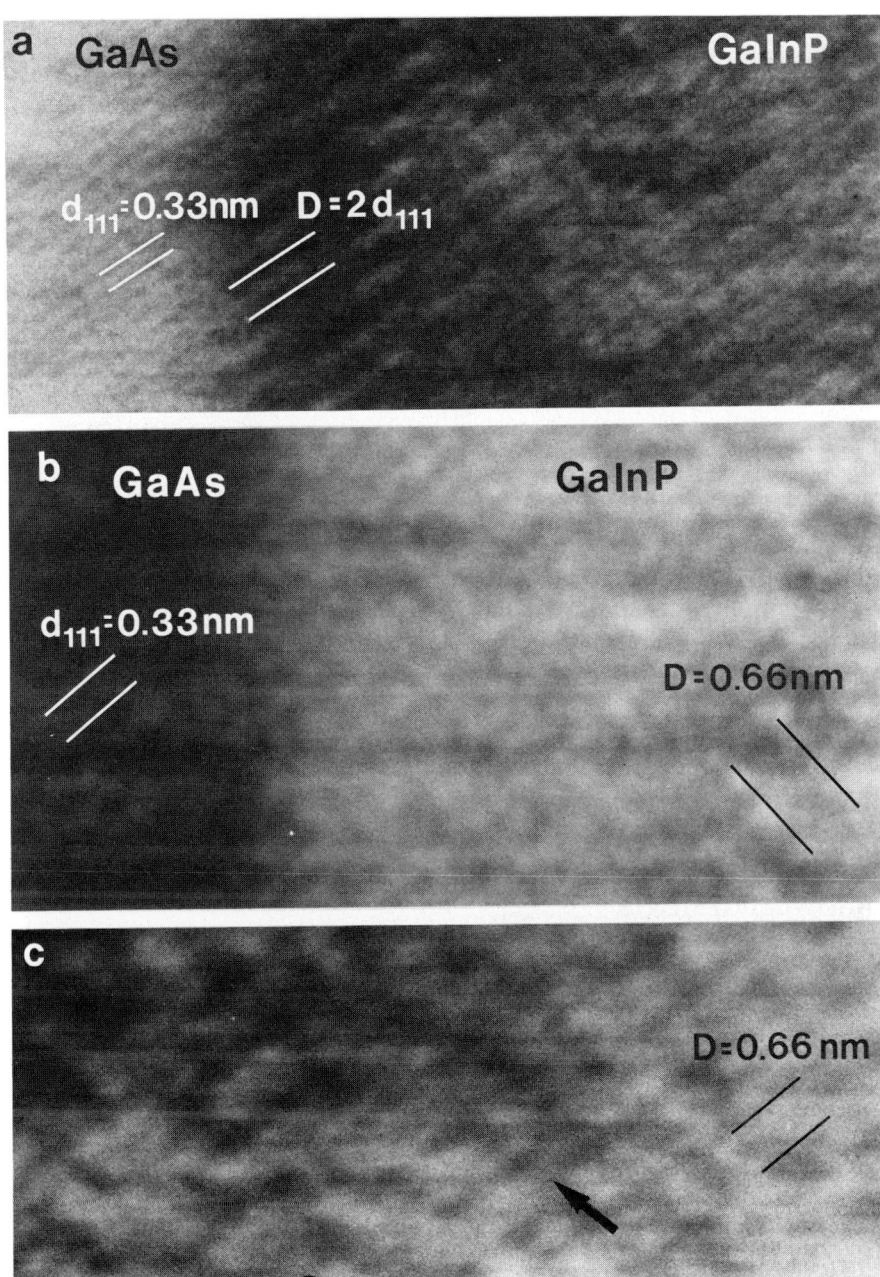

Fig. 3: a) Bright-field image and   b) Z-contrast image of the interface GaAs/GaInP
c) Z-contrast image of GaInP

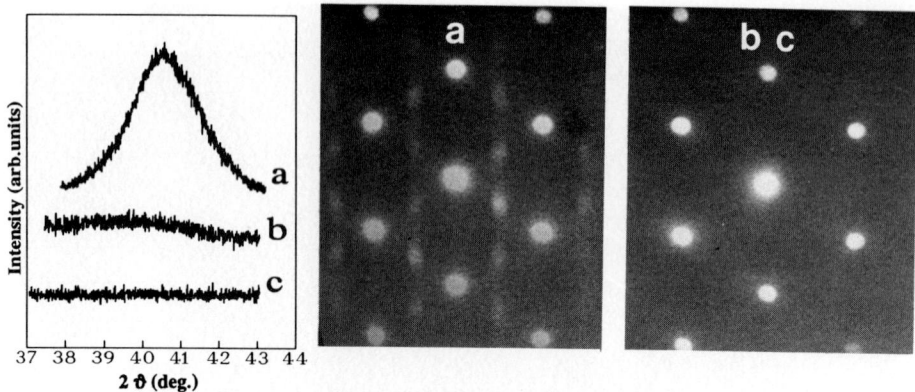

Fig. 4:   XRD-rocking curves and TED-patterns of:
            a) ordered GaInP   b) disordered GaInP   c) GaAs

peak can be explained by the existence of antiphase boundaries in ordered GaInP (Liu et al 1993). Ordering was observed by XRD only in $(\bar{1}, 1, 5)$ and $(1, \bar{1}, 5)$ planes. This result is consistent with TED observations reported previously (Bellon et al 1989, Suzuki et al 1988).

4. CONCLUSIONS

High-resolution STEM-Z-contrast imaging and XRD are suitable tools for the observation of interface and ordering characteristics in GaAs/GaInP heterostructures. With the described application of XRD it is possible to use a non-destructive and time-saving technique for the analysis of ordering in the future.

ACKNOWLEDGEMENTS

The authors like to thank Prof. Silcox (Cornell University) for the provision of the STEM image simulation software. This work was financially supported by the Deutsche Forschungsgemeinschaft and by the Ministerium für Wissenschaft und Forschung des Landes Nordrhein-Westfalen.

REFERENCES

Baxter C S, Stobbs W M and Wilkie J H 1991 Inst. Phys. Conf. Ser. 117 469
Bellon P, Chevalier J P, Augarde E, Andre J P, Martin G P 1989 J. Appl. Phys. 66 (6) 2388
Garcia J C, Maurel P, Bove P and Hirtz J P 1991 Jap. J. of Appl. Phys. 30 (6) 1186
Kirkland E J, Loane R F and Silcox J 1987 Ultramicroscopy 23 77
Lakner H, Maywald M, Balk L J, Kubalek E 1992 Surface and Interface Analysis 19 374
Liu Q, Lakner H, Scheffer F, Lindner A, Prost W 1993 J. Appl. Phys. 73 (6) 2770
Pennycook S J and Jesson D E 1991 Ultramicroscopy 37 14
Suzuki T, Gomyo A and Iijima S 1988 J. Cryst. Growth 93 396

Inst. Phys. Conf. Ser. No 134: Section 7
Paper presented at Microsc. Semicond. Mater. Conf., Oxford, 5–8 April 1993

503

Direct electron-beam fabrication of nanometre scale silicon columns

G S Chen, C B Boothroyd and C J Humphreys

Department of Materials Science and Metallurgy, University of Cambridge, Pembroke Street, Cambridge CB2 3QZ, UK

ABSTRACT: We have developed a novel way of making nanometre diameter silicon columns in a controlled manner. We have found that if silicon dioxide (SiO_2) is irradiated with a high intensity 100 keV electron beam of nanometre scale then a silicon column is formed as small as 2 nm in diameter. If the beam is moved in a straight line then a very thin plate of silicon is formed. These nanometre silicon structures are formed directly under electron irradiation with a dose of $\geq 3 \times 10^9$ Cm^{-2} and no resists or chemical development are required.

1. INTRODUCTION

Conventional electron beam lithography (CEBL) using organic resists has a resolution limit of 5 nm, the limit being a fundamental one set by the range of secondary electrons which ionise the resist and long range coulomb interactions of the primary electron beam with the resist, resulting in ionisation. For example, the smallest silicon features produced using CEBL are 5-7 nm wide (H. Ahmed, private communication). In order to produce structures of a size less than 5 nm a new type of lithographic process is necessary. At present, direct electron beam writing in inorganic materials, such as AlF_3 or Al_2O_3, is the most promising method for producing structures less than 5 nm across (Humphreys et al 1990). In this paper, we show that electron-induced damage in SiO_2 is very different from that induced in AlF_3 or Al_2O_3 and we demonstrate that this effect can be used to fabricate silicon structures significantly smaller than can be made using CEBL.

2. ELECTRON-INDUCED DAMAGE IN SiO_2

Self-supporting amorphous 15 nm thick SiO_2 films were prepared using electron beam evaporation. These were irradiated using an intense 100 keV electron beam of current density typically 1×10^9 Am^{-2} from the field emission gun in a VG HB501 scanning transmission electron microscope. The vacuum in the specimen chamber was $<1 \times 10^{-9}$ Torr. The irradiation process was studied by performing windowless energy dispersive X-ray spectroscopy (EDXS) and electron energy loss spectroscopy (EELS) of the electrons transmitted by the specimen during irradiation. Figure 1 gives X-ray intensities of silicon and oxygen during irradiation, acquired using a stationary probe which had been slightly defocussed in order to reduce the damage rate and hence to facilitate the EDXS study, showing that both silicon and oxygen are removed with oxygen being removed faster. Figure 2(a) shows a series of low loss spectra taken during the irradiation process, acquired using the same stationary defocussed probe. Initially the SiO_2 volume plasmon peak at 22.9 eV dominates the spectrum, but as the dose increases the position and width of the plasmon peak are shifted continuously towards those of silicon. (The volume plasmon peak of silicon is at 16.7 eV.) Figure 2(b) is a series of higher energy loss spectra, showing that the silicon L edge from silicon (99 eV) appears after a dose of only about 2×10^8 Cm^{-2} whereas the silicon L edge from SiO_2 requires a dose of 3×10^9 Cm^{-2} to decrease below detectable levels. The oxygen K edge was also monitored, and its intensity decreases to a level too low to be detected after a dose of 3×10^9 Cm^{-2}. The evidence

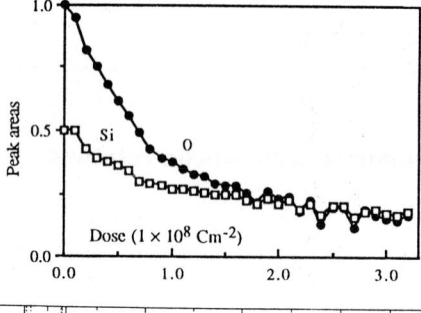

Fig. 1: Si and O EDXS peak areas as a function of dose, acquired using a slightly defocussed stationary probe, showing that both silicon and oxygen are removed with oxygen being removed faster.

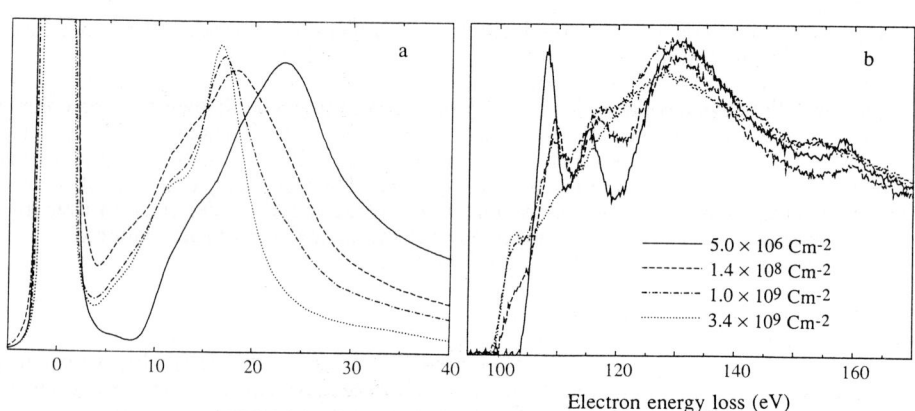

Fig. 2: A series of low loss (a) and Si L edge (b) spectra acquired using a slightly defocussed stationary probe during irradiation, demonstrating that the 15 nm thick SiO_2 is reduced to Si after a dose of 3×10^9 Cm^{-2} of 100 keV electrons. Only 4 of the 12 spectra recorded are shown.

of Figs. 2(a) and (b), together with the oxygen spectrum, suggests that the 15 nm thick SiO_2 is reduced to silicon after a dose of 3×10^9 Cm^{-2} of 100 keV electrons.

The behaviour of amorphous SiO_2 under electron irradiation described above is very different from the behaviour of AlF_3. Figure 3(a) is an aluminium plasmon loss energy-filtered image of an array of regions in AlF_3 which has been irradiated by a slightly defocussed probe of diameter ~10 nm, showing that aluminium is transported to the walls surrounding the irradiated areas. The formation of sheaths of metallic aluminium is believed to be due to the migration of aluminium ions under an intense electric field set up in the specimen as a result of irradiation (Humphreys et al 1985). Figure 3(b) is a silicon plasmon loss energy-filtered image of an array of regions in SiO_2 irradiated, as for the AlF_3 image in Fig. 3(a), by the same defocussed probe of diameter ~10 nm, showing that silicon stays in the centre of the irradiated areas rather than moving to the sides.

In Auger electron analysis, it is well known that if SiO_2 is irradiated by a low energy (typically 1 to 3 keV) electron beam then oxygen is lost from the surface layers (Schwidtal 1978; Chao et al 1987) by a Knotek and Feibelman electron-stimulated desorption mechanism (Knotek and Feibelman 1979; Knotek and Houston 1983). Silicon can also be lost by electron sputtering from the electron exit surface of the specimen (Bradley 1988; Bullough et al 1990). For our case of irradiation by an intense electron beam, windowless energy dispersive X-ray analysis shows that both oxygen and silicon are lost and oxygen is lost faster than silicon. It is suggested that oxygen is desorbed from the surface layers by a Knotek and Feibelman mechanism, and also displaced internally with the assistance of the same mechanism. As oxygen is removed or displaced, the Si-O-Si bonds of amorphous SiO_2 are replaced by Si-Si

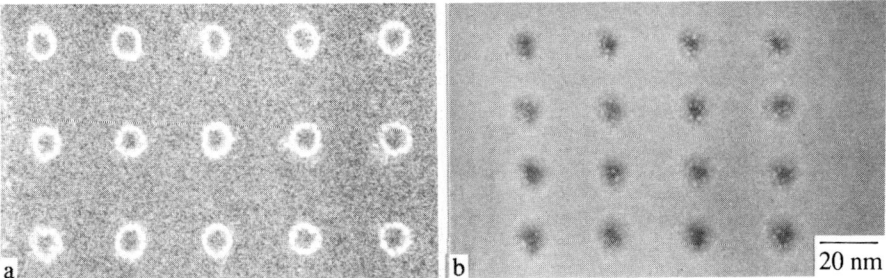

Fig. 3(a): Al plasmon loss energy-filtered image of regions in AlF$_3$ irradiated by a slightly defocussed probe of diameter ~10 nm for one second per area. (b): Si plasmon loss energy-filtered image of regions in SiO$_2$, each area irradiated, as for the AlF$_3$ image on the left, by a slightly defocussed probe of diameter ~10 nm, but for 30 seconds per area.

bonds. Thus particles of silicon can be built up in the centre of the irradiated area where the incident electron beam is most intense.

3. THE NOVEL FABRICATION METHOD

As an application of this effect we have considered the fabrication of nanometer-diameter columns and plates of silicon. It has recently been discovered that porous silicon can emit visible light (Canham 1990). This is an important discovery since it demonstrates that silicon has the potential to be an optoelectronic material. The mechanism of light emission from porous silicon is still not understood, but it may be due to quantum size effects (Takagi et al 1990; Liu et al 1992), and electron microscopy of light-emitting porous silicon reveals that it contains irregular columns of crystalline silicon typically 3 nm in diameter (Cullis and Canham 1991). The direct reduction of SiO$_2$ to silicon using electron irradiation reported in this paper enables silicon columns and plates, which can be as small as 2 nm, to be produced in a controlled manner, and this may allow us to investigate, in a controlled manner, light emission from silicon.

Figures 4(a)-(c) show an array of irradiated areas formed using a 4 second electron beam dwell time, equivalent to $5 \times 10^9 \, Cm^{-2}$, per irradiated spot. The full width at half maximum (FWHM) of the electron beam was ~2nm. Figure 4(a) is an energy-filtered image using electrons which have lost energies from 16 to 17 eV, corresponding to the silicon plasmon loss peak. Bright dots, ~2 nm in diameter, appear in the centre of each irradiated area. Figure 4(b) is an energy-filtered image using electrons which have lost energies from 22 to 24 eV, corresponding to the SiO$_2$ plasmon loss peak. Figure 4(c) is an annular dark field image using electrons of all energies. We identify the 2 nm bright dots in Fig. 4(a) as 2 nm diameter columns of silicon since they are present in this silicon plasmon energy-filtered image, and also in the annular dark-field and bright-field images, but absent from the SiO$_2$ plasmon loss image. Surrounding each bright 2 nm silicon column in Fig. 4(a) is a darker annular region which we identify as a thinned region of SiO$_{2-x}$.

The ability of this silicon column fabrication process to produce silicon plates less than 10 nm in width is illustrated in the silicon plasmon images of Fig. 5. The electron beam was rastered many times along a line in the SiO$_2$ specimen. The focussed beam initially creates a damaged area with a width of ~30 nm [Figs. 5(a) and (b)]. Presumably the width of this damaged area is mainly due to the 'tails' of the electron beam (the beam shape is non-Gaussian: it has a narrow central peak of 2 nm FWHM, and long extended tails). After rastering for 10 minutes an 8 nm wide silicon plate is formed, surrounded by SiO$_{2-x}$, whose width in this case is mainly due to specimen drift [Fig. 5(c)]. After prolonged irradiation the thinned SiO$_{2-x}$ has been completely cut through on one side of the plate and the plate has been pulled towards the

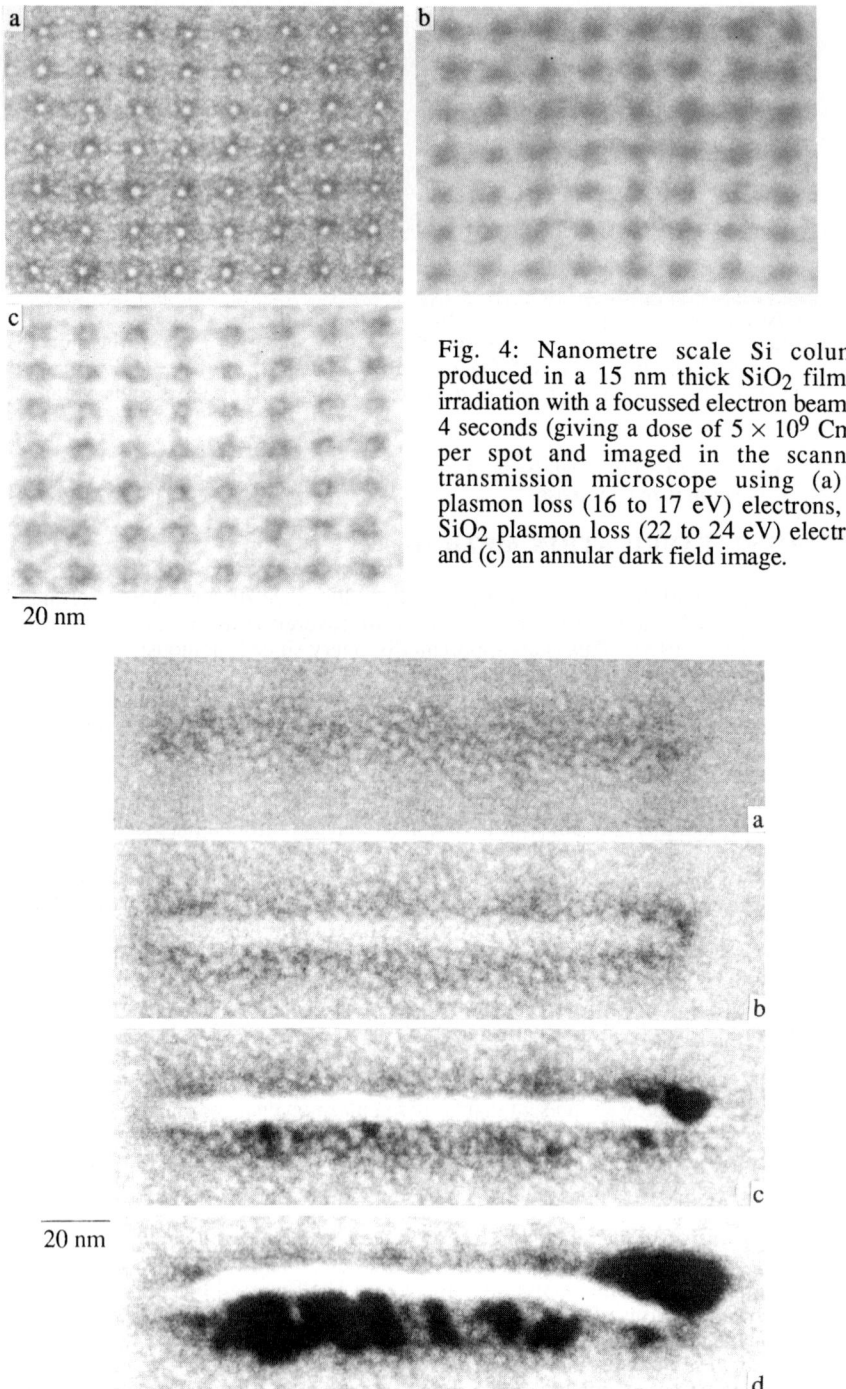

Fig. 4: Nanometre scale Si columns produced in a 15 nm thick SiO_2 film by irradiation with a focussed electron beam for 4 seconds (giving a dose of 5×10^9 Cm^{-2}) per spot and imaged in the scanning transmission microscope using (a) Si plasmon loss (16 to 17 eV) electrons, (b) SiO_2 plasmon loss (22 to 24 eV) electrons and (c) an annular dark field image.

20 nm

20 nm

Fig. 5: Nanometre scale Si plate fabricated by scanning the electron beam across a 15 nm thick SiO_2 film and imaged during fabrication using Si plasmon loss electrons after scanning times of (a) 1 min, (b) 5 mins, (c) 10 mins and (d) 12 mins.

side still attached [Fig. 5(d)]. Figure 6 shows an array of several longer and thinner 4-5 nm silicon plates.

To clarify the composition of these silicon columns and plates, the energy loss spectra from 'undamaged' SiO_2 powder, thermally oxidised SiO_2, SiO powder and ion-beam thinned silicon wafers were acquired using a PEELS in a conventional transmission electron microscope. Figure 7(a) is the low loss spectra of these materials, showing that the composition and width of the peak shift towards lower energies as the composition of oxygen decreases. Figure 7(b) is the higher loss spectra, showing that the threshold of silicon L edge in pure silicon occurs at 99 eV, and the threshold of silicon L edge in SiO_2 is chemically shifted to 106 eV, with a large peak at 108 eV. The spectra of SiO are intermediate between silicon and SiO_2. Figures 8(a) and (b) show the low loss and silicon L edge spectra from a silicon column, the SiO_2 matrix and the thinned SiO_{2-x} region surrounding each silicon column, demonstrating the composition of these regions. It is clear from these spectra (Figs. 7 and 8) that the bright dots in Fig. 4(a) are indeed images of 2 nm diameter columns of silicon and that the matrix in Fig. 4 is SiO_2 (the matrix appears fairly bright in Fig. 4(a) because the broad SiO_2 plasmon peak overlaps the narrower silicon plasmon peak). The dark annular region surrounding each bright silicon column in Fig. 4(a) is produced by the less intense tails of the electron beam, and it has spectra intermediate between silicon and SiO_2. We therefore identify this annular region as SiO_{2-x}.

Diffraction patterns from these 2 nm diameter silicon columns and plates do not show sharp rings or spots, but consist of diffuse rings. Dark-field images, recorded using an objective aperture on the diffuse rings, exhibit 'speckle' characteristic of amorphous materials. We therefore believe that these structures are composed of amorphous silicon.

4. CONCLUSIONS

We have found that 15 nm thick amorphous SiO_2 can be reduced to silicon by electron irradiation. This effect can be used to produce nanometre-diameter columns and nanometre-wide plates of silicon. We are currently attempting to crystallize these structures by annealing them, and plan to study their electrical and optical properties.

ACKNOWLEDGEMENTS

The authors thank C. J. D. Hetherington for useful discussion. This work is supported by the SERC. GSC is grateful to the SERC and University of Cambridge for financial support.

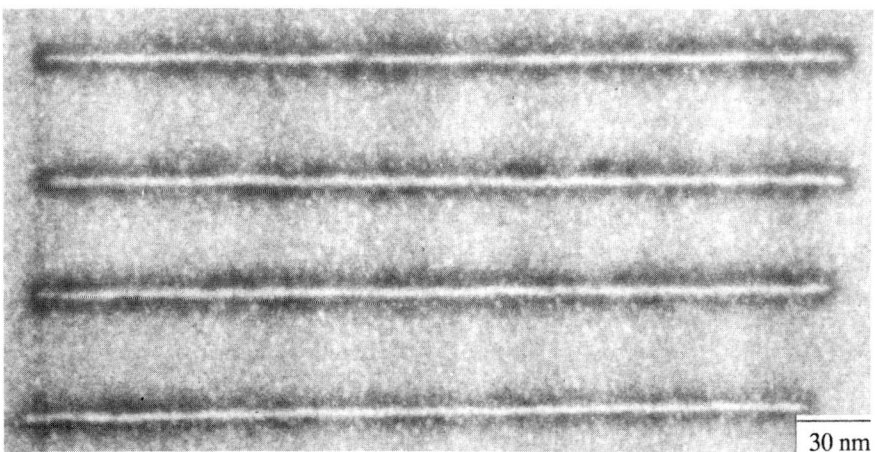

Fig. 6: 4-5 nm Si plates produced by rastering the electron beam for 4 minutes per plate.

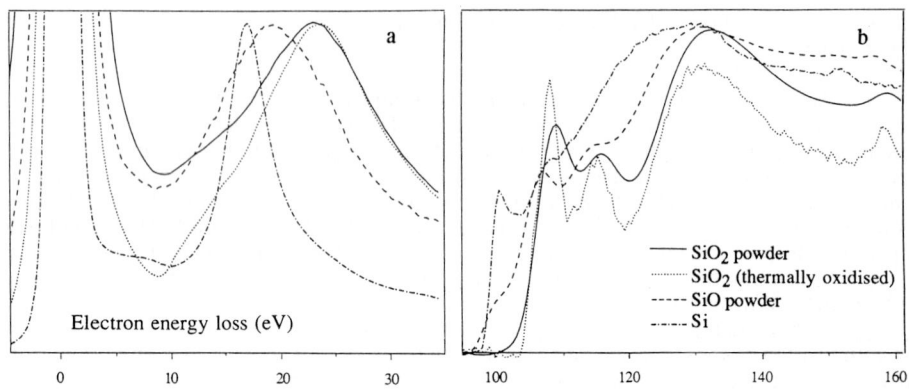

Fig. 7: Low loss (a) and Si L edge (b) spectra from 'undamaged' SiO₂ powder, thermally oxidised SiO₂, SiO powder and ion-beam thinned Si wafers.

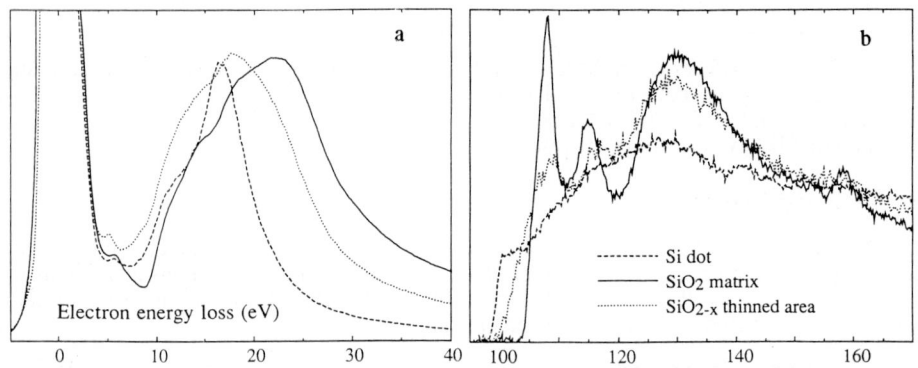

Fig. 8: Low loss (a) and Si L edge (b) spectra from a Si column, the SiO₂ matrix and the thinned SiO$_{2-x}$ region surrounding each Si column. These spectra prove that the columns shown in Figs. 4(a) are indeed Si (compare Figs. 2 and 7).

REFERENCES

Bradley C R 1988 Argonne National Laboratory Report No. ANL-88-48

Bullough T J, Humphreys C J and Devenish R W 1990 Mater. Res. Soc. Symp. Proc. 157 323

Canham L T 1990 Appl. Phys. Lett. 57 1046

Chao S S, Tyler J E, Tsu D V, Lucovsky G and Mantini M J 1987 J. Vac. Sci. Technol. A5 1283

Cullis A G and Canham L T 1991 Nature 353 335

Humphreys C J, Bullough T J, Devenish R W, Maher D M and Turner P S 1990 Scanning Microscopy Supplement 4 (Chicago, Scanning Microscopy International) p185

Humphreys C J, Salisbury I G, Berger S D, Timsit R S and Mochel M E 1985 Inst. Phys. Conf. Ser. 78 1

Knotek M L and Feibelman P J 1979 Surf. Sci. 90 78

Knotek M L and Houston J E 1983 J. Vac. Sci. Technol. B1 899

Liu H I, Maluf N I, Pease R F W, Biegelsen D K, Johnson N M and Ponce F A 1992 J. Vac. Sci. Technol. B10 2846

Schwidtal K 1978 Surf. Sci. 77 523

Takagi H, Ogawa H, Yamazaki Y, Ishizaki A and Nakagiri T 1990 Appl. Phys. Lett. 56 2379

Inst. Phys. Conf. Ser. No 134: Section 7
Paper presented at Microsc. Semicond. Mater. Conf., Oxford, 5–8 April 1993

Strain and defects in Si/Ge superlattices fabricated on a Si substrate

H Matsuhata, K Miki, K Sakamoto and T Sakamoto

Electrotechnical Laboratory, 1-1-4 Umezono, Tsukuba, Ibaraki, 305 Japan

ABSTRACT: Microstructures of MBE-grown Si_{12}/Ge_4 and Si_{18}/Ge_6 strained-layer superlattices (SLSs) were investigated using transmission electron microscopy and X-ray diffraction. Results showed that while the Ge/Si interfaces were flat and abrupt, the Si/Ge interfaces were neither flat nor distinct. In a plan-view image, dislocations in the Si substrate regarded as 60 degree type running along the orthogonal [110] directions were observed. V-shaped misfit islands were observed in the Si_{18}/Ge_6 SLS, but almost not in the Si_{12}/Ge_4 SLS. Shrinkage of the SLSs along the c axis from the ideal SLS due to their expanding along the a and b axes caused by the misfit dislocations was detected using X-ray diffraction.

1. INTRODUCTION

Because of the possibility of altering the optical properties of indirect-type semiconductors Si and Ge by the Brillouin-zone folding effect, fabrication of short-period Si/Ge strained-layer superlattices (SLSs) has been attracting attention (e.g., Pearsall *et al.*, 1989). However, there is disagreement between optical experimental results and theoretically calculated band structures (e.g., Noël *et al.*, 1990; Schmid *et al.*, 1991; Sturm *et al.*, 1991). The theoretical calculations usually assume an ideal structure for the SLSs, and also investigation of the details and actual microstructures in fabricated SLSs has not been satisfactorily carried out. In this study, the microstructures in Si_{12}/Ge_4 and Si_{18}/Ge_6 SLSs were investigated using TEM and X-ray diffraction. Since defects strongly influence properties of fabricated SLSs, our interest focused on the defects, strain, and unit cell size in the SLSs.

2. EXPERIMENTS

A Si wafer substrate with a (001) surface was first chemically polished using H_2O_2 and HCl, and then baked in an MBE chamber. A buffer layer of Si was grown to a thickness of a few hundred nanometers, then either Si_{12}/Ge_4 or Si_{18}/Ge_6 SLS was fabricated on the substrate at 673 K by monitoring the RHEED intensity oscillations. The lattice parameter along the c axis in the Si_{12}/Ge_4 SLS was then measured using the X-ray diffraction. Cross-sectional and plan-view specimens were prepared using the standard technique of ion milling, then examined using a TEM that was operated at 200kV or 150kV to avoid radiation damage.

3. RESULTS

3.1. Dislocations

Figure 1 shows the cross-sectional, high-resolution image of the SLS/Si-substrate interface. The upper part indicates the Si_{18}/Ge_6 SLS, and the lower part the Si substrate. The heterointerfaces of Ge/Si appear smooth and distinct, contrary to that of Si/Ge (Matsuhata *et*

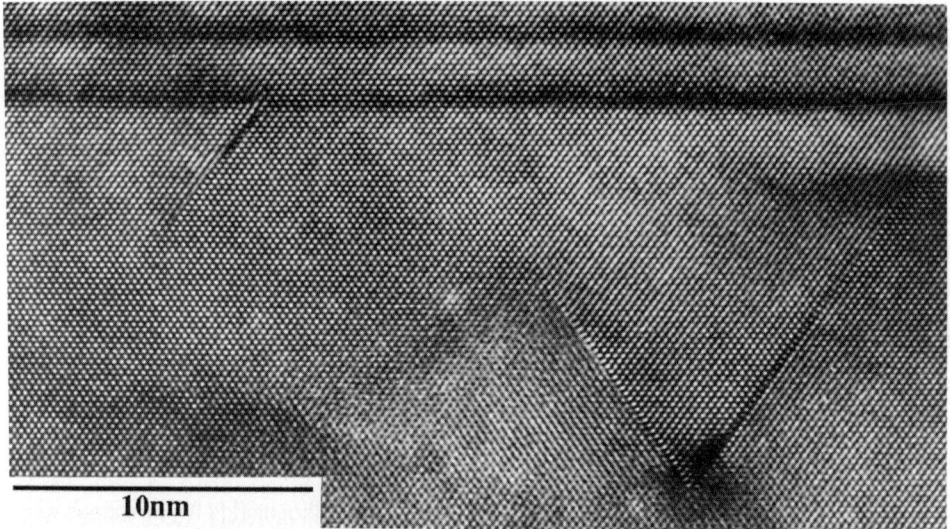

Fig. 1 High-resolution, cross-sectional view of heterointerfaces between the Si18/Ge6 SLS and the Si substrate.

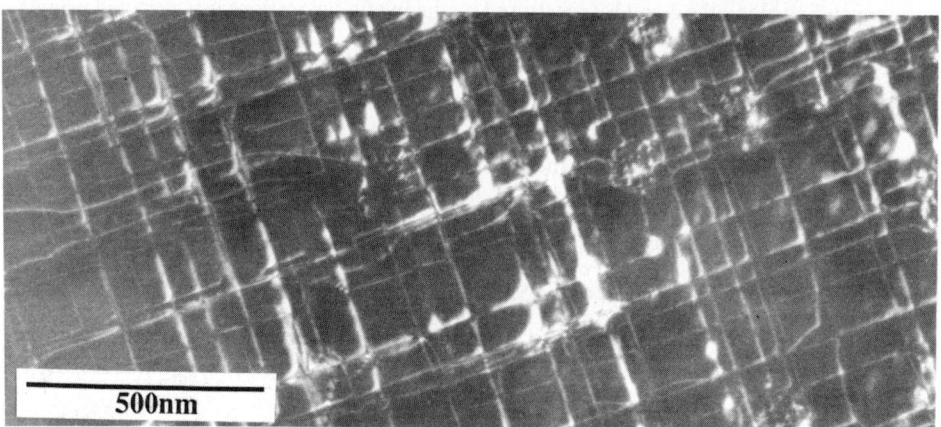

Fig. 2 Dark-field, weak-beam, plan-view image of Si18/Ge6 SLS.

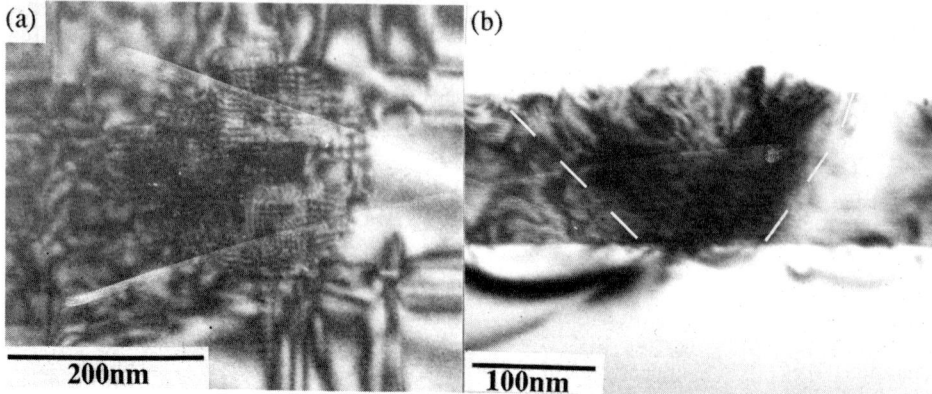

Fig. 3 Plan-view image (a) and cross-sectional image (b) of a misfit island.

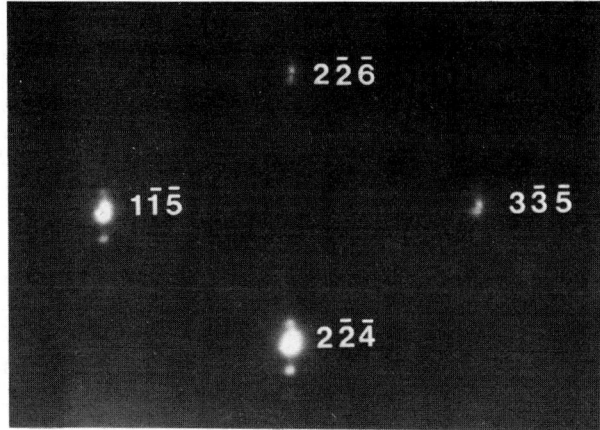

Fig. 4 Diffraction pattern of Si18/Ge6SLS and Si substrate taken at the [110] orientation.

al., 1991). The indistinctness of the Si/Ge interface is due to the segregation of Ge toward the surface of the growing Si layer, and the roughness is due to the Stranski-Krastanov growth mode of Ge layers on the Si surface. Dislocations are seen in the Si substrate near the heterointerface of SLS/Si substrate. The dislocation on the left is a commonly observed 60 degree dislocation, but this dislocation dissociates slightly into a 30 degree and a 90 degree partial dislocation with a small stacking fault. The dislocation on the right is a rarely observed Lomer-type dislocation.

Figure 2 shows a weak-beam, 040 dark-field, plan-view image of the Si18/Ge6 SLS taken at a 080 Bragg position. Dislocations are seen running along the two orthogonal [110] directions and forming a network. Since these dislocations also appear at a Bragg condition of 400, 040, 220, and $2\bar{2}0$, we conclude that they are the 60 degree type at this magnification level. As we have seen in Fig. 1, these 60 degree dislocations dissociate into two types of partial dislocations with small stacking faults. Various dislocation multiplication models have been discussed in Si-base thin films by Hull *et al.* (1991). If we consider the model proposed by Eaglesham *et al.* (1989), incoherent particles observed in the substrate at the interface between the Si buffer layer and original Si surface may be dislocation sources. Misfit dislocation densities obtained from the plan-view images are 1.7 dislocations per 100 nm along the [110] direction in the substrate of the Si12/Ge4 SLS, indicating a 0.67% elongation perpendicular to the c axis, and 2.2 dislocations in the Si18/Ge6 SLS, indicating a 0.84% elongation. These densities were estimated assuming that all misfit dislocations are the 60 degree type, however, some dislocations were actually the Lomer-type, and in the various cross-sectional observations we also noticed that these dislocations had Burgers vectors of various directions to reduce the misfit stress within local region of the heterointerface.

3.2. V-shaped Misfit Islands

In the plan-view observation, misfit islands with moirè fringes were observed in the Si18/Ge6 SLS, but not in the Si12/Ge4 SLS. A misfit island is seen in the bright-field, plan-view image of the Si18/Ge6 SLS shown in Fig. 3a. The cross-section of a misfit island is shown in Fig. 3b. The V-shaped island formed a hump on its top surface. A misfit dislocation is joined at the bottom of the island. These misfit islands are probably one source of the misfit dislocations in a Si substrate. These islands are considered to be due to the agglomerated growth of Ge on Si formed to reduce the elastic stress at the growing surface. The critical thickness of a grown layer for formation of these misfit islands is between 4 monolayers (ML) and 6ML of Ge in these SLSs.

3.3. Strain in the SLS

A selected area electron diffraction pattern taken at the [110] zone axis is shown in Fig. 4. The diffraction spots are from both the Si18/Ge6 SLS and the Si substrate. Fundamental diamond lattice spots and satellites due to the superstructure are both observed. Also seen is splitting of the fundamental spots due to the different lattice parameters for the substrate and the SLS. The 000 beam is located in the lower left of the figure. The inner fundamental spots with satellites are from the SLS and the outer spots are from the Si substrate. The deviation of the spot positions towards the 000 spot indicates that the lattice parameters are elongated in both directions, that is, perpendicular to and along the c axis. The elongation perpendicular to the c axis in the SLS is caused by the incoherent growth at the SLS/Si-substrate heterointerface. By X-ray diffraction one period of Si_{12}/Ge_4 SLS along the c axis is estimated to be 2.19635 nm. This value is shorter than that calculated from macroscopic elastic theory assuming coherent growth on the Si substrate.

4. DISCUSSION AND CONCLUSIONS

As we have seen in the electron micrographs, SLSs contain certain MBE growth defects: the Stranski-Krastanov mode growth of Ge on Si, surface segregation of Ge toward growing surface of Si, misfit islands in the SLS, misfit dislocations in the Si substrate, and the deviation of the unit cell size from the ideal coherent growth model. These growth defects and strain are considered to influence the optical and electrical properties of the SLSs.

According to theoretical calculations, band structures are very sensitive to strain in the SLS and achieving the direct gap critically depends on the lattice parameters along the a and b axes(e.g., Schmid *et al.*, 1991). The calculations indicate that coherent growth of the SLSs on the Si substrate does not have the conduction-band-minimum at the Γ point because of the compressive strain in the SLSs.   The lattices of our SLSs are elongated perpendicular to the c axis due to incoherent growth on the Si substrate; this is an advantage in changing the conduction-band-minimum position. However, the existence of dislocations at the interface disturbs the accuracy of the optical measurements. On the other hand, the Si/Ge SLS fabricated on a Ge substrate, which is interesting from the view point of optical properties because of the tensile stress in the SLS perpendicular to the c axis, showed planar defects in the SLS due to the stress (Wegscheider *et al.*, 1991). Either SLSs on a strain-symmetrised substrate or thin SLS films buried in Ge films, as fabricated by Pearsall *et al.* (1989), are preferable to those on a pure Si or Ge substrate. Asymmetric mixing at the heterointerface reduces the symmetry of the SLSs from *Pmma* to *Pmm2*. Though the transition probabilities may be very small, new optical transitions due to the absence of an inversion center are expected. The mixing at the heterointerface in the SLS due to segregation of Ge reduces the mini-band gap introduced by the superstructure, and also reduces dipole matrix elements. To improve the abruptness at the heterointerfaces, application of surfactant (Sakamoto *et al.*, 1992) is expected to be effective.

ACKNOWLEDGMENT

A part of the high-resolution electron microscopy was carried out at Kyushu University and at JEOL. Prof. Tomokiyo, Mr. Manabe and Dr. Ibe are gratefully acknowledged for their kindness.

REFERENCES

Eaglesham D J, Kvam E P, Maher D M, Humphreys C J and Bean J C, 1989, Philos Mag. A59, 1059.
Hull R, Bean J C, Bahnck D, Bonar J M and Peticolas L J, 1991, IOP series No. 117 p497.
Matsuhata H, Miki K, Unoki S, Sakamoto K and Sakamoto T, 1991, IOP series No. 119 p71.
Noël J P, Rowell N L, Houghton D C and Perovic D D, 1990, Appl. Phys. Lett. 57 1037.
Pearsall T P, Vandenberg J M, Hull R and Bonar J M, 1989, Phy. Rev. Lett. 63 2104.
Sakamoto K, Miki K, Sakamoto T, Yamaguchi H, Oyanagi H, Matsuhata H and Kyoya K, (1992) Thin Solid Films, 222 112.
Schmid U, Christensen N E, Alouani M and Cardona M 1991, Phys. Rev. B43 14597.
Sturm J C, Manoharan H, Lenchyshyn L C, Thewalt M L W, Rowell N L, Noël J P and Perovic D D, 1991, Phys. Rev. Lett. 66 1362.
Wegscheider W, Eberl K and Abstreiter G, 1991, IOP series No 117 p21.

Inst. Phys. Conf. Ser. No 134: Section 7

513

Paper presented at Microsc. Semicond. Mater. Conf., Oxford, 5–8 April 1993

Profiling of Ge_xSi_{1-x}/Si strained-layer superlattices by large-angle convergent beam electron diffraction and electron holography

X F Duan[1,2], N Grigorieff[2], D Cherns[2], J W Steeds[2] and C Sheng[3]

1 Beijing Laboratory of Electron Microscopy, Chinese Academy of Sciences, P.O. Box 2724, Beijing 100080, P.R. China
2 H.H. Wills Physics Laboratory, University of Bristol, Tyndall Avenue, Bristol BS8 1TL, U.K.
3 Surface Physics Laboratory, Fudan University, Shanghai 200433, P.R. China

ABSTRACT: Large-angle convergent-beam electron diffraction (LACBED) is suitable to reveal information on local strain and misfit stress relaxation of Ge_xSi_{1-x}/Si strained-layer superlattices (SLS) because the diffraction lines are sensitively dependent on small changes of the spacing between lattice planes. We shall demonstrate that the line shifts in a cross-sectional specimen caused by the effects of misfit strain and stress relaxation can be separated. An electron hologram taken from a Ge_xSi_{1-x}/Si SLS is also presented as a preliminary result.

1. INTRODUCTION

The band structure of Ge_xSi_{1-x}/Si SLSs is known to be strongly dependent on the state of strain within the epilayers. LACBED has been established as one of the most widely used techniques in materials science. The diffraction lines in the LACBED patterns can be used to investigate the strain in a lattice misfit system because their positions are sensitively dependent on a small change in lattice parameter. Since the probe is slightly defocussed, equivalent to imaging the specimen with a convergent spherical wave, a shadow image of the specimen is superimposed on a LACBED pattern in the back focal plane of the microscope. (Tanaka, Satio, Ueno and Harada, 1980, Fung, 1984, Duan, 1992a) It is important that the probe size, which defines the spatial resolution in LACBED should be as small as possible. Several workers have attempted to measure SLS misfit strains directly. In practice, the possibility of elastic relaxation of strain in the very thin samples used for cross-sectional transmission electron microscopy (XTEM) must be taken into account. In order to interpret the diffraction contrast due to the stress relaxation, some theoretical treatments of free surface stress relaxation at a misfitting interface have been carried out. (Gibson and Treacy, 1984, Perovic and Weatherly, 1991) Recently it has been shown that the relaxation in the cross-sectional specimen of Ge_xSi_{1-x}/Si SLS can be directly measured by the shift of the diffraction lines in LACBED. In this paper we shall demonstrate that the shifts of diffraction lines caused by strain and stress relaxation in XTEM specimen can be separated. The technique can be used to profile strain and misfit stress relaxation in the SLS. With progress in field emission gun transmission electron microscopy, there is interest in using the amplitude and phase information recorded in electron holograms. The composition and elastic strain in Ge_xSi_{1-x}/Si SLS can introduce phase shift which can be evaluated by electron holography. We present a preliminary electron hologram taken from Ge_xSi_{1-x}/Si SLS.

2. EXPERIMENTS

$Ge_{0.2}Si_{0.8}$(10nm)/Si(40nm) SLS was grown on (001) Si by molecular beam epitaxy (MBE). Conventional transmission electron microscopy (CTEM) observation showed that the structure was pseudomorphic with few dislocations at the superlattice-substrate interface or in the superlattice itself. The superlattice-substrate interface and the interfaces between the GeSi and Si layers appeared even and sharp. LACBED was carried out in Philips EM420 and CM12 TEMs at 100kV and a Hitachi HF-2000/FEG TEM at 200kV at room temperature. The HF-2000/FEG electron microscope can give a probe less than 1nm. The spatial resolution of the shadow image obtained is better than 2nm. The electron holography was carried out in the HF-2000/FEG TEM with a biprism between the objective lens and the intermediate lens.

3. RESULTS AND DISCUSSIONS

If a layer has been coherently grown on the substrate, i.e., without misfit dislocations, the strain in the growth direction can be given as $e^{\perp} = -f(1+v)/(1-v)$, where f is the misfit and v is Poisson's ratio. For a [110] XTEM specimen, this strain will be relaxed during preparation. The relaxation along the [110] sample normal is dependent on the specimen thickness, while the strain along the [$\bar{1}$10] interfacial direction in the plane of the sample remains unrelaxed. Thus the inclination of the lattice planes in the strained-layer to the interfacial plane will depend on both the unrelaxed strain and the relaxation σ. This makes the shift of the diffraction lines complex. If we tilt the specimen to a special zone axis, the problem may be simplified. The rotation of the reflection plane due to strain in the (001) plane and the partial relaxation along [110] was discussed by Cherns et al. (1991) and Duan (1992b). In the case of the [110] zone axis LACBED, the relaxation σ in [110] does not affect the zero order diffraction lines because σ is parallel to the reflection planes. If the sample is tilted about the [001] axis, e.g. to the [210] zone axis, the relaxation can be detected since it is no longer parallel to the reflection planes which are inclined to the interfacial plane. To investigate σ experimentally, we choose a reflection plane perpendicular to the interfacial plane, for example ($\bar{4}$80) plane. In this case, the diffraction line $\bar{4}$80 in GeSi layer is shifted from that in the substrate. (Duan, 1992b)

Figure 1a shows a LACBED pattern of the Ge_xSi_{1-x}/Si SLS tilting away from the [110] zone axis a few degree about the [$\bar{1}$10] axis. The shadow image of the SLS is clear. The diffraction lines, e.g. $\bar{2}2\bar{2}$ and $\bar{3}3\bar{3}$, near the interface and the buffer layer between the substrate and the SLS are blurred as discussed in our previous works. It is evident that the $\bar{2}2\bar{2}$ and $3\bar{3}3$ lines are shifted in the GeSi layers from that in the Si layers by the rotation of the ($\bar{1}1\bar{1}$) reflection plane (Fig.1b). The shifts can be used to measure the unrelaxed strains in (001) plane in each GeSi layer. As expected, $\pm$220, $\pm$440 and $\pm$880 lines are not changed near the interface and the buffer layer between the SLS and the substrate. It is interesting that the thickness fringes of $\pm\bar{2}$20 contours are shifted in the GeSi layers. This is a composition effect since the extinction distance in GeSi layers is quite different from that in the Si layers.

Figure 2 shows the [210] LACBED patterns from a thinner area of the same specimen. The diffraction lines near the edge of the specimen are zigzagged very much. This is caused by the relaxation. As mentioned before, the diffraction lines $\pm\bar{4}$80 are very useful to measure the relaxation in [110] without the effect of the strain along [001] because the ($\bar{4}$80) plane is parallel to [001]. Since the shift of the $\bar{4}$80 line is proportional to the relaxation σ, the curve can be used to profile the relaxation in each Si layer. The 242 lines are also very much curved. The shift of the 242 diffraction lines has two components: one corresponding to the relaxation along [110], the other to the strain along [001]. Since the latter is proportional to the tilt angle from the [210] zone axis, it is a small term near the 210 pole and then can be neglected for first approximation. Because of the presence of the thickness fringes even in very thin area, the 242 lines can give much information on the strain relaxation vs the specimen thickness.

The small probe size available on the HF-2000/FEG TEM allows us to obtain high spatial resolution in LACBED patterns since, in the limit of geometrical optics, the resolution is given by the minimum probe size. (Vincent,1989) In order to achieve high spatial resolution we need to have a small specimen defocus to increase the spatial magnification of the shadow image. This required the use of a 1µm selected area aperture specially fabricated by etching a small hole in a thin metal foil. (Vincent private communication) By minimizing the spot size to less than 1nm we can obtain results such as those illustrated in Fig.3. This shows two LACBED patterns taken near the 210 pole showing 242 and 004 contours. The spatial resolution is better than 2nm. Fig. 3a shows that the diffraction lines are evidently curved in the Si and GeSi layers with layer thickness = 40nm and 10 nm respectively. Fig. 3b shows that the presence of an interfacial dislocation splits the diffraction lines into segments in agreement with previous observations (Cherns and Preston, 1986).

Off-axis electron holography was performed in the HF-2000/FEG TEM equipped with both a cold field emission gun and a electron biprism. When a positive potential is applied to the central filament of the electron biprism located below the objective lens, the image of object and a reference beam can be made to overlap each other to form an interference pattern. Fig. 4 shows an electron hologram taken from the edge of the Si layer in the XTEM specimen of the Ge_xSi_{1-x}/Si SLS. An arrow-pair indicates the edge of the specimen. In region A the hologram fringes are from free-space and in region B the lattice image is crossed by the interference fringes from which local phase and amplitude information can be extracted. (Lichte, Volkl and Scheerschmit, 1992)

ACKNOWLEDGEMENT

X. F. Duan was supported by the exchange program between the Chinese Academy of Science and the Royal Society, U.K. and by the National Natural Science Foundation of China

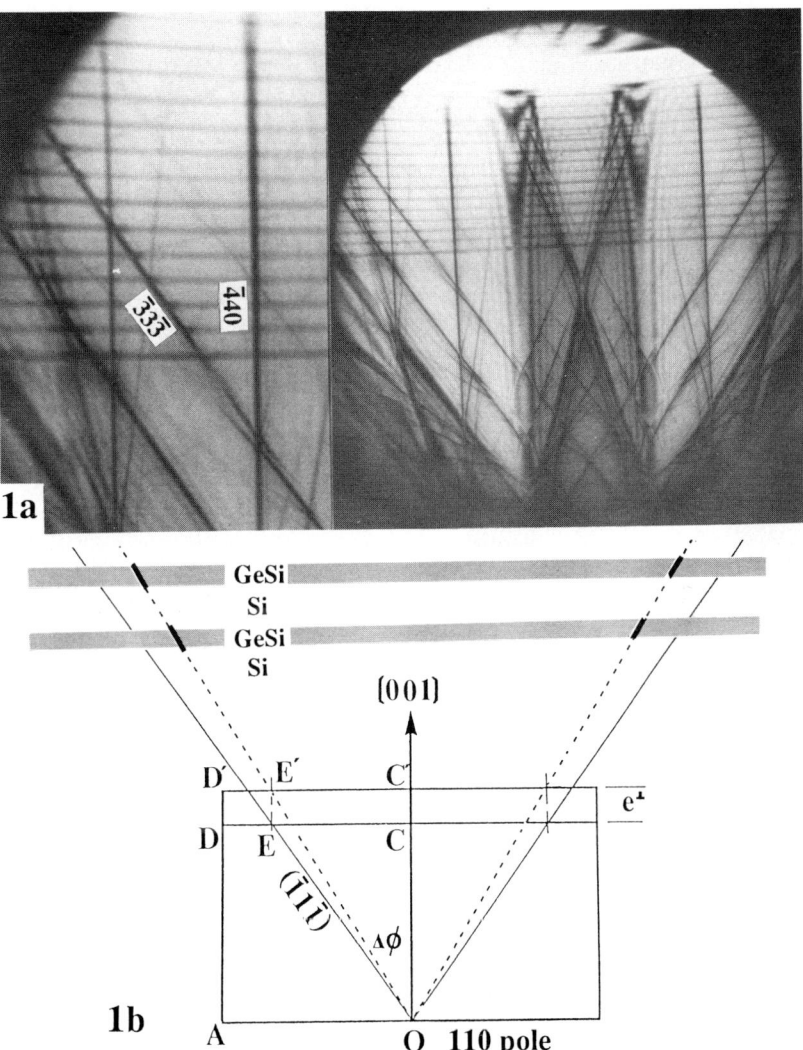

Fig.1 a [110] LACBED pattern of the Ge_xSiB_{1-x}/Si SLS showing the shift of $\bar{3}3\bar{3}$ lines in GeSi layers by the strain $e^{\perp}$ along [001].

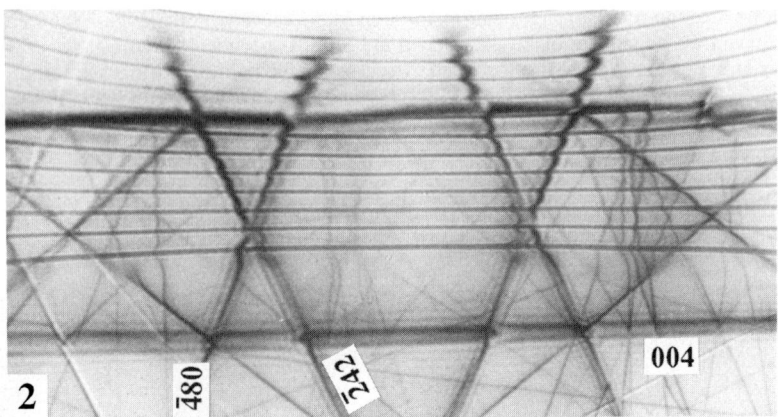

Fig.2 a [210] LACBED pattern from a thinner area of the same specimen.

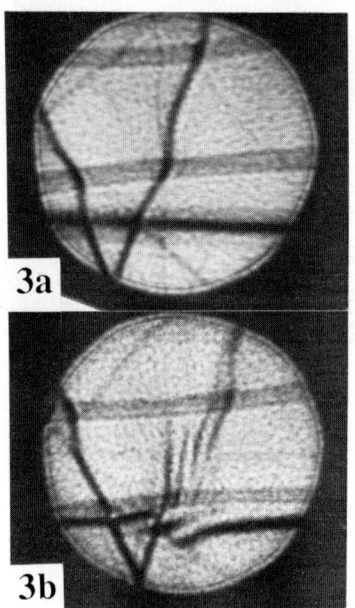

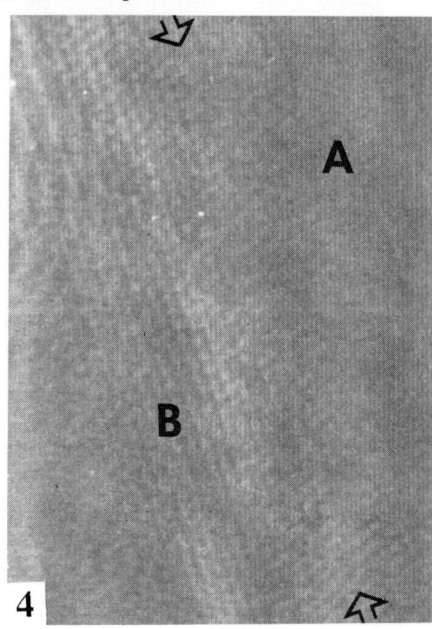

Fig.3 LACBED patterns taken using 1µm SA aperture.   Fig.4 an electron hologram from Si layer.

REFERENCES

Cherns D., Touaitia R., Preston A. R., Rossouw C. J. and Houghton D. C., 1991, Phil. Mag. A64, 597

Cherns D. and Preston A. R., 1989, J. Electron Microscopy Technique, 13, 111

Duan X. F., 1992a, Ultramicroscopy 41, 249, 1992b, Appl. Phys. Lett. 61, 324

Fung K.K., 1984, Ultramicroscopy 12, 243

Gibson J.M., and Treacy M.M.J., 1984, Ultramicroscopy 14, 345

Lichte H., Volkl E. and Scheerschmit K., 1992, Ultramicroscopy 47, 231

Perovic D. D., and Weatherly G. C., 1991, Phil. Mag. A64, 1

Tanaka M., Saito R., Ueno K. and Harada Y., 1980, J. Electron Microscopy 29, 408

Vincent R., 1989, J. Electron Microscopy Technique 13, 40

Inst. Phys. Conf. Ser. No 134: Section 8
Paper presented at Microsc. Semicond. Mater. Conf., Oxford, 5–8 April 1993

517

Defect structures and impurity distribution inhomogeneities in LEC GaAs crystals

C Frigeri

CNR-MASPEC Institute, via Chiavari, 18/A - 43100 Parma, Italy

ABSTRACT: The defect structures, including slip traces, originating from the interaction between dislocations and impurities in n-type LEC GaAs, grown from either As- or Ga-rich melts, have been investigated. The crystals were doped with either Si, S or Te. Independently of the dopant used, the gettering regions around dislocations have been found to be mostly depleted of dopant atoms suggesting that EL2 is generated, rather than gettered, by the dislocations.

1. INTRODUCTION

The behaviour of impurities during growth of LEC GaAs crystals is important in setting the final mechanical and electrical characteristics of these crystals. Migration, or gettering, of impurities to dislocations is particularly important as, in general, it causes retardation of dislocation motion (Sumino 1989). It also produces solution hardening of the lattice. These properties have made it possible to obtain GaAs crystals of very low dislocation density by means of high doping, e. g., with S, Te, Al, N (Seki et al 1978), Si (Fornari et al 1983) or In for semi-insulating (SI) material (Holmes et al 1988). Impurity gettering engineering is also used in device processing to remove unwanted impurities from the "active" regions of the devices (Kang et al 1992).

Despite these great advantages, the interaction between impurities and dislocations during the crystal growth process also has the remarkable drawback of producing a great inhomogeneity of the electrical and optical properties of the crystals, especially in the as-grown state, that requires further processing of the crystals, e. g. annealing, which is usually time-consuming and expensive. Well known for instance is the correlation between dislocations and the deep trap EL2 in SI GaAs (Brozel et al 1983, Stirland et al 1985, Alt and Packeiser 1986). Much attention has been dedicated in the recent years to SI GaAs (see, e.g., Stirland 1991), whose quality is still not satisfactory even in the case of commercially available material (Bassignana and Macquistan 1993). The situation is not much better for semiconducting GaAs that takes nearly 80% of the GaAs consumption, being used as substrate for optical devices such as laser diodes and light emitting diodes (Oda et al 1992). This paper presents a study of the defect structures and impurity distribution inhomogeneities, originating from the interaction between dislocations and impurities, which are usually found in n-type LEC GaAs doped with Si, S or Te.

2. EXPERIMENTAL

The GaAs crystals were grown by the liquid encapsulated Czochralski (LEC) method under As-rich conditions. All crystals were n-type and doped with different dopants, namely Si, Te, S. The doping level ranged between $\sim 1 \cdot 10^{16}$ and $\sim 5 \cdot 10^{17}$ cm^{-3}. Si-doped crystals grown from Ga-rich melts were also investigated. The As atomic fraction $\delta = [As]/[As] + [Ga]$ of the Ga-rich melts was 0.496.

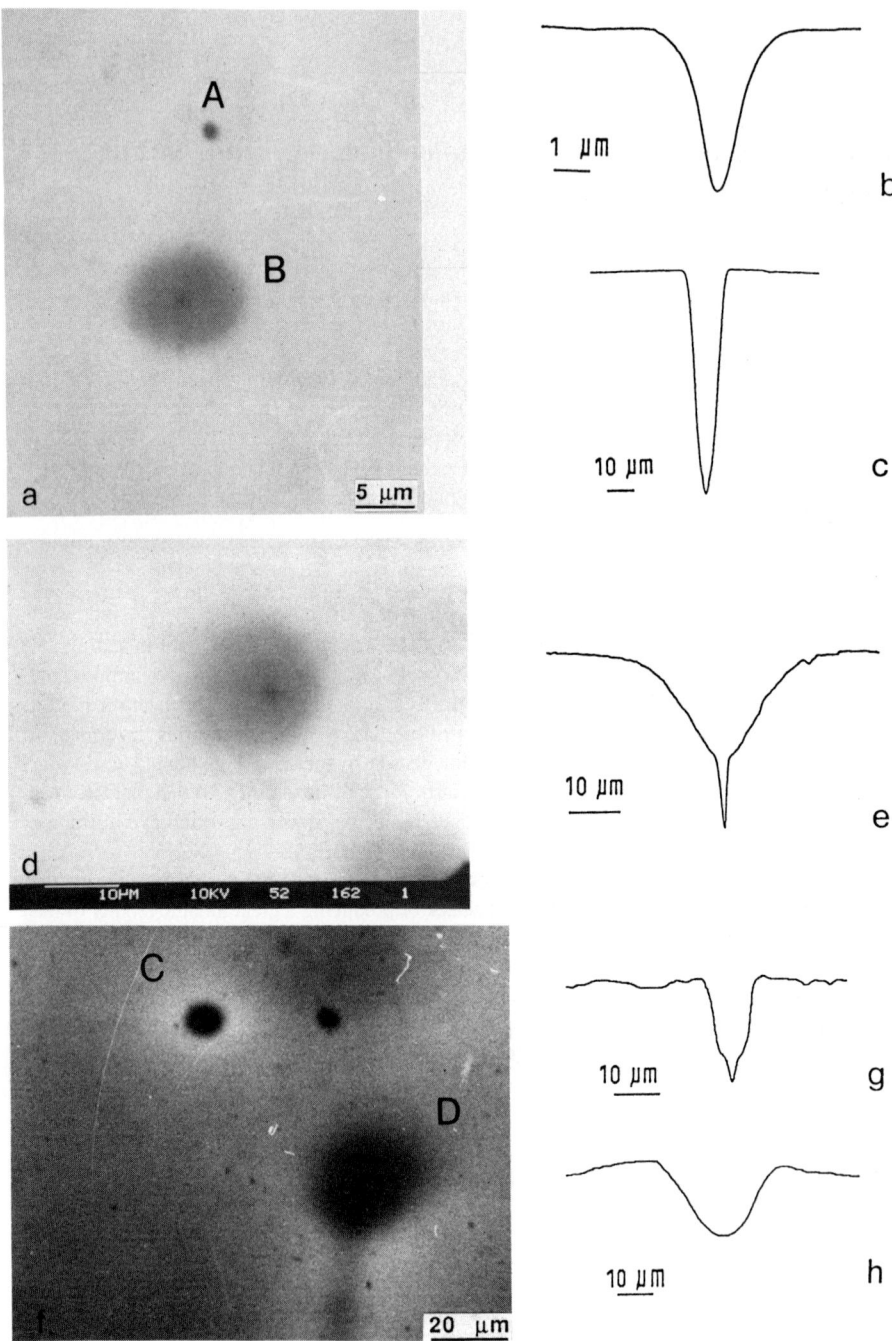

Fig. 1 - EBIC images and linescans of typical defect structures in n-type LEC GaAs. For details see text.

The samples were investigated by EBIC using Au Schottky diodes and Au-Ge ohmic contacts. The diodes had an ideality factor ≤ 1.05. Low injection conditions were fulfilled by using beam currents ≤ 0.1 nA. The energy-dependent method was used. By this method it is possible to assess the behaviour of both the deep impurities, through the measurement of the diffusion length L, and the shallow impurities, through the determination of the net ionized donor density $N = N_D - N_A$ (Frigeri 1987), N_D and N_A being the donor and acceptor density, respectively. By the energy-dependent method, in fact, the width w of the depletion region associated with the EBIC diode can also be measured (Frigeri 1987) thus giving N since $N = k/w^2$, where $k = 1.35 \cdot 10^7$ cm^{-1} for GaAs. TEM in the diffraction contrast mode was also used.

3. RESULTS AND DISCUSSION

3.1 Defect Structures in n-type GaAs Grown from As-rich Melts

Independently of the type of doping, all the investigated LEC GaAs crystals grown from As-rich melts exhibited the same types of defect structures (shown in Figs. 1-2) that can be classified into four groups according to their EBIC contrast:
  i) very small (~1-2 μm) dark dots, i. e. pure glide dislocations;
  ii) very large (~10-20 μm) atmospheres, exhibiting non-radiative recombination, surrounding a smaller but darker dot at their centres, i. e. grown-in dislocations;
  iii) large non-radiative atmospheres as ii) but surrounded by an EBIC bright halo;
  iv) large non-radiative atmospheres without the darker dot at their centres, i. e. without any associated dislocation;
  v) slip traces of moving dislocations.
  In fig. 1a) A is a dislocation of type i), whose EBIC line scan is shown in fig. 1b), whereas feature B is a dislocation surrounded by a larger but less dark area (type ii) defect). The presence of the dislocation (as will be discussed below) in the centre of feature B is clearly evidenced by the EBIC line scan across the central part of B (fig. 1c)). Figure 1c) shows that the EBIC contrast profile narrows at its tip corresponding to the central dot of feature B. Thus feature B consists of two regions with different recombination efficiencies. This is more evident in Figs. 1d-e) showing a defect like B in Fig. 1 a) but with a much larger impurity atmosphere. Its EBIC line scan is shown in fig. 1e). In Fig. 1f) C is a dot-and-halo defect of type iii) formed by a bright halo surrounding a large dark impurity atmosphere which contains a more recombinative defect (i.e. a dislocation, as will be shown below) at its centre as seen from the line scan of fig. 1g). Feature D, instead, is a defect of type iv) in which the absence of the dislocation is confirmed by the corresponding EBIC line scan of Fig. 1h).
  A slip trace, detected in a sample Si-doped to $4 \cdot 10^{17}$ cm^{-3}, is shown in Fig. 2a). It connects the dark dot S to the dot-and-halo feature E. The slip trace was left by a dislocation that has moved from S, where it was sitting as a grown-in dislocation, to E, where now the dislocation can be detected (Weyher and van de Ven 1986, Frigeri and Weyher 1989). The EBIC line scans at S and E are similar to those shown in Fig. 1h) and 1g), respectively, thus confirming that there is no dislocation at S. At S there is only the impurity atmosphere that surrounded the grown-in dislocation before it moved away. Fig. 2b) is an example of a very irregular distribution of dislocations and clusters of non-radiative recombination centres and gettering regions in a GaAs sample Si-doped to $\sim 1 \cdot 10^{16}$ cm^{-3}.
  The EBIC profile of Fig. 1b) for dislocation A has a full width at half maximum of 1.2 μm (at 15 KV). This width agrees well with the width calculated according to Donolato's theories (Donolato 1978/79) for the EBIC contrast profile at dislocations using a diffusion length of 0.7 μm as experimentally measured in these samples.
  It is therefore apparent that in defect structures of type ii) (B in Fig. 1a, Fig. 1d) and type iii) (C in Fig. 1f) only the small non-radiative dot at the centre of the larger cloud represents the dislocation

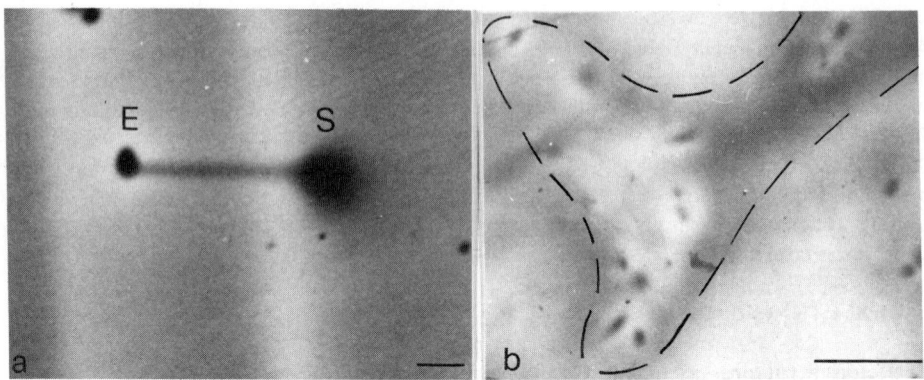

Fig. 2 - EBIC images of a) A slip trace in Si-doped GaAs. Bar is 20 μm. b) Defect structures with irregularly shaped impurity clouds and gettering areas in the surroundings of dislocations. Bar is 100 μm.

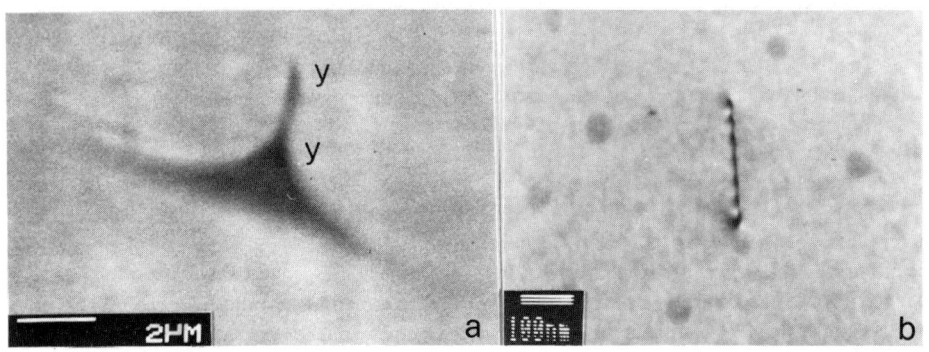

Fig. 3 - a) SEM-SE image of a hillock with tip YY created at a grown-in dislocation by photoetching (sample courtesy of J L Weyher). b) TEM image of a grown-in dislocation exhibiting a photoetched feature as in a). The dislocation turns out to be undecorated.

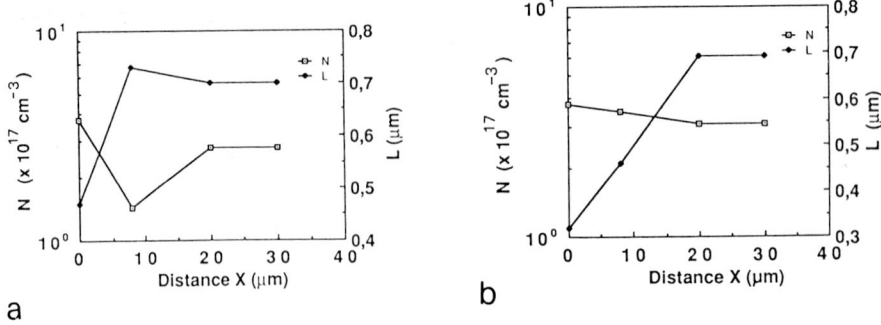

Fig. 4 - Typical plots of L and N for a) defect of type iii), i.e. dot and halo defect; b) defect of type ii) or iv). In a) the bright halo/gettering area is at ~8 μm.

and the related dangling bonds together with possible contaminating impurities. Reconstruction of the narrower part of the profiles shown in Figs. 1c), 1e) and 1g) gives, in fact, widths of these profiles comparable with the width of the profile in Fig. 1b) discussed above (difference of accelerating voltage taken into account).

This conclusion is also confirmed by the expected size of the recombination areas around a dislocation calculated according to the existing models (Read 1954, Labusch and Schrœter 1978) for the recombination of the minority carriers at dislocations. For a doping level of $1 \cdot 10^{17}$ cm^{-3} such size ranges between 25 and 60 nm depending on the model. The EBIC profile width would tend to this low value for beam voltages << 5 kV (Donolato 1978/79) that, however, cannot be used because the electron beam would be totally inside the depletion region of the EBIC Schottky diode.

It is well known that dislocations in LEC GaAs are often decorated by As precipitates in both SI (Cullis et al 1980, Wurzinger et al 1991) and n-type crystals (Fornari et al 1989). The possibility that the more recombinative defect in the centre of the large impurity atmospheres be a precipitate is ruled out, in our samples, by the fact that such defects do not give rise to etch pits after DSL photoetching that is known to produce pits at As precipitates (Weyher and van de Ven 1986, Weyher et al 1991). On the contrary, DSL photoetching of the defect structures of type ii) and iii) produces a surface hillock with a sharp tip at its centre which is higher (less etched) than the surrounding hillock. This is shown in the SEM image of Fig. 3a) that was taken from a deeply photoetched wafer tilted ~80° to the electron beam in the SEM. The long needle YY corresponds to the dislocation and would never have formed this long if there had been precipitates along the dislocation. The absence of precipitates along the dislocations of the defect structures of type ii) and iii) has also been checked by TEM on specimens selected by means of photoetching (Fig. 3 b).

3.2 Impurity Atmospheres and Gettering Areas in GaAs Grown from As-rich Melts

Typical results about N (calculated from EBIC measurements of w) and L are given in Fig. 4 where X is the distance from the centre of the dark atmosphere; the defect free matrix is at X > ~20 μm. Fig. 4 a) refers to a defect structure of type iii), i. e. an impurity atmosphere with a dislocation at its centre and surrounded by a gettering region located at about 8 μm. It is seen that N decreases in the gettering area and increases in the dark atmosphere around the dislocation. On the other hand, L decreases in the dark atmosphere and increases, but only slightly, in the gettering region. Similar results were obtained in all crystals irrespective of the dopant type and density, within the range investigated. This result means that the gettering region is mostlty depleted of dopant atoms, whose density, instead, increases in the atmosphere close to the dislocation. By taking the doping level in the matrix as a reference value, the example of Fig. 4 a) shows that, in this case, in the dark atmosphere N has increased by nearly the same amount by which it has decreased in the gettering area. It can, thus, be concluded that the dislocation has gettered dopant atoms.

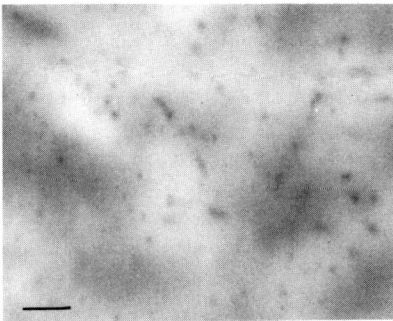

Fig. 5 - Gettering areas round dislocations of type i) in a GaAs crystal S-doped to $2.5 \cdot 10^{16}$ cm^{-3}. Bar is 20 μm.

Even in the absence of a large impurity atmosphere round the dislocation, as for the defects of type i) shown in Fig. 5, the gettering areas have always been found to be depleted of dopant atoms with respect to the matrix. For defect structures like those of Fig. 5 it was not possible to make any N measurement at the dark dots, i.e. dislocations, because they are too much smaller than the interaction volume of the SEM electron beam with the specimen. On the basis of the results for the defect structure of type iii) it is believed, however, that also in such a case the dopant atoms have been captured by the dislocations.

A slight increase of N and a decrease of L in the dark atmospheres with respect to the matrix was also observed for the defect structures of type ii) and iv), i. e. large impurity atmospheres with and without associated dislocation and not surrounded by any bright halo (Fig. 4b). Even such atmospheres, therefore, contain dopant atoms.

As the net ionized dopant density $N = (N_D - N_A)$ is measured, it might also be that not only N_D but also the acceptor density N_A changes in the atmospheres with respect to the matrix. This hypothesis was assumed in Si-doped crystals to explain the fact that sometimes $(N_D - N_A)$ had the same value in the atmospheres as in the matrix (Frigeri and Weyher 1989). Assuming that the atmospheres contain Si atoms gettered by the dislocation, it was argued that in the atmospheres the amphoteric Si dopant might have occupied not only the Ga sublattice, thus behaving as a donor, but also the As sublattice, thus becoming an acceptor. This would require a higher probability of Si occupancy of the As sites. This should be favoured, in the next neighbouring areas around the dislocations, by the simultaneous increase of the density of the Si atoms and decrease of Ga vacancies (V_{Ga}) close to the dislocation. To explain the decrease of $[V_{Ga}]$ one should consider that the occurrence of As precipitates at dislocations is evidence of effective diffusion of As_i to the dislocations whose surroundings should thus be enriched of As_i on their way to the dislocation. This should occur even without As precipitation that takes place, in fact, only when some threshold density of As_i along the dislocation is achieved. An increased density of As_i close to the dislocation might favour the formation of As antisite defects, that have a small energy of formation (van Vechten 1975), through the reaction $As_i + V_{Ga} = As_{Ga}$, so that $[V_{Ga}]$ decreases. Enrichment of As_i close to the dislocations has also been suggested by Miyazawa and Wada (1986).

Investigations of the same defects in the same Si-doped GaAs crystals by scanning photoluminescence at 4 °K with a 2 µm spatial resolution have confirmed the presence of the dopant Si in the atmospheres, although on Ga sites (Visser et al 1989, Weyher et al 1990). No Si_{As} was detected within the sensitivity of the technique. Possible compensation of Si_{Ga} should thus take place, alternatively, by other acceptors either shallow (e.g. C) or deep of unknown origin.

The rate of change with time t of the concentration C of an impurity in the strain field of a dislocation is given by

$$\delta C / \delta t = D \nabla [\nabla C + (C / kT) \nabla E_i] \qquad (1)$$

where D is the diffusion coefficient of the impurity and E_i the energy of elastic interaction between an impurity atom and the dislocation, k and T as usual (Bullough and Newman 1964). The elastic energy is mostly due to the difference in size between the impurity atoms and the available atomic sites in the host lattice (Bullough and Newman 1964, Sumino 1989, Kang et al 1992). The lattice mismatch between the host lattice atoms and the dopant atoms considered here is -7.1% for Si, -11.9% for S and + 11.9% for Te as calculated by Seki et al (1978) by employing Pauling's tetrahedral covalent radii (Phillips 1973).

The dislocations have been seen to getter all three types of dopant with the same efficiency. For a doping level of $2.5 \cdot 10^{16}$ cm^{-3}, in fact, the ratio between N in the bright halo and in the matrix for the three types of dopant ranged between 0.5 and 0.63. The gettering of the dopant atoms seems, therefore, to be independent of their size misfit with respect to the host lattice. The negligible role of the size misfit was also suggested by Yonenaga and Sumino (1989) by studying the influence of the dopant on the retardation of the dislocation motion in GaAs. The gettering of dopant impurities should, therefore, be mostly affected by the diffusivity of the impurities (diffusion coefficient D in eq. (1)). However, despite the fact that Si has a lower diffusivity than S and Te (Yonenaga and Sumino 1989), it is gettered by the dislocations as efficiently as S and Te. At present this is not fully understood. It might be that the Si atoms form complexes that interact very strongly (large E_i in eq. (1)) with the dislocations as suggested by Yonenaga and Sumino (1987) to explain their result that Si has a great ability to lock dislocations, at least in the high temperature range, despite its lower diffusivity in GaAs.

As the diffusion length in n-type LEC GaAs is mainly affected by non-radiative recombination processes at deep traps (Frigeri 1993), the decrease of L in the atmospheres clearly means that such impurity atmospheres also contain non-radiative deep centres. Several deep traps act as lifetime killers in LEC GaAs (Frigeri 1993). All these traps have characteristics that can account for the low value of L measured here. Their relationship, however, with the dislocations and surrounding atmospheres has never been established except for EL2. EL2 clustering around dislocations was observed by Stirland et al (1985) in SI crystals by near IR absorption and by Wosinski and Breitenstein (1986) in n-type plastically deformed Si-doped GaAs by Scanning Deep Level Transient Spectroscopy (SDLTS).

Although L decreases significantly in the dark atmospheres, Fig. 4 a) shows that L increases only slightly (<~10%) in the gettering regions with respect to the matrix. This would indicate that the gettering regions are depleted of non-radiative deep centres only to a small extent. This conclusion has been confirmed by SDLTS mapping (Frigeri and Breitenstein 1990). Combined EBIC and SDLTS investigations of the sample shown in Fig. 2b) (Si-doped to ~$1 \cdot 10^{16}$ cm^{-3}) have shown, in fact, that EL2 is homogeneously distributed all over the area inside the dash line in Fig. 2b) (Frigeri and Breitenstein 1990). *No* depletion of EL2 was detected in the gettering regions, i. e. EBIC bright haloes, that instead were found to be significantly depleted of shallow dopant atoms (Frigeri and Breitenstein 1990). It was also found that the distribution of the lower charge state of EL2 (which is a hole trap) exactly overlaps to that of EL2. These results confirm that, in the n-type GaAs crystals investigated here, the dislocations tend to preferentially getter dopant atoms rather than EL2.

This conclusion could be supported by the low diffusivity of the antisite defect As$_{Ga}$, that is generally considered to be the main constituent of the deep trap EL2, as it is a substitutional impurity (von Bardeleben et al 1986). Moreover, according to Sumino (1989) the interaction energy E_i between a dislocation and an EL2 centre is too small to cause EL2 gettering by the dislocation at distances from the dislocation greater than few atomic distances.

The question, still not fully answered by the GaAs community, arises whether the dislocations getter EL2 or rather generate it, e. g., by climb according to the model of Weber et al (1982). Our results are more in agreement with the hypothesis that the dislocations do not getter EL2. It should be noted, however, that the EL2 cluster around dislocations can be as large as some hundreds μm or even not surrounding any dislocation at all (Frigeri and Breitenstein 1990). It is thus possible that EL2 is created not only at dislocations by climb but also in any other place in the crystal due to inhomogeneities in the distribution and density of the point defects necessary to form EL2 (V$_{Ga}$ and As$_i$) very likely caused by inhomogeneities in the distribution of the dopant itself.

The slip traces (Fig. 2a) were always too small to be reliably analysed by EBIC to establish whether they contain dopant atoms or not. On the other hand, photoluminescence measurements have shown that the traces do not contain dopant atoms but only non-radiative recombination centres (Weyher et al 1992, Visser et al 1990). By studying the photoetching features of these traces in both As- and Ga-rich GaAs crystals it was possible to establish that the only constituent of the traces should be the As$_{Ga}$ defect (Weyher et al 1992). This allows to experimentally confirm that both the climb process (Weber et al 1982) and the climb/glide process (Figielski 1985) can account for the movement of the dislocations in GaAs. In the pure climb model of Weber (1982) the As$_{Ga}$ is always produced because an As_i reacting with a negative climb step -dc, that is defined as the simultaneous emission of a V_{Ga} and a V_{As}, yields a V_{Ga} that then reacts with an As_i producing an As_{Ga} defect:

$$As_i - dc \rightarrow V_{Ga}$$
$$V_{Ga} + As_i \rightarrow As_{Ga}.$$

If Ga vacancies predominate instead of As insterstitials, as for instance when As precipitation consumes many As_i, As_{Ga} can still form but through the reaction between a V_{Ga} and a positive climb step (emission of As_i and Ga_i) (Wurzinger et al 1991).

The climb/glide process requires that the gliding dislocation has jogs that absorb point defects. Four jog configurations have been proposed by Figielski (1985), namely (* denotes jog position): 1) regular jog *Ga_{Ga} - *As_{As} ; 2) irregular jog *Ga_{As} - *Ga_{Ga} ; 3) irregular jog *Ga_{As} - *As_{Ga}; 4) irregular jog *As_{Ga} - *As_{As}. Process 1) is not possible because no defect is created in the matrix so that the trace would not be detectable, contrary to the experimental results. Process 2) has also to be excluded as Ga_{As} is created in the matrix. Since Ga_{As} is an acceptor-like impurity the trace would give rise to grooves upon photoetching. On the contrary, the traces give rise to ridges (Weyher and van de Ven 1986, Frigeri and Weyher 1989, Weyher et al 1992). In process 3) the antistructure pair $(Ga_{As}-As_{Ga})$ is created that is electrically neutral and so would not account for the formation of photoetching features at the traces. Then only process 4) remains with the formation of As_{Ga} that would explain the combined experimental results of EBIC, photoetching and photoluminescence reported by Weyher et al (1992).

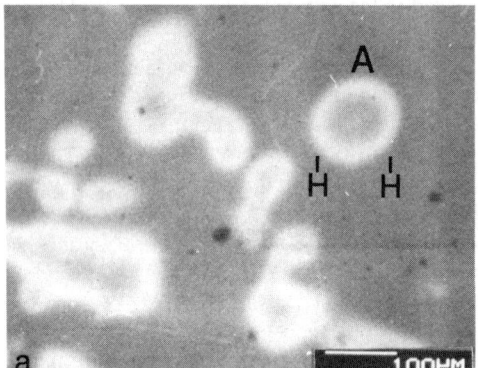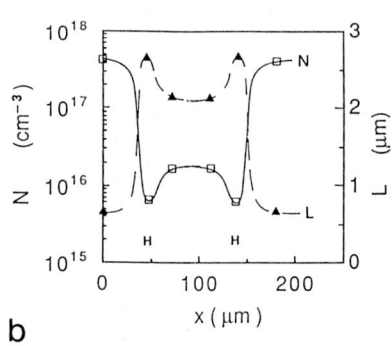

Fig. 6 -   a) EBIC image of typical defect structures in Si-doped GaAs grown from Ga-rich melts.
b) Plot of N and L at defect A. X = 100 μm is the centre of the defect. H = bright halo.

3.3 Defect Structures in Si-doped GaAs Grown from Ga-rich Melts

Fig. 6 a) shows typical large defect structures in Si-doped GaAs crystals grown from Ga-rich melts. They exhibit EBIC bright contrast, contrary to what is observed in crystals grown from As-rich melts. This anomaly was confirmed by photoetching that produces large depressions at such defects instead of the usual hillocks of the As-rich crystals (Weyher et al. 1990). Line scan analysis has shown that in the centre of the annular ring A in Fig. 6 a) there is a dislocation exhibiting the usual EBIC dark contrast. Plots of N and L across defect A are shown in Fig. 6 b). In the bright ring L increases, whereas N decreases but much more than in the As-rich crystals.

Even in GaAs crystals grown from Ga-rich melts, therefore, the gettering regions around dislocations are depleted of dopant atoms. It should be noted that in Ga-rich crystals a great contribution to the decrease of N in the bright haloes could also be due to an increase of the density of the deep double acceptor Ga_{As} that is suggested to be the dominant defect in Ga-rich crystals (Bugajski et al 1989). Ga_{As} should form, upon solidification, in crystals containing a large amount of V_{As} through the reaction $V_{As} \Leftrightarrow V_{Ga} + Ga_{As}$ that also introduces the other acceptor V_{Ga} whereas the strong donor V_{As} is removed (Bugajski et al 1989). Photoluminescence, however, confirmed that the matrix contains more Si, in both As and Ga sites, than the bright halo (Weyher et al 1990), so that the bright haloes are really depleted of shallow Si donors to a significant extent. The increased value of L in the bright halo indicates a smaller density of deep traps in the gettering regions, as also found by PL measurements. This might be related to a decreased content of EL2 due to the fact that fewer As interstitials are available in the Ga-rich crystals.

ACKNOWLEDGMENTS

The author wishes to thank Dr J L Weyher for stimulating discussions as well as for supplying some of the samples investigated here. Thanks are also due to Drs O Breitenstein and P J van der Wel for useful discussions.

REFERENCES

Alt H Ch and Packeiser G 1986 J. Appl. Phys. 60 2954
Bassignana I C and Macquistan D A 1993, Proc. 7th Inter Conf. on III-V SI Materials, Ixtapa (Mexico), April 21-24 1992, in press.
Brozel M R, Grant I, Ware R M and Stirland D J 1983 Appl. Phys. Lett. 42 620
Bugajski M, Ko K H, Lagowski J and Gatos H C 1989 J. Appl. Phys. 65 596
Bullough R and Newman R C 1964 in Progress in Semiconductors, ed A F Gibson and R E Burgess (London: Heywood) vol 8 p 100
Cullis A G, Augustus P D and Stirland D J 1980 J. Appl. Phys. 51 2556
Donolato C 1978/79 Optik 52 19
Figielski T 1985 Appl. Phys. A 36 217
Fornari R, Frigeri C and Gleichmann R 1989 J. Electron. Mat. 18 185
Fornari R, Paorici C, Zanotti L and Zuccalli G 1983 J. Crystal Growth 63 415
Frigeri C 1987 Inst. Phys. Conf. Ser. 87 745
Frigeri C 1993 J. Crystal Growth 126 91
Frigeri C and Breitenstein O 1990 in Defect Control in Semiconductors, ed K Sumino (Amsterdam: North-Holland) pp 685.
Frigeri C and Weyher J L 1989 J. Appl. Phys. 65 4646
Holmes D E, Kuwamoto H, Sandberg C J and Johnston S L 1988 J. Crystal Growth 91 557
Kang N S, Zirkle T E and Schroder D K 1992 J. Appl. Phys. 72 82
Labusch R and Schræter W 1978 in Dislocations in Solids, ed F R N Nabarro (Amsterdam: North Holland) vol 5 ch 20 p 127.
Miyazawa S and Wada K 1986 Appl. Phys. Lett. 48 905
Oda O, Yamamoto H, Seiwa M, Kano G, Inoue T, Mori M, Shimakura H and Oyake M 1992 Semicond. Sci. Technol. 7 A215
Phillips J C 1973 Bonds and Bands in Semiconductors (New York: Academic)
Read W T 1954 Phil. Mag. 45 775
Seki Y, Watanabe H and Matsui J 1978 J. Appl. Phys. 49 822
Stirland D J 1991 Inst. Phys. Conf. Ser. 117 327

Stirland D J, Brozel M R and Grant I 1985 Appl. Phys. Lett. 46 1066

Sumino K 1989 in Point, Extended, and Surface Defects in Semiconductors, ed G Benedek, A Cavallini and W Schrœter (New York: Plenum) p 22

van Vechten J A 1975 J. Electrochem. Soc. 122 423

Visser E P, van der Wel P J, Weyher J L and Giling L G 1990 J. Appl. Phys. 68 4242

von Bardeleben H J, Stiévenard D, Deresmes D, Huber A and Bourgoin J C 1986 Phys. Rev. B 34 7192

Weber E R, Ennen H, Kaufmann U, Windscheif J, Schneider J and Wosinski T 1982 J. Appl. Phys. 53 6140

Weyher J L, Frigeri C and van der Wel P J 1990 J. Crystal Growth 103 46

Weyher J L, Pascal P, Frigerio G and Zanotti L 1990 J. Crystal Growth 106 175

Weyher J L and van de Ven J 1986 J. Crystal Growth 78 191

Weyher J L, van der Wel P J and C Frigeri 1992 Semicond. Sci. Technol. 7 A294

Wosinski T and Breitenstein O 1986 Phys. St. Sol. a) 96 311

Wurzinger P, Oppolzer H, Pongratz P and Skalicky P 1991 J. Crystal Growth 110 769

Yonenaga I and Sumino K 1987 J. Appl. Phys. 62 1212

Yonenaga I and Sumino K 1989 J. Appl. Phys. 65 85

Inst. Phys. Conf. Ser. No 134: Section 8
Paper presented at Microsc. Semicond. Mater. Conf., Oxford, 5–8 April 1993

527

HRTEM study of the {111} planar defect

Y Ohno, S Takeda and S Horiuchi*

Department of Physics, College of General Education, Osaka University, Toyonaka, Osaka 560, Japan
*National Institute for Research in Inorganic Material, 1-1 Namiki, Tsukuba, Ibaraki 305, Japan

ABSTRACT: Planar defects on the {111}-type planes of heavily Si-doped GaAs have been examined by TEM techniques. We have determined the relationship between the defects and Si concentration by nm-probe EDX measurement and 1 MV-HRTEM observation. Considerable amounts of aggregated Si are detected at the defect, especially at the dislocation core. This aggregation of Si causes the compression of the lattice near the defects, which has been observed in the HRTEM images.

1. INTRODUCTION

It is well known that Si is an interesting dopant in a GaAs crystal because it acts as a donor or acceptor by occupying the Ga or As site, respectively. Therefore, the behavior of Si in GaAs is attractive both fundamentally and practically. In a heavily Si-doped and subsequently annealed GaAs crystal, defects of interstitial character were frequently observed (Swaminathan and Copley 1976, Chen and Spitzer 1981). The defects were observed in the area of high Si-concentration, and were characterized as Frank-type loops on the {111}-type planes. Since the defects have never been detected in a pure GaAs crystal, they are considered to be Si-related. Based on an analysis with TEM and EDX techniques, it was suggested that the defects are Si aggregates, precipitated in a GaAs matrix and forming interstitial-type stacking faults (Muto, Takeda, Hirata, Fujii and Ibe 1992).

We report here on the experimental results by HRTEM and nm-probe EDX analysis concerning these defects. The results show that Si is detected on the defect, especially at the core of the Frank partial dislocation. The noticeable compression of the lattice near the defect is observed in the HRTEM images, indicating the aggregation of Si. The structure of the dislocation core is also revealed in the HRTEM image.

2. EXPERIMENTS

Specimens were prepared from Si-doped GaAs crystals grown by the gradient freeze method (Orito, Tsujikawa and Tajima 1984). The average Si concentration around the defects was evaluated by the secondary ion mass spectroscopy (SIMS) as $(2-4) \times 10^{19}$ cm^{-3}. Samples were then heat-treated at 850 °C in an As atmosphere for 20 hours, followed by air cooling to room temperature. It is established that the heat treatment enhances the formation of the defects of interest. TEM images and CBED patterns were taken with a JEOL JEM2000EX operated at 200 kV and HRTEM images were taken with a high voltage TEM operated at 1000kV. A nm-probe energy-dispersive X-ray analysis (EDXA) was performed with a JEOL JEM2010 equipped with a LaB$_6$ filament. The electron probe size was estimated to be 3 nm.

3. RESULTS AND DISCUSSION

3-1 Energy-Dispersive X-Ray Analysis

The internal structure of the GaAs crystal with a high Si concentration, after annealing at 850°C for 20 h, is seen in a dark field TEM image (Fig. 1A). Larger defects on the {111}-type planes exhibit triangular shape, with their edges parallel to the <110> directions. The apex of each triangle points in the same direction, and the inverted triangle has never been observed in the present observation. The phenomenon is common in Frank-type defects in the zinc blende -type structure. This is because the structures have two kinds of core structures and the stability and/or the mobility of them are considered to be different. Smaller defects seen in Fig. 1(A) have the same nature as the larger ones, even though they do not look like triangles.

The larger defect was viewed edge-on as seen in Fig. 1(B), and the EDX measurement was performed. The locations of the probe are indicated by the circles in Fig. 1(B). It is found in Fig. 1 (a)-(d) that Si aggregates at the defect, particularly around the dislocation core. Si is also detected in the central part of the defect. The evidence of Si aggregation around the defect is also found in the HRTEM images as described below.

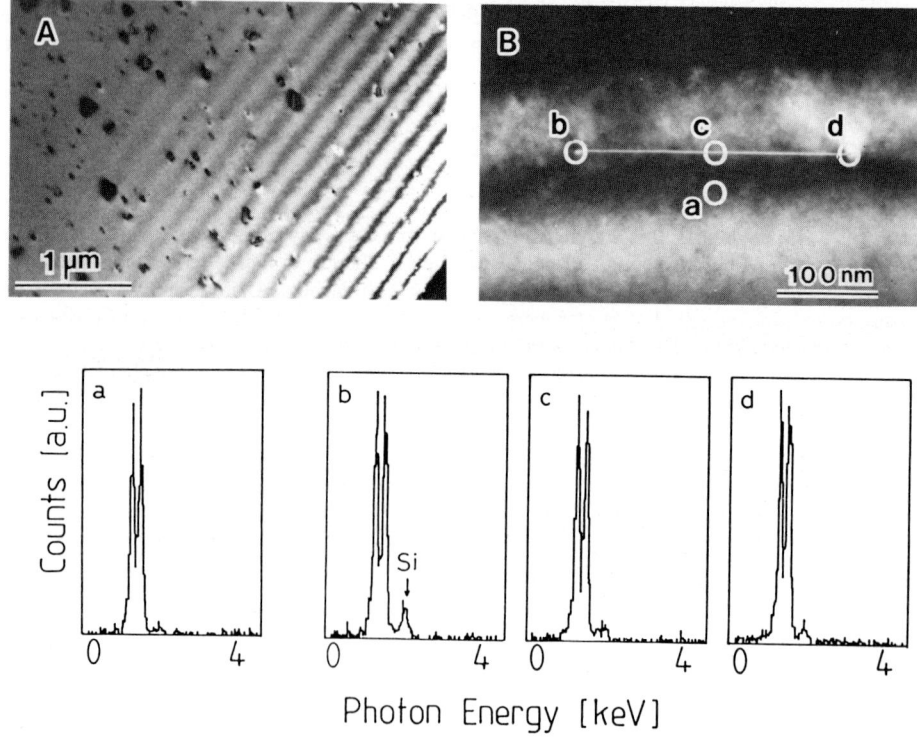

Fig. 1 (A) A weak-beam image of the internal structure of Si-doped GaAs. (B) A dark field image of the {111} planar defect, on which the locations of the electron probe for EDX measurement are marked by the open circles (a)-(d). EDX spectra with the probe as indicated in (B); (a) outside the defect, (b) on an edge, (c) on the center of the defect and (d) on the other edge.

3-2 HRTEM observation

The cross-section of the planar defect was observed by utilizing a 1 MeV-HRTEM with <110> incidence. The image depicted in Fig.2(a) was recorded near the Scherzer defocus condition, and the dark dots in the image definitely correspond to the atomic columns. Based on this assignment, it is found that two {111} net planes are inserted into closely spaced interatomic planes of Ga and As, forming an extrinsic Frank-type loop. Measurement of the (111) lattice spacing in the image in Fig. 2(a) showed that the lattice is compressed by 2 to 10 % near the defect. The elastic energy of several models of Si precipitates in the periodic manner on the {111}-type planes was estimated based on the valence force field treatment (Martins and Zunger, 1984), and the energetically favorable relative displacement of the crystal on both sides of the defect was estimated to be (0.92-0.95)x 1/3[111]a. These facts suggest the aggregation of Si in the planar defect.

Both edges of the defect are shown in the enlarged images of Fig. 2(b) and (c). As mentioned in the previous section, the triangular defect is bounded by a dislocation line which lies parallel to the <110> direction. The polarity of the crystal was determined by the CBED technique as in the

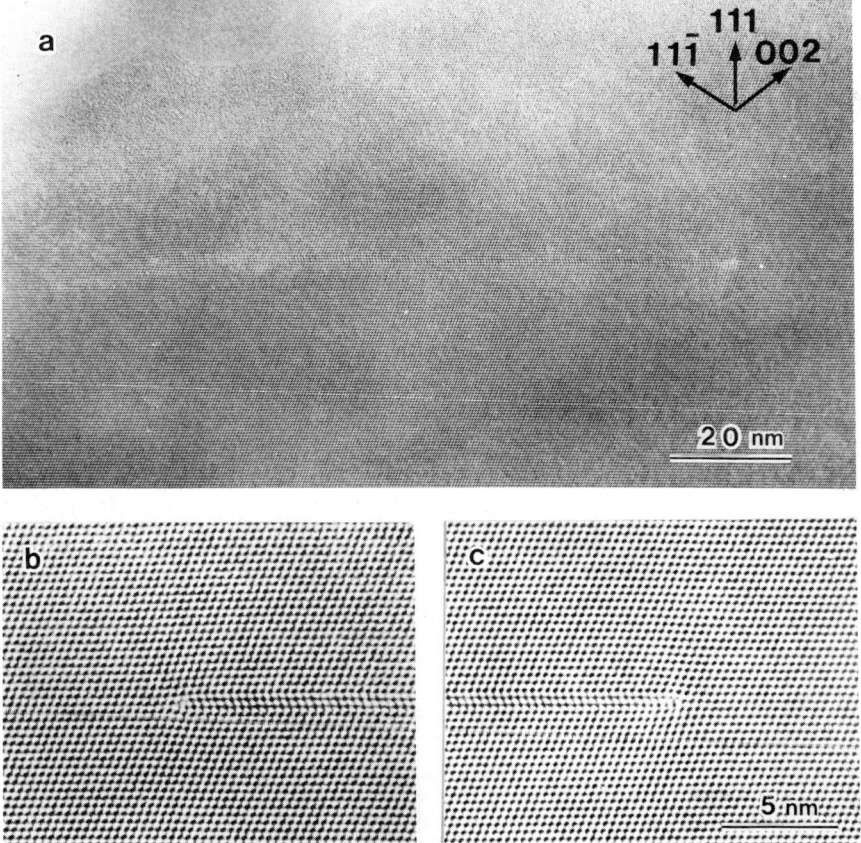

Fig. 2 (a) Cross-section view of the {111} planar defect. The image was taken with the <110> incidence. The images in (b) and (c) show both edges of the defect seen in (a). The dislocation line in (b) is parallel to the incident beam, and that in (c) makes an angle of 30 degrees to the incident beam.

previous paper (Muto et al. 1992). According to the analysis, the dislocation line in Fig. 2(b) is parallel to the electron beam, while that in Fig. 2 (c) makes an angle of 30 degrees to the beam.

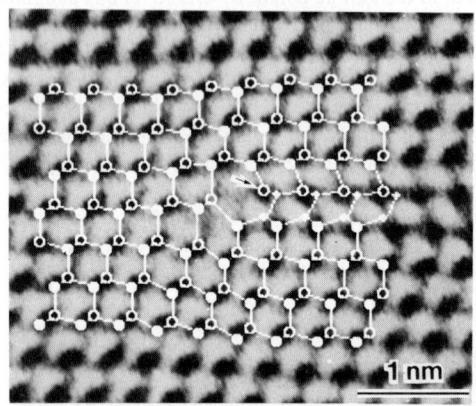

Fig. 3 Magnified image of Fig. 2(b), on which the model of the dislocation core is superimposed. Notice that the displacement of the GaAs crystal on both sides of the planar defect is about $0.25[111]a$.

From the contrast analysis of the image in Fig 2(b), it is possible to construct a model of the core structure. The present model is superimposed on the magnified image shown in Fig. 3. The larger full and open circles represent Ga and As atoms, respectively. It is assumed that the inserted two {111} net planes near the core region consist of Si atoms (the smaller full circles). The atoms in the inserted layers arrange regularly to the edge of the defect. It is considered that As atoms in the edge of the defect (indicated by the arrow in Fig. 3) have a dangling bond in the model. The crystal around the core is very compressed. The relative displacement of the GaAs matrix on both sides of the inserted layer is measured to be about $0.25[111]a$ near the edge, while that at the centre of the defect is measured to be $(0.30\text{-}0.33) \times [111]a$. This fact also indicates that Si is involved in the defect, since the atomic size of the Si atom is smaller than that of Ga or As atoms.

4. CONCLUSION

We found that Si atoms in heavily Si-doped GaAs crystals aggregate in the {111} planar defects, especially around the dislocation core of the defect. A model of the core structure was obtained by utilizing a 1 MeV-HRTEM technique. It was expected that As atoms in the core have a dangling bond. The GaAs matrix was compressed around the core, and it is considered that the compression effect is caused by the aggregation of Si atoms around the core.

REFERENCES

Chen R T, Spitzer W G, 1981 J. Electron Mater. **10**, 1085
Orito F, Tsujikawa Y, Tajima M, 1984 J. appl. Phys. **55**, 1119
Muto S, Takeda S, Hirata M, Fujii K, Ibe K, 1992 Philos. Mag. **A 66**, 257
Martins J L and Zunger A, 1984 Phys. Rev. **B15**, 6217
Swaminathan V, Copley S M, 1976 J. appl. Phys. **47**, 4405

Inst. Phys. Conf. Ser. No 134: Section 8
Paper presented at Microsc. Semicond. Mater. Conf., Oxford, 5–8 April 1993

531

Formation of void/Ga-precipitate pairs during Zn diffusion into GaAs: competition of two thermodynamic driving forces

W Jäger, A Rucki, K Urban, H-G Hettwer*, N A Stolwijk*, H Mehrer* and T Y Tan+

Institut für Festkörperforschung Forschungszentrum KFA Jülich D-5170 Jülich Germany
*Institut für Metallforschung Universität Münster D-4400 Münster Germany +Dept.
Materials Science and Engineering Duke University Durham NC 27706 USA

ABSTRACT: Defect formation and evolution during Zn indiffusion into GaAs at $900^{\circ}C$ has been characterized. Using a pure Zn source, the Zn profile is box-shaped and the Zn diffused region contains dislocations and void/Ga-precipitate pairs, with the void to precipitate volume ratio being essentially constant. Using As also in the source, the Zn profile is of the kink-and-tail type with the Zn diffused region containing the same kinds of defects. However, only in the profile tail region the Ga precipitate to void volume ratio is large, while in the region of the profile at high Zn concentration near the surface only voids are present. This shows that indiffusion of high concentrations of Zn produces Ga-rich GaAs crystals, but in the As-rich diffusion ambient case the surface region has been subsequently converted into being rich in As.

1. INTRODUCTION

Zn is a substitutional-interstitial species in GaAs. It diffuses by the migration of the interstitial species which is subsequently incorporated substitutionally onto Ga sites. It has been found that diffusion of Zn from vapor sources into GaAs to a high concentration is associated with the formation of dislocations and void/Ga-precipitate pairs (Luysberg 1989, 1992). The formation mechanism of these defects has been discussed previously (Luysberg et al. 1989 and 1992, Tan et al. 1991). In this paper we discuss results on Zn diffusion induced defect structures which are significantly different between Ga-rich and As-rich diffusion ambient cases.

2. EXPERIMENTS

Zn was diffused at $900^{\circ}C$ into semi-insulating GaAs wafers with a dislocation density of $< 10^4$ cm^{-2} using sealed quartz ampoules. The source materials were either pure Zn (Ga-rich ambient), or a mixture of Zn and As at a weight ratio of 3:7 (As-rich ambient). The Zn concentration profiles were determined using the spreading resistance technique (SRT) and electron microprobe analysis (EMP). EMP measures the total concentration of Zn whereas SRT monitors only the electrically active substitutional Zn acceptor atom concentration, C_s. Defects in the Zn diffused regions were studied using a transmission electron microscope (TEM, JEOL 4000FX) equipped with an energy dispersive X-ray spectrometer (EDS).

3. RESULTS

Fig. 1a shows box-type Zn concentration profiles as measured by SRT and EMP on the same sample. They are characterized by one plateau and one step which is typical for the Ga-rich ambient cases (Casey 1973, Tuck 1988, Jäger et al. 1992). Fig. 1b shows kink-and-tail type profiles (Casey 1973, Tuck 1988, Luysberg et al. 1992) typical for As-rich ambient conditions (Luysberg et al. 1989, 1992). These profiles are characterized by two plateaus and two steps. The C_s value at the surface is 2×10^{20} cm^{-3} and decreases at the position of the first kink to a second plateau value of 1-2×10^{19} cm^{-3}. A similar behavior is observed beyond the kink position with a sharp decrease of C_s to values below the detection limit of 1×10^{17} cm^{-3}. For the box-profile the C_s value at the surface is smaller by about a factor of 2 ($\tilde{}10^{20}$ cm^{-3}), and, although the diffusion time is more than a factor 4 shorter, the maximum penetration depth is larger.

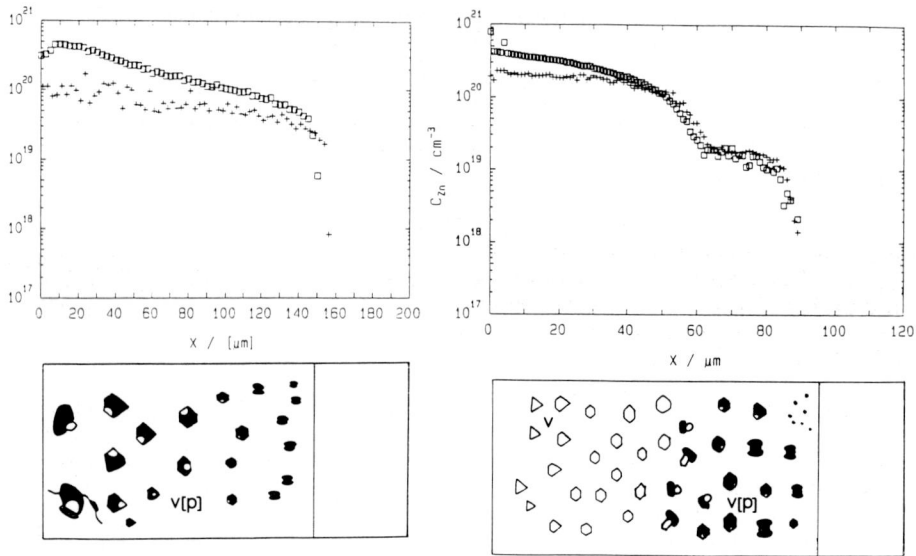

Fig. 1 Zinc indiffusion profiles at 900°C, obtained by SRT (+) and by EMP (□): (a) Ga-rich ambient, 90 min (left) and (b) As-rich ambient, 435 min (right). Schematic diagrams indicating the morphologies and distributions of voids (v) and void/Ga-precipitate pairs (v[p]) are shown below.

Structural investigations showed that extended defects have formed throughout the Zn diffused regions. The defect varieties and distributions constitute a complicated picture and are subject to temporal evolution (for details see Jäger et al 1992). Very close to the diffusion front, only dislocation loops and small precipitates are present. The loops are identified to be perfect interstitial type loops, and EDS analyses showed that the precipitates essentially consist of Ga. Behind the diffusion front dislocations and precipitates with a void volume are present, independent of the diffusion ambient conditions. In this region, the void/Ga-precipitate sizes and the ratios of precipitate and void volume are practically identical for the two samples (Fig.1 schematics). However, a significant difference is observed for the void/precipitate structures in the near-surface regions. For box-like profiles void/Ga-precipitate pairs have an essentially constant precipitate/void volume ratio throughout the Zn diffused region (Fig. 2a). The precipitate structure coarsens with increasing diffusion time, i.e., larger void/Ga-precipitate pairs are observed closer to the surface. Using EDS, it was found that these large Ga precipitates near the surface contain also Zn. For kink-and-tail profiles the tail region contain void/Ga-precipitate pairs with a large precipitate/void volume ratio whereas in the region of high Zn concentration near the surface only voids are present (Fig. 2b).

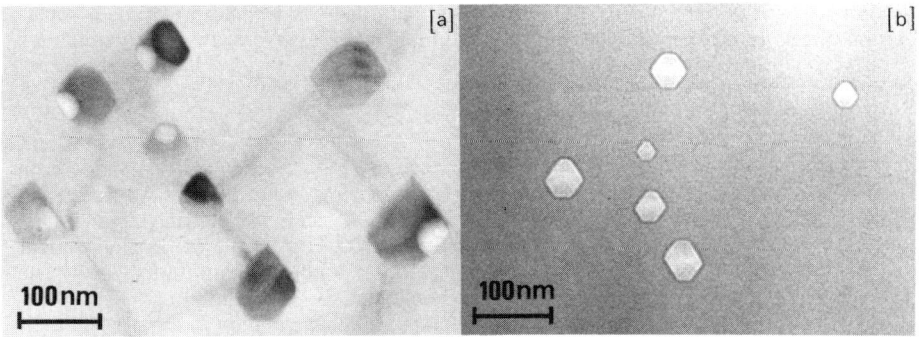

Fig. 2 Bright field TEM micrographs of void/Ga-precipitate pairs: (a) Tail region of the Zn profile (Ga-rich ambient)(b) High concentration kink region of the Zn profile; only voids are left (As-rich ambient).

The defect structures for both the As-rich and the Ga-rich case in a depth zone near the surface are dominated by the presence of small Zn_3As_2 precipitates, which are largely responsible for the higher Zn concentrations measured by EMP, Figs. 1 and 2 (Rucki et al. 1993).

4. DISCUSSION AND CONCLUSIONS

The formation of interstitial-type dislocation loops at the diffusion front is a clear indication that a supersaturation of Ga interstitials, I_{Ga}, develops as a consequence of the incorporation of diffusing Zn interstitials onto substitutional Ga sites. The interstitial-substitutional exchange is governed by the kick-out mechanism (Morehead and Gösele 1981, Tan et al. 1991), which generates the excess I_{Ga}. The I_{Ga} supersaturation decays by formation of interstitial dislocation loops whose growth leads to emission of As vacancies, V_{As}, thus creating a supersaturation of V_{As}. These V_{As} condense to form voids which are partly filled with Ga. A high I_{Ga} supersaturation is present only at the diffusion front (Luysberg et al. 1992, Jäger et al. 1992), see Fig. 3.

The dislocations and void/Ga-precipitate pairs influence the further Zn diffusion by establishing local point defect equilibria. Fig. 3 shows schematically the Zn indiffusion kink-and-tail profile under As-rich conditions, and the corresponding Ga self-interstitial distribution, respectively normalized to thermal equilibrium values for Ga- and As-rich GaAs. The scale shift reflects the dependence of I_{Ga} concentration on the As_4 pressure at 900°C. For both the Ga- and the As-rich cases the defect formation process leads initially to a GaAs crystal which is in equilibrium with the liquid Ga and the vapor phase within the voids, thereby holding the I_{Ga} and V_{As} concentrations close to the appropriate thermal equilibrium values of a Ga-rich GaAs crystal (Ga-rich portion of Fig.3). Diffusion under As-rich conditions for longer time leads to an As-rich GaAs crystal in the near-surface region. In this region only voids are left and the excess Ga atoms, produced by Zn indiffusion and present earlier in the void volume have all disappeared due to efficient outdiffusion of Ga to the surface. In this region the I_{Ga} concentration is close to the corresponding equilibrium value of the As-rich crystal. With reference to local compositions of the GaAs crystal, the local I_{Ga} concentrations may be regarded as the allowed thermal equilibrium values. In a substantial portion of both the profile high concentration and tail regions these values are constants, i.e., those at two different points of the GaAs solidus line. In a transition region the crystal composition changes along the GaAs solidus line and hence also the corresponding I_{Ga} equilibrium concentration value. Therefore, the kink-and-tail profiles can be interpreted as a super-position of two box-like profiles with Zn solubilities and effective diffusivities which are

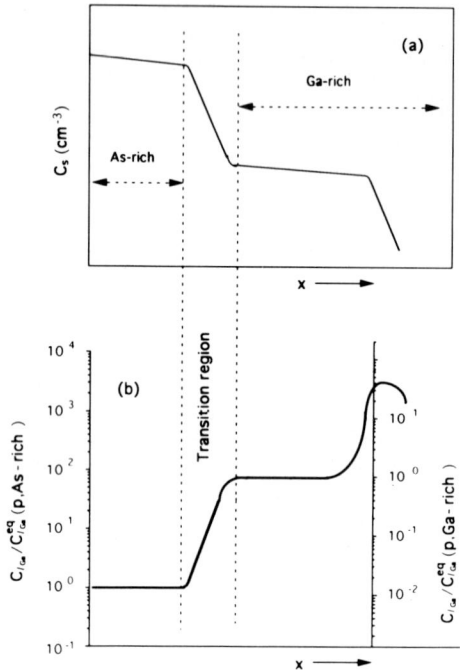

Fig. 3 (a) A schematic Zn indiffusion profile for which the As-rich, the transition, and the Ga-rich regions are indicated. The profile is obtained under As-rich ambient conditions. (b) The (schematic) corresponding Ga self-interstitial distribution in the present thermal equilibrium type interpretation. The left- and right-hand ordinate scale shift is in accordance with the thermal equilibrium Ga self-interstitial concentration dependence on the As_4 pressure value (P_{As4}) at 900°C.

different for the Ga-rich and the As-rich crystal regions. Assuming such local dynamic point defect equilibria and considering the ternary Ga-As-Zn phase diagram the observed Zn solubilities and effective Zn diffusivities were estimated from the dependence of the Ga interstitial concentrations on the Zn and As_4 partial pressure (Fig. 3) and are found to be in good quantitative agreement with the experimental values (Jäger et al. 1992). The fundamental physical reason leading to this situation is that, under the As-rich ambient condition, the vapor phase Zn and As species constitute two competing thermodynamic driving forces for producing GaAs crystals with two extreme compositions: Zn for Ga-rich crystals and As for As-rich crystals. The effective diffusivity of the Zn substitutional species via the kick-out mechanism is very high, while that of As is fairly low. Therefore, during Zn indiffusion, a Ga-rich crystal is first produced, but later on in the near surface region further Zn incorporation onto the Ga sites ceases to be effective which allows the crystal to become As-rich.

REFERENCES

Casey H C 1973 Atomic Diffusion in Semiconductors Ch.7 (London:Plenum Press) p 351
Jäger W, Rucki A, Urban K, Hettwer H-G, Stolwijk N A , Mehrer H and Tan T Y 1992, submitted to J. Appl. Phys.
Morehead F and Gösele U 1981 J. Appl. Phys. 52 4617
Luysberg M, Jäger W, Urban K, Schänzer M, Stolwijk N A and Mehrer H 1989 Mat. Res. Soc. Symp. Vol. 163 659 and 1992 Mat. Sci. Eng. B13 137
Rucki A, Jäger W, Urban K, Hettwer H-G, Stolwijk N A and Mehrer H 1993, to be published
Tan T Y, Gösele U and Yu S 1991 Crit. Rev. Sol. St. Mat. Sci. 17 47
Tuck B 1988 Atomic Diffusion in III/V Semiconductors (Bristol: Adam Hilger)

Inst. Phys. Conf. Ser. No 134: Section 8
Paper presented at Microsc. Semicond. Mater. Conf., Oxford, 5–8 April 1993

535

High-resolution electron microscopy of radiation damage in implanted and laser treated GaAs

N Pashov, M Kalitzova, M Rossi* and G Vitali*

Institute of Solid State Physics, Bulgarian Academy of Sciences, Tzarigradsko Chaussee 72, 1784 Sofia, Bulgaria
*Dipartimento di Energetica, Universita "La Sapienza", via A. Scarpa, 14, 00161 Roma, Italy

ABSTRACT: Single crystals of semi-insulating (100) GaAs implanted with 140 keV Zn^+ at elevated temperature ($110\pm10°C$) to a dose of $10^{14}/cm^2$ were treated with low power laser pulses of 4.5 MW/cm^2. Irradiation induced defect structures in as-implanted GaAs and the degree of restoration after their low power pulsed laser annealing (LPPLA) were investigated by high resolution transmission electron microscopy (HRTEM) of cross-sectional and planar specimens. Digital filtering (by Bragg filters) was applied to experimental HRTEM images to extract additional information about the details in the structure of radiation-induced defect cluster zones.

1. INTRODUCTION

The possibility of annealing ion implanted GaAs by LPPLA treatment at a temperature far below the melting point of the material (Vitali et al 1991a, Vitali et al 1991b, Vitali 1992) has introduced new aspects of interest for electron microscopical study of the fine structural details of the annealed crystals.

When ion implantation doping of GaAs is employed, due to the large energy transfer several thousand atoms are displaced in each collision cascade, thus creating defect zones in the crystal lattice characterized by a vacancy-rich core and a periphery rich in interstitials. The kind of clustering which preferentially develops in such a two-component system far from thermodynamic equilibrium is still a topic for investigation.

In the case of FCC metals it seems now well established that the main effect in the process of implantation is to form dislocation loops of interstitial and vacancy type. The interstitial clusters show an unexpectedly high mobility, as pointed out by Trinkaus and Jäger (1992).

In recent years it has been recognised that HRTEM provides a very valuable, direct insight into the atomic structure of the damage zones and their transformations after annealing (Gibson 1991). Extremely comprehensive information about such structures can be gained by additional digital image processing (De Jong et al 1989, Pashov et al 1992) as will be shown in the present communication regarding HRTEM of Zn^+-implanted and also implanted and subsequently LPPLA treated semi-insulating GaAs.

2. MATERIALS AND METHODS

Single crystal semi-insulating (100) GaAs slices were implanted with Zn^+ ions at an energy of 140 keV, with a fluence of $10^{14}/cm^2$, and at a temperature of $110\pm10°C$, along a random implantation direction. A part of the implanted crystal was then irradiated in air and at room temperature with 30 successive superimposed pulses at $P_o = 4.5$ MW/cm² from a Q-switched ruby laser ($\lambda = 694.3$ nm) with pulse duration $t = 25$ ns. The calculated temperature of the crystal surface during such a treatment does not exceed 700 K (Vitali et al 1991a).

All cross-sectional high resolution transmission electron microscopy (XHRTEM) observations in <110> projection were carried out in a JEOL 100C electron microscope operating at 100 keV. The plan-view specimens were imaged down their [100] axis using a 400 keV JEOL 4000EX instrument. Ion beam and chemically thinned specimens were used for XHRTEM and plan view HRTEM respectively. In both cases, the imaging of the GaAs crystal lattice was performed in bright field many-beam mode, providing "structure images" of the specimens. Digital image processing was performed using a KONTRON system.

3. RESULTS

Figure 1 represents a (110) cross-section image showing the damage depth distribution in (100) implanted GaAs. The left edge of the micrograph corresponds to an irradiated (100) surface plane. Near the edge the GaAs lattice is strongly disturbed as a result of ion beam thinning of the sandwich structure. It should be pointed out that the implantation damage zones are characterized by a specific contrast and appear as dark-centred spots surrounded by grey boundaries (areas marked C in Fig. 1). The concentration of these zones increases with distance from the edge, their mean position being about 85 nm from the top (see the arrow marks on the micrograph). The calculated mean projected ion range and its standard deviation are 61.6 and 27.5 nm, respectively (Kalitzova et al, in press). These values are close to those tabulated for amorphous targets ($R_p = 57$nm and $\Delta R_p = 28$ nm) (Burenkov et al 1986)

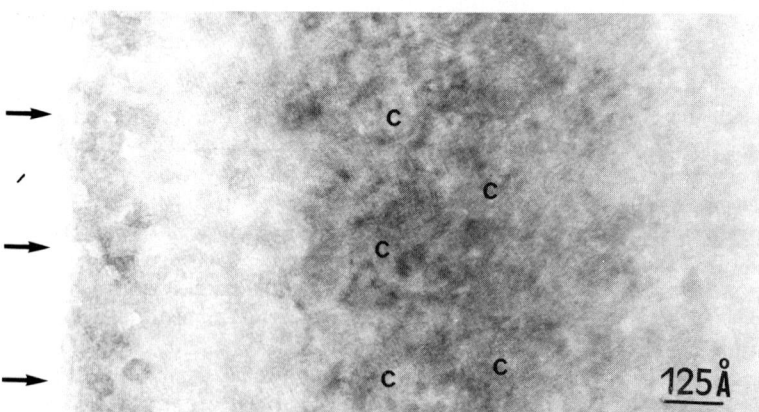

Fig. 1. (110) cross-section, multi-beam image of GaAs. Damage depth distribution is shown.

Due to a poor signal to noise ratio in the inner part of a typical isolated damage zone the lattice fringes in HRTEM micrographs cannot be clearly resolved. In order to obtain more precise information about the kind of localized periodicity disturbances in ion damaged regions in the GaAs lattice, images of isolated damage zones were digitally processed using different types of Bragg filters in Fourier space. After removing background noise, Bragg filters (mask, 20 pixels) using one or two perpendicular pairs of reflections were applied. From experimental results it was concluded that a useful filter configuration was one using a pair of strong reflections corresponding to one set of {220} resolved lattice planes. In this case the reconstructed image of the analysed part of the crystal lattice projection is presented with improved contrast for all periodicity disturbances. Figure 2a shows a filtered image of an isolated damage zone, observed by HRTEM in < 100 > projection, of an as-implanted GaAs crystal. From the filtered image it is evident that this damaged region contains a large number of extended lattice defects, mainly dislocation loops of vacancy and interstitial type (indicated as D_L) and edge dislocations (indicated as D). The number of observable extended defects in this case gave a calculated defect density in the order of $10^{13}/cm^2$. For comparison, a filtered image of an undisturbed part of the sample is shown in Fig. 2b and was obtained using the same processing conditions, in the same (100) HRTEM projection.

Figure 3 shows an isolated damaged zone in an implanted crystal specimen after LPPLA treatment, with exactly the same image filtering procedure as described above. From the filtered image it is clear that no dislocation loops are found in the zone examined, and only very short lattice distortions can be detected. In Fig. 4 a filtered image, generated with a Bragg filter using two pairs of strongly scattering reflections, of a damaged zone in an implanted and LPPLA specimen is presented.

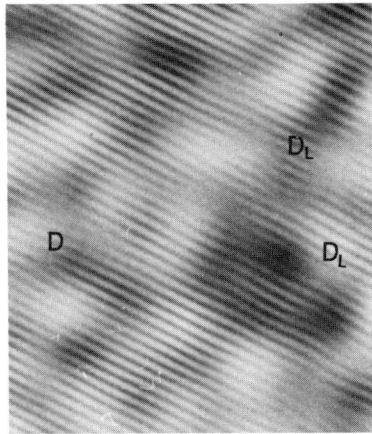

Fig. 2a. Filtered image of an isolated damage zone in as-implanted GaAs.

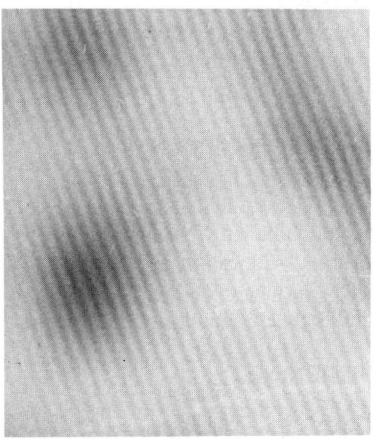

Fig. 2b. Filtered image of an undisturbed part of the GaAs lattice.

Fig. 3. Filtered image of an isolated damage zone after LPPLA.

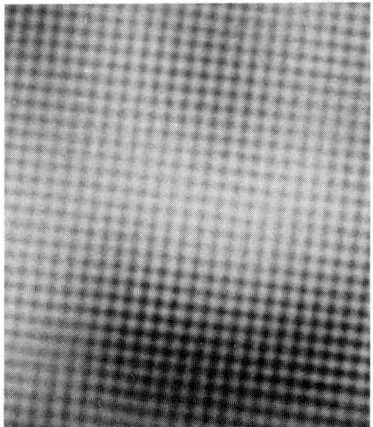

Fig. 4. Filtered image of a damage zone after LPPLA. Two pairs of strongly scattering reflections.

From these examples we note the high efficiency of the applied LPPLA treatment for restoration of the quality of the GaAs lattice disturbed by ion implantation.

It is important to notice that all the above discussed data, concerning the type, concentration and degree of restoration of extended lattice defects in GaAs, relate only to the behaviour of isolated defect zones embedded in an undisturbed matrix. In the case of overlapping defect zones the situation is much more complicated and will not be discussed in this presentation.

4. CONCLUSION

As shown by HRTEM and following image processing, in GaAs crystals implanted at elevated temperature with Zn+ ions, the observed mean depth of the damage distribution is located about 50% deeper than indicated by the tabulated and calculated data. In isolated primary defect zones of as-implanted crystals, we observe clustering of point defects in elongated aggregates as dislocation loops of vacancy and interstitial types and edge dislocations at concentrations in the order of $10^{13}/cm^2$. LPPLA treatment of implanted GaAs leads to nearly complete restoration of the original quality in initially damaged zones.

ACKNOWLEDGEMENTS

This research has been carried out within the framework of the Scientific Agreement between the Italian CNR and the Bulgarian Academy of Sciences. It is financed by the Bulgarian Ministry of Education and Science (Contract F-3). The authors are indebted to Dr R Scholz (Halle, Germany) and to Dr V Yamakov (Sofia, Bulgaria) for their valuable assistance in HRTEM investigations and image processing operations.

REFERENCES

De Jong A F, Coene W and Van Dyck D 1989 Ultramicroscopy 27 53
Burenkov A F, Komarov F F, Kumahov M and Temkin M M 1986 Tables Ion Implantation Spatial Distribution (New York; Gordon and Breach)
Gibson J M 1991 MRS Bull. XVI 3
Kalitzova M, Karpuzov D, Pashov N, Vitali G, Rossi M and Scholz R 1993 Nucl. Instr. Meth. Res. Phys. B (in press)
Pashov N, Kalitzova M, Vitali G, Rossi M and Hofmeister H 1992 Proc. VIIth ISCMP '92, Varna, Bulgaria (Singapore; World Scientific Publ. Co.)
Trinhaus H and Jäger W 1992 IFF Bull. 41 3
Vitali G, Rossi M, Karpuzov D, Budinov H and Kalitzova M 1991a J. Appl. Phys. 69 3882
Vitali G, Rossi M, Karpuzov D, Budinov H, Kalitzova M and Katardjiev I 1991b Nucl. Instr. Meth. Res. Phys. B 50/60 1077
Vitali G 1992 J. Appl. Phys. 31 46

Inst. Phys. Conf. Ser. No 134: Section 8
Paper presented at Microsc. Semicond. Mater. Conf., Oxford, 5–8 April 1993

Defects in Zn-diffused InP single crystals

R H Dixon, W Jäger, A Rucki, K Urban, H-G Hettwer*, N A Stolwijk* and H Mehrer*

Institut für Festkörperforschung Forschungszentrum Jülich D-5170 Jülich Germany
* Institut für Metallforschung Universität Münster D-4400 Münster Germany

ABSTRACT: The defect structure of semi-insulating Fe-doped LEC InP single crystals following Zn ampoule diffusion at 700°C with metallic Zn and P as sources is characterized for different diffusion times ($<$ 3000 min) by analytical transmission electron microscopy of cross-section samples. The observations are correlated with Zn concentration profiles obtained by electron microprobe measurements. A model which describes the formation of the defects is suggested.

1. INTRODUCTION

Zn is the most commonly used p-type dopant for InP-based material in electronic and optoelectronic devices. Diffusion of Zn in InP occurs via an interstitial-substitutional exchange mechanism (Tuck and Hooper **1975**, Serreze and Marek **1986**, Van Gurp et al **1989**). From analyses of the diffusion profiles obtained under various diffusion conditions it has been concluded that the Zn diffusion at high temperatures can be described by models based on the Frank-Turnbull (**1956**) or dissociative mechanism (Tuck **1988**). However, a diffusion mechanism involving self-interstitials at least at high temperatures is indicated by recent experiments of Zn indiffusion into doped InP (Blaauw et al. **1992**) in which enhanced diffusion and subsequent redistribution of dopants has been found and has been explained with the kick-out reaction. This paper compares the results of a transmission electron microscopy investigation of defects formed in Zn-diffused semi-insulating InP crystals with Zn concentration profiles obtained by electron microprobe measurements of the same samples.

2. EXPERIMENTS

Zn was diffused into semi-insulating LEC grown and Fe-doped InP single crystals with a dislocation density of $< 10^5 \text{cm}^{-2}$. The diffusion anneals were carried out in Ar-flushed and sealed quartz ampoules at 700°C for two different diffusion times (700 min, 3000 min) using elemental Zn and P at a weight ratio of 3:2 as the diffusion source. Zn concentration profiles were determined using electron-microprobe depth profiling (EMP) (Hettwer et al **1991**), whereby the total concentration of Zn is measured. The defect structure was investigated in $<110>$ cross-section samples using transmission electron microscopy (TEM) at 400 kV and compared to the microstructure of the non-diffused substrate material before and after annealing at 700°C. The cross-section specimens were prepared by iodine-assisted Ar^+ ion beam thinning to avoid formation of In islands on the specimen surfaces. Precipitates were analysed by energy-dispersive X-ray analyses (EDS) at 200 kV.

3. RESULTS

Fig. 1 shows Zn concentration profiles as measured by EMP after diffusion at 700°C for 3000 min. The profile is characterised by a surface concentration of about 4.10^{19}cm^{-3}, a slow decrease with increasing depth and an abrupt decrease at the front to a value of about 1.10^{17}cm^{-3} which represents the EMP detection limit. The maximum Zn penetration depths observed are 150 μm for t = 3000 min and 70 μm for t = 700 min.

TEM investigations of cross-section samples show that defects are formed only in the diffused samples.

No defects are observed in the annealed substrate material, and only few defects are observed in the unannealed substrate material. However, their different nature and the much lower volume density allow a clear distinction from the defect structure observed in diffused samples. The most important results concerning types and spatial distribution of defects observed at the longer diffusion time will be summarized here.

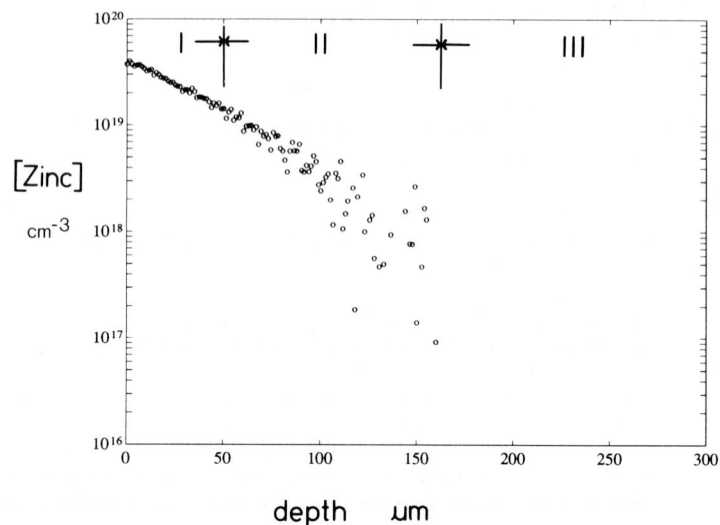

Figure 1 Zn concentration profile in Fe-doped SI InP obtained by EMP after diffusion at 700°C for 3000 minutes.

Fig. 2 shows a typical defect structure near the Zn diffusion front at a depth of about 150 μm consisting of a loose network of dislocations, dislocation loops and a high density of precipitates, imaged as small black dots under the imaging conditions applied. The precipitates are always found to be closely spatially correlated with the dislocations or dislocation loops. Fig. 3 shows a band of precipitates and an isolated dislocation helix that indicates that effective dislocation climb must have occurred. Occasionally, larger precipitates (up to 100 nm) of clearly different contrast are found (Fig.2, the darker precipitates). The defects are observed in a depth region behind the diffusion front extending from about 50 μm to about 150 μm (region II), and beyond the diffusion front to about 500 μm (region III). The high Zn concentration surface region is free of defects up to a depth of 50 μm (region I).

Contrast analyses of the dislocation loops by means of the 'inside-outside' contrast method (Föll and Wilkens 1975) proved the loops to be of interstitial type irrespective of their depth location. All of the loops analysed have {110} habit planes and show no indication of stacking fault contrast when imaged under various

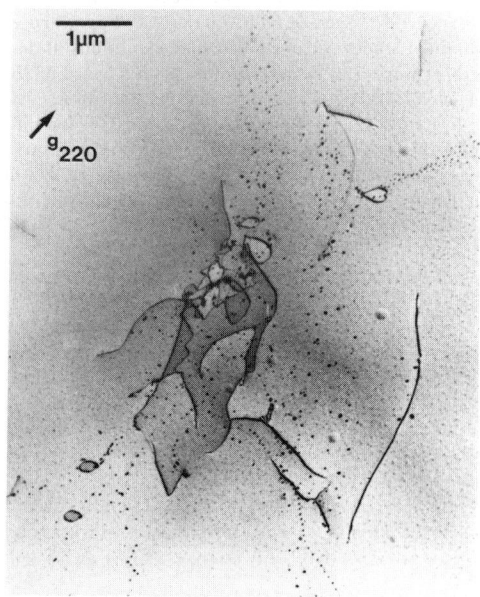

Figure 2 Dislocations, dislocation loops and precipitates in Zn-diffused InP at a depth of about 150μm.

diffraction conditions, indicating that they consist of an extra layer containing both In and P. Contrast analyses of the precipitates using their defocus contrast behaviour show that all precipitates are connected with voids (Fig.4). They are partly facetted and have average diameters of about 25 nm (region II) and 30 nm (region III), at estimated volume densities of about 10^{12}cm^{-3} to 10^{13}cm^{-3}. Precipitate analyses by EDS with electron beam diameters of the same order as the precipitate diameter show in all cases significant enrichment in In with respect to the surrounding InP matrix, irrespective of the precipitate location in the depth profile. This indicates that the precipitates consist predominantly or completely of In. Some of the large precipitates (Fig. 2) are found to be enriched in Fe.

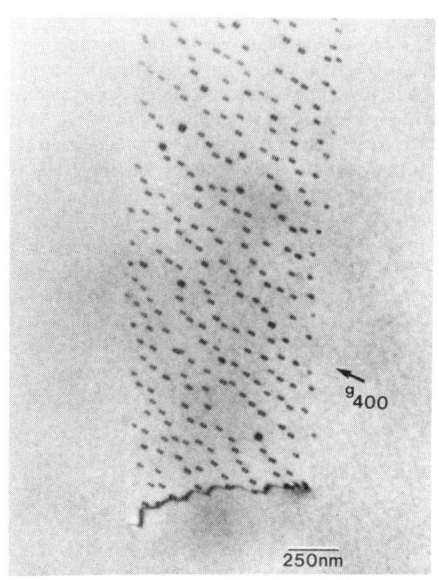

Figure 3 *Dislocation in spatial correlation with a band of precipitates.*

4. DISCUSSION AND CONCLUSIONS

The experimental results show that the diffusion of Zn into the InP lattice leads to concentration profiles (Fig.1) that are similar in shape and surface concentration values to experiments published earlier for comparable diffusion conditions (Tuck and Hooper **1975**, Van Gurp et al. **1989**, Yamada **1983**). A second diffusion tail at low Zn concentrations ($< 5.10^{16}$cm^{-3}) extending to a depth about twice as deep as the penetration depth for the first front has been reported for ampoule diffusion of Zn at elevated temperatures (Yamada et al. **1983**). Whether such a concentration tail is present also under our diffusion conditions is unclear at present because of the EMP detection limit at concentrations of about 10^{17}cm^{-3}.

Comparison of our results from diffused and non-diffused samples show that the defect formation is clearly related to the Zn diffusion. Defects are observed only in depth regions II and III but not in the surface region I (Fig.1), independent of the diffusion time. Based on a comparison of the present results with the Zn diffusion-induced defect formation in GaAs which has been studied extensively for various diffusion conditions (Luysberg et al. **1989**, **1992**, Jäger et al. **1993**), we suggest in the following model the mechanism for defect formation during high concentration Zn indiffusion

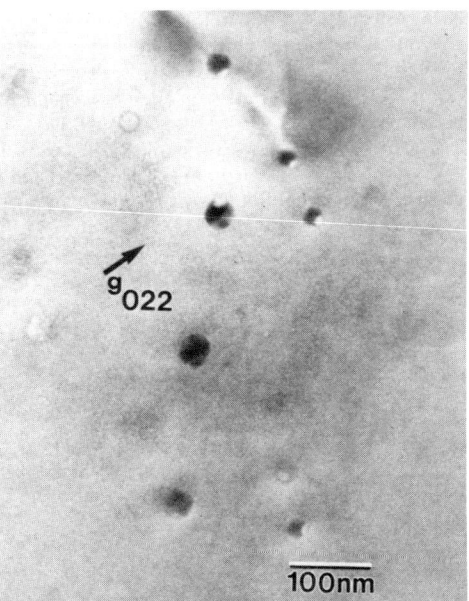

Figure 4 *In-enriched precipitates decorated with voids shown at higher magnification than in figure 2.*

into InP.

The incorporation of Zn on the In sublattice produces a supersaturation of In interstitials by a kick-out reaction. Migrating interstitial In atoms condense into dislocation loops. In order to avoid an energetically unfavourable stacking fault and to maintain stoichiometry P interstitials have to be provided simultaneously. P interstitials can be provided at the loop periphery during loop growth thereby forming free P vacancies. Condensation of these vacancies produces voids which can act as sinks for In interstitials. This explains the observations of voids filled with In. The observed Fe enrichment in larger precipitates could be the result of a segregation of Fe to In-rich precipitates indicating a redistribution of Fe during Zn diffusion. Zn-induced diffusion of Fe has been recently reported also for InP grown by vapor phase epitaxy (Young and Fontijn 1990). The entanglement of loops during growth would result in the formation of dislocation networks. Dislocation climb by absorption of further In interstitials explains the observed spatial correlation of precipitates and dislocations.

The spatial distribution of defects indicates that a point defect supersaturation leading to defect production is effectively maintained only in a depth region near the diffusion front and beyond (up to 500 μm for t = 3000 mins). The observation of a surface zone free of defects can be understood by assuming that defect annealing at long diffusion times at 700°C has taken place.

REFERENCES

Blaauw C, Emmerstorfer B, Kreller D, Hobbs L and Springthorpe A J, 1992 J. Electronic Mat. 21 173
Föll H and Wilkens M, 1975 Phys. Stat. Sol. 31 519
Frank F C and Turnball D, 1956 Phys. Rev. 104 617
Hettwer H-G, Lerch W, Lentfort B, Stolwijk N A and Mehrer H, 1991 Appl. Surf. Sci. 50 470
Jäger W, Rucki A, Urban K, Hettwer H-G, Stolwijk N A and Mehrer H, 1992 Submitted to J. Appl. Phys.
Luysberg M, Jäger W, Urban K, Schänzer M, Stolwijk N A and Mehrer H, 1989 Mat. Res. Soc. Symp. Vol.163 659 and 1992 Mat. Sci. Eng. B13 137
Serreze H B and Marek H S, 1986 Appl. Phys. Lett. 49 210
Tan T Y, Gösele U and Yu S, 1991 Crit. Rev. Sol. St. Mat. Sci. 17 47
Tuck B 1988 Atomic Diffusion in III-V Semiconductors (Bristol Adam Hilger)
Tuck B and Hooper A, 1975 J. Phys. D:Appl. Phys. Vol. 8 1806
Van Gurp G J, van Dongen T, Fontijn G M, Jacobs J M and Tjaden D L A, 1989 J. Appl. Phys. 65 553
Yamada M, Tien P K, Martin R J, Nahory R E and Ballman A A, 1983 Appl. Phys. Lett. 43 594
Young E W A and Fontijn G M, 1990 Appl. Phys. Lett. 56 146

Inst. Phys. Conf. Ser. No 134: Section 8
Paper presented at Microsc. Semicond. Mater. Conf., Oxford, 5–8 April 1993

543

Basal and prismatic defects in α-SiC 6H single crystals irradiated at the GANIL accelerator with 5.5 GeV Xe ions

I Lhermitte-Sebire[1], J Vicens[1], J L Chermant[1], M Levalois[1] and E Paumier[1,2]

[1]LERMAT, URA CNRS No1317, ISMRA, 6 bd du Maréchal Juin, 14050 Caen Cedex, France
[2]CIRIL, Rue Claude Bloch, BP 5133, 14040 Caen Cedex, France

ABSTRACT : Single crystals of α-SiC 6H were irradiated at the GANIL accelerator with 5.5 GeV Xe ions. Irradiated samples were observed in TEM after annealing at 1373K. Dislocation loops were found in the whole irradiated area except close to the implantation zone located at 405μm from the surface. Dislocation loops identified as interstitial faults were observed in $\{10\bar{1}0\}$ prismatic and $\{0001\}$ basal planes. From contrast analyses a geometrical model is proposed for the two types of faults.

1. INTRODUCTION

Silicon carbide is an important material for its thermal, mechanical and semiconducting properties. This is also one of the candidates for the first wall of fusion reactors because of its small neutron capture cross-section and its high chemical stability. Electronic property modifications after irradiation by light particles were measured in SiC: photoluminescence of radiation defects in cubic SiC (Choyke and Patrick, 1971 ; Choyke, 1990) and electrical conductivity modification (Corelli et al., 1983). By optical absorption measurements, Yasuda et al., 1982, showed that the band gap of SiC was narrowed after electron irradiation. The Rutherford back scattering technique was also used on samples implanted with H^+, N^+ or Al^+ ions at 300K (Spitznagel et al., 1986). More recently the crystalline to amorphous transition induced by electron irradiation was studied in high voltage electron microscopy (Inui et al., 1990, 1992 a and b). Observation of structural defects was also performed by transmission electron microscopy (TEM) after neutron irradiation (Price, 1973; Yano et al., 1988). Recently Yano and Iseki (1990) studied by high resolution electron microscopy (HREM) interstitial loops in α and β SiC after neutron irradiation. The aim of this work is to investigate structural defects created in α SiC single crystals by swift Xe ions, using TEM.

2. EXPERIMENTAL

Single crystals of 6H α-SiC, p type (p = 10^{17} cm^{-3})*, were irradiated by 5.5 GeV Xe (44^+) ions supplied by GANIL with the ion beam direction along the <0001> crystal orientation. Total fluences were chosen in the range 10^{13} to 10^{15} Xe.cm^{-2}. The implantation zone was located at 405 μm from the crystal surface as calculated by the TRIM program (Biersack and

* Supplied by Elektroschmelzwerk Company, Kempten (Germany)

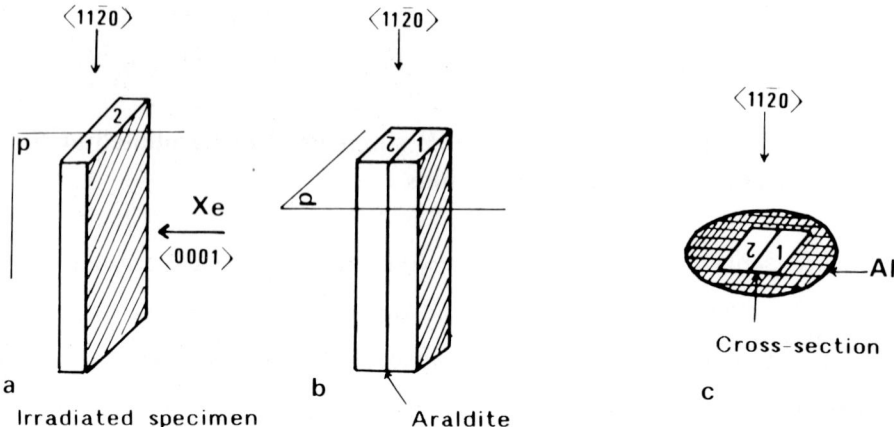

Fig. 1, (a) Single crystal (700 μm thick) irradiated along [0001]. Specimen was cut along p plane, (b) two parts 1 and 2 were stuck with aral-
dite, (c) cross-section of the sample inserted in an aluminum cylinder.

Haggmark, 1980). In order to study the irradiated area, cross-sections
were prepared as shown in figure 1. The disc-shaped samples were mechanically polished (10 μm thick) then ion-milled with argon flux (Gatan,
5kV). Contrast studies of defects were performed with a Jeol 200 CX
microscope operating at 200 kV.

3. RESULTS

As-received 6H SiC crystals have a good crystallographic quality. The
rarely observed defects consist of dissociated Shockley dislocations,
most likely introduced during the specimen preparation. After irradiation
experiments, the colour of the crystal changed and became yellow in irradiated areas (10^{14} Xe.cm^{-2}) while they have a blue grey colour in the
as-received state. However no additionnal defects could be detected by
TEM or HREM in irradiated samples. Thus annealing of the irradiated samples (1h at 1373K under argon flux) was performed is order to coalesce
defects in the crystal.

Observations of a sample irradiated with 10^{14} Xe.cm^{-2} and annealed
reveal the presence of numerous dislocation loops in irradiated zones.
The size and density of loops depend on the depth where they are located.
Loop sizes decrease when approaching the implantation zone and loops
appear as black dots. However, contrary to all expectations, the implantation zone is deprived of any loops. A first example (Fig. 2a) shows
loops nucleated close to the surface of the irradiated crystal (100 μm
deep) while an area located at 350 μm in depth from the surface is illustrated in figure 2b. Near the surface numerous loops are found to be
parallel to the basal plane (Fig. 2a). When the density of loops increases, prismatic {10$\bar{1}$0} and basal {0001} loops are imaged (Fig. 2c). In the
orientation shown both faults are parallel to the electron beam. Prismatic {10$\bar{1}$0} loops are unusual defects by comparison with results already
published on α-SiC or compounds with an equivalent crystallographic description (Yano and Iseki, 1990). It is not the case for basal loops, identified in HREM images as interstitial loops with $\vec{b}=1/6$ [0001] = 0.25nm.

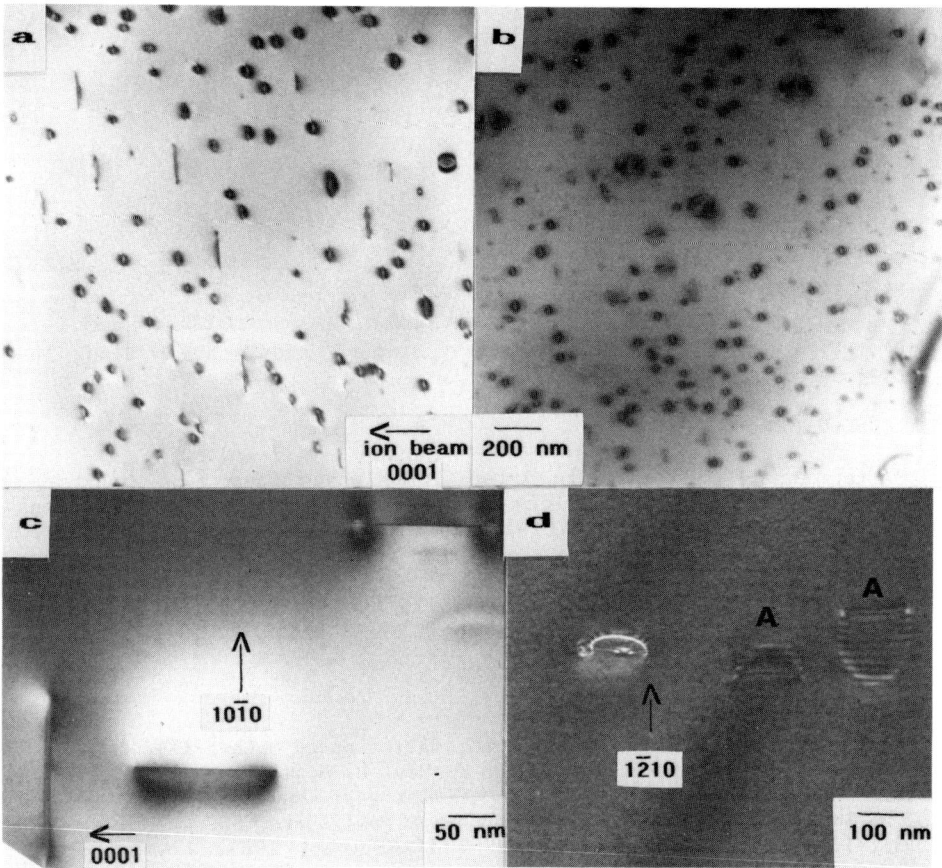

Fig. 2, (a) Loop precipitation in irradiated then annealed SiC sample near the surface. (b) At 350 μm from the surface. (c) Example of prismatic ($10\bar{1}0$) and basal (0001) loops. In this orientation loops are parallel to the electron beam. (d) Weak beam image of prismatic loops with $\vec{g}=1\bar{2}10$. Loops denoted A are out of contrast.

Contrast analyses have been performed on prismatic faults. They are found lying in the three equivalent prismatic planes and fault contrast is imaged in two beam conditions with $\vec{g}$ = 0006. Moreover, as shown in figure 2d, one family (denoted A) is out of contrast with $\vec{g}$ = $1\bar{2}10$. These results indicate that the Burgers vector is of the type 1/n $<10\bar{1}\ell>$. The HREM image (Fig.3) shows also that the fault plane does not belong to the prismatic plane at the atomic level but is periodically shared between the two ($10\bar{1}2$) and ($10\bar{1}2$) pyramidal planes. This gives rise to a zig-zag configuration of the fault plane. All these observations can be interpreted by an interstitial precipitation on {$10\bar{1}2$} planes inducing a displacement $\vec{b}_2$ = 1/18 $<80\bar{8}1>$ = 0.25 nm.

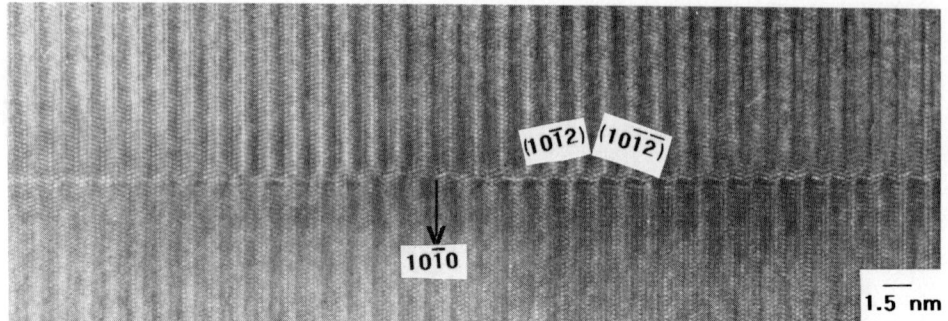

Fig. 3, Prismatic fault showing the "zig-zag" configuration along the
$(10\bar{1}2)$ and $(10\bar{1}2)$ planes.

4. CONCLUSION

After annealing treatments interstitial loops have been found and
analysed in 6H α-SiC single crystals irradiated with 5.5 GeV Xe ions
supplied by GANIL. Two types of interstitial loops have been identified
in basal and prismatic planes. Prismatic faults appear as a new defect
for the 6H polytype and correspond to interstitial precipitation in
$\{10\bar{1}2\}$ pyramidal planes. Prismatic faults have a zig-zag configuration
and a Burgers vector $\vec{b}_2$ = 1/18 <80$\bar{8}$1> = 0.25 nm.

REFERENCES

BIERSACK J.P. and HAGGMARK L.G., 1980, Nucl. Instr. Meth., **174**, 257.
CHOYKE W.J. and PATRICK L., 1971, Phys. Rev. B, **4**, 1843.
CHOYKE W.J., 1990, The Physics and Chemistry of Carbides ; Nitrides and
    Borides, FREER R. (ed), Kluwer Academic Publishers, 563.
CORELLI J.C., HOOLE J., LAZARRO J. and LEE C.W., 1983, J.Amer. Ceram.
    Soc., **66**, 529.
INUI H., MORI H. and FUJITA H., 1990, Phil. Mag. B., **61**, 107.
INUI H., MORI H., SUZUKI A. and FUJITA H., 1992a, Phil. Mag. B, **65**, 1.
INUI H., MORI H. and SAKATA T., 1992b, Phil. Mag. A, 66, 737.
PRICE R.J., 1973, J. Nucl. Mater., **48**, 47.
SPITZNAGEL J.A., WOOD S., CHOYKE W.J., DOYLE N.J., BRADSHAW J. and
    FISHMAN S.G., 1986, Nucl. Instr. Meth., B **16**, 237.
YANO T. and ISEKI T., 1990, Phil. Mag. A, **62**, 421.
YANO T., SUZUKI T., MARUYAMA T. and ISEKI T., 1988, J. Nucl. Mater., **155**,
    311.
YASUDA K., TAKEDA M., MASUDA H. and YOSHIDA A., 1982, Phys. Stat. Sol.
    (a), **71**, 549.

ACKNOWLEDGEMENTS

The authors wish to thank Dr S. Bouffard (CIRIL) for helpful discussion and the technical staff of CIRIL for his help during irradiation
experiments. One author (I.L.S.) would like to thank the CNRS (France)
and SEP (Etablissement de Bordeaux, France) for their financial support.

Ion implantation effects in single crystal CuInSe$_2$

C A Mullan, C J Kiely, M V Yakushev[1], M Imanieh[1], R D Tomlinson[1] and A Rockett[2]

Department of Materials Science and Engineering, The University of Liverpool, Liverpool, UK.
[1]Department of Electronic and Electrical Engineering, The University of Salford, Salford, UK.
[2]Coordinated Science Laboratory, The University of Illinois at Urbana-Champaign, Illinois, USA.

ABSTRACT : A series of CuInSe$_2$ single crystals have been implanted with oxygen, xenon and hydrogen ions. Oxygen (and to a lesser extent Xe) implantation causes an n to p-type carrier conversion and an enhancement of photoconductivity. Proton implants cause a p to n-type change. A combination of Transmission Electron Microscopy (TEM) and Secondary Ion Mass Spectroscopy (SIMS) has been used to characterise such implants and the results are correlated with photoconductive response measurements.

1. INTRODUCTION

CuInSe$_2$ is a I-III-VI$_2$ ternary semiconductor which has the chalcopyrite structure. Its high optical absorption coefficient and direct band gap make it an excellent choice for use as an absorber layer in thin film solar cells (Mitchell, 1989). The electrical properties of CuInSe$_2$ are controlled by the complex interplay between compensating populations of intrinsic point defects (Neumann, 1983). Calculations suggest that vacancies on the cation sublattice and Cu$_{In}$ antisite defects are p-type centres whereas Se vacancies and In$_{Cu}$ antisite defects are n-type centres (Rinçon et al, 1989). In practise, the conductivity type in a thin film device is usually controlled by varying the Cu:In ratio.

Noufi et al (1986) found that the efficiency of polycrystalline CuInSe$_2$ thin film solar cells could be significantly increased by annealing the layer in an oxidizing atmosphere. This prompted Tomlinson et al (1991) to study the effects of introducing controlled amounts of oxygen (and a variety of other elemental species) into the near surface regions of n-type CuInSe$_2$ single crystals by ion implantation. They found that oxygen implantation caused an n to p-type conversion in the surface of the material and also produced a significantly enhanced photoconductive response. In this paper, we present a study of the microstructural changes caused by oxygen, xenon and proton implantation into single crystal CuInSe$_2$ and correlate our observations with the opto-electronic characteristics of the material after implantation.

2. EXPERIMENTAL

Single crystal material was grown by the vertical Bridgman technique (Tomlinson, 1989). Wafers, n-type in character, were cut from the central regions of the melt-textured ingot as close to the (221) orientation as possible. Prior to implantation they were etched in 0.1% Br/methanol solution for 20 seconds. A 40keV ion separator was used for the ion implantation and doses of O, Xe and H in the range 10^{15}-10^{17} cm^{-2} were employed. Cross sectional TEM samples were made using standard preparation techniques so that the damage profile as a function of depth from the implanted surface could be characterised. Considerable care must be taken when ion milling CuInSe$_2$ in order to minimize preferential sputtering artefacts (Mullan et al, 1992). TEM analyses were carried out on JEOL 2000 series electron microscopes operating at 200kV. Chemical depth profiles of the implants were obtained using a Cameca IMS 5f Secondary Ion Mass Spectrometer (SIMS). The relative photoconductivities of samples as a function of wavelength were measured using a chopped light source and a tuned phase sensitive detector.

3. RESULTS AND DISCUSSION

3.1 Oxygen Implantation

Figure 1 shows the variation in photoconductivity with incident wavelength for a series of CuInSe$_2$ samples which have been given different implant doses of oxygen ions. It is clear that these CuInSe$_2$ crystals show a substantially enhanced photoconductive response compared with the as-grown n-type material; the maximum increase occurs for an implant dose of 10^{16}cm^{-2}. A conversion from n to p-type conductivity was also observed in each case. The oxygen serves to remove donor states (possibly by filling Se vacancies) and thereby alters the balance of compensating defects in favour of the acceptor state population. The accompanying optical sensitization is thought to be associated with an overall decrease in carrier recombination caused by the presence of an acceptor state that has a high capture cross-section for holes and a low capture cross-section for electrons (Tomlinson et al, 1991). One oxygen implanted sample (dose 2x10^{16}cm^{-2}) was annealed in forming gas for 1 hr at 200°C. This heat treatment was seen to further increase the photoconductive response as shown in Figure 2.

Figures 3(a),(b) and (c) show cross-sectional TEM micrographs of CuInSe$_2$ samples which have been given oxygen implants of 10^{17}, 5x10^{16} and 10^{16}cm^{-2} respectively. The sub-surface region of the most heavily implanted sample shows three distinct layers, labelled A,B and C in fig 3(a). Layer A extends down about 25nm from the top surface and consists of single crystal CuInSe$_2$ containing only a few microtwin defects on {112} planes. Layer B is a 60nm thick amorphised layer which contains the peak of the implant. Layer C is a heavily damaged crystalline layer containing a high density of microtwins and extrinsic dislocation loops. No planar or line defects were observed deeper than 160nm into the sample. These observations are in good agreement with profiles calculated using the TRIM-89 algorithm which predicts that the peak of the implant should be at a depth of 72nm with a damage tail extending about 170nm into the crystal. The existence of a buried amorphous layer is consistent with previous observations (Tomlinson et al, 1991) that very high dose implanted samples tend to have a peak in the sensitised photoconductivity spectrum at 900nm corresponding to an energy transition at 1.378nm (which is the reported band gap of amorphous CuInSe$_2$ (Neumann et al, 1982)).

The damage caused by the 5x10^{16}cm^{-2} implant is shown in fig.3(b). A 130nm thick layer below the surface is seen to contain a high density of dislocations and stacking faults. A piece of this material which was annealed at 200°C (ie about 0.38T$_m$) in forming gas for 30 minutes was found by TEM to have a similar damage profile. The improvement in photoconductive response after such an annealing step is thought to be primarily due to the redistribution of point defects (in particular allowing oxygen trapped in interstitial sites to diffuse to substitutional positions).

Figure 3(c) shows that the visible lattice damage caused by the 10^{16}cm^{-2} implant is mainly limited to some microtwinning and stacking faults (intrinsic near the surface, extrinsic deeper down) lying on {112} planes. No defects were observed deeper than 100nm in this sample.

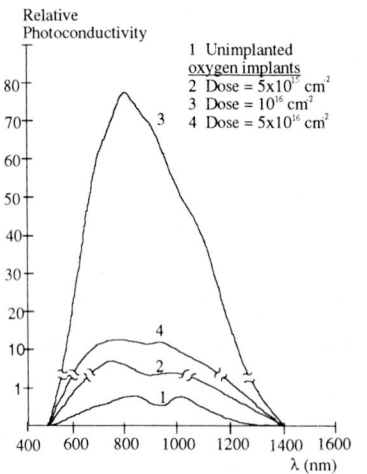

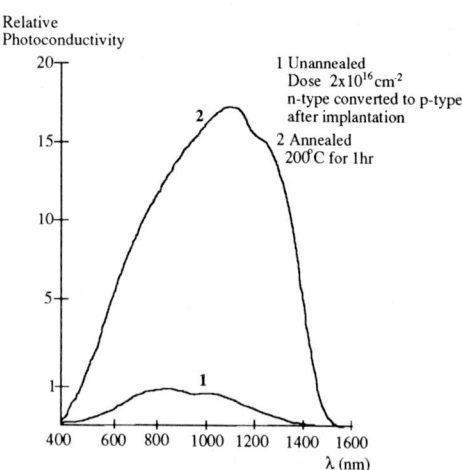

Fig.1 The effect of oxygen implant dose on photoconductive response.

Fig.2 The effect of annealing on the photoconductive response (dose = 2x10^{16}cm^{-2})

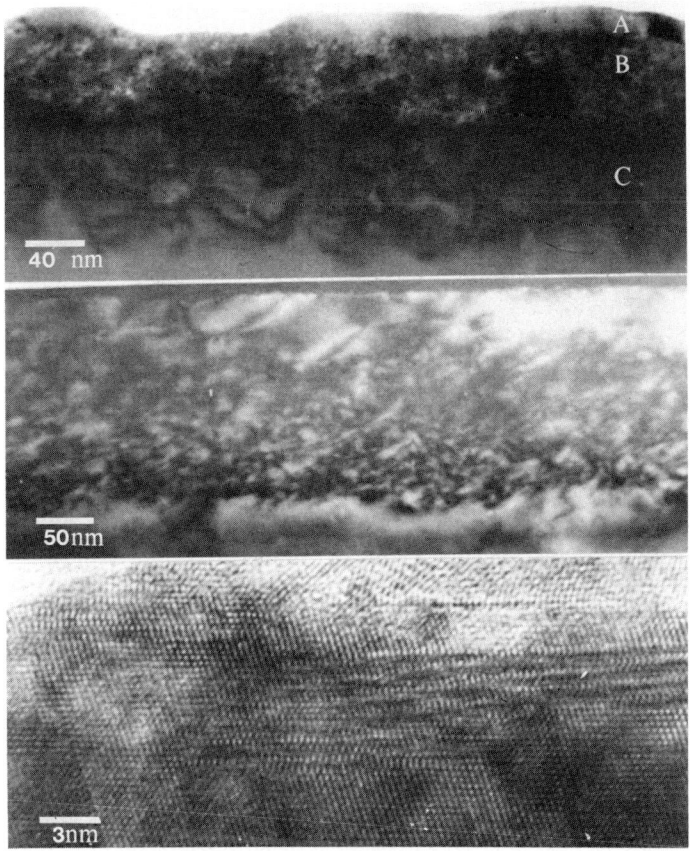

Fig.3 Cross sectional micrographs showing sub-surface damage in a series of oxygen implanted samples. Implant doses were; (a) $10^{17}cm^{-2}$, (b) $5x10^{16}cm^{-2}$ and (c)$10^{16}cm^{-2}$.

Furthermore, no line or planar defect damage has been observed in samples with implant doses below $5x10^{15}cm^{-2}$. This suggests that substitutional oxygen incorporation into $CuInSe_2$ without inducing excessive amounts of near-surface damage is the most beneficial way of improving its photoconductive response.

3.2 Xenon implantation

Implantation of $CuInSe_2$ with inert ions (such as Xe and Ne) has also been found to improve the photoconductive response relative to unimplanted material (Tomlinson et al, 1991). However, the photo-sensitization effect is much less marked than for oxygen implantation (ie, improvements of 5-fold instead of 100-fold are noted). In addition, n to p-type carrier conversion only occurs at implant doses greater than $1x10^{16}cm^{-2}$. Figure 4 shows an axial (021) HREM image of the sub-surface region of a $10^{16}cm^{-2}$ Xe implanted sample in which stacking faults and microtwinning on {112}-type planes is apparent down to a maximum depth of 33nm (again in good agreement with depth ranges suggested by theoretical TRIM profiles). Figure 5 shows a SIMS depth profile of this implant in which it is clear that oxygen is also present in the sub-surface region. The oxygen profile closely follows the Xe damage profile, suggesting that the lattice damage creates fast diffusion paths for oxygen to get into the material after implantation. Hence, we believe that the carrier type conversion and improvement in photoconductive response observed in samples implanted with inert atoms are strongly influenced by the presence of oxygen ions which enter the material by diffusion after implantation.

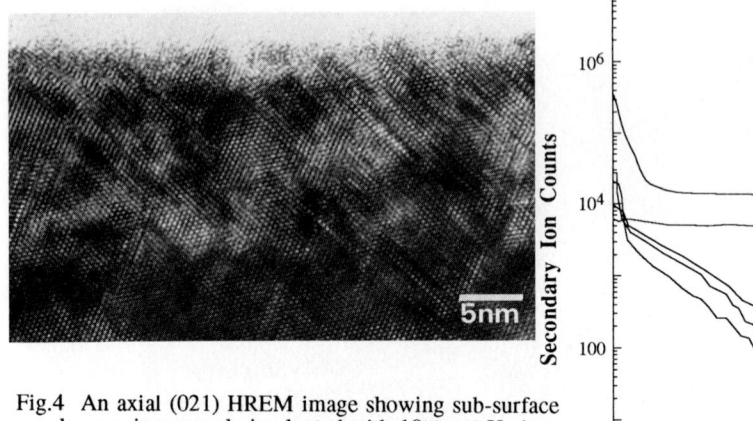

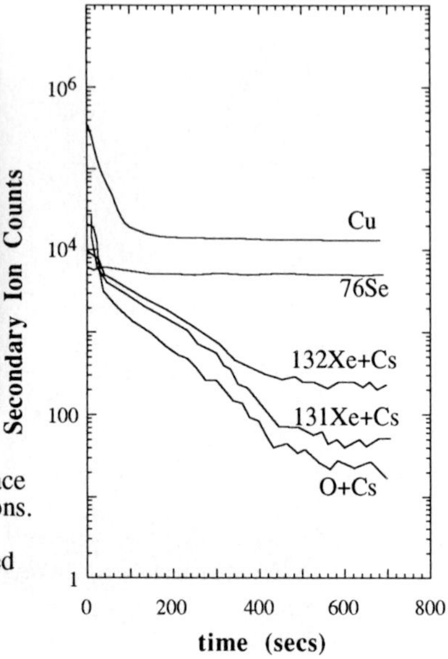

Fig.4   An axial (021) HREM image showing sub-surface
   damage in a sample implanted with $10^{16}cm^{-2}$ Xe ions.

Fig.5   SIMS depth profile of the $10^{16}cm^{-2}$ Xe implanted
   specimen.

3.3 Proton implantation

Several p-type CuInSe$_2$ samples have recently been implanted with 40keV protons to doses up
to $5 \times 10^{15}cm^{-2}$. Contrary to all the other ion-species we have implanted, this causes causes a type
conversion from p to n-type. In addition, definite photo-sensitization and carrier lifetime
improvements have been measured in such specimens. Cross-sectional TEM examination of
proton implanted material has not revealed any line or planar defects down to depths of 400nm.
It is thought that the H+ ions may in this case be filling Cu vacancy sites, thus tipping the balance
of compensating defects in favour of the donor state population.

4. CONCLUSIONS

Oxygen implantation of single crystal CuInSe$_2$ causes an n to p-type carrier conversion and a
strong enhancement of photoconductive response. The largest effect occurs for implant doses of
$10^{16}cm^{-2}$. Above this implant level, a large degree of lattice damage occurs and carrier
recombination effects become important. Low temperature annealing treatments on some samples
are seen to further improve the photoresponse. Implantation with inert xenon ions similarly cause
a weaker increase in photoresponse and an n to p-type conversion at high implant doses. This is
thought to occur as a result of enhanced oxygen in-diffusion through the highly damaged sub-
surface layers of the material. Preliminary work on proton implantation in CuInSe$_2$ shows that
implant doses as low as $10^{15}cm^{-2}$ are effective in causing p to n-type carrier conversions and can
give rise to significant improvements in photoconductive response.

REFERENCES

Mitchell K, 1989, Proc 9th E.C Photovoltaic Solar Energy Conference, Frieburg,, 292.
Mullan CA, Kiely CJ, Rockett A, Imanieh M, Yakushev MV and Tomlinson RD, 1992,
   Proc.Mat.Res.Soc.Symp. 262, 1097.
Neumann H, 1983, Cryst.Res.Technol., 18, 483.
Neumann H, Perlt S, Abdul-Hussein NAK, Tomlinson RD and Hill AE, 1982, Solid State
   Commun 42, 855.
Noufi R, Matson RJ, Powell RC, and Herrington C, 1986 Solar Cells, 16, 479.
Rinçon C, Bellabarba C, Gonzalez J and Sanchez-Perez G, 1989, Solar Cells, 16, 335.
Tomlinson RD, 1989, PVSEC, Sydney, Australia, 467.
Tomlinson RD, Hill AE, Imanieh M, Pilkington RD, Roodbarmohammadi A, Slifkin MA
   and Yakushev MV, 1991, J. Electronic Mats., 20, 659.

Inst. Phys. Conf. Ser. No 134: Section 8
Paper presented at Microsc. Semicond. Mater. Conf., Oxford, 5–8 April 1993

551

Convergent beam electron diffraction analysis of doped indium selenide crystals

D Manno*, R Rella** and P Siciliano**

* Dipartimento di Scienza dei Materiali, Università di Lecce - Via Arnesano 73100 Lecce (Italy); ** Istituto Materiali per l'Elettronica (IME-CNR) Via Arnesano 73100 Lecce (Italy)

ABSTRACT: Structural analysis of InSe, grown from the melt by the Bridgman-Stockbarger method doped with different elements (such as Cd, As, Cl), has been performed by convergent beam electron diffraction techniques in order to study the effect of impurities on the crystal structure. We have observed that Cd and As doped InSe crystals are made of only one structural modification, that is the γ rhombohedral polytype. On the contrary, the Cl doped InSe is made of a superposition of several polytypes in addition to the γ one.

1. INTRODUCTION

Indium selenide is a layered III-VI semiconductor whose crystal structure consists of graphite-like layers bound by Van der Waals forces. The optical and electrical properties of this material are suitable for some applications in solid state devices. In particular InSe is a possible candidate material to form heterojunction devices with a very low density of interface states.

InSe crystals are obtained by the stacking of tightly bound four-atom layers (Lieth 1976). Each layer consists of four two-dimensional sheets of like atoms, which, along the high symmetry axis (c-axis), are in the sequence Se-In-In-Se. These fourfold layers are bound together by Van der Waals forces which give rise to different stackings of layers and so to the polytypism.

Several authors (Celustka et al. 1974, Chevy et al. 1981, De Blasi et al. 1990) have investigated the structure of undoped InSe crystals grown by different methods and have confirmed that this material exhibits several crystallographic modifications, some of which can coexist in the same crystal. On the contrary, as we know, in literature no results are reported about the structure of doped InSe and about the polytype occurrence; only electrical and optical properties are reported.

In order to gain more information about the structure of doped InSe crystals, we have analyzed this material by convergent beam electron diffraction (CBED) techniques. In this paper we report the preliminary results of structural analysis of the InSe crystals doped with Cd, As, and Cl elements.

2. EXPERIMENTAL

The InSe single crystals analyzed in this work have been grown in our laboratory by the Bridgman-Stockbarger method from an indium rich polycrystalline melt in the ratio 52 at %

of indium and 48 at % of selenium. Details of the experimental procedures for crystal growth are reported elsewhere (De Blasi et al. 1982). Cd, As and Cl were introduced into the polycrystalline powder as CdSe, InAs and InCl₃ compounds, respectively.

The analyzed samples have been obtained by cleavage from different doped InSe ingots. Repeated cleavages give samples with regions sufficiently thin for TEM observation without using other thinning methods. Several regions of each sample have been analyzed by a Philips EM 400T transmission electron microscope, operating at a nominal accelerating voltage of 120 kV. Both conventional convergent beam patterns at low camera length and large angle convergent beam patterns have been recorded.

The lattice parameters have been evaluated by using the zero order Laue reflections (hk.0) to calculate the a-axis and the radii R of the higher order Laue zone (HOLZ) rings for the c-axis, according to procedure described by De Blasi et al (1990).

The diffraction patterns, observed in regions of samples made by a mixture of structural modifications, have shown the lowest symmetry of the crystal, therefore the polytypes have been identified by evaluating the lattice parameters.

3. RESULTS AND DISCUSSIONS

Fig. 1 shows typical Tanaka (a) and conventional (b) convergent beam [00.1] zone-axis patterns obtained from As doped InSe. Both fig.1 a and b show 3m symmetry, that corresponds to the symmetry of a rhombohedral material (point group 3m). In addition only three HOLZ rings are present in picture b, the evaluation of 1/H values of each ring allows us to say that the c-axis of the unit cell is 2.5 nm. We can conclude that the three HOLZ rings correspond to first, second, and third order of γ rhombohedral polytype, respectively.

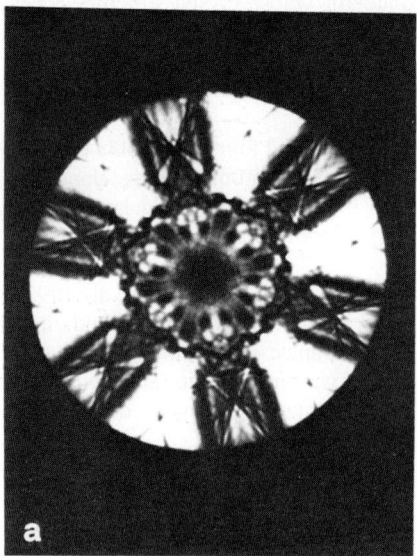

Figure 1 Tanaka (a) and conventional (b) convergent beam [00.1] zone-axis patterns recorded from As doped InSe.

As regards Cd doped InSe, we have observed the material is made by only one structural modification: the γ polytype, in the same way as the As doped InSe.

In a previous paper (De Blasi et al. 1990) it has been observed that the undoped InSe grown from the melt is made of a mixture of four different structural modifications: the 2H-hexagonal ε, the 3R-rhombohedral γ, the 4H-hexagonal δ, and the 9R-rhombohedral polytype. So we can conclude that adding As or Cd in the polycrystalline powder we obtain a crystal characterized by a higher structural order along the c-axis with respect to undoped InSe.

Fig. 2 shows typical Tanaka (a) and conventional (b) convergent beam [00.1] zone-axis patterns obtained from Cl doped InSe. Fig.2a exhibits a slight asymmetry since the observed region is bent (no defect free regions have been observed in Cl doped ingots). Fig.2b shows 3m symmetry and several discontinuous HOLZ rings.

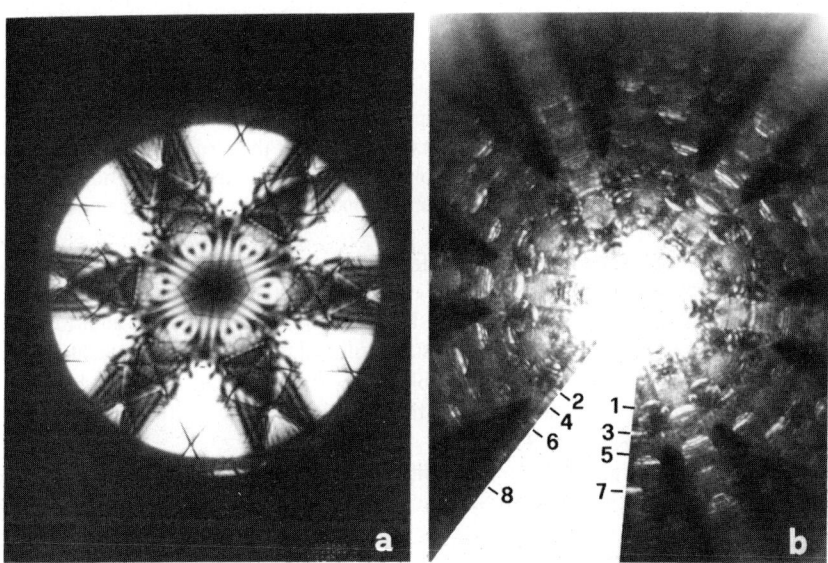

Figure 2 Tanaka (a) and conventional (b) convergent beam [00.1] zone-axis patterns recorded from Cl doped InSe.

TABLE I

Ring	c/n (nm)		HOLZ order	polytype
	As doped InSe	Undoped InSe		
1	7.6	7.6	1	9R
2	5.05		2	12R
3	3.8	3.8	2	9R
4	3.4		3	12R
5	3.3	3.3	1	δ – 4H
6	2.5	2.5	1	γ –3R
			3	9R
			4	12R
7	1.67	1.67	1	ε – 2H
			2	δ – 4H
8	1.11	1.16	3	δ – 4H

Eight rings are evident in fig.2b; the diameters of the rings have been measured and the 1/H values have been determined. As described by De Blasi et al. (1990), for hexagonal lattices (or rhombohedral ones referred to hexagonal unit cells) and [00.1] zone-axis diffraction patterns we have 1/H = c/n (where n is the order of the HOLZ ring). The obtained c/n values are reported in table I together with the corresponding polytype.

By comparison, in table I are also reported the c/n values obtained from undoped InSe (De Blasi et al. 1990). As the 12R structural modification is not present in undoped InSe crystals, we can conclude that the Cl dopant element increases the structural disorder.

An electrical analysis, carried out by Van der Pauw method with the current flowing along the layers, showed the undoped and Cl-doped InSe are n-type while the InSe doped with Cd and As is p-type.

Moreover, deep levels have been detected by DLTS measurements with an ionization energy between 0.45 and 0.6 eV (Micocci et al. 1981, 1992a, 1992b) probably associated with defects or defect complexes due to dopant atoms precipitated in the interlayer regions (Chevy 1984).

4. CONCLUSIONS

Convergent beam electron diffraction has given detailed information about the dependence of InSe structure on dopant elements. In particular the structural analysis showed that the p-doped InSe is made of only one polytype; on the contrary the n-doped InSe is made of a superposition of several polytypes, more than the ones present in undoped InSe (De Blasi et al. 1990).

Nevertheless to make assumptions about the mechanisms determining the InSe structure is difficult because all dopant atoms (p and n type) can precipitate in interlayer sites as excess of indium in undoped InSe as revealed by DLTS measurements. In addition, with regard to the analysed p-type InSe ingots, As is substitutional of Se and Cd is substitutional of In and both the crystals show the same rhombohedral γ structure. On the other hand, in Cl doped InSe, the doping atom replaces Se and the material is made by the superposition of several polytypes (ε, γ, δ, 9R and 12R).

Further analyses are in progress on other different doping atoms in order to understand the function of dopant elements on the InSe structure.

ACKNOWLEDGEMENTS

The authors would like to express their gratitude to Prof. A.Tepore for stimulating discussion, to Prof. G. Micocci for growing crystals and to Mr. L.Monteduro for his technical support.

REFERENCES

Celustka B and Popovic S 1974 J. Phys.Chem. Solids 35 287
Chevy A 1981 J. Cryst. Growth 51 157
Chevy A 1984 J. Appl. Phys. 56 987
De Blasi C, Micocci G, Mongelli S and Tepore A 1982 J. Cryst. Growth 57 482
De Blasi C, Manno D and Rizzo A 1990 J. Cryst. Growth 100 347
Lieth R M A 1976 Crystal Growth of Materials with Layered Structures - Lieth R M A editor
    (Reidel, Dordrecht) p 225
Micocci G, Tepore A, Rella R, Siciliano P, 1981 J. Appl. Phys. 70 6847
Micocci G, Tepore A, Rella R, Siciliano P, 1992a Physica Stat. Sol. (a) 133 421
Micocci G, Tepore A, Rella R, Siciliano P, 1992b Sol. Energy Mater. and Solar Cells 28 223

The complimentary role of XRD to that of TEM in the characterization of semiconductor hetero-epitaxial structures

Mary Halliwell

Philips Analytical X-ray, Lelyweg 1, 7602 EA ALMELO, The Netherlands

ABSTRACT: The complementary nature of X-ray diffraction and electron microscopy goes beyond the obvious difference between a non-destructive macro technique and a destructive micro technique. This paper reviews the structural information available from XRD and TEM studies of single layer and superlattice structures. Although TEM is often required to provide qualitative information, particularly about defects, XRD is usually more quantitative and has a key role to play in monitoring the quality and composition of semiconductor heterostructures.

1. INTRODUCTION

This paper describes XRD data collection techniques using state-of-the-art diffractometers. A few typical applications are discussed illustrating how additional structural information can be obtained by combining TEM and XRD. Finally the relative merits of the two techniques are reviewed.

2. X-RAY DIFFRACTOMETERS

State of the art x-ray diffractometers for materials research are fitted with interchangeable optics for studying a range of sample types. Table 1 shows the options available with the Philips Materials Research Diffractometer. The four reflection monochromator is used as the incident beam optics for most studies of heteroepitaxial layers. The receiving slit is used in the diffracted beam to record rocking curves and the channel cut analyser to record reciprocal space maps (see Section 3.2). The powder collimator together with the parallel plate collimator and graphite monochromator is designed for use with highly textured films. This combination can be used to record pole figures (see for example Sant et al (1991)).

Table 1 - Incident and Diffracted Beam options

OPTICS	OPTION	ANGULAR DIVERGENCE
Incident Beam	Powder collimator	0.3°
	Four reflection monochromator	5" or 12"
Diffracted Beam	Parallel plate collimator/graphite monochromator	0.3°
	Receiving slit	0.2, 0.4, 0.6 or 2.0°
	Channel cut analyser crystal	12"

3. STRUCTURAL ANALYSIS OF HETEROEPITAXIAL LAYERS USING XRD

3.1 The Experimental Method

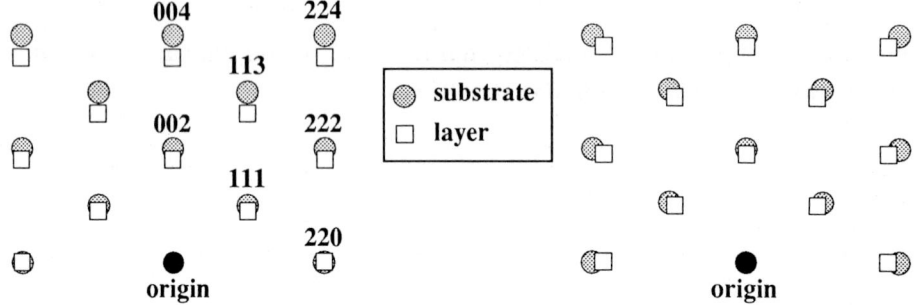

Fig. 1 Reciprocal lattices for fully strained (left) and fully relaxed (right) layers on (001) substrates

The reciprocal lattice for typical layer and substrate combinations is shown in Fig 1. In the figure the substrate surface is (001) and the reciprocal lattice section containing the [001] and [110] vectors has been shown. The layer unit cell is assumed to be larger than that of the cubic substrate. In the fully strained case the layer unit cell is tetragonally distorted to match the substrate unit cell parallel to the interface. In the fully relaxed case the layer unit cell is cubic, with a mismatch dislocation network at the interface. A partially relaxed layer will have an intermediate unit cell distortion which can also be tilted relative to the substrate unit cell.

XRD measurements on semiconductor heterostructures usually employ a Cu X-ray source as CuKα radiation gives a convenient intensity ratio for the layer and substrate features for the majority of samples. If the layer is fully strained it will only be necessary to measure the separation of the layer and substrate reciprocal lattice points for a single reflection (other than a reflection in the [001] zone) to determine the layer unit cell completely. If the layer has mismatch dislocations the state of relaxation can be determined by measuring the separation of the substrate and layer reciprocal lattice points using reflecting planes which are inclined to the interface. If the layer unit cell retains tetragonal symmetry as it relaxes a single reciprocal lattice point separation can be measured. If a tilt develops or the amount of relaxation differs along two orthogonal directions in the interface then at least four reciprocal lattice point separations will need to be measured (Halliwell 1990).

Fig. 2 shows the Ewald Sphere construction with typical dimensions for semiconductor materials: lattice parameter = 5.6Å, Bragg angle = 33° wavelength = 1.54Å. In the diagram the effects of incident beam divergence and angle of acceptance of the detector are indicated. There will also be some divergence of the incident beam normal to the plane of the figure. The resolution achievable in an XRD experiment depends upon the size of the diffractometer sampling volume and the sizes and separations of the reciprocal lattice points. As indicated in Table 1 the dimensions of the sampling volume can be varied from a few seconds to about a degree by changing the incident beam and the diffracted beam optics.

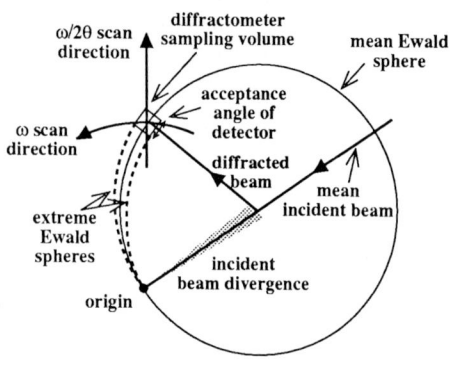

Fig. 2 The Ewald Diagram for a typical XRD measurement

3.2 Data Collection Procedures

Reciprocal space maps are recorded using a combination of ω and $\omega/2\theta$ scans shown in Fig 2. In an ω scan the intensity is recorded as the angle between the incident beam and the surface is increased. The detector is not moved. This is equivalent to moving the diffractometer sampling volume along the circumference of a circle centred on the origin of reciprocal space. In an $\omega/2\theta$ scan the detector is moved at twice the speed with which the sample is moved as intensity is recorded. This is equivalent to moving the diffractometer sampling volume radially away from the origin.

If a number of $\omega/2\theta$ scans are made each with a small offset in ω, then a two dimensional intensity map or reciprocal space map is obtained. These maps are recorded with a small beam divergence and a small detector acceptance angle to maximise the resolution.

Rocking curves are one dimensional scans. They can use either ω and $\omega/2\theta$ scans. If they are to be interpreted using simulation software based on the Takagi-Taupin equations (Halliwell et al 1984) the diffractometer sampling volume used during the recording must be large enough to include all the intensity within the reciprocal lattice points.

If the diffractometer setting is held at the position of a rocking curve feature or a reciprocal space map feature, an image (or topograph) of the regions diffracting can be recorded on a piece of film inserted in the beam path (Fewster 1991). The minimum resolvable defect spacing is limited to a few microns by the photographic process.

The reciprocal lattice point separations within the reciprocal lattice sections shown are measured by having the incident and diffracted x-ray beam lying in the section. The same reciprocal lattice section can be probed by TEM by making a thin section of the sample containing the [001] and [110] vectors and setting this normal to the electron beam. This is the usual method of creating thinned cross sections from semiconductor materials. Thus it is a straightforward procedure to compare x-ray and TEM results. In particular if extra spots, due to ordering for example, are observed by TEM, the position to look for x-ray intensity is easily identified.

4. EXPERIMENTAL RESULTS FOR SINGLE LAYERS

4.1 Studies of layer perfection

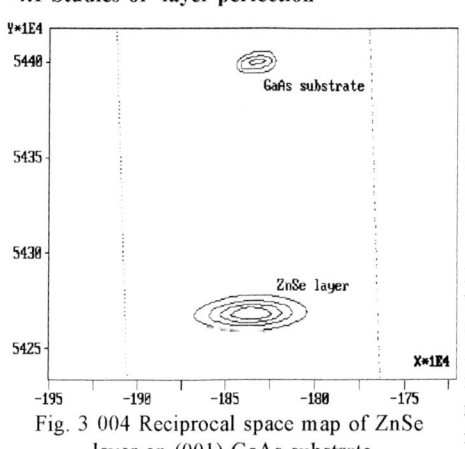

Fig. 3 004 Reciprocal space map of ZnSe layer on (001) GaAs substrate

Electron micrographs show the distribution of defects at the substrate interface and throughout the layer thickness. Frequently it is possible to analyse the source of the defects from the micrographs. X-ray rocking curves and reciprocal space maps imply the presence of defects when the layer peaks are broadened and thickness fringes do not appear.

Some layers have a mosaic structure, possibly arising from an island growth mechanism, where regions of more perfect material are surrounded by regions with a higher defect density. In such cases the lattice continuity parallel to the interface is destroyed leading to an increase in the reciprocal lattice point breadth parallel to the interface. This broadening can be measured using XRD and the average size of the mosaic blocks can be deduced. Fig. 3 shows the 004 reciprocal space map recorded for ZnSe deposited on GaAs. The plot is in units of $\lambda/2d$ and is oriented as Fig. 1. The mosaic structure broadens the reciprocal lattice point parallel to the interface. In Fig 3 both reciprocal lattice points show some broadening due to overall curvature of the sample. The ZnSe reciprocal lattice point shows additional broadening due to mosaic structure. If a lattice parameter variation had also been present this would have extended the spot perpendicular to the interface.

4.2 Relaxation

Partially and fully relaxed layers have mismatch dislocation networks at their interfaces. TEM images these dislocation networks and the average dislocation spacing can be estimated as well as the type of dislocation forming the network (e.g. edge, screw or 60°). An XRD reciprocal space map shows the relative positions of the layer and substrate reciprocal lattice points. Fig.4 is a map of the 224 reciprocal lattice points for $Ga_xIn_{1-x}As$ on (001) GaAs. The orientation of this diagram is identical to the orientation of Fig.1. The dimensions of layer unit cell can be calculated from a set of maps for all the symmetrically related 224 reflections. This allows the composition of the layer to be determined as well as the lattice parameter difference

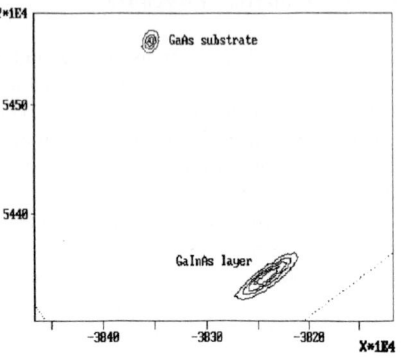

Fig. 4 224 map from a partially relaxed layer of $Ga_xIn_{1-x}As$ on GaAs

parallel to the two orthogonal [110] directions in the interface. If the predominant form of mismatch dislocation is known, then the minimum density of dislocations required to accommodate the measured lattice parameter difference can be calculated (Halliwell et al, 1989).

4.3 Ordering in Gallium Indium Phosphide

$Ga_xIn_{1-x}P$ layers grown on GaAs substrates display CuPt type ordering when grown under certain conditions. On (001) substrates the ordered layers have a complex structure in which domains ordered on two different {111} planes occur as alternating laminae. Each lamina is divided laterally by antiphase boundaries (Baxter et al 1991). The scale of the structure depends upon the growth temperature while the relative percentages of the two domains present depends on the exact orientation of the interface. Fig. 5 is an electron diffraction pattern from a thin cross section of a thick $Ga_xIn_{1-x}P$ grown on an off-axis GaAs substrate which shows extra spots due to ordering. One half of the extra spots are much weaker than the rest indicating that one domain orientation predominates. The intensity of the spots is also dependent on the indium enhancement in group III {111} planes.

To quantify the indium enhancement occurring on the {111} planes it is necessary to measure the relative intensities of these extra reflections. There are problems in determining relative intensities with conventional TEM because of multiple diffraction effects. Multiple diffraction rarely occurs in XRD because of the larger wavelengths employed. Thus intensity measurements are in principle

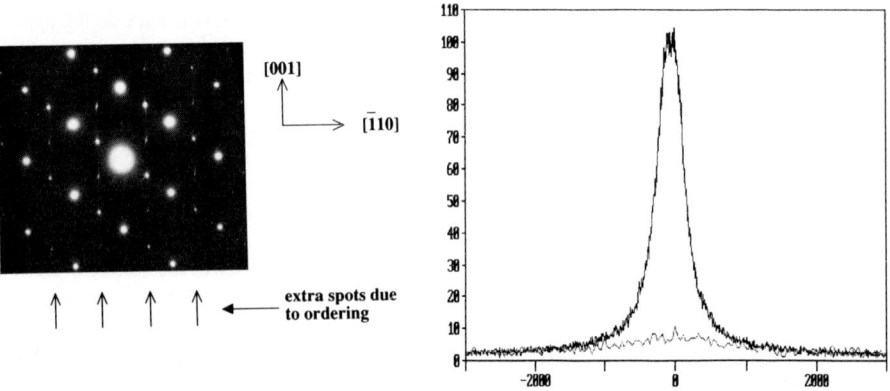

Fig. 5 Electron diffraction pattern for an GaInP layer with CuPt ordering

Fig. 6 Half order 115 peaks recorded from layer in Fig 5 (x-axis in seconds)

much easier with XRD provided sufficient material is present to give an intensity significantly above background. In addition, measurement of the angle between a substrate surface and the nearest (001) plane is straightforward with XRD.

Fig. 6 shows the result of preliminary XRD measurement on the sample used for Fig. 5. The sample was chosen because the layer was $2\mu m$ and showed well separated antiphase domain boundaries using TEM. The thick layer was chosen to increase the volume of ordered material in the x-ray beam while the large domain structure was expected to give peaks with narrower half widths. The peaks shown are for the half order 115 reflection which had the smallest accessible Bragg angle. The full width at half maximum for the reflection from the predominating domain orientation was under 500 seconds. A second sample with more closely space antiphase boundaries gave a full width at half maximum of over 2000 seconds.

5. EXPERIMENTAL RESULTS FOR SUPERLATTICE STRUCTURES

5.1 SiGe superlattice

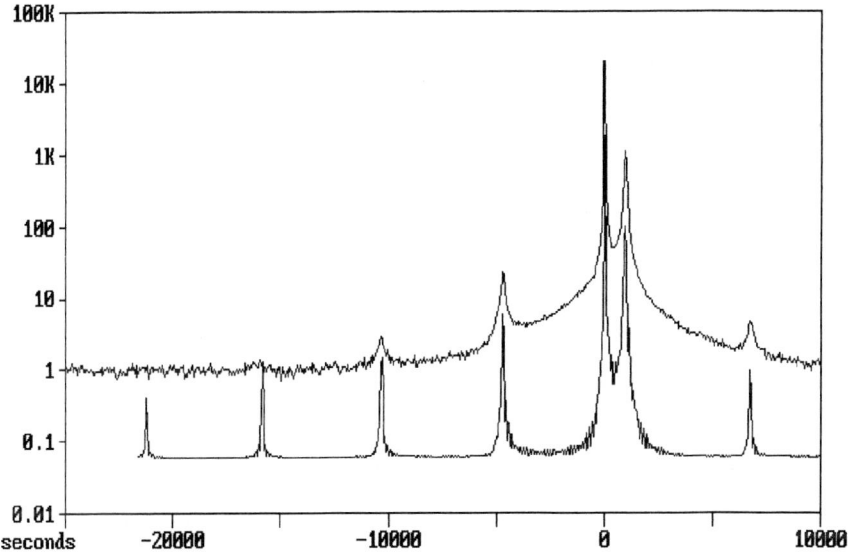

Fig.7 004 Rocking curves for 50x(2ml Si+22ml Ge) on (001) Ge.
Top curve experiment, bottom curve simulation

The top curve in Fig.7 is a rocking curve (logarithmic intensity scale) recorded from an early example of a SiGe superlattice on Ge. The superlattice was intended to have 50 periods consisting of 2 monolayers of Si and 22 monolayers of Ge grown on a (001) Ge substrate. As a monolayer consists of one quarter of the unit cell dimension for Si and Ge the period should have been about 34Å. The 004 rocking curve indicates that the period was 33Å and the mean mismatch was 3800 parts per million. This indicates that the Ge layers had a mean thickness slightly thinner than intended. The bottom curve in Fig.7 is a calculated curve using the period and mismatch values derived from the experimental curve. The calculated curve is for a perfect structure. The experimental curve has broader peaks and more

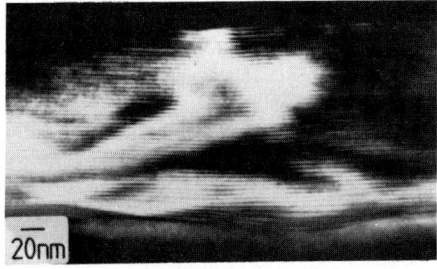

Fig. 8 TEM cross section of the superlattice shown in Fig. 7

diffuse scattering around the central substrate peak indicating a considerable degree of disorder.

The reason for the disorder in the superlattice stacking is not available from the x-ray data. Fig.8. shows a TEM cross section for the same sample where the layers are clearly seen but they are often far from flat. In particular irregularities in the original substrate surface lead to waviness in the layers which decreases towards the top surface of the superlattice. This indicated the need for more careful surface preparation.

Attempts to improve the quality of this type of superlattice can be monitored by either technique. Since a single rocking curve can usually be recorded more easily than producing and examining a TEM cross section this is usually the preferred method for routine checks on all samples. Selected samples showing distinct improvements in quality are looked at in more detail by TEM to identify the source of the improvement.

In principal the superlattice unit cell can be determined by XRD alone when the lattice parameter is dependent on a single variable (Fewster 1988). In practice a large number of satellites must be measured to make the necessary calculations. This is not always possible for samples such as this with less than 0.2µm of material in the superlattice stack. If the lattice parameter variation through the superlattice period is determined the width of each interface can be defined. However it is not possible to differentiate between a flat interface with interdiffusion or a roughened interface which gives the same average compositional variation. To a lesser extent the same problem arises with conventional TEM where an image of an interface is an average over the thickness of the sample used.

5.2 InP based superlattice

Fig. 9 shows a micrograph recorded for a cleavage edge sample (Kakibayashi et al 1985) produced from a 16 well structure grown by atmospheric pressure MOVPE (Nelson et al 1988) on an (001) InP substrate. The intended structure was: InP substrate plus buffer layer + 0.1µm Q1.1 confining layer + (16x55Å Q1.45 wells separated by 15x100Å Q1.1 barriers) + 0.1µm Q1.1 confining layer + 0.2µm InP capping layer. Where Q1.1 and Q1.45 represents the GaInAsP compounds lattice matched to InP which have absorption edge at 1.1 µm and 1.45 µm respectively. The micrograph confirms that the intended layer structure was achieved with abrupt interfaces within the limits of the measurement (about 8Å). Examination of the fringe patterns parallel to the edge of the TEM cleave indicate that all the Q1.1 layer have the same constant composition and all the Q1.45 layers have the same constant composition.

The top curve in Fig.10 is the 004 rocking curve (logarithmic intensity scale) recorded for the structure. The separation of the satellite reflections indicated that the repeat distance within the sixteen period superlattice was 152Å, compared to the intended value of 155Å. From the rocking curve the mean lattice parameter of the superlattice was calculated as 81 parts per million less than indium phosphide.

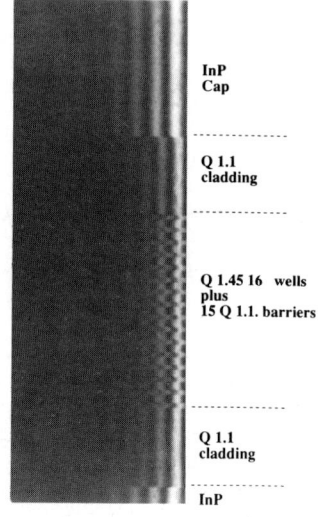

Fig.9 TEM cleavage edge micrograph of a 16 well quaternary superlattice on (001) InP

The shoulder on the high angle side of the central peak was assumed to be due to the two Q1.1 confining layers. This indicated that the Q1.1 layers had a smaller lattice parameter than the Q1.45 layers. The middle curve is a calculated curves where the mismatches of the two quaternary materials have been adjusted such that the mean mismatch is as observed. The individual layer thicknesses have been reduced by a factor of 152/155 the intended values and the interfaces are assumed to be abrupt. The curve shown was around the best fit achieved to the experimental data.

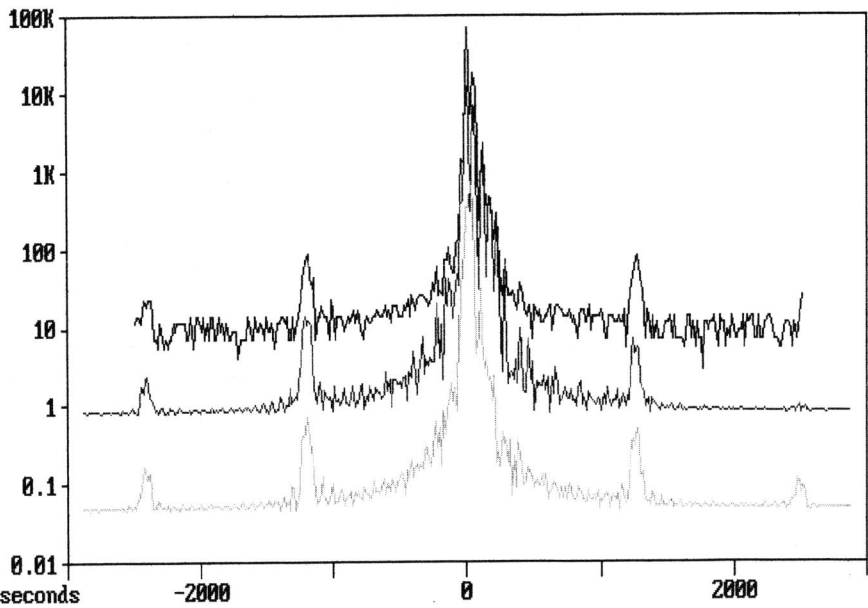

Fig. 10 004 Rocking curves for structure shown in Fig. 9
top curve experiment, lower curves simulations

As the mismatch of the quaternary layers is increased the match between simulation and experiment improves for the satellite peaks and deteriorates for the high angle shoulder.

For this type of superlattice where both the group III and the group V sublattices change at each interface there can be dramatic changes of mismatch at interfaces if there is a lag in altering the gas flows. Within the limits measurement of a TEM cleavage edge micrograph the interfaces look sharp. The bottom curve shows the improved fit achieved by introducing interfaces of 6Å with a mismatch calculated assuming the arsenic flow always stabilises last. The fit is improved.

6. SUMMARY

Table 2 summarises the parameters which can be measured using XRD and conventional TEM. Underlined entries indicate the preferred technique for a particular measurement. A state-of-the-art x-ray diffractometer is around half the cost of a conventional electron microscope. In addition XRD equipment usually requires less maintenance and less accommodation than an electron microscope.

In recent years XRD has become the key technique for routine monitoring in a crystal growth laboratory due to the lower costs and faster turnaround than electron microscopy. However for a full analysis of the potential problems involved in heteroepitaxial growth electron microscopy is essential. This paper has attempted to indicate how they can be used together to maximise the available information.

The examples have illustrated how using both XRD and TEM to study a heteroepitaxial structure more information is available than from only one technique. The indium phosphide based superlattice described in section 5.2 is a particularly good example of the complementary nature of the two techniques. TEM clearly shows that the dimensions of the intended structure have been achieved but is unable to give quantitative estimates of the layer lattice parameters. It would be impossible to derive all the layer thicknesses and lattice parameters from XRD alone as there are too many variables. By using both techniques the layer thicknesses and the layer lattice parameters can all be derived.

Table 2 - Comparison of XRD and TEM

Procedure	XRD	TEM
Sample preparation	None	Cleaving, Making thin section
Data collection - time required	½hr - 2 days	2 hrs - 1 week (including preparation)
Quantifying intensities	straightforward	multiple diffraction effects distort measured values
Imaging defects	topography possible if defect spacing > 1 μm	defects visible on atomic scale
Detecting defects using diffraction patterns	defects produce obvious diffraction spot broadening and peak shifting	diffraction patterns relatively insensitive to defect levels
Detecting interface roughness/ interdiffusion	usually indistinguishable	readily differentiated
Measuring d-spacings and layer composition	d spacings to <20ppm some compositions to <0.1%	contrast changes with d spacing and composition, difficult to quantify
Measuring layer thickness	precise measurements from fringes & satellites if present	direct from micrographs always possible but less precise
Quantifying relaxation	exact dimensions of partially relaxed layer cells can be determined	state of relaxation deduced from estimated defect density

6. ACKNOWLEDGEMENTS

The samples used in this paper were grown at BT Laboratories, Martlesham Heath, Philips Research Laboratories, Redhill and the University of Aachen. Additional experimental data was supplied by Mike Stobbs and Caroline Baxter from the Department of Materials Science and Metallurgy at Cambridge University, Mark Hockly at BT Laboratories, Martlesham Heath and Paul Fewster at Philips Research Laboratories, Redhill.

REFERENCES

Baxter C S, Stobbs W M and Wilkie J H 1991 J. Cryst. Growth **112** 373
Fewster P F 1988 J Appl Cryst **21** 524
Fewster P F 1991 J. Appl. Cryst **24** 178
Halliwell M A G, Lyons M H and Hill M J 1984 J. Cryst. Growth **68** 523
Halliwell M A G, Lyons M H, Davey S T, Hockly M, Tuppen C G and Gibbings C J 1989 Semicond. Sci. Technol. **4** 10
Halliwell M A G 1990 Advances in X-ray Analysis **33** 61 (Plenum Press)
Kakibayashi H and Nagata N 1985 Japan. J. Appl. Phys. **24** L9
Nelson A W, Spurdens P C, Cole S, Walling R H, Moss R H, Wong S, Harding M J, Cooper D M, Devlin W J and Robertson M J 1988 J. Cryst. Growth **93** 792
Sant S B, Weatherly G C and Smith R W 1991 Inst. Phys. Conf. Ser. **117**, 527

Inst. Phys. Conf. Ser. No 134: Section 9
Paper presented at Microsc. Semicond. Mater. Conf., Oxford, 5–8 April 1993

In situ synchrotron X-ray studies of epitaxial strained-layer growth processes

C R Whitehouse[1]*, S J Barnett[1], B F Usher[1+], A G Cullis[1], A M Keir[1], A D Johnson[1], G F Clark[2], B K Tanner[3], W Spirkl[3], B Lunn[4], W E Hagston[4] and J C H Hogg[4]

[1]DRA Malvern, St Andrews Road, Malvern, Worcestershire, WR14 3PS, UK.
[2]SERC Daresbury Laboratory, Daresbury, Warrington, Cheshire, WA4 4AD, UK.
[3]Department of Physics, University of Durham, South Road, Durham, DH1 3LE, UK.
[4]Department of Applied Physics, University of Hull, HU6 7RX, UK.

ABSTRACT: The work described here uses, for the first time, *in situ* synchrotron X-ray topography to study misfit dislocation introduction processes during the growth of strained InGaAs on (001) GaAs. The central importance of GaAs substrate threading dislocations in the initial relaxation phase is highlighted. Following this phase, these studies clearly demonstrate that a significant increase in epilayer thickness is then required to activate additional dislocation sources for further stress-relief. The observation of the different relaxation regimes is described in detail.

1. INTRODUCTION

The past decade has seen a very rapid expansion in research relating to the epitaxial growth of lattice-mismatched semiconducting material combinations, and it is clear that the next-generation of optical and electronic devices will rely increasingly on the wider materials flexibility, device performance and multi-function chip capability thus provided. There are, therefore, continuing extensive studies both of highly mismatched (strain-relieved) material combinations (eg GaAs/Si, CdTe/GaAs, etc), and also of non-relaxed pseudomorphic structures (eg InGaAs/GaAs, SiGe/Si), in which the level of strain (and hence the semiconductor band-structure) is controllably tailored to generate the required new device properties (Dunstan and Adams 1990). Therefore, it is vitally important to generate a much more detailed understanding of the different strain-relief processes involved in order that the design, performance and reliability of the resulting devices can be fully optimised.

Previous post growth techniques used to investigate the relaxation of III-V strained-layer semiconductor structures have included double-crystal X-ray diffractometry, low-temperature photoluminescence, Rutherford back-scattering spectrometry, Hall measurements and Raman spectroscopy. However, major discrepancies in the derived critical thickness values exist, relating primarily to the markedly different strain-relaxation sensitivities of the assessment techniques used (Fritz 1987). Furthermore, none of the above techniques provide direct structural information regarding either the specific initial dislocation generation processes involved, or the subsequent dislocation motion and interaction processes. The use of post-growth dislocation-imaging techniques, including transmission electron microscopy (TEM) (Cullis 1990), photoluminescence microscopy (PLM) (van der Wel et al 1992) and X-ray topography (XRT) is therefore becoming increasingly important. However, PLM still fails to provide the Burgers vector information required to fully characterise and model relaxation processes, and cathodoluminescence studies have already indicated that not all misfit dislocations (MDs) generate detectable dark-line images (Fitzgerald et al 1988). It is, therefore, clear that PLM data alone could be easily misinterpreted. In contrast, TEM clearly provides the required

---

* Now at Department of Electronic and Electrical Engineering, University of Sheffield, PO Box 600, Mappin Street, Sheffield, S1 4DU, UK.
+ Working on temporary secondment from Telecom Australia Research Laboratories, 762-772 Blackburn Road, Clayton North, Victoria 3168, Australia.

high-resolution dislocation physics information, but only in the more advanced stages of strain-relaxation when dislocation densities are relatively high ($\geq 10^3$ or 10^5 cm^{-2} for plan-view and cross-section TEM respectively). At this stage of relaxation, complex dislocation interactions are commonly observed, and it is therefore not possible to confirm the specific dislocation generation mechanisms which initiated relaxation.

It is surprising that, despite early reports of the use of XRT in III-V strain-relaxation studies (Petroff et al 1978, 1980), the technique has only occasionally been used in more recent investigations (eg Eaglesham et al 1988, Green et al 1990). Unlike other techniques, XRT allows the direct observation (and full Burgers-vector characterisation) of the very first dislocations which initiate strain-relaxation, and can also be performed non-destructively on technologically-relevant, full-sized, unthinned substrate wafers. In addition, the technique allows specific dislocation events to be correlated directly with the structural properties of the underlying substrate, and the incident X-ray beam also enables vitally important complimentary selected-area diffraction data to be generated regarding the local composition and thickness of the epitaxial layer under investigation.

Our preliminary *ex situ* post-growth XRT investigations of III-V strained-layer growth processes (Whitehouse et al 1992a, Barnett et al 1993) have already confirmed the vitally important role which the XRT technique can play in gaining an improved understanding of strain-relaxation mechanisms. Indeed these studies have led to the successful design of a unique combined XRT/molecular beam epitaxy (MBE) reactor which is now capable of performing *in situ* XRT observations of dislocation generation processes which occur actually during the growth of III-V strained-layer device structures (Whitehouse et al 1992b). The present paper therefore describes data obtained during the first series of *in situ* XRT studies of (100) InGaAs/GaAs strained-layer growth processes. The results highlight the important role played by the substrate properties in influencing the strain-relaxation of the overlying epitaxial layers, and also, for the first time, allow the determination of values for the *genuine* critical thickness (ie that at which the very *first* strain-relieving MD is formed) for the particular layer combinations and growth conditions studied. The work shows a previously-unreported two-stage contribution to the MD generation process, and provides the first experimental confirmation of theoretical predictions (Spirkl et al 1993) that XRT dislocation image contrast can be used to give sub-surface depth information regarding the imaged dislocations.

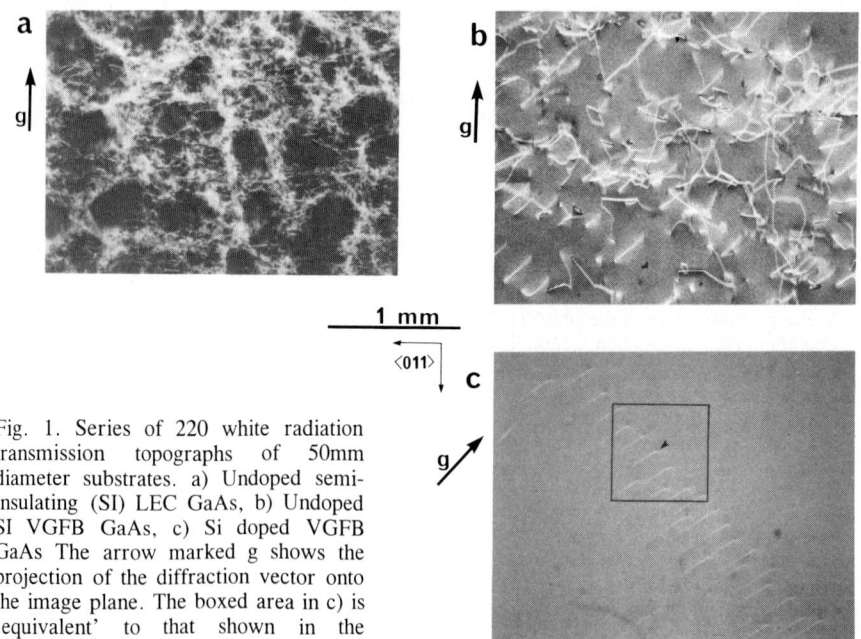

Fig. 1. Series of 220 white radiation transmission topographs of 50mm diameter substrates. a) Undoped semi-insulating (SI) LEC GaAs, b) Undoped SI VGFB GaAs, c) Si doped VGFB GaAs The arrow marked g shows the projection of the diffraction vector onto the image plane. The boxed area in c) is 'equivalent' to that shown in the reflection topographs of fig. 2.

2. EXPERIMENTAL DETAILS

The *in situ* XRT/MBE growth experiments were performed in a new unique purpose-built growth facility specially equipped with a beryllium window assembly which allows both XRT and x-ray diffraction studies to be performed during MBE growth (Whitehouse et al 1992). InGaAs/GaAs epitaxial growth experiments were carried out on (100) GaAs substrates, produced both by liquid encapsulated Czochralski (LEC) and vertical-gradient freeze Bridgmann (VGFB) techniques. In the former case, the substrates were unintentionally doped and semi-insulating (MCP Wafer Technology, UK), whilst samples of both undoped and Si-doped (n = 1.8 - 2.3 x 10^{18} cm^{-3}) VGFB material (AXT, USA) were also investigated. The composition of the InGaAs epitaxial layers grown was calibrated *in vacuo*, using the RHEED oscillation technique, and subsequently confirmed post-growth using the *in situ* selected-area x-ray diffractometry capability. The epitaxial layers described in the present investigation were grown at a substrate temperature of 490°C, calibrated with reference to surface reconstruction changes detected by the in-situ RHEED system, using a target alloy growth rate of one monolayer per second. All InGaAs layers were grown following the deposition of a 200nm GaAs buffer layer.

The *in situ* reflection XRT measurements were performed using the UK SERC Daresbury synchrotron source (station 9.4), with 0.148nm wavelength radiation and a purpose-designed water-cooled double 333-reflection channel-cut silicon monochromator. Typical exposure times for the 422-reflection topographs recorded on Ilford L4 nuclear emulsion plates were of the order of four minutes (synchrotron operating at 2GeV, 150mA; wiggler source at 5T), and the topographs were recorded during short periods of growth interrupt with the sample at the growth temperature. The post-growth *ex situ* white-radiation topographs presented were recorded using a 220-type diffraction vector with 0.04nm wavelength radiation (Daresbury station 7.6, 1.2T dipole magnet).

3. RESULTS AND DISCUSSION

3.1 Initial XRT Comparison of LEC- and VGFB-Grown (100) GaAs Substrate Material

An initial aim of the present study was to investigate the influence of GaAs substrate perfection on the subsequent relaxation processes occurring in the overgrown InGaAs epitaxial layers. Figure 1 therefore illustrates a series of transmission topographs recorded from three different batches of substrate material. The undoped semi-insulating LEC-grown GaAs substrate material, illustrated in Fig. 1a, is seen to possess a highly non-uniform cell-like distribution of threading dislocations, corresponding to an overall average dislocation density of the order of 10^4 cm^{-2}, but with a cell-edge dislocation density as high as 10^5 cm^{-2} (Stirland 1991). In contrast, Fig. 1b shows a corresponding topograph recorded from typical VGFB-grown (100) GaAs substrate material. In this case, the significantly lower thermal stresses associated with the growth process are seen to lead to a very marked reduction in dislocation density (< 3000 cm^{-2}), and a more uniform dislocation distribution.

A further reduction in substrate dislocation density can be achieved by Si doping which has the effect of hardening the GaAs lattice. Figure 2c shows a typical isolated group of dislocations in a Si doped VGFB GaAs substrate for which the average dislocation density over the whole wafer was measured to be < 20cm^2. In this case, the dislocations are arranged in groups of 4 to 100 and, in most cases, a particular group contains dislocations of only one specific ½ < 110 > type of Burgers Vector. There are large areas of the substrate, often > 1cm^2, which are completely free of dislocations. A number of these substrates have now been examined and it is clear that the pattern of threading dislocations is reproduced very closely through a series of several adjacent wafers. For example, the area of substrate shown boxed in Fig 1c is almost identical in its threading dislocation pattern to that shown in the epilayer relaxation stage in Fig. 2. Indeed the individual threading dislocations in these different wafers once formed part of the same single dislocation in the GaAs boule, an example of such a dislocation is shown arrowed in Figs. 1c and 2a,c.

These three types of substrate provide ideal platforms on which to investigate the primary sources of MDs and their effect on the initial relaxation process in strained layers of InGaAs.

3.2 Influence of Substrate Dislocation Density on (100) InGaAs/GaAs Strain-Relaxation Processes

During the growth of InGaAs epilayers on LEC material, earlier XRT results from the present authors have confirmed that the threading dislocations act as primary initial sources of strain-relieving MDs (Green et al 1990, Barnett et al 1993), in accordance with the mechanism proposed in the Matthews and Blakeslee model for strain relaxation (Matthews et al 1974). In this case, the high dislocation density present in the substrates provides numerous potential sources of MDs, resulting in a rapid initial relaxation rate and the generation of a dense array of MDs (Whitehouse et al 1992, Barnett et al 1993). Relaxation proceeds rapidly, therefore, to a point where individual MDs cannot be resolved using X-ray topography.

In contrast, the initial relaxation rate of InGaAs layers grown on the VGFB substrates is expected to be significantly lower. For example, Fig. 2 shows a series of reflection topographs recorded during the growth of a single layer of $In_xGa_{1-x}As$, (x=0.028) on a Si doped VGFB GaAs substrate. The figure shows the same area imaged at two layer thicknesses and using two different diffraction vectors. The latter is necessary since the synchrotron source geometry on station 9.4 results in rather poor spatial resolution in the horizontal direction. In order to image both orthogonal sets of MDs, it is therefore necessary to use two reflections between which there is a 90° rotation of the sample about its surface normal. In these reflection topographs, the 'ends' of the substrate threading dislocations are imaged as white circular contrast features while the bulk of the threading component is not imaged since the reflection topograph only sees the upper 1-3μm of material.

Figure 2a shows the formation of the very first MDs at a epilayer thickness of 110nm. Although all the threading dislocations shown are of the same Burgers vector, at this epilayer thickness they have not all bent over to form MDs in the GaAs/InGaAs interface. As the layer thickness increases, the density of MDs increases sharply but soon reaches a plateau when all the available ½<110> type threading dislocations have been activated. Analysis of the threading dislocations shown here confirms that they give rise to ½<110> 60° MDs, as is commonly observed using TEM. There are some substrate dislocations of other (non <110> type) Burgers vectors but, at this epilayer thickness, they are not observed to be active in the formation of MDs (Barnett et al, in press).

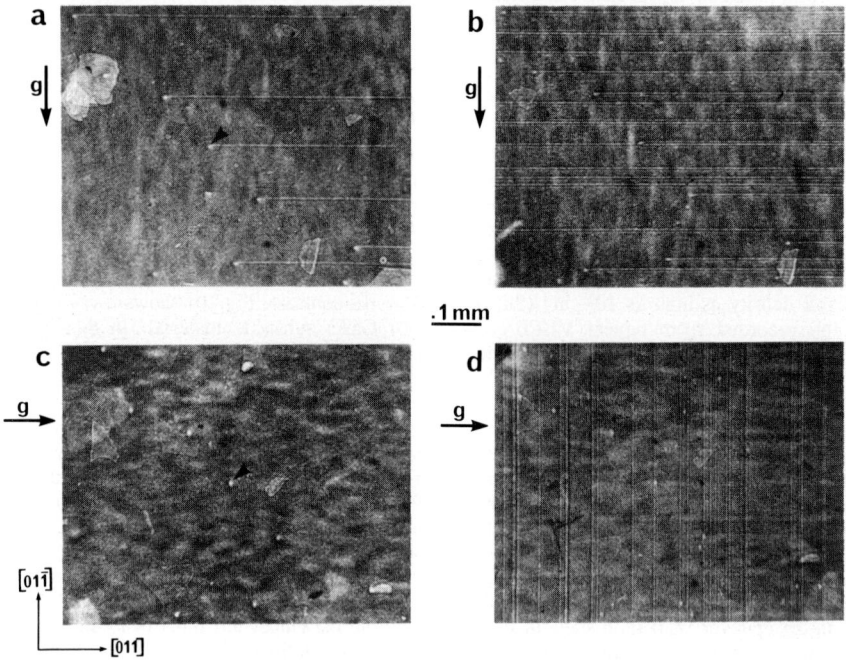

Fig. 2. Series of 422 double crystal reflection topographs recorded *in-situ* at two epilayer thicknesses a,c) 120nm and b,d)268nm. Arrow g indicates the diffraction vector projected on to the image plane.

The variation of the linear MD density with epilayer thickness for each of the inplane <011> directions is shown in Fig. 3. The two data sets represented by closed and open symbols are taken from different areas of the wafer, one close to the group of threading dislocations (closed symbols) shown in Fig. 2, the other remote from the threading dislocations. For each of the <011> directions, following the initial relaxation phase, the height of the plateau region is determined by the local density of threading dislocations, with the required Burgers Vector, which are available to act as sources of

MDs. As the epilayer thickness increases to approximately twice that at which the first MDs were observed there is negligible increase in the MD density as the MDs already present elongate in the plane of the epilayer-substrate interface. Above this thickness, at approximately 240nm, there is a second dramatic increase in the MD density, which is approximately equal in the two $<011>$ directions. Topographs, Figs 2c,d show the MD configuration at a thickness of 268nm. This second relaxation stage, which is an observation unique to the present studies, is believed to be the start of a secondary

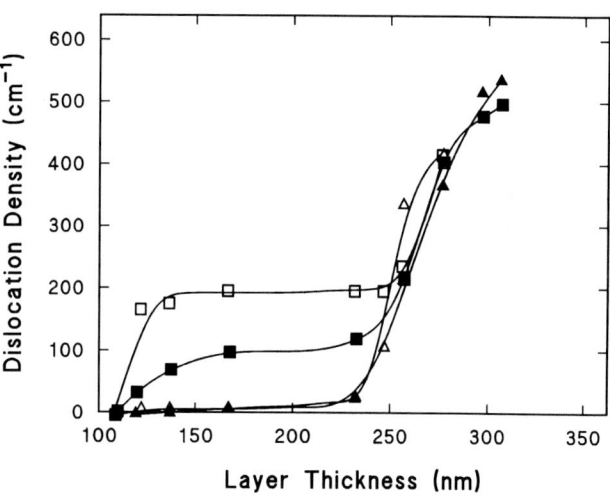

Fig. 3. Graph of MD density versus epilayer thickness for, ■,□ [011] and ▲,△ [0$\bar{1}$1] line direction MDs. See text for more details.

MD generation phase which is not directly controlled by the substrate threading dislocations and is currently under further investigation.

It is important to note that, throughout the present experiments, the MD configuration was observed to be almost invariant during constant temperature annealing periods of no epilayer growth. In almost all cases the elongation of MDs was only observed to occur during an increase in the epilayer thickness.

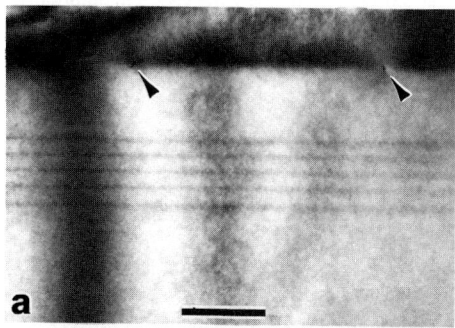

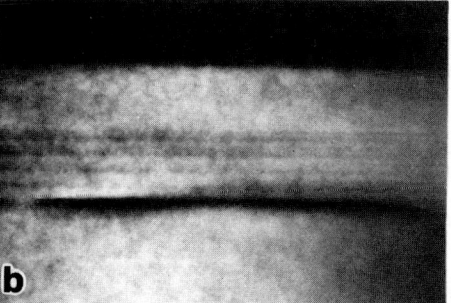

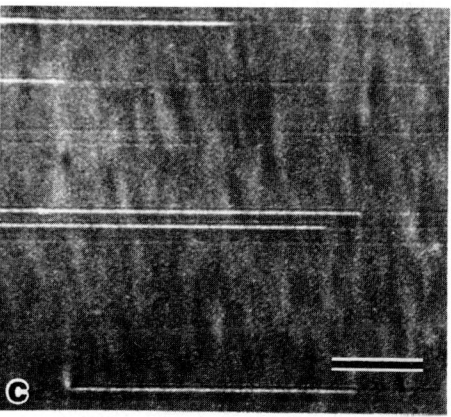

Fig. 4. Dark Field TEM micrographs of a multilayer structure showing MDs lying at a) the top strained layer interface (as arrowed) and b) the base of a buried InGaAs/GaAs superlattice. The scale mark corresponds to 200nm c) Double crystal 422 reflection topograph of the same sample showing weak and strong MD contrast. The scale mark corresponds to 100μm.

3.3 Depth-Dependent Contrast of XRT Images of MDs

Of course, in order to determine the nature of dislocations by analysis of their XRT image contrast, it is necessary to understand the factors which control the contrast. One important parameter is the depth of the dislocations beneath the free surface (Spirkl et al 1993). This has been clearly demonstrated by the growth of test structures with buried superlattices used for calibration purposes. In these cases, when a surface InGaAs layer exceeding the critical thickness was deposited, only a proportion of the MDs introduced resided at the surface-InGaAs/GaAs interface (Fig. 4a): a significant number of the dislocations lay along the lowest stratum of the strain-coupled buried superlattice (Fig. 4b). When this array was imaged *in situ* by XRT (Fig. 4c), it was immediately evident that the deeper dislocations gave very strong diffraction contrast while those at the base of the surface InGaAs layer gave significantly weaker images. Thus, care is needed when imaging dislocations within the thinnest surface layers to take due account of their relatively lower visibility.

4. CONCLUSIONS

The present paper provides the first report of *in situ* XRT for the investigation of MD formation during strained epilayer growth. This work clearly demonstrates the important role of the technique in determining full details of the MD generation process with the observation of the first stress-relieving defects to be introduced. The use of both LEC and VGFB substrates has highlighted the dominant role played by threading dislocations in the first phase of relaxation. After all available ½ < 110 > type dislocations have been activated, a significant increase in the epilayer thickness is required to initiate a second phase of stress-relief resulting from other dislocation generation processes. This plateau region in the MD density versus epilayer thickness curve is of significance, since it defines a window in epilayer thickness / composition space in which the critical thickness can be exceeded without producing more threading dislocations than would be grown in from the substrate. These detailed relationships are reported here for the first time and are currently under further study.

ACKNOWLEDGEMENTS

The authors wish to thank D E J Soley, A D Pitt, J Quarrell, P W Smith, G M Williams, V Stimson and P Moores (DRA Malvern, UK) for their continuing valuable technical assistance. In addition to the strong project support provided by DRA Malvern, additional significant financial support continues to be provided by the UK SERC and is also gratefully acknowledged.

REFERENCES

Barnett SJ, Whitehouse CR, Keir AM, Clark GF, Usher B, Tanner BK, Emeny MT and Johnson AD 1993 J. Phys. D: Appl. Phys. **26** A45
Cullis AG 1990 Defect Control in Semiconductors, ed K Sumino (Amsterdam: North-Holland) pp 1097-1105
Dunstan DJ and Adams AR 1990 Semicond. Sci. Technol. **5** 1202
Eaglesham DJ, Kvam EP, Mayer DM, Humphreys CJ, Green GS, Tanner BK and Bean JC 1988 Appl. Phys. Lett. **53** 2083
Emeny MT, Howard L, Homewood K, Lambkin J and Whitehouse CR 1991 J. Cryst. Growth **111** 413
Fitzgerald EA, Ast DG, Kirchner PD, Petit GD and Woodall JM 1988 J. Appl. Phys. **63** 693
Fritz IJ 1987 Appl. Phy. Lett. **51** 1080
Green G, Tanner BK, Barnett S, Emeny M, Pitt A and Whitehouse CR, 1990 Phil. Mag. Lett. **62** 131
Petroff JF and Sauvage M, 1978 J. Cryst. Growth **43** 628
Petroff JF, Sauvage M, Riglet P and Hashizume H, 1980 Phil. Mag. A **42** 319
Spirkl W, Tanner BK, Whitehouse CR, Barnett SJ, Cullis AG, Johnson AD, Keir AM, Usher B, Clark GF, Hogg CR and Lunn B 1993 Phil. Mag. *in press*
Stirland DJ 1991 Microscopy of Semiconducting Materials 1991 eds AG Cullis and NJ Long (Bristol: Institute of Physics) pp 327-336
van der Wel DJ, te Nijenhuis J, van Eck ERH and Girling LJ 1992 Semicond. Sci. Technol. **7** A63
Whitehouse CR, Barnett SJ, Cullis AG, Keir AM, Johnson AD, Emeny MT, Clark GF, Tanner BK, Cotrell S, Usher BF, Lunn B, Hogg CJ and Hagston W 1992a Proc MRS Spring Meeting
Whitehouse CR, Barnett SJ, Soley DEJ, Quarrell J, Aldridge SJ, Cullis AG, Emeny MT and Johnson AD 1992b Rev. Sci. Instrum. **63** 634

Characterization of InAs on GaAs in the (110) orientation by X-ray diffraction and topography

L Hart and P F Fewster†

Semiconductor Materials IRC, Imperial College, Prince Consort Road, London SW7 2BZ
†Philips Research Laboratories, Cross Oak Lane, Redhill RH1 5HA

ABSTRACT: X-ray diffraction and topography have been used to measure relaxation and tilt in samples of several thicknesses of InAs grown on (110) GaAs by molecular beam epitaxy. Diffraction space mapping enabled accurate measurement of strain and both macroscopic and microscopic tilts. The strain relief was found to be anisotropic: in the [001] direction the relaxation and macroscopic and microscopic tilts increased dramatically with sample thickness, while in the [$\bar{1}$10] direction the relaxation was initially high, increasing only slightly with thickness, and the macroscopic tilt was zero. This was caused by the nucleation of different types of misfit dislocations in the [001] and [$\bar{1}$10] directions.

1. INTRODUCTION

Semiconductor heteroepitaxial growth on (110) orientated surfaces is of considerable interest in studies of growth mode and strain relief behaviour (Zhang et al 1993a). Growth and relaxation of InAs on (110) GaAs has been investigated using grazing-incidence x-ray diffraction by Munekata et al (1987) but there were no measurements of anisotropy of relaxation or tilting. The strain relief mechanisms are affected by the (110) surface geometry since, in the [$\bar{1}$10] direction the {111} slip planes are inclined to the surface, allowing strain relief along [001], while in the [001] direction this is not the case (Pashley 1993). Tilting of the layer with respect to the substrate may also occur as a result of the nucleation and slip of 60° misfit dislocations on inclined {111} planes (Pashley 1991).

2. EXPERIMENTS

The samples were grown by molecular beam epitaxy at a temperature of 420°C with an In:As$_2$ flux ratio of 1:1. The growth mode studied by RHEED and the dislocation microstructure observed by TEM are described by Zhang et al (1993b).

Diffraction space maps were obtained using a low-resolution diffractometer with a graphite analyser (Fewster and Andrew 1992). This enabled measurement of weak layer peaks from the thin, highly relaxed layers. The x-ray generator was run at 50kV, 40mA with a copper target and the beam size was about 5×10mm² on the sample. Maps of the 220, 620 and 331 reflections enabled separation of components of strain and tilt in the [1l0], [$\bar{1}$10] and [00l] directions respectively.

Diffraction space mapping and topography around the substrate peak were carried out using a high-resolution triple-axis diffractometer (Fewster 1989) with a 4-reflection Ge monochromator and a 3-reflection Ge analyser. The dislocation strain field images were separated from the substrate intensity by taking triple-axis topographs on the "wings" of diffuse scattering around the substrate peak (Fewster 1991). Topographs could also be taken on the layer peaks, but the intensities would be low due to the thinness of the layers and only regions of tilts of the same magnitude and sense would be imaged. Ilford L4 nuclear emulsion with a grain size of ~0.25µm was used.

3. RESULTS

The diffraction space maps for the 220 reflections measured along the [$\bar{1}$10] and [001] azimuths for a 400Å InAs layer are shown in Fig. 1.

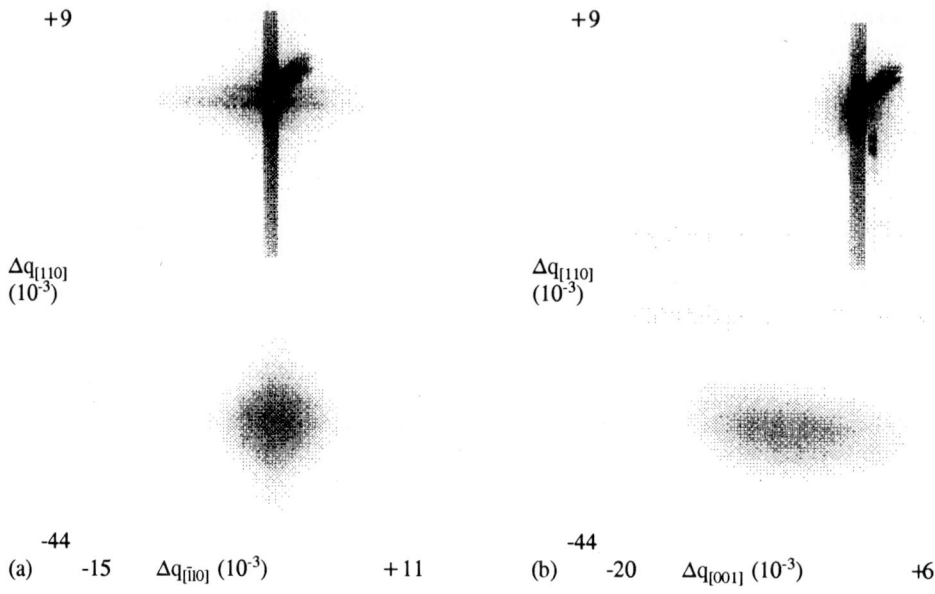

+9   +9

$\Delta q_{[110]}$
(10^{-3}) $\Delta q_{[110]}$
                                             (10^{-3})

-44   -44
(a)   -15 $\Delta q_{[\bar{1}10]}$ (10^{-3}) +11   (b)   -20 $\Delta q_{[001]}$ (10^{-3}) +6

Fig. 1. Diffraction space maps around the 220 reflection for the 400Å sample, along
a) the [$\bar{1}$10] azimuth and b) the [001] azimuth.

The procedure for determining relaxation and tilt is described by Fewster and Andrew (1993). Briefly, from each of the 220 maps (Fig. 1), the surface perpendicular lattice parameter ($a_{[110]}$) is calculated from the difference in $q_\perp$ between the layer and substrate peaks, and the macroscopic tilt is given by the difference in $q_\parallel$. In the 620 and 331 maps, $q_\parallel$ may have components of both tilt and in-plane lattice parameter, so any macroscopic tilts measured from the 220 maps must be subtracted in order to calculate $a_{[\bar{1}10]}$ and $a_{[001]}$. The spread of microscopic tilts is given by the FWHM (full width at half maximum) spread of the layer peak along $q_\parallel$ in the 220 maps.

Table 1. Measured values of lattice parameter, a, and relaxation, R.

thickness (Å)	$a_{[110]}$ (Å)	$a_{[\bar{1}10]}$ (Å)	$R_{[\bar{1}10]}$ (%)	$a_{[001]}$ (Å)	$R_{[001]}$ (%)	$a_{relaxed}$ (Å)
60	6.16910	5.96240	76.2	5.82020	41.1	6.0573
400	6.07808	6.00115	85.7	5.93950	70.5	6.0347

The results calculated from the measured values are given in Table 1. The "relaxed" lattice parameter was calculated using the elastic constants for bulk InAs. These may not be valid for a highly relaxed layer, which would explain the difference between the calculated values and the bulk lattice parameter of InAs (6.05903). It was found that the InAs layer had an orthorhombic unit cell and that the relaxation was anisotropic. The 60Å layer had greater relaxation along [$\bar{1}$10] than along [001] and there was negligible tilting in either direction, while in the 400Å layer, the relaxation was only slightly greater along [Ī10] than [001], but the tilt was much greater in the [001] direction (about the [Ī10] axis). The macroscopic tilt in the [001] direction was 1.0°, with a 1.8° spread of microscopic tilts, compared to a zero average tilt in the [Ī10] direction and a 0.7° spread. A 30Å layer was also grown but this was too thin for these x-ray diffraction studies.

Triple-axis topographs of the 400Å sample, taken on the diffuse scattering near the substrate peak for the 220 reflection along the [Ī10] and [001] azimuths are shown in Fig. 2.

(a)

(b)

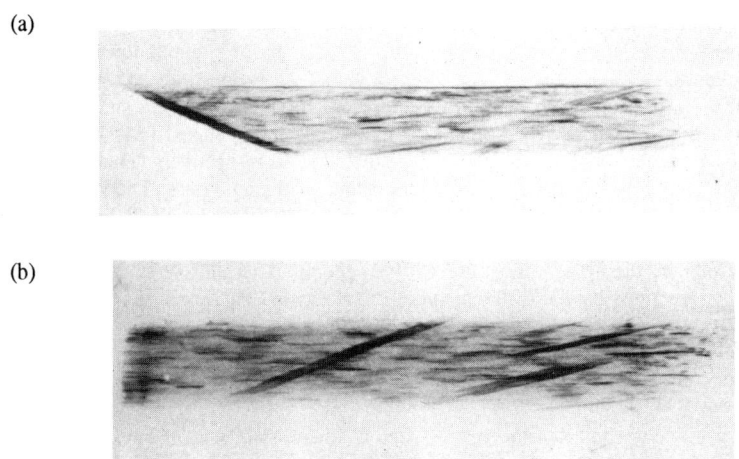

Fig. 2. Triple-axis topographs of the 400Å sample taken close to the substrate peak for the 220 reflection along a) [Ī10] and b) [001]. The width of the topographs is ~ 1mm.

Topographs taken on the diffuse scattering close to the substrate peak show contrast from the long-range strain fields of the dislocations. For relaxed layers grown on (001) substrates, this contrast shows orthogonal arrays of dislocations lying along the < 110 > directions (Fewster, 1991). However, the topographs of the (110) samples showed only short segments of about 300μm in length, which is equal to the resolution of the instrument. The image widths were about 10μm. For the 400Å sample, the average separations of these images were 69μm and 45μm in the [Ī10] and [001] directions respectively. It was unlikely that the images were only of threading dislocations in the substrate, since the separations would then be similar in both directions. It could therefore be deduced that the misfit dislocations in the two orthogonal directions were of different types, relieving different amounts of strain, since the layer was almost fully relaxed in both directions.

High-resolution maps around the 220 substrate peak for the 400Å sample showed different diffuse scattering intensity distributions in the [Ī10] and [001] azimuths, also seen in the low-resolution maps in Fig. 1. The diffuse scattering in the [001] plane had a characteristic "diamond" shape, while in the [Ī10] plane the intensity distribution was more circular. These diffuse scattering intensity distributions can be interpreted in terms of the type and orientation of the dislocations present (Dederichs 1972).

4. DISCUSSION

It has been shown by x-ray diffraction space mapping and topography that when InAs is grown on (110) orientated GaAs, the surface geometry causes anisotropic strain relief. In the [Ī10] direction there is high initial relaxation, increasing only slowly with thickness, and no tilting, while in the [001] direction there is a rapid increase in relaxation with thickness together with the occurrence of large tilts.

The x-ray results tie in well with observations made by TEM (Zhang et al 1993) which showed that the strain in the [$\bar{1}$10] direction was relieved by 90° Lomer-type dislocations, and in the [001] direction by 60° dislocations. At the early stages of growth, before the InAs layer is continuous, Lomer-type dislocations can nucleate at the edges of islands because of the very small (~ monolayer) "critical thickness". This explains why the relaxation in the [$\bar{1}$10] direction is high even for thin layers. However, 60° dislocations can nucleate at the surface because of the high stress, and slip on the inclined {111} planes to the interface. Since the stress levels increase with thickness, the generation of misfit dislocations in the thicker layer is greater. The inclined component of the Burgers vector of the 60° dislocations increases the microscopic tilting, whereas the macroscopic tilt could well arise from differences in the proportions of dislocations slipping down the two non-equivalent {111} planes.

The ratio of the image separations in the [$\bar{1}$10] and [001] topographs (69μm and 45μm) is the same as the ratio of the dislocation separations calculated for a fully relaxed InAs layer on (110) GaAs, which are 60Å and 42Å for Lomer and 60° type dislocations respectively (Zhang et al 1993), although the topographs show bunches of dislocations rather than single ones. The short segments seen in the topographs may be due to interaction between the 60° dislocations slipping down the inclined {111} planes and the Lomer dislocations at the interface. This would also result in extra threading dislocations in the layer along the [$\bar{1}$10] direction. The diffuse scattering intensity distributions can be explained by the different types of misfit dislocations in the two orthogonal directions.

5. CONCLUSIONS

It can therefore be concluded that the strain relief is anisotropic, with the relaxation and tilt increasing with thickness in the [001] direction, and high relaxation but negligible tilting regardless of thickness in the [$\bar{1}$10] direction. These effects are caused by the different types of misfit dislocations generated in the [001] and [$\bar{1}$10] directions. The structural information obtained from x-ray diffraction space mapping and topography is complementary to that obtained by TEM; while TEM shows the detailed microstructure, x-rays give an accurate estimate of the residual strain and relaxation and how it gives rise to macroscopic and microscopic tilts.

ACKNOWLEDGEMENTS

The authors would like to thank Xiaomei Zhang and Don Pashley for valuable discussion, Jim Neave and Peter Fawcett for growth of the samples and Norman Andrew for help with the x-ray measurements.

REFERENCES

Dederichs P H 1972 J. Phys. F. <u>3</u> 471
Fewster P F 1989 J. Appl. Cryst. <u>22</u> 64
Fewster P F 1991 Appl. Surf. Sci. <u>50</u> 9
Fewster P F and Andrew N L 1992 Diffraction from Thin Layers Materials Science Forum, Proc. EPDIC2, eds E J Mittemeijer and R Deldez (TransTech Publications)
Fewster P F and Andrew N L 1993 submitted to J. Appl. Phys.
Munekata H, Segmüller A and Chang L L 1987 Appl. Phys. Lett. <u>51</u> 587
Pashley D W 1991 Processing of Metals and Alloys, Materials Science and Technology <u>15</u>, ed R W Cahn (Weinheim: VCH) pp 289-328
Pashley D W 1993 Phil. Mag. in press
Zhang X, Pashley D W, Neave J H, Fawcett P N and Zhang J 1993a this proceedings
Zhang X, Pashley D W, Neave J H, Fawcett P N and Zhang J 1993b J. Cryst. Growth in press
Zhang X, Pashley D W, Hart L, Neave J H, Fawcett P N and Joyce B A 1993 J. Cryst. Growth in press

Inst. Phys. Conf. Ser. No 134: Section 9
Paper presented at Microsc. Semicond. Mater. Conf., Oxford, 5–8 April 1993

573

X-ray topographic study of the formation of misfit dislocations at the GaAs/Ge(001) interface

N Burle, B Pichaud, N Guelton* and R G St-Jacques*

Laboratoire MATOP ass. au CNRS, Univ. Aix-Marseille III, 13397 Marseille cedex 20, France
* INRS Energie et Matériaux, Varennes CP 1020 Quebec J3X1S2 Canada

ABSTRACT: X-ray topography (XRT) observation of the early stages of misfit dislocation formation allow one to determine the critical thickness of the highly metastable system GaAs/Ge. Heterogeneous nucleation from Ge threading dislocations is clearly observed. Strain and stress in the sample are deduced from the measured deviations from the Bragg angle for the perfect crystal. Tetragonal elastic deformation is found in very good agreement with theoretical calculations.

1. INTRODUCTION

As it leads to a high value of the critical thickness, the low mismatch GaAs/Ge system is a fairly good candidate for the study of the first stages of misfit dislocations nucleation and propagation. So a good control of deposited thickness associated with a suitable observation technique for low dislocation densities, allows their different formation steps to be analysed.

For this purpose we used X-ray transmission and reflection topography which give non destructive imaging of large areas and thus allows the above aim to be fullfiled.

GaAs films are deposited by close-spaced vapour transport (CSVT) onto (001) Ge substrates. GaAs is evaporated from the source at 800°C by (H_2+H_2O) gas flow, and deposited on the germanium substrates at 750°C by the thermal displacement of the reaction :

$$2\,GaAs + H_2O \xrightleftharpoons[750\,°C]{800\,°C} As_2 + Ga_2O + H_2$$

The lattice parameters are respectively $a_{GaAs} = 5.6533$ Å and $a_{Ge} = 5.6576$ Å at room temperature, the lattice mismatch is: $\varepsilon = 7.6\ 10^{-4}$ ($7.26\ 10^{-4}$ at 750°C).

Samples about 1cm x 1cm x 250 μm are observed using X-Ray Transmission topography with rotating Ag anode and using Berg-Barrett setting with standard Cu anode.

2. RESULTS AND DISCUSSION

2.1 Critical Thickness Determination

Topographs were performed on samples with increasing film thickness t_F from 0.3 to 4.0 μm. The first interfacial dislocations were observed for t_F between 1.1 and 1.4 μm. This would give $t_c = 1.3 \pm 0.2$ μm , which is higher by one order of magnitude than the theoretical value calculated from the relation (Hirsch 1991) :

$$t_c = \frac{b}{8\pi\,\varepsilon\,(1+\nu)} \ln\!\left[\frac{e\,t_F}{r_0}\right]$$

(ν : Poisson's ratio, r_0 : core radius). Up to now, we cannot say if the theoretical relation for t_c is flawed, and/or if the cooling rate was too high to allow dislocation development.

Fig. 1 summarizes the various steps of misfit dislocation network formation :
- for $t_F < t_c$ the film is perfectly metastable and no difference can be found between a free Ge

sample and one with a deposited GaAs film (Fig. 1a) ;

- for $t_F \cong t_c$ a few isolated dislocations begin to develop in the interface from Ge threading ones (Fig. 1b)

- for $t_F \geq t_c$ (1.3-1.6 μm) interfacial dislocations are still easily resolved on topographs, but they expand through the whole width of the sample and some interactions can occur (Fig. 1c)

- for $t_F > t_c$ (> 2.0 μm) individual dislocations can no more be resolved (Fig. 1d).

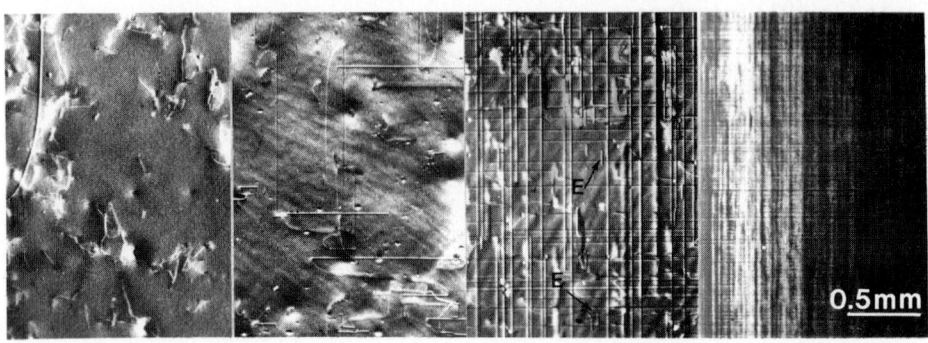

a   b   c   d

Fig. 1 : Different stages of the misfit dislocations formation a : $t_F = 0.5$ μm ;
b : $t_F = 1.2$ μm ; c : $t_F = 1.5$ μm ; d : $t_F = 2.2$ μm

It can be deduced from these observations that misfit dislocations are generated by a heterogeneous mechanism, as expected from the low mismatch.

The number of dislocations outcropping from the substrate in a 1 cm² area sample can be estimated at about 10^3 so that the total length of dislocation lines which can be developed in the interface cannot exceed 10^3 cm if no multiplication occurs.

On the other hand, a rough calculation shows that the complete relaxation of misfit deformation would need about 5.10^4 cm of dislocation lines in a 1 cm² interface so heterogeneous nucleation allows only partial relaxation. Therefore, another mechanism is required to achieve complete relaxation : multiplication, for instance, may be expected from interactions between dislocations of the two orthogonal systems. Homogeneous nucleation also may be invoked regarding the short edge segments labelled "E" in Fig. 1c. One can notice that these segments are not correlated to threading dislocations, but often with long interfacial ones.

Moreover, both mechanisms can simultaneously occur, as observed in In GaAs-GaAsP superlattices by Radzimski et al (1988).

2.2 Evaluation of the Residual Stresses

Assuming that deformation in the two perpendicular directions of the interface, x and y, is related to a biaxial stress in the film:

$$\sigma_{xx} = \sigma_{yy} = \frac{E_F}{1-\nu_F} \varepsilon_F = \sigma_F$$

(E_F, ν_F Young's modulus and Poisson's ratio of the film), the predicted value for σ_F is 94 MPa (resolved stress ≈ 40 MPa).

From this value of σ_F, one can deduce the radius of curvature R_S of the substrate as long as the sample is in elastic state, i.e. as long as dislocations can be resolved on X-Ray topographs. The easiest calculation uses Stoney's formula (Stoney 1909) extended to the biaxial case (Schweitz 1992); a more elaborate one by Chu et al. (1985a) leads to a rather complex expression of R as a function of a_F, a_S, t_F and t_S (respectively the lattice parameters and the thicknesses of the film and the substrate).

On the other hand, experimental values of R_S can be obtained by measurement of the 220 Bragg peak shift along the sample in the [110] direction. Comparison between experimental (R_{exp}) and theoretical values from Stoney's (R_{St}) and Chu's (R_{Ch}) formulas are given in table 1 (All the substrates are $\cong$ 1 cm x 1 cm x 250 μm).

t_F (μm)	0.5	1.0	1.3	1.5	2.0	2.2
R_{St} (m)	17.54	17.55	10.10	11.01	7.69	7.80
R_{Ch} (m)	17.53	17.54	10.09	11.00	7.68	7.79
R_{exp}(m)	18	25	11.5	17	5	9
total dislocation length (cm)	≈ 0	≈ 0	≈ 10	≈ 10	$\approx 10^2$	$> 10^3$

Table 1: Comparison between the measured radius R_{exp} and the theorical ones R_{Ch} and R_{St}.

It can be seen that experimental measurements are in rather good agreement with theoretical predictions; except for the 2.0 μm film-sample in which dislocation density is abnormally low (since the critical thickness is exceeded). Values given by the Stoney formula are quite close to those obtained from Chu's expression.

The stress distribution in the substrate has been given by Chu et al. (1985b) where E_s and v_s are the Young's modulus and the Poisson's ratio of the substrate:

$$\sigma_S = -\frac{t_F}{t_S}\sigma_F + \frac{E_S}{1-v_S}\frac{z}{R} \quad z \in \left[-\frac{t_S}{2}, \frac{t_S}{2}\right]$$

which leads to $\sigma_S \approx 2$ MPa at the interface (film 1.3μm thick). The corresponding stress distribution is shown in Fig. 2b, the neutral surface (NS, $\sigma_S = 0$) is shifted as compared to half the thickness.

Original shapes of dislocations were obtained, which seem to be related to this particular stress distribution. These "hairpin" dislocations were observed for thicknesses about t_c (Fig. 2a).

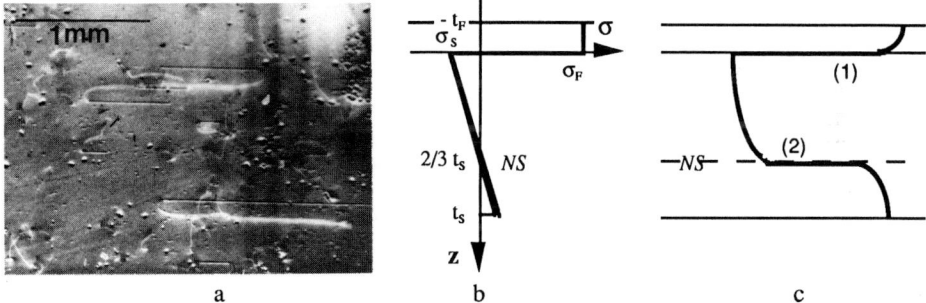

Fig. 2 : Correlation between stress distribution in the sample and shape of hair pin dislocations
a: Hair pin dislocation (XRT); b: Stress distribution through the sample; c: formation of an "hairpin" from a threading dislocation.

In spite of the fact that the interfacial arm (1, Fig 2c) and the internal one (2, Fig 2c) are submitted to largely different stresses σ_F and σ_S the dislocation velocities in Ge (which is higher by one order of magnitude) and in GaAs (extrapolated from Patel and Freeland, 1971, for Ge and Choï, Mihara, Ninomiya, 1977, for GaAs) might produce the same displacements of the dislocation in the substrate and in the film.

2.3 Checking Elastic Character of Deformation in the Film

Since $\sigma_{zz} = 0$, ε_{zz} is easily deduced from the classical Hook's law which leads to:

$$\varepsilon_{zz} = \frac{-2\,v}{1-v}\,\varepsilon_{xx}$$

Writing that the lattice parameter a_F of GaAs is a_S along x and y directions and a_T along the normal z direction, we get :

$$\varepsilon_{xx} = \frac{a_S - a_F}{a_F} \text{ and } \varepsilon_{zz} = \frac{a_T - a_F}{a_F} \quad \text{we found } a_T = 5.6494 \text{ Å.}$$

Such a deformation of the cubic GaAs cell in a tetragonal one implies that planes with the same Miller notation (hkl) (except (001)) are no longer parallel in the layer and in the substrate, and reticular distances d_{hkl} become different for all planes not normal to (001), so that Bragg angles are also different for $(hkl)_F$ and $(hkl)_S$ planes.

As transmission topography uses reflecting planes nearly orthogonal to (001), d_{hkl} in the film are rather close to d_{hkl} in the substrate and so are the Bragg angles. So the substrate and the layer can be imaged on the same topograph. In contrast, Berg-Barrett topography requires inclined planes (on which Bragg conditions in the substrate are far from those in the film). This angular shift can be predicted from the value of a_T.

We present in table 2, a comparison between the calculated angular deviation $\Delta\omega$ deduced from both plane shift and Bragg angle rotations, and the corresponding experimental shift, as measured using a Berg-Barrett setting on a 1.3 μm deposited sample.

g	404, $k\alpha_1$	404, $k\alpha_2$	224 $k\alpha_1$	224 $k\alpha_2$
$\Delta\omega_{cal}(°)$	0.0912	0.0921	0.0858	0.0890
$\Delta\omega_{exp}(°)$	0.088	0.090	0.085	0.095

Table 2: Comparison between measured angular deviations and theorical ones.

Experimental values are found in very good accordance with theoretical ones, which indicates that the elastic hypothesis is fairly valid in $\cong 1$ μm layer with a dislocation density below 10^3 cm.cm^{-2}.

3. CONCLUDING REMARKS

The formation of misfit dislocations in the GaAs/Ge system has been studied by X-Ray techniques. The critical thickness was found to be 1.3 ± 0.2 μm, and strain and stress measurements have been performed and compared with theoretical calculations. We can assume that elastic laws are qualitatively and rather well quantitatively obeyed. Elsewhere, original dislocation configurations, like hairpins, were observed, and if heterogeneous nucleation has been seen, we hope now to elucidate the later stage leading to completely relaxed dislocated structures. Presently, TEM observations are in progress (N. Guelton et al. 1993), as well as new macroscopic studies.

REFERENCES

Choi S, Mimara M and Ninomiya T 1977 Jpn J. Appl. Phys. 16 737
Chu S N G, Macrander A T, Strege K E and Johnston W D 1985a J. Appl. Phys. 57 249
Chu S N G, Macrander A T, Strege K E and Johnston W D 1985b J. Appl. Phys. 60 1238
Guelton N, Feuillet G, Saint-Jacques R G and Dodelet J P (this conference)
Hirsch P B 1991 Polycrystalline Semiconductors II, Springer Proceedings in Physics, eds J M Werner and H P Strunk 54 pp 470-482
Patel J R and Freeland P E 1971 J. Appl. Phys. 42 9 3298
Radzimski Z J, Liang B L, Rozgonyi G A, Humphreys T P, Hamaguchi N and Bedair S M 1988 J. Appl. Phys. 64 5 2328
Schweitz J A 1992 MRS bulletin XVII 7 34
Stoney G G 1909 Proc. Roy. Soc. (London), A82 172

Inst. Phys. Conf. Ser. No 134: Section 9   577

Paper presented at Microsc. Semicond. Mater. Conf., Oxford, 5–8 April 1993

Dislocation structure of heavily strained, MBE-grown GaSb on GaAs and LPE-grown CdHgTe and ZnHgTe on CdTe(ZnSe) revealed by TEM, XRD and XRT

R Kyutt, S Ruvimov, T Argunova, P Kop'ev, V Ratnikov, L M Sorokin, M Scheglov and R Scholz[1]

Ioffe Physical-Technical Institute, St Petersburg 194021, Russia
[1]Max-Planck-Institut für Mikrostrukturphysik, Weinberg 2, 4050 Halle/Saale, Germany

ABSTRACT: TEM methods combined with X-ray diffraction have been applied to the study of MBE-grown, heavily strained $2\mu m$ thick GaSb layers on GaAs and LPE CdHgTe, ZnCdHgTe and ZnHgTe layers on CdTe and ZnCdTe (111) substrates. Triple-crystal X-ray diffractometry was shown to give detailed information about the dislocation structure of the strained heterostructures and the results were in good agreement with those given by TEM

1. INTRODUCTION

GaSb/GaAs and CdHgTe/CdTe heterostructures appear to have considerable promise for microelectronic device fabrication. However, the prospects for using such heterostructures depend upon the structural quality of the material and, indeed, the high misfit levels cause defect generation which can dramatically affect device operation. Transmission electron microscopy (TEM) is widely known to be useful for investigation of the structural quality of heterostructure materials and, in the present work, TEM methods have been combined with X-ray diffraction analyses. In particular, it is shown that triple-crystal diffractometry (TCD) provides detailed information about the dislocations in the strained heterostructures.

2. EXPERIMENTAL DETAILS

Layers of epitaxial GaSb were grown by molecular beam epitaxy (MBE) on (001) GaAs substrates at deposition temperatures of 500-540°C in a Riber system with $\sim 10^{-10}$torr ultimate pressure. CdHgTe layers were grown on (111) CdTe and alloy substrates by closed-system liquid phase epitaxy (LPE) from Te solutions using Cd and Hg.

Using TCD, two dimensional analysis of the diffracted intensity has been carried out. The diffraction curve broadening in directions both normal and parallel to the reciprocal lattice vector **H** have been obtained. Asymmetrical Bragg measurements allow the determination of the deformation components $(\Delta d/d)_\perp$ and $(\Delta d/d)_\parallel$. To reveal the dependence of the broadenings upon the Bragg angle, the series of reflections 200, 400, 422 CuK_α and 200, 400, 800 MoK_α have been measured. The quality of near-surface layers (100-200nm) has been studied by means of grazing incidence diffractometry. X-ray topography observations were used to investigate the defects in the layers and substrates. TEM plan-view images and cross-sections have been obtained by use of JEOL JEM-7A (100kV), JEM 100C (100kV) and JEM 4000EX (400kV) microscopes.

3. RESULTS AND DISCUSSION

3.1 Heavily Strained GaSb/GaAs Heterostructures

For (001) GaSb/GaAs samples, TEM cross-sections showed that the distance between the misfit dislocations (MDs) was 5.5nm (compare previous work by Mallard et al (1989) and Rocher et al (1991)). High resolution electron microscopy (HREM) analysis (Fig. 1.) of the dislocation cores showed that they were of Lomer type, with the Burgers vector $b=a/2<110>$ lying in the interface. From X-ray asymmetrical Bragg measurements it was found that $(\Delta d/d)_\perp = (\Delta d/d)_\parallel$ and, thus, the misfit was compensated completely by MDs. The average MD separation, x, obtained by TEM was equal to that calculated using $x=b/(\Delta a/a)$. The overall dislocation density was reduced at a distance of 200-400nm from the interface. Plan view images (Figs. 2a,b) show that the dislocation density was $\sim 10^{10}\text{cm}^{-2}$ near the interface and $\sim 10^{8}\text{cm}^{-2}$ at the surface. The majority of threading dislocations had Burgers vectors of the type $a/2<110>$ and they lay on their respective $\{111\}$ glide planes. Those dislocations located near the surface were found to be mainly of screw type and, in general, it was concluded that the dislocations were distributed in a random and rather disordered manner. Nevertheless, groups of parallel dislocations with the same Burgers vectors were also observed and may be characteristic of fragments of low-angle boundaries. Near the interface, a three dimensional network of dislocations was found and, since triple nodes were present in close proximity to the interface, dislocation reactions were deduced to occur in the initial stages of growth. Annihilation reactions seem to account for the general reduction in dislocation density with increasing layer thickness.

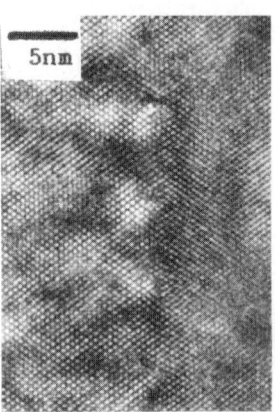

Fig. 1. HREM image of misfit dislocation cores.

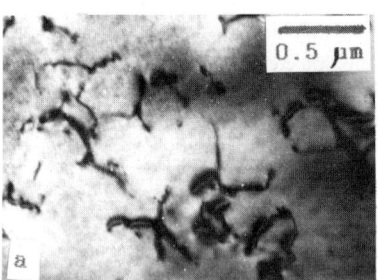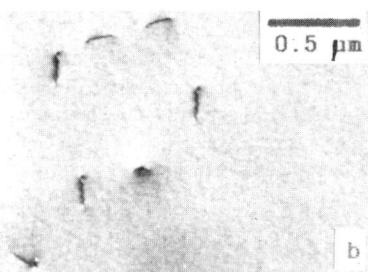

Fig. 2. TEM plan view images of threading dislocations in a GaSb layer: a) near the interface and b) at the layer surface

Diffractometry results were in good agreement with those obtained by TEM. Figure 3 shows typical GaSb layer and GaAs substrate diffraction curves. The main characteristics of these curves are: a) Reflections were broadened both parallel to and normal to the **H** direction, and the normal curve half-widths ($W_\perp$) were always much larger than the parallel half-widths ($W_\parallel$).

b) The $W_\perp$ values did not vary with radiation wavelength nor reflection order, while the $W_\parallel$ values depended on the variation in Bragg angle:

$$W_\parallel \propto \tan\theta_B ,$$ where θ_B is the Bragg angle (Fig. 3b).

These results allow us to determine the dislocation structure of the layers. Since $W_\parallel$ was proportional to $\tan\theta_B$, this shows that the threading dislocations were randomly distributed within

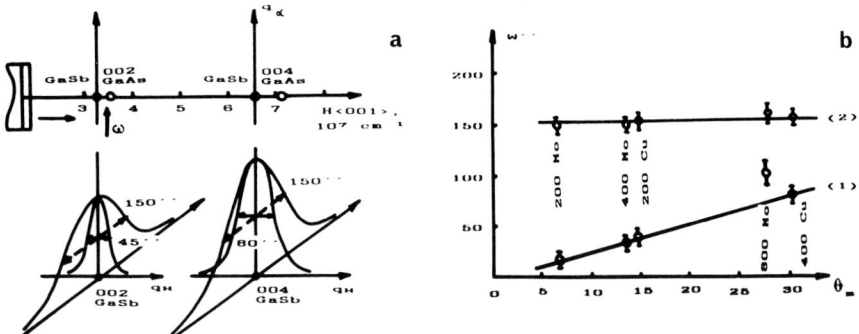

Fig. 3. a) Reciprocal lattice points of GaSb layer and GaAs substrate, and diffraction curves (q_H is parallel to H); b) $W_{\|}$ (1) and $W_{\perp}$ (2) GaSb halfwidths versus Bragg angle θ_B.

the layers. At the same time, large $W_{\perp}$ values were the evidence for misorientations in the crystal lattice. According to the TCD and TEM results, the threading dislocation groups caused the diffraction peaks to broaden. If they are sufficiently small, the size of the coherently scattering areas would affect both $W_{\perp}$ and $W_{\|}$ components and cause them to vary. However, no size contribution to the peak broadening was found, so that the distance between the boundaries was not less than $\sim 1\mu m$. Since the grazing incidence and TCD curves were close together, the average dislocation density in the layer was concluded to be close to that at the surface.

3.2 LPE-grown CdHgTe, CdZnHgTe and ZnHgTe heterostructures

Table 1 presents results of the CdTe-based heterostructure parameter measurements. ($\Delta a/a$) values obtained from microprobe analysis with the application of Vegard's Law have been compared with ($\Delta d/d$) components measured by means of double-crystal diffraction. Since the values are close together, it is clear that the misfit is compensated to a large degree. At the same time, all of the structures have a non-zero curvature and the radius decreases as the misfit increases. $W_{\perp}$ and $W_{\|}$ are the half-widths measured using the ω-scanning and ω-2ω-scanning modes. The dislocation density ρ was calculated from $W_{\|}$ on the assumption that the dislocations were randomly distributed in the

Table 1

	$(\Delta a/a) \times 10^{-2}$	$(\Delta d/d)_{\perp} \times 10^{-2}$	R(m)	W_{DCD}''	$W_{\perp}''$	$W_{\|}''$	$\rho(cm^{-2})$
$Cd_{0.2}Hg_{0.8}Te$ / CdTe	0.247	0.248	200	40	33	15	2.3×10^6
$Zn_{0.17}Cd_{0.08}Hg_{0.75}Te$ / $Zn_{0.115}Cd_{0.885}Te$	0.556	0.565	35	90	85	31	1.0×10^7
$Zn_{0.13}Hg_{0.87}Te$ / $Zn_{0.04}Cd_{0.96}Te$	0.800	0.815	15	100	95	35	1.3×10^7
$Zn_{0.2}Cd_{0.07}Hg_{0.73}Te$ / CdTe	1.400	1.450	6.5	120	115	48	2.5×10^7

layers.   The Burgers vectors were assumed to be a/2 < 110 > (Qadri and Dinan 1985).   The chaotically distributed dislocation model was strengthened by the fact that W_I measured from (111) and (333) Cu was proportional to tan θ_B.   Measured values of $W_\perp$ were independent of the X-ray wavelengths and reflection orders.   Nevertheless, since $W_\perp$ was rather larger than W_I, this indicated conclusively that the dislocations formed the low angle boundaries.

From the X-ray topographs, one can see that the layer has a cellular dislocation structure.   The average dislocation density for the $Cd_{0.2}Hg_{0.8}Te/CdTe$ system can be assumed to equal $5 \times 10^8 cm^{-2}$.

For the $Zn_{0.056}Cd_{0.05}Hg_{0.894}Te/Zn_{0.04}Cd_{0.96}Te$ heterostructure, step-by-step thinning of a 35μm layer was carried out.   The near-surface dislocation density assessed from W_I was $10^7 cm^{-2}$. $W_\perp$ and W_I values were nearly constant until the layer thickness reached 8-10μm.   When the layer was thinner than 5μm, the W_I half-widths drastically increased, at which point the calculated dislocation density was $2 \times 10^8 cm^{-2}$.

4. CONCLUSION

For the heavily strained GaSb/GaAs heterostructure, TEM showed the reduction in threading dislocation density from $10^{10} cm^{-2}$ at the interface to $10^8 cm^{-2}$ at a distance of 200-400nm into the layer. X-ray diffraction analysis allowed us to determine that the threading dislocations were randomly distributed, with the density nearly constant across the layer depth for regions away from the interface.   TEM revealed groups of screw dislocations with the same Burgers vectors.   The large reflection broadening normal to **H** shows that the high dislocation density coexists with low-angle boundaries.

The dislocation structure of the LPE-grown CdTe-based heterostructures showed similar features, except that their dislocation densities were two orders of magnitude lower.   Thus, the dependence of the layer structural parameters on misfit values has been determined.

REFERENCES

Mallard R E, Wilshaw P R, Mason N J, Walker P J and Booker G R 1989 Microscopy of Semiconducting Materials 1989, eds A G Cullis and J L Hutchison (IOP Publishing Ltd, Bristol) pp 331-336
Qadri S and Dinan J 1985 Appl. Phys. Lett. **47**, 1066
Rocher A, Kang J M, H Atmani, Crestou J, Vanderschaeve G, Lassabatère L and Bonnet R 1991 Microscopy of Semiconducting Materials 1991, eds A G Cullis and N J Long (IOP Publishing Ltd, Bristol) pp 509-514

Inst. Phys. Conf. Ser. No 134: Section 9
Paper presented at Microsc. Semicond. Mater. Conf., Oxford, 5–8 April 1993

581

Thickness dependence of the density of threading dislocations in mismatched (001) oriented epilayers

K Durose and H Tatsuoka

Department of Physics, University of Durham, South Road, Durham, DH1 3LE, UK

ABSTRACT: The functional dependence of the threading dislocation density (D_h) with film thickness in mismatched heteroepitaxial films is examined by comparing two theoretical models with published data. A new model is presented which presumes that the residual strain is eliminated by threading dislocations. It is successful in predicting a fall in D_h to a background value, D_∞, via the successive elimination of a fraction (Ab) of the dislocations in each layer. Ayers' half-loop model is also tested for a variety of materials systems.

1. INTRODUCTION

It is generally considered that lattice relaxation in mismatched heteroepitaxial films takes place by the formation of interfacial dislocation arrays, such that $f = \delta + \varepsilon$, where f is the misfit, δ is the strain relieved by the array, and ε the residual strain. The residual strain is due to differential thermal expansion or deviation from the condition $a_{epi}/a_{substrate} = l/m$, an integer ratio. This residual strain is not homogeneous, but is a function of depth, as has been shown experimentally by Tatsuoka et al (1991). The strain is accompanied by a distribution of threading dislocations which must act to relieve it. Such dislocations have been observed by XTEM, and rocking curve data for layers of different thicknesses is available from the literature (see Table 1).

It is the intention of this paper to match models of the thickness dependence of threading dislocation density with experimental data. Two models are described in the literature and these are a) a strain relief model due to Tatsuoka et al (1991) and b) a model based on the equilibrium distribution of half - loops formulated by Ayers et al (1991,1992). Here both are compared with experimental results from a wider range of materials systems and the former is modified since it does not give realistic dislocation densities.

2. DISLOCATION DENSITY MEASUREMENTS

Threading dislocation density ($D_h (cm^{-2})$) and layer thickness data were collected from the literature as shown in Table 1. Most papers gave X-ray rocking curve widths (β) and these values were converted to dislocation densities using the formula of Gay et al (1953):

$D_h = (\beta^2 - B^2)/9b^2$ in which b is the Burgers vector of the dislocations (taken as 60° dislocations here) and B is the rocking curve width in radians with no broadening due to defects (a nominal value of 10" was used throughout this work). In this way data for CdTe/GaAs grown using the hot wall epitaxy (HWE) and MOVPE methods was obtained. The data for annealed and as-grown GaAs/Si was presented by Ayers et al (1992) as D_h values directly. The only data which is not from X-ray measurements is that for ZnTe/GaAs which was obtained by measuring dislocation density as a function of distance from the interface from an XTEM micrograph of a single film. All of the data showed a decrease in D_h with increasing thickness, with that in the thicker layers reaching a constant lower limit.

3. STRAIN RELIEF MODEL

Tatsuoka (1991) presented a model for the relief of the residual strain in the first layer of an epitaxial film (ε_1) by dislocations distributed in the film. Strain in the nth layer (ε_n) was partly relieved by a linear density ($N_{m(n+1)}$) of long straight 'misfit' dislocations in the next layer. The density of dislocations was set as being proportional to the strain in the previous layer so -

$$N_{m(n+1)} = A\varepsilon_n \tag{1}$$

The residual strain in each layer is then given by

$$\varepsilon_{n+1} = \varepsilon_n - N_{m(n+1)}b_{eff} \tag{2}$$

where b_{eff} is the effective Burgers vector of the dislocation. Combining (1) and (2) gives an expression for the strain

$$\varepsilon_n = \varepsilon_1(1 - Ab_{eff})^{n-1} \approx \varepsilon_1(1 - Ab_{eff})^n \tag{3}$$

where each layer has a thickness a/2 for (001) oriented films. Tatsuoka (1991) analysed experimental strain versus thickness data for CdTe/GaAs using this expression and found that a plot of $lg\varepsilon_n$ against thickness (h) was linear yielding $\varepsilon_1 = 2\times10^{-3}$ and $A = 6.5\times10^5$ cm^{-1}. The linear density of complete misfit dislocations from (1) and (3) was

$$N_{m(n)} = A\varepsilon_1(1 - Ab_{eff})^{n-2} \approx A\varepsilon_1(1 - Ab_{eff})^n \tag{4}$$

which yielded (unrealistic) values of between 1 and 10 cm^{-1} for CdTe/GaAs.

The model infers that the long misfit dislocations in each layer arise from the turning over of (vertical) threading dislocations into the (001) planes. This mechanism for the reduction of threading dislocations also gives a massive underestimate of the observed fall-off in dislocation density with increasing thickness.

The model can be modified by describing each of the long misfit dislocations as $1/l$ segments, each of length l in unit length. In this way as an inclined threading dislocation passes through a monolayer its projected length on (001) and effective Burgers vector contribute to the relief of the residual stress. The expression for the number of 'misfit' dislocations (4) may therefore be re-written for an equivalent number of threading dislocation segments, $N_{(n)}$ in the nth layer , each having an effective stress relieving length of l. -

$$N_{(n)} = \frac{A\varepsilon_1}{l}(1 - Ab_{eff})^{n-2} \approx \frac{A\varepsilon_1}{l}(1 - Ab_{eff})^n \tag{5}$$

or as an areal density -

$$D_{(n)} \approx \frac{2A\varepsilon_1}{l}(1 - Ab_{eff})^n \tag{6}$$

This expression indicates that in order to relieve the residual strain ε_1 the threading dislocations in layer n will be present in a density which relieves a fraction Ab of the strain in the previous layer. The fractional reduction in the required threading dislocation density upon passing from one layer to the next is also Ab. This process will continue until, at some thickness, the strain reaches a background level. It is therefore implicit that $2A\varepsilon_1/l$ dislocations will be eliminated when the film reaches a certain thickness. At that thickness any extra dislocations (which did not participate in strain relief) remain as a constant background density which is independent of thickness, the driving force for the annihilation of the dislocations having been strain. If this density of dislocations is $D_{(\infty)}$ then (6) may be rewritten as

$$D_{(n)} = \frac{2A\varepsilon_1}{l}(1 - Ab_{eff})^n + D_{(\infty)} \quad \text{or} \quad D_{(n)} = D_{(1)}(1 - Ab_{eff})^n + D_{(\infty)} \tag{7}$$

where $D_{(1)} = 2A\varepsilon_1/l$. If $D_{(\infty)}$ can be obtained from thick films then this model predicts that plots of the form $lg(D_{(n)}-D_{(\infty)})$ versus n will be linear with intercept $D_{(1)}$ and slope $lg(1-Ab)$.

Figure 1a shows such plots for the two CdTe/GaAs samples, $D_{(\infty)}$ having been found experimentally as 4.27×10^7 and 2.21×10^7 for the HWE and MOVPE material respectively. The value of the coefficient A (see Table 1) for both sets of samples was found to be about an order of magnitude less than that found by Tatsuoka (1991) for the HWE samples. The fraction Ab is about equal for both samples but the dislocation density is reduced slightly faster in the HWE grown samples than in the MOVPE ones. Additionally the density of threading dislocations in the first layer was about 1.7 times greater for the HWE samples. These differences could be due to the differences in the growth techniques, temperatures and substrate preparation for example.

Figure 1b shows similar plots for both the as-grown and annealed GaAs/Si samples. $D_{(\infty)}$ was taken as zero since data for only one very thick sample was reported. This explains the outlying points on the graph. Annealing increases the dislocation density reduction fraction by a factor of 0.7 and decreases D_1 to 0.4 of its original value. It is likely that annealing acts to reduce $D_{(\infty)}$.

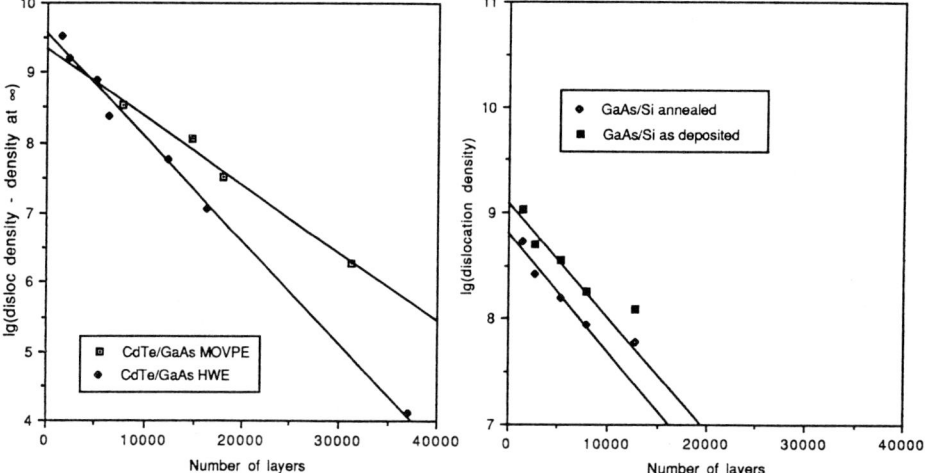

Figure 1. Plots of $lg(D_n - D_\infty)$ versus film thickness in monolayers for a) CdTe/GaAs (left) and b)GaAs/Si (right).

Comparable plots were obtained for other materials systems (see Table 1) and, together with the other data this allowed trends in the Ab values to be sought. It is generally true that Ab increases with increasing misfit. However the trend does not follow a good straight line, the scatter presumably being due to the varying states of equilibrium of the films examined. Examination of sets of annealed films might allow the trend to be confirmed and quantified. The data for ZnTe/GaAs fits the trends very poorly, the D_1 and Ab values being higher than those expected for the misfit. This may be due the method of dislocation density measurement. The observation that Ab increases with misfit might be explained by the fact that for the more highly mismatched layers the threading dislocations are more closely spaced. If they lie on defined crystallographic lines and have a probability of cancelling when they meet, then closely spaced threading dislocations (in highly mismatched systems) will first cancel in fewer monolayers than widely spaced ones (in low misfit systems).

4. HALF LOOP MODEL

Ayers et al (1991,1992) formulated a model for dislocation density as a function of layer thickness which predicted that for thick(>1μm) films with high misfit (f> 0.003) the dislocation density is $D_h = klfl/\alpha bh$ where k is a slowly varying function of D_h and Poisson's ratio and α is a poorly understood empirical factor obtained from experimental plots of D_h versus 1/h. Most such plots are reasonably linear and their slopes for the materials discussed in this work are listed in Table 1. Although these samples were not annealed the data lends modest support to the Ayers' prediction that the slopes should be related to misfit.

Extrapolation of the plots to infinite thickness (see Table 1) and the handling of data for thick samples (CdTe/GaAs here) in this way however indicates that the model does not account well for the thickness independent dislocation densities found in thick layers. The experimental data deviated from linearity and the extrapolations indicate unrealistic and even negative dislocation densities for thick layers.

Table 1. Data from strain relief and half-loop model plots of threading dislocation density versus thickness.

Material	Ref	f	Strain relief model		Loop model		
			D_1 cm^{-2} x10^{-8}	Ab x10^4	A cm^{-1}x10^{-3}	Slope cm^{-2} x μm^{-1}x10^{-8}	D_∞ cm^{-2}x10^{-7}
ZnSe/GaAs MBE	Muggelberg (1992)	-2.77×10^{-3}	3.9	1.5	3.7	1.0	7.3
InP/GaAs MOVPE	Horikawa (1988)	-3.68×10^{-2}	6.8	1.5	3.6	5.0	-2.9
GaAs/Si as grown	Ayers (1992)	-3.93×10^{-2}	1.6	1.9	4.7	3.9	2.8
GaAs/Si annealed	Ayers (1992)	-3.93×10^{-2}	6.3	2.6	6.5	2.0	0.79
ZnTe/GaAs MOVPE	Brown (1993)	-7.38×10^{-2}	1.4	3.8	11.5	43.3	-89.9
CdTe/GaAs HWE	Tatsuoka (1989)	-1.28×10^{-1}	3.7	3.4	7.5	16.9	-26.5
CdTe/GaAs MOVPE	Brown (1988)	-1.28×10^{-1}	2.2	2.2	4.9	10.8	-9.5

5. CONCLUSIONS

The strain relief model provides a function which fits the observed dislocation density versus thickness data for a range of mismatched epitaxial materials. The model predicts that the dislocation density will fall by a fraction Ab in each layer and a background density will be reached when the strain is relieved. Ab is related to the misfit f. This model will be useful in quantifying the efficiency of growth strategies designed to reduce the densities of threading dislocations in mismatched heteroepitaxial films. The half-loop model is reliable but possibly less informative in this context, the parameter α being poorly understood at present.

REFERENCES

Ayers J E, Ghandhi S K and Schowater L J 1991 Mater. Res.Soc Symp. Proc. 209 661
Ayers J E, Schowater L J and Ghandhi S K 1992 Journal of Crystal Growth 125 329
Brown G T, Kier A M, Gibbs M J, Geiss J, Irvine S J C, Astles M G 1989 Electrochem. Soc. Symp. Proc. 89 (5) 171
Brown P D 1993 submitted to Journal of Crystal Growth
Gay P, Hirsch P B and Kelly A 1953 Acta. Metallurgica 1 315
Horikawa H, Ogawa Y, Kawai Y and Sakuta M 1988 Applied Physics Letters 52 (5) 397
Muggelberg C 1992 'Herstellung und Characterisierung von ZnSe und (ZnMn)Se MBE Strichten' diplomarbeit thesis, Humboldt Universitat zu Berlin
Tatsuoka H, Kuwubara H, Fujiyasu H and Nakanishi Y 1989 Journal of Applied Physics 65 (5) 2073
Tatsuoka H, Kuwubara H, Nakanishi Y and Fujiyasu H 1991 Thin Solid Films 201 59

Inst. Phys. Conf. Ser. No 134: Section 9
Paper presented at Microsc. Semicond. Mater. Conf., Oxford, 5–8 April 1993

585

Critical thickness phenomena: the distinction between the existence of interfacial dislocations and significant lattice relaxation

P Kidd[1], P F Fewster[2], N L Andrew[2] and D J Dunstan[3]

[1]Department of Materials Science and Engineering, [3]Department of Physics, University of Surrey, Guildford, GU2 5XH, UK.
[2]Philips Research Laboratories, Cross-Oak Lane, Redhill, UK.

ABSTRACT: We use High Resolution X-ray Diffraction Reciprocal Space Mapping to examine strain relaxation in epitaxial strained layers of $In_{.1}Ga_{.9}As$ on GaAs. We show that the strain-field of an isolated dislocation is confined laterally to within a distance of the order of the layer thickness. For thin layers with a low density of interfacial dislocations this results in the layer containing distinct distorted and non-distorted regions. Only when the dislocation strain-fields overlap does the layer show long-range lattice relaxation.

1. INTRODUCTION

The phenomenon of strain-relaxation in pseudomorphic strained layers, via the introduction of misfit dislocations, has been the subject of much investigation, particularly with regard to a critical thickness, h_{cd}, for the generation of misfit dislocations (For reviews see van der Merwe 1991 and Freund 1992). From a technological viewpoint, with regard to the stability of a strained layer device, it is important to be able to design device structures with layer strain-thickness parameters which keep the structure below h_{cd}. We have recently been interested in the elucidation of strain-relaxation phenomena for layers which exceed h_{cd} in order to develop predictive formulae for relaxation in buffer layer engineering (Dunstan 1991a). We have observed empirically, in the InGaAs/GaAs system, that the residual strain in relaxing single and multilayer structures follows a hyperbolic dependence upon layer thickness, provided that a <u>second</u> critical thickness, h_{cr}, has been exceeded (Dunstan et al 1991b, Kidd et al 1993). This second critical thickness, rather than being identified as the thickness at which dislocations appear in the interface, is identified as the threshold thickness above which long-range lattice relaxation is observed. This phenomenon is illustrated schematically in Fig 1. The relaxation curve e=k/h is a plot of residual strain versus layer thickness and has been obtained empirically using Double Crystal X-ray Diffraction to measure residual strain in $In_xGa_{(1-x)}As$ layers on GaAs, for x = 0.1 and x = 0.2. Despite the observation of interfacial dislocations above thickness h_{cd} lattice strain relaxation is not observed until after h_{cr}. In terms of the dislocation behaviour, there is a sudden increase in interfacial dislocation density at h_{cr} as reported by Dixon et al 1990.

In this paper we present results obtained in a programme to investigate these two critical thicknesses by studying the nature of the dislocation strain-fields in relaxing layers. We have started by examining a series of four samples, each having a single layer of $In_{.1}Ga_{.9}As$ on GaAs with a thickness in the range from h_{cd} up to and beyond h_{cr}. These layers have previously been shown to follow the established hyperbolic strain-thickness dependence regarding lattice relaxation (Kidd et al 1993). Their residual strains as measured by Double Crystal X-ray Diffraction are plotted as filled circles in Fig 1. All the layers have thicknesses exceeding the first critical thickness h_{cd}, i.e. they have dislocations at the interface, however, only the thickest layer shows lattice relaxation, having exceeded h_{cr}. Here we use a new technique which combines High

Resolution X-ray Diffraction (HRD), Reciprocal Space Mapping (RSM), X-ray Topography (XRT), and Simulation of Bragg-case dynamical Diffraction Profiles. We show how, with this technique, we are able to measure the development of dislocation strain-field arrays and the resulting lattice relaxation as the expitaxial layer thickness increases.

2. EXPERIMENTAL ANALYSIS

The experimental details of the eight-crystal high-resolution x-ray diffractometer used for these studies have been presented elsewhere (Fewster 1991a, b, 1992). The samples investigated here were single surface layers of $In_xGa_{(1-x)}As$ grown by MBE at $510^{\circ}C$ on [001] oriented GaAs substrates, with compositions in the range x = 0.08 - 0.09 and nominal thicknesses 70, 100, 140 and 200nm. The 004 Bragg reflection reciprocal space maps for the four samples oriented along the [110] azimuth are shown in Figs 2a-d.

The reciprocal space map is a measure of diffracted intensity measured pixel-by-pixel in a series of step and count scans. A w2w' scan is obtained by coupling the specimen rotation (w) with the point-probe detector rotation (2w') and as such measures the diffracted intensity for a particular crystal lattice spacing

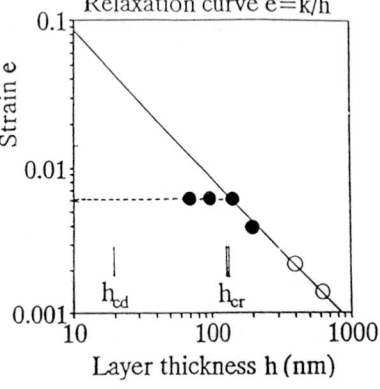

Fig 1. Residual strain versus layer thickness for $In_{.1}Ga_{.9}As/GaAs$.

at a given tilt angle. By decoupling w and rotating it by a fraction dw and recoupling to give the next w2w' scan, the same crystal lattice spacing is being probed for a different lattice tilt angle. The reciprocal space map thus separates lattice parameter measurement (w2w' axis) from lattice tilt (w axis). With the state-of-the-art equipment used here, the instrument function is of the order of the pixel size (Fewster 1991a) thus the diffraction spot shape and intensity can be related directly to the sample reciprocal space lattice without the need to deconvolute for an instrument function.

The full intensity profiles including the Bragg and diffuse scattering can be projected onto the w2w' direction or onto the w direction for each peak. In Fig 2a-d we show the intensity profiles around the layer peaks projected onto the w direction for each layer, and projected onto the w2w' direction for the 70nm and 200nm thick layers. There is thus a wealth of information available in the reciprocal space maps and their resulting intensity profiles. Semi-quantitative information may be obtained, in the first instance, by considering the shapes and positions of peaks and satellites. A more quantitative and detailed analysis may be performed by solution of dynamical diffraction formulae incorporating parameters for coherent (Bragg) diffraction and incoherent (diffuse) diffraction. The complete analysis of these profiles by dynamical diffraction simulation is outside the scope of this paper. In the following section we present a semi-quantitative interpretation of the observed diffraction profiles.

3. INTERPRETATION OF THE RECIPROCAL SPACE MAPS

The reciprocal space maps, Figs 2a-d, are all plotted to the same scale. The diffracted intensity for each uncoupled sample orientation (w) and coupled sample-detector orientation (w2w') is plotted in grey scale contrast. As can be seen clearly, the main layer peak, remains in the same place (accounting for differences in layer composition), except for the 200nm thick layer, where the peak has shifted substantially and the layer has relaxed ~30% of its original strain. Comparison of the 70nm and 200nm layer profiles along w2w' show a change from the profile of a reciprocal lattice streak with thickness fringes for the almost perfect 70nm layer to a diffuse Gaussian-type layer peak resulting from microscopic lattice distortion in the relaxed 200nm layer.

Regarding the intensity distribution around the layer peaks, there is a gradual emergence of two

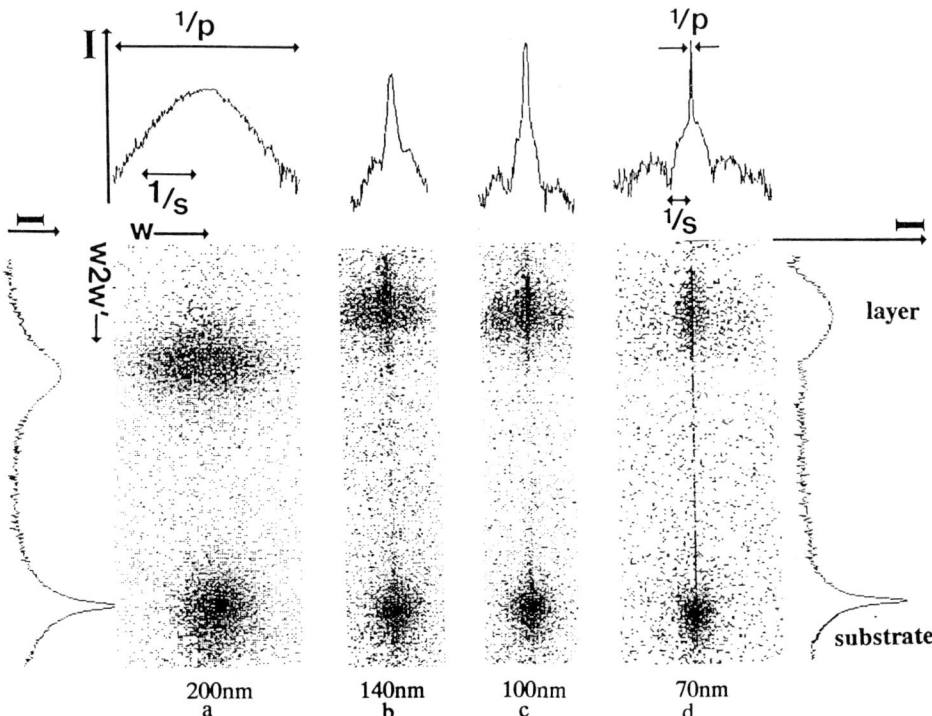

Figs 2a-d. 004 Bragg reflection reciprocal space maps for samples with layer thicknesses 70, 100, 140 and 200nm. Intensity is plotted as a grey scale. The projected diffracted intensity (log scale) along w is plotted for each layer and along w2w' for the 70nm and 200nm thick layers.

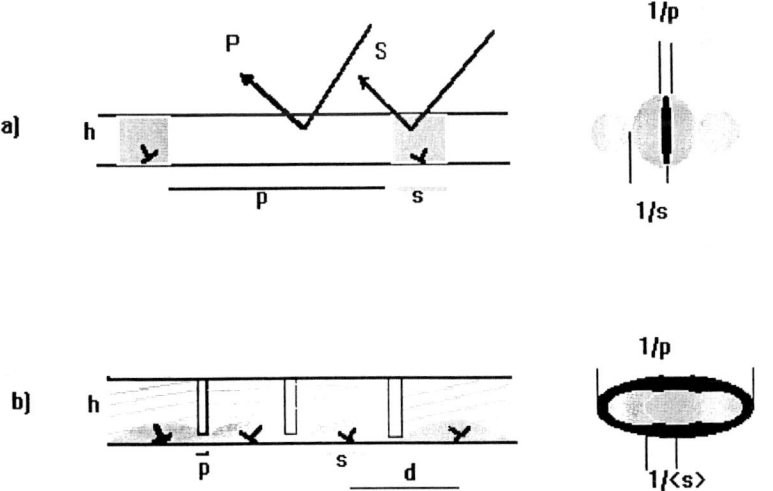

Figs 3a and b. Illustration of the real space components (LHS) contributing to the intensity distribution in the layer peaks in the reciprocal space maps (RHS) for the cases where the strain-field of the dislocation is limited a) by the layer thickness and b) by neighbouring dislocations.

weak lobes either side of the layer peak along the w axis, these are only vaguely visible for the 70nm thick layer, increasing in intensity from 100nm to 140nm and merging into one diffuse peak for the 200nm layer. By taking a topograph of the diffuse scatter from these lobes we have unambiguously identified the scatter as being due to lattice distortion around interfacial dislocations. We have observed (using transmission electron microscopy) that the interfacial dislocation spacings for the 70, 100 and 140nm layers are irregular and all greater that the layer thicknesses, whereas the spacings for the 200nm thick layer, have a range of values with a mean at 10nm (details to be presented elsewhere).

For the 70nm thick layer, the intensity distributions on the reciprocal space maps can be interpreted (see Fig 3a) as consisting of a coherent component from the perfect epitaxial layer, giving rise to the sharp peak, **P**, and an incoherent component, **S**, resulting from diffuse scatter from distorted regions around dislocations. Considering the diffuse scatter component, projected onto w, the simplest shape function whose squared Fourier transform gives an approximation to the observed shape is a simple rectangular function. Using this as a guide we can use the angular separation of the profile minima, **1/s** (1/arc sec.), to calculate a lateral dimension **s** (nm) for the distortion field of the dislocations giving rise to the scatter. For the 100nm and 140nm thick layers the peak **P** becomes less intense and the scatter from distorted regions becomes more irregular. We have calculated the distortion field dimension for the three layer thicknesses to be 68, 88 and 108nm for the 70, 100 and 140nm thick layers respectively. That is, the extent of the distortion field is approximately equal to the layer thickness.

For the 200nm thick layer the w profile is approximately Gaussian. We interpret this as being due to two effects (see Fig 3b): Firstly, the dislocation spacing in the interface has now decreased to a value less than the layer thickness, so that the lateral distortion fields are now dictated by the nearest neighbour dislocation spacings, **d**, which have a range of values about the mean **<s>** =10nm (measured by plan view TEM). This gives rise to an incoherent component in the diffuse scatter with a corresponding profile width **1/<s>**. In addition, the correlation length of perfect crystal, **p**, becomes vanishingly small, resulting in a broadening and weakening of its contribution, **1/p**, to the peak. These two components now combine to give a single diffuse peak.

Thus, the four samples illustrate the transition from the appearance of the first dislocations in thin layers, which only provide localized relaxation (Fig 3a), to the condition where the dislocation density is sufficient to relax all regions of the layer (Fig 3b).

ACKNOWLEDGEMENTS

This work has been supported by the Strained-Layer Structures Research Group at the University of Surrey, and the Royal Society. The samples used in this study were grown by R Grey at Sheffield as part of an SERC funded programme.

REFERENCES

Dixon R H and Goodhew P J, 1990 J Appl. Phys. 68 3163
Dunstan D J, 1991a Semicond. Sci. and Tech. 6 A76-A79
Dunstan D J, Kidd P, Howard L, Dixon R H 1991b Appl. Phys. Lett. 59 3390-3392
Dunstan D J, Young S, Dixon R H, 1991c, J. Appl. Phys. 70 3038-3045
Fewster P F 1992 J Appl. Cryst. 25 714-723
Fewster P F 1991a J. Appl. Cryst. 24 178-183
Fewster P F 1991b Appl. Surf. Sci. 50 9-18
Freund L B, 1992 MRS Bulletin (MRS Pittsburgh PA USA) July p52-60
Kidd P, Dunstan D J, Grey R, David J P R, 1993 Submitted to Appl. Phys. Lett.
Van der Merwe J H, 1991, Critical Reviews in Solid State and Materials Sciences 17 187-209

Determination of the state of deformation in epitaxial layers using X-ray diffraction

P J Dugdale, S J Barnett[1], R C Pond and M Emeny[1]

Department of Materials Science and Engineering, The University of Liverpool,
P O Box 147, Liverpool L69 3BX U K
[1] DRA, St. Andrews Road, Malvern, Worcs U K

ABSTRACT: The state of deformation in epitaxial layers of InGaAs grown by MBE on GaAs substrates has been determined using X-ray diffraction. This method enables the strains and rigid rotations which have occurred in the layers to be measured, and these are described by means of a tensor. Layers of different thicknesses have been grown on substrates exhibiting low and high dislocation density in order to assess the influence of this parameter on layer relaxation.

1. INTRODUCTION

High quality epitaxial films are an essential prerequisite for the fabrication of microelectronic devices. Such layers are generally thinner than the critical thickness (Matthews 1974), and hence do not contain misfit relieving dislocations at the interface but are elastically strained. However, the mechanism and rate of plastic relaxation in thicker layers is of interest (Kavanagh 1988), for example for the production of fully relaxed buffer layers (Fatermi 1991, Dodson 1988). The present work is aimed at extending our understanding of plastic and elastic relaxation processes, particularly the influence of substrate dislocation density. It is now established that dislocations which are initially present in the substrate extend into the growing epilayer and, under the influence of misfit stresses, can bend over to lie in the interface and thereby relieve misfit (Matthews 1974). When the substrate dislocation density is low, this source of defects may be insufficient for complete accommodation of misfit, and other sources, such as half-loop generation at surface features, may operate additionally. The objective of our work is to correlate the state of relaxation with the initial defect density in the substrate, and also to study the rate of relaxation as a function of annealing treatments. Our programme involves the growth of (001) InGaAs by MBE on GaAs substrates with low and high dislocation density. Layers are grown with thicknesses about equal to and greater than the critical thickness, and the state of relaxation is determined using X-ray diffraction (XRD), X-ray topography and transmission electron microscopy (TEM).

In the present paper we report our preliminary results; these include the substrate characterisation, and the state of relaxation of the layers. Emphasis is placed on the XRD method for determining the state of relaxation. Conventional methods (Tanner 1991) concentrate on establishing the extent of relaxation perpendicular to the interfacial plane, and generally assume tetragonal distortion for the parallel relaxation. In the present work, all components of the deformation of epitaxial layers (and hence the state of relaxation) are being determined, and therefore additional information, such as the presence of rigid-body rotations and shears of the overlayer, can be established. Such deformation components may affect the evolution of microstructure in epilayers, for example by modifying the shear stresses acting on the various glide-planes and hence the propensity for dislocation movement on those planes (Beanland and Pond 1989).

2. THE STATE OF DEFORMATION

Deformation is the combination of strains and rotations that occur when a material is subjected to an external stress. It is a homogeneous description and hence ignores localized effects such as the interfacial displacement field. In the case of epitaxy, the source of the stress is the misfit between the substrate and layer. It is generally accepted that misfits of less than about 7% can be accommodated by elastic strain until a critical thickness is reached. In such coherent growth the lattice parameter of the epilayer is constrained to fit that of the substrate parallel to the interface, whilst that normal to the interface will increase or decrease depending on the sign of the misfit. Elastic relaxations have been analysed by Hornstra and Bartels (1978) who defined a two stage process. Firstly the lattice parameter of the epilayer is forced to match that of the substrate in all directions by the application of tractions to its surface. When these are removed the layer relaxes and they showed that all points within the layer move in a particular direction which is normal to the interface in the case of growth on a low index substrate. As a result of the Poisson effect, all planes in the layer rotate by different amounts, except for those parallel or perpendicular to the surface, which remain invariant. Above the critical thickness it is energetically favourable for misfit to be shared between elastic and plastic strain by the introduction of misfit dislocations. If the net Burgers vector content of the misfit dislocation array includes a component perpendicular to the interface rigid-body rotation of the overlayer will arise. In addition, shears parallel to the interface can also arise as a consequence of net screw components.

Using X-ray diffractometry the change in interplanar spacing and the rotation of a set of planes can be found. This information can be used to express the total deformation of the overlayer with respect to the substrate, which is assumed to be rigid, in terms of a tensor. Much of the X-ray work done in the past has concentrated on collecting information from only one set of reflections, usually 004; this enables calculations of either composition or average relaxation. More recent work using {115} reflections has enabled the determination of the layer relaxation in two orthogonal directions parallel to the interface, usually [110] and [$\bar{1}$10] (Leiberich, Levkoft 1990). In this paper we present a description of the state of deformation that requires measurements from more than one set of planes (Beanland 1991). The objective is to express the state of deformation by specification of the elements of a tensor of the form d_{ij}. This can then be split into elastic and plastic parts subject to the restrictions imposed by boundary conditions. Diagonal components relate to the tensile or compressive strains in the layer, for example d_{33} is the change in interplanar spacing for (004) planes and d_{11} and d_{22} give rise to the orthogonal in-plane relaxations. Combinations of the off-diagonal components represent the tilts and shears that have occurred (Beanland 1991). Measurement of the interfacial shears is not directly accessible by our present method, although future work is planned to determine these also.

3. EXPERIMENTAL RESULTS

X-ray topography was used to determine the number of dislocations present and in the case of the lower dislocation density substrate this was found to be 10^3 cm^{-2}. Measuring the number of defects in the poorer quality substrate was more difficult, but a value in excess of 10^6 cm^{-2} was estimated by comparing the dislocation cell size with unpublished experimental data on this subject (Stirland, private communication). InGaAs layers of thickness 30 nm and 150 nm were grown at 520°C in a VG-V80H reactor at a group V:III beam equivalent pressure ratio of 10:1 and an alloy growth rate of 1 micron per hour. These were characterised using X-ray diffraction and topography, reflection topographs of the 150 nm layers were taken using the 620 reflection and CuK$_{\alpha 1}$ radiation. Figs. 1a, b show the topographs of the 150 nm layers.

The X-ray diffraction was carried out on an Apex goniometer equipped with a four crystal monochromator using CuK$_{\alpha 1}$ radiation. The reflections used were 004 and {115} which provided enough data in the form of peak splittings to determine the deformation tensors. These are given in figure 2 for the four layers studied so far.

Having determined the tensors, several useful parameters can be found readily including the natural lattice constant of the InGaAs layer and hence the composition. It is also possible to calculate the relaxations in the <110> directions. For the 30 nm layers these were found to be small. The layer grown on the high dislocation density substrate had relaxed by 7.8% and 5.7% and that grown on the low dislocation density substrate had relaxed by 4.9% and 4.3% in the two orthogonal <110> directions. Their indium compositions were 16.0% and 16.1% respectively and the rigid-body tilt components of both tensors were similar, being of the order of 0.06°.

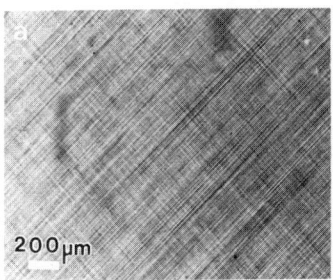 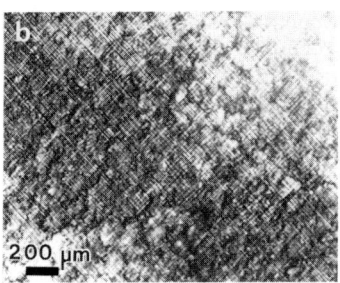

Fig.1a: (620) X-ray topograph of 150 nm layer on low dislocation density substrate, b: on high dislocation density substrate.

High dislocation density substrate.

300 Å   1500 Å

$$10^{-4}\begin{bmatrix} 6.65 & - & 6.07 \\ - & 8.94 & 24.58 \\ 1.79 & 0.00 & 211.07 \end{bmatrix} \qquad 10^{-4}\begin{bmatrix} 39.47 & - & 24.54 \\ - & 24.82 & 13.38 \\ 0.00 & -3.39 & 175.90 \end{bmatrix}$$

Low dislocation density substrate.

300 Å   1500 Å

$$10^{-4}\begin{bmatrix} 4.91 & - & 13.00 \\ - & 5.60 & 13.00 \\ -0.19 & 0.00 & 215.17 \end{bmatrix} \qquad 10^{-4}\begin{bmatrix} 56.91 & - & 10.56 \\ - & 51.48 & 7.38 \\ 1.45 & 5.67 & 166.58 \end{bmatrix}$$

Figure 2. Deformation tensors for the four layers studied so far.

If it is assumed that most of the relief occurs by the glide of 60° dislocations, then the Matthews critical thickness for these layers is about 11 nm, so some plastic relaxation is expected. Both of the thicker layers showed considerably more strain relief, and small rigid-body tilts were recorded. Relaxations of 36.7% and 23.1% were observed for the layer grown on the high density substrate, and 50.3% and 45.5% for the low density case. The significance of this disparity is complicated due to the difference in indium contents of the layers, 15.1% and 15.7% respectively. However there is some evidence from the topographs of more efficient strain relief in the layer grown on the low dislocation substrate. This is apparent from the long straight dislocations in the material on this substrate compared with the shorter, cross hatched array at the interface with the high dislocation density substrate. One possible reason for the poorer relaxation in the high density case could be the larger number of threading dislocations in the epilayer which may impede dislocation motion to the interface.

4. CONCLUSION

The state of deformation in (001) InGaAs layers grown by MBE on GaAs substrates has been determined by X-ray diffraction and expressed in terms of a tensor. Layers 30 and 150 nm thick were grown on substrates with low and high dislocation densities. The extent of relaxation in films grown on the lower density substrates was found to be greater than for corresponding films on the higher density material. In addition to strains, small rigid-body rotations of overlayers were detected.

ACKNOWLEDGEMENTS

The authors wish to thank DRA and SERC for support.

REFERENCES

Beanland R and Pond RC 1989 Inst Phys Conf Ser <u>104</u> 5
Beanland R 1991 Ph D Thesis
Dixon R H and Goodhew P J 1990 J Appl Phys <u>68</u> 7
Dodson B W 1988 Appl Phys Lett <u>53</u>
Fatemi M and Stahlbush R E 1991 Appl Phys Lett <u>58</u>
Hornstra J and Bartels WJ 1978 J Cryst Growth <u>44</u>
Kavanagh K L 1988 J Appl Phys <u>64</u>
Krishnamoorthy V 1991 Appl Phys Lett <u>58</u>
Leiberich A and Levkoff J 1990 J Cryst Growth <u>100</u>
Matthews J W and Blakeslee A E 1974 J Cryst Growth
Tanner B K 1991 - Analysis of Microelectronic Materials and Devices Ed: M
  Grasserbauer and H W Werner (Wiley)
Van der Merwe J H 1962 J Appl Phys <u>34</u>

Inst. Phys. Conf. Ser. No 134: Section 9
Paper presented at Microsc. Semicond. Mater. Conf., Oxford, 5–8 April 1993

Description of the real orientation relationships of epitaxial samples using transformation matrices

P Möck

Humboldt-Universität, Institut für Kristallographie und Materialforschung, Invalidenstrasse 110, D-O-1040 Berlin, Germany

ABSTRACT: A description tool for the real orientation relationships of epitaxial samples (using transformation matrices) is reported. For complete characterization of epitaxial samples, these matrices can be divided into different orientation effects including epitaxial misorientation. Furthermore, information concerning lattice geometry, symmetry, and the admissible line defects at the interface can be derived from these matrices. The matrices can be determined experimentally using a universally applicable diffraction technique (diffraction goniometry). Some experimental details and results of an analysis of epitaxial CdTe on GaAs using X-ray diffraction goniometry are given.

1. INTRODUCTION

There is a lot of experimental evidence in various epitaxial systems that the commonly used laws of overgrowth (i.e., (hkl) of substrate ‖ (h'k'l') of deposit, [uvw] of substrate ‖ [u'v'w'] of deposit) fail in general because epitaxial misorientations exist [see, e.g., Aindow and Pond (1991) or Möck (1993a)]. Up to now, epitaxial misorientation is only partially understood because it is a complex phenomenon including elastic distortion of the deposit lattice, vicinal-step related interfacial dislocations and further crystal defects. On the other hand, Ohki et al (1988) have already shown that engineering of epitaxial misorientation on a phenomenological basis (using substrates with vicinal surfaces) can lead to a significant improvement of the structural quality and of the physical properties of deposits.

For a detailed explanation of epitaxial misorientation the LACBED-technique [see, e.g., Duan (1992)] and high-resolution electron microscopy may be especially suitable. However, the restricted knowledge about epitaxial misorientation can lead to an incorrect image matching of high-resolution electron micrographs of epitaxial interfaces (Knowles 1991). Because an unknown misalignment of the deposit or of the substrate with respect to the electron beam can exist, artefacts may occur in high-resolution images [see, e.g., Smith et al (1983)]. Thus, modelling of the real orientation relationships of epitaxial samples depends on progress in investigations of epitaxial misorientation using other techniques.

The aim of this paper is to describe briefly a universally applicable diffraction technique and a description tool for the real orientation relationships, which allow the complete characterization of epitaxial samples from the lattice geometrical point of view. Besides that, the paper presents some experimental details and results of an analysis of a sample of CdTe (111) on GaAs (001) rotated 2° around [$1\bar{1}0$] using X-ray diffraction goniometry.

2. TRANSFORMATION MATRICES AS A DESCRIPTION TOOL

2.1 Determination of Transformation Matrices

The transformation matrices transform the (average) lattice of the substrate (coordinate system A) to the (average) lattice of the deposit (coordinate system B) and vice versa. Using the notation of Bowles and Mackenzie (1954), the transformation matrix

$$(BNA) = \begin{pmatrix} N_{11} \pm \Delta N_{11} & N_{12} \pm \Delta N_{12} & N_{13} \pm \Delta N_{13} \\ N_{21} \pm \Delta N_{21} & N_{22} \pm \Delta N_{22} & N_{23} \pm \Delta N_{23} \\ N_{31} \pm \Delta N_{31} & N_{32} \pm \Delta N_{32} & N_{33} \pm \Delta N_{33} \end{pmatrix}$$

transforms the direct lattice of the substrate to the direct lattice of the deposit. This matrix can be calculated by the relation

$$(BNA) = (BSE)\ (ETA) = \{(AS^{*}E)\ (ET^{*}B)\}^{t} \ , \qquad [1]$$

where the superscript t means a transposition, and the coordinate system E is a Cartesian coordinate system which is related to the specimen goniometer of the diffraction apparatus. The components ΔN_{ij} are the error limits, which depend on the accuracy and precision of the experimental investigation.

The errors of the goniometer adjustment of the diffraction apparatus determine mainly the accuracy and precision of the transformation matrices. Thus, the use of an X-ray or neutron diffractometer is preferred because these instruments have goniometers with a precision of at least a few hundredths up to a few thousandths of a degree. Using these types of diffractometers, the crystal matrices of the reciprocal lattice $(AS^{*}E)$ and $(ET^{*}B)$ are determined using goniometry of reciprocal lattice vectors. If the deformations of deposit and substrate have to be taken into account for the complete characterization of the epitaxial sample, the use of these two types of diffractometers is absolutely necessary (Möck 1993b).

A common electron microscope (TEM, STEM, SEM) equipped with a high-precision and two-axis specimen goniometer (accuracy of about 0.1°) can also be used for the determination of the transformation matrices. The experimental derivation of the crystal matrices of the direct lattice (BSE) and (ETA) is given by Möck and Hoppe (1992).

2.2 Application of Transformation Matrices

The transformation matrices (BNA) and (AMB) (= $(BNA)^{-1}$) characterize entirely the epitaxial system from the lattice geometrical point of view. For a survey of the different effects of the lattice transformation, these matrices can be divided into several factors using the experimental results and the rotation matrix of the ideal orientation relationship (which can be derived from the law of overgrowth). These factors are for example: the scalar of the volume change $|(BNA)|$, the deformation tensor of the deposit (BDB), the rotation matrix of epitaxial misorientation (BAB), the rotation matrix of the ideal orientation relationship (BIA), and the deformation tensor of the substrate (ACA). For the transformation from the direct lattice of the substrate to the direct lattice of the deposit one obtains

$$(BNA) = \sqrt[3]{|(BNA)|} \; (BDB) \; (BAB) \; (BIA) \; (ACA). \qquad [2]$$

In a series of epitaxy experiments these orientation effects can be correlated to the growth conditions.

From the rotation matrix of epitaxial misorientation the rotation parameters: angle of misorientation

$$\delta = \arccos \{\tfrac{1}{2} \, (\Sigma A_{ii} - 1)\} \qquad [3]$$

and axis of misorientation

$$r_i = (A_{jk} - A_{kj}) \, / \, 2 \sin \delta \quad ; \quad i,j,k \; cyclic \qquad [4]$$

can be derived.

The admissible line defects at the interface can be calculated from the transformation matrices using the theory of Pond (1989). Besides that, information concerning lattice geometry, for example, the angles between prominent low index net planes or lattice directions of substrate and deposit, can be derived from these matrices. Using Curie's and Neumann's principle, the symmetry group(s) of the deformed deposit and of the physical properties of the deposit can also be calculated from the transformation matrices.

For a comprehensive description of transformation matrices as a description tool see Möck (1993a).

3. EXPERIMENTAL EXAMPLE: CdTe (111) on GaAs (001) rotated 2° around [1$\bar{1}$0], grown by MOCVD, 3.3 ± 0.2 μm thick

The transformation matrix of an epitaxial sample of CdTe on GaAs was determined by diffraction goniometry using an X-ray diffractometer equipped with an optical reflection goniometer. The goniometry of reciprocal lattice vectors was performed for 12 reciprocal lattice vectors of substrate $(H_i K_i L_i)_A{}^*$ and deposit $(H_i K_i L_i)_B{}^*$. The Bragg angles were determined with a precision of about 0.02°. The goniometer adjustments can be described by a spherical coordinate system. The two angles of this spherical coordinate system were determined with a precision of about 0.02°. The Cartesian coordinates of the reciprocal lattice vectors $(X_i Y_i Z_i)_E$ under diffraction conditions can be derived from the values read on the goniometer circles.

The correct signs of the indices of the reciprocal lattice vectors (polarity) were determined in the case of the CdTe deposit by a modified anomalous X-ray scattering method (Berger 1987) and in the case of the GaAs substrate by KOH-etching in correlation to polarity determinations using Kossel diffraction (Nolze et al 1990).

The crystal matrices of the reciprocal lattices of substrate and deposit were calculated solving the linear inhomogeneous equation systems

$$(H_iK_iL_i)_A* = (X_iY_iZ_i)_E (AS^*E)^t \qquad [5]$$
$$(X_iY_iZ_i)_E = (H_iK_iL_i)_B* (ET^*B)^t \qquad [6]$$
$$\text{with: } 2\sin\Theta_i / \lambda = (X_i^2 + Y_i^2 + Z_i^2)^{0.5}$$

by means of a least-squares algorithm.

The transformation matrix was calculated using relation [1], and the matrix

$$(BNA) = \begin{pmatrix} -0.1807 \pm 0.002 & 0.6919 \pm 0.003 & -0.4996 \pm 0.003 \\ 0.6918 \pm 0.003 & -0.1795 \pm 0.002 & -0.4999 \pm 0.003 \\ -0.4990 \pm 0.003 & -0.4993 \pm 0.003 & -0.5116 \pm 0.003 \end{pmatrix}$$

was obtained.
This matrix was divided into the scalar of the volume change, the rotation matrix of epitaxial misorientation and the rotation matrix of the ideal orientation relationship. Using the relations [3]-[4], as rotation parameters of epitaxial misorientation the angle $\delta = 0.66 \pm 0.03°$ and the axis [$\overline{7}3$ 68 0] $\pm$ 0.05° were obtained.

Using the theory of Pond (1989) [see also Aindow and Pond (1991)] one can conclude from the relationship of the angle of epitaxial misorientation to the angle of the substrate miscut that in this sample vicinal-step related interfacial dislocations exist. A simple calculation shows that the Burgers vectors of these interfacial dislocations contribute both to the release of the mismatch between the substrate lattice and the deposit lattice, and to the epitaxial misorientation.

The existence of these interfacial dislocations, introduced by a suitable miscut of the substrate, and the related reduction of misfit dislocations inside the bulk of the deposit may be the reasons for a high structural quality of deposits obtained by engineering of epitaxial misorientation.

The author thanks Dr. R. Sukale for providing the epitaxial sample.

REFERENCES

Aindow M and Pond R C 1991 Phil. Mag. A 63 667
Berger H 1987 Cryst. Res. Technol. 22 1101
Bowles J S and Mackenzie J K 1954 Acta Met. 2 129
Duan X F 1992 Ultramicroscopy 41 249
Knowles K M 1991 Proc. Autumn School of the Intern. Centre of
    Electron Microscopy, eds J Heydenreich and W Neumann (Halle:
    Max-Planck-Institute of Microstructure Physics) pp 152-7
Möck P and Hoppe W 1992 Proc. 10th European Cong. on Electron
    Microscopy Vol. I, eds A Ríos, J M Arias, L Megías-Megías
    and A López-Galindo (Granada: University) pp 193-4
Möck P 1993a J. Cryst. Growth, in press
Möck P 1993b Mater. Sci. Eng. B, in press
Nolze G, Geist V, Wagner G, Paufler P and Jurkschat K 1990 Z.
    Krist. 193 111
Ohki A, Shibata N and Zembutsu S 1988 J. Appl. Phys. 64 694
Pond R C 1989 in: Dislocations in Solids 8 1, ed F R N Nabarro
    (Amsterdam: Elsevier)
Smith D A, Saxton W O, O'Keefe M A, Wood G J and Stobbs W M
    1983 Ultramicroscopy 11 263

Inst. Phys. Conf. Ser. No 134: Section 10
Paper presented at Microsc. Semicond. Mater. Conf., Oxford, 5–8 April 1993

597

Work function at a silicon surface atomically resolved by STM

J B Pethica, J Knall [¶] and J H Wilson [‡]

Department of Materials, University of Oxford, Parks Road, Oxford OX1 3PH

ABSTRACT: Aside from atomic topography, three factors influence an STM image: the local density of states, the local work function or potential at an atom, and the shape of the tunnel barrier in the tip-surface gap. We present atomically resolved images of STM barrier height above the Si(113) surface. These show that measurement of the potential above individual atoms is possible, but complicated by topography in practice.

1. INTRODUCTION

Scanning tunnelling microscopy has developed in the last ten years as one of the most powerful techniques of microscopy for semiconductor surfaces. The potential of real space imaging, from atomic scale to hundreds of nm, is now being realised in the study of growth and adsorption (Pashley et al.1991). The combination of atomic resolution with an often strong variation of image contrast as the tunnelling voltage is varied either side of the band gap, can give great insight into atomic structure and bonding characteristics.(Hamers et al.1986, Knall et al. 1991)

Most of the deductions from STM images to date are based on the Tersoff and Hamann model (1984), in which the tip traces out contours of constant charge density above the surface. These contours are in principle given by a combination of the atomic positions and the local density of states (LDOS) at the imaging energy (Feenstra et al. 1987). Image interpretation is simplified if one of these factors dominates. However, in addition to these two, the electron decay length above the surface also strongly influences the STM image. This is affected by the local work function or on-site potential at the atom being imaged, and also by perturbations of the tunnel barrier in the gap, such as electrostatic image potential and the applied voltage.

In a quantitative and energy dependent simulation of imaging of the Si(113) surface, we have shown (Wilson et al.1992) that differing, site dependent decay lengths are necessary to account for the experimental images. In fact, except in cases where the height variation of atoms normal to the surface is small, or where local variations in LDOS are very strong, the values of local decay lengths are likely to control what is seen in the STM image. An understanding of CITS images (Hamers et al. 1986) is completely dependent on knowledge of the decay lengths. It is the purpose of this paper to describe the extent to which the decay length, and hence the local surface potential might be measured by STM.

Many model calculations of atomic and electronic structure of solids use a mathematical scheme to solve a one electron Schrödinger equation in which the interaction with other electrons and the atomic nuclei are accounted for by an effective periodic potential experienced by each single electron. The potential can then be adjusted by imposing self consistency. For example, in tight binding calculations this might be accomplished by assuming local charge neutrality.(Wilson et al.1991) Since there is as yet no way to experimentally measure the amount of charge transfer between individual atoms or the local potential in a crystal, (or indeed to distinguish between them) the basis of the calculations is not experimentally tested. A rigorous test for surface calculations would be to compare calculated local potentials with measured. STM may be able to probe the local potential of the surface with very high spatial resolution by measuring the local tunnel barrier height .

The normal method to measure the barrier is to oscillate the tip surface separation s and measure the resulting changes in tunnel current I_T. In a simple one dimensional model the mean barrier is then given by $0.952 \, (\partial \ln I_T / \partial s)^2$ where s is in Å (Lang 1988). By taking the measurements continuously during image scanning, barrier maps can be built up. However, 3-D topographic structure can change the measured value of $\partial I / \partial s$. Gomez-Rodriguez et al.(1989) tried to separate topographic effects from potential variations by calculating the cosine of the angle between the local gradient vector in the STM topography and the z (surface normal) direction. These images of cos(s,z) were then compared to images of $\partial \ln I_T / \partial s$, and differences between the images were interpreted as changes in chemical composition. Another example where measurements of apparent barrier height has been used to probe potential variations comes from Hamers and Kohler (1989) who studied localisation of defect states on Si(100) surfaces. In these examples the local variation in potential was studied on a lateral scale significantly larger that the atomic. Here we present the first atomic resolution measurements of $\partial I / \partial s$ for semiconductors. Our results demonstrate that there are limits to the ability of STM to probe local potential in topographically complex systems.

2. EXPERIMENTAL

The experiments were performed in an Omicron UHV STM. STM Tips were prepared from 0.2 mm tungsten wire by electrochemical etching in concentrated KOH solution. The tips were subsequently heated by electron bombardment in the vacuum system. The Si samples were 2×10^{18} cm^{-2} Sb doped wafers, which were degreased and cleaned by a wet chemical etch-oxide regrowth procedure (Knall et al. 1991) In the UHV system the samples were degassed and heated to 870 °C for one minute to remove the oxide. STM images of $\partial I / \partial z$ were recorded simultaneously with the topographic images. $\partial I / \partial z$ was measured at each pixel of the topographic image using either a lock-in, or a DC technique. When using the lock-in amplifier the tip height oscillates with an amplitude ∂z of 0.1 Å at a frequency of 3.4 kHz; the value of ∂I was sampled over 5-10 ms for each pixel after a settling time of 5 ms per pixel. The DC measurements were performed by breaking the feedback loop after a settling time of 3 ms, ramping the tip height between -1 and +1 Å with respect to the topographic (closed loop) tip height, and recording I vs z.

3. RESULTS

Figures 1 and 2 show images of a clean Si(113) surface obtained at a tunnelling current I_0 of 2 nA and sample voltages of +2.5 (Fig. 1a) and -2.5 V (Fig. 2a). The simultaneous images of $\partial I / \partial s$ are in Figs. 1b and 2b. Also shown are cross-sections of $\partial I / \partial s$ and topographic height along the dotted lines in the images. The barrier height is similar to the topography, both in the empty and the filled state images. While the tip moved between the B- and A-type sites (see Fig. 3 for structure) along the dotted line in the empty state image of Fig.1a, the tip height increased by 0.4 Å. The corresponding changes in $\partial I / \partial s$ were from 1.4 to 1.8 nA/Å. Similar ranges of topographic corrugation and corresponding $\partial I / \partial s$ variations are observed for the filled state measurements. Note that the topographic images of filled and empty states are completely different, which means that the value of $\partial I / \partial s$ in a particular location above the surface is found to depend on whether electrons are tunnelling into or out of the sample.

Values of $\partial I / \partial s$ can also be obtained from DC measurements of I vs z. Figure 4 shows graphs of I vs z obtained in positions of type A and B as marked in Fig.3. Data from 20 pixels located in crystallographically equivalent positions across the surface were averaged to produce each curve. Positions of type A and B exhibit $\partial I / \partial s$ values of 2 and 1.6 nA/Å respectively at U = +2.5 V and I_0 = 2nA. These values are slightly higher than the values given by the lock-in measurements. These variations between measurement techniques are typical also of measurements performed with the same technique but separated in time, and may be due to microscopic changes of the tip.

To get some feeling for what the $\partial I / \partial s$ values mean we can analyse them using a 1-D trapezoidal barrier of the form

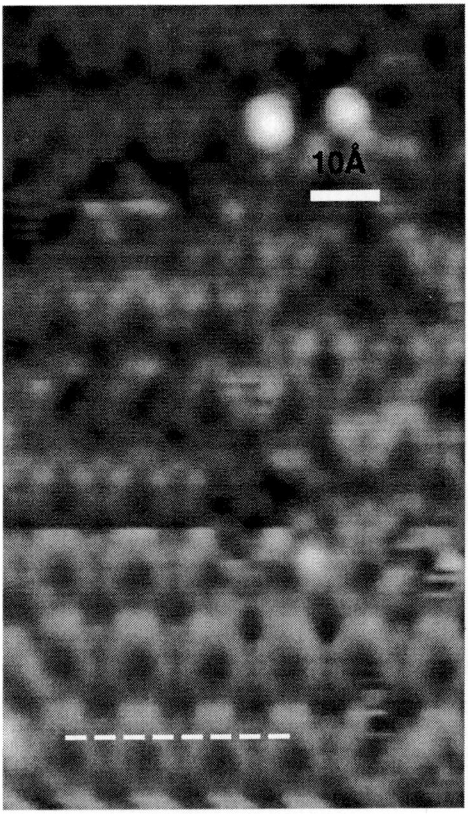

Fig. 1(a) Topography at 2nA tunnel current
Cross sections below

Fig. 1(b) Barrier — Current variation for
0.1Å z modulation

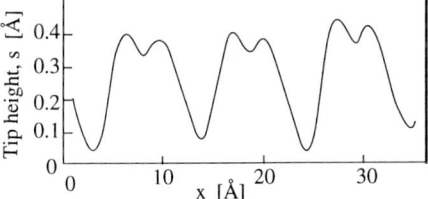

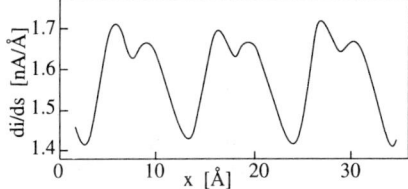

Figure 1. Topography and barrier height images of the same area of surface, simultaneously recorded, for 2.5 V sample bias, giving conduction band (empty) states. Note tip change near middle of image. The line scans are along the three unit cells of the dotted line. The image features are associated with the A-type dangling bonds shown in Fig. 3.

$$\Phi(z) = \Phi_0 - E_v + \frac{zU}{s}$$

where Φ_0 is the work function, E_v is the energy of the state contributing to the current and U is the applied voltage. E_v is zero with respect to tip Fermi level if we consider tunnelling into empty states of the sample and assume that the tunnelling current is dominated by states with energy equal to the Fermi level of the tip. The tunnel current is given by

$$I = A\exp(-B\int_0^s \sqrt{\Phi(z)} \, dz)$$

where A and B are constants. This leads to the following expression:

$$\Phi_0 = \left[(\partial I / \partial z)\left(\frac{3U}{2I_0} \right) \right]^{\frac{2}{3}}$$

Hence with a set current I_0 of 2 nA and a sample bias V of 2.5 V, $\partial I/\partial s$ values of 1.8 and 1.4 correspond to values Φ_0 of 2.3 and 1.9 eV respectively. In Figure 4 we have also plotted the Φ_0 values obtained from the DC measurements. Φ_0 does not appear to vary significantly with distance within he 2Å range that was probed. Note that the above model does not include an image barrier effect.

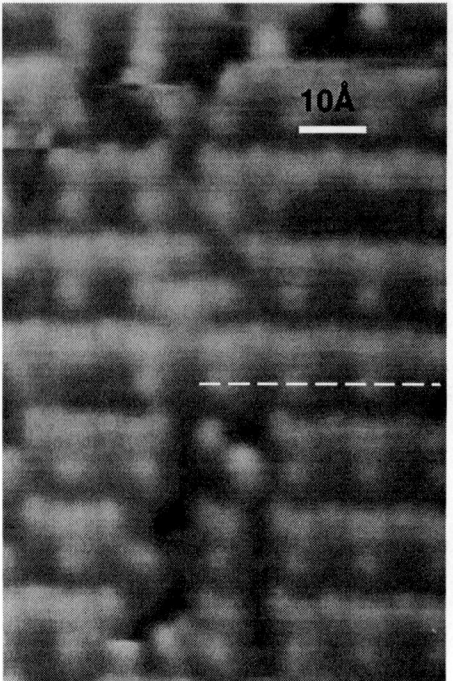

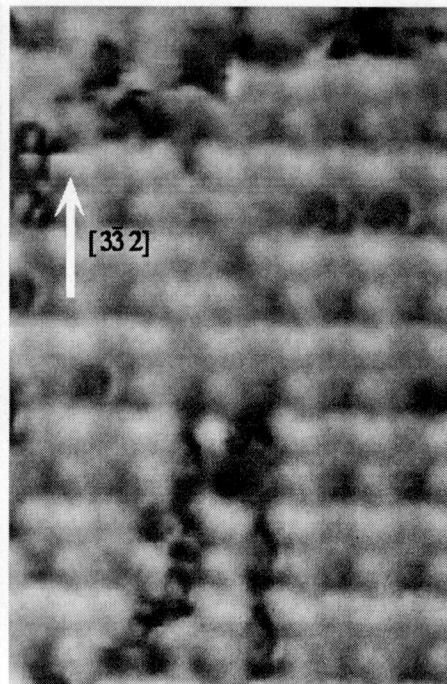

Fig. 2(a) Topography at 2nA tunnel current and -2.5V sample bias (filled states, B-type in Figure 3)

Fig. 2(b) Barrier — Current variation for 0.1Å z modulation, same area as 2(a). Cross-sections below

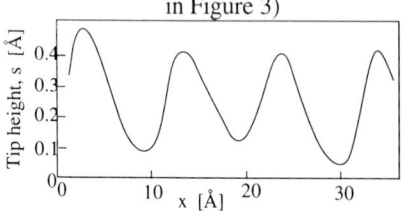

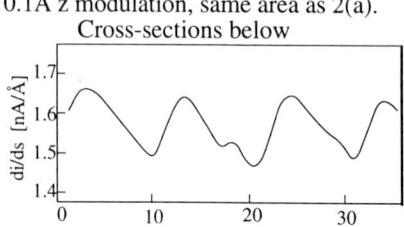

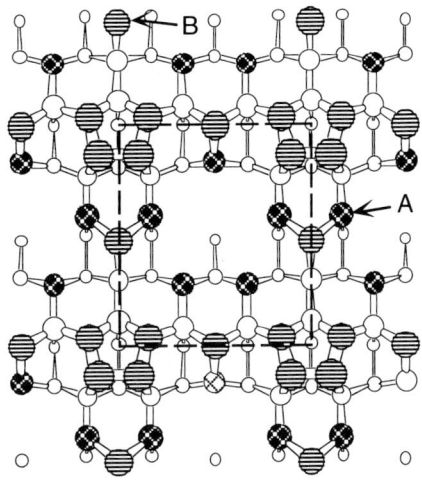

Figure 3. Perspective view of the structure of Si(113). Horizontally shaded atoms (B-type) have dangling bonds appearing in filled state image; diagonally shaded (A-type) in empty state image. 3x2 reconstruction unit cell is dashed outline.

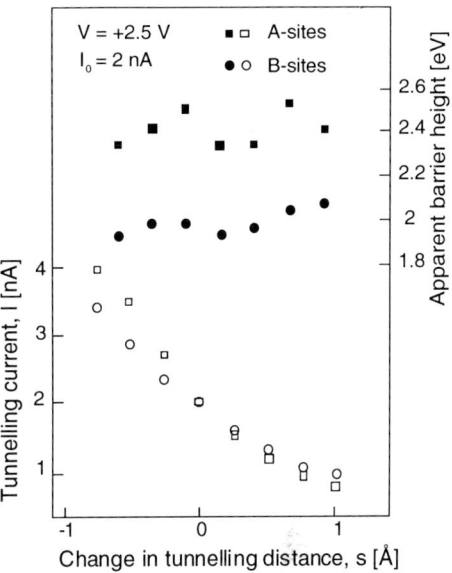

Figure 4. DC current vs. tip position and associated barriers, all for empty state topographic image.

4. DISCUSSION

The difference between empty state and filled state images on the Si(113) surface is related to the mechanism for the stability of the reconstruction.(see Knall et al 1991; Haneman 1961) The dangling bonds in the empty state images have acquired more p-character and their LDOS has risen to the conduction band. The filled state bonds have become more s-like. These dangling bond rehybridisations and associated LDOS changes suggest that charge transfer occurs between dangling bonds on the surface (and hence between atoms,) and thus imply potential shifts at the atoms.

In our tight-binding based simulations of the Si(113) surface (Wilson et al. 1992) we found that LDOS and topography alone were not able to account for the empty state images if, in the usual manner, a single work function was assumed above the whole surface. This is because the empty state bonds are located up to 1.1Å below the filled state bonds. For typically assumed work functions of 4eV, this means that the LDOS difference between adjacent bonds would need to be a factor of ~20 for the lower atoms to outweigh the upper, and so appear in the images as they actually do. This seems unlikely for immediately adjacent and chemically identical atoms. In the simulations, we reduced the barrier by assuming that an electrostatic image potential was present between tip and surface. A reduction of the maximum in barrier height to below 2eV was needed to ensure that LDOS outweighed topography and gave the experimentally observed pictures. The image charge reduction of course depends on tip surface separation, which thus implies a different barrier for each atom, higher for larger separations. However, to achieve the same effect of enhancing the visibility of the lower, empty state atoms (type A), we could also have assumed that the local potential on them was slightly lower than that on the filled state atoms. This would be in accord with the electrostatic result of transfer of charge from the empty to filled state atoms, and would require a rather less drastic image potential reduction in the gap.

Comparing these assertions now with the experimental results reported above we see some agreement. Firstly, the apparent mean barrier does indeed appear to be significantly reduced below normal work functions; Secondly, as shown by Figure 4, the actual value of barrier is indeed higher over the empty state bonds (the A-sites) than the filled state bonds (B-sites, Figure 3) again in agreement with the image potential barrier model used. However, these results are for the empty state images; Fig 2(b) shows that the reverse is true when imaging filled states.

The barrier height images are clearly dominated by the topography. The barrier - topography link follows from Tersoff & Hamann (1984). They treat the wavefunction above the surface as

$$\Psi_s = \Omega_s^{-\frac{1}{2}} \sum_G a_G \exp[-(\kappa^2 + |\kappa_G|^2)^{\frac{1}{2}} z] \exp[i\kappa_G x]$$

The plane wave components with larger surface reciprocal lattice vector G decay faster in the z-direction, which means that the corrugation amplitude of constant charge density falls off with increasing tip surface separation. (Figure 5) To illustrate the effect on $\partial I / \partial s$, we assume that charge density $\rho(r, E_f)$ above the surface varies smoothly, so that only the lowest non-zero Fourier components need to be retained. Following T&H

$$\rho(r, E_f) \propto A_0 \exp(-2\kappa z) + A_1 \exp[-\beta z] \cos(x \cdot G)$$

where r is the position of the tip, x is in a direction parallel to the surface, and $\beta = 2\kappa + \frac{1}{4}\kappa^{-1}G^2$. Since the tunnelling current is proportional to $\rho(r, E_f)$ we can write

$$\frac{\partial I}{\partial z} \propto -2\kappa A_0 \exp(-2\kappa z) - \beta A_1 \exp(-\beta z)\cos(G \cdot x)$$

and as we are following contours of constant current I_0

$$\frac{\partial I}{\partial z} \propto -2\kappa A_0 \exp(-2\kappa z) - \beta[I_0 - A_0 \exp(-2\kappa z)]$$

$$= \frac{A\,G^2}{4\kappa} \exp(-2\kappa z)\ - I_0\beta$$

This tends to a constant far from the surface, as expected. For sufficiently large average z it can be seen that $|\partial I / \partial s|$ is lower over depressions (z low) than over protrusions. The contours result from atomic positions *and* LDOS and barrier effects. This explains the correlation between $|\partial I / \partial s|$ and topography in our images. Note that κ is taken to be a constant in the analysis, which will not be the case if an image potential is present, or if the electrostatic potential varies between adjacent atoms.

Figure 5. Schematic of lines of constant current density above surface, showing decrease in corrugation with increasing z, and site dependence of decay length

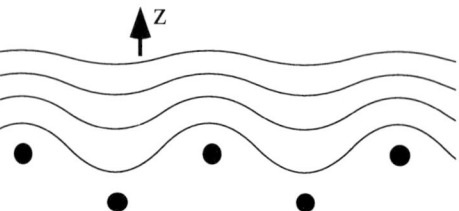

It might be possible in principle to separate out the topological and electronic effects on the measured barrier heights. We do not believe the presently available data is worth a significant quantitative effort in this direction. However, a simple process of image subtraction provides an interesting but tantalisingly imprecise result. If we simply subtract the topographic image from the barrier image, we are, very crudely and arbitrarily, removing the geometric effect within the barrier image — high points in topography correspond to high geometry induced apparent barrier. The result for a few unit cells of the filled states image is shown in Figure 6, along with the barrier image for the empty states. Some similarity can be seen, which suggests that non-topographic effects do indeed give a higher local barrier at sites over the empty state bonds at all imaging voltages. The structure in the unit cell centre, additional to topography (Fig 2(a)), is in fact visible in the barrier image Figure 2(b), and also in the associated cross-section profile. It was interesting to find that in contrast to the filled state images, arbitrary subtraction of topography from barrier for the empty state images was unable to give any patterns recognisably associated with bond types. This suggests that when imaging at a bias which reveals the empty states, all effects on barrier work in the same direction — the topographic effect merely accentuates the already higher barrier above the empty state atoms. Again this asserts higher non-topographic barrier over the A-sites. We conclude that local barrier changes associated with individual atoms can in principle be measured by STM.

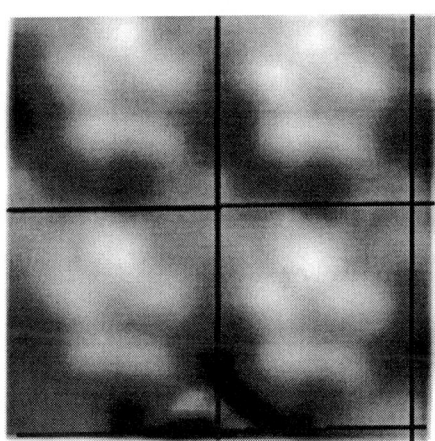 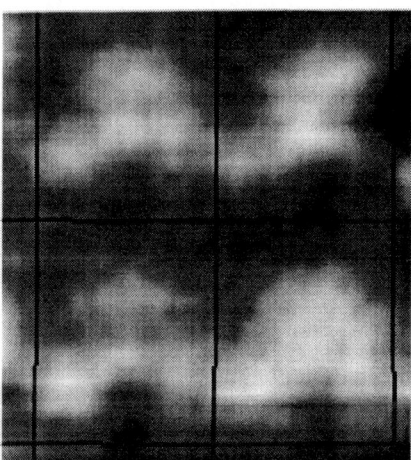

Figure 6. Four unit cells showing barrier image of empty states on left, and on the right the barrier for filled states with topography subtracted.

It is interesting to note that this implies that charge transfer induced effects on local barrier are smaller than image potential effects, as they work in the opposite direction. This in turn suggests that significant screening is occurring.

Some problems should be mentioned. The tunnel barrier is reduced; however, a one-dimensional model shows that the effect of a classical electrostatic image potential on the *measured* value of $\partial \ln I/\partial s$ should be very small, because the barrier also varies as the tip separation is varied (see Coombs et al. 1988 for details.) Three dimensional aspects, and possible deviations from a simple Coulomb image potential need to be taken into account. It is thus not clear at present how much of the barrier reduction is due to image potential effects.

In most of the above discussion, it is assumed that the tip is merely a sampling point; this is likely to be a serious simplification of real tips. The effect of obvious tip distortions can often be seen in images, from regular asymmetries to repeat patterns. When, as here, we are imaging individual atoms and their electron decay to vacuum, it may be difficult to be certain about tip influences. That the tip may interact quite strongly with the surface is clear from recent work by Ruan et al. (1993)

Finally, the Si(113) surface shows strong topographic and LDOS effects even without considering the barrier. This is useful in illustrating the principles involved. However, attempts to quantitatively model local barriers would be best based on images of metal surfaces, so that low applied voltages may be used, a simple LDOS and good screening assumed, and the number of complicating factors kept to a minimum.

Throughout this paper we have considered only the highest resolution features. Variations of apparent barrier also occur over larger lateral scales, which are not so simply related to topography. For example, at the top of figure 1 there are a number of both high and low features in topography which have no clearly associated great changes of barrier. There are unit cells which appear distinctly higher in barrier than their neighbours, in a manner which is not so clear in the topographic image. These effects may be caused by the surface electronic effects described above, but may also be at least in part caused by sub-surface structural differences.

5. CONCLUSIONS

We have presented atomically resolved images of the local apparent barrier height above the Si(113) surface. The tunnel barrier is somewhat reduced from typical work function values and appears to differ from atom site to atom site, in agreement with earlier image simulations. However, the barrier heights are strongly related to the topographic images at constant current, even though the images themselves vary considerably with measuring voltage. The difficulty in separating out the effects of topography from the barrier images means that quantitative measurement of surface potentials or other effects which reduce the barrier at individual atoms, has not yet been made. It is important to investigate these effects, not only to understand STM imaging of regular semiconductor surfaces, but also for adsorbed atoms and molecules where potential changes due to the interaction with the substrate may have a strong influence on what is seen in the STM.

ACKNOWLEDGEMENTS

We thank Adrian Sutton for helpful comments, and the S.E.R.C. for support under grant GR/F 60185.

¶ Now at Xerox Palo Alto Research Centre, 3333 Coyote Hill Rd, Palo Alto, CA94304.
‡ Now at Shell Research Laboratory, Amsterdam, Netherlands.

REFERENCES

Coombs JH, Welland ME and Pethica JB 1988 Surf. Sci. Lett. 198 353
Feenstra RM, Stroscio JA and Fein AP 1987 Phys. Rev. Lett. 58 1192
Gomez-Rodriguez JM, Gomez-Herrero J and Baro A 1989 Surf. Sci. 220 152
Hamers RJ and Kohler U 1989 J. Vac. Sci. Technol.A 7 2857
Hamers RJ, Tromp RM and Demuth JE 1986 Phys. Rev. Lett. 56 1972
Haneman D 1961 Phys. Rev. 121 1093
Knall J, Todd JD, Pethica J and Wilson JH 1991 Phys. Rev. Lett. 66 1733
Pashley MD, Haberern KW and Gaines JM 1991 J. Vac. Sci. Technol.B 9 938
Ruan L, Besenbacher F, Stensgaard I and Laegsgaard E 1993 Phys. Rev. Lett. 70 4079
Tersoff J and Hamann DR 1984 Phys. Rev.B 31 805
Wilson JH, McInnes DA, Knall J, Sutton AP and Pethica JB 1992 Ultramicroscopy 42-44 801
Wilson JH, Todd JD and Sutton AP 1990 J. Phys: Cond. Matt. 2 10259 and 1991 J. Phys: Cond. Matt. 3 1971

Inst. Phys. Conf. Ser. No 134: Section 10
Paper presented at Microsc. Semicond. Mater. Conf., Oxford, 5–8 April 1993

Adsorption of trimethylgallium on semiconductor surfaces: STM observations

A J Mayne, M O Schweitzer, A R Avery*, T S Jones*, C W Smith**, C M Goringe, J B Pethica and G A D Briggs

Department of Materials, Oxford University, Parks Road, Oxford OX1 3PH
*Semiconductor IRC and Department of Chemistry, Imperial College, London SW7 2AY
**Department of Physics and Astronomy, University of Maine, Orono, ME 04469, USA

ABSTRACT: Scanning tunnelling microscopy (STM) has been used to image the adsorption of trimethylgallium (TMGa) firstly, on GaAs(001)-(2×4) surfaces prepared *in situ* by molecular beam epitaxy (MBE) and secondly, on Si(001)-(2×1) surfaces. Filled states images of the clean GaAs surface are dominated by (2×4) unit cells containing only two As dimers. Upon exposure of this surface to TMGa at room temperature, bright oval-shaped features are observed which are centred on the arsenic dimers of the unit cell. Filled and empty states images of the clean silicon surface are dominated by Si dimers and missing dimer trenches running perpendicular to the dimerisation direction. Upon exposure to TMGa at room temperature, no apparent change occurs. However, after annealing the silicon surface to 650 K, large areas of the surface reveal the c(4×4) reconstruction.

1. INTRODUCTION

The interaction of organometallic molecules with semiconductor surfaces plays a crucial role in many semiconductor growth process such as metal organic chemical vapour deposition (MOCVD), metal organic molecular beam epitaxy (MOMBE), chemical beam epitaxy (CBE) and atomic layer epitaxy (ALE). Trimethylgallium (TMGa) has a unique and complex behaviour, since it cannot undergo the facile β-elimination decomposition routes of the higher alkyls (Gibson 1990). Decomposition results in high carbon incorporation levels but also exhibits a self limiting behaviour making TMGa ideal for ALE growth (Nishizawa 1987 and Yu 1993). Recent MOMBE studies on vicinal (001) GaAs surfaces, using reflection high energy electron diffraction (RHEED) critical temperature measurements, have also indicated a significant site selectivity in the growth which may potentially be exploited in the growth of quantum wires (Kaneko 1993).

Scanning tunnelling microscopy (STM) is an ideal technique for studying site specific adsorption since individual atoms and molecules can be identified (Avouris 1992). Recently, we have shown that it is possible to study the adsorption of individual molecules on semiconductor surfaces (Briggs 1992, Mayne 1993, and Avery 1993). The STM study has been expanded to look at TMGa adsorption on the Si(001)-(2×1) surface. The silicon surface was chosen because it is relatively well understood and is stable at elevated temperatures (GaAs begins to lose the As top layer above 650 K). Furthermore, Ga adsorption on silicon, from elemental sources, has been studied by STM (Nogami 1988 and Baski 1990). Thus, a study of TMGa on Si is complementary and, due to the relative simplicity of the silicon surface, will facilitate the understanding of the adsorption mechanism. Also previous studies of TMGa on silicon (Lin 1991) using a range of surface sensitive techniques have indicated possible structures of the TMGa derived species on the surface.

2. EXPERIMENTAL PROCEDURES

The experiments were performed in a two-chamber STM/MBE UHV system (Omicron GmbH) incorporating a fast entry load lock, details of which have been published elsewhere (Mayne 1993). Prior to growth, the oxide layer was removed by heating the GaAs(001) substrates under an arsenic flux at 900 K. All surface temperatures were monitored using an optical pyrometer. GaAs growth was carried out with Ga and As_2 sources, at a substrate temperature of 850 K and with a growth rate of about 0.3ML/s as determined from RHEED intensity oscillation measurements. All samples were Si doped ($n \sim 1 \times 10^{18} cm^{-3}$) up to the last 50-100 layers. A (2×4) RHEED pattern was observed and maintained whilst the As flux and substrate temperature were reduced in a controlled manner before transfer into the STM chamber.

The n-type ($\rho = 0.02 \, \Omega m$) Si(001) substrates were degreased and cleaned by a wet-chemical oxide etch-regrowth procedure prior to insertion in vacuum. Once under UHV, they were degassed by electron bombardment heating at 900 K and finally heated resistively to 1140 K for ~3-4 minutes in order to remove the surface oxide. The clean Si(001) surface was characterised by the appearance of a two domain (2×1) LEED pattern.

TMGa (Epichem, UK) was exposed to the clean semiconductor surfaces at 300K via a gas handling line. Dual bias STM images were recorded on alternate forward and reverse scans to minimise the effects of thermal drift and to facilitate direct comparison of the images.

3. RESULTS

3.1 TMGa on GaAs(001)

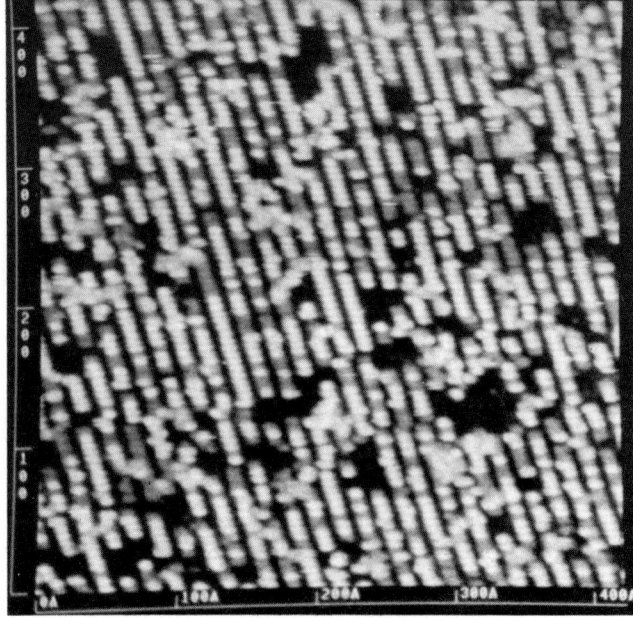

Fig. 1. STM topogragh (410x410Å) of the GaAs(001)-(2×4) surface exposed to 8 L of TMGa. Sample bias -2.8 V and a tunnel current of 0.2 nA.

Fig. 1 shows a GaAs surface exposed to a notional 8 L (1 Langmuir = 1 L = 10^{-6} Torr. s^{-1}) of TMGa at 300 K. The bright, oval shaped features are molecules adsorbed on top of the pairs of arsenic dimers of the unit cell. The images were taken at -2.8 V, which suggests from a comparison with photo-emission data (Claverie 1989), that the features arise from tunnelling from the Ga-C bond to the tip. Occasionally, there are adsorption sites which contain 2 circular features. At low coverages, preferential adsorption on unit cells adjacent to occupied sites along the [$\bar{1}$10] direction is observed.

TMGa adsorption on GaAs has previously been studied using several surface sensitive techniques (Claverie 1989). Unfortunately, confusion abounds as to the exact nature of the molecular species present at the surface. There are several possibilities including intact TMGa, dimethylgallium (DMGa) or monomethylgallium (MMGa). It was not possible to image the empty states (positive sample bias) of either the clean or adsorbate covered surface.

3.2 TMGa on Si(001)

The STM images of the silicon surface after exposure to TMGa at room temperature showed no visible change from the (2×1) structure found on a clean surface. Lin and Masel (1991) have suggested that the TMGa is physisorbed at 300K. It is, therefore, possible that the STM tip 'sweeps' the molecules aside during a scan resulting in the observation of a 'clean' silicon surface.

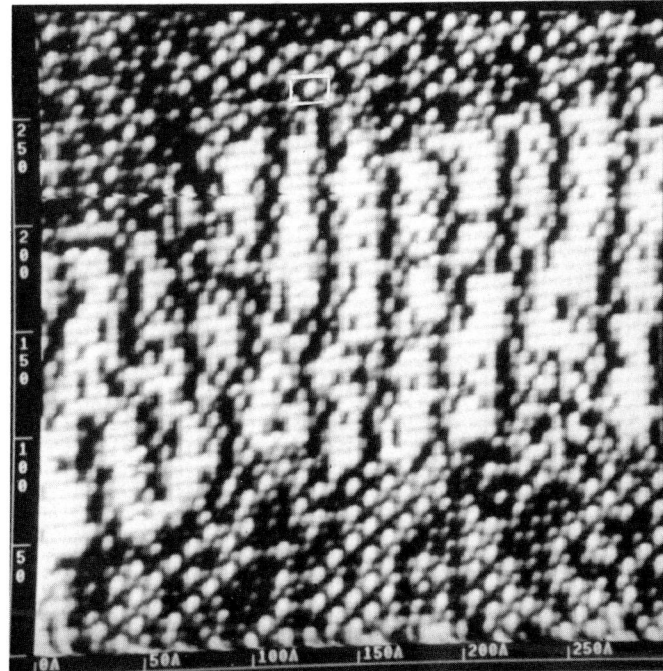

Fig. 2 a) STM topogragh (300x300Å) of the Si(001)-(2×1) surface exposed to 10 L of TMGa showing the Filled states at -2.5 V and 1.0 nA.

Fig. 2 shows filled and empty state images of a silicon surface which has been exposed to 10 L of TMGa and subsequently annealed to 650 K. Lin and Masel also suggest that decomposition of the adsorbed TMGa occurs upon annealing with the loss of methane.

Previous studies of Ga adsorption on Si(001) by Nogami (1988) and Baski (1990) adsorbing Ga from a tungsten filament on to room-temperature substrates which were then annealed to 720-770 K. Their STM studies show that, at low coverages, the Ga atoms line up in rows parallel to the Si dimerisation direction with a 2× periodicity. At higher coverages the rows of Ga atoms are organised into areas which have a 3×2 order. They further suggest that the Ga does not disturb the Si dimer structure and that surface defects have no effect on the bonding of the Ga.

Our STM images show regions of what appear to be clean silicon and regions of a c(4×4) reconstruction. The brightest features appear to be isolated silicon dimers with a feature or features covering the intervening 3 dimers. Analysis of the residual gases using mass spectrometry suggests significant desorption of methane during the annealing process. Therefore it is possible that chemisorption of the TMGa derived species occurs upon annealing and may involve the loss of one or two methane molecules.

4. DISCUSSION AND CONCLUSIONS

Whilst STM gives us good information on site specific adsorption, other techniques, such as HREELS, are required for a more detailed understanding of the adsorbed molecular species. For TMGa adsorption on the GaAs surface, the STM images show bright, oval shaped features occupying sites centred on the two As dimers of the unit cell. These features appear to cluster in the $[\bar{1}10]$ direction and are never found in the missing dimer trenches. Statistical analysis has shown that this is a real effect and not just a feature of random occupation (Goringe (to be published) and Mayne (to be published)).

Room temperature adsorption of TMGa on the silicon surface produced no apparent change in the surface reconstruction. However, annealing the surface to 650 K produced domains of c(4×4) structure. In both systems, the exact nature of the adsorbed organometallic species is not known. In the case of TMGa on silicon, the annealing procedure evolves methane gas, suggesting the probable chemisorption of the TMGa derived species with the subsequent loss of one or two methane molecules.

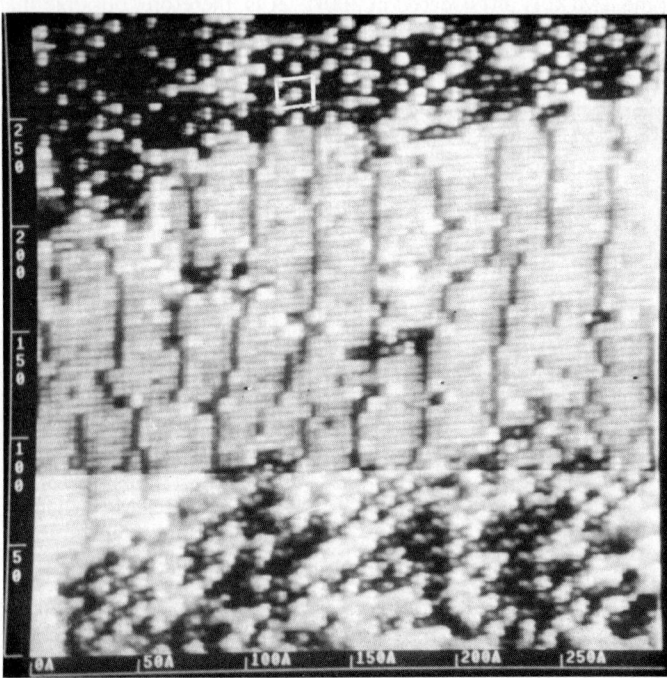

Fig. 2 b).   STM topogragh (300x300Å) of the Si(001)-(2×1) surface exposed to 10 L of TMGa showing the Empty states at +2.5 V and 1.0 nA.

ACKNOWLEDGEMENTS

This work was supported by the SERC as part of a collaboration between Oxford and the Semiconductor IRC in London. CMG is supported by British Gas. Professor Bruce Joyce, Dr Tadaaki Kaneko, Professor Wynand Verwoerd and Dr Andrew Fisher are thanked for many useful discussions.

REFERENCES

Avery A R, Mayne A J, Goringe C M, Owen J H G, Schweitzer M O, Jones T S, Briggs G A D and Weinberg W H, 1993 MRS conference proceedings (in press)
Avouris Ph and Lyo I W 1992 Appl Surf Sci 60/61 426
Baski A A, Nogami J and Quate C F, 1990 J. Vac. Sci. Technol. A 8 245
Briggs G A D, Knall J, Mayne A J, Jones T S, Weinberg W H, Avery A R 1992 Nanotechnology 3 113
Claverie P, Ueyama K, Maeda S, Namba H and Kuroda H 1989 Appl Phys Lett 54 698
Gibson E M, Foxon C T, Zhang J and Joyce B A 1990 J Crystal Growth 105 81
Goringe C M, Smith C W and Briggs G A D (to be published)
Kaneko T, Naji O, Jones T S and Joyce B A 1993 J Crystal Growth 127 1059
Lin R and Masel R I 1991 Surf Sci 258 225
Mayne A J, Avery A R, Knall J, Jones T S, Briggs G A D and Weinberg W H 1993 Surf Sci 284 247
Mayne A J, Goringe C M, Smith C W and Briggs G A D (to be published)
Nishizawa J, Kurabayashi T, Abe H and Nozoe A 1987 Surf Sci 185 249
Nogami J, Park Sang-il and Quate C F 1988 Appl. Phys. Letts. 53 2086
Yu M L 1993 J Appl Phys 73 716

Inst. Phys. Conf. Ser. No 134: Section 10
Paper presented at Microsc. Semicond. Mater. Conf., Oxford, 5–8 April 1993

609

Elastic relaxation by surface rippling of strained $Si_{1-x}Ge_x$/Si heteroepitaxial layers

A J Pidduck, D J Robbins and A G Cullis

Defence Research Agency, St. Andrews Road, Malvern, Worcs, WR14 3PS, UK

ABSTRACT: An undulating surface morphology is developed by in-situ annealing, at 680-750°C, of planar pseudomorphically-strained $Si_{1-x}Ge_x$ epitaxial layers with $0.15 < X < 0.3$ grown on Si by vapour phase epitaxy (VPE) at 610°C. We have used atomic force microscopy (AFM) to measure the amplitude and wavelength of this periodic roughness. The results show that the ratio of undulation amplitude to square of wavelength is constant for a given Ge fraction (proportional to mismatch strain). The observed relationship between this ratio and the mismatch strain is compared with the quantitative predictions of a simple sinusoidal model.

1. INTRODUCTION

The tendency for pseudomorphically-strained $Si_{1-x}Ge_x$ epitaxial layers to form undulations has been reported by Robbins et al. (1991) and Kuan and Iyer (1991). More recently, it has been shown by Cullis et al. (1992) using transmission electron microscopy (TEM) that the undulations are correlated with strain fluctuations and allow a means of partial elastic strain relaxation. Such a rippled surface topography can have lower energy than a corresponding planar uniformly strained layer if the ripple wavelength exceeds a critical value (Asaro and Tiller (1972), Srolovitz (1989)). Pidduck et al. (1992) used AFM to study the development of the undulations as a function of SiGe layer thickness, and concluded that the formation of the rippled topography is enabled by cumulative surface diffusion during growth. In order to obtain a quantitative understanding of how the ripple amplitude and wavelength depends on epilayer mismatch strain, it is useful to decouple the introduction of strain energy from the kinetics of ripple formation. For this reason we have grown planar $Si_{1-x}Ge_x$ epilayers at 610°C and then annealed them in-situ at higher temperatures to develop the rippled topography. Quantitative AFM measurements of the rippled topography will be reported and compared with the predictions of a simplified 1-dimensional sinusoidal model.

2. EXPERIMENTAL

VPE $Si_{1-x}Ge_x$/Si layers were grown and subsequently annealed in an ultrahigh vacuum (UHV) background cold-wall vapour phase epitaxial reactor as described previously by Robbins et al. (1991). Si wafers, 100mm diameter (Shin-Etsu Handotai Ltd) were loaded after brief exposure to UV-ozone or modified RCA cleaning, and the resultant thin passivating surface oxide was decomposed by heating at 880-920°C for a few minutes in 130 Pa H_2. A thin Si buffer layer was grown at 750-850°C before cooling to 610°C for alloy growth. Planar $Si_{1-x}Ge_x$ alloy layers of 12-30nm thickness were then grown by decomposition of SiH_4/GeH_4 mixtures in 18 Pa H_2. Five different SiH_4/GeH_4 partial pressure ratios were used for this work, giving Ge fractions X in the range 0.15-0.27. These layers were finally annealed in-situ by heating to 680-750°C for short periods in either 18 Pa H_2 or UHV. Laser light scattering (LLS) was used to follow the surface roughness changes taking place. The LLS technique uses a 488nm Ar ion laser beam incident on the centre of the wafer, together with a photon-counting photomultiplier to detect the light scattered normal to the surface. Ge fractions X (± 0.02) corresponding to the SiH_4/GeH_4 mixtures used in this work were measured by spectroscopic ellipsometry (Pickering et. al. (1993)), referenced to x-ray diffraction rocking curve results from planar $Si_{1-x}Ge_x$ layers after correction for deviation from Vegard's law. AFM measurements were made in air using a Digital Instruments Nanoscope III contact mode AFM with microfabricated Si_3N_4 stylus at a constant net force in the range of 50-150nN.

3. RESULTS

Figure 1 shows a typical LLS trace obtained during $Si_{1-x}Ge_x$ epitaxy at 610 °C and subsequent in-situ annealing. No rise in LLS signal is observed during epitaxial growth, confirming that the as-grown layer is planar. A pronounced rise in LLS signal then occurs during annealing due to the development of substantial surface roughening.

Figure 2 compares AFM images from annealed layers with different Ge compositions. Pronounced < 100 > surface undulations are apparent in all cases. The topography in fig.2 formed by annealing planar strained layers is similar to that previously observed to develop during growth. This fact shows that the undulations form by surface diffusion, and are not a result of either lateral variations in growth rate or surface evaporation (which is negligible at these temperatures). Fig.2 also shows a clear trend for the scale of the roughness to decrease with increasing Ge fraction X. This is consistent with the general trend for uniform surface roughening

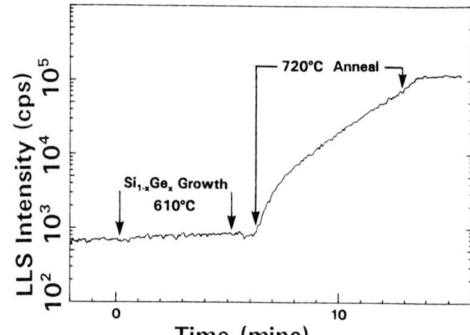

Figure 1. LLS trace obtained during growth of a 25nm-thick planar $Si_{1-x}Ge_x$ layer (X = 0.23) at 610°C followed by in-situ annealing at 720°C.

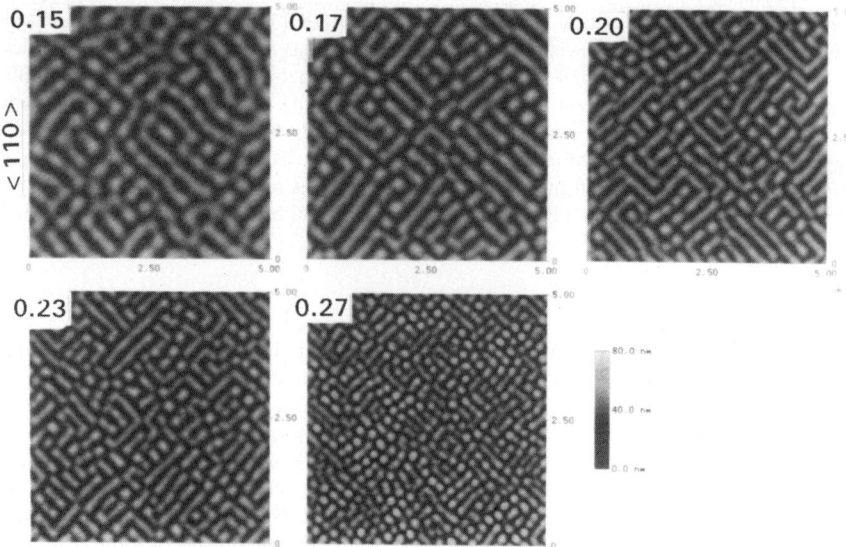

Figure 2. AFM images from $Si_{1-x}Ge_x$ epilayers annealed at 720 °C to a final LLS signal of $\sim 10^5$cps showing dependence on Ge fraction X. All images are 5μm x 5μm, and the white (high) to black (low) height contrast range is 80nm.

during strained-layer epitaxy without misfit dislocations (Pidduck et. al. (1992)).

Figure 3 shows the results of quantitative measurements of roughness period (average < 100 > ridge spacing) and amplitude (rms height calculated over 5μm x 5μm areas) made from the AFM images of annealed layers. A clear pattern which emerges from the graph is that all data points corresponding to a given Ge fraction X lie close to a line of slope ½ (shown dotted). This result is apparently independent of layer thickness (varied in the range 12-30nm), annealing temperature (680-750°C) and annealing ambient (H_2 or UHV). The effect of increased annealing time or temperature is to increase the roughness period and amplitude in such a way as to move the data point from left to right along a line of constant $\lambda/\sqrt{t_{rms}}$ in fig.3.

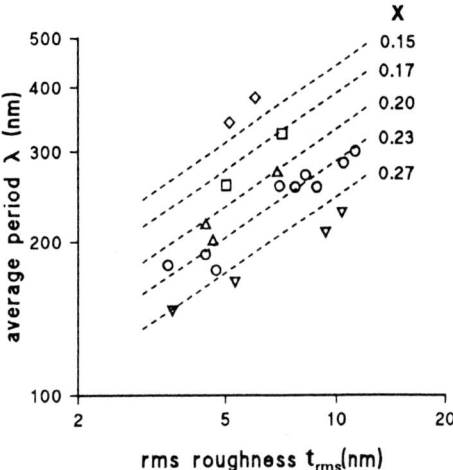

Figure 3. AFM measurements of the mean ridge spacing λ and rms roughness (t_{rms}) of the rippled topography on annealed epilayers. Symbols denote different Ge fractions X.

4. DISCUSSION

Figure 4 is a schematic showing the form of the strain fluctuations associated with the undulations, consistent with transmission electron microscopy (TEM) studies of pseudomorphically strained $Si_{1-x}Ge_x$ undulating layers. Pidduck et. al. (1992) proposed a simple sinusoidal model for the thickness fluctuation t(x) (with t_o, t_1 and λ as defined in fig.4) :

$$t(x) = t_o + t_1 \cos(2\pi x / \lambda) \qquad (1)$$

The strain fluctuation (π out-of-phase) is represented by :

$$\epsilon(x) = \epsilon_o - \epsilon_1 \cos(2\pi x/ \lambda) \qquad (2)$$

where $\epsilon(x)$ is the strain at any position x averaged over the layer thickness at that point, ϵ_o is the mismatch strain between substrate and

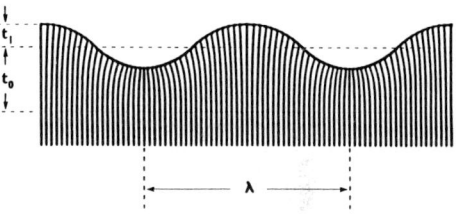

Figure 4. Schematic showing correlation between thickness and strain fluctuations for a compressively strained epilayer of average thickness t_0.

epitaxial layer, and ϵ_1 is the maximum depth-averaged elastic strain modulation (which occurs at the layer crests and troughs). Eqns (1) and (2) are used to calculate the net reduction in elastic energy due to the partially relaxed volume of material at the layer crests. For the undulations to be stable, this must exceed the energy cost of increased roughness and surface area. This is satisfied for wavelengths λ such that :

$$\lambda^2 > 4 \gamma \pi^2 t_1^2 / \{ Y\epsilon_1 \epsilon_o t_o [2t_1/t_o - \epsilon_1/\epsilon_o] \} \qquad (4)$$

where γ is the surface free energy per unit area and Y is Young's modulus of the epilayer material.

For application to the annealed layers, we examine (4) in the limits (a) $t_o \rightarrow t_1$ and (b) $\epsilon_1 \rightarrow \epsilon_o$. Limit (a) is justified by the very thin layers used for this work, and by the experimental observation that, for a given X, the roughness on layers of different thickness annealed under the same conditions is quantitatively similar. The limit is equivalent to considering that any effect of strain variations below the plane of the troughs is small compared with the elastic energy gain due to the relaxed volume of material in the crests. Limit (b) is justified by TEM and X-ray diffraction results (Cullis and Barnett (to be published)) which are consistent with near-complete elastic relaxation occuring in some small volume of material at the crests of uncapped undulating layers.

Eqn. (4) then reduces after rearranging to :

$$t_l / \lambda^2 \quad < \quad Y\epsilon_o^2/4 \, \gamma\pi^2 \qquad (5)$$

Figure 5 shows the results from fig.3 replotted in the form t_{rms}/ λ^2 vs X^2. The data are compared with a line of slope $(0.042)^2 Y/4\sqrt{2}\gamma\pi^2$ (shown dotted) with $Y = 1.66 \times 10^{11} N/m^2$ (Neuberger (1971)) and $\gamma = 2.13 N/m$ (Jaccodine (1963)). This corresponds to eqn.(5) assuming $\epsilon_o = 0.042X$ and $t_l = \sqrt{2} t_{rms}$. The agreement for annealed layers with $X > 0.16$ is consistent with the assumption that the undulations have developed to an extent which allows complete relaxation of some of the material at the layer crests. Once this situation has been reached for a given epilayer mismatch strain, the result of further annealing is to increase both the amplitude t_{rms} and wavelength λ of the roughness at constant t_{rms}/ λ^2.

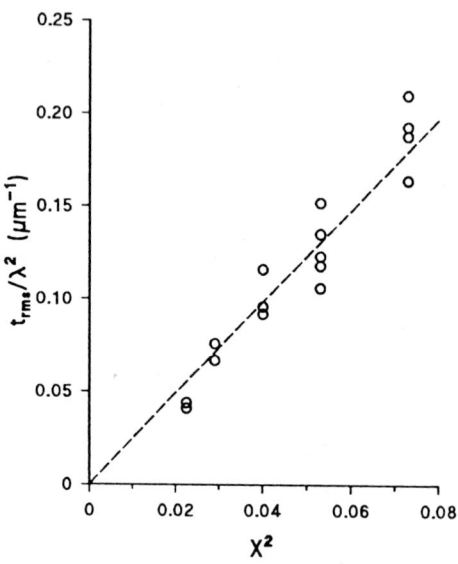

Figure 5. AFM measurements of t_{rms}/λ^2 plotted vs (Ge fraction X)^2.

5. CONCLUSIONS

AFM measurements show that the amplitude/(wavelength)^2 of correlated strain/thickness fluctuations developed by annealing planar uncapped strained $Si_{1-x}Ge_x$ epilayers is a constant value determined by the Ge fraction X. The value of this ratio is found to depend on X^2, and is in reasonable agreement with the predictions of a simple sinusoidal model for the case where there is complete elastic relaxation of a small volume of material at the layer crests.

ACKNOWLEDGEMENTS.

We thank C.Pickering and S.J.Barnett for measurements of Ge composition.

REFERENCES

Asaro R J and Tiller W A 1972 Metall. Trans. 3 1789
Cullis A G , Robbins D J, Pidduck A J and Smith P W 1992 J. Crystal Growth 123 333
Cullis A G and Barnett S J, private communication
Jaccodine R J 1963 J.Electrochem. Soc. 110 524
Kuan T S and Iyer S S 1991 Appl. Phys. Lett. 59 2242
Neuberger M 1971 ed. "Handbook of Electronic Materials",Vol.2 pp46,94 IFI/Plenum,New York
Pickering C, Carline R T, Robbins D J, Leong W Y, Barnett S J, Pitt A D and Cullis A G 1993
    J. Appl. Phys. 73 239
Pidduck A J, Robbins D J, Cullis A G, Leong W Y and Pitt A M 1992 Thin Solid Films 222 78
Robbins D J, Glasper J L, Cullis A G and Leong W Y 1991 J.Appl. Phys. 69 3729
Robbins D J, Cullis A G and Pidduck A J 1991 J.Vac.Sci.Tech. B9, 2048
Srolovitz D J 1989 Acta Metall. 37 621

Inst. Phys. Conf. Ser. No 134: Section 10
Paper presented at Microsc. Semicond. Mater. Conf., Oxford, 5–8 April 1993

613

Characterization of initial stages of growth of CaF$_2$ on Si(111) substrates by atomic force microscopy

J Olivier, G Padeletti*, V Mathet, F Nguyen-Van-Dau and R Bisaro

THOMSON CSF-LCR, Domaine de Corbeville, F-91404 Orsay Cedex, France.
*permanent address: CNR (ICM), Via Salaria Km 29.5 CP 10-00016 Monterotondo Staz (RM), Italy.

ABSTRACT: Initial stages of growth of CaF$_2$ on Si (111) substrates by molecular beam epitaxy (MBE) have been characterized by atomic force microscopy (AFM). We have successively observed pseudomorphic growth for the thinnest layers (e $\approx$ 2 nm), helical growth up to $\approx$ 51 nm and finally, for thick films (100- 300 nm), large flat surfaces striated by straight steps parallel to the three <1-10> directions. A phenomenological model is given to support these observations.

1. INTRODUCTION

In recent years, extensive research has been conducted on growing (III-V, IV-VI...) epitaxial semiconductor compounds on silicon. However, due to the lattice mismatch and the thermal expansion coefficient difference between the overlayer and Si substrate, the quality of the heteroepitaxial layer remains poor (Olivier et al 1992). In order to overcome these problems, epitaxial fluorides are used as a buffer layer. As a matter of fact, CaF$_2$ can lattice match some semiconductor compounds and has the ability to release strain through dislocation motion (Schowalter et al 1986).

In this paper, we present a study of the growth of CaF$_2$ on Si (111) substrates and a phenomenological model is introduced to understand the experimental observations.

2. EXPERIMENTAL

The growth experiments were performed in a RIBER 2300 system. CaF$_2$ molecules have been congruently evaporated from an effusion cell onto silicon (111) substrates heated to 570°C. The substrates were cleaned with a Shiraki modified procedure. Layers of thicknesses e = 2 nm, 5.7 nm, 20 nm, 51 nm, 99 nm and 300 nm have been deposited at a growth rate of 0.16 µm/h.

Our AFM (Nanoscope III) used a microfabricated Si$_3$N$_4$ cantilever with an integrated pyramidal microtip. AFM measurements are performed in air, so due to the capillary forces, they were done in the repulsive mode (Hues et al 1993). The force interaction between the tip and the sample surface is in the range of 2 10^{-8} to 2 10^{-7} N.

3. RESULTS AND DISCUSSION

Though we do not know the surface energy of the chemically-reacted interface layer obtained at the onset of growth (Olmstead 1986, Himpsel 1986), the low Gibbs free surface energy of CaF$_2$ (111) planes (4.76 10^{-5} J x cm^{-2}) (Tasker 1980) is smaller than the corresponding value for Si (111) planes (12.3 10^{-5} J x cm^{-2}) (Jaccodine 1963) and should lead to pseudomorphic layer-by-layer growth. We may assume that such a situation is fulfilled in the case of a 2 nm thick CaF$_2$ layer (Fig. 1) where large plates of several µm^2, separated by narrow depressions, are observed. This result is in good agreement with recent work of Denlinger et al (1993) on the initial stages of CaF$_2$ epitaxy on Si (111) (e < 1.5 nm): for high CaF$_2$ flux rate and medium growth temperature, in the range of our growth parameters (2.7 nm/ min, T = 570°C), they have identified a laminar growth regime by X-ray photoelectron diffraction. On the other hand, low flux, high temperature conditions (0.5 nm/ min, 600°C < T < 775°C) produce island growth, a result explaining the AFM images of Avouris and Wolkow (1989).

Due to the ≈ 2% lattice mismatch at the growth temperature (T_S = 570°C), these first CaF_2 atomic layers are strained to match the Si substrate. However, as the layer thickness increases, the strain energy becomes sufficiently large for misfit dislocations to be introduced. Such a critical thickness h_C was calculated by numerous authors, for example Matthews's theory (Matthews 1975), applied to CaF_2 leads to the value ≈ 4 nm. Beyond this value, some strain relaxation occurs. According to Frank (1949), the observation of growth pyramids on the surface of CaF_2 layers of thickness 5.7 nm (Fig. 2) or greater (Fig. 3), is   direct evidence of stress relaxation via dislocations having a component of displacement vector normal to the crystal face at which they emerge. These dislocations are not necessarily pure screw dislocations characterized by b= a/2<110>. Each of them creates a straight step between the point of emergence which we take as the origin, and the crystal edge. The step advances with the deposition of CaF_2 molecules and since it is pinned at the origin it rotates and winds up into a spiral (Fig. 3).

Fig. 1 AFM image of a 2 nm thick epitaxial CaF_2 layer on Si (111).
The scanned area is 3 μm x 3μm and the total z-range is 2 nm.

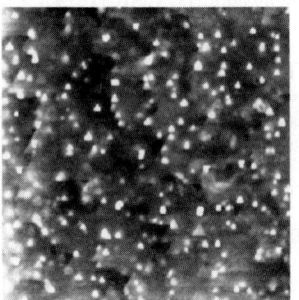

Fig. 2 AFM image of a 5.7 nm thick epitaxial CaF_2 layer on Si (111). Each small triangular pyramid of height in the range 3 nm, corresponds to the emergence of a partly screw dislocation.
The scanned area is 5μm x 5 μm and the total z-range is 6 nm.

The stress relaxation in the pyramids occurs both via the presence of the screw dislocations and the development of large area free surfaces. Luryi et al (1986) have shown that, before coalescence, the elastic energy stored in such pyramids exponentially decreases as their height h increases: thus h can reach values greater than h_C until coalescence of the helical pyramids occurs for e ≥ 20 nm (Mathet et al 1993, see Fig. 1-b).

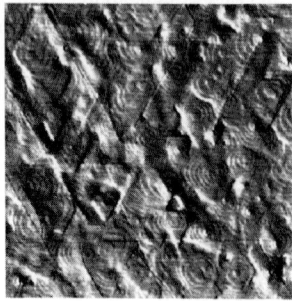

Fig. 3 A highpass filter has been applied to the AFM image of a 51 nm thick epitaxial CaF_2 layer. This treatment highlights edges or areas with rapidly changing height.
The scanned area is 5 μm x 5 μm.

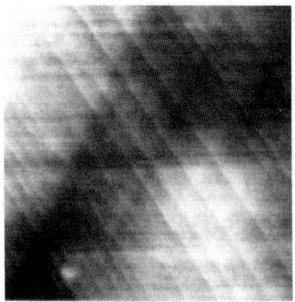

Fig. 4 AFM image of a 99 nm thick epitaxial CaF_2 layer on Si (111). The step height of the terraces lies between 0.3 and 1.2 nm. The scanned area is 1 μm x 1 μm. The total z-range is 10 nm.

Beyond e $\approx$ 51 nm, the coalescence degree is such that the free surface area strongly decreases and we have formation of a highly stressed continuous layer. This situation is favourable for the motion of dislocations (Katz et al 1974): the dislocation lines inside the islands are bent under the stress field and glide towards the edges of the CaF$_2$ layer or the interface CaF$_2$/ Si.

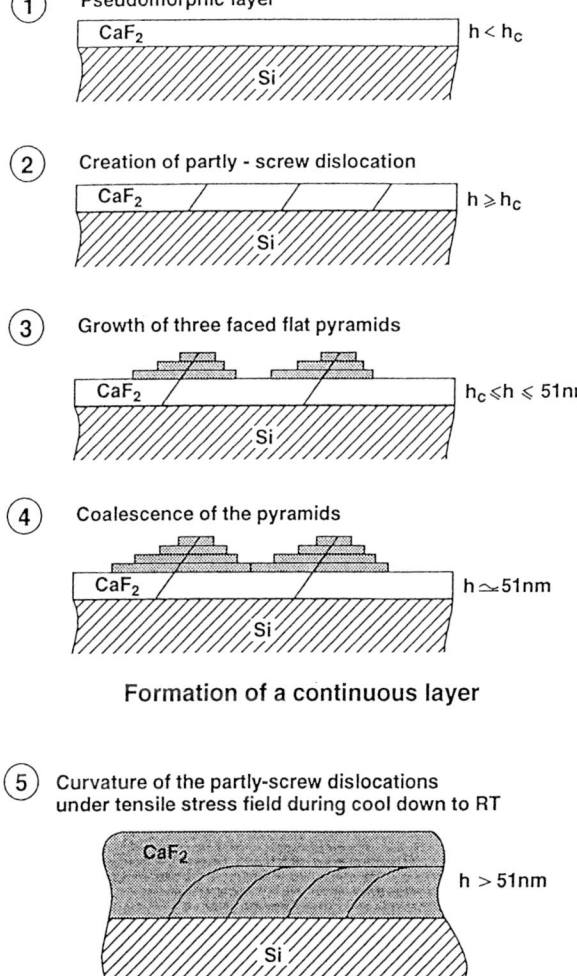

Formation of a continuous layer

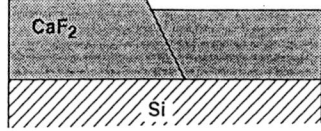

Fig. 5 Phenomenological model used to support all the AFM observations.

Thus, as the primary glide system in CaF_2 is $\{100\}$ $<110>$, dislocations leave straight slip steps parallel to the three $<1\text{-}10>$ directions (Blunier et al 1992), clearly visible for layer thicknesses 99 nm (Fig. 4) and 300 nm but already discernable for e = 51 nm where the tops of some pyramids are joined by $<1\text{-}10>$ ledges (Fig. 3). Some attention will however be focused on the fact that coalescence during growth as well as the cool down procedure may create extra dislocations and other glide systems.

4. CONCLUSION

Initial stages of growth of CaF_2 on Si (111) substrates have been characterized by AFM. We have successively observed large flat plates for the thinnest layers that are likely due to pseudomorphic growth, helical growth up to $\approx$ 50 nm due to the emergence of partly screw dislocations, and large flat surfaces that exhibit three sets of parallel straight steps oriented in the three $<1\text{-}10>$ directions resulting from dislocation glide in the primary $\{100\}$ $<110>$ glide system. The phenomenological model presented in this paper is summarised in Fig. 5. It is worth noting that the modification in the mechanism involved in strain relaxation appears for a CaF_2 layer thickness of about 50 nm, which is in good agreement with recent calculation of Schowalter et al (1993).

REFERENCES

Avouris P and Wolkow R 1989 Appl. Phys. Lett. 55 1074
Blunier S, Zogg H, Maissen C, Tiwari A N, Overnay R M, Haefke H, Buffat P A and Kostorz G 1992 Phys. Rev. Lett. 68 3599
Denlinger J D, Rotenberg E, Hessinger U, Leskovar M and Olmstead M A 1993 to be published in Appl. Phys. Lett.
Frank F C 1949 Disc. Farad. Soc. 5 48
Himpsel F J, Hillebrecht F U, Hughes G, Jordan J L, Karlsson U O, Mc Feely F R, Morar F J, and Rieger D 1986 Appl. Phys. Lett. 48 596
Hues S M, Colton R J, Meyer E and Güntherodt H J 1993 MRS Bulletin vol XVIII 41
Jaccodine R J 1963 J. Electrochem Soc. 110 524
Katz R N and Coble R L 1974 J. Appl. Phys. 45 2382
Luryi S and Suhir E 1986 Appl. Phys. Lett. 49 140
Mathet V, Padeletti G, Olivier J, Galtier P and Nguyen-Van-Dau F 1993 Microscopy of Semiconducting Materials, this issue
Matthews J W 1975 J. Vac. Sci. Technol. 12 126
Olivier J, Bartenlian B, Charasse M N, Bisaro R, Wyczisk F, Chazelas J and Hirtz J P, 1992 Appl. Phys. Lett. 61 766
Olmstead M A, Uhrberg R I G, Bringans R D and Bachrach R Z 1986 J. Vac. Sci. Technol. B4 1123
Schowalter L J, Fathauer R W, Ponce F A, Anderson G and Hashimoto S 1986 Mat. Res. Soc. Proc. 67 125
Schowalter L J, and Li 1993 Appl. Phys. Lett. 62 696
Tasker P W 1980 J. Phys. Paris T41 C6 488

Inst. Phys. Conf. Ser. No 134: Section 10
Paper presented at Microsc. Semicond. Mater. Conf., Oxford, 5–8 April 1993

Top-view construction analysis of 16 Mbit DRAM by atomic force microscopy: a new approach

Evelyne Druet and Paul-Henri Albarède

IBM-France, Dpt 1807-31U, BP 58, 91105 Corbeil-Essonnes, France

ABSTRACT: The knowledge of a device's morphology is commonly thought of in terms of SEM or TEM cross-sections. But TEM has long cycle times (days) and SEM needs heavy chemical etching (hundreds of nanometers) and/or metallization. One powerful alternative technique can be Atomic Force Microscopy (AFM). The advantages of AFM are various and discussed hereafter. For these reasons, and thanks to a convenient specimen preparation method (plan-view polishing with a desired bevel), a complete three dimensional construction analysis can be performed on any kind of chip. The application presented deals with a 16 Mbit DRAM. A first AFM observation is done just after polishing. The interpretation of the images is enhanced by the corresponding TEM view. Then, chemical etching is performed to reveal some specific parts of the device. The valuable AFM information available from this observation is complemented by TEM cross-sections of the same areas.

1. INTRODUCTION

Construction analyses of semiconducting products are currently performed by use of Scanning Electron Microscopy (SEM) and Transmission Electron Microscopy (TEM) cross-sections, leading to two- dimensional information. The high lateral resolution of TEM is balanced by the disadvantages of the technique (expensive and room-consuming) and the cycle time limited by specimen preparation (days). Although SEM has shorter cycle times (hours), it requires severe chemical treatment (hundreds of nanometers) to give a contrasting image.

The originality of the method presented here is to provide complete three dimensional observation of the device by top-view Atomic Force Microscopy (AFM). The interesting layers of the DRAM are situated in the upper few microns. In order to convert the vertical ordering into a lateral change, a very slight bevel is produced on the specimen so that the vertical transition between the five uppermost microns occupies several hundred microns of the free surface. Thus, critical structures become observable in real space. The advantages of the important AFM technique are threefold: workability under air, rapid provision of reliable and reproducible 3D measurements and high vertical and lateral resolutions.

2. METHODOLOGY

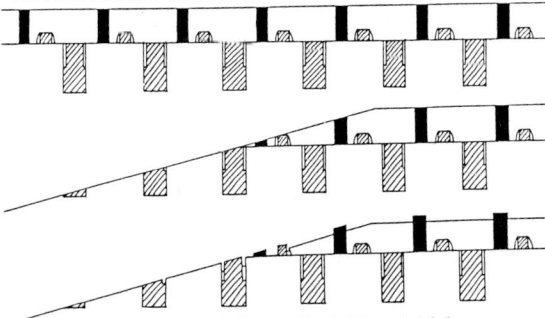

Thanks to the very accurate tripod polisher invented by Stanley Klepeis (Klepeis 1989), the slope of the bevel can be adjusted to any convenient value. In our case, it had to be within two limits: sensitive enough to the thinnest layers of the device, and compatible with the AFM's field of observation. A compromise is found at 20 mrad (figure 1). Another advantage of the Klepeis method is to provide the scratchless surface needed by the AFM, over several times the AFM's field of investigation (hundreds of square microns).

Figure 1: Principle of the method. Top: initial state Center: after polishing. Bottom: after etching.

This step of preparation will give "crude" AFM images. A slight HF/NH4F etching is done on the free surface to give more detailed images: the "etched" AFM images. Since this etching operation is not quantified, the interpretation of the etched AFM images is strengthened by a cross-sectional TEM investigation of the same specimen.

The tools used for analysis are:
- AFM nanoscope II from Digital Instruments
- TEM CM30 Philips (300 keV).

3. OBSERVATIONS

The device is described from bottom to top. Therefore, we start with the deep capacitive trenches where the information will be stored (see schematic cross-section of figure 2).

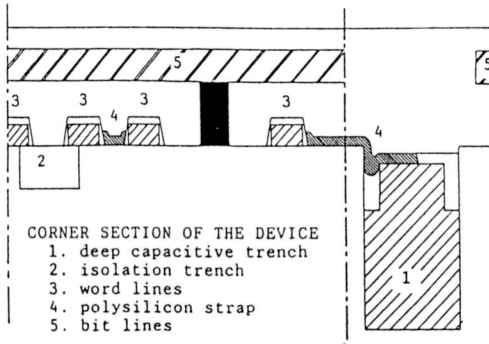

CORNER SECTION OF THE DEVICE
1. deep capacitive trench
2. isolation trench
3. word lines
4. polysilicon strap
5. bit lines

Figure 2: Schematic cross-section of the device.

Figure 3 shows the AFM top-view of the bottom of the deep trench. The polysilicon inside appears humped (5nm above the polished silicon mean surrounding surface). The Oxide-Nitride-Oxide (ONO) at the interface is shown either on top-view or on AFM cross-section (figure 4, Druet 1991). The bright line is the horizontal part of the ONO at the very bottom of the deep trench.

Going up, we cross the collar oxide all around the deep trench. Crude and etched AFM images are given in figures 5 and 6, on which the width of the collar oxide as well as the etched depth (23 nm) can be measured. One complementary advantage of the AFM compared to SEM is its useful computing aid: image improvement and processing. Thanks to a fast Fourier transform (FFT), the periodicity of the device is obtainable. This valuable information tests the mask's quality. In our case, one of the periodicities of the deep trenches is measured with a good accuracy: 1428 nm +/- 30 nm. Figure 7 gives the direct image of the array (7.a) and the unidimensional FFT (7.b).

Switching to the upper structures, we give an overall TEM cross-section, useful for further understanding of the layout (figure 8). One can see the isolation trench (IT) in the center, several word lines and black metallic studs for bit line connection. Embanked between two word lines, small polysilicon straps connect the drain of the access transistor to the neighbouring capacitor. All these active parts are immersed in an isolating Phosphorus Silicon Glass (PSG).

The progressive rise through the isolation trench up to the word lines is presented in the series of figure 9. One can see the lowest part of the isolation trench with the very bottom of the nitride film (9 a and b). The upper part of the IT (9 c and d) shows the appearance of the word lines over the collar oxide valleys. One can interpret the various reliefs in terms of selective wet etching. The central PSG is etched more quickly than the TetraEthylOrthoSilicate (TEOS) of the Isolation Trench (IT), and within the PSG, several steps are evident. The central void, due to early closing of the PSG deposit, is enlarged by the wet etching.

An AFM top view of the word lines (figure 10) enables one to see the appearance of the upper layer (bright zones on the lines). In the same area silicon straps are crossed (dark dashed lines between the word lines). An AFM cross-section of a word line is given on figure 11. The general view (11.a) shows the double layered word line. The detailed view of the spacer (11.b) clearly points out the gate oxide band and the nitride coating.

4. CONCLUSION

Thanks to a very good correlation between AFM and TEM, the reliability of AFM 3D observation of any kind of device is established, even after wet etching. The richness of the three dimensional measurement compares favourably to SEM.

This kind of investigation can be performed on any kind of device, either for construction analysis or for failure analysis. A misalignment or a misprocess in a single cell can be detected, provided the defective device can be located using eg laser shots.

REFERENCES

Druet E and Albarede P-H 1991 Proc. 5th Int. Conf. Quality in Electronic Components vol 2, pp 753-9.
Klepeis S J , Benedict J P , Vandygrift W G and Anderson R
1989 EMSA Bull. vol 19, pp 74-9

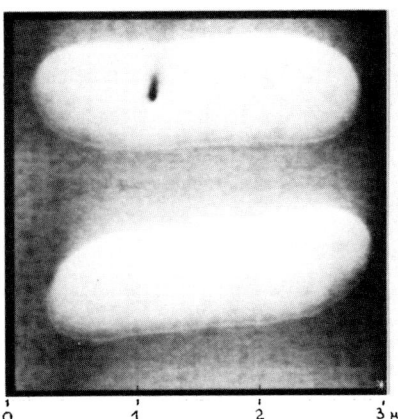

Figure 3: Crude AFM top-view
of the bottom of a deep trench.

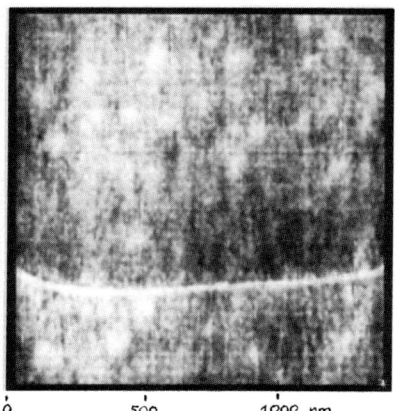

Figure 4: AFM cross-section
over the ONO layer.

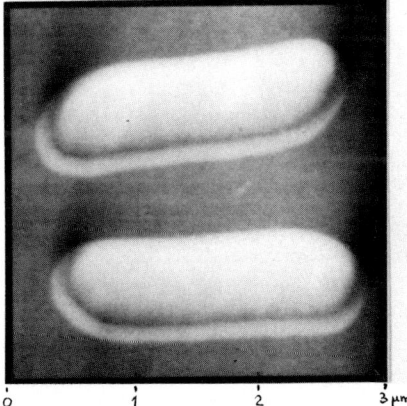

Figure 5: Crude AFM top-view
of the middle of a deep trench.

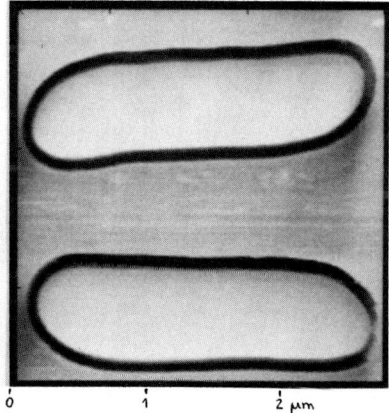

Figure 6: Etched AFM top-view
of the middle of a deep trench.

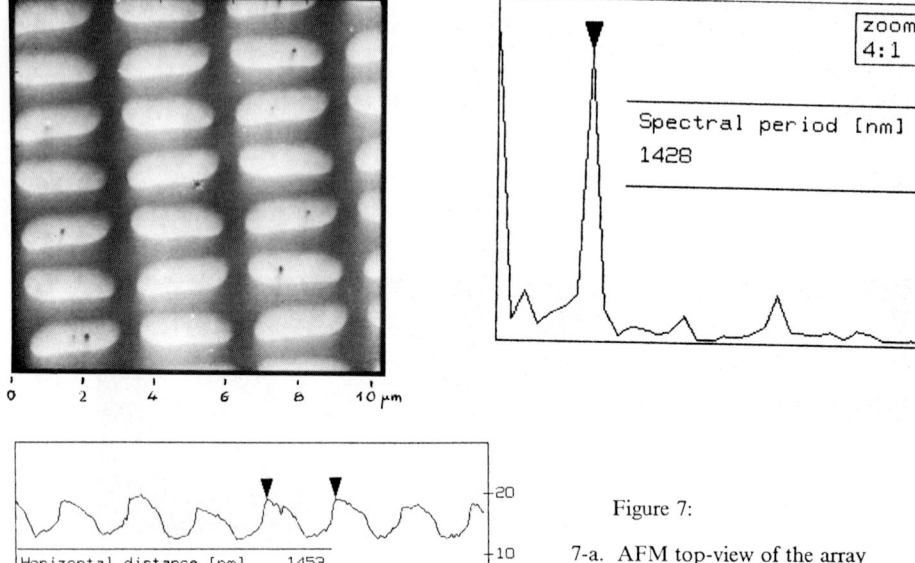

Figure 7:

7-a. AFM top-view of the array
      of deep trenches.
7-b. up: spatial frequency spectrum
      along the vertical direction.
      left: z profile along this direction.

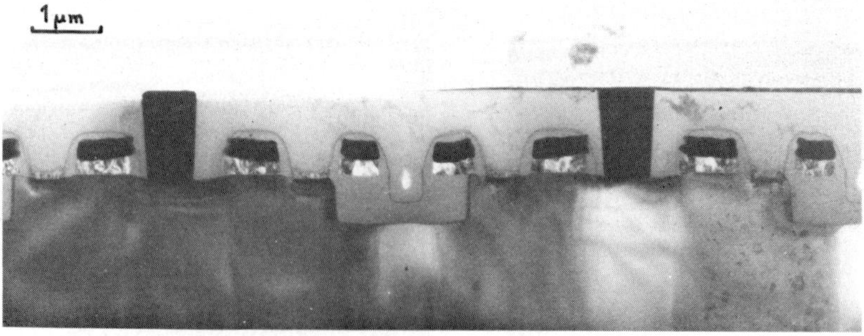

Figure 8: TEM cross-section. General overview on the device.

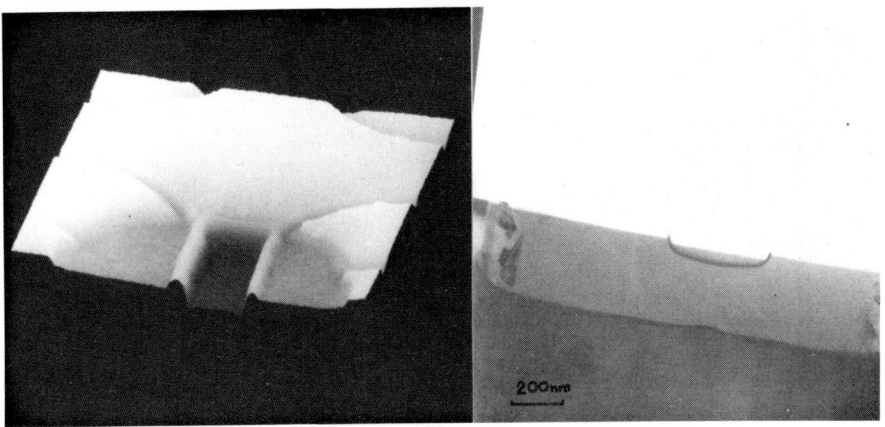

Figure 9-a: AFM 3D view over
the bottom of an isolation trench.

Figure 9-b: TEM cross-section
through the same region.

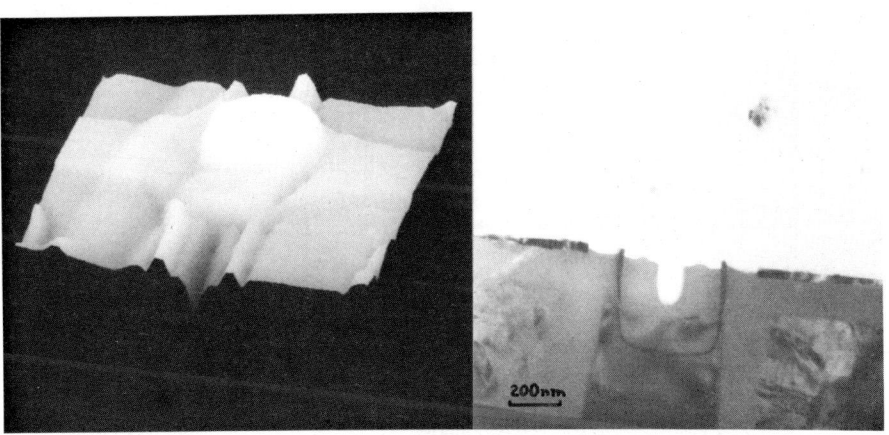

Figure 9-c: AFM 3D view over
the top of an isolation trench.

Figure 9-d: TEM cross-section
through the same region.

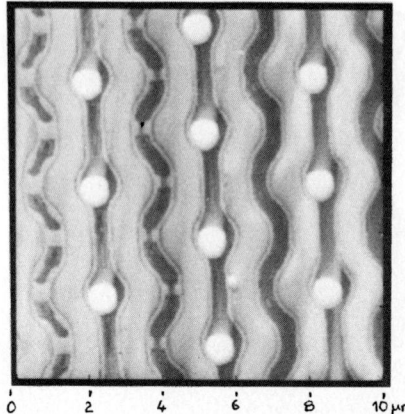

Figure 10:
AFM top-view of word lines.

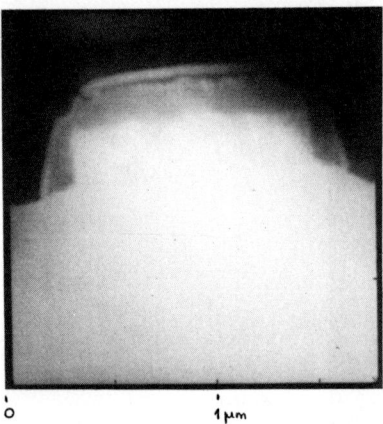

Figure 11-a:
AFM cross-section of a
word line. General view.

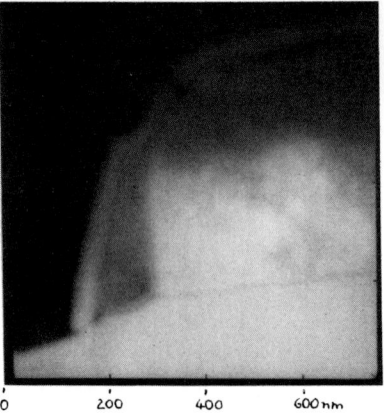

Figure 11-b:
AFM cross-section of a
word line. Detail of
the spacer region.

Inst. Phys. Conf. Ser. No 134: Section 10
Paper presented at Microsc. Semicond. Mater. Conf., Oxford, 5–8 April 1993

623

Scanning tunnelling microscopy in air on as-grown, massive silicon carbide crystals

P Heuell, M A Kulakov, V F Tsvetkov[*] and B Bullemer

Universität der Bundeswehr, München, Fakultät für Elektrotechnik, Institut für Physik, Werner-Heisenberg-Weg 39, D-W 8014 Neubiberg, Germany.

[*]Electrotechnical University of St.Petersburg, Russia.

ABSTRACT: Scanning tunnelling microscopy in air is used to study hexagonal (6H) silicon carbide. The samples were nitrogen doped single crystals of size of about 8x15x5 mm, grown from the vapour phase at a temperature of 2550-2600°C . The (0001)C face revealed extended flat terraces and large facets, running up or down to the next terrace. On such terraces some round islands with a diameter of 2000-4000 Å and height of 200-300 Å were found. They had typical spiral features on their top and are proposed to be growth nuclei.

1. INTRODUCTION

Silicon carbide has recently attracted a great deal of attention due to its excellent high temperature properties which are of great importance for high frequency and high power devices. On the other hand, realisation of these advantages requires reliable and well reproducible growth technology, including polytype control. At present several growth methods are used for high quality layers and crystal production. In general, some of them rely on CVD layer deposition onto SiC or Si substrates at relatively low temperatures of 1200-1400°C (Nishino et al 1983, Liaw and Davis 1985). Others use much higher temperatures 2100-2600°C to grow massive crystals by vapour phase epitaxy (Lely 1955, Tairov and Tsvetkov 1978). This tremendous temperature difference should imply nonequivalent growth mechanisms and, subsequently, different crystal quality and different surface morphologies. Due to peculiarities of these growth processes it is hardly possible to monitor surface properties *in situ*. Therefore all new information obtained *ex situ* is of great importance.

Scanning tunnelling microscopy (STM) has already contributed to studies of SiC films morphology at intermediate (Zheng et al 1988, Chang 1990) and very early (Rogers and Tiedje 1992) stages of CVD growth. It has been shown that faceting of monatomic steps, similar to faceting on silicon in molecular beam epitaxy, occurs in this technology, which shows the important role of accommodation kinetics and indicates the possibility of step flow growth at temperatures of about 1200-1500°C.

In this work we present the first STM observations of massive bulk crystals of hexagonal (6H) silicon carbide, grown by the original Lely method (Lely 1955).

2. EXPERIMENTAL

The crystals were produced from silicon carbide powders which were, in turn, synthesised from Si and C powders of semiconductor purity. The growth proceeded in an argon atmosphere with a certain content of nitrogen, which yielded a doping level of n_a-n_d 2×10^{18}/cm^3. X-ray analysis of the crystals revealed 6H structure. As-grown crystals had the form of a platelet with one well developed flat (0001) and an opposite rounded or stepped face. The flat face can be with equal probability either (0001)C or (0001)Si. Ellipsometry on as-grown crystals showed the presence of a 100-300 Å thick film of SiO$_2$.

The sample preparation procedure for STM study consisted of ultrasonic cleaning in acetone and methanol, both for 5 minutes, 30 second etching in 5% HF, rinsing in boiling ultra pure water for 30-60 seconds and final rinsing in methanol. All the chemicals are of Selectipure quality. Contact to the crystal was made by a small piece of indium, pressed against the surface by a screw.

In order to identify the (0001) face and/or improve the surface quality the crystals were oxidised, as is often done in silicon technology . The oxide films on SiC are known to grow much faster on the (0001)C than on the (0001)Si face. Oxidation was carried out in a commercial laboratory oven at ambient pressure with the use of dry oxygen. The temperatures used were in the range 900-1150°C and typical times were 3-5 hours. The resulting SiO$_2$ films were 200-2000 Å thick, depending upon growth parameters.

Our homemade scanning tunnelling microscope, operated in air, contained a tube scanner which was 14 mm long, 6 mm in diameter, with 0.5 mm wall thickness. Estimated sensitivities were 180 Å for x,y and 60 Å for z axis respectively. The inertial coarse approach unit was able to change the sample position with the precision of about 100 Å. The whole microscope assembly rested on a Viton spaced stack of steel plates, which, in turn, was positioned on a concrete block damped by four springs. No measures were taken to isolate acoustical noise. Feedback loop, scanning and tunnelling parameter control was performed using a digital signal processor DSP 32C of AT&T, working with 60 kHz sampling frequency. The measured height resolution was better than 0.2 Å for metals and 0.4 Å for semiconductor surfaces. Pt-Ir tips were just cut off by clean cutters and normally provided better results than etched ones. All the images were taken with tunnelling currents of 0.2-0.3 nA and biases from -1.8 to -2.4 V with respect to the tip.

3. RESULTS

The crystal face showed a diamond-shaped appearance, typical for crystals grown by the Lely method, fig.1. A network of macro steps in the proximity of the crystal side, which was connected to the crucible, was running mostly parallel to this side. They were easily visible

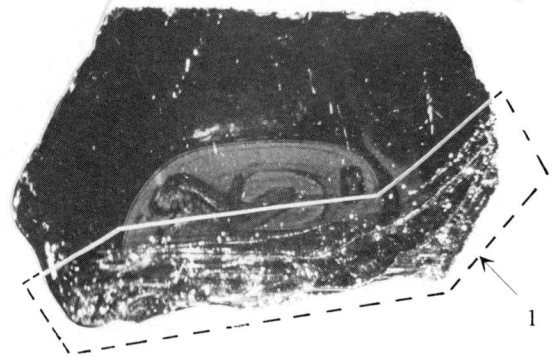

Fig.1: One of the crystals used in the experiments. 1- area containing a network of macrosteps.

optically under low magnification. There was also another network of much smaller steps in the area, free from the first step array. This network was also visible optically but only with the use of differential phase (Nomarski) contrast. This network covered the surface with a density decreasing from one side of the crystal (connected to crucible) to the other. The distances between these steps change from 100-300 μm to 0.8-1 mm . In the middle of the surface a very large rounded closed terrace (island) of approximately 0.5x0.9 mm size was found. Topographically this terrace was at the top of this face.

In areas close to the centre of the crystal face we normally did not see any features which could be attributed to an as-grown morphology. Measured random height corrugations lay in the range 3-5 Å and represented lower stability of the tunnelling current while imaging semiconducting materials in air. Contrary to our expectation of finding a lot of monatomically stepped structures, only once did we hit a 1800 Å wide terrace, separating two 4.7 Å high steps, which approximately corresponds to two monolayers.

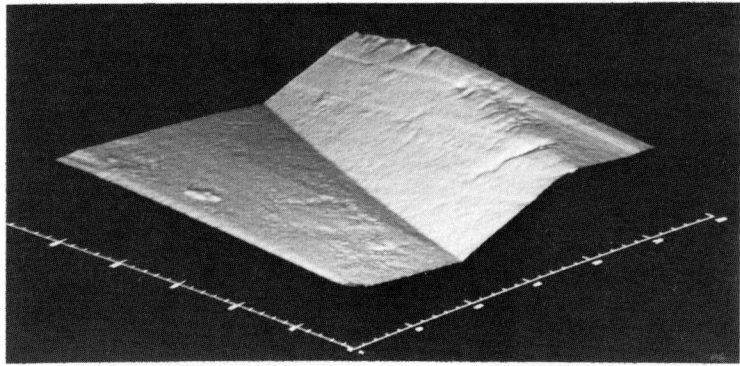

Fig. 2: Two terraces separated by a large facet, 1.4x1.4 μm.

In the part of the surface with a high macro step concentration it was relatively easy to hit occasionally the ledge of a terrace and visualise the transition from a terrace to a facet. An example of such a situation is given in fig. 2 . One can see that both the terrace and the facet are of approximately equal surface perfection, both are flat, the edge between them is sharp

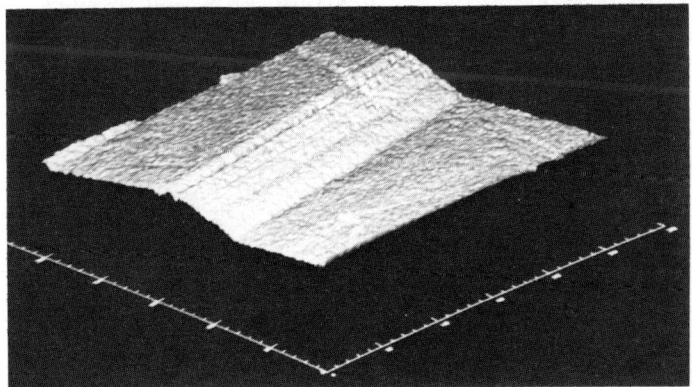

Fig.3. Two terraces separated by a low (~90 Å) step.

and straight. The measured angle between two planes is about 10° and the height of the step is 850 Å. Not all of the transitions revealed such good geometrical quality. A lot of them were blunt and/or not straight and the facets were not always plane. This observation corresponds to an optical picture where we found many rounded features in the step structure.

Another typical example of morphology, revealed by STM, is the step of fig. 3, in the network of low steps, mentioned above. The measured angle between the terrace and a facet is about 7° and the height of the step is 110 Å, which corresponds to 5-6 times the 6H polytype periodicity. The facet in this case is not perfectly flat, it obviously consists of a number of smaller, non equidistant steps. The details of this structure were not resolvable.

Fig. 4 shows two round islands on a flat terrace. The bigger one is about 300 Å high and 4000 Å in diameter, the smaller is 180 Å high 2000 Å in diameter. Both of them have steep sides, their tops on average are plane but contain some structures. Under higher magnification this structure appears as part of a spiral, fig. 5. Three-dimensional representation also reveals in the proximity of these two islands the sharp edges of two intersecting facets.

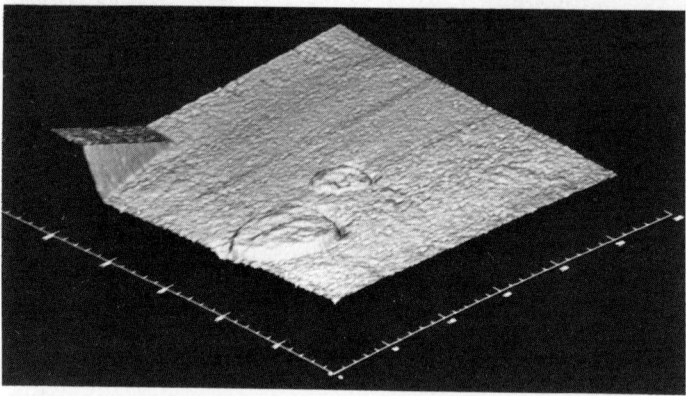

Fig. 4. Two nuclei and the intersection of the two facets on a flat terrace, 1.4x1.4 μm.

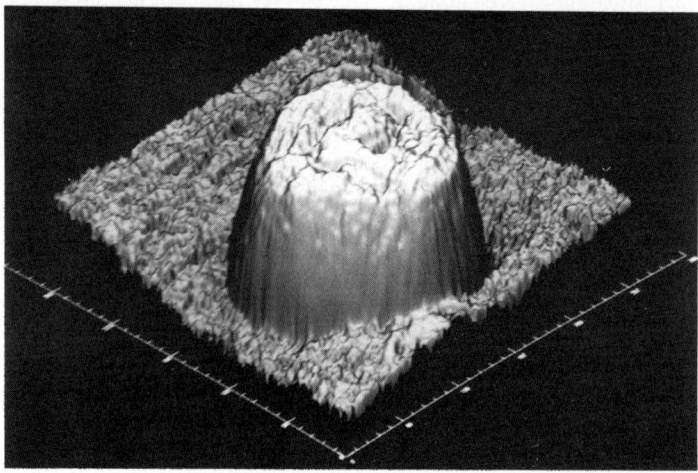

Fig. 5. Smaller nucleus from fig.3 at higher magnification, 5000x5000 Å.

Fig.6 represents a structure consisting of an intersection of three straight lines with the angle between them being 60°. There is very small height variation corresponding to these lines which shows that the contrast in this case is more likely to be electronic than topographical.

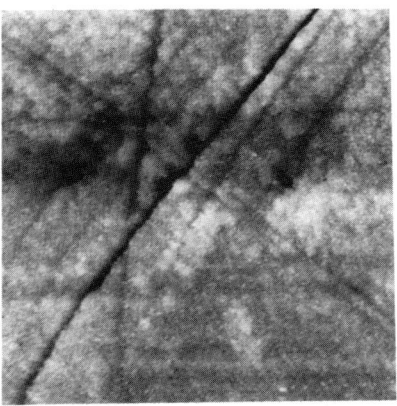

Fig.6: "Star"-like defect, 8000x8000 Å.

Tunelling spectroscopy showed one branch curves, typical for n-doping, in the negative half of the ± 3V ramp.

4. DISCUSSION

The surface morphology of the massive SiC crystals is a very important source of *ex-situ* information about the real growth process. It makes it possible to judge flux and temperature distributions, defect origins, kinetics and thermodynamics.

One of the most important problems in SiC bulk crystal technology is to initiate (induce) and maintain the desirable growth mode. This requires a good understanding of the nucleation mechanism as well as sticking and surface diffusion on the main face and different facets.

Some investigators of silicon carbide argue that on a SiC surface, growth is only possible by means of screw dislocations (Verma and Krishna 1966) and the calculated required density for crystals grown by the Lely method is about 10^2-10^3/cm^2 (Amelincks et al 1960, Tregubova 1972). In our case on both crystals studied we could not find any features which looked like classical screw dislocations (Verma 1966). It is not surprising taking into account the very small field of view of STM. The spiral-looking structures on the tops of both nuclei might be, principally, screw dislocations but at this stage of the work it is too early to discuss it seriously. More data and more thorough study of this point is required.

Experimental estimation of the nuclei density is also rather difficult. In the Lely method the typical supersaturations are about 0.2-1%. This value is very low in comparison with those of CVD or MBE. Therefore, very low nuclei density is to be expected and STM can not be used as a counting instrument. In fact one can only try to count the density of relatively large (1-10 µm) islands with the use of an optical or electron microscope. Only one crystal of the two studied revealed such islands. It is interesting to note that there were approximately equal numbers of hexagonal and round ones. The perfectly flat surface together with the estimated island density of about 0.1-0.5 x 10^2/cm^2 shows that the combination of growth parameters allowed only a very low nucleation rate and that the main growth mechanism on the (0001)C face involved step flow.

The question of the leading mechanism of step generation is still not clear. In practice it is known that steps arise mostly either from natural nuclei on large terraces and crystal edges or from contamination (Chernov 1984). Further investigation with the STM, especially of the atomic structure of crystal and step edges would be very helpful.

The step bunching observed optically in the lower half of the crystal, as shown in fig.1, occurs presumably due to strong flux nonuniformity and higher defect density in the part of the surface close to the area of connection of the crystal to the crucible. Our STM observations show at the same time that beyond the step bunching area steps become much lower. Unfortunately, the very large distances between steps, as compared to the STM field of view, did not allow us to estimate the step height distribution.

The star-like structure shown in fig.5 reflects the sixfold symmetry of the crystal and looks too perfect to be produced artificially by occasional scratching. Such structures might reflect the dopant atom distribution nonuniformity and/or stochiometry variations which are believed to manifest themselves strongly along $[11\bar{2}0]$-like directions (Levin et al 1977, Lilov et al 1976).

5. CONCLUSIONS

Though scanning tunnelling microscopy of semiconducting materials in air is normally associated with a number of difficulties, our study of silicon carbide crystals revealed several morphological features. It was shown that the (0001)C face is a stepped surface, and the smallest observed step height was 4.7 Å, which corresponds to two monatomic layers. The flat and defect free part of the surface was covered with a network of steps which are about 90 Å high. The terrace width in the network of these low steps was very large and beyond the scanning range of the STM. This is in agreement with an optically measured values 0.05-1 mm. The terraces normally did not reveal any features which could be attributed to the as-grown morphology. The facets did not reveal any special orientation. Both STM and optical microscopy showed that depending on their size, surroundings and local growth conditions they can be plane, rounded or stepped. The round islands on flat terraces are believed to be the growth nuclei. Though they have some spiral-like structures on their tops we are not at present confident that these structures could be attributed to screw dislocations.

An STM is proved to be a tool providing very useful morphological information which can be used for control of the growth parameters in SiC bulk crystal production.

REFERENCES

Nishino S, Powell J A and Will H A 1983 Appl.Phys.Lett. 42 460
Liaw H P and Davis R F   1985 J.Electrochem.Soc. 132 642
Lely J A   1978 Ber.Deut.Keram.Ges. 32 229
Tairov Yu M and Tsvetkov V F   1978 J.Crystal Growth 43 208
Zheng N J, Knipping U and Tsong I S T, Petuskey W T, Kong H S and Davis R F 1988
    J. Vac. Sci. Technol. A6 696
Chang C S, Zheng N J and Tsong I S T, Wang Y C and Davis R F 1990 J.Vac.Sci.Technol.B9 681
Rogers D and Tiedje T. 1992 Surf.Sci. 274 L599.
Verma A R Krishna P 1966 Polymorphism and Polytypism in Crystals (New-York:Wiley)
Amelincks S, Strumane J, Webb W W 1960 J.Appl.Phys., 31 1359.
Tregubova A S, Shulpina I L 1972 Soviet Solid State Physics, 14 2670.
Chernov A A 1984 Modern Crystallography v.3 Springer Verlag, .
Levin VI,Pozdnyakova GI,Tairov YuM,Tsvetkov VF,Schaschkov YuM 1977 Inorgan.Mater. 13 212.
Lilov S K, Tairov Yu M, Tsvetkov V F, Shernov M A 1976 Phys.Stat.Sol. A 37 143.

Inst. Phys. Conf. Ser. No 134: Section 11
Paper presented at Microsc. Semicond. Mater. Conf., Oxford, 5–8 April 1993

Cathodoluminescence images of quantum wells and wires

D Bimberg and J Christen

Institut für Festkörperphysik I, Technische Universität Berlin, Hardenbergstr. 36
1000 Berlin 12, Germany

ABSTRACT: Characterization of optical properties of low dimensional structures on a nanometer scale is a key to the understanding of the correlation of their structural and electronic properties. Cathodoluminescence imaging in the cw or time-resolved modes is emerging as an important novel technique for obtaining such insights. Employing high brightness LaB$_6$ cathodes and very low acceleration voltages, a lateral resolution below 100 nm and a temporal resolution of better than 35 ps is achieved. Three examples will demonstrate the versatility of CL: Visualization of the columnar structure of GaAs quantum wells grown with interruption by MBE; identification of highly efficient carrier capture processes into GaAs quantum wires grown on nonplanar substrates and observation of wire-like carrier localization effects in inhomogeneously strained GaAs quantum wells on Si ribs.

1. INTRODUCTION

Semiconductor nanostructures like quantum wells, wires and dots exhibit unique electronic, transport and optical properties not found in 3D structures. Such structures also provide a basis for a whole generation of novel mesoscopic devices. All their properties are a function of one or more dimensional parameters and therefore depend strongly on the spatially varying morphology of their interfaces. Obviously an assessment of the optical properties on a nanometer scale and with picosecond time resolution, to eliminate the influence of carrier diffusion, provides a key to understanding the correlation of their structural and electronic properties.

It is the purpose of this paper to demonstrate that the recently developed technique of cathodoluminescence imaging in the CW or time-resolved modes (Bimberg et al 1986, Christen 1990) is now emerging as probably the most important technique for obtaining such insight(s). Cathodoluminescence (CL) by itself is a technique well established for decades to study e. g. intrinsic recombination processes in phosphors (Bimberg et al 1975) and semiconductors (Bimberg et al 1985) or defects in semiconductors (Petroff et al 1982, Grundmann et al 1991). Breakthroughs have occurred in the hardware configuration of CL initiated by development of miniature cryostats (Steckenborn et al 1980) with variable temperatures down to 5K, high brightness LaB$_6$ cathodes still operating at low acceleration voltages, thus assuring a small Bethe range, and ultra fast electron beam blanking units with ps rise and decay times. Still more important are the breakthroughs in the software for controlling the system functions, rapid data collection, compression, and evaluation, which have enormously extended the capability of CL from a spot technique to a true real space imaging technique similar to scanning tunneling microscopy but different in its applications and lateral resolution. In the next section of this paper we will summarize the present state of the art of CLI, and indicate potential further improvements. The following three sections are devoted to three different recent applications of CLI, studies of the morphology of QW interfaces, the electronic structure of and carrier capture into quantum wires, and the visualization of 1D carrier confinement in inhomogeneously strained QWs.

2. CATHODOLUMINESCENCE IMAGING

The basic setup for cw CL imaging is identical to that for time-resolved experiments. Signal detection and data acquisition, however, are different. We start with a description of CLI.

The CL system is based on a modified commercial scanning electron microscope (SEM) Jeol JSM840 and includes a miniaturized continuous flow He cryostat (5K<T<300K) fitted on the specimen stage of the SEM. The acceleration voltage V_a can be varied from 40 kV down to 500 V, enabling us to adjust the depth of information as well as the lateral resolution of CLI. For V_a = 3 kV, a lateral resolution of $\leq$ 100 nm is experimentally verified.

The Bethe range in GaAs at 3 kV is larger by a factor of three than the diameter of the electron beam for the presently used LaB_6 cathode. We expect that for ultra high brightness field emission cathodes which can be employed in UHV electron microscopes the resolution can be improved by another factor of 5. Depending on the details of the samples studied, the lateral resolution actually reached might be limited by carrier diffusion and/or by the Bohr diameter of the exciton if the recombination is of excitonic origin. Diffusion induced reduction of the resolution can be suppressed by working in the time resolved mode as detailed below at the expense of an increased measurement time.

The scanning of the focused primary electron beam (128 x 100 up to 512 x 400 pixels) as well as all data recording is completely controlled by computer. At each pixel the entire luminescence spectrum emitted from the sample is recorded. For this purpose we use two different optical multichannel analyzer (OMA) systems fitted to a grating spectrometer. This setup presently covers a spectral range from 350 to 1000 nm, which could be easily extended to 1.7 μm by using an InGaAs-based OMA, and has a limiting sensitivity of 5 photons/count.

By choosing appropriate gratings a CL spectrum of $\Delta\lambda$ = 21.7-230 nm width around a preset center wavelength is recorded within 10 ms in 512 channels. Using this pixel dwell time a complete CL image is typically recorded in 8.5 min (256x200 pixels) including online data evaluation. Typically 12 CL images are generated simultaneously from the spectra: four CL intensity maps $I(\lambda_1, x, y)...I(\lambda_4, x, y)$ recorded for four different preselected wavelengths, four CL wavelength images, i. e., $\lambda_{max}(x, y)$ for the wavelength of maximum intensity determined in preselected wavelength intervals, the standard SEM image and three further reference CL integral intensity images.

The principle of CLI applied to structural investigations of quantum well interfaces is outlined in Fig. 1. The inset (a) reviews schematically the origin of the well known QW a/2-triplet splitting (Bimberg et al 1986, Christen 1990) in an idealized QW assuming two interfaces having large growth islands of monolayer height. If the lateral extension of these islands is much larger than the diameter of the QW excition, it senses three discrete QW widths of (n - 1), n, and (n + 1) monolayers with n being integer. This results in three discrete quantization energies and thus in three distinct excitonic luminescence lines in an integral CL spectrum or in a PL spectrum averaged over a large interface area. A GaAs/AlAs heterointerface having huge growth islands of 1 monolayer height is shown in Fig. 1(b). According to the lateral position of the exciting electron beam different CL spectra are observed. Thus, CL imaging directly discloses the width distribution L_z (x, y) of the QW under investigation (its columnar structure). Additionally, the variation of the z coordinate of the upper interface position across the x, y plane is displayed if the lower interface is rough on a scale much smaller than the exciton diameter. For GaAs/AlGaAs and GaAs/AlAs QWs this is the actual case as reported by Köhrbrück et al (1990) and Bimberg et al (1992).

Two different data evaluation approaches are used simultaneously for this application:

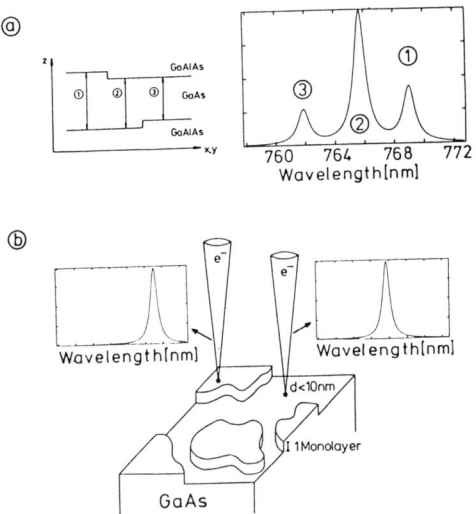

Fig. 1 Principle of CLI: (a) Origin of the a/2 splitting in laterally averaged quantum well spectra. (b) Schematic outline of the lower half of a AlAs on GaAs interface consisting of huge monolayer islands. At different positions of the exciting e-beam in the x-y-plane different CL spectra are observed.

(a) CL intensity maps integrated over various (in Fig. 1) preset wavelength windows around the three a/2 peaks are recorded together with a CLI integrated over the complete QW spectrum for normalization. The CLIs show different QW columns as bright areas in the intensity images obtained for QW widths equals to (n - 1), n, and (n + 1) monolayers.

(b)The CL peak wavelength $\lambda_x(x, y)$ of each single peak is used directly to calculate a histogram of the peak position distribution function.

For time-resolved experiments a beam-blanking unit is employed, located directly below the electron gun. Electron pulses of rise and decay time $\leqq$ 30ps, varying width of $\leqq$0.3ns up to 10µs, and a repetition rate of 1kHz up to 100MHz are used for excitation. An ultrafast Hamamatsu photomultiplier is used for detection in conjunction with a photon counting system operating in an inverted mode, yielding a typical dynamic range of 10^5. The experimentally measured time resolution of the complete system is $\geqq$ 35ps. The data acquisition technique used by us has the advantage that simultaneously up to 14 windows of different widths can be set in order to optimize independently the signal to noise ratio. Thus, up to 14 spectra taken at different time delays with respect to the exciting pulse can be recorded simultaneously and repetitive wavelength scans are not necessary (Bimberg et al 1986).

Three examples shall now demonstrate power and versatility of CLI.

3. STRUCTURE OF GaAs/Al(Ga)As QUANTUM WELL INTERFACES

In general, the two interfaces of a quantum well are inequivalent (Köhrbrück et al 1990). For GaAs/Al(Ga)As/GaAs QWs this inequivalence has been investigated in detail and is now known to account for part of the difficulties with inverted HEMTs. For a standard substrate temperature of 620°C and a growth rate of 1µm/h used to fabricate transistors by molecular beam epitaxy the lower interface (GaAs grown on Al(Ga)As) does not show any formation of islands larger than the exciton diameter, still after an interruption of the growth at the interface of 300s (Köhrbrück etal 1990). The upper interface (Al(Ga)As on GaAs) on the other hand, shows formation of islands much larger than the exciton diameter, and sometimes larger than the exciton diffusion length, after an interruption of as little as 30 s

(Bimberg et al 1986). The step height between such islands is close to or identical to one monolayer, depending on the parameter set used for the modeling (Kopf et al 1991). Consequently, the quantum well can be viewed as having a columnar structure. The number of columns of distinct height (2, 3, . . .) can be directly read from CL spectra taken at low magnification. Each column manifests itself by one luminescence line as schematically shown in Fig. 1. CL images may now be used to reconstruct the complete structure of the QW. Such a reconstruction for a 1.5 nm GaAs/Al$_{0.4}$Ga$_{0.6}$As QW is shown in Fig. 2 (Christen et al 1991). The e-beam was scanned here across an area of 6 µm x 12 µm.

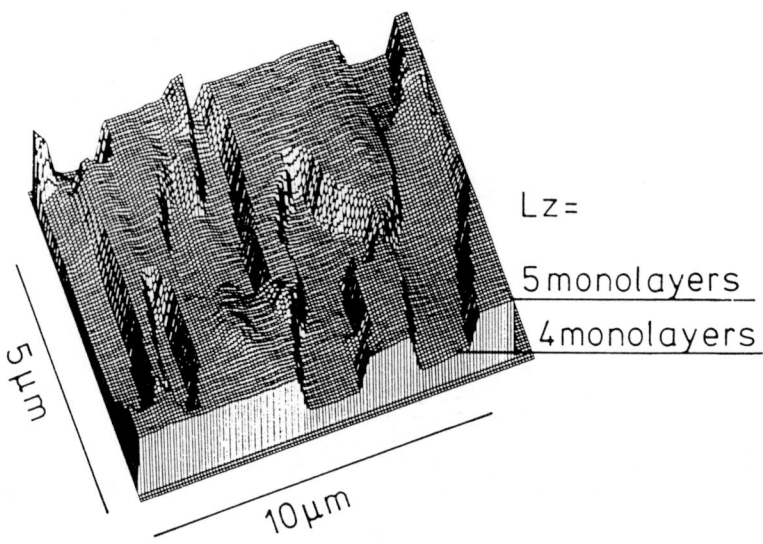

Fig. 2 3D picture of the morphology of the top AlGaAs/GaAs heterointerface of a GaAs QW as sensed by the exciton.

Additional information on the island distribution occurring at a much smaller scale at the lower interface can be obtained by deconvoluting the luminescence line shape and extracting the Gaussian broadening (Christen and Bimberg 1990).

The question how "smooth" such islands at the upper interface (Al(Ga)As on GaAs) can be is presently a subject of controversy (Kopf et al 1991, Warwick et al 1990). In order to avoid ambiguities of the interpretation inherent to ternary barriers we have recently (Bimberg et al 1992 and 1993) investigated the binary/binary system AlAs/GaAs. Composition fluctuations in the ternary barriers (or the ternary well in systems like InGaAs/InP) at the interface lead to the same type of intermixing of atoms in a given lattice plane as structural roughness does. Thus, a structurally flat ternary interface can appear atomically rough due to composition fluctuations. Thus, a precise identification of the interface position and determination of roughness on an atomic scale is not possible in such cases. The AlAs barriers have the additional technical advantage of showing a larger contrast in the chemical lattice imaging mode of transmission electron microscopy (Bimberg et al 1992). Temperature dependent CL-spectra of a 5 nm QW show on a linear scale a doublet structure at any temperature between 5 K and 300 K with varying relative intensities of the components.

The only present meaningful spectroscopic method to quantify "smoothness" is to take histograms of the luminescence peak wavelengths of the components of a multiplet which are characteristic for the columns of a QW (Christen 1990). Fig. 3 shows such a histogram for the 5 nm QW just mentioned. The histogram is constructed from 51200 spectra taken across an area of 12 µm x 6 µm.

Histograms taken at various parts of the sample and various magnifications yield almost identical results. The standard deviation σ_E of the two histograms is identical, 0.3 $\pm$ 0.08 meV, equivalent to 0.02 monolayers. This value presents actually an upper limit of the true standard deviation. Calculation of such histograms is one of the unique software features of our CLI system. High resolution transmission electron micrographs of the same sample confirm the high perfection of the upper interface of the structures investigated here across the full range of view ($\approx$ 110 nm) (Bimberg et al 1992).

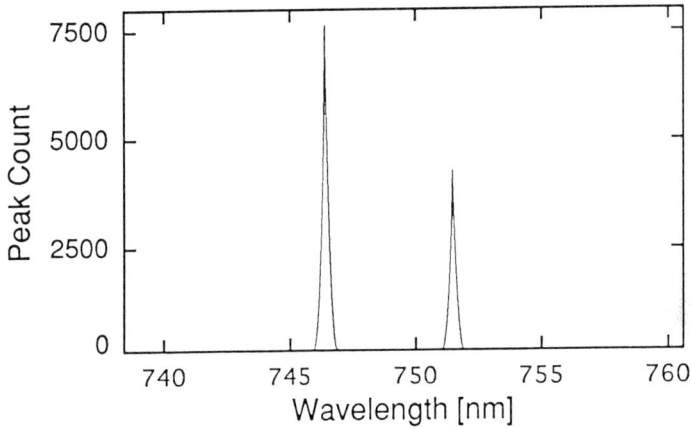

Fig. 3 Histogram of the peak wavelength distributions of a GaAs QW doublet luminescence caused by monolayer steps at the upper interface

4. CARRIER CAPTURE AND RECOMBINATION IN GaAs QUANTUM WIRES

One-dimensional charge carriers in quantum wires (QWRs) are predicted to exhibit unique optical properties as compared to two - or three - dimensional charge carriers. These properties include strongly modified density of states, exciton binding energies, gain spectra etc. attractive for novel photonic devices offering new functions or improved performance (Miller et al 1988, Arakawa and Yariv 1986). An actual QWR structure has to meet a number of requirements in order to really exhibit the desired properties (Kapon et al 1992). Most important is the formation of 1D subbands with intersubband splittings $\Delta E >> kT$ and energy broadening of a single subband $\sigma E << \Delta E$.

Thus, the wires must be sufficiently narrow and uniform. For photonic devices like lasers the wire interfaces should be free of defects for maximum internal quantum efficiency and the carriers injected from the surrounding barriers must be rapidly captured to ensure a large cut-off frequency.

The present most advanced technological concept was proposed 1987 by Kapon et al. This approach is based on in-situ formation of uniform arrays of QWRs as narrow as 10 nm during MOCVD on grooved substrates. Fig. 4 shows a TEM cross-section of such a vertically stacked GaAs/AlGaAs QWR array (Kapon 1992). Carrier capture into the wires after external excitation or injection occurs either via the GaAs QWs on the (111)A side planes of the groove or via a vertical AlGaAs QW having a lower Al content than the neighboring areas. Plan view and edge CLIs show spatially well resolved emissions from the different areas at appropriate wavelengths (Christen et al 1992). Fig. 5 shows time-delayed CL spectra of a single wire structure having a pitch of 240 nm. The wire has a center thickness of 2-8 nm and the confining potential in y-direction for the n = 1 electron state is 14 nm, yielding a n = 1/n = 2 electron subband separation of 23 meV. In this structure the largest distance to travel to the wire for an externally generated carrier is 120 nm.

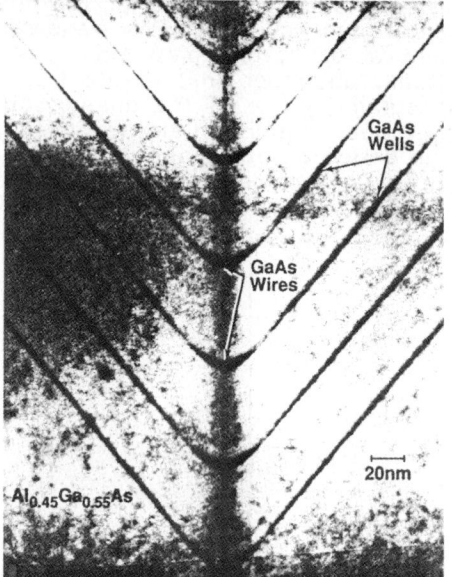

Fig. 4 TEM cross-section of a vertical stack of quantum wires in a V-groove

The CL spectrum is completely dominated by the emission from the wire showing a strong Gaussian broadening. The faint features at 612-618 nm and 640 nm are caused by recombination in the horizontal QWs at the top of the structure and the vertical AlGaAs QWs, respectively. From the intensity ratio I_{wire}/I_{wells} a capture time of < 100 fs and a carrier velocity of 6×10^6 cm/s, close or larger than the drift saturation velocity of holes is estimated.

Fig. 5 Time-delayed spectra (all scaled to the same maximum) visualizing the thermalization of the 1D carriers in QWRs.

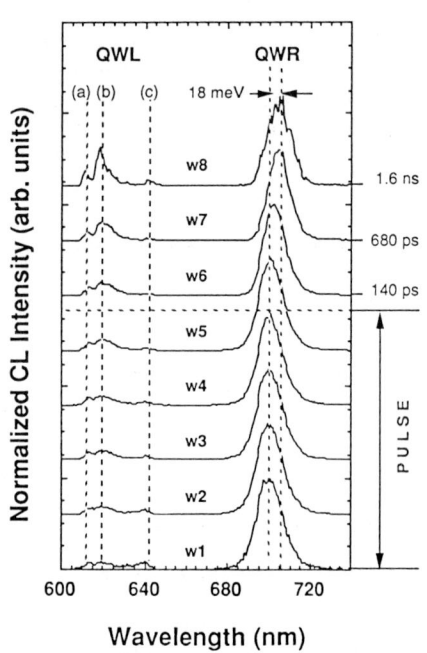

The time delayed spectra show indeed an immediate onset of wire recombination. A minority carrier decay time of 310 ps in the wire, indicative of the high structural perfection is obtained from the transients. This is the largest decay time ever reported for a structure of this size. While the peak position of the QW luminescence remains fixed during the observed time interval a red shift of 18 meV and some substructure is observed for the wire luminescence. The origin of this shift is a relaxation in real space. Due to of geometrical fluctuation, the 1D carriers within a wire thermalize into areas of largest width and lowest energy. CLI top views directly visualize these size fluctuations.

5. IMAGES OF STRAIN-INDUCED ONE-DIMENSIONALLY CONFINED CARRIERS IN "EPITAXIAL LIFT-OFF" QWs

Upon application of inhomogeneous strain on semiconductor structures local potential minima form and zero- or one-dimensional carrier localization (Kash et al 1988, Gershoni et al 1990) occurs.

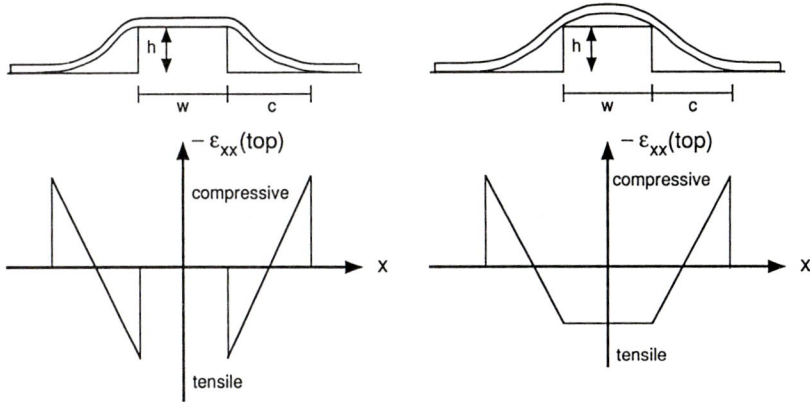

Fig. 6 Schematic view of the flat-top (a) and the arched (b) attachment of a 100 nm thin QW GaAs film to a Si rib and the resulting stress distribution

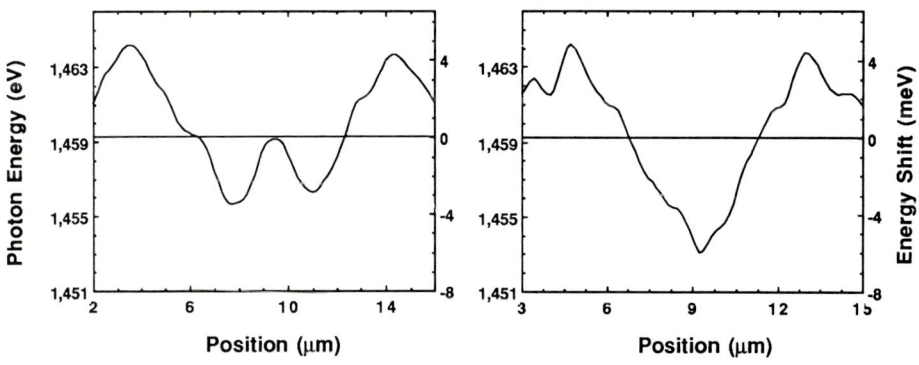

Fig. 7 Linescans of the CL QW peak emission energy for the flat top (left) and arched case (right)

In inhomogeneously strained Ge discs a new type of electron-hole liquid was discovered at low temperatures almost two decades ago (Wolfe et al 1975, Bimberg and Störmer 1976). Most recently Chan et al 1992 reattached a GaAs/AlGaAs single QW structure which was released from the substrate via etching selectively an AlAs release layer (Yablonovitch et al 1987) to a Si substrate having 2 μm wide and 1.6 μm high ribs. The 100 nm thick film bends either archlike across the rib, touching the rib only at its edges or attaches flat on the rib top surface. Cross-sections of the stress distributions for these two fundamentally different cases are shown in Fig. 6. The resulting bandgap variations of the QW in the x-direction vertical to the rib axis leads to the formation of wirelike potential minima along the rib axis as shown in Fig. 7. The extension of the minima in x-direction in the cases investigated until now, however, is presently still too large to lead to subband splittings as big as reported for the wires fabricated on V-grooved substrates. CLI visualize directly the band gap modulation and the wirelike emission. Fig. 8 shows a plan view CL wavelength image for the flat top case.

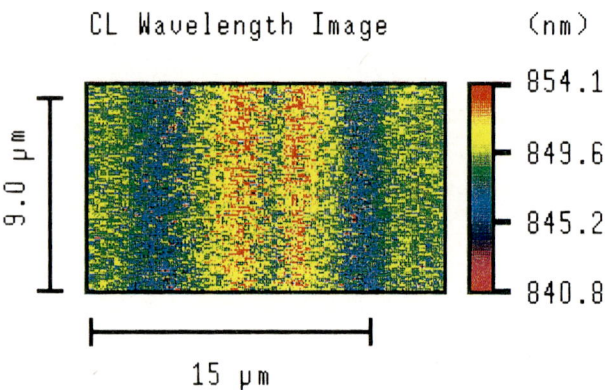

Fig. 8 CLI of the peak emission wavelength for the flat top case

6. CONCLUSION

Cathodoluminescence imaging of semiconductor nanostructures presently provides a lateral resolution $\leq$ 100 nm to detect variations of electronic or structural properties with the prospect of further improvement by a factor 3-5. Time resolution $\leq$ 35 ps is easily obtained. Thus, carrier diffusion does not deteriorate the lateral resolution.

Three recent investigations of nanostructures using CLI demonstrate its power and versatility:

- The top interface of GaAs/AlAs quantum wells grown with growth interruption shows single atomic layer steps across distances of several 100 nm. In between such steps the roughness of the interface is 0.02 monolayers as derived from peak wavelength histograms calculated from more than 50,000 single spectra.
- Time delayed and time resolved spectra of 2.5 nm x 14 nm quantum wires fabricated on V-grooved GaAs substrates show sub ps carrier capture into the wires and long 310 ps decay times demonstrating the superiority of this self-organized in-situ approach to nanofabrication.
- Wavelength images of GaAs quantum wells epitaxially lifted-off from the substrate and redeposited onto Si substrates with 1 μm wide ribs visualize directly locally varying stress, QW ground state energy variations and the formation of wirelike structures.

ACKNOWLEDGMENTS

This work was performed in collaboration with M. Grundmann, M. Joschko, and W. Wittke, TU Berlin. D. E. Mars and J. N. Miller, HP Labs Palo Alto, and E. Kapon, D. M. Hwang, W. K. Chan, T. S. Ravi, T. Gmitter, and L. Florez, Bellcore.

REFERENCES

Arakawa Y and Yariv A 1986 IEEE J. Quant. Electr. QE22 1887
Bauer R K, Oertel D, Mars D E and Miller J N 1986 J. Vac. Sci. Technol B4/1014
Bimberg D, Christen J, Wittke W, Gerthsen D, Stenkamp D, Mars D E
    and Miller J N 1993 Phys. Rev. to be published
Bimberg D, Heinrichsdorff F, Bauer R K, Gerthsen D, Stenkamp D, Mars D E and
    Miller J 1992 J. Vac. Sci. Technol B10 1793 and err. B11 126
Bimberg D, Münzel H, Steckenhorn A, Christen J 1985 Phys. Rev. B 31 7788
Bimberg D and Störmer H L 1976 Il Nuovo Cimento 39B 615
Bimberg D, Robbins D J. Wight D R and Jeser J P 1975 Appl. Phys. Lett. 27 67
Chan W K, Ravi T S, Kash K, Christen J, Gmitter T J, Florez C T and Harbison J P
    1992 Appl Phys. Lett 61 1319
Christen J, Kapon E, Grundmann M, Hwang D M, Joschko M and Bimberg D 1992
    phys. stat. sol. (b)173 307
Christen J, Grundmann M and Bimberg D 1991 J. Vac. Sci. Technol. B9 2358
Christen J 1990 Festkörperprobleme 30 ed. U Rössler (Darmstadt: Vieweg) 239
Christen J and Bimberg D 1990 Phys. Rev. B42 7213
Gershoni D, Weiner J S, Chen S N G, Baraff G A, Vandenberg J M, Pfeiffer L N,
    West K, Logan R and Tanbuk- Ek 1990 Phys. Rev. Lett. 65 1631
Grundmann M, Christen J and Bimberg D 1991 Superlatt. and Microstr. 9 65
Kapon E, Walther M, Christen J, Grundmann M, Caneau C, Hwang D M, Colas E,
    Bhat R, Song G H and Bimberg D 1992 Superl. and Microstr. 12 491
Kapon E, Tamargo M C and Hwang D M 1987, Appl. Phys. Lett. 50 347
Kash K, Worlock J M, Sturge M D, Grabbe P, Harbison J P, Scherer A and Lin P S D
    1988 Appl. Phys. Lett. 53 782
Köhrbrück R, Munnix S, Bimberg D, Mars D E and Miller J N 1990
    J. Vac. Sci. Technol B4 798 and Appl. Phys. Lett. 57 1025
Kopf R F, Schubert E F, Harris T D and Becker R S 1991 Appl. Phys. Lett 58 631
Miller D A B, Chemla D S and Schmitt-Rink S 1988 Appl. Phys. Lett. 52 2154
Petroff P M, Cho A Y, Reinhart F K, Gossard A C and Wiegmann W 1982
    Phys. Rev.Lett. 48 170
Steckenborn A, Münzel H and Bimberg D 1980 BEDO 13 157
Warwick C A, Jan W Y, Ourmazd A and Harris T D 1990 Appl. Phys. Lett 56 2666
Wolfe J P, Hansen W L, Haller E E, Markiewicz R S, Kittel C and Jeffries C D 1975
    Phys Rev. Lett 34 1292
Yablonovitch E, Gmitter T, Harbison J P and Bhat R 1987 Appl. Phys. Lett 51 2222

Inst. Phys. Conf. Ser. No 134: Section 11
Paper presented at Microsc. Semicond. Mater. Conf., Oxford, 5–8 April 1993

639

Cathodoluminescence studies of quantum-well wires grown on patterned substrates

A Gustafsson, L Samuelson, J-O Malm*, G Vermeire** and P Demeester**.

Department of Solid State Physics, Lund University, Box 118, S-221 00 Lund, Sweden. *National Center for HREM, Department of Inorganic Chemistry 2, Lund University, Box 124, S-221 00 Lund, Sweden. **University of Gent-IMEC, Laboratory of Electromagnetism and Acoustics, St.-Pietersnieuwstraat 41, B-9000 Gent, Belgium.

ABSTRACT: We have studied a GaAs/AlGaAs quantum-well wire (QWW) structure grown directly on a substrate with a sub-micron grating. Two cathodoluminescence (CL) peaks are the CL of the planar quantum-well (QW) and the AlGaAs barriers. A third peak is related to the growth on the grating and originates in a thicker region of the QW, forming a QWW. In the same lateral position the barrier forms a vertical quantum well (VQW), resulting in a fourth peak. The identification was confirmed by the cross sectional transmission electron microscope (XTEM) and cathodoluminescence (CL) images.

1. INTRODUCTION

In recent years there has been an increasing interest in low-dimensional semiconductor structures. Novel physical phenomena related to the reduced dimensionality are expected to occur, and have to some extent been demonstrated (Linh 1983). These kinds of structures have a great potential of enhancing the performance of photonic devices. Several approaches to accomplish these structures have been proposed, with varying success (Fukui and Saito 1987, Izrael et al 1990). A special interest has been focussed on growth of QWWs on substrates containing V-grooves (Kapon et al 1989). Multi layer QWW structures with 3.5 µm QWW spacing have previously been characterized by CL imaging and XTEM (Christen et al 1992) and single layer QWW structures on sub-micron spacing have been studied by photoluminescence spectroscopy (Walther at al 1992). Using high resolution XTEM and CL we have studied a QWW structure containing a single QW layer, grown by metalorganic vapour phase epitaxy (MOVPE) on a substrate containing a sub-micron grating.

2. GROWTH

For the growth of the structures a horizontal MOVPE reactor was used at atmospheric pressure. The substrates used for the patterning were (100) GaAs substrates. The gratings were defined by holographic lithography in photoresist. This was followed by wet chemical etching in citric acid, resulting in about 300 nm deep groves. The period of the grating was 765 nm and the grooves were oriented along the [01$\bar{1}$] direction. Maintaining the groove structure, the grating was overgrown by $Al_xGa_{1-x}As$, with a nominal composition of x=0.35. This AlGaAs layer was grown to a thickness of 300 nm and on top of this a thin, 3.4 nm, GaAs QW was deposited. Subsequently, an AlGaAs layer of 250 nm was grown on top of the QW. To serve as a reference, a planar GaAs substrate was simultaneously placed in the reactor during growth.

Due to an increased growth rate of the GaAs at the bottom of the grooves, the QW is significantly thicker in these regions and, thus, the confinement in the growth direction is less. Furthermore, the Ga and Al have different migration properties on the growth surface. Therefore there is a tendency for the barrier to be Ga-rich at the bottom of the groove during growth. This leads to a stripe of material with lower band gap in the growth direction. A schematic picture of the structure is shown in figure 1a. The fabrication of the QWW structures have been described in detail elsewhere (Vermeire et al 1992).

3. CHARACTERIZATION

The XTEM was performed in a JEM-4000EX, operated at 400 kV. The sample was viewed along the [01$\bar{1}$] direction. The CL measurements were performed in a dedicated SEM, with the sample cooled to 25K. The SEM was operated at 6.5 kV using a probe current of about 100 pA. The CL was dispersed by a monochromator and detected by a GaAs photomultiplier. The signal was recorded digitally as spectra, monochromatic images and monochromatic linescans. A more detailed description of the experimental setup has been given previously (Gustafsson 1991)

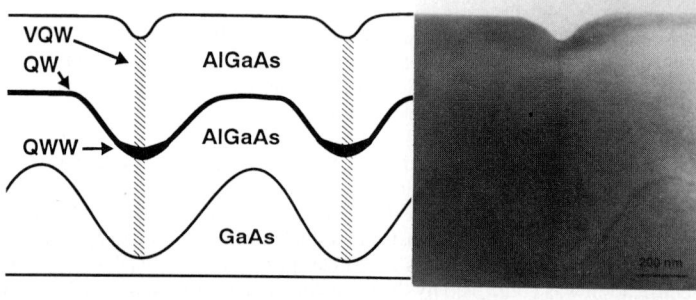

Figure 1. Left) A schematic sketch of the QWW structure. On a GaAs substrate an AlGaAs layer is grown, followed by a thin GaAs QW and a top AlGaAs layer. Due to the different migration of the Al and Ga on the surface during growth of the barrier, the bottom of the groove tends to be Ga-rich, forming a VQW. The growth rate of GaAs is increased at the bottom of the groove and therefore the QW is thicker there, forming a QWW. Right) A low magnification XTEM image of the structure, exhibiting the corrugation of the substrate and a dark stripe, corresponding to the VQW.

4. RESULTS

Figure 1b shows a low magnification XTEM image of the structure. Extending from the bottom of each groove in the barrier, a well defined stripe of material of higher Ga content can be observed. This forms a vertical quantum well (VQW). At this low magnification the QW is barely visible. Figure 2 shows a high resolution image of the QW in the position of the bottom of the groove. Here the

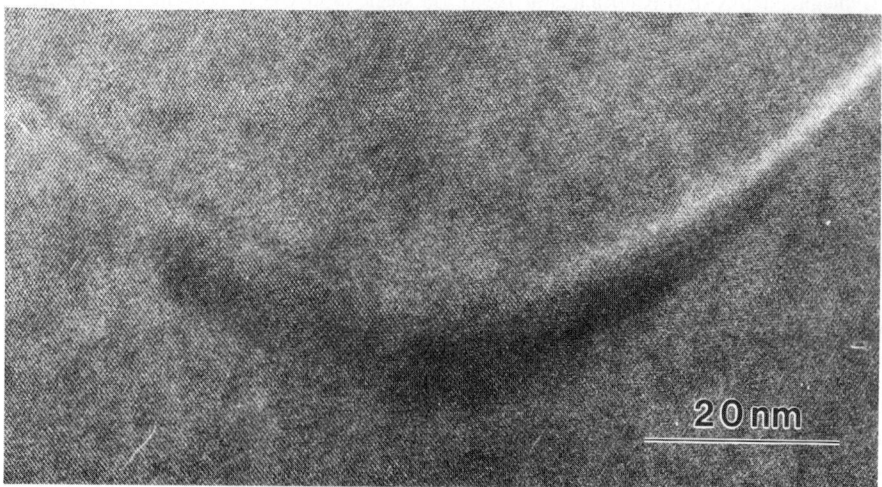

Figure 2. A high resolution XTEM image showing the thicker region of the GaAs QW (dark) at the bottom of the groove, the crescent shaped QWW. The image is formed at an identical groove as shown in figure 1.

QW is thicker, as discussed above, forming a crescent shaped QWW.

Figure 3 displays a typical CL spectrum of the sample and it exhibits four well defined peaks originating in the patterned sample. Two of the peaks were identified by comparing their energy position with those of a sample grown simultaneous on a planar substrate, as their energy positions coincide in the two cases. These are the peaks of the nominal QW and the AlGaAs barriers. A third peak appears at a lower energy than the QW, and is identified as the thicker region of the QW, the QWW. The fourth peak appears at an energy position between the peak of the QW and the peak of the barrier. This is identified as the Ga-rich part of the barrier, i. e. the VQW.

In figure 4, plan view CL images of two of the four peaks originating in the sample are shown for the same area. At the low accelerating voltage used here, 6.5 kV,

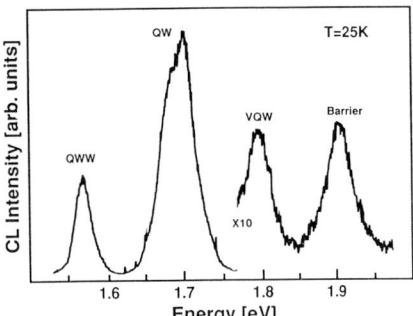

Figure 3. A typical CL spectra of the sample in plan view at 6.5 keV. Four peaks can be observed and identified. At 1.91 eV the peak of the nominal AlGaAs barrier is seen and at 1.70 eV the peak of the nominal QW can be seen. A peak at a lower energy, 1.57 eV, is identified as the QWW. Finally, the peak of the VQW is positioned in between the nominal barrier and QW, at 1.79 eV.

the resolution is high enough to clearly reveal the striped character of the sample for all four peaks involved, even though the resolution is deteriorated by diffusion of the generated carriers. The

Figure 4. The CL images exhibit the same striped character for all of the four peaks originating in the sample. The left image is formed using the emission from the QWW and the right is from the QW. An analysis of the lateral position of intensity variations shows that the variations of the QWW and the VQW coincide and that they are fully out of phase with the variations of the QW and the barrier.

images of the CL from the QWW show the highest intensity variations and they are inverse to the corresponding variations in the images using the CL from the QW. This is illustrated in more detail in figure 5, showing a

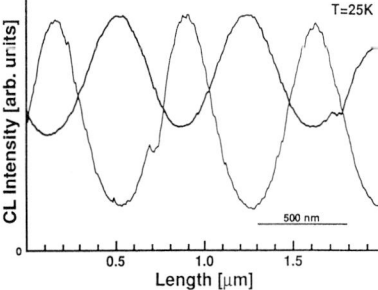

Figure 5. A linescan showing the intensity variations perpendicular to the direction of the grooves. The linescan of the CL from the QW (thick) is fully out of phase with the CL from the QWW (thin). A strong indication of their separate lateral positions.

Figure 6. Cross sectional CL images using the CL from the QWW (left) and the VQW (right) respectively. Taking diffusion into account, the image of the QWW exhibits point like intensity variations, whereas the VQW image shows variations that are more elongated along the growth direction. The latter is an indication that the signal origins in an area that is more elongated, which supports the assumption that the Ga-rich areas of the barrier act as VQWs. The stripes in the images are artifacts of the instrument.

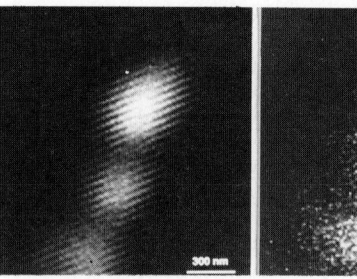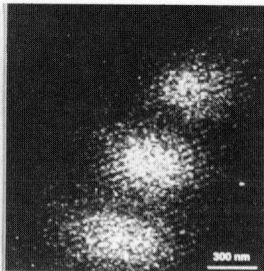

linescan of the intensity variations of the two peaks. Furthermore, the intensity variations of the CL of the QWW and the VQW coincide, thus confirming that their lateral position is the same. Finally, the CL images of the VQW and of the barrier are also found to be complementary. The conclusion of the images is that they support the proposed identification of the different peaks (Vermeire et al 1992).

To further investigate the sample we have performed cross sectional CL imaging. Due to the geometry of the QWW structure, we have been able to image and characterize individual QWWs and VQWs. Figure 6 shows typical images of the QWW and the VQW. The most interesting feature of these images is the shape of the bright areas. The QWW seems to consist of point like luminescence whereas the VQW exhibits luminescence elongated along the growth direction. All in accordance with the interpretation of luminescence and the XTEM images.

5. SUMMARY

We have presented high resolution XTEM and CL measurements of a QWW structure grown on a substrate patterned with a sub-micron grating structure. The CL consists of four well defined and separated peaks. We have been able to identify them as the nominal QW and barrier as well as the additional features of the patterning. A thicker region of the QW in the bottom of the grooves forms a crescent shaped QWW and a well defined Ga-rich stripe of the barrier material, forms a VQW, aligned with the QWW at the bottom of the groove. We have presented a high resolution XTEM image of the crescent shaped QWW. Using CL we have imaged single QWWs and VQWs in cross section. We have also presented plan view CL images of single QWWs and the surrounding QW.

Acknowledgements: The Swedish part of this work was performed within the Nanometer Structure Consortium in Lund with financial support by NUTEK, NFR and TFR. The Belgian part was partly supported by the ESPRIT project BRA 3133 NANSDEV and by IWONL. The authors are thankful to Dr. S. Nilsson and Dr. P. Van Daele for helpful discussions.

REFERENCES

Christen J, Kapon E, Colas E, Hwang D M, Schiavone L M, Grundman M and Bimberg D 1992 Surface Science 267 257

Fukui T and Saito 1987 Appl. Phys. Lett. 50, 824

Gustafsson A 1991 Thesis Lund University.

Izrael A, Sermage B, Marzin J Y, Ougazzaden A, Azoulay R, Etrillard J, Thierry-Mieg V, and Henry L 1990 Appl. Phys. Lett. 56, 830

Kapon E, Hwang D M and Bhat R 1989 Phys. Rev. Lett. 63, 430

Linh N T 1983 in *"Festkörperprobleme XXIII, Advances In Solid State Physics"* (edited by H. J. Queisser, Pergamon press Vieweg

Vermeire G, Yu Z Q, Vermaerke F, Buyens, Van Daele P and Demeester P 1992 J. Crystal Growth 124 513.

Walther M, Kapon E, Hwang D M, Colas E and Nunes L 1992 Phys Rev B45 6333

Inst. Phys. Conf. Ser. No 134: Section 11
Paper presented at Microsc. Semicond. Mater. Conf., Oxford, 5–8 April 1993

Measurements of the diffusion length in GaAs/AlGaAs quantum well (QW) structures by cathodoluminescence: comparison of the bulk and QW diffusion channels

D Araújo[1,3], G Oelgart[2], J-M Bonard[1], J-D Ganière[1] and F K Reinhart[1]

[1] Institut de Micro-et Optoélectronique, Ecole Polytechnique Fédérale de Lausanne,
    1015 Lausanne, Switzerland
[2] Fachbereich Physik der Universität Leipzig, Linnésstr. 5, 7010 Leipzig, Germany
[3] Departamento de Ciencia de Materiales, Universidad de Cádiz, Apartado 40,
    11510 Puerto Real, Spain

ABSTRACT: In a recent paper, we measured the hole diffusion length in a single quantum well (SQW) by a novel method using SEM-cathodoluminescence, assuming that the carriers diffuse only in the SQW. This assumption needs to be checked. Therefore, we present here a model for the carrier transport in the SQW and in the bulk barrier to deduce the lateral hole distribution in both layers. The main lateral transport is shown to be carried by the holes in the QW (2D diffusion) due to their confinement in the well.

1. INTRODUCTION

Investigations of the transport properties of carriers parallel to heterointerfaces have attracted considerable interest, not only because of the related physics but also because of the possible device applications. The transport properties of a semiconductor are usually evaluated by the minority carrier diffusion length L of the material. As L depends directly on the mobility and the lifetime of the carriers, it is possible to deduce the diffusion length by measuring separately both parameters. The mobilities can be obtained by Hall measurements (Guillemot 1987, Gottinger 1988, Sakaki 1987), and the recombination lifetimes by time-resolved cathodoluminescence (Bimberg 1987) and/or time of flight (TOF) measurements (Höpfel 1988, Hattori 1987, Hillmer 1988). A second and far more attractive method consists of the direct determination of the diffusion length, for example by cathodoluminescence (CL) (Zarem 1989, Araújo 1993) or electron-beam induced current (EBIC) (Leamy 1982).

In a previous paper (Araújo 1993), we studied a GaAs/AlGaAs SQW structure, measuring among other parameters the diffusion length. The unintentionally n-doped structure (Si: $\sim10^{16}$cm^{-3}) consisted of a GaAs/Al$_{0.5}$Ga$_{0.5}$As SQW with a 20Å width. The thickness of the GaAs buffer layer and of the AlGaAs barriers were 0.5 and 1.2µm, respectively. The structure contained grown-in dislocations that originated at the GaAs substrate. We analysed the free exciton recombination in the SQW by photoluminescence (PL) and CL. Thickness variations of SQW, due to the growth peculiarities near the dislocations, were observed using CL micrographs at selected wavelengths at 5K. Impurities and native defects were also found around the dislocations. This local change of the optical properties of the material occurred in a region of about 1µm around the dislocation. In this paper, we now compare the contribution of the hole diffusion in the barrier and in the SQW plane near the dislocation region by modelling the lateral carrier diffusion in both regions.

2. EXPERIMENTAL RESULTS AND DISCUSSION

The total diffusion length can be measured by scanning with the beam of the SEM across a dislocation and by measuring the variation of the CL intensity with the position (see Fig. 1). The dislocation region acts then as a local detector (provided that only the CL due to the impurities present around the dislocation is considered for the measurements). Assuming that the diffusion mainly takes place in the SQW, we can compare the measured values to theoretical predictions, and we deduced in (Araújo 1993) $L_{SQW}=1.5\mu m$. This assumption needs to be checked, since the measured diffusion length is obviously the result between two competing processes: (i) the hole diffusion in the SQW plane itself and (ii) the hole diffusion in the barrier followed by diffusion and recombination in the SQW.

Therefore, we estimated the diffusion length in the barrier, by scanning with the beam across the SQW and measuring the variation of the CL of the SQW. By comparing again the measured values to theoretical predictions, we concluded that $0.5\mu m<L_{barrier}<1\mu m$. Since these two diffusion lengths are comparable, our assumption can hold only if we can prove that the 2D diffusion in the SQW (2D because the holes are confined in the SQW) induces a far higher hole concentration than the one induced by the 3D diffusion in the barrier. To compare hole lateral distribution both in the QW and in the barrier, we have to solve the steady state equation for both cases in the Dirac-generation approximation.

2.1   Hole Diffusion in the QW

A simple model of carrier diffusion in a cylindrical system should describe the hole concentration in the SQW plane. For the low injection regime, we can write the steady state equation for the hole concentration, p, in the n-doped SQW:

$$\frac{\partial p}{\partial t} = 0 = D\left(\frac{\partial^2 p}{\partial \rho^2} + \frac{1}{\rho}\frac{\partial p}{\partial \rho}\right) - \frac{p}{\tau} + g(\rho) \qquad (1)$$

where the first term is the hole diffusion term in cylindrical coordinates, the second term the recombination rate, and the third term the generation rate. D is the diffusion constant in $[cm^2/s]$ and τ is the recombination lifetime in [s].

To simplify eq.(1), we approximate the e-h pair generation $g(\rho)$ as a Dirac function. Relating ρ with the diffusion length $L=\sqrt{Dt}$, we introduce a new variable $r=\rho/L$. The exact solution of eq.(1) is then:

$$p(r) = C\ K_0(r) \qquad (2)$$

where $K_0(r)$ is the modified Bessel function of zero order and C a constant. The asymptotic development of the solution is:

$$p(r) \rightarrow \frac{1}{\sqrt{2\pi r}}\ e^{-r} \qquad (3)$$

2.2   Hole Diffusion in the Barrier

To describe the lateral hole distribution in the barrier, we make the following assumptions: (i) all the carriers which reach the well fall in it; this ensures that $p(\rho)=0$ on the QW plane, (ii) the surface recombination is neglected. This last assumption ensures that the estimation of $L_{barrier}$ then becomes an upper value as the surface recombination has the effect of decreasing the diffusion length. In the case of 3D-diffusion, eq. 1 becomes, in spherical coordinates:

$$\frac{\partial p}{\partial t} = 0 = D\left(\frac{\partial^2 p}{\partial \rho^2} + \frac{2}{\rho}\frac{\partial p}{\partial \rho}\right) - \frac{p}{\tau} + g(\rho) \qquad (4)$$

To assure that $p(\rho)=0$ on the SQW plane, we use a "mirror charge" solution:

$$p(x,y,z) = \frac{1}{\rho_1}e^{-\rho_1} - \frac{1}{\rho_2}e^{-\rho_2} \qquad (5)$$

The contribution of the hole diffusion in the barrier to the total hole diffusion should be maximal when the holes are generated at a certain distance from both the (001) surface and the SQW, hence minimising recombination at the surface and at the SQW. We estimate here the lateral concentration of holes in the case of a Dirac generation in the centre of the barrier, at $z_0=0.6\mu m$. Consequently, the obtained lateral hole distribution is an upper estimation. In this geometry, the coordinates ρ_1 and ρ_2 are expressed in cartesian coordinates as:

$$\rho_1 = \frac{\sqrt{x^2 + y^2 + (z-z_0)^2}}{L_{barrier}}$$

$$\rho_2 = \frac{\sqrt{x^2 + y^2 + (z+z_0)^2}}{L_{barrier}} \qquad (6)$$

with the $z=0$ plane defined by the SQW plane, and ρ_2 representing the "mirror" contribution.

2.3 Comparison of the Hole Diffusion in the QW and in the Barrier

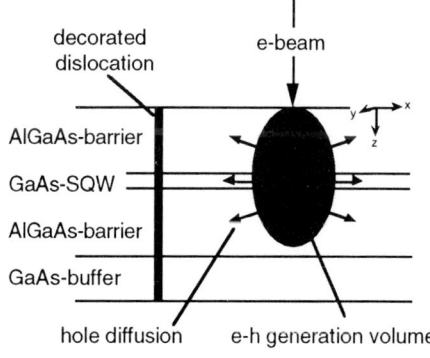

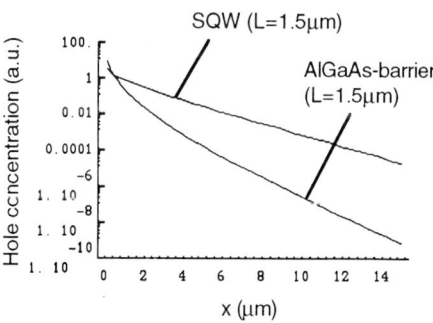

Fig.1: Geometry of the experiment, with the coordinates used in the paper. The different hole diffusion processes, 3D in the barrier and 2D in the SQW, are shown.

Fig.2: Lateral hole concentration in the barrier for $L=1\mu m$ and in the SQW for $L=1.5\mu m$. The abcissa represents the x-axis in fig. 1, with the electron beam on $x=0$.

Using the upper value $L_{barrier}=1\mu m$ and $L_{SQW}=1.5\mu m$, we display in Fig.2 the lateral hole concentration in the barrier and in the SQW, respectively.

The diffusion in the QW follows a $r^{-1/2}exp\{-r\}$ dependence (eq. 3), which leads to a larger extension of holes than the $r^{-1}exp\{-r\}$ dependence in the barrier (solution of eqs 4-6). In spite of the little difference between the hole diffusion length in the SQW and in the barrier, we see that the hole concentration extension is much larger in the case of 2D-diffusion. The calculated hole concentration in the barrier is found to be <0.6% for x > 5 μm of the concentration in the SQW.

3. CONCLUSION

From previous CL measurements, we have estimated the hole diffusion lengths in the barrier and in the SQW, the values being $L_{barrier}<1\mu m$ and $L_{SQW}=1.5\mu m$. To understand the contribution of the hole diffusion in the barrier to the total hole diffusion, we modelled the lateral carrier diffusion in both the AlGaAs-barrier and the GaAs-QW. We showed that the contribution of the diffusion in the barrier is less than 0.6%, for x > 5 μm, of the hole diffusion in the SQW. Therefore the QW is shown to be a very efficient channel for carrier transport compared to the classic bulk diffusion.

ACKNOWLEDGEMENTS

This work was made possible through grants from the Swiss National Science Foundation (n° 2.979-0.88). The authors want to thank G.Peter and B.Garoni for the technical support.

REFERENCES

Araújo D, Oelgart G, Ganière J-D and Reinhart F K 1993 Appl. Phys. Lett. **62**
Bimberg D, Christen J, Fukunaga T, Nakashima H, Mars D E and Miller J N 1987 J. Vac. Sci. Technol. B **5**, 1191
Gottinger R, Gold A, Abstreiter G, Weinmann G and Schlapp W 1988 Europhys. Lett. **6**, 183
Guillemot C, Baudet M, Gauneau M, Regreny A and Portal J-A 1987 Phys. Rev. B **35**, 2799
Hattori K, Mori T, Okamoto H and Hamakawa Y 1987 Appl. Phys. Lett. **51**, 1259
Hillmer H, Hansmann S, Forchel A, Morohashi M, Lopez E, Meier H P and Ploog K 1988 Appl. Phys. Lett. **53**, 1937
Höpfel R A, Shah J, Wolff P A and Gossard A C 1988 Phys. Rev. B **37**, 6941
Leamy H J 1982 J. Appl. Phys. **53**, R51
Sakaki H, Noda T, Hirakawa K, Tanaka M and Matsusue T 1987 Appl. Phys. Lett. **51**, 1934
Zarem H A, Sercel P C, Lebens J A, Eng L E, Yariv A and Vahala K 1989 J Appl. Phys. Lett. **55**, 1647

Inst. Phys. Conf. Ser. No 134: Section 11
Paper presented at Microsc. Semicond. Mater. Conf., Oxford, 5–8 April 1993

Inhomogeneities in multiple quantum wells investigated by SEM-cathodoluminescence

U Jahn, J Menniger, R Hey and C Frigeri*

Paul Drude Institut für Festkörperelektronik, Hausvogteiplatz 5-7, O-1086 Berlin
*MASPEC-CNR, Via Chiavari, 18/A, 43100 Parma

ABSTRACT: The lateral cathodoluminescence distribution of GaAs/Ga$_{1-x}$Al$_x$As-multiple quantum wells with a well thickness of 2nm grown by molecular beam epitaxy is inhomogeneous. The contrast between two neighbouring areas is caused by an average shift of the luminescence spectrum of about 15meV. This shift can be explained neither by well thickness variations nor by fluctuations of the Al-mole fraction of the barriersalone. At selected spots the shift amounts to 50meV or more. We compare the larger spectral splitting with the shift of spectra measured at and around oval defects.

1. INTRODUCTION

Cathodoluminescence (CL) spectroscopy combined with SEM is a powerful tool to image the lateral distribution of luminescence properties in quantum wells and superlattices with a spatial resolution of $\leq 1\mu$m and to determine spectra at selected spots (for instance at defects). In the past several authors established lateral inhomogeneous CL-intensity distributions at single quantum wells (SQWs) with a small well thickness ($L_z < 5$nm) and interpreted this in terms of well thickness fluctuations (see for instance Bimberg et al. (1987), Warwick et al. (1992)).

In the present paper we present results concerning CL-properties of GaAs/Ga$_{0.7}$Al$_{0.3}$As-multiple quantum wells (MQW). These results cannot be explained entirely by layer thickness variations. Therefore we discuss the influence of varying Al-concentration in the barrier.

2. EXPERIMENTAL

The investigated MQWs consisting of 10 periods of a 2nm thick GaAs well and a 18nm thick Ga$_{0.7}$Al$_{0.3}$As barrier were grown by MBE at 590°C with growth interruption on (100)-LEC-GaAs substrates misoriented by 2° towards (111)A. Two samples (S1 and S2) were grown under uniform growth conditions but at two different growth cycles. S1 contains a 3.5nm thick SQW in the midst of the 2nm-wells. S2 is characterized by an essentially higher density of oval defects (about 2×10^4 cm^{-2}) than S1.

The CL-spectra and spectrally resolved CL-images were measured at sample temperatures of 5K and 80K, respectively. The acceleration voltage of the electron beam was chosen to be 10kV and the beam current amounted to 1nA - 20nA, corresponding to an excess carrier density of about 5×10^{16}-10^{18} cm^{-3}. Additionally, TEM investigations were performed to determine thickness fluctuations and interface roughness.

3. RESULTS

The FWHM of the CL-spectra of the 2nm-MQWs amount to 38meV for S1 and 50meV for S2. These linewidths are essentially defined by an effective interface roughness of 1 - 2 monolayers (ML), as revealed by TEM-investigations. Similar to the case of SQWs the lateral distribution of the spectral properties is inhomogeneous on a length scale of 1 to several μm. Fig.1 shows two CL-spectra measured at neighbouring spots (distance about 2μm) of S1. The spectra are shifted by 13meV. Measurements at many spots revealed an average spectral shift between neighbouring areas of 13-15meV. This spatial dependence of the peak energy is responsible for the contrast in the CL-images,

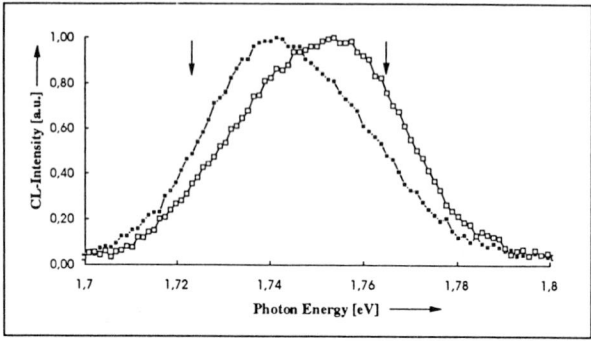

FIG.1 CL-spectra of the 2nm-MQW masured at two neighbouring spots (distance 2μm) of sample S1 at 5K. The arrows mark the photon energies chosen for CL-imaging in Fig.2.

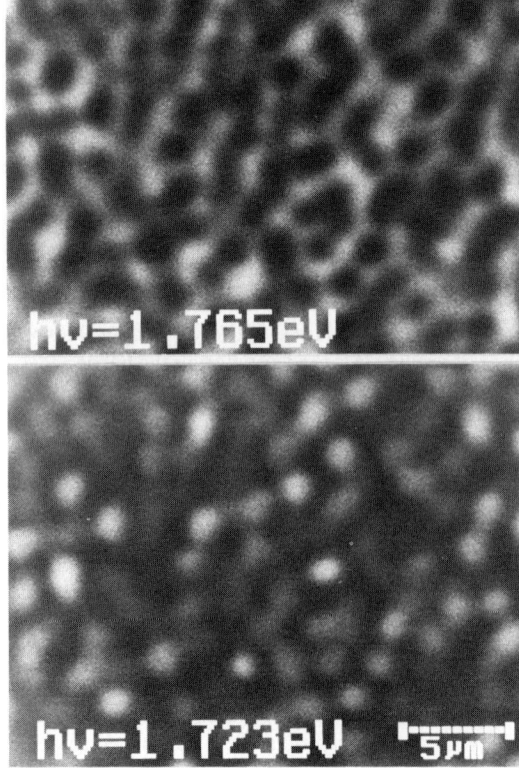

shown for example in Fig.2. The two micrographs image the same specimen region at two different photon energies (arrows in Fig.1). It is remarkable that within an area of about $20 \times 20 \mu m^2$ the bright/dark structures of the CL-intensity distribution at the two different wavelength are exactly complementary. The spectral shift and hence the intensity difference between the dark and bright areas depends strongly on the well thickness L_z. Fig. 3 shows two CL-images reflecting the same region of the sample S1 at the photon energy corresponding to the 2nm-MQW (a) and the 3.5nm-SQW (b). While the image of the 2nm-MQW exhibits clearly the CL-inhomogeneities, the contrast in b) is much lower at comparable signal height.

On S2 we found additionally to the small spectral shift a much larger shift or splitting of the spectra from neigh-

FIG.2. Spectrally resolved CL-images of the 2nm-MQW (sample S1) at two different photon energies (arrows in Fig.1) measured at 5K. The specimen region is the same in both cases.

bouring areas of about 50meV and more at selected spots (see Fig. 4a)).
In Fig.4b) spectra measured close to an oval defect are represented. The peak energies shift by 58meV going from spot 1 to spot 3.

4. DISCUSSION

There is now some evidence, that real interfaces consist of islands with diameters ≥ 1 μm and a substantial roughness on an atomic scale (Warwick 1992). This can cause μm large bright/dark structures in spectrally resolved CL-images of SQWs. In the case of complex MQW layer systems, however the interpretation of μm large CL-intensity inhomogeneities is not so evident, taking into account the above results.

The spectral shift of $\Delta E \simeq 15$meV indicates a fluctuation of the average well width ΔL_z by nearly 0.6ML on a μm-length scale. In the case of the 3.5nm-SQW we expect a smaller spectral shift: $\Delta L_z = 0.6$ML corresponds to $\Delta E = 8$meV. Because of the narrower linewidth of the corresponding spectrum the resulting contrast should be comparable to that of the 2nm-MQW. This conclusion clearly contradicts the experimental result of Fig.3. Hence we are looking for other influences especially those stemming from the whole layer package.

Since $Ga_{1-x}Al_xAs$ is the essential part of this layer system we discuss the influence of Al mole fraction variations, Δx, in the barrier on the spectral properties. The 15meV spectral shift points to $\Delta x \simeq$ 0.03. Provided that the same Δx

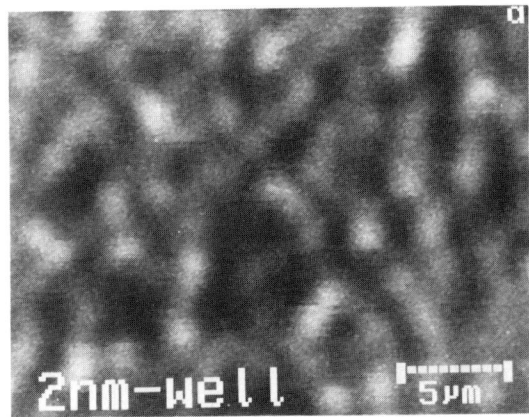
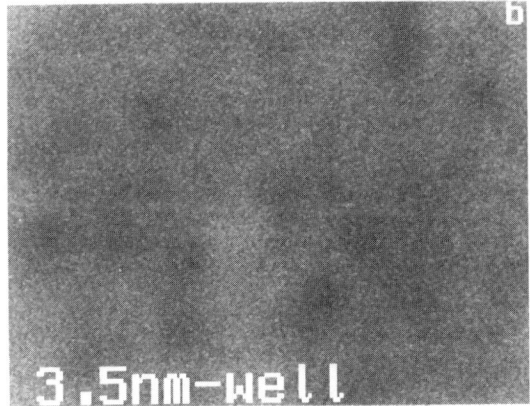

FIG.3. Spectrally resolved CL-images of the sample S1 at two different photon energies: E(a) corresponds to the 2nm-MQW, E(b) to the 3.5nm-SQW.The exciting conditions were chosen in such away, that the signal height was the same in both cases.

affects the 3.5nm-SQW we expect a shift of the corresponding spectrum of 7meV, hence a shift comparable to a thickness fluctuation of 0.6ML. Consequently a direct influence of x-variations on the CL-lineshift cannot explain the result of Fig. 3.

With respect to the large shifts of spectra measured at selected spots of the sample S2, we compare the spectral behaviour with that found close to oval defects. Papadopoulo et al. (1988) and Sapriel et al. (1988) investigated oval defects in $Ga_{0.7}Al_{0.3}As$-MBE layers and found, that the material remains in a rather good crystalline state but its orientation and Al-content is modified. They determine x-values of 0.15-0.18 on the defects and 0.25-0.3 outside. This x-range between inside and outside of the oval defects corresponds to the photon energy differences measured between the split spectra at many selected spots on the sample S2 provided that L_z is constant at 2nm. Hence we conclude, that in this case the

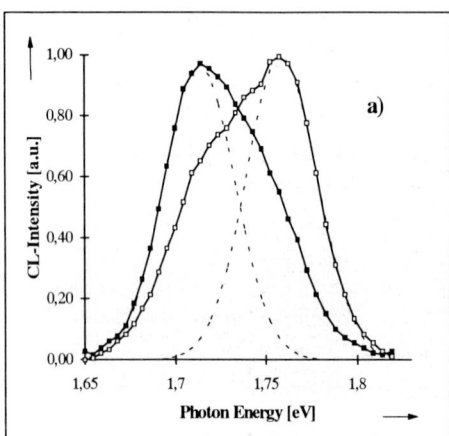 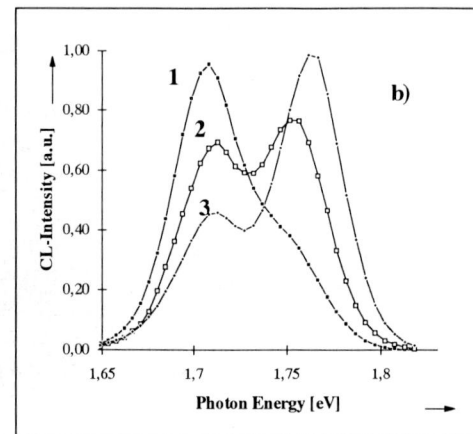

FIG. 4. CL-spectra measured at selected neighbouring spots (distance: few μm) of sample S2 at 80K. a) no oval defect is visible on the surface. b) curve 1: measured on an oval defect, curve 2: 1.5 μm away, curve 3: additional 1.5 μm away. The dashed curves (a) are fitted Gauss functions indicating a spectral shift of 43meV. All solid lines are only a guide to the eye.

spectral shift is due to a large fluctuation of x going from one spot to another. We suppose, even if there is no defect visible on the surface, that the layer package contains an early stage of an oval defect at those selected spots.

In conclusion: the lateral CL-inhomogeneities of the 2nm-MQW are caused by small shifts (15meV) of the spectra of neighbouring areas and by large spectral shifts ($\geq$50meV) at selected spots. The small spectral shift cannot completely be explained by a variation of the barrier height due to lateral x-fluctuations. Other possible mechanisms such as the influence of the carrier capture on the CL-intensity distribution or the influence of extrinsic and intrinsic defect levels on the excitonic recombination are now under investigation.

REFERENCES

Bimberg D, Christen J, Fukunaga T, Nakashina H, Mars D E, Miller J N 1987 J. Vac. Sci. Technol. B5 1191
Papadopoulo A C, Alexandre F, Bresse J F 1988 Appl. Phys. Lett. 52 224
Sapriel J, Chavignon J, Alexandre F 1988 Appl. Phys. Lett. 52 1970
Warwick C A, Kopf R F 1992 Appl. Phys. Lett. 60 386

Inst. Phys. Conf. Ser. No 134: Section 11
Paper presented at Microsc. Semicond. Mater. Conf., Oxford, 5–8 April 1993

651

Cathodoluminescence of strained and relaxed GaAs/AlGaAs p-i-n multi-quantum well structures on patterned Si substrates

C E Norman and R Murray

Interdisciplinary Centre for Semiconductor Materials, Imperial College, London, UK

K Woodbridge

Dept. of Electronic and Electrical Engineering, University College, London, UK

ABSTRACT: Optical emission from MBE-grown AlGaAs/GaAs/Si p-i-n MQW optical modulator structures has been studied using low temperature cathodoluminescence. Growth on patterned Si substrates reduces the degree of microcracking, with microcracks occurring only in the larger islands. The emission energy of the MQW varies with island size and significant strain relaxation has been seen within a few µm of microcracks and island edges. Monochromatic CL imaging successfully reveals relaxed regions near both microcracks and island edges. These results suggest the possibility of virtually strain-free devices on islands below ≈ 10µm in diameter.

1. INTRODUCTION

The growth of GaAs on Si has been of interest for some years because it offers the prospect of integrating GaAs-based optoelectronic devices into Si circuitry to form so called optoelectronic integrated circuits (OEICs). Commercial OEIC production remains hampered by the technological problems involved with growing such highly mismatched systems as GaAs/Si. The defect density resulting from the large (≈4%) mismatch is high and the differences in thermal expansion coefficients of GaAs and Si give rise to profuse microcracking upon cooling from the growth temperature to room temperature. These problems are antagonistic: the defect density at the layer surface decreases as thicker GaAs layers are grown but thicker layers (eg a few µm) are more susceptible to microcracking. The use of deliberately misoriented Si substrates can reduce the number of dislocations threading the GaAs layers [Fischer *et al* 1986]. Biaxial tensile strain in MQW structures grown on GaAs/Si causes a lowering of the energy gap and where it is sufficiently large may induce a near-degeneracy of the heavy and light hole states which are strain-shifted by different amounts. Strain levels can be monitored by observing the shift of the luminescence peaks and such studies have been performed on unpatterned GaAs/Si layers with [Jagannath *et al* 1987] and without [Yacobi *et al* 1987] AlGaAs/GaAs QW structures, stripe [Lee *et al* 1988] and island structures of GaAs/Si [Yacobi *et al* 1988], but never (to our knowledge) on GaAs/AlGaAs MQW device structures on small Si islands.

2. EXPERIMENTAL

Our structures were grown by MBE on high resistivity (100) Si substrates misorientated by 3° towards <110>. The p-i-n optical modulator structures were formed by growth of the following layers, listed from substrate to surface: 6µm n+ GaAs buffer, 500nm n+ AlGaAs, 20nm undoped AlGaAs, a 50 period 9.5nm/6nm (undoped) GaAs/AlGaAs MQW, 20nm undoped AlGaAs,

500nm p⁺ AlGaAs and a 10nm p⁺ GaAs cap. The Al content in all AlGaAs layers was 30% and the Si and Be doping was 10^{18} cm⁻³ in the n⁺ and p⁺ regions respectively. Further details of the growth have been published previously [Woodbridge *et al* 1993]. The CL investigations were performed using the Oxford Instruments MonoCL system with CF302 liquid helium cryostat attached to a JEOL JSM840A SEM.

Structures grown on unpatterned substrates were found to be subject to profuse microcracking along the orthogonal (110) directions, with microcracks occurring at separations of (typically) a few tens of μm. When islands of different sizes were wet-etched into the substrate prior to growth microcracking was found to be eliminated, except in a small number of the largest (460μm x 220μm) islands and, interestingly, between the islands. Figure 1 is a secondary electron micrograph showing the islands after the growth of the modulator structure. PL studies of these islands suggested that the MQW emission energy increases with decreasing island size but the relatively large size of the laser spot (≈ 50μm) meant that areas between the smaller islands were also being excited, signals from which would confuse the PL results [Murray *et al* 1993]. The higher spatial resolution of the CL technique clarifies matters immensely. Figure 2 shows the 4.2K CL spectra obtained with a focussed 15keV beam at the centre of islands of different size whose dimensions are shown in the figure. When one dimension of the island decreases below 60μm a shift to higher energies is observed. As one dimension is decreased below 40μm both a shift and an increase in emission efficiency occurs together with the appearence of a high energy tail. The smallest island studied (25μm x 10μm) presents a knee at higher energy.

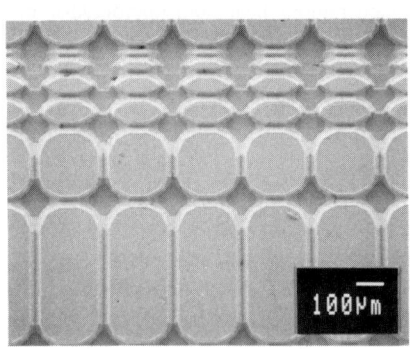

Figure 1.
A secondary electron micrograph showing the island structures after wet etching and MBE growth.

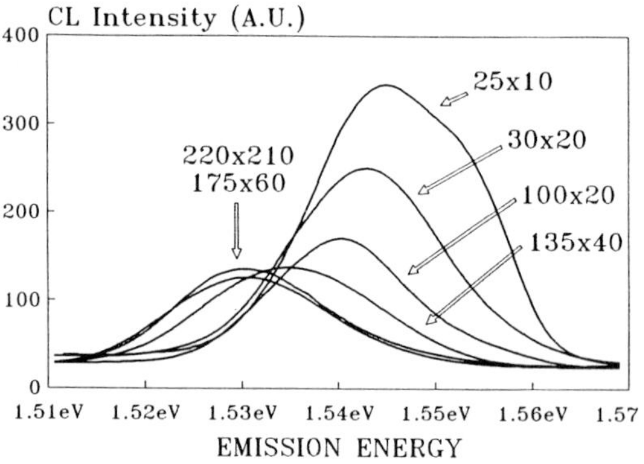

Figure 2.
CL spectra from the island centres. The island sizes are indicated in μm.

To help understand this trend it is useful to examine the optical behaviour with high spatial resolution across one single island. Figure 3 shows the spectra obtained at different points between the centre and the edge of a 60μm-wide island, such as the one shown on the left in figure 4. At the centre the emission occurs at ≈ 1.53eV and has a FWHM of ≈ 19meV. At a point 15μm from the island edge the emission has shifted to a slightly higher energy and has become more efficient. The emission at a distance of 5μm from the edge is noticeably blue-shifted and of much higher efficiency compared to that from the island centre. It also begins to present a high energy tail. The emission at a point 2μm from the island edge is weaker but shows a second peak occurring at ≈ 15meV higher energy which is in good agreement with theoretical predictions of the heavy and light hole energy separation in our MQW under zero biaxial strain. The lower intensity is thought be due to increasing non-radiative surface recombination as the island edge is approached and the second peak is interpreted as being due to light hole states.

It has already been shown that island edges are more efficient at relieving strain than microcracks [Murray *et al* 1993]. The luminescence efficiency is observed to be greatest in a region between 5 and 10μm from both edges and cracks. This behaviour has been seen before in thick GaAs epilayers on Si [Yacobi *et al* 1987] and attributed to the heavy and light hole states reapproaching degeneracy in the relaxed GaAs. In our MQW specimens, however, the opposite effect is occurring: the heavy and light hole states are moving further apart as strain is released. Another possibility is that the edges and cracks act as sinks for defects (point and/or line) giving rise to higher quality (defect denuded) zones nearby. However we observed little qualitative difference between edges and cracks formed during cooling from the growth temperature and those formed at room temperature (eg by cleaving). Assuming the defects to be relatively immobile at room temperature the development of such a denuded zone at freshly cleaved edges seems unlikely. Whilst it is possible that the observed changes in emission efficiency near edges and microcracks could be related to the different rates of strain induced shift of the heavy and light hole states in the MQW, a more likely explanation is that geometrical factors are responsible. As an edge is approached an additional portion of the generated CL can escape from the side wall and be refracted upwards into the collecting mirror. The similar behaviour of pre-existing and cleaved edges is further proof that the peak shifts should be interpreted in terms of strain relaxation and not other factors such as composition or layer thickness variations.

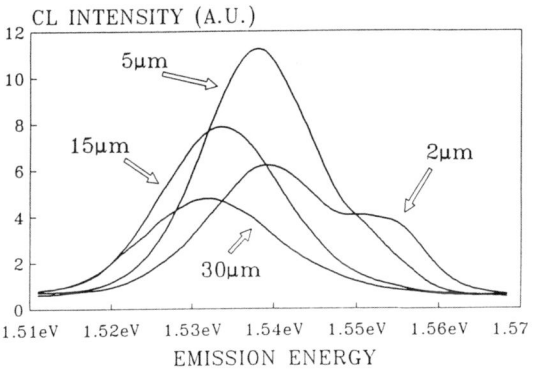

Figure 3.
CL spectra from points across a 60μm wide island. The distance from the island edge in μm is indicated in the figure.

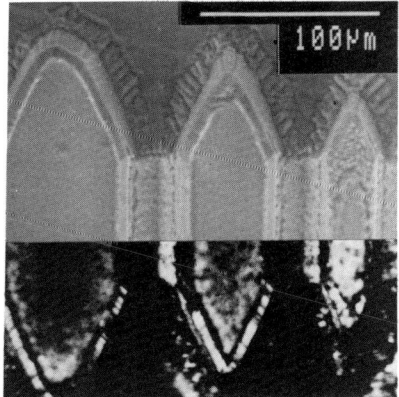

Figure 4.
A composite micrograph showing a secondary electron (upper half) and a monochromatic (λ = 800nm) CL image (lower half) of three small islands.

The significant energy shifts arising due to strain relaxation mean that monochromatic CL imaging can be used to visually display relaxed regions near edges and cracks. The apparent increase in emission efficiency near such features (discussed above) is a slight complication which results in some areas further in from the edge giving strong CL signals even though their emission does not actually peak at the wavelength being used to form the image. The results illustrated in figures 2 and 3 suggest that islands of dimensions of the order of 10μm are largely relaxed. Figure 4 visually confirms this hypothesis by showing a close-up of three small islands in secondary electron mode (upper half of figure) and monochromatic CL (lower half of figure) at λ = 800nm (1.55eV) which corresponds to the emission wavelength of the relaxed MQW structure. The smallest of the islands emits strongly at this wavelength over almost all of its area. Figure 5 shows (a) a secondary electron micrograph of one of the largest islands, and (b) its monochromatic CL image at λ = 800nm (1.55eV). The microcracks which were barely visible in (a) are now clearly delineated by the CL signal from the surrounding relaxed MQW. The CL signal is strongest where the microcracks intersect either each other or the island edges suggesting the reduction of the biaxial strain is greatest in these regions.

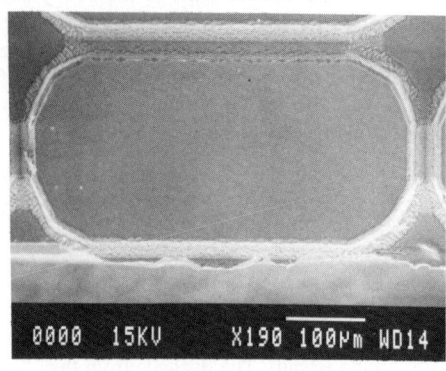

(a)

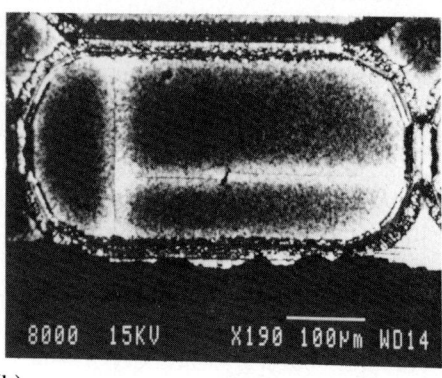

(b)

Figure 5.
Secondary electron (a) and monochromatic CL (b) images of the same large island. Using the wavelength of the relaxed structure to form the image reveals the microcracks.

3. CONCLUSIONS

We have shown that optical devices are a feasible proposition in the AlGaAs/GaAs/Si system when grown on misoriented substrates to lower the threading defect densities, with small island structures to avoid microcracking. Devices grown on Si islands smaller than ≈ 10μm should be relatively free of biaxial strain. In future such structures will be prepared by dry etching techniques since wet etching does not allow adequate control of island size.

ACKNOWLEDGEMENTS

We would like to thank Oxford Instruments Ltd. and the Max Planck Institute für Mikrostrukturphysik in Halle/Salle for the use of their MonoCL systems. CEN acknowledges the EEC Science Plan contract no. SC1-350 and the Max Planck Society for support.

REFERENCES

Fischer R,Neuman D,Zabel H,Morkoç H,Choi C and Otsuka N 1986 Appl. Phys. Lett. 48 1223
Jagannath C,Zemon S,Norris P and Elman BS 1987 Appl. Phys. Lett. 51 1268
Lee HP,Liu X,Lin H,Smith JS,Wang S,Huang Y,Yu P and Huang Y 1988 Appl. Phys. Lett. 53 2394
Murray R,Roberts C,Woodbridge K,Barnes P,Parry G and Norman CE 1993 Appl. Phys. Lett. (to be published)
Woodbridge K,Barnes P,Murray R,Roberts C and Parry G 1993 J. Cryst. Growth (to be published)
Yacobi BG,Zemon S,Norris P,Jagannath C and Sheldon P 1987 Appl. Phys. Lett. 51 2236
Yacobi BG,Jagannath C,Zemon S and Sheldon P 1988 Appl. Phys. Lett. 52 555

Inst. Phys. Conf. Ser. No 134: Section 11
Paper presented at Microsc. Semicond. Mater. Conf., Oxford, 5–8 April 1993

Depth-resolved cathodoluminescence study of GaInP/InP heterojunctions grown by MOCVD

F Cleton, B Sieber, A Bensaada[*], L Isnard[*] and RA Masut[*]

Laboratoire de Structure et Propriétés de l'Etat Solide, URA CNRS n° 234
Bât. C6, Université des Sciences et Technologies de Lille,
59655 Villeneuve d'Ascq Cédex. France.

[*]Groupe de Recherche en Physique et Technologies de Couches Minces, Département de Génie Physique, Ecole Polytechnique, C.P. 6079, Succursale A, Montréal (Québec) H3C 3A7. Canada.

ABSTRACT: $Ga_xIn_{1-x}P/InP$ epilayers (x=0.055 and 0.09) grown by low pressure MOCVD on sulphur doped InP (001) oriented substrates, have been investigated by cathodoluminescence. The epilayers are one micron thick and are under biaxial tensile strain. We give evidence of the electrical activity of dark line defects (DLDs) located within the epilayers and at the interface. The DLDs are preferentially parallel to the [$\bar{1}$10] (β) direction. Segregation of sulphur at these defects is suggested and is found not to be connected to sulphur fluctuations in the substrate.

1. INTRODUCTION

The electrical activity of dislocations or of any kind of extended defect present in semiconducting materials can be easily detected by microscopical techniques based on the injection of excess carriers. In order to characterise the defects produced by strain relaxation in GaInP/InP structures with a direct energy band gap, we have chosen the cathodoluminescence technique (CL) which relies on the emission of light by the specimen bombarded by an electron beam. CL does not to require any specimen preparation, and it gives a fast two-dimensional (in-plane) view of the luminescent properties of the structures with a resolution of about one micron. In-depth luminescent properties of thick epilayers can also be visualised by changing the incident electron beam voltage. In that case, the resolution can be better than one micron. In this paper, we present such a three-dimensional CL characterisation of GaInP/InP heterostructures where the epilayer is under tensile stress. We have examined partially relaxed structures where the epilayer had a thickness of about one micron. We report on the extended defects in two structures that are imaged in the form of dark line defects (DLDs). The CL background of the layers was found to be inhomogeneous.

2. EXPERIMENTAL

$Ga_xIn_{1-x}P$ epilayers were grown at 640°C on (001) oriented sulphur-doped LEC-InP substrates (n $\cong$ 5.10^{18}/cm^3), by Metal Organic Chemical Vapour Deposition (MOCVD) (Bensaada et al 1992). An initial 100 nm InP buffer layer was deposited first, in order to improve the interface quality. High resolution X-ray diffraction was used to determine the epilayer composition by recording (004) symmetric and (115) asymmetric reflections. The asymmetric relaxation with respect to the <110> directions was estimated using a number of diffraction spectra as a function of the azimuthal angle (Bensaada et al. 1993). The composition, thickness and relaxations for the samples analyzed in this paper are listed in Table 1.

| Samples | Composition | Relaxation (%) | | Thickness nm |
		[$\bar{1}$10]	[110]	
A	5.5	4.6	7.3	1.0
B	9.01	16	53	1.3

Table 1

The samples were mounted in a scanning electron microscope (SEM/CL) on a home-made liquid nitrogen specimen holder. Polychromatic CL pictures were obtained with an ellipsoidal mirror and a GaAs photocathode. Backscattered electron (BSE) pictures in the topographic mode were made with a solid state annular detector.

3. RESULTS

The CL images were recorded at various accelerating beam voltages E$_0$. The corresponding penetration depth R of the incident electrons, quoted in table 2, are calculated following the expression of Gruen:

$R = (4.57 \ E_0^{1.75}/ \ 100.\rho)$ where ρ is the material density (g/cm^3).

E$_0$ (kV)	5	9	15	17	20	25
R (μm)	0.15	0.4	0.98	1.22	1.62	2.4

Table 2: Penetration depth of the incident electrons

The CL signal is expected to escape from a depth equal at least to R. Nevertheless, the information depth can be larger, depending on the value of the minority carrier diffusion length. By recording monochromatic images, it was found that luminescence from the substrate became significant for E$_0$ values higher than 12-15 kV. This implies that electron-hole pairs generated in the GaInP layers diffuse to the InP substrate, as it could be expected from the energy band structure.

3.1 Sample A

The BSE picture of the 5.5% specimen did not exhibit any topographic contrast, in agreement with previous observations by means of optical microscopy (Bensaada et al 1992). Cracks were found parallel to the [110] (α) direction. CL images taken at various values of E$_0$ are shown in Figure 1. At an accelerating voltage as low as 5 kV (Fig.1a), the CL image exhibits sets of dark bands (DBs) and of DLDs, the latter

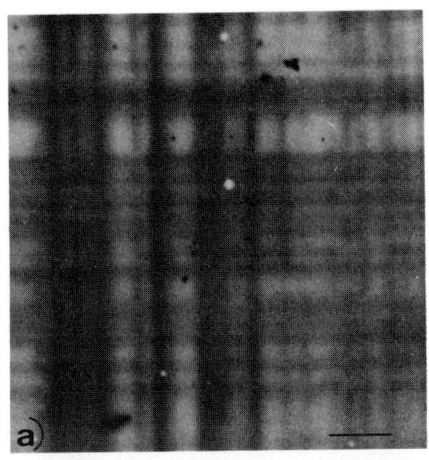

Fig. 1:   5.5% specimen.
Polycromatic CL images.
a - c: same area taken at
a) E₀=5 kV   b) E₀=15 kV:
DLDs are extended defects
within the epilayer. Dark
bands are due to impurity
segregation. DLDs visible
at E₀=25 kV (Fig.1c) are
certainly interface dis-
locations.d) another area
with DLDs only in the β
direction; E₀=9 kV. The
scale of the pictures is
equal to 10 µm.

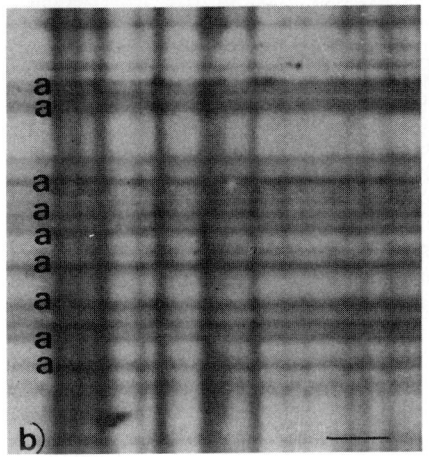

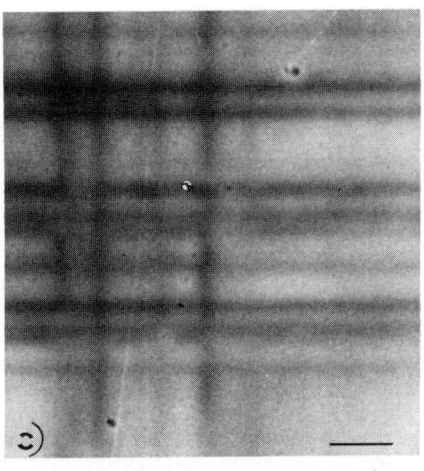

[1̄10]

[110]

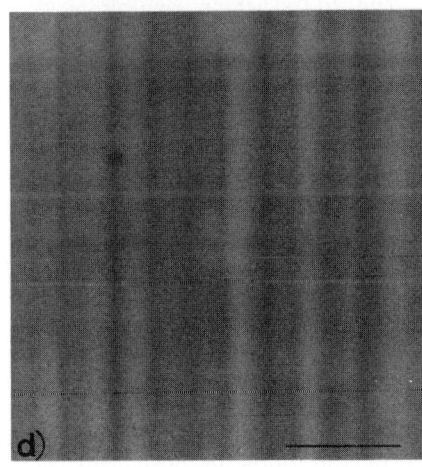

being frequently located within the DBs. Both kinds of defects are parallel to the two <110> directions. The CL contrast of the DLDs parallel to the [$\bar{1}$10] (β) direction is stronger and has a larger width than that of the DLDs parallel to the [110] (α) direction. Futhermore, the CL contrast of the DBs seems also stronger in the β direction. When E₀ increases up to 15 kV, a few DLDs parallel the α direction (a in Fig. 1), exhibit a stronger contrast than the other DLDs parallel to the same direction. Fig. 1c shows that, at 25 kV , only these DLDs are visible in the CL picture, and that the DBs are not visible anymore. In this specimen, the majority of the DLDs were. found parallel to the [$\bar{1}$10] (β) direction (Fig. 1d). Large areas free of defects in the α direction were frequently observed, whereas the largest spacing between series of DLDs in the β direction did not exceed $\simeq$ 20 µm. This gives evidence that relaxation does not proceed homogeneously within the specimen.

3.2 Sample B

Numerous cracks have been observed in the α direction. Fig. 2 shows the BSE and the CL pictures of a typical area of the 9% specimen. The topographic contrast is visible on the BSE image (Fig.2a) in the form of 'step lines' which are more frequently parallel to the [$\bar{1}$10] (β) direction than to the [110] (α) direction. The greater number of DLDs parallel to the β direction is also clearly visible in the CL images taken at E₀ values up to 17 kV (Fig.2b-d). Few of them correspond to the 'step lines'(● and ■ on the CL images). The DLD spacing varies from a few to 20 microns. It is larger for defects along the α direction than in the β direction. At E₀≥20 kV, the strong asymmetry between the two <110> directions is less pronounced, since a few DLDs in the β direction have nearly vanished (□ and ■), while new DLDs appear in the α direction (x). Notice that at E₀≥ 20 kV, a cross-grid in both <110> directions appears in areas nearly free of DLDs at lower values of E₀. In Fig. 2e-f, such an area is located between DLDs n°1 and 3. Background non uniformity is visible in the CL images of this specimen, as shown in Fig. 2 b-c.

4. DISCUSSION AND CONCLUSION

The DLDs which exhibit a CL contrast at all values of E₀, are thought to be extended defects which run through the epilayer. They are more numerous in the 9% specimen than in the 5.5% specimen. Those whose CL contrast disappears at high E₀ should be located within the epilayer and do not reach the interface. Both kinds of DLDs could be dislocations parallel to the surface or twins located in the {111} inclined planes, as observed recently by transmission electron microscopy (Lefebvre et al 1993). The resolution of CL images is such that it is not possible to distinguish between a twin and a dislocation lying in a {111} plane, both defects giving rise to a DLD of similar width. A CL contrast of twins could result from the electrical activity of the partial dislocations or, less probably, from their own electrical activity. It can be assumed that the DLDs visible at 25 kV in the 5.5% specimen are interface dislocations. Since they act as efficient recombination centres, an electrical activity of the partial dislocations in the twins can also be suggested as being at the origin of the CL contrast of the DLDs. Impurity segregation in the twin plane could also give rise to such an image. Such an hypothesis cannot be excluded since the irregular aspect of the DLDs in both specimens and the appeerence of the dark bands in the 5.5% specimen suggest that impurity segregation does occur

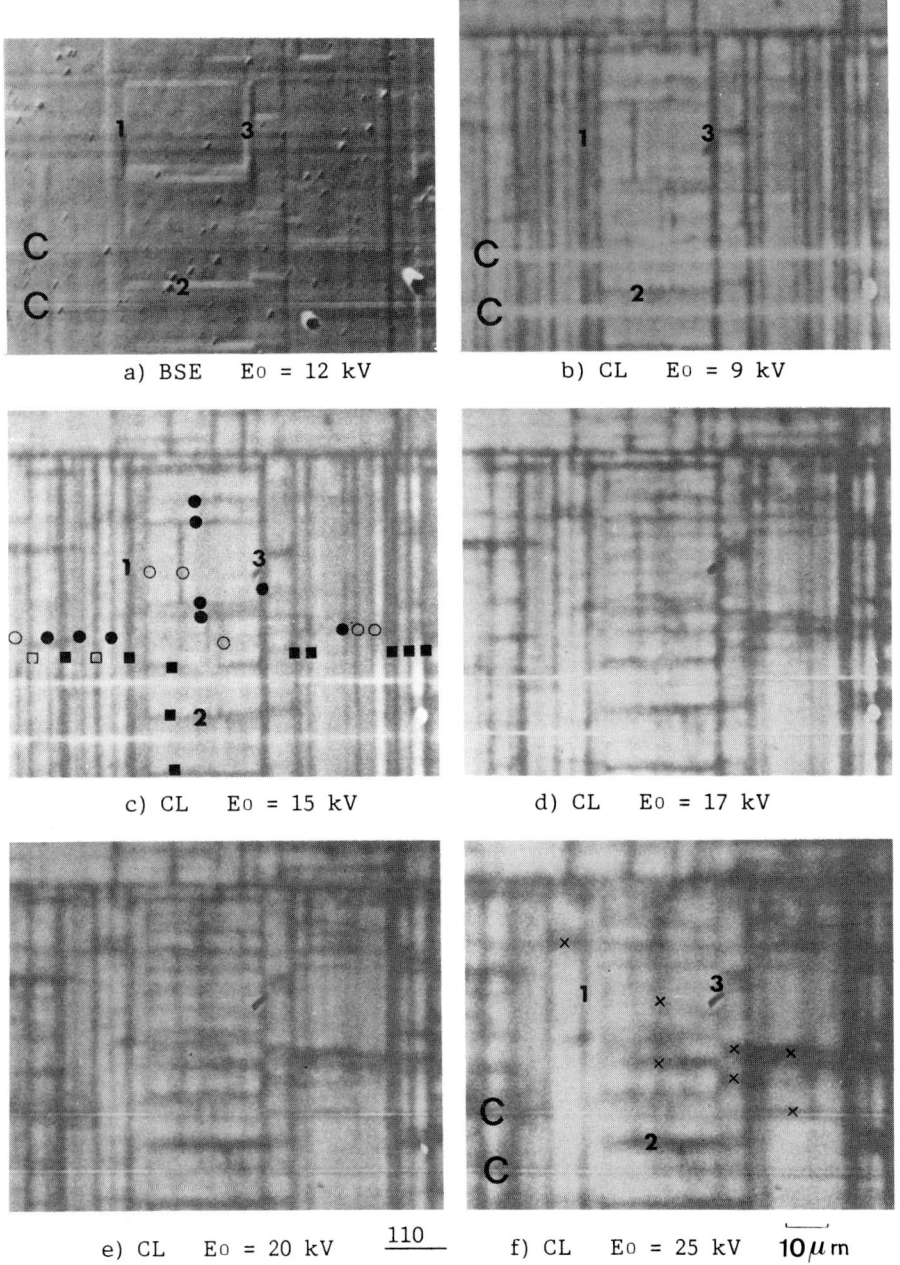

a) BSE   Eo = 12 kV

b) CL   Eo = 9 kV

c) CL   Eo = 15 kV

d) CL   Eo = 17 kV

e) CL   Eo = 20 kV   110

f) CL   Eo = 25 kV   10μm

Fig. 2: 9 % specimen. DLDs are mainly parallel to the [$\bar{1}$10] (β) direction. □ and ■ defects are visible in the CL mode whatever is Eo. □: no BSE contrast. ■: BSE contrast. The CL contrast of ● and ○ defects is higher at low values of Eo. ○: no BSE contrast. ●: BSE contrast. x defects are only visible at high values of Eo. C defects are crack lines at extended defects in the epilayer and at the interface. The CL inhomogeneities in the epilayer do not duplicate those in the substrate (Fig. 1). Even so, sulphur atoms coming from the substrate are, up to now, the best candidates for impurity segregation. STEM/CL experiments are in progress to clarify the origin of the electrical activity of the DLDs. The cross-grids observed in the 9% specimen could correspond to interface dislocations which are not produced by twinning, since they are not related to any CL contrast in the epilayer.

The CL topographic contrast evidenced in the 9% specimen is due to the change of the BSE coefficient with the angle of incidence. Therefore, it corresponds to a variation, across the step lines, of the number of electron-hole pairs created by the electron beam. In the case of the 9% specimen, it cannot be avoided and it has to be distinguished from the usual CL contrast which results from the recombination of electron-hole pairs. The CL contrast of a DLD like n°1 in Fig. 2 becomes quite weak at $E_0 \geq 17$ kV. This suggests that the CL topographic contrast dominates at low E_0 values ($E_0 < 17$ kV for the 9% specimen where the epilayer is 1.3 μm thick), whereas it is the CL recombination contrast which gives rise to the image at high E_0 values. At a certain value of E_0, topographic and recombination contrasts can happen simultaneously, if there are efficient recombination centres underneath the 'step lines'. Examples of this behaviour are given by the DLDs n° 2 and ■ in Fig.2. DLDs which correspond to a CL topographic contrast have been produced by slip of many dislocations in the inclined {111} planes. Steps as high as 0.18 μm, detected by optical confocal microscopy, could be detected in a 13.6% specimen. Such a height could be produced by slip of about 900 partial dislocations. In the 5.5% specimen, the higher and larger contrast of the DLDs in the β direction could result from a higher segregation of impurities. The fact that the dark bands in the 5.5% specimen are darker in the β direction than in the α direction validates this proposal. Indeed a higher segregation could result from a larger number of extended defects which cannot be resolved individually by the CL technique. The different spacing of the DLDs observed in the two specimens, as well as the asymmetry observed in each specimen between the two <110> directions, accounts well for the differences of the relaxation values quoted in Table 1. The result that relaxation is favoured in the [110] direction in GaInP/InP heterostructures where the epilayer is under biaxial tensile stress agrees with those obtained previously in InGaAs/InP structures under tension (Wagner et al 1989).

ACKNOWLEDGMENTS

We would like to thank Dr A. Lefebvre for valuable discussions and Th. Gloriant for the CL images of sample A.

REFERENCES

Bensaada A, Chennouf A, Cochrane RW, Leonelli R, Cova P and Masut RA 1992 J. Appl. Phys $\underline{71}$ 1737
Bensaada A, Cochrane RW, Masut RA, Leonelli R, Cova P and Kajrys N, 1993 J. Cryst. Growth, in the press
Lefebvre A, Cleton F and Sieber B, unpublished results
Wagner G, Paufler P and Rohde G 1989 Zeitschrift für Kristallographie $\underline{189}$ 269

Inst. Phys. Conf. Ser. No 134: Section 11
Paper presented at Microsc. Semicond. Mater. Conf., Oxford, 5–8 April 1993

Beam-induced dislocations and their CL contrast

D B Holt, E Napchan, L Lazzarini*, G Salviati* and M Urchulutegui**

Dept. of Materials, Imperial College of Science, Technology and Medicine, London SW7 2BP
*C.N.R. MASPEC Institute, Via Chiavari 18A, I–43100 Parma, Italy
** permanent address: Universidad Complutense, Ciudad Universitaria, Madrid, Spain

ABSTRACT: Dislocations induced by high SEM beam currents in GaAs were studied by TEM and emission CL. They were found to consist of bowed segments and confirmed to have [001] Burgers vectors. The capacity of different SEMs to induce dislocations varied greatly. Methods for recording CL contrast profiles and a Monte Carlo based program for CL calculations are presented.

1. INTRODUCTION

Beam scanning can induce dark and bright squares visible in several modes of the SEM. Such effects are generally avoided and loosely ascribed to 'contamination' or to 'damage'. A quite different effect was found by Franzosi et al (1988, 1989). Beam currents above a μA induced dense grids of dislocation lines parallel to the surface, running across the scanned area in the two orthogonal <110> directions in (001) GaAs and InP. X–ray topography showed these dislocations to have [001] Burgers vectors. This was ascribed to the production of high densities of vacancies of the volatile constituent under bombardment with subsequent aggregation to produce dislocations. This paper reports TEM and initial ECL contrast studies of these unusual dislocations.

2. EXPERIMENTAL METHODS

At I.C.S.T.M. a JEOL JSM–840A SEM fitted with a Matelect ISM–5 conductive mode detection system was used to take the signal from a Si photodiode mounted above the sample for emission cathodoluminescence (ECL) panchromatic imaging. It was connected to a Kontron image processor for signal averaging, image recording, contrast enhancement and image analysis. At Maspec the TEM work was done using a JEOL 2000FX and the dislocations were induced in a Stereoscan 250 fitted with an EDS (Cuorgné) ECL/TCL (emission and transmission CL) detector. The specimens were cut from wafers grown at Maspec by Dr. R. Fornari.

The ability to induce dislocations was found to vary with the type of SEM. Attempts to beam–induce dislocations using the JEOL 840A on a dozen occasions at magnifications of a few hundred to well over a thousand times using currents up to 4 μA succeeded only once. The Stereoscan 250, however, easily and reliably induced dislocations on going above 1 μA at magnifications of a thousand or more (Fig. 1). The 250 can even produce visible patches showing spectral colours due to the grids of scan lines 'burned' into the material acting as gratings. Reversible changes in luminescent intensity were generally all that was produced by the beam in the 840A. The result of repeated scanning at slightly different magnifications was then to produce concentric areas that could be alternately brighter and darker i.e. show window–frame like contrast in subsequent ECL images (Fig. 2). Which difference or differences in electron optics or scan speeds are responsible for the difference between the two SEMs is not yet known.

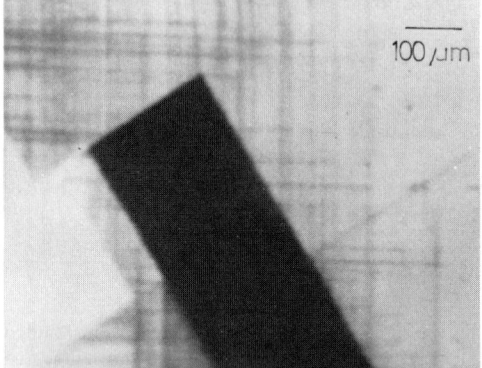

100 μm

Figure 1. ECL micrograph of a number of patches in a GaAs specimen in which dislocations running in the < 110 > directions in the (001) material were previously beam–induced by the Stereoscan 250. The previous scan direction was inclined at 30° to the dislocations and to the scan direction in this micrograph. Part of a dark, damaged area can be seen at lower centre and a bright undamaged area at lower left. The low density grid of dislocations at lower right is the type that was selected for measurement in Fig. 4.

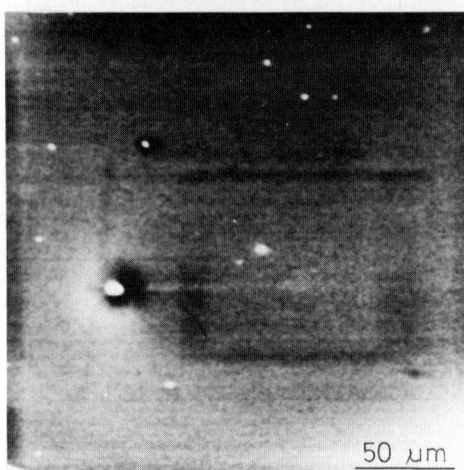

50 μm

Figure 2. SEM ECL micrograph of a GaAs specimen. The effect of scanning at different magnifications was to induce brighter and darker emission (efficiency) in the scanned areas, resulting in window–frame contrast patterns.

3. RESULTS

A part of the scanned area in which dislocations were induced in the 840A is shown in a digitized Kontron 512 x 512 pixel image in Figure 3a. Numerous ECL greyscale contrast line scan profiles could then be extracted along the lines marked in Figure 3a, for examination at leisure (Figure 3b). This method avoids possible further changes due to continued beam scanning and allows rapid checking for changes in dislocation strength along their length due to local impurity decoration or alternating dissociation and constriction (e.g. along d3).

A higher–resolution but slower method for extracting more quantitative ECL line scan profiles for dislocation contrast analyses is provided by the Matelect ISM–5 with computer interface and software. A single isolated 250 beam–induced dislocation was selected from a low defect density area like that at bottom right of Figure 1. A set of line scan profiles at increasing magnification are shown in Figure 4. The shape of the dislocation profile could be made to appear asymmetric by slow, long–range changes of CL intensity (Figure 5) and care was taken to avoid such cases.

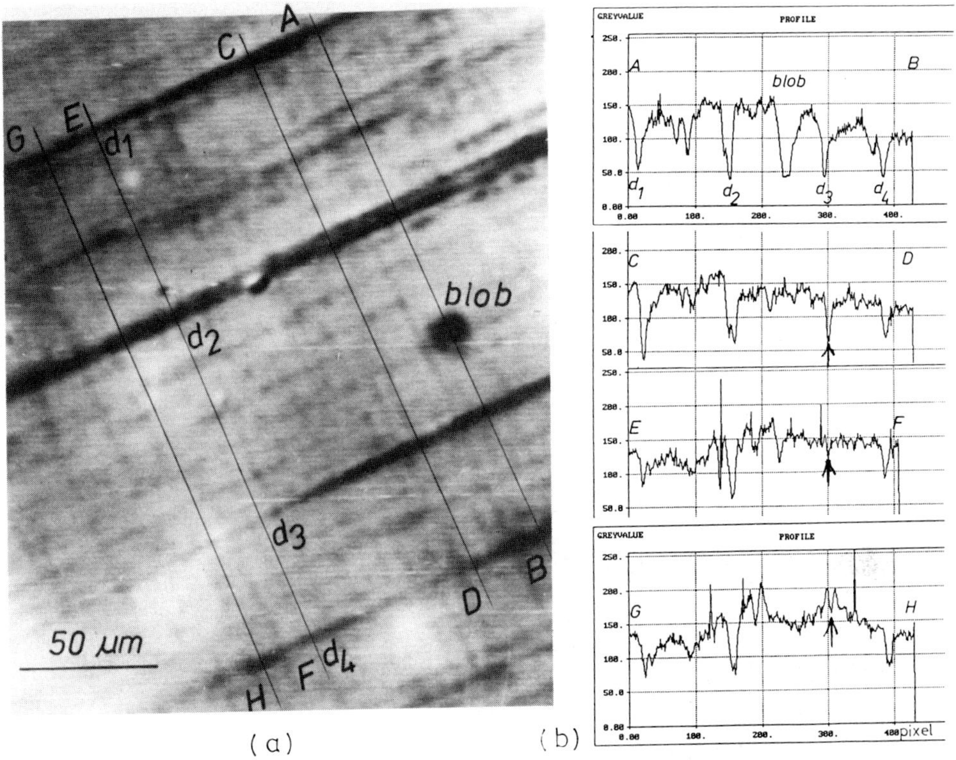

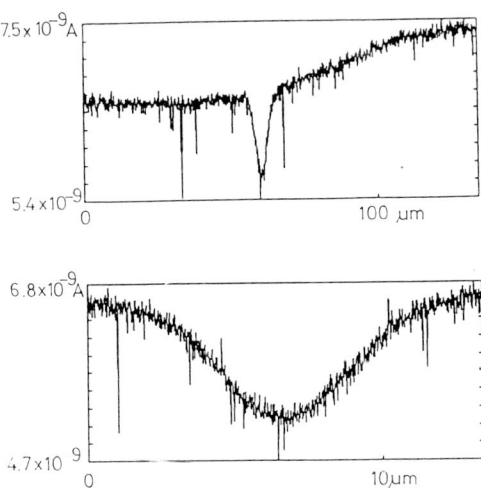

Figure 3. (a) part of a digitized 512×512 pixel ECL image of beam–induced dislocations in GaAs and (b) a number of CL intensity profiles for the scan lines marked in (a).

Figure 4. ECL photodetector current line scan profiles of a single beam–induced dislocation in the specimen of Figure 1 at one and ten thousand times.

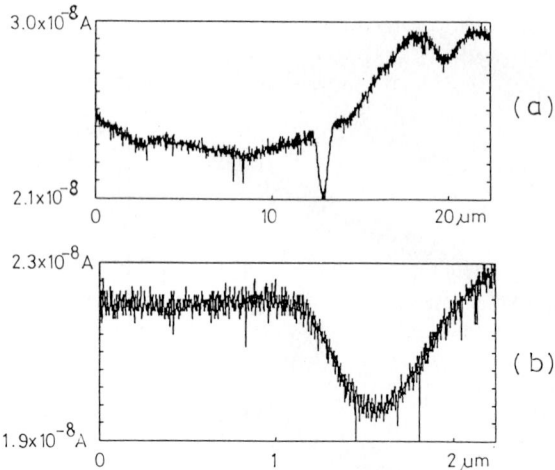

Figure 5. Line scan profiles across another dislocation in an area where the background ECL emission (efficiency) varied across the dislocation resulting in a distorted, asymmetric form of defect contrast. The magnifications were (a) 700× and (b) 6,000×.

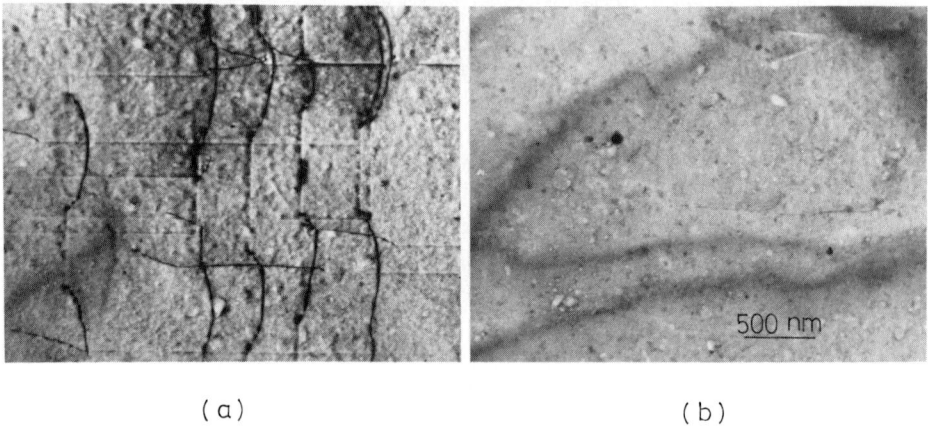

(a)   (b)

Figure 6. TEM plan view images of a beam–affected area crossed by orthogonal grids of beam–induced dislocations taken with reflections (a) g = $\bar{1}3\bar{1}$ and (b) g = 400.

TEM analysis showed the beam–induced material to be crossed by dislocations with [001] Burgers vectors (Figure 6). In such micrographs some dislocations could be seen to run down into the material and then along at a constant depth suggesting that they were introduced from the surface. Segments of the dislocations can be seen to be bowed out, perhaps suggesting the action of shear stresses and the presence of pinning points. This is also reminiscent of the beam activated motion of dislocations previously observed in the TEM (Cockayne et al 1987, Maeda et al 1987).

4. DISCUSSION

A program in C to run on PC compatible microcomputers was written using a simplified model of CL emission. This employs a depth dose distribution of hole–electron pairs derived from a Monte Carlo simulation of electron trajectories program (called MC–SET) adapted to deal with epitaxial multilayer materials and laterally limited device structures. This is an extended and modified version of that described earlier (Napchan and Holt 1987, Napchan 1988) and deriving from the Joy (Joy and Pimentel 1983, Joy 1987) and NBS (Mykleburst et al 1976) programs. The CL program was first tested by simulating the variation of ECL intensity (photodiode current) versus beam voltage for power constant and for current constant and found to give the correct shape of the curve and quantitative fit for reasonable values of the material parameters. It has now been used as the basis for simulation of dislocation ECL line scan profiles using the Donolato phenomenological model, previously applied to CL defect contrast by Lohnert and Kubalek (1983). Work is in progress to use fitting of experimental dislocation contrast data like that in Figures 3 and 4, to determine dislocation depths and strengths. The TEM results show that single ECL dark lines with widths of some μms, may be due to pairs or larger groups of dislocations and variations in effective recombination strength could arise in this way.

It is clear that at the beam powers used to induce the dislocations, preferential evaporation of the volatile constituent can occur since visible grating–like line structures can be produced in severe cases. Moreover, it is well known that the mobility of dislocations in glide can be enhanced by electron bombardment (Cockayne et al 1987, Maeda et al 1987). Hence it appears likely that point defect aggregation and climb and/or beam enhanced glide mobility are involved in introducing beam–induced dislocations.

Numerous line scan profiles can be extracted via image processing from a single ECL stored image thus avoiding inducing further damage during analysis. Higher–resolution quantitative line scans however require extended scanning to select the dislocation and the position(s) along its length for analysis and to record linescans (15 secs each for those in Figures 4 and 5).

Preliminary studies of microcomputer simulations of the strength of the emitted CL and the dislocation contrast profiles showed them to match measured data quite well.

Previous reports of beam damage in CL images were of a reduced CL brightness in ion–implanted Si ascribed to annealing effects (Myhajlenko el al 1983) and of large, slow increases in CL intensity with time in GaAs/AlAs quantum well specimens (Holt et al 1991). Both effects are probably due to electronic 'damage' as contamination films take much longer times to build up effective thicknesses (Wilson et al 1980). The electronic effects appear to involve changes in the numbers of surface states or in the charges on the surface or on point defects.

Dark (bright) patch contrast in ECL pictures is due to a reduction (increase) in radiative recombination efficiency, η_r. This can be written

$$\eta_r = \frac{\tau_{nr}}{\tau_{nr} + \tau_r}$$

where τ_{nr} and τ_r are the non–radiative and radiative recombination times. Hence, dark (bright) contrast implies a reduced (increased) non–radiative recombination time or an increased (decreased) radiative recombination time in the scanned area. Assume in the usual way that

$$\tau = \frac{1}{v_{th}\sigma N_t}$$

where v_{th} is the thermal velocity of the carriers, and σ and N_t are the capture cross–section and density of the non–radiative (radiative) recombination centres (traps) respectively. Suppose, to be specific, that the dark (bright) contrast is due to a reduced (increased) *non–radiative* recombination time. The contrast implies (i) an increase (decrease) in N_{tnr} or (ii) an increase (decrease) in σ_{nr}. Changes in the density of centres, N_{tnr}, are unlikely as knock–on displacements are not possible at the energies (keV) of SEM electron beams. Order of magnitude changes in effective capture cross section, σ_{nr}, however, can be produced by charge state changes in recombination centres (McKeever 1985) which beam irradiation is likely to produce and this seems a probable mechanism for damage effects.

REFERENCES

Cockayne D J H , Lu G and Sikorsky A 1987 Bull. Acad. Sci. USSR Phys. Ser. **51** 87 – 92

Franzosi P , Lazzarini L, Salviati G, Scaffardi M and Fieschi R 1988 in EUREM 88. Conf. Series No. 93 (Inst. Phys.: Bristol) pp. 421 – 422

Franzosi P, Lazzarini L, Salviati G, Scaffardi M and Fieschi R 1989 J. Appl. Phys. **66** 2947 – 2951

Holt D B, Norman C E, Salviati G, Franchi S and Bosacchi A 1991 in Microscopy of Semiconducting Materials 1991. Conf. Series No. 117 (Inst. Phys.: Bristol) pp. 689 – 694

Joy D C and Pimentel C A 1983 in Microscopy of Semiconducting Materials 1983. Conf. Series No. 68 (Inst. Phys.: Bristol) pp. 355 – 360

Joy D C 1988 in EUREM 88. Conf. Series No. 93 (Inst. Phys.: Bristol) pp. 23 – 32

Lohnert K and Kubalek E 1983 in Microscopy of Semiconductors 1983. Conf. Series No. 67 (Inst. Phys.: Bristol) pp. 303 – 314

Maeda K, Kimura K and Takeuchi S 1987 Bull. Acad. Sci. USSR Phys., Ser. 51, 93 – 98

McKeever S W S 1985 Thermoluminescence of Solids (Cambridge University Press) pp. 25 – 40

Myhajlenko S, Ke W and Hamilton B 1983 J.Appl.Phys. **54** 862 – 867

Mykleburst R L, Newbury D E and Yakowitz H 1976 in Use of Monte Carlo Calculations in EPMA and SEM (Ed. K.F.J. Heinrich et al) NBS Special publication 460 pp. 105 – 128

Napchan E and Holt D B 1987 in Microscopy of Semiconducting Materials 1987. Conf. Series No. 90 (Inst. Phys.: Bristol) 355 – 360

Napchan E 1988 Rev. de Phys. Appl. Colloque C6, 15 – 29

Wilson M, Ogden R and Holt D B 1980 J. Mater. Sci. **15** 2321 – 2324

Transmission electron microscopy, cathodoluminescence and secondary ion mass spectroscopy investigations of Si diffusion in GaAs

L Herzog, C E Norman[1], U Egger and O Breitenstein

Max Planck Institute of Microstructure Physics, Weinberg 2, D-O-4050 Halle/Saale, Germany
[1]Semiconducting Materials IRC, The Blackett Laboratory, Imperial College, Prince Consort Road, London SW7 2BZ, U.K.

ABSTRACT: From a thin (50 nm) sputtered film silicon is diffused into undoped semi-insulating and Zn-doped LEC-GaAs in two annealing steps - at 900 °C for 5 h and afterwards at 700 °C for 15 min - under various As pressures. SIMS was used to characterize the silicon diffusion profiles. TEM was carried out on cross-sectional and planar samples to investigate the Si-diffusion induced defect formation. Depth-dependent CL spectra were obtained at helium temperatures on cleaved faces and correlated with the TEM and SIMS results.

1. INTRODUCTION

Silicon is an important n-type dopant in GaAs. Si diffusion can be used for the fabrication of very high-speed transistors and integrated circuits. In addition, the Si impurity enhances layer interdiffusion at $Al_xGa_{1-x}As$/GaAs heterointerfaces (Deppe et al. 1988a). It was assumed by Tan et al. (1991) that owing to a low diffusion coefficient of silicon, thermal equilibrium concentrations of appropriate point defect species are attained and no diffusion-induced defects should be introduced. Microstructural investigations of defects induced by Si diffusion have not yet been performed. In earlier studies of the Si diffusion (Greiner & Gibbons 1984, 1985; Deppe et al. 1987, 1988b; Kavanagh et al. 1985, 1988) only SIMS concentration profiles of Si in GaAs have been investigated and several models of Si diffusion via various point defects were proposed. Some TEM results (Narayanan et al. 1974, Chen et al. 1992) were published for highly (2 x 10^{18} cm^{-3}) Si-doped GaAs. On the system GaAs on Si, CL investigations have been performed (Papadopoulo et al. 1989). This system, however, is essentially different owing to the strain caused by the large lattice mismatch and the difference in the thermal expansion coefficient between the two crystalline materials.

The aim of the present paper is to prove the defect formation during the Si diffusion from a thin sputtered Si film into GaAs by TEM, in correlation with Si concentration depth profiles revealed by SIMS and point defect CL investigations in and below the diffused layer. Though these investigations are still in progress, initial results are presented here.

2. EXPERIMENTAL

Three different wafers - semi-insulating (nominally undoped), Te-doped (1.5 x 10^{18} cm^{-3}), and Zn-doped (1.3 x 10^{18} cm^{-3}) - (100) LEC-GaAs were used as the starting material. A 50 nm thick silicon layer was deposited by HF magnetron sputtering on the wafer surface previously cleaned by sputter etching. The GaAs samples were placed in a capsule of high-purity porous graphite to prevent surface oxidation during annealing. The capsule was then sealed in an evacuated quartz ampoule, with or without pieces of pure elemental As to attain different As pressures during annealing. The annealing was performed in two steps. During the first, i.e. at 900°C for 5h, the actual Si-diffusion occurred.

Afterwards, a second annealing at 700 °C for 15 min was performed to allow defects to form. For all materials diffusion was carried out without As addition (Ga-rich conditions) and with As pressures of 0.1, 1, and 10 Atm (calculated as As_4 at 900°C). After annealing the ampoules were fast-cooled down to room temperature in air.

For the SIMS profiling 7 keV Cs^+ primary ions and the detection of negative secondary ions were used in a CAMECA IMS 3f microprobe. The SIMS profiling was carried out by IFAP GmbH Erfurt. The profiles were calibrated using a Si-implanted GaAs-standard.

TEM was carried out on cross-sectional samples in a JEOL HVEM working at 1 MeV to study the defect structure within the diffused layers. In addition, plan-view specimens were investigated obtained by chemical etching in brommethanol solution. CL experiments were performed in a JSM-840A SEM operating at 15 kV, equipped with a "Mono CL" system from Oxford Instruments. The CL spectra were taken at temperatures between 3.8 and 12 K on surfaces freshly cleaved perpendicularly to the silicon-covered face.

3. RESULTS

3.1 SIMS

A steep diffusion front similar to that obtained by Greiner & Gibbons (1984) and diffusion depths up to 3 μm are obtained for s.i. and p-material. Fig 1 shows the diffusion depths as a function of the As-pressure, showing (contrary e.g. to the results of Vieland (1961)) a non-monotonical dependence on it. The Si depth profiles of Te-doped n-GaAs did not show interpretable diffusion profiles, the reason for which is not yet clear.

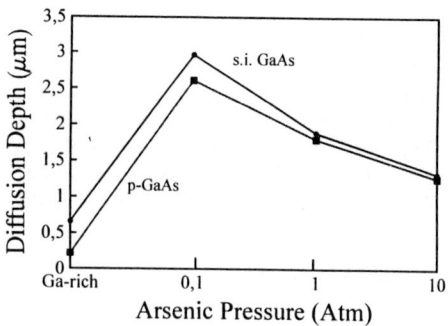

Fig. 1: Si diffusion depth in s.i. and p-material as a function of the arsenic pressure

3.2 TEM results

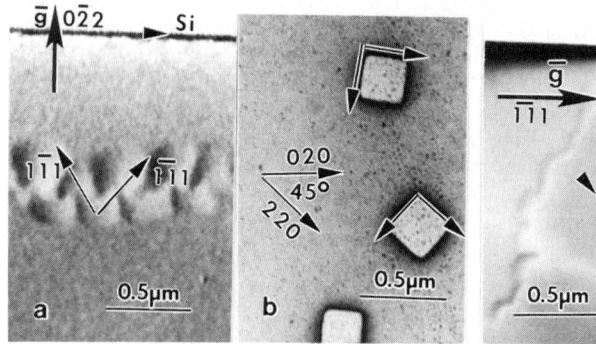

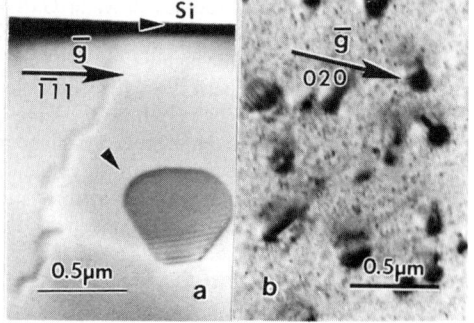

Fig.2: TEM micrographs (bright-field) of diffusion-induced defects in s.i.-GaAs without As overpressure: (a) XTEM; (b) plan-view.

Fig.3: TEM micrographs (bright-field) of diffusion-induced defects in p-GaAs after annealing at 0.1 Atm As_4 pressure: (a) XTEM; (b) plan-view.

Fig. 2 shows a cross-sectional (XTEM) image of s.i.-GaAs annealed without As overpressure and a corresponding plan-view of the same sample. The interface between Si and GaAs after annealing at

900 °C remains planar with no new interfacial phases forming. There is a distinct contrast at a depth of 0.7 - 0.9 μm, approximately corresponding to the diffusion front depth in the SIMS profile under these conditions. The defects in the plan-view are identified as square or rectangular defects having two preferential orientation directions, (200) and (220) at an angle of 45° to each other in the (100) plane. The density of these defects (size ~ 0.2 - 0.3 μm) is about 2×10^9 cm^{-2}.

No spots except regular matrix spots were observed in the SAD pattern of the square defect. The contrast line in the XTEM image shows structures along the {111} planes of GaAs. It might be associated with Ga-precipitation owing to the arsenic loss during annealing without an As overpressure, as they were not found in samples annealed at $P_{As4} \geq 0.1$ Atm.

Fig. 3a is a bright-field image (g = $\bar{1}\,\bar{1}\,1$) of diffusion-induced defects in p-GaAs after annealing at 0.1 Atm As$_4$. A number of defects which can be identified as dislocation loops with typical stacking fault (SF) fringe contrast are clearly visible in the diffused layer. The loops are inhomogeneously distributed, occurring in groups of two or three at depths between 1 and 2.2 μm. The depth of the layer of defects is in excellent agreement with the depth of the Si diffusion front (2.6 μm) detected by SIMS. The loops' radii range from about 300nm to about 600nm; their density is 5.3×10^7 to 6.2×10^8 cm^{-2}, estimated from both XTEM and plan-view micrographs (Fig. 3b). The larger loops tend to assume a triangular shape with their edges parallel to (200) and (0$\bar{2}$2) directions. Stacking faults in the interior of the loops lie on {111} planes and are of extrinsic character. Extrinsic stacking faults were observed in heavily Si-doped GaAs also by Narayanan et al. (1974) and by Muto et al. (1992). The latter suggested that these SFs are associated with an extra layer consisting of Si atoms precipitated between a couple of {111} planes in the GaAs matrix. Our results seem to support this suggestion as the loops with SF contrast were the main type of defect detected after the Si diffusion into both s.i.- and p-type material so that these defects may definitely be Si related.

In Fig. 4, the defect evolution with increasing As pressure is shown on plan-view specimens of s.i. GaAs. The size, shape and density of the defects depend on the As pressure. The largest defect density was observed at 1 Atm As$_4$ for SF loops that have also been obtained in p-type material. Hence, this type of defect seems to be a very important one in the defect formation process during Si-diffusion.

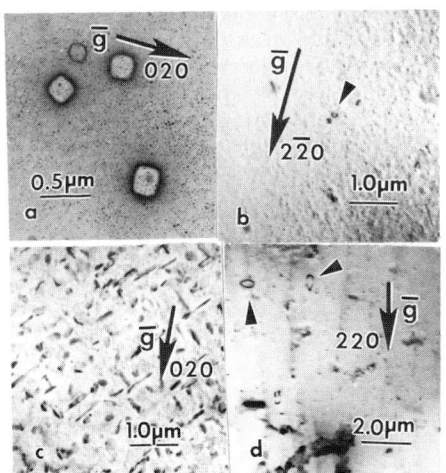

Fig. 4: Defect evolution (plan-view) with increasing As pressure in s.i. GaAs
(a) Ga-rich: square defects (0.2 - 0.3 μm; 2×10^9 cm^{-2})
(b) 0.1 Atm As$_4$: small dislocation loops (60 - 100 nm; 3×10^8 cm^{-2})
(c) 1 Atm As$_4$: high density (1.6 $\times 10^9$ cm^{-2}) of large (⌀ 100 - 600 nm) loops with SF contrast
(d) 10 Atm As$_4$: some loops (⌀ 500 nm; 8×10^6 cm^{-2}), preferentially oriented along (220) and (220) directions

3.3 CL

In the CL investigations on cleaved faces of s.i. material three emission peaks were recorded: one centered around 820 nm (near band edge luminescence, NBE) and two deep-level emissions (DL1 at $\lambda_1 \cong 1040$ nm and DL2 at $\lambda_2 \cong 1280$ nm). Fig. 5 shows a typical example of CL spectra measured on a s.i. sample (0.1 Atm As$_4$) at different depths below the surface. In all s.i. and p-type samples three different regions could be distinguished: 1. a near surface region predominantly showing DL1, DL2, and no NBE luminescence; 2. a region centered approximately around the maximum diffusion depth (2 - 3 μm) with an enhanced NBE luminescence and the DL2 band, and 3. the bulk region showing

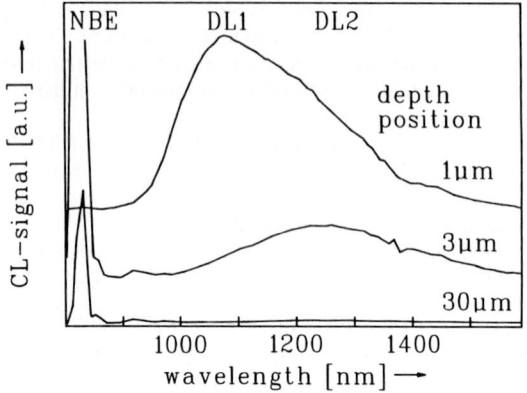

Fig. 5:   3.8 K CL spectra of s.i. sample Si-diffused at 0.1 Atm As$_4$ at different depths below the surface. For better comparison the baselines are shifted.

only intermediate NBE luminescence. Thus, DL1 and DL2 were observed only in the diffusion region, suggesting them to be associated with some defects or complexes within it. Under As pressure neither peaks shift in their positions, but the relative intensities depend on it. DL1 is thought to be due to Si$_{Ga}$-V$_{Ga}$ complexes and DL2 to be V$_{Ga}$ related (Papadopoulo et al. 1989), however some authors (Fujii et al. 1993) have interpreted them as related to Si-cluster defects or to Si$_{Ga}$-Si$_{As}$-pairs.   A   detailed presentation of the CL results and their discussion will follow in a later publication.

4. SUMMARY

Though these are preliminary results of   work still under way, the following conclusions can already be drawn:

- For a 2-step diffusion (900 °C, 5 h and 700 °C, 15 min) of Si into s.i. and p-GaAs steep diffusion fronts are obtained. The diffusion depths vary non-monotonically with the As pressure.
- Si-diffusion induced defects were observed near the diffusion front. The defects were identified in cross-sectional and plan-view images as square defects (perhaps Ga-related precipitates), small ($\varnothing$ 60 - 100 nm) dislocation loops and large ($\varnothing$ 100 - 600 nm) loops with a typical SF contrast. The defect structure depends on the As pressure for both s.i. and p-type material.
- The Si-diffusion affects the NBE luminescence and generates two deep-level emission bands at $\lambda_1 \cong$ 1040 nm and $\lambda_2 \cong$ 1280 nm. The intensities of these bands are As-pressure dependent and show different depth-dependences.

REFERENCES

Chen T P, Chen L J, Huang T S and Guo Y D 1992 Semicond. Sci. Technol. 7 A 300
Deppe et al. 1987 Appl. Phys. Lett. 50 (13) 998
Deppe D G and Holonyak Jr N 1988a J. Appl. Phys. 64 (12) R 93
Deppe et al. 1988b Appl. Phys. Lett. 52 (2) 129
Greiner M E and Gibbons J F 1984 Appl. Phys. Lett. 44 (8) 750
Greiner M E and Gibbons J F 1985 J. Appl. Phys. 57 (12) 5181
Fujii et al. 1993 J. Appl. Phys. 73 (1) 88
Kavanagh et al. 1985 Appl. Phys. Lett. 47 (11) 1208
Kavanagh et al. 1988 J. Appl. Phys. 64 (4) 1845
Muto et al. 1992 Phil. Mag. A 66 (2) 257
Narayanan G H and Kachare A H 1974 Phys. Stat. Sol. (a) 26 657
Papadopoulo et al. 1989 J. Appl. Phys. 66 (8) 3831
Tan et al. 1991 Crit. Rev. in Sol. State and Mat. Sci. 17 (1) 47
Vieland L J 1961 J. Phys. Chem Sol. 21 318

Inst. Phys. Conf. Ser. No 134: Section 11
Paper presented at Microsc. Semicond. Mater. Conf., Oxford, 5–8 April 1993

Cathodoluminescence studies of ZnSe grown on (001) GaAs using laser-assisted MBE

G M Williams, A G Cullis, K Prior*, J Simpson*, B C Cavenett* and S J A Adams*

Defence Research Agency, St Andrews Road, Great Malvern, Worcestershire WR14 3PS, UK
*Department of Physics, Heriot-Watt University, Riccarton, Edinburgh EH14 4AS, UK

ABSTRACT: Zinc selenide based structures grown on gallium arsenide by molecular beam epitaxy (MBE) are of interest for their potential as blue/green light emitting devices. Gallium arsenide is, for technological reasons, a preferred substrate. However, the 0.28% lattice mismatch of this heteroepitaxial system introduces misfit dislocations for layers exceeding the critical thickness. Furthermore, the exposure of the growth surface to laser irradiation during the MBE deposition of zinc selenide can provide advantages. The distribution of optically active defects in this material is of obvious importance and low temperature (17K) scanning cathodoluminescence (CL) is ideally suited to investigate such structures. In this paper we present CL data which reveals the spacial distribution of defect related deep level luminescence in conventional and laser assisted MBE-grown samples. Plan view TEM images of the same series of layers are also presented and the observed defect distribution is correlated with the CL results.

1. INTRODUCTION

Interest in short wavelength (blue or green) laser diodes has increased with their potential to play a major role in the next generation of optical recording systems. The first electrically pumped laser structure emitting at 490nm has been reported fairly recently (Haase et al 1991). This structure and others (Wang et al 1992; Haase et al 1992) based on ZnSe and associated materials are most commonly grown on GaAs substrates using either organometallic chemical vapour deposition (OMCVD) or MBE. Furthermore, during the growth of ZnSe structures using MBE it has been found that the crystalline quality and the doping characteristics of layers can be improved by laser irradiation during deposition (Simpson et al 1992).

In this study we have examined layers of ZnSe grown on (001) GaAs (0.28% lattice mismatch) using both conventional and laser-assisted MBE. Low temperature (17K) scanning CL has been exploited to obtain spectra and also to generate monochromatic images. In particular we have studied the spacial distribution of the defect-related transitions known as I_o° and Y_o (Shahzad et al 1990; Myhajlenko et al 1984; Batstone et al 1985) and identified the specific associated defects using plan view transmission electron microscopy.

2. EXPERIMENTAL METHODS

The ZnSe layers were deposited on semi-insulating (001) GaAs using a VG MB288 MBE system at growth temperatures of 350°C (conventional) and 280°C (laser-assisted), both with a Se:Zn flux ratio of 6:1. In the case of laser irradiated growth, however, the flux ratio at the growth interface was modified by the krypton ion laser radiation used (Simpson et al 1992).

The scanning CL system used to study these layers is described in detail elsewhere (Williams et al 1991). In brief, however, it is based on a Cambridge Stereoscan 150 Mk2 SEM with an LaB$_6$ source operated at 10kV. The samples were cooled to 17K using liquid helium and the light collected by high efficiency DRA-designed optics. The plan view TEM specimens were prepared using sequential mechanical polishing and low voltage argon and iodine ion milling techniques (Cullis et al 1985) and were examined using a JEOL JEM 4000EX microscope operating at 400kV.

3. RESULTS AND DISCUSSION

3.1 ZnSe Grown by Conventional MBE

Figure 1a shows a low temperature (17K) CL spectrum acquired from a 2μm thick layer. The 10kV beam used sampled the top 0.8 μm (Kanaya et al 1972) of the layer grown on GaAs. The spectrum shows a free exciton transition (Fx) and a small donor bound exciton transition (Dx) both near the band edge. The small Dx/Fx value is an indication of high purity material ($< 10^{14}$ cm^{-3}). Also present at 448nm (2.77eV) and 475nm (2.6eV) are the defect related transitions assigned as I$_v^o$ and Y$_o$, respectively, and seen more clearly in Fig. 1b.

The monochromatic CL image in Fig. 2a reveals the distribution of the I$_v^o$ luminescence and clearly shows it to result from a random array of point like features 3 to 4 μm in diameter present with a number density of ~4x10^6 cm^{-2}. In Fig. 2b the same area of material is imaged using the Y$_o$ luminescence. The distribution of Y$_o$ is more uniform than that of the I$_v^o$ signal, although some tendency towards localisation is clearly visible. Overlaying these two images we found some, but not total, coincidence of the two transitions.

Figure 2c shows a bright field 220 plan view TEM image from the same specimen revealing stacking faults, present primarily on one pair of the inclined {111} planes, and threading dislocations. A significant number of stacking faults are clearly associated with threading dislocations. The number density of stacking faults as averaged over many hundreds of square microns of material was found to be ~5x10^6 cm^{-2}.

The Y$_o$ luminescence is known to be associated with dislocations (Myhajlenko et al 1984). We deduce from the results above that the I$_v^o$ luminescence is due to the presence of stacking faults. The association of some stacking faults with threading dislocations gives rise to the observation of the overlapping Y$_o$ signal.

3.2 ZnSe Grown Using Laser-Assisted MBE

Figure 3 is a low temperature (17K) CL spectrum acquired from a ZnSe layer grown at 280°C on (001) GaAs. Laser-assisted MBE layers exhibited a zinc rich C(2x2) reflection high energy electron diffraction pattern during deposition. The growth rate of ZnSe is significantly reduced as a result of laser irradiation and although the growth time for this layer would normally result in ~2 μm of material the layer was in fact <0.5 μm thick. A 10kV SEM beam voltage was employed, sampling ~0.8μm of material, and thus yielded spectral information from the layer and the layer/substrate interface. The free and donor bound exciton transitions (Fx and Dx) are clearly resolved in Fig. 3 and the only other detected transition was the 475nm (2.6eV) defect related Y$_o$ peak as confirmed by Fig. 3b.

Figure 4a shows a monochromatic CL image obtained using the defect related Y$_o$ transition. This luminescence is clearly seen to be emanating from bands ~2μm in width lying predominantly along one [011]-type direction of the crystal and occasionally along the orthogonal direction.

A TEM plan view 400 bright field image of this layer is shown in Fig. 4b. Two set of orthogonal interfacial misfit dislocations are observed. Very occasional threading dislocations are also observed but no stacking faults are present. Clearly the Y$_o$ luminescence in this sample results predominantly from misfit dislocations lying in one particular orientation.

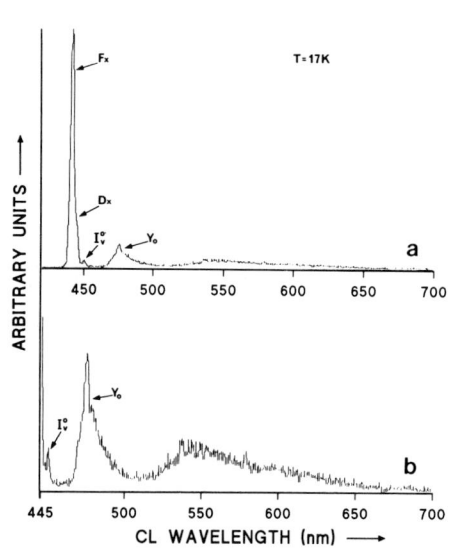

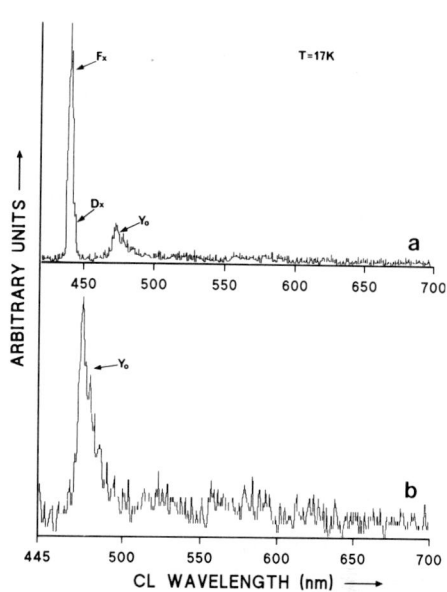

Fig. 1. The CL spectrum shown in (a) was acquired from a 2 μm thick ZnSe layer on GaAs. The two defect-related transitions are seen more clearly in the expanded section of the spectrum shown in (b).

Fig. 3. The CL spectrum shown in (a) was acquired from a 0.5μm thick ZnSe layer on GaAs grown with laser irradiation. Only one defect-related peak was visible (Y_o) as seen more clearly in (b).

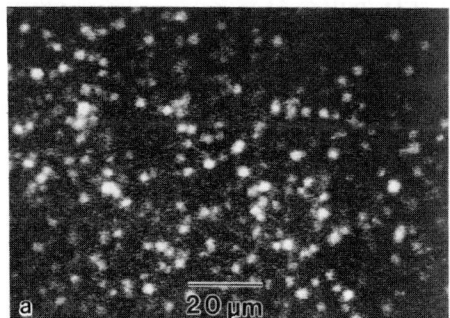

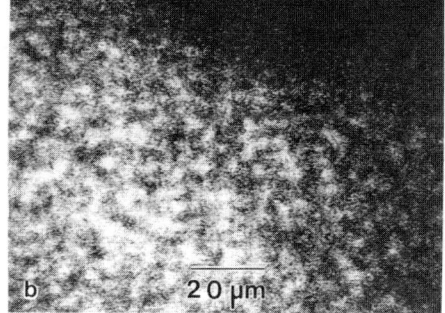

Fig. 2. The monochromatic CL images (a) and (b) reveal the spatial distribution of the defect peaks I_v^o and Y_o respectively. The bright-field TEM image (c) is taken from the same material shown in (a) and (b).

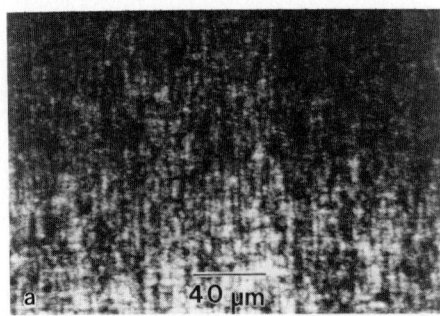 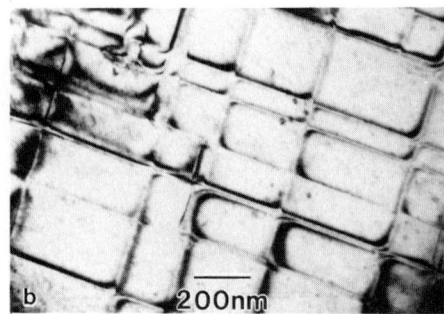

Fig. 4. The monochromatic CL image shown in (a) was acquired using the defect related Y_o peak. The plan view TEM image of the same ZnSe layer on GaAs is shown in (b) and reveals the dislocation array responsible for the CL image.

4. CONCLUSIONS

The powerful combination of low temperature scanning CL and plan view TEM has been used to establish the origin of defect-related transitions in heteroepitaxial ZnSe. Our observations reveal for the first time that the $I_o^?$ luminescence is due to the presence of stacking faults. It has also been shown that the Y_o luminescence can originate from misfit dislocations lying primarily along one [011]-type direction.

ACKNOWLEDGEMENT

The authors wish to acknowledge the assistance of Mr P W Smith in the preparation of TEM samples.

REFERENCES

Batstone J L and Steeds J W 1985 Inst. Phys. Conf. Ser. 76 383
Cullis A G, Chew N G and Hutchison J L 1985 Ultramicroscopy 17 203
Haase M A, Qiu J, DePuydt J M and Cheng H 1991 Appl. Phys. Lett. 59 1272
Haase M A, Qiu J, DePuydt J M and Cheng H 1992 Inst. Phys. Conf. Ser. 120 9
Kanaya K and Okayama S 1972 J. Phys. D Appl. Phys. 5 43
Myhajlenko S, Batstone J L, Hutchison H J and Steeds J W 1984 J. Phys C: Solid State Phys. 17 6477
Shahzad K, Petruzzello J, Olego D J, Commark D A and Gaines J M 1990 Appl. Phys. Lett. 57 2452
Simpson J, Adams S J A, Wallace J M, Prior K A and Cavenett B C 1992 Semicond. Sci. Technol. 7 460
Wang S Y, Hauksson I, Simpson J, Stewart H, Adams S J, Wallace J M, Kawakami Y, Prior K A and Cavenett B C 1992 Appl. Phys. Lett. 61 506
Williams G M, Cullis A G, Sotomayor Torres C M, Thoms S, Beaumont S P, Stanley C R, Lootens D and Van Daele P 1991 Inst. Phys. Conf. Ser. 117 695

Inst. Phys. Conf. Ser. No 134: Section 11
Paper presented at Microsc. Semicond. Mater. Conf., Oxford, 5–8 April 1993

Investigations of high quality Ge_xSi_{1-x} grown by heteroepitaxial lateral overgrowth using cathodoluminescence

A Gustafsson, P O Hansson*, M Albrecht**, H P Strunk** and E Bauser*

Department of Solid State Physics, Lund University, Box 118, S-221 00 Lund, Sweden. *Max-Planck-Institut für Festkörperforschung, Heisenbergstrasse 1, D-7000 Stuttgart 80, Germany. **Institut für Werkstoffwissenschften VII, Universität Erlangen - Nürnberg, Cauerstrasse 6, D-8520 Erlangen, Germany.

ABSTRACT: We present cathodoluminescence studies of relaxed lamellae of Ge_xSi_{1-x}, selectively grown on patterned Si (111) substrates. In the monochromatic images we observe three types of contrast variations. On a 50 μm scale we observe small fluctuations in the composition. On a 20 μm scale the lamellae exhibit a dark band, in the position of the seeding windows. On the scale of the resolution of the measurements we observe dark line defects in some of the lamellae. For lengths less than 250 μm, virtually dislocation-free material is obtained.

1. BACKGROUND

Heteroepitaxial growth of Ge_xSi_{1-x} on Si has a high potential to improve device performance. This has for example been shown in the case heterojunction bipolar transistors (Patton et al 1990). It can also be used to obtain periodic superlattices of Ge_xSi_{1-x}/Si, well suited for tailoring of the band structure (Satpathy et al 1988). It is very important that the structures are grown under conditions where the mismatch between the layers is accommodated by elastic strain, rather than by misfit dislocations. The layers then have to be grown to a thickness which is well below the critical thickness, h_c, for formation of misfit dislocations (Dodson and Tsao 1987). However many of these heteroepitaxial structures need a thickness close to or exceeding h_c. One way to achieve a higher h_c is to grow the structures on a relaxed buffer of Ge_xSi_{1-x} with a suitable lattice constant. The drawback is that part of the dislocations causing the relaxation tend to propagate into the subsequently grown layers. This can seriously degrade the device properties.

2. HETEROEPITAXIAL LATERAL OVERGROWTH

One way to achieve a significant reduction of the dislocation density in a buffer is to use the technique of heteroepitaxial lateral overgrowth (HELO). In this technique the growth starts in seeding windows in the otherwise masked substrate. This initial layer, confined to the seeding window, will grow to a thickness beyond the critical thickness and relax via formation of misfit dislocations. In the later stages of growth the perfectly relaxed layer grows laterally over the masked substrate, with little adhesion to it. The material above the seeding window will contain a relatively high density of defects, whereas the lateral overgrown material will be virtually defect free.

The growth of the HELOs was carried out on patterned Si (111) substrates. First the substrates were thermally oxidized with a 80 nm thick SiO_2 film. Seeding windows were opened up in the film using conventional photolithography and subsequent etching. These seeding windows were 20 μm wide and between 125 and 950 μm long, and they were oriented along several different directions on the substrate. The deposition of the Ge_xSi_{1-x} layer was performed by liquid phase epitaxy (LPE) at 810° C, using a Bi solution (Hansson et al 1990). Three compositions were used, with x=0.80, x=0.90 and x=1.00, nominally. In all cases the thickness of the lamellae was about 15 μm.

3. CATHODOLUMINESCENCE

The CL measurements were performed in a dedicated SEM with the samples cooled to 25 K. The samples were mounted for plan view studies. For all the data presented here the the SEM was operated at 20 kV, with a probecurrent of about 10 nA. The CL was dispersed by a monochromator and detected by a LN_2 cooled Ge diode, using standard lock-in technique by chopping the electron beam. Spectra and images were recorded digitally. The experimental setup has previously been described in more detail (Gustafsson 1991).

4. RESULTS

Figure 1 shows typical CL spectra of three samples with nominal Ge contents of 80, 90 and 100%, respectively. The two samples with the higher Ge content show similar spectra. A strong no-phonon line, with a line width of about 10 meV. From previous studies of the temperature dependence we attribute these peaks to free excitons (Weber and Alonso 1989). The positions of the peaks appear at slightly higher energies than expected from the nominal composition, indicating a 2-3 % higher Si content than expected. The spectra also exhibit phonon replicas. In our case we are unable to resolve the individual phonon peaks, but the replicas appear at 30-35 meV lower energy than the no-phonon peak. This corresponds to the TO_{Ge-Ge} (36 meV) and to LA (30 meV) (Weber and Alonso 1989, Hansson et al 1990). The sample with the 80% Ge exhibits a broader peak, 40 meV, at a lower energy position than expected for the free exciton. This behaviour can either be attributed to a dislocation related luminescence band called L or to multiexciton complexes (Weber and Alonso 1989). The latter is more probable in our case, as we also observe this luminescence in areas which are virtually dislocation free. However, we need higher excitation density in this case to get similar luminescence intensities, which is consistent with the latter case.

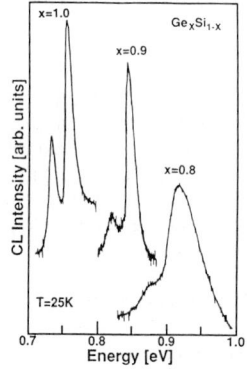

In all three samples the lamellae show similar contrast variations and we will concentrate on the sample with x=0.9. Furthermore, we will also limit this study to the case with the seeding window of different lengths, l, in the <112> direction. Figure 2 shows a CL image of a lamella grown out of a 500 μm long seeding window. Three types of features can be observed in the intensity variations of the monochromatic images. One is a dark stripe in the middle of the lamella, about 20 μm wide. This is in the same lateral position as the seeding window, indicating that there is a higher defect density above the seeding window. It is expected that the interface between the Si substrate and the lamellae will contain a high density of misfit dislocations. However, they will

Figure 1. CL spectra of the three samples, of different Ge content, x=0.8, x=0.9 and x=1.0. The spectra from the samples with x=0.8 exhibits a broad peak with a weak phonon replica. The other two samples exhibits sharper peaks as well as phonon replicas, TO_{Ge-Ge} and LA.

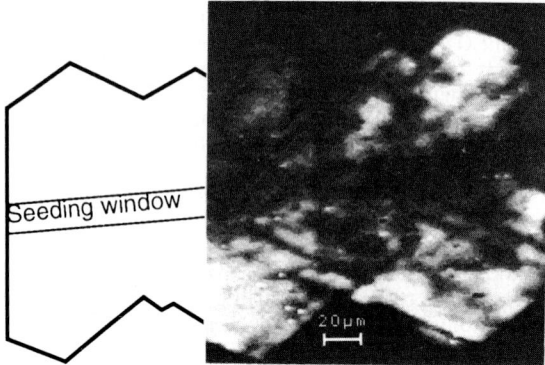

Figure 2. A monochromatic CL image of a lamella grown out of a 500 μm long seeding window. The image shows three different types of contrast variations. On a large scale there are diffuse variations corresponding to variations in the composition. A 20 μm wide band in the lateral position of the seeding window, an area with a relatively high density of threading dislocations. Furthermore, there are dark lines in the <110> directions, corresponding to dislocations.

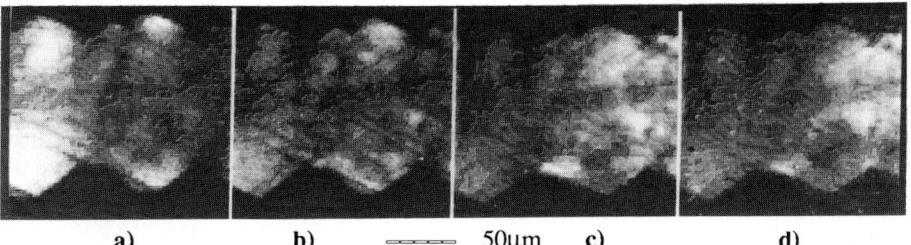

a)   **b)**   ===== 50μm   **c)**   **d)**

Figure 3. A series of monochromatic CL images of the same area of the lamella as in Figure 2, but detecting at different energies. The images correspond to a compositional difference of Δx=0.02, increasing the Ge content in steps of about 0.007 from a to d. The intensity variations on the 50 μm scale are different for the four images and are due to variations in the composition. However, the dark lines of the seeding window and the dislocations, as seen in Figure 2, can be observed in all four images.

not affect the luminescence of the top 5 μm, of the total 15 μm, that can be expected to contribute to the CL signal using 20 keV electrons. Thus all the features seen in the images must lie no more than 5 μm from the surface.

The images also exhibit a series of dark lines, that lie along the <110> directions. In the images they appear to be about 3-5 μm wide, which is the resolution that can be expected at 20 kV. These type of defects are normally attributed to dislocations, or other line defects, in the material. To gain more information about these defects, the accelerating voltage was increased to 30 kV, resulting in an increase in the depth of the excitation. This gave no significant increase in the number of dark lines, but rather a decrease in the contrast of the existing lines. This indicates that the defects giving rise to the dark line contrast are not elongated in the direction normal to the substrate plane.

The third type of contrast variations of the luminescence from the lamellae is on a larger scale and more diffuse than the two discussed previously. The origin of these is identified as fluctuations in the composition of the lamellae. This is illustrated in figure 3, which shows the same area of a lamella, but forming the images at different energies, i.e. detecting different compositions. The four images span a variation in composition of Δx=0.02. This shows that most areas of the lamella are bright in at least one of the images. It is also worth noting that the seeding window appears dark and the dark line contrast can be observed in all four images.

To further investigate the origin of the dark line and seeding window contrast we have investigated the lamellae using transmission electron microscopy (TEM). Figure 4 shows a bright field TEM image of a 500 μm long lamella. As expected the interface between the Ge_xSi_{1-x} lamella and the Si substrate exhibit a high density of misfit dislocations. From the interface into the lamella several threading dislocations can be observed. These types of defects are also present in the TEM images of lamellae grown in shorter seeding windows. The images of the longer lamella, l > 250 μm, exhibit dark line contrast in the laterally overgrown material. This contrast is typical of a dislocation inclined to the electron beam in the TEM.

By comparing the CL images with the TEM images we can conclude that the reduced CL intensi-

Figure 4. A bright field TEM image of a longer lamella reveals that the interface between Si and Ge_xSi_{1-x} in the seeding window contains a high density of misfit dislocations. The area above the

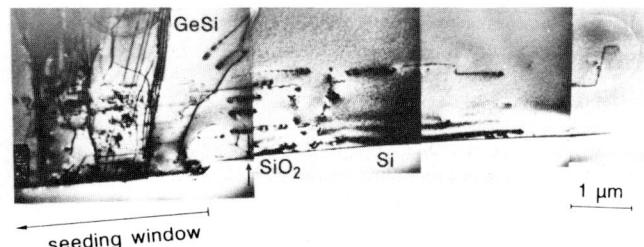

seeding window contains threading dislocations. The laterally overgrown material also exhibits dark line contrast in the image. A contrast that can be attributed to a dislocation inclined to the electron beam in the TEM.

ty in the area above the seeding window is caused by threading dislocations originating from the seeding window. It is also highly likely that the dark line defects in the CL images correspond to the dark lines observed in the TEM images of the overgrown material. This would then account for the dislocations being inclined to the beam.

Figure 5 shows a CL image of a lamella, grown in a shorter window, l=250 μm. This image also shows a dark stripe in the area above the seeding window. The overgrown material also exhibits the intensity variations that we attribute to compositional variations. The most interesting feature is however that the overgrown material is virtually free from the dark line defects that were observed in the longer lamellae. This leads us to conclude that the defects causing the dark lines in CL imaging are introduced after growth. They are probably caused by the larger stress introduced in the lamellae when cooling down from the growth temperature, due to different thermal expansion between the Si substrate, the SiO_2 and the Ge_xSi_{1-x}. In the case of the longer lamellae, the contact area between the SiO_2 and the Ge_xSi_{1-x} is larger, therefore giving rise to more stress. A possible generation mechanism is the gliding of threading dislocations from the area above the seeding window.

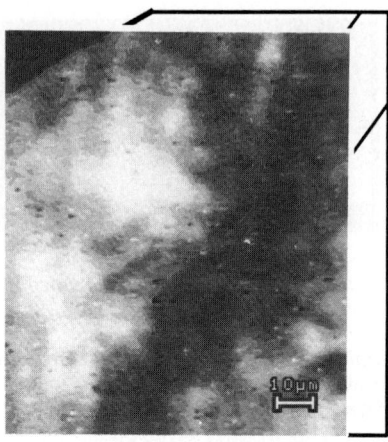

Figure 5. A lamella grown out of a 250 μm long seeding window exhibits the same dark stripe, corresponding to the seeding window. However, it does not show any of the dark line defects, present in the longer lamellae. In this case the overgrown material is virtually defect free.

5. SUMMARY

The contrast variations of the monochromatic images fall into three categories. On a scale of 50 μm there is a fluctuation of the composition, $\Delta x = 0.02$, giving rise to a weak variation of the luminescence intensity. The contrast can be reversed by changing the wavelength of the detected CL. In all the lamellae there is a darker stripe, about 20 μm wide, above the seeding window. This area contains a high density of threading dislocations, affecting the CL signal. In lamellae longer than 250 μm there is an additional feature of dark lines 3-5 μm wide. These are attributed to dislocations and are believed to be caused by the stress introduced by different thermal expansion of the materials involved.

In conclusion, the lamellae grown out of seeding windows longer than 250 μm show defects in an area above the seeding window, as well as in the lateral overgrown material. However, for shorter seeding windows the lateral overgrowth is defect free to within the experimental limit. The conclusion is that for seeding windows shorter than 250 μm, the material is virtually defect free and of high quality.

Acknowledgements: This work was partly financed by the NUTEK and NFR and the CL measurements were performed within the Nanometer Structure Consortium in Lund. The authors are thankful to R. Bergmann and S. Nilsson for valuable discussions and support.

REFERENCES

Dodson B W and Tsao J Y 1987 Appl. Phys. Lett. 51 1325
Gustafsson A 1991 Thesis Lund University.
Hansson P O, Werner J H, Tapfer L P Tilly L P and Bauser E 1990 J. Appl. Phys. 68 2158
Hansson P O, Bergmann R and Bauser E 1991 J. Crystal Growth 114 573
Patton G L, Comfort J H, Meyerson B S, Crabbé E F, Scilla G J, De Frèsart E, Stork j C M, Sun J Y-C, Harame D L and Burghartz J N 1990 IEEE Electron Device Letters EDL-11 171
Satpathy S, Martin R M and Van der Walle C G 1988 Phys. Rev. B 36 13237
Weber J and Alonso M I 1989 Phys. Rev. B 40 5683

Inst. Phys. Conf. Ser. No 134: Section 11
Paper presented at Microsc. Semicond. Mater. Conf., Oxford, 5–8 April 1993

CL imaging of Si/Si$_{1-x}$Ge$_x$/Si quantum wells grown by RTCVD

V Higgs, E C Lightowlers, J C Sturm*, X Xiao* and P J Wright†

Department of Physics, King's College, Strand, London WC2R 2LS, UK
*Department of Electrical Engineering, Princeton University, Princeton, NJ 08544, USA
†Oxford Instruments, Eynsham, Oxon, OX8 1TL, UK

ABSTRACT: Cathodoluminescence (CL) imaging and spectroscopy have been used to characterize fully strained SiGe quantum wells grown on Si. At T≈5 K, CL spectra contain well resolved band edge luminescence features. Monochromatic imaging with the no-phonon line attributed to the bound excitons in the quantum well, shows that the distribution of the luminescence from the wells was not uniform. The thinnest well (33 Å) contained a low density of non-radiative (luminescence reduction up to 100%) areas 40-100 μm in size. The thickest well (500 Å) contained non-radiative areas and also dark line features oriented along the <110> directions. These dark line features are areas of non-radiative recombination (up to 70%) and have been identified by TEM as misfit dislocations.

1. INTRODUCTION

Strained Si$_{1-x}$Ge$_x$ alloy layers and SiGe quantum well (QW) structures grown on Si have become the subject of extensive research interest due to the possibility of improved device performance and capabilities. There is a specific interest for the optoelectronic industry in Si/SiGe heterojunction technology with the potential for both Si based optoelectronic detectors and emitters. Photoluminescence (PL) spectroscopy has been used to characterize Si/SiGe heterostructures and quantum wells grown by molecular beam epitaxy (MBE). Very weak or simply broad features were observed with only a very limited number of reports of well resolved excitonic features.

The application of CVD growth techniques for SiGe epitaxial growth, has demonstrated that high quality fully strained SiGe quantum wells on Si can be grown. In addition, samples grown (growth temperature≈625°C) by a combination of rapid thermal processing and chemical-vapour deposition (RTCVD) show well-resolved band edge luminescence from excitons confined in the fully strained SiGe quantum wells (Sturm et al 1991). These recent advances have revitalized interest in the possibility of Si based optoelectronic devices. Therefore a detailed analysis of the optical properties of such structures is required.

We have recently begun to develop a SEM-CL system for mapping luminescence features both laterally and in depth. We show for the first time that SEM-CL can be used to characterize SiGe quantum wells at low temperatures. We report here the first low temperature CL spectra containing well-resolved band edge luminescence and CL images from SiGe quantum wells.

2. EXPERIMENTAL

CL spectra were recorded from both single (SQW) and multiple (MQW) quantum wells with different well widths grown by RTCVD on Si(100) substrates. In addition, CL measurements were made on SiGe/Si SWQs and MQWs grown by other variants of CVD growth for comparison; these results will be published elsewhere. In this study we concentrate on the CL measurements recorded from two single strained $Si_{1-x}Ge_x$ samples grown by RTCVD. The samples consisted of a single strained epitaxial $Si_{1-x}Ge_x$ quantum well on a Si buffer layer ($\approx 1\mu m$) with a thin ≈ 150 Å Si capping layer. Both samples had a germanium content of 20%, and well widths of 33 Å and 500 Å. These values were determined by high resolution transmission electron microscopy and Rutherford backscattering spectroscopy (Xiao *et al* 1992). PL measurements have also been carried out on these samples at 4.2 K.

CL measurements were made at T$\approx$5 K using a JEOL JSM 35C SEM fitted with an adjustable cold stage, a retractable off-axis paraboloidal collector mirror and a grating monochromator (Mono CL, Oxford Instruments). The detector employed was a North Coast germanium diode detector. CL spectra were recorded using beam energies (E_o) from 10-35 keV with different beam currents (I_b) between 0.1-100 nA. The CL spectra were recorded with a spectral resolution of 2 nm.

3. RESULTS AND DISCUSSION

CL spectra recorded from the 33 Å well sample and the 500 Å well sample are shown in Figures 1a and 1b respectively. The spectral features and band positions are the same as found in the PL spectra (Xiao *et al* 1992), although at much lower resolution. The spectral features are attributed to an exciton bound to a shallow impurity. The highest energy feature is a no-phonon (X_{NP}) transition and at lower energies are the phonon replicas. The transver-

a

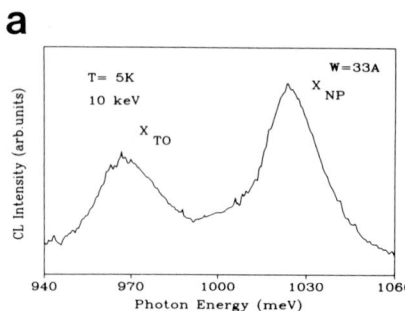

b

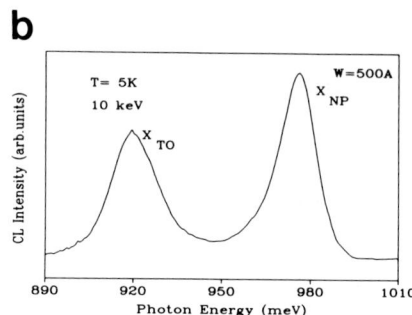

Figure 1.   CL spectra recorded at T$\approx$5 K, E_o=10 keV, I_b=0.1 nA, a) QW $Si_{0.8}Ge_{0.2}$, thickness 33 Å, b) QW $Si_{0.8}Ge_{0.2}$, thickness 500 Å.

se acoustic (TA) phonon replica cannot be resolved in the CL spectra but can be seen on the low energy side of the X_{NP} band. Also the Si-Si transverse optical (TO) phonon assisted band (X_{TO}) can been seen clearly but the other two TO bands (Si-Ge and Ge-Ge) cannot be resolved.

The spectrum of the 500 Å sample shifts to lower energy due to reduced quantum confinement effects. The PL spectra recorded on the 500 Å sample showed wider linewidths compared to the 33 Å sample, whereas in the CL spectra the situation is reversed. These differences could be associated with the different carrier generation rates in the CL spectra and the PL spectra and local heating effects in the later.

CL imaging measurements were made using the X_{NP} feature. Figure 2a shows the CL image of the 33 Å well recorded at T≈5 K. On inspection of the image it is clear that the luminescence from the wells is not uniform, and contains a low density of "blobs" (30-100 μm). Monochromatic line scans showed that these "blobs" are areas of non-radiative recombination with luminescence reduction of up to 100%. Some areas show different levels of non-radiative recombination; the reduction in luminescence can vary from 65% to 100%. An estimate of the spatial resolution can be made assuming that the energy dissipation volume is a sphere of diameter equal to that determined by the Gruen range (Yacobi and Holt 1986). However we have observed previously that in both bulk Si and SiGe alloys the CL spatial resolution can be greatly affected by the large exciton diffusion lengths. CL spectra recorded as a function of beam energy show that bound exciton luminescence from the underlying Si substrate is only observed at $E_o{\geq}15$ keV. The CL images were recorded at $E_o{=}10$ keV, the spatial resolution was estimated to be of the order of 1 μm, indicating that the shape of these "blobs" is not dependent on the spatial resolution.

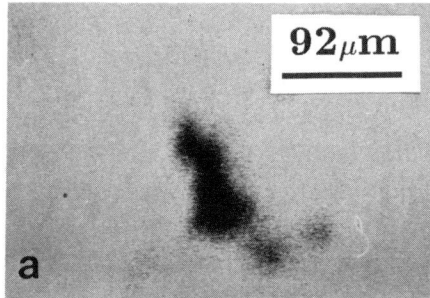 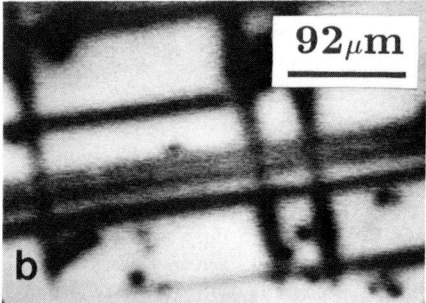

Figure 2.   Monochromatic CL images recorded at T≈5 K, $E_o{=}10$ keV, $I_b{=}0.1$ nA, a) QW $Si_{0.8}Ge_{0.2}$, thickness 33 Å, b) QW $Si_{0.8}Ge_{0.2}$, thickness 500 Å.

On inspection of the CL image of the 500 Å well (see Figure 2b), the "blobs" can be seen but also dark lines oriented along the <110> directions. These dark line features are also areas of non-radiative recombination with a reduction in luminescence intensity by up to 70%. Such CL dark line contrast has been observed in CL imaging experiments on Si/SiGe epilayers containing misfit dislocations (Higgs and Kittler 1993). Transmission electron microscopy was used to examine the areas were the CL measurements were carried out, to establish if there was any correlation with the dark line features and the "blobs". A 1:1 correlation could be established between the dark line contrast and the misfit dislocations. However no unusual structural features could be correlated with the "blobs", suggesting that these areas of non-radiative recombination could be due to point defects trapped by the interfacial strain in the QW. This "blob" type defect has been observed by us with higher densities in CL mapping of QWs grown by other CVD techniques and these results will reported elsewhere.

To investigate the stability of the non-radiative recombination centres, the 500 Å well sample was annealed in the temperature range 100-400°C. CL images were recorded from the sample before and after annealing. The sample was cut into four sections, RCA cleaned then annealed in flowing argon in a RCA cleaned quartz tube. Below 300°C no differences were observed in the CL images. At 300°C and 400°C dramatic changes occurred in the CL images. There were two distinct effects. First the "blobs" became much larger. Figure 3a shows the CL micrograph of a typical region. On closer inspection it was clear there were three distinct regions. As already mentioned there was an area of larger "blobs", a second region with medium sized and small "blobs" (see Figure 3b) and a third region where there were no "blobs" observed.

The second effect which occurred over the whole sample was that the dark line CL contrast at the misfit dislocations had disappeared (see both Figures 3a and 3b). Preferential defect etching revealed that the misfit dislocations were still in the sample. This suggests that the dislocations have been passivated. There have been well documented cases of hydrogen

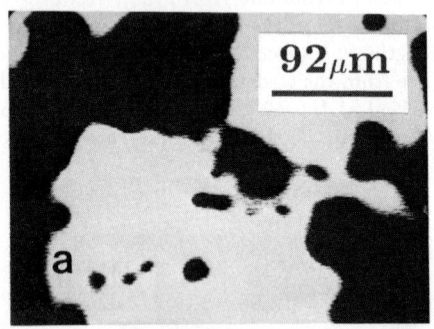 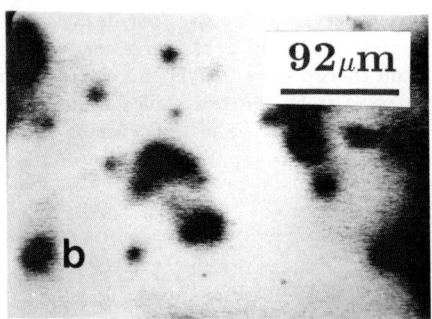

Figure 3.   Monochromatic CL images recorded of the SiGe 500 Å sample,
a) Annealing for 1 hour at T=300°C, b) same treatment as a) different area.

passivation of both deep and shallow levels in Si. It is possible that the passivation could be caused by hydrogen incorporated during layer growth, and that the annealing step has redistributed it. It is difficult to tell if the average density of "blobs" has changed on annealing, but in general there more areas without any "blobs" and dark line contrast. CL spectra were recorded in both regions. It was found that there was a reduction in CL intensity in the areas were the "blobs" became larger, and an increase in the areas were there were no large "blobs". PL spectroscopy revealed that the overall sample luminescence had increased after annealing. This increase in luminescence intensity is caused by a reduction in the non-radiative centres in the material.

This study has demonstrated the feasibility of using SEM-CL to obtain spatial information about the luminescence features and non-radiative processes occurring.

ACKNOWLEDGEMENTS

The authors would like to thank Dr. R. Beanland from the Univesity of Liverpool for carrying out detailed TEM measurements.

REFERENCES

Higgs V and Kittler M 1993 (this conference)
Sturm JC, Manohran H, Lenchyshyn LC, Thewalt MLW, Rowell NL, Noël JP
    and Houghton DC 1991 Phys. Rev. Lett. 66 1362
Xiao X, Liu CW, Sturm JC, Lenchyshyn LC and Thewalt MLW
    1992 Appl. Phys. Lett. 60 1720
Yacobi BG and Holt DB 1986 J. Appl. Phys. 59 R1

Inst. Phys. Conf. Ser. No 134: Section 11
Paper presented at Microsc. Semicond. Mater. Conf., Oxford, 5–8 April 1993

683

3-dimensional mapping of ZnTe by cathodoluminescence spectroscopy

C Trager-Cowan, A Burley, K P O'Donnell, A Naumov[1], K Wolf[1], H Stanzl[1], H P Wagner[1] and W Gebhardt[1]

Physics and Applied Physics Department, Strathclyde University, 107 Rottenrow, Glasgow G4 ONG, Scotland, UK

[1]Institut für Festkörperphysik, Universität Regensburg, Universitätsstrasse 31, D-W-8400 Regensburg, Germany

ABSTRACT: An electron beam of variable energy, has been used to probe the structural properties of epitaxial layers of ZnTe grown on GaAs substrates. The cathodoluminescence intensities of defect and impurity bands relative to the exciton band, were observed to change as a function of position and as a function of depth. In particular, a decrease in defect and impurity luminescence is observed as we probe further from the interface.

1. INTRODUCTION

An electron beam is a versatile and unique probe for studying the microstructure of solids. A tightly focused electron beam can interrogate chosen points of a sample, to provide either topographic information as in scanning electron microscopy, or analytical information as in X-ray spectroscopy or cathodoluminescence spectroscopy (Yacobi and Holt 1990). A variable energy electron beam can be used to interrogate the properties of solids as a function of depth as well as lateral position on the solid's surface, since the mean depth at which an electron beam deposits its energy depends on its kinetic energy. For example in ZnTe, a 10keV electron beam will deposit most energy at a depth of ~110nm, whereas a 30keV electron beam will deposit its maximum energy at a depth of ~0.81μm.

Information on the structure, composition, strain, impurity/doping and defect concentration in solids can be deduced from low temperature luminescence spectra. If cathodoluminescence is excited by an electron beam of variable energy, it should be possible to map sample properties in 3-dimensions (Trager-Cowan et al 1991,1993).

In this paper we describe the use of cathodoluminescence to investigate the structural properties of epitaxial layers of ZnTe, a potential candidate for optoelectronic devices in the green region of the optical spectrum.

2. SAMPLE DESCRIPTION

This paper discusses the properties of two thick (2μm) ZnTe epilayers grown by atmospheric pressure metalorganic vapour phase epitaxy (MOVPE) on (100) GaAs substrates. Layer #35 was grown using diethylzinc (DEZn) and diisopropyltelluride (DiPTe) precursors at a substrate temperature of 350°C; Layer #265 was grown using dimethlyzinc-triethylamine (DMZn-TEN) and DiPTe precursors at a substrate temperature of 340°C. There is a lattice mismatch of 8% between GaAs and ZnTe, (lattice constants of GaAs and ZnTe are 5.653 and 6.101Å respectively at room temperature (Weast and Astle 1982)), that leads to the formation of dislocations in the ZnTe epilayer which may be expected to degrade the optical properties of the ZnTe. Complex defect centres in the dislocation region may act as recombination centres. The observation of defect luminescence bands and their intensity dependence as a function of position and as a function of depth in the ZnTe epilayer are used to characterise the epitaxial layer.

3. DESCRIPTION OF CATHODOLUMINESCENCE SPECTROMETER

The cathodoluminescence apparatus is described in detail by Trager-Cowan et al (1991). In brief, an electron beam of 10-30keV in energy and of moderate current density ($\leq$300μA/cm^2 in a spot size of $\approx$300μm) excites cathodoluminescence from samples mounted on the cold finger of a closed cycle helium cryorefrigerator, cooled to $\approx$15K. The electron beam is normal to the layer and cathodoluminescence detected at right angles to the exciting electron beam, i.e., from the edge of the sample.

4. EXPERIMENTAL RESULTS AND DISCUSSION

4.1 Layer #35

Cathodoluminescence (CL) spectra from layer #35 are shown as a function of position in Fig. 1 and as a function of depth at a fixed position in Fig. 2.

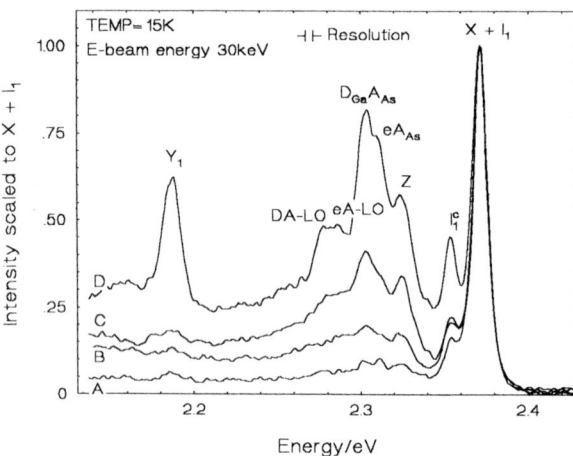

Fig. 1 : CL spectra at separated points A, B, C, D on the surface of layer 35

In both Figs. 1 and 2, the peak labelled X + I$_1$ at 2.371eV includes luminescence from free and bound excitons (Wagner et al 1990, Kuhn et al 1991, Kudlek et al 1992). All spectral intensities are scaled to this peak to allow the relative intensities of the lower energy bands to be compared. The origin of the peak labelled I$_1^c$ is uncertain. Two possible explanations have been advanced: (1) it is an exciton bound to a Si double acceptor (White et al 1974, Kudlek et al 1991) or (2) it is a Zn vacancy, interstitial complex (Kudlek et al 1992). The peak at 2.324eV labelled Z has not previously been reported, we will discuss this peak later. The peak labelled eA$_{As}$ at 2.310eV is attributed to a free to bound acceptor transition, where the acceptor is arsenic diffused from the substrate (Wagner et al 1990) and the peak labelled D$_{Ga}$A$_{As}$ at 2.303eV is attributed to donor acceptor pair recombination, where the acceptor is arsenic and the donor gallium diffused from the substrate (Wagner et al 1990). The peaks labelled eA-LO and DA-LO at 2.285eV and 2.276eV respectively, are phonon replicas of the eA and DA transitions. Finally, the peak labelled Y$_1$ at 2.187eV is thought to originate from the recombination of excitons localised at structural defects such as stacking faults or dislocations and associated defects (Naumov et al 1993).

Spectrum B of Fig. 1 is "typical" of the luminescence from layer #35, at an electron beam energy of 30keV. Most areas of the layer shows strong free/bound exciton luminescence accompanied by weak defect and impurity luminescence; the luminescence appears green to the eye. However, two small regions at the top and bottom of the layer appear "yellow" and show strong defect, eA and DA luminescence. Spectrum D of Fig. 1 was obtained from a "yellow" spot. We tentatively suggest two possible reasons for this "yellow" luminescence. Since higher growth temperatures lead to the generation of defects and the diffusion of impurities, non-uniformity in the substrate temperature during growth may lead to regions of poor growth. Alternatively, regions of poor morphology formed during substrate preparation may lead to degradation of the optical quality of the overgrown epilayer through enhanced diffusion of impurities and defects.

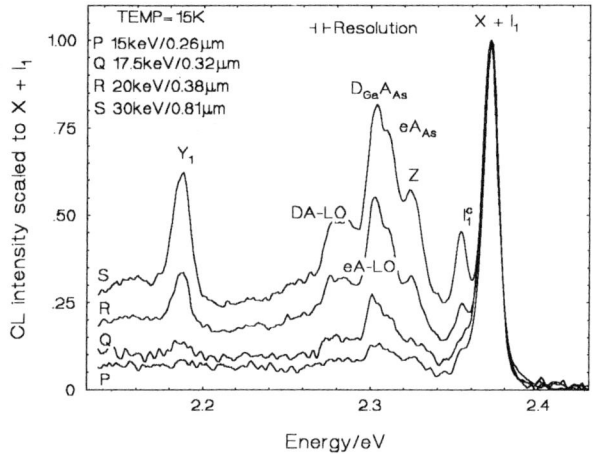

Fig. 2 : CL spectra from "yellow" spot as a function

of depth from layer 35

Fig. 2 shows a depth profile of a "yellow" spot. The penetration depths listed with the beam energies in the figure are calculated using Monte-Carlo simulation of the electron trajectories (Napchan and Holt 1987) and are the depths at which the rate of energy deposition of the electron beam is a maximum. As the electron beam energy decreases and the penetration depth of the electron beam decreases, the luminescence intensity of defect, eA and DA luminescence bands decrease relative to the luminescence intensity of the principal exciton band and the "yellow" spot is observed to turn green. This "improvement" in the optical properties with decreasing depth, is consistent with a reduction of defect and impurity concentration near to the surface of the epilayer as we probe further from the substrate. A reduction in defect density is to be expected, since misfit dislocations are localised at the interface. The reduction in impurity concentration with distance from the substrate is also to be expected, since the main source of impurities is gallium and arsenic diffused from the substrate.

From Fig. 2 it can be seen that there is a marked difference in the relative intensities of the defect and impurity emission with respect to the exciton emission at 15keV and 20keV respectively. The maximum penetration depth of a 15keV electron beam is calculated from Monte-Carlo simulations to be 1.6µm, and the maximum penetration depth of a 20keV beam is 2.4µm. The epilayer is ≈2µm thick, We deduce from this observation, that the defect region is confined to within ≈0.5µm of the interface.

The relative intensity of the Z band scales with those of the I_1^c and Y_1 defect bands, therefore it would appear that this band must also be associated with defects.

4.2 Layer #265

Fig. 3 shows the CL spectra from layer #265 at 10keV and 30keV. The peak labelled X + I_1 at 2.370eV, again includes luminescence from free and bound excitons. In strong contrast to layer #35, luminescence from layer #265 is dominated by the donor acceptor pair emission peaking at 2.206eV. A similar band with a zero phonon line at 2.230 eV, was attributed to recombination of chlorine donors and copper acceptors by Bittebierre and Cox 1986 and Ogawa et al 1991. However, in our case iodine is a more likely candidate for the donor, as epilayers had been intentionally doped with iodine during previous growth runs and some iodine may still have been present in the growth kit when layer #265 was grown. The growth kit is also thought to be the source of copper contamination.

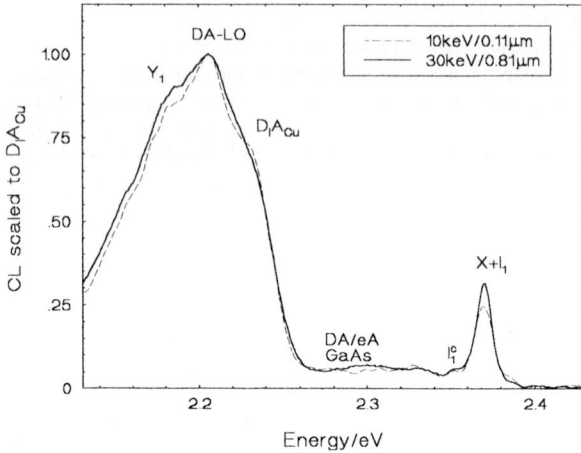

Fig 3 : CL spectra as a function of depth from layer 265

Since the growth kit incorporates copper and iodine at a constant rate, we assume a uniform concentration of iodine and copper in the layer. In this case, therefore, we scale the spectra obtained at 10keV and 30keV respectively, to the DA band. From this scaling, it can be seen that the only features which appear to change appreciably with depth are the exciton band and the shoulder at 2.186eV. The increase in the intensity of the exciton band is tentatively attributed to an increased contribution from the acceptor bound exciton due to arsenic. The shoulder at 2.186eV is due to the defect Y_1 band. The peak at 2.353eV is again the I_1^c band.

The features around 2.3eV, as for layer #35, are attributed to eA and DA recombination due to the presence of gallium and arsenic diffused from the substrate.

ACKNOWLEDGEMENTS

We would like to thank Professor E C Lightlowlers and Dr A T Collins of King's College, London for the loan of the cathodoluminescence system used in this work. We would also like to thank David Clark for his help in the laboratory. A Naumov is supported by a Fellowship from the Alexander von Humboldt Foundation.

REFERENCES

Bittebierre J and Cox R T 1986 Phys Rev B 34 2360

Kudlek G, Presser N, Gutowski J, Hingerl K Abramof E and Sitter H 1991 Semicond. Sci. Technol. 6 A90

Kudlek G, Presser N, Gutowski J, Hingerl K, Abramof E, Pesek A, Pauli H and Sitter H 1992 J. Cryst. Growth 117 290

Kuhn W, Wagner H P, Stanzl H, Wolf K, Wörle K, Lankes S, Betz J, Wörtz M, Lichtenberge D, Leider H, Gebhardt W and Triboulet R 1991 Semicond. Sci. Technol. 6 A105

Naphan E and Holt D B 1987 Inst. Phys. Conf. Ser. 87 eds A G Cullis et al (London: IOP) pp733-6

Naumov A, Wolf K, Reisinger T, Stanzl H and Gebhardt W 1993 Physica B in press

Ogawa H, Nishio M, Ikejiri M and Tuboi H 1991 Appl. Phys. Lett 21 2384

Trager-Cowan C, Parbrook P J, Clark D, Henderson B, O'Donnell K P, Cockayne B and Wright P J 1991 Inst. Phys. Conf. Ser. 117, eds A G Cullis and N J Long (Bristol: IOP) pp715

Trager-Cowan C, Kean A, Yang F, Henderson B and O'Donnell K P 1993 Physica B in press

Wagner H P, Kuhn W and Gebhardt W 1990 J. Cryst Growth 101 199

Weast R C and Astle M J (eds) 1982 CRC Handbook of Chemistry and Physics (Florida: CRC Press) E-98

White A M, Dean P J and Day B 1974 J. Phys C:Solid State Phys. 7 1400

Yacobi B G and Holt D B 1990 Cathodoluminescence Microscopy of Inorganic Solids (New York: Plenum)

A numerical method for simulating cathodoluminescence contrast from localised defects

KL Pey, DSH Chan and JCH Phang

Centre for Integrated Circuit Failure Analysis and Reliability, Faculty of Engineering, National University of Singapore, 10 Kent Ridge Crescent, Singapore 0511

ABSTRACT: A three-dimensional model is given for cathodoluminescence (CL) contrast from localised semiconductor defects. Electron-solid interaction is modelled by a Monte Carlo method and defects are represented by regions of enhanced non-radiative recombination. Optical losses are also taken into account. Differences in the image profiles of localised bulk defects and dislocations which intersect the top surface barrier perpendicularly suggest a method for differentiating between the two types of imperfections. A technique for determining the depth of a subsurface defect is also illustrated.

1. INTRODUCTION

Cathodoluminescence (CL) has been widely employed to investigate electrically active lattice imperfections in luminescent semiconductors. A number of theoretical methods for calculating CL signals from localised defects have been proposed (Löhnert and Kubalek 1984 and Jakubowicz 1986). There have, however, been limited attempts at incorporating a more realistic and accurate carrier generation function into CL models. For example, Löhnert and Kubalek (1984) and Jakubowicz (1986) analysed the contrast of dislocations but only a uniform generation sphere or point source was used. Moreover, the approximations may fail either when the source is sufficiently close to the defects or when the excitation region is comparable to the defect size. In addition, mathematically, it is very difficult to extend these models to investigate CL response in the presence of multiple defects. Other workers (Czyżewski and Joy 1990 and Pasemann and Hergert 1986) presented models mainly to evaluate the ratio of CL contrast to electron beam induced current (EBIC) contrast collected simultaneously over an individual defect.

This paper presents an approach incorporating realistic electron-hole (e-h) pair generation obtained from Monte Carlo calculations into a three-dimensional CL model for a semiconductor having localised defects. The three dimensional carrier diffusion equation is represented by a set of finite difference equations. This approach offers the flexibility of analysing different kinds of defect structures in the bulk by simply reducing the non-radiative recombination lifetime in the region of influence. The contrast profiles of a subsurface localised defect and a dislocation which intersects the top surface perpendicularly in GaAs materials are reported. Arising from the difference in the contrast behaviour in CL images, a method for differentiating dislocations and bulk defects based only on the energy dependent CL contrast is suggested. Additionally, it will be shown that by locating the beam energy at which minimum CL contrast occurs, the depth of a bulk defect can be obtained.

2. THEORETICAL BASIS

The Monte Carlo procedure described by Phang *et al.* (1992) is used to evaluate the rate of energy dissipation of the electron beam on its way through the sample. The sample is assumed to be semi-infinite (Fig. 1), bounded only by the top surface at z=0, and divided into volume elements with dimensions δx, δy and δz and the three dimensional spatial energy dissipation δE of all electrons traversing the sample volume is calculated and stored as a matrix of δE versus x, y and z using the Nearest-Grid-Point method (Hockney and Eastwood 1986).

For a homogeneous semiconductor having minority carrier diffusion coefficient D and lifetime τ, and a surface recombination velocity v_s, a localised defect under the surface can be represented by a bounded region of space F (Fig. 1), where the minority carrier lifetime τ' is lower than that in the rest of the semiconductor. Within F, $\tau(r) = \tau'(r)$, elsewhere $\tau(r)=\tau$. Introducing $L=(D\tau)^{1/2}$ and $L'(r)=(D\tau'(r))^{1/2}$, the three dimensional continuity equation describing the diffusion process of the minority carriers is

$$\nabla^2 p(r) - \frac{1}{D}g(r) + \frac{1}{L^2}p(r) + \gamma(r)e(r)p(r) \tag{1}$$

where   and

$$e(r) - \begin{cases} 1 \text{ for } r \text{ inside } F \\ 0 \text{ elsewhere} \end{cases} \tag{2} \qquad \gamma(r) - \frac{1}{D}\left|\frac{1}{\tau'(r)} - \frac{1}{\tau}\right| - \frac{1}{L'(r)^2} - \frac{1}{L^2} \tag{3}$$

is defined as the 'strength' of the defect (Donolato 1978). If $L'(r) \ll L$ i.e. $\tau'(r) \ll \tau$, $\gamma(r) \approx 1/L'(r)^2$ and the defect strength becomes independent of the diffusion length or the lifetime of the host material.

The top surface boundary is characterised by

$$D\frac{\partial p(r)}{\partial Z}\bigg|_{Z-0} - v_s\, p(r) \tag{4}$$

Eqns. (1) to (4) are discretised using a central-difference quotients scheme and solved by Successive-Over-Relaxation. The light generation and loss in transmission were calculated as described in Phang *et al.* (1992).

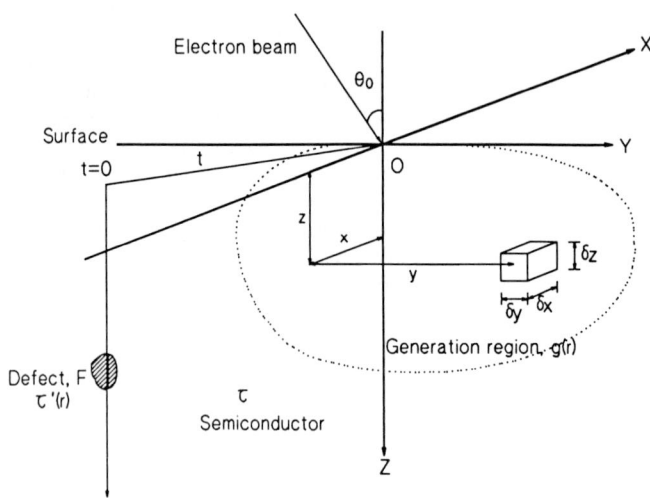

Fig. 1 Physical model used for simulating the CL contrast of a localised lattice imperfection.

3. RESULTS AND DISCUSSION

Fig. 2 presents an example of CL contrast profiles versus distance from the defect centre for a threading dislocation which intersects the top surface barrier at right angles for various beam energies for a GaAs sample. The volume F of the dislocation is represented by a series of nodes with non-zero $e(r)$ and $\gamma(r)$ connected in a form of chain. The contrast $C(t,E)$ is defined as $(I_{CL(t,E)} - I_{CL(\infty,E)})/I_{CL(\infty,E)}$, where $I_{CL(\infty,E)}$ and $I_{CL(t,E)}$ are the CL intensities from the material far away from any defect and at a distance t from the centre of the defect of interest respectively. The full width half minimum $W(E)$ for the contrast profiles which is related to the CL image resolution increases very rapidly with beam energy. However, the minimum contrast $(C(t=0,E))$ $C(E)$ decreases with beam energy.

Fig. 3 gives contrast responses for a bulk point defect located at a depth d of 0.8µm. These results show that the $C(E)$ has a minimum contrast $C(E)_{min}$ at about 12.5keV. Conversely, the $W(E)$ shows only minor variations with respect to beam energy up to about 12.5keV, but for higher beam energies, the curves broaden considerably.

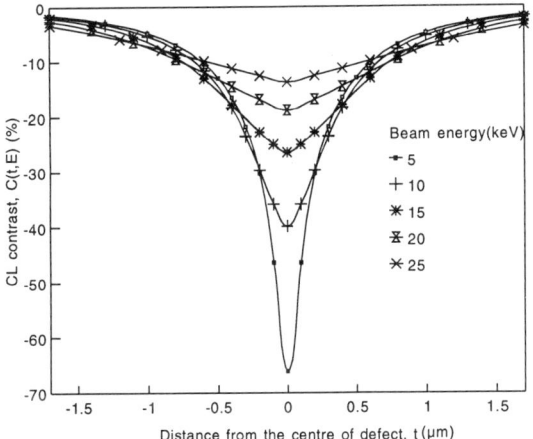

Fig. 2 CL contrast profiles for various values of beam energies for a threading dislocation perpendicular to the surface in GaAs; L=1µm, optical absorption coefficient $\alpha=0.1\mu m^{-1}$ and normalised v_s, $V_s=\infty$, $\theta_o=0°$ and $\gamma(r)=1479\mu m^{-2}$.

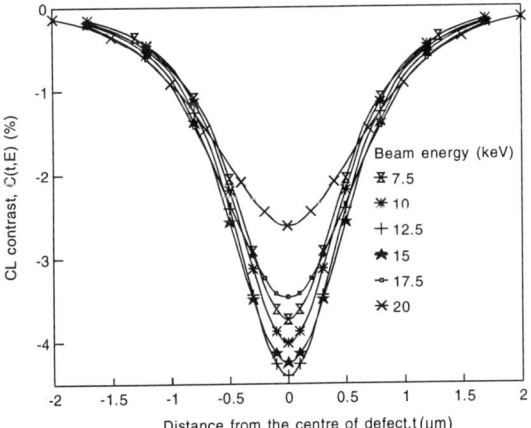

Fig. 3 CL contrast profiles for various values of beam energies for a point defect at a depth of 0.8µm in GaAs; L=1µm, $\alpha=0.1\mu m^{-1}$, $V_s=\infty$, $\theta_o=0°$ and $\gamma(r)=1479\mu m^{-2}$.

The contrast responses of a dislocation (Fig. 2) differ considerably from those of a point defect (Fig. 3). As in the EBIC contrast profiles reported by Donolato (1979), the CL contrast of a threading dislocation intersecting perpendicularly to the top sample surface is a decreasing function of beam energy, whereas the contrast of a subsurface point defect exhibits a contrast reversal phenomenon. Identification of these two imperfections can hence be carried out easily using the different contrast behaviours. For example, one can perform line-scan experiments or simply observe the contrast responses at different beam energies across the defect of interest. If the dark spot is present at all beam energies or contrast features similar to Fig. 2 are observed, it corresponds to a line defect penetrating into the bulk material. If it appears very clearly only at a fixed beam energy or contrast features similar to Fig. 3 are found, it is a subsurface localised bulk defect.

Figs. 4, 5 and 6 show the computational results of the effects of the defect depth, strength and size Wd, respectively, on $C(E)$ and $W(E)$. One of the important features to observe from these graphs is that the beam energy E_{min} at which $C(E)_{min}$ occurs is only a function of d and insensitive to the variation to defect strength and size. Fig. 7 shows the $C(E)$ versus beam energy for selected values of d. All curves exhibit a $C(E)_{min}$. This minimum of each curve corresponds to a particular d.

As an application of these observations, a plot relating the depth to the beam energy at which $C(E)_{min}$ occurs can be constructed. Practically, suppose the $C(E)_{min}$ from the CL image or the line-scanning profiles across a defect is observed at E_{min}. The depth of the defect is then obtained by substituting E_{min} into the plot.

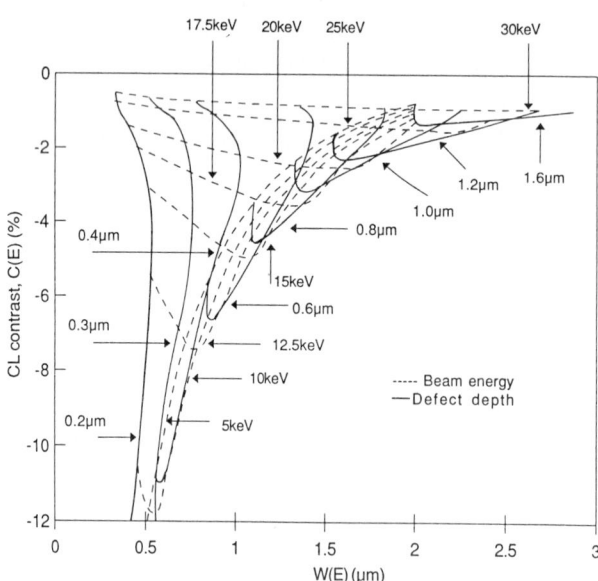

Fig. 4 Plot of $C(E)$ versus $W(E)$ of point defects for different defect depths; $Wd{\rightarrow}0\mu m$ and $\gamma(r)=1479\mu m^{-2}$.

4. CONCLUSION

A three dimensional numerical cathodoluminescence model has been developed. It has been applied successfully to study the contrast of cathodoluminescence images of localised defects in semiconductors. Image contrast of dislocations which intersect the top surface perpendicularly exhibits a decreasing function of the beam energy, whereas for subsurface defects a contrast reversal is observed. This difference in contrast property can be used for distinguishing the above two defects.

The beam energy at which minimum CL contrast occurs is only a function of the depth of the defect and independent of the defect strength and size. A technique for determining the defect depth is also established.

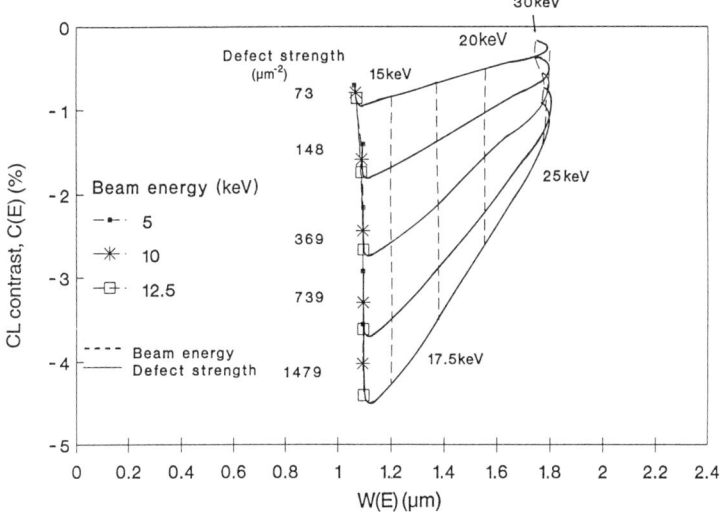

Fig. 5 Plot of **C(E)** versus **W(E)** of point defects for different defect strengths; **Wd→0μm** and **d=0.8μm**.

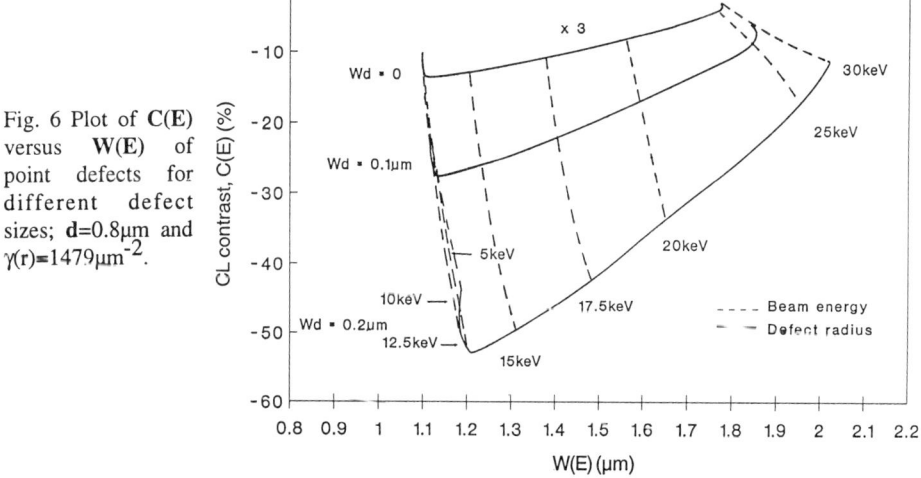

Fig. 6 Plot of **C(E)** versus **W(E)** of point defects for different defect sizes; **d=0.8μm** and γ(r)=1479μm^{-2}.

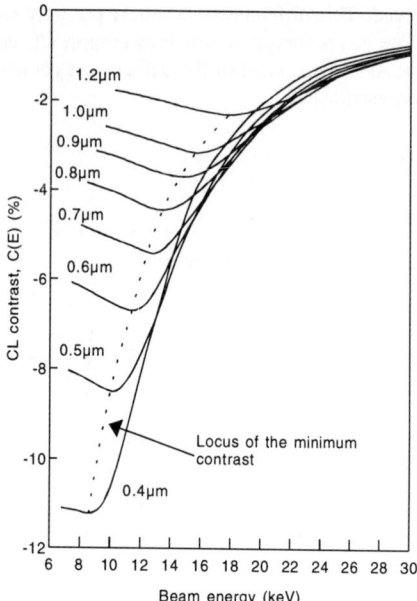

Fig. 7 **C(E)** profiles for point defects located at different depths.

REFERENCES

Czyżewski Z and Joy DC 1990 Scanning 12 5
Donolato C 1978 Optik 52(1) 19
Donolato C 1979 Appl. Phys. Lett. 34(1) 80
Hockney RW and Eastwood JW 1981 Computer simulation using particles (McGraw-Hill:New York)
Jakubowicz A 1986 J. Appl. Phys. 59(6) 2205
Löhnert K and Kubalek E 1984 Phys. Stat. Sol. a83 307
Pasemann L and Hergert W 1986 Ultramicroscopy 19 15
Phang JCH, Pey KL and Chan DSH 1992 IEEE Trans. on Electron Devices 39(4) 782

Inst. Phys. Conf. Ser. No 134: Section 11
Paper presented at Microsc. Semicond. Mater. Conf., Oxford, 5–8 April 1993

693

A calibrated spectroscopic cathodoluminescence system for SEM

E Napchan, D O'Neill and C L M Zanotti–Fregonara

Dept. of Materials, Imperial College, London SW7 2BP

ABSTRACT: A system for cathodoluminescence (CL) mode scanning electron micros-copy which can detect radiation from 400 to 1800 nm is described. This incorporates a collecting mirror inside the SEM, which directs collected light to a periscope attached to a monochromator fitted with a photomultiplier and a liquid nitrogen cooled Germanium detector. The complete system is computer controlled, allowing it to operate in either spectroscopic or imaging mode. Wavelength accuracy and spectral radiance (absolute light intensity) calibration of the system have been carried out, and the results are discussed in this paper. Experimental quantitative CL results using the above system are presented.

1. INTRODUCTION

Cathodoluminescence is the name given to light emitted from a crystal after it has been excited by an electron beam (Yacobi and Holt 1990). If the crystal is a semiconductor then the radiative recombination of excited electron–hole pairs produces CL, and is thus characteristic of the particular band structure of the semiconductor. The commonest radiative transition in lightly doped semiconductors at room temperature is band–to–band. If the SEM is used to provide the excitation (beam voltages of 1 to 40 keV) and a spectroscopic system is used to collect and detect the cathodoluminescence then the bandgap of the material can be measured. Extension of the spectroscopic range to 1.8 μm (near–infrared) allows investigations to be made on devices that operate in the 1–1.6 μm region of the spectrum, which is of high interest for telecommunications.

If the CL system is calibrated absolutely then various other optical and electronic param-eters can be measured (Pankove 1967). A radiance calibrated CL system opens the doors to quantitative CL analysis of semiconducting materials. Uncorrected system output values do not reflect the true emission characteristics of the specimen. Correcting this data yields the true relative spectral intensity distribution of the emission, allowing its comparison with theo-retical expressions. In addition calibration is the key to improvement of the detection system because it provides knowledge of its limitations.

This paper describes and discusses the steps made in the design, construction and testing of such a system.

2. EXPERIMENTAL SYSTEM

An Oxford Instruments CL301 mirror–cryostat assembly mounted on a side port of a JEOL 840A SEM is used for light collection inside the microscope. The light is focused and directed by means of a special turret to a Bentham M300EA monochromator fitted either with a 69×69 mm holographic 1800 lines/mm grating (0–0.9 μm wavelength range, blazed for 0.45 μm) or with a 900 lines/mm grating (0–1.8 μm range, blazed at 1 μm). From the monochromator the light goes either to a photomultiplier tube (PMT) or to a Germanium diode detection system. A schematic diagram of the CL experimental system is shown in Fig. 1.

For the detection of near–infrared CL (in the 0.8–1.8 μm range) the focusing optics were redesigned to allow the insertion of a plane gold mirror, M in Fig. 1, at an angle of 45 degrees in the light path after the exit slit of the monochromator, to direct the light to a pair of quartz lenses (L2 and L3) which focus it on the germanium crystal of a North Coast 817L solid state detector assembly. The signal from the Ge detector can be converted to a pulsed frequency (Hz) signal by a V–f converter and then fed to the microcomputer, to use the same pulse counting features as the PMT detection system.

The mirror and lens system is housed in a special adapter which enabled the mirror M to be retracted from the horizontal light path so that both a PMT and the Ge detector could be used in quick succession. It is also possible to connect the PMT signal directly to a current meter and to the auxiliary channel of the SEM CRT screen, thus enabling mono and pan–chromatic images to be acquired. The infrared system can also be operated in a phase sensitive detection mode, by feeding simultaneously a square wave signal to a beam blanking system (Deben Research) and to the reference channel of a lock–in amplifier. The measured signal from the Ge diode, in this case, is fed to the input channel of the lock–in amplifier.

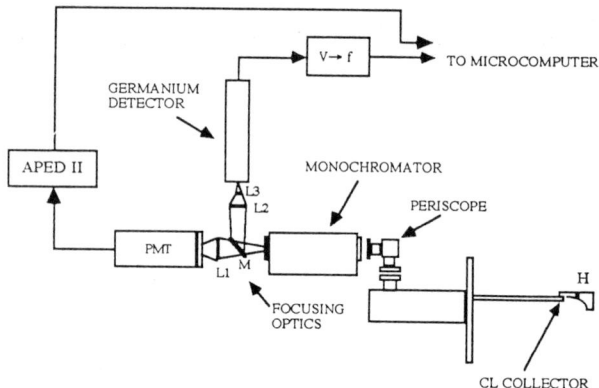

Fig. 1 — Spectroscopic Cathodoluminescence System (L1: crown glass lens, M: mirror, L2,L3: quartz lenses)

The S20, S1 or GaAs PMT is mounted in a Products for Research Inc. cooled housing, together with a BK7 glass lens, L1 in Fig. 1. The PMT output is fed to an Amplifier/Discriminator (APED II), which provides pulses to the data acquisition and display software in a dedicated PC type computer.

The computer system controls the monochromator stepping action and signal acquisition parameters through an IEEE channel, and through dedicated I/O lines. Further details of the system's software are discussed in the following section.

3. ACQUISITION AND ANALYSIS SOFTWARE

The system is controlled by a PC–AT type microcomputer with two internal expansion boards. An IOTech Personal488 board drives the grating and filter stepper motor drives, allowing commands to be sent using high–level languages. The other internal board is an Amplicon PC–14A Programmable I/O and Counter Timer Interface. A 16 bit software controlled timer in this board is used for the measurement of up to 65000 pulses for each acquisition channel (selected wavelength). In addition to the control of the boards the software program can automatically insert order sorting filters (at 0.4 and 0.7 μm wavelength) during spectrum acquisition, for removal of higher order diffraction peaks.

Trigger control for the pulse counter, which determines the dwell time per acquisition channel is performed with one of the I/O lines in the interface board according to the acquisition time measured with the computer main clock. The travel speed of the stepper motors in the monochromator has been taken into consideration in the controlling program, by introducing suitable delays during their operation.

Intensity and wavelength data are displayed during acquisition, with automatic resizing of x and y scales. If a spectrum is to be stored for further analysis, relevant comments related to the experimental conditions and specimen are input, stored and displayed along with the acquired spectrum.

Whenever a previously acquired spectrum is displayed it is possible to overlay on it calibration data built into the program. This includes the higher intensity spectral lines for mercury, xenon and neon, and consists of first and second order diffraction wavelengths. The reference data selected is overlayed on the acquired spectrum as vertical lines, with the line's height adjusted according to the reference intensity in the measured wavelength range. This feature is useful for checking the wavelength calibration of the monochromator.

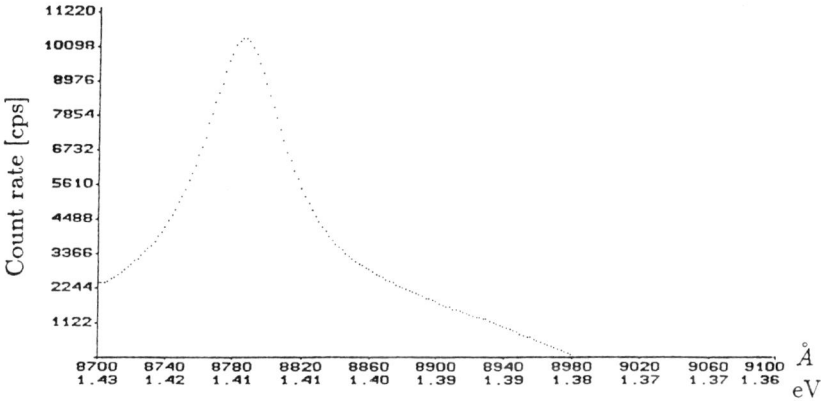

Fig. 2 — Hardcopy System Output of Acquired Spectrum

The software produces two outputs: a graphical plot of intensity against wavelength (on an HP type laser printer in about 2 seconds), and a text file with comments and the measured intensity and wavelength data pairs. An example of a printout is shown in Fig. 2. The data stored in the text file can be easily transferred to other software for further analysis and plotting. A smoothing routine based on Hayden (1987) can be used on-line with the stored data files to remove noise incompatible with the 'instrument (transfer) function'.

4. WAVELENGTH AND SPECTRAL CALIBRATIONS

A spectral calibration was performed on the system to allow for errors in wavelength due to a misaligned monochromator grating. The grating was aligned on its mount, and a standard mercury vapour lamp was used to illuminate the entry slit of the monochromator. A number of emissions were observed in the range 0.4 – 1.8 μm using the S20 PMT and the Ge detector, and these were compared with literature values source (Reader and Corliss 1991). Fig. 3 presents data on the absolute wavelength error ($\lambda_{exp} - \lambda_{lit}$) plotted against the true wavelength (λ_{lit}). It can be seen from this figure that the wavelength error was a maximum of 1.4 nm in the range 0.4 to 0.8 μm (visible), and up to 2.0 nm in the range 0.8 to 1.8 μm (near infrared).

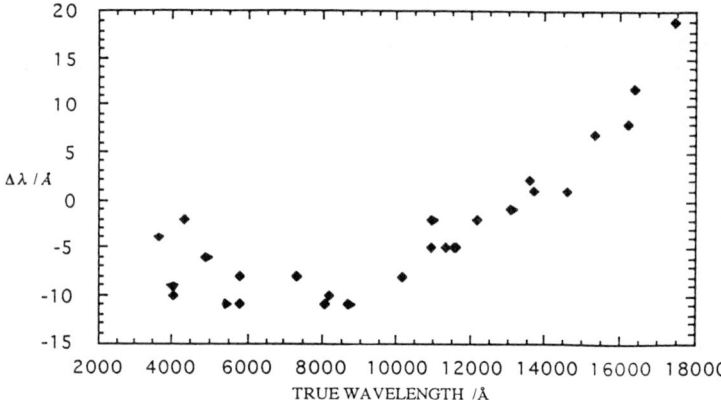

Fig. 3 — Wavelength Calibration $\Delta\lambda$ = error in wavelength, against true wavelength

A radiance calibration was performed to correct for attenuation of the CL intensity by the components of the spectroscopic system prior to detection. The spectroscopic CL system measures R, the number of pulses per second recorded by the detector (Steyn et al. 1976):

$$R_\lambda = Q_\lambda E_\lambda F_\lambda \{1 - A_{B\lambda}\} N_\lambda \qquad (1)$$

where N is the quantum yield (number of photons of wavelength λ generated per second in the material) and $(1 - A_B)$ is the fraction of those photons that escape from the specimen (A_B is the fraction lost internally). F_λ is the fraction of the externally emitted photons that are collected and transmitted to the monochromator, E_λ is the fraction of photons delivered to the input slit of the monochromator that reach the detecting surface and Q_λ is the quantum efficiency of the detector in use. The aim of the radiance calibration was to obtain QEF for all detectors in use.

If a blackbody radiator is used as the source in Eq. 1 instead of the cathodoluminescent solid, then the following equation applies:

$$\frac{R_\lambda}{\Phi_p} = Q_\lambda E_\lambda F_\lambda \qquad (2)$$

where $\Phi_p = L_p \Omega A$ is the photon flux of the radiator, and L_p is the photon sterance of the source as given by Plank's law for a blackbody radiator, Ω is the solid angle of emission, and A is the radiator area.

Thus the correction factor QEF obtained in a radiance calibration will be identical to that given by Eq. 1 provided the experimental geometry of calibration is such that the product ΩA does not differ significantly from the one occurring during normal cathodoluminescence experimentation.

The source used for the calibration was a standard tungsten lamp, with intensity calibration traced to a National Physical Laboratory standard lamp. To account for the fact that in the infrared the emission of such a lamp is not strictly that of a black body, a correction function B_λ was used in Eq. 2, as described by Jones (1970). Separate experiments were used to determine: QE, QEL and $QEFL$ (where L is the attenuation of the signal by the lens used in the calibration to focus the source light), from which the final calibration result QEF was calculated. Fig. 4 shows the main result of the calibration. Each QEF curve represents the efficiency of the system with a different detector. Their relative performance can be derived from this figure. Regions of steep gradients in Fig. 4 are to be avoided because there the true band shape of the detected emission cannot be correctly deduced.

A further feature of Fig. 4 is the region of high fluctuation in the Germanium photodetector efficiency curve at about 1.35 μm. This region can be matched to theoretical absorption curves for O–H radicals in the quartz lenses (here used for the infrared extension).

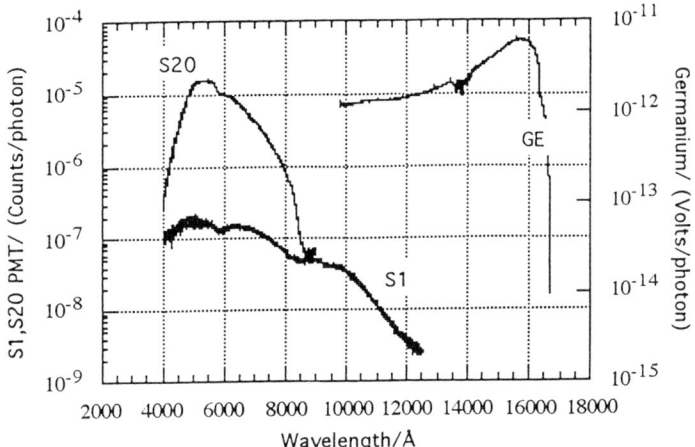

Fig. 4 — $Q_\lambda E_\lambda F_\lambda$ for the S1 and S20 PMT's and for the Ge detector in the CL system

5. EXPERIMENTAL RESULTS

Fig. 5 shows CL emission from a n–type InP substrate material of doping density 10^{19} cm^{-3} acquired with the germanium detector (full line). The specimen was cooled using the continuous flow cryostat technique with liquid nitrogen. The sample temperature is estimated to be about 100 K. The dotted line is a plot of the band to band radiative recombination characteristic for this direct band gap material at this temperature (using equations from Mooradian and Fan 1966). Comparison of two curves indicates that there may be other processes occurring at the band edge, such as self–absorption, that affect the radiative recombination.

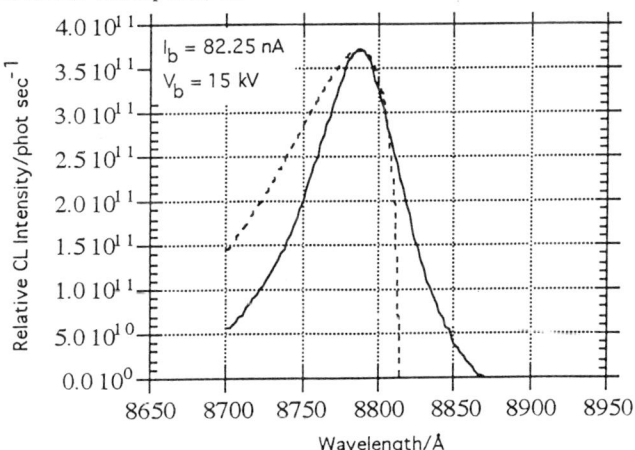

Fig. 5 — Corrected experimental CL emission for InP (solid line) and theoretical characteristic (dotted line) at 102 K

A further example of the application of the system and the radiance calibration presented here is shown in Fig. 6. In this case, the edge emission of InP at room temperature has been measured both with the S1 PMT and the Ge photodetector. The left hand figure shows the uncorrected spectra, while after correction (shown on the right hand side) it can be seen that the bands exhibit the same peak and similar emission band shape.

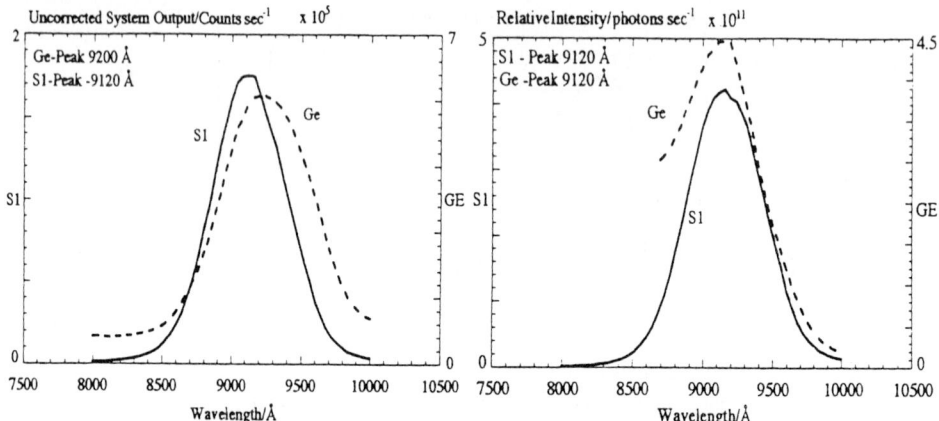

Fig. 6 — Uncorrected and corrected CL emission for InP at room temperature (S1 PMT and Ge photodetector)

6. CONCLUSIONS

This paper discusses the additions and changes to a spectroscopic CL system used with an SEM. In addition to the spectroscopic capabilities the system can also be used for the display of mono and pan–chromatic images.

The main advantages of the present system are its high spectral resolution, the output of which can be corrected for the losses and variations in signal attenuation due to the various components. System operation and data acquisition are by means of a user friendly software, allowing easy operation and data management. A further advantage of this system is its low component cost, which allows a broader access to this powerful technique.

The results presented in the previous figures provide information about the experimental system and the properties of the devices under investigation. Fig. 4 clearly describes the limitations of the detection system, and the applicability of the various detectors for specific wavelength ranges. Fig. 5 shows the comparison of corrected experimental data with quantum mechanically derived expressions for band–to–band emission in semiconductors.

The use of the calibration procedures discussed in this paper is shown in Fig. 6 for the same emission measured with two different detectors. The correction procedures are essential for removing from the measurements the effects of the experimental system, revealing the basic material properties.

REFERENCES

Hayden H C 1987 Computers in Physics **12** pp 74–75
Jones O C 1970 Quantum Metrology Div. Rep. **Qu.** 14
Mooradian A and Fan H Y 1966 Phys. Rev. **148** 873
Pankove J I 1975 Optical Processes in Semiconductors (New York: Dover)
Reader J and Corliss C H 1991 CRC Handbook of Physics and Chemistry E–205
Saba F M 1986 Cathodoluminescence Measurement (PhD Thesis: University of London)
Steyn J B, Giles P and Holt D B 1976 Journal of Microscopy **107** 107
Yacobi B G and Holt D B 1990 Cathodoluminescence Microscopy of Inorganic Solids (New York and London: Plenum Press)

Inst. Phys. Conf. Ser. No 134: Section 11
Paper presented at Microsc. Semicond. Mater. Conf., Oxford, 5–8 April 1993

SEM-CL/EBIC studies of dislocations in semiconductors

J Schreiber, S Hildebrandt and H S Leipner

Martin–Luther–Universität Halle–Wittenberg, Fachbereich Physik, Friedemann–Bach–
Platz 6, D – O – 4020 Halle (Saale), Germany

ABSTRACT: We report on experimental studies of the recombination activity of dislocations in the III–V semiconductors GaAs, GaAsP and GaP. The SEM–CL and EBIC
High Resolution Defect Imaging (HRDI) of configurations of grown–in and introduced
glide dislocations is discussed. By detailed defect contrast measurements the recombination activity of surface–parallel or –perpendicular dislocation lines was characterised
quantitatively. The evaluation of the defect strength for the surface–parallel dislocation
was carried out allowing a realistic generation distribution as well as a large defect
strength in the comprehensive theoretical description.

1. INTRODUCTION AND PROBLEMS

SEM–CL (CathodoLuminescence) and SEM–EBIC (Electron Beam Induced Current)
are suitable tools for recognition and identification of electronically active defects in
semiconductor specimens. The scanning electron microscopy (SEM) in the recombination-
sensitive signal modes allows observations of the defect density, structure and distribution.

The intrinsic CL and EBIC dislocation contrast has a dimension of the order of several
tenths up to a few μm depending on its geometry, e g depth position, the electron beam
accelerating voltage and minority carrier diffusion length. The High Resolution Defect Imaging (HRDI) by CL and EBIC enables studies on microscopic dislocation configurations and
even investigations at selected individual dislocations or chosen parts of single dislocations.

From the SEM micrographs important information on the geometrical defect arrangement in the specimen crystal are obtained which may be correlated with generation, migration and reactions of the defects observed.

In this work several aspects of HRDI by means of CL/EBIC are dealt with in order to
deduce qualitative conclusions about the dislocation contrast. The geometrical dislocation
configuration in epitaxial layers near the specimen surface is analyzed concerning crystallographic directions, depth positions and, especially, the distinction between up and down
threading dislocation parts and the identification of crossing points and inclinations of line
defects in samples of GaAsP and GaP are demonstrated.

Combined SEM–CL/EBIC prove the recombination activity of the imaged defects. The
quantitative consideration of this quantity requires the application of a comprehensive
theoretical description of the signals and the defect contrast behaviour. Results of combined
investigations simultaneously provide information on semiconductor bulk properties, microscopic sample geometry and defect recombination strength.

By detailed defect contrast measurements the recombination activity of line dislocations lying parallel or perpendicular to the specimen surface was characterised quantitatively. A numerical evaluation of the defect strength for the surface–parallel dislocation was carried out taking into account a realistic depth and radial generation distribution of excess carriers. Strongly recombination–active defects cause a noticeable change in the excess minority carrier density around the defect having consequences on contrast and defect strength values. The theoretical description must be able to cover this wide range of non–linear defect strength values.

2. EXPERIMENTAL RESULTS AND DISCUSSION

2.1 High Resolution Defect Imaging (HRDI)

By SEM–CL and EBIC defect imaging dislocations were reliably resolved up to densities of $\sim 10^6 \ldots 10^8$ cm^{-2}. In activated $\{111\}\langle110\rangle$ slip processes the hexagonal loop structures consisting of screw– and edge–type dislocation segments are observed. On selected individual dislocations the defect contrast properties can be investigated. Fig. 1 shows CL micrographs of a (001) GaAs$_{0.62}$P$_{0.38}$ and (001) GaP epilayer specimen, respectively, with point and line contrasts representing the array of recombination–active misfit dislocations in the near–surface region of the epilayer. The images containing a variety of contrast types give a clear information about the geometric arrangement of dislocation segments concerning crystallographic types and directions. In agreement with results of cross–section plane investigations, the point–like contrasts (either symmetrically or slightly comet–shaped) are defined as $\langle001\rangle$ and $\langle112\rangle$ threading dislocations reaching the specimen surface. The line contrasts (see especially Fig. 1b) are due to surface–parallel dislocation lines (in $\langle110\rangle$ direction) or slightly inclined dislocations (inclination angle $\alpha \approx 1 \ldots 4°$ according to the misorientation angle of the substrate in $\langle110\rangle$ direction), all being 60° or 90° dislocation lines. A stronger dot at the end of the contrast line indicates the point where the dislocation penetrates the surface almost perpendicularly after having changed its

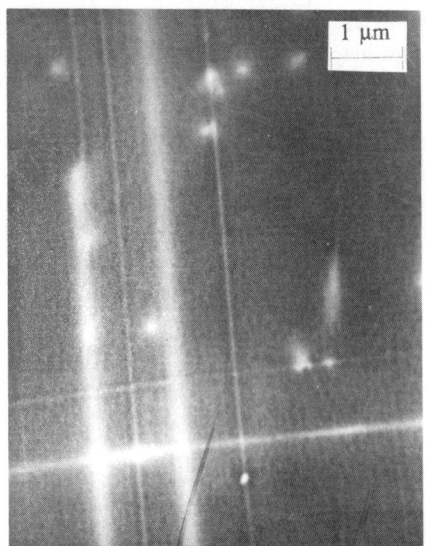

a   b

Fig. 1. CL micrographs (U_b = 20 kV, I_b = 1 nA, T = 300 K, reverse signal) showing typical dislocation contrasts in a (001) epitaxial layer of (a) GaAsP (slightly etched surface) and (b) GaP (as–grown surface)

direction rather abruptly. On the other hand, typical longer comet–shaped contrasts found arise from the inclined dislocations which pass slowly into the depth. Experimental determinations of the depth position of dislocation lines in Fig. 1 yield values between 0.6 and 1.9 μm.

Deviations from the expected intrinsic contrast behaviour, e g dot–and–halo contours or a randomly inhomogeneous line contrast, may be correlated with effects due to dislocation – point defect or impurity interaction.

2.2 Quantitative Studies on the Recombination Activity of Single Dislocations

Crystallographically defined dislocation configurations such as the misfit networks attract much interest in quantitative considerations of the defect recombination activity by means of analysing experimental contrast measurements and using simulation calculations.

In the theoretical description, as an example the EBIC contrast of a surface–parallel dislocation can be written as

$$c_{EBIC}(\xi, U_b) = \lambda \ c^*_{EBIC}(\xi, U_b, z_D, r_D, L) \tag{1}$$

(ξ – electron beam position relative to the defect, U_b – beam voltage, c^* – contrast profile function, z_D – depth position of the dislocation, r_D – dislocation radius, L – minority carrier diffusion length). According to Schreiber et al (1993), λ describes a non–linear defect strength with arbitrary lifetime ratio τ/τ' up to a possible saturation of the contrast:

$$\lambda = \lambda(\tau, \tau', r_D, z_D, L) . \tag{2}$$

In Fig. 2 the CL and EBIC maximum contrast behaviour in dependence on U_b for the case of a surface–parallel line dislocation in GaP and for a surface–perpendicular dislocation in GaAs is represented. The experimental results on the surface–parallel dislocation can be fitted with the theoretical model in the extended region for high defect strength λ (Schreiber et al 1993) using the known depth position. For the surface–parallel dislocation the same defect strength value of $\lambda = 2.42$ describes well both the EBIC and CL contrast. However, the consideration of a realistic generation $g(\mathbf{r})$ in the numerical simulations as done

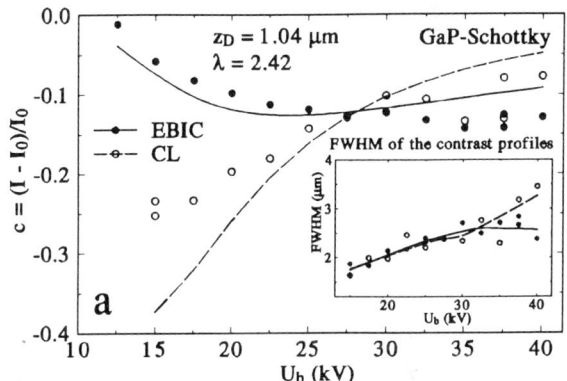

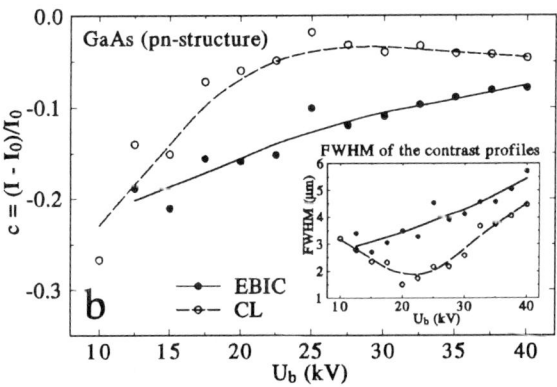

Fig. 2. EBIC and CL contrast of (a) surface–parallel dislocation and (b) surface–perpendicular dislocation as a function of beam voltage U_b

in Fig. 2a using the results of Koch (1987) and Werner et al (1988) is of great importance. Slight deviations of the theoretical U_b dependence from the experimental points may be due to uncertainties in the *radial* generation distribution (which is assumed to be independent from the well–known depth distribution) and its normalisation especially at high U_b. This circumstance recommends use of the contrast profile area analysis procedure (Schreiber et al 1993).

In Fig. 2b the results of defect contrast measurements on a surface–perpendicular dislocation are given showing a qualitatively different behaviour. The diagram shows increasing contrast with lowering of U_b in both CL and EBIC mode. The EBIC contrast dependence has been generally predicted by Pasemann (1991), but for the sphere generation approximation only. On the other hand, an analogous CL contrast theory applicable to surface–perpendicular dislocations is yet an unsolved problem. Especially, compared with the surface–parallel dislocation, here EBIC and CL contrast exhibit different half–widths which originate in the different mechanisms of signal formation and could provide extended information about the recombination properties of the dislocation.

As Pasemann has shown, since the recombination activity of a surface–perpendicular dislocation is characterized by *both* τ/τ' and r_D a defect strength as λ for the surface–parallel dislocation cannot be defined meaningfully. However, this should open the possibility to estimate the defect radius r_D from contrast measurements in the region of low U_b provided that the contrast behaviour in the surface space charge layer may be described appropriately. Moreover, the geometrical situation of the perpendicular dislocation is better defined than that of the parallel one (no depth position required).

3. CONCLUSIONS

Qualitative and quantitative SEM–CL/EBIC studies of the dislocation contrast behaviour in crystalline semiconductor specimens are well–directed to reach a better understanding of the contrast formation in both modes used for the characterization of the dislocation recombination activity. By means of CL and EBIC HRDI essential conclusions on the microscopic geometric–crystallographic configuration and even dislocation dynamics and defect reactions may be deduced.

The dislocation contrasts experimentally observed are described by the theoretical model of defect recombination. The quantitative analysis of the contrast measurements applying numerical fits enables an evaluation of the defect strength which characterises the recombination activity that should depend on the localized electronic defect properties, e g dislocation type and point–defect situation in the vicinity of the dislocation line. This corresponds to our earlier findings of a lower defect strength value for a screw–type glide dislocation line in GaAs and higher ones for edge–type grown–in dislocations in GaAsP and GaP (Schreiber et al 1993).

REFERENCES

Koch F 1987 Thesis (Berlin) pp 39–58
Pasemann L 1991 J. Appl. Phys. 69 6387
Schreiber J, Hildebrandt S and Leipner H S 1993 to be published in phys. stat. sol (a)
Werner U, Koch F and Oelgart G 1988 J. Phys. D 21 116

Inst. Phys. Conf. Ser. No 134: Section 11
Paper presented at Microsc. Semicond. Mater. Conf., Oxford, 5–8 April 1993

Investigation of recombination at misfit dislocations in SiGe CVD structures by 1:1 correlation of EBIC and CL

V Higgs* and M Kittler**

*Department of Physics, King's College London, Strand, London WC2R 2LS, UK
**Institut für Halbleiterphysik GmbH, PSF 409, D-O 1200, Frankfurt (Oder), Germany

ABSTRACT: Misfit dislocations in as-grown and Ni contaminated Si/SiGe epilayers have been characterised by CL spectroscopy and CL imaging and EBIC. Dislocations in the as-grown layers had no radiative recombination (D-bands) and no detectable room temperature EBIC contrast, the dislocationsonly became visible in EBIC at temperatures below 300 K. Following Ni contamination the D-bands were observed and the EBIC contrast increased. The temperature dependence of the EBIC contrast is attributed to shallow levels connected with the dislocations. CL dark line contrast is observed when monochromatic imaging the Si substrate luminescence. The CL dark line contrast and EBIC contrast show a 1:1 correspondence of the non-radiative recombination at the misfit dislocation and also a semi- quantitative agreement with the variation in measured contrast of the individual dislocations.

1. INTRODUCTION

Photoluminescence (PL) spectroscopy studies have long established that the D-bands (D1-D4) are associated with the presence of dislocations. The D1 and D2 bands have been variously attributed to electronic transitions at the stacking fault between the dislocations, transitions at dislocation kinks or point defects trapped in the strain fields around dislocations. Whereas D3 and D4 are considered to be associated with electronic transitions at the dislocation core. From most of these investigations reported, there was no clear evidence that impurities played a role in dislocation related luminescence. However it has been recently demonstrated (Higgs et al 1990) that dislocation related luminescence could not be observed in the absence of transition metal contamination.

For a more detailed understanding about the origins of the D-band luminescence a high resolution scanning technique is required, to gain an insight into the spatial location of the different features. Cathodoluminescence (CL) imaging and spectroscopy has been successfully developed for mapping dislocation structures in both Si and SiGe alloys (Higgs et al 1992). Monochromatic imaging of the dislocations showed that D3 and D4 bands originate on or near the dislocation cores whereas the location of D1 and D2 seem to originate between the dislocations.

The EBIC technique has been routinely used to study the recombination activity of individual dislocations. EBIC contrast has been found to be extremely sensitive to impurity decoration (Kittler and Seifert 1981). It has been generally assumed until recently that defects showing no room temperature contrast were "clean". However, it has been found that the EBIC contrast may very significantly with sample temperature. It is now clear that

EBIC investigations as a function of both temperature and beam current (I_b) are required for a more detailed understanding of the defect properties.

In this paper we report the first combined EBIC and CL investigation of both as-grown and Ni contaminated dislocations in Si/SiGe structures. On comparison with EBIC measurements we show a 1:1 correspondence of the non-radiative recombination activity at the misfit dislocations.

2. EXPERIMENTAL

A wide range of samples, untreated and deliberately transition metal contaminated Si/SiGe layers grown by CVD using SiHCL$_3$ and GeH$_4$ at 1120°C, has been characterized by CL. These layers were supplied by North Carolina State University. These samples consisted of a common structure. A Si capping layer (thickness= 3 μm) grown on top of a Si$_{1-x}$Ge$_x$ alloy layer (thickness= 2 μm) on a Si buffer layer (thickness= 2 μm) on top of a heavily doped Si (100) substrate. The Ni contaminated sample was produced by Ni evaporation on the back face of the wafer followed by rapid transient annealing at 1000°C for a few seconds.

CL measurements were made at T≈5K using a JEOL JSM 35C SEM fitted with an adjustable cold stage, a retractable off-axis paraboloidal collector mirror and a grating monochromator (Mono CL, Oxford Instruments). The detector employed was a North Coast germanium diode detector. The measurements were carried out with a beam energy of 25 keV and a beam current between 0.1 nA-100 nA.

EBIC investigations were performed in a Cambridge Stereoscan S 360 SEM using a Matelect ISM5 EBIC amplifier fitted with a Kontron image processing system. The EBIC signal was collected using electron transparent Schottky barriers, at a beam energy of 30 keV and a beam current below 0.1 nA.

3. RESULTS AND DISCUSSION

CL spectra were recorded from a series of uncontaminated samples, all these samples contained misfit dislocations. All the samples showed the bound exciton features from the heavily doped Si substrate and the majority of these samples showed D-band luminescence. However three samples had no observable D-band luminescence. These samples were further examined with CL imaging. Monochromatic imaging with the Si substrate feature revealed dark line features oriented along the <100> directions. Monochromatic line scans revealed that these dark lines resulted in a reduction of the luminescence intensity. On inspection of exactly the same region with EBIC (T=80 K) it was clear that these dark line features were misfit dislocations. These dark line features were also observed in all the samples containing dislocations both contaminated and as-grown. Similar combined CL and EBIC experiments have often been used to study dislocations in compound semiconductors, and the basic mechanism of defect contrast formation (non-radiative) is the same. However in this study the CL measurements are made close to He temperatures and are dominated by excitonic transitions, therefore the mechanism for non-radiative recombination may be the same but the main species involved will be different (exciton or minority carrier).

The variation of CL intensity observed occurs because the excitons are produced in the upper part of the sample structure, then diffuse towards the Si substrate. The exciton density becomes reduced as the excitons recombine non-radiatively at the misfit dislocations. Therefore the density reaching the Si substrate is reduced and the Si CL intensity is reduced.

The dislocations observed in the as-received layer could not be observed at room temperature (detection limit ≈0.1%) it was only on cooling that the dislocations became visible with a very small contrast (c<0.5%). This behaviour can be attributed to shallow levels connected with misfit dislocations (Kittler *et al* 1993).

CL spectra recorded on the Ni contaminated sample contained only the D1 and D2 bands and the Si substrate features. This is in contrast to previous measurements on SiGe alloys and maybe due to the high temperature annealing that has destroyed or modified the structure of the centres responsible for D3 and D4.

The EBIC measurements showed that the contrast was non-uniform following Ni contamination. As with the as-grown sample there were regions where the dislocations could be observed at temperatures below 300K, however there were also regions where dark spots were observed at room temperature. In these regions following sample cooling to T=250 K the misfit dislocations became visible. The EBIC contrast measured on these samples reached a few percent at T=80 K, much larger than those measured on the as-grown sample. An in depth study has been carried out on these samples (Kittler and Seiffert 1993) and the variation of temperature dependence of the EBIC contrast for the as-grown and Ni contaminated (identified as Ni(low)) can be explained by shallow levels connected with the dislocations. Shockley-Read-Hall recombination simulations show a semi-quantitative agreement of the experimental results. In the areas were the dark spots were dominant (identified as Ni(high)), the corrected contrast of these spots decreases as the sample temperature decreases. This behaviour has been observed for $NiSi_2$ precipitates in FZ Si, in addition TEM measurements have shown that Ni precipitates are observed in these samples.

Figure 1a shows the EBIC micrograph recorded at T=80 K. The corresponding dark line CL image is shown in Figure 1b. It is clear there is a 1:1 correlation of the non-radiative recombination at the dislocation. On inspection of both the EBIC and CL dark line profiles it is clear that there is a semi-quantitative agreement in the relative variation of the contrast.

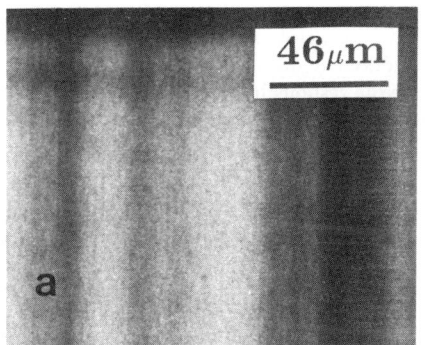 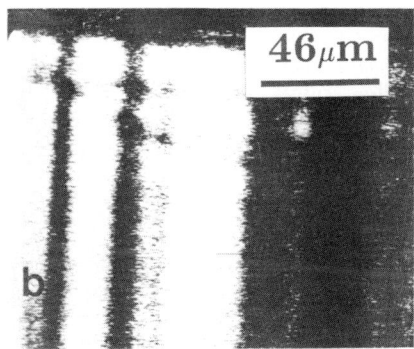

Figure 1. 1:1 correlation of EBIC and CL imaging of Ni contaminated misfit dislocations, a) EBIC image E_o= 30 keV, T=80 K, $I_b \approx 0.1$ nA, b) CL monochromatic image, E_o= 25 keV, T=5 K, $I_b \approx 1$ nA.

Monochromatic CL imaging of the Ni(low) area sample revealed that the D1 and D2 intensity increased at the dislocation and was also distributed between the dislocations. Line scans show that the CL intensity increased by up to 15% at the dislocation. This is in contrast to the previously measured results, where the D1 and D2 intensity was more dominant in between the dislocations. The same effect has been observed on other CVD layers. These differences may be due to competition for the excitons, in these samples only D1 and D2 are present whereas in the previous measurements all four D-bands were present.

The corresponding EBIC and CL micrographs of the Ni (high) area is shown in Figures 2a and 2b respectively. In the strongly Ni contaminated area of the sample (Ni(high)), monochromatic CL line scans revealed that the D-band intensity decreased by up to 30% at the dislocation. This is consistent with previous PL measurements on defects in Si, as the level of transition metal contamination increased there was a decrease in the D-band luminescence. This was thought to be due to the microprecipitates absorbing the centres responsible for the

D-bands (Higgs *et al* 1992). In addition, the dark line CL contrast has increased and so has the EBIC contrast, clearly the precipitates can also act to increase the non-radiative recombination occurring at the dislocation. The competition between non-radiative and radiative transitions for exciton capture will be greatly affected by the presence of metallic impurities.

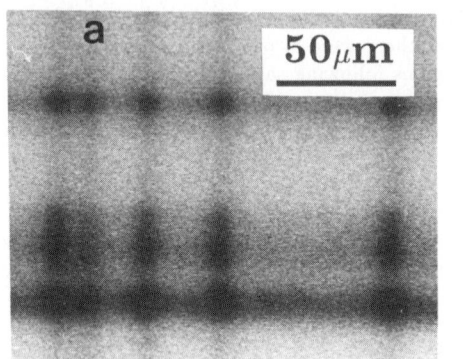 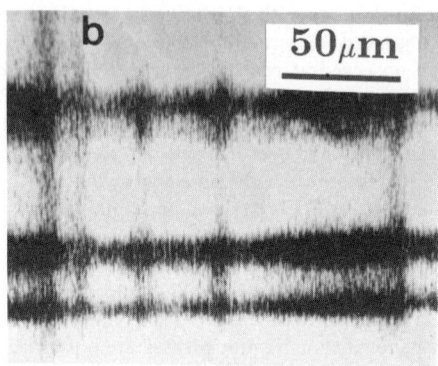

Figure 2. Stronger Ni contaminated misfit dislocations   a) EBIC image E_o= 30 keV, T=80 K, $I_b \approx 0.1$ nA, b) CL monochromatic image, E_o= 25 keV, T=5 K, $I_b \approx 1$ nA.

Although the EBIC and CL measurements were carried out under different conditions of injection level and temperature, also the CL contrast measurements have not been corrected for absorption losses. It is clear that the EBIC and CL contrasts follow a similar behaviour, with the same relative variation in contrast between different dislocations. It is feasible that the shallow centres responsible for EBIC contrast maybe responsible for variations in the CL dark line contrast. These shallow levels maybe introduced during growth and may result from the elastic strain field of the dislocation.

ACKNOWLEDGEMENT

The authors would like to thank Prof. G. A. Rozgonyi from North Carolina State University (USA) for supplying the Si/SiGe epilayer samples.

REFERENCES

Higgs V, Lightowlers EC and Kightley P 1990 Mat. Res. Sci. Soc. Proc. <u>163</u> 57

Higgs V, Lightowlers EC, Tajbakhsh S and Wright PJ 1992 Appl. Phys. Lett. <u>61</u> 1087

Kittler M and Seifert W 1981 Phys. Stat. Sol. (A) <u>66</u> 573

Kittler M and Seifert W 1993 (this conference)

EBIC and AFM studies of reactive ion and reactive ion beam etched silicon for damage depth research

G Jäger-Waldau, H-U Habermeier, G Zwicker * and E Bucher +

Max-Planck-Institut für Festkörperforschung, Postfach 80 06 65, W-7000 Stuttgart 80, Germany
* Fraunhofer-Institut für Mikrostrukturtechnik, Dillenburgerstr. 53, W-1000 Berlin 33, Germany
+ Universität Konstanz,Fakultät für Physik, Postfach 55 60,W-7750 Konstanz, Germany

ABSTRACT: The Planar Electron Beam Induced Current (PEBIC) mode of a Secondary Electron Microscope (SEM) is used to study the active electrical defects in silicon created by Reactive Ion Etching (RIE) and Reactive Ion Beam Etching (RIBE) processes. There is found to be a reduction in the EBIC signal from the etched areas. Because of the differences in the EBIC signals, the EBIC image received from a scan across the sample reflects the geometric etch mask. In conjunction with the exact geometric structure of the etched sample as determined by Atomic Force Microscope (AFM) measurements, the maximum lateral damage depth can be measured. The influence of the etch parameters on the etch damage is discussed.

1 INTRODUCTION

In micro-fabrication of silicon devices, reactive ion etching (RIE) is one of the standard processes for anisotropic etching and also a key component in the production line. A large number of papers have been published which describe the interaction and operation of this etch process and the origin of defects. The control of these defects is critical for proper device operation, especially when the integrated circuits are miniaturised to the sub-micron range. Lattice defects, passivation of donors, amorphisation of the material, insulating layers on the surface etc. were found by different experimental techniques including XPS, DLTS, TEM, Auger, SIMS, RBS, SPV, He-ion channeling, PL, Raman spectroscopy, IV and CV measurements (Misra (1990), Harris et al. (1989), Cerofolini and Ottaviani (1989), Oehrlein and Lee (1987), Oehrlein et al. (1986), Noerstroem et al. (1991), Lee et al. (1989), Kuroda and Iwakuro (1990), Tsang et al. (1985), Spiritoet al. (1986), Misra and Heasell (1987)). The EBIC mode of the SEM can also be used to study the recombination behaviour of point defects (Eckstein et al. (1989), Bode et al. (1988), Jakubowicz and Habermeier (1985)) and to compare the results with those from DLTS investigations from etched and unetched silicon samples (Ikuta (1987)). The reactive ion beam etching (RIBE) is also used for etch processes, since this method allows better seperation of the individual parameters, and therefore better control over them. In the micro-device fabrication industry RIBE is not often used.

The use of the secondary electron microscope (SEM) in the electron beam induced current (EBIC) mode is not very common, although it is a very powerful and helpful tool for the investigation of semiconductors (Holt (1989), Leamy (1982)). The EBIC technique has the advantage of the simultaneous observation of etched and unetched pattern of the devices. This enables the direct observation of the recombination behaviour of semiconductors and therefore the characterisation of active electrical defects. Also, two-dimensional images of EBIC intensities across a sample area are possible and since EBIC is not an integral method, it is possible to get a high lateral resolution in the measurements. This is very efficient for developing microdevices in the sub-micron range and studying the variation in the defects between etched and unetched structures.

In this paper we report on the correlation between the EBIC signal and the etch pattern, and the possibility of obtaining a lateral damage depth due to this effect.

2   EXPERIMENTAL SETUP

The investigations were performed on commercial, dislocation free, about 0.002Ωcm p-type CZ-silicon single crystals doped with boron. First an oxide was grown by thermal oxidation. Then the samples were lithographically patterned and etched in an 8111 Hexode RIE-etcher (Applied Materials) using $CHF_3 : CF_4 : O_2$ (80:40:20 sccm) as etching gas , or for RIBE, in an OXI-150-LLC (Oxford) with an Astex ECR source using CF_4 .

Parameters such as pressure and bias-voltage (DC_{BIAS}) for RIE and U_{Beam} for RIBE were varied for different samples. The silicon oxide protected by the resist during the etch process was lifted off with buffered HF-solution (wet etch), to avoid a second plasma process.

The etched samples were cleaned by the RCA method (Kern and Puotinen (1970)) ($NH_4OH : H_2O_2 : H_2O$ with 1:2:7) and a short dip in buffered HF solution. Aluminum was evaporated on the reverse side of the sample, followed by a diffusion process ($T = 400°C; t = 10$min) to produce ohmic contacts. Titanium contacts with diameters of 1mm and a thickness of about 400Å were evaporated for the semi transparent Schottky contacts.

For measurement we use a Cambridge Stereoscan 250 Mk2 equipped with a cooling stage inside the vacuum chamber and devices for DC EBIC measurements. The cooling stage allows temperature dependent EBIC measurements between $5K - 300K$. A computer supports the control of the whole system and is also applied for data analysis.

For quantitative statements, we use the well known declaration of a contrast between regions with different EBIC signals (Eckstein et al. (1989), Bode et al. (1988), Jakubowicz and Habermeier (1985)).

3   RESULTS AND DISCUSSION

Simple EBIC measurements at room temperature show a different intensity in the EBIC signals between etched and unetched areas. Figure 1 shows the etched pattern of the samples by an optical microscope (sample is over-etched), whereas Figure 2 represents the measurement of the EBIC signal in nearly the same geometric dimension of a RIE etched sample. The brightness of the image corresponds to the intensity of the EBIC signal, although in order to make the differences clearer,   background suppression was done. The dark areas belong to the etched pattern and the bright ones to the unetched patterns, so a more enhanced recombination and thus reduced EBIC signal has to be in the etched areas of the sample. This is due to the fact that the amount of generated electron-hole pairs is the same for all scanned coordinates. The etch process leads to an electrical defect in the sample. The same observations can be done with samples etched by RIBE. The quantitative description of the differences of the signals with an EBIC contrast C ($C = (I_0 - I_1)/I_0; I_1$: EBIC signal of the etched pattern, I_0 : EBIC signal of the non-etched pattern) gives a value of about $0.4\% - 0.9\%$ (RIE), which is not very much. It was because of this small value that the background suppression was done, in order to depict the changes in EBIC signal more clearly.

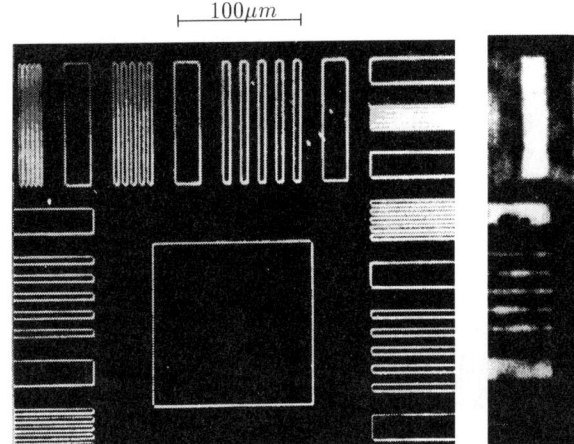

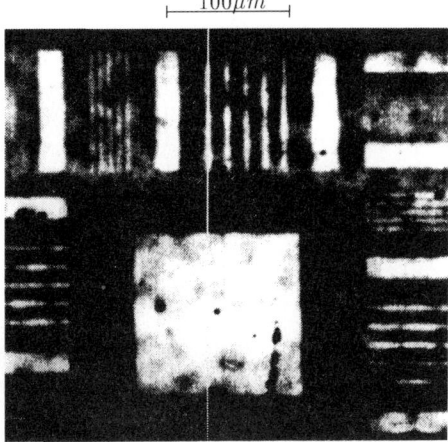

Figure 1: Pattern of the etched sample obtained by an optical microscope.

Figure 2: EBIC signals of etched and non-etched areas.

It is not the aim of this paper to discuss what kind of defects are responsible for the contrast, but rather to obtain some information about the damage depth of the etch processes. As Figure 2 shows, the two dimensional EBIC reflects the etch mask pattern and provides information about the damaged zones. By comparison with an EBIC linescan, a direct correlation from the signal slope and the etch design is given. This can be seen in Figure 3, where both EBIC linescan and etch design are shown. The idea and aim of our studies is to extract information about the maximum lateral damage depth created by RIE and/or RIBE.

From diffusion length measurements of minority carriers we know that the damaged layer has to be near the surface region, because no different values for the diffusion length in etched and unetched areas, when measured by planar EBIC and fitted to the theoretical model could be detected. The EBIC contrast increases with decreasing temperature, with a maximum around $T \approx 40K$. This fact is of interest for defect characterisation. For our studies the increase of contrast is only used to cause a better signal-noise ratio, an advantage for graphical analysis.

The damage is detectable by EBIC because of the lower signal in the etched area, the slope at the etch steps and the shape of the signal in the unetched pattern. In Figure 3 the shape of the EBIC signal shows a lowering in the EBIC signal even in small non-etched plateaus. The geometric analysis of the device obtained using an atomic force microscope (AFM) does not depict differences in the heights or widths of the etch mask, or between the same structures. At the small non etched steps no geometrical changes in the structure could be found. This implies the electrical defects were created without removing material. Because the top of the steps was protected by the photo resist and silicon-dioxide, a lateral defect influences the electrical behaviour of the etch steps. This effect has its maximum on the top of the non-etched plateau, because of the maximum duration of the damaging etch process.

In the case when the lateral damage depth comes to the same geometric dimension as a half of the non-etch step width, the EBIC signal in the non-etched area increases, since there is an overlay of the defect zones. This is the reason why the EBIC signal does not reach the normal maximum value in small etch plateaus.

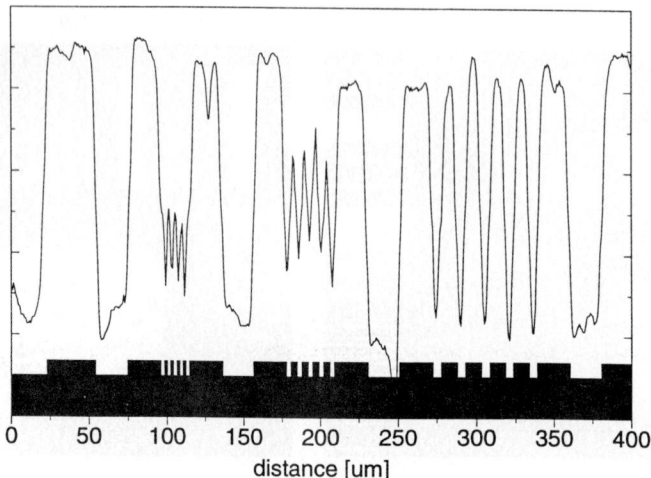

Figure 3: EBIC linescan and etch design in correlation with each other.

A very important consideration in the analysis is the influence of the generation volume, which is related to the penetration range and the lateral dose function, controlled by the accelerating voltage of the electron beam and material data. The geometric expansion of the generation volume changes the EBIC linescan across a signal step from an ideal step function to a wide slope (Figure 4). We took the mathematical description for the lateral dose function given by Shea, Partain and Warter (et al. (1985)):

$$g(x) = \quad exp(\frac{-x}{0.25 \cdot R_G}) \qquad 0.1 R_G < x < 0.6 R_G$$
$$g(x) = \quad 8.6 \cdot exp(\frac{-(x-0.6 \cdot R_G)}{0.13 \cdot R_G}) \qquad 0.6 R_G < x < 1.23 R_G$$
$$g(x) = \quad 0 \qquad x > 1.23 R_G$$

The Grün-range R_G has to be known because of the depth dose function. However, a very small beam voltage does not succeed in minimizing the influence of the generation volume, since the beam voltage is responsible for the amount of generated electron-hole pairs as well as for the density of the generated charge carriers in the volume and the factor for the charge collection efficiency. Therefore a balanced value has to be chosen.

When we start with the ideal EBIC curve obtained by the geometric structure and the knowledge of different signals, we get a non-realistic signal function (Figure 5). It is therefore necessary to subtract the influence of the generation volume. For our chosen accelerating voltage of 11kV a Grün-range of $R_G = 1\mu m$ is calculated by Monte-Carlo simulation. The new, corrected curve is depicted with dashed lines in Figure 5. If we lay the recorded EBIC signal on top of the ideal curve, it should be possible to estimate the maximum lateral damage length.

For representation, we took an EBIC linescan recorded at T = 50K to give a reduction in the signal to noise ratio allowing better analysis. A maximum damage length of d $\approx$ 300nm was graphically measured for a sample etched with RIE and the following parameters: $DC_{BIAS} = 450V, p = 5332mPa, P_{for} = 1400W$ (etch gas see above). A RIBE sample, $U_{Beam} = 600V, U_{acc} = 500V, I_{Beam} = 96mA, I_{acc} = 6mA, P_{for} = 158W, p = 1.7 \cdot 10^{-4}mbar$, was found to have the the maximum lateral damage depth of d $\approx$ 500nm.

The method works well with a relatively large damage length, which means a large EBIC

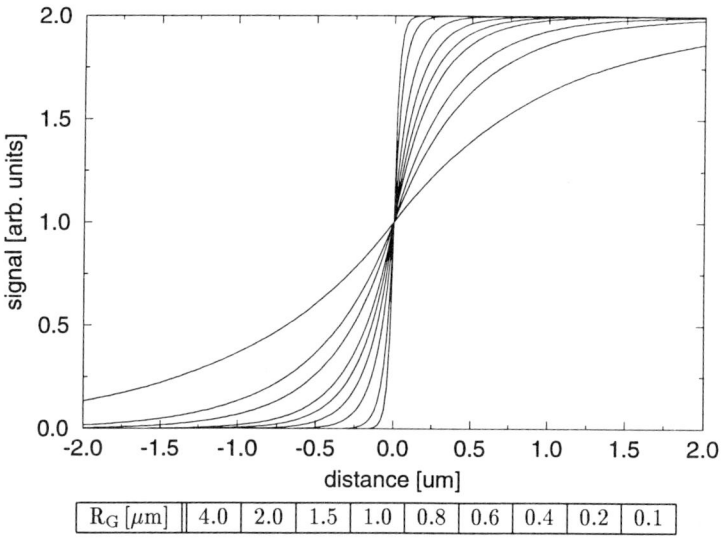

Figure 4: EBIC slopes across a single signal step varied with the dimension of the generation volume or the Grün range, respectively.

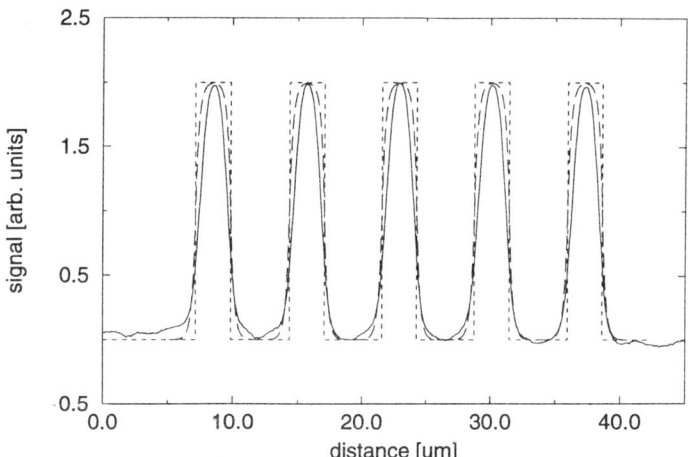

Figure 5: Real EBIC linescan across the ideal etch pattern signal (width w $= 4\mu$m) corrected by the generation volume at the Grün range of $R_G = 1\mu$m .
- - - ideal EBIC signal in correlation with the etched structure
- - ideal EBIC signal corrected due to the generation volume
—— measured EBIC signal at T $= 50$K across the etched steps

contrast. With a variation of the DC_{BIAS} in the RIE process, the damage caused is also varying, detectable by a shift in the maximum EBIC contrast. With smaller EBIC contrast the damage decreases so the signal to noise ratio is getting worse as well. This leads to increased errors, which is one disadvantage of this method.

Whereas it is possible to get different values for the lateral damage depth at RIE etched samples with variation of DC_{BIAS}, no changes in RIBE etched silicon samples were detectable, since the damage depth is dependent on the silicon etch rate and for the range of the RIBE parameter $U_{Beam} = 300V..600V$ the rate is constant (Heath (1982)).

ACKNOWLEDGEMENTS

The authors would like to thank S. Tippmann and B. Hammer for technical assistance in film preparation and F. Schartner for the mask production as well as Y. Kershaw for helpful discussions.
The authors appreciate the financial support by the German Ministry of Research and Technology [contract number NT 2792 B].

REFERENCES

Bode M, Jakubowicz A, and Habermeier H-U (1988). *Scanning*, 10:169.
Cerofolini G and Ottaviani G (1989). *Mater. Sci. & Eng.*, B(4):19, 1989.
Eckstein M, Jakubowicz A, Bode M, and Habermeier H-U (1989). *Appl. Phys. Lett.*, 54:2659.
Harris C, Sawyer W, Konuma M, and Weber J (1989). *Sci. Eng.*, 4(1-4):457.
Heath B A (1982). *J.Electrochem.Soc.*, 129(2):396.
Holt W (1989). *Techniques of Physics*, volume 12. Academic Press, London.
Ikuta K (1987). *Jap. J. Appl. Phys.*, 26:1034.
Jakubowicz A and Habermeier H-U (1985). *J. Appl. Phys*, 58:1407.
Kern W and Puotinen D (1970). *RCA Rev.*, 31:187.
Kuroda T and Iwakuro H (1990). *Jpn. J. Appl. Phys.*, 29(5):923.
Leamy H (1982). *J. Appl. Phys.*, 53(6):R51.
Lee Y, Oehrlein G, and Ransom C (1989). *Radiat. Eff. Defects Solids*, 111-112(1/2):221.
Misra D (1990). *Semicond. Sci. Technol.*, 5(3):229.
Misra D and Heasell E (1987). *J. Electrochem. Sem.: Solid-state science and technology*, :956.
Noerstroem H, Blom H, Ostling M, Lylandsted-Larson L, and Keinonen J (1991). *J. Vac. Sci. Technol.*, B9(1):34.
Oehrlein G, Clabes J, and Spirito S (1986). *J. Electrochem. Soc.: Solid-state science and technology*, 133(5):1002.
Oehrlein and Lee (1987) Oehrlein G and Lee Y (1987). *J. Vac. Sci. Technol.*, A5(4):1585.
Shea S, Partain L, and Warter P (1985). *Scanning electron microscopy*, 1:435.
Spirito P, Ransom C, and Oehrlein G (1986). *Solid State Electronics*, 29(6):607.
Tsang J, Oehrlein G, Haller I, and Custer J (1985). *Appl. Phys. Lett.*, 46(6).

Inst. Phys. Conf. Ser. No 134: Section 11
Paper presented at Microsc. Semicond. Mater. Conf., Oxford, 5–8 April 1993

713

Frequency dependence of the EBIC signal from dislocations in Si measured with a lock-in amplifier

H Mohr[1], H Alexander and J Palm

II Physikalisches Institut, Abteilung für Metallphysik, Universität zu Köln, D-5000 Köln 41, Germany
[1]Department of Physics, University of Surrey, Guildford, Surrey GU2 5XH

ABSTRACT: The lock-in technique in EBIC-imaging is known to improve the signal to noise ratio as well as the resolution. We have studied the dependence of the EBIC amplitude on modulation frequencies up to 200kHz for dislocations in Si. We also observed a frequency dependence of the phase between the modulation signal and the EBIC signal. The amplitude-versus-frequency curves as well as the phase-versus-frequency curves are characteristic of dislocations in certain environments, (i.e. they depend on doping and sample growth etc) and they are also different for grain boundaries. So for example phase can decrease or increase depending on the sample. The cause of this behaviour is explained. The use of amplitude- and phase-versus-frequency curves for the life time determination of dislocation traps is illustrated.

1. INTRODUCTION

With the EBIC-method it is possible to reveal dislocations in semiconductors due to the higher recombination activity at dislocation and defects. The recombination rate affects the amplitude of the EBIC-current. Normally the electron beam is a permanent source to create electron and hole pairs. Only a few EBIC examinations exist which explore the behaviour of dislocations under time dependent stimulation (Ourmazd et al, 1981). If the beam is modulated and detected with a lock-in amplifier the phase between the modulation signal and the EBIC signal will depend on the frequency of the modulation.

2. EXPERIMENTAL

The measurements were made with a 30kV Philips Scanning Electron microscope REM 515, which has an electrostatic Beam Blanking unit. The electron beam was modulated by a square wave function. The EBIC-current was detected as the voltage drop over a 10 ohm resistor with an ITHACO 393 two phase lock-in amplifier. Single crystal n-type silicon with $n=5*10^{14}cm^{-3}$ and $10^{16}cm^{-3}$ and polycrystalline p-type silicon for solar-cells (SILSO) with $p=10^{16}cm^{-3}$ Boron-doped from Wacker-Chemitronic were used. The samples were mechanically polished with Al_2O_3-powder and diamond paste down to 0.25 μm. A chemical etch containing HNO_3, CH_3COOH, HF was used for etching the Si and the oxide layer was removed with HF. Schottky-diodes were made by evaporating a 25nm thick Au layer. The ohmic contacts were made with InGa on the backside of the sample. Amplitude-and phase-line scans were recorded in the frequency range of 10kHz-170kHz. Ground loops between the lock-in amplifier, the beam blanking unit and the microscope have a large influence on the recorded signal. It is very important to avoid such ground loops.

3. RESULTS

3.1 Behaviour of Amplitude and Contrast

Six different samples were examined. The amplitude-versus-frequency curves reveal very different behaviour. For bulk material the curves decreased either with increasing modulation frequency from 10kHz to 170kHz or remained constant or they decreased only at high frequencies. A similar behaviour occurs for the amplitude at dislocations. The contrast (I(background)-I(defect))/I(background) for deformed n-Si decreased from 7% to 3% with increasing frequency (see (a) in table 1). An n-Si sample annealed at 1000°C for 15min showed an increase of the contrast from 24% to 39% (tab.1,b). A dislocation in SILSO showed a contrast of 10% which decreased by about 3% (4%) (tab.1 d,e and fig. 1). A grain boundary in SILSO had a decreasing contrast from 62% to 42% (tab.1 f).

Material		kind of defect	Cont rast	ΔC	Phase backgrou nd	Phase defect	$\Phi_{bulk}-\Phi_{defect}$ over f
a)	n-Si deformed	dislocation	7%	-4%	1..-24°	1..-24°	0(const)
b)	n-Si annealed	dislocation	24%	+15%	1..-26°	+10°..14°	10°..12°
c)	n-Si	no defects			-1..26°		
d)	SILSO	dislocation	10%	-3%	1°..-30°	1°..-36°	0..6°
e)	SILSO	dislocation	10%	-4%	1°..-28°	1°..-28°	0°
f)	SILSO	grain boundary	62%	-20%	0°..-60°	0°..-40°	+4°..-20°

Table 1 - examined samples.

3.2 Behaviour of Phase

For all samples the phase between modulation-signal and EBIC-signal decreased for increasing frequency. At the lowest frequency of 10kHz no phase difference is seen for bulk or defective material. At higher frequencies the phase difference increased. For the deformed sample (tab.1 a) no detectable difference between the bulk and the dislocation behaviour can be seen. The phase in both changes by about 24°. A dislocation in heat treated Si reveals a constant difference of about 11° between bulk and dislocation; here the phase changes by about 26° (tab.1 b). SILSO shows a decrease for the bulk by about 60° and for the grain boundary by about 20° (tab.1 f). Remarkably, there is an overlap of the two curves at 120kHz. One dislocation in SILSO shows no difference from the bulk measurement (tab.1 e). Another shows a difference of up to 6° (tab.1 d and fig.1). It seems that the behaviour of the curve depends strongly on the history of the sample temperature treatment.

3.3. Phase Line Scan

Generally a line scan across a dislocation indicates the amplitude of the EBIC-current along the scanning line of the electron beam. The amplitude has a minimum at the position of the dislocation. A phase line scan only shows the position of the dislocation at higher frequencies. The scan rate must be very slow for this purpose. Only at modulation frequencies higher than 10kHz does the phase line scan show the position of the defect (Fig.2).

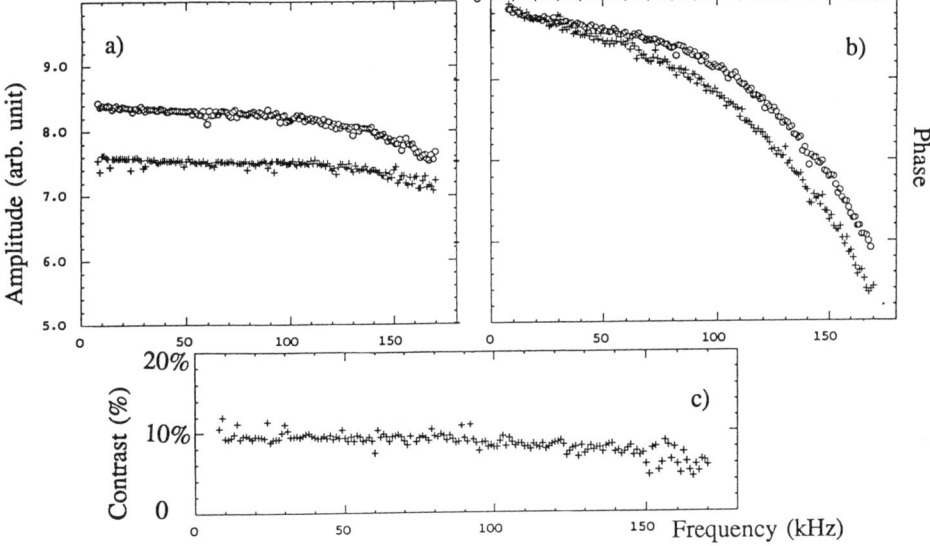

Fig.1 - Behaviour of a dislocation in SILSO at 30kV. a) Amplitude for background and defect, b) Phase for background and defect, c) Contrast of the defect.

Fig.2 - Phase line scan over a defect at several modulation frequencies.

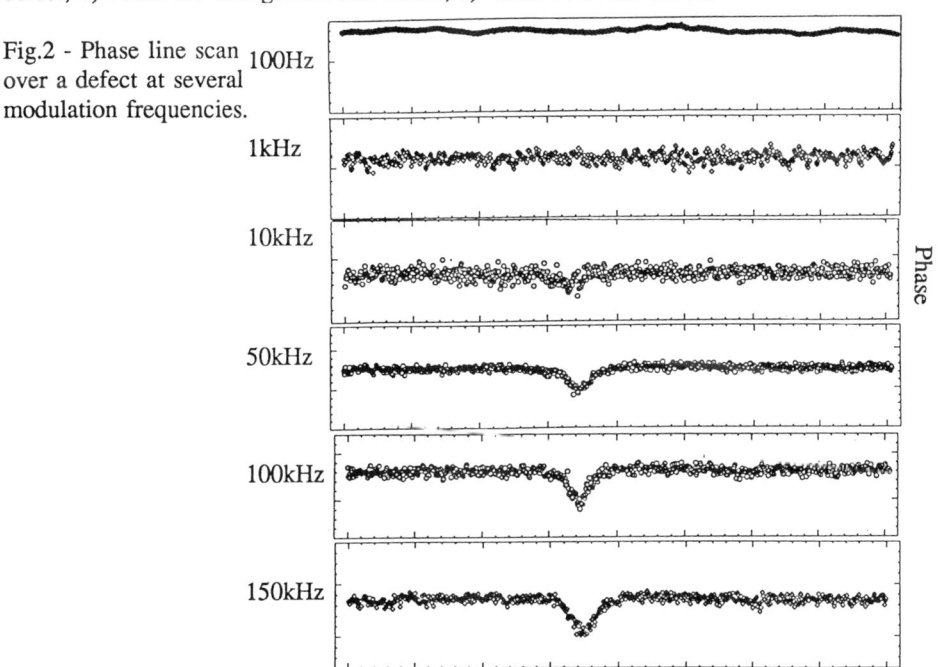

Electron beam position

4. DISCUSSION

The behaviour of the minority carriers is described by the diffusion equation which contains the diffusion length L and the minority lifetime τ. The diffusion equation is solved assuming time dependent carrier generation. By determination of L with a time independent method (Wu, Wittry 1978) and τ as a fit parameter for the measured amplitude or phase curve it is possible to give an estimation of the lifetime for the bulk-material. But the solution depends strongly on the form of the generation volume and is not sensitive enough to give a good correspondence with the measured curve. At a frequency higher than 120kHz there is no correspondence between the theoretical fit and measured curve. At the moment it is not clear if further calculations or more precise measurements could overcome the discrepancy between the theoretical and the measured curve. The frequency dependence of the amplitude is a diffusion effect of the minority carriers from the generation volume to the depletion region. The change of the phase at a defect is greater than in the bulk, this means that the carriers at a defect can travel further and have therefore a longer lifetime. This is in contradiction to the strength of a defect $\gamma = \pi r^2/D$ $(1/\tau_o - 1/\tau')$ (Donolato, 1979), where τ' the lifetime at the defect is shorter than τ_o, the lifetime in the bulk (r is the defect radius and D the diffusion constant). An explanation for this behaviour involves the area around a dislocation. If the defect has the majority of impurities bound, a small area around the dislocation is very clean. The absence of recombination centres around the dislocation leads to a higher diffusion length than in the average dislocation free bulk material. Compared to the size of the defect the scanning probe is large (≈ 10 µm). This means that mainly the area with greater diffusion length is scanned. The carriers produced deeper inside the material now have the chance to reach the Schottky-contact within their lifetime and make a contribution to the current at a later time which accounts for the observed phase shifts. The amplitude of the EBIC-signal will be smaller at a defect because the carriers are always captured and will recombine. The examined samples let us presume that an annealing step leads to an increase in diffusion length in the vicinity of a dislocation as in the case of gettering.

The use of the phase-signal might be a useful tool for examinations of defects. An increase in the modulation frequency up to several MHz could give a higher accuracy and the possibility of measuring lower lifetimes (Pietzsch 1982).

ACKNOWLEDGEMENT

This work has been undertaken as part of a Diploma thesis at Universität zu Köln, Germany supervised by Prof H Alexander.

REFERENCES

Donolato C, 1979, Optik **52** 19
Ourmazd A, Wilshaw P R and Crips R M, 1981, Inst. Phys. Conf.Ser. **60** 63
Pietzsch J, 1982, Solid-St. Electron. **25** 295
Wu C J and Wittry D B, 1978, J. Appl. Phys. **49** 2827

On the origin of recombination activity of extended defects in silicon as studied by the method of the electron beam induced current

M Kittler and W Seifert

Institut für Halbleiterphysik GmbH, PSF 409, D-O 1200 Frankfurt (Oder), Germany

ABSTRACT: Individual NiSi$_2$ particles are found to show a high recombination activity, increasing with temperature. The activity dependence on beam current is influenced by the doping level and shows a negative slope for low doping. 60 ° misfit dislocations increase their activity upon cooling the sample. The findings are discussed in terms of recombination activity controlled by defect charging or shallow centres.

1. INTRODUCTION

EBIC investigations as a function of temperature, T, and beam current, I_b, provide a chance to get a deeper insight into the origin of recombination at defects, see e.g. Kimerling et al. (1977), Ourmazd et al. (1983), Wilshaw et al. (1989,1991), Radzimski et al. (1991), Kusanagi et al. (1992) and Kittler et al. (1993a and b). Analyzing published data on T- and I_b-dependence of defect contrast, two main classes of contrast behaviour can be distinguished (Kittler and Seifert 1992). For type 1 the slope of the c(T) curve is positive and that of c(I_b) negative. Just the opposite is characteristic for type-2 behaviour, negative c(T) and positive c(I_b) slope.

EBIC investigations of misfit dislocations, formed at the interface between Si(2%Ge) layers and (100) Si substrate, and of well-defined NiSi$_2$ particles in Si are reported here. The samples were of n-type conductivity and gold Schottky contacts were used for charge collection. A beam energy of E_o = 30 keV was chosen to allow measurements at defects in the neutral semiconductor beneath the Schottky junction. The contrast values were determined from EBIC line scans using c = 1 - I_d/I_o, with I_d and I_o the collected signal at the defect and far away from the defect, respectively.

2. MISFIT DISLOCATIONS

Si(Ge) layers on (100) Si substrates result in networks of two perpendicular sets of 60° dislocations located at the interface (e.g. Radzimski et al. 1991). When using small Ge contents in the range of 2% the electronic properties of Si(Ge) are practically equal to that of Si. The dislocation density can be controlled by Si(Ge) composition and thickness and so the dislocations form well-defined model defects. The dislocations studied exhibit an increase of EBIC contrast upon cooling the sample from 300 to 80 K. Examples for two typical groups of experimental c(T) dependencies are demonstrated in Fig. 1. One group is characterized by a marked contrast for 300 K and contrast saturation for temperatures between 100 and 80 K. The other group exhibits at 300 K a very small contrast near the detection limit and shows pronounced increase of contrast down to 80 K, without any indication of saturation.

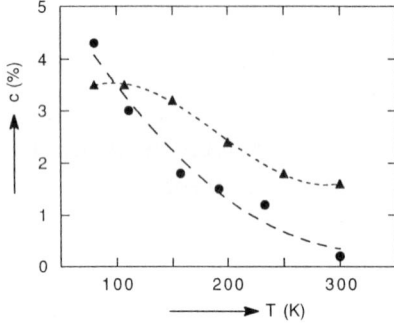

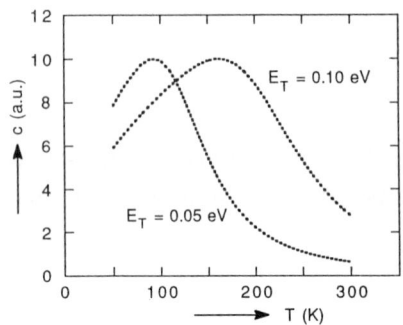

Figure 1. 60° misfit dislocations at the interface Si/Si(2%Ge). Contrast vs temperature measured for as-grown material (▲) exhibiting D-band luminescence (Higgs and Kittler 1993) and for Ni contaminated material (●)

Figure 2. Expected dependence of contrast vs temperature, calculated using $c \propto 1/D\tau$, Shockley-Read-Hall recombination theory, diffusivity and capture cross-section inversely proportional to temperature (doping conc. 10^{16} cm^{-3}, E_T level pos.)

Fig. 2 shows a rough simulation of the c(T) behaviour of a defect, taking $c \propto 1/D\tau$ (Donolato 1978/79) and Shockley-Read-Hall recombination theory for lifetime $\tau(T)$. Minority-carrier capture cross-section and diffusivity D were assumed to be inversely proportional to temperature. According to Fig. 2 the contrast maximum is found at lower temperatures and the ratio between maximum contrast and contrast at 300 K is larger for shallower centres. A comparison of Figs. 1 and 2 demonstrates a qualitative agreement between experimental and calculated dependencies. The absence of contrast maxima in our experimental data is probably a consequence of the temperature limitation of the experimental set-up to a minimum value of 80 K. Indeed, EBIC investigations of dislocations in n-type Si performed by Kusanagi et al. (1992) for temperatures down to 50 K exhibit such maxima and the character of the experimental dependencies is in agreement with Fig. 2 .

3. NiSi$_2$ PARTICLES

Small NiSi$_2$ particles can be formed in Si without generation of secondary defects, see e.g. Seibt and Schröter (1989). Such particles were shown to be very efficient recombination sites despite of their small size (Kittler et al. 1991). According to calculations by Donolato (1992) the minimum defect radius, necessary to produce the large contrast observed, is larger than the radius of the precipitate. In view of the very low strain induced by NiSi$_2$, Schottky barriers around the particles are the most likely explanation of the ´increased˜defect radius. The recombination behaviour of NiSi$_2$ particles (thickness below 20 nm, diameter around 1μm) in n-type Si of different doping levels (4 x 10^{14} cm^{-3} and 3 x 10^{16} cm^{-3}), is reported here. It is found that the contrast is larger in the material with lower doping concentration. This is in qualitative agreement with the picture of NiSi$_2$ particles acting as internal Schottky junctions.

For both doping levels the contrast of the particles decreases when cooling the sample, see Fig. 3a. This is similar to findings of Ourmazd et al. (1983) and Wilshaw et al. (1989,1991) for charged dislocations in Si. Fig. 3b and 3c show $c(I_b)$ dependencies measured for exactly the same particles referred to in Fig. 3a. The contrast decreases with increasing beam current for the sample with lower doping level (Fig. 3 b). This is again similar to charged dislocations in Si (Wilshaw et al. 1989, Bondarenko and Yakimov 1990). In the sample with higher doping level an additional regime of contrast behaviour with positive $c(I_b)$ slope is found at low beam currents, particularly pronounced for T = 80 K (Fig. 3c).

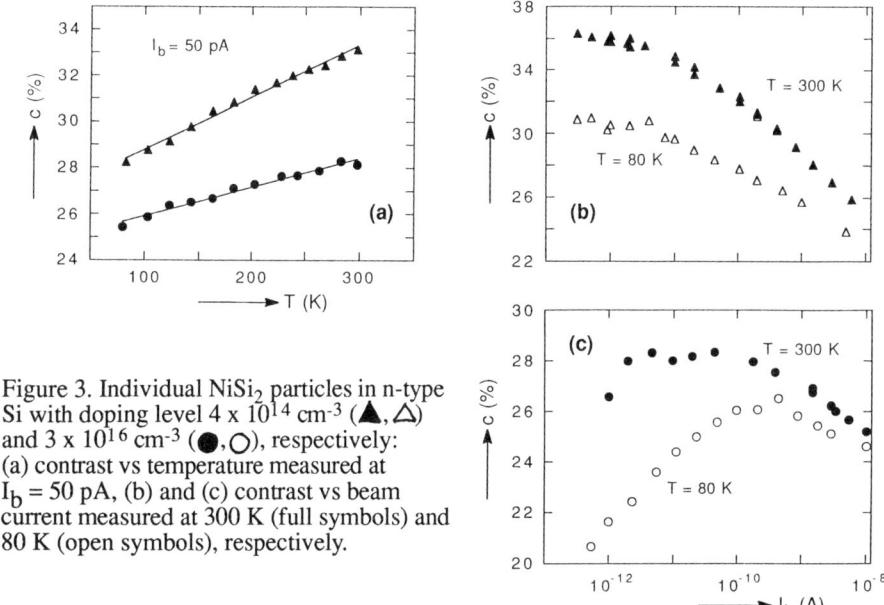

Figure 3. Individual NiSi$_2$ particles in n-type Si with doping level 4 x 10^{14} cm^{-3} (▲,△) and 3 x 10^{16} cm^{-3} (●,○), respectively: (a) contrast vs temperature measured at I$_b$ = 50 pA, (b) and (c) contrast vs beam current measured at 300 K (full symbols) and 80 K (open symbols), respectively.

A similar c(I$_b$) behaviour exhibiting three regimes was also observed for dislocations. In the work of Norman and Holt (1991) the positive slope at low beam current was attributed to a variation of the ratio of EBIC background current I$_o$ and beam current I$_b$, originating from beam-current-dependent screening effects in the electric field of the collecting junction. In our sample no marked variation of the I$_o$/I$_b$ ratio could be observed. Consequently, the positive slope at low current must have a different origin.

It was pointed out by Leamy et al. (1976) that for shallow recombination centres a lifetime reduction/contrast increase is expected when the injection level increases; for simulation see also Kittler and Seifert (1992). We believe that, in accordance with Cavallini and Castaldini (1992), the positive slope at low beam currents might be caused by an atmosphere of shallow levels surrounding the NiSi$_2$ particles. An accumulation of shallow levels could be a consequence of Coulomb interaction with the charged NiSi$_2$ particles. In that case the accumulation is larger for the sample with the higher doping level, which could explain the absence of the positive c(I$_b$) slope in the sample of lower doping.

Another explanation is related to the fact that the level of injection at a certain beam current is lower by two orders of magnitude in case of Fig. 3c as compared to the sample shown in Fig. 3b because of the large difference in doping level. So, one may speculate that a behaviour similar to Fig. 3c is expected also for the sample of lower doping, but at much lower beam currents not realized here experimentally.

4. CONCLUDING REMARKS

Misfit dislocations and NiSi$_2$ particles in n-type Si show opposite behaviour in EBIC. The origin of recombination activity of the dislocations studied here can be attributed to shallow recombination levels. The effects observed for NiSi$_2$ particles could be interpreted, in similarity to Wilshaw´s model (1989) of charged dislocations, as an action of internal Schottky contacts, i.e. negatively charged particles surrounded by a depletion region. Additionally, the contrast behaviour in highly doped material seems to indicate an atmosphere of shallow levels around NiSi$_2$ particles.

The most surprising observation is, that dislocations in n-type Si exhibit either the character presented here for 60° misfit dislocations -also Kusanagi et al. (1992) found the same character for 60° and screw dislocations introduced by plastic deformation- or a behaviour similar to that of NiSi$_2$ particles, as reported for screw and 60° dislocations introduced by plastic deformation (Ourmazd et al. 1983, Wilshaw et al. 1989, Bondarenko and Yakimov 1990). These observations give strong arguments concerning the dominance of extrinsic influences on dislocation recombination activity.

ACKNOWLEDGEMENTS

The authors acknowledge Drs. M. Seibt and F. Riedel for supplying the samples containing NiSi$_2$ particles and Dr. Z.J. Radzimski for samples containing misfit dislocations. Further they thank J. Lärz for technical assistance during EBIC measurements. Prof. W. Schröter and Dr. V. Higgs are acknowledged for stimulating discussions. Parts of this work were done under support of the Volkswagenstiftung.

REFERENCES

Bondarenko I E and Yakimov E B 1990 *Defect Control in Semiconductors*, ed K Sumino
     (Amsterdam: North Holland) pp 1443-6
Cavallini A and Castaldini A 1992 MRS Proc. Vol. 262 223
Donolato C 1978/79 Optik 52 19
Donolato C 1992 Semicond. Sci. Technol. 7 32
Higgs V and Kittler M 1993 this conference
Kimerling L C, Leamy H J and Patel J R 1977 Appl. Phys. Lett . 30 217
Kittler M, Lärz J, Seifert W, Seibt M and Schröter W 1991 Appl. Phys. Lett. 58 911
Kittler M and Seifert W 1992 ESF Conf: *Extended Defects in Semiconductors*, held in
     Holzhau/Germany in August '92, submitted to Phys. Stat. Sol.
Kittler M, Seifert W and Radzimski Z J 1993a Appl. Phys. Lett. in press
Kittler M, Seifert W and Higgs V 1993b Phys. Stat. Sol. in press
Kusanagi S, Sekiguchi T and Sumino K 1992 Appl. Phys. Lett. 61 792
Leamy H J, Kimerling L C and Ferris S D 1976 *Scanning Electron Microscopy*, ed O Johari
     (Chicago: IIT Research Institute) pp 529-538
Norman C E and Holt D B 1991 Inst. Phys. Conf. Ser. 117 755
Ourmazd A, Wilshaw P R and Booker G R 1983 Physica 116B 600
Radzimski Z J, Zhou T Q, Buczkowski A B and Rozgonyi G A 1991 Appl. Phys. A 53 189
Seibt M and Schröter W 1989 Phil. Mag. A 59 337
Wilshaw P R, Fell T S and Booker G R 1989 *Point and Extended Defects in Semiconductors*,
     eds G Benedek, A Cavallini and W Schröter (New York: Plenum) pp 243-256
Wilshaw P R, Fell T S and De Coteau M 1991 J. de Physique IV 1 C6-3

Inst. Phys. Conf. Ser. No 134: Section 11
Paper presented at Microsc. Semicond. Mater. Conf., Oxford, 5–8 April 1993

721

Temperature dependence of gold induced conductivity inversion

Z J Radzimski, A Buczkowski[1], G A Rozgonyi[1], T Sekiguchi[2], S Kusanagi[2] and K Sumino[2]

Analytical Instrumentation Facility, [1] Department of Materials Science and Engineering,
North Carolina State University, Raleigh, NC 27695-7916, USA.
[2] Institute for Materials Research, Tohoku University, Sendai 980, Japan.

ABSTRACT: Defect induced inversion of conductivity type was studied at the surface of heteroepitaxial Si(Ge) structures as a function of sample temperature. The inversion was achieved by controlled contamination with Au introduced by diffusion from a backside evaporated layer. A theoretical explanation is presented. The electrical activity of defects and the inversion effect were evaluated using the electron beam induced current technique in a scanning electron microscope

1. INTRODUCTION

Electrically active defects influence the semiconductor properties in a variety of ways (Kittler et al 1989, Radzimski et al 1991). An important aspect of defect electrical activity is the electric field associated with trap charges. As shown by Wilshaw and Fell (1989), the electric field strongly influences the activity of extended defects. Also, the carrier distribution within the semiconductor may be altered to the point where a conductivity inversion occurs. A surface conductivity inversion has been observed for silicon samples after fluorine (Chu et al 1990) and hydrogen (Zhou et al 1991) implantation. It was explained as the result of a dopant compensation due to the formation of shallow level defects. However, the inversion can be induced also by deep level defects, such as intentionally introduced gold or nickel (Radzimski et al 1991, 1993). The goal of this work is to study defect induced inversion and its temperature dependence. The electrical activity of defects was evaluated using the electron beam induced current (EBIC) technique in a scanning electron microscope (SEM). The theoretical explanation based on calculation of the surface potential resulting from the charge neutrality conditions for surface and space charge region is presented, as previously outlined by Many et al (1965).

2. EXPERIMENT AND RESULTS

Defect induced conductivity inversion was studied using a heteroepitaxial Si(Ge) structure with misfit dislocations confined to two planes at 4 and 6 μm depth (Rozgonyi et al 1987). The epilayer and substrate were n-type with 10^{15} cm^{-3} dopant concentration. The structure was decorated with gold by diffusion from a backside evaporated layer during rapid thermal annealing from 400°C to 1130°C for 30 seconds. An example of the EBIC image for a sample annealed at 1000°C is shown in Fig. 1a. A typical EBIC image of a Schottky contact with a bright area and dark lines corresponding to electrically active misfit dislocations can be observed for this sample. As shown in our previous work (Radzimski et al 1991, 1992), the dislocations yield no EBIC contrast in the as grown sample and they activate in the gold decorated samples after annealing at a temperature higher than 400°C. Following an 1130°C anneal a strong EBIC signal was collected far beyond the 1 mm diameter Schottky contact, as shown in Figs. 1b. For example, at room temperature and at a distance of 500 μm from the contact, which is much larger than a diffusion length, the signal is approximately 5% higher than the signal measured at the contact far away from the defect, see Fig. 2. This indicates that the Au contact operates actually as an ohmic electrode to the inverted p-type surface layer. The signal outside the contact decreases with respect to the signal at the contact when the sample temperature is lowered during EBIC measurements, as shown in Fig. 2. These two signals are approximately equal at the temperature of 150K. Below this temperature the signal outside the Schottky contact is smaller and drastically decreases with temperature and the bright area in the EBIC image is confined only to the Schottky contact.

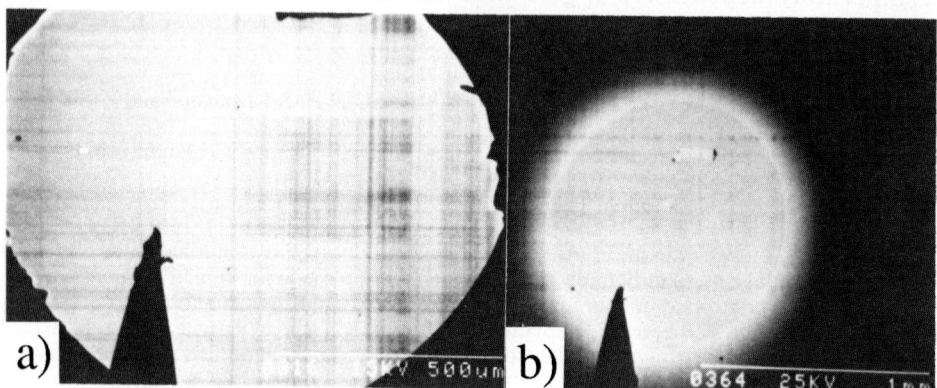

Fig. 1 EBIC/SEM image of the Si(Ge) heterostructure with misfit dislocations following decoration with Au at (a) 1000°C and (b) 1130°C for 30 sec.

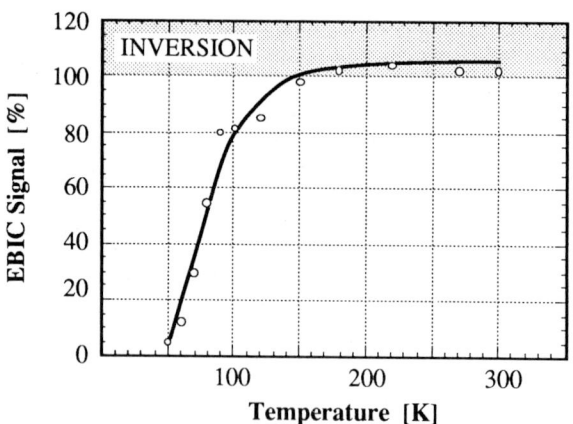

Fig. 2. The EBIC signal measured at a distance of 500 μm from the Schottky contact as a function of temperature. The signal is normalized versus the background signal measured on the Schottky contact.

3. DISCUSSION

The impact of structural defects and impurities on the properties of the surrounding semiconductor material can be analyzed theoretically by calculation of the potential induced by trap charges. In the case of impurities gettered at the surface it would be a surface potential induced by traps associated with contaminants. For acceptor type states the charge is neutral when these states are empty and negative when filled with electrons. If the surface is negatively charged then a positive space charge region is formed beneath it, with the energy bands at the surface bending upward with respect to the Fermi level. Wilshaw and Fell (1989) describe in a similar way the charging mechanism and band bending at the extended line defects. The defect induced surface electric field separates carriers generated by the electron beam in a fashion similar to the electric field of a Schottky contact. A similar effect yielding an anomalous EBIC signal was observed also by Partin and Luk'yanov (1990) at a silicon surface in a poor vacuum. The surface band bending and the barrier height depend on the total charge per unit surface, Q_{surf}. Under equilibrium conditions the Q_{surf} is in balance with the total charge of the space charge region, Q_{scr}. The surface potential or the concentration of free carriers due to the trap presence can be calculated by coupling the equations for Q_{surf} and Q_{scr} in the way described by Many et al (1965) and explained in detail by Buczkowski at al (1993). The concentration of free carriers depends on the trap concentration, their donor or acceptor type nature, and the occupancy status, which is temperature dependent. Theoretical results are given in Fig. 3 for n-type material with acceptor type states of 0.02 eV energy which is a well recognized level for gold (Einspruch and Larrabe 1983). The calculations were done for room temperature and the trap energy is given in respect to the intrinsic level ($E_i = 0$ eV).

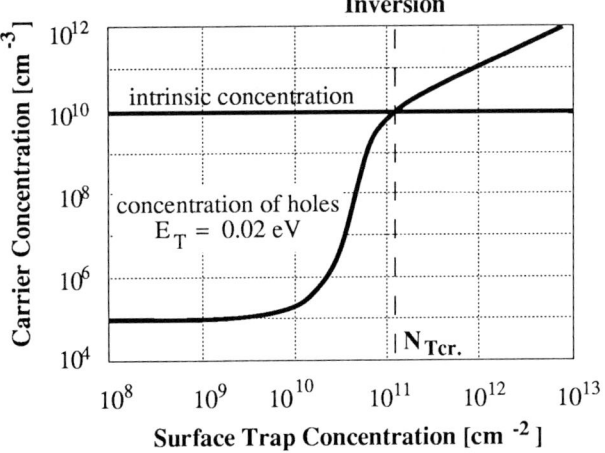

Fig. 3 Carrier concentration of holes at the surface and electrons in the bulk of (a) the n - type silicon as a function of surface concentration of acceptors with $E_T = 0.02$ eV.

An increase of the acceptor trap concentration results in an increase of the hole concentration. However, there is a threshold level of traps below which no significant impact can be observed. This would correspond to the samples annealed in the experiment below 1000°C, where the amount of impurity gettered at the surface was too small to have impact on the background concentration of majority carriers. The significant increase of holes for 0.02 eV traps occurs between 10^{10} and 10^{11}/cm^2 surface concentration. The point where the concentration of holes provided by surface traps is equal to the intrinsic concentration of electrons, defines the critical value of trap density N_{Tcr} for conductivity type inversion. Our earlier qualitative EBIC study of Si(Ge) heteroepitaxial structures (Radzimski et al 1991) showed that such a threshold was reached both at the surface and at the misfit dislocation plane in the structure contaminated by gold and annealed at 1130°C. The electric field observed via the EBIC signal is correlated with regions known to have an enhanced concentration of gold (due to gettering at the surface and the dislocations) and indicates the acceptor type action of the Au precipitates.

The critical value of trap density where conductivity type inversion occurs depends strongly on sample temperature. This is shown in Fig. 4 for acceptor traps with various energy levels in n-type material. As the sample temperature decreases more traps are needed to reach the critical value. For example, at room temperature a 2×10^{11}/cm^2 concentration of acceptor traps with energy level 0.02 eV is required to invert the conductivity of n-type silicon with dopant concentration 1×10^{15}/cm^3, while at 100K a concentration of 1.5×10^{12}/cm^2 is needed. Therefore, in a sample with a constant concentration of traps the inversion effect should vanish if the temperature is decreased as shown in Fig. 2. If the defect level is known one can estimate the density of defects which are responsible for the inversion using the results from Fig. 4. The exact evaluation of defect density would require the accurate measurement of the EBIC signal outside the metal contact which should take into account the signal decrease due to the finite resistance of the inverted surface layer. In the case of a signal measured at the metal contact the electron beam energy losses in the metal layer should be considered.

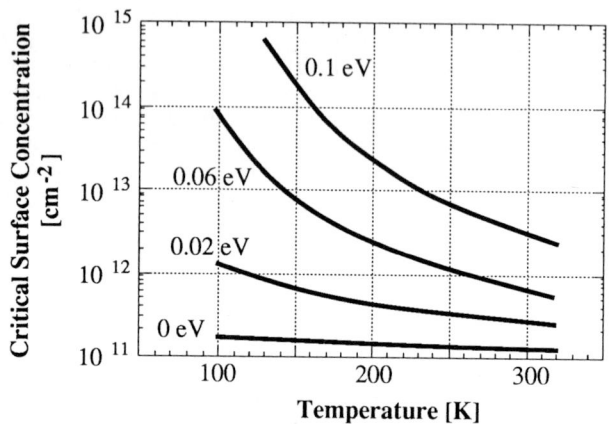

Fig. 4 Critical surface concentration of acceptor traps with energy level as a parameter in n-type Si, as a function of sample temperature. Background doping concentration = 1×10^{15} cm^{-3}.

4. SUMMARY

It has been shown that the inversion of conductivity type may occur when the density of defects is higher than a threshold value even if these defects are known to occupy deep levels in the band gap. Defect induced inversion has been observed at the surface of heteroepitaxial Si(Ge) n-type structures intentionally contaminated with gold. The presence of the non-intentionally introduced electric field was found at the entire surface of the samples subjected to a high temperature decoration process. At this temperature a critical concentration of the gold acceptor type traps in silicon has been reached. The inversion disappeared during EBIC observation when the temperature was lowered below 150K.

REFERENCES

Buczkowski A Rozgonyi G A and Shimura F 1993 Jap. J. Appl. Phys. 32 L218
Chu C H Chen L J and Hwang H L 1990 J. Crystal Growth 103 188
Kittler A 1989 Solid State Phenomena 6 & 7 367
Many A Goldstein Y and Grover N B 1965 Semiconductor Surfaces, North-Holland
    Publishing Company Amsterdam 128
Einspruch N G Larrabe G B 1983 VLSI Electronics-Microstructure Science 6 Academic Press
Patrin A A and Luk'yanov A E 1990 Scanning 12 334
Radzimski Z J Buczkowski A and Rozgonyi G A 1990 ECS Proc. 90-15 436
Radzimski Z J Zhou T Q Buczkowski A and Rozgonyi G A 1991 Appl. Phys. A53 189
Radzimski Z J Zhou T Q Buczkowski A Rozgonyi G A Finn D Hellwig LG and Ross J A
    1992 Appl. Phys. Let. 60 1096
Radzimski Z J Buczkowski A Zhou T Q and Rozgonyi G A 1993 to be published in
    Scanning Microscopy.
Rozgonyi G A Salih A S M Radzimski Z J Kola R R and Honeycutt J 1987 J. Crystal
    Growth 85 300
Wilshaw P R and Fell T 1989 Inst. Phys. Conf. Ser. No. 104 85
Zhou T Q Radzimski Z J Patnaik B Rozgonyi G A Sopori B 1991 Appl. Phys. A58

Electron-beam-induced gate currents in GaAs MESFET

K Kaufmann and L J Balk[*]

Universität Duisburg SFB 254 FB Werkstoffe der Elektrotechnik D 47048 Duisburg Germany
[*]Bergische Universität Wuppertal Lehrstuhl für Elektronik Fuhlrottstr. 10 Wuppertal Germany

ABSTRACT: Measurements of electron beam induced gate currents have been used to investigate GaAs MESFETs. By means of such measurements microscopic inhomogeneities of the electrical properties of the gate contact and of the intermediate areas between gate and ohmic contacts can be detected. In order to achieve quantitative interpretability of the results, numerical simulations of the dependencies of the electron beam induced gate current on different electron beam and device parameters have been carried out.

1. INTRODUCTION

EBIC - (Electron Beam Induced Current) - investigations are an effective method to investigate semiconductor materials and devices (Leamy 1982). Devices such as Schottky-diodes, pn-junctions, bipolar transistors, MOSFET-structures and GaAs MESFETs have been investigated by EBIC. The main aims of these device investigations are to control the position of depletion layers and to detect electrically active crystal defects.

GaAs MESFETs have been investigated both by Drain-EBIC (between drain and source contacts) and by Gate-EBIC (between gate and source/drain contacts) (Kaufmann and Balk 1992). With these techniques inhomogeneities of the electrical properties of the semiconductor material and of the contacts can be easily detected. Whereas an unambigious interpretation of the measuring results seems to be extremely difficult for the case of Drain-EBIC, the Gate-EBIC results may be quantified by simulation of the EBIC dependencies on electron beam parameters and device properties.

But up to now appropriate simulations have been carried out only for relatively simple device structures such as Schottky-Diodes (Donolato 1978) and pn-junctions (Kuhnert 1991). However, these simulation models are based on many simplifications making them unsuitable for more complicated devices like GaAs MESFETs; most importantly the influence of device inherent electrical fields has not yet been considered. Consideration of the microscopic distribution of electrical potentials inside the device structure, however, is necessary for quantification of the results and can only be introduced into the simulation by sophisticated numerical methods.

2. EXPERIMENTAL

2.1 Experimental Set-up

The structure of the investigated GaAs MESFETs and the measuring circuit used is shown in Fig. 1. The GaAs- and AlAs-layers of the GaAs MESFETs have been epitaxially grown by MOVPE (Metal-Organic-Vapour-Phase-Epitaxy). The conducting layer which is n-doped ($n=3\cdot10^{17}\text{cm}^{-3}$) with silicon has a thickness of 100nm. The thickness of the buffer layer is 350nm. The density of free charge carriers in the buffer layer is very low at thermal equilibrium and is not known exactly. Both an n⁻- or p⁻ - charge density is possible. Beneath the buffer layer is a AlAs/GaAs-superlattice serving as growth barrier for crystal defects. On the top of the conducting layer are three contacts. Source and drain are ohmic contacts; whereas the gate is a metal-semiconductor diode with a Schottky barrier and a depletion layer. Within this depletion layer the electrical field is directed towards the semiconductor surface. Thus, excess holes generated by a focused primary electron beam are collected when reaching the depletion layer. The electron beam induced gate current (Gate-EBIC) generated thereby is measured by means of a lock-in amplifier in order to allow the use of very small beam currents I_{PE}.

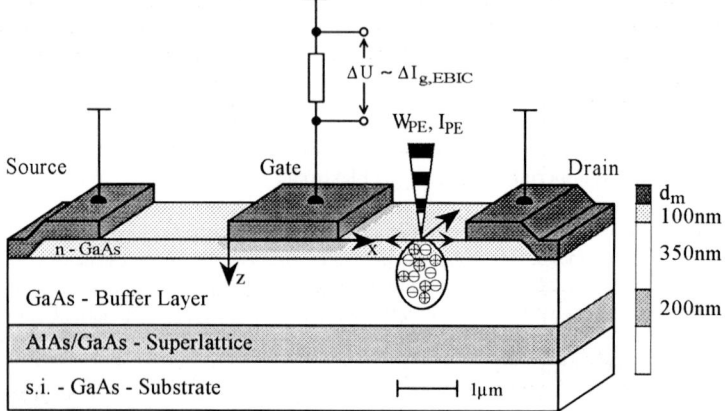

Fig. 1　Structure of the investigated GaAs MESFETs and measuring circuit

2.2 Experimental Results

An illustrative example of the Gate-EBIC method is shown in Fig. 2. The GaAs MESFET has electrical shorts between the gate and the ohmic contacts caused by local electrical break-throughs. These shorts are clearly recognizable in the Gate-EBIC images, whereas they are hardly discoverable in SE/RE images. The contrasts and the attainable resolution depend strongly on the primary electron energy. Nevertheless, the resolution of about 100nm at a primary electron energy $W_{PE}=20keV$ is astonishingly good (compare linescans of Fig. 3 for this statement). Due to the size of the energy dissipation volume, a resolution of 1μm would be expected.

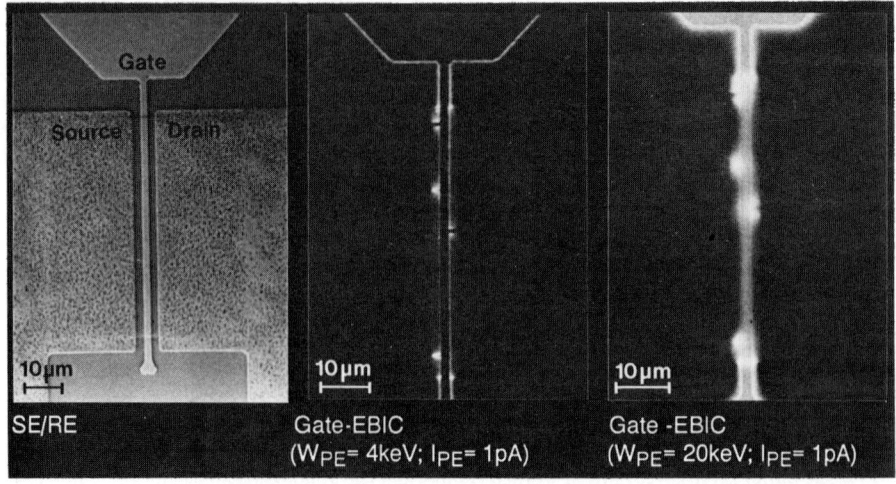

Fig. 2　GaAs-MESFET with electrical connections between the gate and the ohmic contacts

Fig. 3 shows experimental Gate-EBIC signals along a line perpendicular to the gate, extending from source to drain, and taken from a faultless area of Fig. 2. These exemplary results are treated theoretically in detail, subsequently. At W_{PE}= 4keV the Gate-EBIC signal has small values in the area between the contacts and high values nearby the gate edges. At W_{PE}= 20keV the signal has its maximum in the middle of the gate. The signal decreases towards the gate edges and has local minima inside the gate edges. Outside the gate area are sharp local maxima nearby the gate edges. In the area between gate and ohmic contacts the Gate-EBIC signal decreases strongly towards the ohmic contacts. Interestingly enough, the maximum signal height is almost equal at both primary electron energies although at W_{PE}= 20keV the number of generated excess holes is five times higher than at W_{PE}= 4keV.

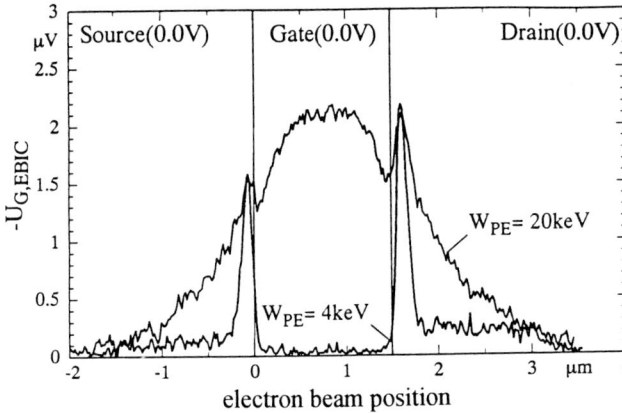

Fig. 3   Linescans of the Gate-EBIC signal for W_{PE}= 4 and W_{PE}= 20keV

3. SIMULATIONS

3.1 Model

In case of GaAs MESFETs with an n-doped conducting layer the Gate-EBIC signal is caused by those excess holes collected in the depletion layer of the gate contact. Thus, to calculate the electron beam induced gate current, the hole current at the semiconductor surface must be known, presupposing knowledge of the total hole density. This simulation of the hole density has been carried out for the low injection case. In this case the disturbance of the electrical potential and of the electron density can be neglected. Under these conditions simulations can be carried out by solving the hole continuity equation only, but in its full form:

$$-\frac{1}{q} \cdot div \, \bar{J}_p + RG_p + G_{PE} = 0$$

For the description of the hole current density $\bar{J}_p$ the drift diffusion approximation is used:

$$\bar{J}_p = -q \cdot \mu_h \cdot p \cdot grad\,\varphi - q \cdot D_h \cdot grad \, p$$

This equation can be simplified using the Einstein relationship $D_h = \mu_h \cdot (k \cdot T/q)$; D_h and μ_h are the hole diffusion constant and the hole mobility, respectively. The recombination rate is assumed to be $RG_p = D_h \cdot p/(L_h \cdot L_h)$; L_h is the hole diffusion length (Sze 1981). Under these assumptions the hole continuity equation to be solved reads:

$$div \, (grad \, p + \frac{q}{k \cdot T} \cdot (p \cdot grad \, \varphi)) - \frac{p}{L_h^2} + \frac{G_{PE}}{D_h} = 0$$

As the continuity equation cannot be solved analytically, the numerical finite differences method has been used. This method requires definition of a simulation region and of a simulation grid, on which the continuity equation is to be solved. Direct discretization of the continuity equation is only possible if the potential difference between two grid lines is smaller than $(k \cdot T/q)/2$ (Selberherr 1984). Thus, to avoid the necessity of a very high number of grid lines, the discretization of the continuity equation given by Selberherr (1984) is used. The dimensions of the simulation region are given in Fig. 4.

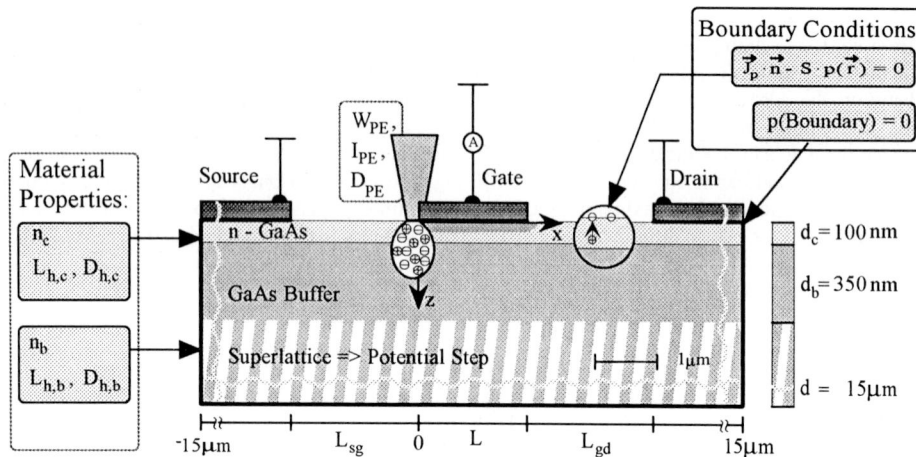

Fig. 4   Two dimensional simulation region, boundary conditions and simulation parameters

The potential distributions φ inside the GaAs MESFETs have been calculated using the device simulation program MINIMOS 5.1 (TU Vienna). The potential distribution depends on the charge carrier densities n_c and n_b at thermal equilibrium in the conducting layer and the buffer layer, respectively. The effect of the AlAs/GaAs-superlattice on the Gate-EBIC signal has also been taken into account. Because the bandgap of AlAs is greater than the bandgap of GaAs, there are steps in the band scheme. Usually, the holes cannot surmount the steps in the valence band. This effect of the superlattice is considered by adding artificial steps to the calculated potential distribution. These artificial potential steps have a height of 0.5V and are at the position of the superlattice. For calculation of the hole generation rate G_{PE} the following main steps were undertaken: for the depth distribution of the generated holes an approach given by Wu and Wittry (1978) is used, taking into consideration the energy loss of the primary electrons in the contacts; the penetration depth of the primary electrons is calculated by the equation given by Kanaya and Okayama (1972); the influence of the finite electron beam diameter D_{PE} is considered by convolution of the lateral electron beam intensity with the lateral extension of the energy dissipation volume; partial shading of the electron beam at the contact edges is taken into account. Based on this model a program has been written allowing the simulation of the Gate-EBIC.

3.2 Simulation Results

The simulated linescans shown in Fig. 5 demonstrate the dependence of the electron beam induced

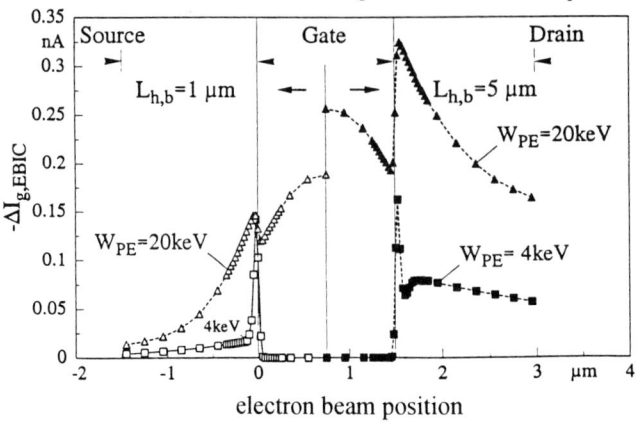

electron beam position

Fig. 5   Simulated Gate-EBIC linescans. Only half linescans are shown as simulation is symmetrical about the middle of the gate. Hole diffusion lengths are for the curves on the left $L_{h,b}=1\mu m$ and on the right $L_{h,b} = 5\mu m$. Parameters used are $W_{PE}=4keV$ and 20keV, $I_{PE} = 1pA$, $D_{PE} = 25nm$, $L_{h,c} = 1\mu m$, $n_c = 3 \cdot 10^{17} cm^{-3}$, $S = \infty$, $n_b = 1 \cdot 10^{14} cm^{-3}$, $d_m = 260nm\,(Au)$.

gate current on the hole diffusion length $L_{h,b}$ for a GaAs MESFET with a superlattice. The calculations of the linescans for $L_{h,b} = 1\mu m$ allow a very good fit to the shape of the measured linescans in Fig. 3, whereas use of a hole diffusion length of $L_{h,b} = 5\mu m$ results in linescans whose shapes show strong deviations from the measured linescans. The strong dependency of the Gate-EBIC signal on the hole diffusion length is also recognizable in Fig. 6a. For comparison, in Fig. 6b corresponding curves of the

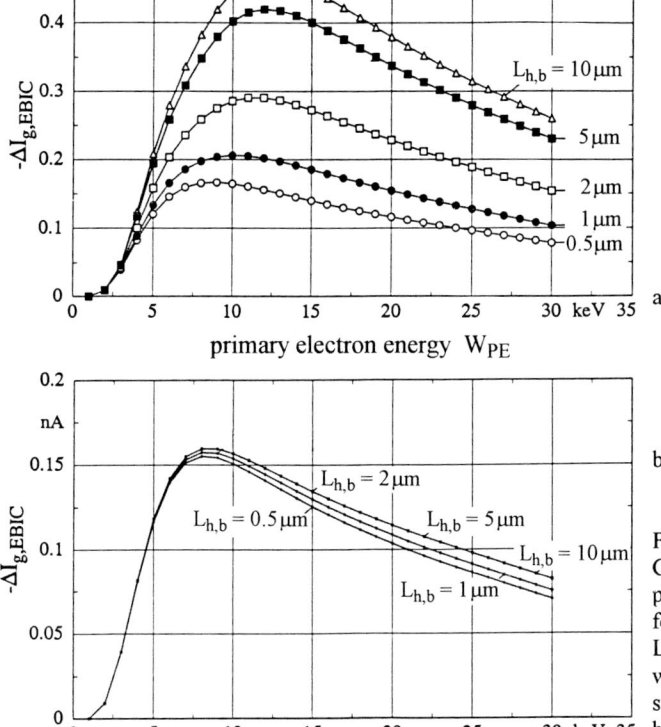

a)

b)

Fig. 6 Dependence of the Gate-EBIC signal on the primary electron energy W_{PE} for different diffusion lengths $L_{h,b}$ for a GaAs MESFET with (a) and without (b) a superlattice for an electron beam position nearby the gate. Parameters are as in Fig. 5.

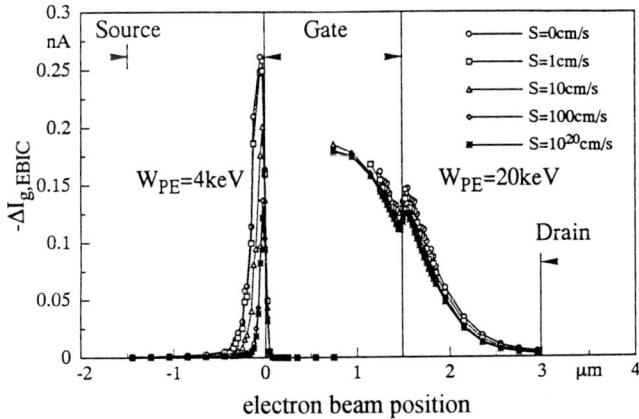

Fig. 7 Linescans of the Gate-EBIC signal for different surface recombination velocities S. Primary electron energies used are $W_{PE} = 4keV$ for the left curves and $W_{PE} = 20keV$ for the right curves. Other parameters are as in Fig. 5.

Gate-EBIC signal calculated for a GaAs MESFET without a superlattice are shown. In this case the Gate-EBIC signal depends only very weakly on the hole diffusion length. This means that in GaAs MESFETs without a superlattice, inhomogeneities of recombination centre distributions are hardly detectable. The physical reason for this is that due to the potential barrier between the conducting layer and the buffer layer the holes drift into deeper regions of the buffer layer. These holes cannot be collected by the gate.

Surface recombination has a significant effect on the Gate-EBIC signal only at low primary electron energies, as shown in Fig. 7. At low primary electron energies, the signal height as well as the width of the contrasts at the gate edges are influenced by the surface recombination velocity S. Finally, in Fig. 8 the influence of the electron beam diameter is demonstrated. Both the obtainable resolution and the Gate-EBIC signal height at the gate edges improve with decreasing electron beam diameter.

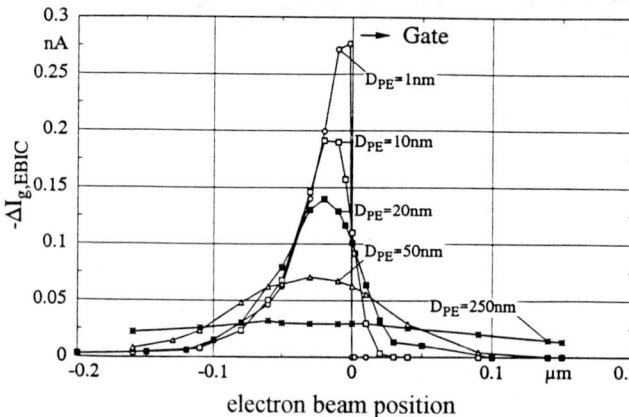

Fig. 8 Influence of the electron beam diameter D_{PE} on the Gate-EBIC signal near a gate edge for the primary electron energy $W_{PE} = 4keV$. Other parameters are as in Fig. 5.

4. CONCLUSIONS AND SUGGESTIONS FOR FURTHER WORK

It has been shown that the electron beam induced currents can be understood quantitatively even for complicated devices. To achieve this, numerical simulations are necessary and the electrical potential distribution inside the devices has to be considered explicitly. The effect of structures like superlattices on the EBIC signal can be taken into account by relatively simply models. In future this model as presented for the case of electron beam induced gate currents in GaAs MESFETs can be extended. These extensions could cover topics as: adaptation to special measuring conditions and other device structures; inhomogeneous distributions of recombination centers (e.g. dislocations); three dimensional simulation domains (in order to simulate edge effects etc.).

ACKNOWLEDGEMENTS

We thank the Department for Solid State Electronics of Duisburg University (Prof. Dr. F.-J. Tegude) for supplying the GaAs MESFETs. The work was financially supported by the Deutsche Forschungsgemeinschaft within Sonderforschungsbereich 254.

REFERENCES

Donolato 1978 Optik 52(1) pp 19-36
Leamy H J 1982 J.Appl.Phys. 53(6) pp R51-R80
Kanaya K and Okayama S 1972 J.Phys.D:Appl.Phys. 5 pp 43-58
Kaufmann K and Balk L J 1992 Microelectronic Engineering 16 pp 513-20
Kuhnert R 1991 J.Appl.Phys. 70(1) pp 476-84
Selberherr S 1984 Analysis and Simulation of Semiconductor Devices Springer-Verlag Wien
Sze S M 1981 Physics of Semiconductor Devices Wiley-Interscience
Wu C J and Wittry D B 1987 J.Appl.Phys. 49(5) pp 2827-36

Inst. Phys. Conf. Ser. No 134: Section 11
Paper presented at Microsc. Semicond. Mater. Conf., Oxford, 5–8 April 1993

731

EBIC in HEMT structures on SI substrates

D B Holt, E Napchan, A Wojcik, M Ammou* and P Gibart*

Dept. of Materials, Imperial College of Science, Technology and Medicine, London SW7 2BP
*Laboratoire de Physique du Solide et Energie Solaire–CNRS, Sophia Antipolis, Rue Bernard Gregory, 065560 Valbonne, France

ABSTRACT: EBIC microscopy using the built–in fields in a HEMT structure showed several types of electrically active defects to occur. Monte Carlo simulations were used to interpret the observations.

1. INTRODUCTION

It was recently found that it is possible to make EBIC observations on HEMT structures on semi–insulating substrates using contacts to the top and bottom of the wafer despite the thick semi–insulating layer between (Holt 1993).

2. EXPERIMENTAL AND MONTE CARLO METHODS

A JEOL JSM–840A SEM, fitted with a Matelect ISM–5 detection system for EBIC and other signals, was used for the measurements. The specimens examined were cut from a slice grown at LPSES (Laboratoire de Physique du Solide et Energie Solaire). This had the layer structure and energy band diagram shown in Figure 1. The specimens were mounted on TO–5 headers with the back in electrical contact with the header via Ag paint and the top connected via Ag paint and a gold wire to one of the insulated posts. The edge of the silver paint of the upper contact runs across the top of the micrographs in Figure 2 and the gold wire appears at the top right.

The MC–SET (Monte Carlo simulation of electron trajectories) suite of programs is written in C with additional enhancements but is otherwise similar to the program previously described (Napchan and Holt 1987, Holt et al 1989). This now has the facility to print out the percentage of the beam energy dissipated in each layer of an epitaxial multilayer structure like that in Figure 1 (Figure 3) and to calculate the total charge collection current for any assigned values of charge collection efficiency (percentage of the carriers that are separated by the built–in field, etc) for each layer.

3. RESULTS

Despite the fact that there are semi–insulating substrates several hundred microns thick underlying the device layers of epitaxial material, EBIC signals can be readily obtained using contacts to the top and bottom of the specimens. This made EBIC microscopy possible over an area hundreds of microns wide around the top contacts (Figure 2) so electrically active defects in the active layers could be imaged (Figure 2b). The sign of the current detected depends on whether the top contact is earthed or connected to the input terminal of the EBIC head amplifier. In this first observation the connections were such that the defects appear bright. Figure 4 shows the Γ shaped defect near the top of Figure 2 at higher magnifications. It could be seen that the roughly vertical line was double as confirmed by the EBIC contrast linescan profile across these lines in Figure 5.

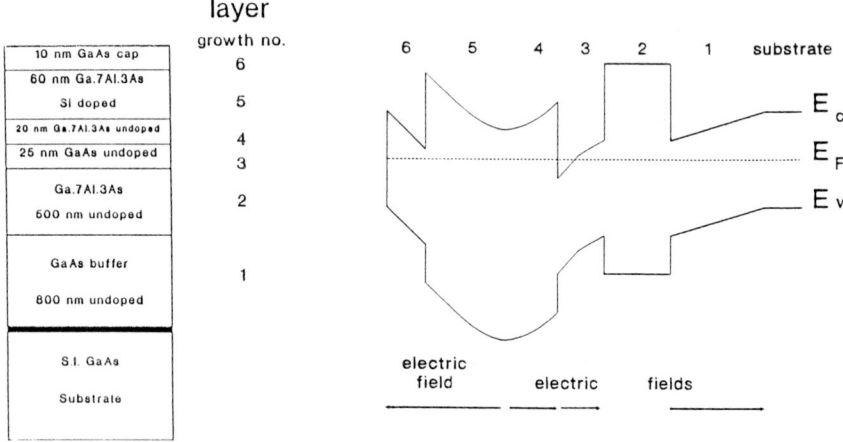

Figure 1. (a) Layer structure and (b) energy band structure of the slices grown at LPSES

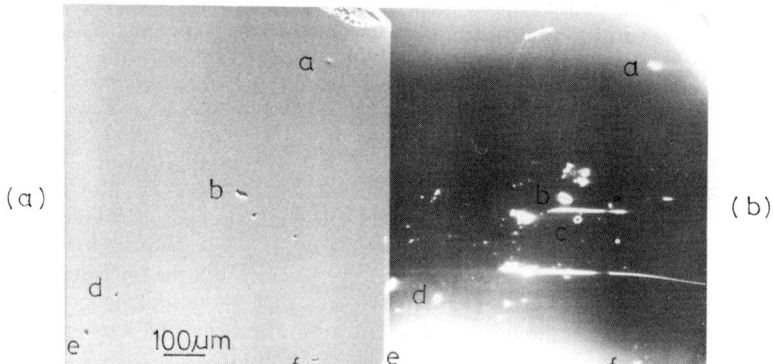

Figure 2. (a) SEI and (b) EBIC image of an area adjacent to the edge of the silver paint contact on the top of one of the specimens. The EBIC brightness varies with distance from the contact and a number of defects can be seen that do not correspond to the surface dust particles labelled a to f.

By varying the beam voltage, layers at different depths can be explored (depth resolution). Dislocations could be seen in one area that were visible at 10 keV but not at 5 keV. They were thus at a deeper level than that affected by lower beam energies. Examination of MC–SET graphics displays (like those in Figure 3) for 5 and 10 keV suggested that the dislocations were in layer 2 or possibly at the 2/1 interface (Figure 1).

The EBIC current was measured as a function of beam energy in keV on a number of occasions and at a number of points on each of the specimens. Two types of result were obtained as shown in Figure 6. The result in 6b is the more commonly observed and is thought to be typical of the material after beam damage. The result in 6a is obtained only when measured at a fresh, undamaged site and provided the beam is carefully cut off except when taking a reading.

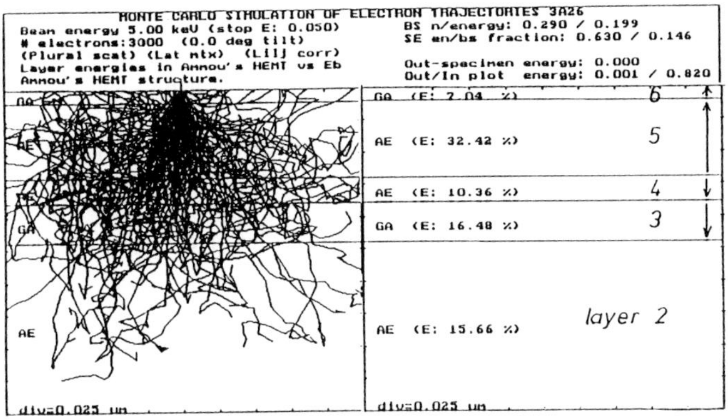

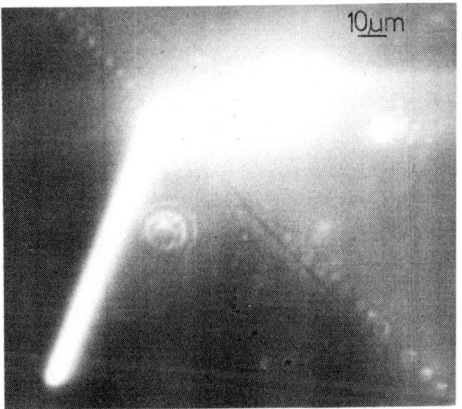

MONTE CARLO SIMULATION OF ELECTRON TRAJECTORIES 3A26
Beam energy 5.00 keV (stop E: 0.050)   BS n/energy: 0.290 / 0.199
electrons:3000   (0.0 deg tilt)   SE en/bs fraction: 0.630 / 0.146
(Plural scat) (Lat ntx) (Lill corr)
Layer energies in Annou's HEMT vs Eb   Out-specimen energy: 0.000
Annou's HEMT structure.   Out/In plot energy: 0.001 / 0.820

GA (E: 7.04 %)   6
AE (E: 32.42 %)   5
AE (E: 10.36 %)   4
GA (E: 16.48 %)   3
AE (E: 15.66 %)   layer 2

d1v=0.025 um   d1v=0.025 um

Figure 3. Graphics output from the MC–SET program showing a plot of trajectories and display of the percentages of total beam energy deposited in the layers of the HEMT structure for 5 keV.

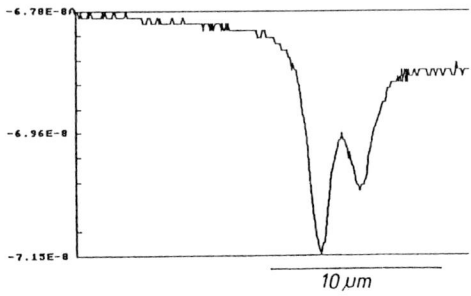

Figure 4. EBIC image of the complex defect near the top of Figure 2b, at higher magnification showing the vertical line to be double and the 45 degree line to be a row of dots.

Figure 5. EBIC linescan across the double vertical line of the defect showing that due to recombination the two line defects reduce the current collected to give 2.4% and 4% contrast.

4. DISCUSSION

It can be seen in Figure 1b that the energy band structure of the specimens is such that there are built–in fields in five of the epitaxial layers. The field in layers 6 and 5 will send holes and, by convention, therefore, the current up to the top contact (Figure 3a). The built–in field in layers 4, 3 and 1 will send holes (current) downwards with different efficiencies. As the beam energy per electron, E_b, in keV goes up, for a constant beam current, at first hole–electron pairs are generated in greater numbers but only in the shallower layers so the upward current first increases with E_b. Due to the shallowness of the layers this applies only up to about 2 to 3 keV as the Monte Carlo results show e.g. in Figure 3. With continuing increases in E_b, the beam penetrates and generates hole–electron pairs more deeply. Thus a downward current will begin to be produced from 4 or 5 keV and will increase relative to the upward component thereafter (Figure 3b). The net EBIC current upward should therefore pass through a maximum and decrease, become zero and then a net downward current that will finally saturate. This reversal of current sign with E_b is observed in Figure 5a. Work is in hand to simulate EBIC current as a function of beam energy using the expression

$$I_{cc} = V_b I_b \sum \eta_{cn} p_n / e_{in} \tag{1}$$

where $V_b I_b$ is the total power in the SEM beam and we sum over the n layers in the structure for η_c the charge collection efficiency for the layer times the number of pairs generated in the layer which is given by p/e_i where p the percentage of the beam energy in eV dissipated in the layer (calculated by the MC–SET program – see Figure 3) and e_i is the hole–electron pair formation energy for the layer given approximately by

$$e_i = 2.1 E_g + 1.3 \tag{2}$$

where E_g is the band gap energy in eV (Ehrenberg and Gibbons 1981). In the case of insulating and semi–insulating materials η_c should include a factor λ/D where λ is the drift length (the distance the carriers move when separated by a built in field and D is the distance between the contacts (Gunn 1964). The small value of λ/D is the reason that the EBIC currents in Figure 5 are relatively small compared to I_b.

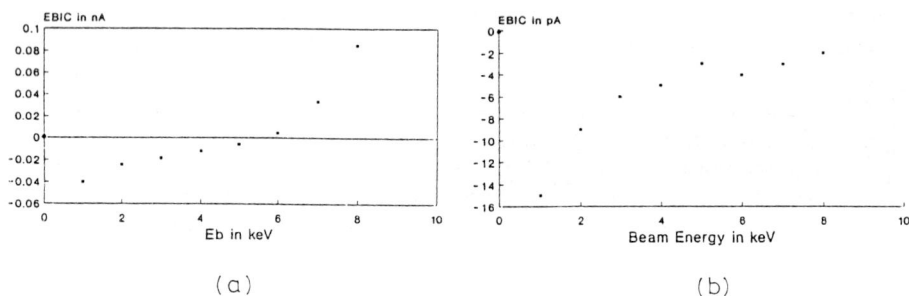

Figure 6. Measured values of charge collection current Icc at points well away from the edge of the top contacts, as a function of beam energy in keV for a constant beam current of $I_b = 2$ nA: (a) in an undamaged area, carefully minimizing beam irradiation during measurements and (b) without special precautions to avoid damage.

5. CONCLUSION

Monte Carlo electron trajectory simulations appeared likely to be able to account for the variation of the EBIC signal with E_b. The EBIC currents, although relatively small compared to those in Si devices, could be used to observe defect structures at different depths in this multilayer structure. At high beam currents, especially at low beam voltages, strong damage effects occurred.

REFERENCES

Gunn J B 1964 Solid–State Electron. **7** 739 – 742
Holt D B 1993 to be published.

Inst. Phys. Conf. Ser. No 134: Section 11
Paper presented at Microsc. Semicond. Mater. Conf., Oxford, 5–8 April 1993

EBIC from δ-doped quantum wires

D J C O'Neill, Y Feng[*], T J Thornton[*] and J J Harris [#]

Department of Materials, Imperial College of Science, Technology and Medicine, London SW7 2BP
[*]Department of Electrical and Electronic Engineering, Imperial College of Science, Technology and Medicine, London SW7 2BT
[#]IRC for Semiconductor Materials, Blackett Laboratory, Prince Consort Road, London SW7 2BZ

ABSTRACT: Test structures which employ a side gating technique to vary the width of a narrow wire made from δ-doped GaAs have been developed by Feng et al (1992). Electron beam lithography and reactive ion etching were used to cut two trenches 0.4 μm deep and 1 μm wide in δ-doped GaAs. The wire thus produced was made to be 1 μm in width, however wire resistance as a function of gate bias (voltage applied across the wire) and magnetoresistance measurements showed that the application of bias causes the wire width to be electrically reduced under very small gate bias (Feng et al 1992). This paper presents electron beam induced current measurements produced in the SEM which show visually the effect of bias. An induced current is also generated when the position of the scanning electron beam is not on the wire but is incident close to it. This is indicative of carriers excited by the electron beam leaking to the δ-doped layer.

1. INTRODUCTION

The device consists of δ-doped MBE GaAs containing a single layer of Si atoms (sheet density of 1×10^{13} m^{-2}) located 0.25 μm below the surface. Electron beam lithography was used to pattern a mask followed by reactive ion etching of the exposed surface in order to cut two trenches 0.4 μm deep and 1 μm wide (see fig. 1). The trenches were cut down through the δ-doped layer and into the semi-insulating GaAs substrate. Therefore semi-insulating GaAs and some GaAs in the buffer layer insulates the wire from the gate terminals.

The physical width of the wire is 1 μm but is reduced electrically by applying a bias to the gate terminals. Magnetoresistance measurements show that wire width decreases linearly with bias but the precise nature of the pinch-off mechanism is unclear.

2. EXPERIMENTAL

The electrical connections made for the EBIC experiment are shown in fig. 1. The δ-layer provides electrical contact to all points in the device. As there is no δ-layer between the wire and the gates an electric field is applied to the wire when the gates are biased.

The EBIC measurements were made using a JEOL JSM-840a SEM equipped with a Matelect ISM-5 EBIC monitor.

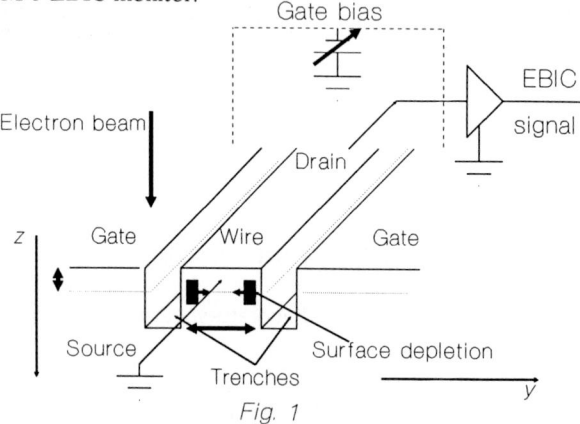

Fig. 1

3. RESULTS AND DISCUSSION

A series of EBIC measurements were recorded at different gate bias. Fig. 2 is an SE image showing the wire at high magnification. The etched trenches and the edges of the split gate are visible too. Fig. 3 is an EBIC micrograph of the same field of view as in fig. 2, taken at a gate bias of -7 V. The dark areas, which could be material depleted by the external electric field, appear to grow when the gate bias is increased. This is consistent with wire width decreasing as gate bias is increased. However bright contrast is present in the EBIC micrograph at locations on the device surface which lie in the trenches close to the wire.

Figures 5 and 6 show linescan profiles of EBIC signal vs. beam position, as the beam scans from one gate to the other, measured at 0 V and -4 V gate bias respectively. A signal two orders of magnitude larger than the beam current is obtained when the gate is not biased. Furthermore the EBIC signal is greater at the centre of the wire than at the edges. However when the gates are biased, in addition to the signal increasing, minima in the linescan profile develop (fig. 6).

Plotted in fig. 4 is the number of electron-hole pairs generated by the electron beam estimated to reach the δ-layer, vs. beam position on the wire. The calculation was done using Monte-Carlo simulations (Napchan 1993). Since the wire width is 1μm across some of the primary electrons penetrating the wire surface can exit from the sides of the wire into the etched trenches. This becomes increasingly more probable as the beam scans from the centre of the wire towards the edge.

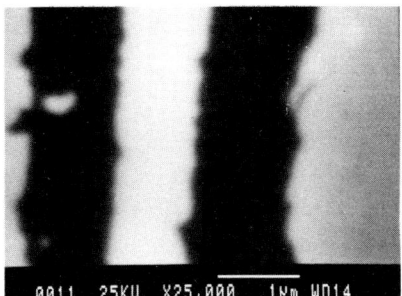

0011 25KV X25,000 1µm WD14

Fig. 2   A SE micrograph showing
the wire and split gates.

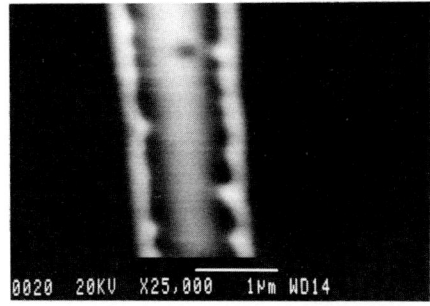

0020 20KV X25,000 1µm WD14

Fig. 3   An EBIC micrograph
recorded at -7 V bias

The electron beam is considered to be incident on the wire in the z-direction (see fig. 1). The number of carriers estimated to reach the δ-layer, N, was calculated using the expression :

$$N = \frac{1}{e_i} \sum_j \frac{dE}{dz} \delta z \; e^{-z_j/L}$$

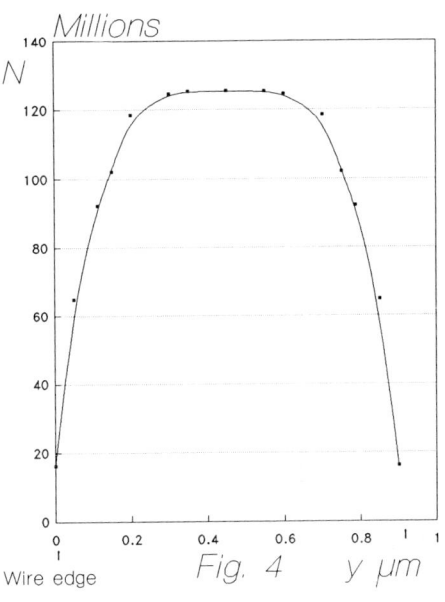

Number of carriers vs. distance scanned

Fig. 4

where dE/dz is the depth dose calculated by the simulation, L is the minority carrier diffusion length and e_i is the electron-hole pair formation energy. The electron beam energy deposited in a layer of thickness δz at a distance z_j from the δ-layer is (dE/dz)δz. The exponential term in the above equation is the probability of an electron-hole pair reaching the δ-layer. This is approximate because the term correctly applies to electron-hole pairs diffusing to an infinite plane, however the distances over which diffusion occurs are of the same order of magnitude as the width of the wire. The Monte-Carlo simulation calculates a depth dose which allows for the energy lost due to primary electrons exiting from the sides of the wire. The simulation was repeated for different beam positions on the wire and the number of carriers to reach the δ-layer was calculated for each position.

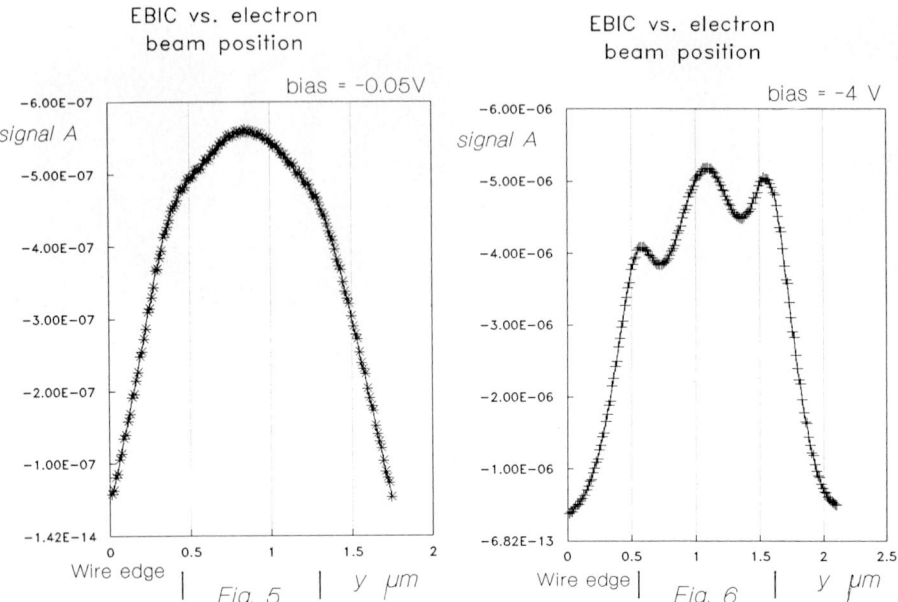

Fig. 5

Fig. 6

4. CONCLUSIONS

The EBIC contrast at different gate bias can be explained qualitatively as a measure of the change in the charge collection by the electric field penetrating the wire, as the electron beam scans across the wire. This being the case, EBIC signal generated when the electron beam is incident on the etched trenches could be due to electron-hole pairs diffusing to the depleted part of the δ-layer. However in order to prove this hypothesis and to understand the contrast quantitatively, further modelling of the electron-hole pair distribution in the wire, which has been induced there by the electron beam, is necessary.

ACKNOWLEDGMENTS

The authors wish to thank Dr E Napchan for use of his Monte Carlo Simulation of Electron Trajectories for Personal Computers.

REFERENCES

Feng Y, Thornton T J, Green M and Harris J J   1992   Superlattices and Microstructures 11 No. 3 281
Napchan E 1993 Simulation of Electron Trajectories for Personal Computers

Inst. Phys. Conf. Ser. No 134: Section 11
Paper presented at Microsc. Semicond. Mater. Conf., Oxford, 5–8 April 1993

Twins in Si InP

D B Holt and G Salviati*

Dept. of Materials, Imperial College of Science, Technology and Medicine, London SW7 2BP
*MASPEC Institute, Via Chiavari 18A, I-43100 Parma, Italy

ABSTRACT: It is shown that the tilt sensitive twin/matrix contrast in semi–insulating InP in SEM emissive mode and ECL images is due to channeling. Remote EBIC produced observable contrast from coherent twin boundaries showing that they are electrically active.

1. INTRODUCTION

Profuse twinning tends to occur in liquid–encapsulated Czochralski growth of InP. The twins were found to exhibit strong contrast in both secondary electron and emission cathodoluminescence (ECL) scanning micrographs of InP:Cr material (Holt et al 1989, Holt and Salviati 1990). This paper reports the results of further studies of this contrast as well as new remote electron beam induced current (REBIC) results.

2. EXPERIMENTAL METHODS

The specimens were cut from polished (111) slices of liquid encapsulated Czochralski InP:Fe and InP:Cr grown by R. Fornari of MASPEC and intended to be semi–insulating.

The emission cathodoluminescence (ECL) technique employs as detector a silicon photodetector looking directly down at the specimen. It is thus a form of panchromatic (all wavelengths) microscopy that produces clear images of GaAs and InP material even of relatively low luminescence efficiency.

For remote EBIC (REBIC) two Ag paint contacts were applied with a relatively wide separation (mm's). No p–n junction or Schottky barrier is necessary. Any contrast present arises from charge collection by the fields around charged defects. It was recently shown that REBIC is also applicable to high resistivity samples (Holt 1993).

A JEOL 840–A SEM and a Matelect ISM — 5 detector for both the REBIC and ECL signals were used at Imperial College. A Stereoscan 250 and an EDS ECL detector were used at MASPEC.

3. RESULTS AND DISCUSSION

Tilt–sensitive contrast was found in secondary and backscattered electron images (SEI and BEI's) and ECL images. That the contrast in the emissive mode is channeling dependent is shown by SEI's taken at low magnification with defocussed beam (Figure 1). These images also show a low resolution channeling pattern. It can then be seen that the contrast between a twin band and the matrix on either side, reverses as the band crosses from one region of the channeling pattern to the next so the contrast is due to channeling. The twin/matrix contrast in BEI's and ECL images varies together, simultaneously reaching maxima and minima at the same values of tilt about an axis normal to the undeflected beam direction (Figure 2), so the ECL contrast is also a channeling effect.

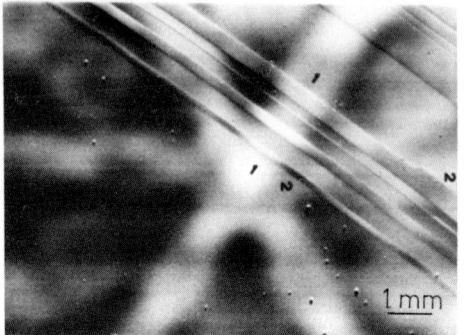

Figure 1. Secondary electron image of InP:Fe at low magnification with the beam defocussed to reveal the basic star–like (111) channeling pattern also. The contrast of twins can be seen to change from darker to lighter than the surroundings on crossing from a bright channeling pattern arm 1 or 2 into the dark 'armpit' area between them.

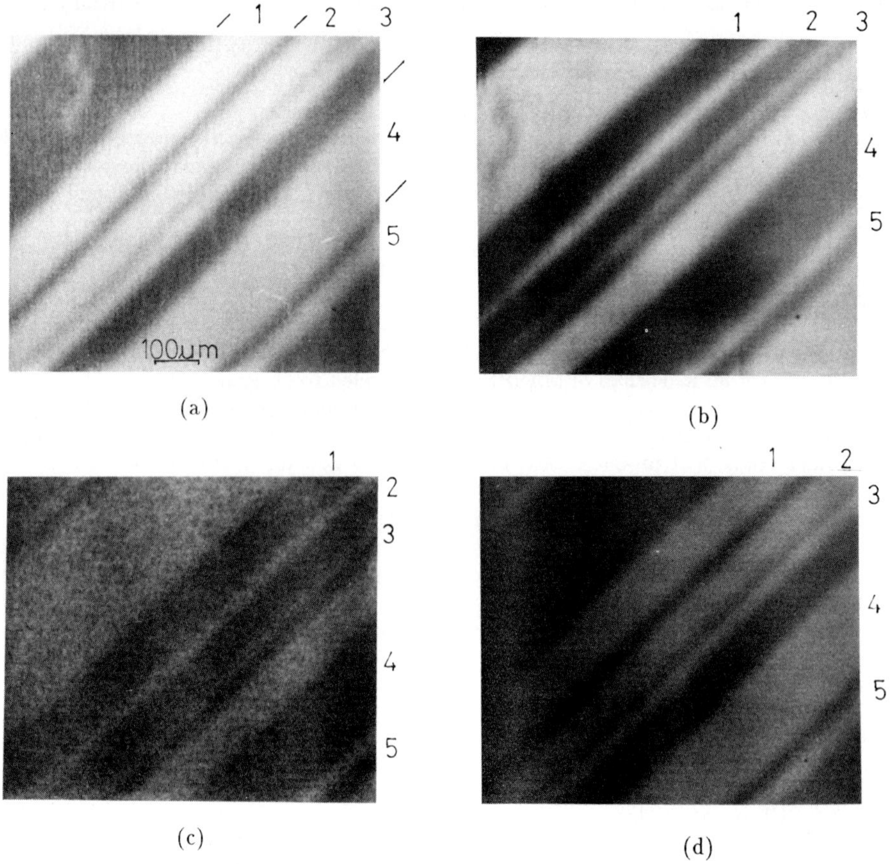

(a)

(b)

(c)

(d)

Figure 2. (a) Backscattered electron (BE) and (b) ECL images of twin bands. (c) BE and (d) ECL images with the specimen tilted through 1.3° relative to the orientation in (a) and (b). Some twin bands are numbered to assist comparison.

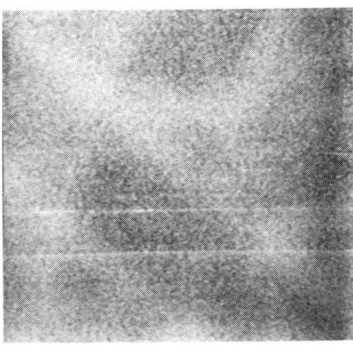

(a)   (b)

Figure 3. Electron channeling patterns recorded using (a) the backscattered electron and (b) ECL signal. The ECL contrast is so low that the pattern is barely visible.

Back scattered electron channeling patterns were recorded from the matrix and twin material and showed the orientations generally to be (111)matrix and (115)twin confirming that the typical, long, straight parallel–sided bands e.g. in Figures 1 and 2, were first order twins. It was found possible also to record ECL ECPs but the contrast was low (Figure 3) as in the only previously published observation of this kind known to us (Schulson et al 1969). The ECL image contrast, however, could be several percent (almost 7 % in Figure 4). It is known that significant CL channeling contrast is possible (Pennycook and Howie 1980). CL spectra were recorded at liquid nitrogen temperature from matrix and twin areas and found to consist of a near–band–edge emission that differed only in intensity between regions of the two orientations.

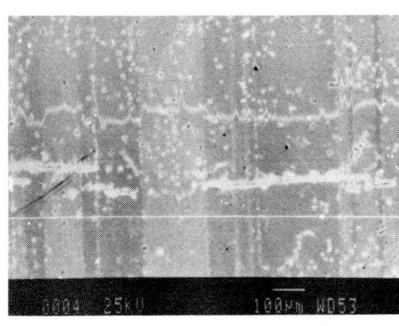
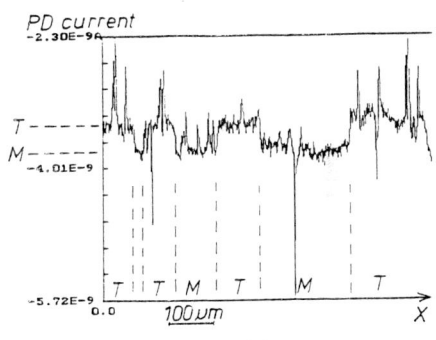

(a)   (b)

Figure 4. InP:Cr (a) ECL image of an InP:Cr specimen tilted to maximise contrast with superimposed linescan (y–modulation) profiles and (b) a quantitative line scan profile of the variation of the ECL photodetector current across this specimen. This is a measure of the ECL intensity (convoluted with the photodetector response). The current is 6.8 % higher in the bright bands in (b) than in the dark areas. The spikes in the trace are not noise but are due to the very strong dot–and–halo contrast arising from Cr segregation to dislocations that can be seen in (a).

Coherent boundaries showed far less ECL contrast than did lateral boundaries (Figure 5) as also reported for CdTe by Durose and Russell (1990). Evidence was previously presented (Holt and Salviati 1990) to show that the bright contrast at scattered points along coherent boundaries is located at steps i.e. at twinning (partial) dislocations so the lateral boundary contrast is also probably associated with impurity segregation to twinning dislocations.

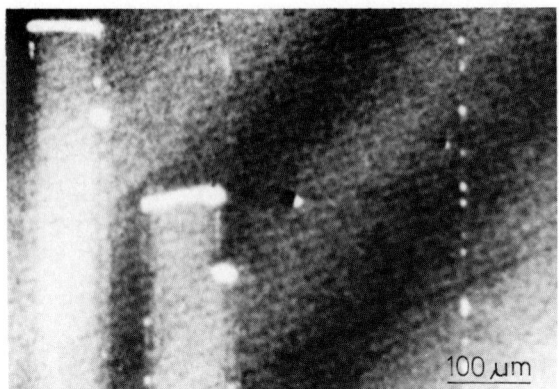

Figure 5. Coherent twin boundaries (the vertical boundaries of the bright twin bands) are much less visible than are lateral twin boundaries (the roughly horizontal lines across the tops of the bright twins) in InP:Fe. There is a vertical row of bright dislocations seen end on at the right of the picture and a few bright twinning dislocations along the vertical coherent boundaries.

Figure 6. (a) REBIC and (b) backscattered electron images of twin boundaries in LEC InP:Fe.

It was found that some coherent twin boundaries showed contrast in REBIC (Figure 6a). The REBIC signals are small (comparable with the beam current rather than hundreds or thousands of times I_b, as in EBIC from Si devices). The REBIC contrast in Figure 6 nevertheless is evidence that coherent twin boundaries have associated electrical fields at room temperature. The non–parallel boundaries of the three regions visible in Figure 6b indicate the presence of something other than a first order twin. These three regions were identified by backscattered electron channeling patterns and one was found to be a second order twin. Such regions were found only infrequently.

Acknowledgements Thanks are due to D.J.C. O'Neill, A. Fernandes and P. Kosmetatos who contributed to the work and to Dr. R. Fornari for supplying the material.

REFERENCES

Durose K and Russell G J 1990 J. Cryst. Growth 101, 246 – 250
Holt D B, Fornari R, Franzosi P, Kumar J and Salviati G 1989 in Microscopy of Semiconducting Materials 1989. Conf. Series No. 100 (Inst. Phys.: Bristol) pp. 421 – 426
Holt D B and Salviati G 1990 J. Cryst. Growth. **100** 497 – 507
Holt D B 1993 to be published
Pennycook S J and Howie A 1980 Phil. Mag. **A41** 809 – 827
Schulson E M, van Essen C G and Joy D C 1969 Scanning Electron Microscopy p. 47 – 55

Inst. Phys. Conf. Ser. No 134: Section 11
Paper presented at Microsc. Semicond. Mater. Conf., Oxford, 5–8 April 1993

SEM–EBIC study of lattice defects in CdTe and CdHgTe subjected to electron and ion bombardment

G N Panin

The Institute of Microelectronics Technology and High Purity Materials, Russian Academy of Sciences, 142432 Chernogolovka, Moscow District, Russia.

ABSTRACT: The effect of low-energy electron and ion bombardments on the electrical activity of extended defects formed during crystal growth and processing of CdTe and CdHgTe crystals has been studied. It has been shown that the electrical properties of these defects and adjacent regions can be changed as a result of tellurium precipitation and bulk acceptor gettering induced by the bombardments.

1. INTRODUCTION

Low-energy electron and ion bombardments have recently been used to adjust the stoichiometric composition and charge carrier concentration in surface layers of compound semiconductors such as $Cd_xHg_{1-x}Te$ (CMT). The main feature of such a bombardment is the formation of an electrically induced layer with a depth much larger than the bombardment range (Blackman et al 1987, Panin and Yakimov 1989). To account for the p to n type conversion induced by ion milling of p-type CMT it was suggested that the extended defects created assist in fast propagation of metal ions, accompanied by annihilation of metal vacancies (Bahir and Finkman 1989). Barbot et al (1990) reported the creation of p-n junctions in p-type CMT ($x=0,7$) by local plastic deformation and electron irradiation. They assumed that the mechanism involved in the p-n junction formation by electron irradiation and local plastic deformation may be of the same kind as that responsible for the ion bombardment-induced conversion.

In this paper the electron beam induced current (EBIC) technique in the scanning electron microscope (SEM) was used to study the effect of lattice defects on the electrical properties of adjacent regions in CdTe and CMT crystals subjected to low-energy electron and ion bombardments.

2. EXPERIMENTAL

Experiments were carried out on undoped samples of solid-state recrystallized CMT and CdTe single crystals grown by the Bridgman technique ($p,n \sim 10^{14}$-10^{15} cm^{-3}). The samples were mechanically lapped and chemically polished in Br:HBr solution. An electron beam of a Superprobe-733 microanalyser with a beam current from 10^{-8} to 10^{-6} A and a beam energy of 25 keV as well as 1 kV Ar$^+$ ions in a sputtering unit at a current density of 0.5 mA/cm were used for the bombardment experiments. To detect the electrical changes expected, two kinds of structures were made: those with the Schottky contact (Au/n-CdTe) for measuring electron beam induced current (EBIC) and those without it using two gold Ohmic contacts plated on the opposite edges of the sample for measuring a remote EBIC (REBIC) (Matare and Laakso 1969, Panin and Yakimov 1992). The REBIC contrast results from the separation of electron beam generated electron-hole pairs in the electric fields of charged defects or local p-n junctions created by the electron or ion bombardment. Electron probe X-ray microanalysis (EPMA) was used for identification of large precipitates. SEM measurements were conducted at electron beam currents of 10^{-10}-10^{-9}A.

3. RESULTS AND DISCUSSION

Subgrain boundaries in $Cd_xHg_{1-x}Te$ (x=0,4) crystals were revealed by the REBIC technique as some extended peculiarities of dark-bright contrast (Fig.1).

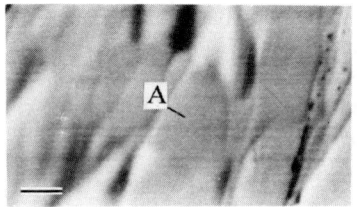
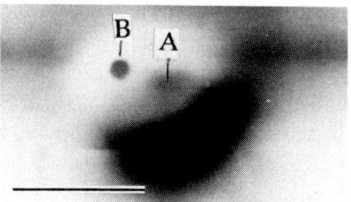

Fig.1   Fig.2

Fig. 1. REBIC image of the subgrain boundaries in the p-type CMT irradiated with a 10^{-7} A fine electron beam for t = 1s (A is the beam incidence point). The bar is 100 μm.

Fig. 2. REBIC image of the n-type region in the p-type CMT crystal created by an electron beam irradiation at a beam current $I_b = 10^{-7}$A for t > 10s. A is the beam incidence point, B is a precipitate-like defect. The bar is 100 μm.

It should be noted that local p-n junctions created by an electron beam irradiation of extended defect free regions in the p-type CMT have earlier been observed as regions of the same REBIC contrast polarity as that of the subgrain boundaries (Panin and Yakimov 1989). The dark-bright REBIC contrast of subgrain boundaries disappears after exposure to a 10^{-7} A electron beam for some seconds (Fig. 1). However a longer exposure (more than 10 s) brings back the REBIC contrast in the region near the beam incidence point. The dark-bright polarity of the REBIC contrast indicated that a p-n junction was formed. Extended defects were revealed which were introduced in this region of the type conversion. Their strong recombination contrast was similar to that of precipitates (Fig. 2, B defect). It should be noted that the REBIC contrast in the region of the subgrain boundary was strongly dependent on the fluence of an electron beam irradiation whereas the REBIC signal in the crystal region containing no extended defects was observed only when the fine electron beam reached its threshold current of about 10^{-7}A.

The recombination EBIC contrast at the subgrain boundaries and twins in CdTe crystals became stronger after an electron beam irradiation with an incident beam current of about 10^{-8}A. The irradiation of a defect free region by a beam of the same current caused no changes in the EBIC (Yegorshev et al 1989). An increase in the electron beam current resulted in an enhanced component desorption from the surface of the crystals as well as in the formation of extended recombination active defects and Te-enriched regions. Figure 3a shows the extended defects revealed by the EBIC technique after a single scan of a 1.5 x 0.9 mm² area of the Au/n-CdTe structure with a 3 x 10⁻⁶ A focused electron beam for a scan time of 100 s.

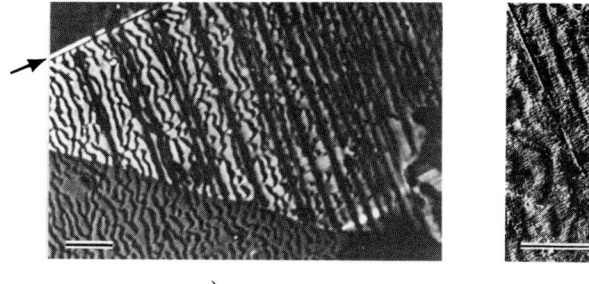 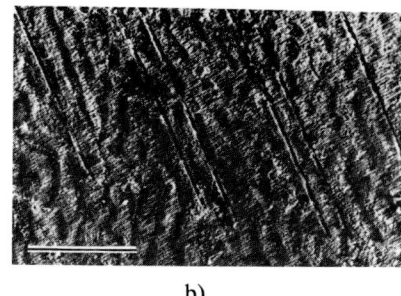

a)   b)

Fig. 3. SEM micrographs of the n-type CdTe crystal region exposed to a single scan of the 1500 x 900 μm² Au-CdTe structure with a fine electron beam at I_b = 3 x 10⁻⁶ A and t = 100 s (the bars are 100 μm):
a) EBIC image (The dark undulating features are Te-enriched regions. The arrow shows the scan direction and the edge of the bombarded region)
b) SE image after selective chemical etching.

A high etch pit density in the Te-enriched region after electron bombardment was revealed with a dislocation etchant which was used by Bagai et al (1988) to display β-dislocations in Te-enriched regions of CdTe. Dark straight strips in the EBIC image after the bombardment correspond well to the etch pattern of twin-like defects (Fig. 3b). They, however, require a more detailed structural examination for reliable identification.

The effect of extended defects decorated by tellurium precipitates on electrical properties of an adjacent crystal zone subjected to ion milling was studied by temperature-dependent REBIC measurements. Figure 4 shows an as-grown laminated tellurium inclusion in the vicinity of the twin boundaries in the p-type CdTe crystals. The upper part of the compound defect and the adjoining crystal zone were milled by Ar⁺ ions through the window in a mask. The bright-dark REBIC contrast in the region subjected to milling indicates that ion bombardment resulted in a p to n conversion (Fig. 4a). The low-temperature REBIC signal of the crystal region far from the inclusion disappears, while remaining in the region adjacent to the defect (Fig. 4b). The change in the REBIC image of the processed region with decreasing temperature implies a p-n junction of variable depth in the vicinity of the extended defects. The efficiency of the charge carrier collection by the junction decreases with temperature owing to a likely decrease of the charge carrier diffusion length in the crystal. The low temperature REBIC image can be interpreted as due to p-n junction formation far from the tellurium inclusion at a greater depth than near it. The shallow p-n junction near the inclusion may be explained by a higher density of acceptor centres diffusing to the tellurium precipitates from the bulk during ion implantation. It should be noted that a decrease in the REBIC contrast at the subgrain boundaries after electron irradiation (see Fig. 1) may also be explained by a bombardment induced increase in the acceptor concentration near the defects. An asymmetrical low temperature REBIC contrast at the tellurium inclusion measured after the ion bombardment (Fig. 4b) may arise because of the specific geometry of the extended defects studied. However, further structural examination of the defects by other techniques is necessary for a better understanding of their effect.

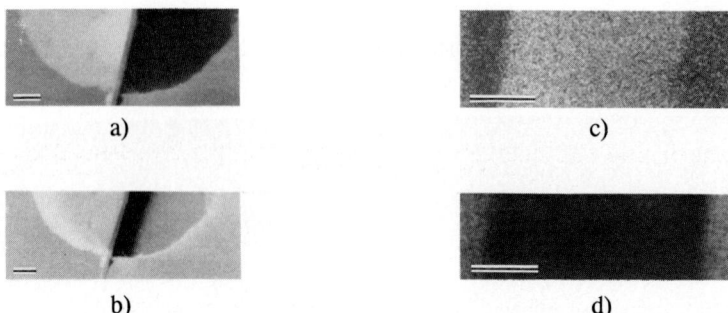

a)

b)

c)

d)

Fig.4. Micrographs of the tellurium inclusion in p-type CdTe bombarded through the window in a mask by Ar⁺ions:
  a) REBIC image at T=300K (a semicircle of the bright-dark contrast corresponds to the exposed region);
  b) REBIC image at T=150K;
  c) EPMA image in Te X-ray;
  d) EPMA image in Cd X-ray.
(The bars are 100 μm (a,b) or 10 μm (c,d).)

4. CONCLUSIONS

The results obtained indicate that extended defects created during crystal growth and by particle bombardment influence the redistribution of point defects induced by the bombardment. A decrease in the REBIC contrast at the subgrain boundaries at the initial stage of the crystal bombardment as well as the results of temperature dependent REBIC examination of extended defects conform to an enhanced acceptor concentration near the defects. Gettering of acceptor centres by tellurium precipitates and by other extended defects is likely to facilitate the type conversion in the crystals at a distance greatly exceeding the bombardment range.

REFERENCES

Bagai R K, Seth G L and Borle W N 1988 Cryst. Growth. **91** 605
Bahir G and Finkman E 1989 J. Vac. Sci. Technol. **A7** 348
Barbot J F, Kronewitz J and Schröter W 1990 Appl. Phys. Lett. **57** 2689
Blackman M V, Charlton D E, Jenner M D, Purdy D R, Wotherspoon J T M, Elliot C T and White A M 1987 Electron. Lett. **23** 978
Matare H F and Laakso C N 1969 J. Appl. Phys. **40** 476
Panin G N and Yakimov E B 1989 Sov. Phys. Semicond. **23** 840
Panin G N and Yakimov E B 1992 Semicond. Sci. Technol. **7** A150
Yegorshev V V, Panin G N and Yakimov E B 1989 Proc. 12th Int. Cong. X-ray Optics and Microanalysis, eds S Jaslenska and L J Maksymowicz (Cracow, Poland) pp 904-907

Inst. Phys. Conf. Ser. No 134: Section 11
Paper presented at Microsc. Semicond. Mater. Conf., Oxford, 5–8 April 1993

747

Diffusion length and surface recombination velocity: injection dose dependence

D Cavalcoli and A Cavallini

Dept. of Physics, University of Bologna, via Imerio 46, I-40126 Bologna, Italy

ABSTRACT: The injection dependence of bulk and surface properties in semiconductors is analyzed on the basis of the Shockley Read (1952) and Hall (1952) recombination mechanism. In the low injection regime the minority carrier diffusion length L is usually assumed to be constant, and the surface recombination velocity s is taken as a constant parameter, too, to define the boundary conditions for the continuity equation that must be solved for obtaining L. A more profound analysis shows that both these quantities depend on the excess carrier concentration, and in particular the s variation can affect the measurement of L.

1. INTRODUCTION

In modelling electron beam interaction with semiconductors for measuring the minority carrier properties, both bulk and surface parameters are assumed to be independent of the variables of the system (the excess minority carrier concentration Δp and the Δp variation with distance). As a matter of fact the diffusion length is obtained by injecting minority carriers into a region of the sample and allowing them to diffuse to a second region where they are collected. Injection may employ light or electron beams while detection may occur at a p-n junction in the bulk or at a surface barrier. It is usual to monitor the detected signal as a function of the distance between injecting source and collector. The problem is reduced to the solution of the continuity equation for the excess carrier concentration with suitable boundary conditions at the semiconductor surface. Both surface recombination velocity and diffusion length are thus assumed to be constant with respect to the injection level. These hypotheses are not always correct: as predicted by the Shockley Read and Hall (1952) (SRH) model, s depends on Δp, and the bulk parameters change with the injection level as well: in particular the lifetime τ of minority carriers depends on Δp and temperature. It is important to take into account both these effects when dealing with diffusion length and surface recombination velocity measurements. Moreover L is often measured using methods based on assumptions of the s value, this can give diffusion lengths strongly dependent on the choice of s, and thus on the s variation with Δp. In this respect an alternative useful method is the moment method (Donolato, 1985), which does not require any assumption of s, thus giving an L value free from the influence of s. By using this model (Cavalcoli et al, 1991) to measure the minority carrier diffusion length in n-type Si in different injection conditions, it has been observed that for an injection level variation of two orders of magnitude L remains constant, while s changes with Δp. In this contribution an analysis of the variation of bulk and surface parameters with the injection level has been carried out, based on the SRH theory, and the theoretical evaluations have been compared with both experimental findings and published results.

2. BULK AND SURFACE PARAMETERS DEPENDENCE ON THE INJECTION LEVEL

2.1 Minority Carrier Lifetime

Let us assume that the recombination occurs through the trapping mechanism, where the trap has an energy E_t measured from the valence band in the energy gap. Following SRH model, the minority carrier lifetime in the neutral region of n-type material can be written as (Cavalcoli et al, 1993):

$$\tau=\tau_{po}\left[1+\frac{1}{h+1}\exp\frac{E_t-E_F}{kT}\right]+\tau_{no}\left[\frac{h}{h+1}+\frac{1}{h+1}\exp\left(\frac{E_i-E_F}{kT}+\frac{E_i-E_t}{kT}\right)\right] \qquad (1)$$

where τ_{no} and τ_{po} are the electron and hole equilibrium lifetime, $h=\Delta n/n_0$ the normalized injection dose ($\Delta n=\Delta p$, in the neutral region) E_F the Fermi energy and E_i the intrinsic energy level, k the Boltzmann constant and T the temperature. From equation (1) the variation of τ with Δn can be obtained for different experimental conditions. In particular τ increases or decreases *vs* Δn depending on the position of the trap level E_t with respect to the Fermi level position E_F. Let the trap energy level E_t lie under the Fermi Level, that is $E_i<E_t<E_F$: since E_t represents the energy level of a trap through which the recombination occurs, it must be an efficient trap, i.e. positioned as near as possible to the mid-gap level. τ/τ_0 is thus an increasing function of Δn, as shown in fig.1a. From eq.1 it is possible to deduce that trend of τ vs Δn depends not only on the trap position, but also on the ratio between τ_{no} and τ_{po}, i.e. from intrinsic properties of the semiconductor material.

2.2 Minority Carrier Diffusion Length.

The minority carrier diffusion length is defined as:

$$L=\sqrt{D\tau} \qquad (2)$$

where D is the diffusion coefficient and τ the minority carrier lifetime. Therefore, to evaluate the dependence of the minority carrier diffusion length on the injection dose two different terms must

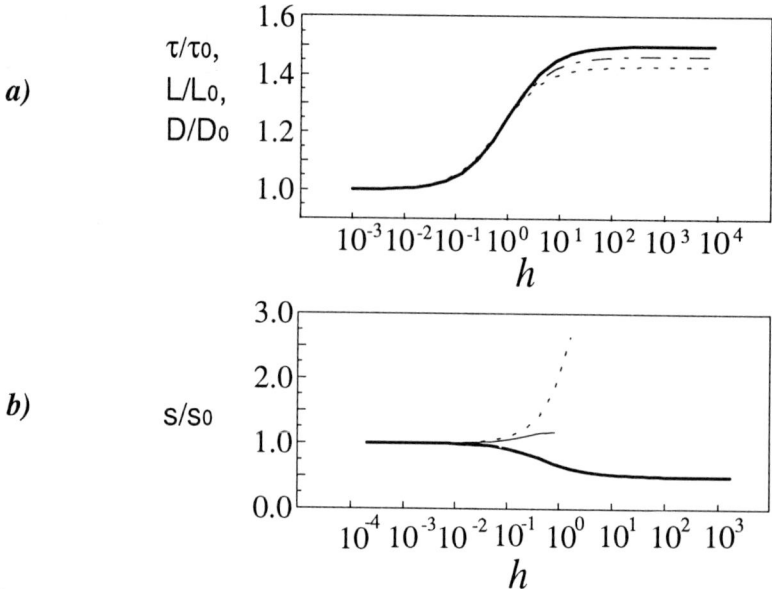

Fig.1 a) $\tau/\tau0$ (—), $L/L0$ (— —), $D/D0$ (- -) vs h. b) s/s0 vs h: strong acc. (— —), strong inv.(—), weak barr. (—).

be analyzed: the contribution of the lifetime, given by eq.1, and that one of the diffusion coefficient. With respect to the latter it is necessary to recall the D variation vs h (Berz et al,1976):

$$D=\frac{D_nD_p(2h+1)}{D_n(1+h)+D_ph} \qquad (3)$$

where D_n and D_p are the electron and hole diffusion coefficients, respectively. The L variation with h can thus be derived from eqs. (1) to (3). The results are shown in fig.1a, where the minority carrier bulk properties are plotted against h for $E_t<E_f$. These quantities are expectedly independent of h only in the low and high injection regions.

2.3 Surface recombination velocity

The recombination rate R at a free semiconductor surface is given by:

$$R=s\Delta p \qquad (4)$$

Equation (4) is often considered as the definition of s, which can be calculated as a function of Δp if R is known in terms of the carrier densities at the surface n_s and p_s. The surface recombination mechanism is a SRH process through surface states. In particular it occurs at these states which can efficiently capture both electrons and holes. The recombination rate depends on the density and distribution of these states, their associated capture coefficient, the equilibrium surface potential v_s (in kT units) and the injection level in a complex way. Following the same approach as Berz (1975) Cavalcoli et al (1993) derived an analytical expression for the recombination rate from which the s variation with h can be obtained. This strongly depends on the values of the surface potential: the following situations are considered: a)strong accumulation, $v_s>4$ i.e. the majority carriers accumulate at the surface $(n_s>>n_0)$, b) weak barriers $|v_s|<<1$, $n_s<n_0$, $p_s<p_0$, c)strong inversion $v_s<-20$, i.e. the minority carriers accumulate at the surface $(p_s>n_0)$. The results are shown in fig.1b. The following features can be noted: s/s_0 remains constant at low injection level in every condition, while it decreases by increasing h and reaches a constant value at high injection only for weak surface barrier. These results show that the assumption of constant surface recombination velocity is not always justified, and the value of the surface potential strongly affects the s behaviour vs h.

3. RESULTS AND DISCUSSION

The surface recombination velocity has been evaluated by using the following procedure:
1) EBIC measurements of minority carrier diffusion length have been performed on n-type Si samples with the moment method;
2) a one-parameter fitting of the induced current profiles has been carried out with the expression obtained by Kuiken et al (1985) in order to obtain the

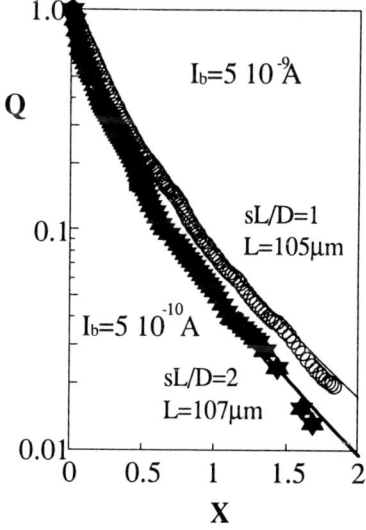

Fig.2. Normalized EBIC current Q as a function of X=x/L, x=beam-junction distance. Experimental points are fitted by theoretical curves.

surface recombination velocity.

This procedure has been applied to EBIC profiles acquired at different injection conditions. An example is shown in fig.2. The results show that the surface recombination velocity doubles for a one order of magnitude decrease of the injection dose, while the diffusion length remains unchanged. In order to understand this behaviour and to compare experimental data with theoretical predictions, a correct evaluation of the injection dose is necessary. In our opinion a different approach for evaluating the injection condition in the bulk (Δn_b) and at the surface (Δn_s) should be used. While Δn_b is evaluated with a model that takes into account the diffusion in the bulk but is based on the assumption of infinite surface recombination velocity, Δn_s should be evaluated taking into account the generation rate in the bulk. The experimental results, and the evaluation of injection at the surface and in the bulk are reported in table 1. The different behaviour of bulk and surface parameters can be understood in terms of the injection condition: for the same value of the generation rate the surface is in high injection condition, while the bulk is in low injection condition. This explain why L does not change, while s halves its value. Furthermore it is worth observing that the surface parameters are more strongly affected by the injection conditions than the bulk ones: for an injection dose value equal to 10% of the majority carrier concentration it is still possible to measure a low injection L value, while s significantly differs from its low injection value.

4. CONCLUSIONS

The injection dose variation of the bulk parameters is strongly affected by the trap energy position E_t, while the s trend vs Δn is related to the value of the surface potential. Both bulk and surface parameters depend on the injection level

Table I: Experimental values of L and s. I_b= electron beam current, E_b= electron beam energy.

I_b (A)	E_0 (keV)	Δn_s (10^{-15}cm^{-3})	s (10^{-3}cm/s)	Δn_b (10^{-15}cm^{-3})	L (μm)
$5\ 10^{-10}$	30	11	3	0.011	107
$5\ 10^{-9}$	35	150	1.5	0.16	105

through a SRH type mechanism, but the resulting trends are different, as already stated. Furthermore great care should be paid in defining the injection conditions as they are quite different in the bulk and at the surface of a semiconductor sample.

REFERENCES

F Berz 1975 "Surface Physics of Phosphorous and Semiconductors", Scott and Reed eds, Academic Press.
F Berz and H Kuiken 1976 Sol. State Electr.19, 437.
D Cavalcoli, A Cavallini, A Castaldini 1991 J.Appl. Phys. 70(4) 2163.
D Cavalcoli A Cavallini, B Fraboni 1993 submitted to Sol. State Electr.
C Donolato, 1985 Sol. State Electr. 7, 717.
R N Hall 1952 Phys. Rev. 87, 387.
H K Kuiken and C van Opdorp 1985 J.Appl. Phys. 57, 2077.
W Shockley and W Read 1952 Phys.Rev.87, 835.

Inst. Phys. Conf. Ser. No 134: Section 11
Paper presented at Microsc. Semicond. Mater. Conf., Oxford, 5–8 April 1993

Simulation of EBIC contrast with account of a realistic generation function and nonzero width of collecting Schottky junction

W Seifert, W Knechtel and M Kittler

Institut für Halbleiterphysik GmbH, PSF 409, D-O 1200 Frankfurt (Oder), Germany

ABSTRACT: A simple simulation procedure based on the linear contrast model of Donolato, but using a Monte Carlo generation function and taking account of the width of the surface junction is introduced. The results of computation are compared to the homogenoeous-generation-sphere case. Contrast of point defects is found to depend strongly on the generation function used while that of line defects is less sensitive to it.

1. INTRODUCTION

The linear contrast model of Donolato (1978/79) with homogeneous generation sphere has been widely used in EBIC studies because of its ability to explain fundamental experimental results and its ease of application. The model neglects the width of the collecting surface junction which may be allowed for defects lying at large depth and for material of high doping level. However, for doping concentrations below 10^{15} cm^{-3} the width of the junction depletion region can reach values of 1 μm or more which is not negligible compared to the electron range R. Another source of error is the assumption of a homogeneous generation sphere. Fig. 1 demonstrates that resolution attained experimentally may be much better than that estimated from the simple model. In recognizing this kind of problem several attempts have been made to improve contrast modelling by more realistic generation functions (e. g. Donolato and Venturi 1982, Holt et al. 1989). In the present paper the effect of both, realistic generation function and of junction width on contrast is studied for charge collection with a surface Schottky barrier.

2. MODEL

The model is essentially that of Donolato (1978/79), but with the following modifications: arbitrary generation distribution and finite width of space-charge region (Fig. 1). As usual, the beam-induced minority-carrier concentration is assumed to vanish at the boundary of the space-charge region. The contrast c caused by a point defect is in linear approximation (in the convention of Kittler and Seifert 1981):

$$c = \gamma \cdot f,$$

with
$$f = \frac{1}{I_0} \cdot \exp\left(\frac{w - z_d}{L}\right) \cdot \iiint_{z_s > w} g(\vec{r_s}) \, G(\vec{r_d}, \vec{r_s}) \, d^3 r_s$$

$$I_0 = \iiint_{0 < z_s < w} g(\vec{r_s}) \, d^3 r_s + \iiint_{z_s > w} g(\vec{r_s}) \exp\left(\frac{w - z_s}{L}\right) d^3 r_s$$

G $(\bar{r}_s, \bar{r})$ the Green's function of the problem, g (r) the distribution of generated carriers (generation function), L the diffusion length and γ the defect strength. To have a realistic as possible distribution of generated carriers data obtained by a Monte Carlo simulation program (Napchan 1989) were used as generation function in our calculation procedure.

The calculation scheme for point defects is used also for extended defects because in the limits of a linear model extended defects can be represented as consisting of a certain distribution of point defects.

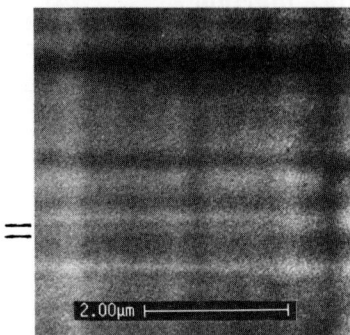

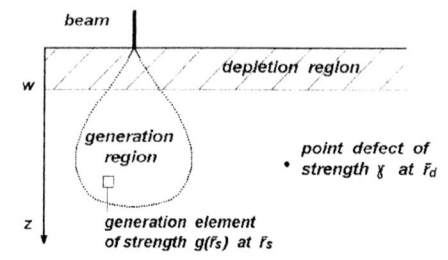

Fig. 2. Schematic view of calculation geometry

Fig. 1. EBIC micrograph of surface-near misfit dislocations in a SiGe/Si structure imaged at E_o = 15 keV with a 2 μm deep p-n junction. Note resolution of 200 nm (see markers).

3. RESULTS

The correction factors f and F (f= c/γ - point defect, F = c/Γ - line defect) and the half width of the contrast profile, hw, were calculated. For comparison, the results obtained with a Monte Carlo (MC) distribution are shown together with those obtained using a homogeneous generation sphere (HGS).

3.1 Point Defects

Pronounced differences between MC and HGS data can be found for both, correction factor and half width (Fig. 3). For surface-near defects (z_d < 0.3 R) stronger contrast and markedly lower half width are obtained in the MC case while contrast is larger and half width smaller in the HGS case for defects at medium depths. At depths exceeding the electron range R both distributions yield similar results (Fig. 3a). Fig. 3b shows the calculated data as a function of beam energy E_0 which is a presentation well suited for comparison with experimental results. It is found that the energy at which maximum contrast occurs is nearly unaffected by the kind of generation function, however the correction factors are different. Further, there are striking changes in the behaviour of half width hw when HGS distribution is replaced by the more realistic MC distribution. The observed differences are essentially due to a very narrow generation distribution near the sample surface in the MC case, compared to a broad distribution in the HGS case. The main effect of non-zero junction width is a reduction of contrast, in particular for defects at low depths. The half width of the contrast profile is reduced, too.

3.2 Line Defects

As expected, the influence of the generation function used is much weaker than for point defects. While surface-parallel line defects show still a dependence on generation function, with trends similar to point defects but less pronounced (Fig. 4a, b), line defects perpendicular to the surface are practically unaffected by the generation function (Fig. 4c). A non-zero junction width leads in both cases to decrease of contrast leaving the half width nearly unchanged.

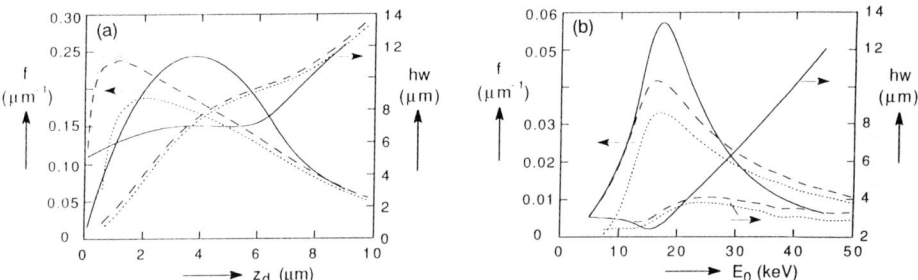

Fig. 3. Correction factor, f, and contrast half width, hw, as calculated for a point defect: a) dependence on defect depth z_d ($E_0 = 30$ keV), b) dependence on beam energy E_0 ($z_d = 2$ μm).

——————— homogeneous generation sphere, junction width w = 0,
- - - - - MC distribution, junction width w = 0,
·········· MC distribution, junction width w = 0.5 μm.

Note: The oscillations in contrast half width at large E_0 in case of MC distribution (Fig.3b) are caused by the roughness of the used MC distribution.

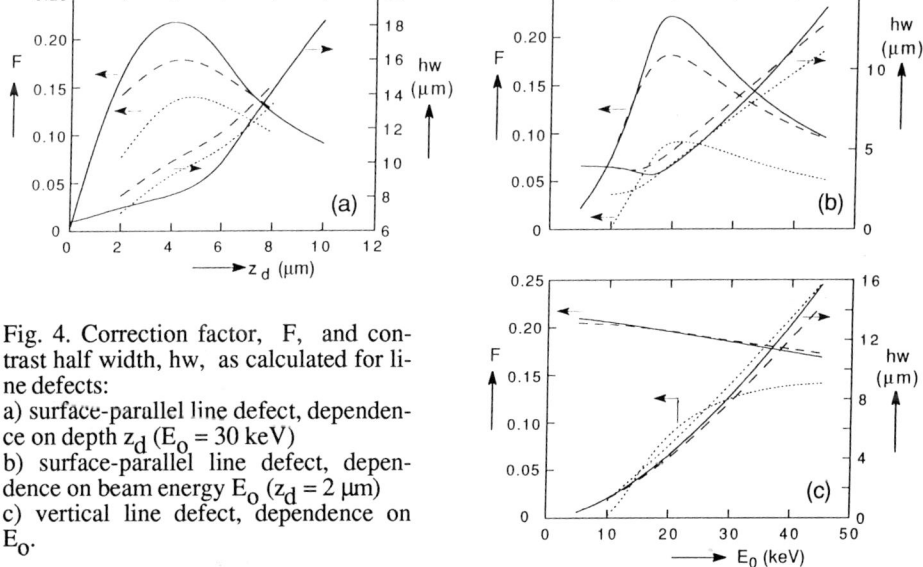

Fig. 4. Correction factor, F, and contrast half width, hw, as calculated for line defects:

a) surface-parallel line defect, dependence on depth z_d ($E_0 = 30$ keV)
b) surface-parallel line defect, dependence on beam energy E_0 ($z_d = 2$ μm)
c) vertical line defect, dependence on E_0.

——————— homogeneous generation sphere, junction width w = 0,
- - - - - MC distribution, junction width w = 0,
············ MC distribution, junction width w = 1 μm.

4. CONCLUSIONS

The developed simulation procedure allows a calculation of defect contrast using a realistic MC generation function and taking account of the width of the Schottky junction. The results obtained are in agreement with experience at least in a qualitative sense. Detailed theoretical and experimental investigations are needed to test the model in a quantitative manner which has not been done so far.

Large differences between cases of MC and HGS generation function are found for point defects in contrast and half width. The point defect assumption is an unrealistic simplification as all defects of interest have a certain size. Moreover, recent calculations (Donolato 1993) have shown that defects need a minimum size in order to produce a certain EBIC contrast. Thus for all real cases a distribution of point defects must be assumed for contrast calculation which will partly remove the differences between MC and HGS case. This is visible for line defects, in particular vertical line defects, very clearly. The action of defects at large depth and of large size seems to be described quite well by the HGS case.

However, defects close to the surface are believed to require a MC generation function for satisfactory description of their EBIC contrast. Another field of application of the proposed simulation procedure is EBIC investigations in cases when the width of the collecting Schottky junction cannot be neglected.

5. ACKNOWLEDGEMENTS

The authors are grateful to Dr. E. Napchan for kindly making available his Monte Carlo simulation program of energy dissipation. Further, they acknowledge technical assistence by P. Hartig and support by the Bundesministerium für Forschung und Technologie under contract No. 03290107 F and by the Volkswagenstiftung under contract No. I/66794.

REFERENCES

Donolato C 1978/79 Optik 52 19
Donolato C and Venturi P 1982 Phys. Stat. Sol. (a) 73 377
Donolato C 1993 Phys. Stat. Sol. (a) 135 K 13
Holt D B, Napchan E and Norman C E 1989 Inst. Phys. Conf. Ser. 104 205
Kittler M and Seifert W 1981 Phys. Stat. Sol. (a) 66 573
Napchan E 1989 Revue Phys. Appl. 24 C6-15

Inst. Phys. Conf. Ser. No 134: Section 11
Paper presented at Microsc. Semicond. Mater. Conf., Oxford, 5–8 April 1993

755

Electron channelling contrast imaging of defects in semiconductors

A J Wilkinson, P B Hirsch, J T Czernuszka and N J Long

Department of Materials, University of Oxford, Parks Road, Oxford OX1 3PH

ABSTRACT: Back scattered electrons can be used to image lattice defects in bulk crystalline specimens in a scanning electron microscope (SEM) with highly efficient detector. Images of individual dislocations lying near the surface of a Si specimen have been obtained using several commercial SEMs. Clusters of dislocations in strained epilayer systems can also be imaged to considerable (>1 μm) depths in a SEM with a LaB_6 filament. The character (i. e. Burgers vector) of the defects can also be assessed with the technique.

1. INTRODUCTION

The idea of using back scattered electrons (BSEs) to image dislocations near the surface of bulk crystalline materials in a scanning electron microscope (SEM) is not a new one. In 1967 Coates demonstrated the strong orientation dependence of BSEs, and shortly after this it was suggested (see Booker 1970) that the local tilting of lattice planes by defects such as dislocations should modulate the BSE signal allowing the defects to be imaged. Theoretical descriptions of the process have been given by Clarke and Howie (1971) and Spencer, Hirsch and Humphreys (1972). These image simulations showed that the incident beam divergence (2α) must be kept small (<4 mrad) while the narrow image width restricts the required spot size to <10 nm. In addition the contrast (C) was calculated to be small (of the order of a few per cent at best) so that a beam current >1 nA would be required to make the image seen above the background noise. Thus near surface dislocations would only be seen in bulk specimens if a source with brightness $B > 10^8$ $Acm^{-2}srad^{-1}$ were used.

Initial experiments (Clarke 1971, and Stern, Ichinokawa, Takashima, Hashimoto, and Kimoto 1972), used tungsten hairpin thermionic emitters operated at 80 keV with $B \cong 5 \times 10^5$ $Acm^{-2}srad^{-1}$ and images of dislocations were obtained in thinned specimens, though not in bulk crystals. Thin specimens were used so as to increase the image contrast by reducing the effects of multiple inelastic scattering. Later, impressive results were obtained by Morin, Pitaval, Besnard and Fontaine (1979) using a cold field emission gun (FEG) with $B \cong 3 \times 10^8$ $Acm^{-2}srad^{-1}$. Again effects of multiple inelastic scattering were suppressed, this time by using a retarding field energy filter so as to collect only those BSEs that had suffered small (<500 eV) energy losses. Good images of dislocations near the surface of bulk specimens were obtained, but the technique was not adopted by other groups presumably due to the complex and bulky detector that was required.

Over the past few years the authors have returned to the electron channelling contrast imaging (ECCI) technique and have shown that dislocations can be imaged in bulk specimens using commercially available SEMs with only minor modifications. This paper describes some of the results obtained with the technique.

2. TILTED SPECIMEN GEOMETRY

Central to the success of this work has been the adoption of a tilted specimen geometry. Tilting the specimen away from the horizontal through large angles (>40°) causes several effects which act to increase the signal to noise ratio. The first effect is that tilting the specimen increases the yield of BSEs thus increasing the number of electrons carrying the signal. Furthermore, BSEs from tilted specimens are preferentially scattered into a forward scattered lobe at low take off angle. This confinement of the majority of BSEs to a restricted range of angles is the second beneficial effect since detection efficiency can be increased simply by assuring the detector is positioned so as to capture electrons in this forward scattered lobe. Thirdly, if one examines the energy distribution of BSEs one finds that with tilted specimens there is a markedly larger proportion of electrons emitted with low energy loss. Thus, the tilted specimen geometry in effect provides some energy filtering, suppresses the multiple inelastic scattering and hence improves the contrast. These factors all combine to make it significantly advantageous to use a tilted specimen and collect electrons emitted at low take off angles. Indeed such a system is efficient enough for images of dislocations in bulk specimens to be produced without the need for additional energy filtering.

3. EXPERIMENTAL SET-UP AND PROCEDURE

Experiments have been undertaken in three microscopes: firstly a V G HB501 STEM, secondly a JEOL 840F FEG SEM, and thirdly a JEOL 6300 SEM with a LaB_6 thermionic emitter. In each case the experimental set-up is similar. The BSE detector simply consists of a YAG scintillator coupled via a light guide to a photo-multiplier tube (PMT). The YAG is coated with a thin layer of Al primarily to prevent charging but this also stops low energy electron contributing to the signal. The detector is mounted on a bellows mechanism so that it can be brought very close to and slightly below the specimen, thus providing a large capture angle. The signal from the PMT is amplified and passed to an external frame store where image processing can be undertaken. The specimen is tilted towards the detector and held at a rather long working distance so as to keep the incident beam divergence small. In the STEM only a single tilt axis can be used to alter the specimen orientation, while in the SEMs more freedom is available with the addition of a rotation axis about the specimen normal. The orientation of the specimen relative to the microscope's optic axis is monitored using selected area channelling patterns (SACP).

4. IMAGING OF INDIVIDUAL DISLOCATION LINES

4.1 Predictions from a Simple Theory

Intensity profiles across dislocation images were calculated using a model based on that of Spencer *et al*, (1972) whose notation will be followed. Two beam dynamical diffraction theory was used with electrons removed from the initial Bloch wave states subsequently treated as two plane waves, one propagating into the specimen and the other backscattered toward the entrance surface. A difference to Spencer *et al's* work was that the backscatter coefficients $p^{(j)}$ were assumed to be proportional to the absorption coefficients $\mu^{(j)}$ (i. e. $p^{(j)} = \sigma \{\mu^{(j)} - A\}$).

The model was used to calculated the image contrast and width expected for a screw dislocation close to the surface of a bulk Si specimen imaged using the (220) reflection for which **g.b**=1 . The effects of finite beam divergence and diameter on the intensity profiles were approximated by assuming a uniform illumination over the range of incident beam angles and positions. The results of these simulations are given in table (1) for some conditions typical of the different microscopes used.

Microscope	Brightness $(Acm^{-2}srad^{-1})$	Divergence (mrad)	spot size (nm)	Contrast (%)	Image Width (nm)	Probe Current (nA)	Signal to noise ratio
FEG-STEM (100 keV)	2×10^9	2.5	8	0.6	30	10	6
FEG-STEM (30 keV)	6×10^8	2.5	12	2.7	20	10	27
FEG-SEM (30 keV)	3×10^8	2.5	20	1.6	35	0.9	5
LaB_6 SEM (30 keV)	2×10^6	2.5	100	0.5	95	3	2

Table (1) : Estimation of signal to noise ratio for ECCI of a screw dislocation near the surface of a bulk Si specimen imaged with the **g**=(220) reflection, in several SEMs.

However, it is not the contrast (C) but rather the signal to noise ratio (S) which determines whether or not the image will be visible. The signal to noise ratio is given by

$$S = C \sqrt{n} = C \sqrt{\frac{I \eta \Omega t}{2 \pi e m}},$$

where I is the probe current and the other parameters and their typical values are as follows: the back scatter fraction η (= 0.5), capture angle of the detector Ω (= $\pi/2$ srad), frame time t (= 60 sec), electronic charge e (= 1.6×10^{-19} C) and the number of picture points m (= 4×10^5). Typical values of the signal to noise ratio that should be obtained in the SEMs used were calculated and are listed in table (1). For the image to be visible S must take a value of about 5 or greater. Thus from table (1) we expect to obtain the best images using the HB501 STEM at 30 keV, while dislocations should also be visible using this machine at 100 keV and the JEOL FEG SEM at 30 keV.

4.2 Experimental Results

Images of dislocations in a deformed Si specimen have been obtained using three microscopes with electron optic parameters similar to those used in the simulations discussed in the previous section. Figure (1) shows two images obtained with the STEM, (a) at 30 keV and (b) at 100 keV. The image obtained at 30 keV appears much cleaner than that obtained at 100 keV in agreement with the calculated signal to noise ratios. The image widths are noticeably broader for the higher beam energy again agreeing with the theory, in which the width varies with the extinction distance of the operating reflection. There is, however, an advantage to using the higher beam energy which is also demonstrated in these two images. The dislocations marked A in the 100 keV image are too deep within the specimen to be seen in the 30 keV image. Indeed, Czernuszka, Long and Hirsch (1991) showed that dislocations in Si could be imaged to a depth of about 100 nm at 30 keV and to about twice this depth at 100 keV.

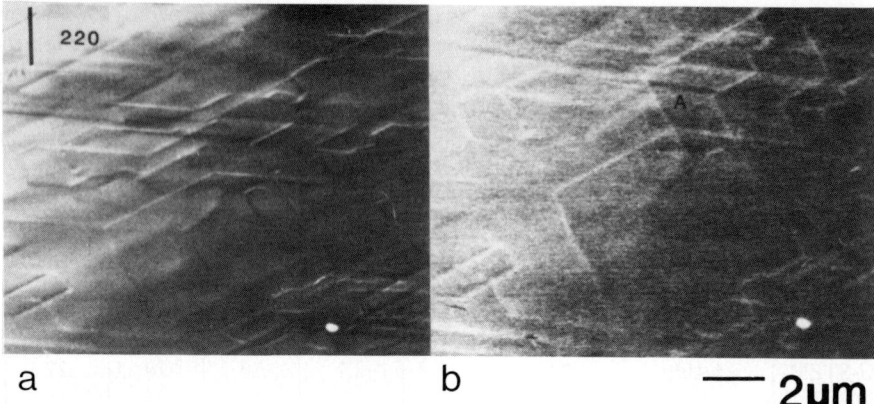

Figure (1) : Dislocations in Si imaged using the HB501 STEM at (a) 30 keV and (b) 100 keV. Dislocations marked A are only visible at the higher beam energy.

Figure (2) shows dislocations in the same specimen imaged using the JEOL 840F FEG SEM at 35 keV. The image is noisier compared to the images obtained using the STEM, again in agreement with the predictions of table (1). Images of this quality are now being obtained routinely and work to improve the signal to noise ratio is in progress. Surprisingly, dislocations were also visible when the same specimen was examined using ECCI in an SEM with a LaB_6 thermionic emitter. The dislocations images were much broader and more diffuse as can be seen in figure (3).

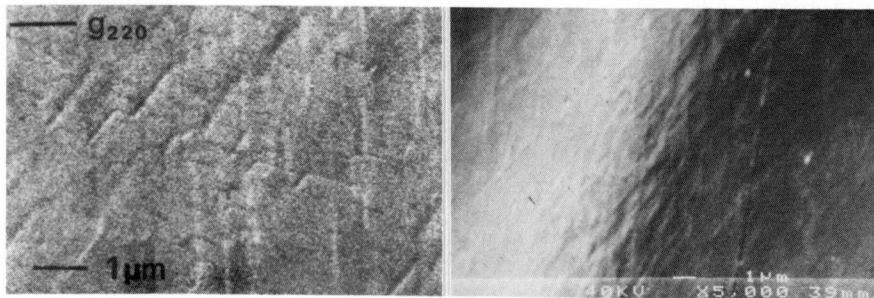

Figure (2) :Dislocations in Si imaged with the JEOL 840F FEG SEM at 35 keV.   **Figure (3)** : Dislocations in Si imaged with the JEOL 6300 LaB_6 SEM at 40 keV.

5. IMAGING OF DISLOCATIONS IN STRAINED EPITAXIAL LAYERS

5.1 Long Range Strain Field of Dislocation Clusters

The contrast observed using ECCI arises within the first few extinction distances from the specimen surface (i. e. within 100 nm or so). The images shown in section 4.2 were undoubtedly of dislocations within this information depth, however, we have also been able to image groups or clusters of dislocations over an order of magnitude deeper than this (i. e. completely outside the information depth). Such dislocation clusters are frequently encountered in strained epitaxial layer systems grown beyond the critical thickness.

When an epitaxial layer is grown pseudomorphically on a substrate with a mismatched lattice parameter dislocations are frequently observed at the interface. These misfit dislocations have been observed using TEM in many systems and for systems with low lattice parameter mismatch the lines tend to be of 60° type. Their distribution is far from uniform with clusters of dislocation often found in which the spacing between lines can be as small as 50 nm (e.g. Eaglesham, Kvam, Maher, Humphreys and Bean 1989).

With the dislocations deep within the specimen the contrast must arise from the long range strain field associated with the dislocation cluster as a whole. Indeed if the dislocations lie close together at the interface between the substrate and the epilayer, which is thick compared to the information depth, then the long range strain field can be obtained by treating the cluster as a single 'superdislocation'. The Burgers vector of the cluster is simply given by the sum the Burgers vectors of all the dislocations in the cluster. The geometry of this situation is compared to that for the case of a near surface dislocation in figure (4). Superposed on these figures are plots showing how the lattice plane bending β (=d($\mathbf{g.R}$)/dz) varies along a scan taken across the defect line at the surface. The near surface dislocation was assumed to be a screw at a depth of 10 nm, while the cluster was taken to consist of 10 screw dislocations 1000 nm below the surface. Obviously β varies much less rapidly for the deep lying defect leading us to expect a much broader image. It is also seen that the variation of β is much smaller for the deep lying cluster indicating that the contrast should be even smaller than that for a near surface dislocation. Fuller calculations using the channelling theory described in section 4.1 (and more completely in Wilkinson, Anstis, Czernuszka, Long, and Hirsch 1993) confirm this prediction of a broad low contrast image.

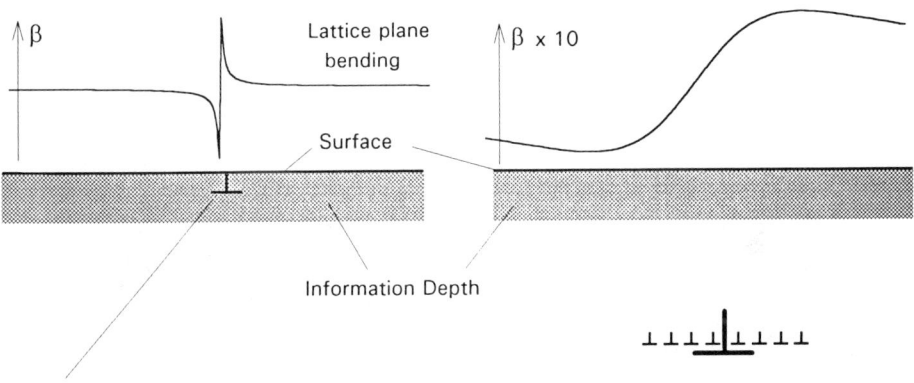

Figure (4) : A comparison of the lattice plane bending (β) caused by a near surface dislocation, and by a cluster of dislocations deeper in the specimen.

Despite the low contrast values, good images can be obtained over a wide range of defect depths (i. e. epilayer thicknesses) using an SEM with a LaB$_6$ (or even W hairpin) filament. Good signal to noise ratios can be achieved since although the contrast is lower the constraint imposed on the beam diameter is now greatly relaxed compared to the case for near surface dislocations. Indeed in this case the electron source brightness is less important than the beam current that can be obtained. Thus, thermionic emitters are more suitable than cold cathode FEGs which although brighter cannot provide sufficient currents. Figure (5) gives examples of typical images obtained from partially relaxed Si$_{1-x}$Ge$_x$ layers grown on Si (001) to different layer thicknesses (h). The ECCI technique affords a rapid assessment of the defect distribution over large areas of bulk specimens with little or no sample preparation required.

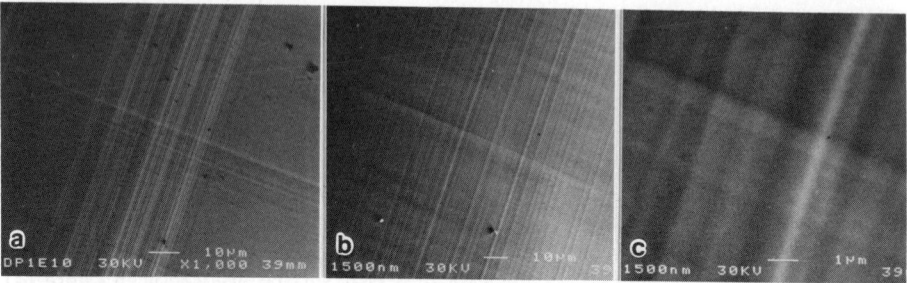

Figure (5) : Misfit dislocation clusters in $Si_{1-x}Ge_x$ layers grown on Si. (a) $x = 0.18$, $h = 0.17\mu m$, (b) and (c) $x = 0.1$, $h = 1.5$ μm at two different magnifications.

5.2 Observation of Dislocations in Mesa Specimens

Some observations of defects in mesa specimens will be given to show the usefulness of the ECCI technique. The mesa specimens consist of an epitaxial layer of $Si_{0.85}Ge_{0.15}$ grown to a thickness of 1 μm on a Si substrate patterned with raised rectangular pads of height 3 μm. It has been suggested that growth of the mismatched epilayer on a substrate of finite extent could result in suppression of the unwanted misfit dislocations. The dislocation suppression might occur due to the lack of suitable nucleation sites over the growth area, but could also be due to elastic relaxation of the lattice mismatch in the upper part of the epilayer, due to the removal of lateral constraint. An analysis of the elastic relaxation has been given by Luryi and Suhir (1986), which predicts a critical mesa width below which the epilayer can be grown to any thickness without creation of misfit dislocations. For the $Si_{0.85}Ge_{0.15}$/Si system this critical width is 5 μm.

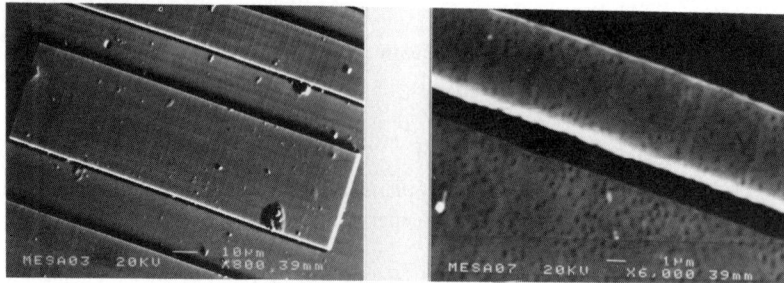

Figure (6) : ECCI of some long rectangular mesas. Dislocations are always present perpendicular to the long axis. Perpendicular to the short axis dislocations are only seen for mesas wider than 5 μm.

ECCI was used in a LaB_6 SEM to image dislocation clusters in mesas of a variety of sizes. The bulk as-grown material could be used and a large number of mesas examined. It should be noted that the height of the mesa structures essentially limits TEM to cross-sectional work, and even then preparation of thin foils is not easy. No dislocation clusters were seen in square mesas below 5 μm in size, while in all larger ones they were clearly seen.

Long rectangular mesas with sides along two <110> directions were also examined (see figure 6). Dislocations were always present perpendicular to the long (>200 μm) sides and for this direction little elastic relaxation is expected. However, perpendicular to the short sides, dislocations were again only seen for mesas wider than 5 μm. These observations agree with the etching experiments of Powell *et al* (1991) and confirm that elastic relaxation can suppress generation of misfit dislocations.

5.3 Characterising the Dislocation Clusters

In addition to imaging defects the ECCI technique also offers a means of characterising them. Just as with diffraction contrast in TEM a screw dislocation of Burgers vector **b** will give no contrast if the operating reflection **g** is chosen such that $\mathbf{g.b} = 0$. For an edge dislocation the additional condition that $\mathbf{g.b_xu} = 0$ must also be satisfied for the dislocation to be out of contrast. Mixed dislocations will always exhibit some contrast though it should be significantly weaker for the reflection where the $\mathbf{g.b} = \mathbf{g.b_xu} = 0$ condition is met. The Burgers vector might then be determined by imaging the same defect using several different reflections in turn until the defects either goes completely out of contrast or gives a much reduced contrast level. Such ideas are routinely applied in TEM, however, more caution is required when using them for Burgers vector analysis in ECCI. Since in ECCI the contrast is always low some consideration must first be made of other factors which could reduce the contrast and hence render a defect invisible.

The first such effect is that the defect contrast decreases rapidly as the incident beam is deviated slightly from the exact Bragg condition for a given set of planes. This is especially true for the deep lying defects where rocking the specimen through angles of the order of 0.1° away from the condition for maximum contrast is sufficient to render it invisible. Good SACPs are therefore crucial for aligning the crystal by bringing the edge of the chosen channelling band (assumed to be at the Bragg condition) to the microscope's optic axis. With practice this is easy to achieve and reproducible results can be obtained.

A second and important effect concerns the angle (γ) between the incident beam and the dislocation line direction. Images obtained with the same reflection operating showed a considerable contrast variation as the specimen was either tilted or rotated so as to change γ (Wilkinson *et al* 1993). The contrast was greatest when the incident beam was normal to the dislocation (γ=90°) and reduced steadily as the beam was brought closer to the line direction. This being due to the dislocation's strain field being independent of the position along the line. Thus when trying to use weak contrast to indicate completion of the $\mathbf{g.b} = \mathbf{g.b_xu} = 0$ condition the operating reflection **g** should be excited in such a way that the line direction is within 10° of perpendicular to the incident beam.

With use of the above conditions useful characterisation of defects can be undertaken. For example, figure (7) shows ECCI of some dislocation clusters in a 170 nm thick $Si_{0.82}Ge_{0.18}$ epilayer on Si (001). The cluster marked E in the g=(04$\bar{4}$) image is out of contrast in the g=(220) image, from which it is deduced that the cluster must have an overall edge character. Other clusters have given similar results, while none have given convincing evidence of being of overall 60° type. If the clusters consist of 60° dislocations it seems that they must contain lines with differing Burgers vectors. The mechanism through which such clusters were created must therefore allow generation of dislocations with different Burgers vectors from the same source.

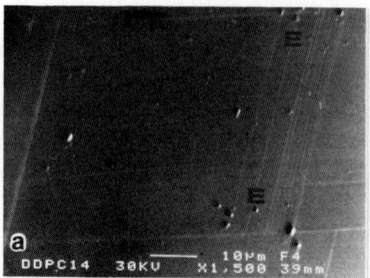 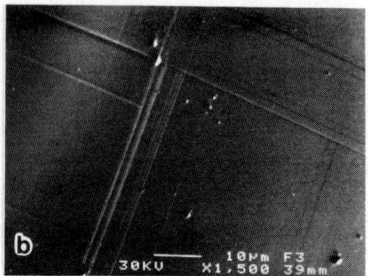

Figure (7) : Dislocation cluster marked E is visible in (a) for g=(04$\overline{4}$), but out of contrast in (b) for g=(220).

6. CONCLUDING REMARKS

The ability to image and characterise individual dislocations near surfaces of bulk specimens in the SEM opens many exciting possibilities. Structures from which thin foil specimens are prohibitively difficult to produce may be examined in the SEM without thinning. Evolution of dislocation structures with annealing or straining might be followed in the same region of a bulk specimen. ECCI could be used to evaluate the defect content and arrangement in a selected structure on a bulk specimen for direct comparison with electronic property measurements. The direct imaging of longer range strain fields from groups of dislocations can be used as a quick quality check for strained epilayer systems and has potential applications in studies of plastic deformation in fracture and fatigue.

7. ACKNOWLEDGEMENTS

This work has be funded by the SERC and the Royal Society (JTC). JEOL (UK) are thanked for the use of SEM facilities.

REFERENCES

Booker G R 1970 Modern Diffraction and Imaging Techniques in Material Science ed. S. Amelinckx (London : North Holland) 647

Clarke D R and Howie A 1971 Phil. Mag. 24 959-971

Clarke D R 1971 Phil. Mag. 24 973-979

Coates D G 1967 Phil. Mag. 16 1179-1184

Czernuszka J T, Long N J, Boyes E D and Hirsch P B 1991 Mat. Res. Soc. Symp. Proc. vol. 209 289-292

Eaglesham D J, Kvam E P, Maher D M, Humphreys C J and Bean J C 1989 Phil. Mag. A 59 1059-1073

Luryi S and Suhir E 1986 Appl. Phys. Lett. 49 140-142

Morin P, Pitaval M, Besnard D and Fontaine G 1979 Phil. Mag. 40 511-524

Powell A R, Kubiak R A, Whall T E, Parker E H C, and Bowen D K 1991 Mat. Res. Soc. Symp. Proc. vol. 220 277-283

Stern R M, Ichinokawa T, Takashima S, Hashimoto H and Kimoto S 1972 Phil. Mag. 26 1495-1499

Spencer J P , Hirsch P B and Humphreys C J 1972 Phil. Mag. 26 193-213

Wilkinson A J, Anstis G R, Czernuszka J T, Long N J and Hirsch P B 1993 Phil. Mag. in press

Inst. Phys. Conf. Ser. No 134: Section 11
Paper presented at Microsc. Semicond. Mater. Conf., Oxford, 5–8 April 1993

763

Novel dislocation contrast effects in scanning-transmission electron microscopy

D D Perovic, A Howie[†] and C J Rossouw[*]

Department of Metallurgy and Materials Science, University of Toronto, 184 College Street, Toronto M5S 1A4 Canada
† Cavendish Laboratory, University of Cambridge, Cambridge, CB3 0HE England
* CSIRO Division of Materials Science and Technology, Clayton, Victoria 3168 Australia

ABSTRACT: We present novel experimental results on the imaging of dislocations in HAADF-STEM. Images of inclined 60° dislocation segments in Si exhibit a number of novel contrast effects which have been shown to depend on the specific position of the dislocation in the foil. The dislocation contrast is initially very dark at the entrant foil surface and eventually exhibits oscillatory contrast at larger foil thicknesses. Ultimately, the dislocation contrast remains bright beyond a certain depth in the crystal towards the exit foil surface. A Bloch-wave scattering description has been formulated to consistently explain the observed contrast features.

1. INTRODUCTION

The use of a high-angle annular dark-field (HAADF) detector in a field-emission gun scanning-transmission electron microscope (STEM) has become an elegant means for acquiring high spatial resolution images of multiphase materials coupled with high compositional sensitivity. With the exclusion of coherent (ie. Bragg) scattering, the HAADF detector allows for truly incoherent imaging at atomic resolution with high atomic number (Z) sensitivity approaching the simple Z^2-dependence of unscreened Rutherford scattering (Pennycook *et al.* 1991). However, recent experimental studies have indicated that HAADF-STEM imaging of 'imperfect' crystals is not always straightforward (Perovic *et al.* 1991a). Here we report on novel specimen thickness-dependent dislocation contrast features in HAADF-STEM imaging. The results are qualitatively explained using Bloch-wave electron scattering theory.

2. EXPERIMENTAL

Several polycrystalline Al and Ni specimens were prepared for electron microscopy using standard electrochemical thinning methods. In addition, specimens of Si-based multilayers grown by Molecular Beam Epitaxy (MBE) were prepared in cross-section with <110> surface normals by mechanical thinning followed by Ar-atom sputtering to electron transparency. The specimens were examined in a Vacuum Generators HB501 dedicated-STEM operating at 100 kV. Bright-field (BF) and HAADF images were recorded directly onto photographic film from a high resolution transparent phosphor screen. Images were taken using various incident beam divergence half-angles in the range, $\alpha = 6\text{-}27$ mrad. The HAADF detector angular collection range was 105-300 mrad.

3. RESULTS

From earlier STEM-electron energy-loss spectroscopy experiments on grain boundary segregation (Perovic *et al.* 1991b), it was observed that lattice dislocation tangles and grain boundary dislocation arrays in polycrystalline Al and Ni appeared predominantly bright relative to the background intensity in HAADF images. Fig. 1 (a,b) compares BF and HAADF images of three low-angle grain boundaries near a triple line in Al which are inclined relative to the electron beam. All three boundaries possess a regularly spaced array of dislocations which accommodate the boundary misorientation. Although the diffraction contrast features (ie. grain boundary extinction fringes) have essentially disappeared in the HAADF image as expected, the grain boundary dislocations are clearly visible with spacings as small as 3 nm. Moreover, some dislocation segments appeared to lose visibility as indicated with arrows in Fig. 1b. In light of these results, it was of interest to study a typical dislocation under specific imaging conditions in order to elucidate the observed contrast behaviour.

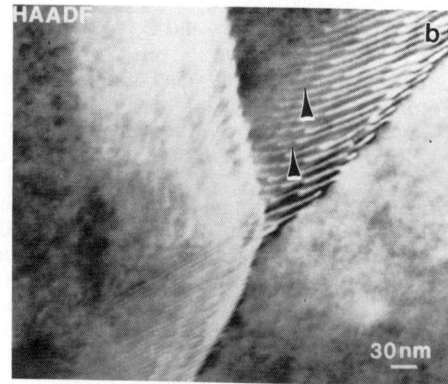

Fig. 1: STEM images of three low-angle grain boundaries in Al possessing interface dislocations to accommodate the lattice misorientation; (a) BF and (b) HAADF.

Fig. 2 (a,b) is a pair of <110> zone-axis STEM images from a B-doped Si multilayer showing two threading dislocations which intersect both thin foil surfaces. Fig 2a shows a BF-STEM diffraction contrast image of the inclined dislocation segments exhibiting the principally dark oscillatory contrast and 'top-bottom' effect which by reciprocity is analogous to conventional transmission electron microscopic (CTEM)-BF images (Cowley 1969). In addition, strong diffraction contrast from the wedge-profile specimen gives rise to <110> thickness extinction contours; the foil thickness increases from the top right to the bottom left in the image. Fig. 2b is a HAADF image of the same area. Although the diffraction contrast features from the matrix have disappeared, the dislocation segments exhibit a number of novel contrast features which have only recently been reported (Perovic *et al.* 1993). At the entrant thin foil surface, the terminating segments give rise to very dark contrast lobes. Further into the crystal, black-white contrast oscillations are observed up to a projected foil depth of ~65 nm. Ultimately, the image contrast remains bright with a monotonically decreasing intensity towards the exit foil surface. As the magnitude of the incident beam divergence (α) was increased under otherwise identical conditions, it was found that the oscillatory contrast eventually disappeared except for the very dark entrant surface lobe contrast features (Perovic *et al.* 1993). In any case it is clear that HAADF dislocation contrast is never complementary with the BF image.

4. DISCUSSION

In light of the experimental results discussed above, it was of interest to formulate a theory based on dynamical Bloch wave scattering effects to explain the observed contrast features. An electron wave field (ψ), built up of a coherent superposition of Bloch waves and generated by low-angle dynamical diffraction events, yields a HAADF signal proportional to the sum of its intensities at all the atomic sites. A given Bloch state may be assigned molecular orbital-type quantum numbers dependent on symmetry. If the first six (ie. high kinetic energy) states are considered in <110> Si, these have *s*- or *p*-type symmetry comprising bonding and antibonding orbital combinations. At the exact zone-axis orientation, 99% of the 100 keV electron flux is contained in the significantly excited states (1), (3) and (6) (ie. 1*s*, 2*p_x* and 2*s*)

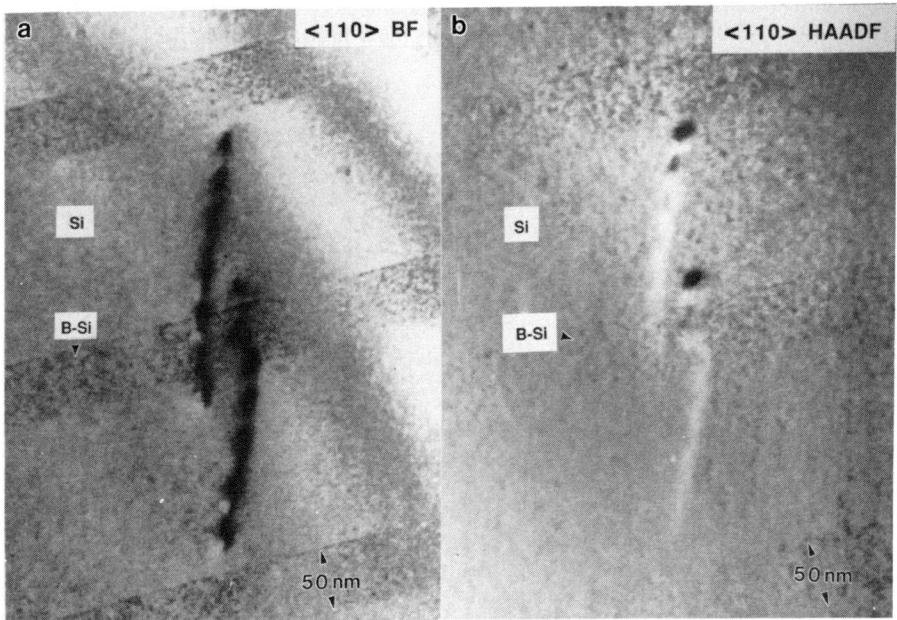

Fig. 2: <110> zone-axis STEM images (α = 13.1 mrad) of a MBE-grown B-doped Si multilayer possessing a pair of inclined threading dislocation segments: (a) BF; (b) HAADF.

with excitation amplitudes ($\alpha^{(i)}$) of 0.685, -0.609 and 0.385 respectively (Fig. 3a). It is worth noting that <110> Si is more complex than <100> Si which has a centre of symmetry on an atomic column and thus only *s*-state excitation is possible on axis. Therefore, in a thin, well-oriented, perfect crystal the scattering is dominated by the *1s*-Bloch wave component of ψ which has a very low dispersion (Pennycook *et al.* 1991). The signal is then simply proportional to the number of atoms in the column multiplied by the mean value of Z^2. Detailed computations of wave propagation and scattering in crystals possessing various frozen-in vibrational states (Loane *et al.* 1992) generally support this picture but also show effects due to the variation of ψ with foil depth (z). Because of anomalous absorption for instance, the *s*-states are heavily attenuated and the HAADF signal falls off. Moreover, in the presence of defects, further complications can arise since transitions between different Bloch-wave components in ψ can occur as a function of z. The influence of these effects on large-angle scattering underlies the technique for imaging defects in the back-scattered electron signal (Czernuszka *et al.* 1990).

The treatment of *high-angle* scattering of an electron wave field (ψ) in terms of Bloch waves can be described by adding a complex component (V_g') into the Fourier components for the elastic lattice potential (V_g) and defined over the angular range of the HAADF detector. V_g' is defined in terms of a thermal diffuse scattering (TDS) absorption potential using an Einstein model of independently vibrating atoms (Allen and Rossouw 1990). Upon writing eigenvalues ($k^{(i)}$) in terms of the elastic ($\gamma^{(i)}$) and absorptive ($\eta^{(i)}$) components (ie. $k^{(i)} = \gamma^{(i)} + i\,\eta^{(i)}$), the corresponding excitation amplitude ($\alpha^{(i)}$) and the effective TDS mean free path ($\lambda^{(i)} = 1/2\eta^{(i)}$) of each partial wave for scattering into the HAADF detector have been calculated as shown in Figs. 3a,b respectively. The HAADF dislocation contrast features of Fig. 2 can be explained with reference to Fig. 3. At the entrant surface of the <110>-oriented crystal, the long-range displacements of the dislocation significantly distort the atomic column symmetry resulting in a change in the initial Bloch wave excitation ($\alpha^{(i)}$) and subsequent *s*-state depletion due to enhanced $s \to p$ interband transitions. The excited p-states are far more dispersive and thus channel through the crystal without being scattered (ie. absorbed) to the HAADF detector (cf. Fig. 3). As a result, the scattering intensity to the HAADF detector from the dislocation is significantly reduced relative to the perfect crystal intensity. Therefore, for very thin crystals (<~ 10 nm), the contrast of inclined dislocations should always appear dark, consistent with the Pennycook and Jesson model.

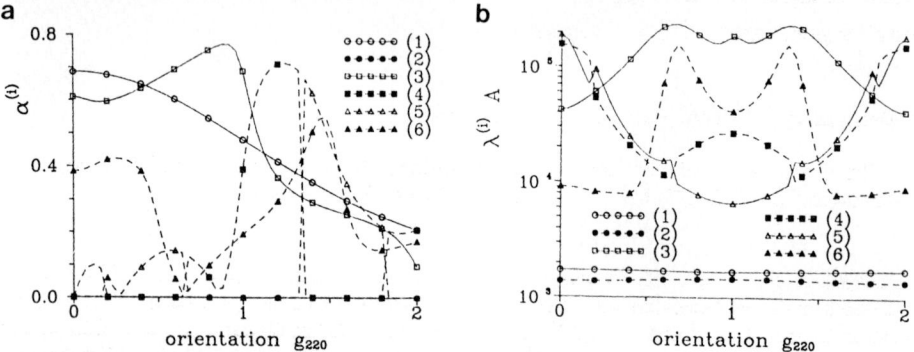

Fig. 3: (a) <110> zone-axis excitation amplitudes $(\alpha^{(i)})$ for the first six Bloch states (i= 1-6) in Si along a g_{220} systematic; (b) the mean free paths $(\lambda^{(i)})$ for TDS scattering into the HAADF detector (100-300 mrad). See text for further details.

At larger thicknesses, the HAADF contrast depends sensitively on the depth of the dislocation in the foil. Analogous to dynamical diffraction in CTEM, the HAADF images exhibit a periodic contrast oscillation indicative of an interference effect among at least two Bloch states contrary to the thin-crystal Pennycook-Jesson single s-state scattering model. At a sufficient depth below the entrant surface, the electrons initially scattered into p-states can subsequently undergo $p \to s$ interband transitions due to the dislocation displacement field. This effect can be understood from Fig. 3a where, for example, the transition from Bloch state (1) to Bloch state (3) gives rise to an interference effect with an extinction length given by the separation of the respective branches at the Brillouin zone boundary. Quantitatively, the measured extinction length (ξ_{110}) of the oscillatory contrast at the dislocation (~ 27.5 nm) agrees best with the wave-vector difference associated with the $B^{(1)}$ to $B^{(3)}$ transition, which from Fig. 3 is calculated $(\xi_g = (\Delta k^{(i)})^{-1})$ to be 23.4 nm. As a result, the visibility or invisibility of a given dislocation will depend on the exact position of the segment in the foil. The Bloch-wave interference effect will continue until the crystal becomes thick enough such that s-states are effectively depleted. Thereafter, the crystal allows only p-state electron channelling without any further interference effects. Thus there is no 'bottom effect' (ie. reverse contrast) where the dislocation intersects the exit face of the crystal since the initial Bloch-wave excitations have been modified by lattice tilting such that $p \to s$ transitions are unlikely.

REFERENCES

Allen L J and Rossouw C J, 1990, *Phys. Rev. B.*, 42 11644.
Cowley J M, 1969 *Appl. Phys. Lett.*, 15 58.
Czernuszka J T, Long N J, Boyes E D and Hirsch P B, 1990 *Phil. Mag. Lett.*, 62 227.
Loane R F, Xu P and Silcox J, 1992 *Ultramicroscopy*, 40 121.
Pennycook S J, Jesson D E and Chisholm M F, 1991 *Inst. Phys. Conf. Ser.*, 117 27.
Perovic D D, Weatherly G C, Paterson J H, Rossouw C J and Jackman T E, 1991a *Inst. Phys. Conf. Ser.*, 117 631.
Perovic D D, McComb D W, McGibbon A J and Brown L M, 1991b *Inst. Phys. Conf. Ser.*, 119 433.
Perovic D D, Howie A and Rossouw C J, 1993 *Phil. Mag. Lett.*, (in press).

Inst. Phys. Conf. Ser. No 134: Section 11
Paper presented at Microsc. Semicond. Mater. Conf., Oxford, 5–8 April 1993

767

Optimization of the imaging conditions for electron- and laser-beam excited scanning DLTS investigations

O Breitenstein, T Heiser[*], J Heydenreich and A Mesli[*]

Max Planck Institute of Microstructure Physics, Weinberg 2, D-O-4050 Halle (Saale), Germany
[*]Laboratoire PHASE CNRS, B.P. 20, F-67037 Strasbourg, Cedex, France

ABSTRACT: Unlike other scanning beam techniques, the spatial resolution of SDLTS strongly depends on the excitation intensity. The optimum excitation conditions can be found by monitoring either the spatial resolution or the SDLTS signal height for different beam currents or filling pulse widths, respectively. Corresponding SDLTS measurements on a GaAs pn junction with a diamond indentation prove the validity of the theoretical predictions. The indentation-induced point defects are proved to be generated mainly in regions outside the dislocation slip planes.

1. INTRODUCTION

Scanning Deep Level Transient Spectroscopy (SDLTS) offers a "missing link" between spectral imaging techniques (CL, scanning PL, or local IR absorption), real microanalytical techniques like scanning SIMS or EDX, and the sensitive detection of electrically active point defect levels provided by standard-DLTS (Lang 1974). SDLTS allows the imaging of energetically well-defined levels in the energy gap with a spatial resolution in the micron range and a detection limit in the 10^{15} cm^{-3} range, well below the detection limit of real microanalytical techniques. Since its introduction by Petroff and Lang (1977), SDLTS and related techniques like Scanning Minority Carrier Transient Spectroscopy (SMCTS) have been further developed (Breitenstein and Heydenreich 1985 and 1989, Woodham and Booker 1987, Heiser et al 1988), with a dedicated SDLTS equipment being now commercially available (Breitenstein and Raith 1991).

The reason why this technique has rarely been used does not only arise from technical difficulties, but also lies in some scepticism with respect to the quantitative interpretation of the results. Therefore, the theoretical modelling of the SDLTS signal generation mechanism and the experimental verification of the results are of great interest. Here, the estimation of the excited area where the SDLTS signal comes from is of particular importance, since the SDLTS signal height amounts to (assuming a homogeneous net doping and $N_t \ll N_D - N_A$):

$$\Delta C = \frac{C_o \, A_e \, N_t}{2A \, (N_D - N_A)} \qquad (1)$$

(C_o = sample capacitance, A_e = excited area, A = sample area, $N_D - N_A$ = net doping concentration, N_t = trap concentration). A detailed theoretical description of the SDLTS signal generation process has been given by Breitenstein et al (1992), so that here only the major final results of these calculations will be provided. The primary aim of the present contribution is to provide new experimental evidence of the validity of these results, which can be summarized as follows.

For long filling pulse widths ($t_f > 1/e_{n;p}$) the extension of the excited area is governed by the local equilibrium between the local capture rate and the emission rate $e_{n;p}$ and does not depend on t_f, hence stationary excitation conditions are established. For short filling pulses ($t_f < 1/e_{n;p}$) the extension of the excited area is dynamically governed by the local capture rate and the filling pulse width (non-stationary excitation conditions), just as for standard-DLTS. Both excitation conditions as well as both electron- and Laser-beam excitations in principle show the same behaviour. The best spatial resolution, which is given by the extension of the beam-induced diffusion current density profile, is attained only if the levels are still partly filled, hence for low beam currents and/or for low filling pulse widths, respectively. In the following, the product of beam current and filling pulse width will be called "excitation intensity" I_e. In this "linear excitation regime" the SDLTS signal linearly depends on I_e. If the traps get saturated with increasing I_e, the SDLTS signal more and more depends sublinearly on I_e. In contrast to standard-DLTS, however, the signal does not really saturate with I_e but finally reaches a logarithmic   dependence, owing to the edges of the diffusion current density profile exponentially falling with the distance. In this "saturation regime" the spatial resolution degrades with the logarithm of I_e. At the junction between the linear and the saturation regimes there are optimum excitation conditions, with the resolution not yet seriously degraded, but the levels almost completely filled within the excited area. Thus, optimum conditions are reached if a further reduction of I_e no more improves the spatial resolution. Though all these considerations hold true of both stationary and non-stationary conditions, non-stationary ones should be preferred as they provide a more time-efficient measurement, a more distinct delineation of the excited area, and the possibility of using a computer-controlled "pulse width scan" to optimize the excitation conditions (Breitenstein and Heydenreich 1989).

2. THE PIXEL SCAN TECHNIQUE

According to eq. (1) the SDLTS signal may be several orders of magnitude smaller than the standard DLTS signal owing to the factor of A_e/A. Therefore, a signal integration time of typically 1s per measuring point is required for SDLTS, leading to considerably long measure times for two-dimensional images. To attain acceptable measure times, the number of points per image is chosen to be low, with the results displayed to consist of more or less extended "pixels". If the image magnification factor of the microscope is chosen sufficiently low, it may happen that the area A_e excited by a focused beam gets smaller than the pixel size. Then the sensitivity is wasted and highly localized features might not at all be detected, since a great part of the area investigated is not excited. To prevent this, instead of a focused beam a well-defined area of beam incidence should be used, most simply realized by defocusing the beam. This, however, is not a very reproducible measure. Alternatively it had been proposed by Woodham and Booker (1987) to quickly scan the focused beam during each excitation pulse across a well-defined pixel area. Thus, a "shaped" beam appears that may be used in a two-dimensional scan to provide a non-overlapping excitation of the whole area investigated pixel by pixel. This "pixel scan" technique can be realized most simply in a computer-controlled beam deflection system. For the pixel scan a meander scan mode is superior to the usual image scan mode, because for a given slew rate of the microscope beam deflection amplifier it provides a faster pixel scan rate.

3. EXPERIMENTAL

The sample used for the following investigations was a Zn-diffused GaAs p^+n junction on <100> oriented epitaxial material (a luminescence diode), where at room temperature a diamond indentation was made at a load of 0.4N for 5 seconds. The investigations were carried out at T = 292 K using a rate window of 1000 s^{-1}. Under electron beam excitation a minority carrier trap appeared having an activation energy of 500 meV. This level probably corresponds to the "Hα"-level investigated e.g. by Wosinski (1985), which had been interpreted to be a second charge state of the EL2 level.

All the results of the present contribution have been measured using the computer controlled "Raith SDLTS" dedicated SDLTS equipment (Breitenstein and Raith 1991), working with a SEM type JSM 840 A operating at 25 kV. This system allows one to carry out both capacitance- and current-based DLTS and SDLTS measurements (line scan and image scan modes) as well as EBIC imaging. Moreover, a number of special techniques are possible, such as SDDLTS (double pulse SDLTS, see Woodham and Booker /5/), the "filling pulse width scan" mentioned above, the "rate window scan" (constant temperature (S)DLTS), and the "pixel scan" technique described above.

4. RESULTS AND DISCUSSION

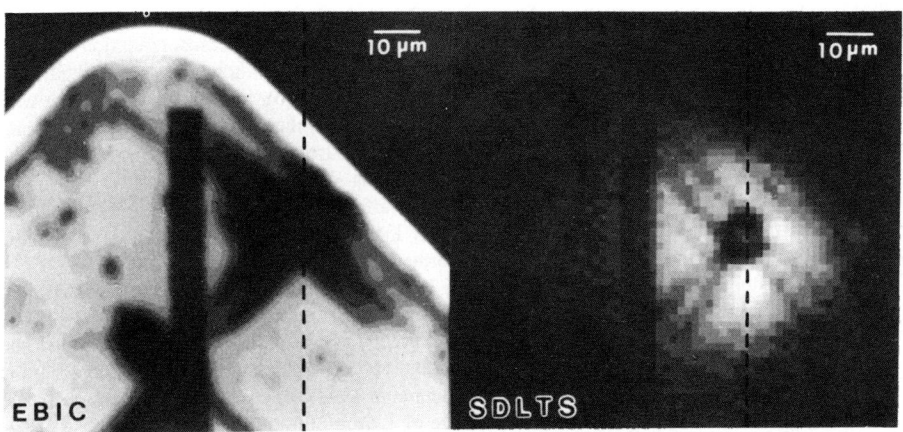

Fig. 1: EBIC image and SDLTS image of the indentation-induced point defects around the indentation. Measuring conditions: $T = 292K$, $e_p = 1000s^{-1}$, $J_{diff} = 3\mu A$. The vertical dashed line is the trace of the SDLTS line scans of Fig. 2.

Fig. 1 shows the EBIC image and the SDLTS image of a sample region containing the edge of the p-diffused region (the bright curved stripe in the EBIC image), a p-contact finger (the vertical dark stripe in the EBIC image) and the indenter impression. From this indentation dislocations are moving in slip bands in <100> directions, yielding a "cross" aligned to the diode edges around a dark contrast region in the actual indentation position in the EBIC image. The additional dark EBIC contrast at the bottom is due to a mechanical surface scratch. The SDLTS image represents the spatial distribution of the deformation-induced point defects (bright contrast correspnds to a high point defect concentration). Along the vertical dashed line in Fig. 1 crossing the indentation position from top to

Fig. 2: SDLTS line scans across the indentation from top to bottom. Measurement conditions: $T = 292K$, $e_n = 1000$ s^{-1}. All results are scaled separately and line shifted.

(a): Exciting diffusion current = $3\mu A$, filling pulse width (from bottom to top): $32\mu s$, $128\mu s$, $512\mu s$, $2ms$.

(b): Filling pulse width = $2ms$, exciting diffusion current (from bottom to top) = $300nA$, $1\mu A$, $3\mu A$, $10\mu A$.

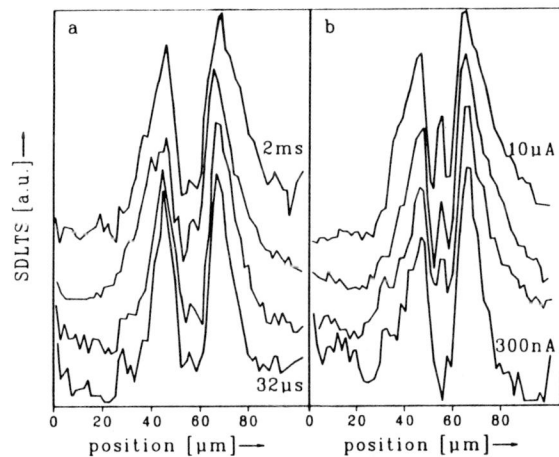

bottom, SDLTS line scans have been measured under different excitation conditions as shown in Fig. 2. On the left (a), for a fixed beam current the filling pulse width was varied in steps by a factor of four from 32 µs to 2 ms. Hence, this series contains the junction between non-stationary (t_f < 1 ms) and stationary excitation conditions (t_f > 1 ms). The predicted spatial resolution degradation appears as a broadening of the two major SDLTS signal maxima above and below the indentation position. In Fig. 2 (b) under stationary excitation conditions (t_f = 2 ms) the beam current was varied over two orders of magnitude in steps by a factor of three. Here, too, the spatial resolution degradation appears, proving the same behaviour of the two excitation conditions. Unlike in (a), however, the local SDLTS maximum in the indentation position at 55 µm, also visible in Fig. 1 (b), increases in Fig. 2 (b) with increasing I_e. The reason for this different behaviour is not yet clear; there are probably different point defect species involved in the image formation. In general, the strongly disturbed region around the very indentation is hard to interpret in detail, as the exciting diffusion current density is strongly reduced, and the net doping concentration, too, is certainly affected here. The four main local SDLTS maxima around the indentation, however, at least partly located in regions outside the strong EBIC contrast, can definitely be interpreted as clouds of deformation-induced point defects.

While for a better comparison all results of Fig. 2 are separately scaled to the same maximum signal height, in Fig. 3 the maximum SDLTS signal height measured of the scans in Fig. 2 (b) has been drawn double-logarithmically over the diffusion current J_{diff} during excitation. The dashed line represents the linear behaviour. A similar measurement but varying the filling pulse   instead of the beam current has been published elsewhere (Breitenstein et al 1992). The junction between linear and saturation regime is found in Fig. 3 to be in the range of J_{diff} = 300 nA for t_f = 2 ms chosen here. Hence, the values of J_{diff} = 3 µA and t_f = 128 µs chosen for measuring Fig. 1 (b) provide nearly the same excitation intensity and can therefore be considered optimum.

In Fig. 1 it is particularly interesting to notice that the point defects are generated predominantly in regions outside the dislocation slip planes. This is in contrast to the usual assumption that during plastic deformation point defects are generated as a result of dislocation motion (see e.g. Wosinski 1985). This result corresponds, however, to earlier studies on silicon, where in regions near an indentation after thermal annealing point defect clusters were revealed by HREM predominantly in regions outside the dislocation slip planes (Werner and Heydenreich, unpublished).

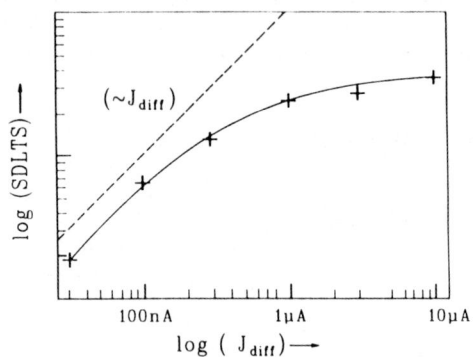

Fig. 3: Maximum SDLTS signal height [a.u.] of Fig. 2 (b) over exciting diffusion current.

The authors acknowledge the experimental cooperation with U. Egger (Halle) in carrying out the indentations.

REFERENCES:

Breitenstein O, Heydenreich J 1985 Scanning 7 273
Breitenstein O, Heydenreich J 1989 SEM Microcharacterization of Semiconductors, eds. D B Holt and
    D C Joy (London: AP) pp 339-371
Breitenstein O, Raith H 1991 J. de Physique IV C6 343
Breitenstein O, Heydenreich J, Heiser T, Mesli A 1992 Optik 92 74
Heiser T, Mesli A, Courcelle E, Siffert P 1988 J. Appl. Phys 64 4031
Lang D V 1974 J. Appl. Phys. 45 3014
Petroff P M, Lang D V 1977 Appl. Phys. Lett. 31 60
Woodham R, Booker G R 1987 Inst. Phys. Conf. Ser. 87 781
Wosinski T 1985 Appl. Phys. A 36 213

Inst. Phys. Conf. Ser. No 134: Section 11
Paper presented at Microsc. Semicond. Mater. Conf., Oxford, 5–8 April 1993

771

A new confocal SIRM incorporating reflection, transmission and double-pass modes either with or without differential phase contrast imaging

P Török, GR Booker, Z Laczik and R Falster[1]

Department of Materials, University of Oxford, Parks Road, Oxford OX1 3PH, UK
[1] MEMC SpA., Via Gherzi 31, 28100 Novara, Italy

ABSTRACT: We have previously used the scanning infra-red microscope (SIRM) in the transmission mode to image individual precipitate particles in bulk semiconductors such as Si, GaAs, InP and CdTe. We have now developed a new confocal SIRM which can be used in either the reflection, transmission or double-pass modes, with or without differential phase contrast imaging, and have applied it to oxide particles in heat treated Cz Si wafers. It has advantages over the previous SIRM, e.g. higher resolution and sensitivity, enabling smaller particles and higher number densities to be assessed.

1. INTRODUCTION

During the last few years we have developed the SIRM (Kidd et al 1987, Booker et al 1992) for operation in transmission so as to image precipitate particles present within semiconductor bulk specimens, e.g. Czochralski (Cz) Si wafers. The beam from a 1.3 μm wavelength 20 mW semiconductor laser is focussed by a microscope objective lens to a probe ~ 2 μm across within the specimen. Most semiconductor specimens transmit this wavelength light without significant absorption. The probe is diffraction or aberration limited, these depending on the numerical aperture (NA) of the lens and the depth at which the probe is focussed into the specimen. The specimen is mechanically either X-Y or X-Z raster scanned, where Z corresponds to the optical axis of the SIRM. The transmitted light is detected by a Ge photodiode and the resulting signal is used to give a bright-field (BF) image on a display screen. The collected signal is built-up in a computer memory where image processing can be performed before the image is displayed. The contrast arises because the individual particles partly scatter and partly absorb the light in the probe as it scans across the particle. For a lens with NA=0.6, the spatial and depth resolutions are ~ 2 and ~ 30 μm respectively. Individual particles can be imaged down to ~ 30 nm across corresponding to a contrast of ~ 0.2%, particles smaller than this not being detected against the background noise. For an individual particle > 2 μm across, the image appears dark and reveals the particle size and shape. For an individual particle < 2 μm across (which is mostly the case for bulk semiconductor specimens), the image appears as a dark spot ~ 2 μm across. For the latter, the contrast decreases as the particle size decreases, and so the contrast is an indication of the particle size. The front and back surfaces of the specimens generally need to be polished flat or some loss in spatial resolution occurs. Precipitate particles in Si, GaAs, InP and CdTe slabs and wafers were investigated using the SIRM in transmission and particle number densities and distributions were determined. The method is non-destructive and non-contaminating.

We have used the SIRM to obtain dark-field (DF) images by placing the detector off the axis of the SIRM (Laczik et al 1989). The contrast arises from scattering and individual particles appear bright. We have used the SIRM in transmission in the confocal mode by using a collector lens and placing a 'pin-hole' in front of the detector. For a lens with NA=0.85, the spatial resolution was improved from ~ 1.5 to ~ 1.0 μm and the depth resolution from ~ 15 μm to ~ 7 μm, compared with the same lens without a pin-hole. We used the SIRM in transmission in the crossed-polariser mode to image local strain fields and to determine the Burgers vectors of individual dislocations (Laczik et al. 1991).

In the present work we obtained new results by modifying the SIRM to produce differential phase contrast images, in addition to amplitude contrast images such as those described above. Such phase contrast images might give better sensitivity. In particular we describe for the first time the imaging of oxide particles within Si wafers using the SIRM in the RC, RC*dpc*, TC*dpc*, DPC and DPC*dpc* modes, where the letters correspond to R - reflection, C - confocal, *dpc* - differential phase contrast, T - transmission and DP - double-pass. It is shown, for example, that TC*dpc* enables additional particles to be imaged and in RC*dpc* mode causes any light simultaneously reflected from the surfaces of the wafer to be reduced in the image. Previous work using visible-light laser scanning microscopes with *dpc* for surface structure examination has been reviewed by Wilson (1991).

2. EXPERIMENTAL

Our modified SIRM is shown schematically in Figure 1. Light from the laser after traversing a collimator lens (CoL) is focussed by the probe forming lens (PfL) to a probe within the Si specimen (Sp). For the TC*dpc* mode, the transmitted light traverses a collector lens (CL) and is incident on a non-polarising beam-splitter (BS). The transmitted and reflected light from the latter are separately focussed by detector lenses (DL) onto pin-hole detectors (P/D). Two half-stops (HS) are arranged so that one detector receives only the one half of the light, and the other detector receives only the other half of the light. Subsequent addition of the two signals gives an amplitude contrast image, denoted here by TC mode, while subtraction of the two signals gives a differential phase contrast image, denoted here by TC*dpc* mode.

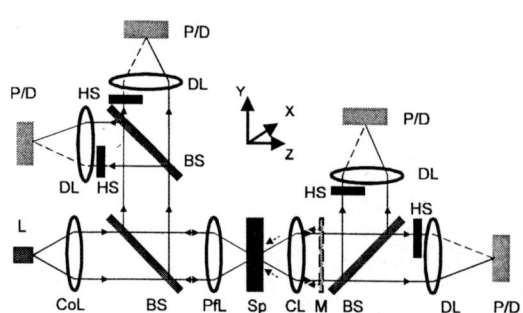

Figure 1. Schematic diagram of the optical system

For the RC*dpc* mode, the light back-scattered by the particles within the specimen is collected by the probe forming lens (PfL) and is incident on a non-polarising beam-splitter (BS). The reflected light from the latter is incident on a second non-polarising beam-splitter (BS). The arrangement is then as for the TC*dpc* mode, two signals being obtained, with addition giving a RC image and subtraction giving a RC*dpc* image. For the DPC*dpc* mode, the light transmitted through the specimen transverses a collector lens (CL) and is incident on a mirror (M). The light reflected from the mirror (M) again traverses the collector lens (CL), the specimen (Sp) and the probe forming lens (PfL). The arrangement is then as for the RC*dpc* mode, signal addition giving a DPC image and signal subtraction giving a DPC*dpc* image.

The new SIRM has been used to image oxide particles in Cz Si wafers heat treated to produce various particle number densities and sizes, these spanning the range typically used for internal oxide gettering of metal impurities during device fabrication.

3. RESULTS

Figures 2a, b and c show RC images for these wafers, which have only the front surface polished and contain oxide particles with number densities of ~ 10^8, 10^9, 10^{10} cm^{-3} respectively. The probe forming lens was of NA=0.85 (X-Y scans are used for all the illustrations in this paper). Each bright spot corresponds to a single oxide particle (or possibly

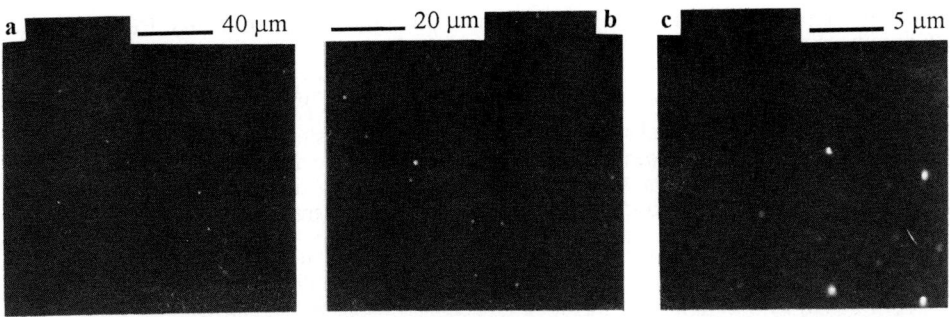

Figure 2. Heat-treated Cz Si specimens, RC SIRM images showing oxide particles with number densities a) $\sim 10^8$ cm^{-3}, b) $\sim 10^9$ cm^{-3} and c) $\sim 10^{10}$ cm^{-3}

in some cases to closely spaced particles). The differences in spot intensity are in general partly due to different particle sizes and partly to some particles being in-focus and others out-of-focus. For the RC mode, those particles which are in-focus can be distinguished either from a focal series obtained on mechanically moving the specimen along the Z axis, or by switching to an X-Z scan image. On using these procedures the in-focus particles gave spots typically 0.7 µm across (FWHM), this corresponding approximately to the spatial resolution, while the depth resolution (FWHM) was 5 to 10 µm. The particle distributions and the differences in the number densities for these specimens are clearly revealed.

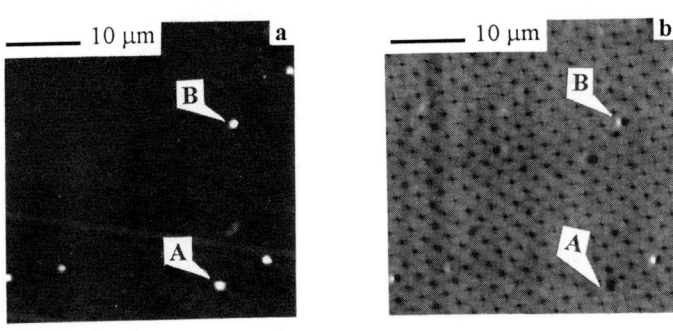

Figure 3. Heat-treated Cz Si specimen, a) RC and b) RC*dpc* images from the same area showing oxide particles. Particle B is in-focus, particle A is out-of-focus.

Figure 3a shows a RC image similarly obtained for the wafer containing $\sim 10^9$ cm^{-3} oxide particles. All of the particles are resolved as bright spots. Figure 3b shows a RC*dpc* image from the same area. Spots corresponding to the same particles of Figure 3a can be seen, but most spots now show an asymmetrical contrast arising because of the two adjacent bright and dark spots that were adjusted with the pinhole alignment to be brought together until they just merged in the *dpc* mode. In Figure 3b, particle B shows the bright and dark sides of the spot to be equally pronounced (in-

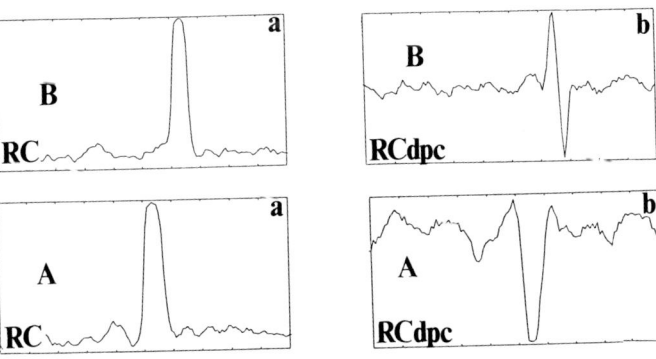

Figure 4. Line traces taken from Figs. 3a and 3b for particles A and B. The asymmetrical RC*dpc* trace for particle B in Fig. 4b indicates that this particle is in-focus.

focus), while particle A shows the dark side more pronounced (out-of-focus). Hence, a single RC*dpc* image can show directly which particles are in-focus.

This effect is illustrated more quantitatively in Figure 4 where intensity line traces for the two particles of Figs. 3a and 3b are given. Figure 4a shows the RC traces and both particles A and B exhibit symmetrical peaks with positive amplitude intensities. Fig. 4b shows the RC*dpc* traces and for particle A the peak is dominated by the negative amplitude intensity, indicating that particle A is out-of-focus. Conversely, the corresponding trace for particle B shows an asymmetrical peak with the positive and negative amplitude intensities closely equal, indicating that particle B is in-focus.

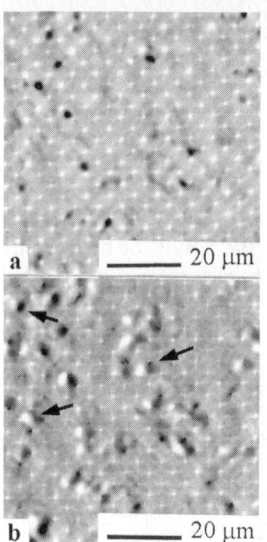

Figure 5a shows a TC image obtained for the wafer containing ~ 10^8 cm^{-3} oxide particles which now has both surfaces polished, and using probe forming and collector lenses of NA=0.6. All of the particles are revealed as dark spots. Figure 5b shows a TC*dpc* image from the same area. Spots corresponding to the same particles of Figure 5a can be seen, but each 'spot' now consists of two separate bright and dark spots. In this case the usual *dpc* adjustment was terminated just before the two images merged so as to illustrate the mode more clearly. Figure 5b shows images for some particles (arrowed) which are not revealed in Figure 5a. This indicates a superior sensitivity of TC*dpc* over TC mode. With these lenses the spatial resolution was ~ 1.5 μm and the depth resolution was 15 to 20 μm.

Images were also obtained using the DPC and DPC*dpc* modes (no images included). Our initial experience suggests that the alignments of the optical components for the DPC modes are more stringent than for the RC and TC modes. However, the DPC modes are less sensitive to, for example, specimen thickness variations, and have better resolutions than the corresponding TC modes.

Figure 5. Heat-treated Cz Si specimen, a) TC and b) TC*dpc* images from same area showing oxide particles. Particles arrowed in b) are not revealed in a)

Important advantages of the new SIRM compared with our previous SIRM include the following. In the TC*dpc* mode, the sensitivity is increased enabling additional particles to be imaged compared with those in the TC mode. In the RC*dpc* mode, the individual particles in-focus can be directly determined without obtaining X-Y scan focal series or X-Z scans. In the RC mode, the lateral and depth resolutions are improved enabling higher particle number densities (up to ~5 x 10^{10} cm^{-3}) to be imaged. Thus, using the SIRM in the RC mode commercial semiconductor wafers can be assessed in a non-destructive and non-contaminating manner.

REFERENCES

Kidd P, Booker GR and Stirland DJ 1987 Inst. Phys. Conf. Ser. No 87 275
Booker GR, Laczik Z and Kidd P 1992 Semicond. Sci. Technol. 7 A110
Laczik Z, Booker GR, Bergholz W and Falster R 1989 Appl. Phys. Lett. 55 2625
Laczik Z, Török P, Booker GR and Falster R 1991 Inst. Phys. Conf. Ser. No 117 785
Wilson T 1991 Confocal Microscopy (London: Academic Press) p 53

Author Index

Subject Index*

*Page numbers refer to the first pages of the papers in which the citations appear